Solutions Manual for

ENGINEERING MECHANICS
DYNAMICS

Arthur P. Boresi
Richard J. Schmidt
University of Wyoming

Brooks/Cole
Thomson Learning™

Australia • Canada • Mexico • Singapore • Spain • United Kingdom • United States

For more information about this or any other Brooks/Cole product, contact:
BROOKS/COLE
511 Forest Lodge Road
Pacific Grove, CA 93950 USA
www.brookscole.com
1-800-423-0563 (Thomson Learning Academic Resource Center)

Printed in the United States of America.

5 4 3 2 1

ISBN 0-534-95163-5

PREFACE

Each homework problem in this book was solved by the authors. We also hired several students to solve the problems and help us achieve three objectives:

- to test the clarity of the problem statements
- to identify the relative difficulty of each problem
- to obtain an approximate solution time for each problem

As a result of this process, we edited and rewrote some of the problems.

In each section of the text, the problems are arranged approximately in order of increasing difficulty, and, for your information and use, the corresponding student solution times are tabulated in Table B, which follows this preface. The solution times for the problems in Statics are in Table A of that manual.

Since students who solved the problems had previously completed a course in dynamics, they might be regarded as experienced problem solvers. However, many of the students had not previously studied a number of topics treated in this book, or had received only a cursory introduction to them. This is particularly true for planetary motion (Chapter 14), energy and momentum principles for rigid bodies (Chapter 19), vibrations (Chapter 20), and three-dimensional kinetics (Chapter 21). So, the solution times that follow should not be regarded as absolute target times for your students, but you might use the times to judge the relative difficulty of the problems and to help estimate how long an assignment might take.

Two students were assigned to each chapter of the book and, for each section of the chapter, were required to study the text material and example problems, then solve all of the homework problems for that section. Each student worked independently to solve each problem and recorded the time required to obtain a solution. The solution times for the two students were averaged for each problem, then rounded to the nearest five minutes (see Table B). These times include the time required to read the problem, plan a method of solution, and solve the problem. For computer problems, time spent working at the computer and obtaining output is also included.

The problems generally are arranged in the order of increasing solution time. Related problems are placed in sequential order. Consequently, a problem relating to a preceding problem could have required less time to solve, because of the student's familiarity with the earlier problem. In general, computer problems and design problems required greater amounts of time and were more difficult to solve. Challenge problems generally required the most time and often, the most insight.

Solutions are not included for many of the review problems, because they usually require discussions of principles or concepts, definitions of terms, or derivations. However, solutions are given for some review problems that require numerical solutions.

As all instructors know, even the best student can start off in the wrong direction to solve a problem and might take longer than ordinarily required. Likewise, an experienced student might see a shortcut to the solution that the novice will not. Our time averages will reflect such deviations, but may or may not be representative of that required by students studying statics for the first time.

We relied on our student problem solvers to help achieve our three objectives, and we believe we were successful. However, in most cases, the solutions presented in this manual are not the work of our students, since generally student work is not sufficiently well organized or complete for presentation in a solutions manual. So we developed our own set of detailed solutions, intended to serve as a teaching aid to your students, after they have attempted to solve the problems. For the most part, we present solutions that are fairly detailed, similar in format to example problems, allowing us to outline the problem solving process, as well as give correct answers. These solutions were rendered into camera-ready form by some of our students with fairly good drafting and lettering skills. Before they were printed, we reviewed the camera-ready sheets to locate and correct transcription errors. Hence, we have developed homework problems that are student-tried and tested, along with solutions that we believe are correct and complete. Nevertheless, errors might exist and we are solely responsible for them.

We welcome any feedback, including corrections or suggestions for improvement, that you might wish to offer. Your comments may be sent to either Arthur P. Boresi or Richard J. Schmidt, Department of Civil and Architectural Engineering, University of Wyoming, Laramie, Wyoming 82071-3295.

Arthur P. Boresi
Richard J. Schmidt
April 2000

TABLE B — APPROXIMATE PROBLEM SOLUTION TIMES

Chapter 13

Problem	Time [min]	Problem	Time [min]	Problem	Time [min]
Sec. 13.2		13.37	25	13.73	35
13.1	10	13.38	25	13.74	35
13.2	10	13.39	25	13.75	35
13.3	10	13.40	25	13.76	35
13.4	10	13.41	25	13.77	35
13.5	10	13.42	25	13.78	40
13.6	10	13.43	30	Sec. 13.4	
13.7	10	13.44	35	13.79	10
13.8	10	13.45	35	13.80	10
13.9	10	13.46	40	13.81	10
13.10	10	13.47	40	13.82	10
13.11	10	13.48	40	13.83	20
13.12	10	13.49	40	13.84	25
13.13	10	13.50	40	13.85	25
13.14	10	13.51	40	13.86	25
13.15	10	13.52	40	13.87	25
13.16	10	13.53	40	13.88	25
13.17	10	13.54	40	13.89	25
13.18	10	13.55	40	13.90	25
13.19	10	13.56	40	13.91	25
13.20	10	13.57	50	13.92	30
13.21	10	13.58	70	13.93	30
13.22	15	13.59	70	13.94	30
13.23	15	13.60	70	13.95	30
13.24	15	13.61	70	13.96	35
13.25	15	13.62	110	13.97	35
13.26	20	13.63	90	13.98	35
13.27	20	Sec. 13.3		13.99	35
13.28	25	13.64	5	13.100	45
13.29	25	13.65	5	13.101	60
13.30	25	13.66	10	13.102	60
13.31	25	13.67	10	13.103	60
13.32	25	13.68	20	13.104	60
13.33	25	13.69	25	13.105	80
13.34	25	13.70	25	13.106	80
13.35	25	13.71	35	13.107	100
13.36	25	13.72	35	13.108	100

13.109	100
13.110	100
Sec. 13.5	
13.111	5
13.112	5
13.113	15
13.114	20
13.115	20
13.116	20
13.117	25
13.118	30
13.119	40
13.120	50
13.121	55
13.122	60
13.123	60–80
Sec. 13.6	
13.124	20
13.125	25
13.126	30
13.127	40
13.128	45
13.129	60
Sec. 13.7	
13.130	5
13.131	10
13.132	20
13.133	25
13.134	25
13.135	25
13.136	35
13.137	35
13.138	40
13.139	45
13.140	45
13.141	50
13.142	50
Sec. 13.8, 13.9	
13.143	5
13.144	10
13.145	15
13.146	20
13.147	25
Sec. 13.10	
13.148	15
13.149	20
13.150	20
13.151	20
13.152	20
13.153	30
13.154	30
13.155	30
13.156	30
13.157	35
13.158	40
13.159	40
13.160	75
Sec. 13.11	
13.161	10
13.162	10
13.163	15
13.164	20
13.165	20
13.166	25
13.167	25
13.168	30
13.169	30
13.170	30
13.171	30
13.172	30
13.173	30
13.174	30
13.175	35
13.176	35
13.177	35
13.178	35
13.179	35
13.180	35
13.181	35
13.182	45
Review	
13.194	5
13.195	5
13.196	5
13.197	5
13.198	10
13.201	10
13.208	10
13.209	5
13.225	10

Chapter 14

Problem	Time [min]
Sec. 14.1	
14.1	10
14.2	10
14.3	10
14.4	10
14.5	10
14.6	10
14.7	15
14.8	15
14.9	15
14.10	15
14.11	20
14.12	25
14.13	25
14.14	25
14.15	30
14.16	50
Sec. 14.4	
14.17	10
14.18	10
14.19	10
14.20	10
14.21	10
14.22	10
14.23	10
14.24	10
14.25	15
14.26	15
14.27	15
14.28	15
14.29	20
14.30	20
14.31	20
14.32	20
14.33	30
14.34	30
14.35	30
14.36	30
14.37	30
14.38	30
14.39	30
14.40	30
14.41	35
14.42	40
14.43	40
14.44	50
14.45	60
14.46	70
14.47	75
14.48	85
14.49	85
14.50	100
14.51	100
14.52	100
14.53	110
14.54	110
Sec. 14.5	
14.55	10
14.56	10
14.57	10
14.58	10
14.59	10
14.60	10
14.61	10
14.62	15
14.63	15
14.64	15
14.65	20
14.66	20
14.67	25
14.68	30
14.69	30
14.70	35
14.71	40
14.72	45
14.73	50
14.74	55
14.75	60
14.76	70
14.77	75
14.78	75
14.79	140
Sec. 14.6	
14.80	10
14.81	10
14.82	15
14.83	15
14.84	20
14.85	25
14.86	30
14.87	35
14.88	35
14.89	35
14.90	40
14.91	50
14.92	80
Sec. 14.7	
14.93	10
14.94	10
14.95	10
14.96	10
14.97	10
14.98	10
14.99	10
14.100	10
14.101	10
14.102	15
14.103	15
14.104	15
14.105	20
14.106	20
14.107	25
14.108	40
Review	
14.110	10
14.122	10
14.124	20

Chapter 15

Problem	Time [min]
Sec. 15.1	
15.1	5
15.2	5
15.3	5
15.4	10
15.5	10
15.6	10
15.7	10
15.8	15
15.9	15
15.10	20
15.11	20
15.12	25
15.13	25
15.14	25
15.15	25
15.16	35
Sec. 15.2	
15.17	5
15.18	10
15.19	10
15.20	10
15.21	10
15.22	10
15.23	10
15.24	10
15.25	10
15.26	10
15.27	10
15.28	10
15.29	10
15.30	10
15.31	15
15.32	15
15.33	15
15.34	15
15.35	15
15.36	20
15.37	25
15.38	25

Problem	Time [min]
15.39	25
15.40	30
15.41	50
15.42	55
15.43	60
15.44	75
15.45	80
15.46	110
15.47	140
15.48	140
Sec. 15.3	
15.49	10
15.50	10
15.51	10
15.52	10
15.53	15
15.54	15
15.55	15
15.56	20
15.57	20
15.58	45
15.59	60
Sec. 15.4	
15.60	10
15.61	10
15.62	10
15.63	10
15.64	10
15.65	10
15.66	15
15.67	25
15.68	25
15.69	25
15.70	30
15.71	30
15.72	45
15.73	95
15.74	100
15.75	110
15.76	120

Problem	Time [min]
Sec. 15.5	
15.77	10
15.78	10
15.79	10
15.80	10
15.81	10
15.82	15
15.83	15
15.84	15
15.85	20
15.86	25
15.87	30
15.88	35
15.89	35
15.90	40
15.91	90
15.92	95
15.93	100
Sec. 15.6	
15.94	10
15.95	15
15.96	20
15.97	20
15.98	20
15.99	25
15.100	30
15.101	30
15.102	30
15.103	35
15.104	40
15.105	40
15.106	80
15.107	120
Review	
15.109	5
15.110	5
15.116	10
15.120	15
15.121	15
15.122	10

Chapter 16

Problem	Time [min]
Sec. 16.1	
16.1	10
16.2	10
16.3	10
16.4	10
16.5	10
16.6	15
16.7	15
16.8	15
16.9	15
16.10	15
16.11	20
16.12	20
16.13	20
16.14	25
16.15	25
16.16	25
16.17	25
16.18	35
16.19	60
16.20	60
16.21	60
16.22	70
Sec. 16.2	
16.23	10
16.24	10
16.25	15
16.26	20
16.27	20
16.28	20
16.29	20
16.30	20
16.31	20
16.32	25
16.33	25
16.34	55
Sec. 16.3	
16.35	5
16.36	5
16.37	5
16.38	10
16.39	10
16.40	10
16.41	10
16.42	10
16.43	20
16.44	25
16.45	25
16.46	45
16.47	45
16.48	45
16.49	55
16.50	55
Sec. 16.4	
16.51	10
16.52	15
16.53	15
16.54	25
16.55	25
16.56	60
16.57	60
Sec. 16.5	
16.58	10
16.59	10
16.60	10
16.61	15
16.62	15
16.63	15
16.64	15
16.65	25
16.66	25
16.67	25
16.68	25
16.69	25
16.70	35
16.71	35
16.72	40
16.73	45
16.74	60
16.75	60
16.76	60
16.77	60
16.78	60
16.79	60
16.80	60
16.81	70
Sec. 16.6	
16.82	10
16.83	10
16.84	15
16.85	20
16.86	20
16.87	25
16.88	25
16.89	25
16.90	35
16.91	35
16.92	35
16.93	35
16.94	35
Sec. 16.7	
16.95	15
16.96	35
16.97	35
16.98	40
16.99	40
16.100	55
Sec. 16.8	
16.101	10
16.102	10
16.103	25
16.104	25

16.105	25
16.106	25
16.107	30
16.108	30
16.109	30
16.110	40
16.111	40
16.112	50
Sec. 16.9	
16.113	15
16.114	15
16.115	15
16.116	20
16.117	20
16.118	20
16.119	20
16.120	20
16.121	20
16.122	20
16.123	30
16.124	35
16.125	35
16.126	35
16.127	60
16.128	90
Review	
16.133	10
16.135	10
16.136	15
16.137	15
16.143	15
16.144	15
16.146	20

Chapter 17

Problem	Time [min]
Sec. 17.1	
17.1	10
17.2	15
17.3	15
17.4	20
17.5	20
17.6	20
17.7	20
17.8	25
17.9	25
17.10	25
17.11	15
17.12	25
17.13	30
17.14	30
17.15	40
17.16	60
17.17	80
Sec. 17.2, 17.3	
17.18	15
17.19	15
17.20	15
17.21	15
17.22	20
17.23	20
17.24	20
17.25	20
17.26	20
17.27	30
17.28	30
17.29	35
17.30	40
17.31	40
17.32	40
17.33	40
17.34	25
17.35	40
17.36	120

Problem	Time [min]
Sec. 17.4	
17.37	10
17.38	10
17.39	15
17.40	15
17.41	15
17.42	15
17.43	15
17.44	20
17.45	20
17.46	20
17.47	25
17.48	25
17.49	30
17.50	30
17.51	35
17.52	40
17.53	40
17.54	45
17.55	45
17.56	90
Sec. 17.5	
17.57	15
17.58	15
17.59	15
17.60	15
17.61	15
17.62	20
17.63	20
17.64	20
17.65	20
17.66	30
17.67	30
17.68	20
17.69	35
17.70	40
17.71	45
17.72	60
17.73	80

Problem	Time [min]
17.74	200
Sec. 17.6	
17.75	15
17.76	15
17.77	20
17.78	30
17.79	30
17.80	30
17.81	40
17.82	40
17.83	40
Sec. 17.7	
17.84	10
17.85	15
17.86	20
17.87	20
17.88	20
Sec. 17.8	
17.89	15
17.90	20
17.91	20
17.92	20
17.93	20
17.94	40
Sec. 17.9, 17.10	
17.95	30
17.96	30
17.97	40
17.98	40
17.99	40
17.100	40
17.101	40
17.102	40
17.103	50
17.104	60
17.105	120
Review	
17.111	15

Chapter 18

Problem	Time [min]
Sec. 18.1	
18.1	10
18.2	10
18.3	10
18.4	15
18.5	15
18.6	15
18.7	20
18.8	25
18.9	30
18.10	15
18.11	30
18.12	30
18.13	30
18.14	45
18.15	60
18.16	60
18.17	60
Sec. 18.2	
18.18	10
18.19	10
18.20	10
18.21	15
18.22	15
18.23	15
18.24	20
18.25	25
18.26	30
18.27	20
18.28	30
18.29	30
18.30	30
18.31	30
Sec. 18.3	
18.32	15
18.33	15
18.34	15
18.35	20
18.36	20
18.37	20
18.38	20
18.39	20
18.40	20
18.41	20
18.42	20
18.43	20
18.44	20
18.45	25
18.46	25
18.47	25
18.48	30
18.49	30
18.50	30
18.51	30
18.52	30
18.53	35
18.54	40
18.55	40
18.56	80
Sec. 18.4	
18.57	20
18.58	20
18.59	20
18.60	20
18.61	25
18.62	25
18.63	30
18.64	40
18.65	50
18.66	75
18.67	60
Sec. 18.5	
18.68	20
18.69	20
18.70	25
18.71	25
18.72	30
18.73	30
18.74	30
18.75	50
18.76	150
Sec. 18.6	
18.77	15
18.78	15
18.79	15
18.80	15
18.81	20
18.82	20
18.83	20
18.84	20
18.85	20
18.86	20
18.87	25
18.88	25
18.89	25
18.90	25
18.91	25
18.92	25
18.93	30
18.94	30
18.95	30
18.96	60
18.97	90
18.98	140
18.99	100
18.100	100
Sec. 18.7	
18.101	15
18.102	15
18.103	15
18.104	15
18.105	20
18.106	20
18.107	20

18.108	20
18.109	20
18.110	20
18.111	20
18.112	20
18.113	20
18.114	25
18.115	25
18.116	25
18.117	25
18.118	25
18.119	25
18.120	25
18.121	25
18.122	30
18.123	30
18.124	30
18.125	35
18.126	60
18.127	60
18.128	90
Sec. 18.8	
18.129	25
18.130	25
18.131	25
18.132	25
18.133	30
18.134	40
18.135	40
Review	
18.139	10
18.140	10
18.141	10
18.142	15
18.143	20
18.148	15
18.153	20

Chapter 19

Problem	Time [min]
Sec. 19.1	
19.1	10
19.2	10
19.3	15
19.4	25
19.5	25
19.6	30
19.7	30
19.8	35
19.9	40
19.10	40
Sec. 19.2	
19.11	10
19.12	15
19.13	15
19.14	20
19.15	30
19.16	30
19.17	30
19.18	45
19.19	45
Sec. 19.3	
19.20	10
19.21	10
19.22	10
19.23	10
19.24	10
19.25	15
19.26	15
19.27	15
19.28	15
19.29	15
19.30	15
19.31	15
19.32	20
19.33	20

Problem	Time [min]
19.34	20
19.35	35
19.36	35
19.37	40
19.38	40
19.39	60
19.40	75
Sec. 19.4, 19.5	
19.41	10
19.42	20
19.43	25
19.44	25
19.45	25
19.46	25
19.47	25
19.48	25
19.49	30
Sec. 19.6	
19.50	10
19.51	10
19.52	10
19.53	15
19.54	15
19.55	15
19.56	15
Sec. 19.7	
19.57	15
19.58	15
19.59	15
19.60	15
19.61	15
19.62	15
19.63	15
19.64	15
19.65	20

Problem	Time [min]
19.66	20
19.67	20
19.68	25
19.69	25
19.70	25
19.71	25
19.72	25
19.73	25
19.74	25
19.75	40
Sec. 19.8	
19.76	15
19.77	20
19.78	20
19.79	20
19.80	20
19.81	20
19.82	20
19.83	20
19.84	20
19.85	25
19.86	25
19.87	30
19.88	35
19.89	35
19.90	35
19.91	40
19.92	40
19.93	40
19.94	40
Review	
19.96	15
19.100	15
19.101	15
19.102	20

Chapter 20

Problem	Time [min]
Sec. 20.1, 20.2	
20.1	15
20.2	15
20.3	15
20.4	15
20.5	15
20.6	15
20.7	15
20.8	15
20.9	25
20.10	35
20.11	50
Sec. 20.3	
20.12	20
20.13	20
20.14	20
20.15	20
20.16	20
20.17	20
20.18	20
20.19	20
20.20	20
20.21	20
20.22	20
20.23	20

Problem	Time [min]
20.24	20
20.25	20
20.26	20
20.27	25
20.28	25
20.29	25
20.30	30
20.31	30
20.32	40
20.33	50
Sec. 20.4	
20.34	15
20.35	15
20.36	20
20.37	20
20.38	20
20.39	20
20.40	20
20.41	20
20.42	25
20.43	35
20.44	40
Sec. 20.5	
20.45	20
20.46	20

Problem	Time [min]
20.47	10
20.48	20
20.49	20
20.50	25
20.51	25
20.52	25
20.53	25
20.54	40
20.55	40
20.56	40
20.57	40
20.58	40
20.59	40
Sec. 20.6	
20.60	20
20.61	20
20.62	20
20.63	20
20.64	20
20.65	20
20.66	20
20.67	20
20.68	40
20.69	50
20.70	60

Chapter 21

Problem	Time [min]
Sec. 21.1	
21.1	10
21.2	15
21.3	20
21.4	20
21.5	20
21.6	25
21.7	25
21.8	25
21.9	25
21.10	25
21.11	30
21.12	35
Sec. 21.2, 21.3	
21.13	20
21.14	20
21.15	20
21.16	20
21.17	20
21.18	20
21.19	20
21.20	25
21.21	35
21.22	50

Problem	Time [min]
Sec. 21.4	
21.23	15
21.24	20
21.25	25
21.26	25
21.27	25
21.28	25
21.29	35
21.30	40
21.31	50
Sec. 21.5	
21.32	15
21.33	20
21.34	30
21.35	30
21.36	35
21.37	35
21.38	40
21.39	50
Sec. 21.6, 21.7, 21.8	
21.40	20
21.41	20
21.42	20
21.43	25
21.44	25
21.45	35

Problem	Time [min]
21.46	35
21.47	40
21.48	40
21.49	40
21.50	45
21.51	45
21.52	50
21.53	60
Sec. 21.9	
21.54	15
21.55	15
21.56	15
21.57	15
21.58	15
21.59	20
21.60	20
21.61	20
21.62	20
21.63	20
21.64	25
21.65	25
21.66	25
21.67	35
21.68	35
21.69	35

Appendix C

Solutions to the problems in Appendix C are contained in the Statics Solution Manual.

13.1

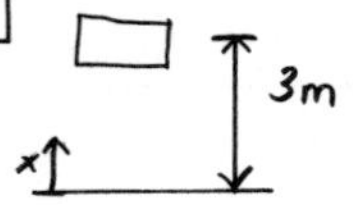

Since hammer is in free fall, $a = -g = -9.81\ m/s^2$

$v = \int a\,dt = -gt + v_0$ ($v_0 = 0$ since dropped from rest)

$v = -9.81\,t$

$x = \int v\,dt = -\frac{1}{2}gt^2 + x_0$ ($x_0 = 3m$, $x_f = 0$)

$x = -4.91t^2 + 3 = 0$ when hammer hits pile

Solving $x(t)$ for t_f: $t_f = 0.782\ s$

$v(0.782) = -9.81(0.782)$ $v = -7.67\ m/s$

or $v = 7.67\ m/s \downarrow$

13.2

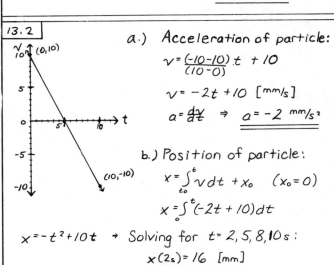

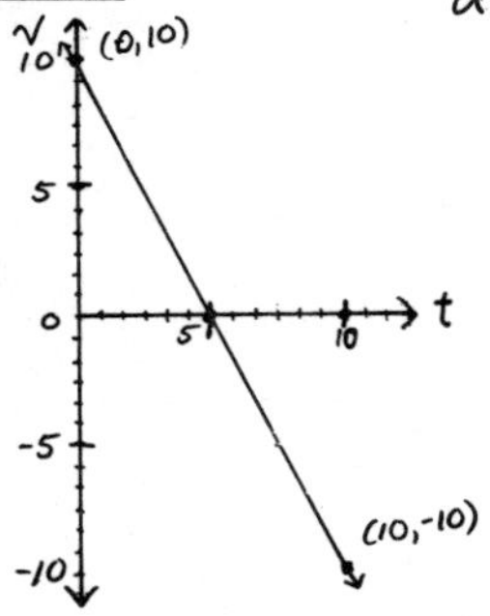

a.) Acceleration of particle:

$v = \frac{(-10-10)}{(10-0)}t + 10$

$v = -2t + 10$ [mm/s]

$a = \frac{dv}{dt} \Rightarrow$ $a = -2\ mm/s^2$

b.) Position of particle:

$x = \int_{t_0}^{t} v\,dt + x_0$ ($x_0 = 0$)

$x = \int_0^t (-2t + 10)dt$

$x = -t^2 + 10t$ → Solving for $t = 2, 5, 8, 10\,s$:

$x(2s) = 16$ [mm]

$x(5s) = 25$ [mm]

$x(8s) = 16$ [mm]

$x(10s) = 0$ [mm]

c.) Distance traveled:

$\Delta S = |\Delta X_1| + |\Delta X_2|$

$\Delta S(2s) = 16$ [mm]

$\Delta S(5s) = 25$ [mm]

$\Delta S(8s) = 25 + |16 - 25| = 34$ [mm]

$\Delta S(10s) = 25 + |0 - 25| = 50$ [mm]

13.3

$v = 4 =$ constant, $x = 0$ when $t = 0$

$a = \frac{dv}{dt}$ → $a(t) = 0$

$v = \frac{dx}{dt} = 4 \Rightarrow x = 4t + C = 0$ for $t = 0$

$\therefore C = 0$ $x = 4t$

13.4

$v = 4t$, $x = 1$ when $t = 1$

$a = \frac{dv}{dt} \Rightarrow$ $a(t) = 4$ (constant)

$v = \frac{dx}{dt} = 4t \Rightarrow x(t) = 2t^2 + C$

For $x = 1$, $t = 1$ $\therefore$ $1 = 2(1)^2 + C \Rightarrow C = -1$

$x(t) = 2t^2 - 1$

13.5

$v = \sin 2t$, $x = 0$ when $t = 0$

$a = \frac{dv}{dt} \Rightarrow$ $a(t) = 2\cos 2t$ (a.)

$v = \frac{dx}{dt} = \sin 2t \Rightarrow x = -\frac{\cos 2t}{2} + C$

$x = 0$ when $t = 0$ $\therefore C = \frac{1}{2}$

$x(t) = -\frac{\cos 2t}{2} + \frac{1}{2}$ (b.)

By Eq. (b). $-x + \frac{1}{2} = \frac{\cos 2t}{2}$

or $t = (\frac{1}{2})\cos^{-1}(-2x + 1)$ (c.)

Substitution of Eq (c.) into Eq (a.) gives:

$a(x) = 2\cos(\frac{1}{2}\cos^{-1}(2 - 4x))$

13.6

$v = e^{-2t}$, $x = 2$ when $t = 0$

$a = \frac{dv}{dt} = -2e^{-2t}$ (a.)

$v = \frac{dx}{dt} = e^{-2t}$ $x = \int v\,dt \Rightarrow x = -\frac{1}{2}e^{-2t} + C$

$x = 2$ when $t = 0$ $\therefore C = 5/2$

$x(t) = \frac{1}{2}(5 - e^{-2t})$ (b.)

Solving Eq (b.) for t: $e^{-2t} = 5 - 2x$

$t = -\frac{1}{2}\ln(5 - 2x)$ (c.)

Substitution of Eq (c) into Eq (a) gives:

$a(x) = -2e^{-2[-\frac{1}{2}\ln(5-2x)]} \Rightarrow$ $a(x) = 4x - 10$

13.7

a.) $\Delta S_f = 2\pi(4000\ \text{miles}) = 25132.74$ miles

$= 132{,}700{,}873$ ft

The displacement of your feet relative to the earth would be zero. $\Delta X = 0$ ft

b.) $\Delta S_h = 2\pi[4000\,\text{mi}(\frac{5280\,\text{ft}}{\text{mi}}) + 6\,\text{ft}]$

$= 132{,}700{,}911.4$ ft

Difference: $132{,}700{,}911.4 - 132{,}700{,}873.7$

$= 37.7$ ft

$\therefore$ Your head travels 37.7 ft more

Alternatively: $2\pi(6\,\text{ft}) = 37.7$ ft

13.8

d - total distance traveled = 2 mi

t - total time over 2 mile stretch

$$t = \frac{1\text{ mile}}{30\text{ mph}} + \frac{1\text{ mile}}{V_{req}}$$

V_{req} = speed required to average 45 mph over 2 mile stretch

$$\frac{d}{t} = 45\text{ mph} \Rightarrow \frac{2\text{ mi}}{\frac{1\text{ mi}}{30\text{ mph}} + \frac{1\text{ mi}}{V_{req}}} = 45\text{ mph}$$

$$V_{req} = \frac{1\text{ mi}}{\frac{2\text{ mi}}{45\text{ mph}} - \frac{1\text{ mi}}{30\text{ mph}}} \Rightarrow \underline{\underline{V_{req} = 90\text{ mph}}}$$

13.9

d - total distance traveled = 2 mi

t - total time around 2 mile stretch

$$t = \frac{1\text{ mile}}{30\text{ mph}} + \frac{1\text{ mile}}{V_{req}}$$

V_{req} = speed required to average 60 mph over 2 mile stretch

$$\frac{d}{t} = 60\text{ mph} \Rightarrow \frac{2\text{ mi}}{\frac{1\text{ mi}}{30\text{ mph}} + \frac{1\text{ mi}}{V_{req}}} = 60\text{ mph}$$

$$V_{req} = \frac{1\text{ mi}}{\frac{2\text{ mi}}{60\text{ mph}} - \frac{1\text{ mi}}{30\text{ mph}}} \Rightarrow \underline{\underline{V_{req} = \infty}}$$

(It is impossible to go fast enough to average 60 mph — there is no time left.)

13.10

a.) $v_0 = (\frac{1}{6}\ \frac{mi}{s})(5280\ \frac{ft}{mi}) = 880\ ft/s$

After power is reduced, $a = -g = -32.2\ ft/s^2$

$\therefore v = v_0 + at = 880 - 32.2t = 0$ at highest pt.

$\therefore t = 880/32.2 \qquad \underline{\underline{t = 27.4\ s}}$

b.) The height the plane rises above the point at which the power is reduced is:

$x = v_0 t + \frac{1}{2}at^2 = 880(27.35) - \frac{1}{2}(32.2)(27.35)^2$

$x = 12,025\ ft = \underline{\underline{2.28\ mi}}$

13.11

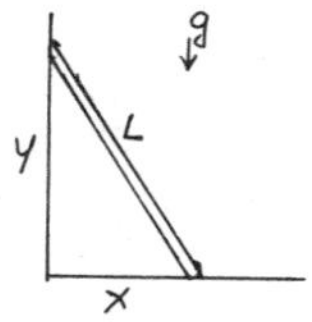

$v_y = \dot{y} \qquad a_y = \ddot{y}$

$L^2 = x^2 + y^2$

Differentiating: $0 = 2x\dot{x} + 2y\dot{y}$

$$\dot{x} = \frac{-y\dot{y}}{x} \Rightarrow \underline{\underline{\dot{x} = \frac{v_y\sqrt{L^2 - x^2}}{x}}}$$

$$\ddot{x} = \frac{-\left(\frac{v_y}{x}\sqrt{L^2-x^2}\right)^2 - v_y^2 + a_y\sqrt{L^2+x^2}}{x}$$

$$\ddot{x} = \frac{a_y\sqrt{L^2-x^2}}{x} - \frac{v_y^2(L^2-x^2)}{x^3} - \frac{v_y^2}{x}$$

$$\underline{\underline{\ddot{x} = \frac{a_y x^2\sqrt{L^2-x^2} - v_y^2 L^2}{x^3}}}$$

13.12

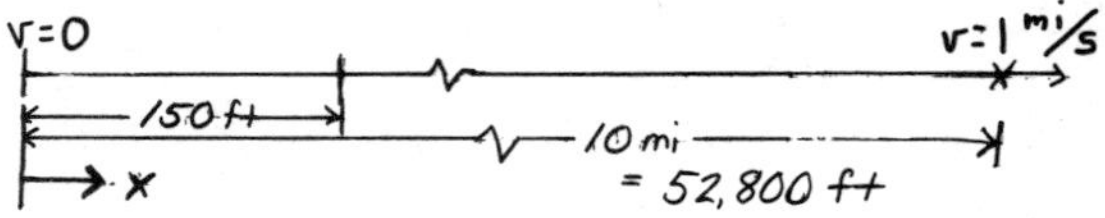

$v = \int a\,dt = at + v_0^{\ 0}$ Since $v = 0$ when $t = 0$

$\therefore t = v/a \quad (A)$

$x = \int v\,dt = \frac{1}{2}at^2 + x_0^{\ 0} = \frac{1}{2}at^2$ since $x = 0$ when $t = 0$

$\therefore x = \frac{1}{2}at^2 \quad (B)$

By Eqns (A) and (B): $\quad x = \frac{1}{2}at^2 = \frac{1}{2}\frac{v^2}{a} \quad (C)$

$$\therefore a = \frac{1}{2}\frac{v^2}{x} = \frac{1}{2}\frac{[(1\ mi/s)(5280\ ft/mi)]^2}{(10\ mi)(5280\ ft/mi)} = 264\ ft/s^2$$

Substituting $a = 264\ ft/s^2$ into (B):

$150\ ft = \frac{1}{2}(264\ ft/s^2)t^2$

$\therefore \underline{\underline{t = 1.066\ s}}$ = time rocket is in contact with guides.

13.13

a.) $a = 32.2\ ft/s^2 \qquad v_0 = 0$

$v = \int a\,dt = at_1 + v_0^{\ 0} \Rightarrow v = 32.2t$

$\left(\frac{5280\ ft}{mi}\right)\left(186{,}000\ \frac{mi}{s}\right) = 32.2\ \frac{ft}{s^2}(t_1)$

$t_1 = 30{,}499{,}378.9\ s \Rightarrow \underline{\underline{t_1 = 0.967\text{ years}}}$

b.) At the end of 0.967 years, the distance traveled is:

$S_1 = \frac{1}{2}at_1^2 = \frac{1}{2}(32.2)(3.0499\times 10^7)^2(\frac{1}{5280})$

$= 2.836 \times 10^{12}\ mi$

Let S_2 be distance remaining to be traveled

$\therefore S_2 = 2.5\times 10^{13} - S_1 = (2.5 - 0.2836)(10^{13})\ mi$

$= 2.216 \times 10^{13}\ mi$

Let t_2 = time to travel distance S_2

$$\therefore t_2 = \frac{2.216(10^{13})}{1.86(10^5)} = 1.191(10^8)\ s = 3.774\text{ years}$$

$t_{total} = t_1 + t_2 = 0.967 + 3.774 = \underline{\underline{4.74\text{ years}}}$

13.14

a.) $v = 16t - t^2\ mi/h = \frac{1}{60}(16t - t^2)\ mi/min$

$\frac{dv}{dt} = \frac{1}{60}(16 - 2t) = 0$ at $t = 8$ min

$\therefore v_{max} = \frac{1}{60}[16(8) - (8)^2] = \frac{64}{60}\ mi/min$

Hence;

$$S = \frac{1}{60}\int_0^8 (16t - t^2)\,dt + \frac{64}{60}(12 - 8)$$

$$= \frac{1}{60}(8t^2 - \frac{1}{3}t^3)\Big|_0^8 + \frac{64(4)}{60} \qquad \underline{\underline{S = 9.96\ mi}}$$

b.)

$$v_{ave}(mi/h) = \frac{9.956\ mi\ (60\ min)}{12\ min\ (1\ h)} = \underline{\underline{49.78\ mi/h}}$$

13.15

$a = \frac{dv}{dt} = \frac{k}{v}$

$\therefore \int k\,dt = \int v\,dv \rightarrow \frac{1}{2}v^2 = kt + C$

when $t=0,\ v=2 \quad \therefore C = 2$

when $t=3,\ v=4 \quad \therefore k=2$

Hence, $v^2 = 4(t+1)$

When $t = 8s,\ v^2 = 36 \rightarrow \underline{|v| = 6\ m/s}$

13.16

$x = c_3 t^3 + c_2 t^2 + c_1$

a.) $v = \frac{dx}{dt} = \underline{3c_3 t^2 + 2c_2 t}$

$a = \frac{dv}{dt} = \underline{6c_3 t + 2c_2}$

b.) Distance traveled from $t=t_0$ to $t=t_1$:

at $t=t_0 \quad x_0 = c_3(t_0)^3 + c_2(t_0)^2 + c_1$

at $t=t_1 \quad x_1 = c_3(t_1)^3 + c_2(t_1)^2 + c_1$

ΔS = distance traveled $= x_1 - x_0$

$\underline{\Delta S = c_3(t_1^3 - t_0^3) + c_2(t_1^2 - t_0^2)}$

13.17

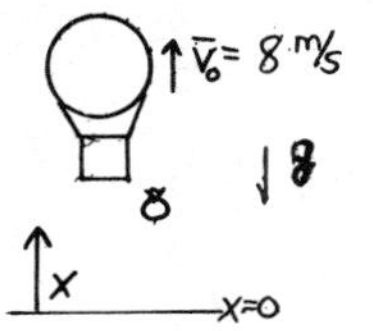

a.) $a = -g = -9.81\ m/s^2$

$v = \int a\,dt = -9.81t + v_0$

$v = -9.81t + 8$

$x = \int v\,dt = -4.91t^2 + 8t + x_0$

$x = 0$ when it hits the ground

$0 = -4.91(10)^2 + 8(10) + x_0 \rightarrow \underline{x_0 = 411\ m}$

b.) $v = -9.81(10) + 8 \qquad v = -90.1\ m/s$ or $\underline{90.1\ m/s \downarrow}$

13.18

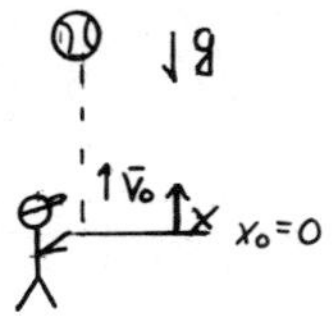

a.) $a = -g = -9.81\ m/s^2$

$v = \int a\,dt = -9.81t + v_0$

$x = \int v\,dt = -4.91t^2 + v_0 t + \cancel{x_0}^{\,0}$

$0 = -4.91(4)^2 + v_0(4) \quad (x=0,\ t=4s)$

$\therefore \underline{v_0 = 19.64\ m/s}$

b.)

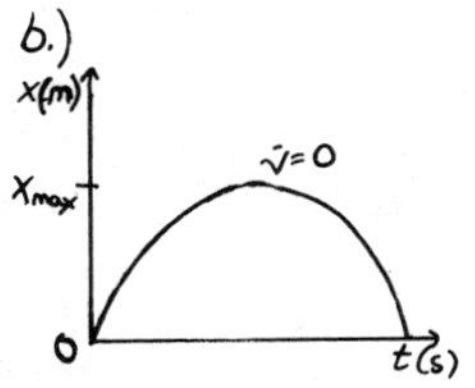

$v = -9.81t + 19.64$

As shown on curve, X_{max} occurs when $\bar{v} = 0$.

$\therefore 0 = -9.81t_m + 19.64$

$t_m = 2.00s$

$x = -4.91t^2 + 19.64t$

$X_{max} = -4.91(2)^2 + 19.64(2)$

$\underline{X_{max} = 19.64\ m}$

13.19

$a = t^3 + 3t\ [ft/s^2] \quad v=0\ @\ t=0,\ x=0$

a.) acceleration for $t = 4s$:

$a = (4)^3 + 3(4) \Rightarrow \underline{a = 76\ ft/s^2}$

velocity for $t=4s$:

$v = \int_0^t (t^3+3t)\,dt \Rightarrow v = \frac{1}{4}t^4 + \frac{3}{2}t^2 + v_0$

Since $v=0$ when $t=0$, $v_0 = 0$

$v(4s) = \frac{1}{4}(4)^4 + \frac{3}{2}(4)^2 \Rightarrow \underline{v = 88\ ft/s}$

displacement for $t=4s$:

$\Delta X = X(4) - X(0) = \int_0^4 v\,dt$

$= \int_0^4 (\frac{1}{4}t^4 + \frac{3}{2}t^2)\,dt = (\frac{1}{20}t^5 + \frac{1}{2}t^3)\Big|_0^4$

$\underline{\Delta X = 83.2\ ft}$

b.) $v = \frac{1}{4}t^4 + \frac{3}{2}t^2$

$v = 186{,}000\ mi/s\ (5280\ \frac{ft}{mi}) = 9.8208 \times 10^8\ ft/s$

$9.8208 \times 10^8 = \frac{1}{4}t^4 + \frac{3}{2}t^2$

Using quadradic equation:

$\underline{t = 250.4s}$

13.20

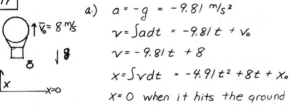

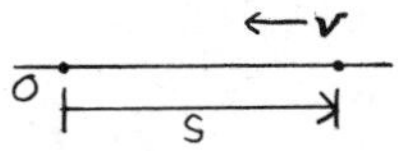

$v = -ks$

a.) $v = \frac{ds}{dt} = -ks \Rightarrow \frac{1}{s}ds = -k\,dt$

$\int_{s_0}^{s} \frac{1}{s}ds = -\int_0^t k\,dt \rightarrow \ln\frac{s}{s_0} = -kt$

$\underline{S = S_0 e^{-kt}}$

b.) $t=0,\ v = 20\ in/s,\ s = 10\ in$

Speed $= |v| = ks$

$20 = k(10) \rightarrow \therefore k = 2s^{-1}$

$10 = S_0 e^{-2(0)} \Rightarrow \therefore S_0 = 10in$

$S = 10e^{-2t} \Rightarrow \therefore t = -\frac{1}{2}\ln\frac{s}{10}$

$t(s=8in) = \underline{0.112s}$

$t(s=6in) = \underline{0.255s}$

$t(s=0in) = \underline{\infty}$

13.21

$v = t^2 + 20t \Rightarrow t = \frac{-20 \pm \sqrt{20^2 - 4(1)(-v)}}{2(1)}$

Negative time does not apply:

$t(v=200) = 7.32s \qquad t(v_2 = 600) = 16.46\ s$

$\Delta s = \int |v|\,dt = \int v\,dt$ since v is positive

$\Delta s = \int_{7.32}^{16.46} (t^2 + 20t)\,dt = [\frac{1}{3}t^3 + 10t^2]\Big|_{7.32}^{16.46}$

$\underline{\Delta S = 3530\ ft}$

13.22

a.) $a = -kv \qquad v = v_0, x = 0$ when $t = 0$

$\int dt = \int \frac{dv}{a(v)} \Rightarrow t - 0 = \int_{v_0}^{v} \frac{dv}{-kv}$

$t = -\frac{1}{k} \ln v \Big|_{v_0}^{v} = -\frac{1}{k} \ln\left(\frac{v}{v_0}\right) \quad (A)$

$\underline{\underline{v = v_0 e^{-kt}}}$

$\int dx = \int v(t) dt \Rightarrow x - 0 = \int_0^t v_0 e^{-kt} dt$

$x = -\frac{v_0}{k} e^{-kt} \Big|_0^t = -\frac{v_0}{k}(e^{-kt} - 1)$

$\underline{\underline{x = \frac{v_0}{K}(1 - e^{-kt})}} \quad (B)$

b.) By Eqn (A), $t = -\frac{1}{k} \ln\left(\frac{v}{v_0}\right)$ and,
By Eqn (B), $x = \frac{v_0}{K}(1 - e^{-kt})$.
Combining (A) and (B) gives:

$x = \frac{v_0}{K}\left(1 - e^{-k\left(\frac{-1}{k} \ln \frac{v}{v_0}\right)}\right)$ or $\underline{\underline{x(v) = \frac{1}{k}(v_0 - v)}}$

13.23

$a = 4, \quad v = -6, \; x = 0$ when $t = 0$

$v = \int a dt = \int 4 dt \Rightarrow v = 4t + v_0$

$v(t=0) = 0 + v_0 = -6 \quad \therefore v_0 = -6 \qquad \underline{\underline{v = 4t - 6}}$

$x = \int v dt = \int_0^t (4t - 6) dt$

$x = 2t^2 - 6t + x_0 \Rightarrow x(t=0) = 0 - 0 + x_0 = 0 \quad \therefore x_0 = 0$

$\underline{\underline{x = 2t^2 - 6t}}$

From $v(t) \Rightarrow t = \frac{v+6}{4}$

Substitution of t into $x(t)$ gives

$x = 2\left(\frac{v+6}{4}\right)^2 - 6\left(\frac{v+6}{4}\right) = \frac{1}{8}v^2 - 4.5$

$\underline{\underline{v = \sqrt{8x + 36} = 2\sqrt{2x + 9}}}$

13.24

$a = 4t; \quad v = 2, \; x = 1$ when $t = 1$

$v = \int_{t_1}^{t} a\, dt + v_1 \Rightarrow v = \int_1^t 4t\, dt + v_1$

$v = 2t^2 - 2 + v_1 \Rightarrow v(t=1) = 2 \quad \therefore v_1 = 2$

$\therefore \underline{\underline{v(t) = 2t^2}} \quad (A)$

$x = \int_{t_1}^{t} v\, dt + x_1 \Rightarrow x = \int_1^t 2t^2 dt + x_1$

$x = \frac{2}{3}(t^3 - 1^3) + x_1 \Rightarrow x(t=1) = 1 = x_1 \quad \therefore x_1 = 1$

$\therefore \underline{\underline{x = \frac{2}{3}t^3 + \frac{1}{3}}} \quad (B)$

By Eqn (A), $t = \sqrt{v/2} \quad (C)$

By Eqn (B) and (C), $x = \frac{2}{3}\left(\frac{v}{2}\right)^{3/2} + \frac{1}{3}$

$\therefore \underline{\underline{v = 2\left(\frac{3}{2}x - \frac{1}{2}\right)^{2/3}}}$

13.25

$a = e^{-2t}; \quad v = 0, \; x = 2$ when $t = 0$

$v = \int a\, dt \Rightarrow v = \int e^{-2t} dt$

$v = -\frac{1}{2}e^{-2t} + v_0 \Rightarrow v(0) = 0 = -\frac{1}{2}(1) + v_0$

$\therefore v_0 = \frac{1}{2} \qquad \therefore \underline{\underline{v(t) = \frac{1}{2}(1 - e^{-2t}) \quad (A)}}$

$x = \int v\, dt \Rightarrow x = \int \left(\frac{1}{2} - \frac{1}{2}e^{-2t}\right) dt$

$x = \frac{1}{2}t + \frac{1}{4}e^{-2t} + x_0 \Rightarrow x(0) = 2 = 0 + \frac{1}{4} + x_0$

$\therefore x_0 = 1.75 \qquad \therefore \underline{\underline{x(t) = \frac{1}{2}t + \frac{1}{4}e^{-2t} + 7/4}} \quad (B)$

By Eqn (A); $t = -\frac{1}{2}\ln(1 - 2v) \quad (C)$

By Eqns (B) and (C):

$x = -\frac{1}{4}\ln(1 - 2v) + \frac{1}{4}e^{\ln(1-2v)} + 7/4$

$\therefore \underline{\underline{x(v) = -\frac{1}{4}[\ln(1 - 2v) - 8 + 2v]}}$

13.26

$a = \sin 2t; \quad v = 0, \; x = 0$ when $t = 0$

$v = \int a\, dt \Rightarrow v = \int \sin 2t\, dt$

$v = -\frac{1}{2}\cos 2t + v_0 \Rightarrow v(t=0) = -\frac{1}{2}(1) + v_0 = 0$

$\therefore v_0 = \frac{1}{2} \qquad \therefore v(t) = \frac{1}{2}(1 - \cos 2t)$

$\underline{\underline{v = \sin^2 t}} \quad (A)$

$x = \int v\, dt \rightarrow x = \int \left(\frac{1}{2} - \frac{1}{2}\cos 2t\right) dt$

$x = \frac{1}{2}t - \frac{1}{4}\sin 2t + x_0 \rightarrow x(0) = 0 - 0 + x_0 = 0$

$\therefore x_0 = 0 \qquad \therefore \underline{\underline{x = \frac{1}{2}t - \frac{1}{4}\sin 2t}} \quad (B)$

By Eqn (A), $t = \sin^{-1}\sqrt{v} \quad (C)$

By Eqns (B) + (C),

$\underline{\underline{x(v) = \frac{1}{2}\sin^{-1}\sqrt{v} - \frac{1}{4}\sin\left[2\left(\sin^{-1}\sqrt{v}\right)\right]}}$

13.27

$a = 32 - 4v; \quad v = 4, \; x = 0$ when $t = 0$

$\int_0^t dt = \int_{v_0}^{v} \frac{dv}{a(v)} \Rightarrow \int_0^t dt = \int_4^v \frac{dv}{32 - 4v}$

Using Integration tables: $\int \frac{dx}{a + bx} = \frac{1}{b}\ln(a + bx)$

$t = -\frac{1}{4}\ln(32 - 4v)\Big|_4^v$

$\therefore \underline{\underline{v(t) = 4(2 - e^{-4t})}} \quad (A)$

By Eqn (A): $x = \int v\, dt = 4\int (2 - e^{-4t}) dt$

or $x = 8t + e^{-4t} + x_0$; $\quad x_0 = -1$ since $x = 0, t = 0$

$\therefore \underline{x = 8t + e^{-4t} - 1} \quad (B)$

By Eqn (A): $t = -\frac{1}{4}\ln\left(\frac{8 - v}{4}\right) \quad (C)$

By Eqns (B) and (C)

$x = 8\left[-\frac{1}{4}\ln\left(\frac{8 - v}{4}\right)\right] + e^{\ln\left(\frac{8-v}{4}\right)} - 1$

Simplifying, $\underline{\underline{x(v) = 3.77 - 2\ln(8 - v) - v/4}}$

13.28 $a = -\frac{4}{x^2}$; $v = 4,\ x = 2$ when $t = 0$ $\quad v > 0$

Find $v(x)$:

$$a = \frac{-4}{x^2} = \frac{dv}{dt}\frac{dx}{dx} = v\frac{dv}{dx} \qquad \therefore \int_4^v v\,dv = -4\int_2^x \frac{dx}{x^2}$$

$$\frac{1}{2}v^2\Big|_4^v = \frac{4}{x}\Big|_2^x \quad \text{or} \quad \frac{v^2}{2} = \frac{4}{x} + 6$$

$$\therefore v = 2\sqrt{\frac{3x+2}{x}} \qquad (A)$$

Find $x(t)$:

$$v = \frac{dx}{dt} = 2\sqrt{\frac{3x+2}{x}} \qquad \therefore \int_2^x \frac{\sqrt{x}\,dx}{2\sqrt{3x+2}} = \int_0^t dt = t \qquad (B)$$

By Integral tables: or computer:

$$\int_2^x \frac{\sqrt{x}\,dx}{\sqrt{3x+2}} = \int_2^x \frac{dx}{\sqrt{\frac{2}{x}+3}} = -2\int_1^{2/x} \frac{dy}{y^2\sqrt{y+3}} \qquad (C)$$

where $y = 2/x$.

By Integration tables or computer:

$$\int_1^{2/x} \frac{dy}{y^2\sqrt{y+3}} = -\frac{\sqrt{y+3}}{3y}\Big|_1^{2/x} - \frac{1}{6}\int_1^{2/x} \frac{dy}{y\sqrt{y+3}}$$

$$= -\frac{\sqrt{y+3}}{3y}\Big|_1^{2/x} - \frac{1}{6}\left[\frac{1}{\sqrt{3}}\ln\frac{\sqrt{y+3}-\sqrt{3}}{\sqrt{y+3}+\sqrt{3}}\right]_1^{2/x} \qquad (D)$$

By Eqns (B), (C), and (D):

$$t = \frac{1}{6}\sqrt{2x+3x^2} + \frac{1}{10.392}\ln\left[\frac{\sqrt{2x+3x^2}-\sqrt{3}\,x}{\sqrt{2x+3x^2}+\sqrt{3}\,x}\right] - 0.413$$

Find $v(t)$:

By Eqn (A): $x = \frac{8}{v^2-12}$ $\qquad (F)$

Hence, $\sqrt{2x+3x^2} = \frac{4v}{v^2-12}$ $\qquad (G)$

So, by substituting (F), (G), and $x(t)$:

$$t = \frac{2v}{3v^2-36} + \frac{1}{10.392}\ln\left(\frac{v-2\sqrt{3}}{v+2\sqrt{3}}\right) - 0.413$$

for $v > 2\sqrt{3}$

13.29 $a = -8x$; $v = 0,\ x = 6$ when $t = 0$

Find $v(x)$:

$$\tfrac{1}{2}\left(v(x)^2 - v(x_0)^2\right) = \int_{x_0}^{x} a(x)\,dx$$

$$\tfrac{1}{2}\left(v(x)^2 - (0)^2\right) = \int_6^x (-8x)\,dx$$

$$\tfrac{1}{2}v^2 = -4x^2\Big|_6^x \quad \rightarrow \quad v^2 = 288 - 8x^2 \qquad (A)$$

$$\therefore v = \sqrt{288-8x^2} \quad \text{or} \quad v = 2\sqrt{72-2x^2}$$

Find $x(t)$:

$$t - t_0 = \int_{x_0}^{x} \frac{dx}{v(x)} \quad \rightarrow \quad t = \int_6^x \frac{dx}{\sqrt{288-8x^2}} + \cancel{t_0}^{\,0}$$

By integration tables:

$$\int \frac{dx}{\sqrt{cx^2+bx+a}} = \frac{1}{\sqrt{-c}}\sin^{-1}\left(\frac{-2cx-b}{\sqrt{b^2-4ac}}\right) \quad \text{for } c < 0$$

Then, $t = \frac{1}{\sqrt{8}}\sin^{-1}\left(\frac{16x}{96}\right) - \frac{1}{\sqrt{8}}\sin^{-1}\left(\frac{16(6)}{96}\right)$

$$t = \frac{1}{\sqrt{8}}\sin^{-1}\left(\frac{x}{6}\right) - \frac{\pi}{2\sqrt{8}}$$

$$t + \frac{\pi}{2\sqrt{8}} = \frac{1}{\sqrt{8}}\sin^{-1}\left(\frac{x}{6}\right) \rightarrow \sin\left(\sqrt{8}\,t + \frac{\pi}{2}\right) = \frac{x}{6}$$

$$\therefore x = \cos(\sqrt{8}\,t)$$

Find $v(t)$:

From Eqn (A): $v^2 = 288 - 8x^2$

and $x = 6\cos(\sqrt{8}\,t)$ so,

$$v^2 = 288 - 8\left(6\cos(\sqrt{8}\,t)\right)^2$$

$$= 288 - 288\cos^2(\sqrt{8}\,t) = 288(1-\cos^2\sqrt{8}\,t)$$

$$v^2 = 288\sin^2(\sqrt{8}\,t) \qquad \therefore v = 12\sqrt{2}\sin(\sqrt{8}\,t)$$

13.30 $a = -kv$; $v = v_0,\ x = 0$ when $t = 0$

Find $v(t)$:

$$a = -kv = \frac{dv}{dt} \rightarrow \int_0^t dt = -\frac{1}{k}\int_{v_0}^{v}\frac{dv}{v}$$

$$t = -\frac{1}{k}\ln(v)\Big|_{v_0}^{v} = -\frac{1}{k}\ln\frac{v}{v_0}$$

$$\therefore v = v_0 e^{-kt} \qquad (A)$$

Find $x(t)$:

$$\frac{dx}{dt} = v \rightarrow \int_0^x dx = \int v\,dt = v_0\int_0^t e^{-kt}\,dt$$

$$x = \frac{-v_0}{k}e^{-kt}\Big|_0^t \quad \text{or} \quad x = \frac{v_0}{k}(1-e^{-kt}) \qquad (B)$$

Find $v(x)$:

By Eqn's (A) + (B): $x = \frac{v_0}{k}\left(1 - \frac{v}{v_0}\right) = \frac{1}{k}(v_0 - v)$

$$v = v_0 - kx$$

13.31 $a = -kv^2$; $v = v_0,\ x = x_0$ when $t = 0$

Find $v(t)$:

$$a = -kv^2 = \frac{dv}{dt} \rightarrow \int_0^t dt = -\frac{1}{k}\int_{v_0}^{v}\frac{dv}{v^2}$$

$$t = \frac{1}{kv} - \frac{1}{kv_0} \qquad \therefore v = \frac{v_0}{1+kv_0t} \qquad (A)$$

Find $x(t)$:

$$v = \frac{dx}{dt} \quad \text{or} \quad \int_0^x dx = \int_0^t v\,dt = \int_0^t \frac{v_0\,dt}{1+kv_0t}$$

$$\therefore x = v_0\int_0^t \frac{dt}{1+kv_0t} = \frac{1}{k}\ln(1+kv_0t) \qquad (B)$$

By Eq'ns (A) and (B):

$$x = \frac{1}{k}\ln\frac{v_0}{v}$$

13.32 GIVEN: $Q\ [in^3/s]$ = constant

Volume of bubble = $\frac{4}{3}\pi r^3$

FIND $v(r)$ and $a(r)$ for particle of soap film

SOLUTION: Time rate of change of Volume is:

$$Q = \frac{dV}{dt} = 4\pi r^2 \frac{dr}{dt} = 4\pi r^2 v$$

where $\frac{dr}{dt} = v$ is the velocity of a particle.

Hence, $v = \frac{Q}{4\pi r^2}$

The acceleration of a particle is given by the chain rule;

$$a = \frac{dv}{dt} = \frac{dv}{dr}\frac{dr}{dt} = v\frac{dv}{dr}$$

$$= \left(\frac{Q}{4\pi r^2}\right)\left(-\frac{2Q}{4\pi r^3}\right) \quad \text{or} \quad a = -\frac{Q^2}{8\pi^2 r^5}$$

13.33 GIVEN

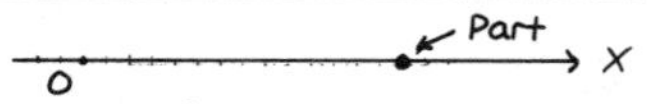

Position of part given by:

$$x = t^4 - 10t^2 + 24 \quad [mm]$$

FIND a) v and a of part

b) x as function of a

SOLUTION a.) $v(t) = \frac{dx}{dt} = 4t^3 - 20t \quad [mm/s]$

$$a(t) = \frac{d^2x}{dt^2} = 12t^2 - 20 \quad [mm/s^2]$$

b.) $t(a) = \sqrt{\frac{a+20}{12}}$ → Sub into given $x(t)$

$$\therefore x(a) = \left(\frac{a+20}{12}\right)^2 - 10\left(\frac{a+20}{12}\right) + 24$$

13.34 GIVEN Body falls 16 ft in first second and falls 32 ft more each second thereafter.

FIND Show this is correct if $g = 32\ ft/s^2$

SOLUTION

For the first second with $x_0 = v_0 = 0$ for $t_0 = 0$, the body falls a distance:

$$x_1 = 16\,ft = \tfrac{1}{2}at^2 = \tfrac{1}{2}a(1)^2 \qquad \therefore a = 32\ ft/s^2$$

At end of first second, $v_1 = at = 32(1) = 32\ ft/s$

Therefore, at the end of the next second, the body falls an additional distance:

$$x_2 = v_1 t_2 + \tfrac{1}{2}at_2^2 = (32)(1) + \tfrac{1}{2}(32)(1) = 48\ ft$$

Since $48\,ft = 32\,ft + 16\,ft$, this verifies that in the second second, the body falls 32 ft more than it fell in the first second (16 ft).

At the end of the 2nd second,

$$v_2 = v_1 + at = 32 + 32(1) = 64\ ft/s$$

So, at the end of the third second, the body falls an additional distance:

$$x_3 = v_2 t + \tfrac{1}{2}at_3^2 = (64)(1) + \tfrac{1}{2}(32)(1)^2 = 80\ ft$$

It can be seen that $80\,ft = 48 + 32$.

Thus, in the third second, the body falls 32 ft more than it fell (48 ft) in the second second, and so on.

13.35 GIVEN $a = 1\ m/s^2$, $v_{max} = 100\ km/hr$

FIND a.) least time [s] required for car to travel 1 km if starting from rest

b.) sketch the velocity-time graph

SOLUTION $v = 100\ km/hr = 27.78\ m/s \qquad v_0 = 0$

a.) The minimum time to travel 1 km is attained if the car accelerates to 100 km/hr and then maintains that speed.

The time required to attain a speed of 100 km/hr is found by: $v = at$; $\quad 27.78\ m/s = (1)t$ or $t_1 = 27.78\,s$

In this time, the car travels the distance

$$x = \tfrac{1}{2}at^2 = \tfrac{1}{2}(1)(27.78)^2 = 385.8\ m = 0.386\ km.$$

In other words, it has not yet traveled 1 km. To travel the remaining distance, the time required is found from the condition that:

$$\Delta s = (1000 - 385.8) = vt_2 = 27.78\,t_2$$

or $t_2 = 22.11\ s$

Therefore, total time is:

$$t = t_1 + t_2 = 27.78 + 22.11 = 49.89\ s$$

b.) Velocity-time graph:

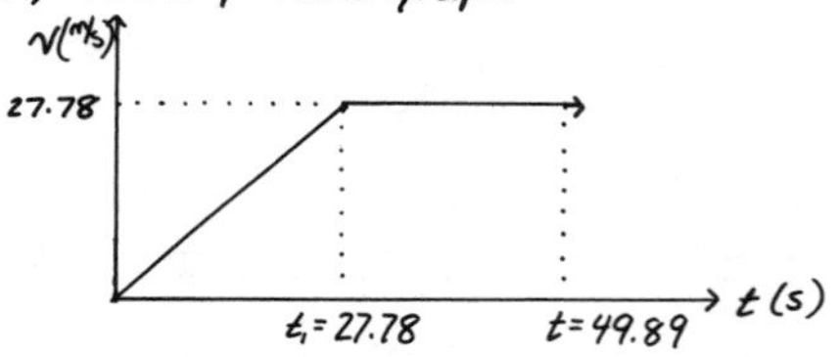

13.36 GIVEN Machine part moves along straight line, has time-velocity graph shown.

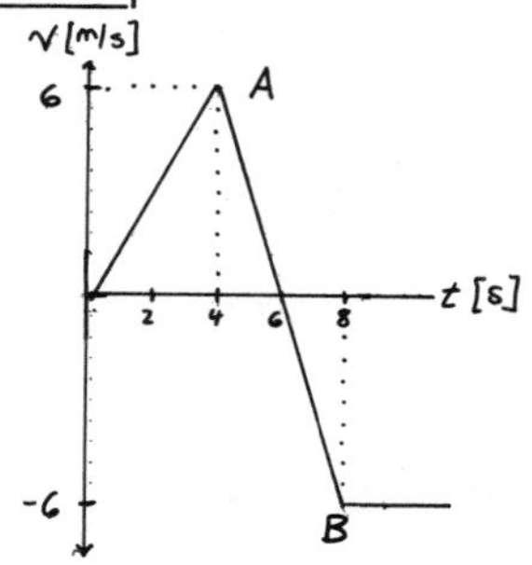

FIND a.) displacement of part in intervals $t = 0-4s$, $t = 4-8s$, and $t = 0-8s$.

b.) distance traveled by particle in same intervals

(Continued)

13.36 cont.

SOLUTION a) Displacement, $t=0$ to $4s$:

$\Delta X_{0-4} = \int_0^4 3/2\, t\, dt = 3/4\, t^2 \Big|_0^4$ $\qquad \underline{\underline{\Delta X_{0-4} = 12\text{ m}}}$

Displacement, $t=4$ to $8s$:

$\Delta X_{4-8} = \int_4^8 (-3t+18)\,dt = (-3/2\, t^2 + 18t)\Big|_4^8$

$\underline{\underline{\Delta X_{4-8} = 0\text{ m}}}$

Displacement, $t=0$ to $8s$:

$\Delta X_{0-8} = \int_0^4 3/2\, t\, dt + \int_4^8 (-3t+18)\,dt$

$\underline{\underline{\Delta X_{0-8} = 12\text{m}}}$

b) Distance traveled, $t=0$ to $4s$:

$\Delta S_{0-4} = |\Delta X_1| + |\Delta X_2| = |0| + |12|$ $\qquad \underline{\underline{\Delta S_{0-4} = 12\text{ m}}}$

Distance traveled, $t=4$ to $8s$:

$\Delta S_{4-8} = |\Delta X_1| + |\Delta X_2| = |18-12| + |12-18| = 6+6$

$\underline{\underline{\Delta S_{4-8} = 12\text{ m}}}$

Distance traveled, $t=0$ to $8s$:

$\Delta S_{0-8} = |\Delta X_1| + |\Delta X_2| + |\Delta X_3| = |12-0| + |18-12| + |12-18|$

$= 12 + 6 + 6$ $\qquad \underline{\underline{\Delta S_{0-8} = 24\text{m}}}$

*** ALTERNATIVE SOLUTION**

Displacement from $t=0$ to $t=4s$ is equal to the area of the v-t diagram. Therefore,

$\Delta X_{0-4} = \frac{1}{2}(4)(6) = \underline{\underline{12\text{ m}}}$

Displacement from $t=4$ to $t=8s$:

$\Delta X_{4-8} = \frac{1}{2}(6-4)(6) - \frac{1}{2}(8-6)(6) = \underline{\underline{0\text{ m}}}$

$\therefore$ Displacement from $t=0$ to $t=8s$ is:

$\Delta X_{0-8} = \Delta X_{0-4} + \Delta X_{4-8} = 12 + 0 = \underline{\underline{12\text{m}}}$

Distance traveled from $t=0$ to $t=4s$ is

$\Delta S_{0-4} = \frac{1}{2}(4)(6) = \underline{\underline{12\text{m}}}$

Distance traveled from $t=4$ to $t=8s$ is

$\Delta S_{4-8} = |\Delta X_{4-6}| + |\Delta X_{6-8}| = |\frac{1}{2}(6-4)(6)| + |\frac{1}{2}(8-6)(6)|$

$\therefore \Delta S_{4-8} = 6 + 6 = \underline{\underline{12\text{ m}}}$

Distance traveled from $t=0$ to $t=8s$ is

$\Delta S_{0-8} = \Delta S_{0-4} + \Delta S_{4-8}$

$= 12\text{m} + 12\text{m}$ $\qquad \therefore \underline{\underline{\Delta S_{0-8} = 24\text{m}}}$

13.37 GIVEN Figure from Problem 13.36

FIND a.) acceleration of part at $t=2s$ + $t=6s$

b.) Discuss Significance of pts A and B

SOLUTION

a.) From v-t diagram:

$$v = \begin{cases} 3/2\, t & 0 \le t < 4 \\ -3t+18 & 4 \le t < 8 \\ -6 & t > 8 \end{cases}$$

$$a = \frac{dv}{dt}$$

$$a = \begin{cases} 3/2 & 0 \le t < 4 \\ -3 & 4 \le t < 8 \\ 0 & t > 8 \end{cases}$$

At $t=2s$, $\underline{\underline{a = 1.5\text{ m/s}^2}}$

At $t=6s$, $\underline{\underline{a = -3\text{ m/s}^2}}$

b.) By plotting an a-t diagram, it can be seen that points A and B are points at which the acceleration changes from 1.5 to -3 m/s^2 and from -3 to 0 m/s^2 instantaneously.

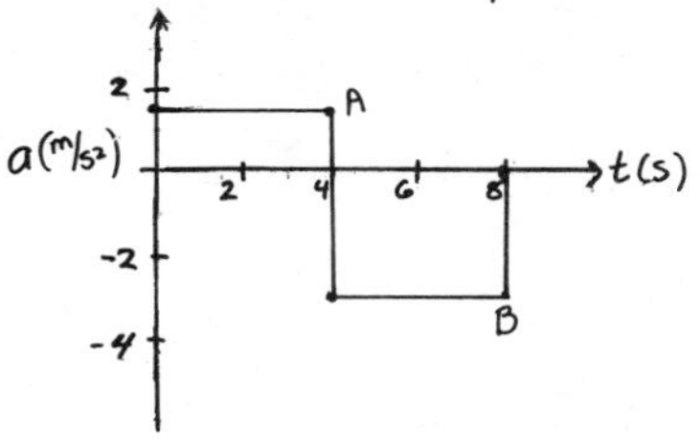

13.38 GIVEN

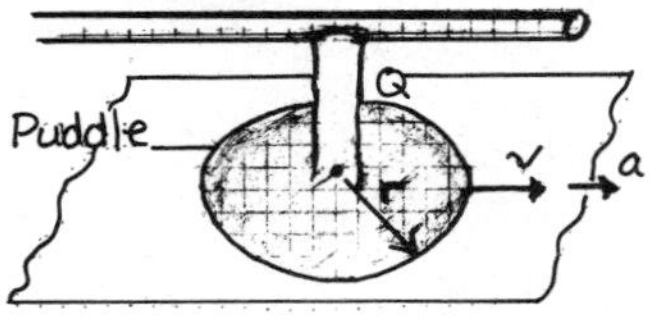

h = thickness of puddle = constant

Q = rate at which water leaks [in³/s]

FIND velocity, v and acceleration, a on edge of puddle as functions of r, Q and h.

SOLUTION

Volume of puddle $\Rightarrow V = \pi r^2 h$

By chain rule, $\frac{dV}{dt} = \frac{dV}{dr}\frac{dr}{dt} = 2\pi r h \frac{dr}{dt} = Q$

$\therefore v = \frac{dr}{dt} = \underline{\underline{\frac{Q}{2\pi r h}}}$

Again, by chain rule:

$a = \frac{dv}{dt} = \frac{dv}{dr}\frac{dr}{dt} = \left(\frac{-Q}{2\pi r^2 h}\right) v = \underline{\underline{\frac{-Q^2}{4\pi^2 h^2 r^3}}}$

13.39 GIVEN Position of body given by:
$x = 3t^2 - 6t$ $\quad x\,[m],\ t\,[s]$ (A)

FIND a.) $v(t)$
b.) Plot of time - velocity graph
c.) Body's displacement for $t=0$ to $t=2s$ and $t=0$ to $t=4s$
d.) Distance traveled for same intervals

SOLUTION

a.) $v = \frac{dx}{dt} = 6t - 6\ [m/s]$ (B)

b.) By Eq (B), the graph of v vs. t is shown in Figure a:

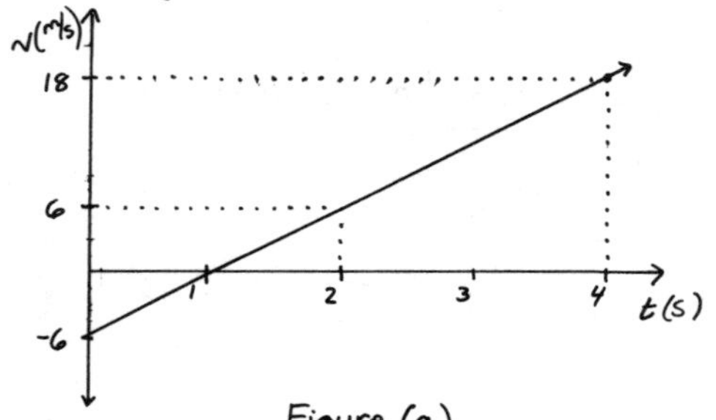

Figure (a)

c.) Displacement: $\Delta X = X_{(final)} - X_{(initial)}$

For $t=0$ to $t=2s$, the displacement is:

$\Delta X_{0-2} = X(2) - X(0)$ (C)

By Eq (A)

$X(0) = 3(0)^2 - 6(0) = 0$

$X(2) = 3(2)^2 - 6(2) = 0$

$\therefore \Delta X_{0-2} = X(2) - X(0) = 0$

For $t=0$ to $t=4s$,

$\Delta X_{0-4} = X(4) - X(0)$

$X(4) = 3(4)^2 - 6(4) = 24\ m$

$\therefore \Delta X_{0-4} = 24 - 0 = 24m$

d.) Distance traveled: $\Delta S = |\Delta X_1| + |\Delta X_2|$

For $t=0$ to $t=2s$, noting by Figure a. that $v=0$ at $t=1s$:

$\Delta S_{0-2} = |\Delta X_{0-1}| + |\Delta X_{1-2}|$

By Eq (A):

$|\Delta X_{0-1}| = |X(1) - X(0)| = |-3-0| = 3m$

$|\Delta X_{1-2}| = |X(2) - X(1)| = |0+3| = 3m$

$\therefore \Delta S_{0-2} = 3 + 3 = 6\ m$

For $t=0$ to $t=4s$;

$\Delta S_{0-4} = |\Delta X_{0-1}| + |\Delta X_{1-4}|$

By Eq (A), and since v is positive from $t=1$ to $t=4s$,

$|\Delta X_{1-4}| = |X_4 - X_1| = |24-(-3)| = 27\ m$

$\therefore \Delta S_{0-4} = 3 + 27 = 30\ m$

ALTERNATIVELY:

By Figure (a.)

$\Delta X_{0-2} = \frac{1}{2}(-6)(1) + \frac{1}{2}(6)(2-1) = 0$

$\Delta X_{0-4} = \frac{1}{2}(-6)(1) + \frac{1}{2}(18)(4-1) = 24\ m$

$\Delta S_{0-2} = |\frac{1}{2}(-6)(1)| + |\frac{1}{2}(6)(2-1)| = 6\ m$

$\Delta S_{0-4} = |\frac{1}{2}(-6)(1)| + |\frac{1}{2}(18)(4-1)| = 30\ m$

13.40 GIVEN First ball thrown upward with velocity v_0. Second ball thrown T seconds later with same velocity.

FIND Derive a formula for time t after the second ball is thrown at which the balls pass each other.

SOLUTION

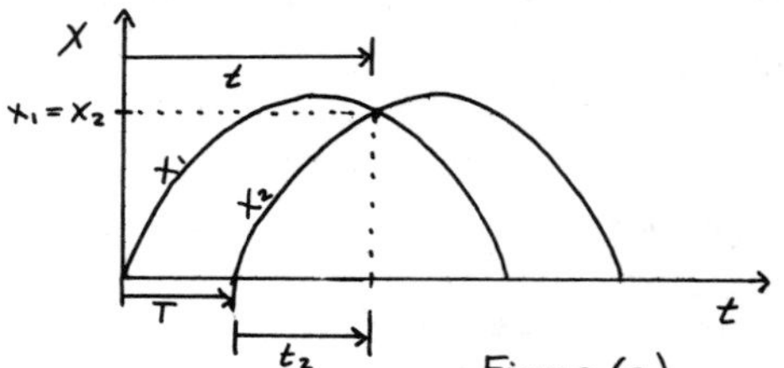

Figure (a.)

Ball one

$a = -g \qquad v_1 = \int a\,dt = -gt + v_0$

$x_1 = \int v_1\,dt = -\frac{1}{2}gt^2 + v_0 t + \cancel{x_0}^0$

$x_1 = -\frac{1}{2}gt^2 + v_0 t$ (A)

Ball two - thrown T seconds later, $\therefore t_2 = (t-T)$ (See Fig. (a.))

$a = -g \qquad v_2 = \int a\,dt = -gt_2 + v_0$

$x_2 = \int v_2\,dt = -\frac{1}{2}gt_2^2 + v_0 t_2 + \cancel{x_0}^0$

$x_2 = -\frac{1}{2}gt_2^2 + v_0 t_2 = -\frac{1}{2}g(t-T)^2 + v_0(t-T)$ (B)

Balls pass each other when $x_2 = x_1$ (See Fig (a.))

Equating (A) + (B) gives:

$-\frac{1}{2}gt^2 + v_0 t = -\frac{1}{2}g(t-T)^2 + v_0(t-T)$

Simplifying gives:

$0 = gTt - \frac{1}{2}gT^2 - v_0 T$ or $t = \frac{1}{2}T + \frac{v_0}{g}$

and $t_2 = t - T$

$\therefore t_2 = \frac{1}{2}T + \frac{v_0}{g} - T \qquad t_2 = \frac{v_0}{g} - \frac{1}{2}T$

13.41 GIVEN Time-velocity graph of racing car as shown (Fig (a))
Note 1in = 10s on time scale,
1 in = 5 ft/s on velocity scale

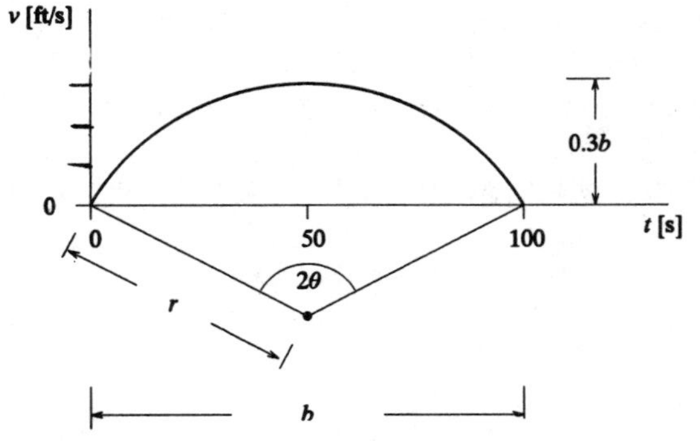

Figure (a.)

FIND Distance traveled by car from $t=0$ to $t=100$ s.

SOLUTION

$b = 100\,s\left(\frac{1in}{10s}\right) = 10\,in$

$0.3b = 0.3(10\,in) = 3\,in$

$V_{max} = 3in\left(\frac{5ft/s}{in}\right) = 15\ ft/s$

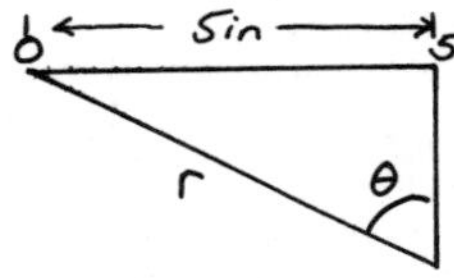

$x = r - 0.3b = (r-3)\,in$

Figure (b.)

By Fig. (b.):

$r^2 = 5^2 + x^2 = 25 + (r-3)^2$

$r = 17/3\ in$

Also by Fig. (b.):

$5 = r\sin\theta \quad \rightarrow \theta = \sin^{-1}\left(\frac{5}{17/3}\right)$

$\theta = 61.93° = 1.0808\ rad$

By Fig. (a.), the area of the circular section is: $A_{cs} = \theta r^2$

$A_{cs} = (1.0808)(17/3)^2 = 34.706\ in^2$

The area of the triangle is:

$A_{Tri} = 2\left[\tfrac{1}{2}(5in)\left(\tfrac{17}{3} - 3\right)\right] = 13.333\ in^2$

The net area between the t-axis and arc is:

$A_{Net} = A_{cs} - A_{Tri} = 34.706 - 13.333 = 21.373\ in^2$

Therefore, the distance traveled by the car is:

$S = (21.373\,in^2)\left(50\frac{ft}{in^2}\right) = \underline{\underline{1069\ ft}}$

13.42 GIVEN From the instant that the ball is dropped from rest 3m above the floor to the instant that it rebounds, 2m, the time elapsed is 1.48 s.
Also, $g = 9.81\ m/s^2$.

FIND Time [s] that ball remains in contact with the floor.

SOLUTION

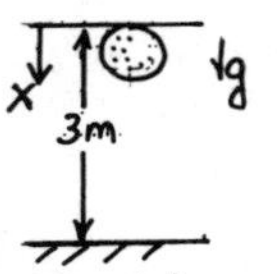

Fig (a.)

Ball is released at time $t=0$ and drops $x=3m$ (Fig (a.)).

With the relationship

$x = \tfrac{1}{2}at^2 + v_0 t + x_0 \qquad (A)$

and t_1 = time the instant before the ball touches the ground,

$3 = \frac{9.81}{2}t_1^2 + 0(t_1) + 0 \qquad t_1 = 0.782\,s$

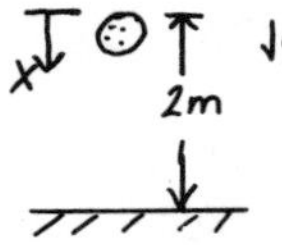

Fig. (b.)

The ball rebounds to a height 2m above the floor (Fig (b.)). Therefore, at $x=2m$ above floor, $v=0$.

The time t_2 that it takes the ball to rise 2m is the same time it takes to fall 2m, starting from rest, ($x_0 = v_0 = 0$). Therefore, by Eq. (A) and Fig (b.):

$2m = \tfrac{1}{2}g t_2^2 = \tfrac{1}{2}(9.81)t_2^2$

OR $t_2 = 0.6385\,s$

Hence, the time that the ball is in contact with the floor is:

$t = 1.48\,s - t_1 - t_2$

$= 1.48\,s - 0.7821\,s - 0.6385\,s$

$\therefore \underline{\underline{t = 0.059\,s}}$

13.43 GIVEN $a = -3t\ [m/s^2]$
at $t=0$, $x = 6m\ (=x_0)$ $v = 4\ m/s\ (=v_0)$

FIND a.) x, v, & a at $t = 2s$
b.) Displacement and distance traveled in intervals $t = 0-2s$ and $t = 0-4s$.

SOLUTION $v(t) = \int a\,dt = \int -3t\,dt = -\tfrac{3}{2}t^2 + v_0$

Since $v_0 = 4$, $\quad v(t) = -\tfrac{3}{2}t^2 + 4 \qquad (A)$

$x(t) = \int v\,dt = \int\left(-\tfrac{3}{2}t^2 + 4\right)dt = -\tfrac{1}{2}t^3 + 4t + x_0$

Since $x_0 = 6m$, $\quad x(t) = -\tfrac{1}{2}t^3 + 4t + 6 \qquad (B)$

a.) By Eqs.(B) and (A),

$x(t=2) = \underline{\underline{10m}}$

$v(t=2) = \underline{\underline{-2\ m/s}}$

$a(t=2) = \underline{\underline{-6\ m/s^2}}$

(Continued)

13.43 cont.

b.) Displacement

For $t=0$ to $t=2s$,

$\Delta X_{0-2} = x(2s) - x(0)$, where $x(0) = x_0 = 6m$

By Eq. (B): $x(2s) = -\frac{1}{2}(2)^3 + 4(2) + 6 = 10m$

$\therefore \Delta X_{0-2} = 10 - 6 = 4m$

For $t=0$ to $t=4s$,

$\Delta X_{0-4} = X(4s) - X_0$

By Eq. (B): $X(4s) = -\frac{1}{2}(4)^3 + 4(4) + 6 = -10m$

$\therefore \Delta X_{0-4} = -10 - 6 = -16\ m$

Distances

Note that, by Eq (A), v is positive from $t=0$ to $t=\sqrt{8/3}$ s $= 1.630$s. From $t=1.630$s to $t=2s$, v is negative. Thus,

$\Delta S_{0-2} = |\Delta X_{0-1.630}| + |\Delta X_{1.630-2}|$

where, By Eq (B),

$\Delta X_{0-1.630} = X(1.630\,s) - X(0) = 10.343 - 6 = 4.343$

$\Delta X_{1.630-2} = X(2s) - X(1.630) = 10 - 10.343 = -0.343$

$\therefore \Delta S_{0-2} = |4.343| + |-0.343| = 4.69\ m$

From $t=0$ to $t=4s$, v is negative from $t=1.630$ s to $t=4s$.

$\therefore \Delta S_{0-4} = |\Delta X_{0-1.630}| + |\Delta X_{1.630-4}|$

where $\Delta X_{0-1.630} = 4.343$ m as before and by Eq (B):

$\Delta X_{1.634-4} = X(4s) - X(1.630) = -10 - 10.343 = -20.343$

$\therefore \Delta S_{0-4} = |4.343| + |-20.343| = 24.69\ m$

13.44 Computer Problem

Given

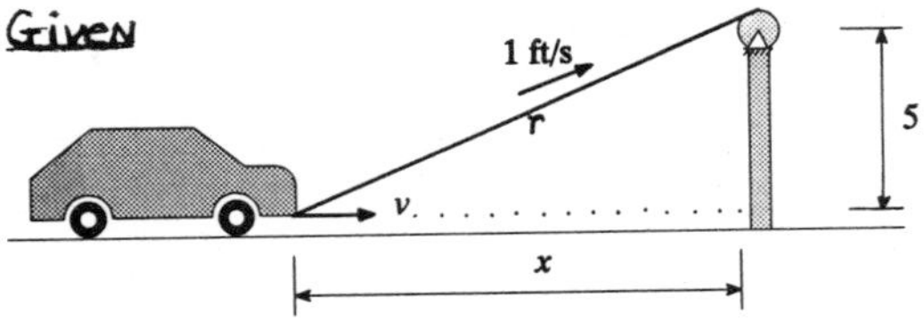

Figure (a)

Find a.) Plot speed, v [ft/s] and acceleration a [ft/s²] of car as function of x for $1 \le x \le 12$ ft.

b.) Is this method of pulling the car a good one? Explain.

Solution

By Fig (a), $r^2 = x^2 + 5^2$

Therefore: $2r\dot{r} = 2x\dot{x}$, OR $rv_R = xv$, ($v_R = 1$ ft/s)

So, $v = \frac{r}{x} = \sqrt{1 + \frac{25}{x^2}}$ [ft/s] (A)

Note that as $x \to 0$, $v \to \infty$ (see plot).

Differentiation of Eq (A) yields the acceleration. By chain rule:

$a = \frac{dv}{dx}\frac{dx}{dt} = v\frac{dv}{dx} = -\frac{25}{x^3}$ [ft/s²] (B)

Again, note that as $x \to 0$, $a \to \infty$ (see plot).

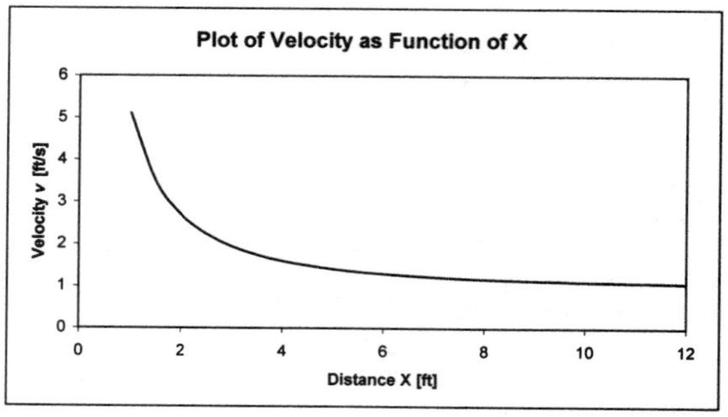

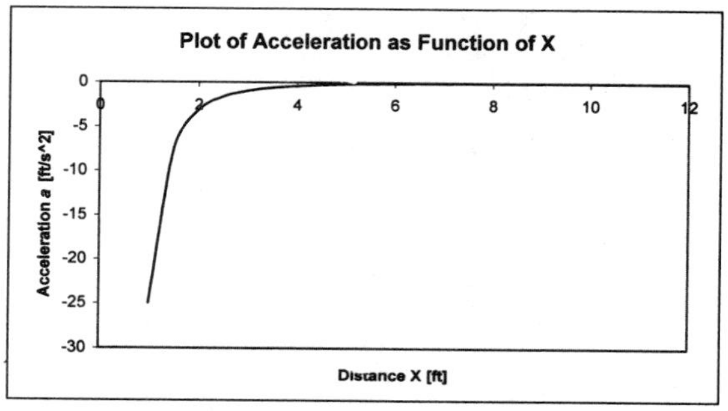

b.) This method is not a good one, since velocity and acceleration $\to \infty$ as $x \to 0$. $\therefore$ The car runs the risk of running into the pulley pole at such a high speed.

13.45 Computer Problem

Given Jet propelled boat moves in straight line such that position is $x = t^3 + 6t^2 + 5$ [ft] where t denotes time in seconds.

Find a.) Plot curves for position, velocity, and acceleration as functions of t for $0 \le t \le 4s$.

b.) V_{AV} + a_{AV} for $0 \le t \le 4s$ and plot on graphs of part a.

c.) Est. time for which $v = v_{AV}$; $a = a_{AV}$.

d.) Verify by calculation.

(Continued)

13.45 cont.

SOLUTION $x = t^3 + 6t^2 + 5$ [ft]

$\therefore v(t) = 3t^2 + 12t$ [ft/s]

$a(t) = 6t + 12$ [ft/s²]

a.) Plotted Together:

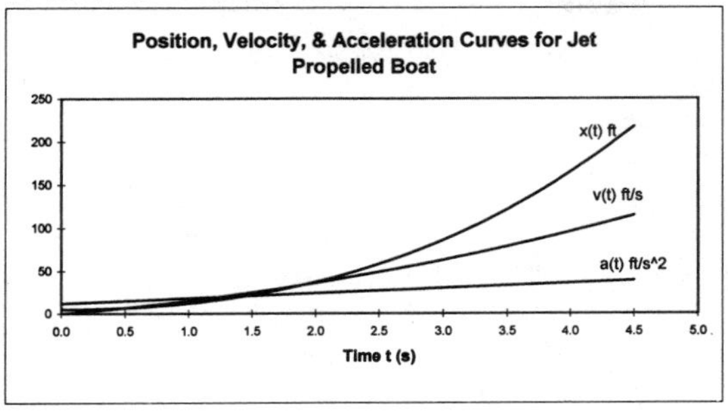

Plotted Seperately:

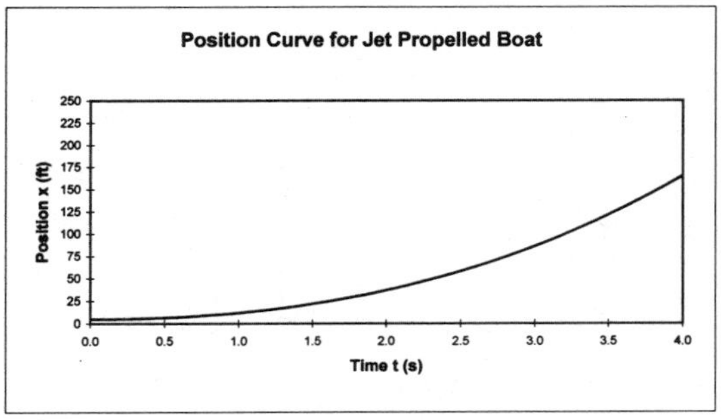

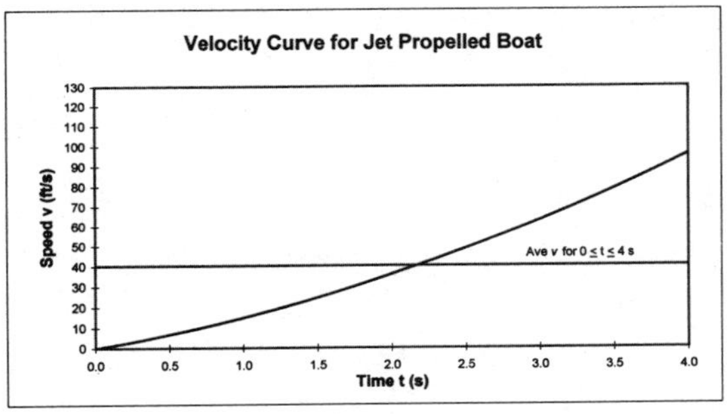

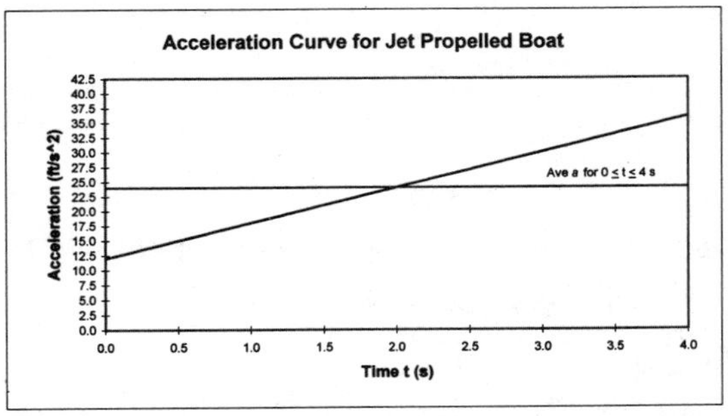

b.) Average speed and acceleration:

$v_{Ave} = \frac{1}{4}\int_0^4 [3t^2 + 12t]\,dt = \underline{\underline{40 \text{ ft/s}}}$

→ plotted on v-t graph above

$a_{ave} = \frac{1}{4}\int_0^4 (6t + 12)\,dt = 24 \text{ ft/s}^2$

→ plotted on a-t graph above

c.) From plots obtained in part a.) and averages obtained in part b.) :

t for instantaneous velocity = ave. velocity $\underline{\underline{\approx 2.15\text{s}}}$

t for instantaneous acceleration = ave acceleration $\underline{\underline{\approx 2\text{s}}}$

d.) These results can be verified by calculation using the average values found in part b.) and solving for t.

$v_{ave} = 40 \text{ ft/s} = 3t^2 + 12t$ $\qquad \underline{\underline{t = 2.16\text{s}}}$

$a_{Ave} = 24 \text{ ft/s} = 6t + 12$ $\qquad \underline{\underline{t = 2\text{s}}}$

13.46 **GIVEN** Particle moves on straight line with velocity shown on time-velocity graph shown in Figure (a.):

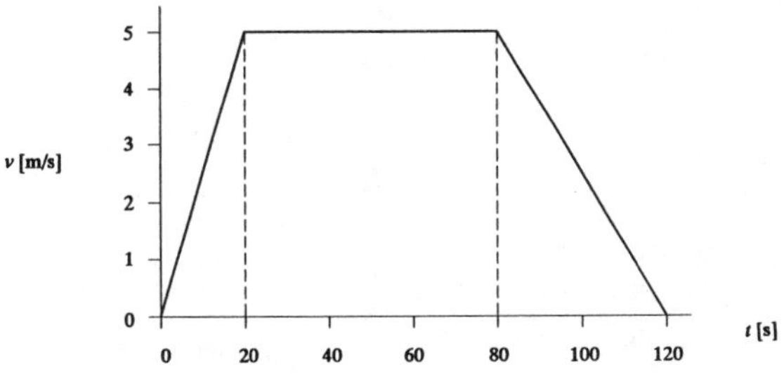

Figure (a)

FIND Plot time-acceleration and time-position graphs and from these plots, determine the acceleration and position for t = 20s, 60s, 80s, and 120s.

SOLUTION By Figure (a.)

$$v(\text{m/s}) = \begin{cases} \frac{1}{4}t, & 0 \le t \le 20\text{s} \\ 5, & 20 \le t \le 80\text{s} \\ -\frac{1}{8}t + 15, & 80 \le t \le 120\text{s} \end{cases}$$

Also by Figure (a.), the acceleration is the slope of the velocity curve (i.e. $a = dv/dt$) Therefore,

$$a(\text{m/s}^2) = \begin{cases} \frac{1}{4}, & 0 \le t \le 20\text{s} \\ 0, & 20 \le t \le 80\text{s} \\ -\frac{1}{8}, & 80 \le t \le 120\text{s} \end{cases}$$

(Continued

13.46 cont.

The displacement, x, is obtained by integration: i.e. $x = \int v\,dt$. Therefore, with Eq (A),

$$x\,(m) = \begin{cases} \frac{1}{8}t^2 & 0 \le t \le 20 \\ 5t - 50 & 20 \le t \le 80 \\ -\frac{1}{16}t^2 + 15t - 450 & 80 \le t \le 120 \end{cases} \qquad (C)$$

With Eqs (B) & (C), the time-acceleration and time-position plots are as follows:

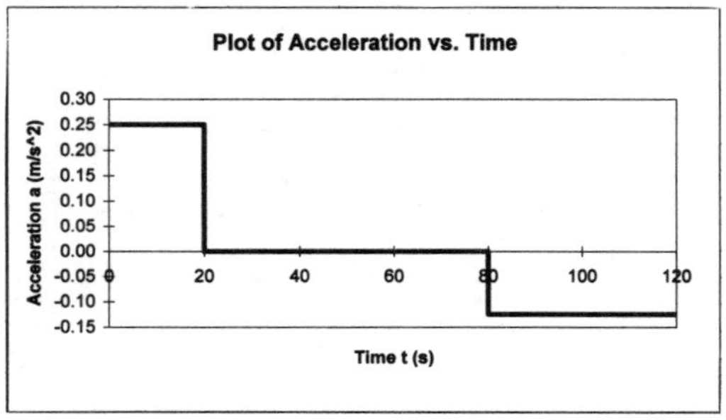

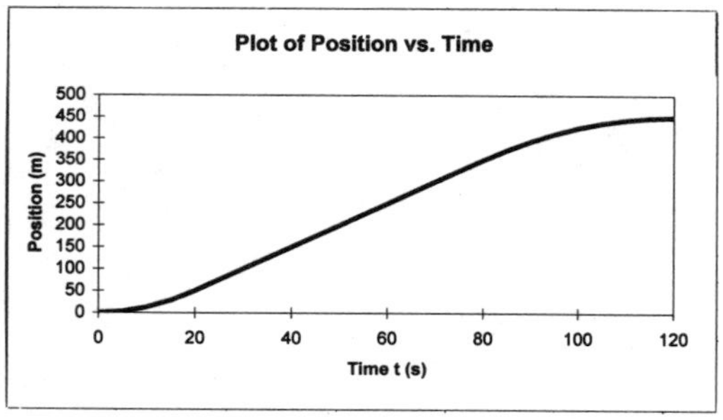

From these plots, it can be determined that:

@ $t = 20\,s$: $a = \frac{1}{4}\ m/s^2$, $x = 50\ m$

@ $t = 60\,s$: $a = 0\ m/s^2$ $x = 250\,m$

@ $t = 80\,s$: $a = -\frac{1}{8}\ m/s^2$ $x = 350\,m$

@ $t = 120\,s$: $a = -\frac{1}{8}\ m/s^2$ $x = 450\,m$

13.47 GIVEN $a = -16x$ (m/s^2) : when $x = 0$, $v = 20\ m/s$

FIND a.) x when $v = 0$

b.) Plot the geometric path in x, v plane

c.) Determine distance traveled from $x = 0$ to position when $v = -10\ m/s$ for first time.

SOLUTION

By the chain rule, $a = \frac{dv}{dt} = \frac{dv}{dx}\frac{dx}{dt} = v\frac{dv}{dx}$

or $\int_0^x a\,dx = \int_0^x (-16x)\,dx = \int_{20}^{v} v\,dv$

Integration yields:

$$-8x^2 = \frac{1}{2}(v^2 - 400) \qquad (A)$$

For $v = 0$, Eq (A) yields: $\underline{x = 5\ m}$ (B)

b.) Geometry of Path in (x, v) plane

By Eq (A): $v^2 + 16x^2 = 400$ (C)

or $\frac{v^2}{20^2} + \frac{x^2}{5^2} = 1$ (D)

Equation (D) is the equation of an ellipse (see Fig. (a.))

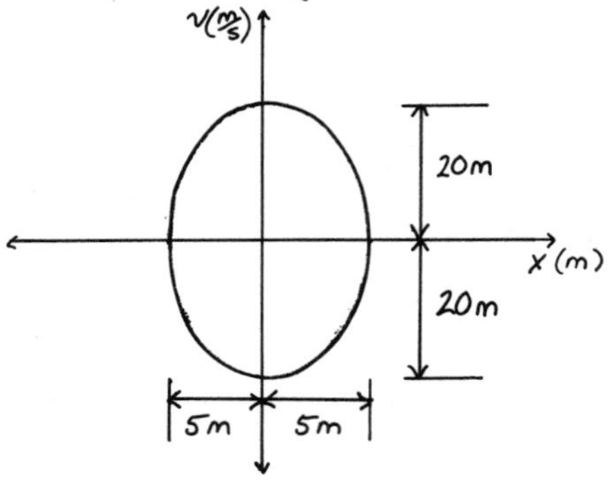

Figure (a.)

c.) The distance traveled by the particle from $x = 0$ to x at $v = -10\ m/s$ for first time:

when $x = 0$, $v = 20\ m/s$ (Given) and,

when $v = 0$, $x = 5$ (by Eq. (D))

Hence, velocity changes signs (direction) at $x = 5\ m$. By Eq. (C):

$$v = \pm\sqrt{400 - 16x^2} \qquad (E)$$

Thus, for $v = -10\ m$, Eq. (E) yields:

$-10 = -\sqrt{400 - 16x^2}$ $\underline{x = 4.33\ m}$

So, the distance traveled from $x = 0$ to position when $v = -10\ m/s$ is

$$\Delta s = |\Delta x_1| + |\Delta x_2|$$
$$= |5 - 0| + |4.33 - 5| \text{ or}$$

$\underline{\Delta s = 5.67\ m}$

13.48 GIVEN $a = \frac{x^3}{2}$; when $t = 1$, $x = 1$, $v = 0$ and $v \ge 0$.

FIND $v(x)$ and $t(x)$ and $v(t)$, $x(t)$, $a(t)$

Sketch graphs of all 5.

SOLUTION

By the chain rule, $a = \frac{dv}{dt} = \frac{dv}{dx}\frac{dx}{dt} = v\frac{dv}{dx}$

$\therefore \int_0^v v\,dv = \frac{1}{2}\int_0^x x^3\,dx$

or $\frac{v^2}{2} = \frac{x^4}{8} \Rightarrow v = \pm\frac{x^2}{2}$

Since $v \ge 0$, $\underline{v = \frac{x^2}{2}}$ (A)

(Continued)

13.48 cont.

Eq (A) expresses v as function of x.

Since $v = \frac{dx}{dt}$, by Eq (A)

$$\frac{dx}{dt} = \frac{x^2}{2} \quad \text{or} \quad 2\int_1^x \frac{dx}{x^2} = \int_1^t dt$$

$$\therefore \; t = 3 - \frac{2}{x} \qquad (B)$$

Eq (B) expresses t as a function of x,

By Eq (B), $$x = \frac{-2}{t-3} = -2(t-3)^{-1} \qquad (C)$$

Eq (C) gives x as a function of t.

Hence, v and a as functions of t are obtained by differentiation as

$$v = dx/dt = 2(t-3)^{-2} \qquad (D)$$

$$a = dv/dt = d^2x/dt^2 = -4(t-3)^{-3} \qquad (E)$$

Graphs of $x(t)$, $v(t)$, $a(t)$, $v(x)$, $a(x)$:

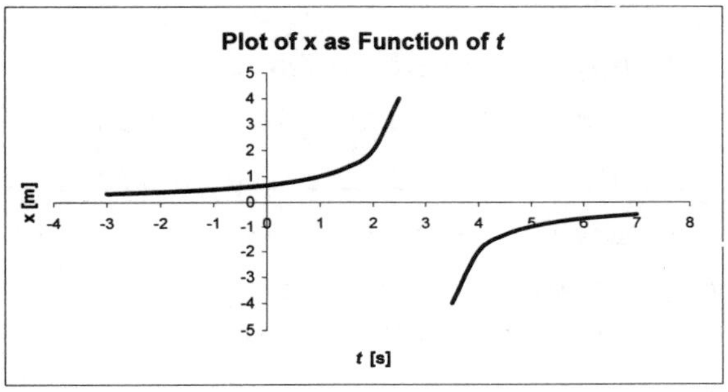

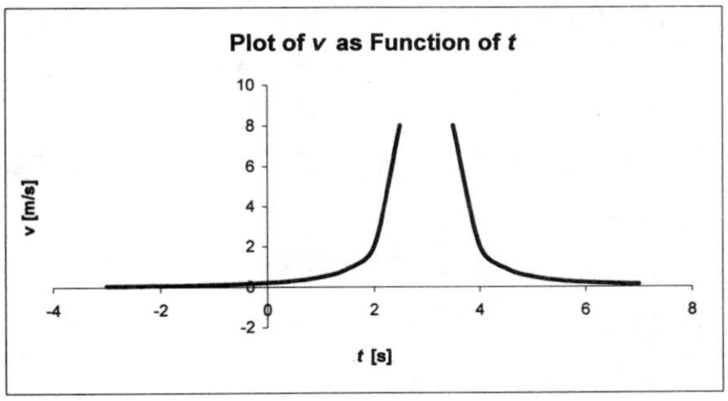

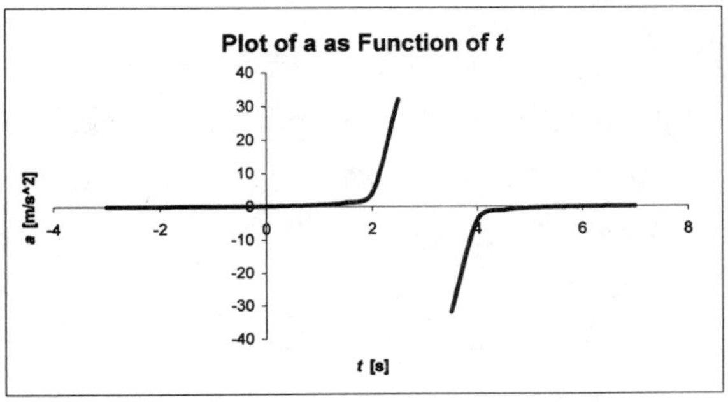

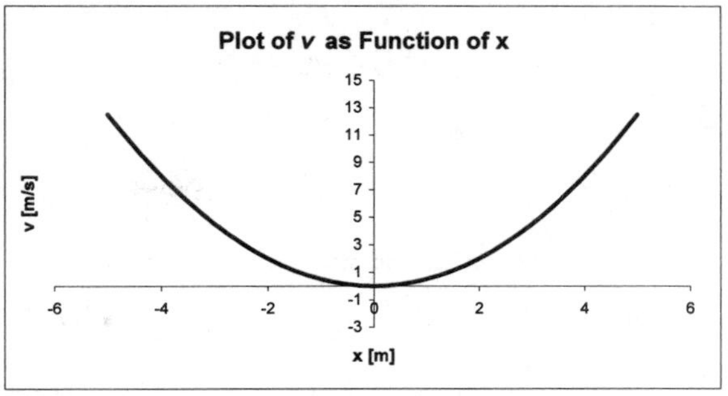

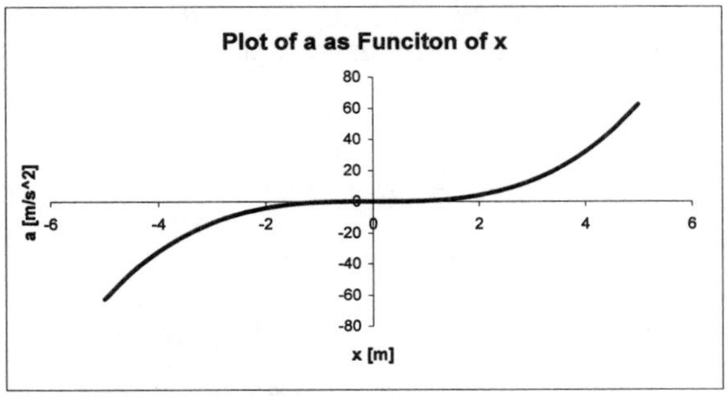

13.49

GIVEN Two cars A and B travel along straight highway. At a given instant, cars A and B have velocities 30 ft/s and 10 ft/s and accelerations 1 ft/s² and 2 ft/s² respectively. At this instant, car B is 40 ft ahead of car A.

FIND a.) Determine the times at which the cars pass one another.

b.) Explain the results in words.

SOLUTION

Given information can be denoted as follows:

at $t=0$:	Car A	Car B
	$a_A = 1$ ft/s²	$a_B = 2$ ft/s²
	$v_{A_0} = 30$ ft/s	$v_{B_0} = 10$ ft/s
	$x_{A_0} = 0$ ft	$x_{B_0} = 40$ ft

Finding velocities and positions of each car as a function of time:

Car A:

$a_A = 1$ ft/s² $\rightarrow$ $v_A = \int a_A \, dt = t + v_{A0}$

$v_A = t + 30$ [ft/s]

$x_A = \int v_A \, dt + x_{A0}$ (with x_{A0} crossed out and replaced by 0)

$$x_A = \tfrac{1}{2} t^2 + 30t \qquad (A)$$

Car B

$a_B = 2$ ft/s² $\rightarrow$ $v_B = \int a_B \, dt + v_{B0}$

$v_B = 2t + 10$

$x_B = \int v_B \, dt + x_0$

$$x_B = t^2 + 10t + 40 \qquad (B)$$

(Continued)

13.49 cont.

Car A passes Car B when $x_A = x_B$

Therefore, by Eqs. (A) and (B),

$\frac{1}{2}t^2 + 30t = t^2 + 10t + 40$

$t = 2.11\,s$ and $37.9\,s$

b.) Car B starts out ahead of Car A, but at a slower speed. Thus, at $t = 2.11\,s$, Car A passed Car B. However, Car B is accelerating faster, thus at $t = 37.9\,s$, Car B passes Car A.

This can be seen clearly by plotting the motion of the cars:

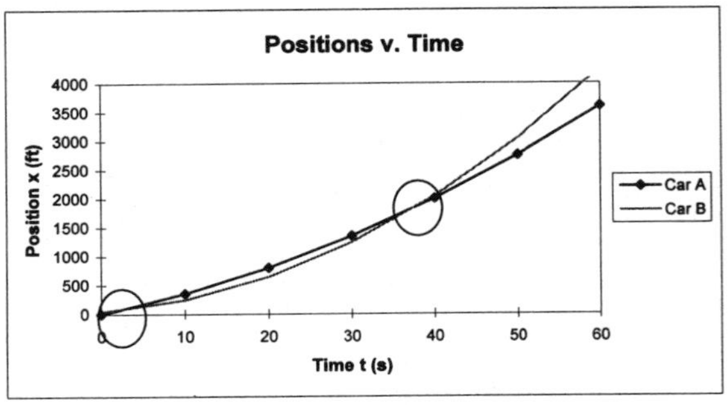

Enlarging each "instant of passing" gives:

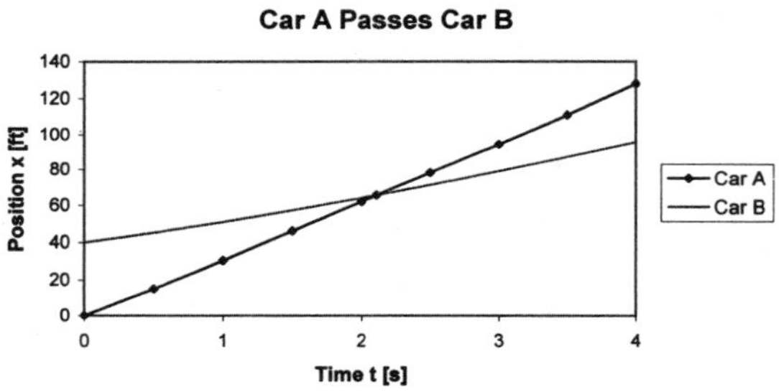

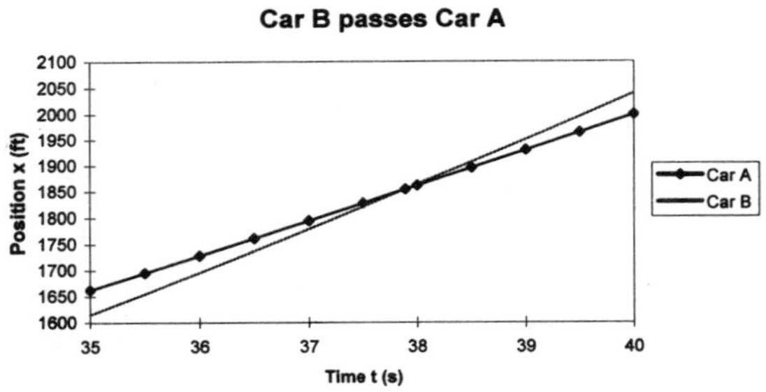

It can be seen from these plots that Car A indeed passes Car B just past $t = 2\,s$ and Car B passes Car A just before $t = 38\,s$.

13.50 GIVEN Problem 13.49

FIND Distance traveled by each of the cars from their initial positions to the times at which they pass one another.

SOLUTION

From the solution to Prob. 13.49, we have:

$x_A = \frac{1}{2}t^2 + 30t, \quad x_B = t^2 + 10t + 40$

and we know the cars pass each other at $t = 2.11\,s$ and $t = 37.9\,s$.

Therefore, the distance traveled by Car A when it passes Car B at $t = 2.11\,s$ is:

$x_A = \frac{1}{2}(2.11)^2 + 30(2.11) = 65.5\ ft$

The distance traveled by Car B at this time:

$x_B = (2.11)^2 + 10(2.11) = 25.5\ ft$

* Note, these are equal, as they should be

When Car B passes Car A at $t = 37.9\,s$, distance traveled by Car A is:

$x_A = \frac{1}{2}(37.9)^2 + 30(37.9) = 1855\ ft$

distance traveled by Car B is:

$x_B = (37.9)^2 + 10(37.9)\ \ 40 = 1815\ ft$

13.51 GIVEN Problem 13.49

FIND Repeat for the case where Car A is 60 ft ahead of Car B.

SOLUTION

a.) From the solution to 13.49, it was found that: $x_A = \frac{1}{2}t^2 + 30t + x_{A0}$

and $x_B = t^2 + 10t + x_{B0}$

In this case, x_{B0} now $= 0$, and x_{A0} now $= 60$

Therefore: $x_A = \frac{1}{2}t^2 + 30t + 60$

$x_B = t^2 + 10t$

Once again, the cars pass when $x_A = x_B$

$\frac{1}{2}t^2 + 30t + 60 = t^2 + 10$

$t = 42.8\,s$

b.) Car B passes Car A at $t = 42.8\,s$. Since Car B is accelerating faster than Car A, Car A will not catch Car B after Car B passes.

13.52 GIVEN Boy flies a kite as shown in Fig (a). The kite moves horizontally from its position away from the boy.

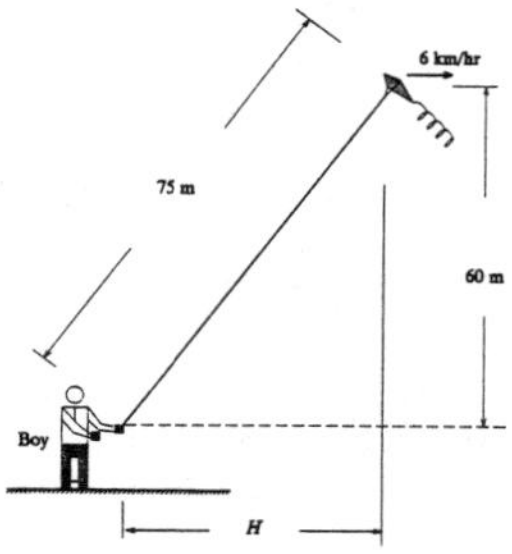

Figure (a)

FIND a.) how fast the cord is being let out at this instant

b.) how fast the cord is being let out at $t = 6\,s$.

SOLUTION

Diagram:

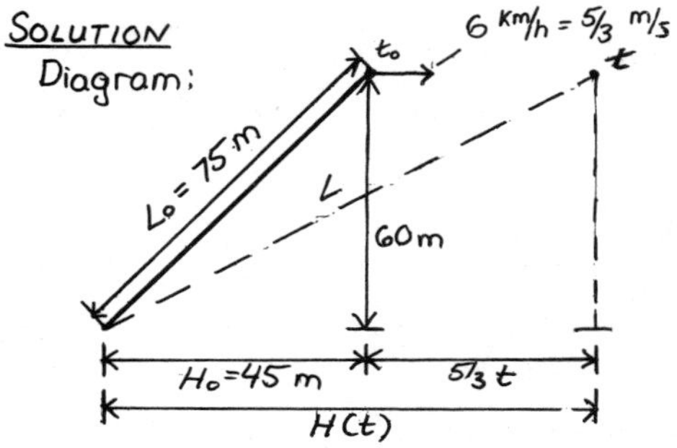

Figure (b)

Figure (b) represents the position of the kite at t_o and at some arbitrary point at time t. By inspection of Fig (a), H_o is found to be 45m (Pythagorean Thm).

$6\,Km/h = 5/3\,m/s$, Therefore, the "additional" horizontal distance moved by the kite is vt or $5/3\, t$ as shown in Fig (b).

Therefore, the total horizontal distance of the kite from the boy is

$$H(t) = 5/3\, t + 45 \quad [m] \qquad (A)$$

The length of the cord is found by triangles as well. The kite moves only horizontally therefore the height remains constant.

The length of the cord is

$$L(t) = \sqrt{60^2 + (5/3\, t + 45)^2} \qquad (B)$$

a.) The time rate of change of L is, by Eq (B),

$$\dot{L}(t) = \frac{5(135 + 5t)}{9L} \qquad (C)$$

at time $t = 0$, by Eq (B), $L = 75\,m$

Therefore, by Eq (C), $\dot{L} = 1.0\ m/s$

b.) At time $t = 6s$, by Eq (B), $L = 81.39\,m$

Therefore, $\dot{L}(6) = 1.126\ m/s$

13.53 GIVEN Cone shaped bottle. Liquid is poured in at a rate $Q = 10\ in^3/s$.

FIND Express the velocity, v and acceleration, a of the surface as functions of radius, r of the surface.

SOLUTION

Diagram of bottle:

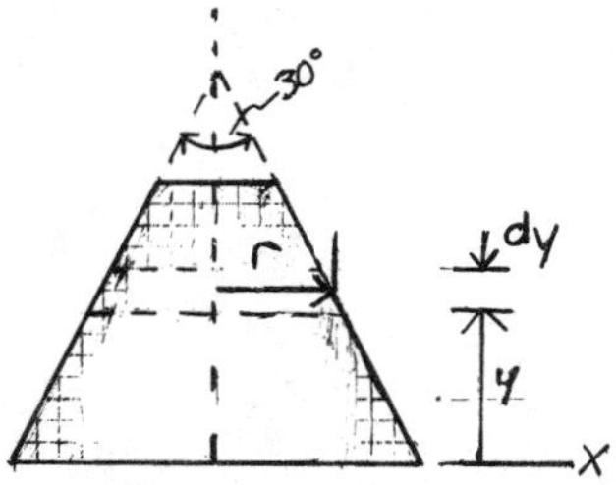

Figure (a.)

Consider a volume element of liquid: $dV = \pi r^2 dy$ in the bottle (Fig a.)

The rate at which the volume changes is:

$$\frac{dV}{dt} = \pi r^2 \frac{dy}{dt} = Q \qquad \therefore \frac{dy}{dt} = \frac{Q}{\pi r^2} = v \qquad (A)$$

is the velocity of the surface of the liquid.

Since $Q = 10\ in^3/s$, Eq (A) yields

$$v = 3.183/r^2 \qquad (B)$$

The acceleration of the surface of the liquid is, with Eq (A) and the chain rule,

$$a = \frac{dv}{dt} = \frac{dv}{dr}\frac{dr}{dt} \qquad (C)$$

By Fig (b),

$$\tan 15° = -\,dr/dy$$

$$\therefore \frac{dr}{dt} = -\frac{dy}{dt} \tan 15°$$

$$= -v \tan 15° \qquad (D)$$

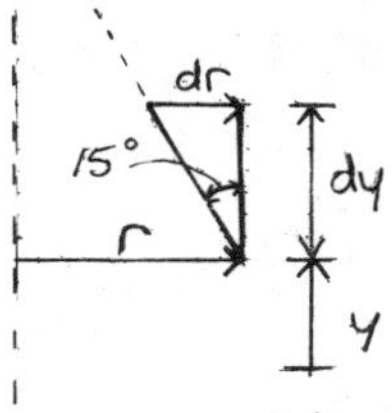

Figure (b.)

Hence, Eqs. (A), (C), and (D) yield

$$a = \frac{2Q^2}{\pi^2 r^5} \tan 15°$$

or, since $Q = 10\ in^3/s$,

$$a = 5.430/r^5$$

13.54 **GIVEN** Position-velocity (x-ν) graph for particle P is shown in Fig (a).

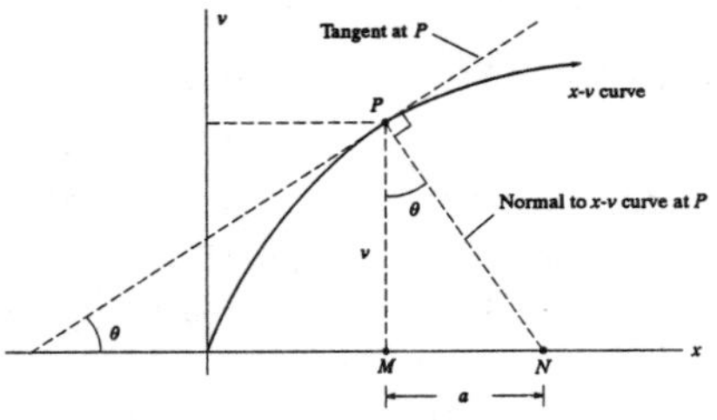

(a)

Consider the x-ν curve of a particle that is a semicircle of radius 4 in shown in Fig (b.)

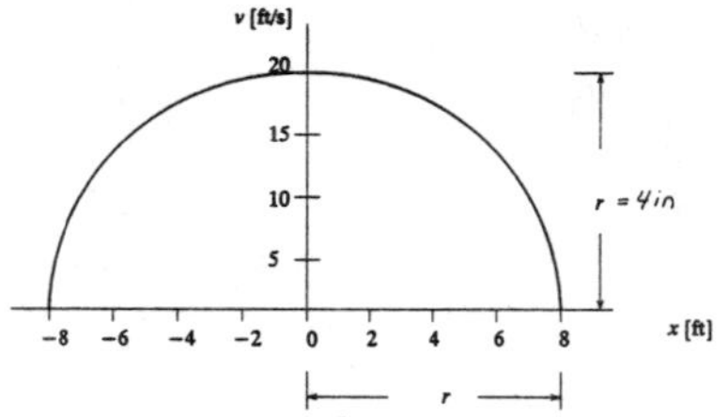

(b)

FIND Determine the acceleration at $x = 4$ ft and $x = -4$ ft.

SOLUTION

Since 1 in = p ft on the x scale,

$$p = \frac{8\text{ ft}}{4\text{ in}} = 2\text{ ft/in}$$

Also, since 1 in = q ft on the ν scale,

$$q = \frac{20}{4} = 5\text{ ft/in}$$

Therefore, 1 in, on the subnormal MN, $= \frac{q^2}{p} = 12.5\ \text{ft/s}^2$

For $x = 4$ ft, N is located at the origin to the left of M (Fig c.)

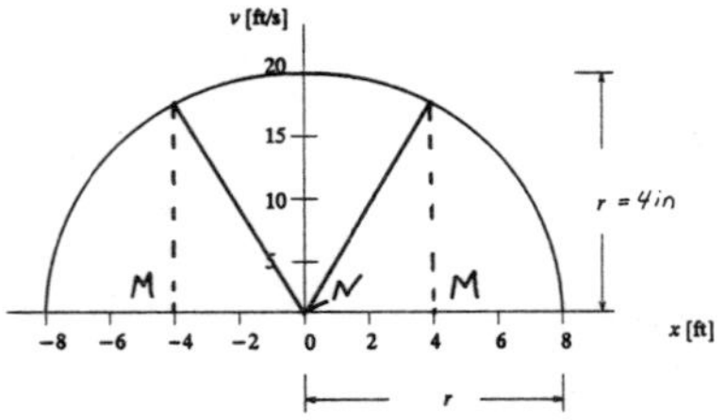

Figure (c)

Hence, MN = -2 in

Therefore, $a = (-2)(12.5) = -25\ \text{ft/s}^2$

For $x = -4$ ft, N is again located at the origin, but to the right of M. (Fig. c).

Therefore, MN = +2 in, and

$$a = (2)(12.5) = 25\ \text{ft/s}^2$$

13.55 **GIVEN** The x-v relation as shown in Fig (a), and for x scale, 1 mm = 0.1 m and for the ν scale, 1 mm = 0.025 m/s.

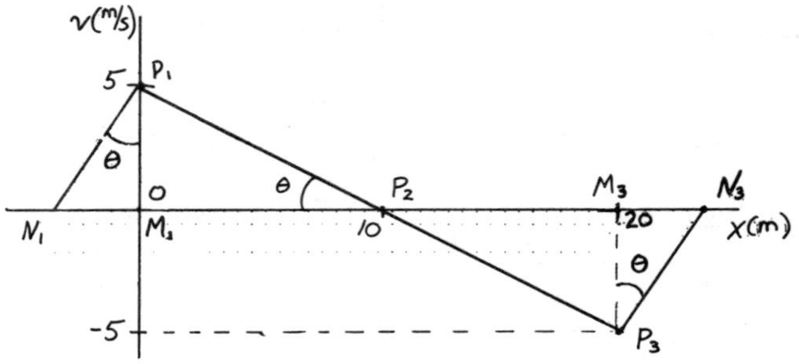

Figure (a)

FIND a.) What acceleration does 1 mm of the subnormal equal?

b.) Sketch a plot of the position-acceleration curve and from this plot, estimate the accelerations at $x = 0$, 10, and 20 m.

HINT: $a = \nu\, d\nu/dx = MN$

SOLUTION

a.) $1\text{ mm} = \frac{1}{10}\text{ m} = p$ for the x scale

$1\text{ mm} = \frac{1}{40}\text{ m/s} = q$ for the ν scale

$$1\text{ mm} = \frac{q^2}{p}\left(\frac{m}{s^2}\right) = \frac{1}{160} = 0.00625\ \text{m/s}^2$$

∴ a on the subnormal = 0.00625 m/s^2

b.) To determine the position-acceleration relationship, we must determine the subnormal for several points on the ν-x graph (Fig a.).

Consider points P_1, P_2, and P_3 shown at $x = 0$, $x = 10$, and $x = 20$ respectively.

Since 5 m/s on the ν scale = 200 mm and 10 m on the x scale = 100 mm,

$$\tan\theta = \frac{200}{100} = 2$$

(Continued)

13.55 cont.

Therefore, for point P_1 (Fig a.), Since N_1 is to the left of M_1,

$$M_1N_1 = -200 \tan\theta = -400 \text{ mm}$$

and $a_1 = -400(0.00625) = -2.5 \text{ m/s}^2$

For point P_2 ($x = 10$), $M_2 = N_2 = 0$

Therefore $a_2 = 0$

For point P_3 ($x = 20$), since N_3 is to the right of M_3, $M_3N_3 = +200 \tan\theta = 400 \text{ mm}$

and $a_3 = (400)(0.00625) = 2.5 \text{ m/s}^2$

Also, since v varies linearly with respect to x, MN (and thus **a**) varies linearly with respect to x. This is shown in the following plot:

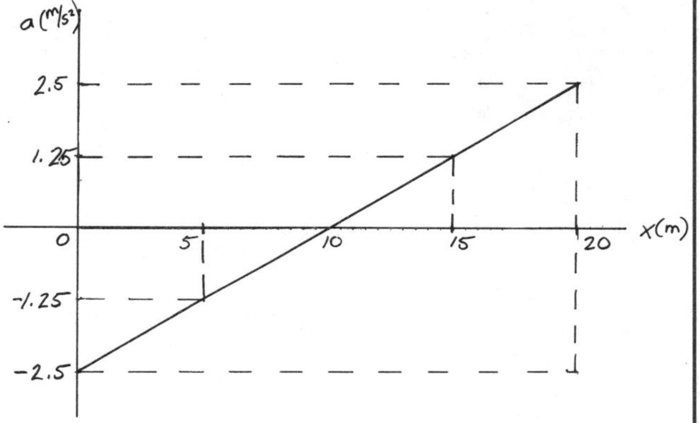

13.56 GIVEN Car passes policeman going 90 mi/h. Policeman chases car, accelerating at constant rate until he reaches 100 mph. Hereafter, the speed of both cars remains constant. Policeman catches car in 6.875 miles.

FIND The policeman's acceleration

SOLUTION

Diagram of motion of each car:

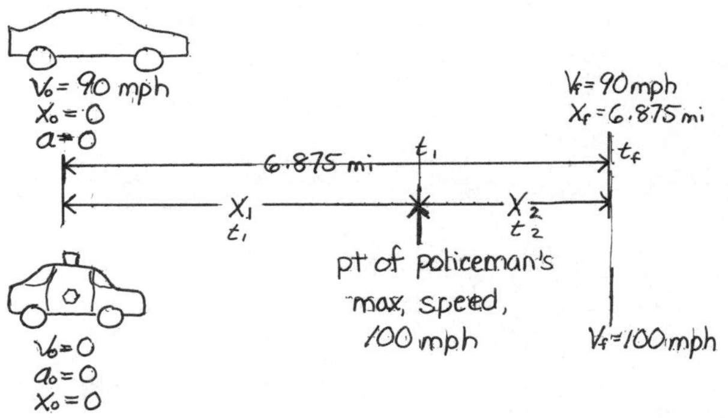

Motion in x given by $x = \frac{1}{2}at^2 + v_0t + x_0$

$\therefore 6.875 \text{ mi} = \frac{1}{2}(0)(t_f)^2 + 90t_f + 0$

$t_f = 0.076389 \text{ h}$

This is the total time it will take the officer to catch the car.

For the Policeman:

$v = at + v_0$ (A)

$100 \text{ mph} = at_1 + 0 \rightarrow t_1 = \frac{100}{a} \text{ [h]}$

Also: $x = \frac{1}{2}at^2 + \cancel{v_0}^{0}t + \cancel{x_0}^{0}$ (B)

$x = \frac{1}{2}at^2$ [mi]

And: $v^2 - v_0^2 = 2a(x - x_0)$ (C)

$100^2 = 2a(x) \rightarrow x = \frac{5000}{a}$ [mi]

From the diagram:

$x = x_1 + x_2$; $x_1 = \frac{1}{2}at_1^2$, $x_2 = v_1t_2$

$v_1 = at_1 = 100 \text{ mph}$

$t_f = t_1 + t_2$ or $t_2 = t_f - t_1 = 0.076389 - t_1$

Therefore:

$x = 6.875 \text{ mi} = \frac{1}{2}at_1^2 + 100(0.076389 - t_1)$

$.5at_1^2 - 100t_1 + 0.7639 = 0$ and $t_1 = \frac{100}{a}$

$.5a\left(\frac{100}{a}\right)^2 - 100\left(\frac{100}{a}\right) + 0.7639 = 0$

$-\frac{5000}{a} = -0.7639$

$$a = 6545.36 \text{ mi/h}^2 = 2.667 \text{ ft/s}^2$$

13.57 GIVEN Hiker drops stone over a cliff and hears it strike the ground 6 s later.

FIND The height of the cliff in meters

SOLUTION

Assume speed of sound is 340 m/s

Diagram

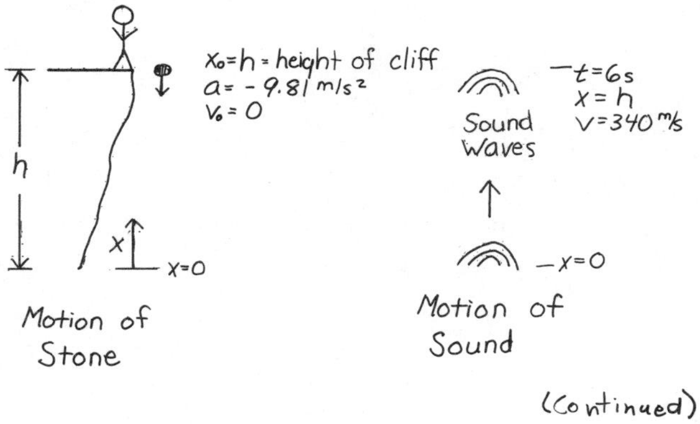

(Continued)

13.57 cont.

When the hiker hears the stone:

$$t_{stone} + t_{sound} = 6s \qquad (A)$$

When stone strikes the ground,

$$x = \frac{-9.81}{2}(t_{stone})^2 + 0(t_{stone}) + h = 0$$

$$t_{stone} = \sqrt{\frac{h}{4.905}} = \text{time for stone to drop} \qquad (B)$$

The speed of the sound wave (due to the stone hitting the ground) is:

$$v = 340\,m = \frac{h}{t_{sound}} \qquad (\text{from } vt = x)$$

where t_{sound} is the time for the sound wave to reach the top of the cliff.

$$\therefore t_{sound} = h/340 \qquad (C)$$

By Eqs (A), (B), and (C),

$$6s = \sqrt{\frac{h}{4.905}} + \frac{h}{340} \rightarrow \underline{\underline{h = 151.35\ m}}$$

13.58 GIVEN Steel ball is dropped from the roof of a building. Passes a window whose vertical height is 3 m. It takes 0.30s to travel from the top of the window to the bottom

FIND Determine the distance from the point at which the ball is dropped to the top of the window.

SOLUTION

We first determine the velocity of the ball at the top of the window:

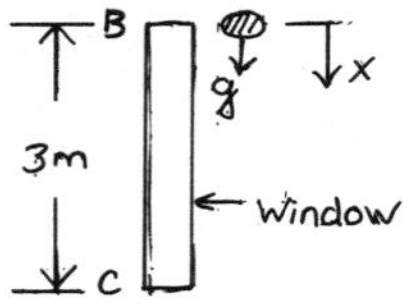

Figure (a)

At top of window, let $x_B = 0$, $t = 0$, $v_0 = v_B$

At bottom of window, $x_c = 3m$, $t = 0.3\,s$

By Fig (a),

$$x_c = \tfrac{1}{2}at^2 + v_B t + \cancel{x_B}^{\,0}$$

When $t = 0.3s$:

$$x_c = \tfrac{1}{2}(9.81)(0.3)^2 + v_B(0.3) + 0 = \mathbf{3m}$$

$$\therefore v_B = 8.5285\ m/s$$

We now find x_B by relating values at the roof to those at point B.

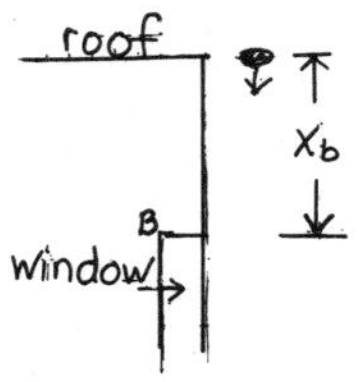

— $x_r = 0$, $v_r = 0$

— x_b = distance from roof to top of window

v_B

Figure (b.)

$$v_B^2 - v_r^2 = 2a(x_B - x_r)$$

$$(8.5285)^2 = 2(9.81)(x_B) \rightarrow \underline{\underline{x_B = 3.71\,m}}$$

ALTERNATIVE SOLUTION

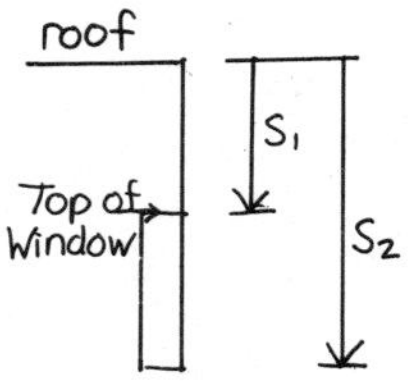

Find S_1

$$S_2 - S_1 = 3m \qquad (A)$$

$$S_1 = \tfrac{1}{2} g t_1^2 \qquad (B)$$

$$S_2 = \tfrac{1}{2} g (t_1 + 0.30)^2 \qquad (C)$$

Subbing (B) and (C) into (A):

$$\tfrac{1}{2}(9.81)(t_1 + 0.3)^2 - \tfrac{1}{2}(9.81)(t_1)^2 = 3m$$

$$t_1 = 0.869s$$

By Eq.(B): $S_1 = \tfrac{1}{2}(9.81)(0.869)^2$

$$\underline{\underline{S_1 = 3.71\ m}}$$

13.59 GIVEN Record 200m race at 19.32s

FIND a.) Speed as Michael crossed finish line

b.) If reaction time of 0.10s, and accelerated at a constant rate for the first 10 m, after which his speed was constant, determine rate of acceleration, his speed at finish, and average speed for the race.

SOLUTION

a.) Since Michael accelerated at a constant rate for the entire 200m, starting from rest, and covered the distance in 19.32s,

$$x = 200m = \tfrac{1}{2}at^2 = \tfrac{1}{2}a(19.32)^2$$

or $\underline{a = 1.072\ m/s^2}$

$$v(19.32s) = 1.072\,(19.32) = 20.70\ m/s$$

$$\underline{\underline{v = 20.70\ m/s = 46.3\ mph}}$$

(continued)

13.59 cont

b.) Since Michael accelerated at a constant rate for only the first 10 m, he covered this distance in t_1 seconds given by:

$x_1 = \frac{1}{2} a t_1^2 = 10\,m \qquad \therefore t_1 = \sqrt{\frac{20}{a}}$ [s] (A)

His speed at the end of t_1 seconds was:

$v_1 = a t_1 = \sqrt{20a}$ (B)

He covered the remaining distance (190 m) running at this speed v_1 in time t_2.

$\therefore 190\,m = v_1 t_2 = \sqrt{20a}\, t_2$ (C)

where $t_1 + t_2 + 0.10 = 19.32$ s
(0.10 = reaction time)

$\therefore t_2 = 19.22 - t_1$ (D)

Hence, by Eqs. (C) and (D),

$\sqrt{20a}\,(19.22 - t_1) = 190$ (E)

Eqs. (A) and (E) are 2 equations and two unknowns. By Eq (E),

$t_1 = 19.22 - \frac{190}{\sqrt{20a}}$ (F)

Then by Eqs (A) and (F),

$19.22 - \frac{190}{\sqrt{20a}} = \sqrt{\frac{20}{a}}$

Multiplying by $\sqrt{20a}$ and simplifying gives:

$\sqrt{20a} = 10.926 \rightarrow \underline{\underline{a = 5.97\ m/s^2}}$ (G)

Then, by Eqs. (B) and (G), the speed at the finish line is; since Michael ran at a constant speed after 10 m:

$v_1 = \sqrt{20a} = \sqrt{20(5.97)} \qquad \underline{\underline{v_1 = 10.93\ m/s}}$

Michael's average speed, v_{Ave}, from the time the starter's gun fired to the finish line is:

$v_{Ave} = \frac{200}{19.32} = 10.35\ m/s \approx \underline{\underline{23.2\ mph}}$

→ Note also, by part a), where Michael ran at a constant acceleration, and finished with a speed of 46.31 mph,

$v_{Ave} = \frac{46.31}{2} = 23.2$ mph

13.60

Computer Problem

GIVEN Car in Problem 13.44, being pulled into repair stall as shown.

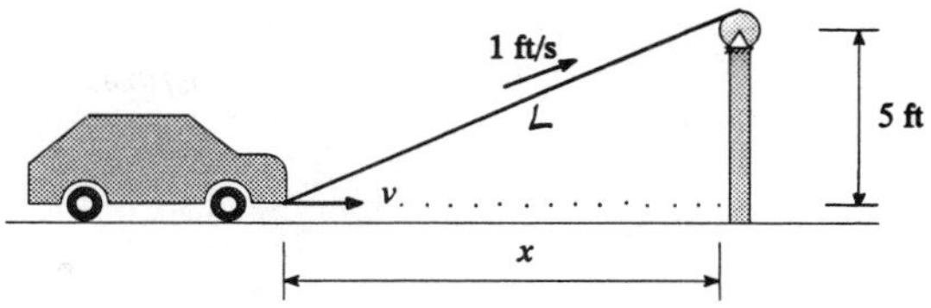

Figure (a)

FIND a.) Calculate and plot the position x, speed v, and acceleration a, of the car as a function of time t as the car moves from $x = 12$ ft to $x = 1$ ft.

b.) Is this method of pulling the car into the repair stall a good one? Explain.

SOLUTION

a.) By Fig (a), we have $L^2 = x^2 + 25$ (A)

Determine $x(t)$

Initially, at $t = 0$, $x = x_0 = 12$ ft, $L = L_0 = 13$ ft

Also (Given) $\frac{dL}{dt} = \dot{L} = -1$ ft/s

Therefore:

$\int_{13}^{L(t)} dL = \int_0^t (-1)\,dt \Rightarrow L(t) = 13 - t$ (B)

Substituting Eq (B) into Eq (A), we have

$\underline{\underline{x(t) = \sqrt{t^2 - 26t + 144}}}$ (C)

Determine $v(t)$

Differentiation of Eq (A) yields $2x\dot{x} = 2L\dot{L}$, or with Eqs (B) and (C):

$\underline{\underline{\dot{x}(t) = v(t) = \frac{L\dot{L}}{x} = \frac{t - 13}{\sqrt{t^2 - 26t + 144}}}}$ (D)

Determine $a(t)$

By Eq (D), after simplification,

$\underline{\underline{a(t) = \dot{v}(t) = \frac{-25}{(t^2 - 26t + 144)^{3/2}}}}$ (E)

Equations (C), (D), and (E) yield plots of $x(t)$, $v(t)$, and $a(t)$ as follows:

b.) Since $v \rightarrow \infty$, $a \rightarrow \infty$, as $t \rightarrow 8$ s (or $x \rightarrow 0$), the design is not good; that is the velocity and acceleration become very large as the car approaches the post (Fig. a).

(Continued)

13.60 cont

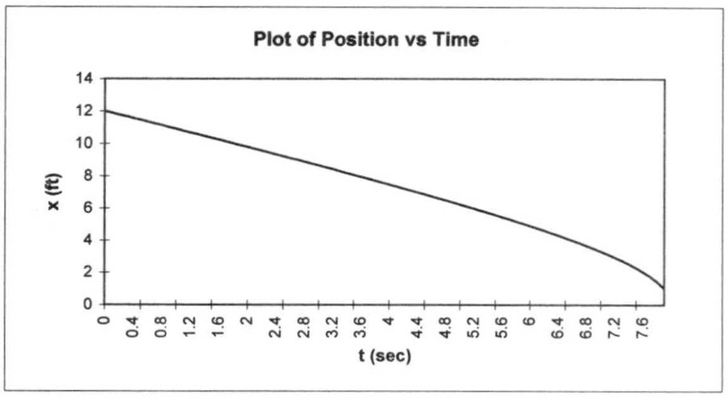

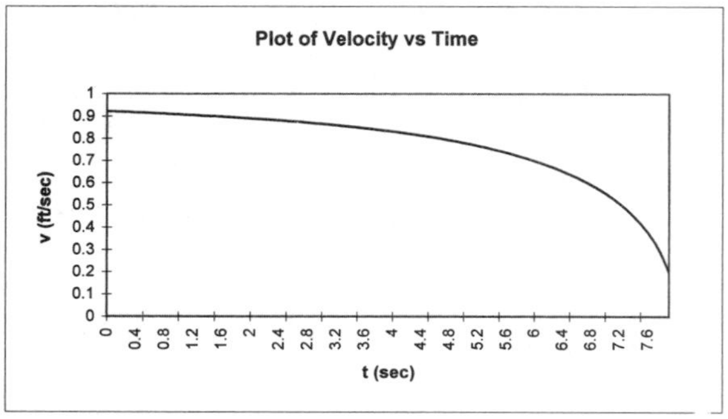

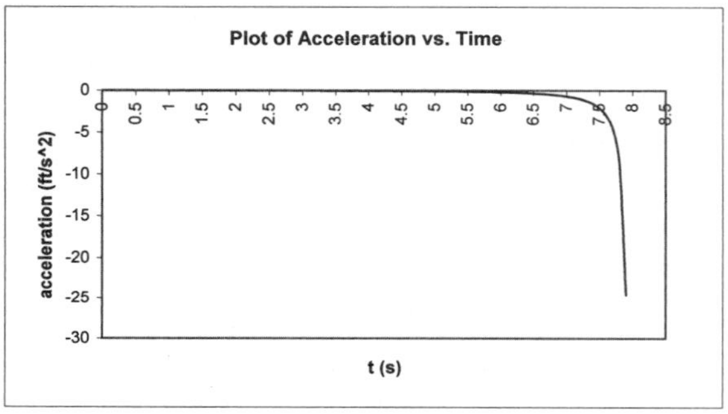

13.61 **Computer Problem**

GIVEN Bag of sand is dropped from hot-air balloon hovering with zero velocity.

Air resistance varies with square of speed v:

$a = g - kv^2$

At $t=0$, $x_0=0$, $v_0=0$

Also, $v_t = \sqrt{g/k}$ is the terminal velocity of the bag.

FIND a.) Derive formulas for v and x (the distance that the bag drops) as functions of g, v_t, and t.

b.) Plot graphs of v [m/s] and x [m] as functions of t [s] over the range $0 \le t \le 4$. Let $g = 9.81$ m/s² and $k = 0.003924$ m⁻¹. On plot of v, indicate the terminal speed v_t and on the plot of x, indicate the terminal slope of $x(t)$.

SOLUTION

a.) $$a = \frac{dv}{dt} = g - kv^2 = k(v_t^2 - v^2) \qquad (A)$$

Therefore, $$\int_0^t dt = \frac{1}{k}\int_0^v \frac{dv}{v_t^2 - v^2}$$

By integration tables or computer software,

$$t = \frac{v_t}{2g}\ln\left(\frac{v_t + v}{v_t - v}\right)$$

or $$\frac{v_t + v}{v_t - v} = e^{2gt/v_t}$$

Solving for v, we obtain,

$$v = v_t\left[\frac{e^{2gt/v_t} - 1}{e^{2gt/v_t} + 1}\right] = v_t\left[\frac{e^{gt/v_t} - e^{-gt/v_t}}{e^{gt/v_t} + e^{-gt/v_t}}\right]$$

or, alternatively,

$$v(t) = v_t \tanh\left(\frac{gt}{v_t}\right) \qquad (B)$$

Also, since $v = \frac{dx}{dt}$, we have

$$\int_0^x dx = \int_0^v v\,dt = v_t\int_0^t \tanh\left(\frac{gt}{v_t}\right)dt$$

By integration tables or computer software,

$$x(t) = \frac{v_t^2}{g}\ln\left[\cosh\left(\frac{gt}{v_t}\right)\right] \qquad (C)$$

b.) With $k = 0.003924$ m⁻¹, $g = 9.81$ m/s², the following plots of $v(t)$ and $x(t)$ are obtained:

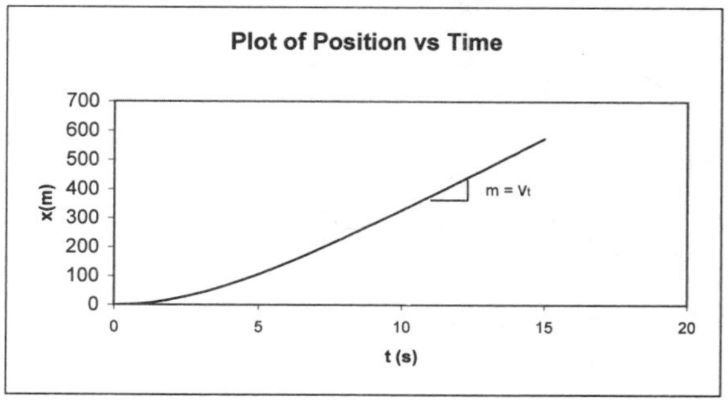

Note: $v_t = \sqrt{g/k} = \sqrt{9.81/0.003924} = 50$ m/s.

(Continued)

13.61 cont

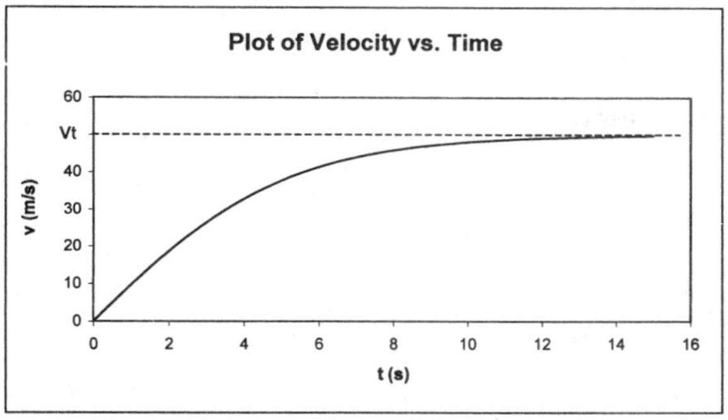

13.62 Computer Problem / Challenge

GIVEN A test car is driven on a straight track. An accelerometer measures its acceleration a. At the same time, its distance s is recorded. Results are as follows: (Table 13.62)

s [m]	0	10	20	30	40	50	60	70	80
a [m/s²]	10	20	30	25	5	-15	-10	0	10

FIND a.) Determine the speed v of the car at $s = 80$ m

b.) Determine the time t [s] that it takes the car to travel the 80 m distance.

SOLUTION

a.) Plotting the tabulated results for s and a and using computer software to find a polynomial function of the best-fit curve through the data gives:

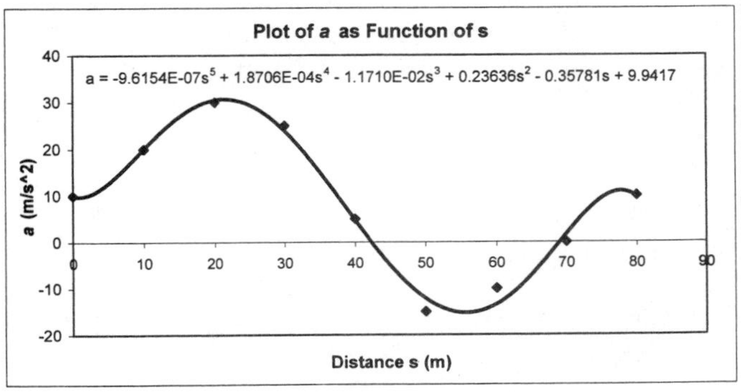

Thus, the polynomial function is found to be:

$$a(s) = -9.615\times10^{-7}s^5 + 1.8706\times10^{-4}s^4 - 1.171\times10^{-2}s^3 + 0.2363s^2 - 0.35781s + 9.9417$$

* NOTE, 4-5 Significant digits must be used in coefficients of polynomial for best accuracy (i.e. s^5 is a large number! The coefficient will have a significant impact.)

With this relationship, values of a can be obtained for any chosen s.

As noted in the problem statement,

$$a = v\frac{dv}{ds} \quad \text{or} \quad \int_{10}^{v} v\,dv = \int_0^s a\,ds$$

and

$$\int_0^{s=80} a\,ds = \tfrac{1}{2}(v^2 - 10^2) \qquad (A)$$

We use the trapezoid rule to evaluate the integral on the left.

The trapezoid rule is given by:

$$\int_a^b f(x)\,dx = \left(\frac{b-a}{n}\right)\sum_{i=1}^{n}\left[\frac{f(x)_{i-1} + f(x)_i}{2}\right]$$

Evaluating from $s=0$ to $s=80$ and using $n=20$, we obtain:

$$\int_0^{80} a\,ds = \left(\frac{80-0}{20}\right)\sum_{i=1}^{20}\left[\frac{a(s)_{i-1} + a(s)_i}{2}\right] \qquad (B)$$

To evaluate this, we divide the interval into n equal segments and find corresponding values of a using the polynomial.

s	a(x) using polynomial
0	9.942
4	11.590
8	16.945
12	23.089
16	28.012
20	30.502
24	30.024
28	26.600
32	20.694
36	13.090
40	4.777
44	-3.168
48	-9.704
52	-13.937
56	-15.244
60	-13.387
64	-8.634
68	-1.875
72	5.258
76	10.274
80	9.704

These results were plugged into Eq. (B)

The result of the Trapezoidal Rule evaluation of the integral was:

$$\int_0^{80} a\,ds = 658.916$$

Now, plugging this result into Eq. (A), we obtain:

$$658.916 = \tfrac{1}{2}(v^2 - 10^2)$$

Therefore, $\underline{v(80) = 37.65 \text{ m/s}}$

(Continued)

13.62 cont.

b.) We note that $v = ds/dt$ or $dt = (1/v)ds$. With that, integration can be used to find t as follows:

$$\int_0^{80} \frac{1}{v}\, ds = \int_0^{t} dt = t \qquad \text{(C)}$$

To evaluate the integral on the left, we use a similar method as in part a.)

Once again, we must find a relationship between v and s to obtain a function $v(s)$. This time however, we do not have a brief table of results. In order to obtain a sample of points from which to plot a curve, we use repeated applications of part a.) to find values for $v(10)$, $v(20)$...

* NOTE: To use the trapezoid Rule to find these values, you must find values for new intervals each time (e.g. 20 equal divisions for each Trapezoid Integral interval.)

With this method, the following results were obtained for $v(10) \rightarrow v(80)$:

Distance s	Trap Result	Velocity	
10	134.66	v(10)=	19.22
20	398.68	v(20)=	29.96
30	685.27	v(30)=	38.35
40	833.37	v(40)=	42.03
50	788.15	v(50)=	40.94
60	645.50	v(60)=	37.30
70	580.35	v(70)=	35.51
80	658.92	v(80)=	37.65

These results can then be plotted to again find a best-fit curve and thus a relationship between v and s. The resulting plot is as follows:

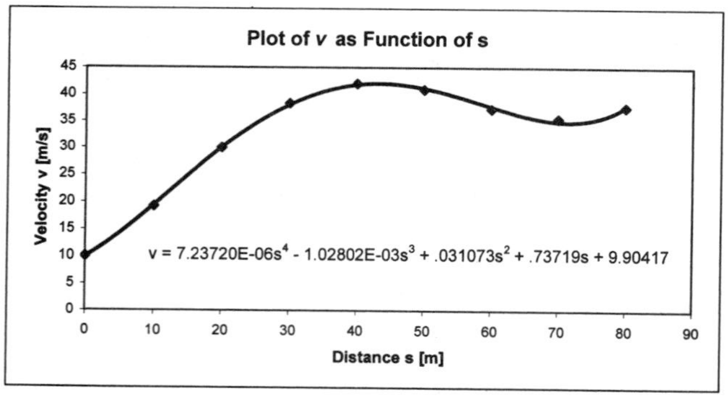

With that, $v(s)$ is found to be:

$$v(s) = 7.237\times10^{-6}s^4 - 1.028\times10^{-3}s^3 + 0.03107s^2 + 0.73719s + 9.904 \qquad \text{(D)}$$

Once again, the interval $s=0$ to $s=80$ is divided into $n=20$ segments to evaluate the integral using the Trapezoid Rule as follows:

$$\int_0^{80} \frac{1}{v(s)}\, ds = t = \left(\frac{80-0}{20}\right)\sum_{i=1}^{20} \frac{1/v(s)_{i-1} + 1/v(s)_i}{2} \qquad \text{(E)}$$

The following results were obtained for v using Eq. (D):

s	v(s) using polynomial
0	9.904
4	13.286
8	17.293
12	21.598
16	25.916
20	30.008
24	33.680
28	36.782
32	39.207
36	40.894
40	41.826
44	42.032
48	41.582
52	40.594
56	39.228
60	37.690
64	36.231
68	35.143
72	34.767
76	35.484
80	37.725

With these results, the Trapezoid Rule evaluation of the integral (Eq. E) was found to be:

$$\int_0^{80} \frac{1}{v(s)}\, ds = t = 2.73 \text{ s}$$

13.63 Computer Problem

GIVEN A particle travels along the x-axis. Its position relative to the origin O is

$$x = t^2 - 8t + 12 \quad [\text{in}] \qquad \text{(A)}$$

FIND a.) Velocity and acceleration as functions of t ($v(t)$ and $a(t)$)

b.) Determine displacement and the distance traveled in the time intervals $t=0$ to $t=3$ s and $t=0$ to $t=5$ s

c.) Determine average velocity and average speed for same intervals.

d.) Plot x and v as functions of time and verify parts b. and c.

SOLUTION

a.) By Eq (A)

$$v = \frac{dx}{dt} = 2t - 8 \quad [\text{in/s}] \qquad \text{(B)}$$

$$a = \frac{dv}{dt} = \frac{d^2x}{dt^2} = 2 \quad [\text{in/s}^2] \qquad \text{(C)}$$

b.) By Eq (A): $x = 12$ when $t=0$
$x = -3$ when $t = 3$ s
$x = -3$ when $t = 5$ s

and by Eq (B): $v = 2t - 8 = 0$ for $t = 4$ s

(Continued)

13.63 cont.

With that, a diagram of particle's movement is:

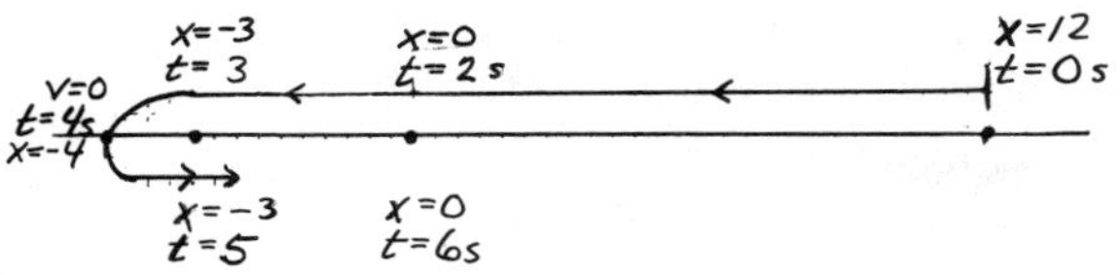

Figure (a)

Displacement

Displacement $(t=0 \text{ to } t=3s) = -12 - 3 = \underline{\underline{-15 \text{ in}}}$

Displacement $(t=0 \text{ to } t=5s) = -12 - 3 = \underline{\underline{-15 \text{ in}}}$

Distance Traveled

Distance $s(t=0 \text{ to } t=3) = 12 + 3 = \underline{\underline{15 \text{ in}}}$

Distance $s(t=0 \text{ to } t=5) = 12 + 4 + 1 = \underline{\underline{17 \text{ in}}}$

c.) Average velocity $(t=0 \text{ to } t=3)$

$$V_{Avg} = \frac{\Delta x}{\Delta t} = \frac{-15}{3} = \underline{\underline{-5 \text{ in/s}}}$$

Average speed $(t=0 \text{ to } t=3)$

$$\left|\frac{s}{\Delta t}\right| - \left|\frac{\Delta x}{\Delta t}\right| = \left|\frac{-15}{3}\right| = \underline{\underline{5 \text{ in/s}}}$$

Average velocity $(t=0 \text{ to } t=5)$

$$V_{Avg} = \frac{\Delta x}{\Delta t} = \frac{-15}{5} = \underline{\underline{-3 \text{ in/s}}}$$

Average speed $(t=0 \text{ to } t=5)$

$$\left|\frac{s}{\Delta t}\right| = \left|\frac{\Delta x}{\Delta t}\right| = \left|\frac{17}{5}\right| = \underline{\underline{3.4 \text{ in/s}}}$$

d.) Plots are as follows:

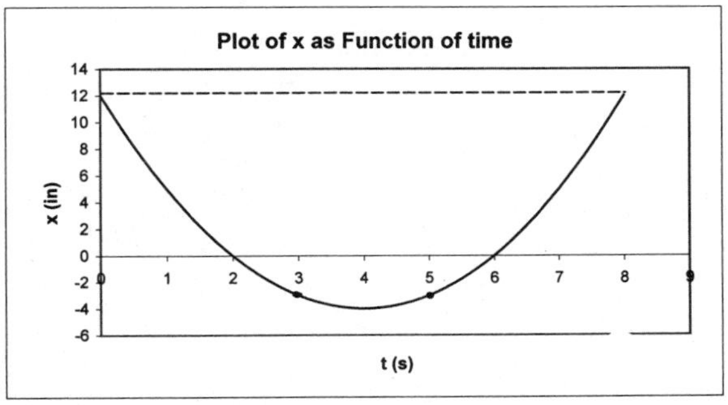

It can be seen on the plot of x vs. t that the displacement for $t=0$ to $t=3$ and $t=0$ to $t=5$ is indeed 12+3 or 15 in.

Similarly, the distances traveled are verified.

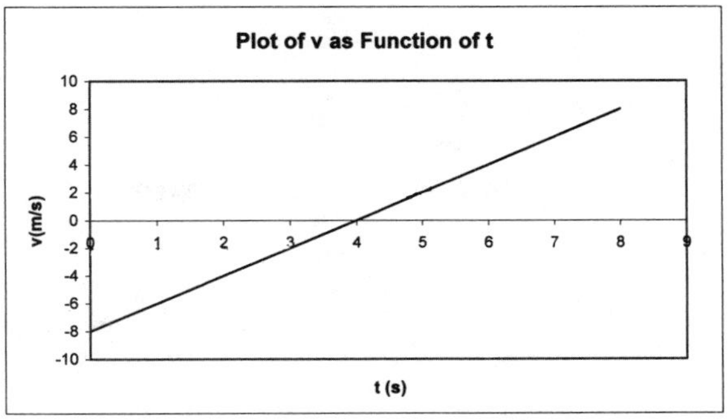

To verify the average velocities, it may be more convenient to visualize the intervals in question. They are distinguished here:

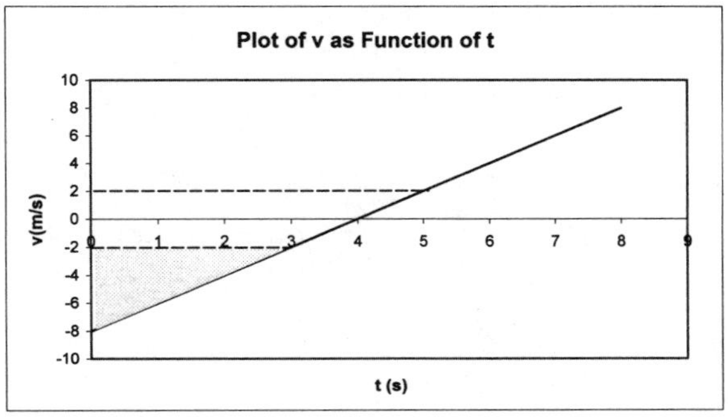

By plot of v vs t:

$$V_{AVG}(t=0 \text{ to } t=3s) = \tfrac{1}{3}\left[\tfrac{1}{2}(-8+2)(3)\right] = \underline{\underline{-3 \text{ in/s}}}$$

$$V_{AVG}(t=0 \text{ to } t=5) = \tfrac{1}{5}\left[\tfrac{1}{2}(-8)(4) + \tfrac{1}{2}(2)(5-4)\right] = \underline{\underline{-3 \text{ in/s}}}$$

avg speed $(t=0 \text{ to } t=3s)$:

$$= \left|\frac{\left[\tfrac{1}{2}(-8+2)(3)\right]}{3}\right| = \underline{\underline{3 \text{ in/s}}}$$

avg speed $(t=0 \text{ to } t=5s)$:

$$= \left|\frac{\tfrac{1}{2}(-8)(4)}{5}\right| + \left|\frac{\tfrac{1}{2}(2)(5-4)}{5}\right|$$

$$= 3.2 + 0.2 = \underline{\underline{3.4 \text{ in/s}}}$$

Thus, answers in parts b.) and c.) are verified.

13.64 GIVEN Motion of a particle is given by parametric equations

$x = t^2 + 4, \quad y = t^3 - 9t \quad [m]$

FIND Slope of the path and acceleration of the particle for $t = 2s$.

Solution

$\bar{r} = (t^2+4)\bar{i} + (t^3 - 9t)\bar{j} \quad [m]$

$\bar{v} = \frac{d\bar{r}}{dt} = (2t)\bar{i} + (3t^2 - 9)\bar{j} \quad [m/s]$

For $t = 2s$:

$\frac{d\bar{r}}{dt}(2) = 2(2)\bar{i} + (3(2)^2 - 9)\bar{j}$

$= 4\bar{i} + 3\bar{j}$

The slope, then is $3/4 = 0.75$

$\bar{a} = \frac{d\bar{v}}{dt} = 2\bar{i} + 6t\bar{j}$

$\bar{a}(2) = 2\bar{i} + 12\bar{j}$

13.65 GIVEN Position vector of a particle is given by $\bar{r} = (c_1 - c_2 t^3)\bar{i} + t^2\bar{j}$

FIND Express the velocity and acceleration vectors in terms of c_1, c_2 and t

SOLUTION

$\bar{v} = \frac{d\bar{r}}{dt} = -3c_2 t^2 \bar{i} + 2t\bar{j} \quad [ft/s]$

$\bar{a} = \frac{d\bar{v}}{dt} = -6c_2 t\,\bar{i} + 2\bar{j} \quad [ft/s^2]$

13.66 GIVEN Motion of a particle given by

$x = 4t^2 + 3t, \quad y = -t^3 + t, \quad z = t^4$

FIND a.) x, y, z projections of the velocity and acceleration for $t = 2s$

b.) Direction cosines of the tangent to the path and the speed of the particle for $t = 2s$.

SOLUTION

a.) $v_x = \frac{dx}{dt} = 8t + 3 \qquad v_x(2) = 19$

$v_y = \frac{dy}{dt} = -3t^2 + 1 \qquad v_y(2) = -11$

$v_z = \frac{dz}{dt} = 4t^3 \qquad v_z(2) = 32$

$a_x = \frac{dv_x}{dt} = 8 \qquad a_x(2) = 8$

$\bar{a}_y = \frac{dv_y}{dt} = -6t \qquad a_y(2) = -12$

$a_z = \frac{dv_z}{dt} = 12t^2 \qquad a_z(2) = 48$

b.) $\cos\theta_x = 0.490$

$\cos\theta_y = -0.283$

$\cos\theta_z = 0.825$

√Check: $\cos^2\theta_x + \cos^2\theta_y + \cos^2\theta_z = 1$

Speed:

$|v| = \sqrt{19^2 + (-11)^2 + (32)^2} = 38.8$

13.67 GIVEN Acceleration projections:

$a_x = 8t^3, \quad a_y = -6t^2 + 1, \quad a_z = 4 \quad [m/s^2]$

For $t = 0$, $v_x = 1\,\frac{m}{s}$, $v_y = 4\,\frac{m}{s}$, $v_z = 0$

$x = 0, \quad y = 0, \quad z = 5\,m$

FIND a.) Velocity projections v_x, v_y, v_z as functions of time

b.) Equations of motion

SOLUTION

a.) $v_x = \int a_x\,dt = 2t^4 + v_{0x} = 2t^4 + 1 \quad [m/s]$

$v_y = \int a_y\,dt = -2t^3 + t + v_{0y} = -2t^3 + t + 4 \quad [m/s]$

$v_z = \int a_z\,dt = 4t + v_{0z} = 4t \quad [m/s]$

b.) $x = \int v_x\,dt = \frac{2}{5}t^5 + t + x_0$

$x = \frac{2}{5}t^5 + t \quad [m]$

$y = \int v_y\,dt = -\frac{1}{2}t^4 + \frac{1}{2}t^2 + 4t + y_0$

$y = -\frac{1}{2}t^4 + \frac{1}{2}t^2 + 4t \quad [m]$

$z = \int v_z\,dt = 2t^2 + z_0$

$z = 2t^2 + 5 \quad [m]$

13.68 GIVEN Equations of motion in (x, y):

$x = r\cos\omega t$, $y = r\sin\omega t$ (A)

r and ω positive constants

FIND a.) Show path is a circle with radius r

b.) Determine time T in which the particle travels once around the circle.

c.) Show particle travels at constant speed

d.) Determine $\bar{v}$ and $\bar{a}$ for $t=0$ and show them accurately on a figure

SOLUTION

a.) By Eqns (A),

$$x^2 + y^2 = r^2(\cos^2\omega t + \sin^2\omega t) = r^2 \quad (B)$$

Equation (B) is the equation of a circle with radius r (Fig a).

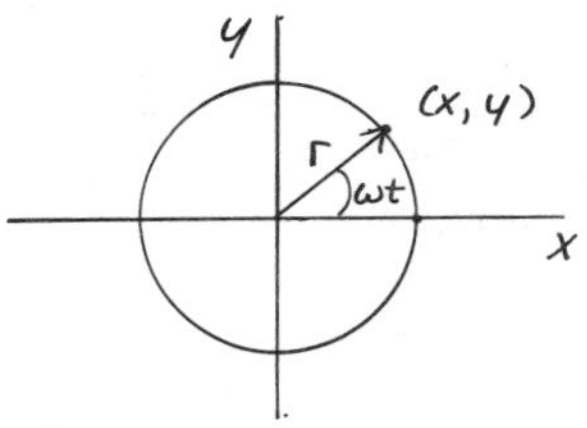

Figure (a)

b.) At time $t=0$, $x=r$, $y=0$ (Fig a.)

For the particle to travel around the circle once, again $x = r = r\cos\omega T$

Therefore, $\cos\omega T = 1$ or $\omega T = 2\pi$

Hence $T = \dfrac{2\pi}{\omega}$

c.) By Eqs. (A)

$$\dot{x} = -r\omega\sin\omega t, \quad \dot{y} = r\omega\cos\omega t \quad (C)$$

By Eqs. (C), the speed is

$$v = \sqrt{\dot{x}^2 + \dot{y}^2} = r\omega\sqrt{\sin^2\omega t + \cos^2\omega t}$$

or, $v = r\omega = \text{constant}$

d.) By Eqs. (C), the velocity vector is

$$\bar{v} = \dot{x}\bar{i} + \dot{y}\bar{j} = r\omega[-(\sin\omega t)\bar{i} + (\cos\omega t)\bar{j}] \quad (d)$$

Differentiation gives

$$\bar{a} = \ddot{x}\bar{i} + \ddot{y}\bar{j} = -r\omega^2[(\cos\omega t)\bar{i} + (\sin\omega t)\bar{j}] \quad (e)$$

For $t=0$, $\bar{v}$ and $\bar{a}$ yield: (Fig b)

$$\bar{v}(0) = r\omega\bar{j} \qquad \bar{a}(0) = -r\omega^2\bar{i}$$

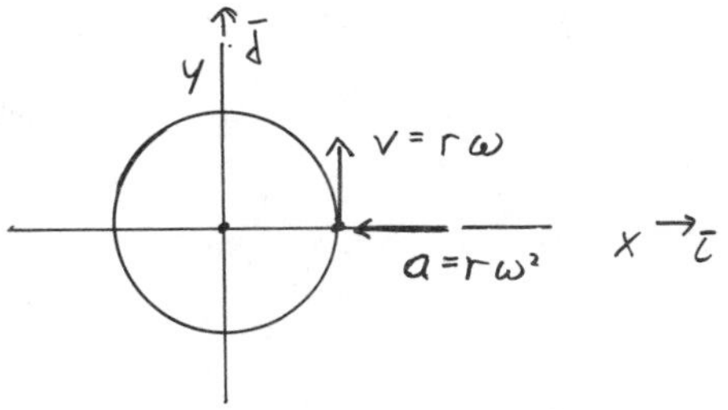

Figure (b)

13.69 GIVEN Data shown in Figure (a)

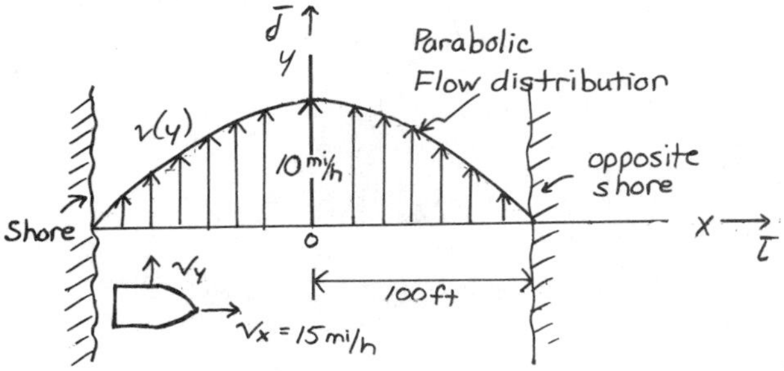

Figure (a.)

FIND Distance boat travels downstream

SOLUTION

v_y at 0 = 10 mi/h = 88/6 ft/s

By equation for a parabola,

$$v_y = \frac{88}{6}\left(\frac{100^2 - x^2}{100^2}\right)$$

or $$v_y = \frac{88}{6}\left(\frac{10{,}000 - x^2}{10{,}000}\right) \quad (A)$$

∴ The velocity vector is

$$\bar{v} = v_x\bar{i} + v_y\bar{j}$$

Also, $v_x = 15$ mi/h = 22 ft/s

$$\therefore\ x = 22t + C_1 = 22t - 100 \quad (B)$$

(since $x = -100$ ft when $t = 0$)

Hence, by Eqs. (A) and (B),

$$v_y = \frac{dy}{dt} = \frac{88}{60{,}000}(4400t - 484t^2) \quad (C)$$

(Continued)

13.69 cont'd

Integration of Eq. (C) yields

$$y = \frac{88}{60{,}000}\left(2200t^2 - \frac{484}{3}t^3\right) + C_2 \qquad (D)$$

at $t=0$, $y=0$, $\therefore C_2 = 0$

Now, by Eq. (B), for $x = 100$ ft, $t = \frac{200}{22}$ s

$\therefore$ By Eq. (D), with $t = 200/22$ s and $C_2 = 0$, the distance the boat moves downstream is: $y = \frac{800}{9} = \underline{\underline{88.9 \text{ ft}}}$

13.70 GIVEN Mechanism vibrates about the equilibrium position E.

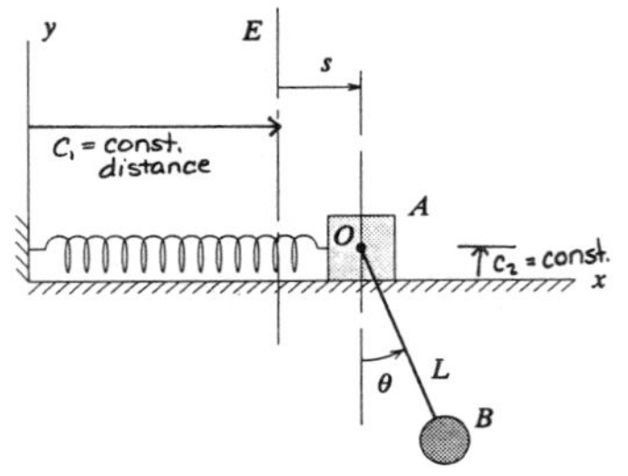

Figure (a.) - Spring - bob system

FIND $\dot{x}, \ddot{x}, \dot{y}, \ddot{y}$ in terms of $L, s, \dot{s}, \ddot{s}, \dot{\theta}, \ddot{\theta}$

SOLUTION

By Fig.(a)
$$x = C_1 + s + L\sin\theta$$
$$y = C_2 - L\cos\theta \qquad (A)$$

By Eqs.(A), differentiation yields,

$$\underline{\underline{v_x = \dot{x} = \dot{s} + L\dot{\theta}\cos\theta}}$$

$$\underline{\underline{v_y = \dot{y} = L\dot{\theta}\sin\theta}}$$

$$\underline{\underline{a_x = \ddot{x} = \ddot{s} + L\ddot{\theta}\cos\theta - L\dot{\theta}^2\sin\theta}}$$

$$\underline{\underline{a_y = \ddot{y} = L\ddot{\theta}\sin\theta + L\dot{\theta}^2\cos\theta}}$$

13.71 GIVEN Coordinates of moving particle:

$$x = t^3 - 3t^2 + 6 \qquad y = t^2 + 3$$

FIND
a.) Position, velocity, and acceleration vectors of particle at time $t = 1$ s.
b.) Average velocity and average acceleration over $t=0$ to $t=1$ s.
c.) Displacement at time $t=2$ s relative to its position at $t=0$.
d.) Speed and magnitude of acceleration at $t=2$ s.

SOLUTION

a.) $\bar{r} = x\bar{i} + y\bar{j}$
$= (t^3 - 3t^2 + 6)\bar{i} + (t^2+3)\bar{j}$

$\underline{\underline{\bar{r}(1) = 4\bar{i} + 4\bar{j} \quad [m]}}$

$\bar{v} = \frac{d\bar{r}}{dt} = (3t^2 - 6t)\bar{i} + (2t)\bar{j}$

$\underline{\underline{\bar{v}(1) = -3\bar{i} + 2\bar{j} \quad [m/s]}}$

$\bar{a} = \frac{d\bar{v}}{dt} = (6t - 6)\bar{i} + 2\bar{j}$

$\underline{\underline{a(1) = 2\bar{j} \quad [m/s^2]}}$

b.) $a_{AVG} = \frac{\Delta V}{\Delta t} \qquad \Delta t = 1$ sec

$\Delta\bar{V} = \bar{V}_{t=1} - \bar{V}_{t=0} = -3\bar{i} + 2\bar{j}$

$\therefore \underline{\underline{\bar{a}_{AVG} = -3\bar{i} + 2\bar{j} \quad [m/s^2]}}$

$V_{AVG} = \frac{\Delta\bar{r}}{\Delta t} \qquad \Delta t = 1$ sec

$\Delta\bar{r} = \bar{r}_{t=1} - \bar{r}_{t=0} = -2\bar{i} + 1\bar{j}$

$\therefore \underline{\underline{\bar{V}_{AVG} = -2\bar{i} + 1\bar{j} \quad [m/s]}}$

c.) For $t = 2$ sec: $x(2) = 2 \quad y(2) = 7$
$t = 0$ sec: $x(0) = 6 \quad y(0) = 3$

displacement $d = \sqrt{(2-6)^2 + (7-3)^2}$

$\underline{\underline{d = 5.66 \quad [m]}}$

d.) speed $|\bar{v}| = \sqrt{(V_x(2))^2 + (V_y(2))^2}$

$V_x(2) = 3(2)^2 - 6(2) = 0$

$V_y(2) = 2(2) = 4$

$\underline{\underline{|\bar{V}| = 4 \; [m/s]}}$

$|\bar{a}| = \sqrt{(6(2)-6)^2 + (2)^2}$

$\underline{\underline{|\bar{a}| = 6.32 \quad [m/s^2]}}$

13.72 GIVEN Particle in (x, y) plane located by

$r = r_o \sin\omega t, \quad \theta = \omega t$

r_o, ω are positive constants

FIND a.) Equation of path and show path is a circle

b.) Radius of circle

c.) Speed of particle and show it is const.

SOLUTION

a.) Polar coordinates give:

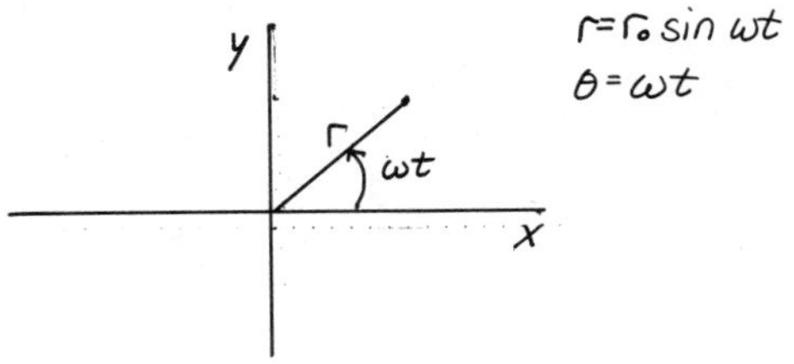

Figure (a.)

By Fig (a), $x = r\cos\omega t$ [ft]

$y = r\sin\omega t$ [ft]

$$\therefore x = r_o \sin\omega t \cos\omega t = \tfrac{1}{2} r_o \sin 2\omega t$$
$$y = r_o \sin\omega t \sin\omega t = r_o \sin^2\omega t \qquad (A)$$

and $x^2 + y^2 = r_o^2 \sin^2\omega t(\cos^2\omega t + \sin^2\omega t)$

$= r_o^2 \sin^2\omega t$

or $x^2 + y^2 - r_o^2 \sin^2\omega t + \frac{r_o^2}{4} = \frac{r_o^2}{4}$

$$\therefore x^2 + \left(y - \frac{r_o}{2}\right)^2 = \frac{r_o^2}{4} \qquad (B)$$

Hence, the path is a circle.

b.) By Eq. (B), the radius of the circle is $\sqrt{\frac{r_o^2}{4}}$ or $\frac{r_o}{2}$

The circle is shown in Figure (b.)

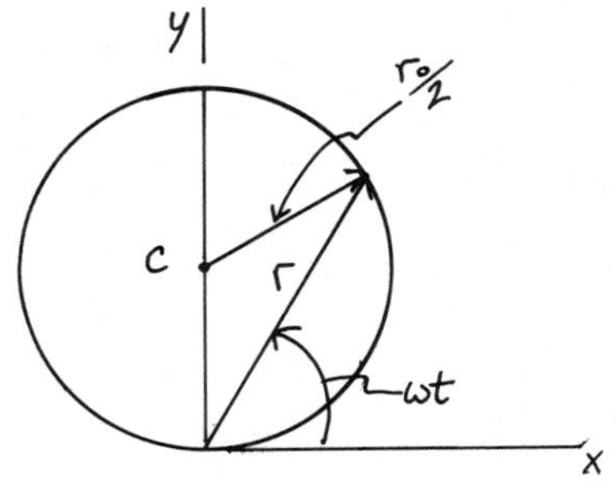

Figure (b)

c.) Differentiation of Eqs. (A) yields

$$\dot{x} = \frac{d}{dt}\left[\frac{r_o}{2}\sin 2\omega t\right] = r_o\omega\cos 2\omega t \text{ [ft/s]}$$

$$\dot{y} = \frac{d}{dt}[r_o \sin^2\omega t] = 2r_o\omega\sin\omega t\cos\omega t$$
$$= r_o\omega\sin 2\omega t \text{ [ft/s]}$$

$$\therefore \text{speed} = |\dot{s}| = \sqrt{\dot{x}^2 + \dot{y}^2} = r_o\omega \text{ [ft/s]} = \text{constant}$$

13.73 GIVEN Equations of motion of a particle are:

$$x = r\cos\omega t, \quad y = r\sin\omega t, \quad z = kt \qquad (A)$$

r, ω, k are positive constants.

FIND a.) Show that path is a helix wound on a cylinder of radius r

b.) Time T in which the particle travels once around the cylinder.

c.) Determine the pitch of the helix

d.) Show the particle travels at constant speed.

e.) Determine velocity and acceleration vectors for $t=0$ and show on sketch of the path

SOLUTION

a.) By Eqs (A):

$$x^2 + y^2 + z^2 = r^2 + k^2t^2 \qquad (B)$$
$$x^2 + y^2 = r^2 \qquad (C)$$

Equation (C) is the equation of a cylinder of circular cross section in (x, y, z) space.

By Eqs. (B) and (C),

$$z = kt \qquad (D)$$

Equation (D) shows that z increases linearly with time t. (Fig a.)

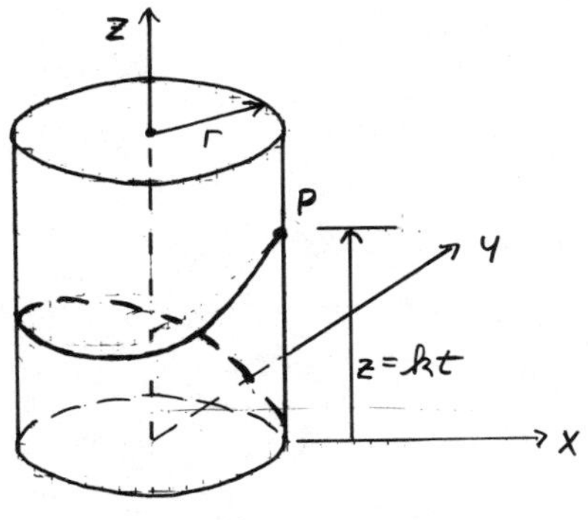

Figure (a.)

Hence, the particle P spirals upward on a circular cylinder at a constant rate k. The curve of the path is a helix.

(Continued)

13.73 cont.

b.) The time required for x to go from $x=r$ at time $t=0$ to $x=r$ at time T for one revolution is given by:

$$x = r\cos\omega T = r \quad \text{or} \quad \omega T = 2\pi$$

Therefore, $T = \dfrac{2\pi}{\omega}$ (E)

c.) The pitch of the helix is, by Eqs. (D) & (E):

$$p = z(T) = kT = \frac{2\pi k}{\omega}$$

d.) Speed of the particle, by Eqs. (A):

$$\dot{x} = -r\omega \sin\omega t, \quad \dot{y} = r\omega\cos\omega t \quad \text{(F)}$$
$$\dot{z} = k$$

Therefore,

$$v = \sqrt{\dot{x}^2 + \dot{y}^2 + \dot{z}^2} = \sqrt{r^2\omega^2 + k^2} = \text{const.}$$

e.) By Eqs. (F)

$$\ddot{x} = -r\omega^2 \cos\omega t, \quad \ddot{y} = -r\omega^2 \sin\omega t$$
$$\ddot{z} = 0 \quad \text{(G)}$$

By Eqs. (F) and (G), for $t=0$:

$$\bar{v} = r\omega\bar{j} + k\bar{k} \qquad \bar{a} = -r\omega^2\bar{i}$$

Shown on sketch:

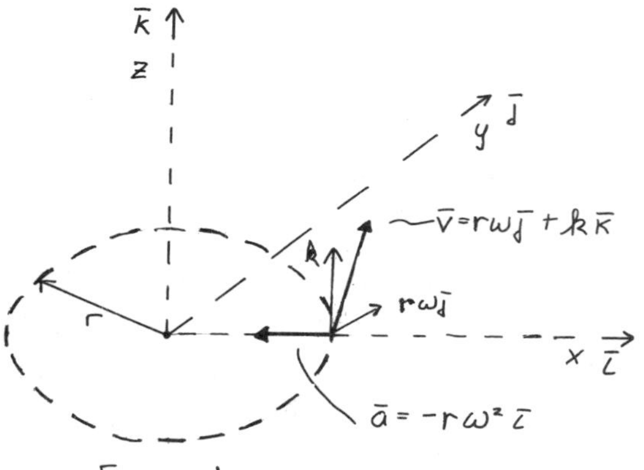

Figure b

13.74 GIVEN The cable AP unwinds at the rate of $\dot{\theta} = 10$ rad/s from the cylindrical drum. (See Fig (a))

For $t=0$, the points B and P coincide.

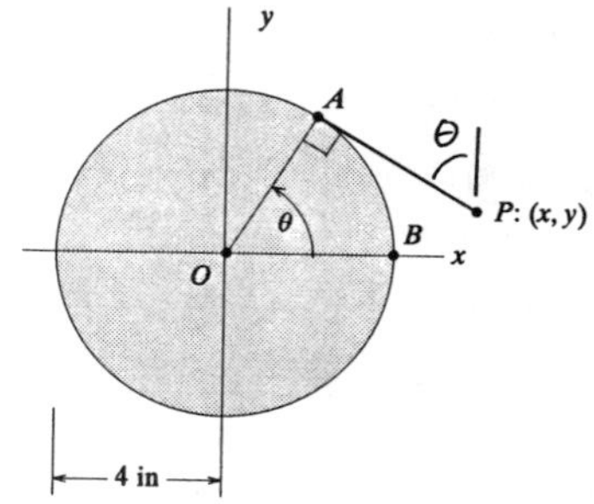

Figure (a) Cable-drum

FIND a.) x, y projections of the velocity and acceleration of a point P for $\theta = 45°$

b.) Speed and magnitude of the acceleration of a point P for $\theta = 45°$

SOLUTION

a.) By Fig (a), $AP = AB = 4\theta$

Also,

$$x = OA\cos\theta + AP\sin\theta = 4\cos\theta + (4\theta)(\sin\theta)$$
$$y = OA\sin\theta - AP\cos\theta = 4\sin\theta - (4\theta)(\cos\theta)$$

Differentiation of the above yields, since $\ddot{\theta} = 0$,

$$\left.\begin{aligned} \dot{x} &= 4\theta\dot{\theta}\cos\theta \quad [\text{in/s}] \\ \dot{y} &= 4\theta\dot{\theta}\sin\theta \quad [\text{in/s}] \\ \ddot{x} &= 4(\dot{\theta})^2(\cos\theta - \theta\sin\theta) \quad [\text{in/s}^2] \\ \ddot{y} &= 4(\dot{\theta})^2(\theta\cos\theta + \sin\theta) \quad [\text{in/s}^2] \end{aligned}\right\} \text{(A)}$$

For $\theta = 45° = \pi/4$ rad, with $\dot{\theta} = 10$ rad/s, Eqs (A) yield:

$$\left.\begin{aligned} \dot{x} &= 22.2 \ [\text{in/s}] & \dot{y} &= 22.2 \ [\text{in/s}] \\ \ddot{x} &= 60.7 \ [\text{in/s}^2] & \ddot{y} &= 505 \ [\text{in/s}^2] \end{aligned}\right\} \text{(B)}$$

b.) The speed and magnitude of acceleration of P are, with Eqs (B):

$$|\dot{s}| = v = \sqrt{(\dot{x})^2 + (\dot{y})^2} = \sqrt{(22.2)^2 + (22.2)^2}$$
$$= 31.4 \text{ in/s}$$

$$a = \sqrt{(\ddot{x})^2 + (\ddot{y})^2} = \sqrt{(60.7)^2 + (505)^2}$$
$$= 509 \text{ in/s}^2$$

13.75 Computer Problem

GIVEN $v_x = 2t$, $v_y = 20t^3$

When $t=1$, particle is at the point $(1, 7)$

FIND a.) (x, y) coordinates of the particle as functions of t

b.) y coordinate as a function of x

c.) Plot y versus x for $0 \le x \le 2$ m and show the velocity vector at the points $x = 0, 1, 2$.

SOLUTION

a.) $\bar{v} = \dfrac{d\bar{r}}{dt} = 2t\,\bar{i} + 20t^3\,\bar{j}$ (A)

$\bar{r} = (t^2 + C_1)\bar{i} + (5t^4 + C_2)\bar{j}$

$\therefore x = t^2 + C_1 \rightarrow 1 = (1)^2 + C_1 \Rightarrow C_1 = 0$

$y = 5t^4 + C_2 \rightarrow 7 = 5(1)^4 + C_2 \Rightarrow C_2 = 2$

Then, $x = t^2$ $\quad y = 5t^4 + 2$ (B)

b.) By Eqs. (B), $y = 5x^2 + 2$ (C)

c.) By Eq. (C), Plot of y vs. x:

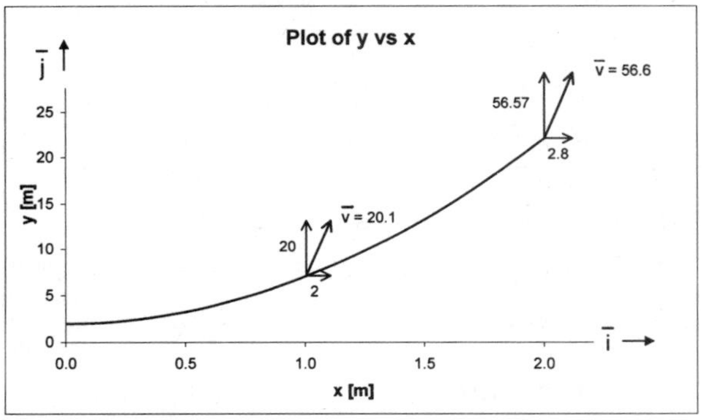

By Eqs. (A) and (B),

$\bar{v} = 0$ for $x = 0$, $(t = 0)$

$\bar{v} = 2\bar{i} + 20\bar{j}$ for $x = 1$, $(t = 1)$

$\bar{v} = 2.828\,\bar{i} + 56.568\,\bar{j}$ for $x = 2$, $(t = \sqrt{2})$

Velocity vectors shown in plot above.

13.76 Computer Problem

GIVEN The coordinates of a particle for the interval $-4 \le t \le 4$ are given by

$x = t, \quad y = [(16 - t^2)^{3/2}]/4$ (A)

FIND a.) x, y projections of the velocity and acceleration as functions of time t.

b.) Plot the path of the particle

c.) Show the velocity vector and the acceleration vector for $t = 3$ s.

SOLUTION

a.) $v_x = \dfrac{dx}{dt} = 1$ [m/s]

$v_y = \dfrac{dy}{dt} = -\dfrac{3}{4}t(16 - t^2)^{1/2}$ [m/s]

$a_x = \dfrac{dv_x}{dt} = 0$ [m/s²]

$a_y = \dfrac{dv_y}{dt} = \dfrac{3t^2}{4(16 - t^2)^{1/2}} - \dfrac{3}{4}(16 - t^2)^{1/2}$ [m/s²]

b.) By Eqs. (A), $y = [(16 - x^2)^{3/2}]/4$ (B)

See plot of path for $0 \le x \le 4$:

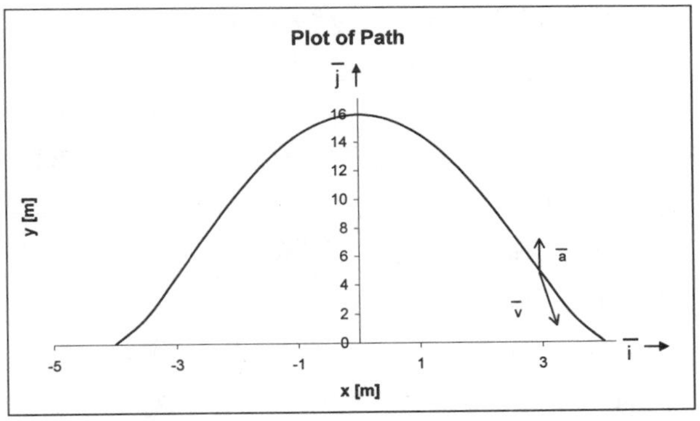

c.) For $t = 3$ s, with Eqs. (A),

$\bar{v} = 1\bar{i} - 3/4\,t\,(16 - t^2)^{1/2}\,\bar{j}$

$\bar{v} = 1\bar{i} - 5.95\bar{j}$ [m/s]

$\bar{a} = 3/4\left[\dfrac{t^2}{(16 - t^2)^{1/2}} - (16 - t^2)^{1/2}\right]\bar{j}$

$\bar{a} = 0.57\,\bar{j}$ [m/s²] at $x = t = 3$

$\bar{a}$ and $\bar{v}$ shown on plot above

13.77 Computer Problem

GIVEN Particle path given by $y = 5x^2$

The x projection of the velocity is 3 ft/s

For $t = 0$, $x = y = 0$

FIND a.) x, y, v_x, v_y, a_x, a_y as functions of t

b.) Speed of particle as a function of t

c.) Plot the path of the particle, and show the velocity and acceleration vectors for the point $x = 1/6$ ft

SOLUTION

a.) By $v_x = \frac{dx}{dt} = 3$, $x = 0$ for $t = 0$,

we find $x = 3t + C = 3t$ [ft] (A)

Therefore, with Eq (A)

$y = 5x^2 = 45t^2$ [ft] (B)

Hence, $v_y = \frac{dy}{dt} = 90t$ [ft/s] (C)

$a_x = \frac{dv_x}{dt} = 0$ $a_y = \frac{dv_y}{dt} = 90$ [$\frac{ft}{s^2}$] (D)

b.) The speed is

$v = \sqrt{v_x^2 + v_y^2} = \sqrt{3^2 + (90t)^2}$

$= 3\sqrt{1 + 900t^2}$ [ft/s] (E)

c.) For $x = 1/6 = 3t$, $t = 1/18$ s Therefore,

$x = 0.167$ ft, $y = 5/36 = 0.139$ ft

$v_x = 3$ ft/s $v_y = 5$ ft/s

$a_x = 0$ ft/s² $a_y = 90$ ft/s²

$\bar{v} = 3\bar{i} + 5\bar{j}$ → $v = \sqrt{34} = 5.83$ ft/s

$\theta = \tan^{-1} 5/3 = 59.04°$ $\bar{a} = 90\bar{j}$

See plot:

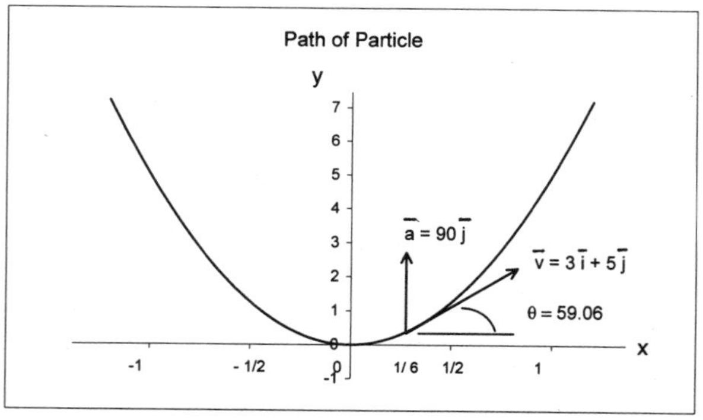

13.78 Computer Problem

GIVEN The Equations of motion of a particle that moves in (x, y) plane are

$x = 3\cos 5t$, $y = 4\sin 5t$ [in, sec.]

FIND a.) Show the path is an ellipse with principle radii 3 and 4.

b.) Time T in which the particle travels once around the ellipse.

c.) Velocity and acceleration vectors for $t = 1/5$ s, and show them accurately on a figure.

SOLUTION

a.) By the given equations,

$\frac{x}{3} = \cos 5t$, $\frac{y}{4} = \sin 5t$ (A)

$\therefore \frac{x^2}{9} + \frac{y^2}{16} = \cos^2 5t + \sin^2 5t = 1$ (B)

Equation (B) is the equation of an ellipse with major radius 4 in. and minor radius 3 in. (See plot below)

b.) Let the particle start at $x = 3$ when $t = 0$. The particle first returns to $x = 3$ (after traveling once around the ellipse) when $x = 3 = 3\cos 5T$

Therefore, $5T = 2\pi$

or $T = 0.4\pi = 1.257$ [s]

c.) For $t = 1/5$ s, Eqs (A) yield

$x = 3\cos 1 = 3\cos 57.296° = 1.621$ in

$y = 4\sin 1 = 4\sin 57.296° = 3.366$ in (C)

Differentiation of Eqs (A) yields, for $t = 1/5$ s,

$\dot{x} = -15\sin 5t = -12.622$ in/s

$\dot{y} = 20\cos 5t = 10.806$ in/s (D)

$\ddot{x} = -75\cos 5t = -40.523$ in/s²

$\ddot{y} = -100\sin 5t = -84.147$ in/s² (E)

By Eqs (D),

$\bar{v} = -12.62\bar{i} + 10.81\bar{j}$ [in/s]

or $v = \sqrt{(-12.62)^2 + (10.81)^2} = 16.62$ in/s

$\theta_x = \tan^{-1}\left(\frac{10.81}{-12.61}\right) = 139.43°$

(continued)

13.78 cont.

By Eqs (E),

$\bar{a} = -40.52\,\bar{i} - 84.15\,\bar{j}$

or $a = \sqrt{(-40.52)^2 + (-84.15)^2} = 93.39$ in/s²

$\theta_x = 244.28°$ or $-115.72°$

The path and velocity and acceleration vectors are shown on the plot below.

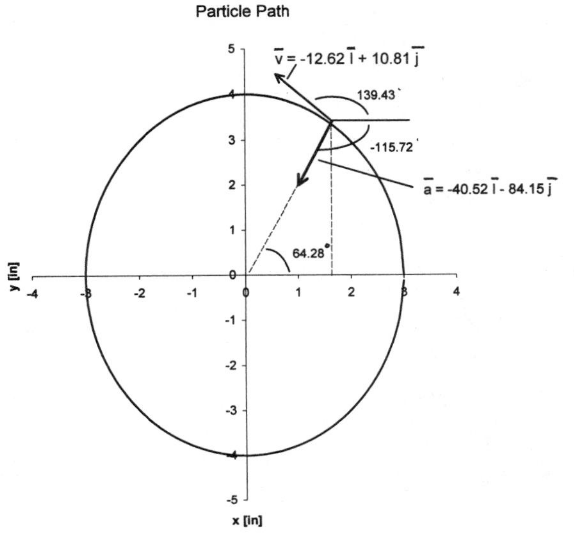

13.79 **GIVEN** Angular speed, ω, of an electric motor changes at a constant angular acceleration α from 10 rpm to 1800 rpm in 4s.

FIND α [rad/s²] and θ [rad] for 4s. interval

SOLUTION

Since α = constant,

with $\omega_o = 10 \text{ rpm} = 10\frac{(2\pi)}{60} = \pi/3$ rad/s,

$\omega = \alpha t + \omega_o = \alpha t + \pi/3$ (rad/s) (A)

For $t = 4$s, $\omega = 1800\frac{(2\pi)}{60} = 60\pi$ rad/s

Therefore, by Eq (A)

$\alpha = \frac{(\omega - \omega_o)}{t} = \frac{(60\pi - \pi/3)}{4} = 46.86 \frac{\text{rad}}{s^2}$

Also, by Eq (A),

$\omega = \frac{d\theta}{dt} = \alpha t + \omega_o$

Therefore, $\int_0^\theta d\theta = \int_0^t (\alpha t + \omega_o)\, dt$

or $\theta = \frac{1}{2}\alpha t^2 + \omega_o t$

$= \frac{1}{2}(46.86)(4)^2 + (\pi/3)(4)$

so $\theta = 379.1$ rad $= 379.1/2\pi$ rev

or $\theta = 60.3$ revolutions.

13.80 **GIVEN** A DC electric motor with an armature diameter of 15 in is rotating at 1800 rpm when it is short-circuited. It speeds up with angular acceleration of 100 rad/s².

The armature will fly apart at 10,000 rpm.

FIND a.) Time (seconds) that the operator has to shut off the current.

b.) If the armature explodes, with what speed do the fragments leave its rim?

SOLUTION

a.) Note that the initial angular velocity is, $\omega_o = 1800 \text{ rpm} = 1800\frac{(2\pi)}{60} = 60\pi \frac{\text{rad}}{s}$

and the final angular velocity is,

$\omega_f = 10{,}000 \text{ rpm} = 10{,}000\frac{(2\pi)}{60} = \frac{1000}{3}\pi \frac{\text{rad}}{s}$

Hence, $\omega_f = \alpha t + \omega_o$

$\frac{1000}{3}\pi = (100)t + 60\pi$

So, $t = 8.587$ s

b.) The tangential speed v is given by

$v = \omega_f r = \frac{1000}{3}\pi\left(\frac{15}{2}\right)\left(\frac{1}{12}\right)$

$v = 654.5$ ft/s

13.81 **GIVEN** The angular acceleration of a turbine rotor is $\alpha = k t^{-1/3}$ where k is a constant.

When $t = 0$, $\omega = 10$ rad/s and $\theta = 0$

When $t = 1$s, $\theta = 28$ rad.

FIND α, ω, and θ for $t = 8$s

(Continued)

13.81 cont.

SOLUTION

By definition, $\alpha = \frac{d\omega}{dt} = kt^{-1/3}$ (A)

so $\int_{10}^{\omega} d\omega = \int_0^t kt^{-1/3}\,dt$

or $\omega = 3/2\, kt^{2/3} + 10$ (B)

Also, $\omega = \frac{d\theta}{dt} = \frac{3}{2} kt^{2/3} + 10$

so $\int_0^\theta d\theta = \int_0^t (\frac{3}{2} kt^{2/3} + 10)\,dt$

or $\theta = \frac{9}{10} kt^{5/3} + 10t$ (C)

For $t = 1s$, $\theta = 28$ rad. Therefore, by Eq (C), $k = 20$.

Hence, by Eqs (A), (B), and (C) at $t = 8s$,

$\alpha = 10$ rad/s² $\omega = 130$ rad/s $\theta = 656$ rad

13.82

GIVEN Wheel is initially rotating at 1200 rpm and is braked so that the angular deceleration is proportional to the square root of the time that the brake acts. After 36 s, $\omega = 600$ rpm.

FIND In how many more seconds will the wheel stop?

SOLUTION

$\omega_0 = 1200$ rpm $= 40\pi$ rad/s at $t=0$

$\alpha = -k\sqrt{t}$

and at $t=36$, $\omega = 600$ rpm $= 20\pi$

The time it takes for the wheel to stop from $\omega = 20\pi$:

$\alpha = \frac{d\omega}{dt} = -k\sqrt{t} = -kt^{1/2}$

or $\int d\omega = \int \alpha\, dt = -\int kt^{1/2}\,dt$

so $\omega = -2/3\, kt^{3/2} + \omega_0$ (A)

at $t=0$, $\omega_0 = 40\pi$

$\therefore \omega = 40\pi - 2/3\, kt^{3/2}$ (B)

at $t = 36s$, $\omega = 20\pi$.

Therefore, $k = 0.436$.

Hence, by Eq (B), with $\omega = 0$ and $k = .436$,

$t = 57.2s$

Then, $\Delta t = 57.2 - 36 = 21.2s$

13.83

GIVEN A spotlight rotating at a constant rate $\omega = 90$ rpm casts a horizontal beam on a straight wall that is 30 m from the light.

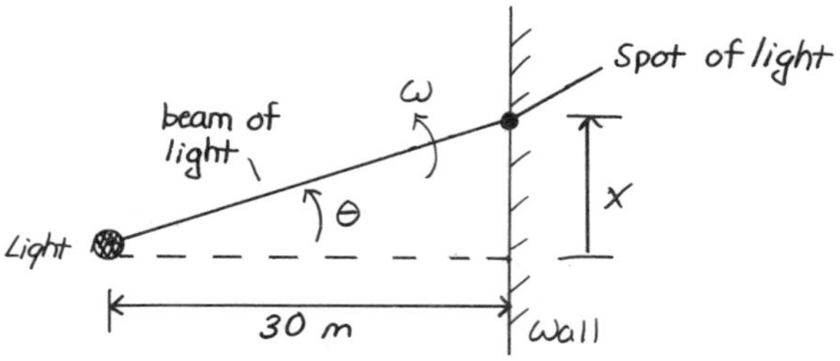

Figure (a.)

FIND The velocity, $\dot{x}$ and acceleration $\ddot{x}$ of the spot light on the wall for the following angles between the beam and the perpendicular to the wall: 0°, 30°, 60°, 70°, 80°

SOLUTION

By Fig (a.),

$x = 30 \tan\theta$ [m]

$\dot{x} = 30\dot{\theta}\sec^2\theta$ [m/s]

$\ddot{x} = 60\dot{\theta}^2(\sec^2\theta)\tan\theta$ [m/s²]

where $\omega = \dot{\theta} = 3\pi$ rad/s and $\ddot{\theta} = 0$

Hence,

$x = 30\tan\theta$ [m]

$\dot{x} = 90\pi\sec^2\theta$ [m/s]

$\ddot{x} = 540\pi^2(\sec^2\theta)\tan\theta$ [m/s²]

The values of $\dot{x}$ and $\ddot{x}$ are tabulated in Table (a.) for $\theta = 0, 30°, 60°, 70°, 80°$

Table (a)

θ°	$\dot{x}$ [m/s]	$\ddot{x}$ [m/s²]
0	282.74	0
30	376.99	4102.72
60	1130.97	36,924.4
70	2417.06	125,176.8
80	9376.77	1.002×10^6

13.84 GIVEN The wheel shown in Fig (a) rolls at constant angular velocity ω along the x axis.
A point P on the rim generates the cycloid $x = r(\theta - \sin\theta)$, $y = r(1-\cos\theta)$

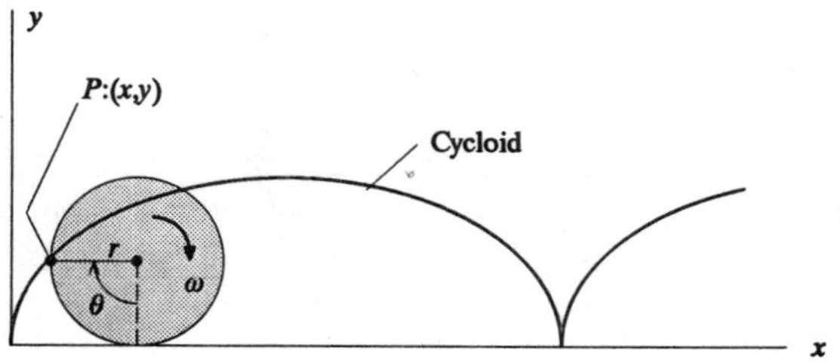

Figure (a.)

FIND a.) Speed and acceleration vector of point P as functions of θ.

b.) Calculate the speed and the acceleration that the particle P possesses when it is at the highest point of its path.

SOLUTION

a.) By differentiation with respect to time:

$\dot{x} = r\dot{\theta}(1-\cos\theta)$

$\dot{y} = r\dot{\theta}\sin\theta$

$\ddot{x} = r(\ddot{\theta} - \ddot{\theta}\cos\theta + \dot{\theta}^2\sin\theta)$

$\ddot{y} = -r(\ddot{\theta}\sin\theta - \dot{\theta}^2\cos\theta)$

But $\dot{\theta} = \omega$ = constant, therefore $\ddot{\theta} = 0$

Hence,

$$|\bar{v}| = \sqrt{\dot{x}^2 + \dot{y}^2} = r\omega\sqrt{2(1-\cos\theta)}$$

and,

$a_x = \ddot{x} = r\omega^2\sin\theta$

$a_y = \ddot{y} = r\omega^2\cos\theta$

$$\bar{a} = a_x\bar{\imath} + a_y\bar{\jmath} = r\omega^2[(\sin\theta)\bar{\imath} + (\cos\theta)\bar{\jmath}]$$

b.) Point P is at its highest point when $\theta = \pi$. For $\theta = \pi$, $\sin\theta = 0$ and $\cos\theta = -1$

Hence,

$$v = 2r\omega$$

$$a_x = 0, \quad a_y = -r\omega^2$$

$$\bar{a} = -r\omega^2\bar{\jmath}$$

13.85 GIVEN $\omega = \dot{\theta} = 2\sqrt{\theta}$ rad/s
For $t=0$, $\theta=0$ (t in seconds)
Positive ω is counterclockwise.

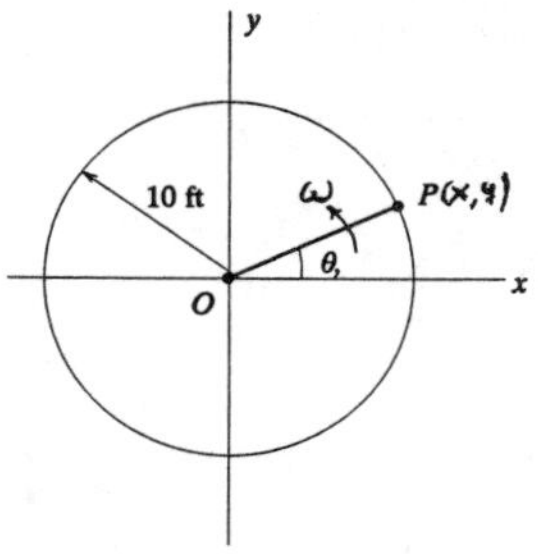

Figure (a.)

FIND x, y components of acceleration for point P for $\theta = \pi/2$ rad.

SOLUTION

By Fig (a.),

$x = 10\cos\theta, \quad y = 10\sin\theta \qquad (A)$

Differentiation of Eqs (A) yields

$\dot{x} = -10(\sin\theta)\dot{\theta}, \quad \dot{y} = 10(\cos\theta)\dot{\theta}$

$$\begin{aligned}\ddot{x} &= -10(\cos\theta)(\dot{\theta})^2 - 10(\sin\theta)\ddot{\theta} \\ \ddot{y} &= -10(\sin\theta)(\dot{\theta})^2 + 10(\cos\theta)\ddot{\theta}\end{aligned} \qquad (B)$$

By $\omega = 2\sqrt{\theta} = \dot{\theta} = \frac{d\theta}{dt}$, we may write

$2\,dt = \frac{d\theta}{\sqrt{\theta}} = \theta^{-1/2}d\theta$

Integration yields

$2\theta^{1/2} = 2t + C = 2t$, since $\theta = 0$ for $t = 0$

$\therefore \theta = t^2, \quad \dot{\theta} = 2t, \quad \ddot{\theta} = 2$

For $\theta = \pi/2$ rad, $t = \sqrt{\pi/2}$. Hence,

$\theta = \pi/2, \quad \dot{\theta} = \sqrt{2\pi}, \quad \ddot{\theta} = 2 \qquad (C)$

Also, for $\theta = \pi/2$, Eqs. (B) become

$\dot{x} = -10\dot{\theta}, \quad \dot{y} = 0$

$\ddot{x} = -10\ddot{\theta}, \quad \ddot{y} = -10\dot{\theta}^2 \qquad (D)$

Substitution of Eqs (C) into Eqs (D) yield the x and y components of acceleration:

$$\ddot{x} = -20 \text{ ft/s}^2 \qquad \ddot{y} = -20\pi \text{ ft/s}^2$$

13.86 GIVEN The data of Example 13.16

FIND
a.) Show if $a=b$ and $\dot{\theta}$ = constant, the arm OS travels at constant $\dot{\phi}$.
b.) Determine $\dot{\phi}$ in this case.

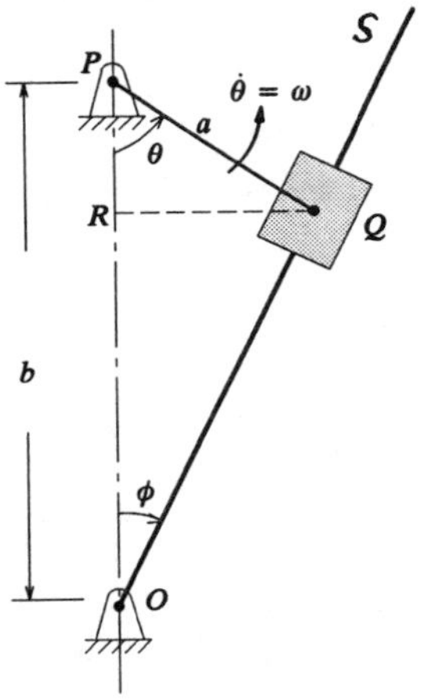

Figure (a.)

SOLUTION

a.) By Eq (D) of Example 13.16, $\ddot{\phi} = 0$. Therefore

$\dot{\phi}$ = constant

b.) By Eq (b) of Example 13.16,

$$\dot{\phi} = -\frac{a\omega(a - b\cos\theta)}{a^2 + b^2 - 2ab\cos\theta} \qquad (A)$$

With $a=b$, Eq (A) yields

$$\dot{\phi} = \frac{-a^2\omega(1-\cos\theta)}{2a^2(1-\cos\theta)}$$

or $\dot{\phi} = -\omega/2$

13.87 GIVEN The data of Example 13.16.

FIND a.) Show that arm **OS** attains its maximum speed ($\dot{\phi}$ = max) for $\theta = 0$ and minimum speed for $\theta = \pi$ provided $\dot{\theta}$ = const.

b.) Ratio of $\dot{\phi}_{max}$ to $\dot{\phi}_{min}$ if $a/b = 1.1$

SOLUTION

a.) For $\dot{\phi}$ = max (or min), $\ddot{\phi} = 0$. Therefore, by Eq (D) of Example 13.16,

$$ab(a^2 - b^2)\omega^2 \sin\theta = 0$$

Therefore, for $a \neq b$, $\sin\theta = 0$

or $\theta = 0$ or π rad $\qquad (A)$

With $\theta = 0$, by Eq (b) of Example 13.16

$$\dot{\phi}_{max} = -\frac{a\omega}{a-b} \qquad (B)$$

and with $\theta = \pi$

$$\dot{\phi}_{min} = \frac{-a\omega}{a+b} \qquad (C)$$

b.) With $a/b = 1.1$, By Eqs (B) and (C),

$$\frac{\dot{\phi}_{max}}{\dot{\phi}_{min}} = \frac{a+b}{a-b} = \frac{a/b + 1}{a/b - 1}$$

or $\dfrac{\dot{\phi}_{max}}{\dot{\phi}_{min}} = 21$

13.88 GIVEN The velocities and accelerations of blocks 1 and 3 are indicated in Fig. (a.):

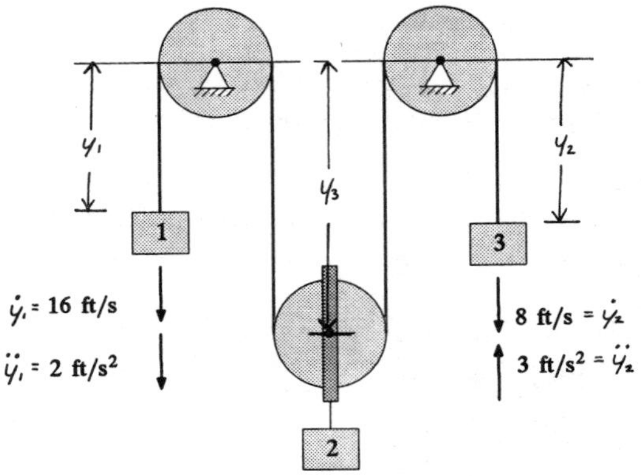

Figure (a.)

FIND Velocity and acceleration of block 2.

SOLUTION

Kinematics: Let C equal the length of the cord wrapped around the pulleys

L (length of cord) = $y_1 + y_2 + 2y_3 + C$ = constant $\qquad (A)$

Differentiation of (A) with respect to time yields

$$\dot{y}_1 + 2\dot{y}_3 + \dot{y}_2 = 0 \qquad (B)$$

or $\dot{y}_3 = -\frac{1}{2}(\dot{y}_2 + \dot{y}_1) = -\frac{(16+8)}{2} = -12$ ft/s

or $\dot{y}_3$ = velocity of block 2 = 12 ft/s ↑

Differentiation of Eq (B) yields

$$\ddot{y}_1 + 2\ddot{y}_3 + \ddot{y}_2 = 0$$

or $\ddot{y}_3 = -\frac{1}{2}(\ddot{y}_1 + \ddot{y}_2) = -\frac{(2-3)}{2} = 0.5$ ft/s²

or $\ddot{y}_3$ = acceleration of block 2 = 0.5 ft/s² ↓

13.89

GIVEN Block A oscillates in the vertical slot: $y = 4\sin 5t$ (A)

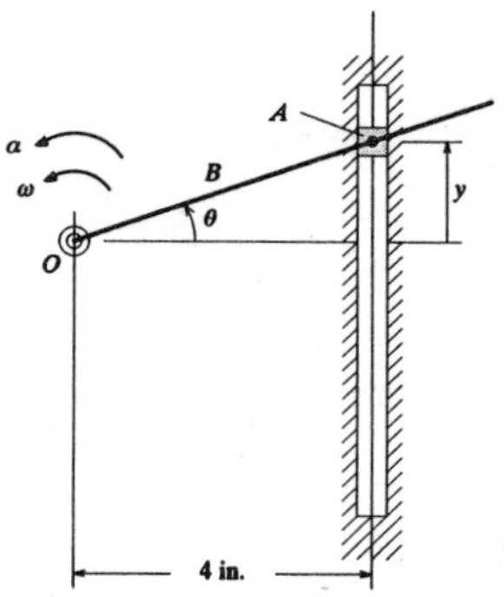

Figure (a.)

FIND The angular velocity $\omega = \dot{\theta}$ and angular acceleration $\alpha = \ddot{\theta}$ for $y = 3$ in.

SOLUTION

For $y = 3$ in, Eq (A) yields: $3 = 4\sin 5t$

or $t = \frac{1}{5}\sin^{-1}(0.75) = 0.1696$ s (B)

By Fig (a), for $y = 3$ in,

$y = 4\tan\theta = 4\sin 5t = 3$ (C)

Therefore (see Fig (a)):

$\tan\theta = 3/4$, $\sec\theta = 1.25$ (D)

Also, by Eq (C), $\tan\theta = \sin 5t$ (E)

Differentiation of Eq (E) yields:

$\omega\sec^2\theta = 5\cos 5t$ (F)

or, with Eqs (B) and (D),

$\omega = \frac{5\cos 5t}{\sec^2\theta} = \frac{5\cos(5\times 0.1696)}{(1.25)^2}$ (F)

Therefore, for $y = 3$ in,

$\omega = 2.117$ rad/s

By Eq (F): $\omega = 5(\cos 5t)\cos^2\theta$

Differentiation yields (with $\dot{\theta} = \omega$)

$\alpha = \dot{\omega} = -25(\sin 5t)(\cos^2\theta) - 10(\cos 5t)(\cos\theta)(\sin\theta)\omega$

For $y = 3$ in, $\omega = 2.117$ rad/s, and $t = 0.1696$ s,

By Fig (a), $\sin\theta = 0.6$ and $\cos\theta = 0.8$ for $y = 3$ in.

$\therefore\ \alpha = -18.72$ rad/s²

13.90

GIVEN $y = 4\sin 5t$

Length of bar B = 8 in

FIND Speed of upper end of Bar B for $\theta = \tan^{-1}(1/2)$

SOLUTION

By Fig (a),

$y = 4\sin 5t = 4\tan$

For $\tan\theta = 1/2$,

$\sin 5t = 0.5$

Hence,

$t = \frac{1}{5}\sin^{-1}(0.5) = 0.1047$ s

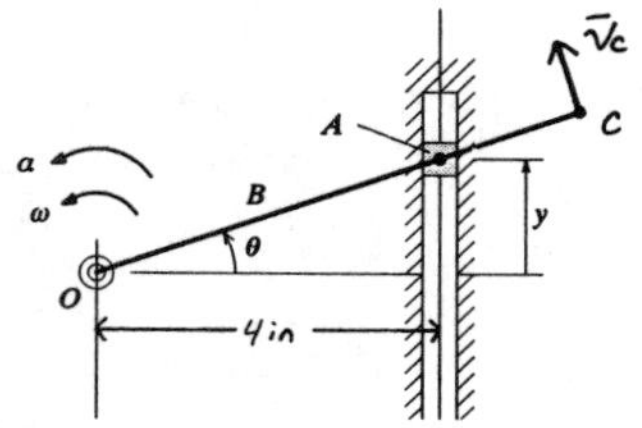

Figure (a)

Differentiation yields, with $\omega = \dot{\theta}$,

$\omega\sec^2\theta = 5\cos 5t$

or $\omega = 5(\cos 5t)(\cos\theta)^2$ (A)

For $\tan\theta = 1/2$ $\cos\theta = \frac{2}{\sqrt{5}}$ (B)

With Eqs (A) and (B) and $t = 0.1047$ s,

$\omega = 3.464$ rad/s

Then, by Kinematics

$v_c = r\omega = (\overline{OC})\omega$

or $v_c = (8)(3.464) = 27.71$ in/s

13.91

GIVEN Links A and B are parallel.

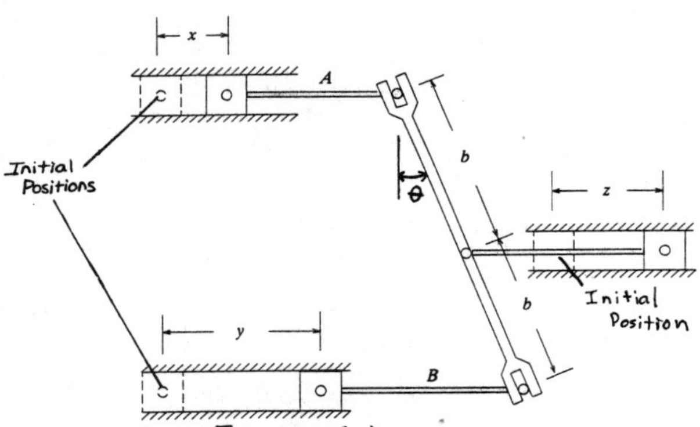

Figure (a.)

FIND Show that $z = (x+y)/2$

SOLUTION

By Fig (a.), $z = x + b\sin\theta$ (A)

Also, by Fig (a), if Links A and B remain approximately horizontal

$\sin\theta \approx (y-x)/(2b)$ (B)

(Continued)

13.91 cont.

By Eqs (A) and (B), $z = x + b\left(\frac{y-x}{2b}\right)$

or $z = \frac{x+y}{2}$

→ ALTERNATIVELY,

By Fig (a), $z = x + b\sin\theta$
$= y - b\sin\theta$

$2z = x + y \rightarrow z = \frac{x+y}{2}$

13.92 GIVEN A plane mirror rotating with angular velocity ω and angular acceleration α is struck by a ray of light directed at a right angle to the mirror's axis of rotation (Fig (a))

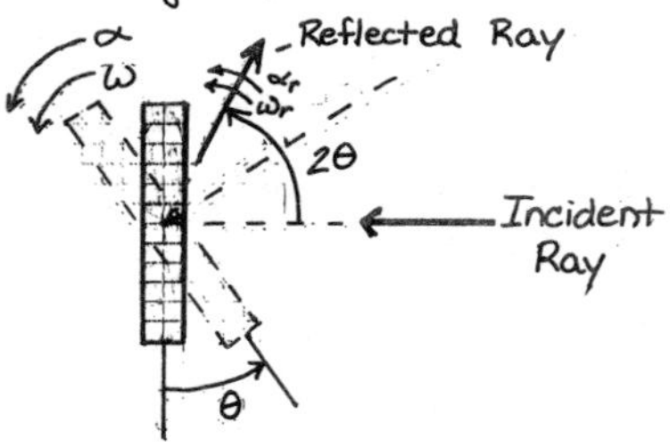

Figure (a)

FIND The angular velocity ω_r and angular acceleration α_r of the reflected ray.

SOLUTION

Let $\dot\theta = \omega$ and $\ddot\theta = \alpha$ be the angular velocity and angular acceleration of the mirror.

The angular displacement of the reflected ray is 2θ.

$\therefore \omega_r = 2\dot\theta = 2\omega$ and $\alpha_r = 2\ddot\theta = 2\alpha$

13.93 GIVEN Ray of light passes from air into water according to Snell's Law: $\sin\theta = \eta \sin\phi$ with $\dot\theta = \text{const} = 12$ rad/s.

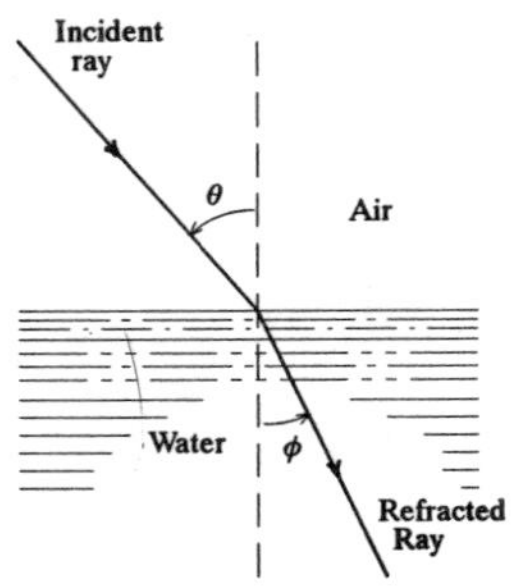

Figure (a.)

FIND Angular velocity $\omega = \dot\phi$ and angular acceleration $\alpha = \ddot\phi$ of refracted wave.

SOLUTION

Differentiation of Law yields (η = constant)

$\dot\theta\cos\theta = \eta\,\dot\phi\cos\phi$

$\ddot\theta\cos\theta - (\dot\theta)^2\sin\theta = \eta\,\ddot\phi\cos\phi - \eta(\dot\phi)^2\sin\phi$

Solving these equations for $\dot\phi$ and $\ddot\phi$, we obtain, since $\ddot\theta = 0$,

$$\dot\phi = \frac{\dot\theta\cos\theta}{\eta\cos\phi} \quad (A)$$

$$\ddot\phi = \frac{\eta(\dot\phi)^2\sin\phi - (\dot\theta)^2\sin\theta}{\eta\cos\phi} \quad (B)$$

For $\theta = 60°$, and $\eta = 1.3$, By Snell's Law:

$\sin\phi = (\sin 60°)/1.3 = 0.6662$

$\therefore \cos\phi = 0.7458$

Hence, by Eqs (A) and (B)

$\dot\phi = 6.188$ rad/s, $\theta = 60°$

$\ddot\phi = -94.41$ rad/s², $\theta = 60°$

For $\theta = 0°$, $\phi = 0°$, $\sin\phi = 0$, $\cos\phi = 1$

By Eqs (A) & (B),

$\dot\phi = 9.231$ rad/s, $\ddot\phi = 0$; $\theta = 0°$

For $\theta = \pi/2$ rad, $\cos\theta = 0$, $\sin\phi = 0.7692$

$\cos\phi = 0.6390$. By Eqs (A) and (B),

$\dot\phi = 0$, $\ddot\phi = -173.3$ rad/s²; $\theta = 90°$

13.94 GIVEN Elliptic trammel (Fig. 13.94) with distance c = constant. (See schematic, Fig. (a))

FIND

a.) Show point P describes an ellipse as B slides (Fig a)

b.) Express v_A, a_A, v_B, a_B in terms of θ, $\dot\theta$, $\ddot\theta$.

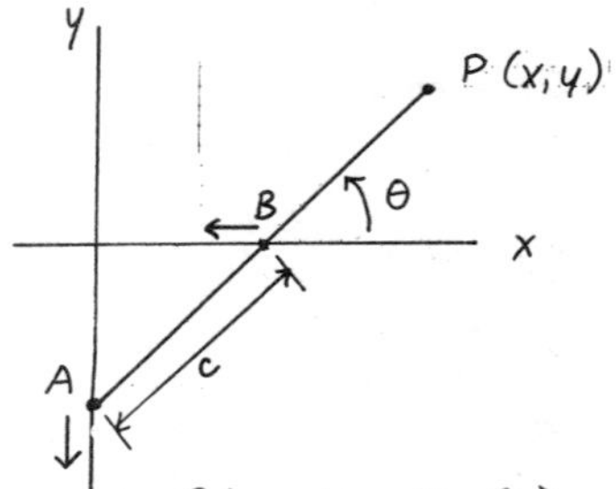

Schematic Fig (a.)

SOLUTION

By Fig (a): $x_P = c\cos\theta + BP\cos\theta$ (A)
$y_P = BP\sin\theta$

By Eqs (A): $\frac{x_P}{a} = \cos\theta$, $\frac{y_P}{b} = \sin\theta$ (B)

(Continued)

13.94 cont.

where $a = c + BP, \quad b = BP$ (C)

By Eqs (B),

$$\left(\frac{x_P}{a}\right)^2 + \left(\frac{y_P}{b}\right)^2 = 1 \quad = \text{equation of ellipse} \qquad (D)$$

b. By Fig (a) and differentiation,

$x_B = c\cos\theta$

$\dot{x}_B = v_B = -c\dot{\theta}\sin\theta$

$\ddot{x}_B = a_B = -c\ddot{\theta}\sin\theta - c(\dot{\theta})^2\cos\theta$

$y_A = -c\sin\theta$

$\dot{y}_A = v_A = -c\dot{\theta}\cos\theta$

$\ddot{y} = a_A = -c\ddot{\theta}\cos\theta + c(\dot{\theta})^2\sin\theta$

13.95 GIVEN Elliptic trammel (Fig. P13.94) with AB = 4 in and BP = 3 in (See schematic) and ω = 2 rad/s = constant

FIND

a.) Velocity and acceleration vectors of P for θ = 30°

b.) Speed of P for θ = 30°

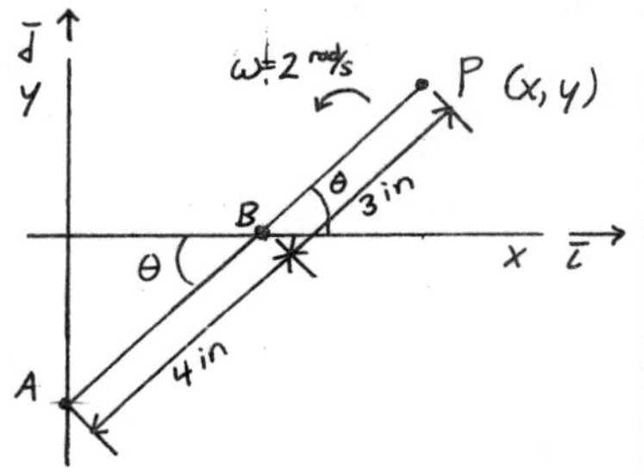

Schematic Fig (a)

SOLUTION

a.) By Fig (a)

$x_P = 4\cos\theta + 3\cos\theta = 7\cos\theta$

$y_P = 3\sin\theta$ (A)

Differentiation of Eqs (A) yields, with $\omega = \dot{\theta} = 2$ rad/s and $\theta = 30°$,

$\dot{x}_P = -7\dot{\theta}\sin\theta = -7$ in/s $\qquad \dot{y}_P = 3\dot{\theta}\cos\theta = 5.196$ in/s

$\therefore \bar{v}_P = -7\bar{i} + 5.20\bar{j}$ in/s

$\ddot{x}_P = -7(\dot{\theta})^2\cos\theta = -24.25$ in/s²

$\ddot{y}_P = -3(\dot{\theta})^2\sin\theta = -6$ in/s²

$\therefore \bar{a}_P = -24.25\bar{i} - 6\bar{j}$ in/s²

b.) Speed of P is:

$s = |\bar{v}| = \sqrt{(-7)^2 + (5.196)^2}$

or $s = 8.72$ in/s

13.96 GIVEN Eccentric Cam (Fig (a.))

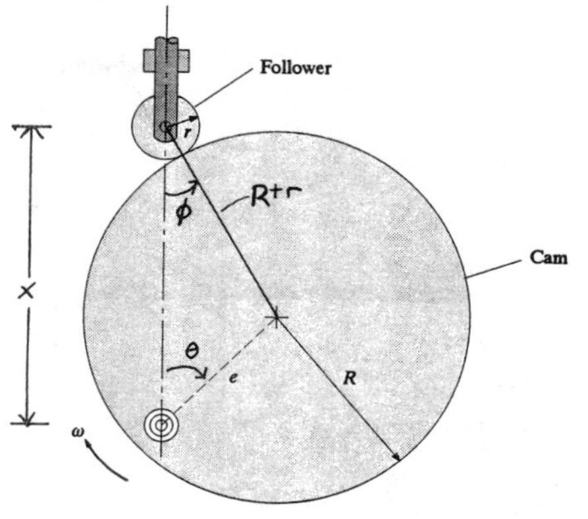

Figure (a.)

FIND Show that motion of follower is same same as motion of the piston of the slider-crank mechanism

SOLUTION

By Fig (a), $x = e\cos\theta + (R+r)\cos\phi$ (A)

For the piston, see Fig (b):

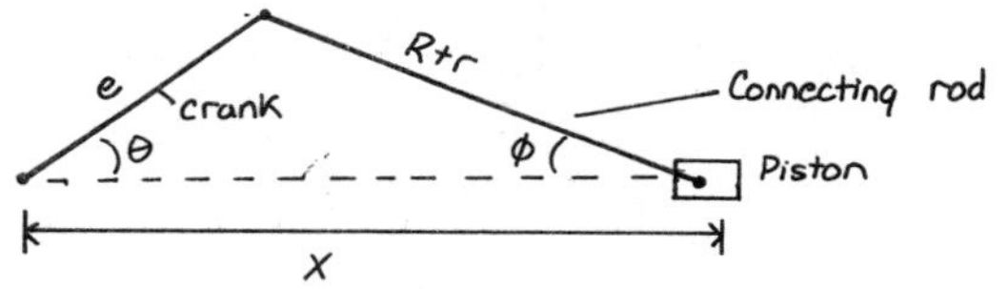

Figure (b)

By Fig (b), $x = e\cos\theta + (R+r)\cos\phi$ (B)

Equations (A) and (B) are identical.

13.97 GIVEN Schematic of a slider-crank mechanism (Fig (a))

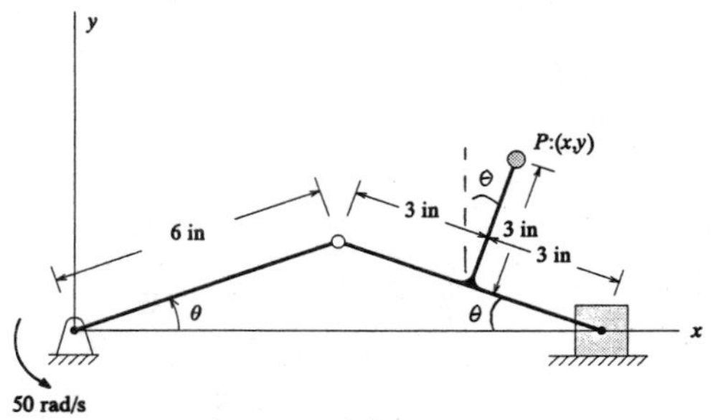

Figure (a).

FIND $\dot{x}$, $\dot{y}$, and $\ddot{x}$, $\ddot{y}$ of P in terms of θ. ($\omega = \dot{\theta} = 50$ rad/s = constant)

(Continued)

13.97 cont.

By Fig (a), $x = 12\cos\theta - 3\cos\theta + 3\sin\theta$

or

$$\left.\begin{aligned} x &= 9\cos\theta + 3\sin\theta \\ y &= 3\sin\theta + 3\cos\theta \end{aligned}\right\} (A)$$

Differentiation of Eqs (A) yields

$\dot{x} = -9\dot{\theta}\sin\theta + 3\dot{\theta}\cos\theta$

$\therefore \underline{\dot{x} = -450\sin\theta + 150\cos\theta \quad [\text{in/s}]}$

$\dot{y} = 3\dot{\theta}\cos\theta - 3\dot{\theta}\sin\theta$

$\therefore \underline{\dot{y} = 150\cos\theta - 150\sin\theta \quad [\text{in/s}]}$

$\ddot{x} = -9(\dot{\theta})^2\cos\theta - 3(\dot{\theta})^2\sin\theta$

$\therefore \underline{\ddot{x} = -22{,}500\cos\theta - 7500\sin\theta \quad [\text{in/s}^2]}$

$\ddot{y} = -3(\dot{\theta})^2\sin\theta - 3(\dot{\theta})^2\cos\theta$

$\therefore \underline{\ddot{y} = -7500\sin\theta - 7500\cos\theta \quad [\text{in/s}^2]}$

13.98 GIVEN Analog mechanism (Fig P13.98) Shown schematically in Fig (a).

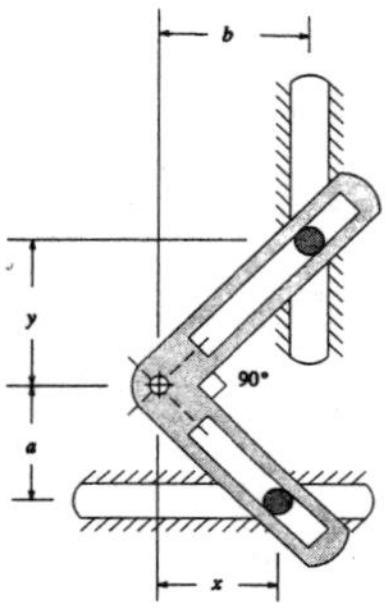

Figure P13.98

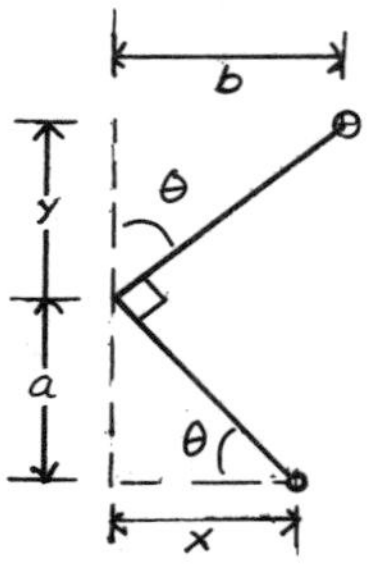

Schematic, Figure (a)

FIND a.) Show that $y = kx$, k = constant
b.) Find k in terms of a and b

SOLUTION

a.) By Fig (a),

$$\frac{b}{y} = \tan\theta = \frac{a}{x}$$

$$\therefore \frac{b}{y} = \frac{a}{x} \quad \text{or} \quad \underline{y = \frac{b}{a}x}$$

b.) Thus, with $y = kx$, $\quad \underline{k = \frac{b}{a}}$

13.99 GIVEN Disk-and-wheel integrator shown in Fig (a),

FIND Show that when disk D undergoes rotations from θ_0 to θ, the wheel undergoes rotation $\beta = \frac{1}{a}\int_{\theta_0}^{\theta} x\,d\theta$

SOLUTION

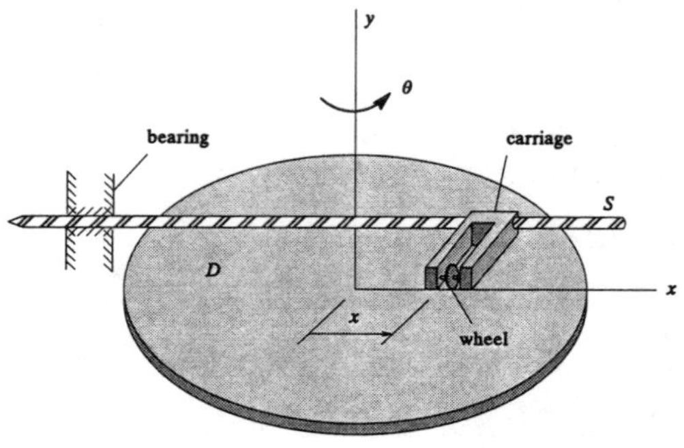

Figure (a)

Let the disk undergo a rotation $d\theta$. Then the wheel travels on the disk a distance:

$$ds = a\,d\beta = x\,d\theta$$

$$\therefore d\beta = x/a\,d\theta \quad \text{or} \quad \underline{\beta = 1/a \int_{\theta_0}^{\theta} x\,d\theta}$$

13.100 GIVEN Rocking block linkage shown in Fig (a).

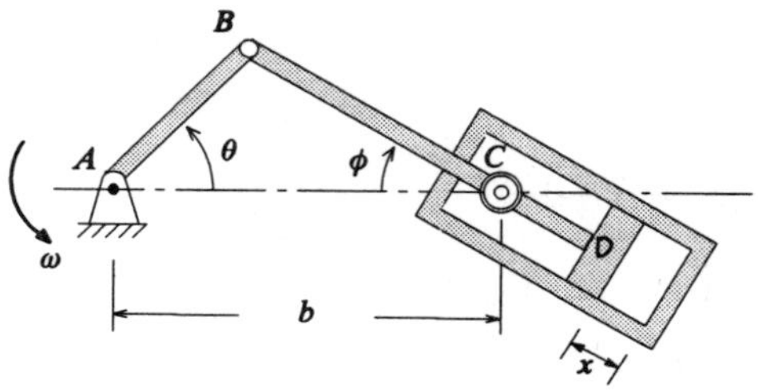

Figure (a)

FIND Show that $\dot{x} = b\omega\sin\phi$

SOLUTION

By Fig (a) (note length $\overline{BC}$ changes as the crank AB rotates)

$$b = \overline{AB}\cos\theta + \overline{BC}\cos\phi = \text{constant} \quad (A)$$

$$\overline{AB}\sin\theta = \overline{BC}\sin\phi \quad (B)$$

Differentiation of Eqs (A) and (B) yields:

$$\overline{AB}\,\dot{\theta}\sin\theta = -\overline{BC}\,\dot{\phi}\sin\phi + \dot{\overline{BC}}\cos\phi \quad (C)$$

$$\overline{AB}\,\dot{\theta}\cos\theta = \overline{BC}\,\dot{\phi}\sin\phi + \dot{\overline{BC}}\sin\phi \quad (D)$$

Also, by geometry (see Fig (b))

Figure (b)

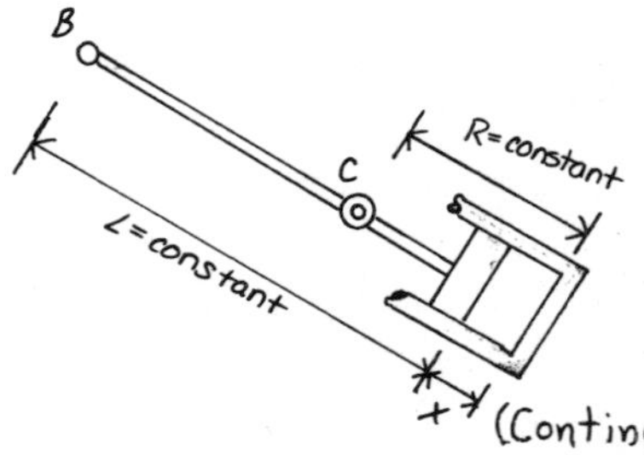

(Continued)

13.100 cont.

By Fig (b), $L - \overline{BC} = R - x$

$L - \overline{BC} + x = R$

Differentiation yields $\dot{x} = \dot{\overline{BC}}$ (E)

Equations (C), (D), and (E) yield

$\dot{x}\cos\phi = \overline{BC}\,\dot{\phi}\sin\phi + \overline{AB}\,\dot{\theta}\sin\theta$ (F)

$\dot{x}\sin\phi = -\overline{BC}\,\dot{\phi}\cos\phi + \overline{AB}\,\dot{\theta}\cos\theta$ (G)

Multiplying Eq (F) by $\cos\phi$, Eq (G) by $\sin\phi$, and adding, we obtain, with Eqs (A) and (B),

$\dot{x} = \dot{\theta}(\overline{AB}\sin\theta\cos\phi + \overline{AB}\cos\theta\sin\phi)$

$\dot{x} = \dot{\theta}(\overline{BC}\cos\phi + \overline{AB}\cos\theta)\sin\phi$

$\dot{x} = \dot{\theta} b \sin\phi = b\omega\sin\phi$

13.101 GIVEN Michelson experiment described in problem 13.101 (See Fig P13.101).

ω = constant = 15,420 rpm

$MM' = 625$ m, $\beta = 0.77146°$

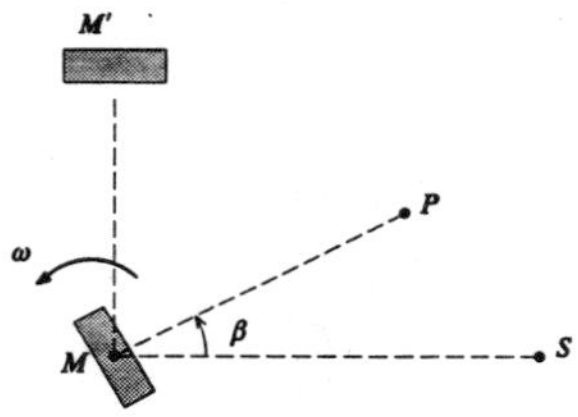

Figure P13.101

FIND Calculate the speed of light.

SOLUTION

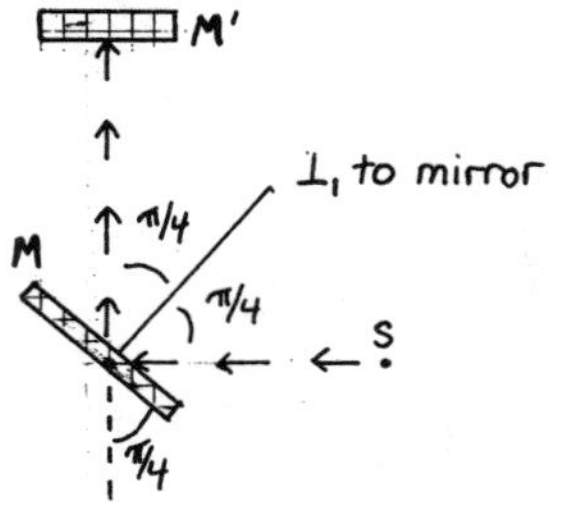

Figure (a)

Reflection of light from pt. S to M':
MM' is ⊥ to MS. Therefore, mirror M is at an angle of $\pi/4$, when light beam strikes M and reflects to mirror M'.
(See Fig (a))

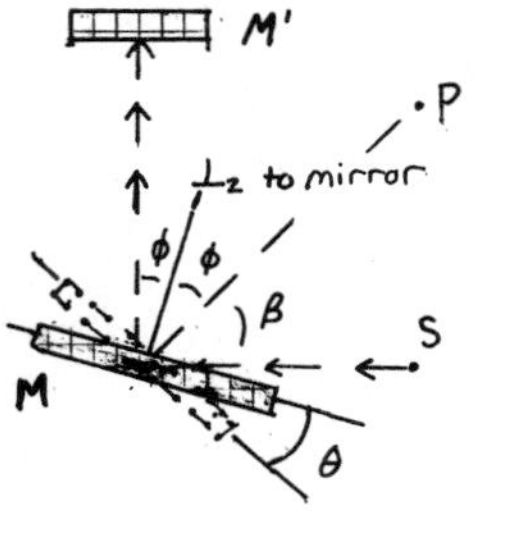

Figure (b)

Mirror M rotates through angle θ as light travels from M to M' and back. The light is then reflected to pt. P

By Figure (b),

$\beta + 2\phi = \pi/2 = 90°$

Combine Figs (a) and (b) to determine relationships between the angles.

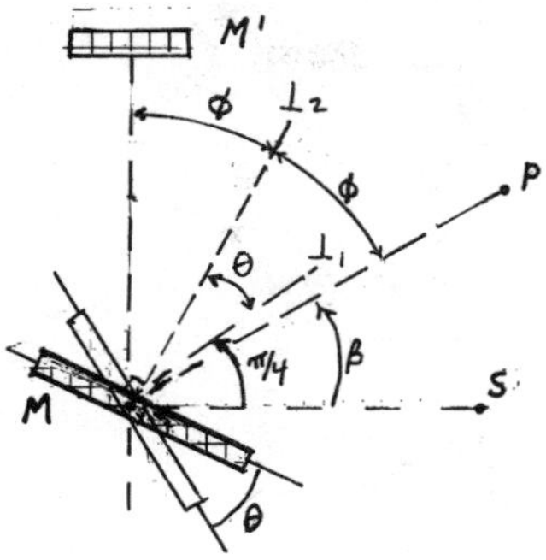

$\beta + 2\phi = \frac{\pi}{2} \Rightarrow \phi = \frac{\pi}{4} - \frac{\beta}{2}$

$\frac{\pi}{2} = \frac{\pi}{4} + \theta + \phi$

$\frac{\pi}{4} = \theta + \frac{\pi}{4} - \frac{\beta}{2}$

$\hookrightarrow \beta = 2\theta$

We know $\beta = 0.77146°$
$= 0.0134645$ rad

Therefore,

$\theta = 0.00673225$ rad

$\omega = \frac{d\theta}{dt} \Rightarrow \theta = \omega t$

$\omega = 15{,}420\ \frac{\text{rev}}{\text{min}} \cdot \frac{2\pi\ \text{rad}}{60\ \text{sec}} = 1614.779$ rad/s

The time t it takes for the mirror to rotate θ rad is equal to the time for the light to travel from M to M' and back.

This distance is $d = 2(625\text{m}) = 1250$ m $= 1.250$ km

$\theta = \omega t \Rightarrow t = \theta/\omega$ (A)

speed of light $= \frac{d}{t} = \frac{\omega d}{\theta}$ (B)

speed of light $= \frac{1.250(1614.779)}{0.00673225} = 299{,}800$ km/s

ALTERNATIVE SOLUTION

From Problem 13.92, we know that $\beta = 2\theta$.

$\beta = 0.77146° = 0.0134645$ rad

Once again, use Eqs (A) and (B) to get

speed of light = 299,800 km/s

13.102 GIVEN Figure (a), where right end of bar slides to right horizontally with speed v.

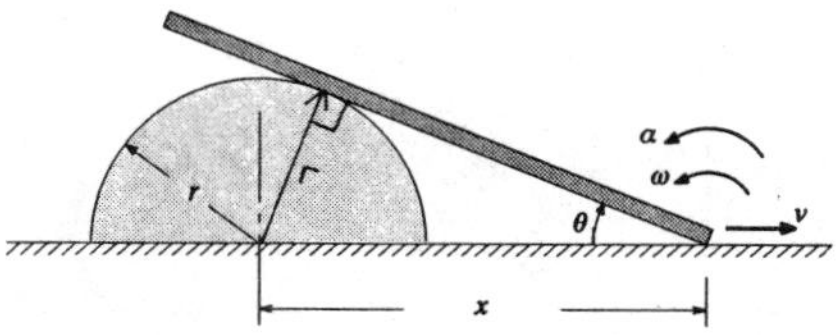

Figure (a)

FIND ω and α in terms of v, x, and r

(Continued)

13.102 cont.

By Fig. (a):

$$x = r/\sin\theta, \quad \omega = -\dot{\theta} \qquad (A)$$

$$\sin\theta = \frac{r}{x}, \quad \cos\theta = \frac{\sqrt{x^2 - r^2}}{x} \qquad (B)$$

By Eq (A),

$$v = \dot{x} = -\frac{r\dot{\theta}\cos\theta}{\sin^2\theta} = \frac{\omega r \cos\theta}{\sin^2\theta}$$

$$\therefore \omega = \frac{v \sin^2\theta}{r\cos\theta} = \frac{vr}{x\sqrt{x^2 - r^2}} \qquad (C)$$

Differentiation of Eq (C) yields:

$$\alpha = \dot{\omega} = vr\left[\frac{-\dot{x}}{x^2\sqrt{x^2-r^2}} - \frac{2x\dot{x}}{2x(x^2-r^2)^{3/2}}\right]$$

or

$$\alpha = \frac{-rv^2(2x^2 - r^2)}{x^2(x^2-r^2)^{3/2}}$$

13.103

GIVEN From Example 13.17,

$$\sin\phi = \frac{r}{R}\sin\theta \qquad (A)$$

$$\dot{x} = -\frac{r\omega\sin(\theta+\phi)}{\cos\phi} \qquad (B)$$

$$\ddot{x} = -\frac{r\omega^2}{\cos\phi}\left[\cos(\theta+\phi) + \frac{r}{R}\frac{\cos^2\theta}{\cos^2\phi}\right] \qquad (C)$$

FIND Plot $\dot{x}$ and $\ddot{x}$ versus θ $(0 \le \theta \le 360°)$ for $r/R = 0.1$, 0.5, and 1.0.
(Note Eq (A) gives ϕ as a function of θ.)
Label maximum values of $\dot{x}$ and $\ddot{x}$ on the plots, and verify that

$$(\ddot{x})_{\theta=0} = -r\omega^2(1 + r/R)$$

$$(\ddot{x})_{\theta=\pi} = r\omega^2(1 - r/R)$$

SOLUTIONS

Results for $r/R = 0.1$, 0.5, and 1.0 are listed in table (a), with $r = 250$ mm, $\omega = 10\pi$ rad/s

Table (a)

r/R	$(\ddot{x})_{\theta=0}$ [m/s²]	$(\ddot{x})_{\theta=\pi}$ [m/s²]
0.1	-271.4	222.1
0.5	-370.1	123.4
1.0	-493.5	0

Plots are shown as follows:

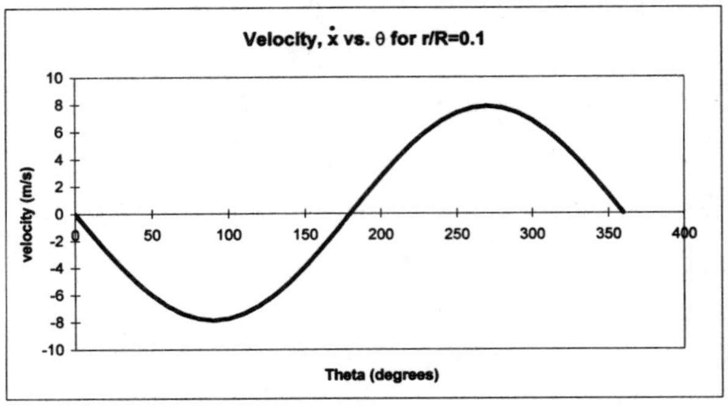

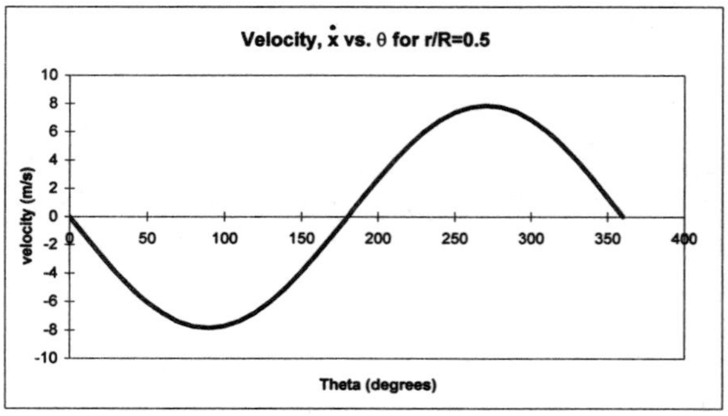

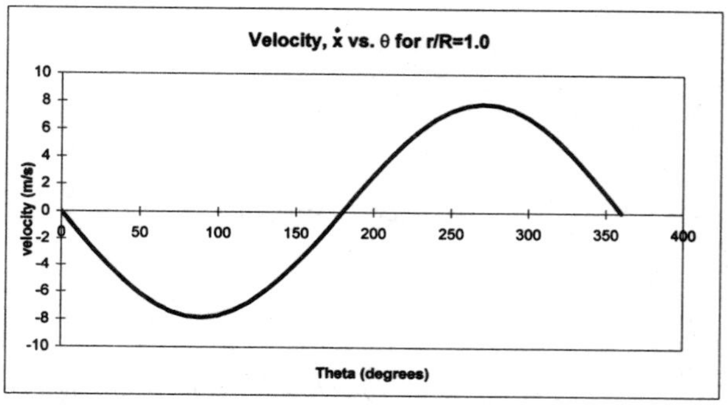

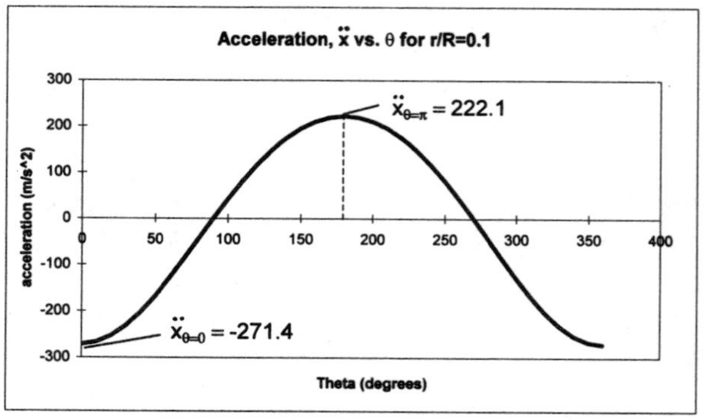

(Continued)

13.103 cont.

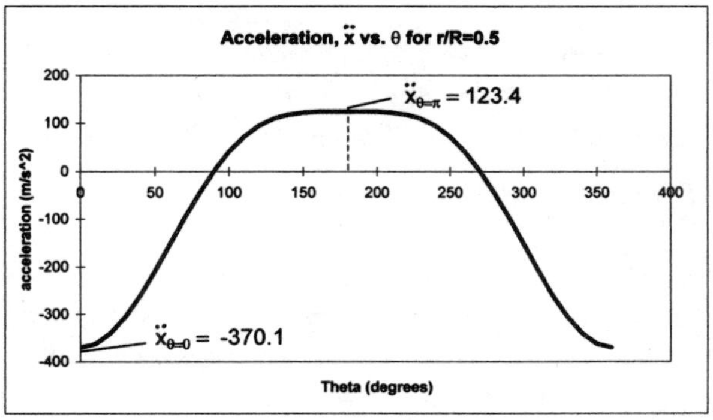

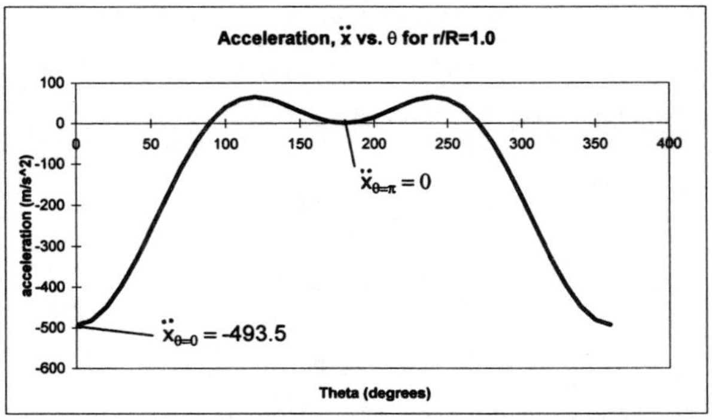

13.104 Computer Problem

GIVEN Data shown in Fig (a)

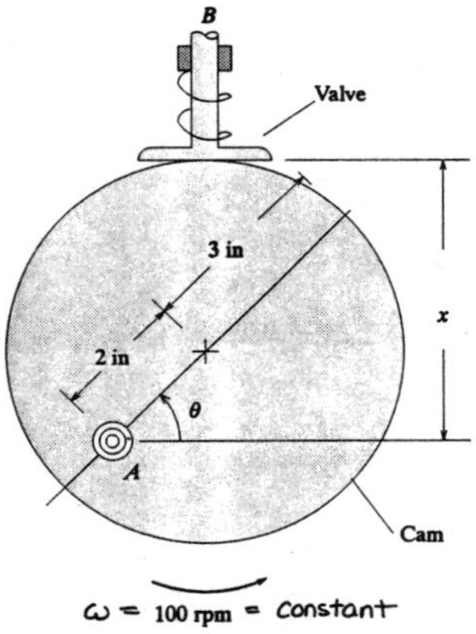

Figure (a) Cam-valve mechanism

FIND a.) Express x, $\dot{x}$, $\ddot{x}$ of valve B in terms of θ.

b.) Construct graphs of x, $\dot{x}$, $\ddot{x}$ for $0 \le \theta \le 360°$

Check graphs to verify that
$\dot{x} = 0$ for $\ddot{x}$ = max and
$\ddot{x} = 0$ for $\dot{x}$ = max

SOLUTION

a.) By Fig (a), $x = 3 + 2\sin\theta$ (A)

Differentiation of Eq (A), with $\omega = \dot{\theta} = 100$ rpm $= 10(\pi/3)$ rad/s,

$\dot{x} = 2\dot{\theta}\cos\theta = 20.944\cos\theta$ [in/s] (B)

$\ddot{x} = -2(\dot{\theta})^2\sin\theta = -219.32\sin\theta$ [in/s²] (C)

b.) By Eqs (A) and (B) and (C), plots are constructed as shown.

From the plots,
for $\ddot{x}$ = max, $\theta = 270°$ and $\dot{x} = 0$
for $\dot{x}$ = max, $\theta = 0°$ (also $\theta = 360°$) and $\ddot{x} = 0$

These results verify as desired

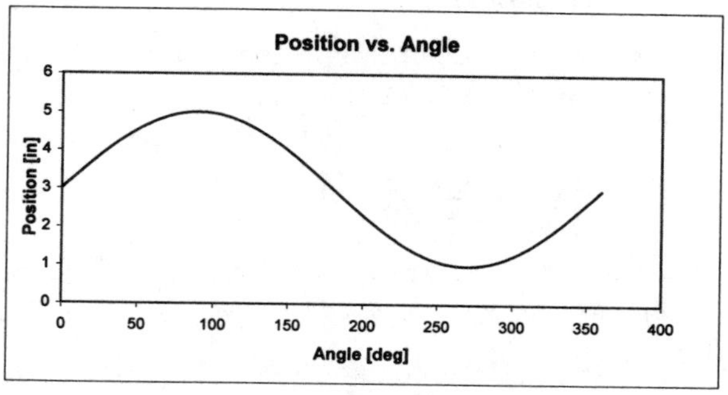

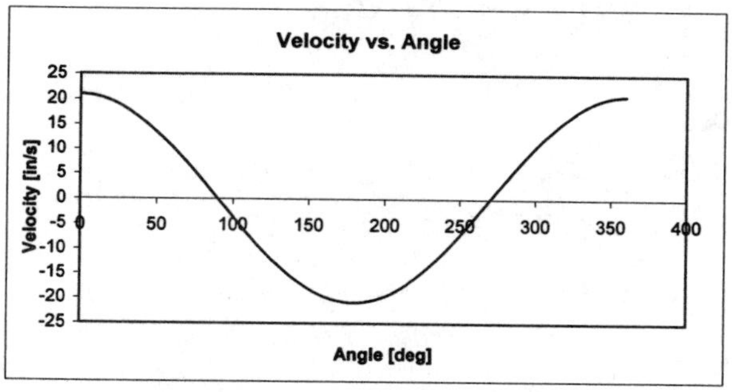

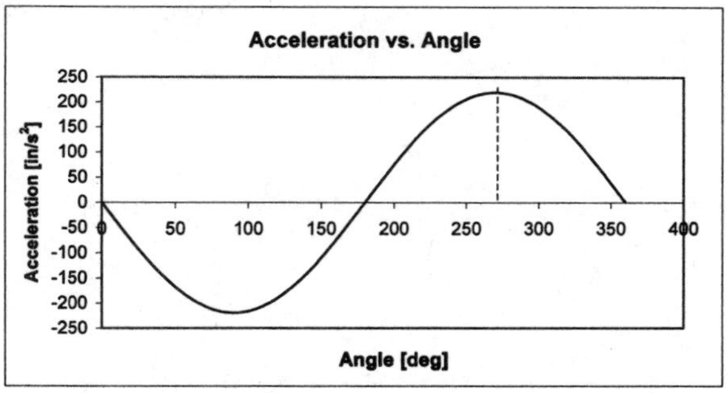

13.105

Computer / Design Problem

GIVEN Slider - crank mechanism shown:

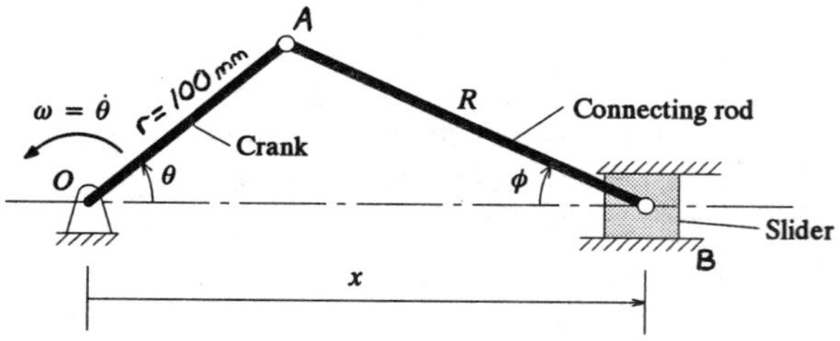

Figure (a)

FIND

a.) Select length $\overline{AB}$ so that the angle ϕ lies in the range $-45° \le \phi \le 45°$ and $x \ge 50$ mm for $0 \le \theta \le 360°$

b.) Discuss your choice of length. Would another value meet the design requirement? If yes, why did you chose the value you did?

c.) With your value of $\overline{AB}$, plot $v = \dot{x}$ and $a = \ddot{x}$ for $0 \le \theta \le 360°$, where $\omega = \dot{\theta} = 2\pi$ rad/s is the constant angular velocity of the crank.

d.) Redesign the slider so that the maximum acceleration of the slider does not exceed $6\ m/s^2$.

SOLUTION

a.) By Fig (a), $x = r\cos\theta + R\cos\phi$ (A)

$r\sin\theta = R\sin\phi$

or $\phi = \sin^{-1}\left(\frac{r\sin\theta}{R}\right)$ (B)

The conditions $\phi = \pm 45°$ are shown in Fig. (b). Since $r = 100$ mm:

$R = \frac{100}{\sin 45°} = 141.4$ mm

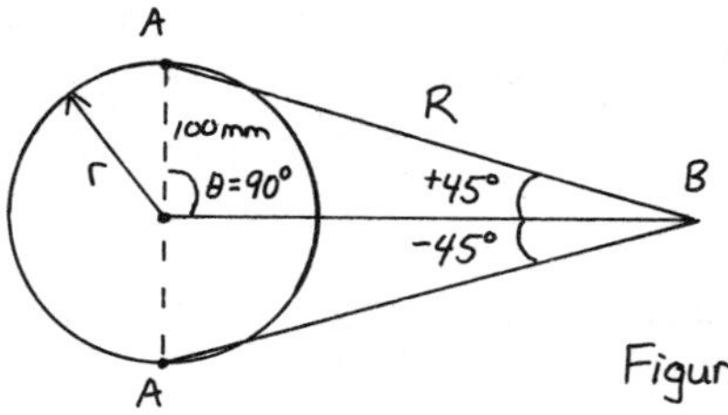

Figure (b)

The condition $x = 50$ mm is shown in Fig (c):

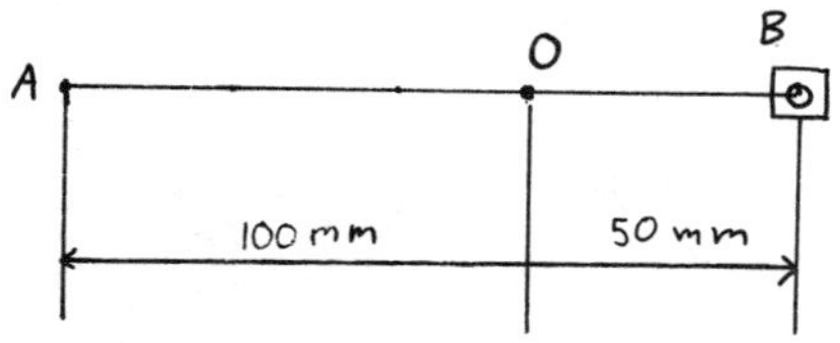

Figure (c)

By Fig (b), $x = \overline{AB} - r = R - 100 = 50$ mm

or $R = 150$ mm

Thus, any length $R \ge 150$ mm will meet requirements $-45° \le \phi \le 45°$, $x \ge 50$ mm

Choose $\underline{R = 150\ \text{mm}}$

b.) As noted in part a, any value $R \ge 150$ mm will meet the requirements. However, the larger R, the more space required and the more energy required to move the slider.

c.) By Example 13.17, with $\theta = \omega t$

$$v = \dot{x} = -r\omega\sin\omega t - R\dot{\phi}\sin\phi \quad (C)$$

$$a = \ddot{x} = \frac{-r\omega^2}{\cos\phi}\left[\cos(\omega t + \phi) + \frac{r}{R}\frac{\cos^2\omega t}{\cos^2\theta}\right] \quad (D)$$

where $\sin\phi = \frac{r}{R}\sin\theta = \frac{r}{R}\sin\omega t$ (E)

By Eq (E), $\dot{\phi} = \frac{r\omega}{R}\frac{\cos\omega t}{\cos\phi}$ (F)

and $\phi = \sin^{-1}\left(\frac{r}{R}\sin\omega t\right)$ (G)

With Eqs (C), (D), (E) and (F), v and a are defined for given values of r, R and ω. See plots:

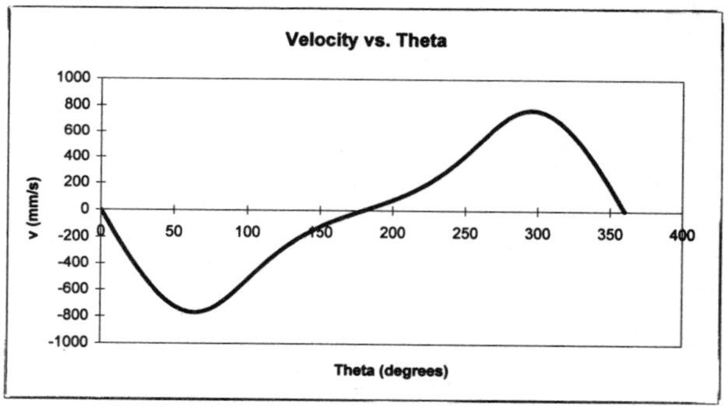

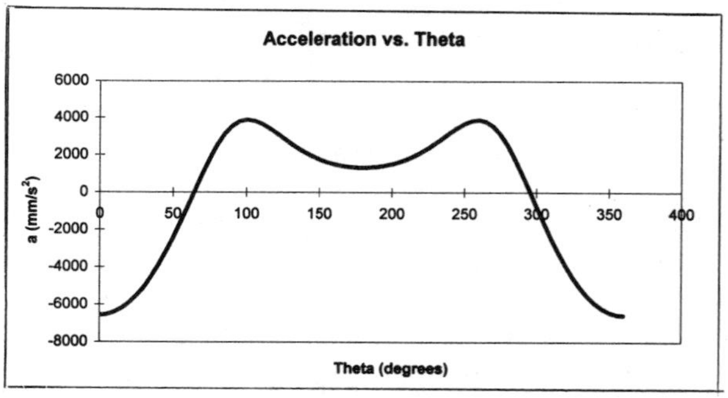

d.) By Example 13.17, Eq (K), the maximum acceleration of the slider is given for $\theta = 0$ (or 360°). (Also see plot of acceleration.)

(Continued)

13.105 cont.

$$a_{max} = r\omega^2(1 + r/R)$$

$$\therefore\ r\omega^2(1 + r/R) \le 6\ m/s^2 \qquad (H)$$

With $r = 0.1\ m$ and $\omega = 2\pi$ rad/s, Eq (H) yields $R \ge 192.4$ mm or minimum $R = 192.4$ mm, which is an awkward length to manufacture. Therefore, use <u>R = 200 mm</u>

13.106 Computer Problem

GIVEN The square cam shown in Fig (a) rotates with constant angular velocity.

FIND
a.) Determine the displacement, x as a function of θ $(0 \le \theta \le 180°)$

b.) Determine velocity, $\dot{x}$ and acceleration, $\ddot{x}$ of the follower in terms of θ

c.) Plot $x, \dot{x}, \ddot{x}$ for $(0 \le \theta \le 180°)$

d.) By plots, explain why cam would hammer badly.

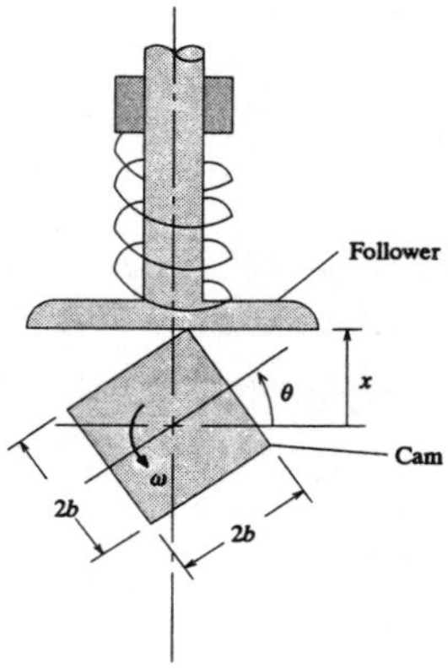

Figure (a)

SOLUTION

a.) By Fig (a),

for $0 \le \theta \le 90°$; $\quad x = \sqrt{2}\, b \sin(\theta + 45°)$ (A)

for $90° \le \theta \le 180°$; $\quad x = \sqrt{2}\, b \sin(\theta - 45°)$ (B)

b) Differentiation of Eqs (A) and (B) yields

for $0 \le \theta \le 90°$ $\quad \dot{x} = \sqrt{2}\, b\omega \cos(\theta + 45°)$ (C)

for $90° \le \theta \le 180°$ $\quad \dot{x} = \sqrt{2}\, b\omega \cos(\theta - 45°)$ (D)

for $0 \le \theta \le 90°$ $\quad \ddot{x} = -\sqrt{2}\, b\omega^2 \sin(\theta + 45°)$ (E)

for $90° \le \theta \le 180°$ $\quad \ddot{x} = -\sqrt{2}\, b\omega^2 \sin(\theta - 45°)$ (F)

c.) With Eqs. (A) through (F), plots of $x, \dot{x}, \ddot{x}$ are obtained, with $\omega = b = 1$ for simplicity.

Plots are shown as follows:

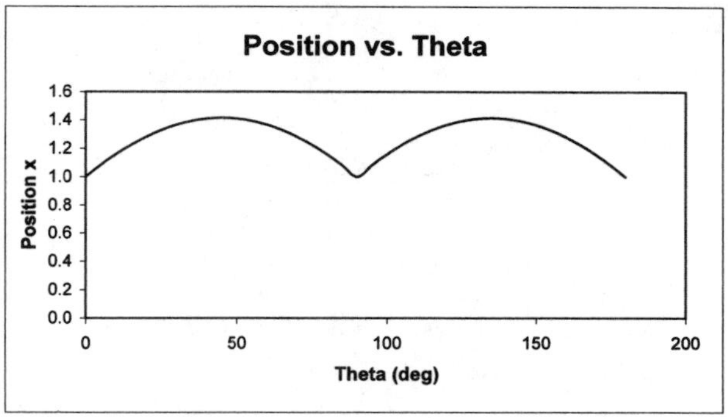

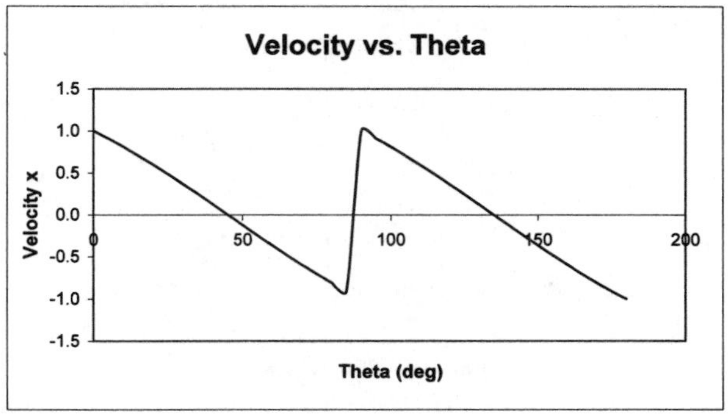

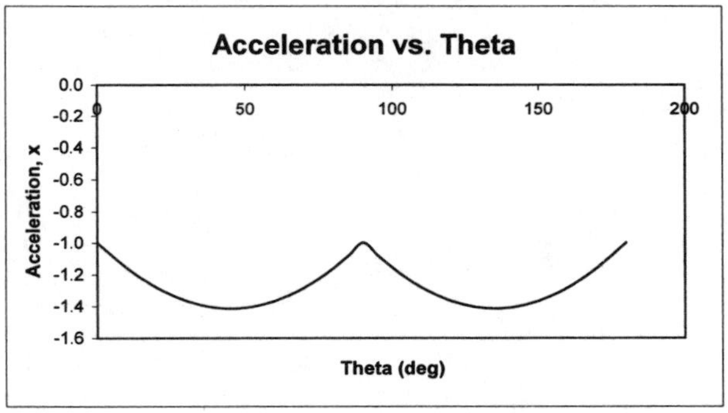

d.) By the plots, it is seen that at $\theta = 90°$, there is a sudden jump in velocity from -1 to $+1$. This jump results in the follower banging into the cam.

13.107 Computer Problem

GIVEN $\tan\phi = \dfrac{\sin\theta}{\frac{b}{a} - \cos\theta}$ (A)

FIND a.) Determine ratio ω_2/ω_1 where $\omega_2 = \dot{\phi}$ and $\omega_1 = \dot{\theta}$

b.) Plot $|\omega_2/\omega_1|$ for $-\frac{180°}{n} \le \theta \le \frac{180°}{n}$ for $n = 3, 4,$ and 6

c.) Discuss significance of $\omega_2/\omega_1 > 1$

(Continued)

13.107 cont.

SOLUTION

a.) Differentiation of Eq (A) with respect to time yields:

$$\dot{\phi}\sec^2\phi = \frac{(\frac{b}{a}-\cos\theta)(\cos\theta)\dot{\theta} - (\sin\theta)(\sin\theta)\dot{\theta}}{(\frac{b}{a}-\cos\theta)^2}$$

or $$\omega_2(1+\tan^2\phi) = \frac{\omega_1[\frac{b}{a}\cos\theta - 1]}{(\frac{b}{a}-\cos\theta)^2}$$

Substitution for $\tan^2\phi$ by Eq (A) yields:

$$\frac{\omega_2}{\omega_1}\frac{[(b/a-\cos\theta)^2+\sin^2\theta]}{(b/a-\cos\theta)^2} = \frac{(b/a\cos\theta-1)}{(b/a-\cos\theta)^2}$$

or $$\frac{\omega_2}{\omega_1} = \frac{(b/a)\cos\theta - 1}{1+(b/a)^2-(2b/a)\cos\theta}$$

b.) See plots of $|\omega_2/\omega_1|$ versus θ. (Values of b/a given by Table P13.107)

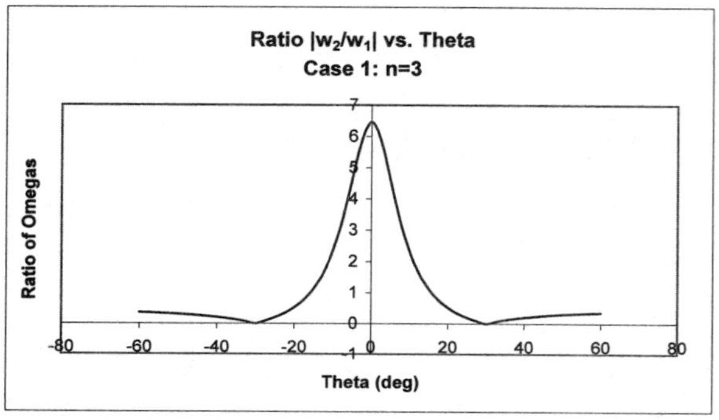

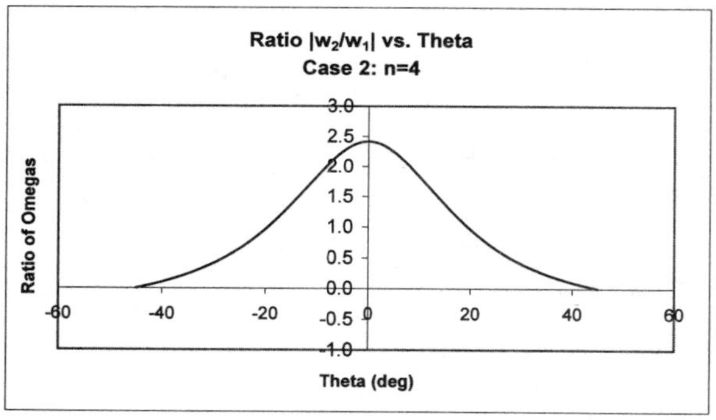

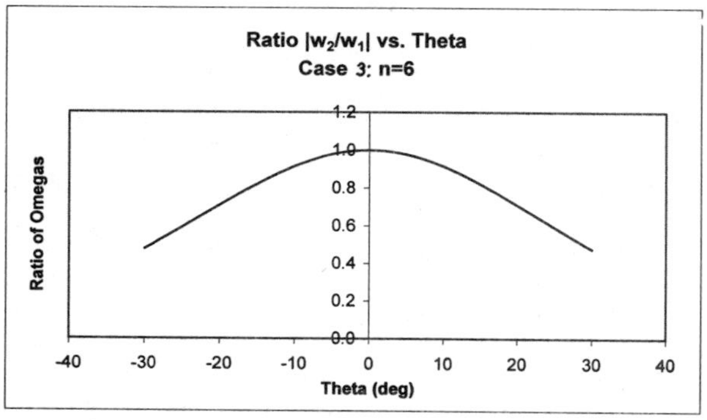

c.) When $\omega_2 > \omega_1$ (n=3 and 4), the angular velocity of the Geneva wheel exceeds the angular speed of the crank. This results in a high rate of angular acceleration of the wheel and its shaft, with corresponding high torques.

13.108 Computer Problem

GIVEN $$\frac{\omega_2}{\omega_1} = \frac{(b/a)\cos\theta - 1}{1+(b/a)^2-2(b/a)\cos\theta} \quad (A)$$

ω_1 = constant

FIND

a) Show that

$$\frac{\dot{\omega}_2}{\omega_1^2} = (b/a\sin\theta)\frac{1-(b/a)^2}{[1+(b/a)^2-2(b/a)\cos\theta]^2}$$

b.) Plot $\dot{\omega}_2/\omega_1^2$ for n=3, 4, and 6 vs. θ for $-180°/n \le \theta \le 180°/n$

c.) Explain instantaneous jump in $\dot{\omega}_2/\omega_1^2$ at $\theta = -180°/n$ and drop at $\theta = 180°/n$

SOLUTION

a.) Differentiation of Eq (a) yields:

$$\frac{\dot{\omega}_2}{\omega_1} = \frac{-(b/a)\omega_1\sin\theta}{1+(b/a)^2-2(b/a)\cos\theta} - \frac{[(b/a)\cos\theta-1][2(b/a)\omega_1\sin\theta]}{[1+(b/a)^2-2(b/a)\cos\theta]^2}$$

or simplifying:

$$\frac{\dot{\omega}_2}{\omega_1^2} = \left(\frac{b}{a}\sin\theta\right)\frac{1-(b/a)^2}{[1+(b/a)^2-2(b/a)\cos\theta]^2}$$

b.) See plots of $\dot{\omega}_2/\omega_1^2$ versus θ (Values of b/a given by Table P13.107)

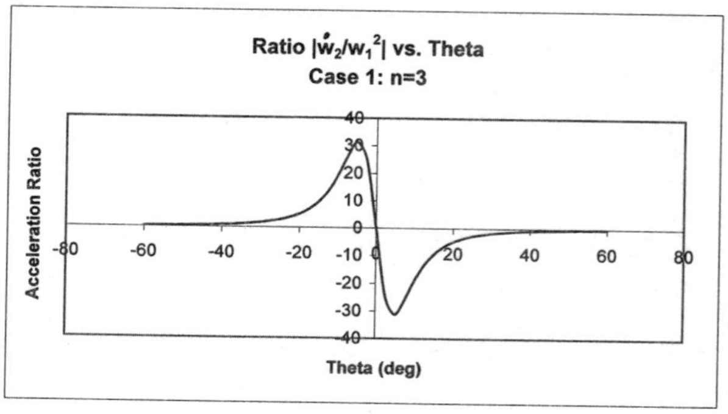

(Continued)

13.108 cont.

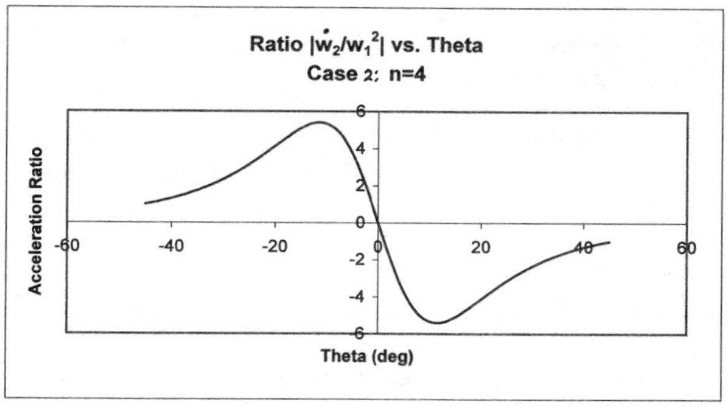

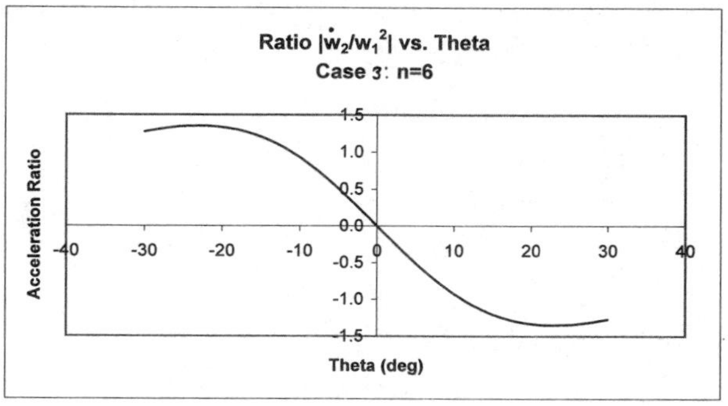

c.) The instantaneous rise (or drop) in acceleration of the wheel at $\theta = -180°/n$ (or at $180°/n$) is due to the fact that the wheel is given a sudden acceleration when the pin engages the slot (or a sudden deceleration as the pin leaves the slot). This results in a suddenly applied force on the wheel shaft (an impact load), when the pin engages a slot. This impact cannot be avoided in the Geneva wheel. Hence, it is used only in slow-speed applications.

13.109 Computer Problem

GIVEN Mechanism shown in Fig (a)

$y = 100\,(5t - \pi/2)$

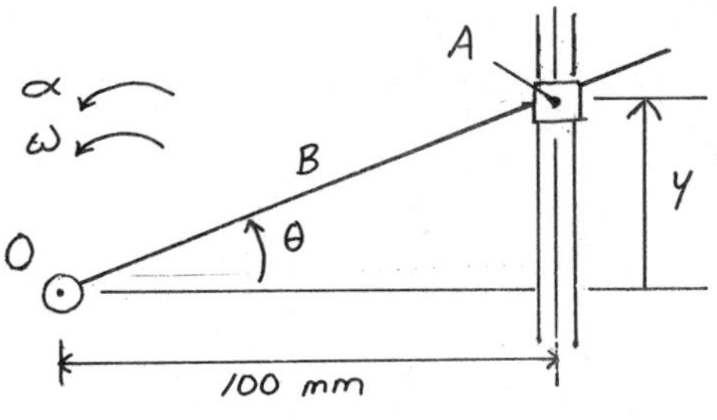

Figure (a)

FIND a.) Formulas for $\omega = \dot{\theta}$ and $\alpha = \ddot{\theta}$ of bar B as functions of time t.

b.) Plots of ω and α as functions of time t, for $-45° \le \theta \le 45°$

c.) From plots, estimate max. and min. values of ω and α and corresponding times. Calculate angles at which these values occur.

SOLUTION

a.) By Fig (a), $y = 100\tan\theta = 100\sin(5t - \pi/2)$

or $\tan\theta = -\cos 5t$ (A)

$\therefore \quad \theta = \tan^{-1}(-\cos 5t)$ (B)

Differentiation of Eq (A) yields (noting that $\frac{d(\tan^{-1}x)}{dt} = \left(\frac{dx}{dt}\right)/(1+x^2)$)

$$\omega = \dot{\theta} = \frac{5\sin 5t}{1+\cos^2 5t} \qquad (C)$$

Then, by differentiation of Eq (C),

$$\alpha = \ddot{\theta} = \frac{75\cos 5t - 25\cos^3 5t}{(1+\cos^2 5t)^2} \qquad (D)$$

b.) To plot ω and α [Eqs (C) and (D)] as functions of t for $-45° \le \theta \le 45°$; note that by Eq (B)

$$t = \tfrac{1}{5}\cos^{-1}(\tan\theta) \qquad (E)$$

Therefore, for $\theta = -45°$

$t = \tfrac{1}{5}\cos^{-1}[\tan(-45°)]$
$= 0.6283$ s

Similarly, for $\theta = 0°$, $t = 0.31416$ s
and, for $\theta = 45°$, $t = 0$

Hence, by Eqs (C) and (D),

$\omega(0) = 0 \qquad \alpha(0) = 12.5$ rad/s²

$\omega(0.314) = 5$ rad/s $\qquad \alpha(0.314) = 0$

$\omega(0.6283) = 0 \qquad \alpha(0.6283) = -12.5$ rad/s²

These values can be used as a check on the plots.

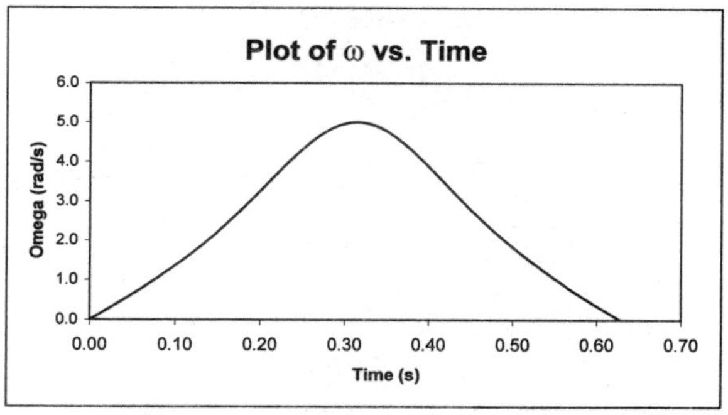

(Continued)

13.109 cont.

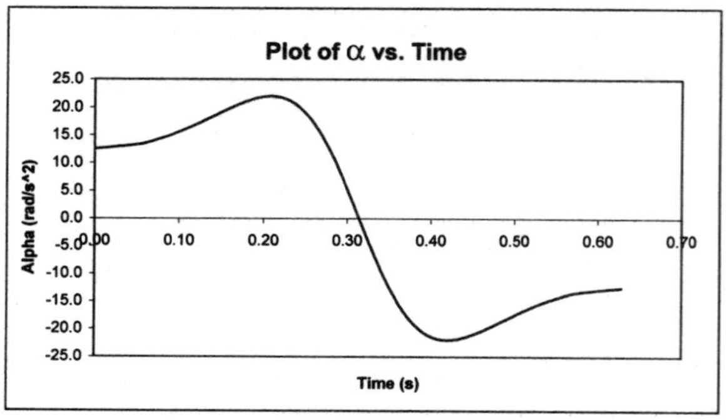

c.) By the plots,

$\omega_{max} = 5$ rad/s at $t = 0.31$ s $\therefore \theta \approx 0°$

$\alpha_{max} = 22$ rad/s² at $t = 0.20$ s $\therefore \theta \approx -27°$

13.110

Design Problem

GIVEN An eccentric circular cam is to produce a rise and a fall of distance h of a follower, in a simple harmonic motion of the cam with respect to time t during a counter clockwise rotation of 180° at constant angular velocity $\omega = \dot{\theta}$

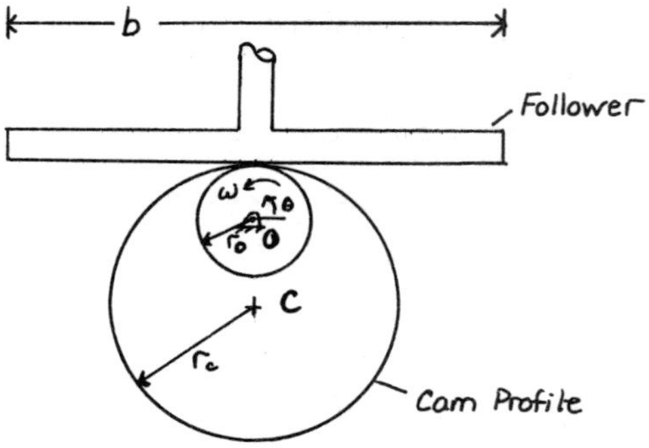

Figure (a) - Initial Position $\theta = 0$

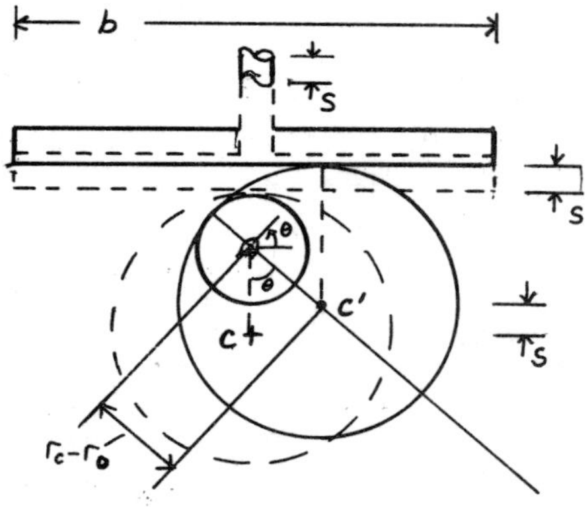

Figure (b) - Displaced Position θ

FIND a.) A design that will meet requirements

b.) Reasonable values of r_o and r_c for $h = 100$ mm

c.) A reasonable value of the width b of the follower to ensure contact with the cam.

SOLUTION

a.) Fig (a) shows the cam at $\theta = 0°$

Fig (b) shows the cam at $\theta > 0°$

By Fig (b), the displacement of the follower is
$$s = (r_c - r_o) - (r_c - r_o)\cos\theta$$
$$= (r_c - r_o)(1 - \cos\theta) \qquad (A)$$

where $\theta = \omega t$. Therefore, the displacement s of the follower is simple harmonic motion with respect to time t.

Also, by Fig (b), when $\theta = 180°$, $s = h$. Therefore, by Eq (A),
$$h = (r_c - r_o)(1 - \cos 180°) = 2(r_c - r_o)$$
or
$$h/2 = (r_c - r_o) \qquad (B)$$

b.) For $h = 100$ mm, by Eq (B)
$$r_c - r_o = 50 \text{ mm} \qquad (C)$$

Any pair of values of r_c, r_o with a difference of 50 mm will work. However, the larger the value of r_c, the more energy required to rotate the cam. Therefore, r_c should be kept to a minimum. Hence, take $r_c = 100$ mm and $r_o = 50$ mm, for a reasonable size cam.

c.) The follower must maintain contact with the top of the cam for any position of the cam. (See Fig (c))

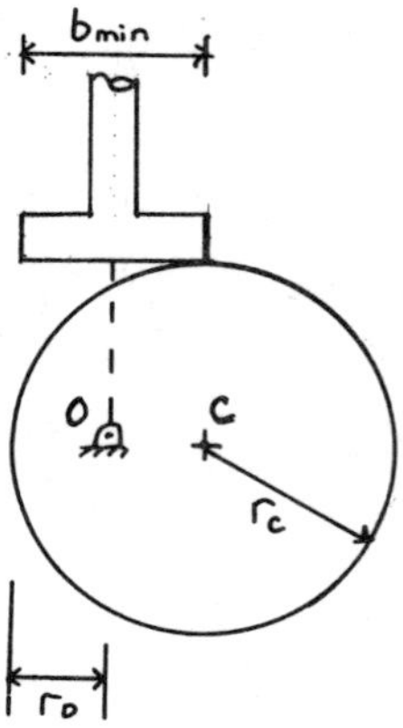

Figure (c) - Minimum b, $\theta = 90°$

(Continued)

13.110 cont.

The critical position is when $\theta = 90°$ or $\theta = 270°$ (Fig (c))

Therefore, the minimum value of b is given by: $\frac{b}{2} = r_c - r_o = \frac{h}{2}$

or $b_{min} = h = 100\ mm$

To ensure that contact is maintained (since in manufacturing of the cam, slight errors might occur), it would be wise to have $b_{min} > 100\ mm$, say 110 mm.

13.111

GIVEN Particle undergoes simple harmonic motion with Period $T = 2s$ and Amplitude $A = 6\ in$

FIND Maximum velocity and maximum acceleration of the particle.

SOLUTION

$T = \frac{2\pi}{\omega}$, $\therefore \omega = \frac{2\pi}{T} = \frac{2\pi}{2} = \pi\ rad/s$

$x = A\cos(\omega t - \beta) = 6\cos(\pi t - \beta)$

$\dot{x} = -A\omega\sin(\omega t - \beta)$

$\dot{x}_{max} = v_{max}$ occurs at $\sin(\omega t - \beta) = 1$

$\therefore v_{max} = -6\pi(1) = -18.85\ in/s$

$\ddot{x} = -A\omega^2\cos(\omega t - \beta)$

$\ddot{x}_{max} = a_{max}$ occurs at $\cos(\omega t - \beta) = 1$

$\therefore a_{max} = -6\pi^2(1) = -59.2\ in/s^2$

13.112

GIVEN A vibrating engine mount with $a_{max} = 40g$, and $f = 50$ cycles/s

FIND Amplitude of vibration

SOLUTION

$a_{max} = 40g = 40(9.81) = 392.4\ m/s^2$

$\omega = 2\pi f = 2\pi(50) = 314.16\ rad/s$

$a = \ddot{x} = -A\omega^2\cos(\omega t - \beta)$;

a_{max} occurs at $\cos(\omega t - \beta) = -1$

$\therefore a_{max} = 392.4 = -A(314.16)^2(-1)$

Thus, $A = 0.003976\ m = 3.98\ mm$

13.113

GIVEN A simple harmonic motion
$x = 3\cos(2t - 0.45) + 5\cos(2t - 0.15)$

FIND The amplitude, A, the frequency f, the period T, and the phase β, of the resultant motion.

SOLUTION

Expanding the above, we have

$x = 3[\cos 2t\cos 0.45 + \sin 2t\sin 0.45]\ldots$
$\ldots + 5[\cos 2t\cos 0.15 + \sin 2t\sin 0.15]$

$x = (3\cos 0.45 + 5\cos 0.15)\cos 2t\ldots$
$\ldots + (3\sin 0.45 + 5\sin 0.15)\sin 2t$ (A)

By Eq (A),

$A^2 = (3\cos 0.45 + 5\cos 0.15)^2 + \ldots$
$\ldots + (3\sin 0.45 + 5\sin 0.15)^2$
$= 58.449 + 4.211 = 62.66$

or $A = 7.92\ in$

Also by Eq (A), $\omega t = 2t$. or $\omega = 2\ rad/s$

$\therefore f = \frac{\omega}{2\pi} = \frac{2}{2\pi} = 0.3183\ Hz$

and $T = 1/f = \pi\ (s)$

The phase β is obtained (with Eq (A)) from

$\tan\beta = \frac{A_1\sin\beta_1 + A_2\sin\beta_2}{A_1\cos\beta_1 + A_2\cos\beta_2}$

$= \frac{3(\sin 0.45) + 5(\sin 0.15)}{3(\cos 0.45) + 5(\cos 0.15)}$

$= 0.2684$

Thus, $\beta = \tan^{-1} 0.2684 = 0.262\ rad = 15.02°$

13.114

GIVEN A simple harmonic motion is defined by the equation
$x = -2\cos(3t - \pi/6) + 5\sin(3t + \pi/4)\ldots$
$\ldots - 4\sin(3t - \pi/3)$ (A)

FIND Amplitude A, the frequency f, and the phase of the resultant motion

SOLUTION

First, we transform Eq (A) into cosine terms by the substitution $\sin\theta = \cos(\theta - \pi/2)$ to obtain:

$x = -2\cos(3t - \pi/6) + 5\cos(3t - \pi/4) - 4\cos(3t - \frac{5\pi}{6})$

We expand the above to obtain

$x = (-2\cos\pi/6 + 5\cos\pi/4 - 4\sin\pi/3)\cos 3t\ldots$
$\ldots + (-2\sin\pi/6 + 5\sin\pi/4 - 4\sin 5\pi/6)\sin 3t$ (B)

(Continued)

13.114 cont.

or $x = 5.268\cos 3t + 0.5355\sin 3t$ (C)

By Eq (C), the amplitude squared is

$A^2 = 5.268^2 + 0.5355^2$

$\therefore A = 5.295 \text{ m}$

Also, by Eq (C), $\omega t = 3t$, or $\omega = 3$ rad/s

So the frequency is

$f = \frac{\omega}{2\pi} = \frac{3}{2\pi} = 0.4775$ Hz

It follows that the period is : $T = \frac{1}{f} = 2.094$ s

With Eq (B), the phase β is obtained from

$$\tan\beta = \frac{A_1\sin\beta_1 + A_2\sin\beta_2 + A_3\sin\beta_3}{A_1\cos\beta_1 + A_2\cos\beta_2 + A_3\cos\beta_3}$$

$$= \frac{-2(\sin\pi/6) + 5(\sin\pi/4) + (-4)(\sin 5\pi/6)}{-2(\cos\pi/6) + 5(\cos\pi/4) + (-4)(\cos 5\pi/6)}$$

$$= \frac{0.5355}{5.2676} = 0.1017$$

$\therefore \beta = \tan^{-1}(0.1017) = 0.101 \text{ rad} = 5.81°$

13.115

GIVEN A simple harmonic motion with frequency $f = 20$ cycles/min ($= 1/3$ Hz) and amplitude $A = 75$ mm

FIND The maximum velocity v_{max} and the maximum acceleration a_{max}

SOLUTION

For a simple harmonic motion

$x = A\sin\omega t$

$\therefore v = \dot{x} = A\omega\cos\omega t$

$a = \ddot{x} = -A\omega^2\sin\omega t$

where $\omega = 2\pi f = 2\pi(1/3) = 2\pi/3$ rad/s

Therefore:

$v_{max} = A\omega = (75)(2\pi/3) = 157.1$ mm/s

$a_{max} = A\omega^2 = (75)(2\pi/3)^2 = 329.0$ mm/s²

13.116

GIVEN A simple harmonic motion with $v_{max} = 15$ ft/s and $a_{max} = 45$ ft/s²

FIND The magnitudes of velocity and acceleration when the displacement x is 80% of the amplitude A.

SOLUTION

For a simple harmonic motion,

$x = A\sin\omega t$ (A)

$\therefore v = \dot{x} = A\omega\cos\omega t$ (B)

$a = \ddot{x} = -A\omega^2\sin\omega t$ (C)

Hence, $v_{max} = A\omega = 15$ ft/s (D)

$a_{max} = A\omega^2 = 45$ ft/s² (E)

Solving Eqs (D) and (E) for A and ω, we find: $A = 5$ ft, $\omega = 3$ rad/s (F)

When $x = 0.80A$, Eq (A) yields

$\sin\omega t = 0.80$

$\therefore \cos\omega t = 0.60$

Then, Eqs (B) and (C) yield

$|v| = (15)(0.6) = 9$ ft/s

$|a| = (45)(0.8) = 36$ ft/s²

13.117

GIVEN Figure E13.17 with constant angular velocity ω.

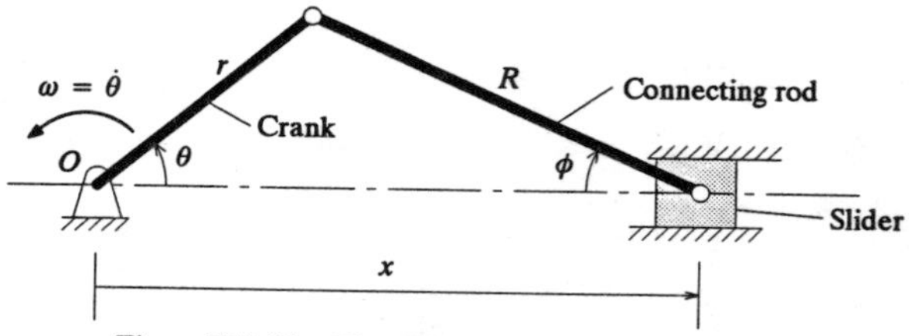

Figure E13.17 – The slider-crank mechanism

FIND a.) Show that if $r = R$, the piston executes simple harmonic motion

b.) Find the amplitude of the oscillation

SOLUTION

a.) If $r = R$, then $\theta = \phi$ (Fig E13.17)

Then, $x = r\cos\theta + R\cos\phi$

$= 2R\cos\theta$

For $\omega = \dot{\theta} = \text{constant}$, $\theta = \phi = \omega t$

$\therefore x = 2R\cos\omega t$ (A)

b.) Equation (A) represents a simple harmonic motion of the piston, with amplitude

$A = 2R = 2r$

13.118 GIVEN Voltage of generator given by

$E = E_o \sin \omega t$ — E_o is max voltage

$f = \omega / 2\pi$

FIND a.) Show that $|\bar{E}| = 2E_o/\pi$ for any whole number of cycles

b.) Show that $(E^2)_{mean}$ for same period is $E_o^2/2$

SOLUTION

a.)

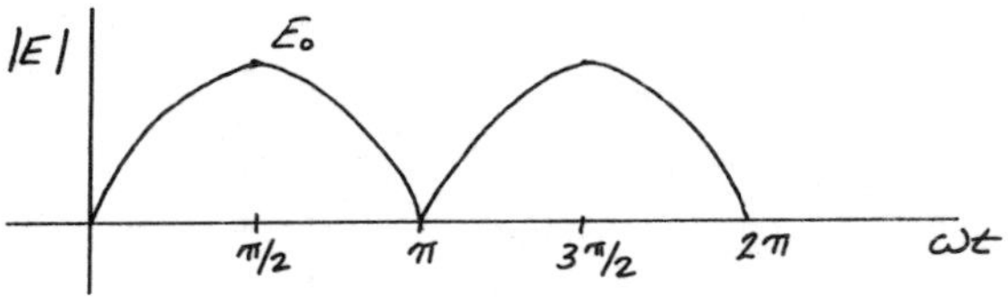

Figure (a)

By Fig (a), the value $|\bar{E}|$ is given by

$$|\bar{E}| = 1/\pi \int_0^{\pi} E_o \sin(\omega t)\, d(\omega t)$$

$$= \frac{E_o}{\pi}\left[-\cos(\omega t)\right]_0^{\pi}$$

or $$\underline{\underline{|\bar{E}| = \frac{2E_o}{\pi}}}$$

b.) $E^2 = E_o^2 \sin^2(\omega t)$

$$\therefore (E^2)_{mean} = \frac{1}{\pi}\int_0^{\pi} E_o^2 \sin^2(\omega t)\, d(\omega t)$$

or, by trig:

$$(E^2)_{mean} = \frac{E_o^2}{\pi}\int_0^{\pi} \frac{1-\cos 2\omega t}{2}\, d(\omega t)$$

$$= \frac{E_o^2}{\pi}\left[\frac{\omega t}{2} - \frac{\sin 2\omega t}{4}\right]_0^{\pi}$$

$$\therefore \underline{\underline{(E^2)_{mean} = 1/2\, E_o^2}}$$

13.119 GIVEN Figure (a) with $\omega = \dot{\theta}$ = constant

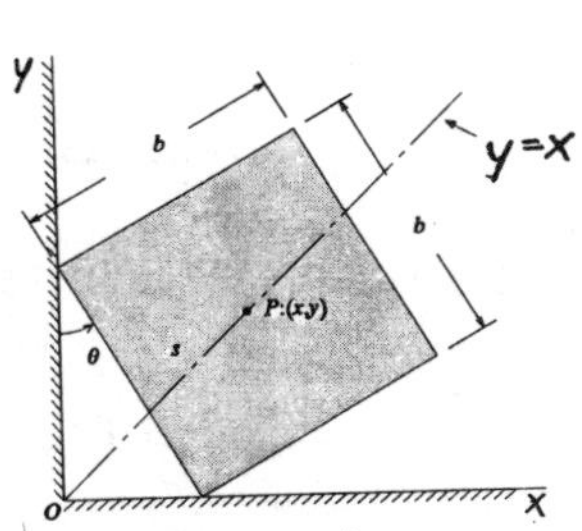

Figure (a)

FIND a.) Show that center P:(x, y) of block performs simple harmonic motion

b.) Determine A and β where $s = A\cos(\omega t - \beta)$ is the distance OP (Fig (a)) and $\theta = 0$ for $t = 0$

SOLUTION

a.) By Fig (a)

$$x_p = \frac{b}{2}(\sin\theta + \cos\theta) \qquad (A)$$

$$y_p = \frac{b}{2}(\sin\theta + \cos\theta) \qquad (B)$$

By Eqs (A) and (B),

$x_p = y_p$

$\ddot{x}_p + (\dot{\theta})^2 x_p = 0$ (harmonic motion)

$\ddot{y}_p + (\dot{\theta})^2 y_p = 0$ (harmonic motion)

Therefore, since $x_p = y_p$ and each coordinate of P undergoes same harmonic motion,

P performs simple harmonic motion along the line $x = y$

b.) When $\theta = 45°$,

$s = OP = s_{max} = \underline{\underline{b = A}}$

$\therefore s = b\cos(\omega t - \beta)$ (C)

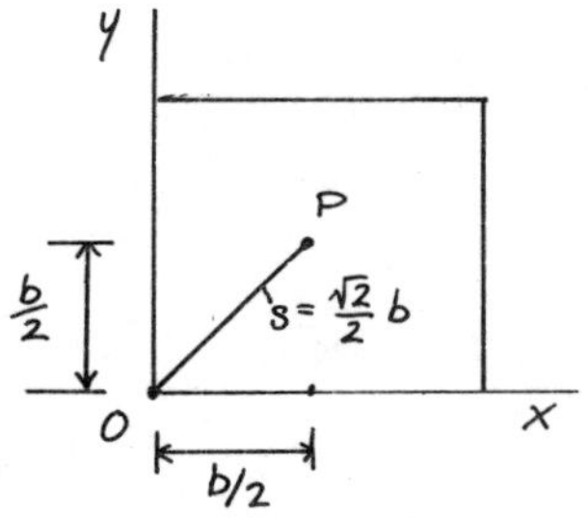

Figure (b)

For $t = 0$, $\theta = 0$ and $s = \frac{\sqrt{2}}{2} b$ (See Fig (b))

Then, by Eq (C),

$\frac{\sqrt{2}}{2} b = b\cos(-\beta)$

$= b\cos\beta$

or $\underline{\underline{\beta = 45°}}$

13.120 GIVEN Figure (a) with $\omega = \dot{\theta}$ and β = constant

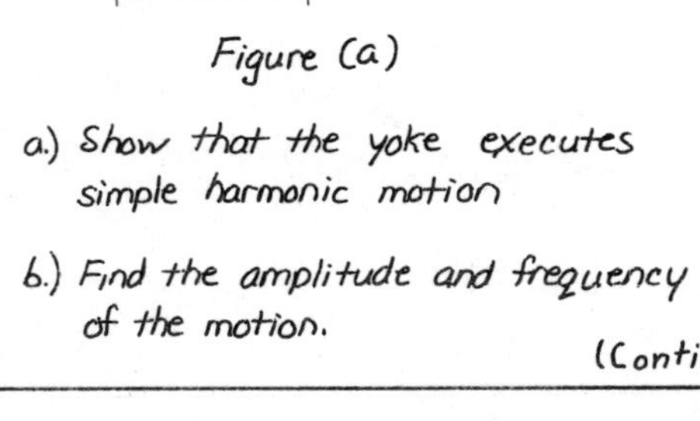

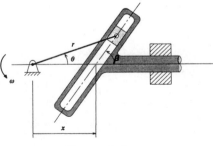

Figure (a)

FIND a.) Show that the yoke executes simple harmonic motion

b.) Find the amplitude and frequency of the motion.

(Continued)

13.120 cont.

c.) Express v_{max} and a_{max} in terms of r, ω, and β

SOLUTION

a.) By Fig (a), since β = constant;

$$180° - \beta = \text{constant}$$

Also, by Fig (b),

$\alpha = \beta - \theta = \beta - \omega t$ (A)

or

$\alpha + \theta = \beta = \text{constant}$

$\therefore \dot{\alpha} + \dot{\theta} = 0$

or

$\dot{\alpha} = -\dot{\theta} = -\omega = \text{constant}$ (B)

and

$\ddot{\alpha} = -\ddot{\theta} = 0$ (C)

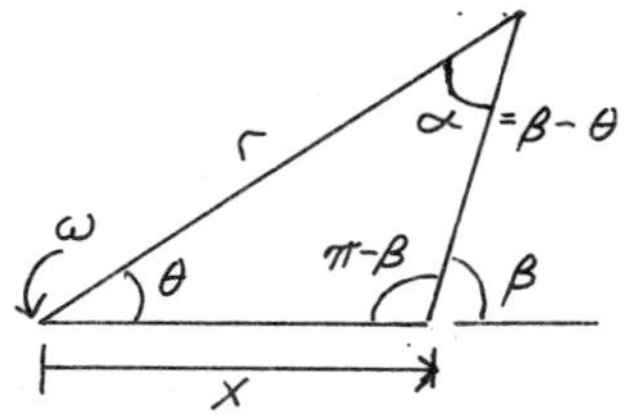

Figure (b)

By law of sines and Fig (b)

$$\frac{x}{\sin\alpha} = \frac{r}{\sin(\pi-\beta)} = \frac{r}{\sin\beta}$$

or (with Eq (A)),

$$x = \left(\frac{r}{\sin\beta}\right)\sin\alpha = \left(\frac{r}{\sin\beta}\right)\sin(\beta - \omega t) \quad (D)$$

Since β = constant, Eq (D) is simple harmonic motion

b.) By Eq (D), The amplitude of motion is

$$A = \frac{r}{\sin\beta}$$

Also, by Eq (D),

$$\dot{x} = -\frac{r\omega}{\sin\beta}\cos(\omega t - \beta) \quad (E)$$

$$\ddot{x} = \frac{r\omega^2}{\sin\beta}\sin(\omega t - \beta) \quad (F)$$

$$= -\omega^2 x$$

or $\ddot{x} + \omega^2 x = 0$

Hence, the frequency of motion is

$$f = \omega/2\pi$$

c.) By Eqs (E) and (F)

$$v_{max} = \dot{x}_{max} = \frac{r\omega}{\sin\beta}$$

$$a_{max} = \ddot{x}_{max} = \frac{r\omega^2}{\sin\beta}$$

13.121 GIVEN Figure (a), with ω = constant

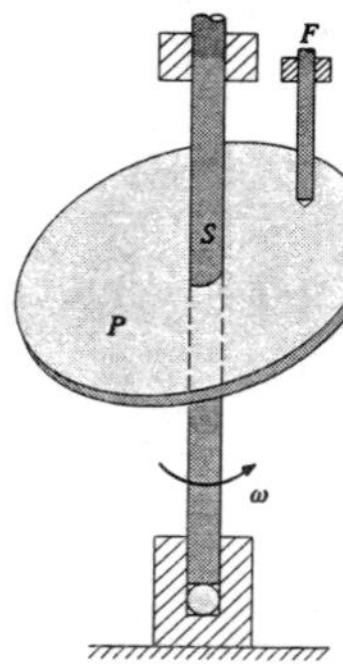

Figure (a)

FIND Show that the follower F executes simple harmonic motion.

SOLUTION

The path of the follower on the disk is the intersection of a cylinder with radius r and an inclined plate (cam plate P). The equation of the cylinder is, by Fig (b):

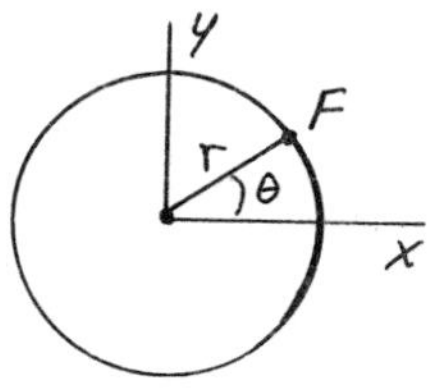

Figure (b)
(Top view of Fig (a))

$x_F = r\cos\theta$
$y_F = r\sin\theta$ (A)
where $\theta = \omega t$

The equation of the plane is, by Fig (c)

$z = x\tan\beta$ (B)

$\therefore$ For $z = z_F$ and

$x = x_F = r\cos\theta$

$z_F = r\cos\theta\tan\beta$

or

$z_F = (r\tan\beta)\cos\omega t$ (C)

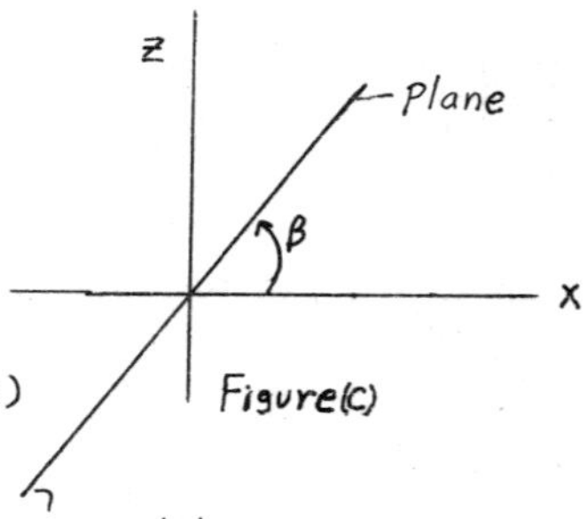

Figure(c)

Since $r\tan\beta = A$ is constant, the equation of motion is harmonic.

That is, $z_F = A\cos\omega t$

13.122 **Computer Problem**

GIVEN Fig (a) depicts motion of center Q given by $y = A\cos 2\omega t$, A = constant, ω = constant

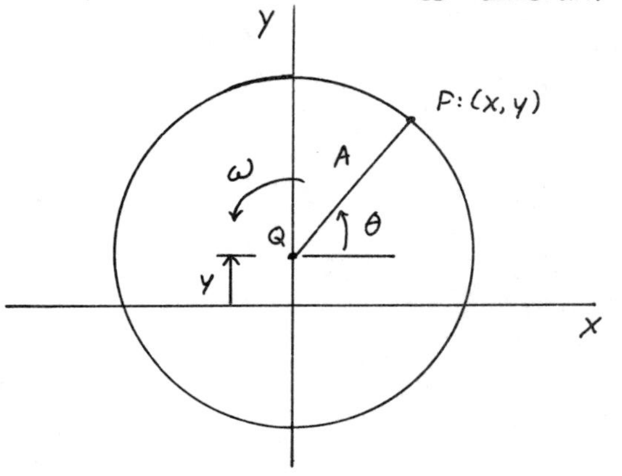

Figure (a)

FIND a.) Determine equations for coordinates (x, y) of particle P

b.) Determine equations for $\dot{x}$, $\dot{y}$

c.) Plot path of P in (x,y) plane with $A = 2$ in, $\omega = \pi$ rad/s, for $0 \le t \le 2$ s. Show $\dot{x}$ and $\dot{y}$ on the plot for $t = 0$, 0.5 s, & 1.0 s.

SOLUTION

a.) By Fig (a),

$$x_P = A\cos\omega t \qquad (A)$$

$$y_P = A\sin\omega t + A\cos 2\omega t \qquad (B)$$

b.) Differentiation of Eqs (A) and (B) gives

$$\dot{x}_P = -A\omega\sin\omega t \qquad (C)$$

$$\dot{y}_P = A\omega\cos\omega t - 2A\omega\sin 2\omega t \qquad (D)$$

c.) For $A = 2$in, $\omega = \pi$ rad/s, Eqs (A), (B), (C), and (D) yield

$$x_P = 2\cos\omega t \qquad y_P = 2(\sin\omega t + \cos 2\omega t)$$

$$\dot{x}_P = -2\pi\sin\omega t \qquad \dot{y}_P = 2\pi(\cos\omega t - 2\sin 2\omega t)$$

See the plot for y_P vs. x_P

Path of P, Y_P vs X_P

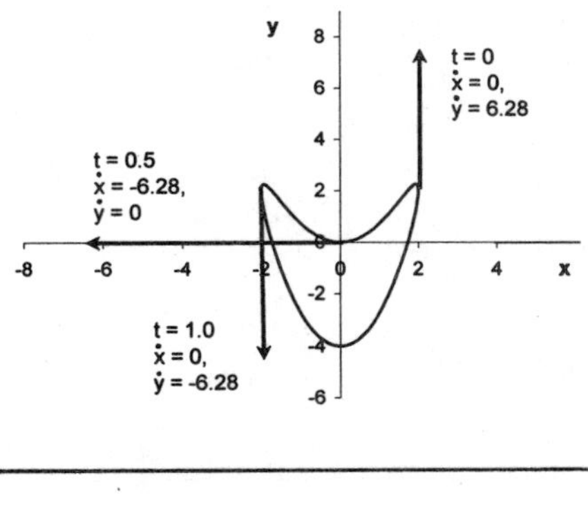

13.123 **Computer / Design Problem**

GIVEN The need to design a cam to produce fine harmonic oscillations of a follower. The cam must be contained in a cylindrical housing of radius 75 mm, and the magnitude of the acceleration must not exceed 500 mm/s². The minimum and maximum radial lengths of the cam are r_1 and r_2 respectively.

FIND a.) Select r_1 and r_2 to meet the design requirements and justify your choice.

b.) With your choices of r_1 and r_2, plot the profile of the cam.

SOLUTION

a.) The equation of the profile of the cam may be written as [See Eq (e), Example 13.18]

$$R = r_1 + (r_2 - r_1)\cos 5\theta \qquad (A)$$

Hence, the displacement of the follower is:

$$x = (r_2 - r_1)\cos 5\theta \qquad (B)$$

Differentiation of Eq (a) yields

$$v = \dot{x} = -5(r_2 - r_1)\sin 5\theta$$

$$a = \ddot{x} = -25(r_2 - r_1)\cos 5\theta \qquad (C)$$

So, the magnitude of the acceleration of the follower is

$$|a| = 25(r_2 - r_1) \qquad (D)$$

Since $|a| \le 500$ mm/s², Eq (D) yields

$$25(r_2 - r_1) \le 500 \qquad (E)$$

However, to contain the cam in its housing: $r_2 < 75$ mm $\qquad (F)$

Hence, Eqs (E) and (F) yield

$$r_1 \ge r_2 - 20 \ge 75 - 20 = 55$$

To facilitate the manufacturing of the cam system (to ensure sufficient clearance between the cam and its housing), say, take $r_2 = 70$ mm. Then,

$$r_1 \ge r_2 - 20 = 50$$

If we take $r_1 = 50$ mm, $r_2 = 70$mm, $|a| = 500$ mm/s².

To take into account manufacturing inaccuracies, it may be wise to take r_1 somewhat larger, say $r_1 = 52$ mm.

(Continued)

13.123 cont.

Then, with $r_1 = 52$ mm,

$$|a| = (25)(70-52) = 450 \text{ mm/s}^2$$

b.) With $r_1 = 52$ mm and $r_2 = 70$ mm, Eq (A) yields the profile equation

$$R = 52 + 18\cos 5\theta \qquad (G)$$

See the plot of **R** as a function of **θ**:

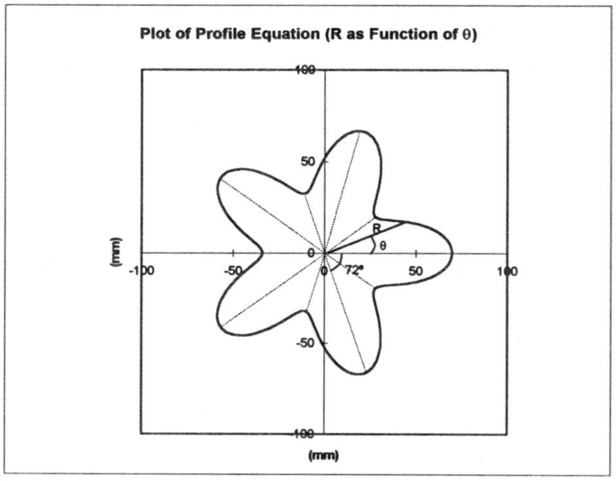

13.124 GIVEN A curve defined by

$$x = t^2, \quad y = t^3 \qquad (A)$$

FIND

a.) Sketch the curve in the (x, y) plane.

b.) Determine the radius of curvature at $x = y = 1$.

SOLUTION

a.) By Eqs (A):

$$y^2 = t^6$$
$$x^3 = t^6$$

or $y^2 = x^3 \qquad (B)$

See the plot of the curve $y^2 = x^3$ →

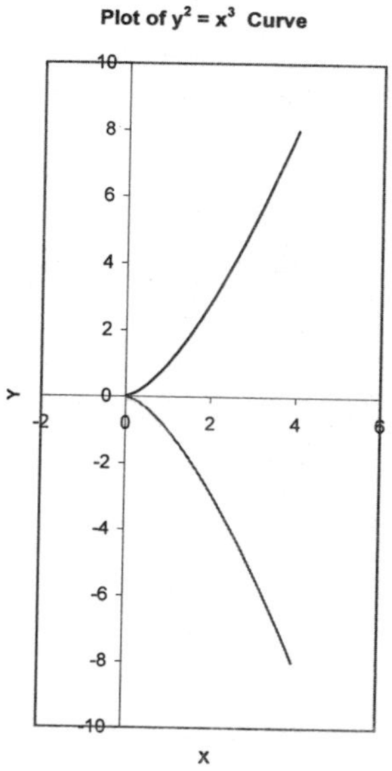

As $t \to \infty$, $x \to \infty$ and $y \to \infty$ (faster than x).

By Eqs (A),

$$\dot{x} = 2t, \quad \dot{y} = 3t^2 \qquad (C)$$
$$\ddot{x} = 2, \quad \ddot{y} = 6t \qquad (D)$$

By Eq 13.56

$$\frac{1}{r} = \left| \frac{\dot{x}\ddot{y} - \dot{y}\ddot{x}}{[(\dot{x})^2 + (\dot{y})^2]^{3/2}} \right| \qquad (E)$$

Substitution of Eqs (C) and (D) into Eq (E) yields

$$\frac{1}{r} = \left| \frac{12t^2 - 6t^2}{[4t^2 + 9t^4]^{3/2}} \right|$$

$$= \left| \frac{6}{t[4 + 9t^2]^{3/2}} \right| \qquad (F)$$

as $t \to \infty$ (see sketch),

$$\underline{\underline{\frac{1}{r} \to 0}}$$

b.) By Eqs (A), when $x = y = 1$, $t = 1$.

∴ For $t = 1$, Eq (F) yields

$$\frac{1}{r} = \frac{6}{[13]^{3/2}} = 0.1280$$

or $\underline{\underline{r = 7.81}}$

13.125 GIVEN Railroad track transition curve defined by $y = x^3/12{,}000{,}000$; x and y in ft.

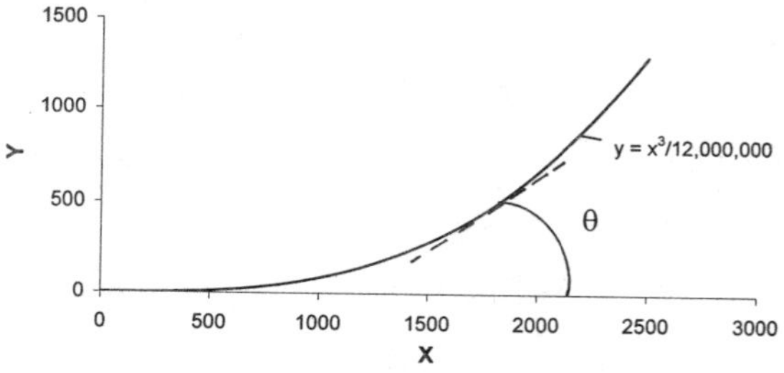

Figure (a)

FIND Time rates that an engine traveling at 40 ft/s changes direction at $x = 500$ ft and $x = 2000$ ft.

SOLUTION

The time rate of changing direction is (Fig.(a))

$$\omega = \frac{d\theta}{dt} = \frac{d\theta}{ds}\frac{ds}{dt} = \frac{1}{r} v \qquad (A)$$

Where, by definition:

$$\frac{1}{r} = \left| \frac{d^2y/dx^2}{[1 + (\frac{dy}{dx})^2]^{3/2}} \right| = \frac{d\theta}{ds} \qquad (B)$$

and

$$v = \frac{ds}{dt} \qquad (C)$$

(Continued)

13.125 cont

With $y = \frac{x^3}{12{,}000{,}000}$, we have

$$\frac{dy}{dx} = \frac{x^2}{(4)(10^6)} \qquad (D)$$

$$\frac{d^2y}{dx^2} = \frac{x}{(2)(10^6)} \qquad (E)$$

at $x = 500$ ft,

$$\frac{dy}{dx} = 0.0625 \text{ ft/s}, \quad \frac{d^2y}{dx^2} = 0.000250 \text{ ft/s}^2$$

So, by Eq (B),

$$\frac{1}{r} = \frac{0.000250}{(1.003906)^{3/2}} = 0.0002485$$

With $v = 40$ ft/s, Eq (A) yields

$$\omega = \frac{1}{r} v = (0.0002485)(40)$$

or $\omega = 0.00994$ rad/s at $x = 500$ ft

at $x = 2000$ ft,

$$\frac{dy}{dx} = 1 \text{ ft/s}, \quad \frac{d^2y}{dx^2} = 0.001 \text{ ft/s}^2$$

So, by Eq (B)

$$\frac{1}{r} = \frac{0.001}{(2)^{3/2}} = 0.0003536$$

With $v = 40$ ft/s, Eq (A) yields

$$\omega = \frac{1}{r} v = (0.0003536)(40)$$

or $\omega = 0.01414$ rad/s at $x = 2000$ ft

13.126 GIVEN Equation of catenary

$$wy = H \cosh\left(\frac{wx}{H} + C\right) + K \qquad (A)$$

FIND Show that w/H represents the curvature at the lowest point of the curve.

SOLUTION

Since $\cosh\theta = \frac{e^{\theta} + e^{-\theta}}{2}$ is symmetric in θ (i.e. $\cosh\theta = \cosh(-\theta)$),

$$y = \frac{H}{w}\cosh\left(\frac{wx}{H} + C\right) + \frac{K}{w}$$

is symmetric in x. (See Fig (a)).

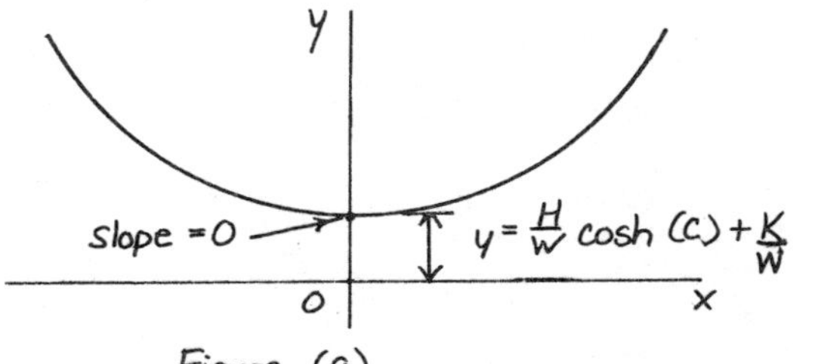

Figure (a)

By Eq (a), for $x = 0$

$$\frac{dy}{dx} = \sinh\left(\frac{wx}{H} + C\right) = \sinh C = 0$$

$$\therefore C = 0 \text{ and } \frac{dy}{dx} = \sinh\left(\frac{wx}{H}\right) \qquad (B)$$

$$\therefore \frac{d^2y}{dx^2} = \frac{w}{H}\cosh\left(\frac{wx}{H}\right) \qquad (C)$$

By Eq 13.58, the curvature is

$$\frac{1}{r} = \left| \frac{d^2y/dx^2}{[1 + (dy/dx)^2]^{3/2}} \right| \qquad (D)$$

Substitution of Eqs (B) and (C) into Eq (D) yields at $x = 0$, the lowest point of the curve,

$$\frac{1}{r} = \left| \frac{\frac{w}{H}\cosh(0)}{[1 + (\sinh(0))^2]^{3/2}} \right|$$

By definition of $\cosh\theta$ and $\sinh\theta$, $\cosh(0) = 1$, $\sinh(0) = 0$. Therefore,

$$\frac{1}{r} = \frac{w}{H}$$

13.127

FIND

a.) Show that $x = a\cosh t$, $y = b\sinh t$, where a and b are positive constants, represents a hyperbola

b.) Curvature of hyperbola at vertex

SOLUTION

The equation of a hyperbola is given by

$$\frac{x^2}{a^2} - \frac{y^2}{b^2} = 1$$

a.) Write $\frac{x}{a} = \cosh t$, $\frac{y}{b} = \sinh t$ (A)

Then $\frac{x^2}{a^2} - \frac{y^2}{b^2} = \cosh^2 t - \sinh^2 t = 1$ (see Appendix B)

so, $\frac{x^2}{a^2} - \frac{y^2}{b^2} = 1$ (Eq. of hyperbola)

b.) at vertex ($x = a$, $y = 0$; see Fig (a))

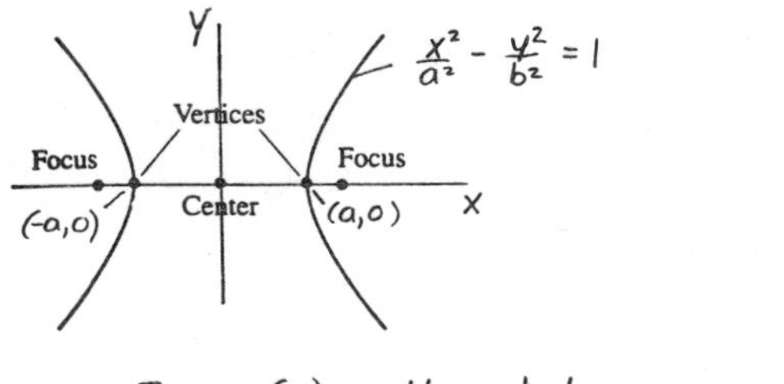

Figure (a) - Hyperbola

(Continued)

13.127 cont

For $x=a$, $y=0$, $t=1$

Therefore, by Eq (A)

$\dot{x} = a\sinh t = 0 \qquad \ddot{x} = a\cosh t = a$

$\dot{y} = b\cosh t = b \qquad \ddot{y} = b\sinh t = 0$

$$\therefore \quad \frac{1}{r} = \left| \frac{\dot{x}\ddot{y} - \dot{y}\ddot{x}}{(\dot{x}^2 + \dot{y}^2)^{3/2}} \right| = \frac{a}{b^2}$$

13.128 GIVEN Plane curve defined by

$$x = x(s), \quad y = y(s) \qquad (A)$$

where s is arc length on the curve (Fig (a))

PROVE The curvature is $\frac{1}{r} = \frac{y''}{x'} = -\frac{x''}{y'}$ (B)

where primes denote derivatives with respect to s, then show that

$$\frac{1}{r^2} = (x'')^2 + (y'')^2 \qquad (C)$$

SOLUTION

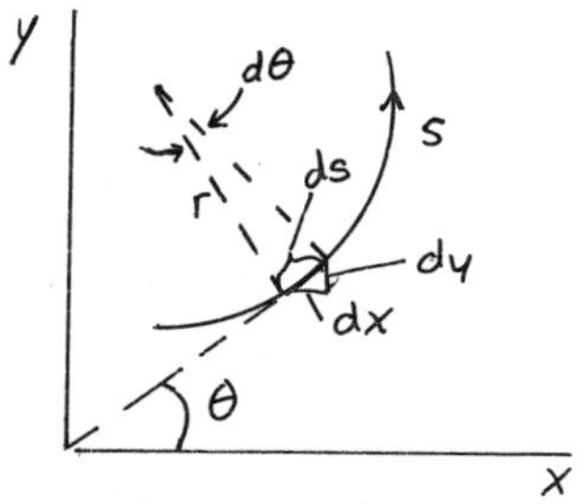

Figure (a)

By definition,

$$\frac{1}{r} = \frac{d\theta}{ds} \qquad (D)$$

By Fig (a),

$$\left.\begin{aligned} \cos\theta &= \frac{dx}{ds} = x' \\ \sin\theta &= \frac{dy}{ds} = y' \end{aligned}\right\} (E)$$

Differentiation of Eqs. (E) yields:

$$\left.\begin{aligned} x'' &= -\theta'\sin\theta \\ y'' &= \theta'\cos\theta \end{aligned}\right\} \quad (F)$$

where primes denote derivatives with respect to s.

Eqs. (D), (E), and (F) yield

$$\frac{1}{r} = \theta' = \frac{y''}{x'} = -\frac{x''}{y'} \qquad (G)$$

Also, by Eqs. (E):

$$(x')^2 + (y')^2 = 1 \qquad (H)$$

Hence, Eqs. (G) and (H) yield

$$(x')^2 + (y')^2 = (ry'')^2 + (-rx'')^2 = 1$$

or

$$\frac{1}{r^2} = (x'')^2 + (y'')^2$$

13.129 Computer Problem

GIVEN The parabola $y = 3 - 2x^2$ (A)

FIND

a.) The radius of curvature and coordinates of the center of curvature of the curve at P, the point it cuts the x axis. $(x>0)$

b.) Plot the curve and show the circle of curvature at P.

SOLUTION

By Eq (A): $\frac{dy}{dx} = -4x, \quad \frac{d^2y}{dx^2} = -4$ (B)

at $y=0$, $3-2x^2=0$

or $x = \sqrt{3/2}$, $(x>0)$ (C)

$\therefore \frac{dy}{dx} = \sqrt{24}, \quad \frac{d^2y}{dx^2} = -4 \quad \left(\frac{dy}{dx}\right)^2 = 24$ (D)

Hence, with Eqs (D),

$$\frac{1}{r} = \left| \frac{d^2y/dx^2}{\left[1 + \left(\frac{dy}{dx}\right)^2\right]^{3/2}} \right| = 0.032$$

or $r = 31.3$ (E)

To determine the coordinates of the center of curvature, consider the sketch shown in Fig (a):

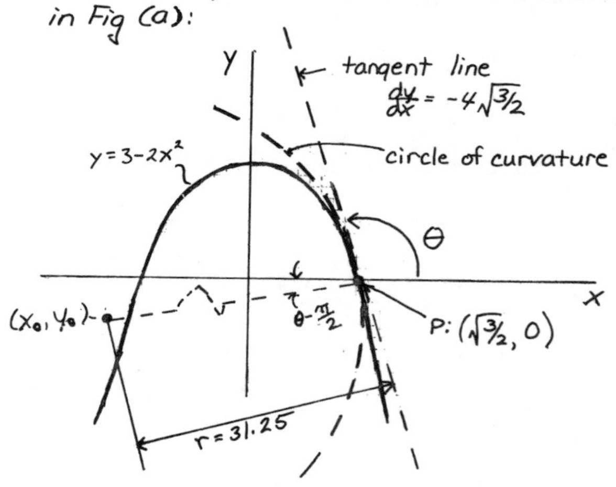

Figure (a)

at $y=0$, $x = \sqrt{3/2}$,

$\tan\theta = \frac{dy}{dx} = -4\sqrt{3/2} = -4.899$

$\therefore \theta = 180° - \tan^{-1}(-4.899)$

$= 180° - 78.46° = 101.54°$

By Fig (a), $x_0 = \sqrt{3/2} - 31.25\cos(\theta - \pi/2)$

$y_0 = -31.25\sin(\theta - \pi/2)$

$\therefore \quad x_0 = -29.29, \quad y_0 = -6.248$

(Continued)

13.129 cont.

b.) See plot of $y = 3 - 2x^2$ and trace of circle of curvature.

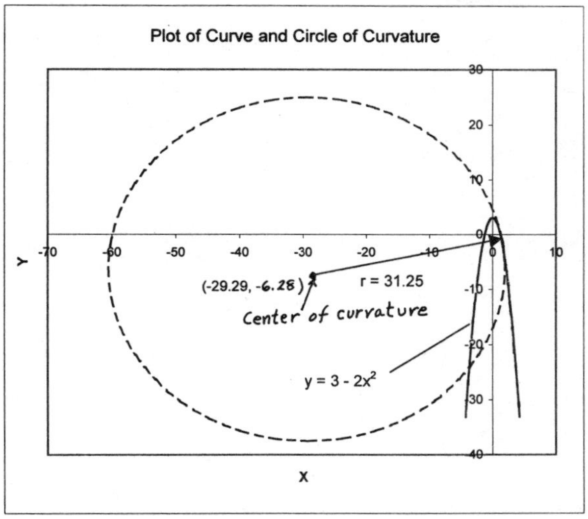

13.130 GIVEN An airplane makes a banked horizontal turn at a constant speed of $v = 240$ m/s

FIND Minimum allowable radius (m) of the turn, if $a_{max} \le 4g$

SOLUTION

Since the plane flies at a constant speed v in the turn,

$$a_t = 0, \quad a_n = \frac{v^2}{r} \le 4g$$

Therefore, $r \ge \frac{v^2}{4g} = \frac{(240)^2}{4(9.81)}$

or $r_{min} = 1467.9$ m

13.131 GIVEN Particle moves with constant speed v along the curve $y = 3x^2$ (A)

FIND a.) Centripetal acceleration as a function of v and x

b.) Tangential acceleration

SOLUTION

a.) By Eq. (A), $\frac{dy}{dx} = 6x$, $\frac{d^2y}{dx^2} = 6$

$$\therefore \frac{1}{r} = \left| \frac{d^2y/dx^2}{[1 + (dy/dx)^2]^{3/2}} \right| = \frac{6}{(1+36x^2)^{3/2}}$$

and $a_n = \frac{v^2}{r} = \frac{6v^2}{(1+36x^2)^{3/2}}$

b.) The tangential acceleration is

$$a_t = \frac{dv}{dt} = 0 \qquad \text{since } v \text{ is constant}$$

13.132 GIVEN Fig (a) with ω = constant and length of oscillating arm OS equal to L.

FIND Expressions for the speed v and tangential and centripetal components a_t and a_n of the tip of arm OS in terms of a, b, L, θ, and ω.

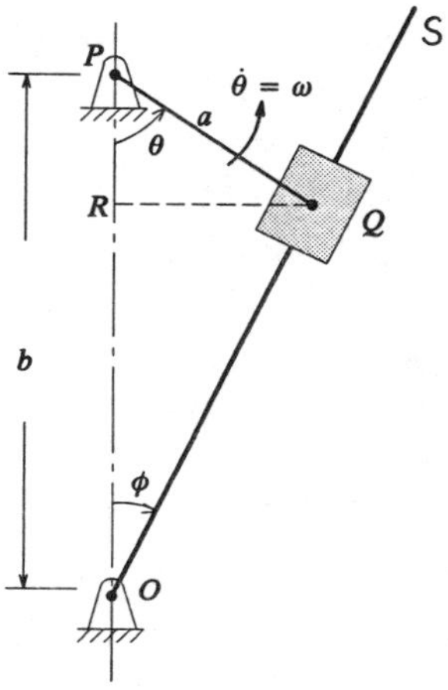

Figure (a)

SOLUTION

Since arm OS rotates about point O, the speed v of its tip is:

$$v = L|\dot{\phi}| \quad (A)$$

By Eq (b) of Example 13.16,

$$\dot{\phi} = -\frac{a\omega(a - b\cos\theta)}{a^2 + b^2 - 2ab\cos\theta}$$

∴ By Eq. (A): $v = \frac{aL\omega(a - b\cos\theta)}{a^2 + b^2 - 2ab\cos\theta}$ (B)

The tangential acceleration a_t of its tip is

$$a_t = L\ddot{\phi} \quad (C)$$

By Eq. (c) of Example 13.16,

$$\ddot{\phi} = \frac{ab(a^2 - b^2)\omega^2 \sin\theta}{(a^2 + b^2 - 2ab\cos\theta)^2}$$

∴ By Eq. (C), $a_t = \frac{abL(a^2 - b^2)\omega^2 \sin\theta}{(a^2 + b^2 - 2ab\cos\theta)^2}$

The centripetal acceleration is given by

$$a_n = \frac{v^2}{L} \quad (D)$$

Hence, by Eqs. (B), and (D)

$$a_n = L\left[\frac{a\omega(a - b\cos\theta)}{a^2 + b^2 - 2ab\cos\theta}\right]^2$$

13.133 GIVEN Ocean wave traveling with constant wave length L and speed

$$c = \sqrt{gL/2\pi} \quad (A)$$

A particle of water performs a circular orbit of radius r (Fig (a)) with constant speed as waves advance a distance L.

SHOW That centripetal acceleration of a water particle at the surface is

$a_n = \pi g h / L$

where $h = 2r$ is the total wave height from trough to crest (see Fig (b)).

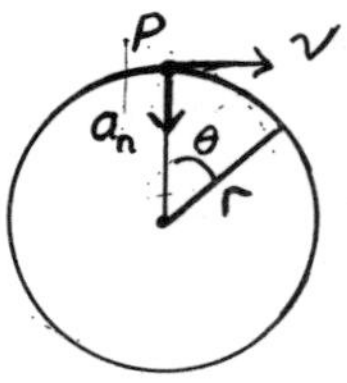

Figure (a)

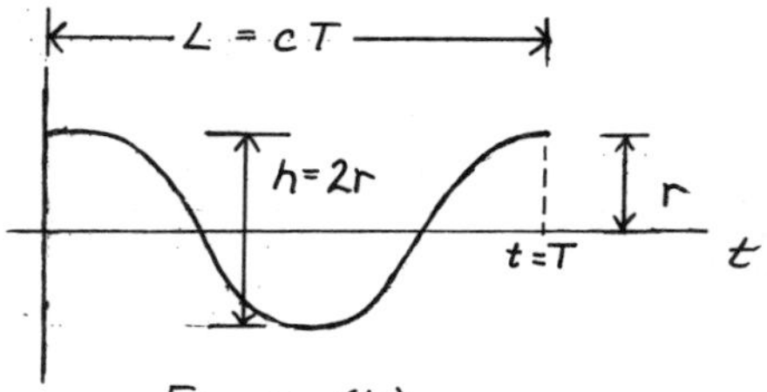

Figure (b)

SOLUTION

Let T = period for a wave to travel distance L at speed c (Fig (b)).

$$\therefore L = cT \quad (B)$$

The particle travels the distance (Fig.(a))

$d = 2\pi r$ during time T.

$$\therefore d = 2\pi r = vT$$

or with Eq (B), $$v = \frac{2\pi r}{T} = \frac{2\pi r c}{L} \quad (C)$$

The centripetal acceleration is (Fig (a))

$$a_n = \frac{v^2}{r} \quad (D)$$

By Eqs (A), (C), and (D),

$$a_n = \frac{4\pi^2 r}{L^2}\left(\frac{gL}{2\pi}\right)$$

or $$a_n = \frac{2\pi r g}{L}$$

13.134 GIVEN A particle moves along a circle of radius r such that it travels an arc distance

$$s = \pi r t^2/4 \quad (A)$$

FIND The magnitudes of the tangential and centripetal components of velocity and acceleration of the particle in terms of t.

SOLUTION

By Eq (A), $$v_t = \frac{ds}{dt} = \frac{\pi r t}{2} \quad (B)$$

Since the velocity is tangent to the circle always, $v_n = 0$

By Eq (B), the tangential acceleration is

$$a_t = \frac{dv}{dt} = \frac{d^2 s}{dt^2} = \frac{\pi r}{2}$$

Also, the centripetal acceleration is

$$a_n = \frac{v_t^2}{r} = \frac{\pi^2 r t^2}{4}$$

13.135 GIVEN Angular displacement

$$\theta = t^3 + 2t + 3 \text{ (rad)} \quad (A)$$

of a rigid body about the z axis (point O); Fig.(a).

FIND Speed of point P and its tangential and centripetal components of acceleration for $t = 2$ s. OP = 10 in.

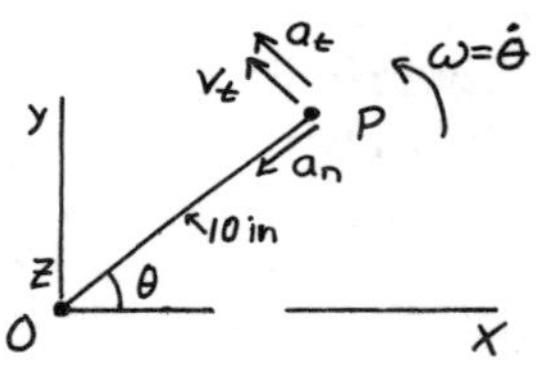

Figure (a)

SOLUTION

By Eq. (A)

$$\dot{\theta} = 3t^2 + 2 \quad [\text{rad/s}]$$
$$\ddot{\theta} = 6t \quad [\text{rad/s}^2] \quad (B)$$

For $t = 2$s, by Eqs (A) and (B)

$\theta = 15$ [rad], $\dot{\theta} = 14$ [rad/s]

$\ddot{\theta} = 12$ [rad/s²]

Hence,

$$v_P = \overline{OP}\,\dot{\theta} = (10)(14) = 140 \text{ in/s}$$

$$a_t = \overline{OP}\,\ddot{\theta} = (10)(12) = 120 \text{ in/s}^2$$

$$a_n = \frac{v_P^2}{\overline{OP}} = \frac{19600}{10} = 1960 \text{ in/s}^2$$

or $$a_n = \overline{OP}\,(\dot{\theta})^2 = (10)(196) = 1960 \text{ in/s}^2$$

13.136 *GIVEN* Water flowing around a cylindrical body (Fig (a)) At P, $v=\sqrt{2gh}$

FIND

a.) a_t and a_n of a particle of water as a function of θ

b.) a_t and a_n for $\theta = 45°$ and $\theta = 90°$

c.) Angle at which water leaves the cylinder. The effects on part b.

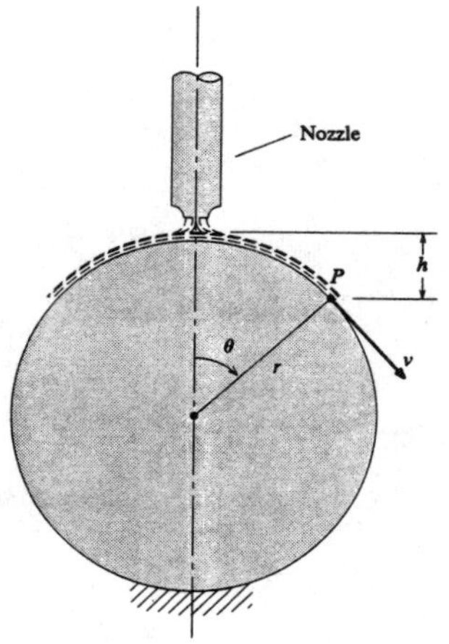

Figure (a)

SOLUTION

a.) By Fig (a)

$h = r(1-\cos\theta)$ (A)

Hence, at P, with Eq (A),

$v=\sqrt{2gh} = \sqrt{2gr(1-\cos\theta)}$ (B)

$\therefore a_n = \dfrac{v^2}{r} = 2g(1-\cos\theta)$ (C)

$a_t = \dfrac{dv}{dt} = \dfrac{dv}{d\theta}\dfrac{d\theta}{dt}$ (D)

By Eq (B),

$\dfrac{dv}{d\theta} = \dfrac{g\,r\sin\theta}{\sqrt{2gr(1-\cos\theta)}}$ (E)

Also, $v = r\dot{\theta} = r\dfrac{d\theta}{dt}$ or $\dfrac{d\theta}{dt} = \dfrac{v}{r}$ (F)

Therefore, by Eqs (B), (D), (E), and (F),

$a_t = g\sin\theta$ (G)

b.) With Eqs (C) and (G),

For $\theta = 45°$; $a_n = 0.586g$ $a_t = 0.707g$

For $\theta = 90°$; $a_n = 2g$ $a_t = g$

→ Does $a_n = 2g$ make sense? See part c.

c.) The water leaves the cylinder when, for free fall, $a = \sqrt{a_n^2 + a_t^2} = g$

Substitution for a_n and a_t yields

$a = \sqrt{4g^2(1-\cos\theta)^2 + g^2\sin^2\theta} = g$

or $3\cos^2\theta - 8\cos\theta + 4 = 0$

$\therefore \theta = 48.19°$ <continued>

Thus, the value $a_n = 2g$ of part b. is nonsense, since at $90°$, the water is no longer following a circular path.

13.137 *GIVEN* Angular acceleration for a toppling pole of length L is

$\alpha = \ddot{\theta} = k\sin\theta$ (A)

Where θ = angle between vertical and pole and k is a constant. See Fig (a). Also, for $\theta = 0$, $\dot{\theta} = 0$.

FIND a_t and a_n of the upper end of the pole in terms of k, L, and θ.

SOLUTION

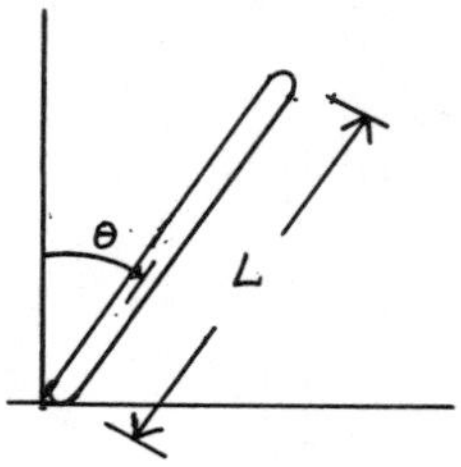

Figure (a)

By Eq (A),

$\alpha = \ddot{\theta} = \dfrac{d\omega}{dt} = k\sin\theta$ (B)

$\therefore a_t = L\ddot{\theta} = Lk\sin\theta$

By Eq (B)

$d\omega = k(\sin\theta)\,dt$

$= k(\sin\theta)\dfrac{dt}{d\theta}\,d\theta$

$= k(\sin\theta)\dfrac{d\theta}{\omega}$

or $\omega\,d\omega = k(\sin\theta)\,d\theta$

Integration yields

$\omega^2 = -2k\cos\theta + C$

For $\theta = 0$, $\dot{\theta} = \omega = 0$; $\therefore C = 2k$

$\therefore \omega^2 = 2k(1-\cos\theta)$

and, $a_n = L\omega^2 = 2kL(1-\cos\theta)$

13.138 *GIVEN* Particle P traveling on a circle with decreasing speed (Fig (a)). As shown, P has acceleration of 52 ft/s²

FIND Speed v of the particle, the sense in which it moves, the angular velocity ω, and the angular acceleration α for the conditions as shown.

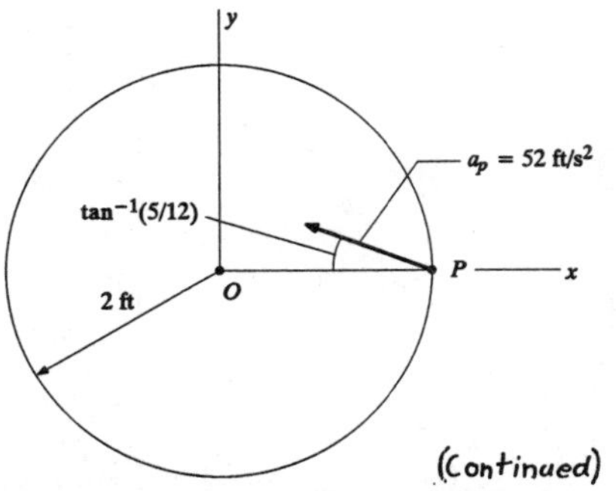

Figure (a)

(Continued)

13.138 Cont

SOLUTION

By Fig (a), for the position shown,

$$a_x = -a_n = -\frac{12}{13}(52) = -48 \text{ ft/s}^2$$

$$a_y = a_t = \frac{5}{13}(52) = 20 \text{ ft/s}^2$$

Since a_t is positive (directed upward), the angular acceleration is directed counterclockwise ↺ and has a magnitude

$$\alpha = a_t/r = 20/2 = 10 \text{ rad/s}^2$$

Also, since particle is traveling on the circle,

$$a_n = 48 \text{ ft/s}^2 = \frac{v^2}{r}$$

the speed then is $v = \sqrt{48(2)} = 9.798$ ft/s

Since the particle is slowing down and the acceleration a_t is directed upward, the particle is traveling downward at P, with speed v, that is, it is traveling in the clockwise sense.

The magnitude of the angular velocity ω is

$$|\omega| = \frac{v}{r} = \frac{9.798}{2} = 4.899 \text{ rad/s}$$

and ω is directed clockwise ↻

13.139

GIVEN A flywheel of radius r rotates counterclockwise with angular velocity ω_0 (rad/s) at time $t = 0$. It has angular acceleration $\alpha = 60t^2$ (rpm/s) where t denotes time in seconds. See Fig (a)

FIND a.) a_t and a_n of a point P on the rim in terms of r, ω_0, and t.

b.) a_x and a_y of P in terms of r, ω_0, and t. The radius to point P forms angle θ with the x axis. For $t = 0$, $\theta = 0$.

c.) Calculate the values of a_x and a_y, for $\omega_0 = 10$ rad/s, $r = 1$ m, and $t = 2$ s

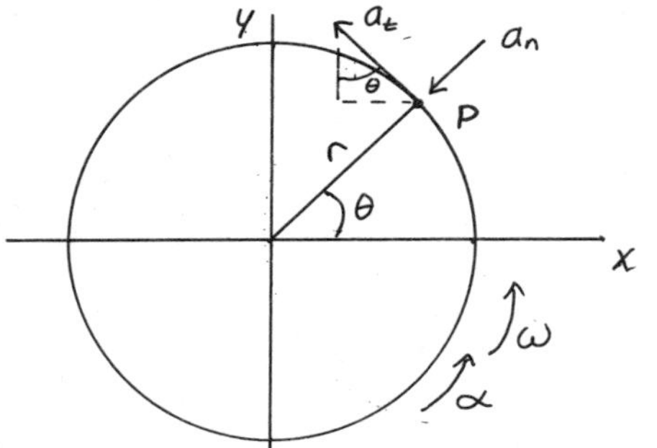

Figure (a)

SOLUTION

a.) Convert $\alpha = 60t^2 \left[\frac{\text{rev}}{\text{min}\cdot\text{s}}\right]$ to [rad/s²]

$$\alpha = 60t \frac{\text{rev}}{\text{min}\cdot\text{s}}\left(2\pi \frac{\text{rad}}{\text{rev}}\right)\left(\frac{1 \text{ min}}{60 \text{ s}}\right)$$

or $\alpha = 2\pi t^2$ [rad/s²] (A)

By Eq (A)

$$\alpha = \ddot{\theta} = \frac{d\omega}{dt} = 2\pi t^2 \quad \text{(B)}$$

or $$\omega = \frac{d\theta}{dt} = \frac{2}{3}\pi t^3 + \omega_0 \quad \text{(C)}$$

$$\therefore \quad v = r\omega = \frac{2\pi r t^3}{3} + \omega_0 r \quad \text{(D)}$$

and, By Eqs (A) and (D); see Fig (a),

$$\left.\begin{aligned} a_t &= r\alpha = 2\pi r t^2 \\ a_n &= \frac{v^2}{r} = r\left(\frac{2\pi t^3}{3} + \omega_0\right)^2 \end{aligned}\right\} \text{(E)}$$

b.) By Eq (C), with $\theta = 0$ for $t = 0$,

$$\theta = \frac{\pi}{6}t^4 + \omega_0 t \quad \text{(F)}$$

By Fig (a),

$$\left.\begin{aligned} a_x &= -a_n \cos\theta - a_t \sin\theta \\ a_y &= -a_n \sin\theta + a_t \cos\theta \end{aligned}\right\} \text{(G)}$$

Equations (E), (F), and (G) yield

$$a_x = -r\left(\frac{2\pi t^3}{3} + \omega_0\right)^2 \cos\left(\frac{\pi}{6}t^4 + \omega_0 t\right) - 2\pi r t^2 \sin\left(\frac{\pi}{6}t^4 + \omega_0 t\right) \quad \text{(H)}$$

$$a_y = -r\left(\frac{2\pi t^3}{3} + \omega_0\right)^2 \sin\left(\frac{\pi}{6}t^4 + \omega_0 t\right) + 2\pi r t^2 \cos\left(\frac{\pi}{6}t^4 + \omega_0 t\right) \quad \text{(I)}$$

c.) With $\omega_0 = 10$ rad/s, $r = 1$ m, and $t = 2$ s, Eqs (H) and (I) yield

$a_x = 714.6$ m/s² $a_y = 48.78$ m/s²

13.140

GIVEN Motion of a particle in (x, y) plane is (t in seconds):

$$x = 3t^2, \quad y = 2t^3 \quad \text{(m)} \quad \text{(A)}$$

FIND a_t and a_n at $t = 1$ s and show them on the plot of the path for the range $-1.5 \le t \le 1.5$ (s)

SOLUTION

By Eq (A) $$y^2 = \frac{4x^3}{27}, \quad x \ge 0 \quad \text{(B)}$$

See the plot of particle path.

(Continued)

13.140 cont.

To determine a_t and a_n, by Eq. (A):

$x = 3t^2$ $\quad y = 2t^3$

$\dot{x} = 6t$ $\quad \dot{y} = 6t^2$

$\ddot{x} = 6$ $\quad \ddot{y} = 12t$

at $t = 1\,s$:

$x = 3\,m$ $\quad y = 2\,m$

$\dot{x} = 6\,m/s$ $\quad \dot{y} = 6\,m/s$

$\ddot{x} = a_x = 6\,m/s^2$ $\quad \ddot{y} = a_y = 12\,m/s^2$

By the plot, the slope of the path at $t = 1\,s$ is: $\tan\theta = \frac{\dot{y}}{\dot{x}} = 1$ or $\theta = 45°$

But, $a_t = a_x \cos\theta + a_y \sin\theta$

$a_n = -a_x \sin\theta + a_y \cos\theta$

or $a_t = (6+12)(0.7071) = 12.73\ m/s^2$

$a_n = (-6+12)(0.7071) = 4.24\ m/s^2$

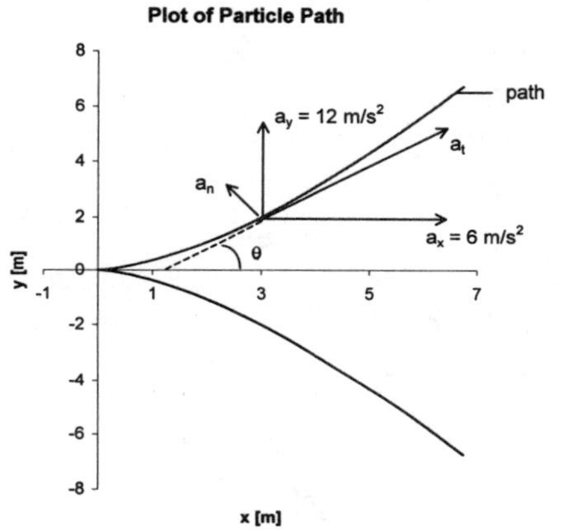

13.141 GIVEN A particle moves in the (x, y) plane such that, with t in seconds,

$a_x = -32\ cm/s^2$, $\quad a_y = 10t\ cm/s^2$ (A)

For $t = 0$, the particle is at $x = y = 0$ and its velocity is 92 cm/s in the +x direction.

FIND a.) Position and speed of the particle at $t = 3\,s$

b.) Plot the particle's path for $-2\,s \le t \le 6\,s$

c.) Show and label on the plot a_n and a_t for the points where $t = 0$ and x is a max.

SOLUTION

a.) By Eqs (A), $\quad a_x = -32 = \frac{dv_x}{dt}$, $\quad a_y = 10t = \frac{dv_y}{dt}$ (B)

Therefore, by integration, with $v_x = 92\ cm/s$ and $v_y = 0$, for $t = 0$,

$v_x = \frac{dx}{dt} = 92 - 32t$ [cm/s]

$v_y = \frac{dy}{dt} = 5t^2$ [cm/s] (C)

Again, by integration and with $x = y = 0$, $t = 0$

$x = 92t - 16t^2$

$y = 5/3\ t^3$ (D)

For $t = 3\,s$, by Eqs (C) and (D)

$x = 132\ cm$, $\quad y = 45\ cm$

and, $v_x = -4\ cm/s$, $\quad v_y = 45\ cm/s$

Therefore, the speed is

$s = |v| = \sqrt{(-4)^2 + (45)^2} = 45.18\ cm/s$

b.) The path is plotted with Eqs (D) for the range $-2\,s \le t \le 6\,s$ (See plot)

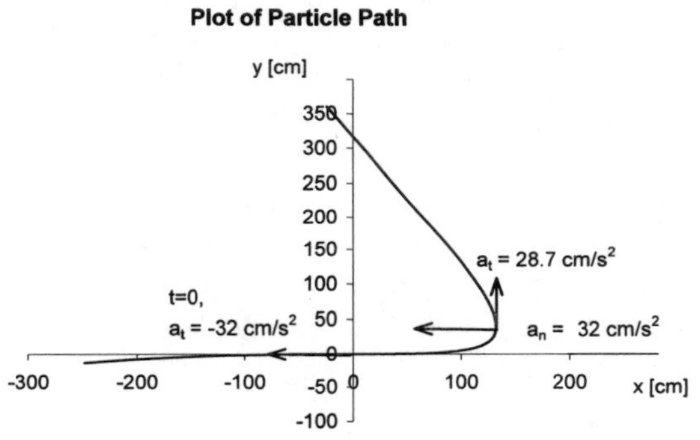

c.) To determine a_n and a_t, note that for $t = 0$, $x = y = 0$. And by Eqs (A) and (C):

$a_x = -32\ cm/s^2$ $\quad a_y = 0$

$v_x = 92\ cm/s$ $\quad v_y = 0$

The slope of the path is

$\tan\theta = \frac{v_y}{v_x} = 0$

Therefore, as shown on plot:

$a_t = a_x = -32\ cm/s^2$ $\quad a_n = a_y = 0$

The time for which x is a maximum occurs when $dx/dt = v_x = 0$, or $t = 92/32\ s$ and $x = x_{max} = 132.25$

For $t = 92/32\,s$,

$v_x = 0$, $\quad v_y = 41.33\ cm/s$

Therefore, the slope of the path is

$\tan\theta = \frac{41.33}{0} = \infty$

or $\theta = 90°$

(Continued)

13.141 cont.

Also, by Eqs (A), for $t = 92/32\ s$,

$a_x = -32\ cm/s^2 \qquad a_y = 10\left(\frac{92}{32}\right) = 28.75\ cm/s^2$

Since $\theta = 90°$ (see plot),

$a_n = -a_x = 32\ cm/s^2 \qquad a_t = a_y = 28.7\ cm/s^2$

13.142 **GIVEN** Motion in (x, y) plane of a particle defined as

$$x = 24t, \quad y = (4-t^2)^2/32 \qquad (A)$$

where t is in seconds, x & y in feet.

FIND a.) Slopes of tangent and normal of the particle's path for $t = 2\ s$.

b.) The speed v and a_t and a_n for $t = 2s$.

c.) Show the acceleration vector for $t = 2s$ on a plot of the particle path for $-3s \le t \le 3s$.

SOLUTION

a.) By Eq (A),

$x = 24t \qquad y = (4-t^2)^2/32$

$\dot{x} = 24 \qquad \dot{y} = -1/8\,(4-t^2)t$

$\ddot{x} = 0 \qquad \ddot{y} = 3/8\,t^2 - 1/2$

When $t = 2s$,

$\dot{x} = 24, \quad \dot{y} = 0 \quad \therefore v = \sqrt{\dot{x}^2 + \dot{y}^2} = 24\ ft/s$

The slope of the tangent is:

$\tan\theta = \frac{\dot{y}}{\dot{x}} = 0; \qquad \theta = 0$

The slope of the normal is:

$\tan\theta = \frac{\dot{x}}{\dot{y}} = \infty, \qquad \theta = 90°$

b.) Since the tangent to the curve is horizontal ($\theta = 0$) and the normal of the curve is vertical ($\theta = 90°$), for $t = 2s$,

$a_t = a_x = 0 \qquad a_n = a_y = 1\ ft/s^2$

(speed, $v = 24\ ft/s$ found in part a.)

c.) The plot may be generated by Eq (A), expressing y as a function of x, $y = \left[4 - \left(\frac{x}{24}\right)^2\right]^2/32$

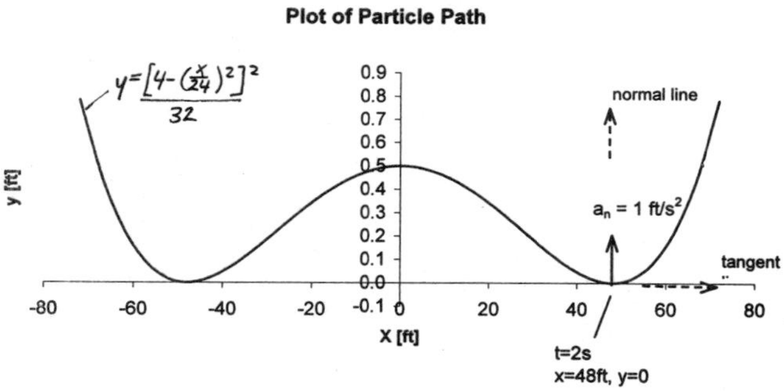

13.143 **GIVEN** Centrifuge with radius $r = 12\ ft$ rotates at constant angular speed ω. A pilot is subjected to acceleration of $a_n = 11g$

FIND The rotational speed ω (rpm).

SOLUTION:

With $r = 12\ ft$, $a_n = 11g$, $a_n = r\omega^2$

$$\omega^2 = \frac{a_n}{r} = \frac{(11)(32.2)}{12}$$

$\therefore \omega = 5.433\ rad/s$

or $\omega = (5.433\ \frac{rad}{s})\left(\frac{1\ rev}{2\pi\ rad}\right)\left(\frac{60s}{1\ min}\right)$

$\omega = 51.88\ rev/min$

13.144 **GIVEN** Particle moving in (x, y) plane such that the coordinates of the particle are (x and y in meters, time t in seconds)

$$x = 3\sin(2t-5), \quad y = 2\sin(4t+1)$$

FIND a.) The (x, y) components of velocity and acceleration for $t = 1s$.

b.) Speed of particle for $t = 1s$.

SOLUTION

a.) From the given equations:

$\dot{x} = v_x = 6\cos(2t-5) \qquad \dot{y} = v_y = 8\cos(4t+1)$

$\ddot{x} = a_x = -12\sin(2t-5) \qquad \ddot{y} = a_y = -32\sin(4t+1)$

Hence, for $t = 1s$,

$v_x = 6\cos(-3) = -5.940\ m/s$

$v_y = 8\cos(5) = 2.269\ m/s$

$a_x = -12\sin(-3) = 1.693\ m/s^2$

$a_y = -32\sin(5) = 30.68\ m/s^2$

b.) The speed at time $t = 1s$ is

$v = \sqrt{v_x^2 + v_y^2} = \sqrt{(-5.940)^2 + (2.269)^2}$

or $v = 6.359\ m/s$

13.145 **GIVEN** Source of light emits horizontal ray that strikes a cylindrical body at P (Fig (a)). S moves upward with speed u.

FIND a_t and a_n of P in terms of u, r, and θ

(Continued)

13.145 cont.

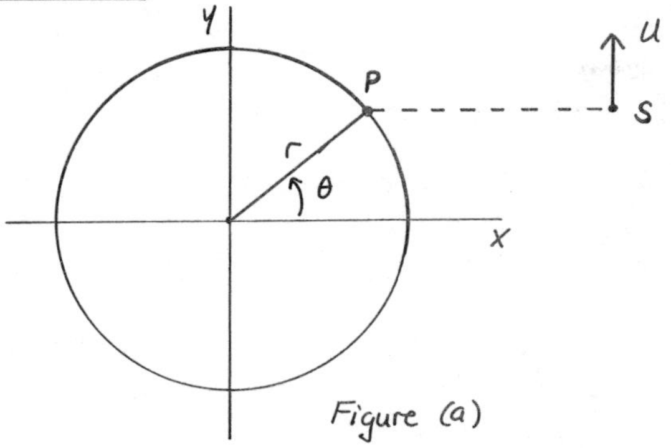

Figure (a)

SOLUTION

By Fig (a), $y = r\sin\theta$

Therefore, differentiation with respect to time t yields: $\dot{y} = r\dot{\theta}\cos\theta = u$

$\therefore \dot{\theta} = \omega = \frac{u}{r}\sec\theta$

$\ddot{\theta} = \alpha = \frac{u}{r}\dot{\theta}\sec\theta\tan\theta$

Hence, $a_t = r\alpha = u\dot{\theta}\sec\theta\tan\theta$

or, in terms of u, r, θ;

$$a_t = \frac{u^2\sec^2\theta\tan\theta}{r}$$

$$a_n = r\omega^2 = \frac{u^2\sec^2\theta}{r}$$

13.146

GIVEN A particle travels on a circular path $x^2 + y^2 = 100$ (A)

When $x = 6$ mm and $y = 8$ mm,

$v_x = 40$ mm/s and $a_x = -150$ mm/s²

FIND a.) The velocity, speed, and (x, y) components of acceleration of the particle for $x = 6$ mm, $y = 8$ mm

b.) The normal and tangential accelerations (a_n and a_t) of the particle for $x = 6$, $y = 8$ mm

SOLUTION

a.) By Eq (A), differentiation with respect to t yields: $x\dot{x} + y\dot{y} = 0$ (B)

or $\frac{\dot{y}}{\dot{x}} = -\frac{x}{y}$

For $x = 6$, $y = 8$

$\dot{y} = v_y = -\frac{3}{4}\dot{x} = -\frac{3}{4}v_x$

$\therefore v_y = -\frac{3}{4}v_x = -\frac{3}{4}(40) = -30$ mm/s

Hence, the velocity of the particle is,

$$\bar{v} = 40\bar{i} - 30\bar{j} \text{ [mm/s]}$$

and the speed is: $|v| = \sqrt{40^2 + 30^2} = 50$ mm/s

To determine (x, y) components of acceleration, differentiation of Eq (B) yields

$\dot{x}^2 + x\ddot{x} + \dot{y}^2 + y\ddot{y} = 0$

or, with $\dot{x} = v_x = 40$, $\dot{y} = v_y = -30$, $\ddot{x} = a_x = -150$, and $(x = 6, y = 8)$

$\ddot{y} = \frac{-(\dot{x}^2 + \dot{y}^2) - x\ddot{x}}{y} = \frac{-2500 - 6(-150)}{8}$

or $\ddot{y} = a_y = -200$ mm/s²

$\therefore a_x = -150$ mm/s², $a_y = -200$ mm/s² (C)

b.) To determine a_n and a_t, see Fig. (a) + (b)

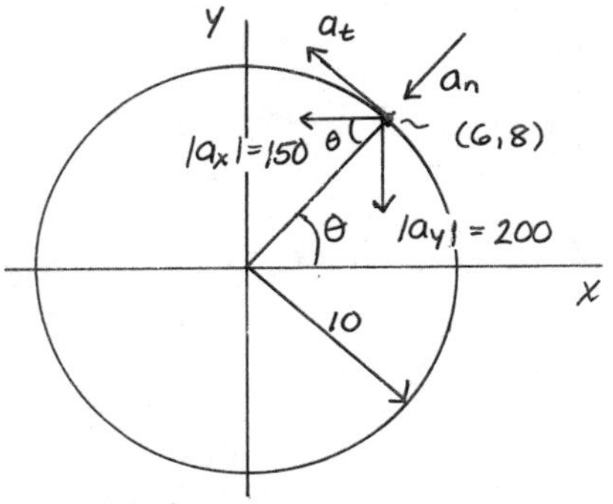

Figure (a)

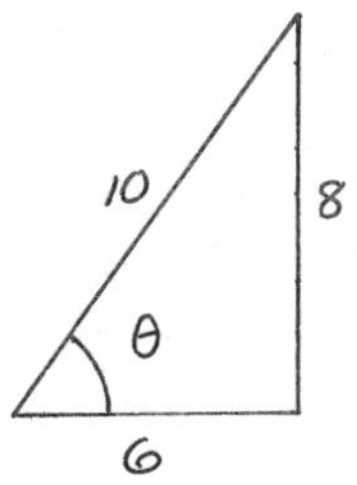

Figure (b)

By Fig (a),

$a_n = |a_x|\cos\theta + |a_y|\sin\theta$

$a_t = |a_x|\sin\theta - |a_y|\cos\theta$ (D)

where, by Fig (b),

$\cos\theta = 6/10 = 0.6$, $\sin\theta = 8/10 = 0.8$

Hence, by Eqs. (C) and (D),

$a_n = (150)(.6) + (200)(.8) = \underline{250 \text{ mm/s}^2}$

$a_t = (150)(0.8) - (200)(0.6) = \underline{0}$

13.147

GIVEN A 3-D Lissajous figure generated by a particle with path defined by

$$\left.\begin{aligned} x &= 9\cos(3t - 5) \\ y &= 6\cos(4t + 6) \\ z &= 3\cos(8t - 2) \end{aligned}\right\} \quad (A)$$

FIND a.) The speed of the particle in terms of t

b.) The magnitude of acceleration directed toward the origin in terms of t.

(Continued)

13.147 cont

SOLUTION

a.) By Eqs (A)

$$\left.\begin{aligned}\dot{x} &= -27\sin(3t-5)\\ \dot{y} &= -24\sin(4t+6)\\ \dot{z} &= -24\sin(8t-2)\end{aligned}\right\} \quad (B)$$

$$\therefore\ v = \sqrt{(\dot{x})^2 + (\dot{y})^2 + (\dot{z})^2}$$

or
$$v = \left[\begin{aligned}&729\sin^2(3t-5)\\ &+576[\sin^2(4t+6)+\sin^2(8t-2)]\end{aligned}\right]^{1/2} \quad (C)$$

b.) The acceleration directed toward (0,0,0) may be calculated as follows: By Eq (B),

$$\left.\begin{aligned}\ddot{x} &= -81\cos(3t-5)\\ \ddot{y} &= -96\cos(4t+6)\\ \ddot{z} &= -192\cos(8t-2)\end{aligned}\right\} \quad (D)$$

Therefore, the acceleration vector is

$$\bar{a} = \ddot{x}\bar{i} + \ddot{y}\bar{j} + \ddot{z}\bar{k} \quad (E)$$

By Eq (A), the position vector of the particle is

$$\bar{r} = x\bar{i} + y\bar{j} + z\bar{k} \quad (F)$$

The magnitude of acceleration toward the origin is, with Eqs (A), (D), (E), and (F),

$$a_o = \left|\frac{\bar{r}}{|\bar{r}|}\cdot\bar{a}\right|$$

$$= \left|\frac{(x\ddot{x}+y\ddot{y}+z\ddot{z})}{\sqrt{x^2+y^2+z^2}}\right|$$

or

$$a_o = \frac{\{729\cos^2(3t-5)+576[\cos^2(4t+6)+\cos^2(8t-2)]\}}{[81\cos^2(3t-5+36\cos^2(4t+6)+9\cos^2(8t-2)]^{1/2}}$$

13.148 GIVEN Particle P moves in (x, y) plane. Distance from origin O is [m]

$$r = bt^2 \quad (A)$$

b = constant, t = time [s]

Line from O to P forms angle with x axis $\theta = ct$ (B)

c = constant

FIND v_r, v_θ, a_r, a_θ in terms of b, c, and t

SOLUTION

By Eq (A), $v_r = \dot{r} = 2bt$

By Eqs (A) and (B), $v_\theta = r\dot{\theta} = (bt^2)(c)$

or $v_\theta = bct^2$

By definition,

$$a_r = \ddot{r} - r(\dot{\theta})^2 \quad (C)$$

$$a_\theta = 2\dot{r}\dot{\theta} + r\ddot{\theta} \quad (D)$$

By Eqs (A), (B), (C), and (D)

$$a_r = 2b - bc^2t \qquad a_\theta = 4bct$$

13.149 GIVEN Particle's path is a spiral, in polar coordinates (r, θ):

$$r = k\theta, \quad k = \text{constant} \quad (A)$$

$$\alpha = \ddot{\theta} = \text{constant} \quad (B)$$

FIND The most general expressions for v_r, v_θ, a_r, a_θ, in terms of t, k, and α.

SOLUTION

By Eqs (A) and (B)

$$\dot{r} = k\dot{\theta} \quad (C)$$

$$\ddot{r} = k\ddot{\theta} = k\alpha \quad (D)$$

$$\ddot{\theta} = \frac{d\dot{\theta}}{dt} = \alpha \text{ or } \dot{\theta} = \int\alpha\, dt = \alpha t + \omega_0 \quad (E)$$

where $\omega_0 = \dot{\theta}$ for $t=0$

By Eq (E), $\theta = \frac{1}{2}\alpha t^2 + \omega_0 t + \theta_0$ (F)

where $\theta = \theta_0$ for $t=0$

$$\therefore\ v_r = \dot{r} = k\dot{\theta} = k(\alpha t + \omega_0)$$

$$v_\theta = r\dot{\theta} = k\theta(\alpha t + \omega_0)$$

or
$$v_\theta = k(\tfrac{1}{2}\alpha t^2 + \omega_0 t + \theta_0)(\alpha t + \omega_0)$$

Now, $a_r = \ddot{r} - r(\dot{\theta})^2$, $a_\theta = 2\dot{r}\dot{\theta} + r\ddot{\theta}$

Hence,
$$a_r = k\alpha - k(\tfrac{1}{2}\alpha t^2 + \omega_0 t + \theta_0)(\alpha t + \omega_0)^2$$

$$a_\theta = 2k(\alpha t + \omega_0)^2 + k\alpha(\tfrac{1}{2}\alpha t^2 + \omega_0 t + \theta_0)$$

13.150 GIVEN Train enters a circular track (r = 600 m) at speed v = 175 km/h and deceleration of 4 km/h/s for a period of 10 s.

FIND a_θ, a_r of the engine for the instant (t = 0) after it enters the curve, at t = 1 s, and t = 6 s.

SOLUTION

Since $a_\theta = -4$ km/h/s $= -1.111$ m/s² = const and 175 km/h = 48.611 m/s = v_0 at t = 0, the speed v of the train is

$$v = v_0 + at \quad \text{or} \quad v(t) = 48.611 - 1.111t \quad (A)$$

(Continued)

13.150 cont.

Since r is constant ($r = 600$ m)

$$a_r = \ddot{r} - r(\dot{\theta})^2 = -r(\dot{\theta})^2 = -\frac{v^2}{r}$$

or, with Eq (A):

$$\left.\begin{aligned} a_r &= -\frac{(48.611 - 1.111t)^2}{600} \\ a_\theta &= -1.111 \ \text{m/s}^2 \end{aligned}\right\} \quad (B)$$

By Eqs (B), since the deceleration remains constant for 10 s,

For $t = 0$, $a_r = -3.938$ m/s²

$a_\theta = -1.111$ m/s²

For $t = 1$ s, $a_r = -3.760$ m/s²

$a_\theta = -1.111$ m/s²

For $t = 6$ s, $a_r = -2.932$ m/s²

$a_\theta = -1.111$ m/s²

See Figure (a): (a_θ is constant and negative for all t)

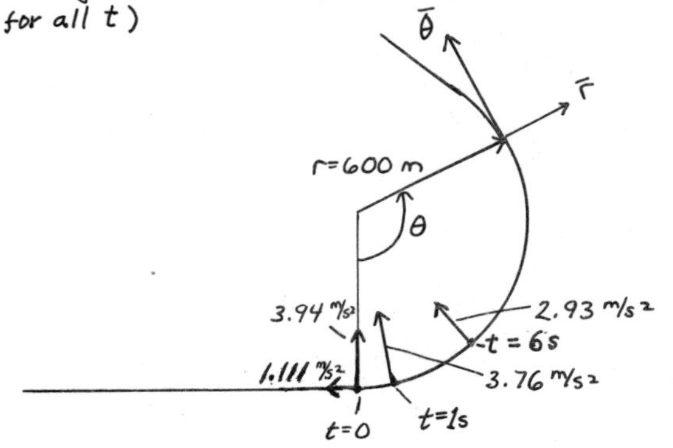

Figure (a)

13.151 GIVEN Fig. (a) where bead B slides along the rod OC and is constrained to travel on the circular track.

FIND Expressions for v, a_n, and a_t in terms of θ

SOLUTION

By Fig (a),

$\overline{OB} = r = (2)(8)\sin\theta$ (A)

Since $\dot{\theta} = 2$ rad/s is constant, $\ddot{\theta} = 0$.

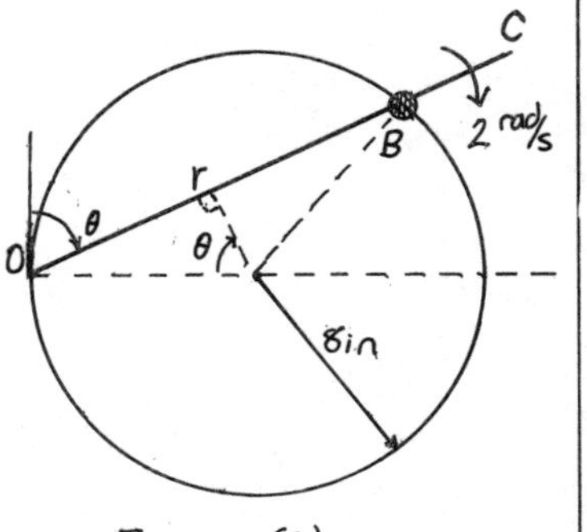

Figure (a)

Therefore, by Eq (A),

$$v_r = \dot{r} = 16\,\dot{\theta}\cos\theta = 32\cos\theta$$

$$v_\theta = r\dot{\theta} = 32\sin\theta$$

or $v = |\bar{v}| = \sqrt{v_r^2 + v_\theta^2} = 32$ in/s

Since v is constant, $a_t = 0$

and $a_n = \dfrac{v^2}{r} = \dfrac{(32)^2}{8} = 128$ in/s²

13.152 GIVEN Fig (a), Race car enters circular portion of track (point P) with $v = 120$ km/h and $a_\theta = 5$ m/s²

FIND

a.) a_r and $\bar{a}$ (relative) to x axis.

b.) v_x, v_y, a_x, a_y at point P

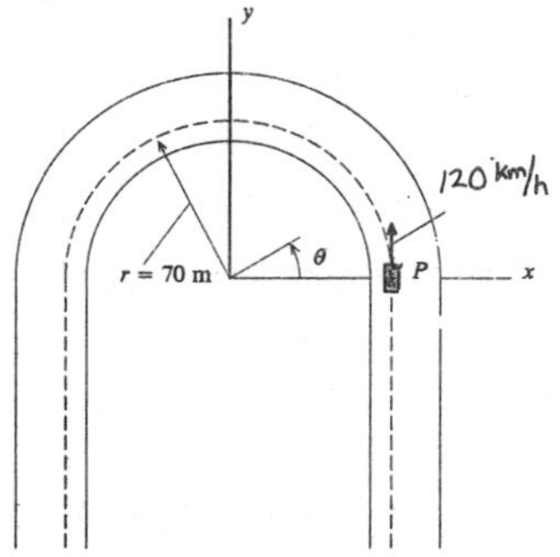

Figure (a)

SOLUTION

a.) Since $v = 120$ km/h $= 33.33$ m/s

and $r = 70$ m $=$ const.

$$a_r = \ddot{r} - r(\dot{\theta})^2 = 0 - \frac{v^2}{r}$$

$$\therefore a_r = -\frac{(33.33)^2}{70}$$

$$= -15.87 \ \text{m/s}^2$$

By the given data, $a_\theta = 5$ m/s². Therefore, the magnitude of the acceleration $\bar{a}$ is:

$$\bar{a} = \sqrt{a_n^2 + a_\theta^2} = \sqrt{(-15.87)^2 + 5^2} = 16.64 \ \text{m/s}^2$$

The direction of $\bar{a}$ is (See Fig (b).)

$$\theta_x = \tan^{-1}\frac{a_\theta}{a_r}$$

$$= \tan^{-1}\frac{5}{-15.87}$$

or $\theta_x = -17.5°$

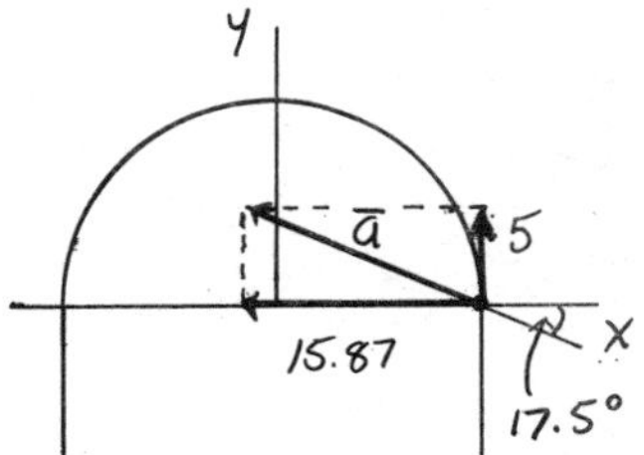

Figure (b)

b.) By Fig (a), at P the (x, y) components of velocity are

$v_x = 0$, $v_y = 120$ km/h

By Fig (b), at P,

$a_x = a_r = -15.87$ m/s² $\quad a_y = a_t = 5$ m/s²

13.153 GIVEN: Data of Problem 13.152 for $t = 3s$ after car enters the circular portion of the track (See Figure in Problem 13.152)

FIND a.) a_r and $\bar{a}$ (relative to X axis)
b.) v_x, v_y, a_x, a_y

SOLUTION

a) For $t = 3s$, the distance the car has traveled is, with $v_0 = 120$ km/h $= 33.33$ m/s and $a_\theta = 5$ m/s²,

$$s = v_0 t + \tfrac{1}{2} a_\theta t^2 = (33.33)(3) + \tfrac{1}{2}(5)(3)^2$$

or $s = 122.5$ m

Therefore, the car has traveled an arc that subtends the angle (Fig (a))

$$\theta = \frac{122.5}{70} = 1.75 \text{ rad}$$

or $\underline{\theta = 100.3°}$

Since $r = 70\text{m} = \text{const}$

$$a_r = -\frac{v^2}{r}$$

and the speed v of the car at $t = 3s$ is

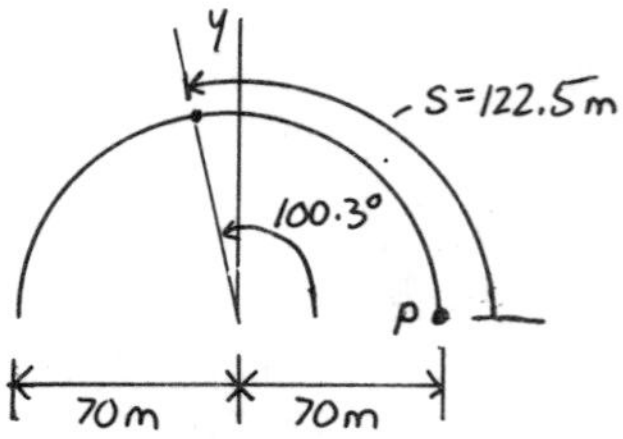

Figure (a)

$$v = a_\theta t + v_0 = (5)(3) + 33.33 = 48.33 \text{ m/s}$$

$$\therefore a_r = -\frac{(48.33)^2}{70} = -33.37 \text{ m/s}^2$$

By given data, $a_\theta = 5$ m/s²

$$\therefore a = |\bar{a}| = \sqrt{(-33.37)^2 + 5^2} = 33.74 \text{ m/s}^2$$

By Fig (b),

$$\phi = \tan^{-1}\frac{5}{33.37} \text{ or}$$

$\phi = 8.52°$

Hence,

$$\theta_x = -90° - 8.52° + 10.3°$$

or $\underline{\theta_x = -88.2°}$

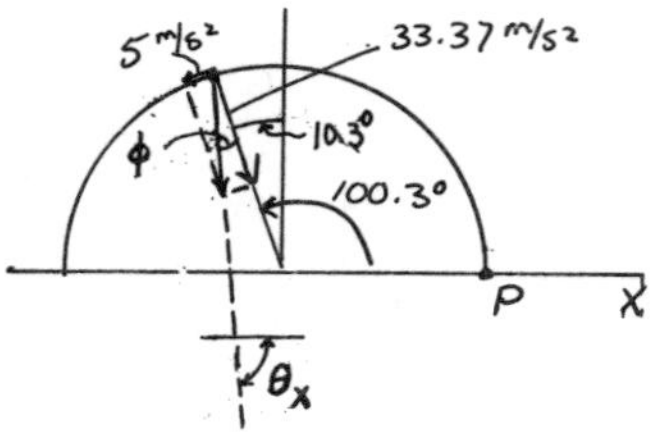

Figure (b)

b.) By Fig (c),

$$v_x = (-48.33)\cos 10.3° = -47.55 \text{ m/s}$$

$$v_y = (-48.33)\sin 10.3° = -8.64 \text{ m/s}$$

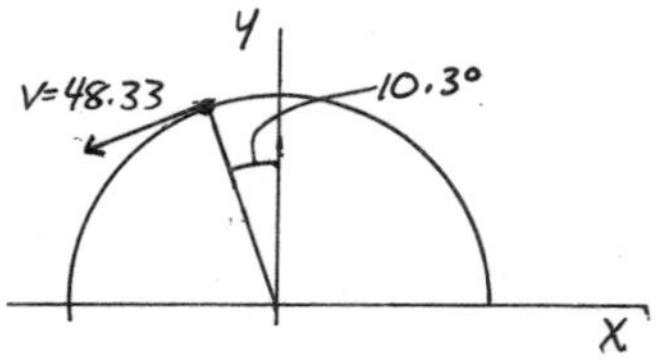

Figure (c)

By Fig (d),

$$a_x = (-5)\cos 10.3° + (33.37)\sin 10.3°$$

or $\underline{a_x = 1.047 \text{ m/s}^2}$

$$a_y = (-5)\sin 10.3° - (33.37)\cos 10.3°$$

or $\underline{a_y = -33.73 \text{ m/s}^2}$

See Fig (e).

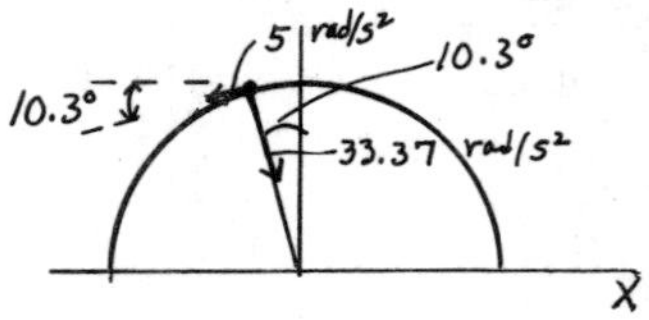

Figure (d)

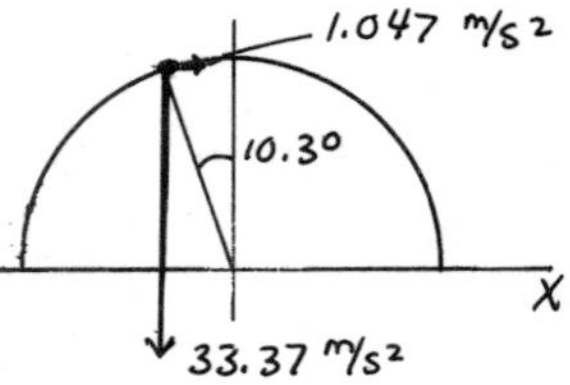

Figure (e)

13.154 GIVEN A frictionless bead slides out along a rigid rod that rotates in a horizontal plane with constant angular acceleration α about a fixed pivot.

SHOW If the rod is at rest for $t = 0$, the distance r of the bead from the pivot at time t is defined by the equation

$$\ddot{r} - \alpha^2 t^2 r = 0 \qquad (A)$$

SOLUTION

In polar coordinates, the radial acceleration is given by

$$a_r = \ddot{r} - r\dot{\theta}^2 \qquad (B)$$

and since $\alpha = \ddot{\theta}$ is constant,

$$\dot{\theta} = \alpha t + \dot{\theta}_0$$

However, since the rod is at rest when $t = 0$, $\dot{\theta} = 0$ for $t = 0$, and thus $\dot{\theta}_0 = 0$.

$$\therefore \dot{\theta} = \alpha t \qquad (C)$$

Substitution of Eq (C) into Eq (B) yields

$$a_r = \ddot{r} - \alpha^2 t^2 r \qquad (D)$$

So, since the bead is frictionless, there is no radial force acting on it. Therefore

$$\Sigma F_r = m a_r = 0 \quad \text{or} \quad a_r = 0 \qquad (E)$$

Equations (D) and (E) yield

$$\ddot{r} - \alpha^2 t^2 r = 0 \qquad (F)$$

as the differential equation that determines r.

13.155 GIVEN Jerk $\bar{J} = \dot{\bar{a}}$ and in polar coordinates:

$$\bar{a} = (\ddot{r} - r\dot{\theta}^2)\bar{e}_r + (2\dot{r}\dot{\theta} + r\ddot{\theta})\bar{e}_\theta \qquad (A)$$

FIND $\bar{J}$ in terms of r, θ, $\bar{e}_r$, $\bar{e}_\theta$ and derivatives.

SOLUTION

Differentiation of Eq (A) yields

$$\dot{\bar{a}} = (\dddot{r} - \dot{r}\dot{\theta}^2 - 2r\dot{\theta}\ddot{\theta})\bar{e}_r + (\ddot{r} - r\dot{\theta}^2)\dot{\bar{e}}_r + (2\ddot{r}\dot{\theta} + 2\dot{r}\ddot{\theta} + \dot{r}\ddot{\theta} + r\dddot{\theta})\bar{e}_\theta + (2\dot{r}\dot{\theta} + r\ddot{\theta})\dot{\bar{e}}_\theta$$

But, $\dot{\bar{e}}_r = \dot{\theta}\bar{e}_\theta$, $\dot{\bar{e}}_\theta = -\dot{\theta}\bar{e}_r$

Hence,

$$J = \dot{\bar{a}} = (\dddot{r} - \dot{r}\dot{\theta}^2 - 2r\dot{\theta}\ddot{\theta} - 2\dot{r}\dot{\theta}^2 - r\dot{\theta}\ddot{\theta})\bar{e}_r + (2\ddot{r}\dot{\theta} + 3\dot{r}\ddot{\theta} + r\dddot{\theta} + \ddot{r}\dot{\theta} - r\dot{\theta}^3)\bar{e}_\theta$$

$$\therefore J = (\dddot{r} - 3\dot{r}\dot{\theta}^2 - 3r\dot{\theta}\ddot{\theta})\bar{e}_r + + (3\ddot{r}\dot{\theta} + 3\dot{r}\ddot{\theta} - r\dot{\theta}^3 + r\dddot{\theta})\bar{e}_r$$

13.156 GIVEN Jerk $= \bar{J} = \dot{\bar{a}}$ and

$$\bar{a} = \ddot{s}\bar{e}_t + \frac{v^2}{r}\bar{e}_n = \ddot{s}\bar{e}_t + \frac{\dot{s}^2}{r}\bar{e}_n \qquad (A)$$

SHOW $\bar{J} = \left(\dddot{s} - \frac{\dot{s}^3}{r^2}\right)\bar{e}_t + \left(\frac{3\dot{s}\ddot{s}}{r} - \frac{\dot{r}\dot{s}^2}{r^2}\right)\bar{e}_n$

SOLUTION

By differentiation of Eq (A),

$$\dot{\bar{a}} = \dddot{s}\bar{e}_t + \ddot{s}\dot{\bar{e}}_t + \left(\frac{2\dot{s}\ddot{s}}{r} - \frac{\dot{s}^2\dot{r}}{r^2}\right)\bar{e}_n + \frac{\dot{s}^2}{r}\dot{\bar{e}}_n$$

But, $\dot{\bar{e}}_t = \dot{\theta}\bar{e}_n$, $\dot{\bar{e}}_n = -\dot{\theta}\bar{e}_t$

Hence,

$$\dot{\bar{a}} = \dddot{s}\bar{e}_t - \frac{\dot{s}^2\dot{\theta}}{r}\bar{e}_t + \left(\frac{2\dot{s}\ddot{s}}{r} - \frac{\dot{s}^2\dot{r}}{r^2} + \ddot{s}\dot{\theta}\right)\bar{e}_n$$

Also, $r\dot{\theta} = \dot{s}$

$$\therefore \bar{J} = \dot{\bar{a}} = \left(\dddot{s} - \frac{\dot{s}^3}{r^2}\right)\bar{e}_t + \left(\frac{3\dot{s}\ddot{s}}{r} - \frac{\dot{r}\dot{s}^2}{r^2}\right)\bar{e}_n$$

13.157 Challenge Problem

GIVEN A drug runner in a fog at sea is detected at point P 10 miles from a Coast Guard boat and immediately starts off on a straight course at 10 mi/h. Coast Guardmen wait 1 hr and begin to spiral around P at 50 mi/h with a radial component of 10 mi/h

FIND The maximum time that can elapse after the drug runner is detected until he is caught.

SOLUTION

Set up (x, y) axes at P with the Coast Guard boat at C (Fig (a))

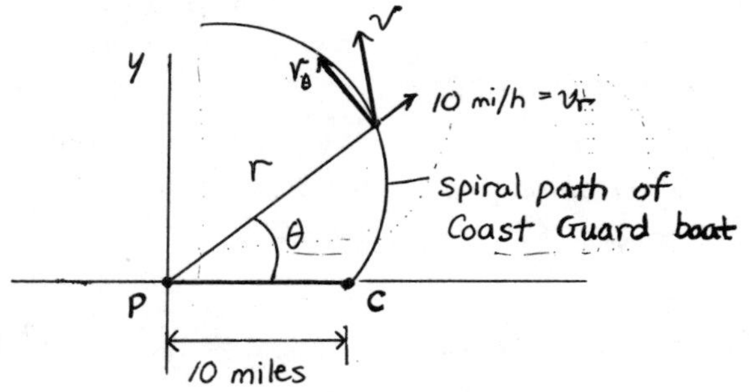

Figure (a)

By Fig (a), on the spiral path,

$v_r = 10$ mi/h $\qquad v = 50$ mi/h

$\therefore v_r^2 + v_\theta^2 = 50^2$

or $v_\theta = 20\sqrt{6}$ mi/h

Also, $v_\theta = r\dot{\theta}$ $\qquad \therefore \dot{\theta} = \frac{20\sqrt{6}}{r}$ $\qquad (A)$

Also by Fig (a), $\dot{r} = 10$ mi/h

$\therefore r = 10t + 10 \qquad (B)$

since $r = 10$ for $t = 0$.

Equations (A) + (B) yield

$$\dot{\theta} = \frac{2\sqrt{6}}{1+t}$$

Integration yields, with $\theta = 0$ for $t = 0$,

$$\theta = 2\sqrt{6}\ln(1+t) \qquad (C)$$

The maximum time T that the Coast Guard boat can travel until it intercepts the drug runner occurs at $\theta = 2\pi$ (one revolution).

Therefore, by Eq (C),

$$2\pi = 2\sqrt{6}\ln(1+T)$$

or $T = 2.606$ h

$\therefore$ The maximum time after the message is recieved is

$$T_{max} = 2.606 + 1 = 3.606 \text{ h}$$

13.158 *Computer Problem*

GIVEN Particle P travels along the path

$r = \theta$ (A)

(r in meters, θ in radians), where (r, θ) are polar coordinates.

FIND

a.) For $r = 2t^2$ (B) t in seconds, find velocity $\bar{v}$ of the particle as a function of t.

b.) Plot the path of the particle in the (x, y) plane for $0 \le \theta \le 360°$ (2π rad)

c.) Find $\bar{v}$ for $\theta = 30°$ ($\pi/6$ rad)

SOLUTION

a.) By Eqs (A) and (B), $\theta = 2t^2$ (C)

The velocity, in polar coordinates is

$\bar{v} = \dot{r}\,\bar{e}_r + r\dot{\theta}\,\bar{e}_\theta$ (D)

By Eqs. (B), (C) and (D),

$\bar{v} = 4t\,\bar{e}_r + 8t^3\,\bar{e}_\theta$ (E)

b.) To plot the path in the (x, y) plane for $0 \le \theta \le 360°$, express x and y as functions of θ:

$x = r\cos\theta = \theta\cos\theta$

$y = r\sin\theta = \theta\sin\theta$

See plot:

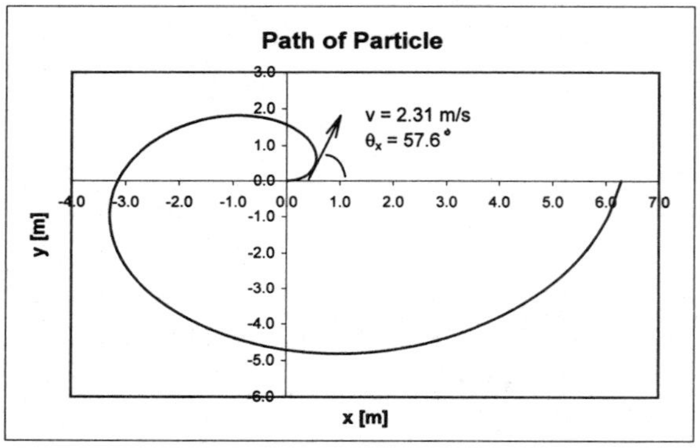

c.) For $\theta = 30° = 0.5236$ rad, by Eq (C), $t = 0.5117$ s.

Hence by Eq (E), $\bar{v} = 2.047\,\bar{e}_r + 1.072\,\bar{e}_\theta$

or (see Fig (a))

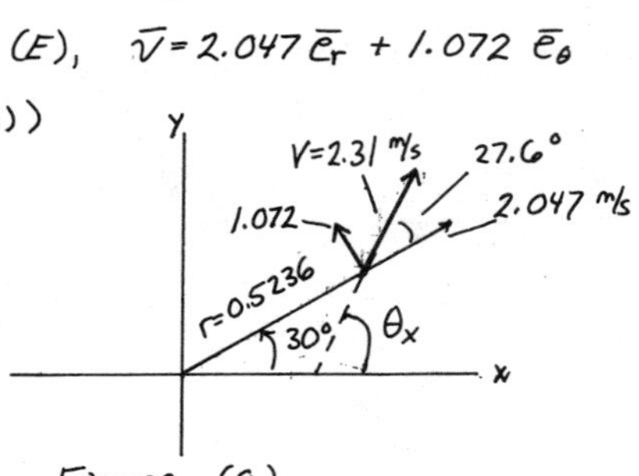

Figure (a)

By Fig (a), $v = |\bar{v}| = 2.31$ m/s

$\theta_r = \tan^{-1}\dfrac{1.072}{2.047} = 27.6°$

or $v = 2.31$ m/s

$\theta_x = 30° + 27.6° = 57.6°$

See plot above, where $\bar{v}$ is shown at $x = \theta\cos\theta = 0.453$ m, $y = \theta\sin\theta = 0.262$ m

13.159 *Computer Problem*

GIVEN A particle's path is

$R = b\cos 3\theta$ (A)

b = constant, and $\dot{\theta} = \omega$ = constant.

FIND a.) Plot the path in the (x, y) plane

b.) Find v for $\theta = 15°$

c.) Find a_n for $\theta = 15°$

SOLUTION

a.) The plot is obtained with Eq (A):

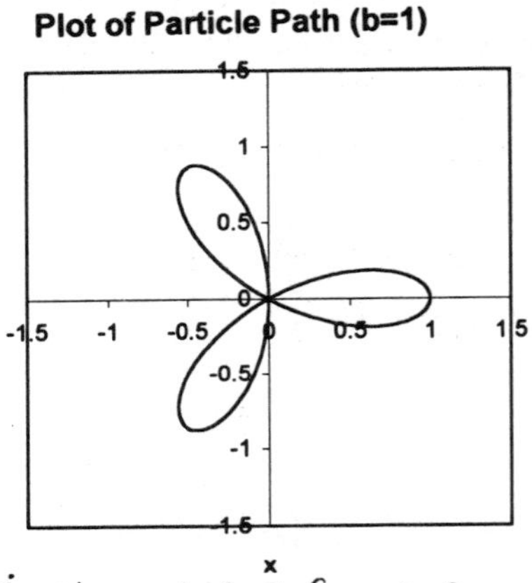

b.) Since $\dot{\theta} = \omega$ and $\theta = 0$ for $t = 0$,

$\theta = \omega t$ (B)

By Eqs (A) and (B), since $\ddot{\theta} = 0$,

$R = b\cos 3\omega t$

$\therefore\ \dot{R} = -3b\omega\sin 3\omega t$

$\ddot{R} = -9b\omega^2\cos 3\omega t$

Now,

$v_r = \dot{R} = -3b\omega\sin 3\omega t$

$v_\theta = R\dot{\theta} = b\omega\cos 3\omega t$

For $\theta = 15° = \pi/12$ rad $= \omega t$

$v_r = -3b\omega\sin(\pi/4) = -2.121\,b\omega$

$v_\theta = b\omega\cos(\pi/4) = 0.7071\,b\omega$

$\therefore v = \sqrt{v_r^2 + v_\theta^2} = \sqrt{(-2.121)^2 + (0.7071)^2}\ b\omega$

or $v = 2.236\,b\omega$ (C)

(Continued)

13.159 cont.

c.) The centripetal acceleration is

$$a_n = \frac{v^2}{r} \qquad (D)$$

where $$\frac{1}{r} = \frac{|R^2 + 2(R')^2 - RR''|}{[R^2 + (R')^2]^{3/2}} \qquad (E)$$

By Eqs (A) and (E), for $\theta = 15°$,

$$\frac{1}{r} = \frac{14}{5^{3/2} b} = \frac{1.252}{b} \qquad (F)$$

∴ For $\theta = 15°$, with Eqs (C), (D), (E), and (F),

$$a_n = \frac{(2.236\, b\omega)^2 (1.252)}{b}$$

or $a_n = 6.26\, b\omega^2$

13.160 Design Problem

GIVEN A test track is to be designed at 30° N latitude for missiles with maximum speed of 600 m/s. Maximum horizontal acceleration perpendicular to the track must not exceed 0.06 m/s^2.

FIND A suitable design for the track

SOLUTION

a.) Let the track be laid along the 30° N latitude and let the missile travel from West to East (Fig (a))

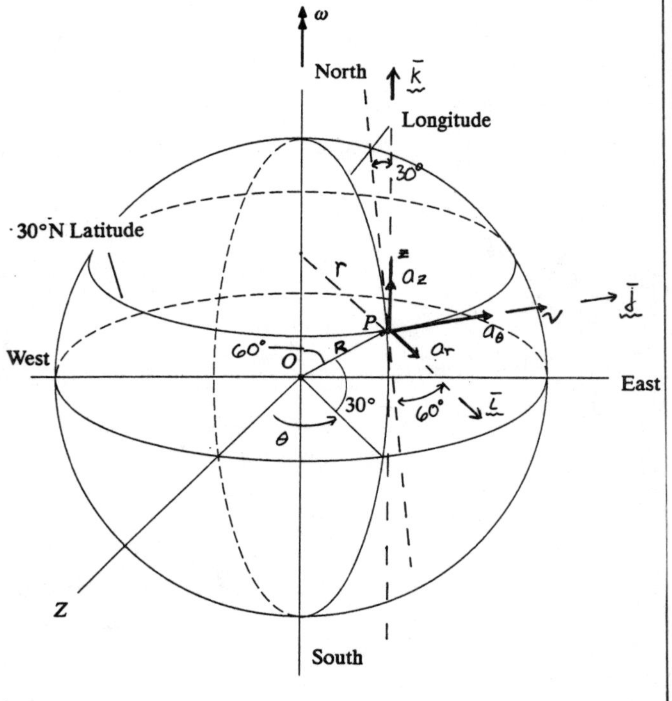

Figure (a)

Use cylindrical coordinates, r, θ, z (Fig.(a)).
By Eq (13.72)

$$a_r = \ddot{r} - r(\dot{\theta})^2$$
$$a_\theta = 2\dot{r}\dot{\theta} + r\ddot{\theta} \qquad (A)$$
$$a_z = \ddot{z}$$

By Fig (a), with the track along the 30°N Latitude and the missile traveling from West to East, the horizontal acceleration perpendicular to the track is

$$a_\perp = a_z \cos 30° - a_r \cos 60° \qquad (B)$$

But, by Eq (A), $a_z = \ddot{z} = 0$, since the missile travels in a circular path of constant radius r perpendicular to the z axis; that is, z = constant.

Also, since r = constant,

$$a_r = \ddot{r} - r\dot{\theta}^2 = 0 - r\left(\frac{v}{r}\right)^2 = -\frac{v^2}{r} \qquad (C)$$

where $v = 600 + |\omega \times r| = 600 + \omega r$

and where ω is the angular velocity of the earth, that is

$$\omega = \omega k$$

$$\omega = \frac{2\pi}{(24)(60)(60)} = 7.2722 \times 10^{-5} \text{ rad/s}$$

and $r = R\cos 30° = (6.437 \times 10^6)(\sqrt{3}/2)$
$= 5.5746 \times 10^6$ m

Therefore,

$$v = 600 + 7.2722 \times 10^{-5} \times 5.5746 \times 10^6$$
$$= 600 + 405.4 = 1005.4 \text{ m/s}$$

Hence, $$a_r = -\frac{(1005.4)^2}{5.5746 \times 10^6} = -0.18133 \text{ m/s}^2$$

and $a_\perp = -a_r \cos 60° = 0.18133 \times 0.50$
$= 0.0907$ m/s^2

Since $a_\perp = 0.0907$ $m/s^2 > 0.06$ m/s^2, this configuration is not acceptable.

b.) Let the missile travel from East to West. Then, $v = -600 + 405.4 = -194.6$ m/s

and $$a_n = \frac{-(-194.6)^2}{5.5746 \times 10^6} = -0.006793 \text{ m/s}^2$$

$$a_\perp = -a_r \cos 60° = 0.00340 \text{ m/s}^2$$

and $a_\perp = 0.00340$ $m/s^2 < 0.06$ m/s^2

Thus, if the track is laid along the 30°N latitude and the missile travels from East to West, the design is acceptable.

(Continued)

13.160 cont

c.) Let the track be laid along a longitude and let the missile travel from North to South. Then, the horizontal acceleration perpendicular to the track is a_θ (Fig (a)).

By Eqs (A): $a_\theta = 2\dot{r}\dot{\theta} + r\ddot{\theta}$
$= 2\dot{r}\omega + 0$

where (see Fig (b))
$\dot{r} = v\cos 60 = 600 \times 0.50 = 300$ m/s

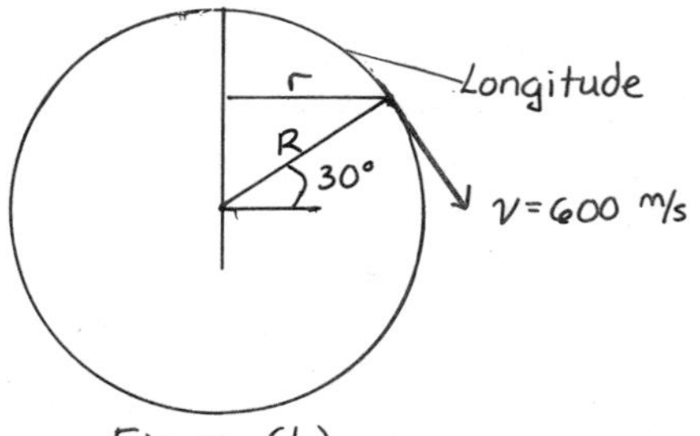

Figure (b)

Therefore,
$a_\perp = a_\theta = 2(300)(7.2722 \times 10^{-5}) = 0.0436$ m/s²

Hence, since $|0.0436 \text{ m/s}^2| < 0.06$ m/s² this design is satisfactory.

Likewise, if the missile travels from South to North, the design is also satisfactory.

RECOMMENDATION:
Placing the track along a longitude is satisfactory regardless of whether the missile travels from South to North or from North to South.
If the track is laid along the 30°N Latitude, the design is satisfactory provided the missile travels from East to West.
The margin of safety (0.00340 < 0.06) is greater in this case than in the case where the track is laid along a longitude (0.0436 < 0.06). Other factors may enter, for example, the availability of a site (either along the latitude or longitude)
There are also other directions (non-longitudinal, non-latitudinal) along which the track can be laid. The analysis of these directions is more complex.

→ From the data above, you reccommend that, if possible, the track be laid along the 30°N Latitude and the missile travel from East to West.

13.161

GIVEN Highway crosses a railroad at 90°. A car and a train approach the crossing. Initially the car, traveling at 60 mi/h, is 3/4 mile from the crossing, and the engine, traveling at 90 mi/h, is 1 mile from the crossing.

FIND The rate [mi/h] that the distance between the car and the engine is decreasing.

SOLUTION

By Fig (a), the rate of decrease of S is:

$\dot{S} = 60\cos\theta_1 + 90\cos\theta_2$
$= 60\left(\frac{3}{5}\right) + 90\left(\frac{4}{5}\right)$

or

$\dot{S} = 180$ mi/hr

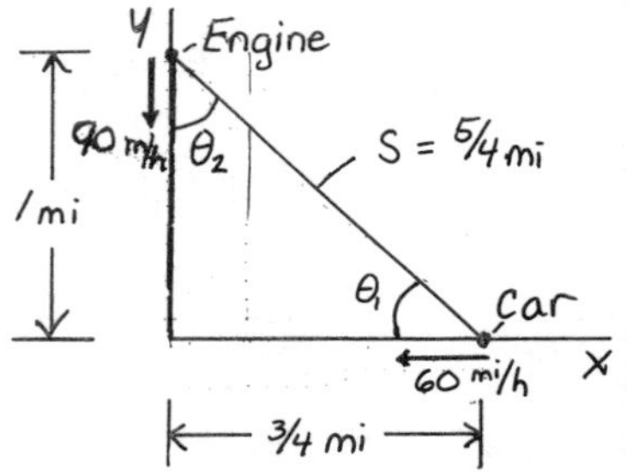

Figure (a)

13.162

GIVEN Collar B is raised by pulling the cord (Fig (a))

FIND The rate of seperation $\dot{S} = \dot{PB}$ of collar B and point P.

SOLUTION

By Fig (a),
$v = v_B \cos\theta$, or
$v_B = \dfrac{v}{\cos\theta} = v\sec\theta$ (A)

Therefore, by Fig (a)
$\dot{S} = v_B\cos\phi + v\cos\phi$ (B)

Eqs. (A) and (B) yield:
$\dot{S} = v(1+\sec\theta)\cos\phi$

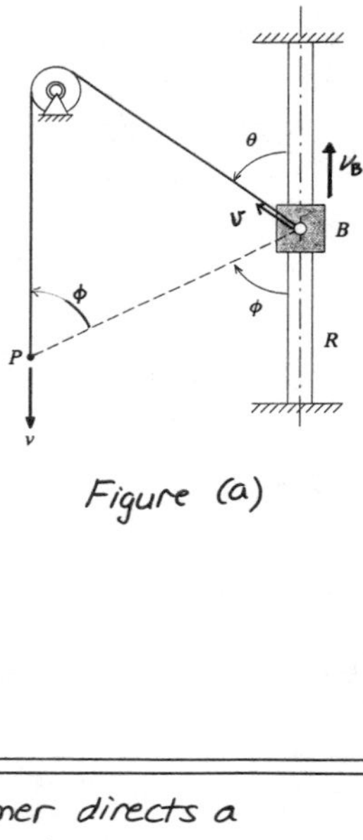

Figure (a)

13.163

GIVEN An astronomer directs a telescope at angle θ_1 to measure the altitude angle θ_2 of a star S (Fig (a)) the velocity c of light and the velocity v of the earth, in a Newtonian reference frame relative to the star, are known.

(Continued)

13.163 cont

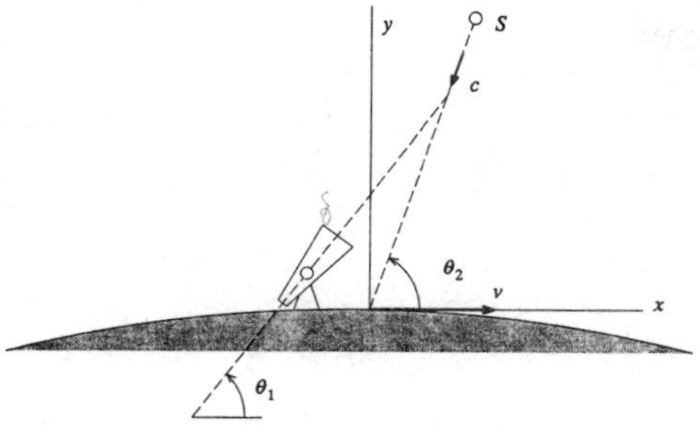

Figure (a)

FIND Formula for θ_2 in terms of θ, v, c. Then simplify the formula for $\theta_2 - \theta_1 << 1$

SOLUTION

By Figs (a) and (b)

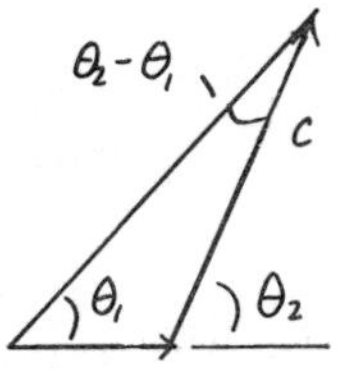

Figure (b)

$$\frac{c}{\sin\theta_1} = \frac{v}{\sin(\theta_2-\theta_1)}$$

or

$$\sin(\theta_2-\theta_1) = \frac{v}{c}\sin\theta_1 \quad (A)$$

$$\therefore \theta_2-\theta_1 = \sin^{-1}\left(\frac{v}{c}\sin\theta_1\right)$$

or $$\theta_2 = \theta_1 + \sin^{-1}\left(\frac{v}{c}\sin\theta_1\right) \quad (B)$$

If $\theta_2-\theta_1 << 1$, $\sin(\theta_2-\theta_1) \approx \theta_2-\theta_1$ (C)

So, by Eqs. (A) and (C),

$$\theta_2-\theta_1 \approx \frac{v}{c}\sin\theta_1$$

$$\theta_2 \approx \theta_1 + \frac{v}{c}\sin\theta_1$$

13.164 GIVEN Refer to Example 13.32, and assume the relative speed of water at exit equals its relative speed of approach.

FIND v_x', v_y' in terms of v, V, θ, ϕ

SOLUTION

By Fig (a),

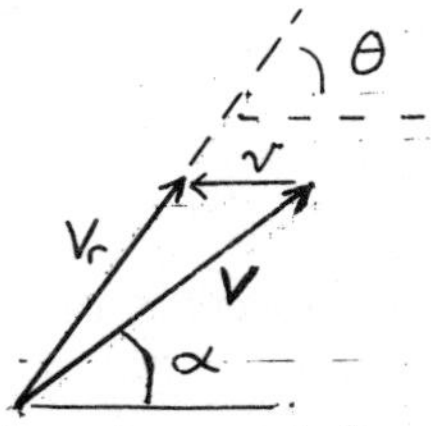

Figure (a)

$$V^2 = v^2 + V_r^2 + 2vV_r\cos\theta$$

or

$$V_r^2 + 2vV_r\cos\theta + (v^2 - V^2) = 0$$

Therefore,

$$V_r = -v\cos\theta + \sqrt{V^2 - v^2\sin^2\theta} \quad (A)$$

By Fig (b),

$$\bar{V}' = V_x'\bar{i} + V_y'\bar{j}$$

where

$$V_x' = v - V_r\cos\phi \quad (B)$$

$$V_y' = V_r\sin\phi \quad (C)$$

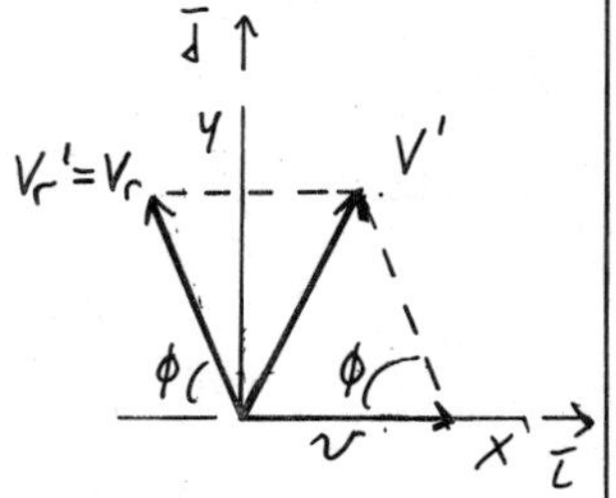

Figure (b)

Equations (A), (B) and (C) yield:

$$V_x' = v + v(\cos\theta)(\cos\phi) - \left(\sqrt{V^2 - v^2\sin^2\theta}\right)(\cos\phi)$$

$$V_y' = -v(\cos\theta)(\sin\phi) + \left(\sqrt{V^2 - v^2\sin^2\theta}\right)(\sin\phi)$$

13.165 GIVEN Water flowing steadily through a uniform pipe bend that is moving at 30 ft/s. The water enters the pipe at 60 ft/s (Fig a)

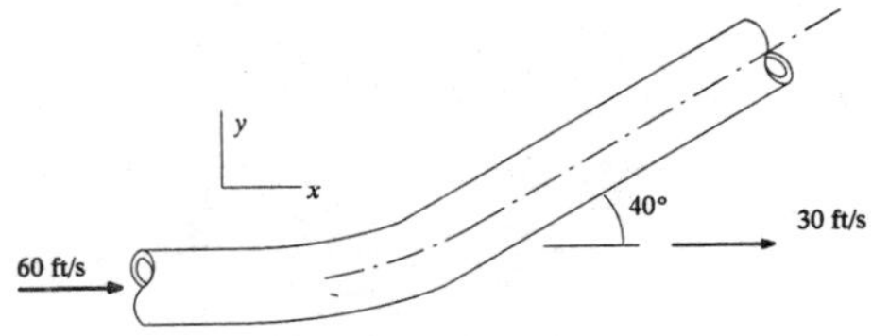

Figure (a)

FIND The velocity of the water with which it exits the bend.

SOLUTION

Let the velocity of the pipe be v_p.
Let the velocity of the water at the exit be v_w and the velocity relative to the pipe be $v_{w/p}$. (Fig b)

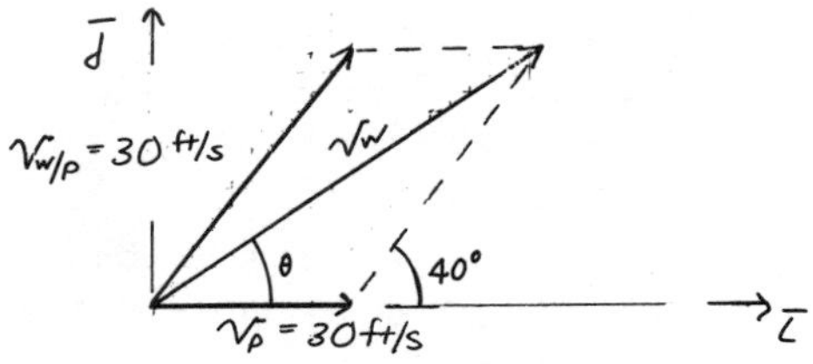

Figure (b)

By Fig (b),

$$\bar{v}_w = \bar{v}_p + \bar{v}_{w/p}$$

$$= (30 + 30\cos 40°)\bar{i} + (30\sin 40°)\bar{j}$$

or $$\bar{v}_w = (52.98)\bar{i} + 19.28\bar{j} \quad [\text{ft/s}]$$

13.166 GIVEN Fig (a) and data shown.

FIND $v_{A/G}$, $v_{A/B}$ and $a_{A/G}$, $a_{A/B}$

SOLUTION

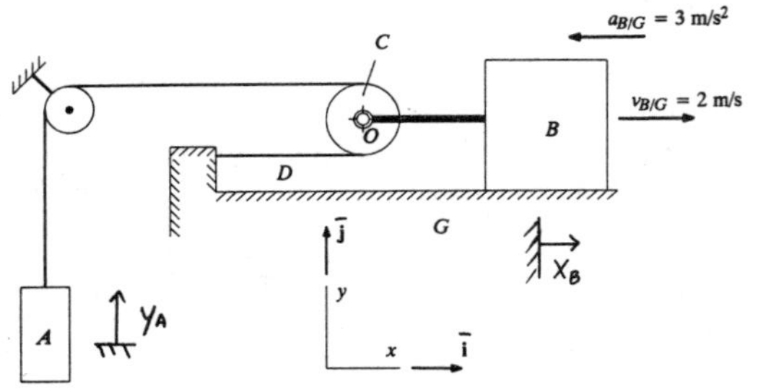

Figure (a)

Let the displacement of B be $X_{B/G}$ (Fig a.) Then, the displacement of block A is

$$y_{A/G} = 2X_{B/G}$$

$$\therefore\ v_{A/G} = \dot{y}_A = 2\dot{X}_{B/G} = 2v_{B/G}$$

$$a_{A/G} = \ddot{y}_A = 2\ddot{X}_{B/G} = 2a_{B/G}$$

Since,
$$\left.\begin{aligned}\bar{v}_{B/G} &= 2\bar{i}\ \text{m/s}\\ \bar{a}_{B/G} &= -3\bar{i}\ \text{m/s}\end{aligned}\right\}\ (A)$$

$$\left.\begin{aligned}\bar{v}_{A/G} &= 2(2)\bar{j} = 4\bar{j}\ \text{m/s}\\ \bar{a}_{A/G} &= 2(-3)\bar{j} = -6\bar{j}\ \text{m/s}^2\end{aligned}\right\}\ (B)$$

Also, since
$$\left.\begin{aligned}\bar{v}_{A/G} &= \bar{v}_{B/G} + \bar{v}_{A/B}\\ \bar{a}_{A/G} &= \bar{a}_{B/G} + \bar{a}_{A/B}\end{aligned}\right\}\ (C)$$

With Eqs (A) and (B), Eqs (C) yield

$$\bar{v}_{A/B} = -2\bar{i} + 4\bar{j}\ \text{m/s}$$

$$\bar{a}_{A/B} = 3\bar{i} - 6\bar{j}\ \text{m/s}^2$$

13.167 GIVEN Airplane with propeller 3 m in diameter rotating at 1800 rpm, is flying 500 km/h = 138.89 m/s

FIND The speed (m/s) of the tip of the propeller, relative to the plane. Relative to the ground.

SOLUTION

The angular velocity of the propeller is

$$\omega = 1800\ \text{rpm} = 60\pi\ \text{rad/s}$$

Therefore, the speed of the tip relative to the plane is:

$$v_{P/A} = \omega r = (60\pi)(1.5) = 282.74\ \text{m/s}$$

The velocity of the tip relative to the ground is: $\bar{v}_{P/G} = \bar{v}_{P/A} + \bar{v}_{A/G}$ where $\bar{v}_{P/A}$ is the velocity of the tip relative to the plane and $\bar{v}_{A/G}$ is the velocity of the airplane relative to the ground.

Since $\bar{v}_{P/A} \perp \bar{v}_{A/G}$, the speed of the tip relative to the ground is:

$$v_{P/G} = |\bar{v}_{P/G}| = \sqrt{|\bar{v}_{P/A}|^2 + |v_{A/G}|^2}$$

$$\therefore\ v_{P/G} = \sqrt{(282.94)^2 + (138.89)^2}$$

or $v_{P/G} = 315.0$ m/s

13.168 GIVEN Figure (a.) where

$$\left.\begin{aligned}u &= -V + a^2 V r^{-4}(X^2 - Y^2)\\ v &= 2a^2 V r^{-4} XY\end{aligned}\right\}\ (A)$$

where $r = X^2 + Y^2$, V = speed at $X = \infty$ and B = moving reference frame with speed V in −X direction.

FIND

a.) The (x, y) velocity projections u′, v′ relative to B in terms of X, Y.

b.) tan θ as a function of (X, Y); θ = angle between x axis and vector (u′, v′)

SOLUTION

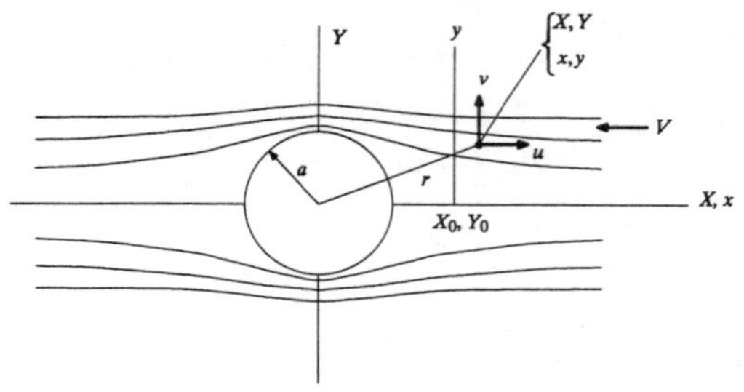

Figure (a)

a.) By Fig a, since $Y_0 = 0$,

$$X = X_0 + x \qquad Y = Y_0 + y = y \qquad (B)$$

By Eqs. (B),

$$\dot{X} = \dot{X}_0 + \dot{x} = u \qquad \dot{Y} = \dot{y} = v$$

$$\therefore\ \dot{x} = u' = u - \dot{X}_0 = u + V$$

$$\dot{y} = v' = v$$

(Continued)

13.168 cont

or, by Eqs (A),

$$u' = u + V = a^2 V r^{-4}(x^2 - y^2)$$
$$v' = v = 2a^2 V r^{-4} xy \qquad \text{(C)}$$

b.) By Eqs (C),

$$\tan\theta = \frac{v'}{u'} = \frac{2xy}{x^2 - y^2}$$

13.169 GIVEN Particle motion in (x, y) plane

$$x = 4\cos 2t, \quad y = 3\sin 2t \qquad \text{(A)}$$

where t denotes time and (x, y) axes are attached to frame B that translates with respect to frame A and the origin of (x, y) moves according to

$$X_0 = 2t^2, \quad Y_0 = 1 - 3t \qquad \text{(B)}$$

X, Y attached to frame A and parallel to (x, y)

FIND The velocity $\bar{v}_{P/A}$ and the acceleration $\bar{a}_{P/A}$ relative to frame A as functions of t.

SOLUTION

Since axes (x, y) translate relative to axes (X, Y), the coordinates of P in the X, Y frame are:

$$X = X_0 + x, \qquad Y = Y_0 + y$$

Hence, with Eqs (A), and (B),

$$X = 2t^2 + 4\cos 2t$$
$$Y = 1 - 3t + 3\sin 2t$$

$$\therefore \quad \dot{X} = 4t - 8\sin 2t = v_x$$
$$\dot{Y} = -3 + 6\cos 2t = v_y$$
$$\ddot{X} = 4 - 16\cos 2t = a_x$$
$$\ddot{Y} = -12\sin 2t = a_y$$

So, the velocity $\bar{v}_{P/A}$ and the acceleration $\bar{a}_{P/A}$ of particle P relative to frame A are

$$\bar{v}_{P/A} = (4t - 8\sin 2t)\bar{i} - (3 - 6\cos 2t)\bar{j}$$

$$\bar{a}_{P/A} = (4 - 16\cos 2t)\bar{i} - (12\sin 2t)\bar{j}$$

13.170 GIVEN Figure (a.), water with velocity v_1 (constant) strikes horizontal vane moving with velocity $= v_2$ = constant

FIND a.) Express the exit speed v_2 of water in terms of u, v_1, and θ. Neglect friction.
b.) Check your equation from part a. for θ = 0° and θ = 180°

SOLUTION

a.) Since friction is negligible, v_1 remains constant as water flows around the vane. ∴ By the velocity diagram (Fig b) and the law of cosines:

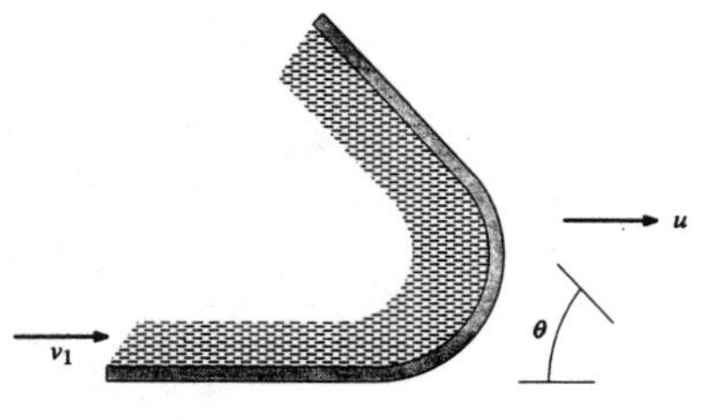

Figure (a)

$$v_2 = \sqrt{(v_1 - u)^2 + u^2 - 2(v_1 - u)(u)\cos\theta} \qquad \text{(A)}$$

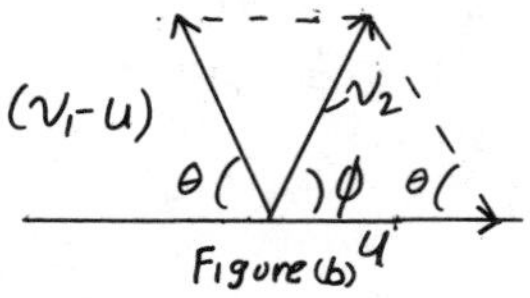

Figure (b)

b.) For θ = 0°, Equation (A) yields (φ = 180°; see Fig. (b))

$$v_2 = v_1 - 2u \qquad \text{(B)}$$

Equation (B) is intuitively correct.

For θ = 180°, Eq (A) yields (φ = 0°; see Fig. (b))

$$v_2 = v_1 \qquad \text{(C)}$$

Equation (C) is also intuitively correct.

13.171 GIVEN Position of cam shown in Fig (a):

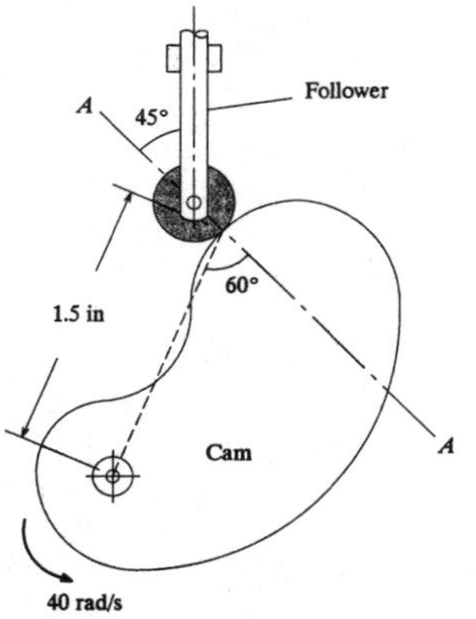

Figure (a)

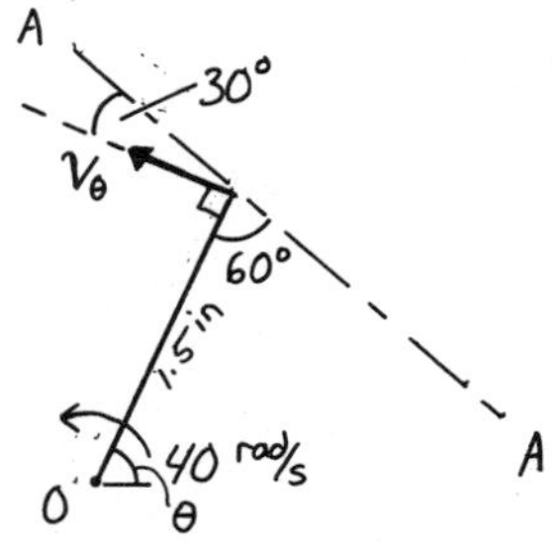

Figure (b)

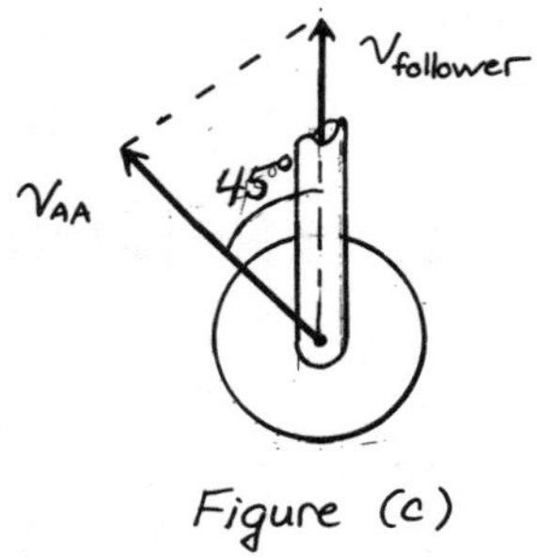

Figure (c)

(Continued)

13.171 cont

SOLUTION

By Fig (b), $v_\theta = (1.5)(40) = 60$ in/s

The component of v_θ along the direction AA is

$v_{AA} = v_\theta \cos 30° = 51.96$ in/s

By Fig (c), $v_{follower} = \dfrac{v_{AA}}{\cos 45°}$

or $v_{follower} = 73.48$ in/s

13.172 GIVEN A woman rows a boat from point A on the East bank of a river at a speed of 8 km/h relative to the water, to a point B directly opposite on the West bank (Fig (a)). The river flows South at 5 km/h and is ½ km wide

FIND

a.) The direction she must head the boat

b.) The time required to cross the river.

SOLUTION

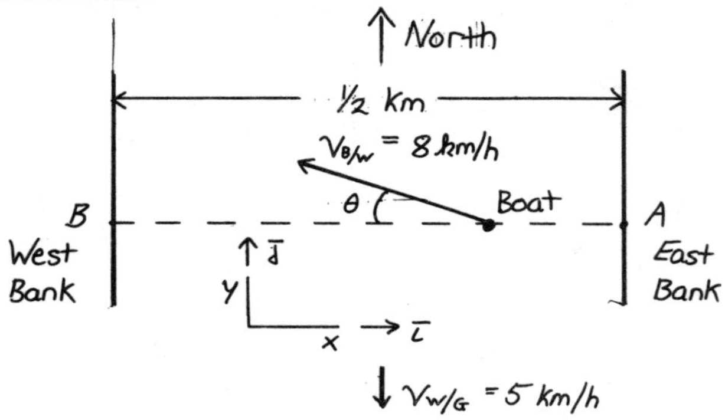

Figure (a)

a.) By Fig (a), with $v_{B/w} = 8$ km/h and $v_{w/G} = 5$ km/h :

$\bar{v}_{B/G} = \bar{v}_{B/w} + \bar{v}_{w/G}$, directed along AB.

$\therefore \quad \bar{v}_{B/w} = -(8\cos\theta)\bar{i} + (8\sin\theta)\bar{j}$ km/h (A)

$\bar{v}_{w/G} = -5\bar{j}$ km/h (B)

$\therefore \quad \bar{v}_{B/G} = -(8\cos\theta)\bar{i} + (8\sin\theta - 5)\bar{j}$ (C)

For the boat to travel from point A to point B, the $\bar{j}$ component of $\bar{v}_{B/G}$ must be zero.

$\therefore \sin\theta = \frac{5}{8}$; or $\theta = 38.68°$

b.) The time required to cross the river is determined by:

$|\bar{v}_{B/G}|t = \frac{1}{2}$ km

By Eq. (C), $|\bar{v}_{B/G}| = 8\cos\theta = 8(0.7806)$
$= 6.245$ km/h

$\therefore t = \dfrac{0.5 \text{ km}}{6.245 \text{ km/h}} = 0.08\text{ h} = 4.8$ min

13.173 GIVEN Data shown in Fig (a) (Length in inches, time t in seconds) Initially (t=0) both blocks are at rest.

FIND $v_{B/A}$, $a_{B/A}$

SOLUTION

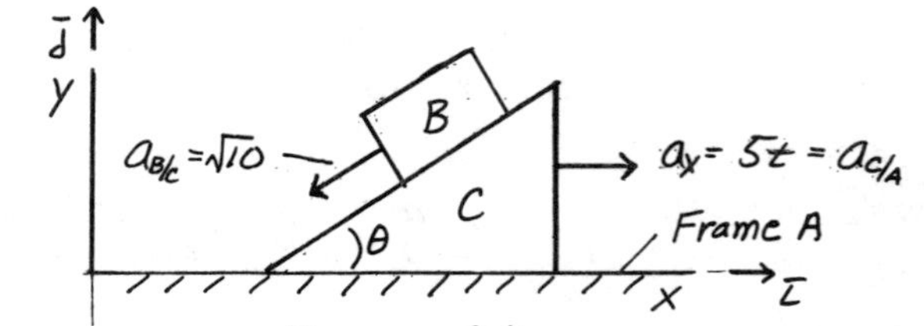

Figure (a)

Since the blocks are initially at rest;

$v_{B/c} = a_{B/c}t = \sqrt{10}\,t$ in/s

$v_{c/A} = \frac{1}{2}5t^2 = 2.5t^2$ in/s

$\bar{v}_{B/c} = \sqrt{10}\,t\,[(-\cos\theta)\bar{i} + (-\sin\theta)\bar{j}]$

$\bar{v}_{c/A} = (2.5t^2)\bar{i}$

$\therefore \bar{v}_{B/A} = \bar{v}_{B/c} + \bar{v}_{c/A}$

or $\bar{v}_{B/A} = (2.5t^2 - \sqrt{10}\,t\cos\theta)\bar{i} - (\sin\theta)(\sqrt{10}\,t)\bar{j}$

Hence, $v_{B/A} = |\bar{v}_{B/A}| = \sqrt{6.25t^4 + 10t^2 - 5\sqrt{10}\,t^3\cos\theta}$

Also, $\bar{a}_{B/A} = \bar{a}_{c/A} + \bar{a}_{B/c}$

or $\bar{a}_{B/A} = [(5t) - \sqrt{10}(\cos\theta)]\bar{i} - \sqrt{10}(\sin\theta)\bar{j}$

Hence, $a_{B/A} = |\bar{a}_{B/A}| = \sqrt{25t^2 + 10 - 10\sqrt{10}\,t\cos\theta}$

13.174 GIVEN Figure (a). When plane A shoots a rocket at plane B with a speed $v_{R/A} = 2000$ ft/s.

FIND The required direction of the rocket guides with respect to the flight path of plane A for a hit (neglect gravity)

SOLUTION

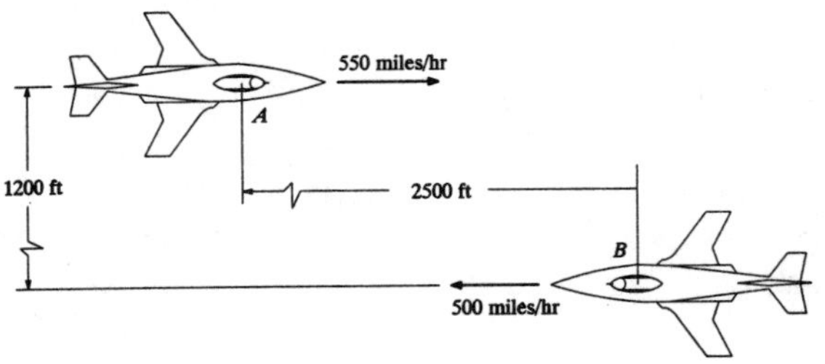

Figure (a)

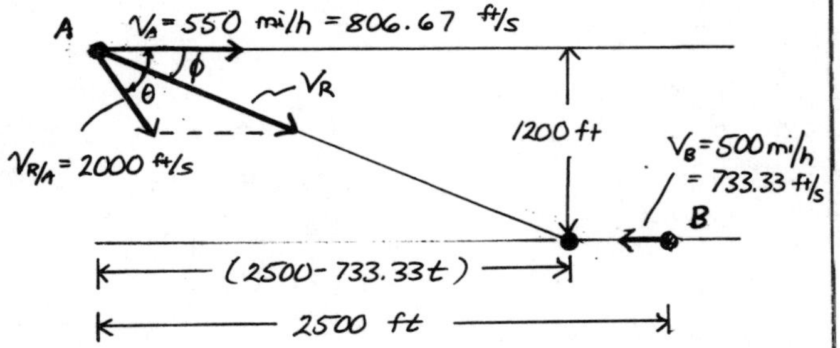

Figure (b)

By Fig (b),

$2500 - 733.33t = v_R(\cos\phi)t$ (A)

(t = flight time of rocket)

and $v_R(\sin\phi)t = 1200$ (B)

$\therefore\ t = \dfrac{1200}{v_R \sin\phi} = \dfrac{2500}{733.33 + v_R\cos\phi}$ (C)

Also, $v_R \sin\phi = 2000 \sin\theta$ (D)

$v_R\cos\phi = 806.67 + 2000\cos\theta$ (E)

Therefore, by Eqs. (C), (D), and (E),

$$\frac{1200}{v_R \sin\phi} = \frac{1200}{2000\sin\theta} = \frac{2500}{1540 + 2000\cos\theta}$$

$\therefore\ 1848 + 2400\cos\theta = 5000\sin\theta$ (F)

The solution of Eq. (F) is

$\theta = 45.105°$

13.175 GIVEN Refer to Fig (a). The sound wave travels from S to R and back to S as the platform moves with velocity v directed as shown.

FIND The time t required for the sound wave to travel from S to R back to S.

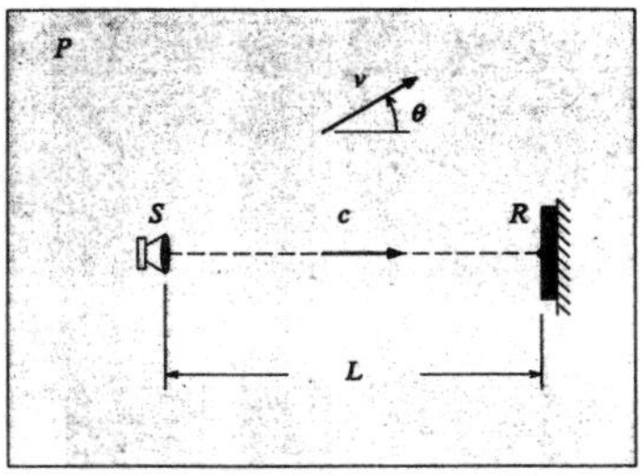

Figure (a)

SOLUTION

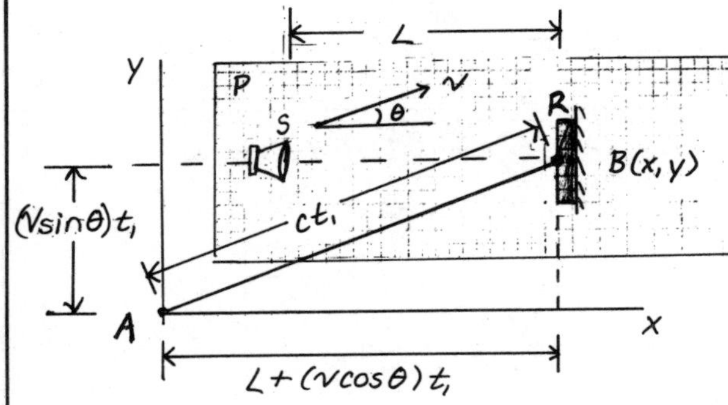

Figure (b)

By Fig (b), the distance traveled from A (initial position of S) to B (position of R when sound wave arrives after t_1 seconds is)

$$(\overline{AB})^2 = x^2 + y^2$$

or $(ct_1)^2 = \left[L + v(\cos\theta)t_1\right]^2 + \left[v(\sin\theta)t_1\right]^2$

$\therefore\ (c^2 - v^2)t_1^{\,2} - 2vL(\cos\theta)t_1 - L^2 = 0$

or $t_1 = \dfrac{L\left[v\cos\theta + \sqrt{c^2 - v^2\sin^2\theta}\,\right]}{c^2 - v^2}$ (A)

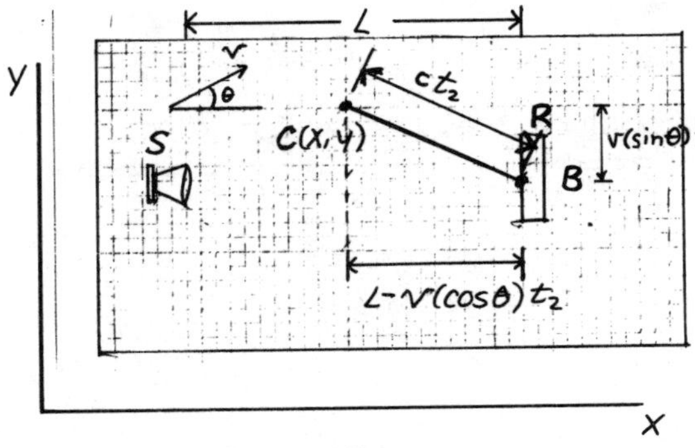

Figure (c)

(Continued)

13.175 cont.

By Fig. (c), the distance from B to C (the position of the source when the sound wave returns to the source) is:

$$(\overline{BC})^2 = (ct_2)^2 = [L - v(\cos\theta)t_2]^2 + [v(\sin\theta)t_2]^2$$

or

$$(c^2 - v^2)t_2^2 + 2vL(\cos\theta)t_2 - L^2 = 0$$

Hence,

$$t_2 = \frac{L[-v\cos\theta + \sqrt{c^2 - v^2\sin^2\theta}\,]}{c^2 - v^2} \qquad (B)$$

By Eqs. (A) and (B), the total time is:

$$t = t_1 + t_2 = \frac{2L\sqrt{c^2 - v^2\sin^2\theta}}{c^2 - v^2}$$

13.176 GIVEN Figure (a) and data shown

FIND: ω_{AB}, α_{AB}, ω_{BC}, and α_{BC}.

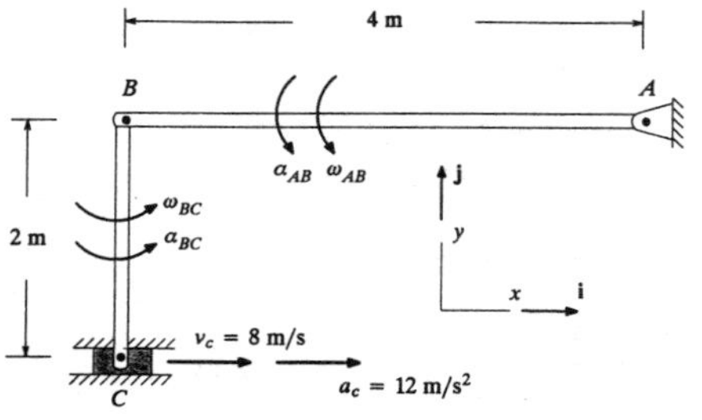

Figure (a)

SOLUTION

By Fig (a), $\bar{v}_C = \bar{v}_B + \bar{v}_{C/B}$

or $8\bar{i} = -4\omega_{AB}\bar{j} + 2\omega_{BC}\bar{i}$

$\therefore \omega_{AB} = 0, \quad \omega_{BC} = 4 \text{ rad/s} \qquad (A)$

Also by Fig (a), $\bar{a}_C = \bar{a}_B + \bar{a}_{C/B}$

or $12\bar{i} = -4\alpha_{AB}\bar{j} + 4\omega_{AB}^2\bar{i} + 2\alpha_{BC}\bar{i} + 2\omega_{BC}^2\bar{j}$

Hence, with Eq. (A),

$12\bar{i} = 2\alpha_{BC}\bar{i} + (32 - 4\alpha_{AB})\bar{j}$

$\therefore \quad \alpha_{BC} = 6 \text{ rad/s}^2, \quad \alpha_{AB} = 8 \text{ rad/s}^2$

13.177 GIVEN Figure (a) with

$$\left.\begin{aligned} X_0 &= A\cos\omega t + \text{constant} \\ Y_0 &= \text{constant} \\ \theta &= \theta_0 \sin\omega t + \text{constant} \end{aligned}\right\} (A)$$

t denotes time, A, θ_0, and ω are constants, and axes (x, y) translate relative to axes X, Y

FIND

a.) Velocity and acceleration of bob B, relative to axes (x, y) in terms of L, A, ω, t, and θ_0

b.) Velocity and acceleration of B relative to axes (X, Y).

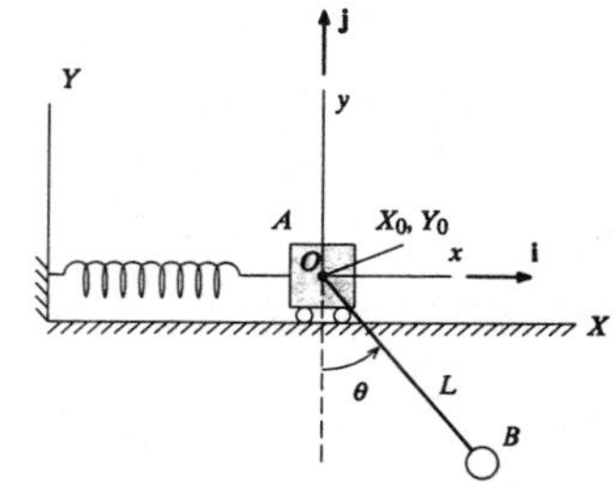

Figure (a)

SOLUTION

a.) By Fig (a), and by differentiation,

$$\left.\begin{aligned} x &= L\sin\theta, & y &= -L\cos\theta \\ \dot{x} &= L\dot{\theta}\cos\theta, & \dot{y} &= L\dot{\theta}\sin\theta \end{aligned}\right\} (B)$$

$\therefore \bar{v}_{B/xy} = L\dot{\theta}[(\cos\theta)\bar{i} + (\sin\theta)\bar{j}]$

or, with Eqs. (A),

$\bar{v}_{B/xy} = L\theta_0\omega\cos\omega t\,[\cos(\theta_0\sin\omega t + \text{constant})\bar{i} + \sin(\theta_0\sin\omega t + \text{constant})\bar{j}]$

In a similar manner,

$$\left.\begin{aligned} \ddot{x}_B &= -\omega^2 L\theta_0\cos(\theta_0\sin\omega t + \text{constant})\sin\omega t \\ &\quad -\omega^2 L\theta_0^2\sin(\theta_0\sin\omega t + \text{constant})\cos^2\omega t \\ \ddot{y}_B &= -\omega^2 L\theta_0\sin(\theta_0\sin\omega t + \text{constant})\sin\omega t \\ &\quad +\omega^2 L\theta_0^2\cos(\theta_0\sin\omega t + \text{constant})\cos^2\omega t \end{aligned}\right\} (C)$$

$\therefore \bar{a}_{B/xy} = \ddot{x}_B\bar{i} + \ddot{y}_B\bar{j}$

where $\ddot{x}_B$, $\ddot{y}_B$ are given by Eqs. (C).

b.) By Fig (a), the (X, Y) coordinates of B are

$X = X_0 + x, \quad Y = Y_0 + y$

$$\left.\begin{aligned} \therefore \dot{X} &= \dot{X}_0 + \dot{x} & \dot{Y} &= \dot{y} \\ \ddot{X} &= \ddot{X}_0 + \ddot{x} & \ddot{Y} &= \ddot{y} \end{aligned}\right\} (D)$$

where, by Eqs. (A)

$\dot{X}_0 = -A\omega\sin\omega t \qquad \ddot{X}_0 = -A\omega^2\cos\omega t$

$\dot{Y}_0 = \ddot{Y}_0 = 0$

and $(\dot{x}, \dot{y})$, $(\ddot{x}, \ddot{y})$ are given by Eqs. (B) & (C)

(Continued)

13.177 cont

Hence,
$$\bar{v}_{B/XY} = \dot{X}\bar{i} + \dot{Y}\bar{j}$$
$$\bar{a}_{B/XY} = \ddot{X}\bar{i} + \ddot{Y}\bar{j}$$

13.178 GIVEN Airplane with air speed of 225 km/h is flying in a gale of 80 km/h directed 110° clockwise from North. (Fig (a))

FIND
a.) Direction plane must be pointed if it is to fly a course with azimuth 40° relative to the earth
b.) Speed of airplane relative to earth
c.) Check your computations graphically

SOLUTION

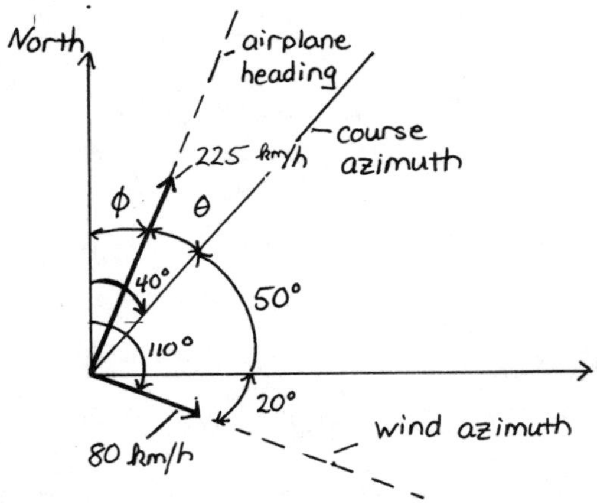

Figure (a)

a.) Since the plane is to fly an azimuth of 40°, the component of its velocity ⊥ to the 40° azimuth must cancel the ⊥ component of the wind velocity.

Hence, by Fig (a),

$$225 \sin\theta = 80 \sin 70°$$

or $\theta = 19.578°$

$\therefore$ $\phi = 40° - \theta = 20.48°$

East of North (Fig (a)) is the required plane heading (course azimuth).

b.) The speed of the plane P relative to earth E is (see Fig a.)

$$v_{P/E} = 225 \cos\theta + 80 \cos 70°$$

or $v_{P/E} = 212.07 + 27.36 = 239.43$ km/h

c.) By Fig (b),

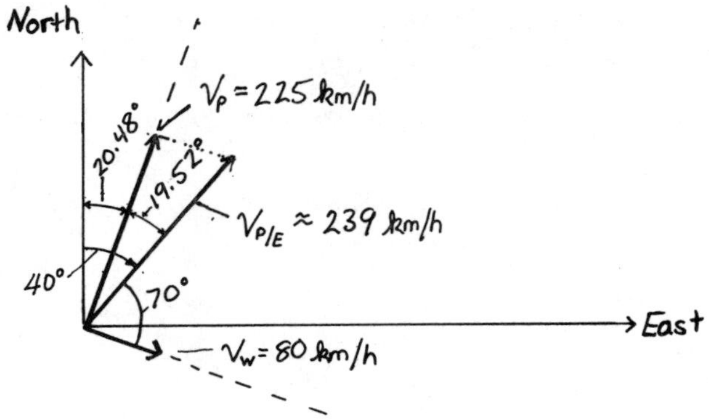

Figure (b)

$$\bar{v}_{P/E} = \bar{v}_P + \bar{v}_w$$

or, by measurement,

$\bar{v}_{P/E} \approx 239$ km/h @ 40° East of North

13.179 GIVEN At time $t = 0$, plane A is 640 km East of plane B. Plane A is flying N.E. at 560 km/h and plane B is flying on a straight course at 800 km/h to catch plane A.

FIND
a.) Direction (from North) that plane B must fly and rate at which distance between planes is decreasing.
b.) Time required for plane B to catch plane A

SOLUTION

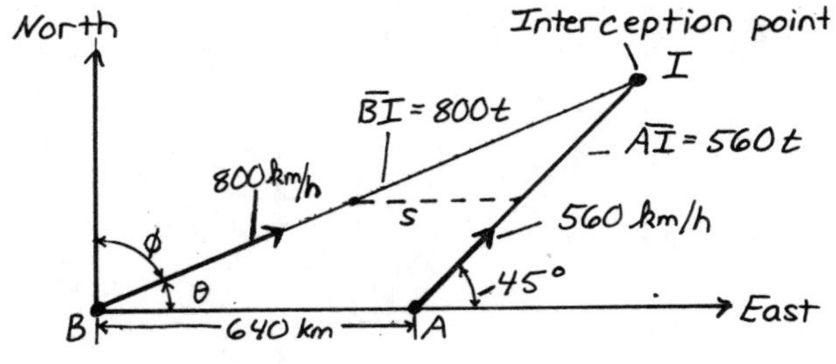

Figure (a)

a.) Consider Fig (a). With that,

$$(800t)(\sin\theta) = (560t)\sin 45°$$

(Continued)

13.179 cont

$\therefore \sin\theta = 0.4950, \qquad \theta = 29.668°$

Hence, plane B must fly at azimuth (East of North)

$\phi = 90° - \theta = 60.33°$

Also, by Fig (a), the rate $\dot{s}$ at which the distance between A and B decreases is

$\dot{s} = (800)\cos\theta - (560)\cos 45°$

or $\dot{s} = 695.126 - 395.980 = 299.15$ km/s

b.) The time required for B to catch A is obtained from Fig (a)

$(800t)\cos\theta = 640 + (560t)\cos 45°$

$\therefore t = \dfrac{640}{(800)\cos(29.668°) - (560)\cos 45°}$

or $t = 2.139$ h

13.180 GIVEN Figure (a). The speed of end B of rigid link AB is 40 ft/s, and end A moves as shown.

FIND v_x and v_y for the position shown. Use the fact that $\dot{\overline{AB}} = \dot{s} = 0$

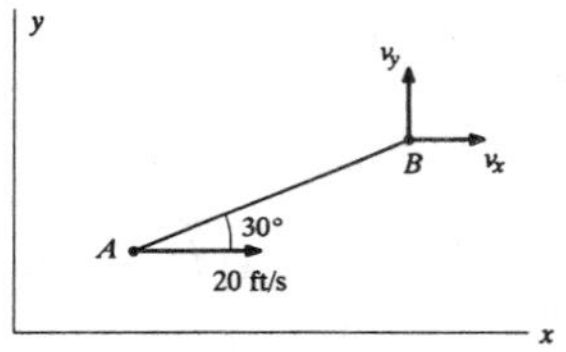

Figure (a)

SOLUTION

By Fig (a),

$\dot{s} = v_x \cos 30° + v_y \cos 60° - (20)\cos 30°$

or $v_x = 20 - \dfrac{1}{\sqrt{3}} v_y \qquad (A)$

Also, $v_x^2 + v_y^2 = (40)^2 = 1600 \qquad (B)$

By Eq. (B), $v_x = \pm\sqrt{1600 - v_y^2} \qquad (C)$

By Eqs. (A) and (C),

$v_y^2 - 10\sqrt{3}\, v_y - 900 = 0$

$v_y = 5\sqrt{3}\,(1 \pm \sqrt{13}) \qquad (D)$

or

$v_y = 39.88$ ft/s or -22.56 ft/s

Hence, by Eqs (A) and (D)

$v_x = -3.028$ or 33.03 ft/s

Therefore, the two solutions are

$v_x = -3.03$ ft/s, $v_y = 39.88$ ft/s

and $v_x = 33.03$ ft/s, $v_y = -22.56$ ft/s

13.181 GIVEN Figure (a). As the platform swings the ball rolls on it.

FIND The speed of the center of the ball relative to the earth in terms of $\dot{x}$ and $\dot{\theta}$

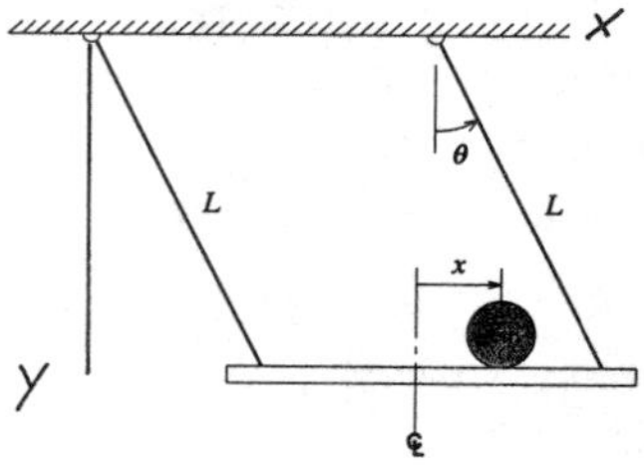

Figure (a)

SOLUTION

Select axes X, Y fixed to the supports (shown in Fig. a). Then the coordinates of the center of the ball are:

$X = L\sin\theta + x + \text{constant}$

$Y = L\cos\theta$

and $\dot{X} = L\dot{\theta}\cos\theta + \dot{x}$

$\dot{Y} = L\dot{\theta}\sin\theta$

$\therefore v^2 = \dot{X}^2 + \dot{Y}^2 = L^2(\dot{\theta})^2 + 2L\dot{x}\dot{\theta}\cos\theta + (\dot{x})^2$

or $v = \sqrt{L^2(\dot{\theta})^2 + 2L\dot{x}\dot{\theta}\cos\theta + (\dot{x})^2}$

13.182 Computer Problem

GIVEN Figure (a) with plane at height h [ft] and speed v (mi/h).

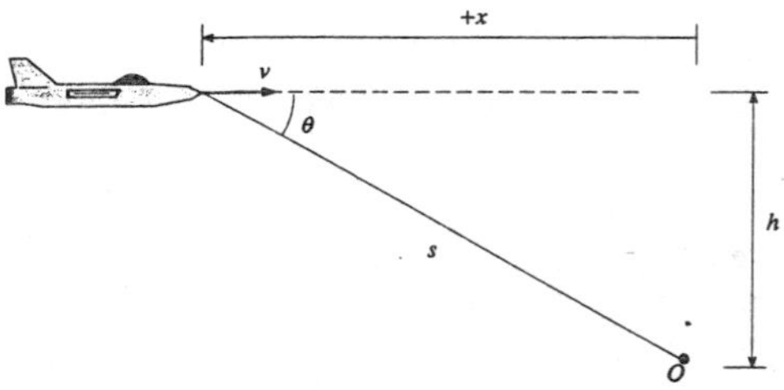

Figure (a)

(Continued)

13.182 cont.

FIND

a.) Formula for time rate $\dot{s}$ the distance s between the plane and station O is decreasing, in terms of v, x, and h.

b.) For $h = 36{,}000$ ft, plot ratio $\dot{s}/v$ as a function of x for $80{,}000 \text{ ft} \geq x \geq -80{,}000$ ft

c.) For $h = 36{,}000$ ft and $v = 600$ mi/h plot $\dot{s}$ [ft/s] as function of time t [s], for $0 \leq t \leq 182$ s. For $t=0$, $x = 80{,}000$ ft.

SOLUTION

a.) By Fig (a),

$$\dot{s} = v\cos\theta = \frac{vx}{s} = \frac{vx}{\sqrt{x^2+h^2}} \qquad (A)$$

b.) See plot as follows:

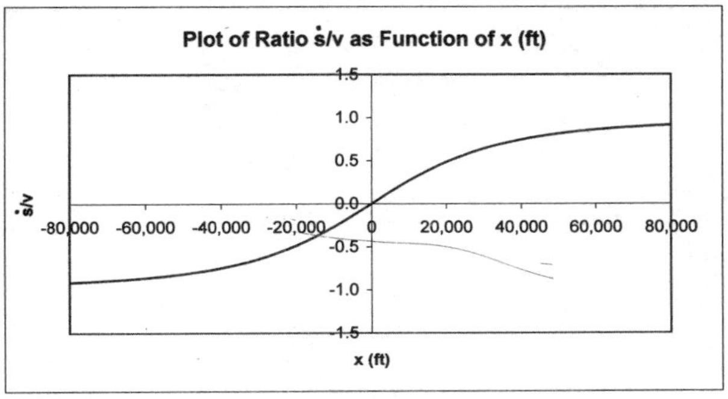

c.) By Fig (a),

$$x = x_0 - vt = 80{,}000 - (600)\left(\frac{88}{60}\right)t$$

or $x = 80{,}000 - 880t$ [ft] (B)

∴ By Eqs (A) and (B),

$$\dot{s} = \frac{880\,(80{,}000 - 880t)}{\sqrt{(80{,}000-880t)^2 + (36{,}000)^2}}$$

See plot:

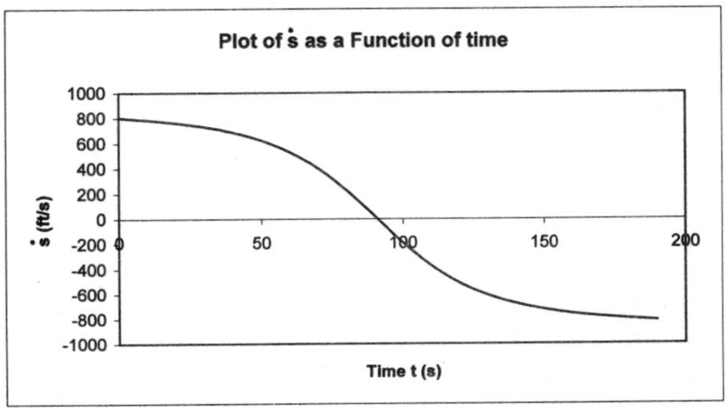

d.) Explain plots of b.) and c.):

In plot of $\dot{s}/v$ versus x, when x is positive (plane is left of point O; Fig a.), $\dot{s}$ is positive, that is, s is decreasing. For x negative, (plane is right of point O), $\dot{s}$ is negative, that is, s is increasing. The plane is flying away from point O.

In plot of $\dot{s}$ versus time t, for $0 \leq t \leq 80{,}000/880 = 90.91$ s, $\dot{s}$ is positive (distance is decreasing). At $t = 90.91$ s, the plane passes over point O. Hence, for $t > 90.91$ s, $\dot{s}$ is negative (s is increasing).

REVIEW QUESTIONS

13.194 GIVEN a = constant; at $t=0$, $x=v=0$

SHOW $v = \sqrt{2ax}$

SOLUTION

$$a = \frac{dv}{dt} = \frac{dv}{dt}\frac{dx}{dx} = \frac{dx}{dt}\frac{dv}{dx} = v\frac{dv}{dx}$$

∴ $v\,dv = a\,dx$

or $\frac{1}{2}v^2 = ax + \text{constant} = ax$, since $x=v=0$ for $t=0$.

∴ $v = \sqrt{2ax}$

13.195 GIVEN Particle moves on x axis. a = constant; for $t=0$, $v=v_0$ and $x=x_0$

Find a.) Express v as a function of time t

b.) Express x as function of t.

SOLUTION

$a = \frac{dv}{dt}$ → ∴ $v = at + v_0$

But, $v = \frac{dx}{dt}$ → ∴ $x = \frac{1}{2}at^2 + v_0 t + x_0$

13.196 GIVEN Acceleration of particle traveling along x axis, $a = 20t^3$ [m/s²]

For time $t = 1$ s, $v = 3$ [m/s]

For $t = -1$ s, $x = 5$ m

FIND: x for $t = 2$ s

SOLUTION

$a = 20t^3 = \frac{dv}{dt}$ ∴ $v = 5t^4 + v_0 = 5t^4 - 2$

Also, $v = \frac{dx}{dt}$ ∴ $x = t^5 - 2t + x_0$

Since $x=5$ for $t=-1$, $x_0 = 4$ ∴ $x = t^5 - 2t + 4$ [m]

For $t = 2$, $x = 32$ m

13.197 GIVEN Acceleration along x axis; $a = 2/x$ For $x=1$, $v=2$

FIND v as function of x

SOLUTION

$$a = \frac{dv}{dt} = v\frac{dv}{dx}$$

$$\therefore \tfrac{1}{2}v^2 = 2\ln|x| + \text{constant}$$

Since $v=2$ for $x=1$, constant $=2$

$$\therefore v = 2\sqrt{(1+\ln|x|)}$$

13.198 GIVEN Particle velocity $v = x+1$ along x axis. For $t=0$, $x=0$

FIND a.) $t = t(x)$

b.) time t when $x = e-1$; e = base of natural logarithm

SOLUTION

a.) $v = x+1 = \frac{dx}{dt}$ $\quad\therefore t = \ln|x+1|$

b.) For $x = e-1$,

$t = \ln|e-1+1| = \ln|e| = 1$ $\quad\therefore t=1$

13.201 GIVEN Plane flies from A to B, then back to A with mean speeds 300 mi/h to B and 200 mi/h back to A.

SHOW Mean speed for round trip is 240 mi/h.

SOLUTION

Time to fly A to B is $d = 300\,t_{AB}$ where d is distance between A and B and t_{AB} is time taken. $\therefore t_{AB} = \frac{d}{300}$ (A)

Similarly, from B to A, time taken is $d = 200\,t_{BA}$ $\therefore t_{BA} = \frac{d}{200}$ (B)

Let mean speed for round trip be v_{mean}. Then $2d = v_{mean}(t_{AB}+t_{BA})$

$$\therefore t_{AB} + t_{BA} = \frac{2d}{v_{mean}} \quad (C)$$

By Eqs. (A), (B), and (C):

$$\frac{2d}{v_{mean}} = \frac{d}{300} + \frac{d}{200}$$

or $v_{mean} = \frac{120{,}000}{500} = 240$ mi/h

13.208 GIVEN Angular acceleration of rotating line is $\alpha = 2\omega$. For $t=0$, $\omega=1$.

FIND $\omega = \omega(t)$, t = time and interpret ω + α for the limiting cases $t\to 0$ and $t\to\infty$.

SOLUTION

By definition, $\alpha = 2\omega = \frac{d\omega}{dt}$

or $\int_0^t 2\,dt = \int_1^\omega \frac{d\omega}{\omega}$

or $2t = \ln|\omega|$ $\rightarrow$ $|\omega| = e^{2t}$

as $t\to\infty$, $\omega\to\infty$, $\alpha\to\infty$

as $t\to 0$, $\omega\to 1$, $\alpha\to 2$

13.209 GIVEN Angular velocity of rotating line is $\omega = 2\theta^2$

FIND Angular acceleration, α as function of θ

SOLUTION $\alpha = \frac{d\omega}{dt} = \omega\frac{d\omega}{d\theta} = 8\theta^3$

13.225 GIVEN Particle travels along a curved path in one sense. Arc length $s = t^3$ (s in meters, t in seconds)

For $t = 3$ s, $a_n = 27$ m/s²

FIND v_t, v_n, a_t, $\frac{1}{r}$ (curvature) at $t=3$s

SOLUTION

The particle's velocity is tangent to the path

$\therefore v_n = 0$ for all times

$$v_t = \frac{ds}{dt} = 3t^2\Big|_{t=3} = 27 \text{ m/s}$$

$$a_t = \frac{d^2s}{dt^2} = 6t\Big|_{t=3} = 18 \text{ m/s}^2$$

$$a_n = \frac{v^2}{r} = \frac{v_t^2}{r} = 27 = \frac{27^2}{r}$$

$$\therefore \frac{1}{r} = \frac{1}{27}\ \text{m}^{-1}$$

14.1

Given: Newton's law of gravitation and Fig. 14.1

$$F = G\frac{m_1 m_2}{r^2}$$

Verify: Ratios ω/ω_1 for $r/r_1 = 2$ and 4

Solution:

For $r/r_1 = 2$, by Fig. 14.1, $\omega/\omega_1 \simeq \underline{0.25}$

By Eq. (14.1),

$$\frac{\omega}{\omega_1} = \frac{G m m_E/r^2}{G m m_E/r_1^2} = \left(\frac{r_1}{r}\right)^2 = \frac{1}{4} = \underline{\underline{0.25}}$$

For $r/r_1 = 4$, by Fig. 14.1, $\omega/\omega_1 \simeq \underline{0.06}$

By Eq. (14.1),

$$\frac{\omega}{\omega_1} = \frac{G m m_E\ r^2}{G m m_E\ r_1^2} = \left(\frac{r_1}{r}\right)^2 = \left(\frac{1}{4}\right)^2 = \underline{\underline{0.0625}}$$

14.2

Given: Mass of moon $= \frac{1}{81}$ Mass of earth; $m_{moon} = \frac{1}{81} m_{earth}$

r_1 = distance from the center of the earth to the mass m.

r_2 = distance from the center of the moon to the mass m.

$r_1 + r_2 = 240{,}000$ miles (a)

Find: The point of equal attraction,

Solution:

For equal attraction (Fig a)

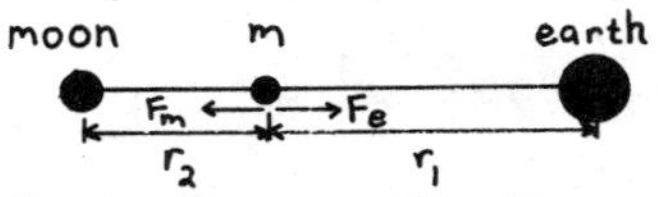

figure a

$$F_e = G\frac{m\, m_{earth}}{r_1^2} = F_m = G\frac{m\, m_{moon}}{r_2^2}$$

or $$\frac{m_{earth}}{m_{moon}} = \left(\frac{r_1}{r_2}\right)^2 = 81$$

$\therefore\ r_2 = \frac{1}{9} r_1$ (b)

By eqs. (a) and (b), $\frac{10 r_1}{9} = 240{,}000$ miles

$\therefore\ \underline{r_1 = 216{,}000 \text{ miles}}$, $\underline{r_2 = 24000 \text{ miles}}$

14.3

Given: A body near the surface of the sun

Find: a) g near the surface of the sun.

b) the weight of a body near the sun's surface, if the body weighs 700 N on earth

Solution:

a) $F = G\frac{m\, m_s}{(r_s)^2} = m g_s = W_s$, where subscript s denotes the sun

$$\therefore\ g_s = \frac{(6.672\times 10^{-11})(1.989\times 10^{30})}{(6.960\times 10^{8})^2} \quad \text{(See Table 14.1)}$$

or $\underline{\underline{g_s = 274\ \frac{m}{s^2}}}$

b) $W_e = 700\ N = m(9.81)$, or $m = 71.4$ kg

$W_s = m g_s = (71.4)(274) = \underline{\underline{19.55\ kN}}$

14.4

Given: The mass of each lead ball is $m_L = 1000$ Kg; of each brass ball is $m_B = 10$ Kg (Fig. a).

The distance between the brass balls is 2 m; between a brass and a lead ball is 0.4 m.

The wire is twisted $\theta = 1°$ (see Figure a)

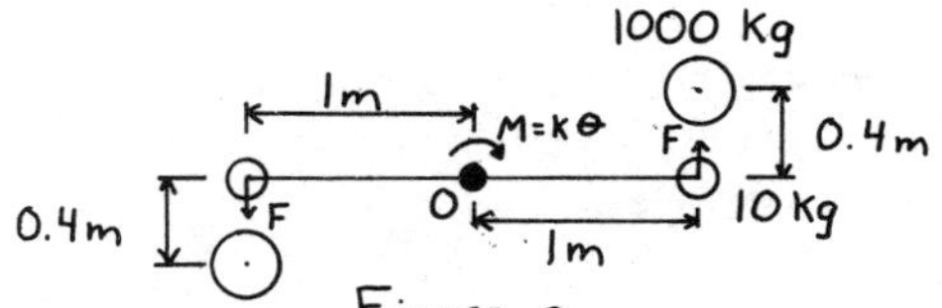

Figure a

Find: The torsional spring constant k of the wire.

Solution:

By Eq. (14.1), with the given data,

$$F = G\frac{m_B m_L}{r^2} = \frac{(6.672\times 10^{-11})(1000)(10)}{(0.4)^2} = 4.17\times 10^{-6}\ N$$

By Fig. (a), with $\theta = 1° = \pi/180$ rad,

$$\Sigma M_O = 2F(1) - M = 0$$

or $M = k\left(\frac{\pi}{180}\right) = 2(4.17\times 10^{-6})$

$$\underline{\underline{k = 4.778\ \frac{N\cdot m}{rad}}}$$

14.5

Given: Mass m_s of the sun in Astronomical units; $m_s = 4\pi^2 [L^3 T^{-2}]$

Find: Mass m_e of earth in Astronomical units.

Solution: By Table 14.1,

$m_s = 1.989\times 10^{30}$ kg $\qquad m_e = 5.974\times 10^{24}$ kg

$$\therefore\ \frac{m_s}{m_e} = 332{,}900 = \frac{4\pi^2 [L^3 T^{-2}]}{m_e}$$

$$\therefore\ m_e = \frac{4\pi^2 [L^3 T^{-2}]}{332{,}900} = \underline{\underline{1.186\times 10^{-4}\ [L^3 T^{-2}]}}$$

Alternatively, for a mass m on earth

$$F = mg = G\frac{m\, m_e}{(r_e)^2} \qquad \therefore\ m_e = \frac{g(r_e)^2}{G}$$

where $g = \left(9.81\ \frac{m}{s^2}\right)\left(\frac{1\ AU}{1.4960\times 10^{11} m}\right)\left(\frac{3.1558\times 10^7 s}{1\ yr}\right)$

$g = 6.5306\times 10^4\ AU/(yr)^2$

$r_e = (6378\times 10^3 m)\left(\frac{1\ AU}{1.4960\times 10^{11} m}\right) = 4.263\times 10^{-5}$ AU

$G = 1 \qquad \therefore\ m_e = \left(6.5306\times 10^4 \frac{AU}{(yr)^2}\right)(4.263\times 10^{-5} AU)^2$

or $\underline{\underline{m_e = 1.187\times 10^{-4} \frac{(AU)^3}{(yr)^2} = 1.187\times 10^{-4}\ [L^3 T^{-2}]}}$

14.6

Given: Figure a

Find: Resultant gravitational force acting on the 4-kg ball.

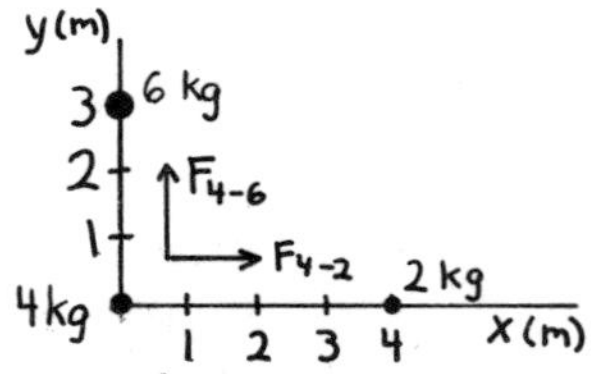

Figure a

Solution: By Eq.(14.1) and Fig. a,

$$F_{4-6} = \frac{G m_1 m_2}{r^2} = \frac{(6.672\times10^{-11})(4)(6)}{(3)^2} = 1.779\times10^{-10}\ N$$

$$F_{4-2} = \frac{(6.672\times10^{-11})(4)(2)}{(4)^2} = 3.336\times10^{-11}\ N$$

Since F_{4-2} is $\perp$ to F_{4-6} (Fig. b)

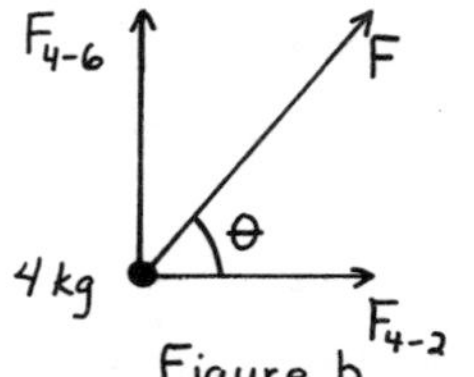

Figure b

$$F = \sqrt{F_{4-2}^2 + F_{4-6}^2} = \underline{\underline{1.810\times10^{-10}\ N}}$$

$$\theta = \tan^{-1}\frac{F_{4-6}}{F_{4-2}} = \tan^{-1}(5.3327)$$

$$\underline{\underline{\theta = 79.38^\circ}}$$

14.7

Given: Figure a

Find: Resultant gravitational force acting on the 2-kg ball.

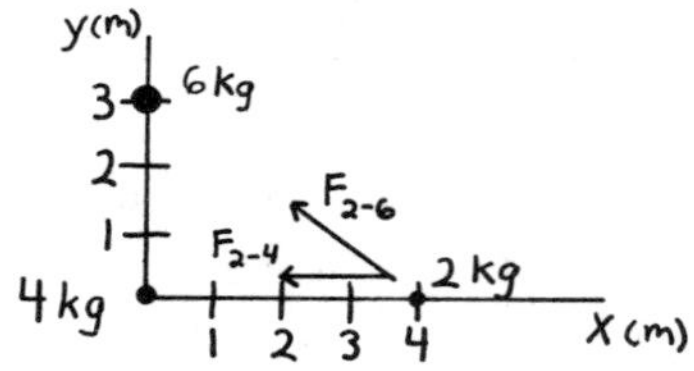

Figure a

Solution: By Eq.(14.1) and Fig. a,

$$F_{2-4} = \frac{G m_1 m_2}{r^2} = \frac{(6.672\times10^{-11})(4)(2)}{(4)^2} = 3.336\times10^{-11}\ N$$

$$F_{2-6} = \frac{(6.672\times10^{-11})(6)(2)}{(5)^2} = 3.203\times10^{-11}\ N$$

Resolve F_{2-6} into (x,y) projections (Fig a)

$$(F_{2-6})_x = -F_{2-6}\left(\tfrac{4}{5}\right) = -2.562\times10^{-11}\ N$$

$$(F_{2-6})_y = F_{2-6}\left(\tfrac{3}{5}\right) = 1.922\times10^{-11}\ N$$

Hence,

$$\sum F_x = (F_{2-6})_x - F_{2-4} = -5.898\times10^{-11}\ N$$

$$\sum F_y = (F_{2-6})_y = 1.922\times10^{-11}\ N$$

Hence, resultant gravitational force acting on the 2 kg mass is (Fig. b)

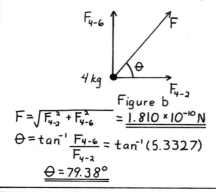

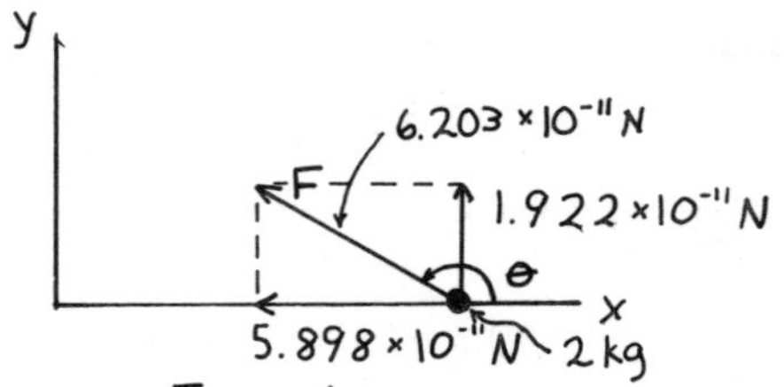

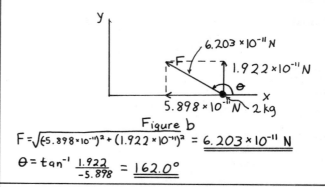

Figure b

$$F = \sqrt{(-5.898\times10^{-11})^2 + (1.922\times10^{-11})^2} = \underline{\underline{6.203\times10^{-11}\ N}}$$

$$\theta = \tan^{-1}\frac{1.922}{-5.898} = \underline{\underline{162.0^\circ}}$$

14.8

Given: Distance between moon and earth is $d = 384{,}000$ km. Radius of earth is $r_e = 6{,}400$ km and mass of moon is $m_m = 7.35\times10^{22}$ kg

Find: Gravitational attraction of moon on a mass m_p

a.) on side of earth nearest moon

b.) on side of earth farthest from moon

Find: c.) Differences of forces in a and b

d.) Corresponding change in g

Solution:

a) By Eq. (14.1), near force is

$$F_n = \frac{G m_p m_m}{r^2}, \quad r = d - r_e$$

$$F_n = \frac{(6.672\times10^{-11})\, m_p\, (7.35\times10^{22})}{(384{,}000{,}000 - 6{,}400{,}000)^2} = \underline{\underline{3.44\times10^{-5}(m_p)\ N}}$$

b) By Eq. (14.1), far force is

$$F_f = \frac{G m_p m_m}{r^2}, \quad r = d + r_e$$

$$F_f = \frac{(6.672\times10^{-11})\, m_p\, (7.35\times10^{22})}{(384{,}000{,}000 + 6{,}400{,}000)^2} = \underline{\underline{3.22\times10^{-5}(m_p)\ N}}$$

c) The difference of near and far forces is

$$\Delta F = F_n - F_f = \underline{\underline{2.22\times10^{-6}(m_p)\ N}}$$

d) The net gravitation forces on m_p due to earth and moon at near and far sides are

$$F_n' = \frac{G m_p m_e}{(r_e)^2} - F_n, \quad F_f' = \frac{G m_p m_e}{(r_e)^2} + F_f \qquad (e)$$

The change Δg in gravity is obtained from

$$m_p \Delta g = F_f' - F_n' \qquad (f)$$

By Eqs. (e) and (f)

$$\Delta g = \frac{F_f + F_n}{m_p} \qquad (g)$$

By Eq. (g) and results from a) and b)

$$\underline{\underline{\Delta g = 6.66\times10^{-5}\ \frac{m}{s^2}}}$$

14.9

Given: A mass $m = 1\ kg$ weighs 9.81 N when the mass is 6,378 Km from the center of the earth.

Find: a) The mass m_e of the earth
b) Average specific gravity sg_E of the earth.

Solution:

a) By Eq. (14.1) and the weight of the mass,

$$F = \frac{G\, m\, m_e}{r^2}$$

$$9.81\ N = \frac{(6.672 \times 10^{-11} \frac{N \cdot m^2}{Kg^2})(1\ Kg)(m_e)}{(6,378,000\ m)^2}$$

$$m_e = 5.981 \times 10^{24}\ Kg$$

b) The mass density of the earth is

$$\rho_e = \frac{m_e}{Vol.}\ ;\ vol. = \frac{4\pi(6,378,000)^3}{3} = 1.0868 \times 10^{21}\ m^3$$

$$\therefore \rho_e = \frac{5.981 \times 10^{24}}{1.0868 \times 10^{21}} = 5,504\ Kg/m^3$$

The mass density of water is

$$\rho_w = \left(\frac{1\ g}{cm^3}\right)\left(\frac{1\ Kg}{1000\ g}\right)\left(\frac{100\ cm}{1\ m}\right)^3 = 1000\ Kg/m^3$$

$$\therefore sg_E = \frac{\rho_e}{\rho_w} = 5.50$$

14.10

Given: Figure a, with data

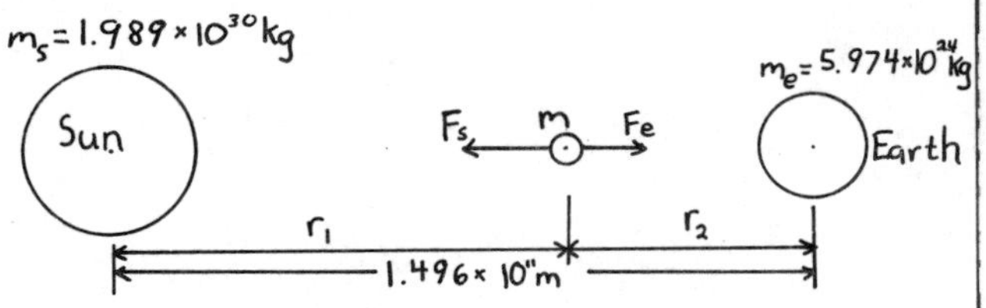

Figure a

Find: Value of r_1 at which attraction F_s of sun on m equals attraction F_e of earth on m.

Solution:

$$F_s = \frac{G\, m\, m_s}{r_1^2} \qquad F_e = \frac{G\, m\, m_e}{r_2^2}$$

$$F_s = F_e \longrightarrow \frac{m_s}{r_1^2} = \frac{m_e}{r_2^2} \qquad \text{But } r_1 + r_2 = 1.496 \times 10^{11}\ m$$

$$\therefore \frac{m_s}{r_1^2} = \frac{m_e}{(1.496 \times 10^{11} - r_1)^2} \text{ or } \frac{1.989 \times 10^{30}}{r_1^2} = \frac{5.974 \times 10^{24}}{(1.496 \times 10^{11} - r_1)^2}$$

The solution for r_1, between the sun and the earth, is

$$r_1 = 1.492 \times 10^{11}\ m$$

14.11

Given: A space craft weighs 100 kN on earth

Find: a) The mass of the space craft
The weight of the space craft
b) at 12,000 Km above the earth's surface
c) at 24,000 Km above the earth's surface
d) The mass of the space craft at 12,000 Km and 24,000 Km above the earth's surface.

Solution:

a) By Eq. (14.2),

$$W = m g_e \quad \text{or} \quad 100\ kN = m(9.81\ \tfrac{m}{s^2})$$

$$\therefore m = 10,190\ kg$$

b) at 12,000 Km,

$$W = \frac{G\, m_1 m_2}{r^2} = \frac{(6.672 \times 10^{-11})(10,190)(5.974 \times 10^{24})}{[(6.378 \times 10^6) + (1.2 \times 10^7)]^2}$$

$$\therefore W = 12.03\ kN$$

c) at 24,000 Km,

$$W = \frac{(6.672 \times 10^{-11})(10,190)(5.974 \times 10^{24})}{[(6.378 \times 10^6) + (2.4 \times 10^7)]^2}$$

$$\therefore W = 4.40\ kN$$

d) The mass of a body is a constant. Therefore, the mass of the space craft is 10,190 Kg at any distance from the earth's surface.

14.12

Given: Figure a, with θ_{12}, θ_{13}, θ_{23} Known; i.e, relative direction of m_1, m_2, m_3 known

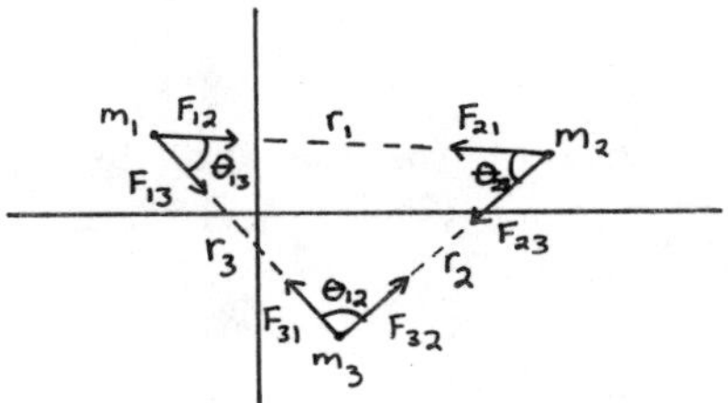

Figure a

Find: Number of quantities needed to solve for gravitational forces $F_{12} = F_{21}$, $F_{23} = F_{32}$ and $F_{13} = F_{31}$

Solution:

$$F_{12} = \frac{G\, m_1 m_2}{r_1^2}, \quad F_{23} = \frac{G\, m_2 m_3}{r_2^2}, \quad F_{13} = \frac{G\, m_1 m_3}{r_3^2}$$

G is known. Therefore, need 5 quantities m_1, m_2, m_3, r_1, r_2 (r_3 can be found by trigonometry in terms of r_2 and r_1).

14.13

Given: Mass systems with different numbers of masses

Find: The number of quantities required to calculate the gravitational attractions between each pair of masses.

Solution:

a) For 4 masses:

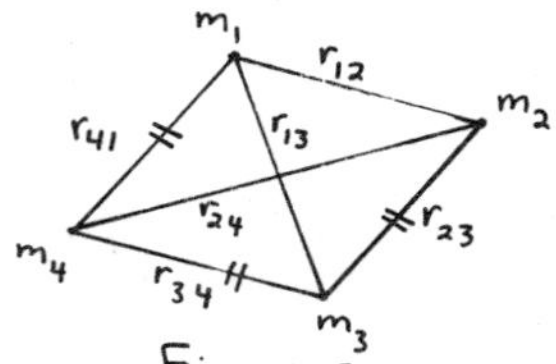

Figure a

With 3 lengths (marked in Fig. a), the other lengths can be calculated by trigonometry.

∴ 7 quantities are needed: 4 masses, 3 lengths

b) For 5 masses:

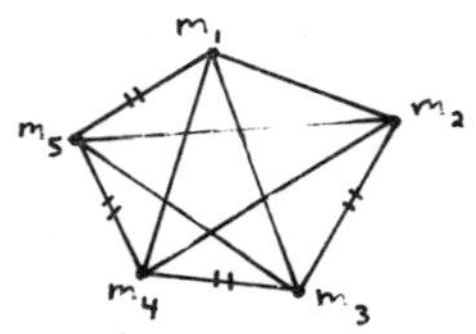

Figure b

With 4 lengths, (marked in Fig. b), all other lengths can be found by trigonometry

∴ 9 quantities are needed: 5 masses, 4 lengths

c) For n masses:

By analogy, n-1 lengths connecting n masses are needed.

∴ (2n − 1) quantities are needed, n masses, (n-1) lengths.

14.14

Given: A mass on the surface of various planets

Find: The acceleration of the mass due to gravity on the surfaces of Venus, Mars, Jupiter, Saturn, and Pluto

Solution:

By Eq. (14.1) and (14.2)

$F = m_1 g = \frac{G m_1 m_2}{r^2}$ or $g = \frac{G m_2}{r^2}$

and $G = 6.672 \times 10^{-11}\ N \cdot m^2/kg^2$

masses in kg, radii in meters

Venus ($m_v = 4.869 \times 10^{24}$, $r_v = 6.052 \times 10^6$)

$$g_V = \frac{G m_V}{r_V^2} = 8.87\ \frac{m}{s^2}$$

Mars ($m_m = 6.419 \times 10^{23}$, $r_m = 3.397 \times 10^6$)

$$g_m = \frac{G m_m}{r_m^2} = 3.71\ \frac{m}{s^2}$$

Jupiter ($m_J = 1.899 \times 10^{27}$, $r_J = 7.149 \times 10^7$)

$$g_J = \frac{G m_J}{r_J^2} = 24.8\ \frac{m}{s^2}$$

Saturn ($m_s = 5.685 \times 10^{26}$, $r_s = 6.027 \times 10^7$)

$$g_s = \frac{G m_s}{r_s^2} = 10.44\ \frac{m}{s^2}$$

Pluto ($m_p = 1.500 \times 10^{22}$, $r_p = 1.151 \times 10^6$)

$$g_P = \frac{G m_P}{r_P^2} = 0.755\ \frac{m}{s^2}$$

14.15

Given: Radial hole to center of earth. Mass particle is in hole

Find: a) Formula for force of attraction earth exerts on particle as a function of distance r between the particle and the center of the earth

b) What force the earth exerts on the particle when it is at the center of the earth.

c) Give a physical argument for results found in part b.

Solution:

a) The force that acts on the particle of mass m is due to only the mass M_E' of the Earth that lies within a sphere of radius r (Fig. a).

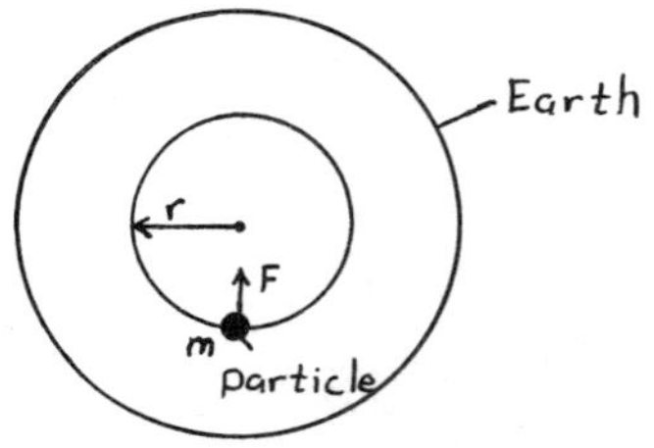

Figure a

The part of the earth that lies outside this sphere does not exert a net force on the particle.

Mass M_E' is given by

$$M_E' = \rho V_E' = \rho \frac{4\pi r^3}{3},$$

Where ρ is the assumed uniform mass per unit volume and $V_E' = 4\pi r^3/3$ is the volume of the sphere of radius r.

The force that acts on the particle is

$$F = -\frac{G m M_E'}{r^2} = -\frac{G m \rho 4\pi r^3}{3 r^2} = -\left(\frac{4}{3}\pi m G \rho\right) r$$

or $F = -kr$ where $k = \frac{4}{3}\pi m G \rho$

(continued)

14.15 Cont.

b.) When the particle is at the center of the earth, $r=0$

$\therefore$ $F=0$

c.) at $r=0$, the mass particle is attracted equally in all directions. Therefore, the resultant force acting on it is zero.

14.16

Given: Figure a, with data

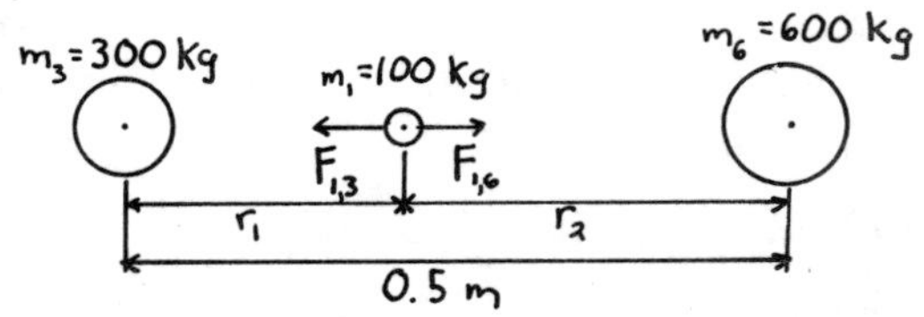

Figure a

Find: a) Resultant gravitational force R exerted on m_1

b) Position of m_1 at which $R=0$

Solution:

a) $F_{1,3} = \dfrac{G m_1 m_3}{r_1^2} = \dfrac{(6.672\times10^{-11})(100)(300)}{(0.25)^2} = 3.2026\times10^{-5}$ N

$F_{1,6} = \dfrac{G m_1 m_6}{r_2^2} = \dfrac{(6.672\times10^{-11})(100)(600)}{(0.25)^2} = 6.4051\times10^{-5}$ N

$R = F_{1,6} - F_{1,3} = 3.2026\times10^{-5}$ N

b) For $R=0$, $F_{1,3}=F_{1,6}$

$\therefore \dfrac{m_3}{r_1^2} = \dfrac{m_6}{r_2^2}$ or $2r_1^2 = r_2^2$ (a)

Also

$r_1 + r_2 = 0.50$ (b)

Equations (a) and (b) have the solutions

$r_1 = 0.20711$ m, $r_2 = 0.29289$ m

and $r_1 = -1.20711$ m, $r_2 = 1.70711$ m

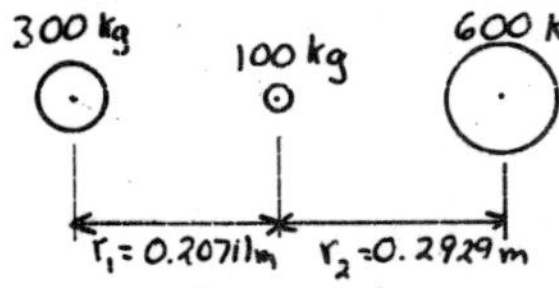

Figure b

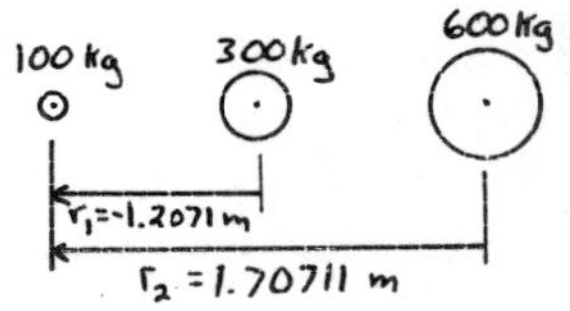

Figure c

Note: Also, $F_{13} \to F_{16}$ as $|r_1|$ and $|r_2| \to \infty$

14.17

Given: Figure a, with data

$r_{em} = 240{,}000$ mi

$r = \frac{1}{4} r_{em}$

$r = 60{,}000$ mi

$= 3.168\times10^8$ ft

$m_e = 4\times10^{23}$ slugs

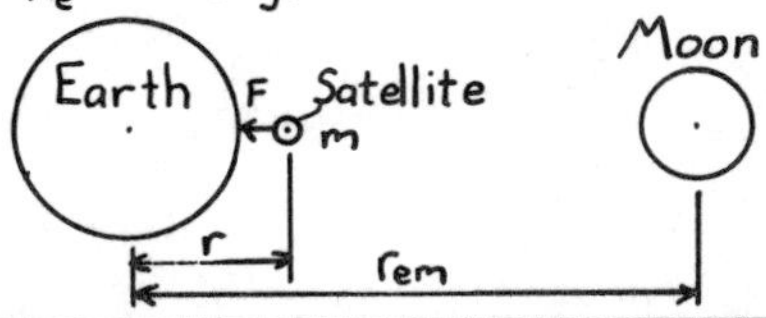

Find: a) The satellite's speed

b) The satellite's period.

Solution:

a) By Newton's gravitational law,

$$F = \frac{G m m_e}{r^2} = \frac{(3.439\times10^{-8})(m)(4.0\times10^{23})}{(3.168\times10^8)^2} = 0.1371m$$

Since the satellite travels on a circle,

$F = \dfrac{mV^2}{r}$ [see Eq. (14.6)]

or $V = \sqrt{\left(\dfrac{rF}{m}\right)} = \sqrt{\dfrac{(3.168\times10^8)(0.1371m)}{m}}$

$V = 6590$ ft/s

b) The circumference of the satellite's circular path, $C = 2\pi r$, is the distance traveled in one revolution. The period T is:

$$T = \frac{C}{V} = \frac{(2\pi)(3.168\times10^8)}{6590\times10^3} = 3.02\times10^5 \text{ s}$$

or $T = 3.496$ days

14.18

Given: Two satellites orbit the earth in the same circular path. The mass m_1 of one satellite is 4 times the mass m_2 of the other. The satellite of mass m_1 circles the earth at a speed of 6000 ft/s.

Find: The speed of the satellite whose mass is $m_2 = \frac{1}{4} m_1$.

Solution:

The gravitational force acting on masses m_1 and m_2 are, respectively

$F_1 = \dfrac{G m_1 m_{earth}}{r^2}$, $F_2 = \dfrac{G m_1 m_{earth}}{4r^2}$ (a)

Since the satellites travel the same circular path of radius r,

$F_1 = \dfrac{m_1 V_1^2}{r}$, $F_2 = \dfrac{m_1 V_2^2}{4r}$ (b)

or by Eqs. (a) and (b)

$V_1^2 = \dfrac{G m_{earth}}{r}$, $V_2^2 = \dfrac{G m_{earth}}{r}$ (c)

or $V_1 = V_2 = 6000$ ft/s

14.19

Given: A 10 lb box set on a conveyor belt that has a speed of 20 ft/s; $\mu_K = 0.4$ (Fig. a)

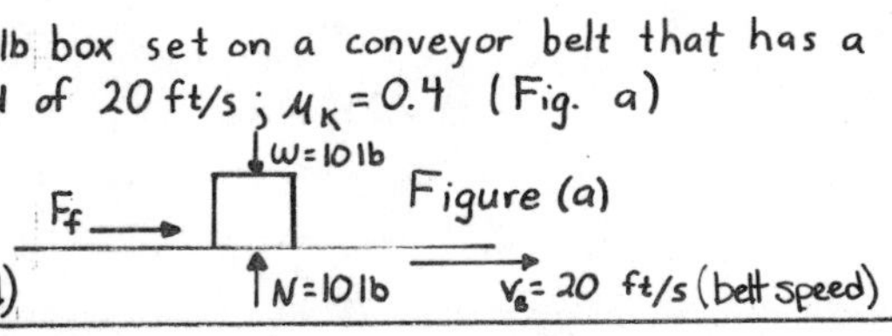

Figure (a)

(continued)

14.19 Cont.

Find: The time needed for the box to accelerate to the belt's speed.

Solution:

$F_f = \mu_K N = (0.4)(10) = 4\,lb$

By Fig. a and $F = ma = m\,dv/dt$,

$F_f = 4 = \frac{10}{32.2}\frac{dv}{dt}$

Integration yields

$12.88\int_0^t dt = \int_0^{20} dv$ or $12.88t = 20$

$\therefore\ t = 1.553\ s$

14.20

Given: Driver claims his car had not exceeded 70 km/h before an accident. A police officer measured the skid marks to be 15 m long.

Find: Discuss the driver's claim from the view of mechanics

Solution:

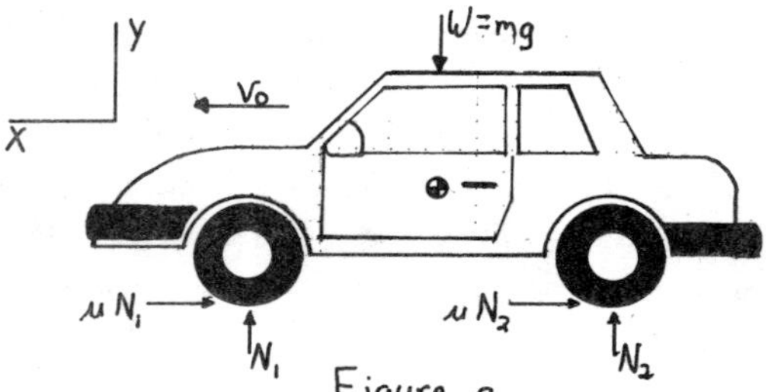

Figure a

By Fig. a, $\Sigma F_y = N_1 + N_2 - mg = 0$, $N_1 + N_2 = mg$

$\Sigma F_x = -\mu(N_1 + N_2) = ma = mv\frac{dv}{dx}$

or $v\,dv = -\mu g\,dx$, Integration yields

$\int_{v_0}^{0} v\,dv = -\mu g\int_0^d dx$ or $\mu g d = \frac{v_0^2}{2}$

$\therefore\ \mu = \frac{v_0^2}{2gd}$

With $v_0 = 70\,km/h = 19.44\,m/s$, $d = 15\,m$, and $g = 9.81\,m/s^2$, the coefficient of kinetic friction required to stop the car is $\mu_K = 1.28$

It is difficult to prove or disprove the driver's claim. Depending on the surface of the road, μ_K can be as high or higher than 1.28 (see Table 10.1).

14.21

Given: Car traveling at 72 km/h (20 m/s) when driver jams on the brakes. Car skids on all four tires to a stop. $\mu_K = 0.70$, $\mu_S = 0.80$

Find: The difference in distance to stop if the driver had applied the brakes so that the tires were on the verge of skidding

Solution:

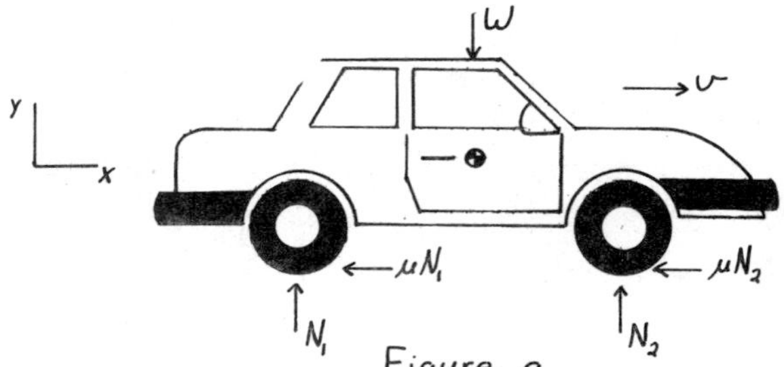

Figure a

By Fig (a)

$\Sigma F_y = -W + N_1 + N_2 = 0\ \therefore\ N_1 + N_2 = W = mg$

$\Sigma F_x = -\mu(N_1 + N_2) = ma = mv\frac{dv}{dx}$

or $v\,dv = -\mu g\,dx$

Integration yields

$\int_v^0 v\,dv = -\mu g\int_0^d dx$ or $d = \frac{v^2}{2\mu g}$ (a)

with $\mu = \mu_K = 0.70$ Eq. (a) yields

$d_K = \frac{(20)^2}{2(0.7)(9.81)} = 29.12\ m$

with $\mu = \mu_S = 0.80$ Eq. (a) yields

$d_S = \frac{(20)^2}{2(0.8)(9.81)} = 25.48\ m$

The difference is

$\Delta d = d_K - d_S = 3.636\,m$

14.22

Given: Earth orbits sun in circular path in 1 year, Fig. a

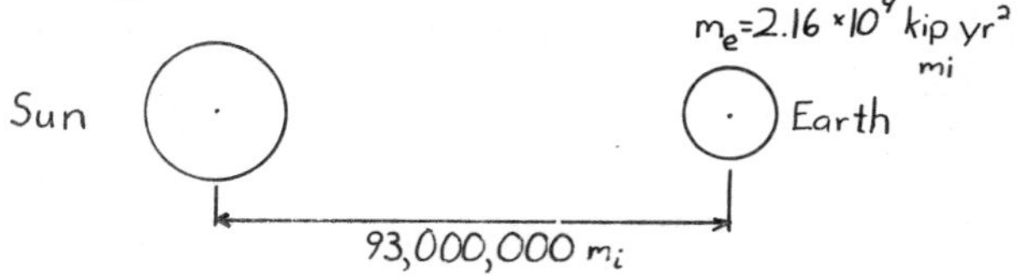

Figure a

Find: The diameter of a cable that can replace the sun's gravitational pull on the earth if the cable can withstand 2×10^{11} kips tension per square mile of cross section.

Solution:

The speed of the earth is

$v = \frac{2\pi(93,000,000)}{1\ day} = 5.843\times10^8\ mi/day$

(continued)

14.22 Cont.

The force required to keep it in a circular orbit of 93,000,000 mi is

$$F = \frac{mv^2}{r} = \frac{(2.16\times10^9)(5.843\times10^8)^2}{(9.3\times10^7)} = 7.930\times10^{18} \text{ kips}$$

The minimum area of a cable that will withstand this force is

$$A = \frac{\pi d^2}{4} = \frac{F}{2\times10^{11}}\ (\text{mi})^2 \quad \text{or} \quad d = \sqrt{\frac{(4)(7.93\times10^{18})}{(2\pi)(10^{11})}} = \underline{\underline{7105 \text{ mi}}}$$

14.23

<u>Given:</u> Jet plane of mass m is brought to rest on aircraft carrier by an arrestor cable ($F_1 = k_1 t^2$) and wheel brakes ($F_2 = k_2 t$), k_1, k_2, constants. The landing speed is v_0 relative to the carrier.

<u>Find:</u> Show that time t_s required to stop the plane is a root of

$$2k_1 t^3 + 3k_2 t^2 - 6 m v_0 = 0$$

<u>Solution:</u>

The plane may be represented as a body subjected to forces F_1 and F_2 (Fig. a).

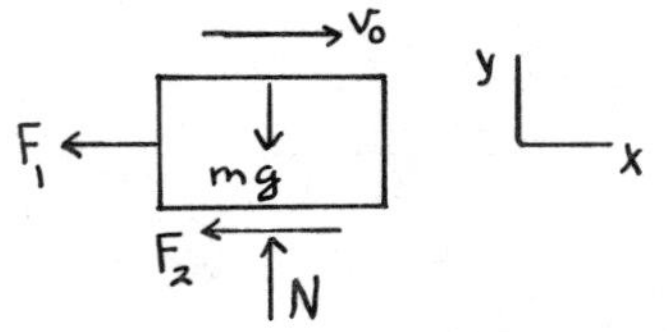

Figure a

By Fig. a,

$$\Sigma F_x = -F_1 - F_2 = m a_x = m\frac{dv_x}{dt}$$

Hence, $m\,dv_x = -(F_1 + F_2)dt \longrightarrow m\int_{v_0}^{0} dv_x = -\int_0^{t_s}(k_1 t^2 + k_2 t)dt$

$$\therefore -m v_0 = -\frac{k_1 t_s^3}{3} - \frac{k_2 t^2}{2} \quad \text{or} \quad \underline{\underline{2k_1 t_s^3 + 3k_2 t_s^2 - 6 m v_0 = 0}}$$

14.24

<u>Given:</u> A man pushes horizontally on a crate until it slides on the floor.
$\mu_s = 0.50$ and $\mu_k = 0.40$

<u>Find:</u> The acceleration of the crate (m/s²) when it begins to slide.

<u>Solution:</u>

At impending motion, the man exerts the force

$$F = \mu_s mg = 0.50\,mg \quad \text{(Fig. a)}$$

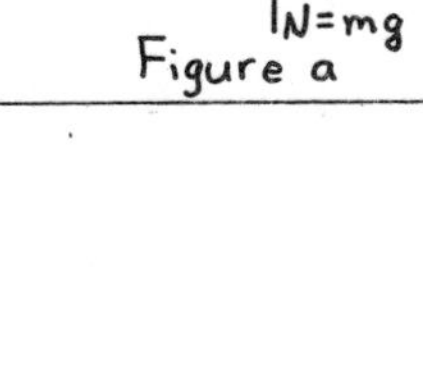

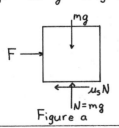

Figure a

When sliding begins, the frictional is $\mu_K mg = 0.40\,mg$ but F remains at $0.50mg$

Hence, the unbalanced force is

$$\Delta F = (0.50 - 0.40)\,mg = ma$$

$$\therefore a = 0.10g = (0.10)(9.81) \text{ or } \underline{\underline{a = 0.981 \text{ m/s}^2}}$$

14.25

<u>Given:</u> Figure a. mass of yoke is 1.50 lb·s²/ft
Neglect friction.

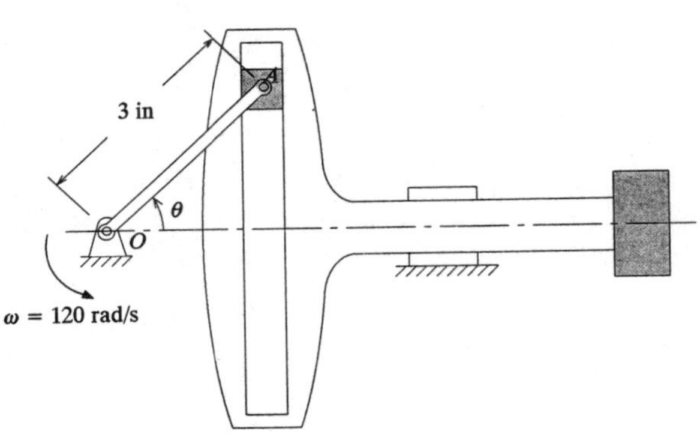

Figure a

<u>Find:</u> The maximum normal force that the crank exerts on the yoke

<u>Solution:</u>

Let x be the distance that the yoke moves as the crank rotates. Then,

$$x = 3\cos\theta \text{ (in)} \qquad (a)$$

Let N be the normal force exerted on the yoke by the slider (Fig. b)

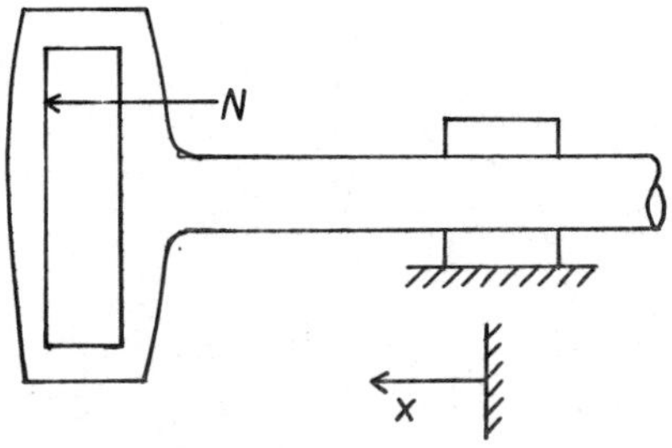

Figure b

Then, $\Sigma F_x = N = m\ddot{x}$ (b)

By Eq. (a), $\dot{x} = -3\dot{\theta}\sin\theta$

and since $\dot{\theta} = \omega = \text{constant}$

$$\ddot{x} = -3(\dot{\theta})^2\cos\theta = -3\omega^2\cos\theta$$

The maximum value of $\ddot{x}$ is $3\omega^2 = 3(120)^2 = 43{,}200$ in/s

or $\ddot{x}_{max} = 3600$ ft/s²

$\therefore N_{max} = (1.50)(3600) = \underline{\underline{5400 \text{ lb}}}$

This maximum value occurs at $\theta = 0°$ and $\theta = 180°$ (Fig. a)

14.26

Given: Refer to Fig. P14.25, where now friction between the slider A and the yoke is 0.30. Other friction is negligible.

Find: a) The angle θ for which the torque of the crank shaft O is a maximum.

b) The maximum crank torque

Solution:

a) From the solution of Problem 14.25,

$$\ddot{x} = -\frac{3}{12}\omega^2\cos\theta = -3600\cos\theta \text{ ft/s}^2$$

and, $N = m\ddot{x} = (1.5)(-3600\cos\theta)$

or, $N = -5400\cos\theta$ (lb)

is the normal force exerted by the slider on the yoke (directed to the left). Hence, the equal normal force exerted on the slider is directed to the right (Fig. a).

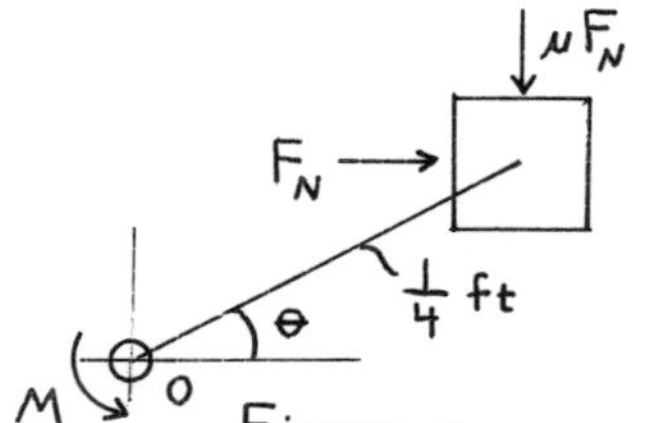

Figure a

Hence, the friction force is

$\mu F_N = 0.3F_N$ Then, by Fig. a,

$$\Sigma M_o = M - F_N(\tfrac{1}{4})\sin\theta - \mu F_N(\tfrac{1}{4})\cos\theta = 0$$

$$\therefore M = 1350\sin\theta\cos\theta + 405\cos^2\theta$$

or, by trigonometry identities

$$M = 675\sin 2\theta + 202.5(1+\cos 2\theta)$$

For maximum M, $dM/d\theta = 0$.

$$\therefore 1350\cos 2\theta - 450\sin 2\theta = 0$$

or $\tan 2\theta = 10/3$ $\therefore \theta = \frac{1}{2}\tan^{-1}\left(\frac{10}{3}\right)$

$\underline{\underline{\theta = 36.65^\circ}}$ (is the angle when $M = M_{max}$)

b) Since $\theta = 36.65^\circ$

$\sin 2\theta = 0.9578$, $\cos 2\theta = 0.2873$

$$\therefore M_{max} = 675(0.9578) + 202.5(1+0.2873)$$

$$M_{max} = 907.20 \text{ lb·ft}$$

or $\underline{\underline{M_{max} = 10,890 \text{ lb·in}}}$

14.27

Given: After brakes are applied, a car skids 65 m before crashing into another car. $\mu_s = 0.50$

Find: Minimum speed (km/h) that car could have had when brakes were first applied.

Solution:

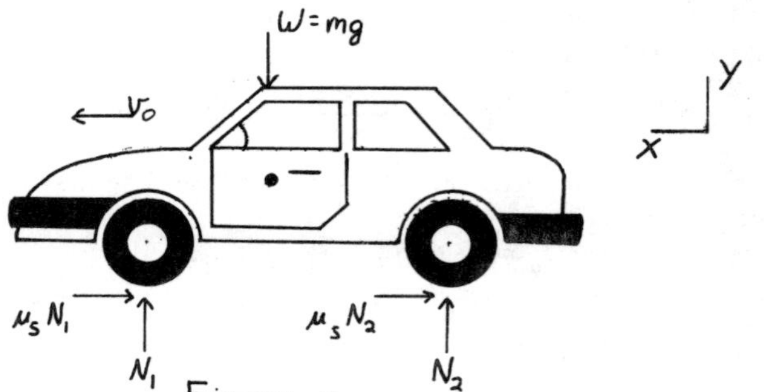

Figure a

By Fig. a,

$\Sigma F_y = N_1 + N_2 - W = 0$ or $N_1 + N_2 = mg$

$\Sigma F_x = -\mu_s mg = ma$ $\therefore a = -\mu_s g = -0.5(9.81) = -4.905 \text{ m/s}^2$

$\therefore$ Since a is constant, $v = at + v_0 = -4.905t + v_0$

when $v = 0$, $t = \dfrac{v_0}{4.905}$

Then, $d = \frac{1}{2}at^2 + v_0 t$ or $65 = -\dfrac{4.905}{2}\left(\dfrac{v_0}{4.905}\right)^2 + \dfrac{v_0^2}{4.905}$

$\therefore v_0 = \sqrt{(130)(4.905)} = 25.25 \text{ m/s}$

$\therefore \underline{\underline{v_0 = 90.9 \text{ km/h}}}$

14.28

Given: Figure a, with $F = 20250/x^2$ (N)

The magnets are pushed slowly toward each other until they suddenly jump together.

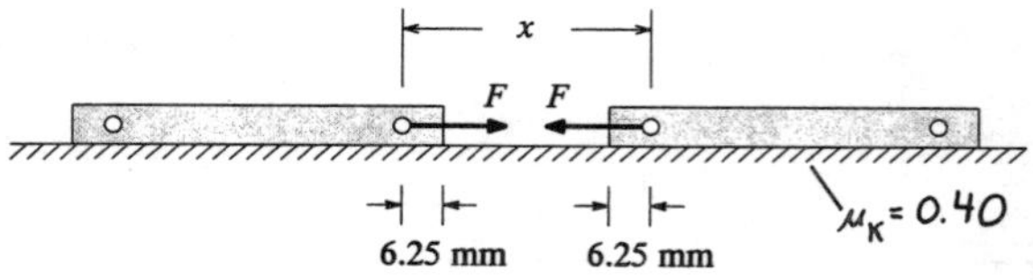

Figure a

Find: The relative speed (m/s) with which they collide.

Solution:

The frictional force is $\mu_K N = (0.4)(9) = 3.6$ N

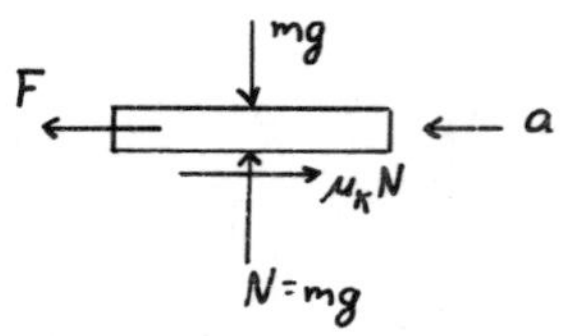

Figure b - FBD of right magnet

By the FBD of a magnet (Fig. b),

$$\Sigma F_x = F - \mu_K N$$

(Continued)

14.28 Cont.

Hence, when the jump occurs $x = x_0$

$\therefore F = \dfrac{20250}{x_0^2} = 3.6$ or $x_0 = 75$ mm

Also, by Fig a and b, once sliding occurs,

$\Sigma F_x = F - \mu_k N = -m\dfrac{\ddot{x}}{2}$ (a)

since $-\ddot{x}/2$ is the absolute acceleration of a magnet, $m = 9/g$ and $-\dot{x} = v$ where v is the relative velocity of approach of the magnets. Hence, by the chain rule,

$$-\ddot{x} = \frac{dv}{dt} = \frac{dv}{dx}\frac{dx}{dt} = \frac{dv}{dx}\dot{x} = -v\frac{dv}{dx} \quad \text{(b)}$$

Hence, by Eqs.(a) and (b),

$$\frac{20250}{x^2} - 3.6 = -\left(\frac{4.5}{g}\right) v \frac{dv}{dx}$$

$$\therefore v\,dv = \left(0.8 - \frac{4500}{x^2}\right) g\,dx$$

or

$$\int dv = g\int \left(0.8 - \frac{4500}{x^2}\right) dx$$

Integration yields, $\dfrac{v^2}{2} = \left(0.8x + \dfrac{4500}{x}\right) g + C$

For $x = x_0 = 75$ mm, $v = 0$ $\therefore C = -120g$

and $\dfrac{v^2}{2} = \left(0.8x + \dfrac{4500}{x} - 120\right) g$ (c)

The magnets collide when $x = 12.5$ mm. For this value of x Eq. (c) yields

$v^2 = 500g = (500)(9810)$ $\therefore$ $\underline{v = 2.215 \text{ m/s}}$

14.29

Given: Two balls of the same material are dropped from a balloon. One ball is 25 mm in diameter, the other is 300 mm. The drag on the ball is kv^2, k is proportional to the square of the diameter of a ball (Fig. a); i.e., $k = cd^2$, where c is a constant.

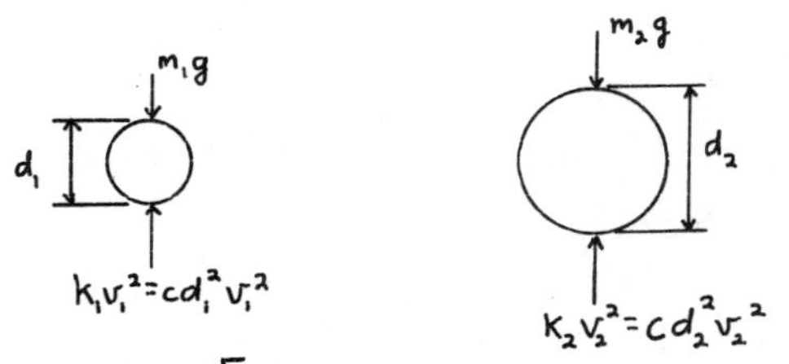

Figure a

Find: The ratio of the terminal velocity of the larger ball to the terminal velocity of the smaller ball.

Solution:

In general for a ball, $\Sigma F = mg - cd^2 v^2 = ma$ (a)

The terminal velocity is reached when $a = 0$. Therefore, by Eq. (a),

$$v = \sqrt{\frac{mg}{cd^2}} \quad \text{(b)}$$

where the mass of a ball is,

$$m = \rho\,\text{Vol} = \rho\left(\tfrac{4}{3}\pi r^3\right) \quad \text{(c)}$$

where ρ is the mass density

Then, by Eqs. (b) and (c),

$$v_1 = \sqrt{\frac{\rho\left(\frac{4}{3}\pi\right) r_1^3}{c\,(2r_1)^2}} = \sqrt{\frac{\pi\rho r_1}{3c}}, \quad v_2 = \sqrt{\frac{\pi\rho\, r_2}{3c}}$$

Hence,

$$\frac{v_2}{v_1} = \sqrt{\frac{r_2}{r_1}} = \sqrt{\frac{300}{25}} = \underline{\underline{3.46}}$$

14.30

Given: Identical balls ($m = 0.10$ lb·s²/ft) are released from rest and slide on the frictionless hoop (Fig. a).

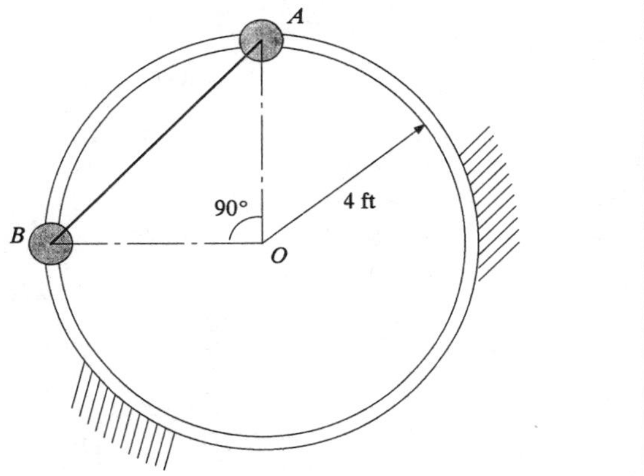

Figure a

Find: Tension in cord connecting balls at instant balls are released.

Solution:

The two balls have the same acceleration at the instant they are released. The accelerations are tangent to the hoop, since the initial velocity is zero.

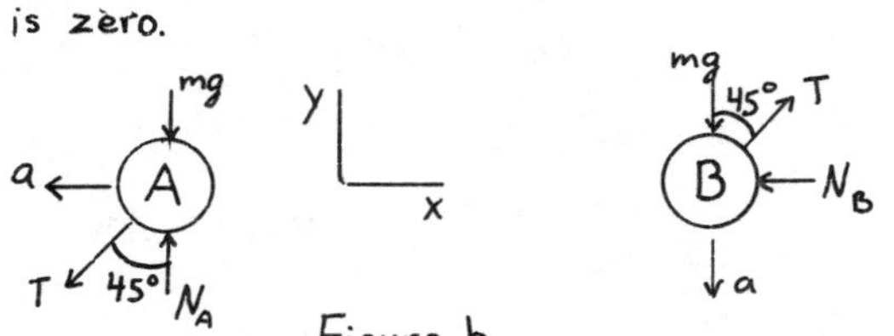

Figure b

From the FBD's, Fig. b,

Ball A: $\Sigma F_x = T\sin 45° = ma$

Ball B: $\Sigma F_y = mg - T\cos 45° = ma$

$\therefore T\dfrac{\sqrt{2}}{2} = mg - T\dfrac{\sqrt{2}}{2}$

Since $m = 0.10$, this equation yields

$\underline{T = 2.28 \text{ lb}}$

14.31

Given: Test sled travels on a straight track at 320 ft/s, and decelerates at a rate of 80 t (ft/s²); t in seconds.

Find: Distance sled travels until it is stopped.

Solution:

The sled may be represented as a body of mass m (Fig. a)

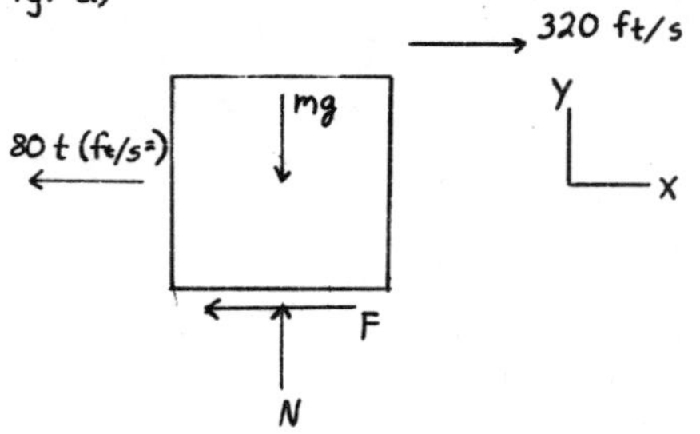

Figure a

Therefore, $a = -80t = \frac{dv}{dt}$ (a)

or, $\int_{320}^{0} dv = -\int_{0}^{t_s} 80t\,dt$ $\therefore t_s = \sqrt{320/40} = 2.828\,s$

where t_s is time required for sled to come to rest. Also by Eq. (a),

$v = -40t^2 + v_0$ or $v = \frac{dx}{dt} = -40t^2 + 320$

$\therefore \int_0^{x_s} dx = \int_0^{2.828} (-40t^2 + 320)\,dt$

or, $x_s = -\frac{40}{3}(2.828)^3 + 320(2.828)$

$\therefore x_s = 603.4$ ft is the distance s d travels

14.32

Given: Particle of mass m moves along x-axis and is attracted toward the origin by $F = kx$, k = constant. At $x = A$, it is at rest (Fig. a)

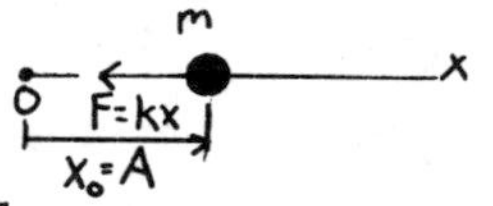

Figure a

Find: Show that it move according to the equation $x = A\cos(t\sqrt{K/m})$

Solution:

By Fig. a, $\Sigma F_x = -kx = ma = mv\frac{dv}{dx}$

$\therefore -\int_A^x kx\,dx = \int_0^v mv\,dv$

Integration yields, $k(A^2 - x^2) = mv^2$

$\therefore v = \frac{dx}{dt} = \sqrt{\frac{k}{m}(A^2 - x^2)}$

or, $\int_0^t dt = \int_A^x \sqrt{\frac{m}{K}} \frac{dx}{\sqrt{A^2 - x^2}}$

Integration yields, since $\cos^{-1}(1) = 0$,

$t\sqrt{\frac{k}{m}} = [\cos^{-1}(\frac{x}{A}) - \cos^{-1}(1)] = \cos^{-1}(\frac{x}{A})$

$\therefore \frac{x}{A} = \cos(t\sqrt{\frac{k}{m}})$ or $x = A\cos(t\sqrt{\frac{k}{m}})$

14.33

Given: A particle (mass = m) slides on a frictionless cylinder of radius r (Fig. a). It starts from rest at $\theta = 0°$

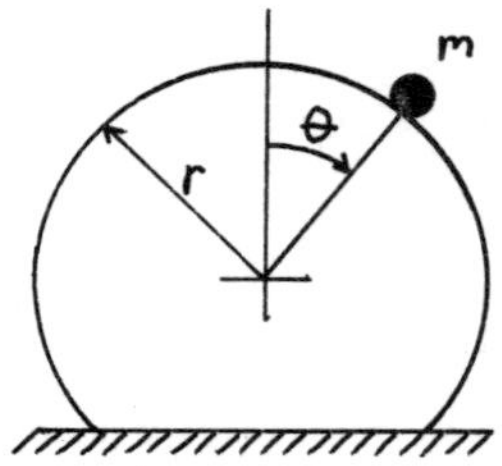

Figure a

Find: The value of θ at which it leaves the cylinder.

Solution:

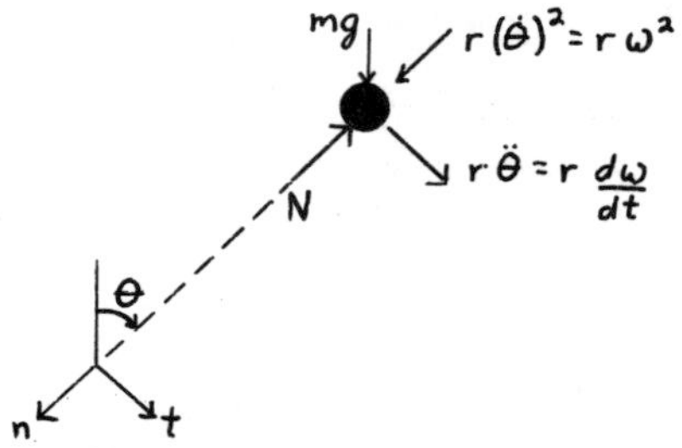

Figure b

By Fig. b, $\Sigma F_n = mg\cos\theta - N = mr\omega^2$

$\Sigma F_t = mg\sin\theta = mr\frac{d\omega}{dt} = mr\omega\frac{d\omega}{d\theta}$

$\therefore N = mg\cos\theta - mr\omega^2$

and $\int_0^\omega r\omega\,d\omega = \int_0^\theta g\sin\theta\,d\theta$

so, $r\omega^2 = 2g(-\cos\theta)\Big|_0^\theta = 2g(1 - \cos\theta)$

and, $N = mg(3\cos\theta - 2)$

The particle leaves the cylinder when $N = 0$. Hence,

$\cos\theta = \frac{2}{3} \implies \theta = 48.19°$

14.34

Given: Meteor falls to the earth from a height h, with zero initial speed relative to earth. Neglect air resistance.

Find: a) a formula for its speed v in terms of h and r (the meteor's distance from the center of the earth (Fig. a)). Take the radius of the earth as $r_e = 4000$ mi.

b) Calculate the meteor's speed (mi/hr) for $h = \infty$ and $r = r_e = 4000$ mi.

Solution:

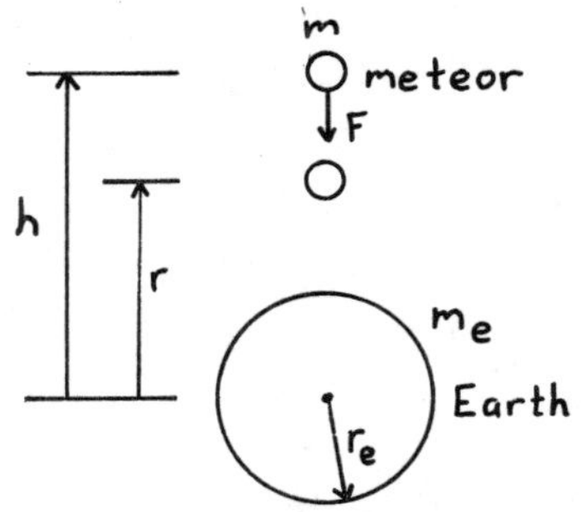

Figure a

a) The force acting on the meteor at r is

$$F = \frac{kmm_e}{r^2} \qquad \therefore \frac{kmm_e}{r^2} = -m\ddot{r} = -mv\frac{dv}{dr}$$

Integration yields

$$\frac{v^2}{2} = \frac{km_e}{r} + C \qquad \text{For } r = h,\ v = 0,\ \therefore C = -\frac{km_e}{h}$$

and $v^2 = 2km_e\left(\frac{1}{r} - \frac{1}{h}\right)$

on the earth's surface, $r = r_e$ and

$$F = mg_e = \frac{kmm_e}{r_e^2} \quad \text{or} \quad km_e = g_e r_e^2$$

where g_e is the acceleration of gravity on the earth's surface.

$$\therefore v^2 = 2g_e r_e^2\left(\frac{1}{r} - \frac{1}{h}\right)$$

b) With $g_e = 32.2\ \text{ft/s}^2 = 32.2/5280\ \text{mi/s}^2$, $h = \infty$, and $r = r_e = 4000$ mi, we obtain

$$v = 6.98\ \text{mi/s} = 175{,}500\ \text{mi/h}$$

14.35

Given: Two masses $m_1 > m_2$ are released from rest from position shown in Fig. a.

Find: The time that elapses before the masses pass each other, t.

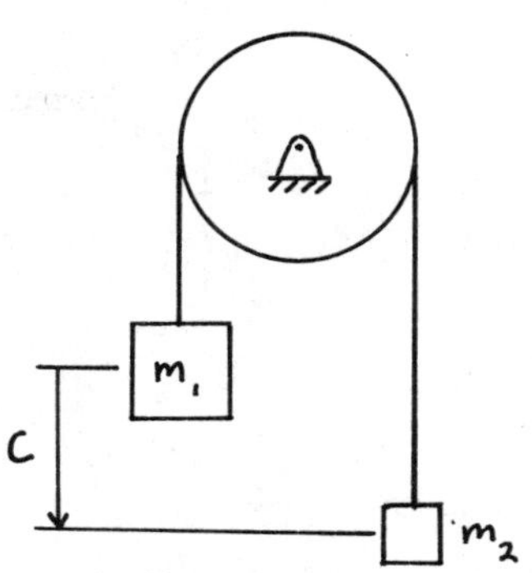

Figure a

Solution:

Free body diagrams of m_1 and m_2 are shown in Figs. b and c.

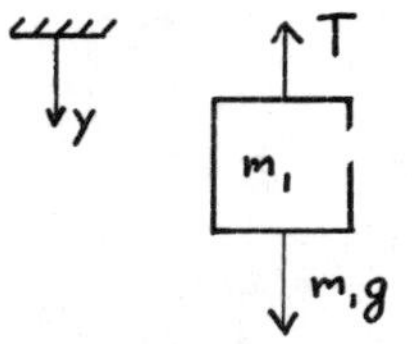

Figure b

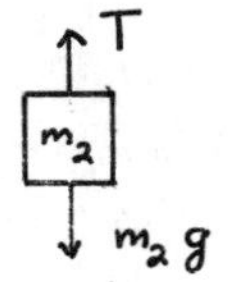

Figure c

By Fig. b, $\Sigma F_y = m_1 g - T = m_1 a_1$

or $T = m_1(g - a_1)$ (a)

By Fig. c, $\Sigma F_y = m_2 g - T = m_2 a_2$

or $T = m_2(g - a_2)$ (b)

Equating Eqs. (a) and (b) and noting that $a_2 = a_1$, we find

$$m_1(g - a_1) = m_2(g + a_1) \quad \text{or} \quad a_1 = \left(\frac{m_1 - m_2}{m_1 + m_2}\right)g \qquad \text{(c)}$$

Since $a_1 = \frac{d^2y}{dt^2}$, $y = dy/dt = 0$ at $t = 0$, and $y = C/2$ when the masses pass each other, Integration of Eq. (c) yields

$$t = \sqrt{\left(\frac{m_1 + m_2}{m_1 - m_2}\right)\left(\frac{C}{g}\right)}$$

14.36

Given: Figure a with data shown.

$W_A = 9\text{N}$
$W_B = 18\text{N}$
$\mu = 0.20$
$r = 0.6\text{m}$

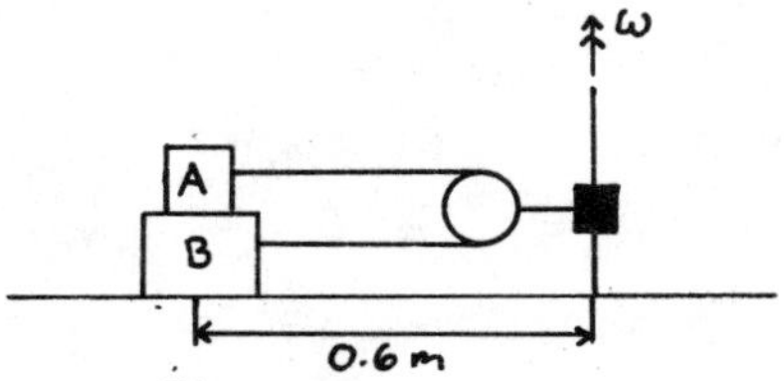

Figure a

Find: Angular velocity ω for which blocks A and B slip.

(continued)

14.36 Cont.

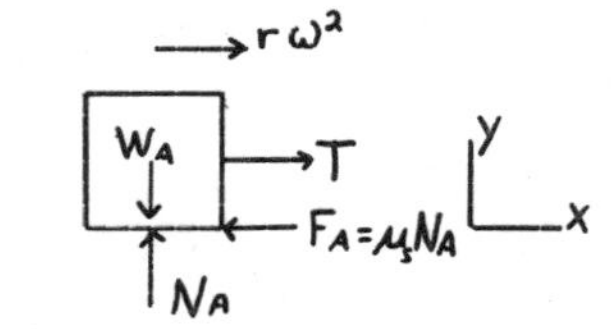

Figure b

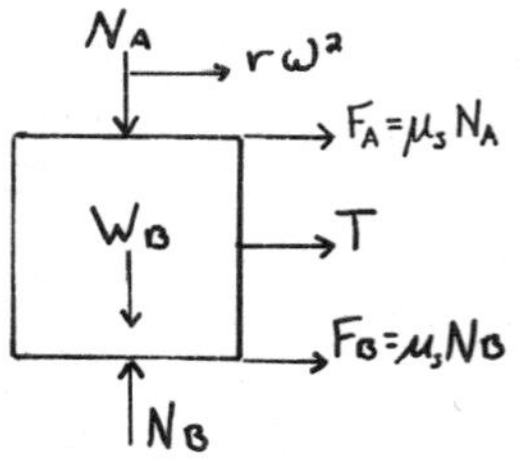

Figure c

By the FBD of A (Fig. b),

$\Sigma F_y = N_A - W_A = 0;\quad N_A = W_A = m_A g$

$\Sigma F_x = m_A a = T - 0.2 m_A g = m_A r\omega^2$

$\therefore T = 0.550\,\omega^2 + 1.8$ (N) (a)

By FBD of B (Fig. c),

$\Sigma F_y = N_B - W_B - N_A = 0;\quad N_B = W_B + N_A = m_B g + m_A g$

$\Sigma F_x = T + \mu_s (N_A + N_B) = m_B r\omega^2$

$\therefore T + 0.2(2m_A + m_B)g = m_B r\omega^2$

or $T = 1.101\,\omega^2 - 7.2$ (N) (b)

Equating Eqs. (a) and (b), we find

$0.551\,\omega^2 = 9.0$ or $\underline{\omega = 4.04 \text{ rad/s}}$

14.37

Given: Figure a. Block on inclined plane that has acceleration a.

Figure a

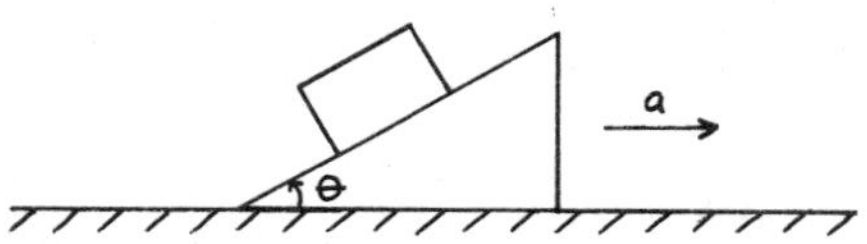

Find: Show that the block will slide on plane if $a > g\tan(\phi - \theta)$ where $\tan\phi = \mu_s$

Solution:

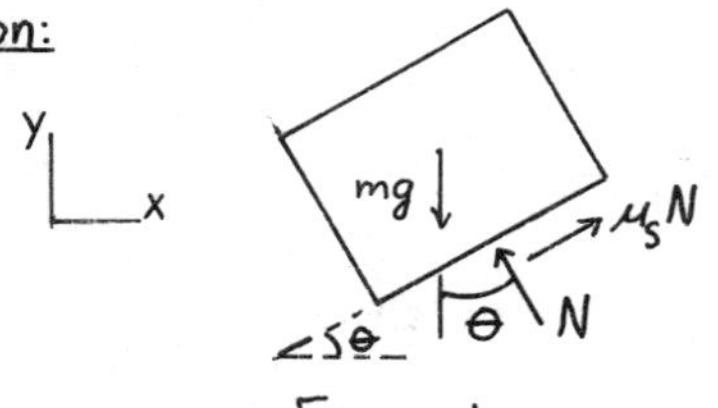

Figure b

By the FBD of the block (Fig. b), impending sliding, $\Sigma F_x = \mu_s N\cos\theta - N\sin\theta = ma$

$\Sigma F_y = N\cos\theta + \mu_s N\sin\theta - mg = 0$

Eliminating N, we get

$$a = \frac{g(\mu_s\cos\theta - \sin\theta)}{(\cos\theta + \mu_s\sin\theta)} = g\left(\frac{\mu_s - \tan\theta}{1 + \mu_s\tan\theta}\right)$$

with $\mu_s = \tan\phi$,

$$a = g\left(\frac{\tan\phi - \tan\theta}{1 + \tan\phi\tan\theta}\right)$$

or, by trigonometric identity,

$$a = g\tan(\phi - \theta)$$

This is the value of a for impending sliding. Hence, for sliding,

$$\underline{a > g\tan(\phi - \theta)}$$

14.38

Given: Figure a of Problem 14.37, but with acceleration directed to the left.

Find: Acceleration a inequality for the sliding of the block.

Solution:

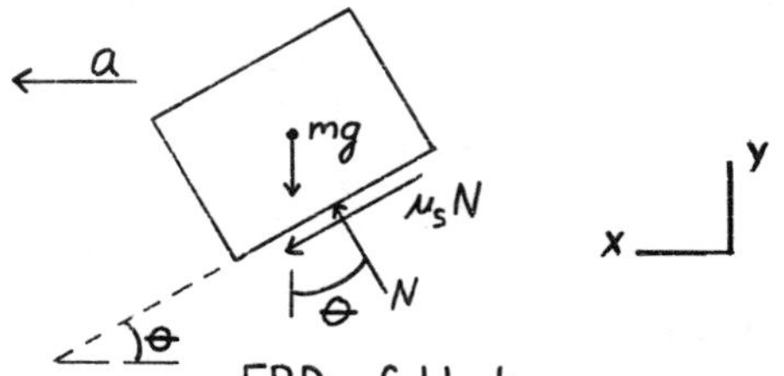

FBD of block

By the FBD of the block,

$\Sigma F_x = \mu_s N\cos\theta + N\sin\theta = ma$

$\Sigma F_y = N\cos\theta - \mu_s N\sin\theta - mg = 0$

Following the procedure of Problem 14.37, we obtain for sliding

$$\underline{a > g\tan(\phi + \theta)}$$

14.39

Given: A particle of mass $0.100\ \text{lb}\cdot\text{s}^2/\text{ft}$ is propelled along the x-axis by a force $F = 2\pi\sin 10\pi t$ (lb); t in seconds. At time $t = 0$, its velocity is -80 ft/s.

Find: The particle's velocity at $t = 0.100$ s.

Solution:

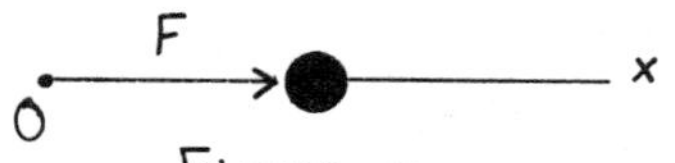

Figure a

By Fig. a, $\Sigma F_x = F = ma = m\dfrac{dv}{dt}$ or $F\,dt = m\,dv$

$$2\pi\int_0^{0.100}(\sin 10\pi t)\,dt = m\int_{-80}^{v} dv$$

Therefore, $2\pi\left(-\dfrac{\cos 10\pi t}{10\pi}\right)_0^{0.100} = (0.100)(v + 80)$

Hence, $\underline{v = -76 \text{ ft/s}}$

14.40

Given: A body of mass m, starts from rest and sinks slowly in liquid of resistance $F=kv$; k = constant, v = speed

Find: Show that, in time t, the body drops a distance s given by,

$$s=\frac{m^2}{k^2}g\left(\frac{kt}{m}-1+e^{-kt/m}\right)$$

Solution:

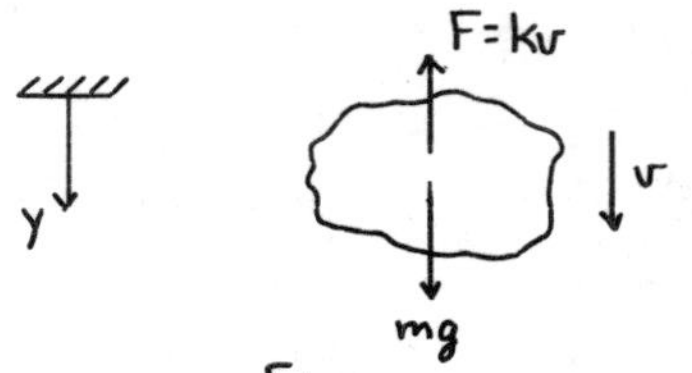

Figure a

By the FBD of the body (Fig. a),

$$\Sigma F_y = mg - kv = m\frac{dv}{dt} \quad \text{or} \quad \frac{dv}{g-\frac{k}{m}v} = dt$$

and $\displaystyle\int_0^v \frac{dv}{g-\frac{k}{m}v} = \int_0^t dt$

By integration,

$$-\frac{m}{k}\ln\left(g-\frac{k}{m}v\right)\Big|_0^v = t \quad \text{or} \quad t=-\frac{m}{k}\left[\ln\left(g-\frac{kv}{m}\right)-\ln g\right]$$

$$\therefore t=\frac{m}{k}\ln\left(\frac{g}{g-\frac{kv}{m}}\right) \quad \text{Hence,} \quad \frac{g}{g-\frac{kv}{m}} = e^{kt/m}$$

or $v=\dfrac{mg}{k}\left(1-e^{-kt/m}\right)$ (a)

Since $v=dy/dt$, integration of Eq.(a) yields

$$y=\frac{mg}{k}\left(t+\frac{m}{k}e^{-kt/m}\right)+C$$

where C is a constant of integration.

Since $y=0$, for $t=0$, $C=\dfrac{m^2g}{k^2}$

$\therefore$ The distance that the body drops in time t is

$$s=y=\frac{m^2g}{k^2}\left[\frac{kt}{m}-1+e^{-kt/m}\right]$$

14.41

Given: Figure a, with block and sheet initially at rest. Then, the sheet is given acceleration $a=32\ ft/s^2$; $\mu=0.50$, $g=32\ ft/s^2$

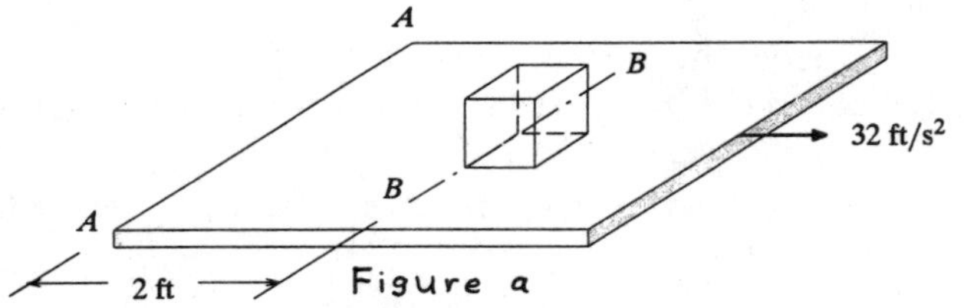

Figure a

Find: Absolute acceleration, velocity, and displacement of the blocks when the edge B-B of the block reaches the edge A-A of the sheet

Solution:

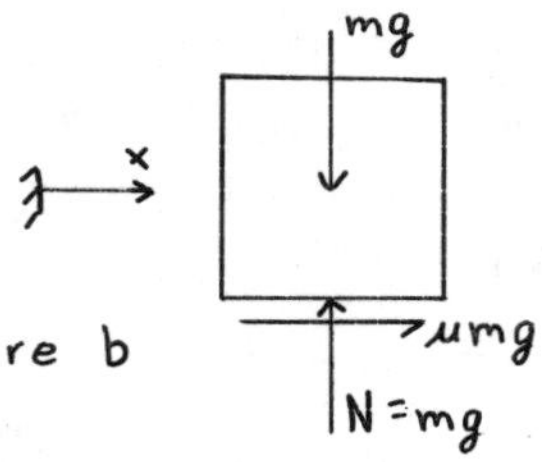

Figure b

See the FBD of the block (Fig. b).
Let x be the absolute displacement of the block and s be the absolute displacement of the sheet. By Fig. b,

$\Sigma F_x = \mu mg = m\ddot{x}$ or $\ddot{x}=\mu g=(0.5)(32)$

$\therefore \ddot{x}=16\ ft/s^2$ (a)

Hence, the block slides, since otherwise $\ddot{x}$ would be $32\ ft/s^2$. Integration of Eq. (a) yields, the block starts from rest ($x_0=v_0=0$),

$\dot{x}=16t$, $x=8t^2$ (b)

Also, $\ddot{s}=32\ ft/s^2$

$\therefore \dot{s}=32t$, $s=16t^2$ (c)

By geometry, $s=2+x$

Hence, by Eqs. (b) and (c), the edge B-B coincides with edge A-A when

$16t^2=2+8t^2$ or $t=0.5\ s$

and by Eq. (b), $\dot{x}=8\ ft/s$; $x=2\ ft$

14.42

Given: Figure a; A weighs 400 lb. Each block of B weighs 50 lb and is held against the guides by two springs (each exert a force of 100 lb). μ between the blocks and guides is 0.20.

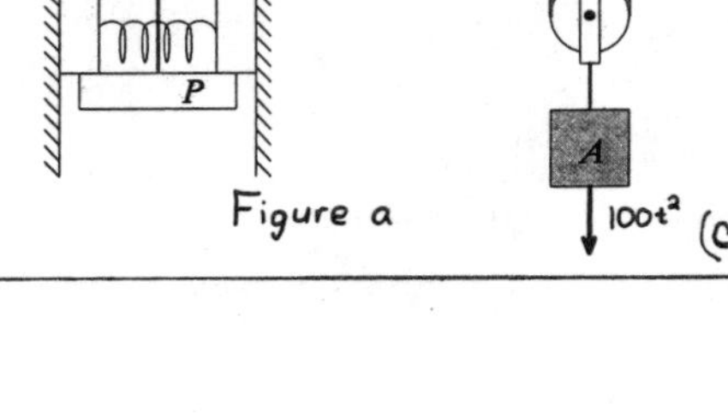

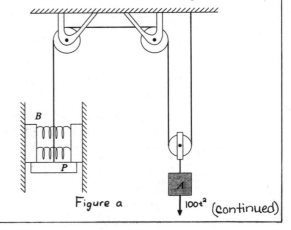

Figure a

(continued)

14.42 Cont.

Find: The tension T in the cable and the acceleration of A in terms of t. Use $g = 32\ ft/s^2$

Solution:

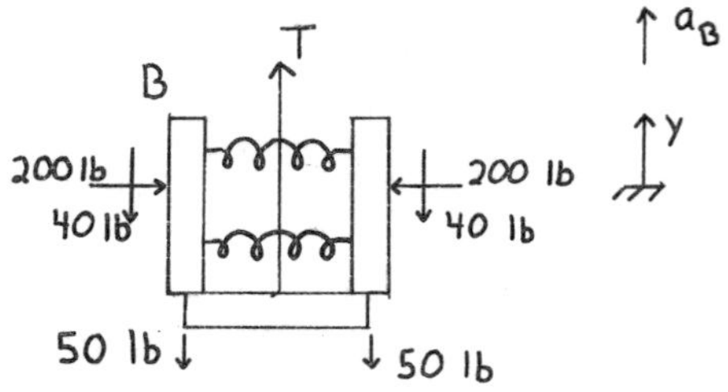

Figure b

When the brake B is sliding upward, the frictional force on each block is (Fig. b)

$\mu N = (0.2)(200) = 40\ lb$

By Fig. b, $\Sigma F_y = T - 180 = \frac{100}{g} a_B$ (a)

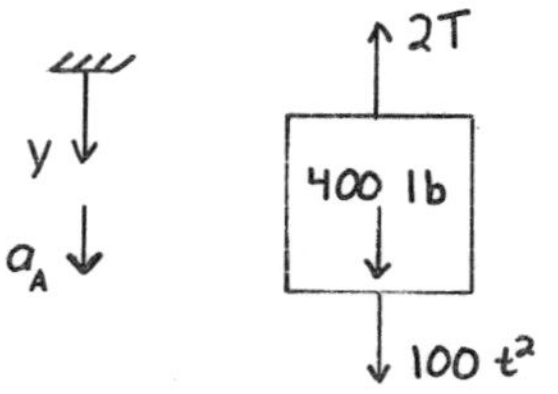

Figure c

By the FBD of block A (Fig. c),

$\Sigma F_y = 100t^2 + 400 - 2T = \frac{400}{g} a_A$ (b)

Let $a_A = a$. By kinematics, $a_B = 2a$ (c)

Hence, by Eqs. (a), (b), and (c)

$a_A = a = 4t^2 + 1.6\ ft/s^2$ (d)

and by Eqs. (a), (c), and (d)

$T = 25t^2 + 190\ (lb)$ (e)

Note that Eq. (e) is valid for all t, since at t = 0, T = 190 lb, and by Fig. (b) and Eq. (a), at time t = 0, the brake B is accelerating upward. If the spring forces are increased so that initially at t = 0 the brake does not move, the tension T would have to increase with time t until motion begins (see Problem 14.43).

14.43

Given: System of Problem 14.42, with compressive force in each spring equal to 200 lb.

Find: a) The time (seconds) at which motion begins.
b) Tension T in the cable and acceleration of A in terms of t, t ≥ 0.

Solution:

As in Problem 14.42,

$a_A = a$, $a_B = 2a$

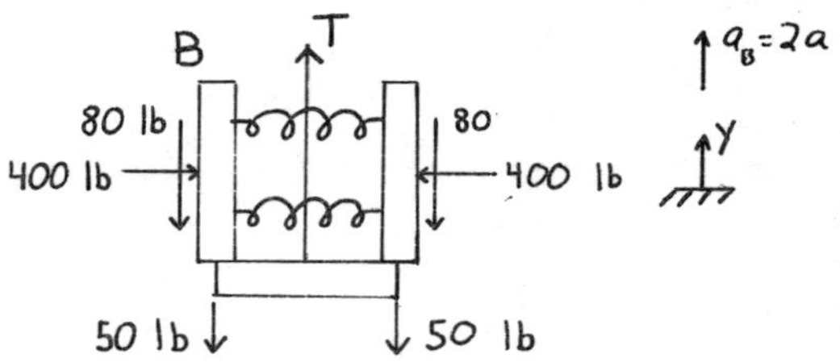

Figure a

By the FBD of brake B (Fig. a), if a > 0,

$\Sigma F_y = T - 260 = \frac{100}{32}(2a)$ or $T = 260 + 6.25a$ (a)

Equation (b) of the solution of Problem 14.42 remains valid ∴

$T = 50t^2 + 200 - 6.25a$ (b)

Equating Eqs. (a) and (b), we obtain

$50t^2 - 60 = 12.5a$, $a > 0$ (c)

By Fig. (a) [or by Eq. (a)], it is seen that motion impends when T = 260 lb. Hence, by Eq. (b) at impending motion (a = 0)

$50t^2 - 60 = 0$

or at impending motion,

t = 1.095 s

For $t \le 1.095$ s, $a_A = a = 0$

and by Eq. (b)

$T = 50t^2 + 200$ (lb); $t \le 1.095$ s

For t > 1.095 s, Eq. (c) applies. Hence,

$a_A = a = 4t - 4.8\ (ft/s^2)$; t > 1.095 s

Then, by Eq. (a),

$T = 25t + 230$ (lb); t > 1.095 s

14.44

Given: Jet plane weighs 10,000 lb. In horizontal flight, it is subected to air resistance (drag) of $D = 0.005v^2$ (lb); v = speed (ft/s). The engine thrust is increased suddenly by 3,000 lb, when the plane's speed is 400 mi/hr.

Find: The time required for the speed to increase to 500 mi/h.

Solution:

Let F_0 be the initial thrust.

Hence, after the 3000 lb increase in thrust

$\Sigma F = F_0 + 3000 - 0.005v^2 = m\frac{dv}{dt}$ (a)

Initially, at the constant speed of $v_0 = 400$ mi/h (586.67 ft/s)

$\Sigma F = F_0 - 0.005v^2 = 0$ (b)

∴ $F_0 = 0.005(586.67)^2 = 1720.9$ lb (continued)

14.44 Cont.

Therefore, the net engine thrust is, after the 3000 lb increase,

$T = F_0 + 3000 = 4720.9$ lb

Now Eq. (a) may be written as

$$\frac{m\,dv}{T - kv^2} = dt \quad , \quad k = 0.005$$

Therefore, the time required to attain $v_f = 500$ mi/h (733.33 ft/s) is given by integration. Thus

$$\frac{m}{k}\int_{v_0}^{v_f} \frac{dv}{\left(\frac{T}{k} - v^2\right)} = \int_0^t dt = t$$

After integration,

$$t = \frac{m}{2\sqrt{kT}} \ln\left[\frac{\sqrt{T/k} + v}{\sqrt{T/k} - v}\right]\Bigg|_{v_0}^{v_f}$$

with $k = 0.005$ lb·s²/ft², $T = 4720.9$ lb, $v_f = 733.33$ ft/s, $v_0 = 586.67$ ft/s, and $m = 10{,}000/32.2 = 310.6$ lb·s²/ft, we obtain

$$\underline{\underline{t = 18.20 \text{ s}}}$$

14.45

Given: A small steel ball is dropped into a straight shaft that passes through the center of the earth. (Fig. a)

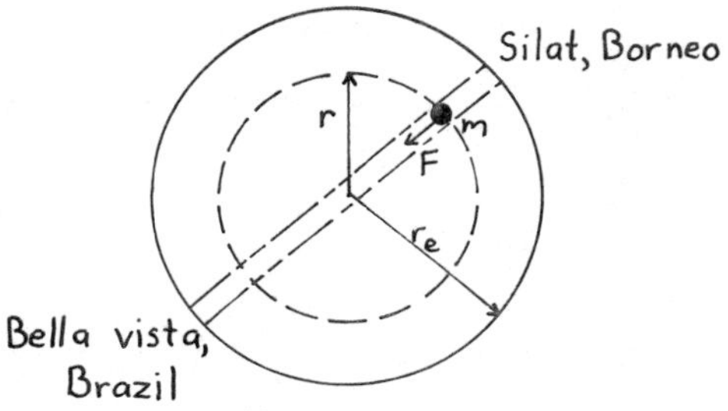

Figure a

Find: The equation of motion of the ball. Neglect air resistance.

Solution:

The gravitational force that acts on the ball is due to the earth matter inside radius r.

$$\therefore \Sigma F_r = -F = ma = m\frac{d^2r}{dt^2} \qquad \text{(a)}$$

where,

$$F = \frac{Gmm_e}{r^2} \qquad \text{(b)}$$

where in Fig. (a), the effective mass of the earth is

$$m_e = \rho\frac{4\pi}{3}r^3 ; \quad \rho = \text{mass density} \qquad \text{(c)}$$

Therefore, by Eqs. (a), (b), and (c)

$$\underline{\underline{\frac{d^2r}{dt^2} + \left(\frac{4\pi}{3}\rho G\right)r = 0}}$$

14.46

Given: A body projected upward with speed v_0 is subjected to air drag $D = Kv^2$; K = constant, v = speed. Assume g = constant

Find: Show that the speed of the body when it returns to earth is

$$v_1 = \frac{v_0}{\sqrt{1 + Kv_0^2/mg}}$$

Solution:

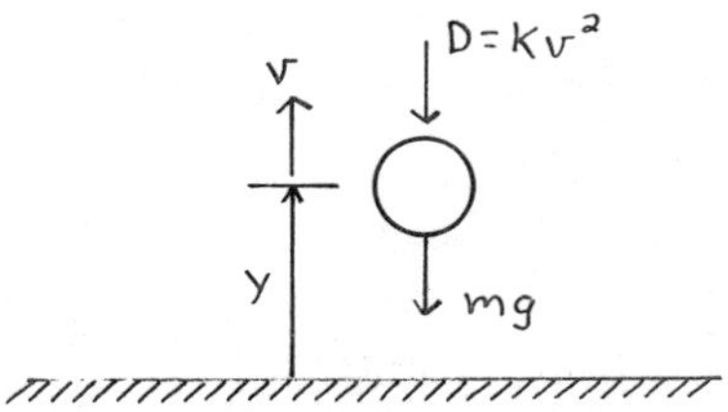

Figure a

Let positive sense be up (Fig. a). By Fig. a,

$$\Sigma F_y = -mg - Kv^2 = m\frac{dv}{dt} = mv\frac{dv}{dy}$$

Hence, going up,

$$\frac{v\,dv}{v^2 + \frac{mg}{K}} = -\frac{K}{m}dy$$

Integration yields

$$\frac{1}{2}\ln\left(v^2 + \frac{mg}{K}\right) = -\frac{K}{m}y + C$$

For $y = 0$, $v = v_0$ $\therefore C = \frac{1}{2}\ln\left(v_0^2 + \frac{mg}{K}\right)$

$$\therefore \frac{K}{m}y = \ln\sqrt{\frac{v_0^2 + mg/K}{v^2 + mg/K}}$$

For $v = 0$, $y = y_{max} = y_1$, the height to which the body rises.

Hence,

$$y_1 = \frac{m}{K}\ln\sqrt{1 + \frac{Kv_0^2}{mg}} \qquad \text{(a)}$$

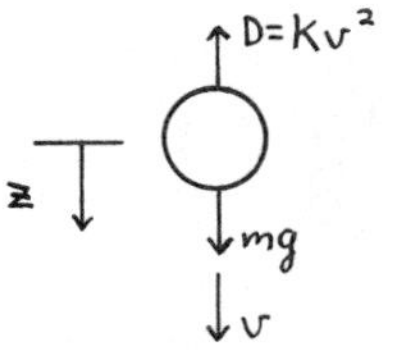

Figure b

For the falling body, let positive sense be down (Fig b). By Fig. (b)

$$\Sigma F_z = mg - Kv^2 = mv\frac{dv}{dz} \quad \text{or} \quad \frac{v\,dv}{\frac{mg}{K} - v^2} = \frac{K}{m}dz$$

Integration yields (with $v = 0$ for $z = 0$)

$$\frac{K}{m}z = \frac{1}{2}\ln\left[\frac{mg}{mg - Kv^2}\right] \qquad \text{(b)}$$

When body returns to earth $z = y_1$ and $v = v_1$. Then, by Eqs. (a) and (b),

$$2\ln\sqrt{1 + \frac{Kv_0^2}{mg}} = \ln\left[\frac{mg}{mg - Kv_1^2}\right] \qquad \text{(c)}$$

(continued)

14.46 Cont.

or $\ln\left(\frac{mg + kv_0^2}{mg}\right) = \ln\left(\frac{mg}{mg - kv_1^2}\right)$

$$\therefore \quad \frac{mg + kv_0^2}{mg} = \frac{mg}{mg - kv_1^2}$$

and

$$v_1 = \frac{v_0}{\sqrt{1 + kv_0^2/mg}}$$

14.47

Given: Thrust (lb) vs time (s) in Table a. Thrust acts on spacecraft to lift it vertically from the moon's surface

Table a

t (s)	0	0.5	1.0	1.5	2.0	2.5	3.0	4.0	5.0
T (lb)	0	220	360	450	510	550	560	560	560

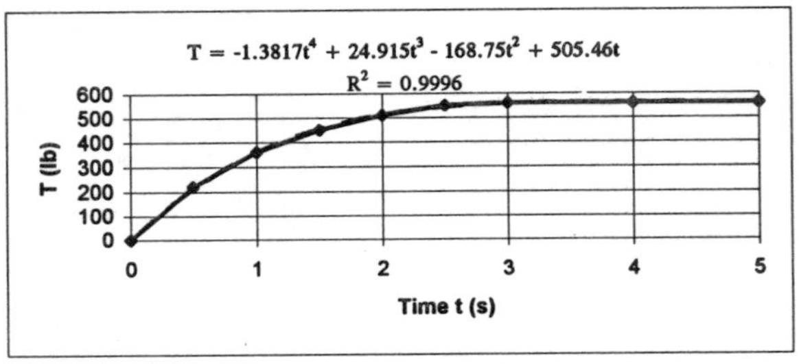

Find: velocity of the craft at $t = 5$ s after ignition

Solution

By a polynomial fit (using a software package), see plot, $T(t) = -1.3817t^4 + 24.915t^3 - 168.75t^2 + 505.46t$ (a)

The spacecraft does not lift off, until

$T = mg = (75+3)(5.36) = 418.08$ lb

By Eq. (a), when $T = 418.08$ lb, $t = 1.275$ s

By Fig. (a), for $t \geq 1.275$ s,

$\Sigma F_y = T(t) - mg = m\frac{dv}{dt}$

so, with $T(t)$ given by Eq. (a)

$\int_{1.275}^{5.0} (T(t) - 418.08)\,dt = 78v$

Integrating and solving for v, we find,

$v = 5.73$ ft/s at $t = 5$ s

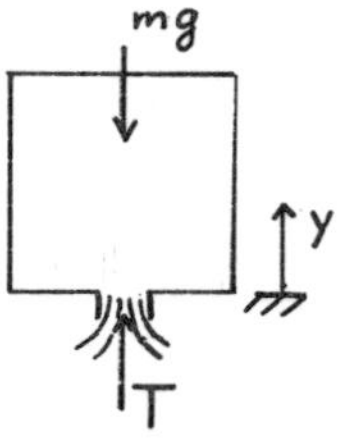

Figure a

14.48

Given: A bathysphere that in time t (s) sinks from rest in the ocean a distance

$$S = \frac{m^2 g}{k^2}\left(\frac{kt}{m} - 1 + e^{-kt/m}\right) \qquad (a)$$

Find: a) Formulas for speed v and the magnitude a of the craft as functions of t

b) For $k = 1\ \text{s}^{-1}$, and $g = 9.81\ \text{m/s}^2$ plot s, v, and a vs t, $0 \leq t \leq 10$ s.

c) From the plots, verify that $v = ds/dt$ and $a = dv/dt$ for $t = 2$ s; $t = 4$ s

d) From plots determine t for $v = 50\%$ and 90% of its maximum speed and depths.

Solution:

a) By Eq. (a)

$$v = \frac{ds}{dt} = \frac{mg}{k}\left(1 - e^{-kt/m}\right) \qquad (b)$$

$$a = \frac{dv}{dt} = g\,e^{-kt/m} \qquad (c)$$

b) Following plots

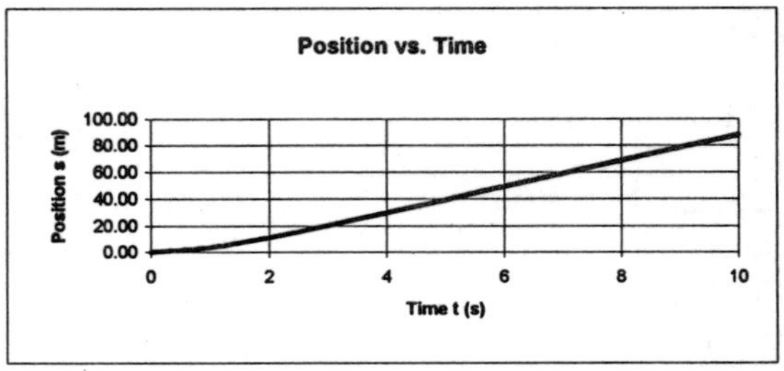

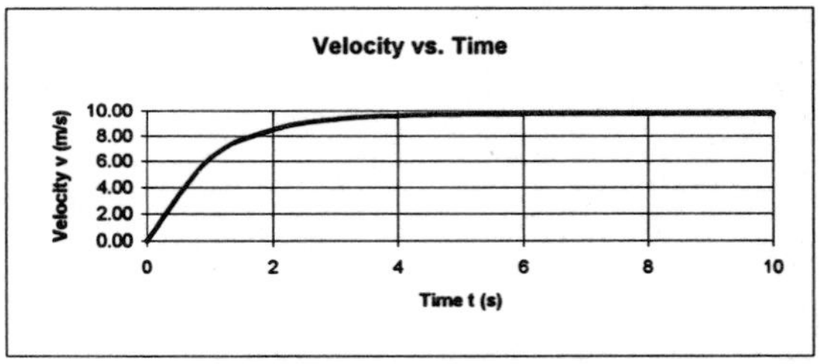

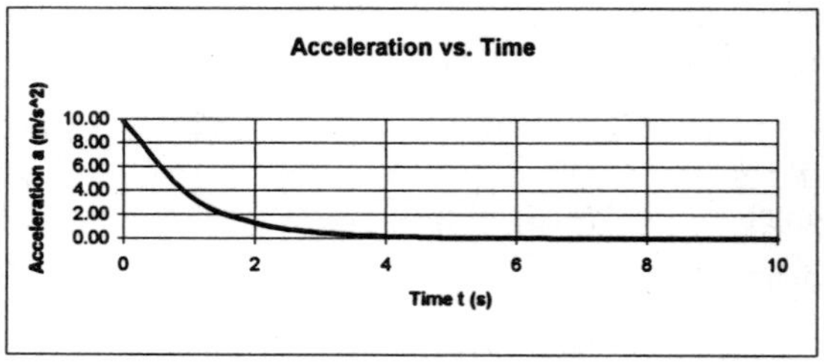

c) By slope of s plot at $t = 2$ s, $t = 4$ s, respectively,

$v \approx 8.5$ m/s and $v \approx 9.6$ s

By the slope of v plot at $t = 2$ s and $t = 4$ s

$a \approx 1.3\ \text{m/s}^2$ and $a \approx 0.2\ \text{m/s}^2$

(continued)

14.48 Cont.

As a check,

By the plot of v, $v(2) \approx 8.5$ m/s and $v(4) \approx 9.6$ m/s.

By the plot of a, $a(2) \approx 1.4$ m/s² and $a(4) \approx 0$.

Note: These values were obtained from larger plots than shown here.

d) By Eq. (b) $v_{max} = 9.81$ m/s.

By plot of v, at $0.5\,v_{max}$, $t \approx 0.75$ s;
at $0.9\,v_{max}$, $t \approx 2.2$ s

By Eq. (b), $t = 0.693$ s and $t = 2.30$ s

14.49

Given: Refer to Problem 14.44 and data shown in Fig. a below.

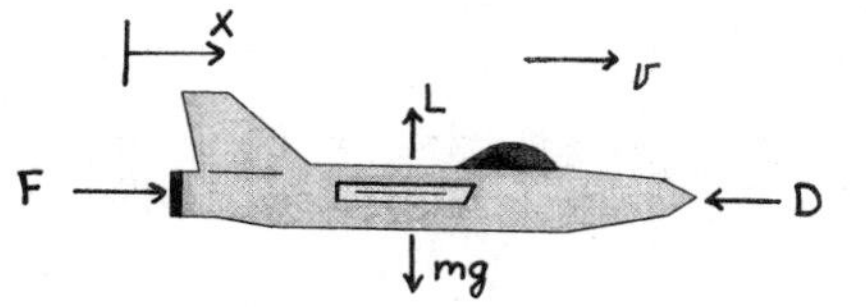

$F = 4721$ lb
$D = 0.005\,v^2$
$m = 310.6$ slugs
$v \geq 586.7$ ft/s
$v \leq 880$ ft/s

Figure a

Find: Provide plots of v vs. t and v vs. s

Solution:

By Fig. a, $\Sigma F_x = F - D = m\,a_x$

$$4721 - 0.005\,v^2 = m a_x = 310.6\,\frac{dv}{dt} \qquad (a)$$

or $$\int_0^t dt = \int_{586.7}^{v} \left[\frac{310.6\,dv}{4721 - 0.005\,v^2}\right]$$

Using computer software to solve for $v(t)$, we get

$$v(t) = \frac{3.03\,(2.624 \times 10^7 \times e^{0.03129t} - 6.482 \times 10^6)}{81874\,e^{0.03129t} + 20229}\ \frac{ft}{s} \qquad (b)$$

Alternatively, using the integration method of Problem 14.44, we find

$$v(t) = \frac{971.7\,e^{t/31.965} - 240}{e^{t/31.965} + 0.247} \qquad (c)$$

Equations (b) and (c) agree closely.

Equation (b) is used to obtain plot of $v(t)$ below,

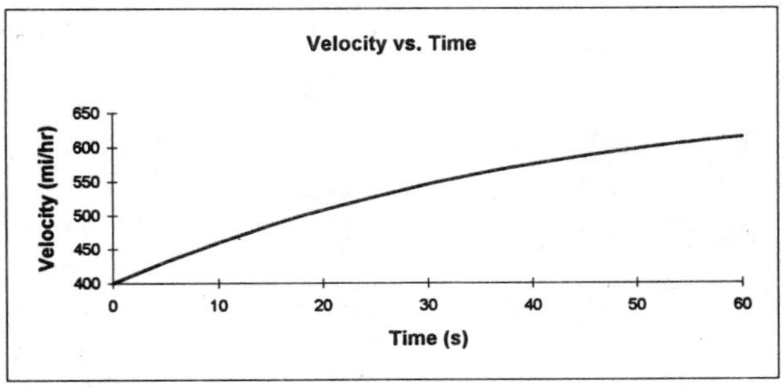

Rewriting Eq. (a) we have

$$4721 - 0.005\,v^2 = 310.6\,v\,\frac{dv}{ds}$$

or, $$\int_0^s ds = \int_{586.7}^{v} \left[\frac{310.6\,v\,dv}{4721 - 0.005\,v^2}\right]$$

Using computer software to solve for $v(s)$, we get,

$$v(s) = 4.714\,e^{(-1.61 \times 10^{-5})s} \sqrt{(42490\,e^{(3.22 \times 10^{-5})s} - 27000)}\ \frac{ft}{s}$$

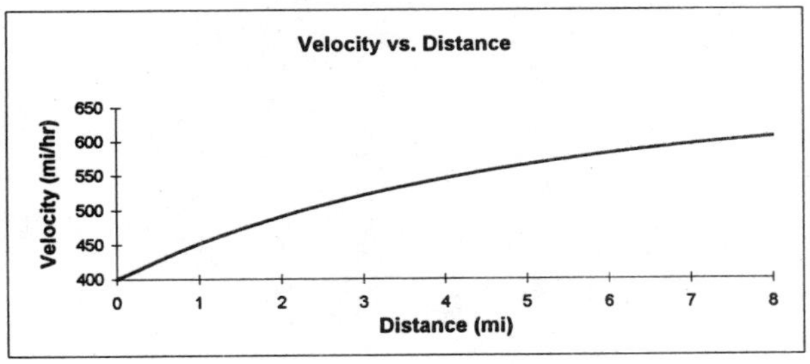

14.50

Given: Figure a

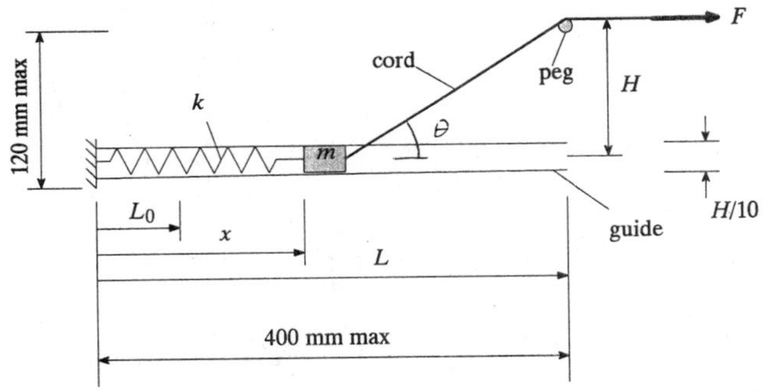

Find: Figure a

a) Show that the acceleration of m is

$$a = \ddot{x} = \frac{F\,(L - x)}{m\sqrt{(L - x)^2 + H^2}} - \frac{K}{m}(x - L_0)$$

b) Design the system so that

$a_{max} < 10$ m/s²
L_0 (unstretched length) = 100 mm
$L \geq 3H$
$L_{max} \leq 400$ mm
vertical dimension ≤ 120 mm

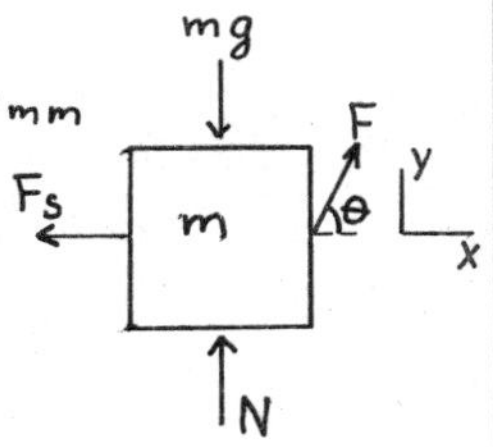

Figure b

Solution:

a) By Figs. a and b,

$$\Sigma F_x = -F_s + F\cos\theta = ma \qquad (a)$$

where, $F_s = K\,(x - L_0)$

(continued)

14.50 Cont.

$$\cos\theta = \frac{L-x}{\sqrt{(L-x)^2 + H^2}} \qquad (b)$$

Therefore, by Eqs. (a) and (b),

$$a = \frac{F(L-x)}{m\sqrt{(L-x)^2 + H^2}} - \frac{k}{m}(x - L_o) \qquad (c)$$

b) Many possibilities exist. One solution can be obtained by selecting geometrical parameters with in the constraints. For example, select $L = 300$ mm, $H = 100$ mm. With $L_o = 100$ mm.

Then Eq. (a) yields

$$a = \frac{F(0.30 - x)}{m\sqrt{(0.30-x)^2 + 0.10^2}} - \frac{k}{m}(x - 0.10) \qquad (d)$$

By inspection of Eq. (d) and Fig. a,

$a = a_{max}$ when $x = L_o = 0.10$ m

Then, Eq. (d) yields

$$a = 0.9487\frac{F}{m} \ (m/s^2) \qquad (e)$$

In general, the smaller m, the less energy required to move the part. But by Eq. (e), the smaller the mass, the larger the acceleration of the part for a given force F. As a compromise, Let's take $m = 4$ kg < 6.75 kg. For $a < 10\,m/s^2$,

$$a = 0.9487\frac{F}{4} < 10\,m/s^2$$

or

$$F < 42.16 \text{ N}$$

Thus, we have defined all the parameters, except the spring stiffness k.

Again, we see by Eq. (d) that at $x = L = 0.30$ m,

$$a = -\frac{k}{m}(0.30 - 0.10)$$

Hence, we must have $0.20\frac{k}{m} < 10\,m/s^2$

or with $m = 4$ $\quad k < 200$ N/m

In other words, we need a very soft spring.

Thus, the design specifications are met with the following values

$L = 300$ mm, $L_o = 100$ mm, $H = 100$ mm, $F < 42.16$ N, $m = 4$ kg, and $k < 200$ N/m

Also to ensure that $L_o \le x \le L$, we need to place stops at $x = L_o$ and $x = L$.

Note: The final selection of parameters may depend on other factors, such as economics.

14.51

Given: Refer to textbook problem statement. A company wins a contract to design a new pulley drive system. The company uses an analysis that has worked on previous designs. In tests of the new system, the belts keep failing. You are brought in to solve the reason for the belt failures.

Find: Why the belts fail and give your recommendations for the design.

Solution:

The previous belt drive operates successfully at $\omega = 120$ rpm. Examining the required specifications for the high-speed drillpress (Table 14.51) and checking the company engineers' calculations you find that there seems to be nothing unusual and that the calculations are correct. However, you do notice that $\omega = 120$ rpm for the companies pulley drive and $\omega = 3000$ rpm (314.16 rad/s) for the new system.

Referring to the discussion of pulley-belt tensions in your mechanics book (e.g., see section 10.5 Footnote 1) you discover that at high speeds, centrifugal forces can significantly affect belt tension.

In particular, at high speeds, Eq. (a) relating the belt tensions T_1 and T_2 may not be sufficiently accurate. Rather, T_1 and T_2 are related by the equation,

$$T_1 - mv^2 = (T_2 - mv^2)e^{\mu\theta} \qquad (a)$$

where m is the mass per unit length of belt in contact with the pulley and v is the belt speed. With the data of Table 14.51, you find that,

$$m = \left(\frac{\gamma}{9.81}\right)h w$$

$$\therefore m = \frac{10000}{9.81}hw = 1019.4\,hw \ (kg/m)$$

where h and w are the belt thickness and width, respectively.

Hence, since $v = r\omega = (0.075)(314.16) = 23.56$ m/s and $mv^2 = 5.659(10^5)hw$. Using $w = 6.25$ mm and $h = 1.5$ mm (as did the engineer), you obtain $mv^2 = 5.659(10^5)(0.00625)(0.0015) = 5.30$ N, which is not negligible compared to $T_1 = 17.84$ N and $T_2 = 9.52$ N. Therefore, you continue and find by Eq. (a),

$$T_1 = [T_2 - 5.659(10^5)hw]e^{0.20\pi} + 5.659(10^5)hw$$

$$\therefore T_1 = 1.874T_2 - 4.949(10^5)hw \ (N) \qquad (b)$$

or,

$$T_2 = 0.5335T_1 + 2.640(10^5)hw \ (N) \qquad (c)$$

Since power requirement is

$$E = [(T_1 - T_2)r]\omega = 196 \text{ N m/s} \qquad (d)$$

Equations (b), (c), and (d) yield

$$10.99T_1 - 6.220(10^6)hw = 196 \text{ N}\cdot\text{m/s} \qquad (e)$$

And since the allowable stress in the belt is $\sigma = 2\times10^6$ N/m² and $T_1 = A\sigma = \sigma h w$ or $T_1 = 2(10^6)hw$.

(continued)

14.51 Cont.

Hence, by Eq. (e),

$15.76 \times 10^6 hw = 196 \;\; N\cdot m/s$

so, the minimum requirement for

$hw = 196/[15.76(10^6)] = 12.436(10^{-6})m^2 = 12.436\ mm^2$

selecting $h = \frac{1}{5}w$, you find

$w = 7.88\ mm, \quad h = 1.58\ mm$

As a safety margin, you recommend that

$w = 8.75\ mm, \quad h = 1.75\ mm \qquad (f)$

In your report, you will point out the reason you believe the belts are failing and recommend the belt size (Eq. (f)). You also suggest that although a belt with allowable stress of $2 MN/m^2$ is satisfactory, a belt with a 5% increase in allowable stress would increase the margin of safety.

14.52

Given: The brake system of Problem 14.42 (Fig. a). The weight of block A must remain at 400 lb. However, the weight W of each block in the brake B, the spring force F_s of each spring, and the coefficient of friction μ may be changed.

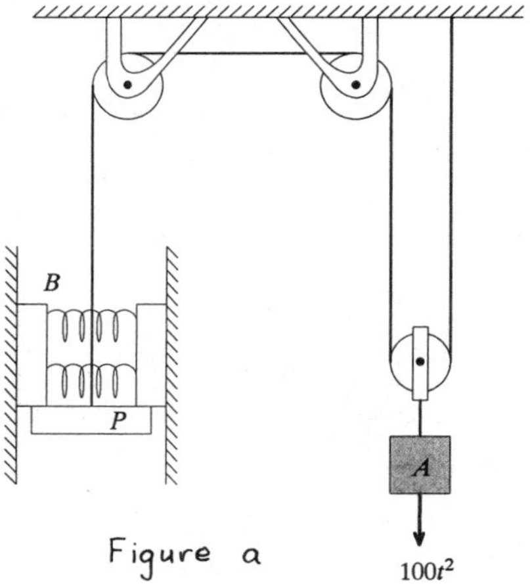

Figure a

Find: Design the system so that the acceleration a of block A lies in the range $10\ ft/s^2 \le a \le 12\ ft/s^2$ at time t=2s and such that motion is impending up at t=0. Use $g = 32\ ft/s^2$

Solution:

Since W, F_s, and μ may be selected to meet the design requirements, let's derive a formula for the acceleration a in terms of W, F_s, and μ. Let's start with Block A and work back to brake B. Thus, by the FBD of A (Fig. b), where T is the tension in the cable,

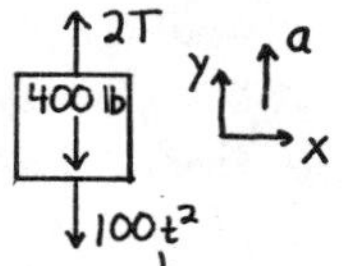

Figure b

$\Sigma F_y = 100t^2 + 400 - 2T = \frac{400}{g} a$

$\therefore a = \frac{1}{4} g t^2 + g - \frac{gT}{200} \qquad (a)$

Since motion impends up at t=0, a=0 and T = 200 lb at t=0.

Next, consider the FBD of one of the blocks of brake B, say the left block (Fig. c), where Fp is the force exerted on the block by the plate. By Fig. c,

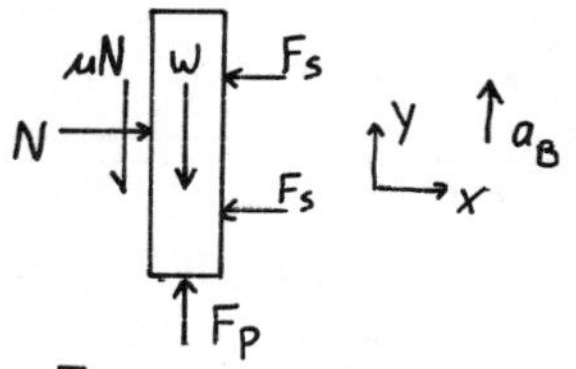

Figure c

$\Sigma F_x = N - 2F_s = 0$

$\Sigma F_y = F_p - W - \mu N = \frac{W}{g} a_B$

Hence, $N = 2F_s$ and

$\therefore F_p = 2\mu F_s + W + \frac{W}{g} a_B \qquad (b)$

The FBD of the plate is shown in Fig. d. By Fig. d, since the mass of the plate is negligible,

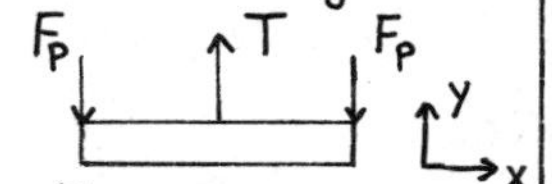

Figure d

$\Sigma F_y = T - 2F_p = 0$

or

$F_p = \frac{1}{2} T \qquad (c)$

By Eqs. (b) and (c),

$a_B = \frac{gT}{2W} - \frac{2g\mu F_s}{W} - g \qquad (d)$

At t=0, motion is impending, T = 200 lb, and $a_B = 0$. Therefore, by Eq. (d),

$\mu F_s = 50 - 0.5W \quad (lb) \qquad (e)$

Then, by Eqs. (a) and (d), with $g = 32\ ft/s^2$ and $a_B = 2a$ (by kinematics), we find,

$a = \frac{800t^2}{100 + 2W} \qquad (f)$

For t=2s, Eq. (f) yields

$a = \frac{3200}{100 + 2W} \qquad (g)$

or

$W = \frac{1600 - 50a}{a} \qquad (h)$

For $a = 10\ ft/s^2$, W = 110 lb

For $a = 12\ ft/s^2$, W = 83.33 lb

By Eq. (e),

For W = 110 lb, $\mu F_s = -5$ lb

and motion of the brake is impending down. Then, the force $100t^2$ would have to increase until motion impends upward, before the block A would move. Hence, a given by Eq. (a) is invalid, since a would be zero until $100t_1^2$ (i.e. $t = t_1$) is large enough to cause impending motion upward. The largest value that W can have for motion of B impending up at time t=0 is, by Eq. (e), W = 100 lb, with $\mu F_s = 0$; i.e., either $\mu = 0$, or $F_s = 0$ (no springs) (continued)

14.52 Cont.

Then, by Eq. (g), $a = 10.67 \text{ ft/s}^2 > 10 \text{ ft/s}^2$.
For $W = 83.33$ lb, $a = 12 \text{ ft/s}^2$, and $\mu F_s = 8.33$ lb.
Hence, the design requirements are met provided

$$83.33 \text{ lb} \le W \le 100 \text{ lb}$$

with corresponding values of μF_s

$$0 \le \mu F_s \le 8.33 \text{ lb}$$

As determined by Eq. (e) for corresponding values of W.

14.53

Given: A baseball and a length of rope. Also, see detailed problem statement.

Find: A design of a low-cost pitching machine with the baseball and rope. The ball is to travel at a speed of 35 mph, pass through a strike zone 18 in and 42 in above the ground, and have a path that is horizontal.

Solution:

Assumptions

- The free end of the rope is held at a point (the center of rotation) h (ft) above the ground.
- The ball, attached to the other end of the rope, is swung about the rotation center with constant angular velocity ω (rad/s).
- The center of rotation is fixed.
- The mass of the rope is negligible.
- The center of mass of the ball is at the other end of the rope; that is, the length L (ft) of rope is measured from the center of rotation to the center of mass of the ball.
- The ball is a point mass
- Air resistance is negligible.
- The tensile strength T of the rope will not be exceeded for the required angular velocity.

Analysis

A sketch of the rope-ball system is shown in Fig. a.

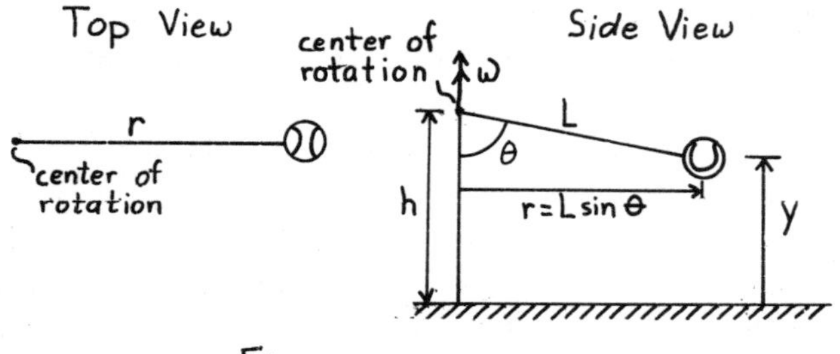

Figure a

Since $v = 35$ mph $= 51.33$ ft/s, by Fig. a,

$$r\omega = L(\sin\theta)\omega = 51.33 \text{ ft/s} \qquad (a)$$

Also, since the ball travels in a circular path of radius r, the acceleration of the ball is directed toward the center O of the circular path and is

$$a = r\omega^2 = L(\sin\theta)\omega^2 \qquad (b)$$

The FBD of the ball is shown in Fig. b.
By Fig. b,

$$\Sigma F_r = T\sin\theta = ma$$

$$\Sigma F_y = T\cos\theta - mg = 0$$

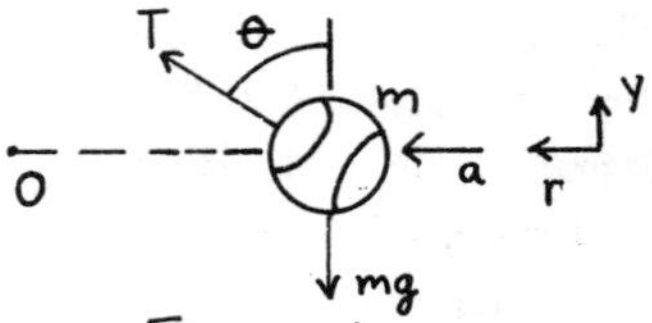

Figure b

$$\therefore \; a = g\tan\theta \qquad (c)$$

By Eqs. (b) and (c)

$$\omega^2 = \frac{g}{L\cos\theta} \qquad (d)$$

The height y of the circular path of the ball is, by Fig. a,

$$y = h - L\cos\theta \qquad (e)$$

The design parameters L, h, y, θ, and ω are governed by Eqs. (a), (d), and (e). Since we have three equations and 5 parameters, we must specify two parameters.

Let's say that h is about 6.5 ft (78 in). Since y must be in the range $18 \text{ in} \le y \le 42 \text{ in}$, let's take the mid range $y = 2.5$ ft (30 in). Then, by Eq. (e),

$$L\cos\theta = h - y = 4 \text{ ft } (48 \text{ in}) \qquad (f)$$

Now by Eqs. (d) and (f)

$$\omega^2 = \frac{32}{4} = 8 \quad \text{or} \quad \omega = 2\sqrt{2} \text{ rad/s} \qquad (g)$$

Hence, by Eqs. (a) and (g),

$$L\sin\theta = 18.15 \text{ ft} \qquad (h)$$

and by Eqs. (f) and (h), $L^2 = 345.4$

or $\quad L = 18.6$ ft $\qquad$ (i)

Also, by Eqs. (f) and (h) $\quad \tan\theta = \frac{18.15}{4}$

or $\quad \theta = 77.57°$ $\qquad$ (j)

Therefore, the design requirements are satisfied with $L = 18.6$ ft, $h = 6.5$ ft, $y = 2.5$ ft, $\theta = 77.57°$, and $\omega = 2\sqrt{2} = 2.83$ rad/s.
However, many other combinations would be adequate.

Note that if air resistance and the weight of the rope were included, for a given angular velocity the rope would not lie in a straight line and the ball would lag behind the original radial line r (Fig. a). Also, the pitcher probably can not maintain the free end of the rope at a fixed center of rotation.

14.54

Given: Boeing 777-300 Specifications

Find: a) List two or more design items that Boeing engineers had to address in the design of the Boeing 777-300.

b) Discuss a few factors that engineers had to consider in establishing values of design parameters for the design items selected in part a.

Solution:

a) The engineers had to select hundreds of design parameters. Some of these envolved the following parts or systems.

- Overall size
- General interior configuration - including seating options and cabin layout.
- General exterior configuration - including wingspan, tail, engine positions, etc.
- Mechanical systems
- Electrical systems
- Communication systems
- Structural strength (of Wings, of pressurized fuselage, of landing gear, etc.)
- Impact strength and durability of tires
- etc.

b) As examples, consider structural, exterior configuration, and electrical wiring.

Structural strength

Engineers had to determine the loads that the wings would be subjected to. These loads include the weight of the wings and aerodynamic loads. With these loads, engineers had to specify sizes and configurations of the wings and the interior structural support systems.

Exterior Configuration

Engineers had to decide on the position and configuration of the wings and tail, since the lift depends on them. Also, they affect the plane's landing characteristics. (The 777 wings are closer to the ground than earlier planes, such as the 747, thus making it easier to control during landing.)

Electrical Wiring

Engineers had to consider several back-up wiring systems through different parts of the plane, so that if an explosion occurred in one part of the plane, the plane could still be controlled. Also, the wiring had to be shielded to prevent spurious electrical currents from affecting the control of the plane.

14.55

Given: A stream of water issues horizontally from a nozzle 0.60m above level ground and strikes the ground at 3m in front of the nozzle (Fig. a) Neglect air resistance

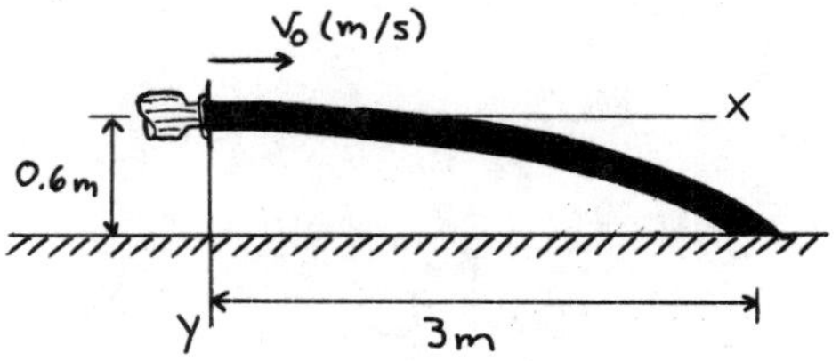

Figure a

Find: v_0 (m/s)

Solution:

Since air resistance is neglected, the stream of water is subjected to gravity only. Hence,

$\ddot{y} = g$, $\ddot{x} = 0$ $\therefore$ $\dot{x} = v_0$, $\dot{y} = gt$

since water is ejected horizontally

$$x = v_0 t \quad , \quad y = \tfrac{1}{2} g t^2 \quad (a)$$

Elimination of t in Eq. (a) yields

$$y = \frac{1}{2} g \frac{x^2}{v_0^2} \quad (b)$$

For $x = 3m$, $y = 0.6m$

$\therefore$ By Eq. (b), $v_0^2 = \frac{x^2}{2y} g = \frac{(3^2)(9.81)}{2(0.6)}$

or $\underline{v_0 = 8.578 \text{ m/s}}$

Alternatively, by Eq. 14.11, with +y downward

$$y = \frac{g x^2}{2 v_0 \cos^2 \theta_0} - x \tan \theta_0$$

with $\theta_0 = 0$, $y = \frac{g x^2}{2 v_0}$ as before [Eq. (b)]

14.56

Given: A stream of water issues from a nozzle at 40 ft/s and strikes a target 30 ft away at the same elevation as the nozzle. (Fig. a)

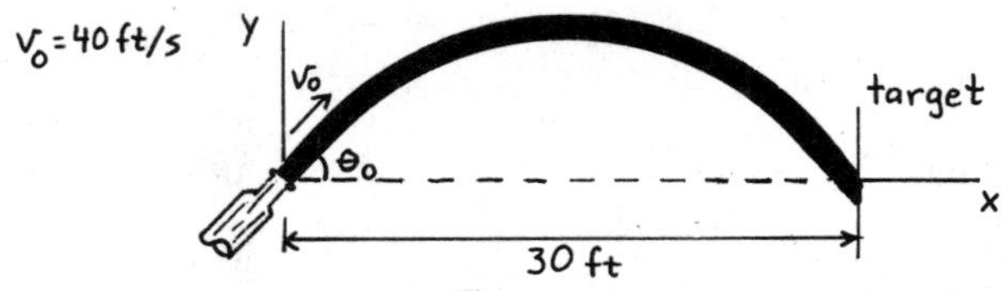

Figure a

Find: a) Angle θ_0 (two solutions)

b) The solution for which the least time is required for water to reach the target.

Solution:

As in previous problem (14.55) Neglect air resistance.

(continued)

14.56 Cont.

a) By Fig. a, the time required to reach the target is given by the equations

$x = 40t\cos\theta_0 = 30$ ft (a)

$y = -\frac{1}{2}gt^2 + 40t\sin\theta_0 = 0$ (b)

By Eqs. (a) and (b),

$\cos\theta_0 = \frac{0.75}{t}$

$\sin\theta_0 = \frac{gt}{80} = \frac{32t}{80} = 0.4t$

$\therefore \cos\theta_0 = \frac{(0.75)(0.4)}{\sin\theta_0}$

or $2\sin\theta_0\cos\theta_0 = 0.6$

or $\sin 2\theta_0 = 0.6$

$\therefore 2\theta_0 = 36.870°,\ 180° - 36.87° = 143.14°$

$\theta_0 = 18.435°,\ 71.56°$

b) By Eq. (a)

$t = \frac{0.75}{\cos\theta_0}$

For $\theta_0 = 18.435°$, $t = 0.79$ s

For $\theta_0 = 71.565°$, $t = 2.37$ s

$\therefore$ The least time is taken when $\theta_0 = 18.435°$

14.57

Given: A stream of water from a nozzle reaches a height h above the nozzle and strikes a target at the same elevation as the nozzle. The nozzle is directed at a higher angle and rises to a height 2h above the nozzle and again strikes the target (Fig. a)

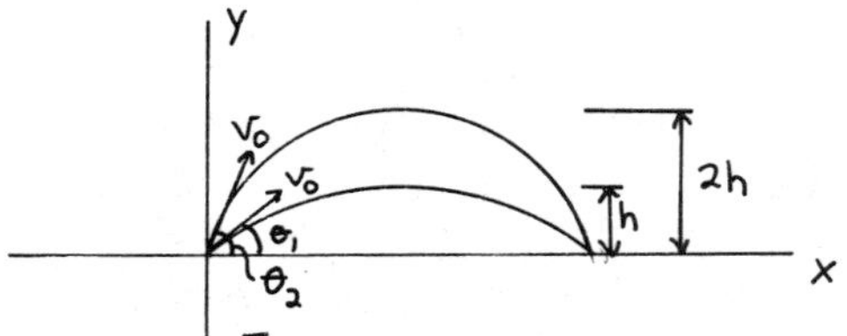

Figure a

Find: The angle of the nozzle for
a) elevation h
b) elevation 2h

Solution:

By Eq. (14.12),

For height h,

$x_1 = \frac{V_0^2}{2g}\sin 2\theta_1$

$y_1 = h = \frac{V_0^2}{2g}\sin^2\theta_1$

For height 2h,

$x_2 = \frac{V_0^2}{2g}\sin 2\theta_2$

$y_2 = 2h = \frac{V_0}{2g}\sin^2\theta_2$

$\therefore x_1 = x_2 \Rightarrow \sin 2\theta_1 = \sin 2\theta_2$ (a)

$y_2 = 2y_1 \Rightarrow \sin^2\theta_2 = 2\sin^2\theta_1$ (b)

By Eq. (a),

$2\theta_1 = \pi - 2\theta_2$

or $\theta_1 = \frac{\pi}{2} - \theta_2$ (c)

By Eqs. (b) and (c)

$2\sin^2\theta_1 = \sin^2(\frac{\pi}{2} - \theta_1) = 1 - \cos^2(\frac{\pi}{2} - \theta_1)$

or $2\sin^2\theta_1 = 1 - \sin^2\theta_1$

$3\sin^2\theta_1 = 1$

$\therefore \theta_1 = 35.26°$, $\theta_2 = 57.74°$

14.58

Given: Astronaut on moon throws ball of mass m = 0.145 kg up to a height of 20 m. The same ball, with the same initial speed is thrown up on earth. Ignore air resistance and note that $g_m = \frac{1}{6}g_{earth}$

Find: a) How high the ball rises on earth.
b) The initial speed of the ball to rise 20 m on earth.

Solution:

a) $g_{moon} = \frac{1}{6}g_{earth} = \frac{1}{6}(9.81) = 1.635$ m/s²

$\therefore$ By Eq. (14.12), with $\theta = 90°$,

$y = \frac{v_0^2}{2g_{moon}}\sin^2\theta = 20$ m or $v_0^2 = 40g_{moon} = 65.4\ \frac{m^2}{s^2}$

or $v_0 = 8.087$ m/s

On earth, $y = \frac{v_0^2}{2g}\sin^2\theta = \frac{(8.087)^2}{(2)(9.81)}$

or $y = 3.33$ m

b) To rise 20 m on earth

$y = 20 = \frac{v_0^2}{2g}\sin^2\theta = \frac{v_0^2}{(2)(9.81)}$

or

$v_0 = 19.81$ m/s

14.59

Given: A baseball ($m = 0.145$ kg) is thrown vertically up with speed $v_0 = 16$ m/s

Find:
a) Height to which ball rises, neglecting air resistance (drag).
b) Due to drag, ball rises only 12 m. Find drag force D.
c) Time it takes ball to rise 12 m.
d) Time it takes ball to drop 12 m from rest.

Solution:

a) Without drag, ball rises

$$y = \frac{v_0^2}{2g} = \frac{16^2}{(2)(9.81)} = \underline{\underline{13.05 \text{ m}}}$$

b) By FBD of ball as it goes up (Fig. a),

$$\Sigma F_y = -mg - D = m\ddot{y}$$

$$\therefore \dot{y} = v = -\frac{(mg+D)t}{m} + v_0$$

$$y = -\frac{1}{2}\frac{(mg+D)t^2}{m} + v_0 t + \cancel{y_0}^{\,0}$$

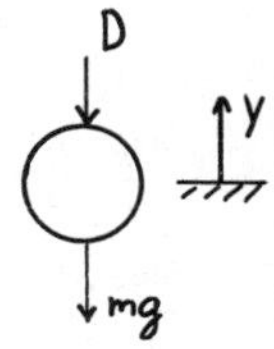

Figure a

$y = y_{max} = 12$ m when $\dot{y} = 0$

or when $t = \frac{m v_0}{mg + D}$ (a)

$$\therefore 12 = \frac{1}{2}\frac{m v_0^2}{mg+D} + \frac{m v_0^2}{mg+D}$$

Solving for D, we obtain

$$D = m\left(\frac{v_0^2}{24} - g\right) = [0.145]\left[\frac{256}{24} - 9.81\right]$$

$$\therefore \underline{D = 0.124 \text{ N}} \quad \text{(b)}$$

c) With Eqs. (a) and (b), the time to rise 12 m is

$$t = \frac{m v_0}{mg + D} = \frac{(0.145)(16)}{(0.145)(9.81) + 0.124}$$

$$\therefore \underline{\underline{t_{up} = 1.50 \text{ s}}}$$

d) By the FBD of the ball as it falls from a height of 12 m (Fig. b),

$$\Sigma F_y = mg - D = ma = m\ddot{y}$$

$$\therefore \dot{y} = \frac{(mg - D)t}{m} + \cancel{\dot{y}_0}^{\,0}$$

$$y = \frac{1}{2m}(mg - D)t^2 + \cancel{y_0}^{\,0} \quad \text{(c)}$$

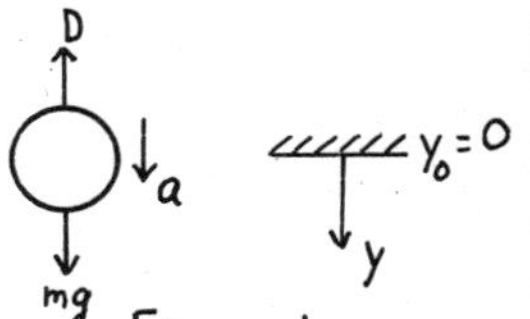

Figure b

For $y = 12$, Eq. (c) yields

$$t = \sqrt{\frac{24m}{mg - D}} = \sqrt{\frac{(24)(0.145)}{(0.145)(9.81) - 0.124}}$$

or

$$\underline{\underline{t_{down} = 1.64 \text{ s}}}$$

14.60

Given: A missle fired from the surface of the earth at an angle θ_0, is subjected to drag $D = kv^2$

Find: Derive the horizontal and vertical differential equations of motion of the missile

Solution:

Let x and y be the horizontal and vertical axes (Fig. a)

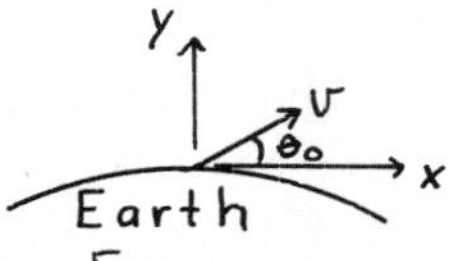

Figure a

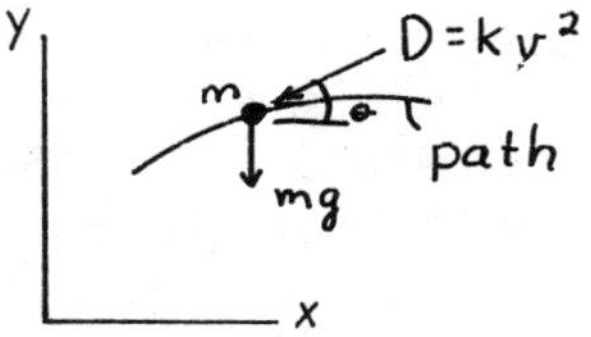

Figure b

By the FBD of the missle (Fig. b),

$$\Sigma F_x = -k v_x^2 = m\frac{d^2x}{dt^2} \quad \text{(a)}$$

$$\Sigma F_y = -mg - k v_y^2 = m\frac{d^2y}{dt^2} \quad \text{(a)}$$

where

$$v_x = v\cos\theta, \quad v_y = v\sin\theta$$

By Eqs. (a),

$$\underline{\frac{d^2x}{dt^2} + \frac{k}{m}\left(\frac{dx}{dt}\right)^2 = 0}$$

$$\underline{\frac{d^2y}{dt^2} + \frac{k}{m}\left(\frac{dy}{dt}\right)^2 + g}$$

14.61

Given: A bullet, fired horizontally, is subjected to drag $D = kv^2$. The muzzle speed is v_0. Trajectory of bullet is flat, so that $D = kv_x^2$, where v_x is horizontal velocity of bullet.

Find: Speed v of bullet as a function of v_0, k, t, and m (mass of bullet).

Solution:

By the FBD of the bullet (Fig. a)

$$\Sigma F_x = -k v_x^2 = m\frac{dv_x}{dt}$$

$$\Sigma F_y = -mg = m\frac{dv_y}{dt} \quad \text{(a)}$$

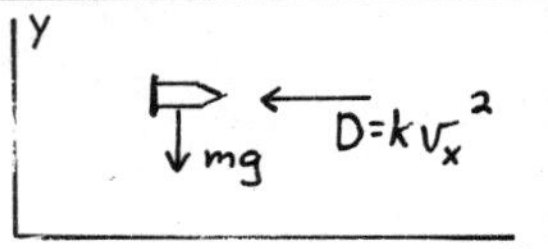

Figure a

By Eqs. (a),

$$\int_{v_0}^{v_x}\frac{dv_x}{v_x^2} = -\int_0^t \frac{k}{m}dt \quad \text{(b)}$$

$$\int_0^{v_y} dv_y = -\int_0^t g\,dt \quad \text{(c)}$$

(continued)

14.61 Cont.

By Eq. (b), after integrating and simplifying,

$$v_x = \frac{v_0}{1+\frac{k}{m}v_0 t} \qquad (d)$$

and by Eq. (c),

$$v_y = -gt \qquad (e)$$

$$\therefore v = \sqrt{v_x^2+v_y^2} = \sqrt{\left(\frac{m v_0}{m+k v_0 t}\right)^2 + (gt)^2}$$

14.62

Given: A missle of mass $m = 14.59$ kg is fired horizontally with a speed $v_0 = 600$ m/s. Horizontal air resistance is $D = 450/(1+t)$ [N], t is in seconds. Neglect vertical air resistance. The missle strikes a target 3 s after it is fired.

Find: a) Elevation of the target and its horizontal distance from point at which the missile was fired.

b) Speed v of missile as it hits target.

Solution:

a) Since vertical air resistance is neglected, only gravity acts in the vertical direction (fig. a).

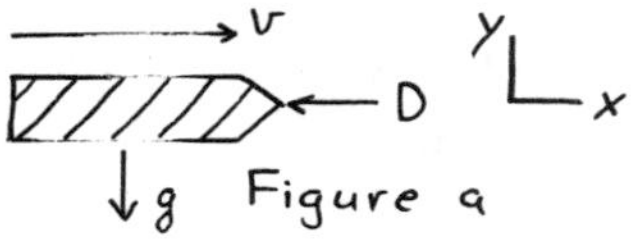

Figure a

∴ Therefore, in the 3 s, the missile drops the distance

$$y = -\tfrac{1}{2}gt^2 = -\tfrac{1}{2}(9.81)(3)^2$$

or $y = -44.14$ m

∴ The elevation of the target below the point at which the missile was fired is

Elevation $= -44.14$ m

By Fig. a,

$$\Sigma F_x = -D = m a_x = m\frac{dv_x}{dt}$$

$$\therefore dv_x = -\frac{(450)}{(1+t)}\left(\frac{1}{14.59}\right)dt$$

or

$$\int_{600}^{v_x} dv_x = -30.84\int_0^t \frac{dt}{1+t}$$

Integration yields

$$v_x - 600 = -30.84\ln(1+t)$$

or

$$v_x = \frac{dx}{dt} = -30.84\ln(1+t) + 600 \ (\text{m/s}) \qquad (a)$$

Integration yields

$$\int_0^{x_t} dx = -30.84\int_0^3 \ln(1+t)\,dt + 600\int_0^3 dt$$

$$\therefore x_t = -(30.84)(2.545) + 1800$$

or

$x_t = 1721.5$ m is the horizontal distance of the target

b) The vertical velocity of the missile when it hits the target is

$$v_y = \frac{dy}{dt} = -gt = -(9.81)(3) = -29.43 \text{ m/s}^2 \qquad (b)$$

By Eq. (a), with t = 3 s, the horizontal velocity of the missile is

$$v_x = 557.2 \text{ m/s}$$

$$\therefore v = \sqrt{v_x^2+v_y^2} = 558.0 \text{ m/s}$$

14.63

Given: Airplane flies at speed V and elevation h along a horizontal path. A projectile is fired at the plane when it is overhead. Neglect air resistance.

Show:

a) the minimum initial speed v of the projectile to hit the plane is

$$v = \sqrt{V^2 + 2gh}$$

b) the corresponding angle of the gun barrel is

$$\theta = \tan^{-1}\sqrt{2gh}/V$$

Solution:

a) The horizontal velocity of the missile must be $v_x = V$

By Eq. (14.13), the vertical velocity of the projectile to reach the height h is the same as the velocity attained by the missile if dropped from rest at a height h (Fig. a)

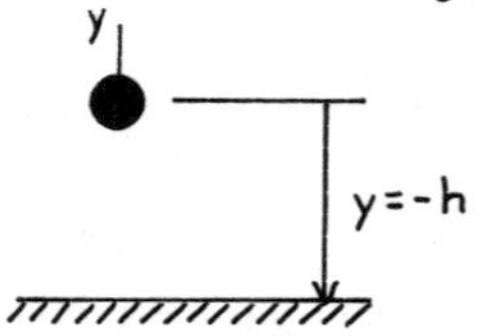

Figure a

Therefore, by Eq. (14.13)

$$v_y^2 = v_0^2 - 2gy = 0 - 2g(-h)$$

or $v_y^2 = 2gh$

Hence,

$$v = \sqrt{v_x^2+v_y^2} \quad \text{or} \quad v = \sqrt{V^2+2gh}$$

b) By Fig. (b),

$$\tan\theta = \sqrt{2gh}/V$$

or

$$\theta = \tan^{-1}\frac{\sqrt{2gh}}{V}$$

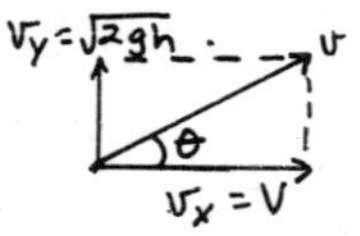

Figure b

14.64

Given: A charged particle of mass m is projected along the positive y axis (Fig. a), with velocity v_0. A uniform electric field exerts force F on the particle in the x direction

a) Derive equation of the path $y = f(x)$

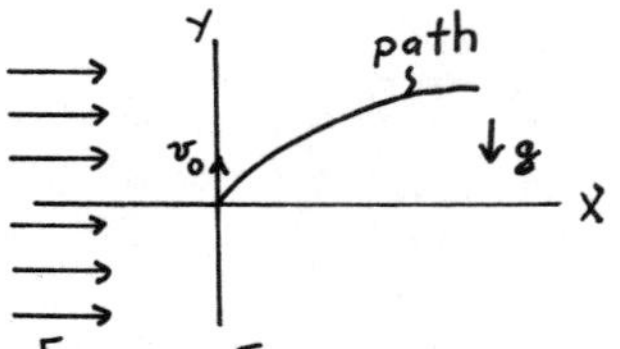

Figure a

b) Determine intercepts of the path with the x axis

c) Determine value of x at $y = y_{max}$

Solution:

a) By Fig. (b)

$$\Sigma F_x = F = m\ddot{x}$$

$$\Sigma F_y = -mg = m\ddot{y}$$

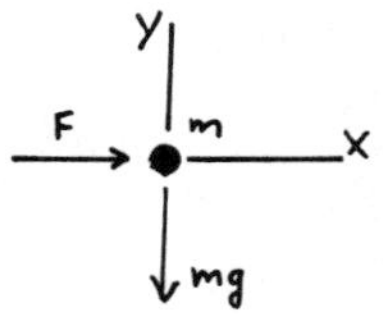

Figure b

$\therefore m\dot{x} = Ft, \quad \dot{y} = -gt + v_0$

$mx = \frac{1}{2}Ft^2, \quad y = v_0 t - \frac{1}{2}gt^2$

$$\therefore t^2 = \frac{2mx}{F} \quad \text{so} \quad \underline{\underline{y = v_0\sqrt{\frac{2mx}{F}} - \frac{gmx}{F}}} \quad (a)$$

b) For $y = 0$, Eq. (a) yields, for the intercepts,

$$v_0\sqrt{\frac{2mx}{F}} = \frac{gmx}{F}$$

$$\therefore \underline{\underline{x = 0, \quad x = \frac{2Fv_0^2}{mg^2}}}$$

c) For $y = y_{max}$, by Eq. (a),

$$\frac{dy}{dx} = v_0\sqrt{\frac{m}{2Fx}} - \frac{gm}{F} = 0$$

$\therefore$ at $y = y_{max}$,

$$\underline{\underline{x = \frac{Fv_0^2}{2mg^2}}}$$

14.65

Given: A 20 lb block at rest on a horizontal surface is subjected to a horizontal force P = 10 lb. $\mu_s = 0.25$; $\mu_k = 0.20$.

Find:

a) if the block moves.

b) if it moves, its speed for instant it travels 10 ft.

Solution:

a) By FBD of block (Fig. a)

$$\Sigma F_x = P - F_f = ?$$

$$\Sigma F_y = N - W = 0; \quad N = W = 20 \text{ lb}$$

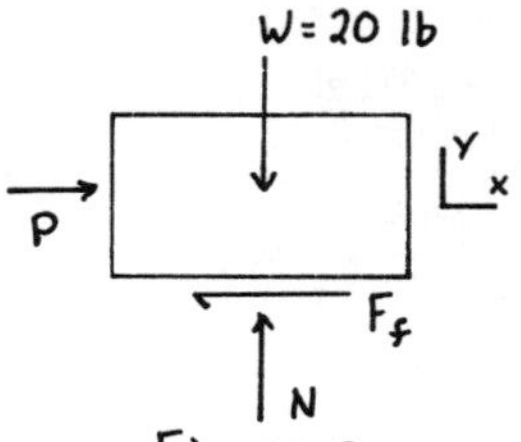

Figure a

If the block moves $P > (F_f)_{max} = \mu_s N$

Since $\mu_s N = (0.25)(20) = 5$ lb, $P = 10 \text{ lb} > 5 \text{ lb}$

$\therefore$ Block moves

b) Once the block moves,

$$F = \mu_k N = (0.20)(20) = 4.0 \text{ lb}$$

Hence, by Fig. a,

$$\Sigma F_x = 10 - 4 = \frac{20}{32.2} a = \frac{20}{32.2} v \frac{dv}{dx}$$

or $\int_0^{10} 9.66\, dx = \int_0^{v_{10}} v\, dv$

$$\therefore 96.6 = \frac{1}{2} v_{10}^2$$

or $\underline{\underline{v_{10} = 13.90 \text{ ft/s}}}$

14.66

Given: Given Fig. a, with $\mu_s = 0.25$, $\mu_k = 0.20$. The system of blocks is released from rest. The mass and friction of the pulley are negligible.

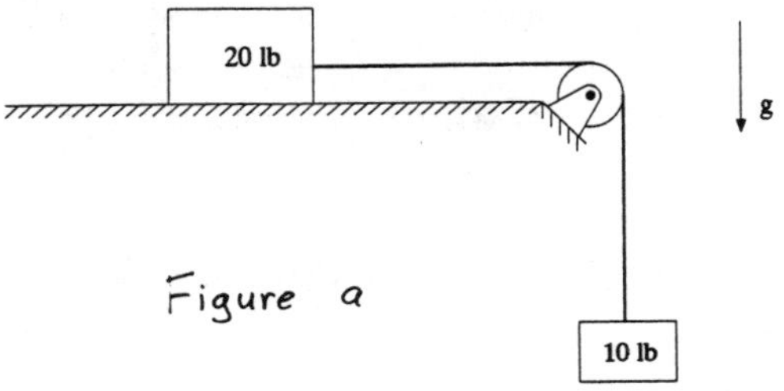

Figure a

Find: a) If the system moves.

b) Speed of blocks.

c) Reason speed of blocks is less than blocks of Problem 14.65.

Solution:

a) By FBD of 20 lb block (Fig. b), for the block to move

$$\Sigma F_x = T - (F_f)_{max} > 0$$

$$\therefore T_1 > (F_f)_{max} = \mu_s N = (0.25)(20)$$

or $T_1 > 5$ lb

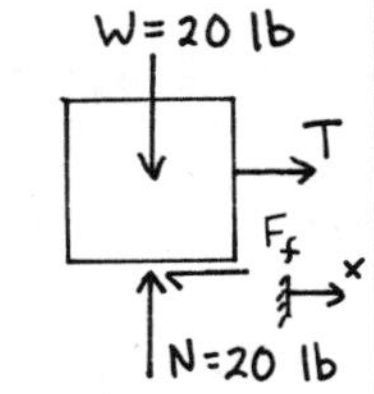

Figure b

By FBD of 10 lb block (Fig. c), the value of T_2 for equilibrium of the block is

$$\Sigma F_y = W - T_2 = 0; \quad T_2 = W = 10 \text{ lb}$$

so, initially, $T_1 = T_2 = 10 \text{ lb} > 5 \text{ lb}$

$\therefore$ system moves

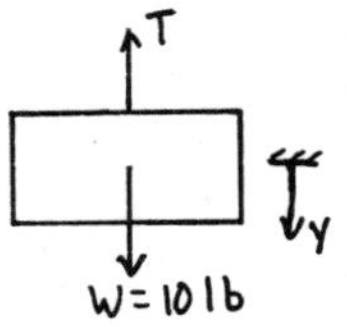

Figure c

b) Now by FBD of Fig. (b), since motion occurs,

$$\Sigma F_x = T - (0.20)(20) = \frac{20}{32.2} a \quad (a)$$

and by Fig. c

$$\Sigma F_y = 10 - T = \frac{10}{32.2} a \quad (b)$$

(Continued)

14.66 Cont.

Adding Eqs. (a) and (b) and solving for a, we obtain

$$a = v\frac{dv}{dx} = 6.44 \text{ ft/s}^2$$

or

$$6.44\int_0^{10} dx = \int_0^{v_{10}} v\, dv$$

$$\therefore 64.4 = \frac{1}{2}v_{10}^2$$

$$v_{10} = 11.35 \text{ ft/s}$$

c) In Problem 14.65, $v_{10} = 13.90$ ft/s > 11.35 ft/s. In Problem 14.65, the force P = 10 lb accelerated a mass of 20/32.2 = 0.621 slugs. In Problem 14.66, the force W = 10 lb (of hanging block) accelerates a mass of 30/32.2 = 0.932 slugs. Therefore, the acceleration is less and so speed gained in 10 ft is less.

14.67

Given: A weight of 54 N is placed on a scale in an elevator. The maximum scale reading is 60 N when the elevator accelerates up. The minimum scale reading is 40 N when the elevator accelerates down.

Find:

a) the maximum upward acceleration of the elevator.

b) the maximum downward acceleration of the elevator.

Solution:

a) By the F.B.D. of the weight as the elevator accelerates up (Fig. a)

$$\Sigma F_y = 60 - 54 = \frac{54}{9.81}a_{up}$$

$$\therefore a_{up} = 1.09 \text{ m/s}$$

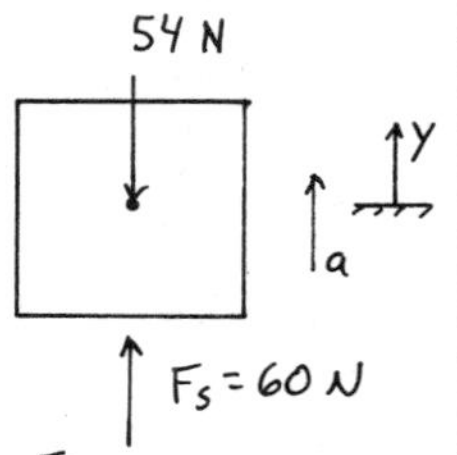

Figure a

b) By the F.B.D. of the weight as the elevator accelerates down (Fig. b),

$$\Sigma F_y = 54 - 40 = \frac{54}{9.81}a_{down}$$

$$\therefore a_{down} = 2.54 \text{ m/s}$$

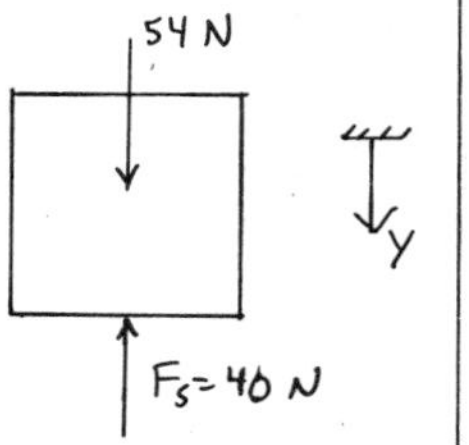

14.68

Given: Missile of mass m = 163 kg is fired horizontally from a plane that travels at 360 m/s.

Find: The average horizontal force required to increase the horizontal speed of the missile from 360 m/s to 600 m/s in 2 s. Neglect drag and mass loss due to fuel burn.

Solution:

By the FBD of the missile (Fig. a)

$$\Sigma F_x = F = m a_x = m\frac{dv_x}{dt}$$

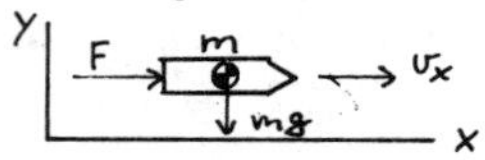

or

$$\int_{v_0}^{v_x} dv_x = \int_0^t \frac{F}{m}\, dt$$

$$\therefore m(v_x - v_0) = Ft$$

or

$$F = \frac{m(v_x - v_0)}{t} = \frac{163(600-300)}{2}$$

$$F_{avg} = 24.45 \text{ kN}$$

14.69

Given: Batter hits a ball (mass m) at height d. Ball's initial velocity is v_0 at angle θ_0 (Fig. a). The ball is subjected to drag $D = kv^2$

Show that the differential equations of motion of the ball are

$\ddot{x} + \lambda v\dot{x} = 0$, $\ddot{y} + \lambda v\dot{y} + g = 0$; $\lambda = k/m$

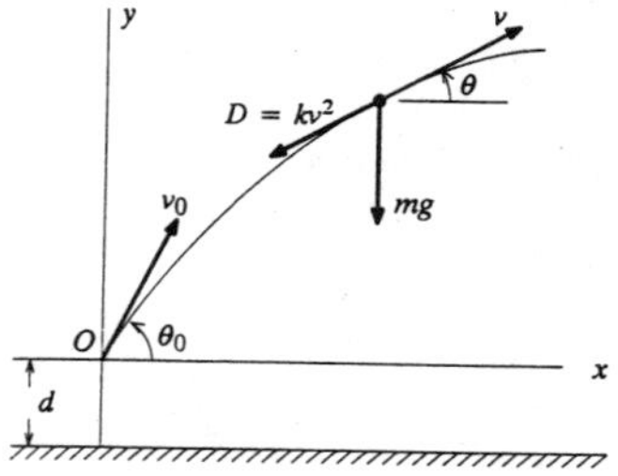

Figure a

Solution:

By FBD of the ball (Fig. b),

$$\Sigma F_x = -kv^2\cos\theta = ma_x = m\ddot{x} \quad (a)$$

$$\Sigma F_y = -kv^2\sin\theta - mg = ma_y = m\ddot{y} \quad (b)$$

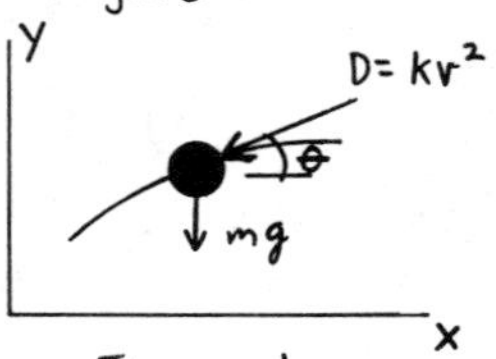

Figure b

But,

$$v_x = \dot{x} = v\cos\theta \quad (c)$$

$$v_y = \dot{y} = v\sin\theta \quad (d)$$

By Eqs. (a) and (c),

$$\ddot{x} + \frac{k}{m}v(v\cos\theta) = 0$$

or

$$\ddot{x} + \lambda v\dot{x} = 0$$

Similarly, by Eqs. (b) and (d),

$$\ddot{y} + \frac{k}{m}v(v\sin\theta) + g = 0$$

or

$$\ddot{y} + \lambda v\dot{y} + g = 0$$

14.70

Given: A home-run ball clears the grand stand 420 ft from home plate at height of 60 ft. When hit the ball was 4 ft above home plate, and left the bat at an angle of 40°.

Find: The speed of the ball as it left the bat.

Solution:

Select (x, y) axes as shown in Fig. a.

By Eq. 14.11,

$$y = -\frac{gx^2}{2v_0^2\cos^2\theta_0} + x\tan\theta_0$$

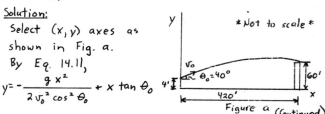

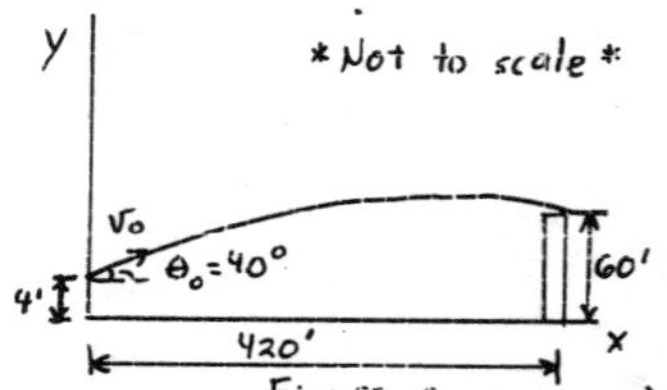

Figure a (Continued)

14.70 Cont.

At $x = 420$ ft, $y = 60 - 4 = 56$ ft

Also $\theta_0 = 40°$. Therefore,

$$56 = -\frac{(32.2)(420)^2}{2(v_0^2)(0.7660)^2} + (420)(0.8391)$$

Solving for v_0, we obtain

$$\underline{v_0 = 127.8 \text{ ft/s}}$$

14.71

Given: A projectile is launched on level ground with initial speed v_0.

a) Show that the projectile can strike a target at $x = R$, $y = 0$ if it is fired at an angle θ_0 or angle $90° - \theta_0$.

b) Verify part (a) for $v_0 = 50$ m/s and $\theta_0 = 15°$; $90° - \theta_0$.

Solution:

a) By the first of Eqs. (14.12)

$$2x_1 = R = \frac{v_0^2 \sin 2\theta_0}{g} \qquad (a)$$

Now note that

$$\sin 2(90° - \theta_0) = (\sin 180°)(\cos 2\theta_0) - (\sin 2\theta_0)(\cos 180°)$$

or $\sin 2(90° - \theta_0) = \sin 2\theta_0$. Hence, the range R is achieved for the angles θ_0 and $90° - \theta_0$.

(b) For $\theta_0 = 15°$ (low aim), Eq. (a) yields

$$R = \frac{(50)^2 \sin 30°}{9.81} = 127.4 \text{ m}$$

For $90° - \theta_0 = 75°$ (high aim),

$$R = \frac{(50)^2 \sin 150°}{9.81} = 127.4 \text{ m}$$

14.72

Given: African warrior throws his spear at a lion with an initial speed v_0 at an angle θ_0 (Fig. a)

a) show that

$$s = \frac{2v_0^2[\sin(\theta_0 + \alpha)]\cos\theta_0}{g\cos^2\alpha}$$

b) For $\alpha = 20°$, $v_0 = 14$ m/s, $s = 30$ m, find angle θ_0 for a hit.

c) If in part b, two values of θ_0 exist, which value of θ_0 would you recommend to the warrior?

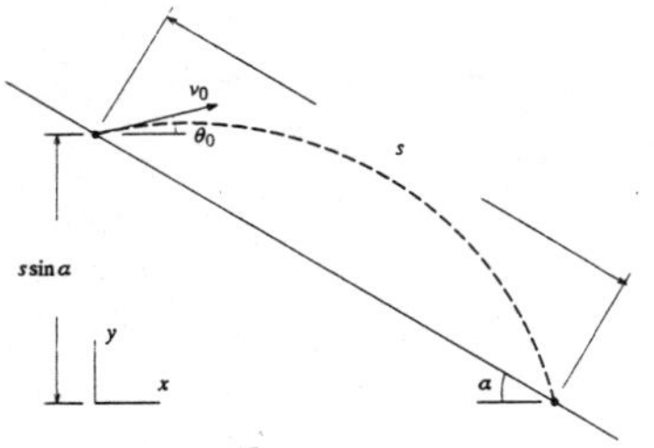

Figure a

Solution:

a) By the FBD of the spear (treated as a particle), Fig. b:

$$\Sigma F_x = 0 = ma_x = m\frac{dv_x}{dt} \qquad (a)$$

$$\Sigma F_y = -mg = ma_y = m\frac{dv_y}{dt} \qquad (b)$$

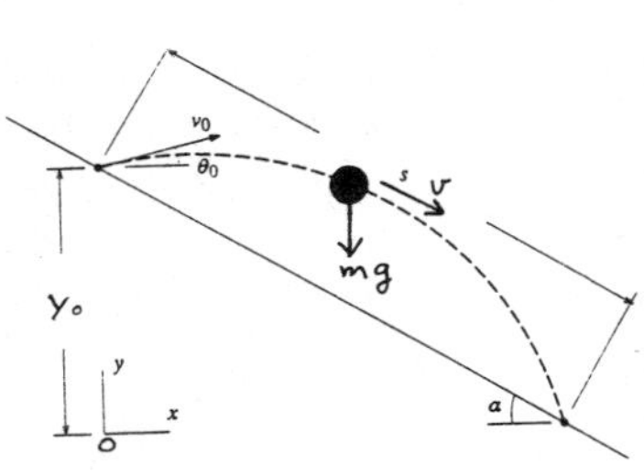

Figure b

By Eqs. (a) and Fig. b,

$v_x = v_0 \cos\theta_0 = \text{constant}$

$\therefore x = (v_0 \cos\theta_0)t$; $x_0 = 0$ (c)

By Eq. (b),

$v_y = -gt + v_0 \sin\theta_0$

$$\therefore y = -\frac{1}{2}gt^2 + v_0 t \sin\theta_0 + s\sin\alpha \qquad (d)$$

when $x = s\cos\alpha$, the spear hits the lion. Then, by Eq. (c)

$$t_s = \frac{s\cos\alpha}{v_0\cos\theta_0} \qquad (e)$$

Substitution of Eq. (e) into Eq. (d), with $y = 0$, yields,

$$\underline{\underline{s = \frac{2v_0^2[\sin(\theta_0 + \alpha)]\cos\theta_0}{g\cos^2\alpha}}} \qquad (f)$$

b) with $\alpha = 20°$, $v_0 = 14$ m/s, $s = 30$ m, Eq. (f) yields

$$0.6629 = \sin(\theta_0 + 20°)\cos\theta_0$$

The solutions of this equation are

$$\underline{\theta_0 = 29.8°}, \ \underline{\text{and } 40.2°} \qquad (g)$$

c) By Eqs. (e) and (g),

For $\theta_0 = 29.8°$, $t_s = 2.32$ s

For $\theta_0 = 40.2°$, $t_s = 2.64$ s

<u>Recommend to warrior to use $\theta_0 = 29.8°$, since spear hits lion sooner.</u>

14.73

Given: That (see Problem 14.72 and Fig. P 14.72)

$$s = \frac{2v_0^2[\sin(\theta_0 + \alpha)]\cos\theta_0}{g\cos^2\alpha} \qquad (a)$$

Find:

a) the value of θ_0 for which $s = s_{max}$ is a maximum, given values of v_0 and α

b) formula for s_{max} in terms of v_0 and α.

Solution:

a) The maximum value of s occurs when $ds/d\theta_0 = 0$. Therefore, by Eq. (a),

$$\frac{ds}{d\theta_0} = \frac{2v_0^2}{g\cos^2\alpha}\left[\cos(\theta_0 + \alpha)\cos\theta_0 - \sin(\theta_0 + \alpha)\sin\theta_0\right] = 0$$

(Continued)

14.73 Cont.

Hence, by trigonometry,

$$\cos(2\theta_0 + \alpha) = 0$$

$$\therefore 2\theta_0 + \alpha = \pi/2$$

or $$\theta_0 = \frac{\pi}{4} - \frac{\alpha}{2} = 45° - \frac{\alpha}{2} \qquad \text{(b)}$$

b) By Eqs. (a) and (b),

$$s_{max} = \frac{2v_0^2}{g\cos^2\alpha}\left[\sin\left(45° + \frac{\alpha}{2}\right)\cos\left(45° - \frac{\alpha}{2}\right)\right]$$

or by trigonometry

$$s_{max} = \frac{v_0^2}{g\cos^2\alpha}(1 + \sin\alpha)$$

14.74

Given: In Problem 14.72, the lion is the distance s uphill from the warrior (Fig. a)

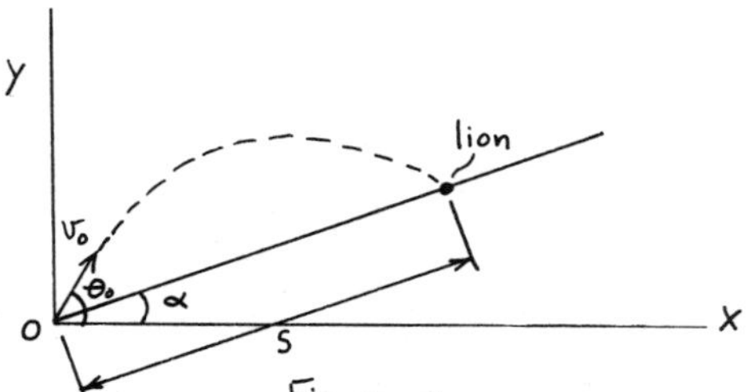

Figure a

Show: a) that

$$s = \frac{2v_0^2[\sin(\theta_0 - \alpha)]\cos\theta_0}{g}$$

b) that for $\theta_0 = 45° + \alpha/2$, the maximum value of s is, for given values of v_0 and α,

$$s_{max} = \frac{v_0^2(1 - \sin\alpha)}{g\cos^2\alpha}$$

Solution:

a) By the FBD of the spear, treated as a particle, Fig. b:

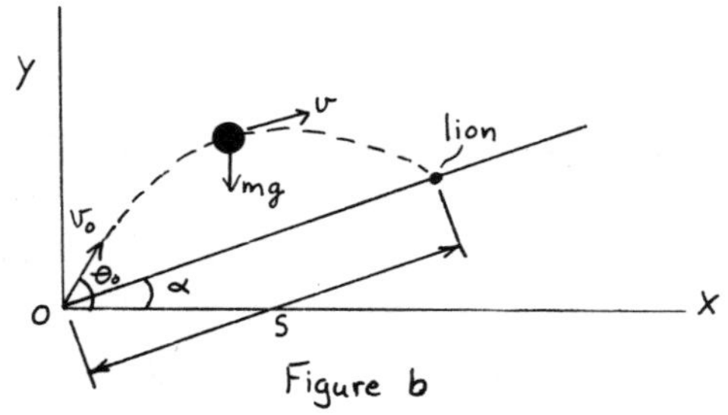

Figure b

$$\Sigma F_x = 0 = m a_x = m\frac{dv_x}{dt} \qquad \text{(a)}$$

$$\Sigma F_y = -mg = m a_y = m\frac{dv_y}{dt} \qquad \text{(b)}$$

By Eq. (a) and Fig. b,

$$v_x = v_0\cos\theta_0 = \text{constant}$$

$$\therefore x = (v_0\cos\theta_0)t\ ;\ x_0 = 0 \qquad \text{(c)}$$

By Eq. (b) and Fig. b,

$$v_y = -gt + v_0\sin\theta_0$$

$$\therefore y = -\frac{1}{2}gt^2 + v_0 t\sin\theta_0 \qquad \text{(d)}$$

when $x = s\cos\alpha$, the spear hits the lion. Then, by Eq. (c),

$$t_s = \frac{s\cos\alpha}{v_0\cos\theta_0} \qquad \text{(e)}$$

Substitution of Eq. (e) into Eq. (d) yields, with $y = s\sin\alpha$,

$$s = \frac{2v_0^2[\sin(\theta_0 - \alpha)]\cos\theta_0}{g\cos^2\alpha} \qquad \text{(f)}$$

b) The maximum value of s occurs when $ds/d\theta_0 = 0$. Therefore, by Eq. (f),

$$\frac{ds}{d\theta_0} = \frac{2v_0^2}{g\cos^2\alpha}[\cos(\theta_0 - \alpha)\cos\theta_0 - \sin(\theta_0 - \alpha)\sin\theta_0] = 0$$

Hence, by trigonometry,

$$\cos(2\theta_0 - \alpha) = 0$$

$$\therefore 2\theta_0 - \alpha = \pi/2$$

or $$\theta_0 = \frac{\pi}{4} + \frac{\alpha}{2} = 45° + \frac{\alpha°}{2} \qquad \text{(g)}$$

By Eqs. (f) and (g),

$$s_{max} = \frac{2v_0^2}{g\cos^2\alpha}\left[\sin\left(45 - \frac{\alpha}{2}\right)\cos\left(45° + \frac{\alpha}{2}\right)\right]$$

or by trigonometry,

$$s_{max} = \frac{v_0^2(1 - \sin\alpha)}{g\cos^2\alpha} \qquad \text{(h)}$$

c) For $\alpha = 20°$, $s = 30$ m, $v_0 = 14$ m/s

Eq. (h) yields

$$s_{max} = 14.89\text{ m} < 30\text{ m}$$

Hence, the spear does not reach the lion. The warrior should wait until the lion is closer than 14.89 m before throwing the spear. (or run!)

14.75

Given: Figure a, in which seat and dummy have combined mass m = 35 slugs. Seat is propelled up guide by thrust that varies as shown in Fig. a.

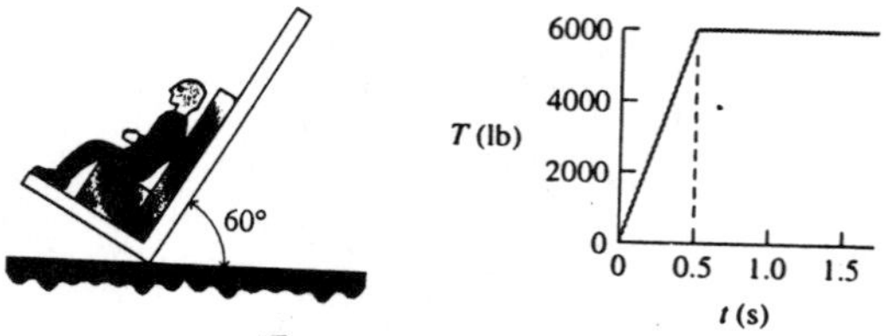

Figure a

(Continued)

14.75 Cont.

Find: a) Maximum acceleration of dummy and seat.

b) Speed of dummy at burnout ($t = 1.5$ s)

Solution:

a) Consider free-body diagram of seat and dummy considered as a particle (Fig. b)

$\Sigma F_x = T - mg\cos 60° = ma$ (a)

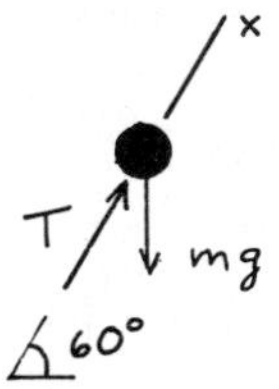

Figure b

check to see when lift off occurs

By Fig. a, for $t \le 0.5$ s

$T = 12{,}000\, t$.

at lift off $a = 0$ and Eq. (a) yields,

$$12\,000\, t = (35)(32.2)(\cos 60°)$$

or

$$t = 0.0813 \text{ s}$$

Therefore, lift off occurs before $t = 0.5$ s. Hence, the seat-dummy system does not move until $t = 0.0813$ s. This fact will be used in part b. However, since lift off occurs before $T = 6{,}000$ lb by Eq. (a), the maximum acceleration occurs when $T = 6000$ lb or

$$a = \frac{6000 - (35)(32.2)(\cos 60°)}{35} = 143.54 \text{ ft/s}^2$$

This value of $a = 4.46\, g$

b) At time $t = 0.0813$ s (lift off), the speed is $v = 0$. Therefore, by Eq. (a)

$$T - mg\cos 60° = m\frac{dv}{dt}, \text{ or}$$

$$dv = \left(\frac{T}{m} - g\cos 60°\right)dt$$

Hence, in the range $0.0813 \text{ s} \le t \le 0.5 \text{s}$,

$$\int_0^{v(0.5)} dv = \int_{0.0813}^{0.50} \left[\frac{12000t}{35} - (32.2)(0.866)\right] dt$$

Integration yields,

$$v(0.5) = 30.0 \text{ ft/s}$$

At $t = 1.5$ s, v is given by

$$v(1.5) = v(0.5) + at\Big|_{0.5}^{1.5}$$

$$v = 30 + (143.54)(1.5 - 0.5)$$

$$v(1.5) = 173.5 \text{ ft/s}$$

14.76

Given: Trajectories of shells in air (Table a)

Table a: Trajectories of shells in air.

45° angle		75° angle	
x (km)	*y* (km)	*x* (km)	*y* (km)
0.00	0.00	0.00	0.00
0.80	0.80	0.80	3.22
1.61	1.61	1.61	4.83
3.82	3.06	3.82	9.33
4.83	4.26	4.83	11.26
6.44	4.75	6.44	8.69
8.04	4.83	8.04	0.80
9.65	4.34	8.21	0.00
11.26	2.90	---	---
12.87	2.65	---	---
13.68	0.00	---	---

a) Plot actual trajectories of shells (table a) with x as abscissa and y as ordinates.

b) Calculate and plot, to the same scale as actual trajectories, shell trajectories without drag

c) Estimate the differences in the maximum horizontal and vertical distances of parts a and b for 45° and for 75°

d) Comment on the effect of air resistance on gun accuracy

Solution:

a)

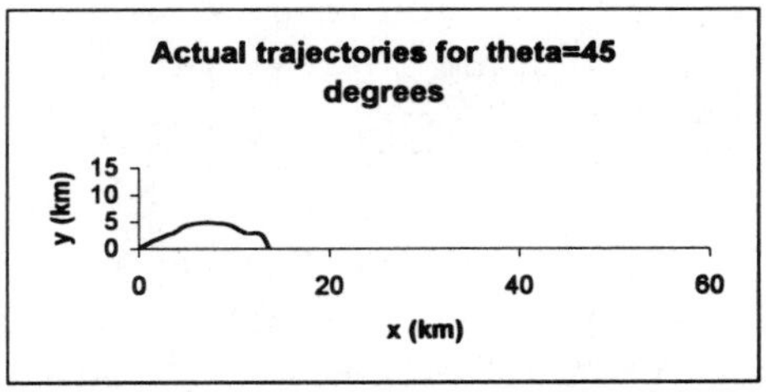

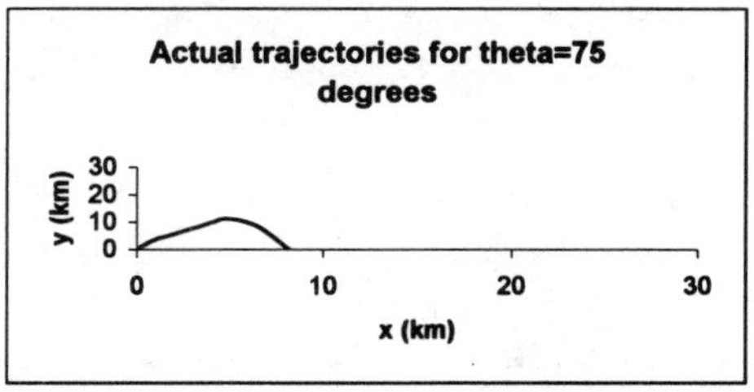

b) see plots for no drag, where

$$y = \frac{-g x^2}{2 v_0^2 \cos^2\theta_0} + x\tan\theta_0$$

(Continued)

14.76 Cont.

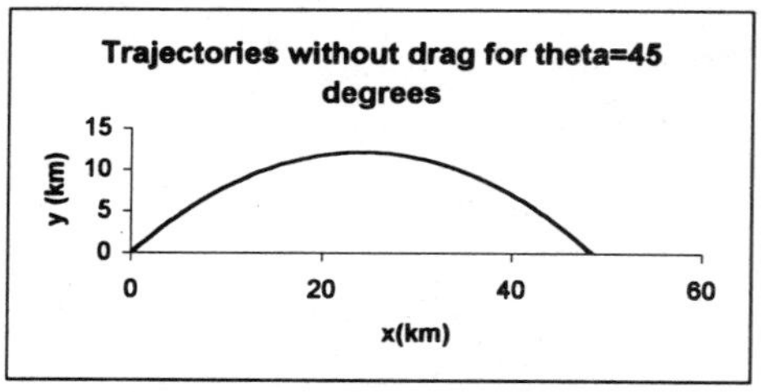

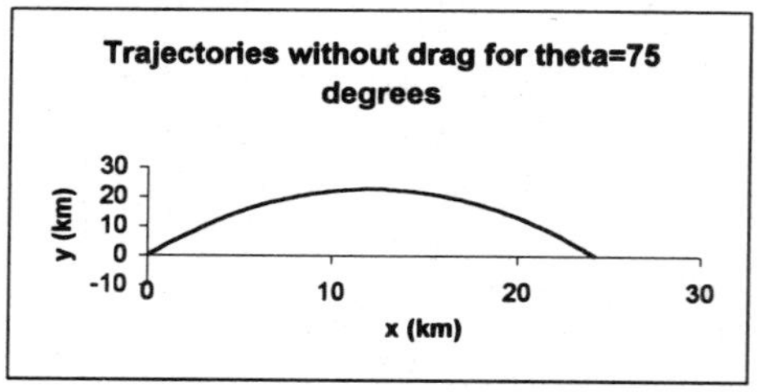

c) From data of plots or estimates directly from plots

For $\theta_0 = 45°$,

$\Delta x \approx 48.5 - 13.7 = 34.8$ km

$\Delta y \approx 12.1 - 4.8 = 7.3$ km

For $\theta_0 = 75°$

$\Delta x \approx 24.2 - 8.2 = 16$ km

$\Delta y \approx 22.5 - 11.3 = 11.2$ km

d) Air resistance has large effect on the trajectories of projectiles, especially at high speeds and large ranges.

14.77

Given: A stock car is tested on a level track, at maximum possible acceleration over a range of speeds. The speed v at varius times t are listed in Table P 14.77. Represent the speed v as function of t by a 4th degree polynomial.

Find:

a) Time required for car to travel first 1200 ft.

b) Maximum incline car can climb at a steady speed of 62.5 mi/h.

c) Maximum acceleration as car descends grade of 1/20 at speed of 100 mi/h.

Solution:

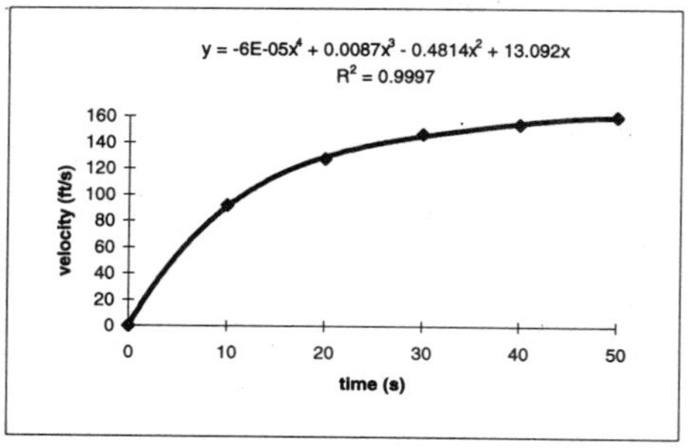

a) By a curve fitting program (see the plot)

$$v = -6\times10^{-5}t^4 + 0.0087t^3 - 0.4814t^2 + 13.092t \text{ (ft/s)} \quad (a)$$

and $$x = -1.2\times10^{-5}t^5 + 0.002175t^4 - 0.1605t^3 + 6.546t^2 \text{ (ft)} \quad (b)$$

at $x = 1200$ ft, and Eq. (b) yields

$t = 16.43$ s

b) At 62.5 mi/h = 91.667 ft/s, Eq. (a) yields $t = 10$ s (see also Table P 14.77). By Eq. (a) (or estimated from the plot),

$dv/dt = a = 5.83$ ft/s²

$\therefore \Sigma F = ma = 5.83\ m$ (lb)

is the force exerted by the car.

Consider the car on an incline θ (Fig. a)

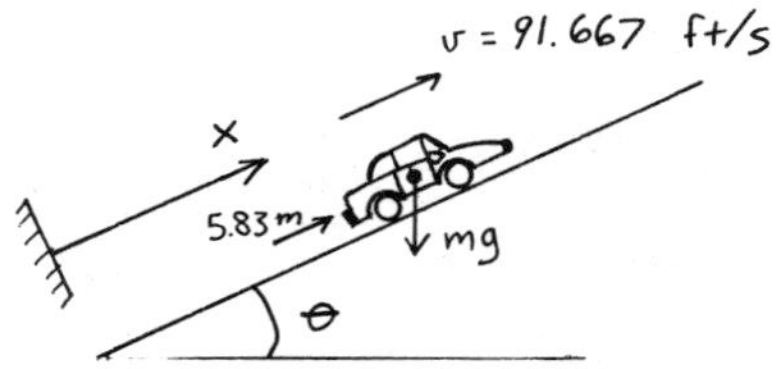

Figure a

By Fig. a, for constant speed

$\Sigma F_x = 5.83\ m - mg \sin\theta = 0$

or $\sin\theta = \dfrac{5.83\ m}{m\ 32.2} = 0.181$

or

$\theta = 10.43°$

c) when $v = 100$ mi/h = 146.7 ft/s; (see Table P14.77). By Eq. (a) or Table P14.77, $t = 30$ s. Also by Eq. (a) (or estimated from the plot),

$dv/dt = a = 1.22$ ft/s²

$\therefore \Sigma F = ma = 1.22\ m$ (lb)

is the force exerted by the car.

Consider the car descending an incline of 1/20 or angle $\tan^{-1}(1/20) = 2.86°$ (Fig. b)

Figure b

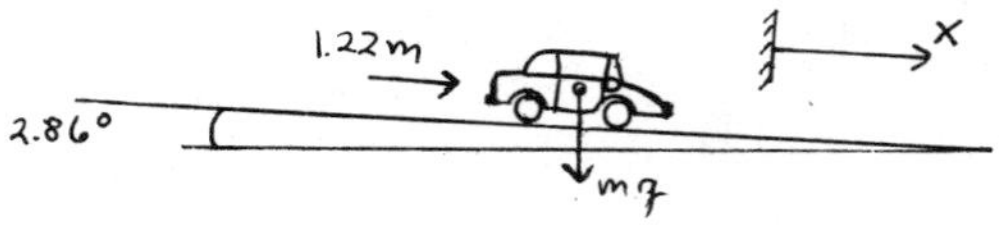

$\Sigma F_x = 1.22\ m + mg \sin 2.86° = ma$

or

$a = 1.22 + (32.2)(0.05) = 2.83$ ft/s²

14.78

Given: Refer to Example 14.17, where formulas for x and y are derived, namely (for $\lambda = k/m \neq 0$),

$$x = \frac{v_0 \cos\theta_0}{\lambda}(1 - e^{-\lambda t}) \quad (a)$$

$$y = \left(\frac{g}{\lambda^2} + \frac{v_0 \sin\theta_0}{\lambda}\right)(1 - e^{-\lambda t}) - \frac{g}{\lambda}t \quad (b)$$

where t denotes time.
For no drag, by Eq. (14.10),

$$x = v_0(\cos\theta_0)t \quad (c)$$

$$y = -\frac{1}{2}gt^2 + v_0 \sin\theta_0 t \quad (d)$$

with the data given in Example 14.17
a) Plot x vs. y for no drag [Eqs. (c) and (d)]
b) Plot x vs. y for $k = 0.001$ lb·s/ft [Eqs. (a) and (b)]
c) Plot x vs y for $k = 0.004$ lb·s/ft
d) Discuss effects of drag on range, max. height, and angle of trajectory at impact with ground.

Solution: see plots for parts a, b, and c

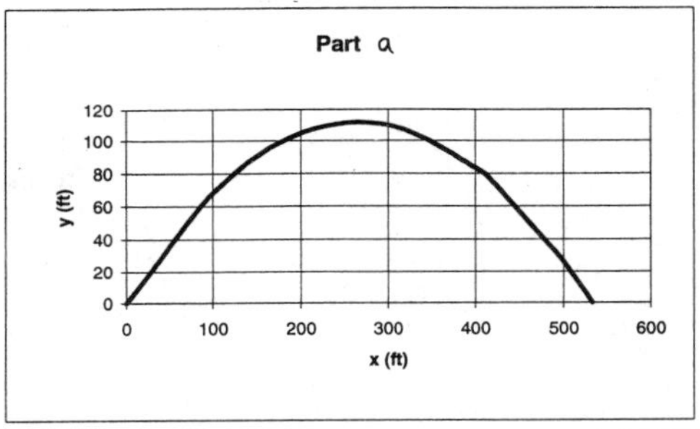

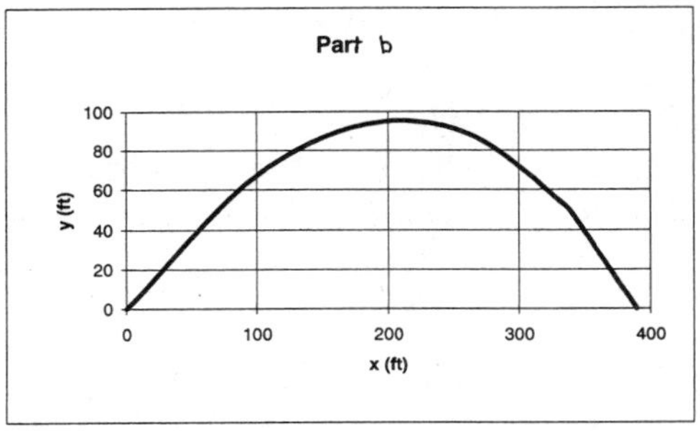

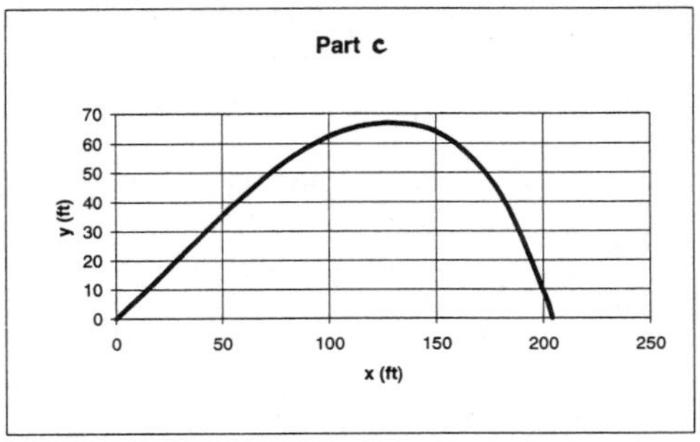

d) As drag (k) increases, max height and range decrease. The angle of impact increases.

14.79

Given: Man stands on a cliff 200 ft above the sea, and he is just able to throw a stone to the edge of the water. On level ground he can throw the stone 250 ft. Neglect air resistance and the height of the man

Find:
a) The angle θ_0 the man must throw the stone for maximum range.
b) The horizontal distance of the man from the water.

Solution:
a) By Eq. (14.11),

$$y = \frac{-gx^2}{2v_0^2\cos^2\theta_0} + x\tan\theta_0 \quad (a)$$

Since his range on level ground is $2x_1$ [see Eq. (14.12)], the maximum distance he can throw on level ground is

$$2x_1 = v_0^2/g = 250 \text{ ft}$$

or $v_0^2 = 250g$ (b)

So, Eqs. (a), and (b) yield (for $y = -200$ ft)

$$-200 = \frac{-x^2}{500\cos^2\theta_0} + x\tan\theta \quad (c)$$

Solving Eq. (c) for x, we get

$$x = 125\sin 2\theta_0 + 10\cos\theta_0\sqrt{1312 - 312.5\cos 2\theta_0} \quad (d)$$

Using a computer solver, determine θ_0 by the conditions $dx/d\theta_0 = 0$, for $x = x_{max}$. This yields $\underline{\theta_0 = 30.81°}$ Then, substitution of $\theta_0 = 30.81°$ into Eq. (d) yields

$$\underline{x_{max} = 403.1 \text{ ft}}$$

Alternatively, we could plot Eq. (d) as a function of θ_0 and determine x_{max} (and θ_0) from the plot.

14.80

Given: The orbits of the planets are concentric circles and that the sun lies at the center.
a) Prove that the square of period T of a planet is proportional to the cube of the distance r from the sun
b) Find the distance (miles) from saturn to the sun given $T = 29.46$ earth years and that the earth is about 93,000,000 miles from the sun.

(Continued)

14.80 Cont.

Solution:

By the FBD of a planet (Fig. a),

$\Sigma F_n = F + I = 0$

$I = -m_p v^2/r$

$F = \frac{G m_p m_s}{r^2}$

$\therefore \frac{G m_s}{r^2} = \frac{v^2}{r}$ (a)

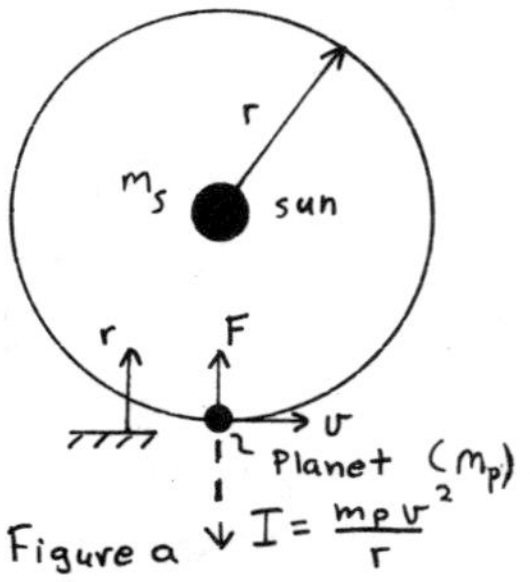

Figure a

The period T of the planet times the speed is the distance traveled in one orbit:

$Tv = 2\pi r$ or $v = \frac{2\pi r}{T}$ (b)

Equations (a) and (b) yield

$T^2 = kr^3 ;\ k = \frac{4\pi^2}{Gm_s}$ (c)

b) Let the earth be the planet. Then, by Eq.(c),

$(1 \text{ earth yr})^2 = k(9.3\times 10^7 \text{ mi})^3$

or $k = (1.243)(10^{-24}) \frac{(\text{earth yr})^2}{(\text{mi})^3}$

Now let Saturn be the planet. Then, by Eq.(c),

$(29.46 \text{ earth yr})^2 = (1.243\times 10^{-24}) r_s^3$

or

$r_s^3 = 6.982\times 10^{26} \ (\text{mi})^3$

or

$r_s = 887{,}000{,}000$ mi

14.81

Given: Artificial satellite circles earth at 1000 mi above the earth's surface (Fig. a)

Find: Period T(h) of the satellite. Take the earth's radius $r_E = 4000$ mi.

Solution:

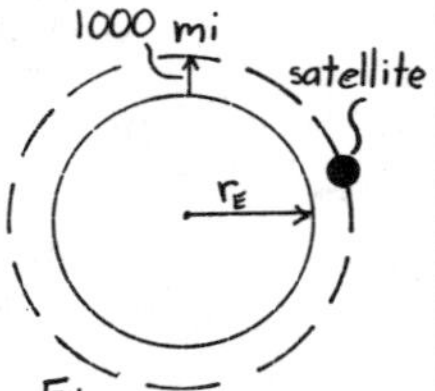

Figure a

Consider the FBD of the satellite (Fig. b) and

$\Sigma \vec{F}_r = \vec{F} + \vec{I} = 0$ (a)

By Fig. b,

$\vec{F} + \vec{I} = 0$

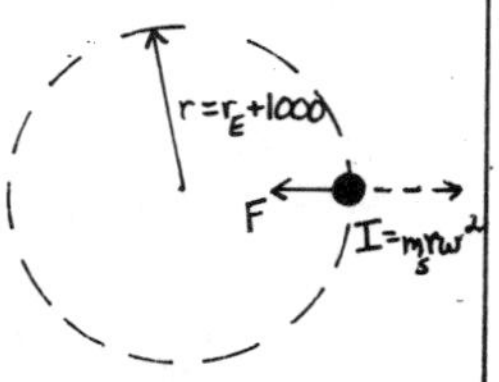

Figure b

where with respect to r

$\vec{F} = -F = \frac{-G m_s m_E}{(r_E + 1000)^2}$

$\vec{I} = +I = +m_s (r_E + 1000)\omega^2$

By Eq. (a),

$\frac{G m_s m_E}{(r_E + 1000)^3} = m_s (r_E + 1000)\omega^2$

or

$\omega^2 = \frac{G m_E}{(r_E + 1000)^3} = \frac{(3.439\times 10^{-8})(4.092\times 10^{23})}{(5000)^3 (5280)^3}$

$\therefore \omega = 8.745\times 10^{-4}$ rad/s

In a period T, the satellite travels a distance of $2\pi r$. Therefore,

$Tv = Tr\omega = 2\pi r$

or $T = \frac{2\pi}{\omega} = \frac{2\pi}{(8.745\times 10^{-4} \frac{\text{rad}}{\text{s}})(3600 \frac{\text{s}}{\text{h}})}$

or

$T = 2.00$ h

14.82

Given: A horizontal force F is applied to the block of mass m (Fig. a) causing it to accelerate from rest. When the acceleration is a, the mass is on the verge of sliding up the plane. $\mu_s = 0.80$ between mass m and inclined face.

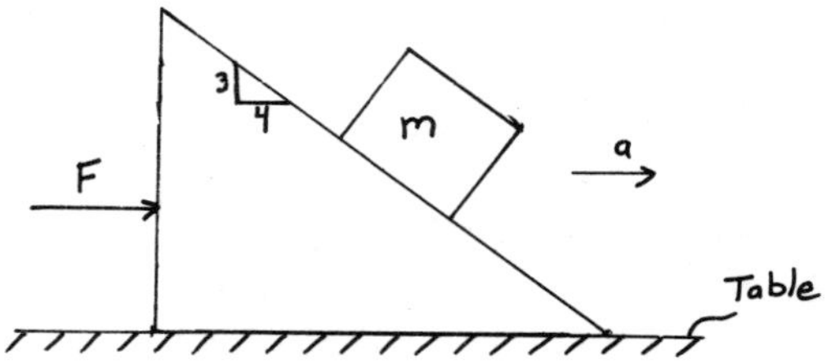

Figure a

Find:

(a) Draw a FBD of the mass, in the inertial frame fixed to the table, for that instant.

b) Draw a FBD of the mass, in the frame fixed to the block, for that instant.

c) Determine the acceleration (m/s²) at which the mass begins to slide.

Solution:

a) The FBD of m in the inertial frame is shown in Fig. b.

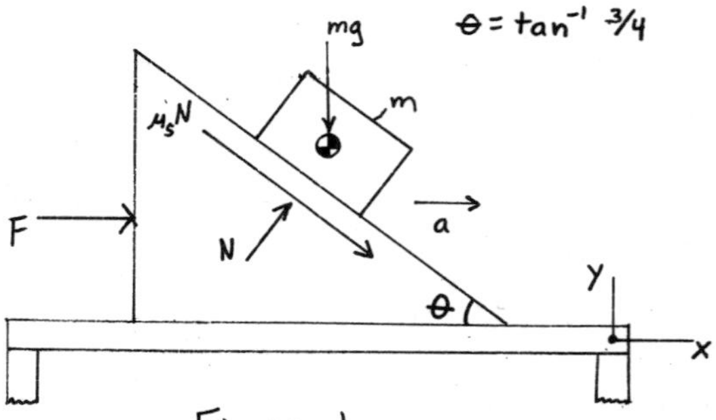

Figure b

(Continued)

14.82 Cont.

b) The FBD of m in the noninertial frame attatched to m is shown in Fig. C.

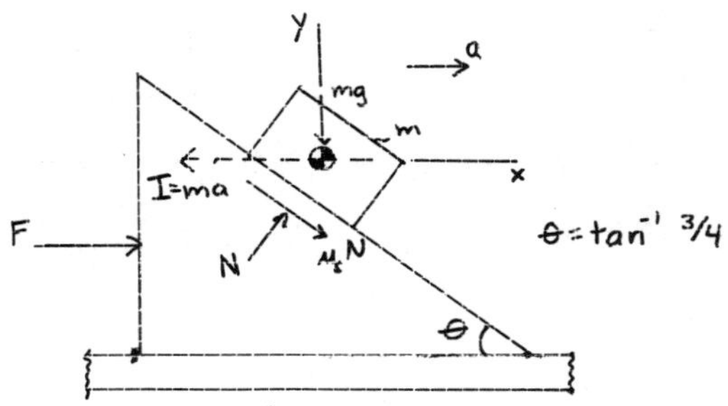

Figure c

c) By the FBD of m (Fig. c),

$$\vec{F} + \vec{I} = 0; \quad \vec{I} = -m\vec{a} \qquad (a)$$

In terms of axes (x, y),

$$\Sigma F_x = \mu_s N \cos\theta + N \sin\theta - ma = 0$$

$$\Sigma F_y = N\cos\theta - \mu_s N \sin\theta - mg = 0$$

$$\therefore \quad N = \frac{mg}{\cos\theta - \mu_s \sin\theta}$$

Hence,

$$a = \frac{\mu_s g \cos\theta + g\sin\theta}{\cos\theta - \mu_s \sin\theta}$$

so,

$$a = \frac{(0.80)(9.81)(0.80) + (9.81)(0.60)}{0.80 - (0.8)(0.60)}$$

$\therefore$ $a = 38.01$ m/s² at impending sliding

For sliding, a must be slightly (at least) greater than 38.01 m/s².

14.83

Given: The force F causes acceleration a of the block and mass m, at which instant mass m is on the verge of sliding down the plane (Fig. a). $\mu_s = 0.80$.

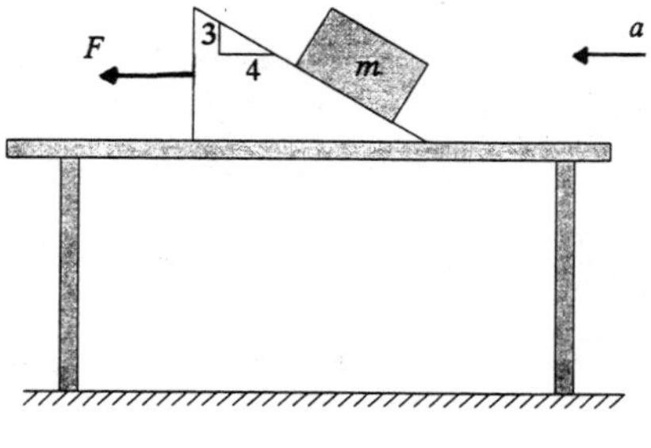

Figure a

a) Draw a FBD of m, in the inertial frame fixed to the table, for that instant.

b) Draw a FBD of m, in the noninertial frame fixed to m, for that instant.

c) Determine the acceleration (m/s²) at which the mass begins to slide.

Solution:

a) The FBD of m is shown in Fig. (b),

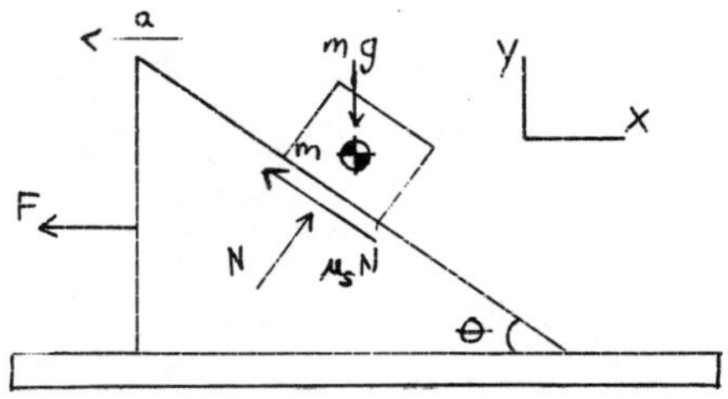

Figure b

b) The FBD of m is shown in Fig. (c),

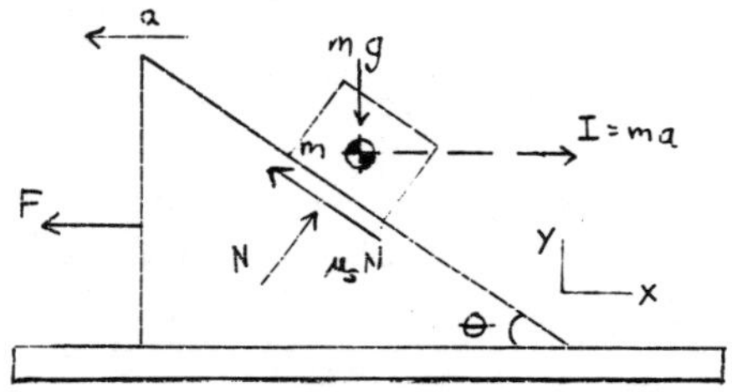

Figure c

c) By the FBD of m (Fig. c),

$$\vec{F} + \vec{I} = 0, \quad \vec{I} = -m\vec{a} \qquad (a)$$

In terms of axes (x, y),

$$\Sigma F_x = -\mu_s N \cos\theta + N\sin\theta + ma = 0$$

$$\Sigma F_y = N\cos\theta + \mu_s N \sin\theta - mg = 0$$

$$\therefore \quad N = \frac{mg}{\cos\theta + \mu_s \sin\theta}$$

Hence,

$$a = \frac{g(\mu_s \cos\theta - \sin\theta)}{\cos\theta + \mu_s \sin\theta}$$

or

$$a = \frac{(9.81)((0.80)(0.80) - 0.60)}{(0.80 + (0.80)(0.60))}$$

$\therefore$ $a = 0.3066$ m/s² at sliding impending

For sliding, a must be at least slightly greater than 0.3066 m/s²

14.84

Given: A highway curve with a 400 ft radius.

Find: Tangent of angle that highway must be banked to prevent a car rounding the curve at 45 mi/h to have no tendency to skid.

Solution:

The FBD of the car on the banked turn is shown in Fig. a.

Since there is no tendency to skid, there is no frictional force.

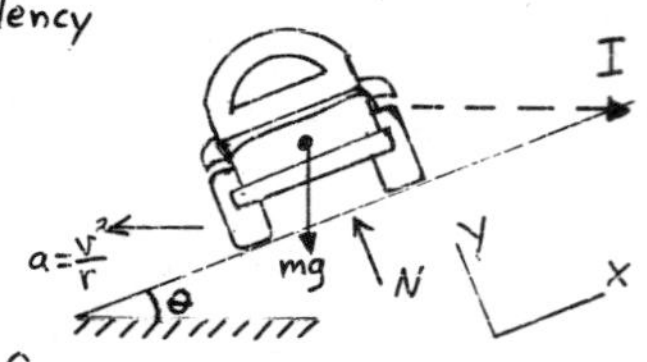

Figure a

By Fig. a,

$$\Sigma F_x = I\cos\theta - mg\sin\theta = 0$$

$$I = \frac{mv^2}{r}\ ; \quad v = 45 \text{ mi/h} = 66 \text{ ft/s}\ ; \quad r = 400 \text{ ft}$$

$$\therefore \tan\theta = \frac{I}{mg} = \frac{v^2}{rg} = \frac{(66)^2}{(400)(32.2)}$$

so,

$$\tan\theta = 0.338$$

$$\therefore \underline{\theta = 18.68°}$$

14.85

Given: Masses m_1 and m_2 hang from a string over a frictionless pulley, of negligible mass that accelerates upward with a magnitude a (Fig. a).

Derive formulas for the tension T and the accelerations a_1 and a_2 of the masses.

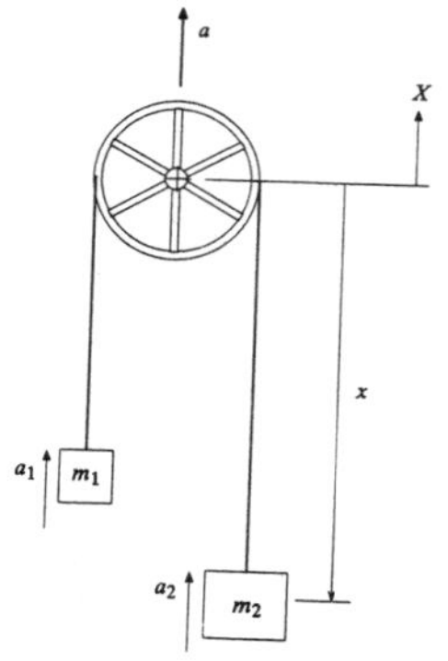

Figure a

Solution:

By Fig. (a), the absolute accelerations of the masses are

$$a_1 = a + \ddot{x} \qquad \text{(a)}$$
$$a_2 = a - \ddot{x}$$

Where $\ddot{x}$ is the acceleration of masses m_1 and m_2 relative to the center of the pulley.

The FBD's of m_1 and m_2, with reference to the noninertial axis x (Fig. a) are given in Figs. (b) and (c).

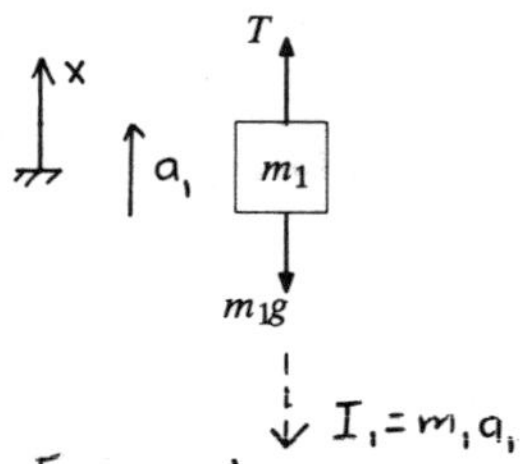

Figure b

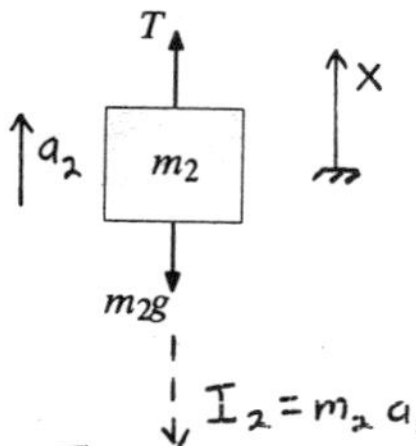

Figure c

Hence, for mass m_1,

$$\Sigma F_x = T - m_1 g - I_1 = 0 \qquad \text{(b)}$$

and for mass m_2,

$$\Sigma F_x = T - m_2 g - I_2 = 0 \qquad \text{(c)}$$

Hence by Eqs. (a), (b), and (c)

$$T = m_1 g + m_1(a + \ddot{x}) = m_2 g + m_2(a - \ddot{x}_1) \qquad \text{(d)}$$

Solving Eq. (d) for $\ddot{x}$ and T, we find

$$\ddot{x} = \frac{(m_2 - m_1)(a+g)}{m_1 + m_2} \qquad \text{(e)}$$

$$T = \frac{2m_1 m_2 (a+g)}{m_1 + m_2} \qquad \text{(f)}$$

Then, by Eqs (a) and (e),

$$a_1 = \frac{2m_2 a}{m_1 + m_2} + \frac{(m_1 + m_2)g}{m_1 + m_2}$$

$$a_2 = \frac{2m_1 a}{m_1 + m_2} + \frac{(m_1 - m_2)g}{m_1 + m_2}$$

14.86

Given: Olympic record for the 16-lb hammer throw is 230 ft. The hammer is whirled in a circle of radius 6.5 ft before being flung.

Find: The force F that the thrower must resist in whirling the hammer.

Solution:

To attain the maximum distance, the hammer must be flung at $\theta_0 = 45°$ (Fig. a)

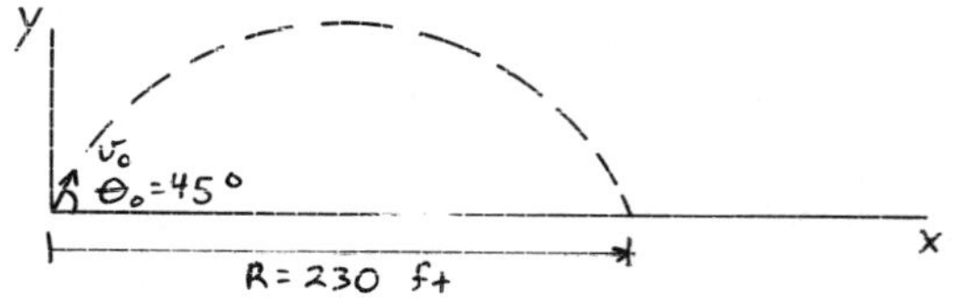

Figure a

By Eq. (14.1),

$$y = -\frac{gx^2}{2v_0^2\cos^2\theta_0} + x\tan\theta_0 \qquad \text{(a)}$$

∴ For $y = 0$, $x = R = 230$ ft, Eq. (a) yields

$$v_0^2 = \frac{(32.2)(230)^2}{(2)(230)(\tan 45°)(\cos^2 45°)}$$

or

$$v_0 = 86.06 \text{ ft/s}$$

Now, by the FBD of the hammer (Fig. b)

$$\Sigma F_r = \vec{F} + \vec{I} = 0 \quad \text{or} \quad F - \frac{mv_0^2}{r} = 0$$

$$\therefore F = \frac{mv_0^2}{r} = \left(\frac{16}{32.2}\right)\left[\frac{(86.06)^2}{6.5}\right] =$$

or $\underline{F = 566 \text{ lb}}$

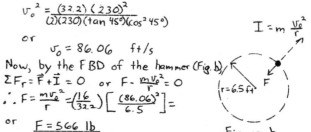

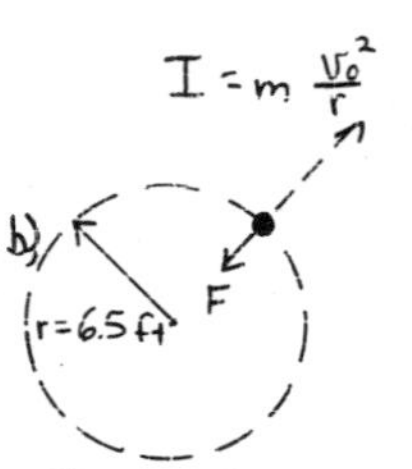

Figure b

14.87

Given: Figure a, with $m = 0.0099\ \text{lb}\cdot\text{s}^2/\text{ft}$ and speed v of ball as it rotates is $v = 55\ \text{mi/h} = 80.667\ \text{ft/s}$.

Find:

a) Height h and tension T in the string.

b) Angular velocity ω of the pole.

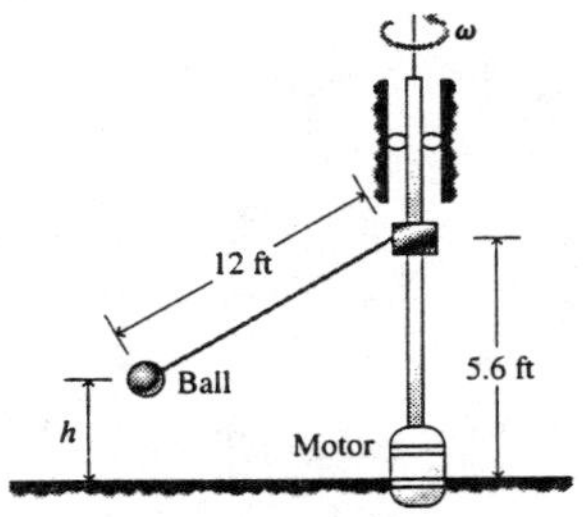

Figure a

Solution:

a) By the FBD of the ball (Fig. b)

$$\Sigma\vec{F} = \vec{F} + \vec{I} \quad (a)$$

By Eq. (a), for the x direction,

$$\Sigma F_x = T\cos\theta - ma = 0 \quad (b)$$

For the y direction,

$$\Sigma F_y = T\sin\theta - mg = 0 \quad (c)$$

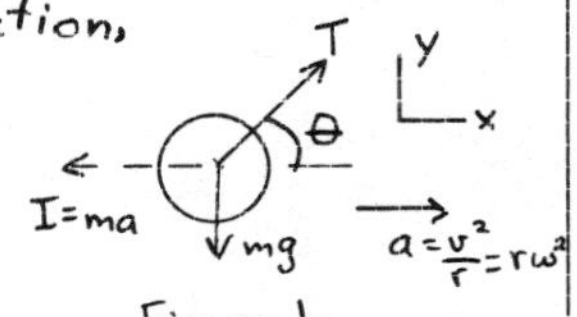

Figure b

By Eq. (b)

$$T = \frac{ma}{\cos\theta} = \frac{mv^2}{r\cos\theta} \quad (d)$$

Now by Fig. a, $r = 12\cos\theta$.

By Eqs. (c) and (d),

$$T = \frac{mv_0^2}{12\cos^2\theta} = \frac{mg}{\sin\theta} \quad (e)$$

$$\therefore \cos^2\theta = 16.840\sin\theta \quad (f)$$

The solution of Eq. (f) is

$$\underline{\underline{\theta = 3.440°}}$$

Hence by Eq. (e),

$$T = \frac{mg}{\sin\theta} = \frac{(0.0099)(32.2)}{0.06} = \underline{\underline{5.310\ \text{lb}}}$$

Also, by Fig. a,

$$\sin\theta = \frac{5.6 - h}{12}$$

or

$$h = 5.6 - 12\sin\theta = 5.6 - (12)(0.06)$$

$$\therefore \underline{\underline{h = 4.88\ \text{ft}}}$$

b)

$$\omega = \frac{v}{r} = \frac{80.667}{12\cos(3.440°)}$$

or

$$\underline{\underline{\omega = 6.73\ \text{rad/s}}}$$

14.88

Given: Angular speed of the pole is adjusted so that $h = 1.6$ ft (see Problem 14.87).

Find:

a) speed (mi/h) of the ball

b) the tension in the rope

Solution:

a) By Fig. (a) of Problem 14.87,

$$\sin\frac{5.6 - 1.6}{12} = 0.3333$$

or

$$\theta = \sin^{-1} 0.3333 = 19.47°$$

By the FBD of the ball (Fig. a)

$$\Sigma F_x = T\cos\theta - \frac{mv^2}{r} = 0$$

$$\Sigma F_y = T\sin\theta - mg = 0$$

$$\therefore \tan\theta = \frac{rg}{v^2} = 0.3535 \quad (a)$$

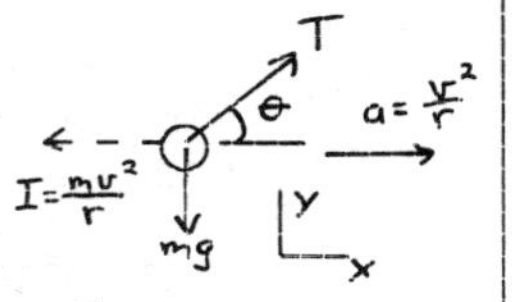

Figure a

and by Fig. a, problem 14.87

$$r = 12\cos\theta = (12)(0.9428)$$

or

$$r = 11.31\ \text{ft} \quad (b)$$

By Eqs. (a) and (b),

$$v^2 = 1030.55\ ;\ \underline{\underline{v = 32.10\ \text{ft/s}}}$$

b) By the equation $\Sigma F_y = T\sin\theta - mg = 0$,

$$T = \frac{mg}{\sin\theta} = \frac{(0.0099)(32.2)}{0.3333}$$

or

$$\underline{\underline{T = 0.956\ \text{lb}}}$$

14.89

Given: Figure a, with $m = 0.0099\ \text{lb}\cdot\text{s}^2/\text{ft}$ and speed of ball $v = 30\ \text{mi/h} = 44\ \text{ft/s}$.

Find:

a) Height h and tension T in string.

b) Angular speed ω of the pole.

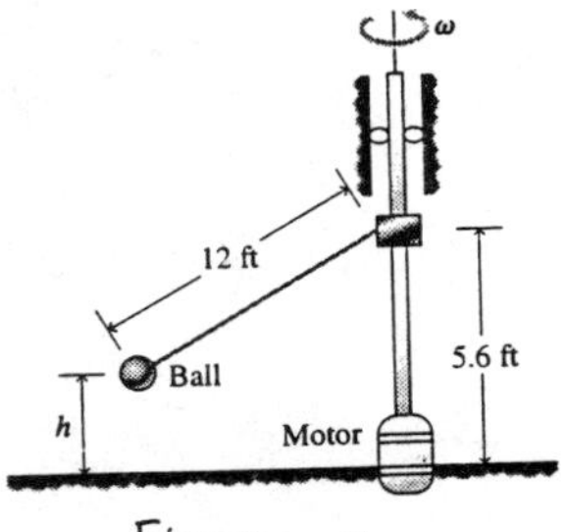

Figure a

Solution:

a) By the FBD of the ball (Fig. b)

$$\Sigma\vec{F} = \vec{F} + \vec{I} = 0 \quad (a)$$

By Eq. (a) for the x-direction,

$$\Sigma F_x = T\cos\theta - ma = 0 \quad (b)$$

For the y-direction,

$$\Sigma F_y = T\sin\theta - mg = 0 \quad (c)$$

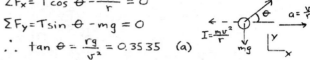

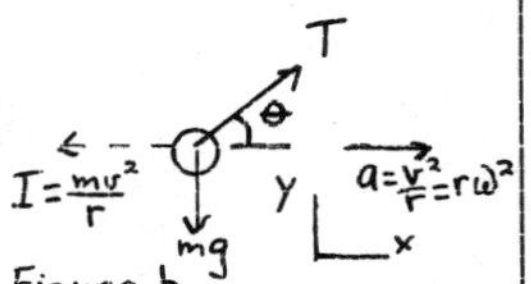

Figure b

(Continued)

14.89 Cont.

By Eq. (b),

$$T=\frac{ma}{\cos\theta}=\frac{mv^2}{r\cos\theta} \qquad (d)$$

Now, by Fig. a, $r=12\cos\theta$ and by Eqs. (c) and (d),

$$T=\frac{mv_0^2}{12\cos^2\theta}=\frac{mg}{\sin\theta} \qquad (e)$$

$$\therefore \cos^2\theta = 5.010\sin\theta \qquad (f)$$

The solution of Eq. f is

$$\underline{\theta = 11.09°}$$

Hence, by Eq. (e)

$$T=\frac{mg}{\sin\theta}=\frac{(0.0099)(32.2)}{0.1924}=1.66 \text{ lb}$$

Also, by Fig. a,

$$\sin\theta=\frac{5.6-h}{12}$$

or

$$h=5.6-12\sin\theta=5.6-12(0.1924)$$

$$\therefore \underline{h=3.29 \text{ ft}}$$

b) $\omega = v/r = 44/[12(\cos 11.09°)]$

or

$$\underline{\omega = 3.74 \text{ rad/s}}$$

14.90

Given: A steel ($\gamma = 0.28$ lb/in³) fly wheel with thin rim of mean radius 30 in will burst if the tension T exceeds 60,000 lb/in².

Find: Angular speed ω (rpm) at which wheel will burst.

Solution:

By the FBD of an element of the rim (Fig. a)

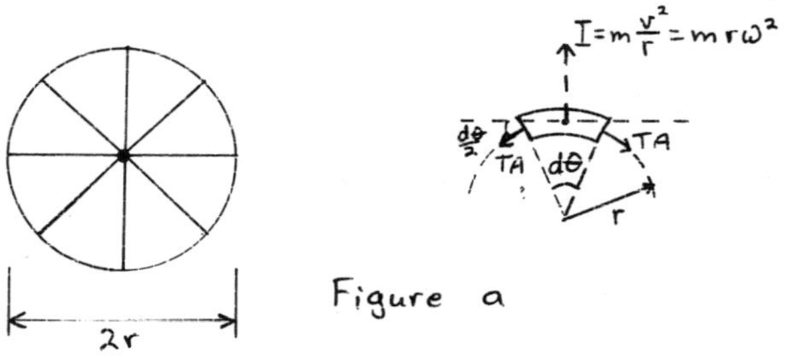

Figure a

$$\vec F + \vec I = 0, \qquad 2TA\sin\left(\frac{d\theta}{2}\right)-mr\omega^2=0 \qquad (a)$$

where

$$m=\frac{\gamma A r\, d\theta}{g}$$

and A = cross-sectional area of the rim.

Then, $\sin\frac{d\theta}{2}=\frac{d\theta}{2}$, Eq. (a) yields (with T=60,000 $\frac{\text{lb}}{\text{in}^2}$)

$$\omega^2=\frac{(60{,}000 \frac{\text{lb}}{\text{in}^2})(386.4 \text{ in/s}^2)}{(30 \text{ in})^2(0.28 \text{ lb/in}^3)}; \quad \omega = 303 \frac{\text{rad}}{\text{s}}$$

or

$$\omega = (303 \text{ rad/s})(1 \text{ rev}/2\pi \text{ rad})(60\text{s}/1 \text{ min})$$

$$\therefore \underline{\omega = 2896 \text{ rpm}}$$

14.91

Given: A light frictionless pulley is attached to the ceiling of an elevator that is accelerating downward with magnitude $a = g/3$ ($m_1 < m_2$); see Fig. a

a) Derive formulas for the absolute accelerations a_1 and a_2 of masses m_1 and m_2 in terms of m_1, m_2, and g.

b) Calculate a_1 and a_2 for $m_2 = 2m_1$; for $m_2 = 3m_1$

c) Derive a formula for the force extered on the pulley by the rod in terms of m_1, m_2, and g.

d) What are the accelerations of m_1 and m_2 relative to the elevator for $m_2 = 3m_1$, in terms of g.

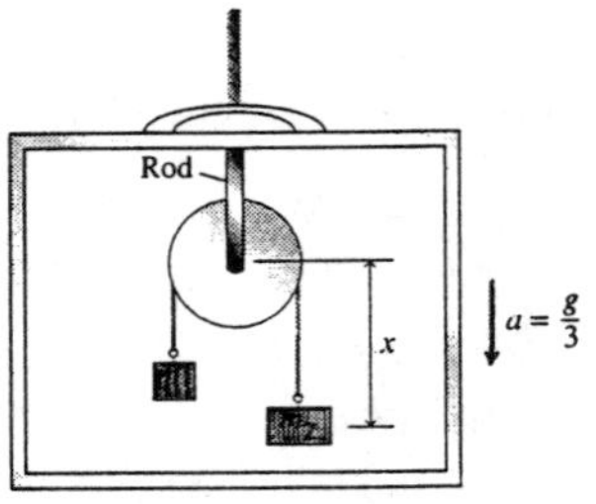

Figure a

Solution:

a) Let x be the coordinate from center of pulley to m_2 (Fig. a). Then, the FBDs of m_1 and m_2 are shown in Figs. (b) and (c)

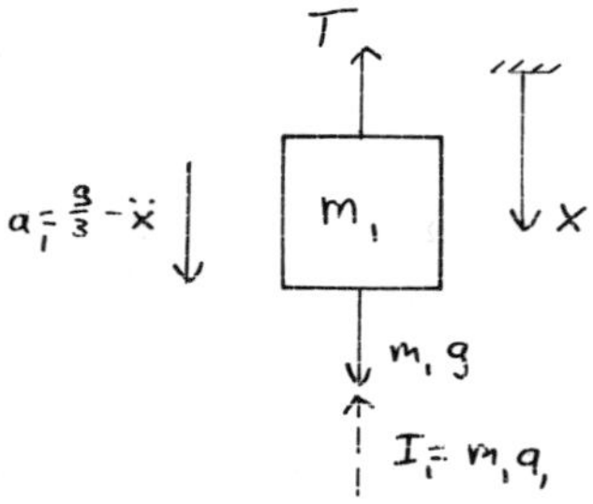

Figure b

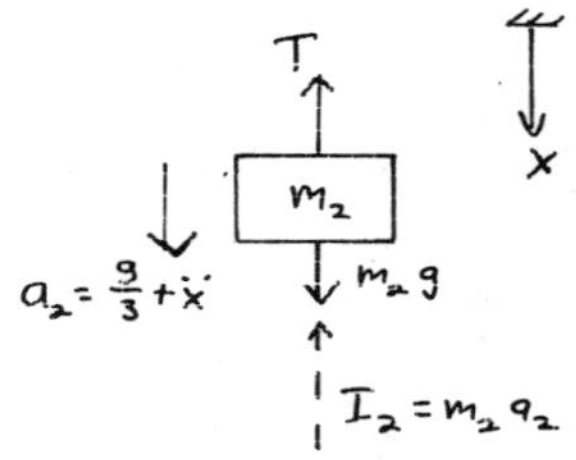

Figure c

By Fig. b,

$$\Sigma \vec F_x = \vec F + \vec I_1 = 0$$

or

$$m_1 g - T - m(\tfrac{g}{3}-\ddot x)=0$$

$$\therefore T = m_1 g - m_1(\tfrac{g}{3}-\ddot x) \qquad (a)$$

By Fig. c,

$$\Sigma \vec F_x = \vec F + \vec I_2 = 0$$

or

$$m_2 g - T - m_2(\tfrac{g}{3}+\ddot x)=0$$

$$\therefore T = m_2 g - m_2(\tfrac{g}{3}+\ddot x) \qquad (b)$$

By Eqs. (a) and (b)

$$m_1 g - m_1(\tfrac{g}{3}-\ddot x) = m_2 g - m_2(\tfrac{g}{3}+\ddot x)$$

or

$$\ddot x = \frac{2(m_2-m_1)g}{3(m_1+m_2)} \qquad (c)$$

(Continued)

14.91 Cont.

Therefore,

$$a_1 = \frac{g}{3} - \ddot{x} = \frac{(3m_1 - m_2)g}{3(m_1+m_2)} \quad (d)$$

$$a_2 = \frac{g}{3} + \ddot{x} = \frac{(3m_2 - m_1)g}{3(m_1+m_2)} \quad (e)$$

b) For $m_2 = 2m_1$, Eqs. (d) and (e) yield

$$a_1 = \frac{1}{9}g \downarrow \; ; \quad a_2 = \frac{5}{9}g \downarrow$$

For $m_2 = 3m_1$,

$$a_1 = 0 \; , \quad a_2 = \frac{2}{3}g \downarrow$$

c) By the FBD of the pulley (Fig. d)

$\Sigma F_x = 2T - T_p = 0$ (f)

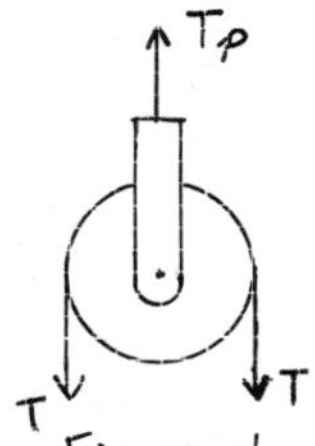

Figure d

Now by Eqs. (a) or (b) and Eq. (c),

$$T = \frac{4m_1 m_2 g}{3(m_1+m_2)} \quad (g)$$

Eqs. (f) and (g) yield the formula for T_p as

$$T_p = \frac{8m_1 m_2 g}{3(m_1+m_2)}$$

d) For $m_2 = 3m_1$, Eq. (c) yields $\ddot{x} = g/3$.
Therefore,

the acceleration of m_1 relative to the center of the pulley is

$$a_1/\text{pulley} = \frac{g}{3} \uparrow$$

and for m_2

$$a_2/\text{pulley} = \frac{g}{3} \downarrow$$

14.92

Given: Figure a; the 2-ton load hangs from a hoist supported by a trolley that travels at 3 ft/s on a circular track.

a) Derive a formula for the distance r that the load moves radially in the circular track.

b) Determine the magnitude of r, for $v = 3$ ft/s and $R = 1.5$ ft.

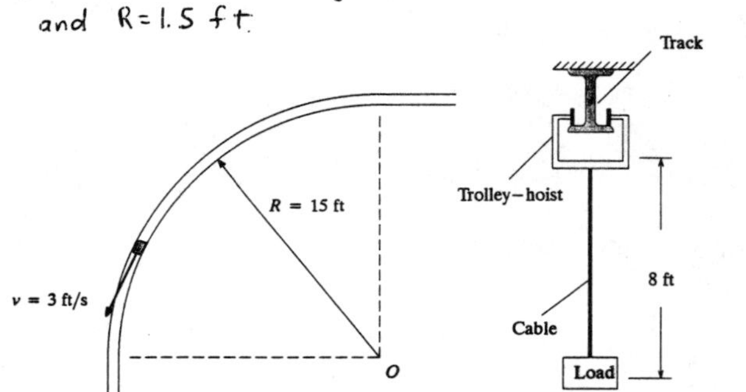

Figure a

Solution:

a) The FBD of the load is shown in Fig. b.
By Fig. b,

$\Sigma \vec{F} = \vec{F} + \vec{I} = 0$ (a)

For the x-direction,
Eq. (a) yields

$\Sigma F_x = T\cos\theta - ma = 0$ (b) ← $I = ma$

For the y-direction

$\Sigma F_y = T\sin\theta - mg = 0$ (c)

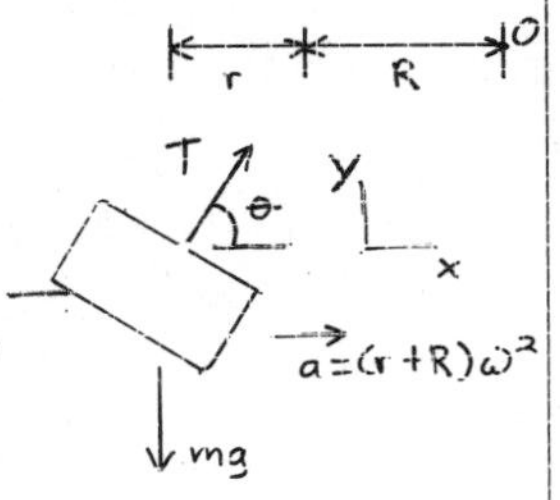

Figure b

By Eqs. (b) and (c),

$$\tan\theta = \frac{g}{a} = \frac{g}{(r+R)\omega^2}$$

or,

$$r = \frac{g}{\omega^2 \tan\theta} - R \quad (d)$$

(b) By Figs. a and b,

$r = 8\cos\theta$ (e)

with $\omega = v/R = \frac{3}{15} = 0.2$ rad/s and $R = 15$ ft,
Eqs. (d) and (e) yield,

$$\tan\theta = 53.667 - 0.80\sin\theta$$

$\therefore \theta = 1.55188$ rad $= 88.916°$

and $r = 12\cos\theta = 0.151$ ft

14.93

Given: A satellite circles the earth in a circular orbit.

Find:
a) The period as a function of the distance d from the surface of the earth.
b) The orbital speed as a function of d.

Solution:

a) Figure a shows the path of the satellite.
By Fig. a,

$a = R + d$ (a)

where R is the radius of the earth. The period T is related to the distance a from the center of the earth by Kepler's third law

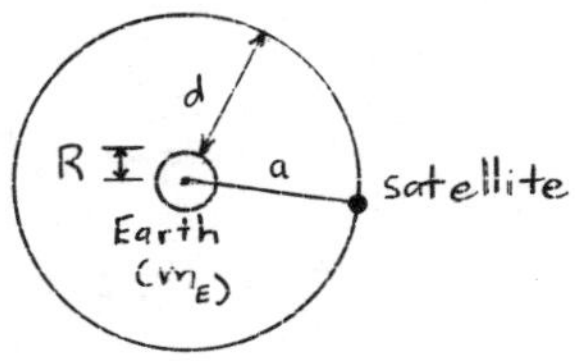

Figure a

$$\frac{T^2}{a^3} = \frac{4\pi^2}{Gm_E} \quad (b)$$

By Eqs. (a) and (b),

$$T = 2\pi\sqrt{\frac{(d+R)^3}{Gm_E}}$$

b) By Eq. (14.38), the orbital speed v_0 is, with $r_0 = a$,

$$v_0 = \sqrt{\frac{gR^2}{a}} = \sqrt{\frac{gR^2}{(d+R)}}$$

14.94

Given: Let T be the period in seconds and a be the mean distance (m) of an orbit (see Table 14.1).

Find: the value of T^2/a^3

a) for planets of the sun.

b) for satellites that orbit Earth

c) for satellites that orbit Mars.

Solution:

By Table 14.1,

$G = 6.672 \times 10^{-11}$ $N \cdot m^2/kg^2$

mass of sun: $m_s = 1.989 \times 10^{30}$ kg

mass of earth: $m_E = 5.974 \times 10^{24}$ kg

mass of mars: $m_m = 6.419 \times 10^{23}$ kg

a) For planets of the sun, by Kepler's third law

$$\frac{T^2}{a^3} = \frac{4\pi^2}{G m_s} = \frac{4\pi^2}{(6.672)(10^{-11})(1.989)(10^{30})}$$

or

$$\frac{T^2}{a^3} = 2.975 \times 10^{-19} \frac{s^2}{m^3}$$

b) For satellite orbiting earth

$$\frac{T^2}{a^3} = \frac{4\pi^2}{(6.672)(10^{-11})(5.974)(10^{24})} = 9.905 \times 10^{-14} \frac{s^2}{m^3}$$

c) For satellites that orbit mars

$$\frac{T^2}{a^3} = \frac{4\pi^2}{(6.672)(10^{-11})(6.419)(10^{23})} = 9.218 \times 10^{-13} \frac{s^2}{m^3}$$

14.95

Given: The period of the moon's orbit around the earth is $T = 2.36055 \times 10^6$ s.

Estimate the distance between the earth and the moon.

Solution:

By the solution of Problem 14.94

$$\frac{T^2}{a^3} = 9.905 \times 10^{-14} \frac{s^2}{m^3}$$

$$\therefore a = \sqrt[3]{\frac{T^2}{9.905 \times 10^{-14}}} = \sqrt[3]{\frac{(2.36055 \times 10^6)^2}{(9.905 \times 10^{-14})}}$$

or

$$a = 3.832 \times 10^5 \text{ km}$$

14.96

Given: NASA plans to send up an earth satellite with a period T = 4h

Find:

a) the required altitude of the satellite

b) the escape velocity of the satellite in its orbit.

Solution:

a) The period T = 4h = 14,400 s.
For a satellite orbiting earth (see Problem 14.94),

$$\frac{T^2}{a^3} = 9.905 \times 10^{-14} \ s^2/m^3$$

$$\therefore a^3 = \frac{(14,400 s)^2}{[9.905 \times 10^{-14} s^2/m^3]}$$

or

$$a = 1.279 \times 10^4 \text{ km}$$

The required altitude is

altitude = a − R = 12,790 − 6400 km

or

$$\text{altitude} = 6390 \text{ km}$$

b) The critical speed (escape velocity) is given by Eq. (14.39) as,

$$v_c = \sqrt{2}\, v_0$$

where v_0 is the orbital speed of the satellite.

Since T = 14,400 s and $a = 1.279 \times 10^4$ km,

$$v_0 = \frac{2\pi a}{T} = \frac{2\pi (1.279 \times 10^4)}{14,400} = 5.5807 \text{ km/s}$$

$$\therefore v_c = \sqrt{2}\, v_0 = 7892 \text{ m/s}$$

14.97

Determine T^2/a^3 for the moon as it orbits the earth; T in seconds and a in feet.

Solution:

By Kepler's third law, for the moon,

$$\frac{T^2}{a^3} = \frac{4\pi^2}{G m_{EARTH}} \qquad (a)$$

By Table 14.1,

$$G = 6.6720 \times 10^{-11} \frac{Nm^2}{kg^2}\left(\frac{1 \text{ lb}}{4.448 \text{ N}}\right)\left(\frac{1 \text{ ft}}{0.3048 m}\right)^2\left(\frac{14.6 \text{ kg}}{1 \text{ slug}}\right)^2$$

or

$$G = 3.442 \times 10^{-8} \text{ lb} \cdot ft^2/(slug)^2 \qquad (b)$$

and

$$m_{EARTH} = 5.974 \times 10^{24} \text{ kg} \left(\frac{1 \text{ slug}}{14.6 \text{ kg}}\right)$$

or

$$m_{EARTH} = 4.092 \times 10^{23} \text{ slugs} \qquad (c)$$

By Eqs. (a), (b), and (c),

$$\frac{T^2}{a^3} = 2.803 \times 10^{-15} \frac{s^2}{ft^3}$$

14.98

Given: The mean distance a of the GOES 4 earth geosynchronous satellite is approximately 42,200 km.

Find: Its period T.

Solution:

In general, by keplers third law,

$$\frac{T^2}{a^3} = \frac{4\pi^2}{G m_{EARTH}} = \frac{4\pi^2}{(6.672\times10^{-11})(5.974\times10^{24})}$$

or

$$\frac{T^2}{a^3} = 9.905\times10^{-14}\ s^2/m^3$$

Hence,

$$T^2 = (9.905\times10^{-14})(4.220\times10^7)^3$$

or

$$T = 8.6277\times10^4\ s = 0.9986\ \text{days}$$

Note: For GOES 4 to be exactly geosynchronous, the period should be exactly 1 day (or 86,400s). (The approximation $a = 42,200$ km may account for the difference.)

14.99

Given: Viking I satellite had a period around Mars of 1655 minutes.

Find:

a) The mean radius of its orbit

b) The average radius of Mars, knowing that at it's closest point to Mars' surface it was approximately 500 km away and at its farthest point it was approximately 35,800 km away. Compare with Table 14.1.

Solution:

a) $T = 1655$ min. $= 99,300$ s

By Problem 14.94, for Mars,

$$\frac{T^2}{a^3} = 9.218\times10^{-13}\ s^2/m^3$$

$$\therefore\ a^3 = \frac{(99,300)^2}{9.218\times10^{-13}}$$

or $a = 2.203\times10^7$ m

b) With given data and Fig. 14.18,

$2a = 1500 + \text{diameter of Mars} + 35,800$ (km)

$\therefore$ 2 × radius of Mars $= 2a - 37,300$

or r_{mars} = radius of Mars $= 2.203\times10^7\ m - 1.865\times10^7\ m$

or

$$r_{mars} = 3.380\times10^6\ m$$

By Table 14.1,

$$r_{Table} = 3.397\times10^6\ m$$

$$\%\ error = \left(\frac{3.397-3.380}{3.397}\right)\times100 = 0.50\ \%$$

14.100

Given: A satellite in an elliptical orbit passes over the North pole at the aphelion of its orbit, 1600 km above the earth's surface and over the South pole at the perihelion, 500 km above the earth's surface. Earth's radius is $R = 6378$ km.

Find:

a) The eccentricity of the orbit.

b) The speeds of the satellite at the aphelion and the perihelion.

c) The speed at the minor semi-axes of the orbit.

Solution:

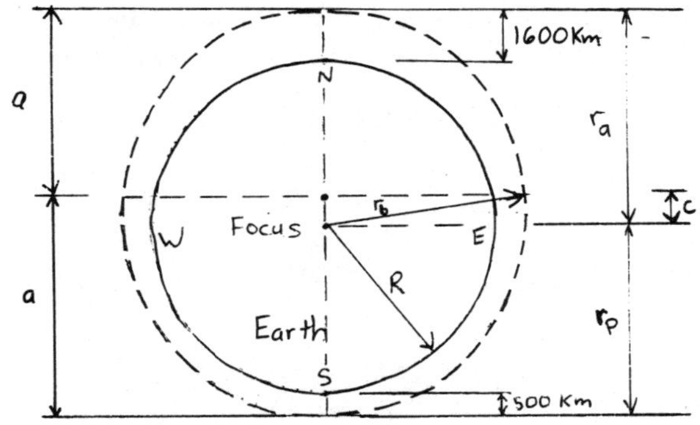

Figure a

a) Figure a shows the elliptical orbit (see also, Fig. 14.18)

By Fig. a,

$r_p = 500 + R$ $\quad\therefore\ r_p = 500 + 6378 = 6878$ km

and

$r_a = 1600 + 6378 = 7978$ km

$$\therefore\ \xi = \frac{r_a - r_p}{r_a + r_p} = 0.0740 \qquad (a)$$

b) The speed v_a at the aphelion is given by

$$v_a = R\sqrt{\frac{2g}{r_a} - \frac{g}{a}} \qquad (b)$$

where by Fig. (a), $a = \frac{1}{2}(r_a + r_p)$

or $a = 7428$ km

$$\therefore\ v_a = 6378\times10^3\sqrt{\frac{2(9.81)}{7978\times10^3} - \frac{9.81}{7428\times10^3}}$$

or

$$v_a = 6805.6\ m/s$$

Likewise,

$$v_p = R\sqrt{\frac{2g}{r_p} - \frac{g}{a}} = 7894.0\ m/s$$

c) At the minor semi-axes,

$r_b = \sqrt{c^2+b^2}$, where, $c = a\xi$, $b = a\sqrt{1-\xi^2}$

$\therefore\ r_b = a = 7428$ km

and, $v_b^2 = gR^2\left(\frac{2}{r_b} - \frac{1}{a}\right)$

or

$$v_b = R\sqrt{\frac{2g}{r_b} - \frac{g}{a}} = 7329.6\ m/s$$

14.101

Given: A missile is fired horizontally from the surface of the moon.

Find: It's escape velocity from the moon and compare it to the escape velocity from the earth.

Solution:

In general, the escape velocity of a body (say a missile) relative to a planet (say the moon) is

$$v_c = \sqrt{\frac{2Gm}{r}} \qquad (a)$$

where m is the mass of the planet (moon) and r is the distance between the body (missile) and the center of the moon. By Table 14.1, the mass and mean radius of the moon are,

$$m_m = 7.348 \times 10^{22} \text{ kg}$$

$$r_m = 1.738 \times 10^{6} \text{ m}$$

also, $G = 6.6720 \times 10^{-11} \text{ N·m}^2/\text{kg}^2$

Thus, the escape velocity of the missile from the moon is

$$v_{cm} = \sqrt{\frac{(2)(6.6720\times10^{-11})(7.348\times10^{22})}{(1.738\times10^{6})}}$$

or

$$v_{cm} = 2375 \text{ m/s}$$

Similarly, the escape velocity of the missile from the surface of the earth is,

$$v_{CE} = \sqrt{\frac{(2)(6.6720\times10^{-11})(5.974\times10^{24})}{(6.378\times10^{6})}}$$

or

$$v_{CE} = 11\ 180 \text{ m/s}$$

Hence,

$$\frac{v_{CE}}{v_{cm}} = 4.71$$

14.102

Given: A geosynchronous satellite is to orbit Saturn. Saturn's period is $T = \frac{2\pi}{\omega} = 10.5$ h.

Find:

a) the required altitude relative to the surface of Saturn.

b) the required orbital speed of the satellite.

Solution:

By Table 14.1,

$G = 6.672 \times 10^{-11} \text{ m}^3/(\text{kg·s}^2)$

mass of Saturn: $m_s = 5.685 \times 10^{26}$ kg

also period $T = 10.5\text{h} = 37{,}800$ s.

mean radius of Saturn: $r_s = 6.027 \times 10^{7}$ m

a) By Kepler's third law,

$$\frac{T^2}{a^3} = \frac{4\pi^2}{Gm_s}$$

or

$$a^3 = \frac{(37{,}800)^2(6.672\times10^{-11})(5.685\times10^{26})}{4\pi^2}$$

or

$$a = 1.111 \times 10^{8} \text{ m}$$

$$\therefore \text{altitude} = a - r_s = 5.086 \times 10^{7} \text{ m}$$

b) $T = \frac{2\pi}{\omega} = \frac{2\pi}{(v/r)} = \frac{2\pi r}{v}$; $r = a$

$$\therefore v = \frac{2\pi a}{T} = \frac{2\pi(1.111\times10^{8})}{3.78\times10^{4}}$$

or $v = 18\ 470$ m/s

14.103

Given: Hyperion revolves about Saturn at 1.481×10^{6} km. Enceladus revolves about Saturn at 2.380×10^{5} km. The period of Enceladus is 1.37 days.

Find: The period of Hyperion.

Solution:

By Kepler's third law,

$$\frac{T^2}{a^3} = \frac{4\pi^2}{Gm_{saturn}} \qquad (a)$$

Equation a applies for all of Saturn's 18 moons. Therefore,

$$\frac{T^2_{HYPERION}}{a^3_{HYPERION}} = \frac{T^2_{ENCELADUS}}{a^3_{ENCELADUS}}$$

and

$$\therefore T^2_{HYPERION} = \frac{(1.37\text{ days})^2(1.481\times10^{6})^3}{(2.380\times10^{5})^3}$$

or

$$T_{HYPERION} = 21.27 \text{ days}$$

14.104

Given: Assume that the moon circles the earth in a circular orbit with a period T = 27.32 earth days.

Find: The moon's sectorial speed.

Solution:

Sectorial speed is defined as

$$\frac{1}{2}\,g = \frac{\pi a^2}{T}\sqrt{1-\xi^2} \qquad (a)$$

where a is the mean radius of the orbit and ξ is the orbit's eccentricity. The sectorial speed is the instantaneous rate that the position vector of a satellite (the moon) sweeps out area in the plane of motion.

(Continued)

14.104 Cont.

For a circular orbit, $\xi = 0$. Therefore by Eq. (a),

$$\frac{1}{2}\mathscr{g} = \frac{\pi a^2}{T} \qquad (b)$$

Now,
$$T = (27.32 \text{ days})(24 \tfrac{h}{day})(\tfrac{3600 s}{h})$$

or,
$$T = 2.36 \times 10^6 \text{ s}$$

and since (see solution of problem 14.95)

$$a = 3.832 \times 10^8 \text{ m}$$

$$\frac{1}{2}\mathscr{g} = 1.955 \times 10^{11} \frac{m^2}{s}$$

14.105

Find: the values of T^2/a^3 for satellites orbiting Saturn. For orbitino Jupiter.

Solution:

For Saturn, by Table 14.1 and kepler's third law,

$$\frac{T^2}{a^3} = \frac{4\pi^2}{Gm_s} = \frac{4\pi^2}{(6.672\times10^{-11})(5.685\times10^{26})}$$

or $$\frac{T^2}{a^3} = 1.041 \times 10^{-15} \text{ s}^2/\text{m}^3$$

For Jupiter,

$$\frac{T^2}{a^3} = \frac{4\pi^2}{Gm_J} = \frac{4\pi^2}{(6.672\times10^{-11})(1.899\times10^{27})}$$

or $$\frac{T^2}{a^3} = 3.116 \times 10^{-16} \text{ s}^2/\text{m}^3$$

14.106

Given: Jupiter's period $T = 0.41354$ earth days.

Find:

a) Mean distance a of a geosynchronous satellite of Jupiter

b) The escape velocity v_c of the satellite from its orbit.

Solution:

a) By solution of Problem 14.105, for Jupiter,

$$\frac{T^2}{a^3} = 3.116 \times 10^{-16} \text{ s}^2/\text{m}^3$$

$$\therefore a^3 = \frac{T^2}{(3.116\times10^{-16})} = \frac{[(0.41354)(3600)(24)]^2}{(3.116\times10^{-16})}$$

or $$a = 1.600 \times 10^8 \text{ m}$$

b) The orbital speed v_0 of the satellite is obtained from

$$v_0 T = 2\pi a, \quad T = 35730 \text{ s}$$

or $$v_0 = \frac{2\pi\,(1.600\times10^8 \text{ m})}{35730 \text{ (s)}} = 28136 \text{ m/s}$$

$$\therefore v_c = \sqrt{2}\, v_0 = 39790 \text{ m/s}$$

14.107

Given: The mean radius of Mars is $r_{Mars} = 3.397\times10^6$ m and its mass is 6.419×10^{23} kg.

Find: The acceleration of gravity on Mars.

Solution:

$$F = \frac{Gm\, m_{MARS}}{r_{MARS}^2} = m\, g_{mars}$$

$$\therefore g_{mars} = \frac{Gm_{mars}}{r_{mars}^2} = \frac{(6.672\times10^{-11})(6.419\times10^{23})}{(3.397\times10^6)^2}$$

or $$g_{mars} = 3.71 \text{ m/s}^2$$

14.108

Given: Data for the Viking I satellite in Problem 14.99 and $g_{mars} = 3.71$ m/s², $r_{mars} = 3.397\times10^6$ m

Find:

a) Eccentricity ξ of its orbit

b) The speed v_p at the perihelion

c) The speed v_a at the aphelion

d) The speed at minor semi-axes.

Solution:

a) $r_p = 1500 + r_{mars} = 4897$ km

$r_a = 35{,}800 + r_{mars} = 39{,}197$ km

$$\xi = \frac{r_a - r_p}{r_a + r_p} = \frac{39{,}197 - 4897}{39{,}197 + 4897}$$

or $$\xi = 0.7779$$

b) $$v_p^2 = g_{mars} r_{mars}^2 \left(\frac{2}{r_p} - \frac{1}{a}\right) = (3.71)(3.397\times10^6)^2\left(\frac{2}{4.897\times10^6} - \frac{1}{2.203\times10^7}\right)$$

or $$v_p = 3942 \text{ m/s}$$

c) $$v_a^2 = g_{mars} r_{mars}^2 \left(\frac{2}{r_a} - \frac{1}{a}\right) = (3.71)(3.397\times10^6)^2\left(\frac{2}{3.9197\times10^7} - \frac{1}{2.203\times10^7}\right)$$

or $$v_a = 491 \text{ m/s}$$

d) $$v_b^2 = g_{mars} r_{mars}^2 \left(\frac{2}{r_b} - \frac{1}{a}\right)$$

where by Fig. 14.18 and Eqs. (14.25) and (14.27)

$$r_b^2 = c^2 + b^2 = a^2\xi^2 + a^2(1-\xi^2) = a^2$$

or $$r_b = a$$

$$\therefore v_b^2 = (3.71)(3.397\times10^6)^2\left(\frac{1}{2.203\times10^7}\right)$$

or $$v_b = 1394 \text{ m/s}$$

14.110

(a) For a mass m on the earth's surface, Newton's gravitational law yields the force of attraction on m as $F = \frac{Gmm_E}{r_E^2}$, where m_E is the mass of the earth and r_E is the radius of the earth. Also, on the earth's surface, the weight of m is $W = mg = F$. Therefore,
$$W = \frac{Gmm_E}{r_E^2}.$$

b) If $r \le r_E$, the force of attraction on m is
$$F = \frac{Gm\bar{m}_E}{r^2}$$
where $\bar{m}_E$ is the mass of the earth contained in the sphere of radius r. Assume that ρ_E = mass density = m_E/vol_E is constant. Then, by Table 14.1,
$$\bar{m}_E = \rho_E \frac{4\pi r^3}{3} = \frac{5.974\times10^{24}\ kg}{\frac{4\pi r_E^3}{3}\ (m^3)} \frac{4\pi r^3\ (m^3)}{3}$$
or,
$$\bar{m}_E = \frac{5.974\times10^{24}}{(6.378\times10^6)^3} r^3 = 23025\, r^3$$
$$\therefore W = F = \bar{W} = \frac{Gm\bar{m}_E}{r^2} = 23{,}025\, Gmr\ ;\ r \le r_E \quad (a)$$
For $r \ge r_E$
$$F = W = \frac{Gmm_E}{r^2} \quad (b)$$
at $r = r_E$, $W = mg = \frac{Gmm_E}{r_E^2}$

With Eqs. (a) and (b), W is given as a function of r (see Fig. a)

at $r = 2r_E$, Eq (b) yields
$$F = \frac{Gmm_E}{4r_E^2} = \frac{1}{4}mg$$

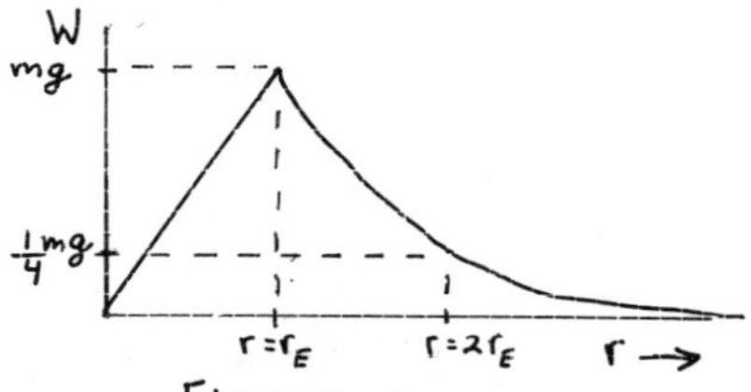

Figure a

14.122

Show, for consistency with the equation F=ma, that the weight of a unit mass is g units of force.

Solution:

If we set m=1 in the equation W=mg, we obtain W=g. If Newton's equation F=ma is applied to a body that falls freely in a vacuum, the force F is the weight W, and the acceleration a is the acceleration of gravity g. Therefore, F=ma yields W=mg, and as noted above for m=1, W=g.

This condition is independent of the local acceleration of gravity, for if g varies, the weight W of the body varies proportionately.

Thus, this relation is universally true.

14.124

Show that a particle that moves under the action of a constant gravity describes a parabolic path.

Solution:

Let the force mg that acts on the particle lie in the (x, y) plane and be directed downward (Fig. a).

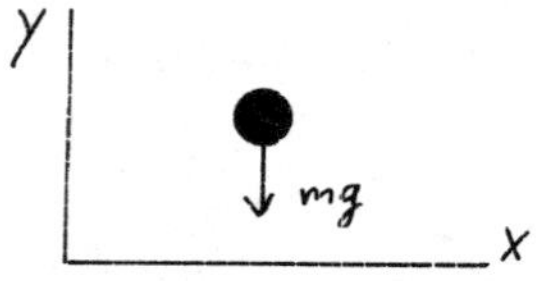

Then, by Newton's second law
$$\Sigma F_x = 0,\ \Sigma F_y = -mg$$
Hence,
$$a_x = \frac{dv_x}{dt} = 0,\quad a_y = \frac{dv_y}{dt} = -g$$
and
$$v_x = C_1,\quad v_y = -gt + v_{y0}$$
$$\therefore v_x = \frac{dx}{dt} = C_1 \Rightarrow x = C_1 t + x_0 \quad (a)$$
and,
$$v_y = \frac{dy}{dt} = -gt + v_{y0} \Rightarrow y = -\frac{1}{2}gt^2 + v_{y0}t + y_0 \quad (b)$$
Solving Eq. (a) for t and substituting for t in Eq. (b), we obtain, where x_0, y_0 are constants,
$$y = -\frac{1}{2}g\left(\frac{x-x_0}{C_1}\right)^2 + v_{y0}\left(\frac{x-x_0}{C_1}\right) + y_0$$
Expanding and collecting terms, we find
$$y = C_2x^2 + C_3x + C_4 \quad (c)$$
where C_2, C_3, and C_4 are constants in terms of C_1, x_0, y_0 and v_{y0}.

Equation (c) is the general equation of a parabola

15.1

GIVEN: The body in Fig. a. is lowered slowly from the unstretched position to the stretched position where the body is in equilibrium.

FIND: Spring constant k [F/L] in terms of m, g, and d.

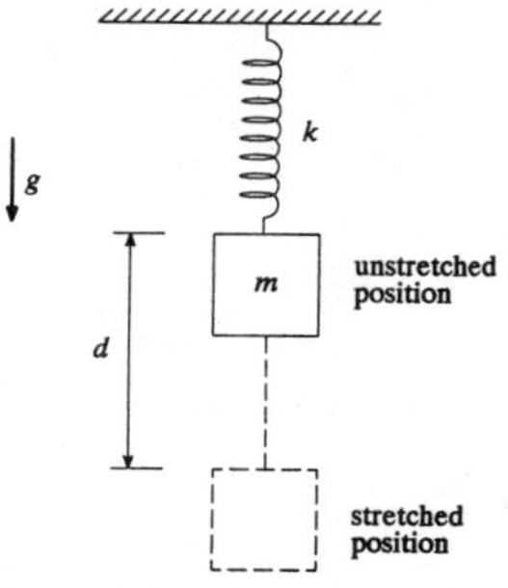

Figure (a)

SOLUTION

Conservative System:

$$T_1 + V_1 = T_2 + V_2 \quad (T_1 = 0,\ V_1 = 0,\ T_2 = 0)$$

$$V_2 = -mgd + \tfrac{1}{2}k(d^2 - 0^2) = 0$$

$$mgd = \tfrac{1}{2}kd^2$$

$$k = \frac{2mgd}{d^2} \rightarrow \underline{\underline{k = \frac{2mg}{d}}} \quad [F/L]$$

15.2

GIVEN: Body in Fig (a) from Problem 15.1 is dropped from rest from the unstretched position.

FIND:
a.) Speed of the body at d.
b.) Total distance it drops before it comes to rest.

SOLUTION

a.) $T_1 + V_1 = T_2 + V_2 \quad (T_1 = 0,\ V_1 = 0)$

$$0 = \tfrac{1}{2}mv^2 + \tfrac{1}{2}k(d^2 - 0^2) - mgd$$

$$\tfrac{1}{2}mv^2 = mgd - \tfrac{1}{2}kd^2$$

$$\underline{\underline{v = \sqrt{2gd - \frac{kd^2}{m}}}}$$

b.) $T_1 + V_1 = T_2 + V_2 \quad (T_1 = 0,\ V_1 = 0,\ T_2 = 0)$

$$V_2 = \tfrac{1}{2}kd^2 - mgd = 0 \ ; \quad \underline{\underline{d = \frac{2mg}{k}}}$$

15.3

GIVEN: Graph of propellor force exerted on plane during take off.

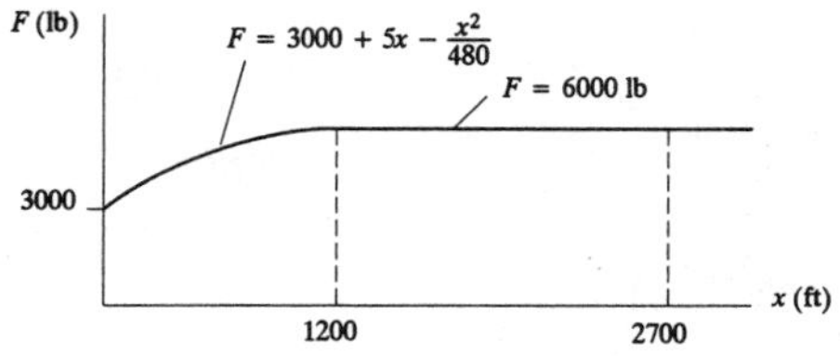

Figure (a)

FIND: Work done by propellors during take off

SOLUTION

$$U_{1-2} = \int_{x_1}^{x_2} F \cdot dx$$

$$U_{1-2} = \int_0^{1200} \left(3000 + 5x - \frac{x^2}{480}\right) dx + \int_{1200}^{2700} 6000\, dx$$

$$U_{1-2} = \left(3000x + \tfrac{5}{2}x^2 - \frac{x^3}{1440}\right)\Big|_0^{1200} + 6000(2700 - 1200)$$

$$\underline{\underline{U_{1-2} = 15{,}000{,}000 \text{ (ft·lb)} = 15{,}000 \text{ (kip·lb)}}}$$

15.4

GIVEN: Work required to lift six people 100 meters = 408 KJ (Fig. a)

FIND: The average mass per person and the average weight per person.

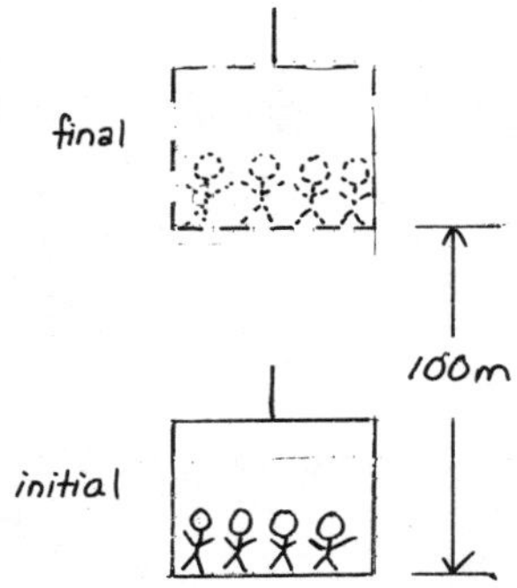

Figure (a)

SOLUTION

$$F = m_{all}\, g$$

$$U_{1-2} = \int_{r_1}^{r_2} F \cdot dr$$

$$U_{1-2} = 408 \text{ KJ}$$

$$= \int_0^{100} m_{all}\, g\, dr$$

$$408 \text{ KJ} = m_{all}\, g\, (100 - 0) \ ; \quad g = 9.81 \text{ m/s}^2$$

$$m_{all} = \frac{408 \text{ kN·m}}{9.81(100)\text{m}} = \frac{408000 \text{ N·m}}{9.81 \text{ m/s}^2 (100\text{m})} = 415.9 \text{ kg}$$

$$m_{ave} = \frac{m_{all}}{6 \text{ people}} = \underline{\underline{69.3 \text{ kg}}}$$

$$W_{ave} = m_{ave}\, g = 69.3(9.81) = \underline{\underline{680 \text{ N}}}$$

15.5

GIVEN

$\mu_{KA} = 0.20$

$\mu_{KB} = 0.30$

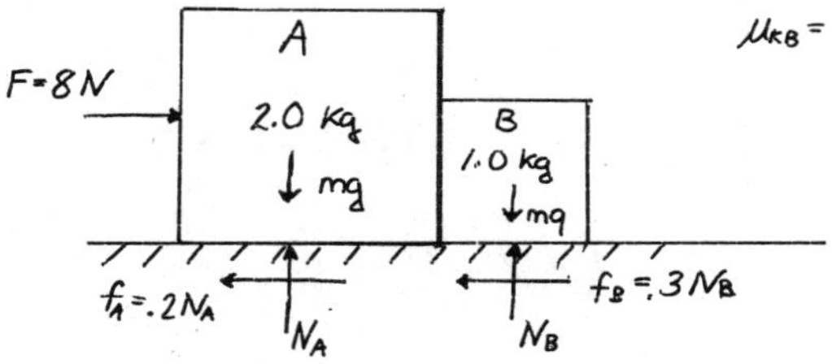

FIND

a.) Total work done on crates when both slide 3m to the right.

b.) Speed of crates at 3m.

c.) Power generated by force F when crates are slid 3m.

(continued)

15.5 cont.

SOLUTION

a.) $U_{1-2} = \int_{x_1}^{x_2} F \cdot dx$

Only forces that are parallel to the motion do work.

$U_{1-2} = (8x - .2N_{Ax} - .3N_{Bx}) \Big|_0^3$

For block A:

$\Sigma F_y = 0 \qquad N_A = 2.0\,kg\,(9.81\,m/s^2) = 19.62\,N$

For block B:

$\Sigma F_y = 0 \qquad N_B = 1.0\,kg\,(9.81\,m/s^2) = 9.81\,N$

$U_{1-2} = 8(3) - 0.2(19.62)(3) - 0.3(9.81)(3)$

$U_{1-2} = 3.40\ J$

b.) $U_{1-2} = T_2 - T_1$ ($T_1 = 0$)

$U_{1-2} = 3.40\ N\cdot m, \quad T_2 = \frac{1}{2}(m_A + m_B)v^2 = \frac{1}{2}(3)v^2$

$3.40\ N\cdot m = 1.5v^2 \qquad v = 1.505\ m/s$

c.) $E = F \cdot v$

$E = (8N)(1.505\ m/s) = 12.04\ N$

15.6 SHOW that the work done on a piston by an expanding gas is $\int_{V_0}^{V_1} p\,dv$

SOLUTION

$U_{1-2} = \int_{r_1}^{r_2} \vec{F} \cdot d\vec{r}$

Volume $= V = Ar$, $\quad A$ = cross-sectional area of piston (Fig a)

$\therefore dV = A\,dr$

or $dr = \frac{dV}{A}$

also, $p = \frac{F}{A}$, and

$U_{1-2} = \int_{V_0}^{V_1} \frac{F \cdot dV}{A}$

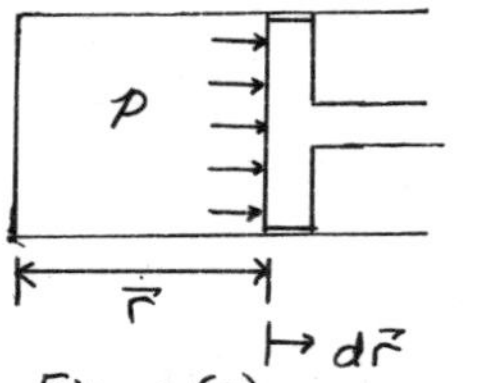

Figure (a)

Thus yielding: $U_{1-2} = \int_{V_0}^{V_1} p\,dV$

15.7 GIVEN Block at rest connected to spring with constant k

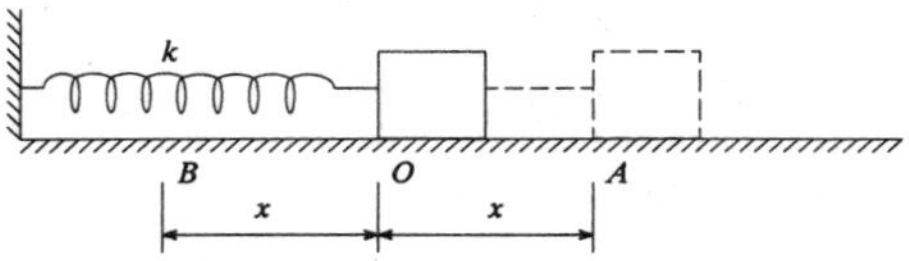

FIND

a.) Work done by spring when block moves from O to A

b.) Work done by spring when moves from O to B

c.) Work done by spring when block moves from O, to A, to B, and back to O.

SOLUTION

a.) $U_{1-2} = -\frac{kx^2}{2}$

b.) $U_{1-2} = -\frac{kx^2}{2}$

c.) $= \underbrace{\left(-\frac{kx^2}{2}\right)}_{O\ to\ A} \underbrace{\left(+\frac{kx^2}{2} - \frac{kx^2}{2}\right)}_{A \to O \quad O\ to\ B} + \underbrace{\left(\frac{kx^2}{2}\right)}_{B \to O} = 0$

15.8 Derive an expression for the work done by a piston to compress an ideal gas. The pressure-volume relation is: pV = constant

SOLUTION

$U_{1-2} = -\int_{V_0}^{V_1} p\,dV \qquad$ (refer to problem 15.6)

Since $pV = C$ (constant), $\quad p = \frac{C}{V}$

$\therefore U_{1-2} = -\int_{V_0}^{V_1} \frac{C\,dV}{V} = -C\int_{V_0}^{V_1} \frac{dV}{V}$

$U_{1-2} = -C\ln(V)\Big|_{V_0}^{V_1} = -pV\ln(V)\Big|_{V_0}^{V_1}$

$U_{1-2} = -p_1V_1\ln V_1 + p_0V_0\ln V_0$

But; $p_1V_1 = p_0V_0 \qquad (pV = \text{constant})$

$\therefore U_{1-2} = p_0V_0\ln\left(\frac{V_0}{V_1}\right) = p_1V_1\ln\left(\frac{V_0}{V_1}\right)$

15.9 GIVEN A 500 N force acts on the block;

$W_B = 150\ N$

$\mu_k = 0.40$

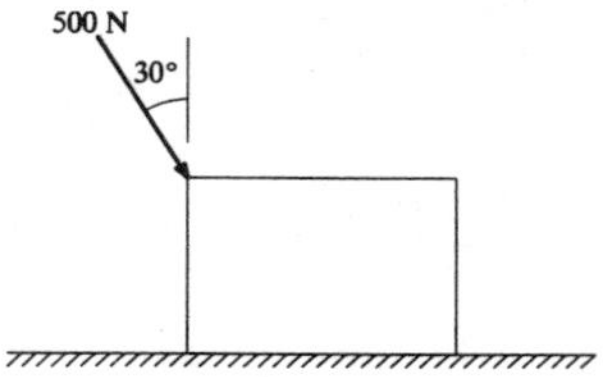

Figure (a)

(Continued)

15.9 cont

FIND

a.) Work performed on the block by all the forces.
b.) Speed of the block, when it has moved 2.5 m
c.) Power generated by 500 N force at 2.5 m.

SOLUTION

FBD

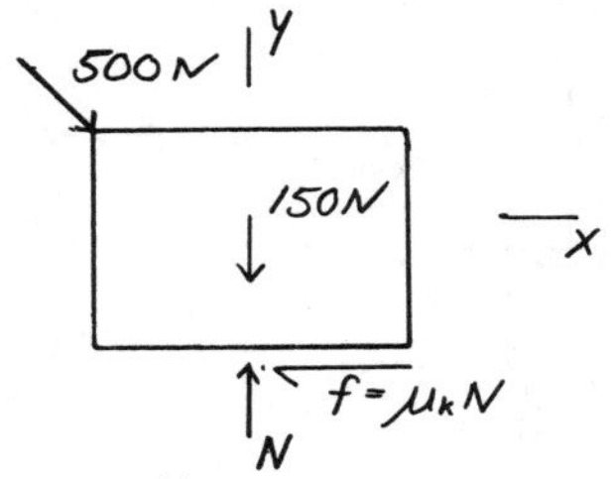

Figure b

a.) From freebody diagram,

$\Sigma F_y = N - 150 - 500 \cos 30° = 0$ [N]

$N = 583$ [N]

$U_{1-2} = \int_{r_1}^{r_2} \vec{F} \cdot d\vec{r}$

$U_{1-2} = \int_0^{2.5} (500 \sin 30°) dr - \int_0^{2.5} (0.40)(583) dr$

$U_{1-2} = 250(2.5-0) - 233.2(2.5-0)$

$U_{1-2} = 42$ N·m $= \underline{\underline{42 \text{ J}}}$

b.) $U_{1-2} = T_2 - T_1$

$U_{1-2} = \frac{1}{2} m v_2^2 - 0$ $\qquad m = \frac{150 \text{ N}}{9.81 \text{ m/s}^2} = 15.29$ kg

42 N·m $= \frac{1}{2}(15.29) v^2$; $\underline{\underline{v = 2.34 \text{ m/s}}}$

c.) $E = \vec{F} \cdot \vec{v}$

$E = 500 \sin 30° (2.34) = \underline{\underline{585 \text{ J/s}}}$

15.10 **GIVEN** Figure (a); 30 N block slides down ramp; $\mu_k = 0.50$

FIND

a.) Work done on crate as it slides down 3 m.
b.) Speed of crate at 3 m if $v_0 = 0$
c.) Power generated by force of gravity at 3 m.

SOLUTION

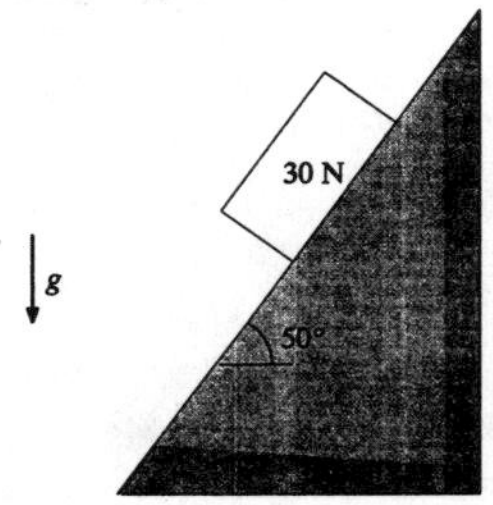

Figure (a)

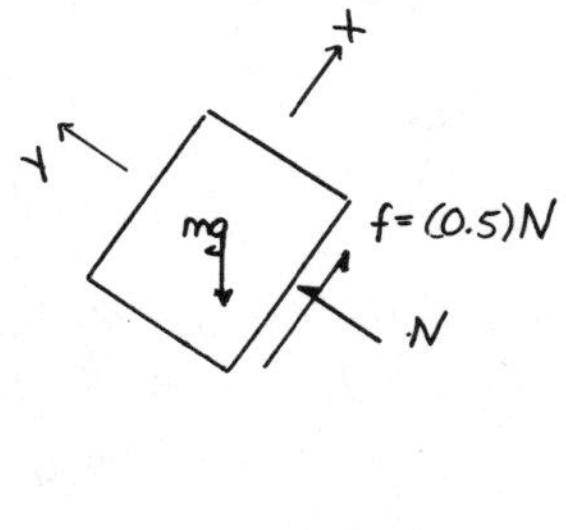

Figure (b) - FBD

a.) By the FBD of the block (Fig b.)

$\Sigma F_y = 0$: $N = 30 \cos 50° = 19.28$ [N]

$f = 0.50(19.28) = 9.642$ [N]

Work is only done by forces parallel to motion.

$U_{1-2} = \int_{x_1}^{x_2} \vec{F} \cdot d\vec{x}$

$U_{1-2} = -(19.642)x + (30)(\sin 50)(x) \Big|_0^{3m}$

$U_{1-2} = -9.64(3) + 22.98(3)$

$U_{1-2} = 40.02$ N·m $= \underline{\underline{40.02 \text{ J}}}$

b.) $U_{1-2} = T_2 - T_1$ ($T_1 = 0$)

$40.02 = \frac{1}{2} m v^2$ $\qquad m = \frac{30 \text{ N}}{9.81 \text{ m/s}^2} = 3.06$ kg

$40.02 = 1.529 v^2$; $\underline{\underline{v = 5.12 \text{ m/s}}}$

c.) $E = \vec{F} \cdot \vec{v}$

$E = 22.98$ [N] $(5.12$ m/s$) = \underline{\underline{117.6 \text{ J/s}}}$

15.11 **GIVEN** Figure a, with $W_A = W_B = 9$ N, $\mu_K = 0.40$, and $F = \frac{22.5}{x^2}$ kN

Magnet A is pushed slowly until it suddenly slides and strikes magnet B.

FIND

a.) Work in Joules performed on Magnet A by Magnet B during slide.
b.) Work done on magnet A by friction during slide
c.) Speed of Magnet A when it strikes Magnet B.
d.) Power generated by F at instant Magnet A strikes Magnet B.

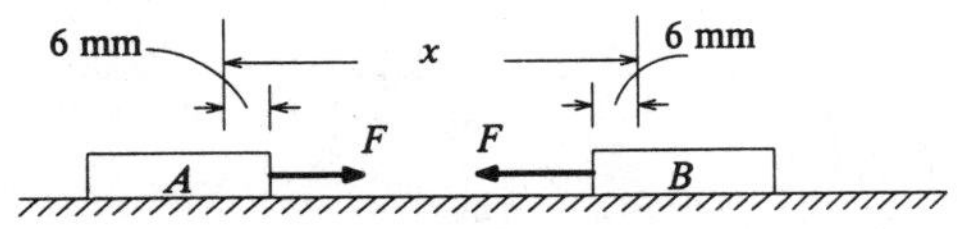

Figure (a)

SOLUTION

a.)

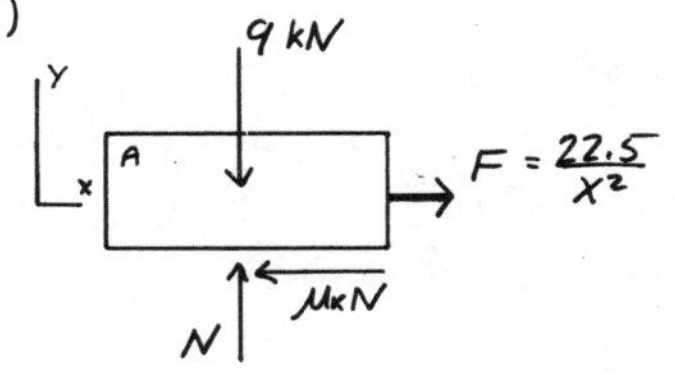

Figure (b) - FBD

(Continued)

15.11 cont

At beginning of slide (Fig b),

$\Sigma F_y = N - 9\text{ kN} = 0$

$\Sigma F_x = \frac{22.5}{X_1^2} - (0.40)(N) = 0$

-or- $X_1^2 = \frac{22500}{(0.4)(9)}$; $X_1 = 79.057$ mm

$\therefore U_{1-2}$ = Work done on A by Magnet B

$= -\int_{X_1}^{X_2} \frac{22500}{X^2}\,dx$

where $X_1 = 79.057$ mm and $X_2 = 12$ mm

Thus,

$$U_{1-2} = -\int_{79.057}^{12} \frac{22500}{X^2}\,dx = \left.\frac{22500}{X}\right|_{79.057}^{12}$$

-or- $U_{1-2} = 1590.4$ N·mm $= \underline{1.5904\text{ J}}$

b.) $U_{1-2} = -(3.6)(79.057 - 12) = -241.4$ N·mm

-or- $\underline{U_{1-2} = -0.2414\text{ J}}$

c.) $U_{1-2} = T_2 - T_1^{\,0}$

$U_{1-2} = 1.5904 - 0.2414 = \frac{1}{2}\left(\frac{9}{9.81}\right)v^2$

-or- $\underline{v = 1.715\text{ m/s}}$

d.) $E = \vec{F}\cdot\vec{v} = \frac{22500}{(12)^2}(1.715)$

-or- $\underline{E = 268.0\text{ Watts}}$

15.12 GIVEN Force - distance diagram (Fig a)

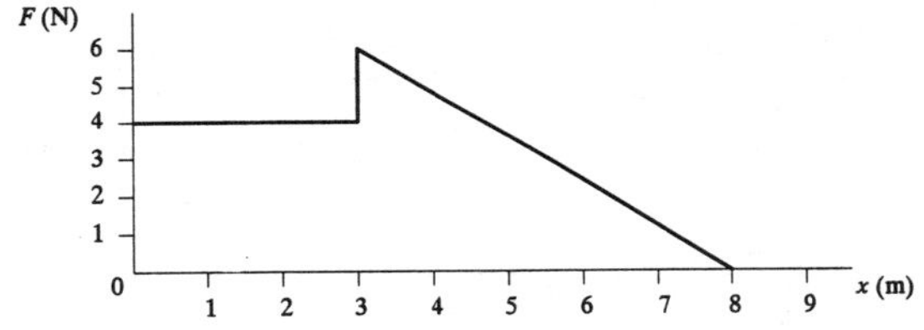

Figure (a)

FIND

a.) Work done by force in moving the body from $X=0$ to $X=8$m.

b.) The speed of the body ($W_b = 10$ N) at $X = 3$m. Body starts from rest and $\mu_K = 0.20$

c.) Speed of body at $X = 8$m

d.) Total distance that block slides.

SOLUTION

a.) By Fig (a), $U_{1-2} = \int_{x_1}^{x_2} \vec{F}\cdot d\vec{x}$

$U_{1-2} = 4(3) + \frac{1}{2}(6)(8-3) = 12 + 15$ [N·m]

$\underline{U_{1-2} = 27\text{ N·m} = 27\text{ J}}$

b.) FBD

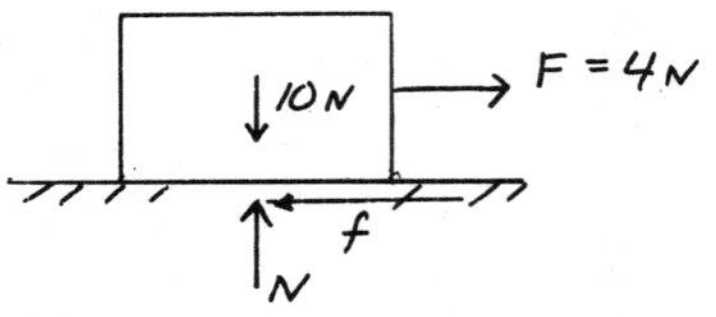

Figure (b)

By Fig b.,

$\Sigma F_y = N - 10 = 0$; $N = 10$N

$f = \mu_K N = (0.2)(10\text{N}) = 2$N

$U_{1-2} = T_2 - T_1$

$4(3) - 2(3) = \frac{1}{2}mv^2 - 0$;

$m = \frac{10\text{N}}{9.81\text{ m/s}^2} = 1.019$ kg

$\therefore 6 = \frac{1}{2}(1.019)v^2$; $\underline{v = 3.43\text{ m/s}}$

c.) $U_{1-2} = T_2 - T_1$

Use $x = 0$m as point 1 and $x = 8$m as point 2, with $T_1 = 0$:

$U_{1-2} = 27 - 2(8) = \frac{1}{2}mv^2$

$11 = \frac{1}{2}(1.019)v^2$ $\underline{v = 4.65\text{ m/s}}$

d.) $U_{1-2} = T_2 - T_1$; $T_1 = T_2 = 0$

$U_{1-2} = 27 - 2x = 0 - 0$; $\underline{x = 13.5\text{ m}}$

15.13 GIVEN Figure a. ; rocket of mass m travels directly away from earth and sun.

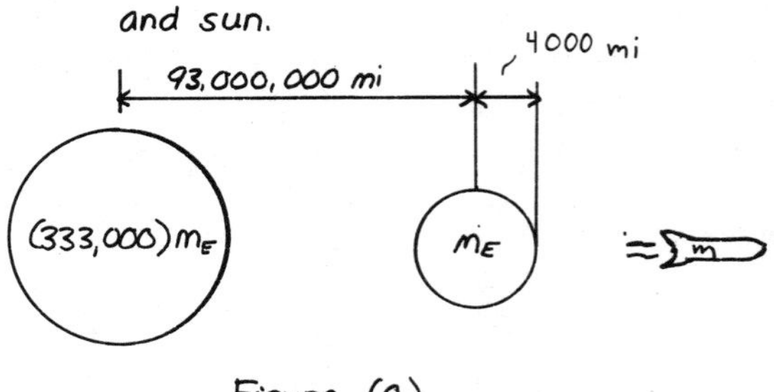

Figure (a)

FIND The ratio U_S/U_E

where U_S = work by gravity of sun on the mass,

U_E = work by gravity of earth on the mass.

(Continued)

15.13 cont.

SOLUTION

Let r be the distance of the rocket from the center of the earth. Then, the distance of the rocket from the center of the sun is $R = 93{,}000{,}000 + r$

The attraction of the earth is (Fig. b)

$$F_E = \frac{G m m_E}{r^2}$$

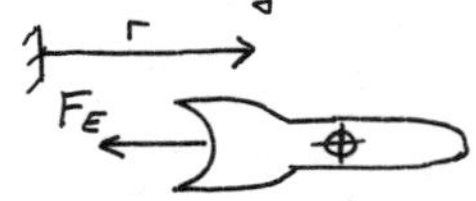

Figure (b)

$$\therefore U_E = -\int_{4000}^{\infty} F_E\, dr = \left.\frac{G m m_E}{r}\right|_{4000}^{\infty}$$

$$\therefore U_E = \frac{-G m m_E}{4000} \qquad (A)$$

The attraction of the sun is (Fig. c)

$$F_S = \frac{G m m_S}{R^2}$$

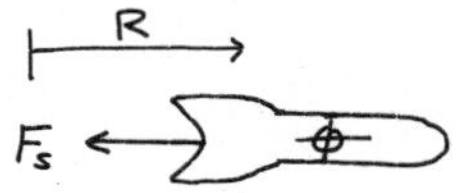

Figure (c)

$$\therefore U_S = -\int_{93{,}004{,}00}^{\infty} \frac{G m m_S}{R^2}\, dR = \left.\frac{G m (333{,}000\, m_E)}{R}\right|_{93{,}004{,}000}^{\infty}$$

$$U_S = -\frac{333{,}000\ G m m_E}{93{,}004{,}000} \qquad (B)$$

By Eqs (A) and (B),

$$\frac{U_S}{U_E} = \frac{(333)(4)}{93.004} = 14.32$$

15.14 GIVEN Figure a.

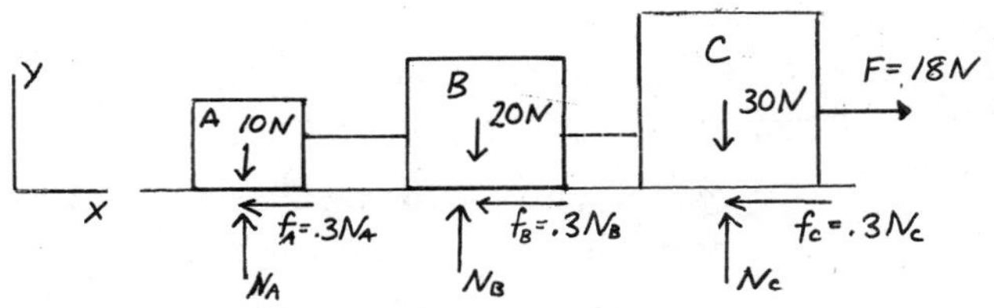

Figure (a)

FIND

a.) The work done on each crate as they slide 4m to the right
b.) The total work done on the crates
c.) Change in speed of the crates
d.) Net power generated by all forces.

SOLUTION

a.) By Fig (a),

$$\Sigma F_y = N_A + N_B + N_C - 10 - 20 - 30 = 0$$

$$N_A + N_B + N_C = 60$$

$$\Sigma F_x = 18 - (0.3)(60) = 0$$

$\therefore$ Crates move at constant velocity.

By Fig (b): FBD A;

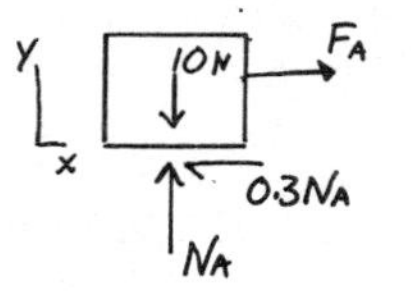

Figure (b)

$$\Sigma F_y = N_A - 10 = 0$$

$$N_A = 10N$$

$$\Sigma F_x = F_A - (0.3)(10) = 0$$

$$F_A = 3N$$

For crate A,

$$(U_{1-2})_A = F_A(4) - (0.3)N_A(4)$$

$$= (3)(4) - (0.3)(10)(4)$$

$$\therefore (U_{1-2})_A = 0\ J$$

By Fig (c): FBD of B;

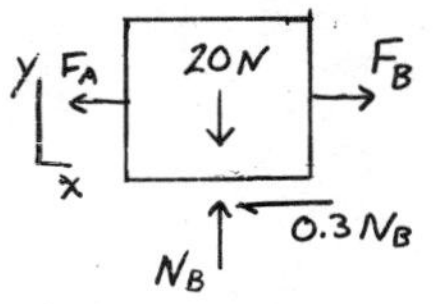

Figure (c)

$$\Sigma F_y = N_B - 20 = 0$$

$$N_B = 20N$$

$$\Sigma F_x = F_B - F_A - (0.3)N_B$$

$$F_B = 3 + (0.3)(20) = 9N$$

For crate B:

$$(U_{1-2})_B = (F_B - F_A)(4) - (0.3)(N_B)(4)$$

$$= (9-3)(4) - (0.3)(20)(4)$$

$$\therefore (U_{1-2})_B = 0\ J$$

By Fig (d): FBD of C;

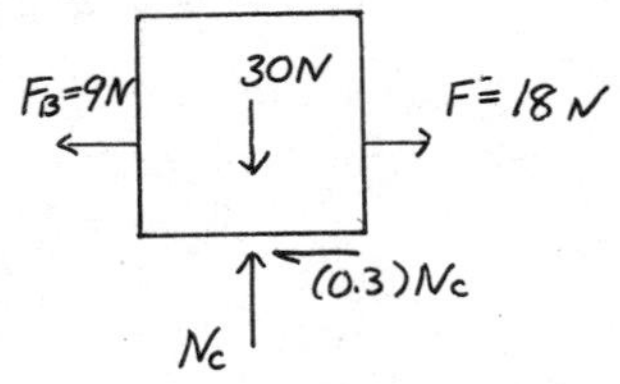

Figure (d)

$$\Sigma F_y = N_C - 30 = 0$$

$$N_C = 30N$$

$$\Sigma F_x = 18 - 9 - (0.3)(30) = 0$$

For crate C,

$$(U_{1-2})_C = (18-9)(4) - (0.3)(30)(4)$$

$$\therefore (U_{1-2})_C = 0\ J$$

b.) $(U_{1-2})_{Total} = (18)(4) - [(0.3)(10+20+30)](4)$

$$\therefore (U_{1-2})_{Total} = 0\ J$$

(Continued)

15.14 cont.

c.) Since velocity is constant

$$U_{1-2} = T_2 - T_1 = \tfrac{1}{2}mv^2 - \tfrac{1}{2}mv^2 = 0$$

$\Delta V = 0$ m/s

d.) The net power generated is

$E = \vec{F} \cdot \vec{v}$

By Fig (a),

$\vec{F} = 18 - (0.30)(10+20+30) = 0$

$\therefore E = 0 \cdot \vec{v} = 0$ watts

15.15 GIVEN Figure a, pressure-volume diagram for an engine

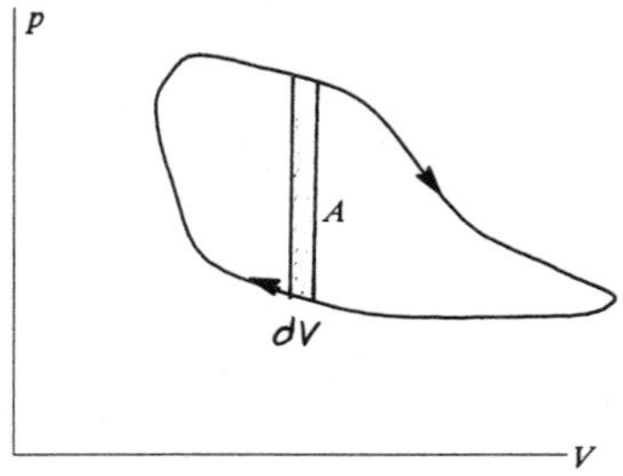

Figure (a)

FIND

a.) Show that the work is $U = KA$, where K is a constant and A is the area enclosed.

b.) The scale for the p axis is 1 in : 150 psi
The scale for the V axis is 1 in : 800 in³
Determine the factor K so that U is expressed in ft-lb

SOLUTION

a.) $U = \oint p\,dV$ if integration is clockwise.
Graphically, this integral is represented by the area A of the indicator diagram.
Hence, $U = KA$

b.) With given scales,

$p = 150y$ and $V = 800x$; $dV = 800\,dx$

$$\therefore U_{1-2} = \int p\,dV = \int 150y \cdot 800\,dx = 120000 \int y\,dx$$

where $\int y\,dx$ = Area.

Hence, $U_{1-2} = 120000\,A$ in·lb $= 10{,}000\,A$ ft·lb

$\therefore K = 10{,}000$

15.16

Show that, if $\gamma \neq 1$, the work performed by a piston in compressing a gas is

$$U = \frac{1}{\gamma - 1}(p_1 V_1 - p_0 V_0)$$

where pV^γ = constant; γ is a constant.

$$U_{1-2} = -\int_{V_0}^{V_1} p\,dV \qquad \text{(Refer to problem 15.6)}$$

$$pV^\gamma = c; \quad p = \frac{c}{V^\gamma}$$

$$\therefore U_{1-2} = -\int_{V_0}^{V_1} \frac{c}{V^\gamma}\,dV \qquad (\gamma \neq 1)$$

$$U_{1-2} = -c\int_{V_0}^{V_1} \frac{dV}{V^\gamma} = -c\int_{V_0}^{V_1} V^{-\gamma}\,dV$$

$$U_{1-2} = -c\left(\frac{V^{(-\gamma+1)}}{(-\gamma+1)}\right)\Bigg|_{V_0}^{V_1}$$

$$U_{1-2} = -\left(\frac{cV_1^{1-\gamma}}{1-\gamma} - \frac{cV_0^{1-\gamma}}{1-\gamma}\right)$$

Now, $c = pV^\gamma$

$$U_{1-2} = -\left(\frac{p_1 V_1^\gamma V_1^{1-\gamma}}{1-\gamma} - \frac{p_0 V_0^\gamma V_0^{1-\gamma}}{1-\gamma}\right)$$

$$U_{1-2} = -\left(\frac{p_1 V_1}{1-\gamma} - \frac{p_0 V_0}{1-\gamma}\right)$$

$$U_{1-2} = \frac{1}{\gamma - 1}(p_1 V_1 - p_0 V_0)$$

15.17 GIVEN A bead slides on a frictionless parabolic wire ($y = 2x^2$) (Fig. a) The bead starts from rest at $x = 1$, $y = 2$ [ft].

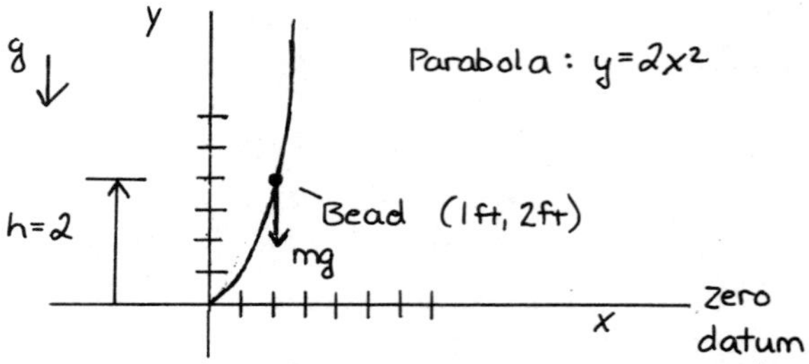

Figure a.

(Continued)

15.17 cont.

FIND The speed at $x=0,\ y=0$

SOLUTION:

$$U_{1-2} = T_2 - \cancel{T_1}^{\,0},$$ since the bead starts from rest

$$U_{1-2} = (mg)(2\text{ ft}) = \tfrac{1}{2} m v_2^2 - 0$$

$$\therefore\ v_2^2 = 4g = (4)(32.2)$$

or $$\underline{\underline{v_2 = 11.349\ \text{ft/s}}}$$

15.18

GIVEN Liquid flows uniformly through a pipe and the pressure drop in 100 ft is 5 lb/in² (Fig. a)

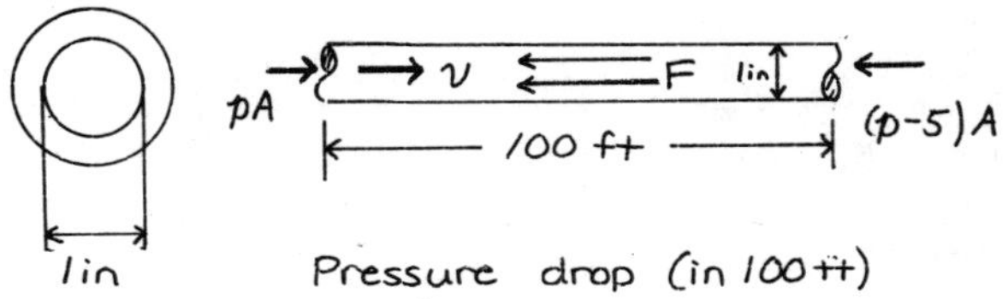

Pressure drop (in 100 ft) = 5 lb/in²

Velocity of liquid along wall = 0.200 ft/s

Figure a

FIND The work done by friction per minute.

SOLUTION:

By Fig. a,

$$\Sigma F = pA - F - (p-5)A = 0$$

(Net force is zero because of uniform flow)

$$\therefore\ F = 5A,$$

where $A = \pi(0.5)^2$ = cross-sectional area of fluid

The work of friction per second is

$$U_{friction} = F\cdot v = \left(5\tfrac{lb}{in^2}\right)\left[\pi(0.5)^2 in^2\right](0.200\ ft/s)$$

$$\underline{\underline{U_{friction} = 0.7854\ \text{ft}\cdot\text{lb/s} = 47.12\ \frac{\text{ft.lb}}{\text{min}}}}$$

15.19

GIVEN A box (8 lb) slides 20 ft on a horizontal pavement; $\mu_K = 0.20$ (Fig. a).

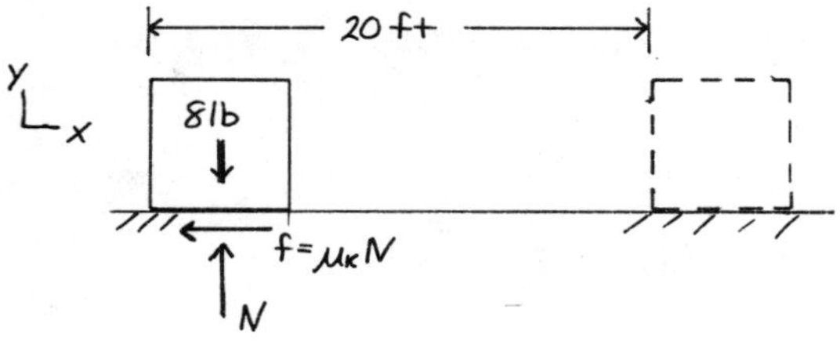

Figure a.

FIND:

a.) The work done by friction

b.) The box's speed at $x = 20$ ft; at $x=0$, $v = 20$ ft/s.

SOLUTION:

a.) $\Sigma F_y = N - 8\text{lb} = 0;\qquad N = 8\text{ lb}$

$\therefore\ f = 0.20(8\text{lb}) = 1.6\text{ lb}$

$$U_{1-2} = \int_{x_1}^{x_2} F\,dx$$

$$U_{1-2} = -\int_0^{20\text{ ft}} 1.6\,dx;$$

$$\underline{\underline{U_{1-2} = -32.0\ \text{ft}\cdot\text{lb}}}$$

b.) $U_{1-2} = T_2 - T_1$

$$\therefore\ -32\ \text{lb}\cdot\text{ft} = \tfrac{1}{2}\left(\frac{8\text{lb}}{32.2\ \text{ft/s}^2}\right)v^2 - \tfrac{1}{2}\left(\frac{8\text{lb}}{32.2\ \text{ft/s}^2}\right)(20\ \text{ft/s})^2$$

$0.1242\,v^2 = 17.689;\qquad \underline{\underline{v = 11.93\ \text{ft/s}}}$

15.20

GIVEN A constant force, $\vec{F} = 3\hat{\imath} + 2\hat{\jmath} + 3\hat{k}$ [N], acts on a particle of mass $m = 1$ gram

FIND

a.) The work done by force as the particle goes from (0,0,0) m to (0,2,4) m.

b.) The speed of particle at (0,2,4) assuming the particle starts from rest.

(Continued)

15.20 cont.

SOLUTION

a.) $U_{1-2} = \int F_x\,dx + \int F_y\,dy + \int F_z\,dz$

$U_{1-2} = \int_0^0 3dx + \int_0^2 2dy + \int_0^4 3dz$

$U_{1-2} = 0 + 2(2) + 4(3)$

$U_{1-2} = 16$ J

b.) $U_{1-2} = T_2 - T_1$, since particle starts from rest. ($T_1 = 0$)

$16\text{ J} = \frac{1}{2}mv^2 = \frac{1}{2}\left(1g \times \frac{1kg}{1000g}\right)v^2$

$16\text{ J} = \frac{v^2}{2000}$

$v = 178.9$ m/s

15.21

GIVEN A crate slides down a 60° plane (Fig. a). $\mu_K = 0.50$

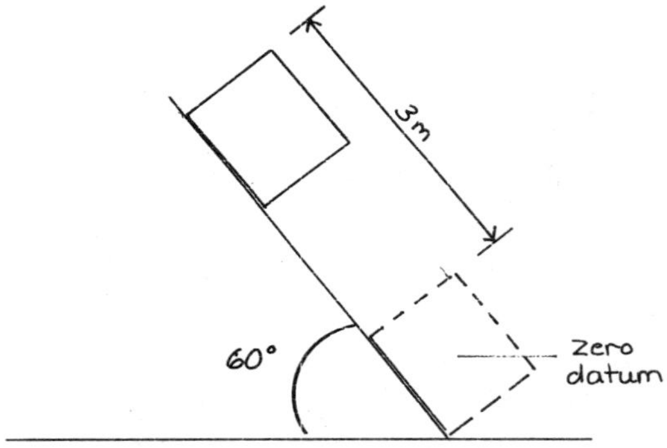

Figure a

FIND By the principle of kinetic energy, find the speed of the crate when it has moved 3m from rest.

SOLUTION

By the free-body diagram of the crate (Fig b).

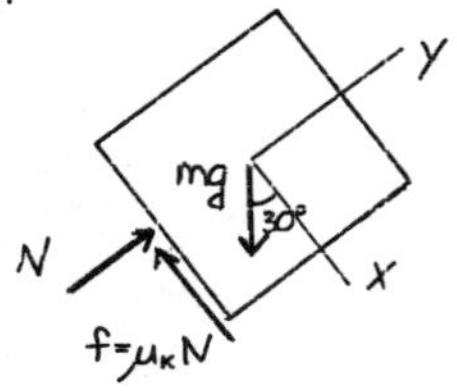

Figure b

$\Sigma F_y = N - mg\sin 30° = 0;$

$N = mg\sin 30°$

$\therefore\ f = 0.50\,(mg\sin 30) = 0.25\,mg$

Since the crate starts from rest,

$U_{1-2} = -(f)(3) + mg(3\sin 60°) = \frac{1}{2}mv^2$

or $U_{1-2} = -0.25\,mg(3) + mg\,(3\times\sqrt{3}/2) = \frac{1}{2}mv^2$

or $v^2 = (3.696)g = (3.696)(9.81)$

$v = 6.02$ m/s

15.22

GIVEN A particle moves along the path $x=t$, $y=t^2$, $z=t^3$ [m], and is resisted by a force $F_y = -3\frac{dy}{dt}$, $F_z = -3\frac{dz}{dt}$, $F_x = -3\frac{dx}{dt}$

FIND: The work done by the resisting force when t goes from 1 to 3 s.

SOLUTION

Since $x=t$, $y=t^2$; and $z=t^3$;
$dx = dt$ $dy = 2t\,dt$ and $dz = 3t^2dt$

The work done by the force is

$U_{1-2} = \int_{t_1}^{t_2} (F_x V_x + F_y V_y + F_z V_z)\,dt$

$\therefore\ U_{1-2} = \int_1^3 [(-3)(1) + (-6t)(2t) + (-9t^2)(3t^2)]\,dt$

$U_{1-2} = \int_1^3 (-3 - 12t^2 - 27t^4)\,dt$

$U_{1-2} = \left(-3t - 4t^3 - \frac{27}{5}t^5\right)\Big|_1^3$

$U_{1-2} = -1429.2 + 12.4 = -1417$ J

15.23

GIVEN: The uniform bar AB weighs 40 lbs and is hinged at A (Fig a.) The disk weighs 18 lb.

FIND: The work done by gravity as the system drops to the vertical position

(Continued)

15.23 cont.

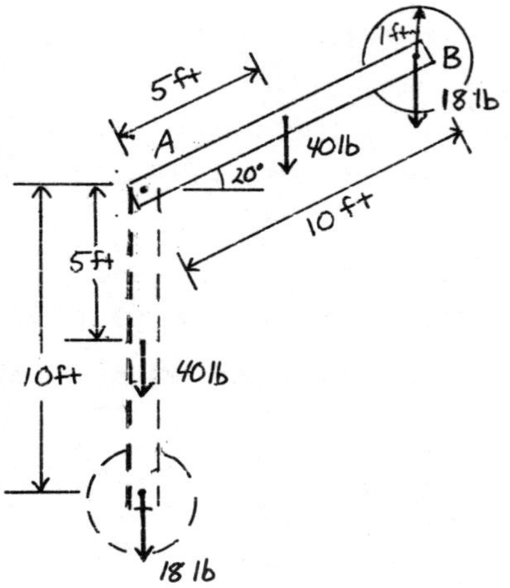

Figure a.

SOLUTION

Let the initial position of the system be the zero datum.

$$U_{1-2} = -W \Delta \bar{z}$$

$$U_{1-2} = -[40(-5\sin 20) + 18(-10\sin 20)] - [40(-5) + 18(-10)]$$

$$U_{1-2} = 130.0 + 380.0 \ (\text{lb}\cdot\text{ft})$$

$$\underline{\underline{U_{1-2} = 510.0 \ (\text{lb}\cdot\text{ft})}}$$

15.24

GIVEN: Roller coaster speed is 16 km/h at top of 30 m drop (Fig. a) During the drop, 20% of work of gravity is dissipated by friction.

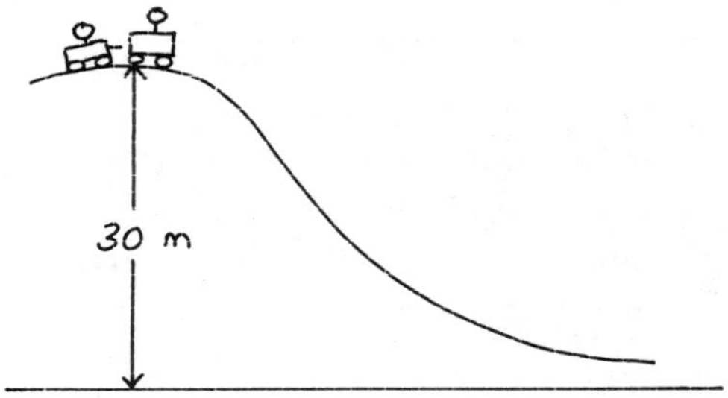

Figure a.

FIND: The speed of the coaster at bottom of drop (Neglect rotational K.E. of wheels).

SOLUTION

By law of kinetic energy and Fig. a,

$$U_{1-2} = T_2 - T_1$$

$$(0.80)(\cancel{m}g)(30) = \tfrac{1}{2}\cancel{m}v_2^2 - \tfrac{1}{2}\cancel{m}v_1^2$$

$$v_1 = 16 \text{ km/h} = 4.444 \text{ m/s}$$

$$\therefore \ v_2^2 = 48g + (4.444)^2 = (48)(9.81) + (4.444)^2$$

or

$$\underline{\underline{v_2 = 22.15 \text{ m/s} = 79.74 \text{ km/h}}}$$

15.25

GIVEN A force $\vec{F} = -3.75t\hat{i} + 5\hat{j}$ [N], (t = time in seconds), acts on a particle that moves in the (x, y) plane such that its coordinates are $x = 2.0 - 0.5t^3$, $y = 2t^2$ [m]

FIND: a.) The work done by the force during the time interval $t = 0$ to $t = 3$ s.

b.) The mass of the particle

c.) The particle's speed at $t = 3$ s.

SOLUTION:

a) $$U_{1-2} = \int_{t_1}^{t_2} F \cdot v \, dt = \int_{t_1}^{t_2} (F_x V_x + F_y V_y)\, dt$$

Differentiation of x and y yields

$$v_x = \frac{dx}{dt} = -1.5t^2 \qquad \text{and}$$

$$v_y = \frac{dy}{dt} = 4t$$

$$\therefore \ U_{1-2} = \int_0^3 [(-3.75t)(-1.5t^2) + 5(4t)]\, dt$$

$$= \int_0^3 (5.625t^3 + 20t)\, dt$$

$$= 1.406t^4 + 10t^2 \Big|_0^3$$

$$\therefore \ \underline{\underline{U_{1-2} = 203.9 \text{ J}}}$$

(Continued)

15.25 cont.

b.) A second differentiation of x and y yields:

$$\frac{d^2x}{dt^2} = -3t, \qquad \frac{d^2y}{dt^2} = 4$$

or $\vec{a} = (-3t)\hat{\imath} + 4\hat{\jmath}$

$$|a| = \sqrt{(-3t)^2 + 4^2} = \sqrt{9t^2+16}$$

$$|F| = \sqrt{(-3.75t)^2 + 5^2}$$

By F = ma

$$\sqrt{(-3.75)^2 t^2 + 25} = m\sqrt{9t^2+16}$$

$\therefore$ $\underline{\underline{m = 1.25 \text{ kg}}}$

c.) $U_{1-2} = T_2 - T_1$ ($T_1 = 0$) since $\frac{dx}{dt} = \frac{dy}{dt} = 0$ at $t = 0$.

$\therefore 203.9 = \tfrac{1}{2}(1.25)v^2$

$\underline{\underline{v = 18.06 \text{ m/s}}}$

CHECK: $\frac{dx}{dt} = -1.5t^2, \quad \frac{dy}{dt} = 4t$

$\therefore \vec{v} = -1.5t^2\hat{\imath} + 4t\hat{\jmath}$

$|v| = \sqrt{(1.5t^2)^2 + (4t)^2}$

$v = \sqrt{2.25t^4 + 16t^2}\,\Big|_{t=3s}$

$v(3) = 18.06$ m/s ✓

15.26

GIVEN: A block of mass m = 6 kg on a frictionless horizontal surface is attached to a nonlinear, unstretched spring (Fig. a), whose force-extension relation is F = 400 sin 10x. The block is moved 0.10 m to the right and released from rest.

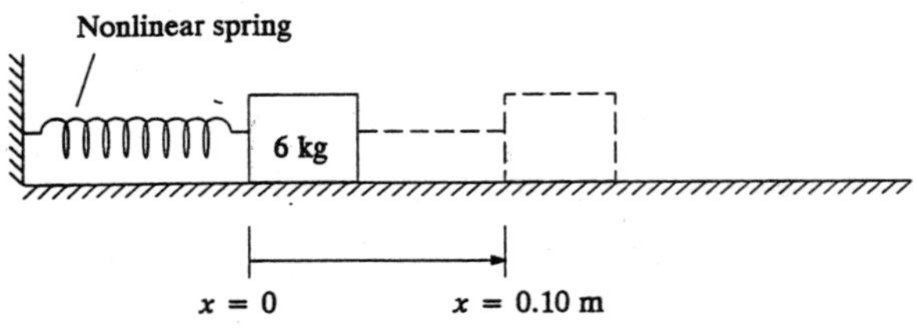

Figure a.

FIND: a.) The work done by the spring on the block during the time it moves back to the equilibrium position (x = 0).

b.) The speed of the block as it passes through its equilibrium position

c.) The work done by the spring as the block moves from 0 to −0.10 m

d.) The speed of the block at x = 0.10 m

SOLUTION:

a.) $U_{1-2} = \int_{x_1}^{x_2} F_x\,dx$

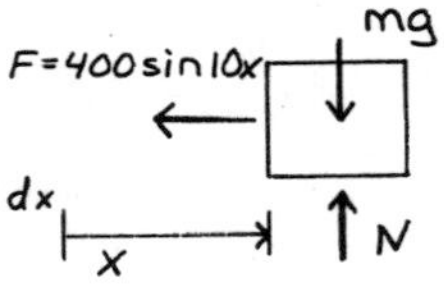

Figure b

By Fig. b, $U_{1-2} = -\int_{0.10}^{0} 400\sin 10x\,dx$

$U_{1-2} = 40\cos 10x\,\Big|_{0.10}^{0}$

$\underline{\underline{U_{1-2} = 40 - 21.612 = 18.39 \text{ J}}}$

b.) $U_{1-2} = T_2 - T_1$ ($T_1 = 0$) since the mass is released from rest

$18.39 = \tfrac{1}{2}mv^2 = 3v^2;$

$\therefore$ $\underline{\underline{v = 2.48 \text{ m/s}}}$

c.) By Fig. c,

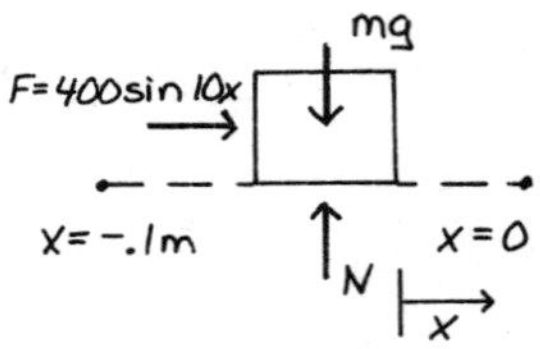

Figure c

$U_{1-2} = \int_0^{0.1} 400\sin 10x\,dx$

$U_{1-2} = -40\cos 10x\,\Big|_0^{0.1}$

$= -40 + 21.612$

$\underline{\underline{U_{1-2} = -18.39 \text{ J}}}$

d.) $U_{1-2} = T_2 - T_1$

$\therefore 18.39 = \tfrac{1}{2}(6)v^2 - \tfrac{1}{2}(6)(2.48)^2$

$v = 0$ m/s

15.27

GIVEN: Water (62.4 lb/ft³) is pumped from a cylindrical well, (dia = 20 ft) into a tank (dia. = 30 ft). The bottom of the tank is 80 ft above the water surface in the well, initially (Fig. a).

(Continued)

15.27 cont.

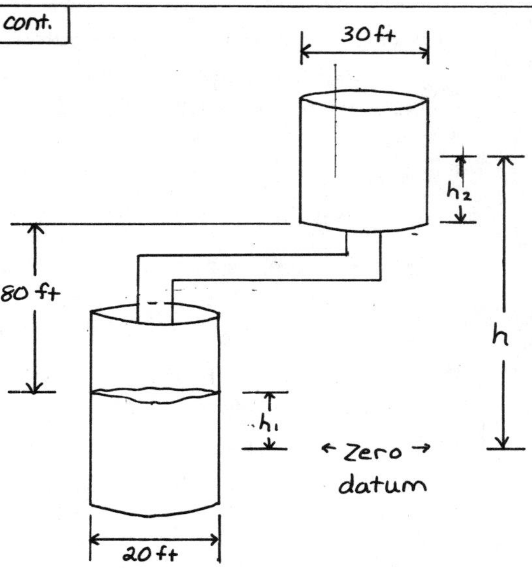

Figure a.

FIND: The total work performed against gravity to lift 10,000 ft³ water from the well into the tank. The well is not refilled during pumping.

Solution:

Average height of the water in the well is:

$$h_1 = \frac{V}{A_w}\cdot\frac{1}{2} = \frac{10{,}000\ ft^3}{2\pi\,(10\,ft)^2} = 15.92\,ft$$

Average height of water in the tank is:

$$h_2 = \frac{V}{A_T}\cdot\frac{1}{2} = \frac{10{,}000\,ft^3}{2\pi\,(15\,ft)^2} = 7.07\,ft$$

Weight of 10,000 ft³ of water is

$$W = (10{,}000)(62.4) = 624{,}000\ lb$$

Hence, work required to lift the water is:

$$U_{water} = Wh = (624{,}000)(15.92+80+7.07)$$

or

$$U_{water} = 64.3\times10^6\ ft\cdot lb = 64.3\times10^3\ ft\cdot kip$$

15.28

GIVEN: A column of mercury (270 N) is held in a bent tube (Fig a.). The mercury is released from rest.

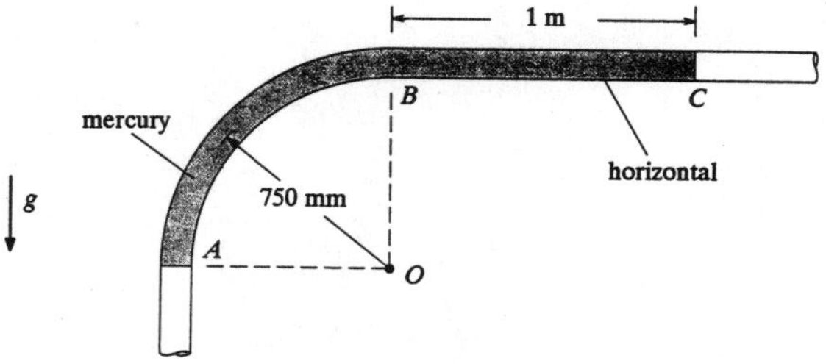

Figure a.

FIND: a.) The work done by gravity from the time the mercury is released until the right end of the column passes B.

b.) The speed when the right end is at B. Neglect friction.

SOLUTION:

a.) The part BC is effectively transferred to the vertical leg. The total length of the column of mercury is $(\pi/2)(0.75) + 1$ m. Hence, the weight of the horizontal part is

$$W = (270)\left(\frac{1}{\left(\frac{\pi}{2}\right)\left(\frac{3}{4}\right)+1}\right) = 123.96\ N$$

This part drops $d = 0.75 + 0.5 = 1.25$ m. Hence, work performed by gravity is,

$$U_{weight} = Wd = (123.96)(1.25) = 154.95\ N\cdot m$$

b.) The speed of the mercury when the right end of the column passes B is obtained from

$$U_{weight} = T_2 - \cancel{T_1}^{\,0},$$ since mercury is initially at rest

$$\therefore\ 154.95 = \tfrac{1}{2}mv^2 = \tfrac{1}{2}\left(\frac{270}{9.81}\right)v^2$$

or $v = 3.356$ m/s

15.29

GIVEN The block weighs 96 lb (See Fig. a.) $\mu_k = 0.50$. The spring constant $k = 8$ lb/ft. The block's speed is 24 ft/s, when it strikes the spring.

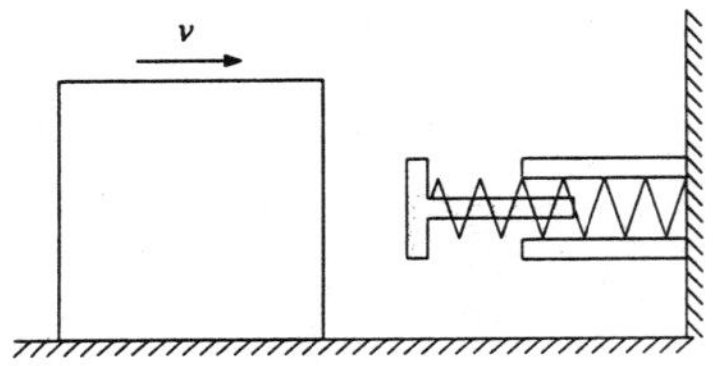

Figure a.

FIND: The max compression of the spring in ft.

SOLUTION:

By Fig. b,

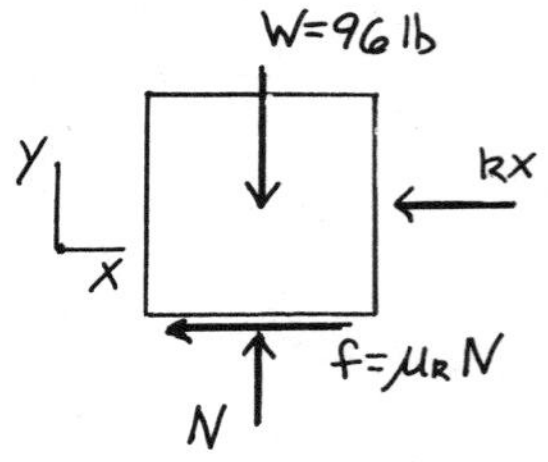

Figure b.

$$\Sigma F_y = N - 96\text{ lb} = 0$$

$N = 96$ lb

$\therefore f = 0.50(96\text{ lb}) = 48$ lb

Let the maximum compression of the spring be e_{max}. Then, the law of kinetic energy yields

$$U_{1-2} = T_2 - T_1 = -T_1$$

since at maximum compression, the block is stopped.

By Fig. b,

$$U_{1-2} = -\int_0^{e_{max}} kx\,dx - \int_0^{e_{max}} f\,dx = -\tfrac{1}{2}mv_1^2$$

or $$U_{1-2} = -\tfrac{1}{2}(8)e_{max}^2 - 48e_{max} = -\tfrac{1}{2}\left(\tfrac{98}{32}\right)(24)^2$$

$$\therefore e_{max}^2 + 12e_{max} - 216 = 0;$$

$$\underline{\underline{e_{max} = 9.87\text{ ft}}}$$

15.30

GIVEN: The spring in Problem 15.29 is replaced by a bumper that exerts a resisting force $F = 16x^2 + 32x$ (lb) where x is the distance (ft) the bumper is compressed.

FIND: The max compression of bumper.

SOLUTION

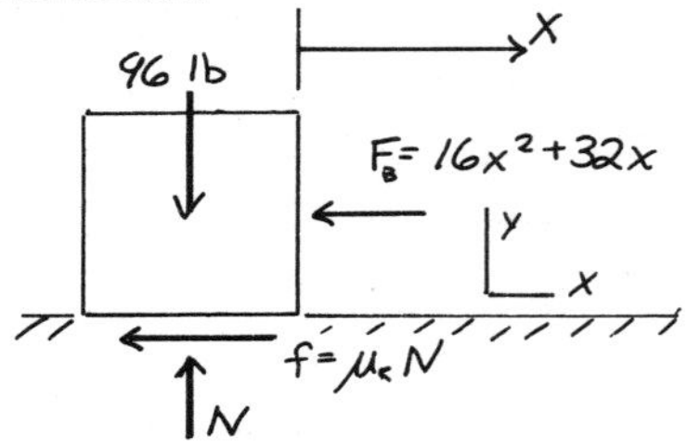

Figure b

By Fig. a,

$$\Sigma F_y = N - 96 = 0; \qquad N = 96\text{ lb}.$$

$$\therefore f = 0.50(96) = 48\text{ lb}$$

The work that the bumper performs on the block is

$$(U_{1-2})_B = -\int_0^{e_{max}} (16x^2 + 32x)\,dx$$

$$= -\tfrac{16}{3}e_{max}^3 - 16e_{max}^2$$

The work of friction is

$$(U_{1-2})_f = -\int_0^{e_{max}} f\,dx = -48e_{max}$$

$$\therefore U_{1-2} = -\tfrac{16}{3}e_{max}^3 - 16e_{max}^3 - 48e_{max}$$

$$= \tfrac{1}{2}\left(\tfrac{96}{32}\right)(24)^2$$

or $$e_{max}^3 + 3e_{max}^2 + 9e_{max} - 162 = 0$$

$$\therefore \underline{\underline{e_{max} = 4.167\text{ ft}}}$$

15.31

GIVEN Particle P constrained to move on frictionless path of radius r (Fig a.). The particle is acted on by gravity and a tangential force

$$F = as + b$$

(Continued)

15.31 cont.

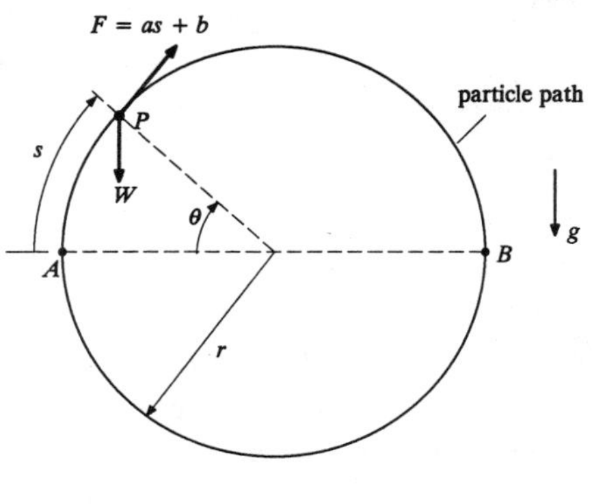

Figure a.

FIND: a.) The total work done on the part as it moves from A to B, in terms of a, b, W, and r. Neglect friction.

b.) The speed of particle at B if $v = 0$ at A.

SOLUTION

a.) Work of gravity:

$$U_{1-2_G} = -W\Delta z$$

$$U_{1-2_G} = -W(0-0) = 0\ \text{J};$$ → W does no work.

Work of Force F:

$$U_{1-2_F} = \int_{x_1}^{x_2} F \cdot dx$$

By Fig. a, $s = r\theta$, $ds = r\,d\theta$

$$\therefore U_{1-2} = \int_A^B (as+b)\,ds = \int_0^{\pi} [(ar\theta + b)(r\,d\theta)]$$

$$U_{1-2_F} = ar^2\frac{\theta^2}{2} + br\theta \Big|_0^{\pi} = \tfrac{1}{2}ar^2\pi^2 + br\pi$$

$$\therefore U_{1-2_F} = \tfrac{1}{2}a\pi^2r^2 + b\pi r \quad [\text{J}]$$

$$\therefore \underline{U_{1-2_{Tot}} = \tfrac{a}{2}\pi^2r^2 + b\pi r \quad [\text{J}];}$$

b.) $U_{1-2} = T_2 - T_1^{\,0}$

$$U_{1-2} = \tfrac{a}{2}\pi^2r^2 + b\pi r = \tfrac{1}{2}\left(\tfrac{W}{g}\right)v^2$$

$$\therefore v = \sqrt{\frac{g\pi r}{W}(a\pi r + 2b)}$$

15.32

GIVEN: A bead (1N) is released at A and it slides down the frictionless wire (Fig. a). The wire is straight from A to P. At P, the wire forms a circular arc of radius 1m.

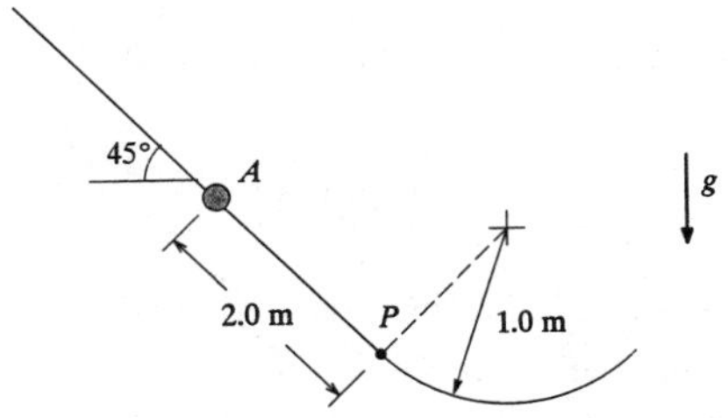

Figure a.

FIND: The change in force that the bead exerts on the wire as it passes point P.

SOLUTION

By Fig b, along the straight part,

$$\Sigma F_y = N_s - 1\cos 45° = 0$$

$$N_s = 0.707\ [\text{N}]$$

$$\therefore \underline{N_s = 0.7071\ \text{N} \text{ from A to P}}$$

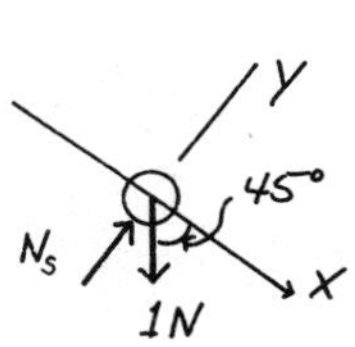

Figure b.

In the circular arc, the normal acceleration (Fig. c) is:

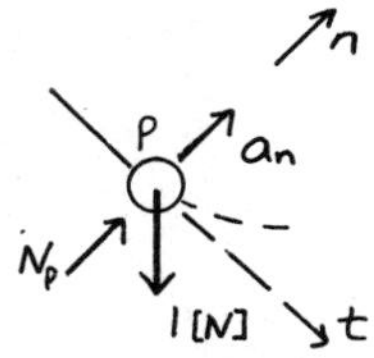

Figure c

$$a_n = \frac{v^2}{r} = v^2;$$

$$r = 1\text{m}$$

To find velocity at P, use the law of kinetics with zero datum at P (Fig. a)

$U_{1-2} = T_2 - T_1 = T_2$, since bead starts from rest

$$U_{1-2} = (mg)(2\sin 45°) = 1.414\ [\text{N}\cdot\text{m}]$$

$$T_2 = \tfrac{1}{2}mv^2 = \tfrac{1}{2}(1/9.81)v^2$$

$$\therefore 1.414 = \tfrac{1}{2}(1/9.81)v^2; \quad v = 5.268\ \text{m/s}$$

(Continued)

15.32 cont.

By Fig. c, at P;

$\sum F_n = N_P - (1)(\cos 45°) = mv^2$

$\therefore\ N_P = \cos 45° + (1/9.81)(5.268)^2$

$N_P = 3.536$ [N]

$\therefore\ \underline{\Delta N = N_P - N_S = 2.829 \text{ [N]}}$

15.33

GIVEN: A 50 N part slides on a circular rod (Fig. a.), under the action of gravity and a resistive magnetic force $F = 18\cos\theta$ [N]

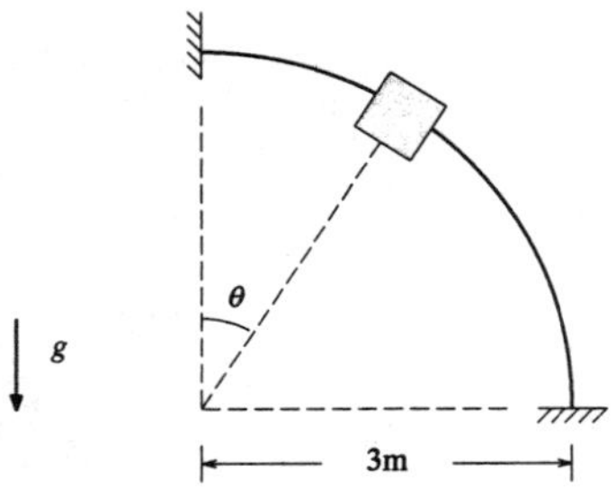

Figure a.

FIND: a.) The total work done on the part as θ goes from 30° to 90°

b.) The speed of the particle at θ = 90°

SOLUTION

a.) Work by gravity:

$(U_{1-2})_G = -W\Delta\bar{z}$ (Theorem 15.6)

$(U_{1-2})_G = -50(z_2 - z_1) = -50(0 - 3\cos 30°)$

$(U_{1-2})_G = 129.9$ J

by magnetic force:

$(U_{1-2})_f = \int_{s_1}^{s_2} F \cdot ds$

$(U_{1-2})_f = \int_{s_1}^{s_2} (-18)(\cos\theta)ds$; $ds = r\,d\theta$

$(U_{1-2})_f = -54\sin\theta \Big|_{30°}^{90°} = -27$ J

$\underline{(U_{1-2})_{Tot} = (U_{1-2})_G + (U_{1-2})_f = 102.9 \text{ J}}$

b.) At θ = 30°, $v = 0$ m/s

$\therefore\ U_{1-2} = T_2 - T_1$

or $102.9 = \tfrac{1}{2}mv^2 = \tfrac{1}{2}\left(\frac{50\text{ N}}{9.81\text{ m/s}^2}\right)v^2$

$\underline{v = 6.35 \text{ m/s}}$

15.34

GIVEN: A uniform cubical crate, set on edge at 30°, is released from rest (Fig. a)

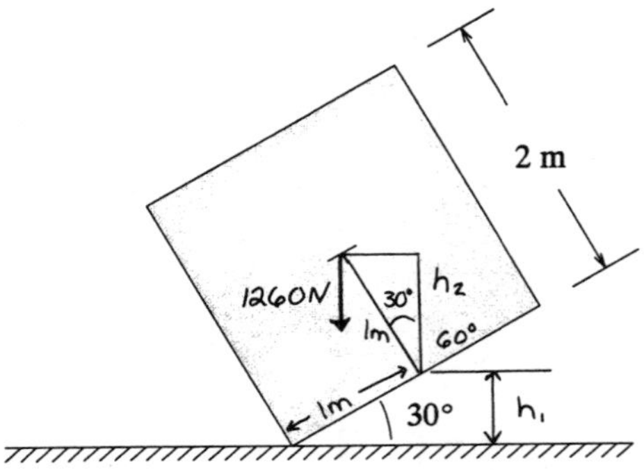

Figure a.

FIND: The work done by gravity from when crate is released to when its face strikes the floor.

SOLUTION:

$U_{1-2} = -W\Delta\bar{z}$ (Theorem 15.6)

Let $\bar{z}$ be the height of the center of the block above the floor. Initially, $\bar{z} = \bar{z}_1 = h_1 + h_2$ where

$h_1 + h_2 = 1\cdot\sin 30° + 1\cdot\cos 30° = 1.366$ m

When the face strikes the floor, $\bar{z} = \bar{z}_2 = 1$ m

$\therefore\ \Delta\bar{z} = 1 - 1.366 = -0.3660$ m

$\therefore\ U_{1-2} = -(1260\text{ N})(-0.3660)$

$\underline{= 461.2 \text{ J}}$

15.35

GIVEN: A block (45 N) rests on a horizontal plate (Fig a.). The plate is given horizontal acceleration $a = 12/(t+1)$ $[m/s^2]$; t in seconds.

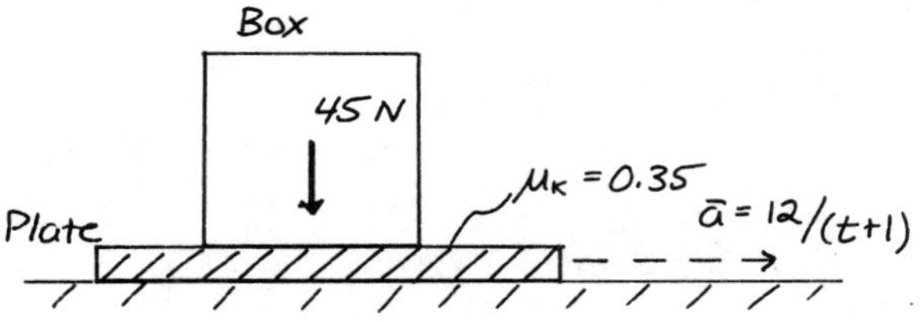

Figure a.

FIND: a.) The work done on the block by friction for $0 \le t \le 2$ [s]

b.) The block's speed at $t = 2s$.

SOLUTION:

a.) By Fig. b,

$\Sigma F_y = N - 45N = 0;$

$N = 45\ N$

$\therefore f = 0.35(45N)$

$= 15.75\ N$

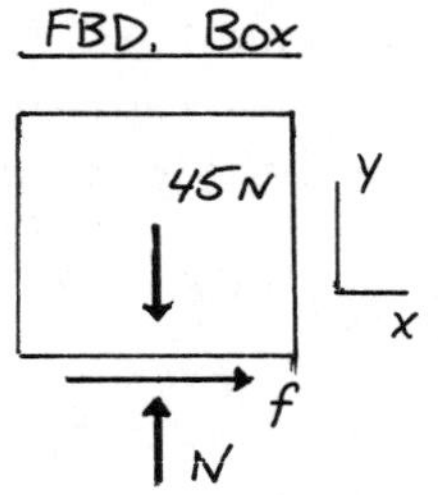

Figure b.

Initially, at $t=0$,

$a_{plate} = 12\ m/s^2$

$\therefore \Sigma F_x = 15.75 = \left(\frac{45}{9.81}\right) a_{Box}$

$a_{Box} = 3.434\ m/s$

$\therefore$ Box slides at $t=0$ and since $3.434 < 12/t+1$ for $0 \le t \le 2$ [s], box slides during interval $0 \le t \le 2$ s.

$\therefore V_{Box} = 3.434\,t$

$(U_{1-2})_{Box} = \int_0^2 (15.75)(3.434\,t)\,dt$

$= 108.16\ N{\cdot}m$

b.) $U_{1-2} = T_2 - T_1^{\,0}$ since block starts at rest.

$108.16\ N{\cdot}m = \frac{1}{2}\left(\frac{45}{9.81}\right) v^2;$

$v = 6.867\ m/s$

Check: By $v = 3.434t$, at $t = 2$ s $v = 6.868$ m/s

15.36

GIVEN: Sleeve B weighs 10 N and can slide along the frictionless bar OA (Fig. a)

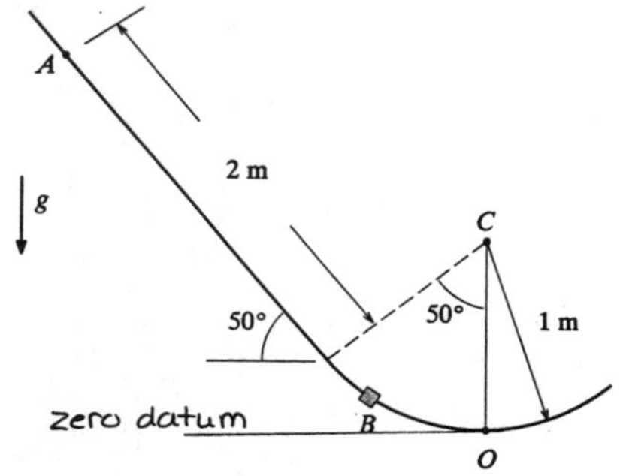

Figure a.

FIND: a.) The work required to raise the body from O to A.

b.) The speed of B at O, if it is released from rest at A.

c.) The force that the bar exerts on the body at O

SOLUTION

a.) The work required to raise B from O to A is: $U_{12} = W\Delta h$

$U_{1-2} = 10\left[[(1-\cos 50°) + 2\sin 50°] - 0\right]$

$U_{1-2} = 18.89\ J$

b.) $U_{1-2} = T_2 - T_1^{\,0}$ since B is released from rest at A

$18.89 = \frac{1}{2}\left(\frac{10\ N}{9.81\ m/s^2}\right) V^2$

$v = 6.09\ m/s$

c.) At O, the acceleration of the sleeve is

$a_n = \frac{V^2}{r} = \frac{(6.09)^2}{1} = 37.1\ m/s^2$

Therefore, by Fig. (b),

$\Sigma F_n = F - W = \frac{W}{g} a_n$

$\therefore F = W + W/g\ a_n$

$= 10 + 10/9.81\ (37.1)$

or $F = 47.82\ N$

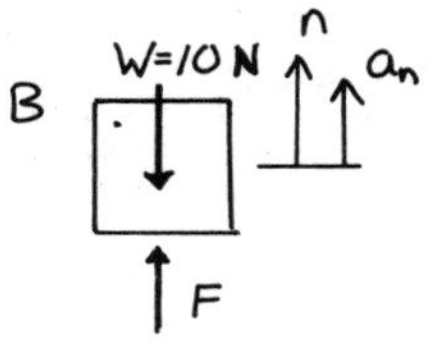

Figure b

15.37

GIVEN: Figure a; linear spring, $k = 175$ N/m, attached to fixed point C and to sleeve B, with unstretched length of 900 mm. Bar OA is frictionless.

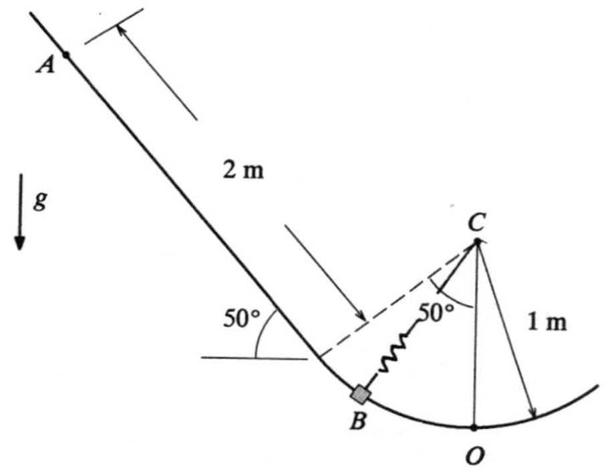

Figure a.

FIND: a.) The work needed to lift the sleeve B from O to A.

b.) The speed at O when the sleeve is released from rest at A.

c.) The force the bar exerts on the sleeve B at O.

SOLUTION

a.) Work required to lift the sleeve against gravity from O to A is (by Prob. 15.31)

$$U_{weight} = 18.89 \text{ J}$$

Work required to stretch spring from O to A is:

$$U_{spring} = \tfrac{1}{2} k \left(e_2^2 - e_1^2\right)$$

$$e_2 = \sqrt{5} - 0.90 = 1.336 \text{ m}$$

$$e_1 = 1 - 0.90 = 0.10 \text{ m}$$

$$\therefore \quad U_{spring} = \tfrac{1}{2}(175)(1.336^2 - 0.10^2) = 155.3 \text{ J}$$

Hence, the total work required to raise the sleeve from O to A is:

$$U_{Total} = U_{weight} + U_{spring}$$

or

$$U_{Total} = 18.89 + 155.3 = 174.2 \text{ J}$$

b.) The speed at O is given by (releasing the sleeve from rest at A)

$$U_{1-2} = T_2 - T_1$$

where

$$U_{1-2} = 174.2 \text{ J}$$

$$T_2 = \tfrac{1}{2} m v_0^2$$

$$T_1 = 0 \quad \text{(since sleeve released from rest at A)}$$

$$\therefore \; 174.2 = \tfrac{1}{2}\left(\frac{10}{9.81}\right) v_0^2;$$

$$v_0 = 18.487 \text{ m/s}$$

c.) Consider the FBD of the sleeve at O (Fig. b). Since the sleeve is moving on a circular path of radius 1 m, and there are no tangential forces acting, the acceleration is

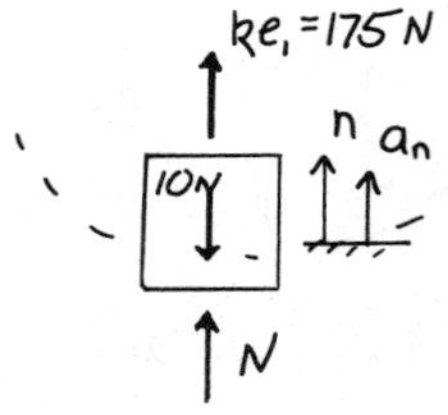

Figure b.

$$a_n = \frac{v_0^2}{r} = \frac{(18.487)^2}{1} = 341.78 \text{ m/s}^2$$

Hence, by Fig. b,

$$\Sigma F_n = N + 175 - 10 = m a_n = \left(\frac{10}{9.81}\right)(341.78)$$

or

$$N = 183.40 \text{ (N)}$$

15.38

GIVEN: Two water tanks (A and B) are connected by a pipe (Fig. a). They are 6 ft long (perpendicular to plane of figure). Valve C is opened to allow water to flow from tank A to tank B.

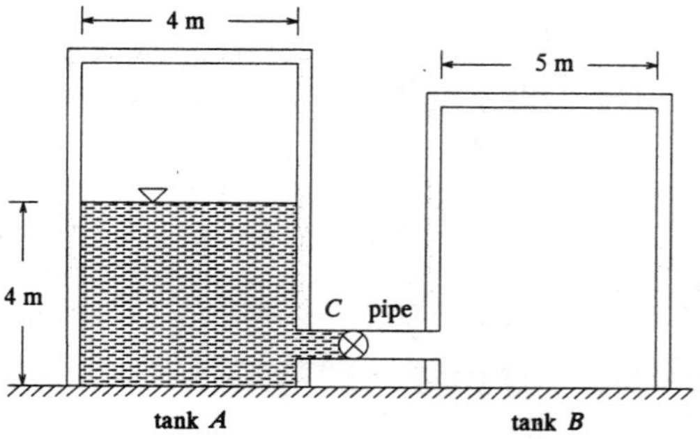

Figure a.

15.38 cont.

FIND: The work performed by gravity to the time the water reaches the same elevation in the tanks.

SOLUTION:

The volume of water is

$$V = (4\,m)(4\,m)(6\,m) = 96\,m^3$$

The final height, h_f of the water in the tank is given by (Fig. a)

$$(4m)(6m)h_f + (5m)(6m)h_f = 96\,m^3$$

$$\therefore\ h_f = 1.778\ m.$$

By Theorem 15.6,

$$U_{1-2} = -W\Delta\bar{z}$$

Effectively, the center of gravity of $(4-1.778)(4)(6) = 53.328\ m^3$ of water is lowered $\left(\frac{4-1.778}{2}\right) + \frac{1.778}{2} = 2m$ from tank A to tank B. Hence

$$U_{1-2} = -(53.328\ m^3)\left(9.80\ \frac{kN}{m^3}\right)(2m)$$

or $U_{1-2} = 1045\ kN\cdot m = 1045\ kJ$

15.39

GIVEN: Tank B in Problem 15.38 (Fig P15.38) is lowered 1m (Fig a.). Refer to data in Problem 15.38.

FIND: The work performed by gravity from the time the valve C is opened to the time that the surface of the water reaches the same elevation in both tanks.

SOLUTION:

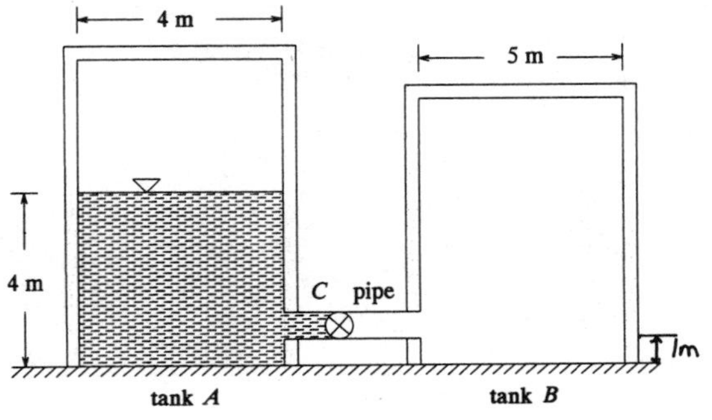

Figure a

Volume of water is

$$(4m)(4m)(6m) = 96\ m^3$$

By Fig a, the final height h_f of the water surface above the base of tank A is given by (Fig. a)

$$(4m)(6m)h_f + (5m)(6m)(1+h_f) = 96\ m^3$$

$$\therefore\ h_f = 1.222\ m$$

By Theorem 15.6,

$$U_{1-2} = -W\Delta\bar{z}$$

Effectively, the center of gravity of $(4-1.222)(4)(6) = 66.667\ m^3$ of water is lowered $\left(\frac{4-1.222}{2}\right) + \left(\frac{1+1.222}{2}\right) = 2.5\,m$ from tank A to tank B. Hence,

$$\therefore\ U_{1-2} = -(66.667\,m^3)\left(9.80\ \frac{kN}{m^3}\right)(-2.5\ m)$$

or $U_{1-2} = 1633\ kN\cdot m = 1633\ kJ$

15.40

GIVEN: A particle P moves on a circle under the action of a force $\bar{F} = (F_x = -ky,\ F_y = kx)$, k = constant and (x, y) are the coordinates of P (Fig. a)

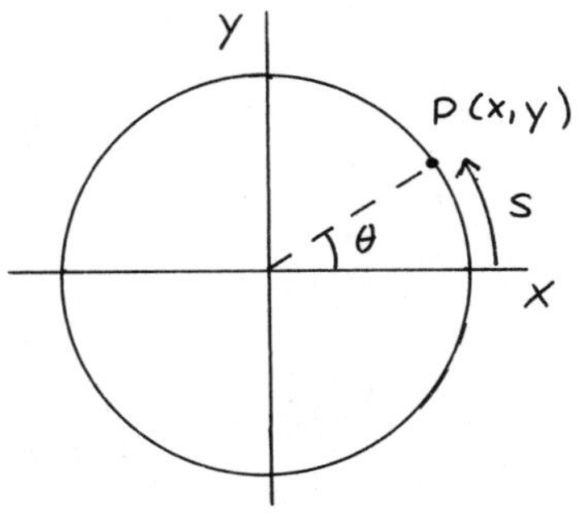

Figure a

FIND: Show that when the particle travels once around the circle (counter clockwise),

The work of the force F is:

$$U = 2kA$$

where A = area of circle.

(continued)

15.40 cont.

Solution:

By Fig. a, $x = r\cos\theta \qquad dx = -r\sin\theta\, d\theta$

$y = r\sin\theta \qquad dy = r\cos\theta\, d\theta$

The work of $\vec{F}$ is

$$U_{1-2} = \int F_y\, dy + \int F_x\, dx$$

$$U_{1-2} = \int kx\, dy + \int (-ky)\, dx$$

$$U_{1-2} = k\int_0^{2\pi} r\cos\theta(r\cos\theta)\, d\theta - k\int_0^{2\pi} r\sin\theta(-r\sin\theta)\, d\theta$$

$$U_{1-2} = k\int_0^{2\pi} [r^2(\sin^2\theta + \cos^2\theta)]\, d\theta$$

$$= kr^2\int_0^{2\pi} d\theta$$

$\therefore\ U_{1-2} = kr^2(2\pi)$

But, $A = \pi r^2$

$\therefore\ \underline{\underline{U_{1-2} = 2kA}}$

15.41

GIVEN: Particle P of weight W moves on a frictionless ring and is attracted toward A by $F = kr$, directed toward A (Fig. a) at B, $F = F_B$

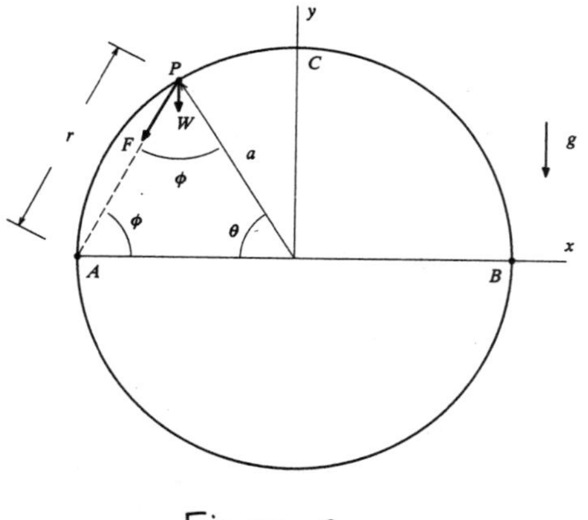

Figure a.

FIND: a.) The work done on a particle by force F in terms of F_B and a as it moves from A to B.

b.) The minimum clockwise speed at $\theta = 0$, in terms of F_B, a, and W, for the particle to reach point C.

SOLUTION:

a.) At A, $r = 0 \quad \therefore F = F_A = 0$

At B, $r = 2a \quad \therefore F = F_B = 2ka$

$$\therefore\ U_{1-2} = -\int_0^{2a} kr\, dr$$

$$= -\tfrac{1}{2}kr^2\Big|_0^{2a} = -2ka^2$$

In terms of F_B, $\underline{\underline{U_{1-2} = -F_B a}}$

b.) At C, $r = \sqrt{2}\, a$

$$U_{1-2} = -\int_0^{\sqrt{2}a} kr\, dr - Wa$$

So, $U_{1-2} = -2ka^2 - Wa$

Now,

$U_{1-2} = T_2 - T_1$

For min. v, $T_2 = 0$, $T_1 = \tfrac{1}{2}\left(\frac{W}{g}\right)v^2$

$\therefore\ -2ka^2 - Wa = -\tfrac{1}{2}\left(\frac{W}{g}\right)v^2$

$$\underline{\underline{v = \sqrt{\frac{4gka^2}{W} + 2ga} = \sqrt{\left(\frac{F_B}{W} + 1\right)(2ga)}}}$$

15.42

GIVEN: The thrust of a space rocket ceases at a distance of 500 miles above the earth's surface.

FIND: The minimum speed $v = v_{escape}$ that the rocket must have at this altitude to escape from the earth. Take the radius of the earth to be 4000 miles (Fig. a)

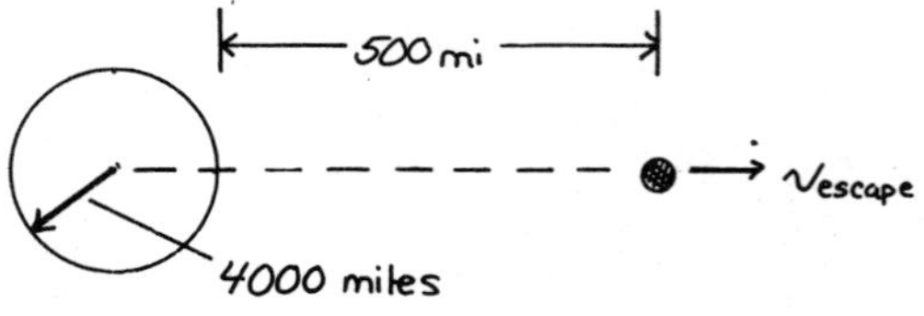

Figure a

(Continued)

15.42 cont.

SOLUTION

Let $r_E = 4000$ mi be the radius of the earth and its mass be m_E.

Let the mass of the rocket be m (Fig b.) Then, at a distance r from the center of the earth, the gravitational force acting on the rocket is

$$F = \frac{G m m_E}{r^2} \qquad (a)$$

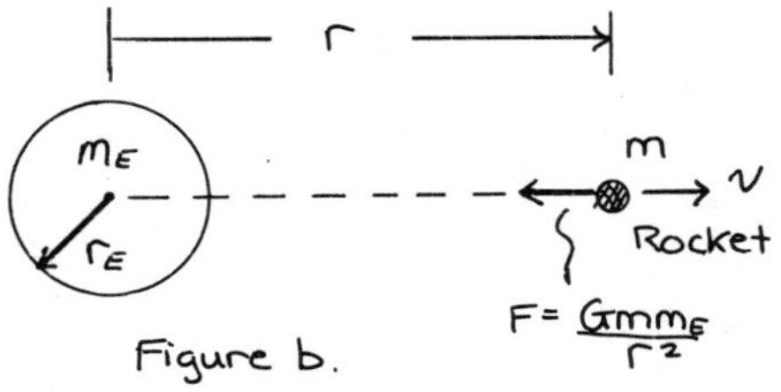

Figure b.

For the minimum escape speed v_{escape} at $r_1 = 4000 + 500 = 4500$ mi, the speed of the rocket goes to zero as r goes to infinity.

Hence, by the law of kinetic energy,

$$U_{1-2} = T_2 - T_1 \qquad (b)$$

where

$$U_{1-2} = \int_{r_1}^{\infty} F\,dr = -\int_{r_1}^{\infty} \frac{G m m_E}{r^2}\,dr$$

$$T_2 = 0$$

$$T_1 = \tfrac{1}{2} m v_{escape}^2$$

Therefore, by Eq. (b)

$$v_{escape}^2 = \frac{2Gm_E}{r_1} \qquad (c)$$

Now, on the surface of the earth,

$$F = \frac{Gmm_E}{r_E^2} = mg$$

or

$$G m_E = g r_E^2 \qquad (d)$$

Equations (c) and (d) yield

$$v_{escape}^2 = 2g\frac{r_E^2}{r_1} \qquad (e)$$

Thus, with $r_E = 4000$ mi $= 2.112\times10^7$ ft, $r_1 = 4000 + 500 = 4500$ mi $= 2.376\times10^7$ ft, and $g = 32.2$ ft/s^2, Eq (e) yields

$$v_{escape}^2 = 1.209\times10^9$$

So, $v_{escape} = 34{,}770$ ft/s $= 23{,}700$ mi/hr

15.43

GIVEN: A box (450 N) starts from rest and slides down an inclined plane (Fig. a), and strikes a spring after sliding 3 m.

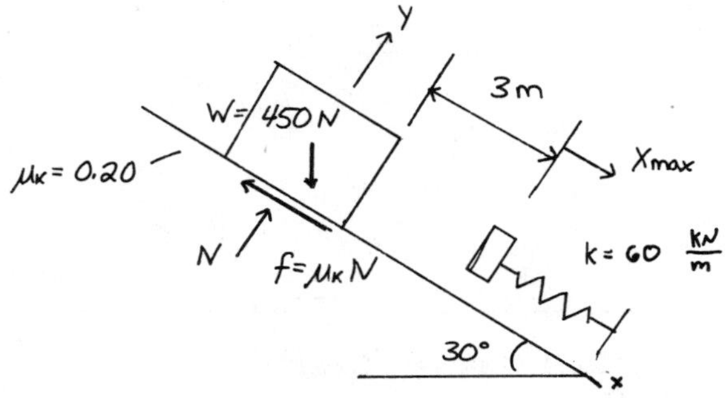

Figure a.

FIND: a.) The maximum compressive force (N) developed in the spring

b.) The distance the box rebounds.

SOLUTION:

a.) At maximum compression X_{max} the box is brought to rest.

$$\therefore U_{1-2} = T_2 - T_1 = 0$$

since the box starts and ends at rest. By Fig. a,

$$U_{1-2} = W(3 + X_{max})\sin 30° - f(3 + X_{max}) - \tfrac{1}{2}kX_{max}^2 = 0$$

or $X_{max}^2 - 0.004902\,X_{max} - 0.014706 = 0$

$$\therefore X_{max} = 0.1237 \text{ m}$$

Hence

$$F_{max} = kX_{max} = \left(60{,}000\ \frac{N}{m}\right)(0.1237\text{ m})$$

or

$$F_{max} = 7424 \text{ N}$$

b.) To determine the rebound distance X,

$$U_{1-2} = T_2 - T_1 = 0$$

since again, the box starts and ends at rest. Assume that the rebound distance is greater than 0.1237 m.

(Continued)

15.43 cont.

$$\therefore\ U_{1-2} = -W(x)\sin 30^\circ - fx + \tfrac{1}{2}k(0.1237)^2 = 0$$

$$\therefore\ -450(x)(0.50) - 77.94(x) + \tfrac{1}{2}(60{,}000)(0.01530) = 0$$

or $302.94\,x = 459$

$\therefore\ x = 1.515\ \text{m}$

15.44

GIVEN: A particle P moves in the (x, y) plane, while it is attracted to the origin O by the force $F = k/x$ (Fig. a)

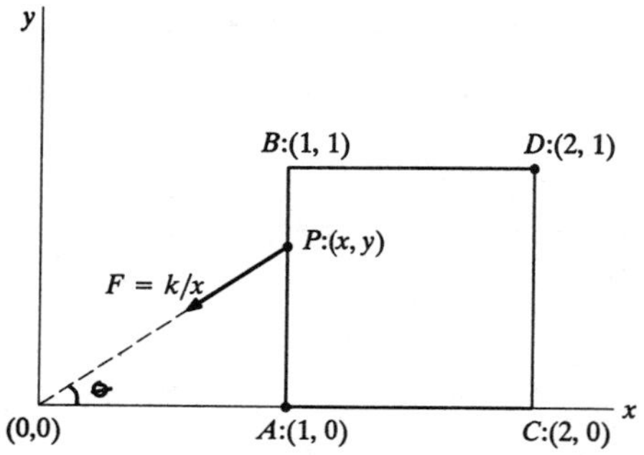

Figure a.

FIND:
a.) The work done by F as particle P moves from (1, 0) to (1, 1) to (2, 1).
b.) The work done by F as particle P moves from (1, 0) to (2, 0) to (2, 1)
c.) Whether or not the work done by the force depends upon the path.

SOLUTION:

a.) $U_{1-2} = \int_{x_1}^{x_2} F_x\,dx + \int_{y_1}^{y_2} F_y\,dy$

where

· From (1,0) to (1,1):

$$U_{1-2} = -\cancel{\int_0^1 \frac{k}{\sqrt{x^2+y^2}}\,dx}\to 0 \ - \int_0^1 \frac{ky}{\sqrt{1+y^2}}\,dy$$

Let $u = 1+y^2$, then $du = 2y\,dy$

$$\therefore\ U_{1-2} = -k/2 \int \frac{du}{u^{1/2}}$$

so,

$$U_{1-2} = -k\,u^{1/2} = -k\sqrt{1+y^2}\,\Big|_0^1 = k - \sqrt{2}\,k$$

· From (1,1) to (2,1):

$$U_{1-2} = -\int_1^2 \frac{k}{\sqrt{x^2+1}}\,dx - \cancel{\int_1^1 \frac{ky}{x\sqrt{x^2+y^2}}\,dy}\to 0$$

$$U_{1-2} = -\int_1^2 \frac{k}{\sqrt{x^2+1}}\,dx = -k\ln\left[x+\sqrt{1+x^2}\right]\Big|_1^2$$

$$U_{1-2} = -k\ln\left(\frac{4.236}{2.414}\right) = -0.562\,k$$

· From (1,0) to (2,1)

$$U_{1-2} = k - \sqrt{2}\,k - 0.562\,k$$

$$U_{1-2} = -0.9765\,k$$

b.) Work done as P moves from (1,0) to (2,0) to (2,1)

· From (1,0) to (2,0):

$$U_{1-2} = -\int_1^2 \frac{k}{\sqrt{x^2+y^2}}\,dx - \cancel{\int_0^0 \frac{ky}{x\sqrt{x^2+y^2}}\,dy}\to 0$$

$$U_{1-2} = -\int_1^2 \frac{k}{x}\,dx$$

$$U_{1-2} = -k\ln x\,\Big|_1^2 = -k\ln(2) = -0.693$$

· From (2,0) to (2,1):

$$U_{1-2} = -\cancel{\int_2^2 \frac{k}{\sqrt{x^2+y^2}}\,dx}\to 0 \ - \int_0^1 \frac{ky}{x\sqrt{x^2+y^2}}\,dy$$

$$U_{1-2} = -\int_0^1 \frac{ky}{2\sqrt{2^2+y^2}}\,dy$$

$$U_{1-2} = -\frac{k}{2}\int_0^1 \frac{y}{\sqrt{4+y^2}}\,dy \qquad u = 4+y^2,\ du = 2y\,dy$$

$$U_{1-2} = -\frac{k}{4}\int_0^1 \frac{du}{u^{1/2}}$$

(Continued)

15.44 cont.

$$U_{1-2} = -\frac{k}{4}\left(\frac{u^{1/2}}{1/2}\right) = -\frac{k}{2}\sqrt{4+y^2}\,\Big|_0^1$$

$$U_{1-2} = \frac{\sqrt{5}}{2}k + k = -0.118\ k$$

$\therefore$ From (1,0) to (2,0) to (2,1)

$$U_{1-2} = -0.693\,k - 0.118\,k = -0.811\,k$$

c.) Since the work performed in parts a and b differs, the work does depend on the path of the particle.

15.45

GIVEN: A horizontal trough (80 ft × 2 ft) contains liquid (γ = 70 lb/ft³) that has a wavy surface profile:

$$z = h + 1.5\sin(\pi x/10)\ \text{[ft]}$$

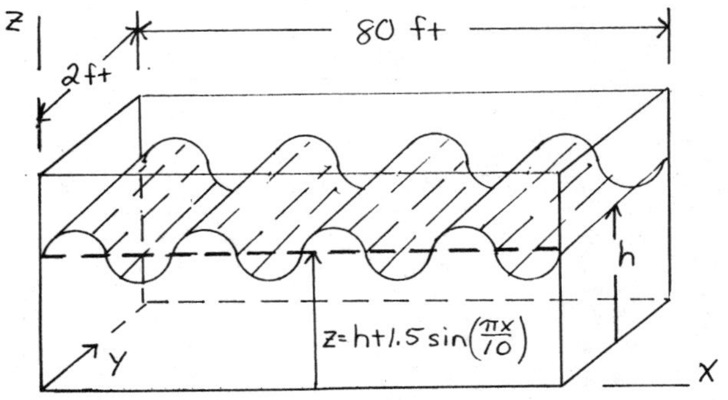

Figure a.

FIND: The work done by gravity as the waves die out.

SOLUTION:

The mean depth of the liquid is h. To solve this problem, determine the initial height of the center of gravity (or the centroid, since the liquid is of constant density) by the formula [See Eq. (8-16)]

$$z_{cg} = \frac{Q_x}{A} \qquad \text{(a)}$$

where Q_x is the moment of the shaded about the x axis (Fig. b) and A is the area of the shaded region.

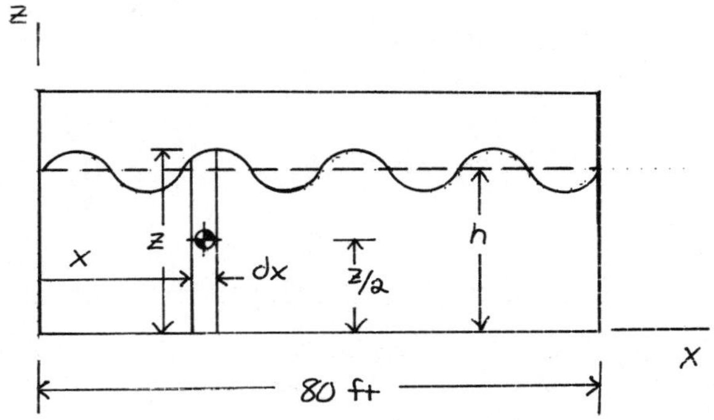

Figure b.

By Figure b.,

$$Q_x = \frac{1}{2}\int_0^{80} z^2\,dx$$

$$= \frac{1}{2}\int_0^{80}\left(h + 1.5\sin\frac{\pi x}{10}\right)^2 dx$$

or $\quad Q_x = 40h^2 + 45 \qquad$ (b)

The shaded area is

$$A = \int_0^{8} z\,dx = 80\,h \qquad \text{(c)}$$

By Eqs. (a), (b), and (c),

$$z_{cg} = \frac{h}{2} + \frac{45}{80h} \qquad \text{(d)}$$

After the waves die out, the height of the center of gravity is:

$$z_{cg}^{*} = \frac{h}{2}$$

Hence, the descent of the center of gravity is:

$$|\Delta z_{cg}| = z_{cg} - z_{cg}^{*} = \frac{45}{80h}$$

The total weight of the liquid is

$$W = (80)(2)h\gamma$$
$$= (80)(2)(h)(70) = 11{,}200\,h$$

(Continued)

15.45 cont.

By Theorem 15.6,

$$U_{1-2} = -W(\Delta z_{cg})$$

$$= -(11{,}200\,h)\left(\frac{-45}{80h}\right)$$

or

$$U_{1-2} = 6300 \text{ ft}\cdot\text{lb}$$

15.46

GIVEN A collar weighs 50 N. It is constrained to move along a smooth, circular ring of radius r = 200 mm (Fig. a)
The collar is moved slowly until it slides under the action of gravity and a tangential resistive force $F = 18\cos\theta$ [N]

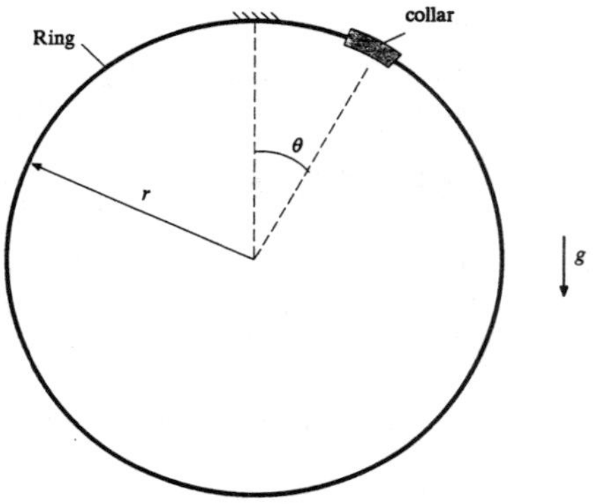

Figure a.

FIND: a.) The angle at which the collar begins to slide

b.) The speed as a function of θ and plot V for $\theta_s \le \theta \le 90°$

c.) The normal force exerted on the ring by the collar and the magnitude of the acceleration of the collar at 90°.

SOLUTION:

a) The FBD of the collar is shown in Fig. b, at the instant sliding impends.

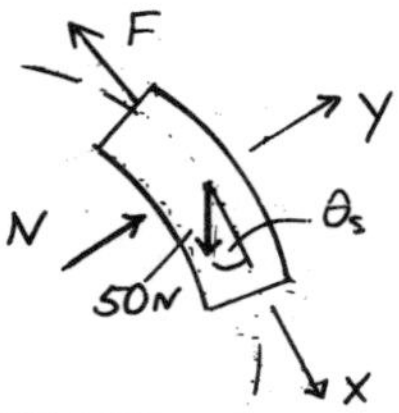

$\Sigma F_x = 0$ when sliding impends

$$\therefore \Sigma F_x = 50\sin\theta_s - 18\cos\theta_s = 0$$

or $\tan\theta_s = \frac{18}{50}$;

$$\theta_s = 19.80°$$

b.) By the law of kinetic energy,

$U_{1-2} = T_2 - T_1$ (with $T_1 = 0$) since initially $v = 0$

$$U_{1-2} = \int_{19.80}^{\theta} (50\sin\theta - 18\cos\theta)\,ds$$

$$ds = r\,d\theta = 0.2\,d\theta$$

$$\therefore U_{1-2} = 0.2\int_{19.80}^{\theta} (50\sin\theta - 18\cos\theta)\,d\theta$$

$$U_{1-2} = 0.2(-50\cos\theta - 18\sin\theta)\Big|_{19.8}^{\theta}$$

or $U_{1-2} = 10.628 - 10\cos\theta - 3.6\sin\theta$

Also, $T_2 = \frac{1}{2}\left(\frac{50\,N}{9.81\,m/s^2}\right)v^2 = 2.548\,v^2$

$\therefore U_{1-2} = T_2$

or $10.28 - 10\cos\theta - 3.6\sin\theta = 2.548\,v^2$

$$v = \sqrt{4.170 - 3.924\cos\theta - 1.413\sin\theta}$$

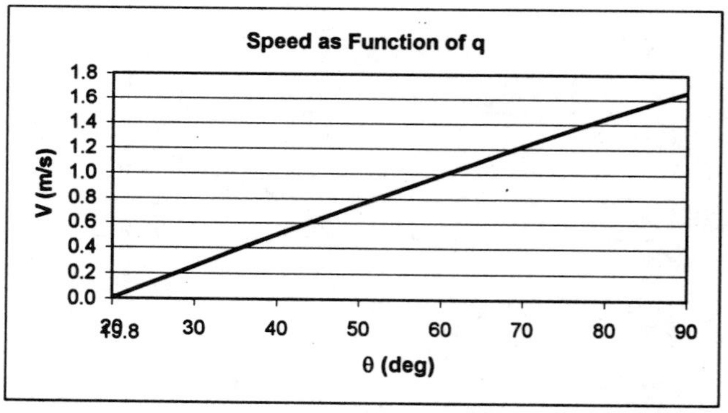

c.) For θ = 90°, the FBD of the collar is shown is Fig. c

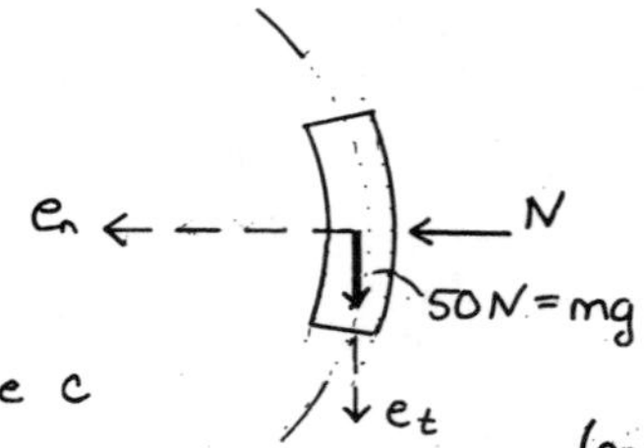

Figure c

(continued)

15.46 cont.

$$\Sigma F_n - N = ma_n$$

$$\Sigma F_t = mg = ma_t$$

$$\therefore \quad a_n = \frac{v^2}{r} \; ; \qquad v = 1.66 \text{ m/s @ } 90°$$

(off graph from b)

$$\therefore \quad a_n = \frac{(1.66)^2}{0.20 \text{ m}} = 13.77 \text{ m/s}^2$$

So, $N = \left(\frac{50 \text{ N}}{9.81 \text{ m/s}^2}\right)(13.77 \text{ m/s}^2)$

$$N = 70.2 \text{ [N]}$$

With $a_n = 13.77 \text{ m/s}^2$,

$a_t = 9.81 \text{ m/s}^2 \qquad (a_t = g)$,

$|a| = \sqrt{(13.77)^2 + (9.81)^2}$;

$$a = 16.91 \text{ m/s}^2$$

15.47

Given: A satellite of mass m_s passes through P_1 at height h_1 and at a later time through P_2 at height h_2 (Fig. a)

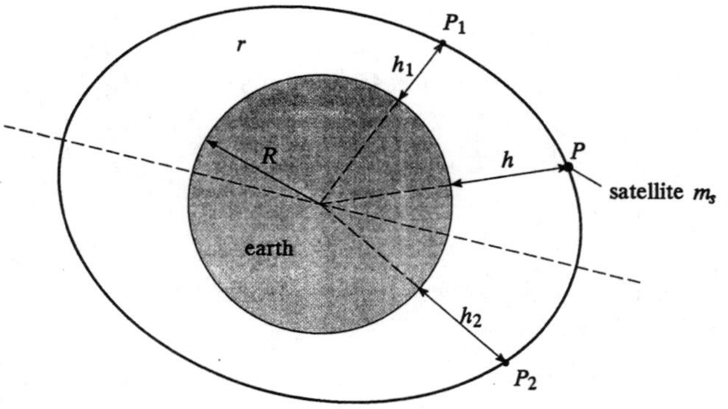

Figure a.

FIND: a.) Show that the work done by gravity on satellite as it goes from P_1 to P_2 is

$$U_G = m_s[g_2(R+h_2) - g_1(R+h_1)]$$

where g_1 and g_2 are the accelerations of gravity at P_1 and P_2, respectively, and R is radius of earth.

b.) Derive a formula for its speed v as a function of h. Plot v for $625 \text{ mi} \le h \le 5000 \text{ mi}$, with $R = 4000$ mi and $g = 32.2 \text{ ft/s}^2$ on the surface of the earth.

c.) Use the plot (part b) to estimate height of satellite at which $v = 15{,}000$ ft/s; and at which $v = 20{,}000$ ft/s. Check the estimates by formula for v (part b).

SOLUTION:

a.) By Eq. (14.1),

$$F = \frac{G m_s m_E}{r^2} = m_s g \qquad \text{(a)}$$

where m_E is the mass of the earth, r is the distance from the center of the earth to the center of mass of the satellite, and g is the value of the acceleration of gravity at r. Therefore

$$\cancel{m_s} g_1 = \frac{G \cancel{m_s} m_E}{(R+h_1)^2}$$

$$\cancel{m_s} g_2 = \frac{G \cancel{m_s} m_E}{(R+h_2)^2} \qquad \text{(b)}$$

The work done on the satellite by gravity is:

$$U_{1\text{-}2} = -\int_{R+h_1}^{R+h_2} F\,dr = -G m_s m_E \int_{R+h_1}^{R+h_2} \frac{dr}{r^2}$$

Integration yields

$$U_{1\text{-}2} = G m_s m_E \left[\frac{1}{R+h_2} - \frac{1}{R+h_1}\right] \qquad \text{(c)}$$

By Eqs. (b) and (c), we obtain

$$W = m_s[g_2(R+h_2) - g_1(R+h_1)] \qquad \text{(d)}$$

b.) By the law of kinetic energy,

$$U_{1\text{-}2} = T_2 - T_1 \; ;$$

or

$$m_s[g_2(R+h_2) - g_1(R+h_1)] = \tfrac{1}{2} m (v_2^2 - v_1^2) \qquad \text{(e)}$$

(continued)

15.47 Cont.

With the given data and noting that [by Eq (a)] on the earth's surface $Gm_E{}^2 = gR_1$ Eq. (b) yields

$$g_1 = \frac{gR^2}{(R+h_1)^2}, \qquad g_2 = \frac{gR^2}{(R+h_2)^2} \qquad (f)$$

where $g = 32.2$ ft/s². By Eqs. (e) and (f), with $h_2 = h$ and v_2 = the speed of the satellite at height h, we find

$$v^2 = 2gR^2\left[\frac{1}{R+h} - \frac{1}{R+h_1}\right] + v_1{}^2 \qquad (g)$$

with $R = 4000$ mi $= 2.112 \times 10^7$ ft, $h_1 = 5000$ mi $= 2.64\times10^7$ ft, $g = 32.2$ ft/s², and h is in feet, Eq (g) yields

$$v = \sqrt{\frac{2.8726\times10^{16}}{2.1120\times10^7 + h} - 4.355\times10^8} \qquad (h)$$

See graph of v versus h:

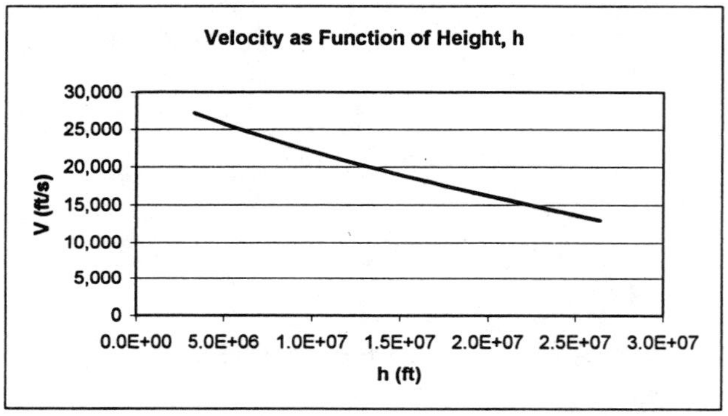

c.) By the plot of part b:

$v \approx 15{,}000$ ft/s at $h \cong 2.25\times10^7$ ft

$v \approx 20{,}000$ ft/s at $h \cong 1.35\times10^7$ ft

By Eq (h),

$v = 15{,}000$ ft/s when $h = 2.237\times10^7$ ft

$v = 20{,}000$ ft/s when $h = 1.326\times10^7$ ft

15.48

GIVEN: A body ($m = 0.06211$ lb·s²/ft) is constrained to slide along the frictionless bar AD (Fig. a). The spring constant is $k = 1.0$ lb/in $= 12$ lb/ft.

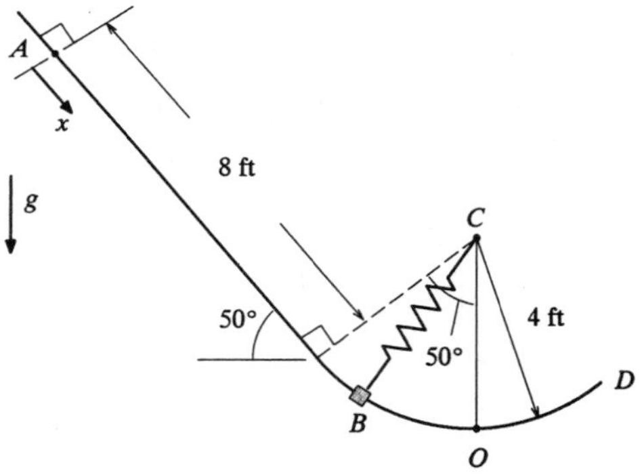

Figure a.

The unstretched length of the spring is 3ft.

FIND a.) The work required to lift the body from point O to point A

b.) The maximum speed attained by the body in traveling from rest at point A to point D.

c.) A formula for the speed v of the body as a function of the distance x along rod AD, from point A where is is released from rest to point O, and plot v as a function of x.

SOLUTION

a.) The work required to stretch the spring from O to A (Fig. a)

$$(U_{1-2})_{spring} = \int_{4-3}^{\sqrt{8^2+4^2}-3} ke\,de$$

$$= \frac{1}{2}(12)e^2\Big|_1^{\sqrt{80}-3} = 206.01 \text{ ft·lb}$$

The work required to overcome gravity is

$$(U_{1-2})_{gravity} = mgh$$

(continued)

15.48 cont.

where by Fig. a, h, the height from point O to point A, is

$$h = 8\sin 50° + 4(1-\cos 50°) = 7.557 \text{ ft}$$

$$\therefore (U_{1-2})_{gravity} = (0.06211)(32.2)(7.557)$$

or $(U_{1-2})_{gravity} = 15.11$ ft·lb

Hence,

$$(U_{1-2})_{Total} = (U_{1-2})_{spring} + (U_{1-2})_{gravity} = 221.12 \text{ ft·lb}$$

b.) The maximum speed occurs at point O. By the law of kinetic energy.

$U_{1-2} = T_2 - T_1$ (with $T_1 = 0$) since body starts at rest at A.

or $U_{1-2} = \frac{1}{2}mv^2$

Hence:

$221.12 = \frac{1}{2}(0.06211)v^2$; $v = 84.38$ ft/s

c.) To determine v as a function of distance x along the rod from A to O, two formulas are required; one for $0 \le x \le 8$ ft and one for $8 \text{ ft} \le x \le 8+4(50° \times \pi/180) = 11.49$ ft.

For $0 \le x \le 8$ ft ;

$U_{1-2} = T_2 - T_1$ (with $T_1 = 0$) yields, where (see Fig. b)

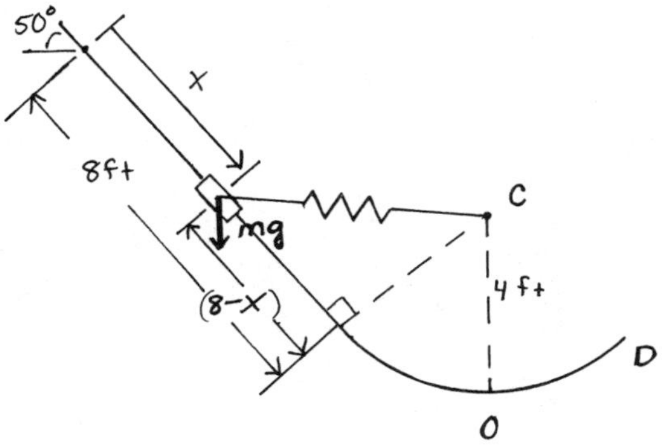

Figure b.

$$U_{1-2} = mgx\sin 50° + \frac{1}{2}k(e_A^2 - e_x^2)$$
$$= mgx\sin 50° + \frac{1}{2}k[\sqrt{8^2+4^2}-3]^2 \ldots$$
$$\ldots + \frac{1}{2}k[\sqrt{(8-x)^2+4^2}-3]^2$$

$T_2 = \frac{1}{2}mv^2$

or

$$-6x^2 + 97.532x - 321.99 + 36\sqrt{80-16x+x^2} = 0.031055v^2$$

$\therefore$ For $0 \le x \le 8$ ft,

$$v = \sqrt{-10{,}368.4 - 193.2x^2 + 3140.6x + \ldots}$$
$$\overline{\ldots + 1159.2\sqrt{80-16x+x^2}}$$

For $8 \text{ ft} \le x \le 11.49$ ft (Fig c)

$U_{1-2} = T_2 - T_1$

where, since the spring length remains constant at 4 ft;

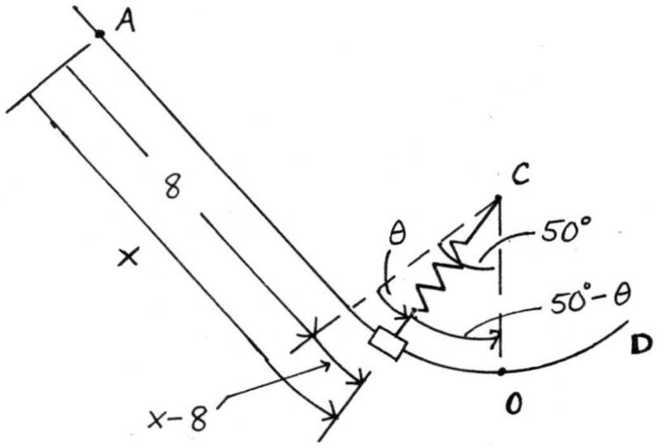

Figure c.

$$U_{1-2} = 4[\cos(50°-\theta) - \cos 50°]mg$$

or since $x - 8 = r\theta$,

$$\theta = \frac{x-8}{r} = \frac{x-8}{4} = 0.25x - 2 \text{ (rad)}$$

In degrees:

$$\theta = (0.25x-2)\frac{180}{\pi} = 14.3239x - 114.59, \text{ x in ft}$$

$$\therefore U_{1-2} = 8[\cos(164.59 - 14.329x) - 0.6428] \text{ ft·lb}$$

and

$$T_1 = \frac{1}{2}mv_1^2 = \frac{1}{2}(0.06211)(84.38)^2 = 221.12 \text{ ft·lb}$$

(continued)

15.48 cont.

$$T_2 = \tfrac{1}{2} m v_2^2$$
$$= \tfrac{1}{2}(0.06211) v_2^2 = 0.31055\, v_2^2 \text{ ft}\cdot\text{lb}$$

where v_1 = speed at $x = 8$ ft, and v_2 = speed at point O. Therefore, $U_{1-2} = T_2 - T_1$ yields, for $8\text{ ft} \le x \le 11.49$ ft,

$$8[\cos(164.59 - 14.3239x) - 0.6428] =$$
$$= 0.03105\, v_2^2 - 221.12$$

or

$$v_2 = \sqrt{6954.68 + 257.61 \cos(164.59 - 14.33x)}$$

where $164.59 - 14.33x$ is in degrees.

or

$$v_2 = \sqrt{6954.68 + 257.61 \cos(2.8726 - 0.250x)}$$

where $2.8726 - 0.250x$ is in radians.

SUMMARY:

$$v = \begin{cases} 0 \le x \le 8 \text{ ft}; \\ v = \sqrt{-10{,}368.4 - 193.2x^2 + 3140.6x + \ldots} \\ \quad \overline{\ldots + 1159.2\sqrt{80 - 16x + x^2}} \\ 8 \le x \le 11.49 \text{ ft}; \\ v = \sqrt{6954.68 + \ldots} \\ \quad \overline{\ldots + 257.61\cos(2.8726 - 0.250x)} \end{cases}$$

See graph:

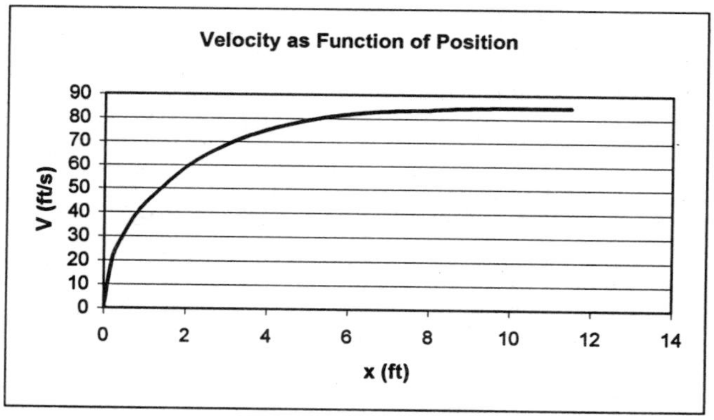

15.49

GIVEN: Air resistance on jet plane is $F = 0.005V^2$ where V is flight speed (ft/s). The engine can deliver 4000 hp.

FIND The maximum speed the airplane can maintain in horizontal flight.

SOLUTION

$$\text{Power} = E = \bar{F}\cdot\bar{V}$$

$$\therefore\ 4000 \text{ hp} = (0.005V^2)(V) = 0.005V^3$$

$$4000 \text{ hp}\left(550 \frac{\text{ft}\cdot\text{lb/s}}{\text{hp}}\right) = 0.005V^3$$

$$V = 760.6 \text{ ft/s}\left(\frac{1 \text{ mi}}{5280 \text{ ft}}\right)\left(\frac{3600 \text{ s}}{\text{hr}}\right)$$

$$V = 518.6 \text{ mi/hr}$$

15.50

GIVEN: Super boy weighs 50 lb and runs up 40-ft hill in 3 s. Super mom weighs 120 lb and runs up the hill in 6 s.

FIND: Whether super boy or super mom exerts more power in climbing the hill.

SOLUTION:

$$E_{boy} = \frac{\text{work}}{\text{time}} = \frac{(50)(40)}{3} = 666 \text{ ft}\cdot\text{lb}$$
$$= 1.21 \text{ hp}$$

$$E_{mom} = \frac{\text{work}}{\text{time}} = \frac{(120)(40)}{60} = 800 \text{ ft}\cdot\text{lb}$$
$$- 1.45 \text{ hp}$$

$\therefore$ super mom exerts more power

15.51

GIVEN: 900-kg car accelerates from rest to 100 km/h in 10 seconds

FIND: a.) Work performed
b.) Mean power required
c.) Power generated when $v = 100$ km/h

SOLUTION:

a.) $U_{1-2} = T_2 - T_1$

But, $T_1 = 0$ so;

$U_{1-2} = T_2 = \frac{1}{2}mv^2$

where $v = 100$ km/h $= 27.778$ m/s

$\therefore \quad U_{1-2} = \frac{1}{2}(900)(27.778)^2 = 347.22$ kJ

b.)

$$E_{mean} = \frac{work}{time} = \frac{347.22}{10}$$

$= 34.72$ kW $(46.54$ hp$)$

c.) The power in general is

$$E = Fv$$

where F is the force exerted by friction on the drive tires.

Since

$$F = ma$$

and

$$a = \frac{v}{t} = \text{constant}$$

at $v = 100$ km/h $= 27.778$ m/s and $t = 10$ s,

$$a = 2.778 \text{ m/s}^2$$

$$\therefore \quad F = (900)(2.778) = 2500 \text{ N}$$

Hence, at $v = 27.778$ m/s

$E = (2500)(27.778)$
$= 69.44$ kW $(93.09$ hp$)$

15.52

GIVEN: A 2000-lb car accelerates at constant rate from rest to 60 mi/h (88 ft/s) on a level road. The car is rated at 100 hp.

FIND: The minimum time required for the car to attain 60 mi/hr with 20% of its horsepower used to overcome air resistance and friction.

SOLUTION

First, calculate the work, U_{1-2} required

$$U_{1-2} = T_2 - T_1$$

But, $T_1 = 0$; so,

$$U_{1-2} = T_2 = \frac{1}{2}\left(\frac{2000}{32.2}\right)(88)^2$$

or $U_{1-2} = 240{,}500$ ft·lb

The maximum horsepower available is $(0.80)(100) = 80$ hp. Therefore, the the maximum useful energy output of the car is

$$U_{car} = (80 \text{ hp})\left(550 \ \frac{\text{ft}\cdot\text{lb}}{\text{hp}\cdot\text{s}}\right)\Delta t$$

where Δt is the time required for the car to reach 60 mi/h.

$$\therefore \ U_{1-2} = (80)(550)\,\Delta t$$

or $\Delta t = \dfrac{240{,}500}{(80)(550)} = 5.46$ s

15.53

GIVEN: 1000-kg car travels up a road inclined at 15° for a distance of 100 m, in 10 s.

FIND: The mean power required if the car:
a.) travels at constant speed
b.) accelerates at a constant rate starting from rest.

SOLUTION

a.) First calculate the work U_{1-2} performed by the law of kinetic energy:

(continued)

15.53 cont.

$$U_{1-2} = T_2 - T_1$$

But, $T_1 = T_2$ (since car at constant v)

$$\therefore U_{1-2} = mgh = (1000)(9.81)(100 \sin 15°)$$

or $U_{1-2} = 253,900$ N·m

Hence, the mean power is:

$$E_{mean} = \frac{U_{1-2}}{\Delta t} = \frac{253.90}{10} = 25390 \frac{N \cdot m}{s}$$

or $\underline{\underline{E_{mean} = 25.39 \text{ kW} \quad (34.04 \text{ hp})}}$

b.) First, calculate the work U_{1-2} performed to accelerate the car a distance of 100 m in 10 s.

$$(U_{1-2})_{accelerate} = T_2 - T_1$$

But, $T_1 = 0$, $T_2 = \frac{1}{2} m v^2$,

To determine v, note that:

$$s = 100 m = \frac{1}{2} a t^2 = \frac{a}{2} (10)^2;$$

or $a = 2 \text{ m/s}^2$

$\therefore v = at = (2)(10) = 20$ m/s

and

$$(U_{1-2})_{accelerate} = \frac{1}{2}(1000)(20)^2 = 200,000 \text{ N·m}$$

In addition, the work required to raise the car $h = 100 \sin 15° = 25.882$ m is

$$(U_{1-2})_{raise} = mgh = (1000)(9.81)(25.882) = 253900 \text{ N·m}$$

Hence,

$$(U_{1-2})_{total} = 200,000 + 253,900 = 453,900 \text{ N·m}$$

Therefore, the mean power is

$$E_{mean} = \frac{453,900 \text{ N·m}}{10 \text{ s}} = 45,390 \frac{N \cdot m}{s}$$

or $\underline{\underline{E_{mean} = 45.39 \text{ kW} \quad (60.84 \text{ hp})}}$

15.54

GIVEN: Linear spring, $k = 4$ kip/ft, is extended 6 in from its unstretched length.

FIND: a.) Work done on the spring
b.) Mean power generated for extension time of 0.5 s, for extension time of 2 s.

SOLUTION:

a.) $U_{1-2} = \frac{1}{2} k x^2 = \frac{1}{2}(4)(\frac{1}{2})^2$

or $U_{1-2} = 0.5$ ft·hp

$\therefore$ $\underline{\underline{U_{1-2} = 500 \text{ ft·lb}}}$

b.) For $t = 0.5$ s

$$\underline{\underline{E_{mean}(0.5) = \frac{500}{0.5} = 1000 \frac{ft \cdot lb}{s}}}$$

or $\underline{\underline{E_{mean} = 1.818 \text{ hp}}}$

For $t = 2$ s

$$E_{mean}(2) = \frac{500}{2} = 250 \frac{ft \cdot lb}{s}$$

or $\underline{\underline{E_{mean} = 0.454 \text{ hp}}}$

15.55

GIVEN: A 20,000 lb airplane requires 300 hp for level flight at 200 mi/h. Its engine can develop 900 hp. The propellar is 70% efficient.

FIND: The maximum possible rate of climb if the speed of airplane is constant.

SOLUTION

Power = $E = F \cdot v$

The excess is $900 - 300 = 600$ hp

$$\therefore 600 \text{hp} \left(500 \frac{ft \cdot lb / s}{hp}\right)(0.7) = 20,000 v$$

where v is the rate of climb.

$$\therefore v = 11.55 \text{ ft/s} (3600 \text{ s/hr})(1 \text{ mi}/5280 \text{ ft})$$

$$\underline{\underline{v = 7.88 \text{ mi/hr}}}$$

15.56

GIVEN: An automobile that weighs 3800 lb can develop 120 hp when traveling 30 mi/h.

FIND: How rapidly it can accelerate if at 30 mi/h it requires 15 hp.

SOLUTION

Power $= \dot{E} = F \cdot v$

$$\left(550 \frac{\text{ft}\cdot\text{lb}/\text{s}}{\text{hp}}\right)(120-15)\,\text{hp} = F\cdot(30\,\text{mi/hr})\left(\frac{5280}{3600}\right)$$

$$57750\ \frac{\text{ft}\cdot\text{lb}}{\text{s}} = (44\ \text{ft/s})F;$$

$$F = 1312.5$$

By $F=ma$;

$$1312.5 = \left(\frac{3800\ \text{lb}}{32.2\ \text{ft/s}}\right)a;$$

$$\underline{\underline{a = 11.12\ \text{ft/s}^2}}$$

15.57

GIVEN: Combined power of engines to dig the Suez Canal was 10,000 hp. The specific weight of sand was 150 lb/ft and the average height of a lift was 25 ft.

FIND: The cubic yards that could have been excavated in an 8 hr. day. Neglect friction losses in machinery.

SOLUTION

Power $= \dot{E} = \frac{\Delta U}{\Delta t}$

$$\Delta U = 10{,}000\,\text{hp}\left(550\ \frac{\text{lbft/s}}{\text{hp}}\right)(8\,\text{h})\left(3600\ \frac{\text{s}}{\text{h}}\right)$$

$$= 1.584\times 10^{11}\ \text{lb}\cdot\text{ft}$$

$$F = 6.336\times 10^{9}\ \text{lb}$$

$$\forall = 6.336\times10^{9}\ \text{lb}\cdot\frac{\text{ft}^3}{150\,\text{lb}} = 4.224\cdot 10^{7}\ \text{ft}^3$$

$$\forall = 4.224\times10^{7}\ \text{ft}^3\left(\frac{\text{yd}^3}{27\,\text{ft}^3}\right)$$

$$\underline{\underline{= 1.564\times 10^{6}\ \text{yd}^3}}$$

15.58

GIVEN: p-V diagram (Fig. a) for an engine that operates at 280 rpm, where $p = a/(V-b)$

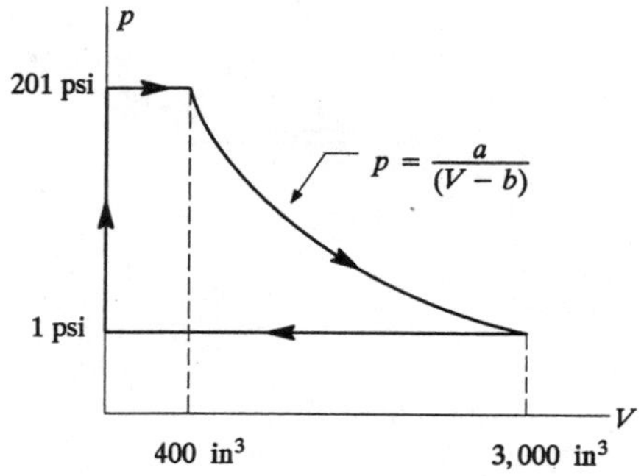

Figure a.

FIND: a.) The constants a and b

b.) The horsepower output of the engine. Neglect friction.

SOLUTION:

a.) By Fig a., and $p = a/(V-b)$,

$$201\ \text{lb/in}^2 = \frac{a}{(V-b)} = \frac{a}{(400\,\text{in}^3 - b)}$$

$$1\ \text{lb/in}^2 = \frac{a}{3000\,\text{in}^3 - b}$$

$$80400 - 201b = a,$$

$$\underline{\underline{a = 2613\ \text{lb/in}}}, \qquad \underline{\underline{b = 387\ \text{in}^3}}$$

Hence $p = \frac{2613}{V-387}$.

b.) Power $= \dot{E} = F\cdot v$

The work performed during expansion from 400 in³ to 3000 in³ is

$$U_{exp} = \int_{400}^{3000} p\,dV = 2613\int_{400}^{3000} dV/(V-387)$$

$$= 2613\ \ln(V-387)\Big|_{400}^{3000}$$

$$\therefore\ U_{exp} = 13{,}860\ \text{in}\cdot\text{lb}$$

The work performed returning to $p = 201$ psi is:

$$U_{201} = (201)(400) - (1)(3000)$$

$$\therefore\ U_{201} = 77{,}400\ \text{in}\cdot\text{lb}$$

(continued)

15.58 cont.

Hence, the work per cycle is

$$U_{cycle} = 13,860 + 77,400 = 91,260 \ \frac{in \cdot lb}{cycle}$$

or $$U_{cycle} = 7605 \ ft \cdot lb/cycle$$

Hence, the horsepower is

$$E = \left(7605 \frac{ft \cdot lb}{cycle}\right)\left(280 \frac{cycles}{min}\right)\left(\frac{1 hp}{33,000 \frac{ft \cdot lb}{min}}\right)$$

or $$E = 64.53 \ hp$$

15.59

GIVEN: The p-V diagram (Fig. a) undergoes a cycle 1-2-3-4-1 for one cylinder of a diesel engine. The engine has 6 cylinders and each cylinder performs the cycle for each revolution of the crank shaft.

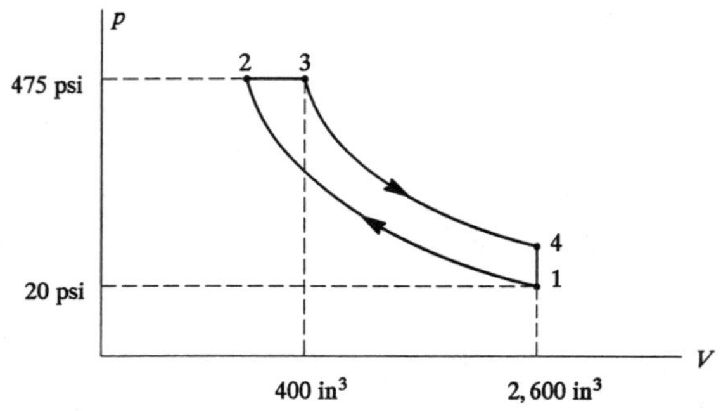

Figure a.

FIND: The horsepower output of the neglecting friction. (See problems 15.6 and 15.16)

SOLUTION:

For compression from 1 to 2,

$$p_1 V_1^{1.2} = p_2 V_2^{1.2}$$

$$(20)(2600)^{1.2} = (475) V_2^{1.2}$$

$$\therefore V_2 = 185.60 \ in^3$$

For expansion from 3 to 4,

$$p_3 V_3^{1.2} = p_4 V_4^{1.2}$$

$$(475)(400)^{1.2} = p_4 (2600)^{1.2}$$

$$\therefore p_4 = 50.257 \ psi$$

For the compression stroke from 1 to 2, (by Problem 15.16)

$$U_{1-2} = \frac{1}{\gamma - 1}(p_1 V_1 - p_2 V_2)$$

$$= \frac{1}{1.2-1}[(20)(2600) - (475)(185.6)]$$

or $$U_{1-2} = -180,800 \ in \cdot lb$$

For the expansion stroke from 3 to 4,

$$U_{3-4} = \frac{1}{1.2-1}[p_3 V_3 - p_4 V_4]$$

$$= 5[(475)(400) - (50.257)(2600)]$$

or $$U_{3-4} = 296,660 \ in \cdot lb$$

For the injection period from 2 to 3.

$$U_{2-3} = (475)(V_3 - V_2)$$

$$= 475(400 - 185.60)$$

or $$U_{2-3} = 101,840 \ in \cdot lb$$

Hence, the work per cycle is

$$U_{cycle} = 296,660 + 101,840 - 180,800$$

or $$U_{cycle} = 217,700 \ in \cdot lb/cycle$$

The power output is:

$$E = \left(217,700 \frac{in \cdot lb}{cycle}\right)\left(\frac{400 \ cycle}{min}\right)\left(\frac{1 min}{60 s}\right)\left(\frac{1 hp}{550 \frac{ft \cdot lb}{s}}\right)\left(\frac{1 ft}{12 in}\right)$$

$$E = 219.90 \ hp$$

15.60

GIVEN: Uniform bar ABC rotates in a vertical plane about the frictionless hinge A (Fig. a) The spring constant is $k = 9$ N/m and the unstretched length of the spring is 90 mm.

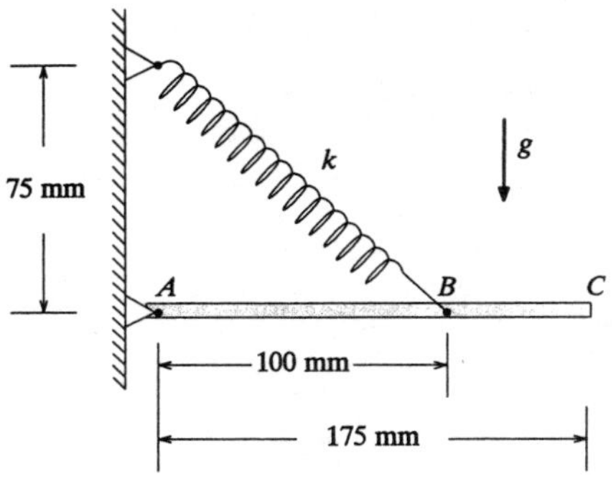

Figure a.

FIND: a.) The increase in potential energy of the spring as the bar swings from the horizontal to vertical position (straight down)

b.) The total work of all the forces acting on the bar as it undergoes the rotation.

SOLUTION

a.) $V_2 = \frac{1}{2} k e^2$

$= \frac{1}{2}(9 \text{ N/m})(0.175\text{m} - 0.090\text{m})^2$

$= 0.03251 \text{ N·m}$

$V_1 = \frac{1}{2} k e^2$

$= \frac{1}{2}(9 \text{ N/m})(\sqrt{0.100^2 + 0.075^2} - 0.090)^2$

$= 0.00551 \text{ N·m}$

$\Delta V = V_2 - V_1 = 0.03251 - 0.00551$

$\Delta V = 0.0270 \text{ N·m}$

b.) $U_{1-2} = -\Delta V$

$\Delta V_{spring} = 0.027 \text{ N·m}$ (From part a.)

$\Delta V_{gravity} = -\frac{0.175}{2}(10\text{N}) - 0 = -0.875 \text{ N·m}$

$U_{1-2} = -0.0270 + 0.875 = 0.848 \text{ N·m}$

$U_{1-2} = 0.848 \text{ N·m}$

15.61

GIVEN Rigid rod of weight w and length L swings in a vertical plane about frictionless pin O (Fig. a). A body of weight W is attached to one end of a spring, the other end of which is attached to the frictionless rod. Let zero datum of potential energy be $\theta = 0$, $x = 0$. Neglect the mass of the spring.

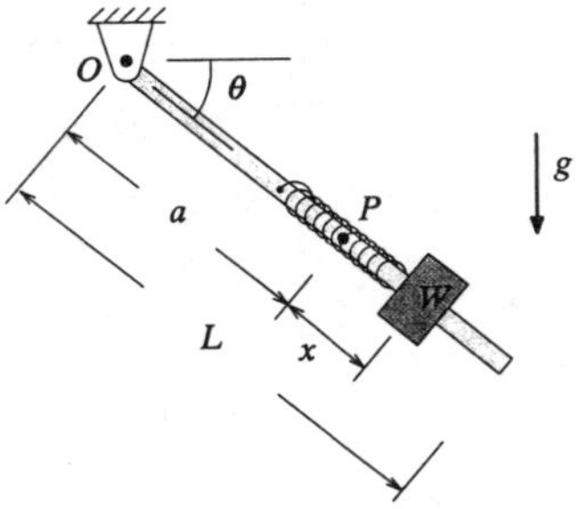

Figure a.

FIND: The potential energy of the system as a function of x and θ.

SOLUTION:

$V = 0 \quad (\theta = 0, \; x = 0)$

$V = -W(a+x)\sin\theta - w(\frac{L}{2}\sin\theta) + \frac{1}{2}kx^2$

$(\theta \neq 0, \; x \neq 0)$

$V = \frac{1}{2}kx^2 - W(a+x)\sin\theta - w\frac{L}{2}\sin\theta$

15.62

GIVEN: The earth and sun attract each other with a force $F = k/r^2$ (Fig. a), where k is a constant.

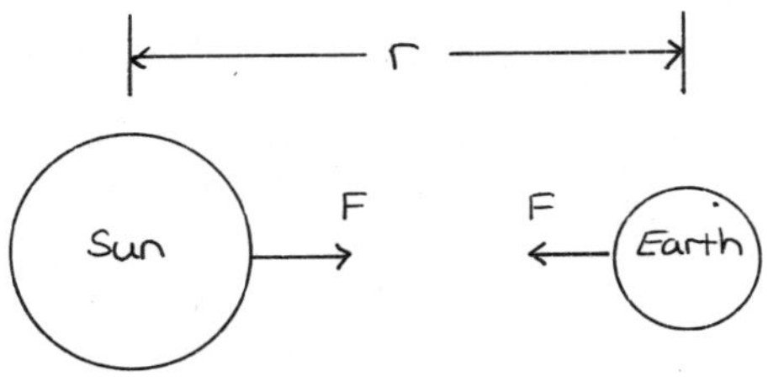

Figure a

(continued)

15.62 cont.

FIND: A formula for the potential energy of the system.

SOLUTION:

Assume the sun is fixed. Then, by Fig. (a)

$$U_{1-2} = -\int F\,dr = -\int \left(-\frac{k}{r^2}\right) dr$$

or $U_{1-2} = -\frac{k}{r} + \text{constant}$

So $V = -U_{1-2} = \frac{k}{r} + \text{constant}$

as $r \to \infty$, $V \to 0$. Therefore, constant $= 0$. Hence,

$$V = \frac{k}{r}$$

15.63

GIVEN: A 120-N block is suspended above a helical spring (Fig a.)

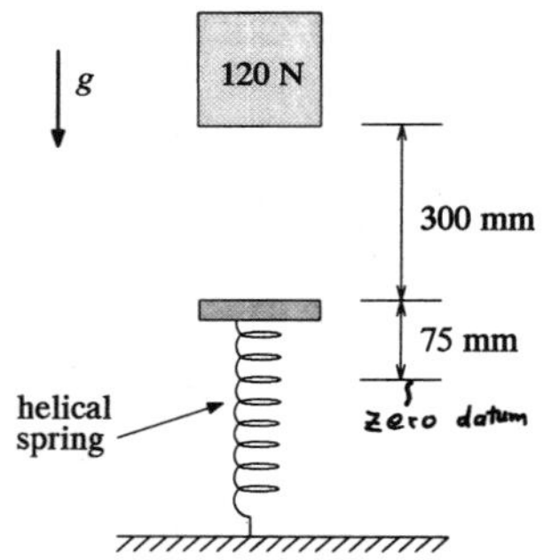

Figure a

FIND The spring constant, k [N/mm] assuming energy is conserved.

SOLUTION

$$\cancel{T_1}^{\,0} + V_1 = \cancel{T_2}^{\,0} + V_2 \qquad (a)$$

Let the compressed position of the spring be the zero datum of potential energy (Fig a)

$\therefore$ $V_1 = mgh = 120\text{ N}(375\text{ mm})$
$= 45000\text{ N·m}$

$V_2 = \frac{1}{2}ke^2 = \frac{1}{2}k(75\text{mm})^2$
$= 2812.5k\text{ N·mm}$

By Eq (a),

$45000\text{ N·mm} = 2812.5k$

$\therefore$ $k = 16\text{ N/mm}$

15.64

GIVEN: The rigid uniform bar is 3 ft long, weighs 30 lb, is hinged at one end and attached to a spring at the other end (Fig a). The free-length of the spring is 3 ft and the spring constant is k = 1 lb/in.

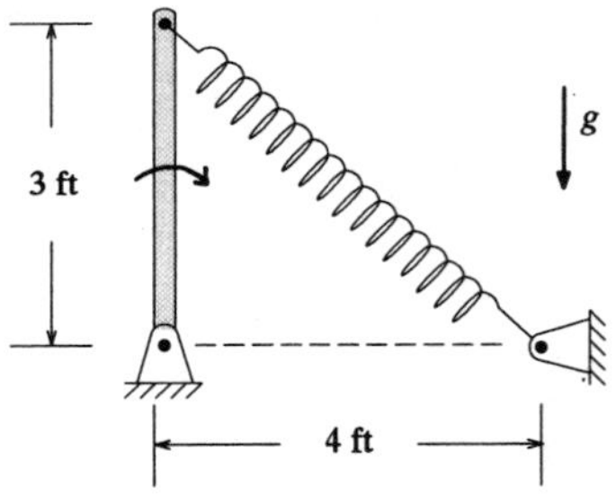

Figure a.

FIND a.) The change in potential energy of the spring under a clockwise rotation to the horizontal position.

b.) The total work of all forces acting on the bar under this rotation.

(continued)

15.64 cont.

SOLUTION:

a.) $\Delta V = V_2 - V_1$

$$V_2 = \tfrac{1}{2} k e^2 = \tfrac{1}{2}(1\ \text{lb/in})(1-3\text{ft})^2(12\ \text{in/ft})$$
$$= 24\ \text{ft}\cdot\text{lb}$$

$$V_1 = \tfrac{1}{2} k e^2 = \tfrac{1}{2}(1\ \text{lb/in})(\sqrt{4^2+3^2}-3)^2(12\ \text{in/ft})$$
$$= 24\ \text{ft}\cdot\text{lb}$$

$$\underline{\underline{\Delta V = 24\ \text{ft}\cdot\text{lb} - 24\ \text{ft}\cdot\text{lb} = 0}}$$

b.) $U_{1-2} = -\Delta V = -\Delta V_{spring} - \Delta V_{gravity}$

$\Delta V_{spring} = 0$ (Part A)

$$\Delta V_{gravity} = -mgh$$
$$= -(30\ \text{lb})(1.5\text{ft}) = -45\ \text{ft}\cdot\text{lb}$$

$$\underline{\underline{U_{1-2} = -0 + 45\ \text{ft}\cdot\text{lb} = 45\ \text{ft}\cdot\text{lb}}}$$

15.65

GIVEN Solve Problem 15.64 for the case where the bar rotates counter clockwise to the horizontal position (Fig. a)

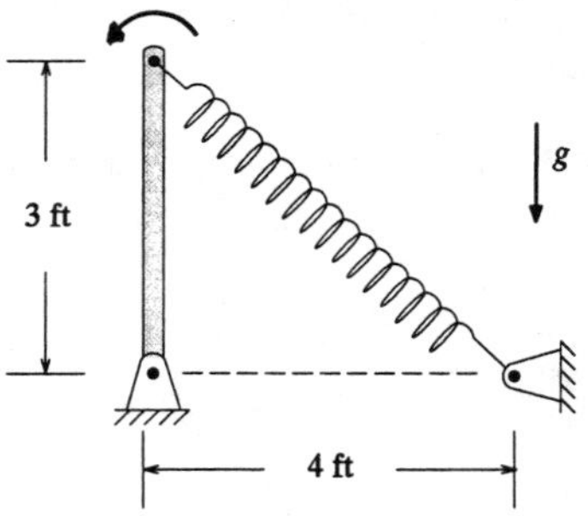

Figure a

FIND a.) The change in potential energy of the spring under a counterclockwise rotation to the horizontal.

b.) The total work of *all* forces acting on the bar under this rotation.

Solution:

a.) $\Delta V = V_2 - V_1$

$$V_2 = \tfrac{1}{2} k e^2 = \tfrac{1}{2}(1\ \text{lb/in})(12\ \text{in/ft})((4\text{ft}-3\text{ft})-3\text{ft})^2$$
$$= 96\ \text{ft-lb}$$

$$V_1 = \tfrac{1}{2} k e^2 = \tfrac{1}{2}(12\ \text{lb/ft})(\sqrt{4^2+3^2} - 3\text{ft})^2$$
$$\doteq 24\ \text{ft-lb}$$

$$\underline{\underline{\Delta V = 96 - 24 = 72\ \text{ft-lb}}}$$

b.) $U_{1-2} = -\Delta V = -\Delta V_{spring} - \Delta V_{gravity}$

$\Delta V_{spring} = 72\ \text{ft-lb}$ (Part A)

$\Delta V_{gravity} = -1.5\ \text{ft}\,(30\ \text{lb}) = -45\ \text{ft-lb}$

$$\underline{\underline{U_{1-2} = -72 + 45 = -27\ \text{ft-lb}}}$$

15.66

GIVEN: The free length of each spring (Fig. a) is 10 in, and their constant is 4 lb/in. The system is released from rest from the position shown.

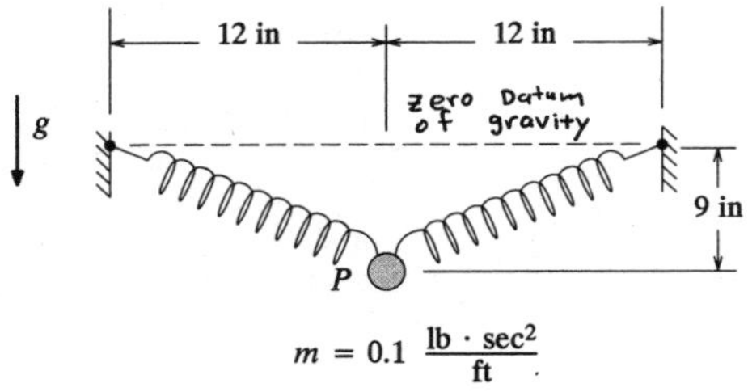

Figure a

FIND: The speed of the mass when the springs are horizontal.

SOLUTION

By conservation of mechanical energy (since the system is released from rest),

$$\overset{0}{\cancel{T_1}} + V_1 = T_2 + V_2 \qquad (a)$$

For the zero datum (Fig. a) of gravity,

$$V_1 = -mgh + 2\left(\tfrac{1}{2} k e^2\right)$$

(continued)

15.66 cont.

$$V_1 = -(0.1 \tfrac{lb\cdot s^2}{ft})(32.2\ ft/s^2)(9in) + \ldots$$
$$\ldots (2)\left(\tfrac{1}{2}(4\ lb/in)(\sqrt{12^2+9^2}-10)^2\right)$$

$V_1 = 71.02$ in · lb

$T_2 = \tfrac{1}{2} m v^2 = 0.05 v^2$

$V_2 = 2(\tfrac{1}{2} k e^2)$
$= 2[\tfrac{1}{2}(4\ lb/in)(12-10)^2] = 16$ in-lb

∴ By Eq. (a),

$71.02 = 0.05 v^2 + 16$ in-lb

$v = 33.2$ in/s $= 2.76$ ft/s

15.67 GIVEN: The narrow uniform tube (Fig. a) lies in a vertical plane. The liquid in the tube oscillates and its speed at $x=0$ is v_0. The length of the column of liquid is L.

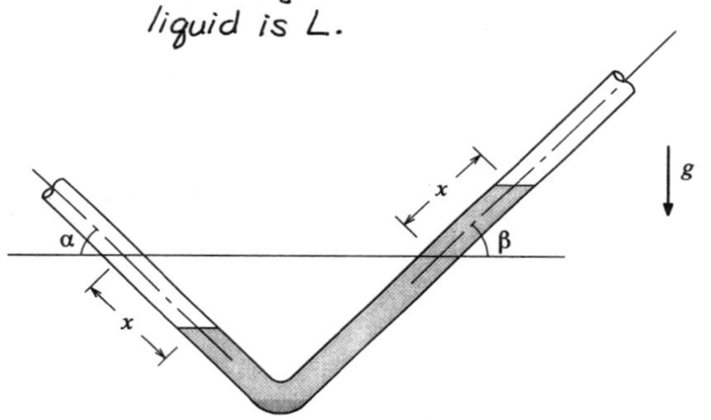

Figure a.

FIND: Show that the maximum value of x is:

$$x_{max} = v_0 \sqrt{\frac{L}{g(\sin\alpha + \sin\beta)}}$$

SOLUTION:

Starting with $x=0$ (configuration 1), in Fig. a (configuration 2) can be obtained by transferring liquid from the left leg to the right leg. The work we perform in doing this is the potential energy in Configuration 2. Hence, by conservation of mechanical energy,

$$T_1 + V_1 = T_2 + V_2 \qquad (a)$$

Taking the zero datum of potential energy at $x=0$, we have

$T_1 = \tfrac{1}{2} m v_0^2$

$V_1 = 0$

$T_2 = \tfrac{1}{2} m v^2 \qquad (b)$

$V_2 = (A x \rho g)(\tfrac{x}{2}\sin\alpha + \tfrac{x}{2}\sin\beta)$

$= \tfrac{1}{2}\tfrac{mg}{L} x^2 (\sin\alpha + \sin\beta)$

where A is the cross-sectional area of the liquid, ρ is the mass density of the liquid, and m is the mass of the liquid.

Equations (a) and (b) yield:

$$\tfrac{1}{2} m v_0^2 = \tfrac{1}{2} m v^2 + \tfrac{1}{2}\tfrac{mg}{L} x^2 (\sin\alpha + \sin\beta)$$

For $x = x_{max}$, $v=0$. Therefore,

$$x_{max} = v_0 \sqrt{\frac{L}{g(\sin\alpha + \sin\beta)}}$$

15.68 GIVEN: A satellite goes around the earth in an elongated orbit.

FIND: a.) Show that its speed v is given by the formula

$$v^2 = 2G m_E/r + \text{constant}$$

where r is its distance from the center of the earth, m_E is the mass of the earth, and G is the gravitational constant.

b.) Show that

$$v^2 = 2 g_0 r_0^2/r + \text{constant}$$

SOLUTION

a.) The potential energy of the satellite is

$$V = \int F\,dr = \int \frac{G m_s m_E}{r^2}\,dr$$

(continued)

15.68 cont.

or $V = -\frac{G m_s m_E}{r} + \text{constant}$

The kinetic energy of the satellite is

$T = \frac{1}{2} m_s v^2$

Hence, since $T + V = \text{constant}$,

$\frac{1}{2} m_s v^2 - \frac{G m_s m_E}{r} = \text{constant}$

or $v^2 = 2\frac{G m_E}{r} + \text{constant}$

b.) On the earth's surface,

$F = \frac{G m_s m_E}{r_0^2} = W = m_s g_0$

Hence,

$G m_s = g_0 r_0^2$

So, by

$v^2 = \frac{2 G m_s}{r} + \text{constant}$ (part a.)

$v^2 = \frac{2 g_0 r_0^2}{r} + \text{constant}$

15.69

GIVEN A particle ($m = 10$ kg) is attached to two springs, each of undeformed length 75 mm and spring constant $k = 2$ N/mm. The particle is dropped from rest at point O.

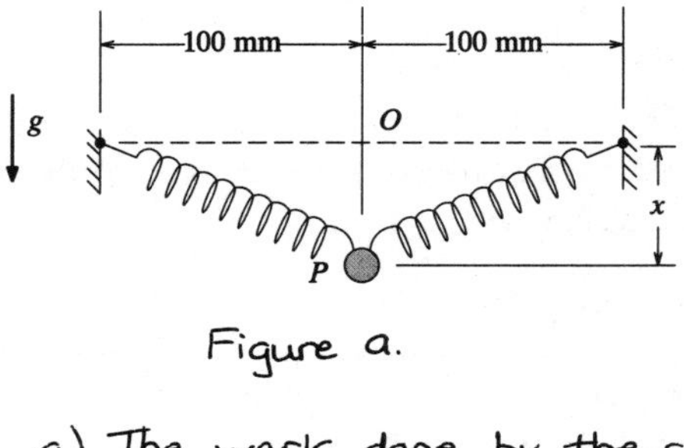

Figure a.

FIND: a.) The work done by the springs as a function of x.

b.) Speed of the particle at $x = 100$ mm

c.) The distance the particle falls before again coming to rest.

SOLUTION:

a.) $U_{1-2} = -\Delta V = V_1 - V_2$

$V_1 = 2(\frac{1}{2} k e^2) = k e^2 =$
$= (2 \text{ N/mm})(100\text{mm} - 75\text{mm})^2$

$\therefore V_1 = 1250 \text{ N}\cdot\text{mm}$

$V_2 = 2(\frac{1}{2} k e^2) = k e^2$
$= (2 \text{ N/mm})(\sqrt{x^2 + 100^2} - 75)^2$
$= 2(x^2 + 10000 - 150\sqrt{x^2 + 100^2} + 5625)$

$U_{1-2} = V_1 - V_2$
$= 1250 - 2x^2 - 20000 + 300\sqrt{x^2 + 100^2} - 11250$

$U_{1-2} = -3000 - 2x^2 + 300\sqrt{x^2 + 10000}$ (N·mm)

b.) $T_1 + V_1 = T_2 + V_2$ ($T_1 = 0$) since particle starts from rest

$V_1 = 1250$ N·mm (part a.)

$T_2 = \frac{1}{2} m v^2 = 5 v^2$

$V_2 = 2[\frac{1}{2}(2 \text{ N/mm})(\sqrt{100^2 + 100^2} - 75)^2] - \ldots$
$- 10 \text{ kg}(9.81 \text{ m/s}^2)(100 \text{ mm})$

$V_2 = -986.4$ N·mm

$1250 = 5v^2 - 986.4$

$\therefore 5v^2 = 2236.4$ $v = 21.15$ mm/s

c.) $T_1 + V_1 = T_2 + V_2$ ($T_1 = 0$, $T_2 = 0$) since particle starts and ends at rest.

$V_1 = 1250$ N·mm (part A)

$V_2 = 2(\frac{1}{2}(2 \text{ N/mm})(\sqrt{100^2 + x^2} - 75)^2 - 98.1x$

$V_2 = 2(10000 + x^2 - 150\sqrt{10000 + x^2} + 5625) \ldots$
$- 98.1x$

$\therefore 1250 = 31250 + 2x^2 - 300\sqrt{10000 + x^2} - 98.1x$

So, $x = 118.9$ mm

15.70

GIVEN: Solve Problem 15.69 for the case where each spring has the nonlinear load-extension relationship $T = 8\sqrt{e}$ (N), where e is the extension (mm) of the spring.

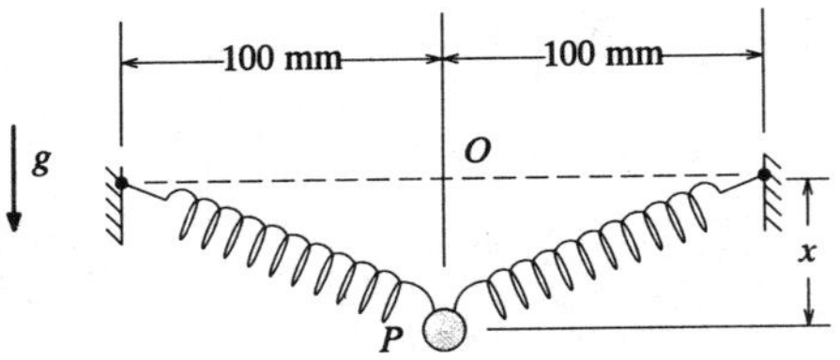

Figure a.

FIND: a.) The work done by the springs as a function of x

b.) The speed of the particle at $x = 100$ mm.

c.) The distance the particle falls before again coming to rest.

SOLUTION:

a.) $V = 2\int 8\sqrt{e}\, de$

$= 2(16/3\, e^{3/2}) = \frac{32}{3} e^{3/2}$

$U_{1-2} = -\Delta V = V_1 - V_2$

$V_1 = \frac{32}{3}(100 - 75)^{3/2} = 1333.3$ N·mm

$V_2 = \frac{32}{3}(\sqrt{x^2 + 100^2} - 75)^{3/2}$

$\underline{\underline{U_{1-2} = 1333.3 - 10.667(\sqrt{x^2 + 10000} - 75)^{3/2} \text{ N·mm}}}$

b.) $\cancel{T_1}^{0} + V_1 = T_2 + V_2$ since the particle starts from rest

$V_1 = 1333.3$ N·mm (part a)

$T_2 = \frac{1}{2} m v^2 = 5v^2$

$V_2 = \frac{32}{3}(\sqrt{20{,}000} - 75)^{3/2} - 98.1\,(100 \text{ mm})$

$= -4036$ N·mm

$1333.3 = 5v^2 - 4036;$

$\underline{\underline{v = 32.77 \text{ mm/s}}}$

c.) $\cancel{T_1}^{0} + V_1 = \cancel{T_2}^{0} + V_2$ since the particle starts and ends at rest

$V_1 = 1333.3$ N·mm (part a)

$V_2 = \frac{32}{3}(\sqrt{x^2 + 100^2} - 75)^{3/2} - 98.1\,x$

$\therefore\ 1333.3 = \frac{32}{3}(\sqrt{x^2 + 100^2} - 75)^{3/2} - 98.1x$

so, $\underline{\underline{x = 219.62 \text{ mm}}}$

15.71

GIVEN: A weight W is hung from a spring (constant = k). A magnet pulls down on the weight with force $F = c/(a-e)^2$, where c and a are constants and e is the extension of the spring (Fig a.)

FIND:

a.) Derive a formula for the potential energy of the system, in terms of c, a, e, W, and k.

b.) Determine the speed of the weight for $e = 3$ in, $k = 2$ lb/in, $W = 4$ lb, $C = 72$ lb·in², and $a = 12$ in, if W is released from rest at $e = 0$

c.) Determine how far the weight drops before it again comes to rest.

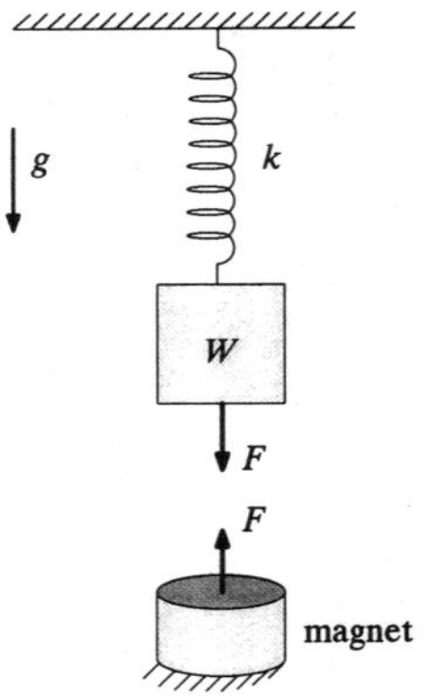

Figure a.

SOLUTION:

a.) The potential energy stored in the spring as the weight drops a distance e is

$V_{spring} = \frac{1}{2} k e^2$

The potential energy of the external forces W and $F = c/(a-e)^2$ are the negative of the work of these forces as the weight drops a distance e.

(continued)

Therefore,

$$V_{\substack{external \\ forces}} = -We - \int \frac{C}{(a-e)^2}\, de$$

$$= -We - \frac{C}{a-e} + constant$$

Hence, the potential energy of the system is:

$$V = \tfrac{1}{2}ke^2 - We - \frac{C}{a-e} + constant$$

Let $V=0$ for $e=0$, Then, constant $= \frac{C}{a}$ and;

$$\underline{\underline{V = \tfrac{1}{2}ke^2 - We - \frac{C}{a-e} + \frac{C}{a}}}$$

b.) For $e=3\,in$, with $k = 2\,(lb/in)$, $W = 4\,lb$, $C = 72\,lb\cdot in^2$, and $a = 12\,in$, the speed is determined from (W dropped from rest when $e=0$)

$$\cancel{T_1}_0 + \cancel{V_1}_0 = T_2 + V_2$$

where

$$T_2 = \tfrac{1}{2}mv^2 = \tfrac{1}{2}\left(\frac{4\,lb}{32.2\,ft/s^2}\right) = 0.06211\,v^2$$

and

$$V_2 = V = \tfrac{1}{2}(2\,{}^{lb}\!/\!in)(3\,in)^2 - (4\,lb)(3\,in) - \ldots$$

$$\ldots - \frac{72\,lb\cdot in^2}{(12-3)\,in} + \frac{72\,lb\cdot in^2}{12\,in}$$

or $V_2 = -5\,in\cdot lb$

$$\therefore\ 0.06211\,v^2 = 5; \quad \underline{\underline{v = 8.972\ in/s}}$$

c.) To determine how far the weight drops from rest at $e=0$,

$$\cancel{T_1}_0 + \cancel{V_1}_0 = \cancel{T_2}_0 + V_2$$

or $V_2 = 0 = \tfrac{1}{2}(2)(e)^2 - 4e - \frac{72}{12-e} + \frac{72}{12}$

or $e^2 - 4e - \frac{72}{12-e} + 6 = 0$

$$\therefore\ e = 0;\quad e = 4.838\,in$$

Thus, the weight drops 4.838 in. before coming to rest again.

GIVEN: When the liquid is at rest ($x=0$), the lengths of the columns of liquid in tubes 1 and 2 are both L. (Fig. a) The liquid oscillates in the tubes and for $x=0$, $\frac{dx}{dt} = v_0$.

FIND: By means of the principle of conservation of energy, dx/dt in terms of x, L, g, and v_0.

SOLUTION:
Starting with $x=0$ (configuration 1, the zero datum for potential energy), the configuration shown in Figure a (Configuration 2) can be obtained by transferring liquid from the left leg to the right leg.

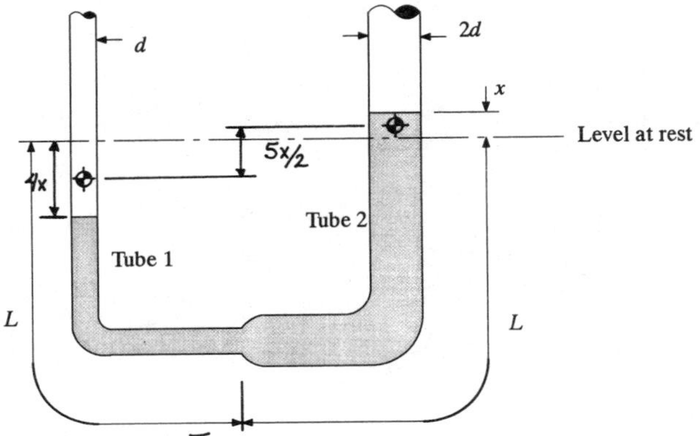

Figure a

The work required to do this is the potential energy in configuration 2.
Hence, by Fig. a, with mass density ρ,

$$V = \left(\frac{\pi}{4}d^2\right)(4x)\left(\frac{5}{2}x\right)\rho g$$

or

$$V = \frac{5}{2}\pi d^2 \rho g\, x^2 \qquad (a)$$

The kinetic energy of the liquid is

$$T = \tfrac{1}{2}(L+x)\frac{\pi}{4}(2d)^2 \rho\left(\frac{dx}{dt}\right)^2$$
$$+ \tfrac{1}{2}(L-4x)\frac{\pi}{4}d^2\rho\left(4\frac{dx}{dt}\right)^2$$
$$\therefore\ T = \frac{5\pi\rho d^2}{2}(L-3x)\left(\frac{dx}{dt}\right)^2 \qquad (b)$$

For $x=0$, $\frac{dx}{dt} = v_0$ and $T = T_1$. Hence, by Eq. (b),

$$T_1 = \frac{5\pi}{2}\rho d^2 L v_0^2 \qquad (c)$$

Also, by Eq. (a), for $x=0$,

$$V_1 = V = 0 \qquad (d)$$

and for $x\neq 0$, by Eqs. (a) and (b),

$$V_2 = V = \frac{5\pi}{2}d^2\rho g\, x^2 \qquad (e)$$

$$T_2 = T = \frac{5\pi}{2}\rho d^2(L-3x)\left(\frac{dx}{dt}\right)^2 \qquad (f)$$

(continued)

15.72 (cont.)

Hence, by the principle of conservation of energy and Eqs. (c), (d), (e), and (f).

$$T_1 + V_1 = T_2 + V_2$$

$$L v_0^2 = (L-3x)\left(\frac{dx}{dt}\right)^2 + g x^2$$

Therefore,

$$\frac{dx}{dt} = \pm\sqrt{\frac{L v_0^2 - g x^2}{L-3x}}$$

15.73

GIVEN: A mass m that is hung from a vertical spring with constant k is dropped from rest when the spring is undeformed. (Fig. a) The magnet pulls down on the mass with a force $F = C/(a-e)^2$; C and a are constants and e is the extension of the spring.

FIND:

a) Derive a formula for the speed v of m in terms of e, k, m, g, C, and a.

b) Let $k = 2$ lb/in , $mg = 4$ lb , $C = 72$ lb·in², and $a = 12$ in. With the formula for v (part a), determine the value of $e = e_{max}$ (inches) at which the mass again comes to rest.

c) Plot v versus e for $0 \le e \le e_{max}$

d) From the plot estimate the maximum value of v and the corresponding value of e.

e) Check your estimates in part d by the formula for v derived in part a.

SOLUTION:

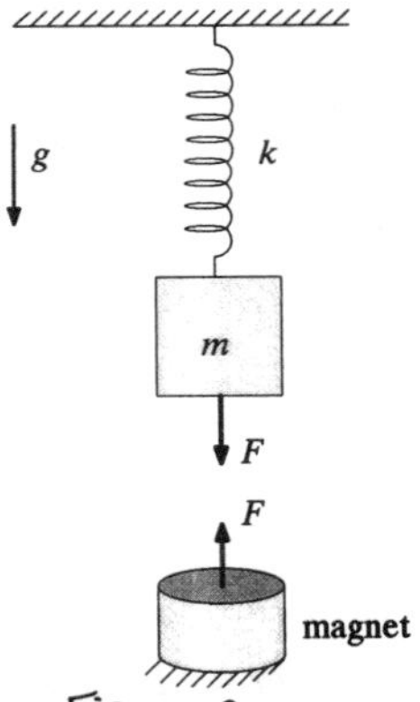

Figure a

a) Let the zero datum of potential energy be the position of the unstreched spring. Consider the free-body diagram of the mass (Fig. b).

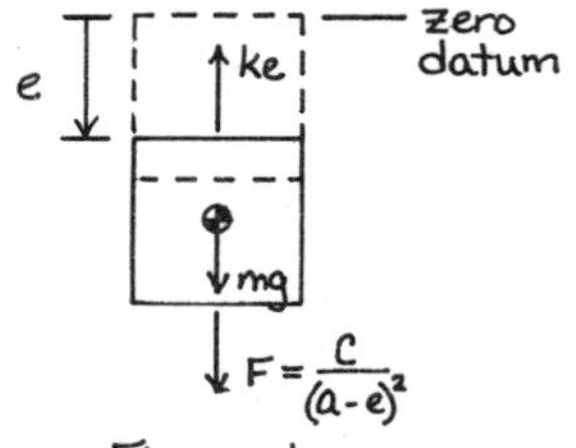

Figure b

The potential energy of the system, after the mass has dropped the distance e, is

$$V = \tfrac{1}{2} ke^2 - mge - \int F de$$

or

$$V = \tfrac{1}{2} ke^2 - mge - \int \frac{C de}{(a-e)^2}$$

Integration yields

$$V = \tfrac{1}{2} ke^2 - mge - \frac{C}{a-e} + \text{constant}$$

Let $V_1 = V = 0$ for $e = 0$; therefore, constant $= \frac{C}{a}$.

and $V = V_2 = \tfrac{1}{2} ke^2 - mge - \frac{C}{a-e} + \frac{C}{a}$

Now by Eq. (15.27),

$$T_1 + V_1 = T_2 + V_2 \qquad (a)$$

where $T_1 = V_1 = 0$

$T_2 = \tfrac{1}{2} mv^2$

$V_2 = \tfrac{1}{2} ke^2 - mge - \frac{C}{a-e} + \frac{C}{a}$

Hence, Eq. (a) yields

$$\tfrac{1}{2} mv^2 = \frac{C}{a-e} - \frac{C}{a} + mge - \tfrac{1}{2} ke^2$$

or

$$v = \sqrt{\frac{2}{m}\left[\frac{C}{a-e} - \frac{C}{a} + mge - \tfrac{1}{2} ke^2\right]} \qquad (b)$$

b) Let $k = 2$ lb/in², $mg = 4$ lb, $C = 72$ lb·in², and $a = 12$ in. The extension e reaches a maximum value e_{max} when v is again zero. Therefore, by Eq. (b):

$$\frac{C}{a-e_{max}} - \frac{C}{a} + mge_{max} - \tfrac{1}{2} ke_{max}^2 = 0$$

or

$$\frac{72}{12-e_{max}} - \frac{72}{12} + 4e_{max} - e_{max}^2 = 0 \qquad (c)$$

Hence, $e_{max} = 4.838$ in

c)

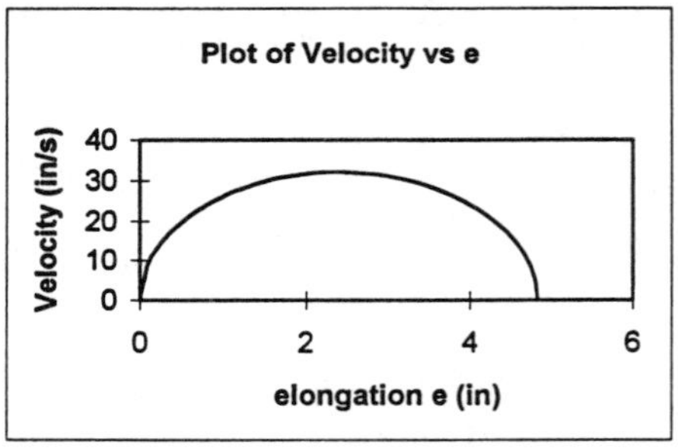

d) By the graph of v,

$v_{max} \approx 32$ in/s at $e \approx 2.4$ in

e) To check the estimates of part d, by Eq. (b), the value of e at which $v = v_{max}$ is obtained from

$$\frac{dv}{de} = \frac{1}{2m} \frac{\frac{C}{(a-e)^2} + mg - ke}{\sqrt{\frac{2}{m}\left[\frac{C}{a-e} - \frac{C}{a} + mge - \tfrac{1}{2} ke^2\right]}} = 0$$

(continued)

15.73 (cont.)

or $\frac{C}{(a-e)^2} + mg = ke$

$$\therefore \frac{72}{(12-e)^2} + 4 = 2e$$

Hence, $\underline{e = 2.390 \text{ in}}$

and, by Eq. (b),

$$v = \sqrt{\frac{2(386.4)}{4}\left[\frac{72}{(12-2.390)} - 6 + 4(2.390) - (2.390)^2\right]}$$

or $\underline{\underline{v = 32.12 \text{ in/s}}}$

15.74

GIVEN: A particle ($m = 0.6$ slugs) is attached to two identical springs ($k = 8$ lb/in). The mass is released at $x = 0$ from rest (Fig. a).

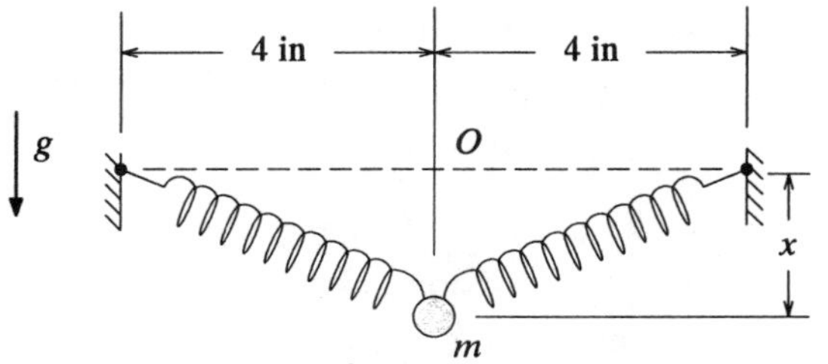

Figure a

FIND:
(a) Derive a formula for speed as a function of x
(b) Find the max displacement of the mass
(c) Plot v versus x.
(d) From the plot, estimate the max speed and corresponding x value
(e) Verify the estimates of part d by the formula for v from part a.

SOLUTION:

(a) By Eq. (15.27),

$$T_1 + V_1 = T_2 + V_2 \qquad \text{(a)}$$

where $T_1 = 0$

$$V_1 = 2(\tfrac{1}{2}kx^2) = kx^2 = 8 \text{ lb/in}(4'' - 3'')^2 = 8 \text{ in·lb}$$

$$T_2 = \tfrac{1}{2}mv^2 = \tfrac{1}{2}(0.6)v^2\left(\tfrac{1\text{ ft}}{12\text{ in}}\right) = 0.025v^2 \text{ (in·lb)}$$

$$V_2 = -mgx + 2(\tfrac{1}{2}kx^2) = -0.6(32.2)x + 8\left(\sqrt{4^2+x^2} - 3\right)^2$$

$$V_2 = -19.32x + 8\left(16 + x^2 - 6\sqrt{16+x^2} + 9\right)$$

$$V_2 = -19.32x + 128 + 8x^2 - 48\sqrt{16+x^2} + 72$$

$$V_2 = 8x^2 - 19.32x + 200 - 48\sqrt{16+x^2} \text{ (in·lb)}$$

Therefore, by Eq. (a),

$$8 = 0.025v^2 + 8x^2 - 19.32x + 200 - 48\sqrt{16+x^2}$$

$$\underline{\underline{v = \sqrt{-320x^2 + 772.8x - 7680 + 1920\sqrt{16+x^2}}}} \qquad \text{(b)}$$

(b) At x_{max}, $v = 0$. Therefore, by Eq. (b)

$$0 = -320x^2 + 772.8x - 7680 + 1920\sqrt{16+x^2}$$

$$1920\sqrt{16+x^2} = 320x^2 - 772.8x + 7680$$

$$\left(\sqrt{16+x^2}\right)^2 = \left(\tfrac{x^2}{6} - 0.4025x + 4\right)^2$$

$$16 + x^2 = \left(\tfrac{x^2}{6} - 0.4025x + 4\right)^2; \therefore \underline{x = 5.46 \text{ in}}$$

(c)

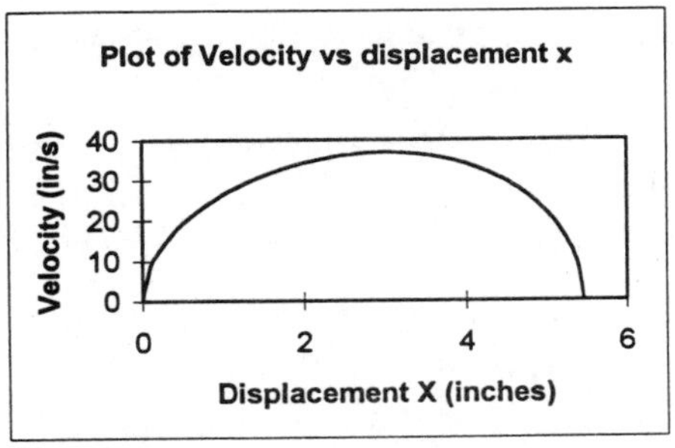

(d) By the graph of v versus x

$$\underline{v_{max} \approx 37 \text{ in/s at } x \approx 3.00 \text{ in.}}$$

(e) By Eq. (b),

$$\frac{dv}{dx} = \frac{1}{2}\,\frac{-640x + 772.8 + \frac{1920}{2}\frac{(2x)}{\sqrt{16+x^2}}}{\sqrt{-320x^2 + 772.8x - 7680 + 1920\sqrt{16+x^2}}} = 0$$

$$\therefore (-640x + 772.8)\sqrt{16-x^2} + 1920x = 0$$

or, $(-640x + 772.8)^2(16 + x^2) = 1920^2x^2$

$\therefore \underline{x = 3.012 \text{ in.}}$

$\therefore \underline{\underline{v_{max} = 36.86 \text{ in/s}}}$

15.75

FIND: Design a bungee cord to prevent jumpers, whose masses lie between 3.0 slugs and 7.0 slugs, from dropping no more than 240 ft (Fig. a).

a) Assume that the cord acts like a linear spring, with constant k (lb/ft) and an unstretched length L_0 (ft). Neglect air resistance and the weight of the rope.

b) With your design, determine how high the jumper rebounds.

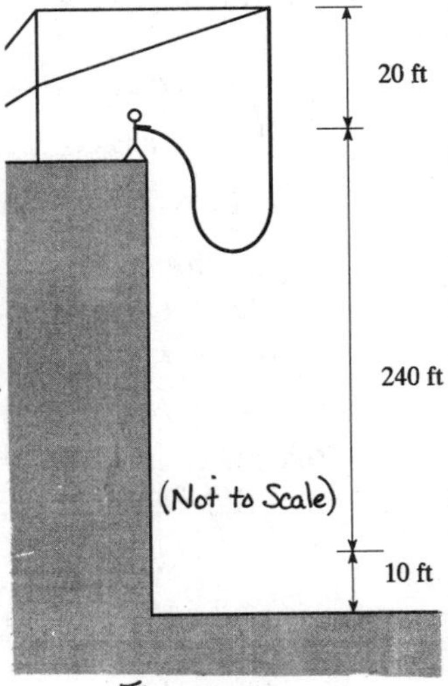

Figure a

(continued)

15.75 (cont.)

SOLUTION:

(a) Since air resistance is neglected, the system is conservative. Therefore, by conservation of energy,

$$T_1 + V_1 = T_2 + V_2 \qquad (a)$$

Since the jumper starts from rest and, at the end of the fall, is again at rest $T_1 = T_2 = 0$.

At the top of the jump, the potential energy of the system is

$$V_1 = mg(240 \text{ ft}) \qquad (b)$$

where the zero datum of potential energy is taken at the end of the fall (Fig. a).

At the end of the fall, the potential energy of the system is that of the extended cord. Thus,

$$V_2 = \tfrac{1}{2}ke^2 = \tfrac{1}{2}k(260-L_0)^2 \qquad (c)$$

where $e = 260 - L_0$ is the stretch of the cord at a drop of 240 ft (Fig. a).

Equations (a), (b), and (c) yield

$$240mg = \tfrac{1}{2}k(260-L_0)^2 \qquad (d)$$

Equation (d) relates the mass m of the jumper to the spring constant k of the cord and the unstretched length L_0 of the cord.

Consider a jumper whose mass is 7 slugs. Then Eq. (d) yields, with $g = 32.2$ ft/s²,

$$k(260-L_0)^2 = 108,192 \qquad (e)$$

Decisions must now be made on appropriate values of k and L_0. For example, a free fall of more than 100 ft may not be desirable for most jumpers. Hence, take

$$L_0 = 100 + 20 = 120 \text{ ft}$$

Then, Eq. (e) yields $k = 5.52$ lb/ft for a jumper whose mass is 7 slugs.

For a jumper whose mass is 3 slugs, Eq. (d) yields $k(260-L_0)^2 = 46,368$

For $L_0 = 120$ ft, this yields $k = 2.36$ lb/ft.

Thus, for $k = 5.52$ lb/ft, both the 7 slug and 3 slug jumper would not drop more than 240 feet.

In fact, the 3 slug jumper would drop a distance $d < 240$ ft, with $k = 5.52$ lb/ft and $L_0 = 120$ ft. For example, for the 3 slug jumper,

$$V_1 = mgd = (3)(32.2)d = 96.6d$$
$$V_2 = \tfrac{1}{2}k(d+20-L_0)^2 = \tfrac{5.52}{2}(d-100)^2 \qquad (f)$$

Hence, by Eqs. (a) and (f),

$$d^2 - 235d + 10,000 = 0$$

or $d = 179.2$ ft

Hence, take $\underline{k = 5.52 \text{ lb/ft}, \; L_0 = 120 \text{ ft}}$

(b) Take $k = 5.52$ lb/ft and $L_0 = 120$ ft. Then for the 7 slug jumper, on the rebound,

$$V_1 = \tfrac{1}{2}(5.52)(260-120)^2 = 54,096 \text{ ft·lb}$$
$$V_2 = (7)(32.2)h$$

or

$$h = \frac{54,096}{(7)(32.2)} = 240 \text{ ft}$$

This result should have been expected since the system is conservative.

Likewise for the 3 slug jumper,

$$V_1 = \tfrac{1}{2}(5.52)(179.2-100)^2 = 17,312.5$$
$$V_2 = (3)(32.2)h \quad \text{and}$$
$$h = \frac{17,312.5}{(3)(32.2)} = 179.2 \text{ ft}$$

Because of air resistance (drag) and damping in the cord the drops and rebounds would be somewhat less than the values computed neglecting drag and damping.

15.76

GIVEN: In Problem 15.75, add the condition that the maximum acceleration be at most 2g.

FIND: Design the bungee cord (Fig. a).

SOLUTION:

Let L_0 = length of unstretched cord

k = spring constant of cord

m = mass of jumper (3-7 slugs)

$h = L_0 - 20$ = distance of free fall

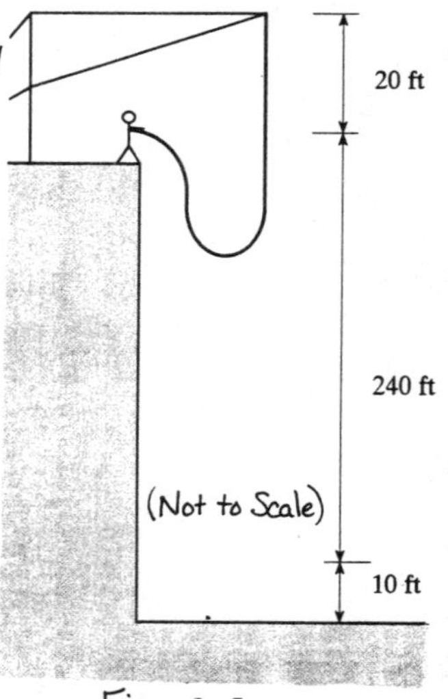

Figure a

Until $h = L_0 - 20$, the jumper is in a free fall (acceleration = g). At the end of the free fall the speed v_f of the jumper is given by

$$U_{1-2} = T_2 - T_1$$

where $U_{1-2} = mgh = mg(L_0 - 20)$

(continued)

15.76 (cont.)

$T_1 = 0$

$T_2 = \frac{1}{2} m v_f^2$

$\therefore\ v_f = \sqrt{2g(L_0 - 20)}$ (a)

The speed v_r during the restrained fall (when the cord is stretched) is given by

$$U_{1-2} = T_2 - T_1$$

where here

$$U_{1-2} = mgx - \frac{1}{2} kx^2$$

$$T_1 = \frac{1}{2} m v_f^2$$

$$T_2 = \frac{1}{2} m v_r^2$$

where x is the distance that the jumper falls from the end of the free fall.

Hence,

$$mgx - \frac{1}{2} kx^2 = \frac{1}{2} m v_r^2 - \frac{1}{2} m v_f^2 \quad (b)$$

Since v_f is a constant [see Eq. (a)], differentiation of Eq. (b) with respect to time yields

$$mg\dot{x} - kx\dot{x} = m v_r \dot{v}_r$$

or since $v_r = \dot{x}$ and $\dot{v}_r = a_r$, the acceleration,

$$a_r = g - \frac{k}{m} x \quad (c)$$

The maximum acceleration of the jumper occurs at the end of the fall ($v = 0$) and is directed upward. Thus, Eq. (c) yields with $a_r = -2g$,

$$kx = 3mg \quad (d)$$

Since the drop is restricted to 240 ft, $L_0 + x \le 240$ ft. Hence, $x = 240 - L_0$, and Eq. (d) becomes

$$k(240 - L_0) = 3mg \quad (e)$$

Also, see Problem 15.75, at the top of the jump,

$$V_1 = mg(240 \text{ ft})$$

and at the bottom of the jump,

$$V_2 = \frac{1}{2} k e^2 = \frac{1}{2} k (260 - L_0)^2$$

$\therefore$ By $T_1 + V_1 = T_2 + V_2$ and since $T_1 = T_2 = 0$,

$$k(260 - L_0)^2 = 480 mg \quad (f)$$

Equations (e) and (f) relate k, L_0, and m, subject to the conditions $|a_{max}| = 2g$ and the fall $\le$ 240 ft. Eqs. (e) and (f) yield

$$(260 - L_0)^2 = 160(240 - L_0)$$

or $L_0^2 - 360 L_0 + 29200 = 0$

or $\underline{L_0 = 123.43 \text{ ft}}$

Hence, by Eq. (e),

$$\underline{\underline{k = \frac{(3)(7)(32.2)}{240 - 123.43} = 5.80 \text{ lb/ft}}}$$

For $L_0 = 123.43$ ft and $k = 5.80$ lb/ft, the distance d that the 3 slug jumper drops is given by

$$V_1 = V_2$$

$$V_1 = mgd = 3(32.2)d$$

$$V_2 = \frac{1}{2} k (d + 20 - L_0)^2 = \frac{5.8}{2} (d - 103.43)^2$$

or

$$d^2 - 240.17d + 10{,}697.8 = 0$$

or

$$\underline{\underline{d = 181.1 \text{ ft}}}$$

Thus, the bungee cord with

$\underline{\underline{k = 5.80 \text{ lb/ft and } L_0 = 123.43 \text{ ft}}}$

satisfies the requirements that $|a_{max}| \le 2g$ and the drop $\le$ 240 ft for 3 slugs $\le m \le$ 7 slugs.

15.77

GIVEN: A body of mass m is hung from a spring (constant k) and it is lowered a distance x from the unstretched position of the spring (Fig. a).

FIND: Internal and External potential energies of the system.

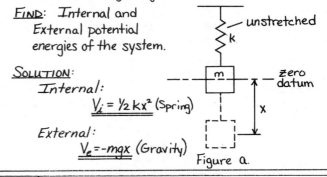

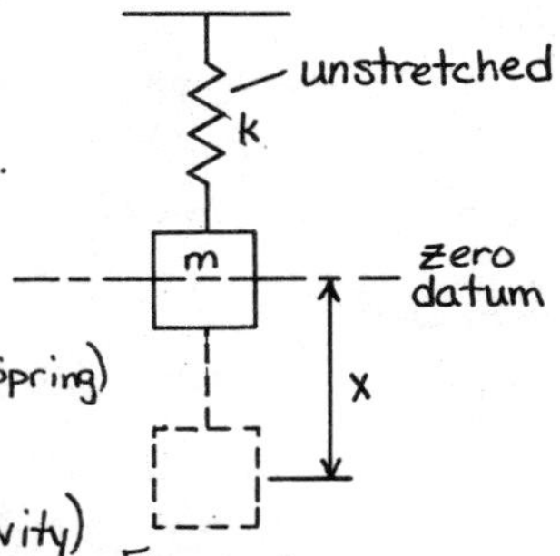

Figure a.

SOLUTION:

Internal:

$$\underline{\underline{V_i = \frac{1}{2} kx^2}} \text{ (Spring)}$$

External:

$$\underline{\underline{V_e = -mgx}} \text{ (Gravity)}$$

15.78

GIVEN: A potential energy function is given by $V = C/x^3$, where C is a positive constant.

FIND:

a) Derive a formula for the conservative force F as a function of x.

b) Does F point toward or away from the origin?

c) Does V increase or decrease as $|x|$ increases?

SOLUTION:

a) By Eq. (15.30), with $V = \frac{C}{x^3}$

$$F = -\frac{dV}{dx} = -\frac{d}{dx}(Cx^{-3}) = 3Cx^{-4}$$

or

$$\underline{\underline{F = \frac{3C}{x^4}}}$$

(continued)

15.78 (cont.)

b) Since F is positive, it points <u>away</u> from the origin (x=0).

c) Since $V=\frac{C}{x^3}$, as $|x|$ increases $V=\frac{C}{x^3}$ <u>decreases</u> for $x>0$ and <u>increases</u> for $x<0$. (see Fig. a)

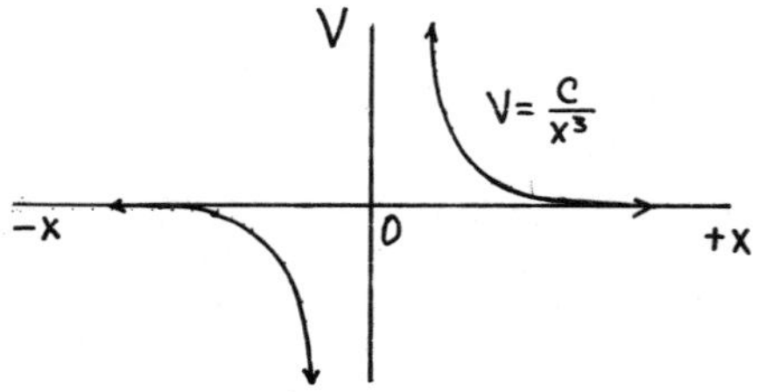

Figure a.

15.79

GIVEN: In a spring mass system, the potential energy of the spring (constant k) is $V=\frac{1}{2}kx^2$, where x is the extension of the spring.

FIND:

a) Determine the force that the spring exerts on the mass.

b) Explain the negative sign.

SOLUTION:

a) For one-dimensional problems

$$F=-\frac{dV}{dx} \quad \text{[see Eq. (15.30)]}$$

Hence, with $V=\frac{1}{2}kx^2$

$$\underline{F=-kx}$$

b) As shown in figure a., the negative sign comes from the fact that the force acting on the mass is opposite in direction to the extension of the spring.

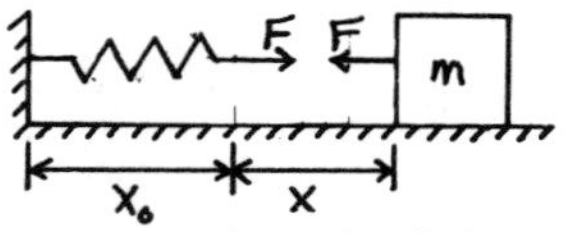

x_o = unstretched length

x = extension

Figure a

15.80

FIND:

a). Show that a constant force (say $\vec{F}=6\hat{\imath}$) is conservative.

b). Determine the potential function V for $\vec{F}=6\hat{\imath}$; V=0 at x=0.

SOLUTION:

a) To show that $\vec{F}=6\hat{\imath}$ is conservative, we may show that the work performed by $\vec{F}$ around a closed path is zero. For example, consider a circular path (Fig. a).

By Fig. a,

$$U=\oint 6dx$$

$$=\int_0^{2\pi} 6d(r\cos\theta)$$

$$=6r\int_0^{2\pi}(-\sin\theta)d\theta$$

or

$$U=6r\cos\theta\Big|_0^{2\pi}=0 \quad \text{Q.E.D.}$$

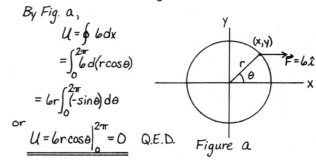

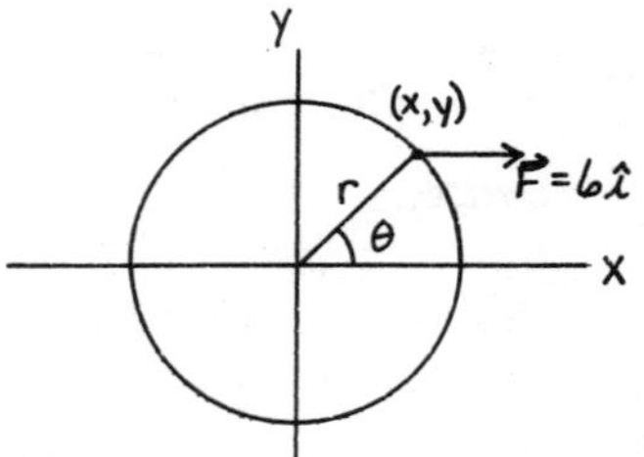

Figure a

b) $V=-\int F\cdot dx=-6x+\text{constant}$

Since V=0 for x=0, constant = 0

$\therefore \underline{V=-6x}$

15.81

GIVEN: A conservative force $\vec{F}=(Ay^2-By)\hat{\jmath}$, where A and B are constants and y is in meters.

FIND:

a). Potential energy function of F; V=0 at y=0.

b). Change of potential energy and change in kinetic energy of particle of mass m as it moves from y=1m to y=3m.

c) The speed v of the particle at y=3m in terms of A, B, and m.

SOLUTION:

a) $V=-\int F dr \qquad V=-\int(Ay^2-By)dy$

$$V=-\frac{Ay^3}{3}+\frac{By^2}{2}+\text{constant}$$

V=0 for y=0. ∴ constant = 0

$$\underline{V=-\frac{Ay^3}{3}+\frac{By^2}{2}}$$

b) The change in potential energy is

$$\Delta V=\int_1^3 -(Ay^2-By)dy+mg(3)-mg(1)$$

$$\underline{\Delta V=-\frac{26}{3}A+4B+2mg}$$

By Eq. (3.26),

$$\underline{\Delta T=-\Delta V=\frac{26}{3}A-4B-2mg}$$

(continued)

15.81 (cont.)

c.) By conservation of energy with the particle starting from rest at y=1m,

$$\cancel{T_1}^{0} + V_1 = T_2 + V_2$$

where $V_1 = -\frac{A}{3} + \frac{B}{2} + mg(1)$

$T_2 = \frac{1}{2} m v^2$

$V_2 = -9A + 4.5B + mg(3)$

$\therefore \frac{1}{2} m v^2 = \frac{26}{3}A - 4B - 2mg$

or $$v = \sqrt{\frac{52}{3}\frac{A}{m} - \frac{8B}{m} - 4g}$$

15.82

GIVEN: In Problem 15.77 the mass is lowered slowly until it is in equilibrium (Fig. a), and then it is lowered an additional amount d.

FIND: The potential energy of the spring-mass system, taking the zero state of potential energy at the equilibrium position.

SOLUTION:

Relative to the equilibrium position, the potential energy of the mass m is (Fig. a)

$$V_{mass} = -mgd \qquad (a)$$

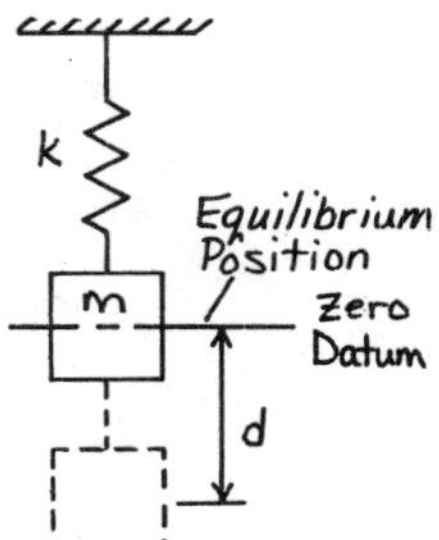

Figure a.

In the equilibrium position (Fig. b),

$$\Sigma F_y = mg - kd_o = 0$$

or $d_o = \frac{mg}{k}$

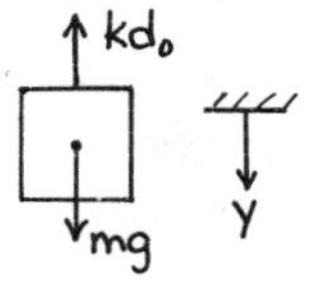

Figure b

where d_o is the stretch in the spring (Fig. c).

Hence, the potential energy of the spring relative to the equilibrium position is

$$V_{spring} = \frac{1}{2}k(d_o+d)^2 - \frac{1}{2}kd_o^2$$
$$= \frac{1}{2}k\left(\frac{mg}{k}+d\right)^2 - \frac{1}{2}\frac{(mg)^2}{k} \qquad (b)$$

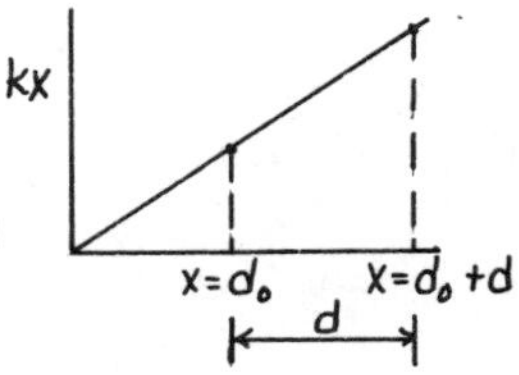

Figure c

Hence, by Eqs. (a) and (b),

$V = V_{mass} + V_{spring}$

$$V = -mgd + \frac{1}{2}k\left(\frac{mg}{k}+d\right)^2 - \frac{1}{2}\frac{(mg)^2}{k}$$

15.83

GIVEN: The force of repulsion between two electric charges is $F = kee'/r^2$, where k is a constant and r is the distance between the charges. A thin wire in the form of a circle of radius a carries a constant density ρ of electric charge e; de = charge on length ds of wire. So $de = \rho ds = \rho a d\theta$ (Fig. a)

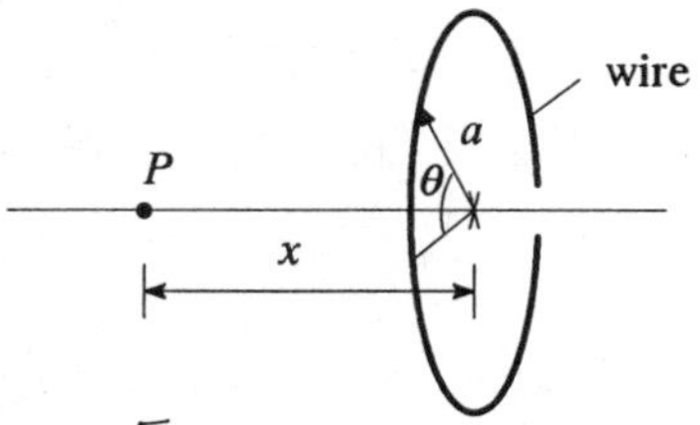

Figure a

FIND:

a) A formula for the potential energy of a unit point charge e' that lies at P (Fig. a) in terms of k, ρ, a, and x.

b) A formula for the force F that acts on the unit charge e'

SOLUTION:

a) By Eq. (15.35), since e' = 1,

$$V = \frac{ke}{r}$$

$$\therefore dV = \frac{k\,de}{r} \qquad (a)$$

By Fig. a, $r = \sqrt{x^2+a^2}$. Also, $de = \rho ds = \rho a d\theta$

Integration of Eq. (a) yields

$$V = \frac{k}{\sqrt{x^2+a^2}}\int_0^{2\pi} \rho a d\theta = \frac{2\pi k \rho a}{\sqrt{x^2+a^2}} \qquad (b)$$

b) By Eq. (15.30),

$$F(x) = -\frac{dV(x)}{dx} \qquad (c)$$

Hence, by Eqs. (b) and (c)

$$F = \frac{2\pi k \rho a x}{(x^2+a^2)^{3/2}}$$

15.84

GIVEN: A suspension system of two masses and two springs (Fig. a). The masses m_1 and m_2 are given displacements y_1 and y_2 from the unstretched positions of the springs

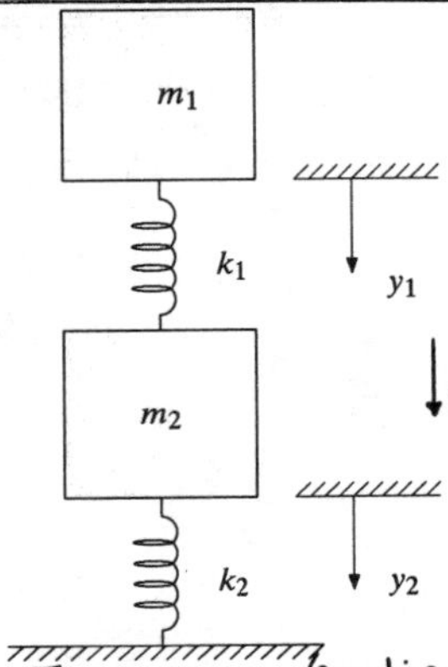

Figure a

(continued)

15.84 (cont.)

FIND: the potential energy of the system relative to its initial configuration.

SOLUTION:

By Fig. a, the potential energy of mass m_1 and spring k_1 is

$$V_1 = \tfrac{1}{2}k_1(y_1-y_2)^2 - m_1g_1(y_1)$$

and of mass m_2 and spring k_2 is

$$V_2 = \tfrac{1}{2}k_2y_2^2 - m_2gy_2$$

Therefore, the total potential energy of the system is

$$V = V_1 + V_2 = \tfrac{1}{2}k_1(y_1-y_2)^2 - m_1g_1y_1 + \tfrac{1}{2}k_2y_2^2 - m_2gy_2$$

or $$\underline{\underline{V = \tfrac{1}{2}k_1(y_1-y_2)^2 + \tfrac{1}{2}k_2y_2^2 - (m_1y_1 + m_2y_2)g}}$$

15.85

GIVEN: A part of a suspension system consists of a mass m supported by a spring (Fig. a). The mass is suspended by a cord so that the spring is unstretched.

FIND:

a.) Determine in terms of m, k, y, and g, the potential energy of the system when m has dropped a distance y.

b.) What value of y is the speed a maximum (v_{max})

c.) Determine v_{max}

d.) Determine y_{max}, the maximum displacement

e.) Evaluate v_{max} and y_{max} for $m = 160\ kg$ and $k = 16\ N/mm$

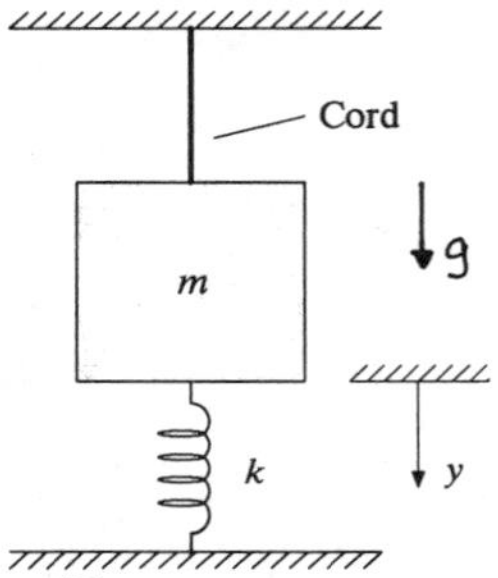

Figure a

SOLUTION:

a) Relative to $y=0$,

$$\Delta V = V_2 - V_1$$

where, $V_1 = 0$ and $V_2 = -mgy + \tfrac{1}{2}ky^2$

$$\therefore\ \underline{\Delta V = \tfrac{1}{2}ky^2 - mgy}$$

b) Since the mass falls from rest,

$$\cancel{T_1}^{\,0} + V_1 = T_2 + V_2$$

or $$T_2 = V_1 - V_2 = -\Delta V$$

$$\therefore\ \tfrac{1}{2}mv^2 = mgy - \tfrac{1}{2}ky^2$$

$v = v_{max}$, when $\frac{dv}{dy} = 0$

$$mv\cancel{\frac{dv}{dy}}^{\,0} = mg - ky = 0$$

$$\underline{\underline{y = \frac{mg}{k}}}$$

c.) Since the mass drops from rest at $y=0$, Conservation of energy yields

$$T_2 = V_1 - V_2$$

where, by parts a and b,

$$V_1 - V_2 = mgy - \tfrac{1}{2}ky^2$$

or $$V_1 - V_2 = \frac{(mg)^2}{k} - \tfrac{1}{2}\frac{(mg)^2}{k}$$

$$T_2 = \tfrac{1}{2}mv_{max}^2$$

$$\therefore\ v_{max} = \sqrt{\left(\frac{2m}{k} - \frac{m}{k}\right)g^2}$$

or $$\underline{\underline{v_{max} = g\sqrt{\frac{m}{k}}}}$$

d.) y_{max} occurs when $v = 0$.

Therefore, by part b,

$$\tfrac{1}{2}m\cancel{v^2}^{\,0} = mgy_{max} - \tfrac{1}{2}ky_{max}^2$$

or $$\underline{\underline{y_{max} = \frac{2mg}{k}}}$$

e.) With $m = 160\ kg$ and $k = 16\ N/mm$,

$$v_{max} = g\sqrt{\frac{m}{k}} = 9.81\ m/s^2\sqrt{\frac{160\ kg}{16\ N/mm \cdot \frac{1000\ mm}{m}}}$$

$$\underline{\underline{v_{max} = 0.981\ m/s}}$$

and $$y_{max} = \frac{2mg}{k} = \frac{2(160\ kg)(9.81\ m/s^2)}{(16\ N/mm)\left(\frac{1000\ mm}{m}\right)}$$

$$\underline{\underline{y_{max} = 0.1962\ m}}$$

15.86

GIVEN: A mass-pulley system of masses m_1 and m_2 connected by a light inextensible cord (Fig. a) such that when released from rest m_2 moves down. Mass m_1 is a smooth cylindrical body.

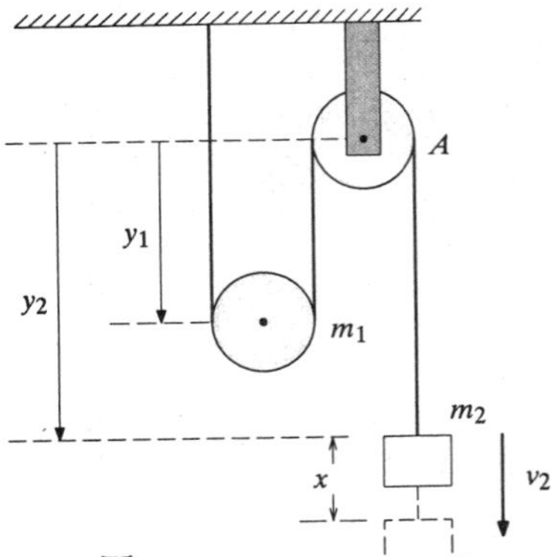

Figure a

FIND: Show that

$$v_1 = \sqrt{\frac{(2m_2 - m_1)gx}{4m_2 + m_1}}\uparrow, \quad v_2 = 2v_1\downarrow$$

(continued)

15.86 (cont.)

SOLUTION:

By geometry, $2y_1 + y_2 = \text{constant}$.
Differentiation yields

$$2v_1 + v_2 = 0;\quad v_2 = -2v_1 = 2v_1 \downarrow$$

By conservation of energy, since the system starts from rest,

$$\cancel{T_1}^{0} + V_1 = T_2 + V_2$$

or $T_2 = V_1 - V_2$ (a)

Take the zero datum for potential energy of m_1 at y_1 and m_2 at y_2. Then $V_1 = 0$, and

$$V_2 = -m_2 g x + m_1 g\left(\tfrac{x}{2}\right)$$

also $T_2 = \tfrac{1}{2} m_1 v_1^2 + \tfrac{1}{2} m_2 (4 v_1^2)$

∴ Equation (a) yields

$$\tfrac{1}{2}(m_1 + 4m_2) v_1^2 = \tfrac{1}{2}(2m_2 - m_1) g x$$

or $$v_1 = \sqrt{\frac{(2m_2 - m_1) g x}{(4m_2 + m_1)}} \uparrow$$

15.87

GIVEN: Two masses are connected by a light, inextensible cord that passes over a frictionless, massless pulley (Fig. a).

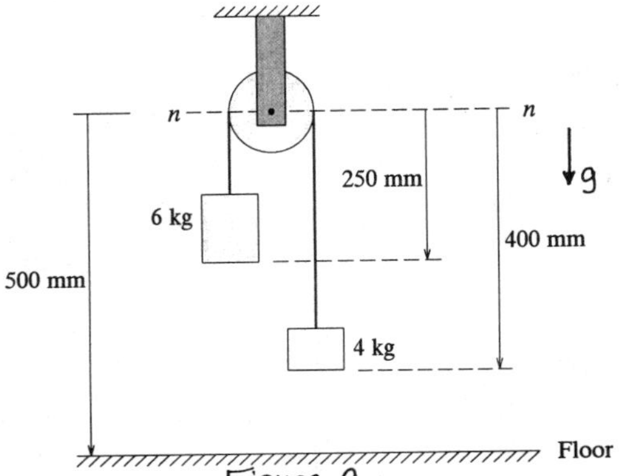

Figure a

FIND:
a.) Determine the potential energy of the external forces for the position shown.
b.) Determine the potential energy of the internal forces for the position shown.
c.) Determine the speed v of the 6 kg. block when it strikes the floor.
d.) Find the height that the 4 kg. mass rises above the floor.

SOLUTION:

a.) $V_e = m_1 g h_1 + m_2 g h_2$

$V_e = (6\text{ kg})(9.81\text{ m/s}^2)(-0.25\text{ m}) + (4\text{ kg})(9.81\text{ m/s}^2)(-0.4\text{ m})$

$V_e = -30.4\text{ N-m}$

b.) Since the cord is inextensible, it stores no energy. Therefore,

$V_i = 0$

c.) Initially, the system is at rest. Therefore,

$$\cancel{T_1}^{0} + V_1 = T_2 + V_2$$

Hence, when the 6-kg. mass hits the floor,

$$-30.4\text{ N·m} = \tfrac{1}{2}(4\text{ kg} + 6\text{ kg})v^2 - 6\text{ kg}(9.81\text{ m/s}^2)(0.5\text{ m}) - (4\text{ kg})(9.81\text{ m/s}^2)(0.15\text{ m})$$

∴ $5v^2 = 4.916$; $v = 0.992\text{ m/s}$

d.) When the 6-kg mass strikes the floor, the 4kg mass is 0.35m above the floor and has velocity of 0.99 m/s upward (Fig. b). Taking this position as the datum for zero potential energy, by conservation of energy,

$$T_1 + \cancel{V_1}^{0} = \cancel{T_2}^{0} + V_2$$

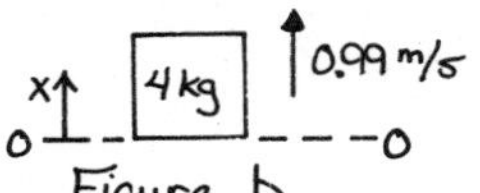

Figure b

Hence,

$$\tfrac{1}{2}(4\text{ kg})(0.99\text{ m/s})^2 = (4\text{ kg})(9.81\text{ m/s}^2)(x)$$

$x = 0.05$ m.

So, the total height above the floor is

$h = 0.35 + 0.05 = 0.40\text{ m}$

15.88

GIVEN: The system shown in Figure a. In the position shown, the spring is unstretched.

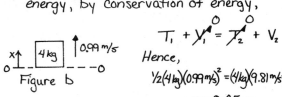

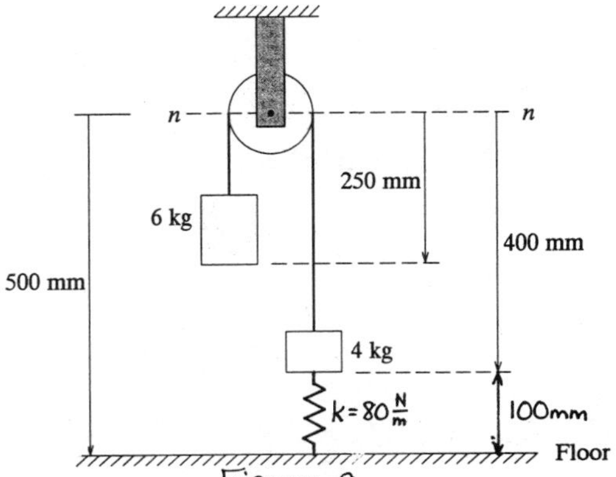

Figure a

FIND:
a.) The potential energy of the external forces relative to the datum n-n, when the 6-kg mass is lowered 50 mm. (continued)

15.88 (cont.)

b.) The potential energy of the internal forces in the position in part a.

c.) The system is released from rest in the unstretched position of the spring. Find the speed of the masses for the instant that the 6-kg mass hits the floor.

d.) The height above the floor that the 4-kg mass rises.

SOLUTION:

a.) $V_e = -(6\,kg)(9.81\,m/s^2)(0.3m) - (4kg)(9.81\,m/s^2)(0.35m)$

$\underline{V_e = -31.4\ N\cdot m}$

b.) $V_i = \frac{1}{2}(80)x^2 = \frac{1}{2}(80)(0.05)^2$

$\therefore \underline{V_i = 0.10\ N\cdot m}$

c.) Since the system is released from rest, $T_1 = 0$, and by conservation of energy,

$$\cancelto{0}{T_1} + V_1 = T_2 + V_2 \quad (a)$$

where, $V_1 = -(6)(9.81)(0.25) - 4(9.81)(0.4) = -30.411$

$V_2 = -6(9.81)(0.5) - 4(9.81)(0.15) + \frac{1}{2}(80)(0.25)^2$

$V_2 = -32.816$

$T_2 = \frac{1}{2}(4+6)v^2 = 5v^2$

By Eq. (a),

$$-30.411 + 32.816 = 5v^2$$

$$\underline{v = 0.695\ m/s}$$

d.) When the 6-kg mass hits the floor, the 4-kg mass is 0.35m above the floor and has a velocity of 0.695 m/s upward. With $T_2 = 0$, by the conservation of energy,

$$T_1 + V_1 = \cancelto{0}{T_2} + V_2 \quad (a)$$

where,

$T_1 = \frac{1}{2}(4\,kg)(0.695\,m/s)^2 = 0.966\ N\cdot m$

$V_1 = \frac{1}{2}(80)(0.25)^2 = 2.50\ N\cdot m$

$V_2 = 4(9.81)(x) + \frac{1}{2}(80)(0.25+x)^2$

$= 39.24x + 40(x^2 + 0.5x + 0.0625)$

or $V_2 = 40x^2 + 59.24x + 2.50$

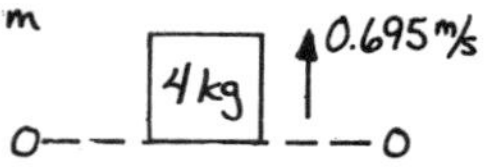

Figure b

By Eq. (a),

$$0.966 + 2.50 = 40x^2 + 59.24x + 2.50$$

$40x^2 + 59.24x - 0.966 = 0$; $\quad x = 0.0161\ m$

Hence, the height above the floor is

$$\underline{h = 0.35 + 0.0161 = 0.366\ m}$$

15.89

GIVEN: In the position shown in Figure a, the spring is unstretched. The spring constant k (lb/in) is such that when the system is released from rest, the speed of the masses is zero as the 0.6-slug mass touches the floor. The mass and friction of the pulley are negligible.

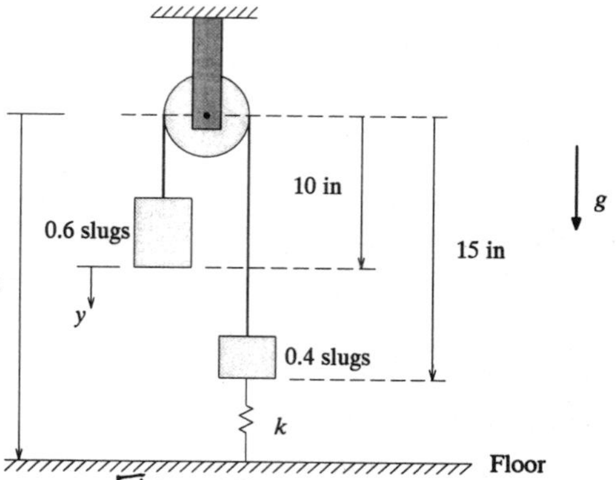

Figure a

FIND:

a.) the value of the spring constant k

b.) Derive a formula for the speed v and plot v versus y.

c.) From the plot, estimate v_{max} and the corresponding value of y.

d.) Check the estimate of part c, using the formula for v of part b.

SOLUTION:

a.) Since the system is initally and finally at rest, and with the zero datum for potential energy in the position of the unstretched spring, Conservation of energy yields

$$\cancelto{0}{T_1} + \cancelto{0}{V_1} = \cancelto{0}{T_2} + V_2 \quad (a)$$

where,

$V_2 = -(0.6\,slugs)(32.2\,ft/s^2)(10\,in) + (0.4\,slugs)(32.2\,ft/s^2)(10\,in) + \frac{1}{2}(k)(10\,in)^2$

When the 0.6-slug mass touches the floor $v_2 = 0$. Therefore, $0 = -64.4 + 50k$; $\underline{k = 1.288\ lb/in}$

b.) By Conservation of energy, since the system starts from rest,

$$\cancelto{0}{T_1} + \cancelto{0}{V_1} = T_2 + V_2 \quad (b)$$

where, $T_2 = \frac{1}{2}(0.6)v^2 + \frac{1}{2}(0.4)v^2 = 0.5v^2$

$V_2 = (0.4)(32.2)y - (0.6)(32.2)y + \frac{1}{2}(1.288)y^2$

By Eq. (b),

$$0 = 0.5v^2 - 6.44y + 0.644y^2$$

(continued)

15.89 (cont.)

Hence, $\underline{\underline{v = \sqrt{12.88y - 1.288y^2}}}$ (c)

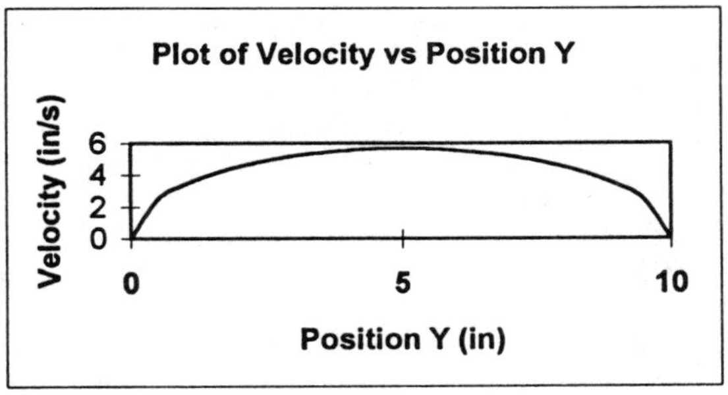

c.) From the plot of v data,

$\underline{v_{max} \approx 5.67 \text{ in/s at } y \approx 5 \text{ in.}}$

d.) By Eq. (c), the maximum value of v occurs when $dv/dy = 0$; that is,

$$\frac{dv}{dy} = \frac{1}{2}\frac{(12.88 - 2.756y)}{\sqrt{12.88y - 1.288y^2}} = 0$$

So, $\underline{y = 5.0 \text{ in.}}$

With $y = 5.0$ in, Eq. (b) yields

$\underline{\underline{v_{max} = 5.67 \text{ in/s}}}$

The results agree with the estimates.

15.90

GIVEN: A spherical shell carries a constant density ρ of electric charge on its inner surface (Fig. a)

FIND:

a.) Prove that the potential energy of a unit point charge e' within the shell is independent of its location.

b.) What conclusion can be drawn concerning the force on the point charge?

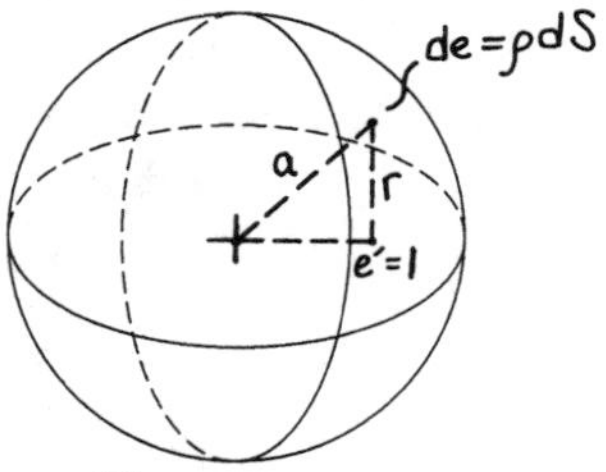

Figure a

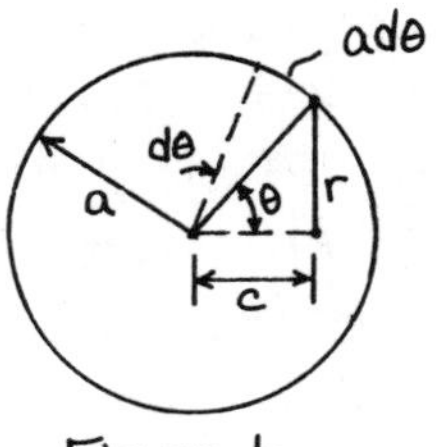

Figure b

SOLUTION:

a.) By Fig. b, an element of surface dS can be represented as

$$dS = 2\pi(a\sin\theta)(a\,d\theta)$$

or $dS = 2\pi a^2 \sin\theta\, d\theta$ (a)

Since $e' = 1$, Eq. (15.35) yields

$$dV = \frac{k\,de}{r} = \frac{k}{r}\rho\, dS \quad (b)$$

or by Eq. (a),

$$dV = \frac{2\pi a^2 \rho}{r}\sin\theta\, d\theta$$

Integration yields

$$V = 2\pi a^2 \rho \int_0^{\pi} \frac{\sin\theta}{r}\, d\theta \quad (c)$$

By Fig. (b), $r^2 = a^2 + c^2 - 2ac\cos\theta$

Therefore, $2r\,dr = 2ac\sin\theta\, d\theta$

or $\sin\theta\, d\theta = \frac{r\,dr}{ac}$

Also, for $\theta = 0$, $r = a - c$ and for $\theta = \pi$, $r = a + c$.

Hence,

$$\underline{\underline{V = \frac{2\pi a\rho}{c}\int_{a-c}^{a+c} dr = 4\pi a\rho}}$$

Since V is constant, the potential energy of the unit charge is independent of location c.

b.) Since V is constant, by Eq. (15.30),

$$\underline{\underline{F = -\frac{dV}{dr} = 0}}$$

15.91

GIVEN: A collar weighs 25 N and is attached to a spring whose unstretched length is 3m (Fig. a). The collar, under the action of gravity and the spring, slides on a smooth vertical rod. The collar is released from rest in the position shown.

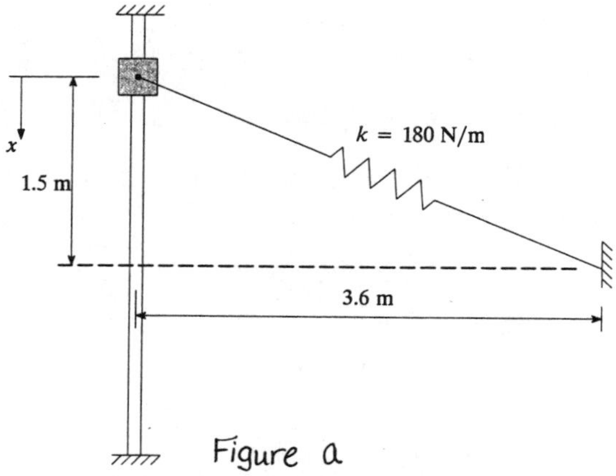

Figure a

FIND:

a.) the speed of the collar for $x = 1.5$ m.

b.) the maximum distance $x = x_{max}$ that the collar drops.

c.) a formula for the speed v of the collar in terms of x; A plot of v versus x for $0 \le x \le x_{max}$.

15.91 (cont.)

d.) from the plot of part c, estimates of the results of parts a and b. Also, an estimate of the maximum speed $v = v_{max}$ and the corresponding x.

e.) A verification of the estimates of part d, using the formula for v of part c.

SOLUTION:

a.) Since the system is released from rest, by conservation of energy, with the zero datum for potential energy at x = 1.5 m,

$$T_1 + V_1 = T_2 + V_2 \qquad (a)$$

where

$$T_1 = 0$$

$$V_1 = mgx + \tfrac{1}{2}ke^2 = (25\,N)(1.5\,m) + \tfrac{1}{2}(180)\left(\sqrt{3.6^2+1.5^2}-3\right)^2$$

or $V_1 = 110.4\ N\cdot m$

$$T_2 = \tfrac{1}{2}mv^2 = \tfrac{1}{2}\left(\tfrac{25}{9.81}\right)v^2 = 1.2742v^2\ (N\cdot m)$$

$$V_2 = \tfrac{1}{2}ke^2 = \tfrac{1}{2}(180)(3.6-3)^2 = 32.4\ N\cdot m$$

Hence, by Eq. (a),

$$110.4 = 1.2742v^2 + 32.4$$

or

$$\underline{\underline{v = 7.824\ m/s}}$$

b.) At $x = x_{max}$, $v = 0$. Take the zero datum for potential energy of the collar at x = 0. Then, by the law of conservation of mechanical energy, at x,

$$T_1 + V_1 = T_2 + V_2$$

where

$$T_1 = T_2 = 0$$

$$V_1 = \tfrac{1}{2}ke^2 = \tfrac{1}{2}(180)\left(\sqrt{3.6^2+1.5^2}-3\right)^2$$

or $V_1 = 72.9\ N\cdot m$

$$V_2 = -mgx + \tfrac{1}{2}ke^2 = -25x + \tfrac{1}{2}(180)\left(\sqrt{3.6^2+(1.5-x)^2}-3\right)^2$$

So, $72.9 = -25x + 90\left[12.96 + (1.5-x)^2 - 6\sqrt{12.96+(1.5-x)^2} + 9\right]$

or

$$90x^2 - 295x + 2106 - 540\sqrt{12.96+(1.5-x)^2} = 0$$

Solving for x, we obtain

$$\underline{\underline{x = 4.007\ m = x_{max}}}$$

c.) By part b,

$$T_1 + V_1 = T_2 + V_2$$

$$T_1 = 0$$

$$V_1 = 72.9\ N\cdot m$$

$$V_2 = 90x^2 - 295x + 2178.9 - 540\sqrt{x^2-3x+15.21}$$

Now, $T_2 = \tfrac{1}{2}mv^2 = \tfrac{1}{2}\left(\tfrac{25}{9.81}\right)v^2 = 1.2742v^2$

Hence,

$$72.9 = 1.2742v^2 + 90x^2 - 295x + 2178.9 - 540\sqrt{x^2-3x+15.21}$$

or

$$\underline{\underline{v = \sqrt{423.8\sqrt{x^2-3x+15.21} - 70.63x^2 + 231.5x - 1652.8}}} \quad (b)$$

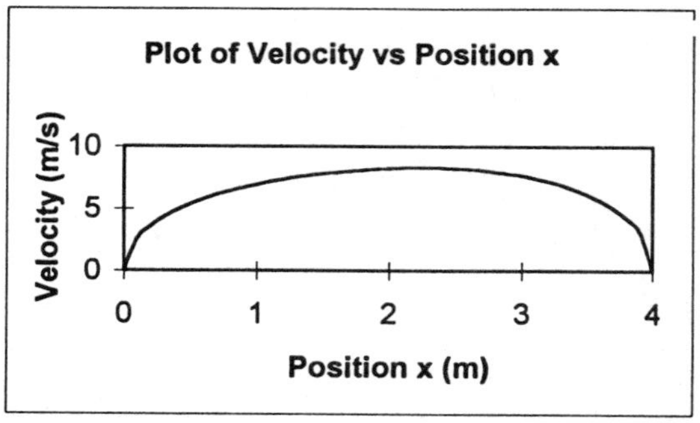

d.) From the plot, we estimate

$$\underline{\underline{v = v_{max} \approx 8.3\ m/s}}$$

$$\underline{\underline{x \approx 2.25\ m}}$$

e.) The maximum value of v occurs when

$$\frac{dv}{dx} = 0$$

Hence, by Eq. (b),

$$\frac{dv}{dx} = \left(\frac{1}{2}\right)\frac{\dfrac{423.8(2x-3)}{2\sqrt{x^2-3x+15.21}} - 141.26x + 231.5}{\sqrt{423.8\sqrt{x^2-3x+15.21} - 70.63x^2 + 231.5x - 1652.8}} = 0$$

or $423.8(2x-3) + 2(231.5 - 141.26x)\sqrt{x^2-3x+15.21} = 0$

$$\therefore\ \underline{x = 2.254\ m}$$

So, by Eq. (b),

$$v = v_{max} = \sqrt{423.8\sqrt{2.254^2 - 3(2.254) + 15.21} - 70.63(2.254)^2 + 231.5(2.254) - 1652.8}$$

or $\underline{\underline{v_{max} = 8.304\ m/s}}$

These results agree with part c.

15.92

GIVEN: You are asked to design a ski hill (Fig. a) for novice skiers so that they do not experience an acceleration $a \geq 2g$ and do not lose contact with the snow, assuming that the skiers start from rest at the top of the run. (Point A). Neglect Friction.

FIND: Recommend appropriate dimensions that will meet these requirements.

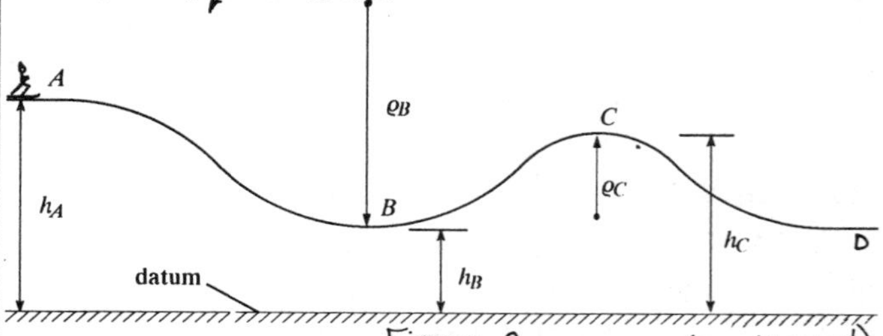

Figure a

(continued)

15.92 (cont.)

SOLUTION:

Since the maximum acceleration will occur at B, let's calculate the velocity v_B at B. Then, the acceleration is $a_B = v_B^2/\rho_B$. Let the ground be the zero datum for potential energy. Then,

$$T_A = 0, \quad V_A = mgh_A \qquad (a)$$

$$T_B = \tfrac{1}{2}mv_B^2, \quad V_B = mgh_B \qquad (b)$$

and by conservation of energy,

$$T_A + V_A = T_B + V_B$$

or

$$mgh_A = \tfrac{1}{2}mv_B^2 + mgh_B$$

$$\therefore \; v_B^2 = 2g(h_A - h_B) \qquad (c)$$

and

$$a_B = \frac{2g(h_A - h_B)}{\rho_B} \le 2g \qquad (d)$$

If we take $\rho_B = h_A - h_B$, $a_B = 2g$. Therefore, the minimum radius of curvature at B should be

$$\rho_B \ge h_A - h_B.$$

For the skier not to leave the snow at C, the normal reaction N to the skier at C (Fig. b) must be $N \ge 0$. The critical case is $N = 0$.

By figure b,

$$\Sigma F_N = mg - N = ma_n$$

$$mg - N = \frac{mv_C^2}{\rho_C} \qquad (e)$$

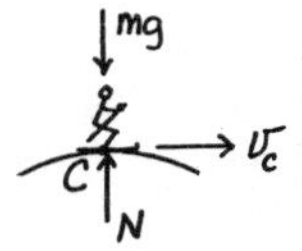

Figure b

To compute v_C, consider the law of conservation of energy between points A and C.

$$T_A = 0, \quad V_A = mgh_A$$

$$T_C = \tfrac{1}{2}mv_C^2, \quad V_C = mgh_C$$

$\therefore$ $T_A + V_A = T_C + V_C$ yields

$$mgh_A = \tfrac{1}{2}mv_C^2 + mgh_C$$

or

$$v_C^2 = 2g(h_A - h_C) \qquad (f)$$

With $N = 0$, Eqs. (e) and (f) yield

$$v_C^2 = g\rho_C = 2g(h_A - h_C) \qquad (g)$$

Thus, if we make $\rho_C = 2(h_A - h_C)$, this is the minimum allowable radius of curvature at C when $N = 0$. Hence, if we take $\rho_C \ge 2(h_A - h_C)$, the skiers will maintain contact with the snow. Friction will reduce these requirements somewhat. However, for safety reasons, they probably should not be reduced.

Hence, the design requirements are

$$\underline{\rho_B \ge h_A - h_B, \quad \rho_C \ge 2(h_A - h_C)}$$

The length of the run ABCD and the heights h_A, h_B, and h_C need to be determined, probably on the basis of experience and/or common sense, as well as available resources.

15.93

GIVEN: A collar mechanism (Fig. a) must be designed so that, when released from rest, it drops a distance x and comes to rest, where $1.0\,m \le x \le 1.2\,m$. Friction is negligible. The mass m of the collar may vary in the range $2.0\,kg \le m \le 2.5\,kg$. The mass of the linear spring is negligible. The spring constant is $k\left(\frac{N}{m}\right)$. The unstretched length of the spring is L (Fig. a). The mechanism must be contained in a region 2.5 m wide and 2.0m high. The depth d perpendicular to the plane of the figure is not specified.

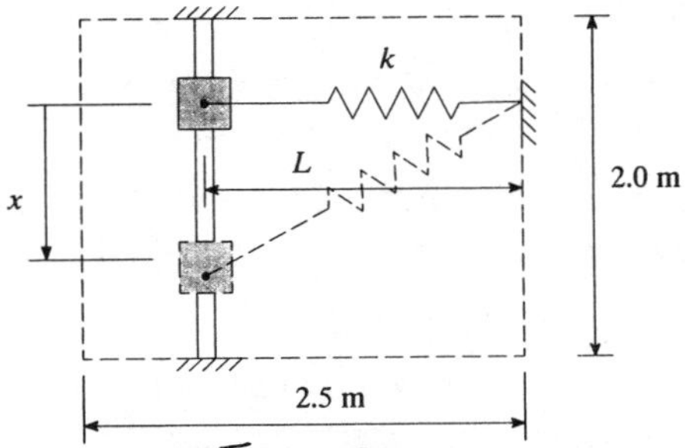

Figure a

FIND: Design the collar mechanism to meet the design requirements.

SOLUTION:

To derive a formula for x in terms of m, k, and L, we may use conservation of energy. Thus, relative to the initial position $x = 0$,

$$T_1 + V_1 = T_2 + V_2 \qquad (a)$$

where

$$V_1 = T_1 = T_2 = 0$$

and

$$V_2 = -mgx + \tfrac{1}{2}ke^2$$

or

$$V_2 = -mgx + \tfrac{1}{2}k\left(\sqrt{L^2 + x^2} - L\right)^2$$

Hence, by Eq. (a),

$$mgx = \tfrac{1}{2}k\left(\sqrt{L^2 + x^2} - L\right)^2 \qquad (b)$$

Equation (b) contains 4 unknown quantities, m, k, x, and L, where $2.0\,kg \le m \le 2.5\,kg$, $1.0\,m \le x \le 1.2\,m$, and $L < (2.5\,m)$ minus the additional width required for the column and collar. Let's first select the maximum values of m and x, and a reasonable value for L, as follows

$$m = 2.5\,kg, \quad x = 1.2\,m, \quad L = 2.0\,m$$

(Continued)

15.93 (cont.)

Then, by Eq. (b),

$$(2.5)(9.81)(1.2) = \tfrac{1}{2}k\left(\sqrt{4+1.44}-2\right)^2$$

or $k = 266.4$ N/m

Check Eq. (b) for m = 2.0 kg, L = 2.0 m, and k = 266.4 N/m. Thus,

$$(2.0)(9.81)x = \tfrac{1}{2}(266.4)\left(\sqrt{4+x^2}-2\right)^2$$

or $133.2x^2 - 19.62x + 1065.6 - 532.8\sqrt{4+x^2} = 0$

$\therefore\ x = 1.431 > 1.2\,m$

So design does not satisfy requirement that $x \le 1.2$. Try increasing k to k = 600 N/m. Then for m = 2.0 kg, L = 2m, Eq. (b) yields

$$(2.0)(9.81)x = 300\left(\sqrt{4+x^2}-2\right)^2$$

or $300x^2 - 19.62x + 2400 - 12\sqrt{4+x^2} = 0$

$\therefore\ x = 1.059\,m$

This is in the range $1.0m \le x \le 1.2m$.
Check for m = 2.5 kg. Then,

$$300x^2 - 24.525x + 2400 - 1200\sqrt{4+x^2} = 0$$

or $x = 1.149\,m$

This is in the range $1.0m \le x \le 1.2m$. Thus, the design requirements are met with

$\underline{k = 600\ \text{N/m},\ L = 2m}$

There are many other possible solutions.

15.94

GIVEN: A truck, weight 5,000 lb, travels at a constant speed of 45 mi/hr up an incline of 10% (Fig. a). The resistive force acting on the truck is 250 lb. The truck is 20% efficient.

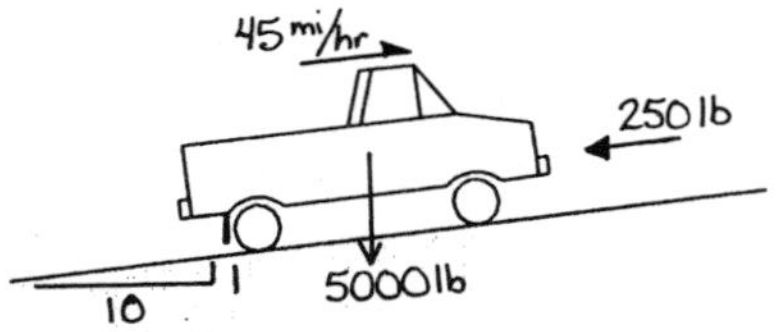

Figure a

FIND:

a.) Determine the horsepower the truck's engine must produce.

b.) Give reasons that the truck's efficiency is only 20%.

SOLUTION:

a.) Power is given by $E = Fv$
where (Fig. a),

$$F = 250 + 5000\sin\theta$$
$$= 250 + 5000\left(\tfrac{1}{\sqrt{101}}\right)$$
$$F = 747.519\text{ lb.}$$

$v = 45$ mi/hr $= 66$ ft/s

$\therefore\ E = (747.519)(66) = 49{,}336$ ft·lb/s

or $E = \left(49{,}336\ \tfrac{\text{ft·lb}}{\text{s}}\right)\left(\tfrac{1\text{ hp}}{550\ \frac{\text{ft·lb}}{\text{s}}}\right) = 89.7$ hp

Since the truck is only 20% efficient, the engine must produce

$$\underline{E_{engine} = \frac{89.7}{0.20} = 448.5\text{ hp}}$$

b) Much of the internal energy loss in the engine is due to friction and the fact that by thermodynamics the engine is not 100% efficient. Also, there are losses in the drive train (gears and shafts, etc.).

15.95

GIVEN: Larry Walker hits a pop-up straight up with an initial speed of 66 mi/hr (96.8 ft/s) from 3 ft above the ground to an altitude of 120 ft (Fig. a). The mass of the ball is 0.00931 slugs.

FIND:

a.) the initial kinetic energy of the ball.

b.) the work that air resistance does on the ball.

c.) the speed of the ball when it returns to its initial position 3 ft above the ground, assuming that air resistance does 4/5 as much work on the way down as on the way up.

SOLUTION:

a.) The initial kinetic energy is

$$T_1 = \tfrac{1}{2}mv_1^2 = \tfrac{1}{2}(0.00931)(96.8)^2$$

or $\underline{T_1 = 43.62\text{ ft·lb}}$

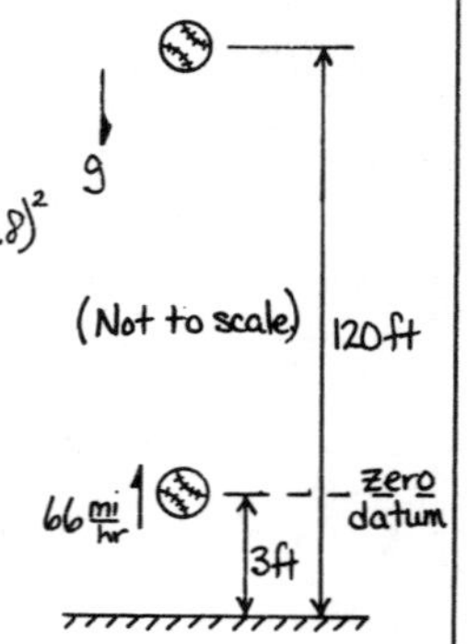

Figure a

b.) Take the zero datum for potential energy to be the initial position of the ball (Fig. a). Then, $T_1 = 43.62$ ft·lb, $V_1 = 0$, $T_2 = 0$ (at altitude of 120 ft) and

$V_2 = mgh = (0.00931)(32.2)(117) = 35.074$ ft·lb (at altitude of 120 ft)

Hence, by

$$T_1 + V_1 + U_{1\text{-}2}^{nc} = T_2 + V_2$$
$$43.62 + 0 + U_{1\text{-}2}^{nc} = 0 + 35.074$$

or $\underline{U_{1\text{-}2}^{nc} = -8.546\text{ ft·lb}}$

where $U_{1\text{-}2}^{nc}$ is the work done by air resistance on the ball.

(continued)

15.95 (cont.)

c) To determine the ball's speed when it returns to 3 ft above the ground from 120 ft altitude, use

$$T_1 + V_1 + U_{1-2}^{nc} = T_2 + V_2 \quad \text{(a)}$$

where $T_1 = 0$ (at altitude of 120 ft)
$V_1 = 35.074$ ft·lb
$V_2 = 0$
$T_2 = \frac{1}{2}mv^2 = 0.004655v^2$

Also in Eq. (a), $U_{1-2}^{nc} = \frac{4}{5}(-8.546) = -6.837$ ft·lb

Hence, Eq. (a) yields

$$0 + 35.074 - 6.837 = 0.004655v^2 + 0$$

or $\underline{v = 77.88 \text{ ft/s} = 53.10 \text{ mi/hr}}$

15.96

GIVEN: A ball of putty of radius r, whose center of gravity is h above the floor, is dropped from rest (Fig. a).

FIND:

a) the work done on the ball from the time it is dropped to the instant it hits the floor (neglect air resistance).

b) the kinetic energy of the ball at the instant it hits the floor

c) the work done by the contact forces exerted on the putty by the floor (Assuming that the putty is squashed flat by the impact and that it does not rebound).

SOLUTION:

a.) By Fig. a, the work done on the ball from the instant it is dropped to the instant it hits the floor is

$$\underline{U_{1-2} = mg(h-r)}.$$

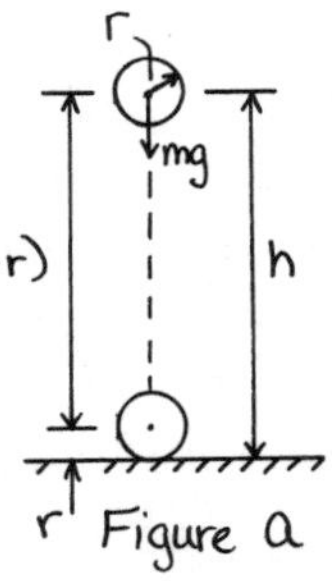

Figure a

b.) By the law of conservation of mechanical energy

$$T_1 + V_1 = T_2 + V_2 \quad \text{(a)}$$

where $T_1 = 0$, $V_1 = mgh$ (relative to the floor)
$T_2 = \frac{1}{2}mv^2$, $V_2 = mgr$

Therefore, by Eq. (a),

$$0 + mgh = \frac{1}{2}mv^2 + mgr$$

Hence, the kinetic energy at impact is

$$\underline{T_2 = \frac{1}{2}mv^2 = mg(h-r)}$$

c.) As the putty ball is squashed flat, non-conservative forces (friction) act. Hence, from the instant that the ball contacts the floor to the time that it is squashed flat (Fig. b).

$$T_1 + V_1 + U_{1-2}^{nc} = T_2 + V_2 \quad \text{(b)}$$

where $T_1 = mg(h-r)$; by part b
$V_1 = mgr$
$T_2 = 0$
$V_2 = 0$ (since the putty is squashed flat)

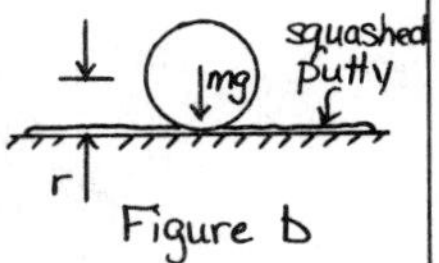

Figure b

Therefore, Eq. (b) yields

$$\underline{U_{1-2}^{nc} = -mgh}$$

where U_{1-2}^{nc} is the work of friction on the putty.

15.97

GIVEN: You must run up a flight of stairs to a height h above the point from which you start (Fig. a). Neglect air resistance.

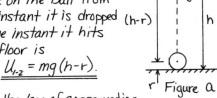

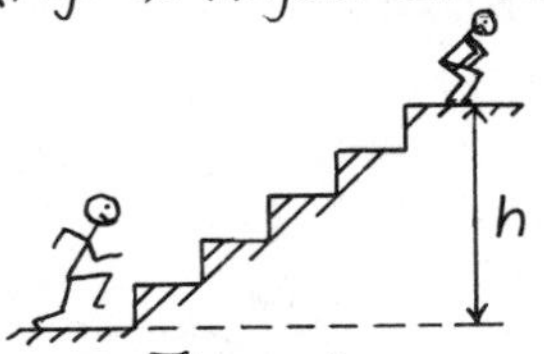

Figure a

FIND:

a) the work done on you by external forces as you run up the stairs

b) the amount of energy that you expend in running up the stairs, assuming that your body is 15% efficient.

c) the number of Calories (kilocalories; a calorie is 4.186 J) you use in running up the stairs.

SOLUTION:

a.) The only force that does work on you (since air resistance is neglected) is gravity. Hence, the work done on you is

$$\underline{U_{gravity} = -mgh}$$

b.) Since your body is 15% efficient, the amount of energy you expend is

$$\underline{U_{body} = \frac{|U_{gravity}|}{0.15} = \frac{mgh}{0.15} = 6.667mgh \text{ (J)}}$$

c.) Since a calorie in 4.186 J,

$$U_{body} = [6.667mgh \text{ (J)}]\left(\frac{1 \text{ calorie}}{4.186 \text{ J}}\right)\left(\frac{1 \text{ Calorie}}{1000 \text{ calories}}\right)$$

or $\underline{U_{body} = 0.001593 \, mgh \text{ (Calories)}}$

used by you in running up the stairs.

15.98

GIVEN: A 400-lb crate is pulled by a tractor at a speed of 6 mi/hr up a 60-ft long asphalt ramp, inclined at 30° with the horizontal (Fig. a). The tractor is 15% efficient and the coefficient of sliding friction is $\mu_k = 0.40$.

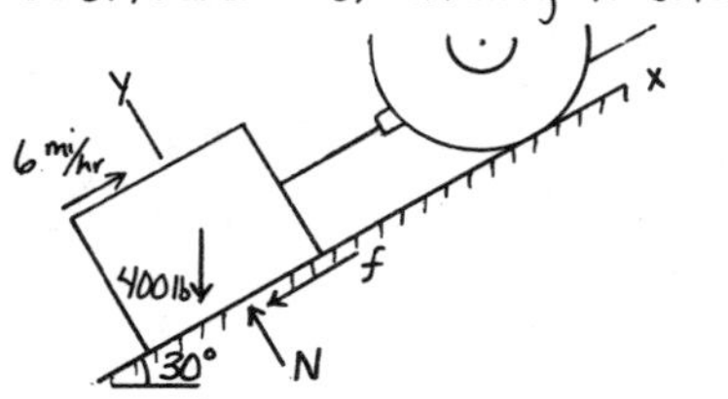

Figure a

FIND:

a.) the horsepower exerted by the tractor to move the crate up the ramp.

b.) the energy expended by the engine in moving the crate up the 60-ft ramp.

SOLUTION:

a.) Consider the FBD of the crate (Fig. b)

$\Sigma F_y = N - 400\cos 30° = 0$

$\therefore\ N = 346.4$ lb

and

$\mu_k N = 0.40(346.4)$

$\mu_k N = 138.56$ lb

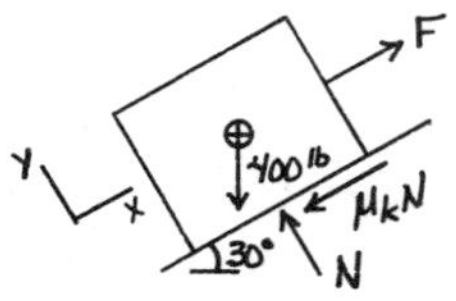

Figure b

The power required to move the crate is given by

$E = Fv$ (a)

where by $\Sigma F_x = 0$ (Fig. b),

$F = 138.56 + 400\sin 30° = 338.56$ lb.

With $v = 6$ mi/hr $= 8.8$ ft/s, Eq. (a) yields

$E = (338.56\ \text{lb})(8.8\ \text{ft/s})\left(\frac{1\ \text{hp}}{550\ \frac{\text{lb}\cdot\text{ft}}{\text{s}}}\right)$

or $E = 5.417$ hp

Since the tractor is 15% efficient, it must expend

$E_{tractor} = \frac{E}{0.15} = \frac{5.417}{0.15} = 36.11$ hp

b.) Traveling at a constant speed of 6 mi/hr = 8.8 ft/s, it will take $t = (60\ \text{ft})/(8.8\ \text{ft/s}) = 6.818$ s to move the crate 60 feet up the ramp. Hence, the energy expended by the engine is

$U = (E_{tractor})(t) = (36.11\ \text{hp})\left(\frac{550\ \frac{\text{ft}\cdot\text{lb}}{\text{s}}}{1\ \text{hp}}\right)(6.818\ \text{s})$

or $U = 135{,}400$ ft·lb $= 135.4$ ft·kip

15.99

GIVEN: A sailor hoists a 40-kg crate with the block and tackle shown in Fig. a by exerting a constant force F. The pulleys are seized so that they do not rotate; the rope slides around the pulleys so that the tension in the rope is reduced 8 N by friction as it passes around each pulley.

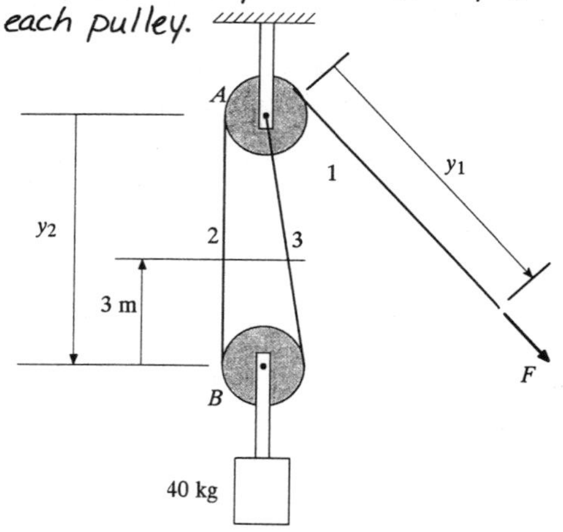

Figure a

FIND:

a.) the work that the sailor does in raising the crate 3m, neglecting the slight angle formed by the rope from A to B with the vertical and the masses of the rope and pulley B. The speed at which the crate is raised is negligible.

b.) the average force that the sailor exerts.

SOLUTION:

a.) By Fig. (a),

$y_1 + 2y_2 = \text{constant}$ (a)

Let Δy_1 and Δy_2 be the changes in y_1 and y_2 as the crate is raised. By Eq. (a),

$y_1 + \Delta y_1 + 2(y_2 + \Delta y_2) = \text{constant}$

Hence,

$\Delta y_1 + 2\Delta y_2 = 0$

or $\Delta y_1 = -2\Delta y_2$

When the crate is raised 3m, $\Delta y_2 = -3$m. Therefore,

$\Delta y_1 = 6$m (b)

By the free-body of pulley B and the crate (Fig. b),

$\Sigma F_y = 392.4 - 2F + 24 = 0$

or $F = 208.2$ N.

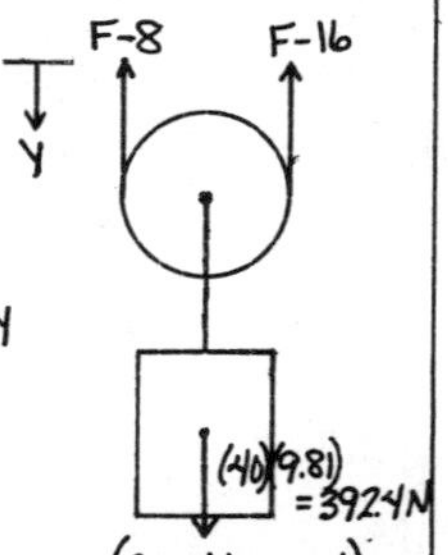

Hence, the work performed by the sailor is

$U = F\cdot\Delta y_1 = (208.2)(6)$

or $U = 1249.2$ N·m

(continued)

15.99 (cont.)

b.) The average force exerted by the sailor is $\underline{F = 208.2\ N}$

15.100

GIVEN: A pulley mechanism is shown in Fig. a. The system is released from rest; mass m_2 moves down and mass m_1 moves up. Mass m_2 is a smooth cylindrical body. Mass m_1 slides in guides with a coefficient of sliding friction μ_k. Pulley A is frictionless and has negligible mass. The light cord is inextensible. Forces F press the guides against m_1.

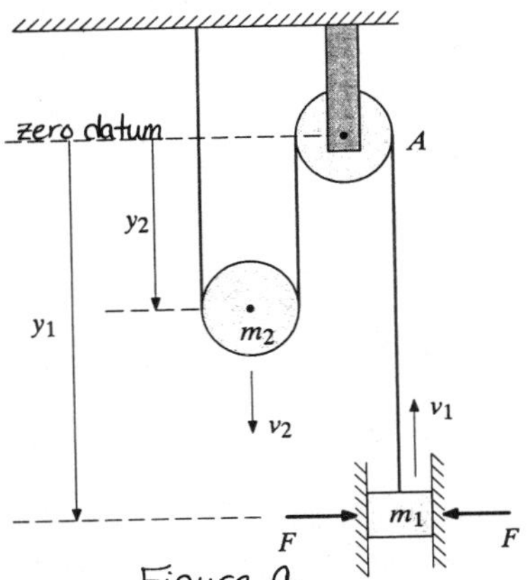

Figure a

FIND: the speeds of the masses for the instant that m_2 has dropped a distance d; express results in terms of $d, m_1, m_2, \mu_k, F,$ and g.

SOLUTION:

By Fig. a,

$$y_1 + 2y_2 = \text{constant} \quad (a)$$

$$\therefore\ \dot{y}_1 + 2\dot{y}_2 = 0$$

$$\text{or}\quad v_1 + 2v_2 = 0;\quad v_1 = -2v_2 \quad (b)$$

Thus, since v_2 is directed down, v_1 is directed up (Fig. a). Since both conservative and nonconservative forces are involved

$$T_1 + V_1 + U_{1\text{-}2}^{nc} = T_2 + V_2 \quad (c)$$

Since the system starts from rest $T_1 = 0$. Let the zero datum for potential energy be at $y_1 = y_2 = 0$. Then, with Eq. (b),

$$V_1 = -m_1 g y_1 - m_2 g y_2$$

$$T_2 = \tfrac{1}{2} m_1 v_1^2 + \tfrac{1}{2} m_2 v_2^2$$

$$= \tfrac{1}{2} m_1 (-2v_2)^2 + \tfrac{1}{2} m_2 v_2^2$$

$$= \tfrac{1}{2}(4m_1 + m_2) v_2^2$$

$$V_2 = -m_1 g(y_1 + \Delta y_1) - m_2 g(y_2 + \Delta y_2)$$

Also by Eq. (a),

$$y_1 + \Delta y_1 + 2(y_2 + \Delta y_2) = \text{constant}$$

Hence,

$$\Delta y_1 + 2(\Delta y_2) = 0$$

Then, since m_2 drops a distance d, $\Delta y_2 = d$ and $\Delta y_1 = -2d$. Therefore,

$$V_2 = -m_1 g(y_1 - 2d) - m_2 g(y_2 + d)$$

$$\text{or}\quad V_2 = -m_1 g y_1 - m_2 g y_2 + (2m_1 - m_2) g d$$

The work $U_{1\text{-}2}^{nc}$ of nonconservative forces is (Fig. b)

$$U_{1\text{-}2}^{nc} = -2\mu_k F(2d)$$

$$= -4\mu_k F d$$

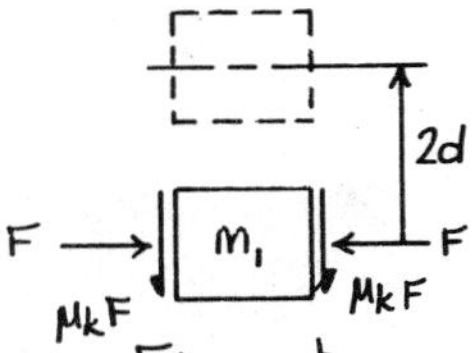

Figure b

Hence, Eq. (c) yields

$$0 - m_1 g y_1 - m_2 g y_2 - 4\mu_k F d = \tfrac{1}{2}(4m_1 + m_2) v^2 - m_1 g y_1 - m_2 g y_2 + (2m_1 - m_2) g d$$

or

$$\underline{v_2 = \sqrt{\frac{2d[(m_2 - 2m_1)g - 4\mu_k F]}{4m_1 + m_2}}\ \downarrow}$$

so

$$\underline{v_1 = 2\sqrt{\frac{2d[(m_2 - 2m_1)g - 4\mu_k F]}{4m_1 + m_2}}\ \uparrow}$$

15.101

GIVEN: The mechanism shown in Fig. a. The spring constant is k, and the pulley is frictionless and of negligible mass. The coefficients of static and kinetic friction are μ_s and μ_k. The block is in equilibrium, with motion impending down the plane and the spring is stretched a distance e.

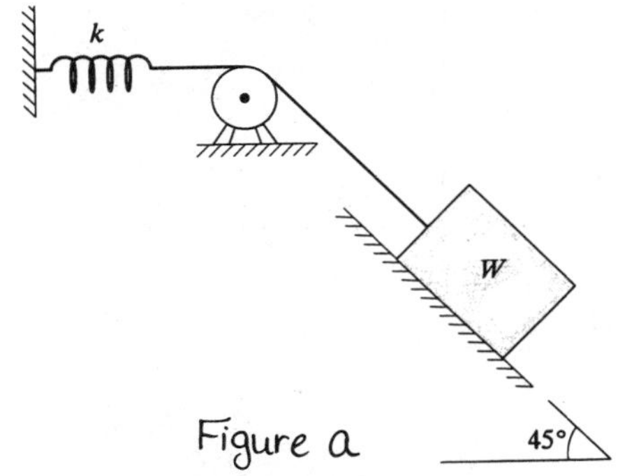

Figure a

FIND:

a.) the weight of the block, with $k = 480$ lb/ft, $\mu_s = 0.30$ and $e = 1$ ft.

(continued)

15.101 (cont.)

b.) the distance the body travels up the plane, when it is pulled down the plane 2 ft from the position shown and released from rest. Take $\mu_k = 0.20$.

c.) What happens immediately after the body again comes to rest?

SOLUTION:

a.) Consider the FBD of the block for motion impending down the plane (Fig. b).

$\Sigma F_x = N - W\cos 45° = 0$

$N = W\cos 45°$

$\Sigma F_y = ke + \mu_s N - W\sin 45° = 0$

$\therefore\ W\sin 45° - \mu_s W\cos 45° = ke$

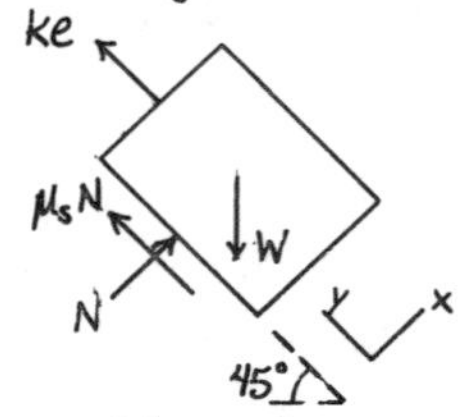

Figure b.

Hence,

$$W = \frac{ke}{\sin 45° - \mu_s \cos 45°} = \frac{(480)(1)}{0.7071(1-0.30)}$$

or $\underline{\underline{W = 969.8\ \text{lb}}}$

b.) First, we check to see if the block moves up the plane. Consider the FBD of the block after it has been pulled down the plane 2 ft.

$\Sigma F_x = N - W\cos 45°$

$N = W\cos 45°$

$\Sigma F_y = ke - \mu_s N - W\sin 45°$

$= 480(3) - (0.30)(969.8)(0.7071)$

$-(969.8)(0.7071)$

$= 548.5 > 0$

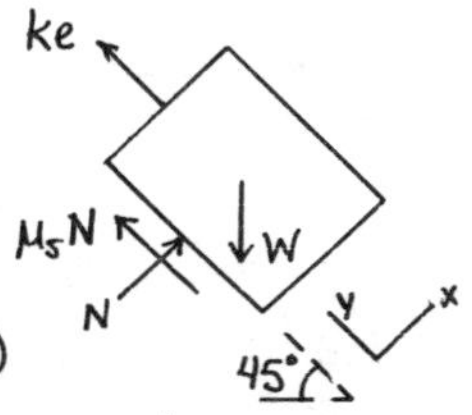

Figure c

$\therefore$ Block moves up the plane.

Since both conservative and nonconservative forces act, to determine the distance y that the block moves up the plane, we use

$$T_1 + V_1 + U_{1-2}^{nc} = T_2 + V_2 \quad (a)$$

where

$T_1 = 0$

$V_1 = \tfrac{1}{2}ke^2 = \tfrac{1}{2}(480)(3)^2 = 2160\ \text{ft}\cdot\text{lb}$

$T_2 = 0$

$V_2 = \tfrac{1}{2}k(e-y)^2 + (W\sin 45°)y$

$= \tfrac{1}{2}(480)(3-y)^2 + (969.8)(0.7071)y$

$= 240y^2 - 754.2y + 2160$

$U_{1-2}^{nc} = -\mu_k N y = -0.20(969.8)(0.7071)y$

$= -137.15y$

Hence, Eq. (a) yields

$0 + 2160 - 137.15y = 0 + 240y^2 - 754.2y + 2160$

or $240y^2 - 617.1y = 0$; $\therefore\ \underline{\underline{y = 2.571\ \text{ft}}}$

c.) To see what happens after the body comes to rest, consider the FBD of the block at $y = 2.571$ ft above its release (Fig. d), assuming that friction acts down the plane.

$\Sigma F_y = ke - F - W\sin 45° = 0$

or $F = ke - W\sin 45°$.

The extension of the spring is

$e = 3 - 2.571 = 0.429\ \text{ft}$

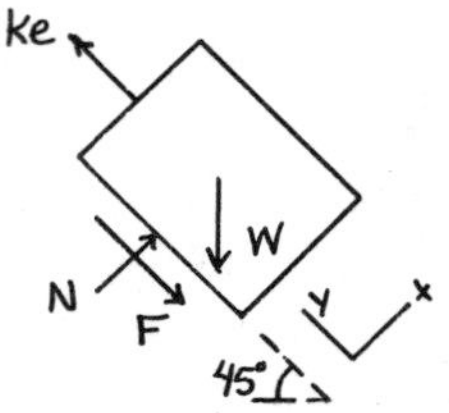

Figure d

Hence,

$F = (480)(0.429) - (969.8)(0.7071) = -479.8\ \text{lb}$

$\therefore$ For equilibrium, F must act up the plane with magnitude 479.8 lb.

The maximum possible magnitude of the friction force is

$F_{max} = \mu_s N = \mu_s W\cos 45°$

or $F_{max} = (0.30)(969.8)(0.7071) = 205.7\ \text{lb}$

$\therefore\ F_{max} = 205.7\ \text{lb} < 479.8\ \text{lb}$

Hence, $\underline{\underline{\text{the block starts to slide down the plane.}}}$

15.102

GIVEN: The model of a tuned mass damper (Fig. a). The springs are initially unstretched when a force F is applied to the mass, displacing it a distance d.

FIND:

a.) work of internal and external forces. Neglect friction.

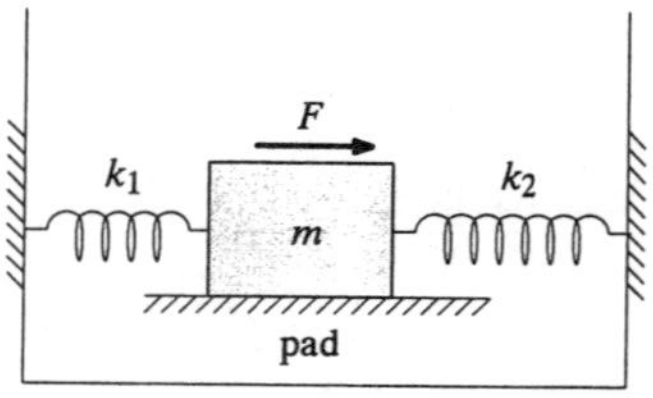

Figure a

b.) the speed of the mass as it passes through its initial position, after the force F is removed and the mass is released from rest at d. Neglect friction.

c.) Repeat part b for sliding friction equal to μ_k. Assume sliding occurs

d.) Calculate the speed (ft/s) of the mass in parts b and c for $k_1 = k_2 = 300$ lb/in, $m = 200$ slugs, $d = 24$ in, and $\mu_k = 0.04$; $\mu_s = 0.10$. (Verify that sliding occurs)

SOLUTION:

a.) The internal forces are the spring forces. Their work is

$$\underline{\underline{U_{springs} = \tfrac{1}{2}k_1 d^2 + \tfrac{1}{2}k_2 d^2 = \tfrac{1}{2}(k_1 + k_2)d^2}}$$

The work of the external force F is

$$\underline{\underline{U_F = Fd}}$$

(continued)

15.102 (cont.)

b.) Since the system is conservative

$$T_1 + V_1 = T_2 + V_2$$

where
$T_1 = 0$
$V_1 = \frac{1}{2}(k_1 + k_2)d^2$
$T_2 = \frac{1}{2}mv^2$
$V_2 = 0$

$$\therefore\ 0 + \tfrac{1}{2}(k_1+k_2)d^2 = \tfrac{1}{2}mv^2 + 0$$

or $$v = d\sqrt{\frac{k_1+k_2}{m}} \qquad (a)$$

c.) When friction is included,

$$T_1 + V_1 + U_{1\text{-}2}^{nc} = T_2 + V_2$$

where $U_{1\text{-}2}^{nc} = -\mu_k(mg)(d)$

With the values of T_1, T_2, V_1, and V_2 from part b,

$$0 + \tfrac{1}{2}(k_1+k_2)d^2 - \mu_k mgd = \tfrac{1}{2}mv^2 + 0$$

or $$v = \sqrt{\frac{(k_1+k_2)d^2}{m} - 2\mu_k g d} \qquad (b)$$

d.) To verify that sliding occurs, consider the free-body diagram of the mass in the displaced position (Fig. b).

$$\Sigma F_x = 2kd - \mu_s mg$$
$$= 2(300)(24) - (0.1)(120)(32.2)$$
$$= 14{,}000 \text{ lb} > 0$$

$\therefore$ Block Slides

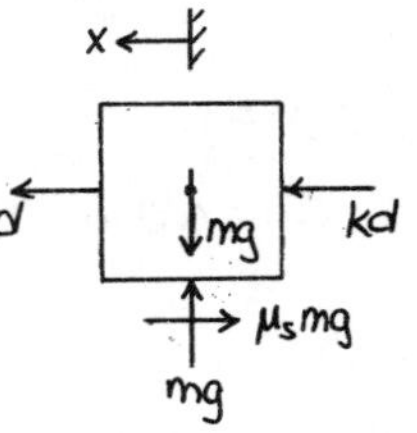

Figure b

Hence, for no friction, by Eq. (a),

$$v = \frac{24 \text{ in}}{12 \text{ in/ft}}\sqrt{\frac{(600 \text{ lb/in})(12 \text{ in/ft})}{120 \text{ lb}\cdot\text{s}^2/\text{ft}}}$$

or $v = 15.49$ ft/s

With friction, by Eq. (b),

$$v = \sqrt{\frac{(600 \text{ lb/in})(12 \text{ in/ft})(24 \text{ in})^2\left(\frac{1 \text{ ft}}{12 \text{ in}}\right)^2}{120 \text{ lb}\cdot\text{s}^2/\text{ft}} - 2(0.04)(32.2)(24)\left(\frac{1 \text{ ft}}{12 \text{ in}}\right)}$$

or $v = 15.32$ ft/s

So, the friction reduces the speed only slightly, a desirable condition for a tuned mass damper.

15.103

GIVEN: A child weighs 250 N and sits on a snow sled that weighs 50 N (Fig. a). Her mother pushes the sled to give it an initial speed of 1 m/s at Point A. $\mu_k = 0.15$

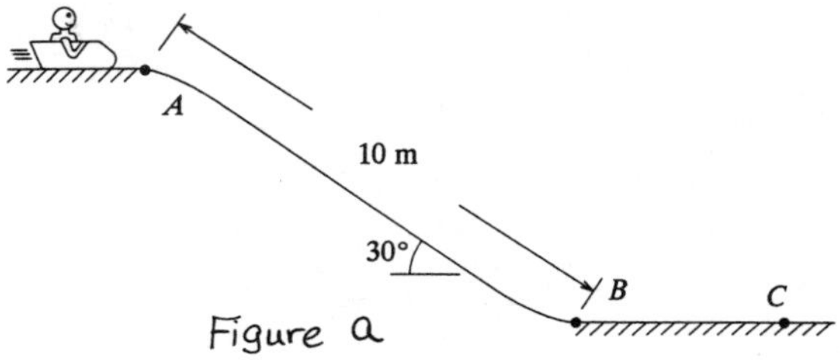

Figure a

FIND:

a.) the distance ABC that the sled slides until it stops.

b.) the maximum speed the sled attains

SOLUTION:

a.) Consider the FBD of the sled in the interval AB (Fig. b).

$$\Sigma F_y = N - 300\cos 30° = 0$$
$$\therefore\ N = 259.8 \text{ N}$$
$$\therefore\ \mu_k N = (0.15)(259.8) = 38.97 \text{ N}$$

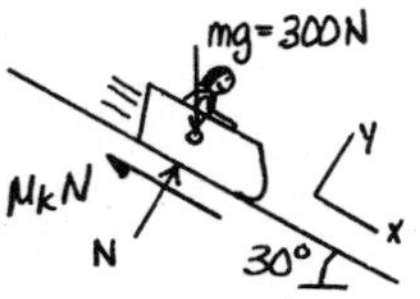

Figure b

For the interval from A to B, with zero datum at B,

$$T_A + V_A + U_{A\text{-}B}^{nc} = T_B + V_B \qquad (a)$$

$T_A = \frac{1}{2}mv_A^2 = \frac{1}{2}\left(\frac{300}{9.81}\right)(1)^2 = 15.291$ N·m
$V_A = (300)(10)(\sin 30°) = 1500$ N·m
$T_B = \frac{1}{2}mv_B^2 = \frac{1}{2}\left(\frac{300}{9.81}\right)v_B^2 = 15.291 v_B^2$ N·m
$V_B = 0$
$U_{A\text{-}B}^{nc} = -(38.97)(10) = -389.7$ N·m

Therefore, by Eq. (a),

$$15.291 + 1500 - 389.7 = 15.291 v_B^2 + 0$$

or $v_B^2 = 73.616$; $v_B = 8.580$ m/s

For the interval from B to C, by Fig. c,

$$\Sigma F_y = N - 300 = 0$$
$$N = 300 \text{ N}$$
$$\therefore\ (0.15)N = (0.15)(300) = 45 \text{ N}$$

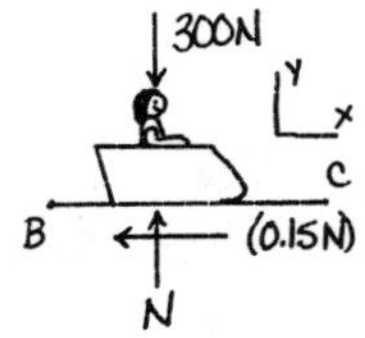

Figure c

Then, by

$$T_B + V_B + U_{B\text{-}C}^{nc} = T_C + V_C$$
$$T_B = \tfrac{1}{2}mv_B^2 = \tfrac{1}{2}\left(\tfrac{300}{9.81}\right)(73.616)$$
$$= 1125.6 \text{ N}\cdot\text{m}$$
$$V_B = V_C = T_C = 0$$
$$U_{B\text{-}C}^{nc} = -45x$$

(continued)

15.103 (cont.)

Hence,

$$1125.6 + 0 - 45x = 0 + 0$$

or $x = 25.01\ m$

So, the total distance is

$$\underline{ABC = 10 + 25.01 = 35.01\ m}$$

b.) The maximum speed of the sled will occur at point B.

Thus,

$$\underline{v_{max} = v_B = 8.580\ m/s}$$

15.104

GIVEN: Refer to Problem 15.101. Let the body be pulled down the plane 2 ft. from its initial position of impending motion ($e = 1$ ft). Then, the extension of the spring is 3 ft (Fig. a).

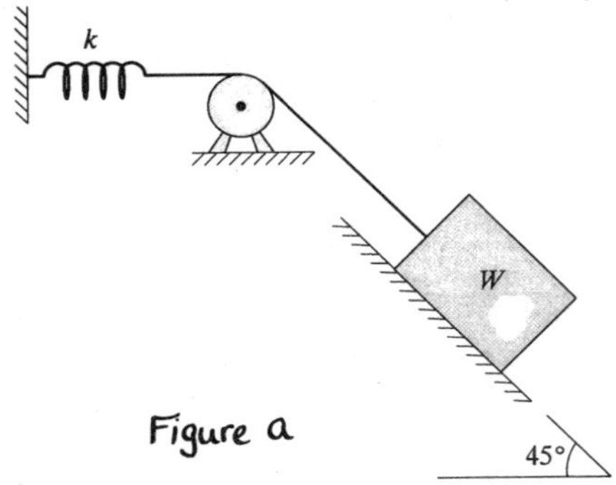

Figure a

Find:

a.) Derive a formula for v the speed of the weight W as a function of x, the distance the weight moves up the plane from the point of release from rest ($e = 3$ ft). Plot v as a function of x, until v again comes to zero.

b.) From the plot of v, determine the maximum distance X_{max} that the weight moves up the plane.

c.) From the plot, estimate the maximum speed v_{max} and the corresponding x.

d.) Verify the estimates obtained in part c, by the formula derived in part a.

SOLUTION:

a.) Since conservative and nonconservative forces act,

$$T_1 + V_1 + U_{1-2}^{nc} = T_2 + V_2 \quad (a)$$

where, since the mass is released from rest with $e = 3$ ft.

$$T_1 = 0$$
$$V_1 = \tfrac{1}{2}ke^2 = \tfrac{1}{2}(480)(3)^2 = 2160\ ft \cdot lb$$
$$T_2 = \tfrac{1}{2}mv^2$$
$$V_2 = (W \sin 45°)x + \tfrac{1}{2}k(3-x)^2$$
$$U_{1-2}^{nc} = -(\mu_k N)x$$

To determine W and N, consider the free-body diagram of the block when $e = 1$ ft (Refer to Problem 15.101), and motion impends down the plane (Fig. b).

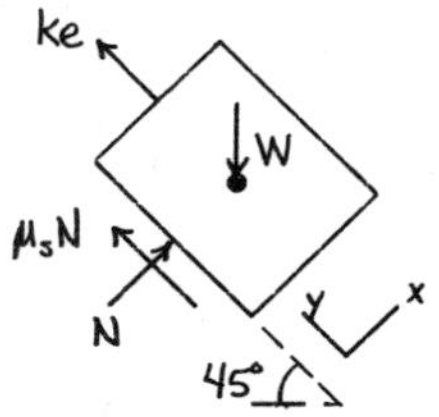

$$\Sigma F_x = N - W\cos 45° = 0$$
$$\therefore N = W\cos 45°$$
$$\Sigma F_y = ke + \mu_s N - W\sin 45° = 0$$
$$\therefore ke + \mu_s W\cos 45° - W\sin 45° = 0$$

or $W = \dfrac{ke}{\sin 45° - \mu_s \cos 45°} = \dfrac{480(1)}{0.7071(1-0.30)}$ Figure b

$$W = 969.8\ lb$$

and $N = W\cos 45° = 685.71$ lb.

and $\mu_k N = (0.20)(685.71) = 137.1$ lb

Hence,

$$T_2 = \tfrac{1}{2}\left(\frac{969.8}{32.2}\right)v^2 = 15.06v^2$$
$$V_2 = (969.8)(0.7071)x + \tfrac{1}{2}(480)(9 - 6x + x^2)$$
$$= 240x^2 - 754.29x + 2160$$
$$U_{1-2}^{nc} = -(0.20)(685.71)x = -137.14x\ ft \cdot lb.$$

Therefore, by Eq. (a),

$$0 + 2160 - 137.14x = 15.06v^2 + 240x^2 - 754.28x + 2160$$

or $\underline{v = \sqrt{41.0x - 15.93x^2}}$ (b) (see plot)

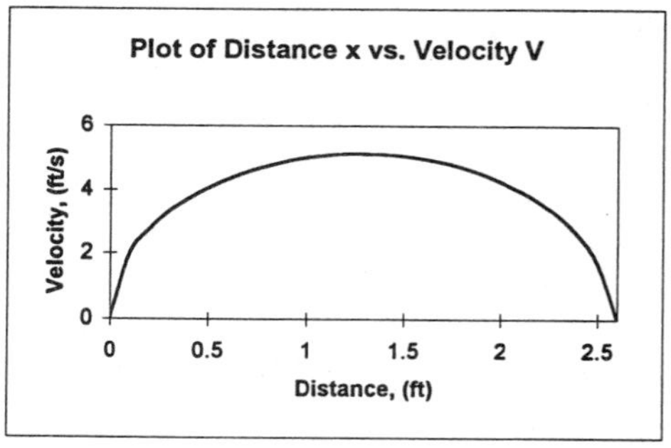

b) By the plot,

$$\underline{X_{max} \approx 2.6\ ft}$$

c) By the plot,

$$\underline{V_{max} \approx 5.2\ ft/s}$$
$$\underline{X \approx 1.25\ ft}$$

d) By Eq. (b), $v = v_{max}$ at x given by $\frac{dv}{dx} = 0$.

(Continued)

15.104 (cont.)

Therefore

$$\frac{dv}{dx} = \frac{(41.0 - 31.86x)}{2\sqrt{41.0x - 15.93x^2}} = 0$$

or $x = 1.29\ ft$

$$\therefore\ v = v_{max} = \sqrt{(41.0)(1.29) - (15.93)(1.29)^2}$$

or $v_{max} = 5.14\ ft/s$

The results of parts c and d agree reasonably well.

15.105

GIVEN: A car is subjected to a drag force $D = C_D(\frac{1}{2}\rho v^2)A$, when traveling at a speed v, where

A = the projected area on a plane normal to the direction of motion,

ρ = mass density of air ($\rho = 1.2256\ kg/m^3$)

C_D = is a dimensionless coefficient (the drag coefficient).

FIND:

a.) For $A = 1.5\ m^2$, $\rho = 1.2256\ kg/m^3$, and $C_D = 0.40$, plot D for $0 \le v \le 108\ km/hr = 30\ m/s$.

b.) Plot the power (kW; 1 kW = 1.34 hp) required to overcome the drag, for $0 \le v \le 108\ km/hr = 30\ m/s$

c.) Discuss the effect of speed on fuel consumption.

d.) Discuss how the design of the car and habits of the driver can reduce fuel consumption.

SOLUTION:

a.) With $A = 1.5\ m^2$, $\rho = 1.2256\ kg/m^3$, and $C_D = 0.40$,

$$D = (0.40)(\tfrac{1}{2})(1.2256)(1.5)v^2$$

or $D = 0.368v^2\ (N)$ (a)

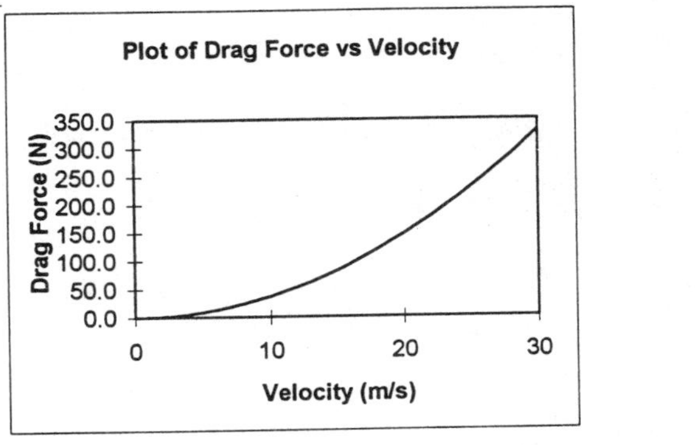

b.) Power E to overcome drag is given by (with v in m/s)

$$E = Fv = Dv\ \ (N\cdot m/s)\ \text{or}\ (\text{Watts})$$

or $E = 3.68 \times 10^{-4} v^3\ (kW)$ (b)

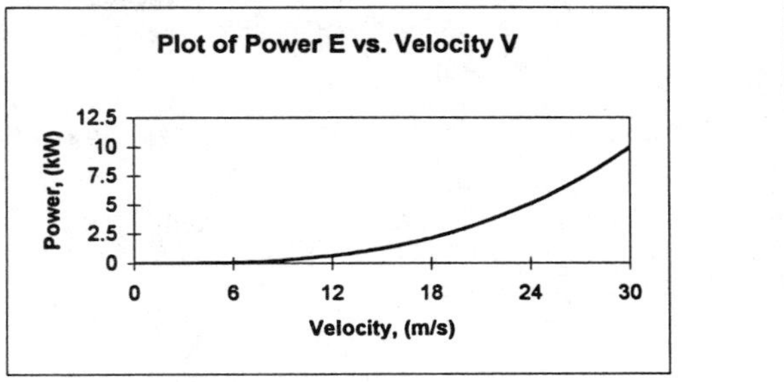

c.) By Eq. (b); the power to overcome drag is proportional to v^3. Thus, as v is increased energy consumption is increased as v^3.

d.) Relative to fuel consumption due to drag, the designer should attempt to reduce C_D further as well as the projected area A, since ρ is fairly constant. Relative to driver habits, drivers should not speed excessively and stay within posted speed limits.

15.106

GIVEN: the system shown in Fig. a. Initially the spring (constant k) is unstretched when the system is released from rest. The coefficients of static and kinetic friction between mass m_A and the plane are μ_s and μ_k respectively.

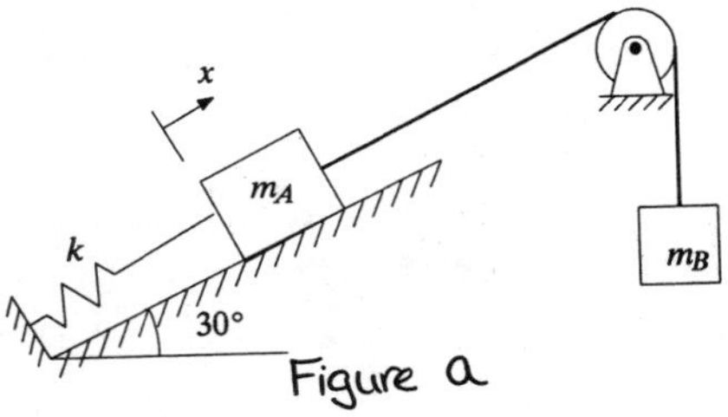

Figure a

FIND:

a.) Assume slip occurs, with mass m_A moving up the plane. Derive a formula for the speed v of masses m_A and m_B in terms of x, μ_k, m_A, m_B, and k.

b.) Let $\mu_s = 0.15$, $\mu_k = 0.10$, $m_A = 60\ kg$, $m_B = 90\ kg$, and $k = 1\ kN/m$. Verify that the masses move from their initial positions.

c.) Determine the distance x_r that the masses move until they again come to rest.

(continued)

15.106 (cont.)

d) Plot v as a function of x for $0 \le x \le x_r$, and verify that the plot agrees with the result of part c.

e) From the plot estimate the maximum value of v and the corresponding value of x. Check the estimates by the formula derived in part a.

SOLUTION:

a) By Fig. b,

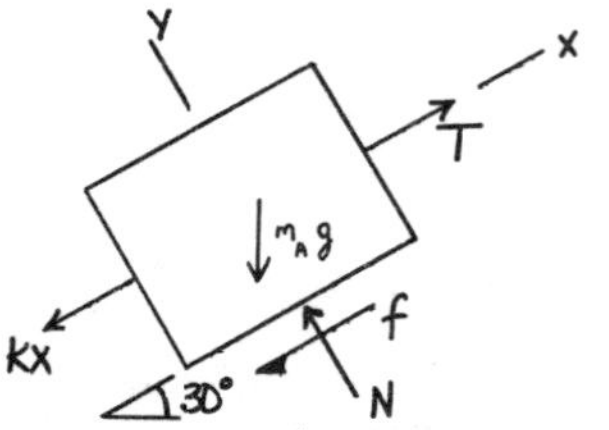

Figure b

$$\Sigma F_y = N - m_A g \cos 30° = 0$$

$$N = m_A g \cos 30°$$

$$\therefore f = \mu_k N = \mu_k m_A g \cos 30°$$

By Eq. (15.38),

$$T_1 + V_1 + U_{1-2}^{nc} = T_2 + V_2 \qquad (a)$$

where

$$U_{1-2}^{nc} = -\mu_k m_A g (\cos 30°)(x)$$

$$T_1 = V_1 = 0$$

$$T_2 = \tfrac{1}{2} m_A v^2 + \tfrac{1}{2} m_B v^2 \qquad (b)$$

$$V_2 = \tfrac{1}{2} kx^2 + m_A g (x \sin 30°) - m_B g x$$

By Eqs. (a) and (b),

$$-\mu_k m_A g (\cos 30°)(x) = \tfrac{1}{2}(m_A + m_B) v^2 + \tfrac{1}{2} kx^2 + m_A g (x \sin 30°) - m_B g x.$$

Hence,

$$v^2 = \left(\frac{2}{m_A + m_B}\right)\left[g(m_B - m_A \sin 30° - \mu_k m_A \cos 30°)x - \tfrac{1}{2}kx^2\right]$$

or

$$v = \sqrt{\left(\frac{2}{m_A + m_B}\right)\left[g(m_B - m_A \sin 30° - \mu_k m_A \cos 30°)x - \tfrac{1}{2}kx^2\right]} \qquad (b)$$

b) Consider the FBD of m_A (Fig. b). In the initial at rest position, $x = 0$ and therefore $kx = 0$. The masses are on the verge of sliding. Then, $T = m_B g$ (Fig. c) and $f = \mu_s N$ (Fig. b).

By Fig. b with $T = m_B g$, $kx = 0$, and $f = \mu_s N$,

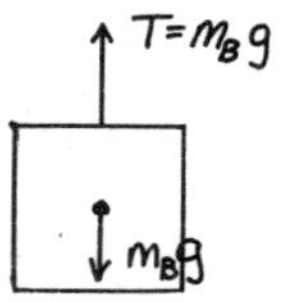

Figure c

$$\Sigma F_y = N - m_A g \cos 30° = 0$$

$$\therefore N = m_A g \cos 30°$$

$$f = \mu_s m_A g \cos 30°$$

$$\Sigma F_x = m_B g - \mu_s m_A g \cos 30° - m_A g \sin 30° \qquad (c)$$

For $m_A = 60$ kg, $m_B = 90$ kg, and $\mu_s = 0.15$, Eq. (c) yields

$$\Sigma F_x = 512.1 \text{ N} > 0$$

Therefore, the masses move; m_A up the plane and m_B down.

c) When m_A and m_B come to rest again, $v = 0$. Then, Eq. (b) yields

$$x = x_r = \frac{2g}{k}\left[m_B - m_A(\sin 30° + \mu_k \cos 30°)\right]$$

With the given data,

$$x_r = \frac{2(9.81)}{1000}\left[90 - 60[0.50 + (0.1)(0.8660)]\right]$$

or $x_r = 1.075$ m

d) With the given data and Eq. (a),

$$v = \sqrt{\frac{2}{150}\left\{(9.81)[90 - 60(0.50 + (0.1)(0.866))]x - 500x^2\right\}}$$

or

$$v = \sqrt{7.168x - 6.667x^2} \qquad (d)$$

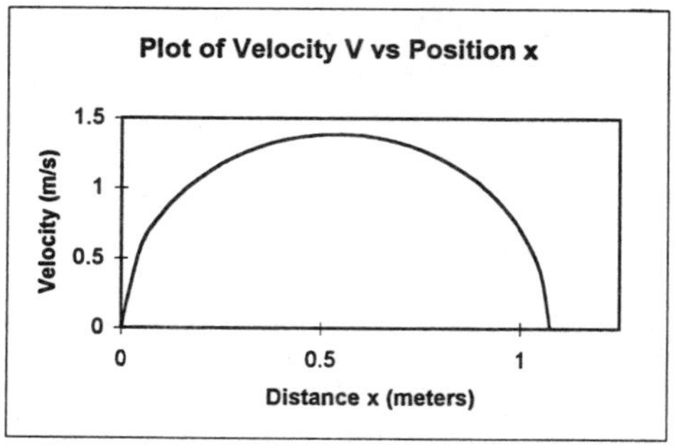

For $x = 1.075$, $v = 0.000975 \approx 0$ to within the accuracy of the calculation.

e) By the plot of Eq. (d),

$$v_{max} \approx 1.39 \text{ m/s}$$

$$x \approx 0.54 \text{ m}$$

By Eq. (b), v_{max} is obtained by the condition $\frac{dv}{dx} = 0$. Thus,

$$\frac{dv}{dx} = \frac{1}{2}\,\frac{7.168 - 13.333x}{\sqrt{7.168x - 6.667x^2}} = 0$$

$$\therefore x = 0.5376 \text{ m}$$

and

$$v_{max} = \sqrt{(7.168)(0.5376) - (6.667)(0.5376)^2}$$

or $v_{max} = 1.388$ m/s at $x = 0.5376$ m

15.107

GIVEN: The data of Problem 15.91 with the addition of friction between the collar and the column. Now, $\mu_s = 0.40$ and $\mu_k = 0.20$.

FIND: Rework Problem 15.91. (see Fig. a)

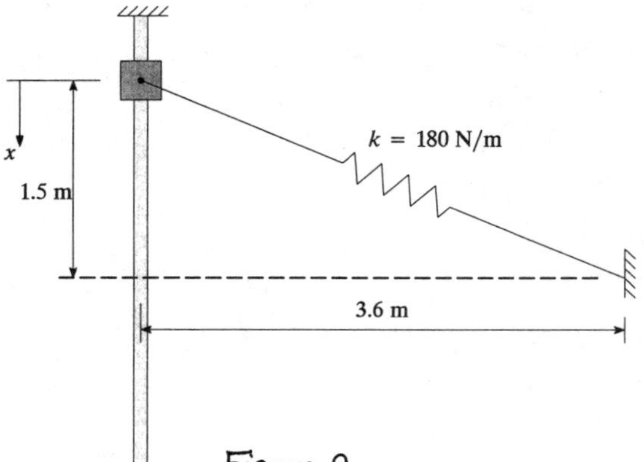

Figure a

SOLUTION:

a.) First, we must ascertain if the collar will slide at $x=0$. So, consider the free-body of the collar at $x=0$. (Fig. b)

$\Sigma F_x = 25 + ke\cos\theta - \mu_s N$

$\Sigma F_y = ke\sin\theta - N = 0$

$N = ke\sin\theta$

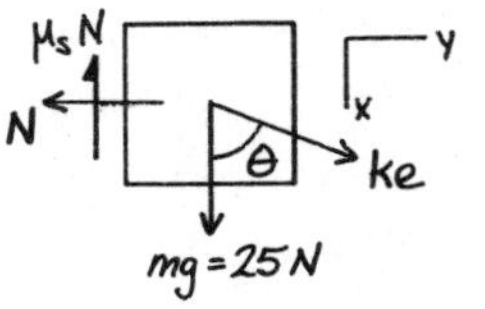

Figure b

By Fig. a,

$\sin\theta = \dfrac{3.6}{\sqrt{1.5^2+3.6^2}} = 0.9230$

$\cos\theta = \dfrac{1.5}{\sqrt{1.5^2+3.6^2}} = 0.3846$

Hence, $N = (180)\left(\sqrt{1.5^2+3.6^2}-3\right)(0.9230)$

$= (180)(0.9)(0.9230) = 149.526\ N$

$\therefore\ \Sigma F_x = 25 + (180)(0.9)(0.3846) - (0.40)(149.526)$

$= 27.89\ N > 0$

Hence, the collar will slide.

To determine the speed of the collar at 1.5m,

$$T_1 + V_1 + U_{1-2}^{nc} = T_2 + V_2 \quad (a)$$

where, since the collar falls from rest,

$T_1 = 0$

$V_1 = \tfrac{1}{2}ke^2 = \tfrac{1}{2}(180)\left(\sqrt{1.5^2+3.6^2}-3\right)^2$

or $V_1 = 72.9\ N\cdot m$

$T_2 = \tfrac{1}{2}mv^2 = \tfrac{1}{2}\left(\tfrac{25}{9.81}\right)v^2 = 1.274v^2$

$V_2 = \tfrac{1}{2}ke^2_{x=1.5} - 25(1.5) = \tfrac{1}{2}(180)(0.6)^2 - 37.5$

or $V_2 = -5.1\ N\cdot m$

Since friction varies with x (Fig. c)

$U_{1-2}^{nc} = -\int_0^x \mu_k N\,dx$

$= -\int \mu_k ke\sin\theta\,dx$

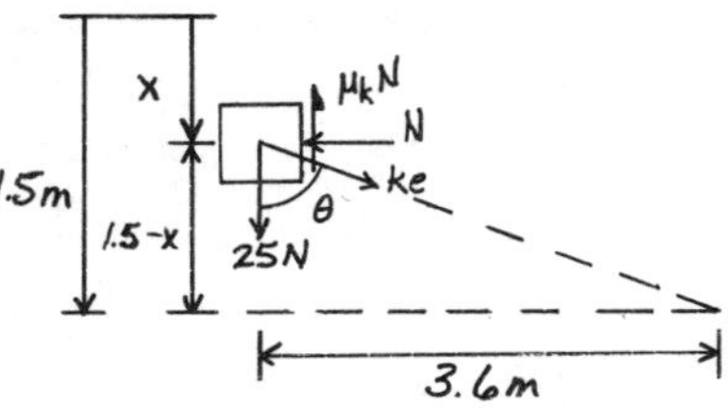

Figure c

By Fig. b,

$e = \sqrt{(1.5-x)^2+3.6^2} - 3$

$\sin\theta = \dfrac{3.6}{\sqrt{(1.5-x)^2+3.6^2}}$

Hence, with $k = 180\ N/m$

$$U_{1-2}^{nc} = -\int_0^x \left[129.6 - \frac{388.8}{\sqrt{12.96+(1.5-x)^2}}\right]dx$$

Integration yields

$$U_{1-2}^{nc} = 655.52 - 129.6x - 388.8\ln\left[1.5-x+\sqrt{12.96+(1.5-x)^2}\right]$$

For $x = 1.5$,

$U_{1-2}^{nc} = -36.91\ N\cdot m$

Hence, by Eq. (a)

$72.9 - 36.91 = 1.274v^2 - 5.1$

$\therefore\ \underline{v(x=1.5) = 5.679\ m/s}$

b.) For $x = x_{max}$,

$T_1 = T_2 = 0$

$V_1 = 72.9\ N\cdot m$

$V_2 = \tfrac{1}{2}ke^2 - 25x = \tfrac{1}{2}(180)\left(\sqrt{12.96+(1.5-x)^2}-3\right)^2 - 25x$

and from part a

$U_{1-2}^{nc} = 655.52 - 129.6x - 388.8\ln\left[1.5-x+\sqrt{12.96+(1.5-x)^2}\right]$

Therefore, Eq. (a) yields

$728.4 - 129.6x - 388.8\ln\left[1.5-x+\sqrt{12.96+(1.5-x)^2}\right]$

$= 90\left(\sqrt{12.96+(1.5-x)^2}-3\right)^2 - 25x$

$\therefore\ \underline{x = 3.02\ m}$

c.) To derive v as a function of x, use Eq. (a), where now T_1, V_1, U_{1-2}^{nc} and V_2 are the same as in part b; but $T_2 = \tfrac{1}{2}mv^2 = 1.274v^2$ (as in part a).

Therefore, by Eq. (a),

$728.4 - 104.6x - 388.8\ln\left[1.5-x+\sqrt{12.96+(1.5-x)^2}\right]$

$= 90\left(\sqrt{12.96+(1.5-x)^2}-3\right)^2 + 1.274v^2$

Therefore,

$$v = \sqrt{\left\{571.74 - 82.10x - 305.18\ln\left[1.5-x+\sqrt{12.96+(1.5-x)^2}\right] - 70.64\left(\sqrt{12.96+(1.5-x)^2}-3\right)^2\right\}}$$

(continued)

15.107 (cont.)

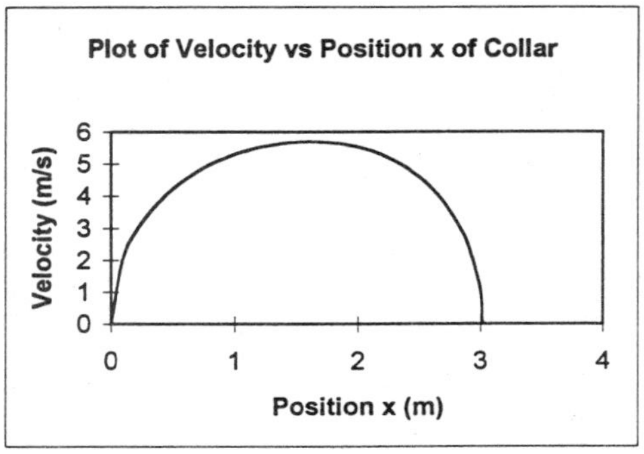

d.) By the plot, at $x = 1.5\,m$, $v \approx 5.7\ m/s$.
Also by the plot, $x_{max} \approx 3\,m$.
These results agree with parts a and b.
Hence, the plot appears correct.

15.109

If $\vec{F}$ is constant, the work performed by $\vec{F}$ is $U = \vec{F} \cdot \Delta\vec{r}$, where $\Delta\vec{r}$ is the displacement vector $\vec{r}_2 - \vec{r}_1$. On a force-displacement graph $|\vec{F}|$ is constant from $\vec{r}_1$ to $\vec{r}_2$ (Fig. a), where the shaded area represents U. [See Eq. (15.17) and Theorem 15.5]

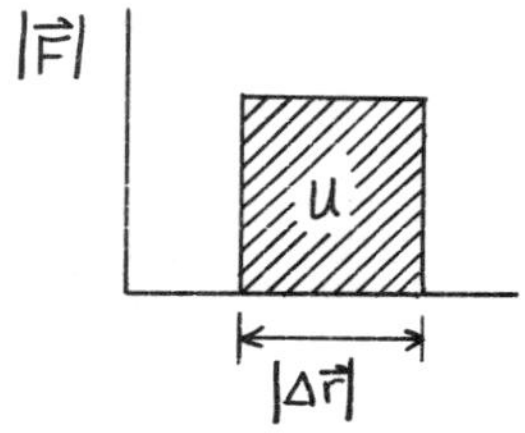

Figure a

15.110

GIVEN: The restoring force of a nonlinear spring is $F = 3x^2$, where F (lb) is the tension in the spring and x is the extension (in).

FIND: The work U that we perform to stretch the spring from $x = 2\,in$ to $x = 5\,in$.

SOLUTION:

$$U = \int F\,dx = \int_2^5 3x^2\,dx = x^3\Big|_2^5 = 125 - 8$$

$$\therefore\ U = 117\ in \cdot lb = 9.75\ ft \cdot lb$$

15.116

GIVEN: a weight W suspended by a spring (constant k)

FIND:
a.) Derive a formula for the total potential energy of the system, relative to the position of the unstretched spring.
b.) Separate the potential energy into potential energies of internal and external forces.

SOLUTION:
a.) Consider Fig. a, where e is the extension length of the spring in the suspended state. Hence, the total potential energy is

$$V = \tfrac{1}{2} k e^2 - We$$

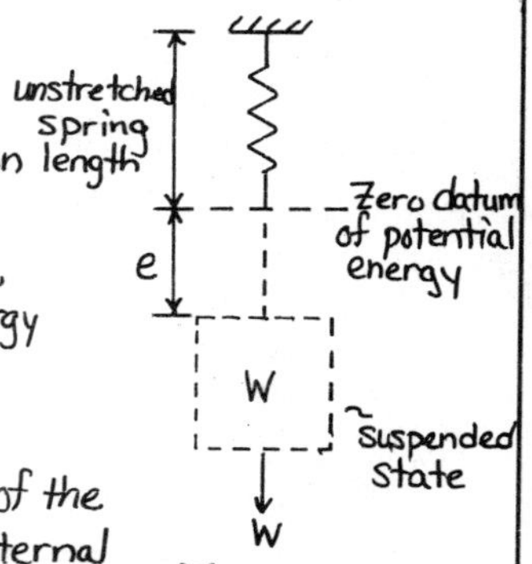

Figure a

b.) The potential energy of the spring results from internal force (the tension in the spring). Hence,

$$V_{internal} = \tfrac{1}{2} k e^2$$

The potential energy of the weight is due to the external gravitational force W. Hence,

$$V_{external} = -We$$

15.120

GIVEN: An archer pulls a bow string back 450 mm with a force of 130 N. The arrow weighs 2.25 N

FIND: the speed with which the arrow leaves the bow, assuming that the force applied is proportional to the deflection and 20% of the strain energy of the bow is lost by internal friction.

SOLUTION:

Since the force is linearly proportional to the displacement,

$$130\ N = k(0.450\,m)$$

or $k = 288.89\ N/m$

Hence, by the law of kinetic energy,

$$U_{1\text{-}2} = T_2 - T_1 \quad (a)$$

where, since 20% of the energy is lost

$$U_{1\text{-}2} = (0.8)\int_0^{0.45} (288.89)\,x\,dx = 115.556\,x^2\Big|_0^{0.45}$$

or $U_{1\text{-}2} = 23.40\ N \cdot m$

(continued)

15.120 (cont.)

$T_1 = 0$

$T_2 = \frac{1}{2} m v^2 = \frac{1}{2}\left(\frac{2.25}{9.81}\right) v^2 = 0.1147 v^2$

Hence, Eq. (a) yields

$23.40 = 0.1147 v^2; \quad \underline{\underline{v = 14.28 \text{ m/s}}}$

15.121

GIVEN: A river delivers 50,000 ft^3 of water per second. There is a hydroelectric power plant at the foot of a dam in the river. The water level upstream from the dam is 400 ft above the level of the power plant.

FIND: the maximum power the plant can deliver. Neglect losses.

SOLUTION:

Assume that the water weighs 62.4 lb/ft^3. The potential energy of 50,000 ft^3 of water relative to the plant is

$$V = (62.4 \text{ lb/ft}^3)(50{,}000 \text{ ft}^3)(400 \text{ ft})$$
$$= 1.248 \times 10^9 \text{ ft}\cdot\text{lb}$$

This amount of energy is delivered to the plant each second. Hence, the maximum power is

$$E = 1.248 \times 10^9 \frac{\text{ft}\cdot\text{lb}}{s} = 1.248 \times 10^9 \text{ Watts}$$

or $\underline{\underline{E = 1.248 \times 10^6 \text{ kW}}}$

15.122

GIVEN:

1 Btu = 778 ft·lb of work

FIND: the number of Btu in a kilowatt-hour (kW·h).

SOLUTION:

$$1 \text{ kW}\cdot\text{h} = 1000 \text{ W}\cdot\text{h} = \left(1000 \frac{\text{N}\cdot\text{m}}{s}\right)(h)$$
$$= \left(1000 \frac{\text{N}\cdot\text{m}}{s}\right)(h)\left(\frac{1 \text{ lb}}{4.448 \text{ N}}\right)\left(\frac{1 \text{ ft}}{0.3048 \text{ m}}\right)\left(\frac{3600 s}{h}\right)$$
$$= 2.655 \times 10^6 \text{ ft}\cdot\text{lb}$$

$$\therefore \quad 1 \text{ kW}\cdot\text{h} = (2.655 \times 10^6 \text{ ft}\cdot\text{lb})\left(\frac{1 \text{ Btu}}{778 \text{ ft}\cdot\text{lb}}\right)$$

or $\underline{\underline{1 \text{ kW}\cdot\text{h} = 3413 \text{ Btu}}}$

16.1

GIVEN: A force that acts on a body increases at a constant rate from 0 to F then decreases at same rate to zero. The total time of impulse is t.

FIND: In terms of F and t, derive formula for impulse of the force in the time interval 0 to t.

SOLUTION:

The force time diagram is shown in Fig. a.

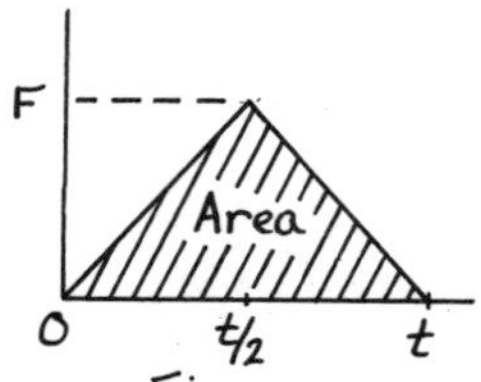

Figure a

The impulse is equal to the area under the F-t curve where Area $= \frac{1}{2}(F)(t)$. Therefore, the impulse is

$$\underline{I = \tfrac{1}{2} F t}$$

16.2

GIVEN: A blow by a hammer imparts a final velocity $v_f = 10$ ft/s to a ½ ounce nail, initially at rest. The impact time is 0.001 s. Assume that the force exerted by the hammer varies as in Problem 16.1.

FIND: The maximum magnitude of force.

SOLUTION:

By the law of impulse-momentum,

$$I = F\Delta t = mv_f - mv_i$$

From Problem 16.1, $I = \frac{1}{2} F_{max} t$

The mass of the nail is

$$m = (\tfrac{1}{2}\ \text{oz})\left(\frac{1\ \text{lb}}{16\ \text{oz}}\right)\left(\frac{1}{32.2\ \text{ft/s}^2}\right) = 9.7(10^{-4})\ \text{slug}$$

$$\therefore\ \tfrac{1}{2} F_{max} t = m(v_f - \overset{0}{\cancel{v_i}}) \quad \text{or} \quad F_{max} = \frac{2mv_f}{t}$$

$$\therefore\ \underline{\underline{F_{max} = \frac{2(9.7)(10^{-4})\ \text{slug}\ (10\ \text{ft/s})}{0.001\ \text{s}} = 19.41\ \text{lb.}}}$$

16.3

GIVEN: A machine gun fires 840 bullets/min. Each bullet weighs 1/20 lb and has a horizontal muzzle velocity of 3220 ft/s.

FIND: The average horizontal force exerted on the gun by its supports.

SOLUTION:

Let the mass of bullets fired per second be m.

$$\therefore\ m = \left(840\ \frac{\text{bullets}}{\text{min}}\right)\left(\frac{1\ \text{min}}{60\ \text{s}}\right)\left(\frac{1}{20}\ \frac{\text{lb}}{\text{bullet}}\right)\left(\frac{1}{32.2\ \text{ft/s}^2}\right)$$

or $m = 0.02174$ slug/s

The momentum imparted to the bullets in Δt seconds is

$$(mv)\Delta t = (0.02174\ \tfrac{\text{slug}}{\text{s}})(3220\ \text{ft/s})(\Delta t\ \text{s}) = 70(\Delta t)\ \text{lb·s}$$

Therefore, by the impulse-momentum principle,

$$F_{avg}(\Delta t) = 70(\Delta t) \quad \text{or} \quad \underline{\underline{F_{avg} = 70\ \text{lb.}}}$$

16.4

GIVEN: A U.S. Navy plane weighing $W_p = 72$ kN and traveling with absolute initial speed $v_{Pi} = 180$ knots, lands on a carrier traveling at $v_c = 15$ knots, in the same direction as the plane (Fig. a)

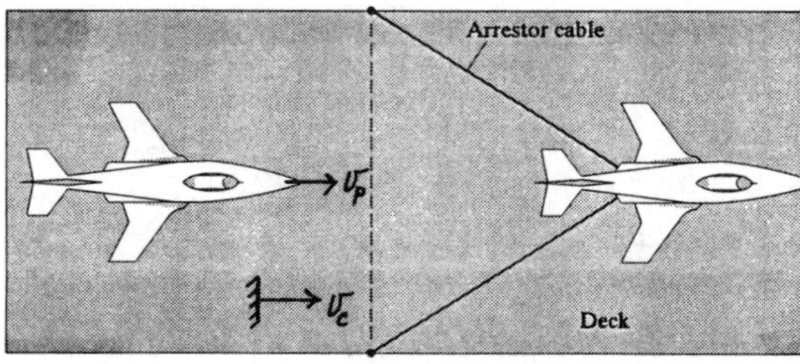

Figure a

The plane is brought to rest on the deck in 2 seconds by the arrestor cable.

FIND: The average force exerted on the plane by the cable.

SOLUTION:

The rotation of the plane is small. Therefore, treat it as a particle. In general, the velocity of the plane relative to the deck is (Fig. a)

$$v_{P/c} = v_p - v_c \qquad (a)$$

When the plane is brought to rest on the deck, $v_{P/c} = 0$, and $v_c = 15$ knots. Therefore, Eq. (a) yields the final velocity of the plane as

$$v_p = v_{Pf} = v_c = (15\ \text{knots})(0.5144\ \tfrac{\text{m}}{\text{knot·s}})$$

$$v_{Pf} = 7.716\ \text{m/s} \qquad (b)$$

The mass of the plane is

$$m = (72\ \text{kN})/(9.81\ \text{m/s}^2) = 7{,}339\ \text{kg} \qquad (c)$$

By the principle of impulse and momentum,

$$F_{avg}\Delta t = m(v_{Pf} - v_{Pi}) \qquad (d)$$

where

$$v_{Pi} = (180\ \text{knots})(0.5144\ \tfrac{\text{m}}{\text{knot·s}}) = 92.59\ \text{m/s} \qquad (e)$$

Equations (b), (c), (d), and (e) yield, with $\Delta t = 2$ s,

$$\underline{F_{avg} = \frac{7339}{2}(7.716 - 92.59) = -311\ \text{kN}}$$

The negative sign indicates that the cable pulls back on the plane (Fig. a).

16.5

GIVEN: An airplane, weighing $W_p = 80$ kN and flying with absolute initial speed of $v_{P_i} = 180$ km/h (50 m/s), lands on a carrier traveling at $v_c = 20$ km/h (5.556 m/s) in the same direction as the plane (see Fig. a, Problem 16.4). Air resistance and brakes exert 10 kN force on the plane. The plane is brought to rest on the carrier deck in 2 seconds by the arrestor cable, air resistance, and brakes.

FIND: The average force exerted on the plane by the cable.

SOLUTION:

As in Problem 16.4, treat the plane as a particle. The velocity of the plane relative to the carrier is

$$v_{P/c} = v_P - v_c \qquad (a)$$

When the plane is brought to rest on the deck $v_{P/c} = 0$, and since $v_c = 5.556$ m/s, Eq. (a) yields

$$v_P = v_{Pf} = v_c = 5.556 \text{ m/s} \qquad (b)$$

The mass of the plane is

$$m = (80{,}000 \text{ N})/(9.81 \text{ m/s}^2) = 8154.9 \text{ kg} \qquad (c)$$

By the principle of impulse and momentum,

$$F_{avg}\,\Delta t = m(v_{Pf} - v_{Pi})$$

or

$$2F_{avg} = (8154.9)(5.556 - 50)$$

$$\therefore \quad \underline{F_{avg} = -181.2 \text{ kN}}$$

The negative sign indicates that the cable pulls back on the plane.

16.6

GIVEN: A long, open railroad tank car travels horizontally at 30 ft/s. It passes under a vertical nozzle that discharges water ($\gamma = 62.4$ lb/ft³) into it (Fig. a) at the rate of 5 ft³/s.

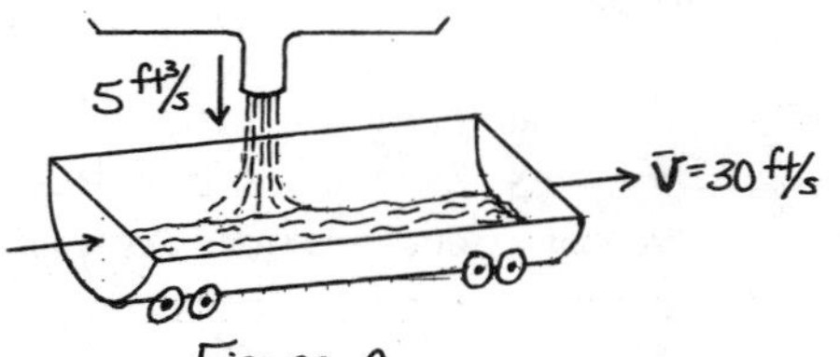

Figure a

FIND: The force required to maintain the car's speed at 30 ft/s. Neglect friction.

SOLUTION:

Use Newton's Second Law as it applies to momentum: namely,

$$F = \frac{d}{dt}(mv) = \frac{dp}{dt}$$

Since the velocity is constant,

$$F = v\frac{dm}{dt} \qquad (a)$$

where

$$\frac{dm}{dt} = \left(\frac{62.4 \text{ lb/ft}^3}{32.2 \text{ ft/s}^2}\right)(5 \text{ ft}^3/\text{s}) = 9.689 \frac{\text{slugs}}{\text{s}}$$

By Eq. (a),

$$\underline{\underline{F = (30 \text{ ft/s})(9.689 \text{ slugs/s}) = 290.7 \text{ lb.}}}$$

16.7

GIVEN: A ship that weighs 90 MN is brought to rest from a speed of 2 km/h in 25 seconds by a constant pull by a hawser.

FIND:

a.) The constant force exerted on the ship

b.) The energy dissipated by friction between the hawser and bollard.

SOLUTION:

a.) For a constant force, by Fig. a and the impulse-momentum principle,

$$F\Delta t = m(v_{sf} - v_{si}) \qquad (a)$$

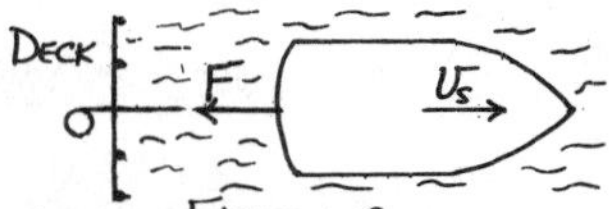

Figure a

where

$$m = \frac{90\times10^6 \text{ N}}{9.81 \text{ m/s}^2} = 9.174\times10^6 \text{ slugs}$$

$$v_{sf} = 0; \quad v_{si} = 2 \text{ km/h} = 0.5556 \text{ m/s}$$

Hence, by Eq. (a), with $\Delta t = 25$ s,

$$F = \frac{1}{25}(9.174\times10^6)(0 - 0.5556)$$

or

$$\underline{F = -203.9 \text{ kN}}$$

The minus sign indicates that the pull is to the left (opposite to v_s), (Fig. a).

b.) By the work-kinetic energy principle,

$$T_1 + U_{1\text{-}2} = T_2 \qquad (a)$$

where T_1, T_2 are the initial and final kinetic energies of the ship, and $U_{1\text{-}2}$ is the work done by friction. Now,

$$T_1 = \tfrac{1}{2} m v_{s_i}^2 = \tfrac{1}{2}(9.174\times10^6)(0.5556)^2 = 1.416\times10^6 \text{ N·m}$$

$$T_2 = 0$$

Hence, by Eq. (a)

$$U_{1\text{-}2} = -1.416\times10^6 \text{ N·m}$$

and the energy dissipated is

$$\underline{E = -U_{1\text{-}2} = 1416 \text{ kJ}}$$

16.8

GIVEN: A rocket burns fuel at the rate of $R = 80$ N/s. Burned fuel is exhausted with a velocity $v_E = 30$ km/s.

FIND: The thrust of the rocket.

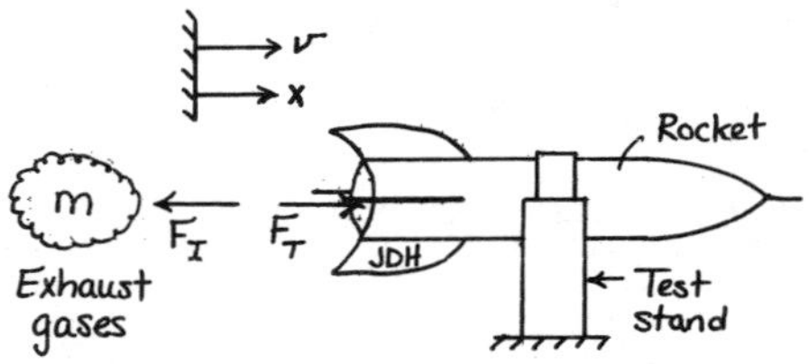

Figure a

SOLUTION:

By impulse-momentum,

$$\Sigma F_x = -F_I = \frac{d(mv)}{dt}$$

Since v is constant,

$$F_I = -v\frac{dm}{dt} \qquad (a)$$

where $\frac{dm}{dt} = \frac{(80 \text{ N/s})}{(9.81 \text{ m/s}^2)} = 8.155$ kg/s = mass rate of exhaust

The impulsive force F_I on the exhausting gas results in an equal and opposite thrust F_T on the rocket.

By Eq. (a) and Fig. a,

with $v = -30{,}000$ m/s (gases exhaust to left)

$\therefore \; F_T = F_I = -(-30{,}000 \text{ m/s})(8.155 \text{ kg/s}) = 245$ kN

or $\vec{F}_T = 245 \text{ kN} \rightarrow$

16.9

GIVEN: A golf ball weighs approximately 1/10 lb. A golfer drives the ball off a tee with a speed of 150 ft/s. The impact time between the ball and the clubhead is 0.01 s.

FIND: By the principle of impulse and momentum,

a) Find the average force F_{avg} that the club exerts on the ball.

b) The average acceleration a_{avg}.

c) Assume that the force exerted on the ball varies as shown in Fig. a. Determine F_{max}, the maximum acceleration a_{max}, and the average acceleration of the ball during the impact.

d) Is the assumption in part c reasonable?

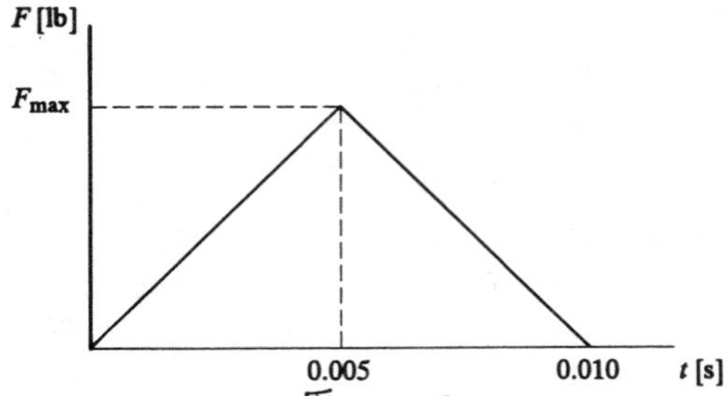

Figure a

SOLUTION:

a.) By impulse-momentum,

$$I = F_{avg}\Delta t = m(v_f - v_i) \qquad (a)$$

where

$m = (1/10 \text{ lb})/(32.2 \text{ ft/s}^2) = 0.003106$ slugs

$v_f = 150$ ft/s ; $v_i = 0$

$\Delta t = 0.01$ s

By Eq. (a),

$$F_{avg} = \left(\frac{1}{0.01}\right)(0.003106)(150-0) = 46.59 \text{ lb.}$$

b.) $$a_{avg} = \frac{F_{avg}}{m} = \frac{46.59}{0.003106} = 15{,}000 \text{ ft/s}^2$$

c.) By Fig. a and impulse-momentum

$$I = \tfrac{1}{2}F_{max}\Delta t = m(v_f - v_i)$$

or $$F_{max} = \frac{2m}{\Delta t}(v_f - v_i)$$

$$= \frac{2(0.003106)}{(0.01)}(150-0)$$

$$\therefore \; F_{max} = 93.18 \text{ lb.}$$

Hence,

$$a_{max} = \frac{F_{max}}{m} = \frac{93.18}{0.003106} = 30{,}000 \text{ ft/s}^2$$

By Fig. a, the average force is $F_{avg} = \tfrac{1}{2}F_{max}$. Therefore, $F_{avg} = 46.59$ lb, and

$$a_{avg} = \frac{F_{avg}}{m} = \frac{46.59}{0.003106} = 15{,}000 \text{ ft/s}^2$$

(Note also that $a_{avg} = \frac{v_f}{\Delta t} = \frac{150}{0.01} = 15{,}000 \text{ ft/s}^2$)

d.) The assumption in part c is reasonable. It gives the same average force and average acceleration as parts a and b.

16.10

GIVEN: A machine gun fires 480 bullets/min. Each bullet weighs 0.275 N and has a horizontal muzzle velocity of 1,000 m/s.

FIND: The average horizontal force exerted by the gun on its supports.

SOLUTION:

Let the mass of the bullets fired per second be m.

$$\therefore \quad m = \left(480 \ \frac{\text{bullets}}{\text{min}}\right)\left(\frac{1 \text{ min}}{60 \text{ s}}\right)\left(\frac{0.275 \text{ N}}{\text{bullet}}\right)\left(\frac{1}{9.81 \text{ m/s}^2}\right)$$

or $\quad m = 0.2243 \text{ kg/s}$

The momentum imparted to the bullets in Δt seconds is

$$(mv)(\Delta t) = (0.2243 \tfrac{\text{kg}}{\text{s}})(1000 \text{ m/s})(\Delta t \text{ s})$$
$$= 224.3(\Delta t) \ (\text{N·s})$$

Therefore, by impulse-momentum principle,

$$F_{avg}(\Delta t) = 224.3(\Delta t)$$

or $\quad \underline{F_{avg} = 224.3 \text{ N}}$

This is the horizontal force that the gun pushes back on its supports.

16.11

GIVEN: A baseball that weighs 1.35 N travels at $v_i = 25$ m/s, strikes the ground at an angle of 30°, and rebounds at an angle of 20° with a speed of $v_f = 15$ m/s (Fig. a)

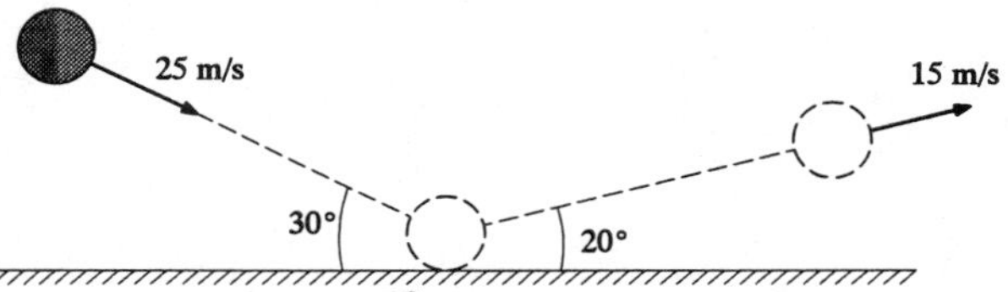

Figure a

FIND:

a.) The (x,y) projections P_{x_i}, P_{y_i} of momentum of the ball for the instant before it strikes the ground.

b.) The projections P_{x_f}, P_{y_f} for the instant after the ball rebounds.

c.) The average force during the time $\Delta t = 0.012$ s of impact.

SOLUTION:

a.) The mass of the ball is

$$m = (1.35 \text{ N})/(9.81 \text{ m/s}^2) = 0.1376 \text{ kg}$$

The (x,y) projections v_i are

$$v_{x_i} = v_i \cos 30° = 25(0.8660) = 21.65 \text{ m/s}$$

$$v_{y_i} = -v_i \sin 30° = -25(0.500) = -12.50 \text{ m/s}$$

Therefore, before impact,

$$\underline{\underline{P_{x_i} = m v_{x_i} = (0.1376)(21.65) = 2.979 \text{ N·s}}}$$

$$\underline{\underline{P_{y_i} = m v_{y_i} = (0.1376)(-12.50) = -1.720 \text{ N·s}}}$$

b.) Similarly, after impact,

$$\underline{\underline{P_{x_f} = m v_{x_f} = (0.1376)(15 \cos 20°) = 1.940 \text{ N·s}}}$$

$$\underline{\underline{P_{y_f} = m v_{y_f} = (0.1376)(15 \sin 20°) = 0.706 \text{ N·s}}}$$

c.) By the principle of impulse-momentum, with $\Delta t = 0.012$ s,

$$(F_{avg})_x \Delta t = P_{x_f} - P_{x_i}$$

$$(F_{avg})_y \Delta t = P_{y_f} - P_{y_i}$$

Hence,

$$(F_{avg})_x = (1.940 - 2.979)/(0.012) = -86.58 \text{ N}$$

$$(F_{avg})_y = [0.706 - (-1.720)]/(0.012) = 202.17 \text{ N}$$

Therefore, the average force that acts on the ball is

$$\underline{\underline{\vec{F}_{avg} = -86.58\mathbf{i} + 202.17\mathbf{j} \ \text{(N)}}}$$

or (see Fig. b)

$$F_{avg} = 219.93 \text{ N @ } \theta = 113.18°$$

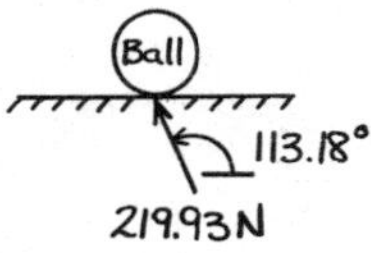

Figure b

16.12

GIVEN: Figure a: Powdered coal is loaded onto a conveyor belt at the rate of 14 kN/s. The speed of the belt is 2 m/s, and the tension in the belt at point A is 100 N.

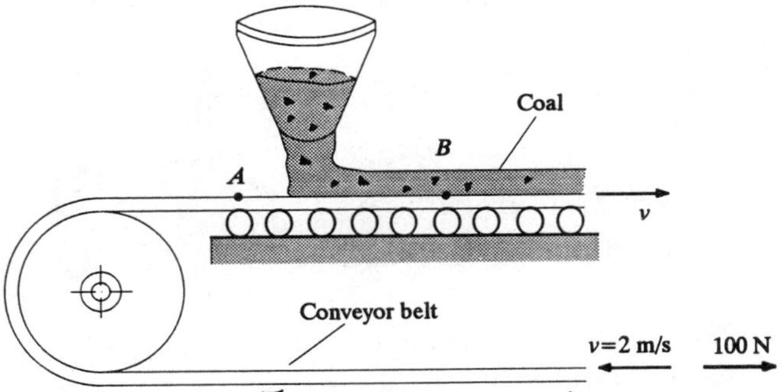

Figure a

FIND: The tension T_B in the belt at point B.

(continued)

16.12 (cont.)

SOLUTION:

Assume that the mass of the pulley is negligible.

By the momentum form of Newton's second law,

$$F = \frac{d(mv)}{dt}$$

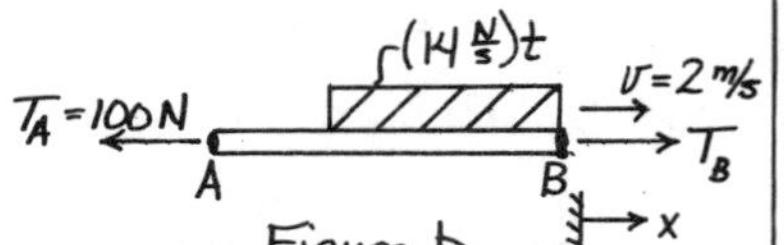

Figure b

where

$$m = \left(\frac{14\ \frac{kN}{s}}{9.81\ m/s^2}\right)t = 1427.11t\ (kg)$$

or

$$mv = (1427.11t)(2\ m/s) = 2854.23t\ (kg\cdot m/s)$$

$$\therefore\ \frac{d(mv)}{dt} = 2854.23\ N$$

Hence, by Fig. b,

$$\Sigma F_x = T_B - T_A = \frac{d(mv)}{dt}$$

or $\underline{\underline{T_B = 2854.23 + 100 = 2.95\ kN}}$

16.13

GIVEN: Figure a: a bullet of weight 0.60 N strikes and enters a block that weighs 450 N. When the bullet enters the block, the block and bullet slide on the table for 0.256 s. The coefficient of sliding friction is $\mu_k = 0.40$.

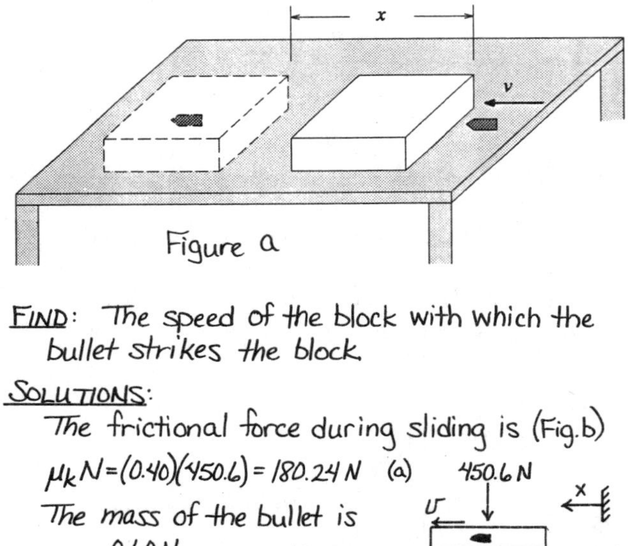

Figure a

FIND: The speed of the block with which the bullet strikes the block.

SOLUTIONS:

The frictional force during sliding is (Fig. b)

$$\mu_k N = (0.40)(450.6) = 180.24\ N \quad (a)$$

The mass of the bullet is

$$m = \frac{0.60\ N}{9.81\ m/s^2} = 0.06116\ kg$$

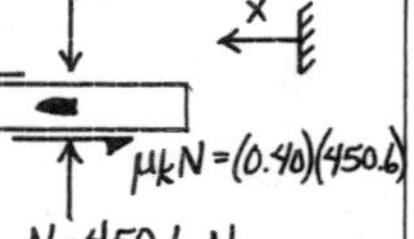

Figure b

Therefore, the initial momentum of the system is

$$P_{x_i} = mv_b = 0.06116 v_{b_i} \quad (b)$$

where v_{b_i} is the initial speed of the bullet.

The final momentum of the system is

$$P_{x_f} = 0 \quad (c)$$

Since the velocities of the bullet and the block are zero at the end of sliding. Therefore, by the principle of impulse-momentum,

$$F_x \Delta t = P_{x_f} - P_{x_i} \quad (d)$$

By Fig. b and Eqs. (a), (b), (c), and (d), with $\Delta t = 0.256s$,

$$-(180.24)(0.256) = 0 - 0.06116 v_{b_i}$$

or $\underline{\underline{v_{b_i} = 754.4\ m/s}}$

16.14

GIVEN: A car pulls a trailer with a constant force of 600 N at a constant speed of 80 km/h.

FIND:

a.) the impulse exerted by the car on the trailer as the car travels 100m.

b.) the impulse exerted by the car on the trailer as the car travels 1 km.

c.) the impulse exerted by the car on the trailer as the car travels for 1s.

d.) Compare the impulse per second exerted by the car in parts a, b, and c. Explain the results.

SOLUTION:

By the definition of impulse,

$$I = \int_0^t F dt = \int_0^t 600 dt = 600t \quad (a)$$

a.) The time required to travel 100m is

$$t = \frac{0.100\ km}{80\ km/h} = \frac{1}{800}h = \frac{3600}{800} = 4.5\ s$$

$$\therefore\ \underline{\underline{I_a = (600)(4.5) = 2700\ N\cdot s}}$$

b.) The time required to travel 1 km is

$$t = \frac{1.0\ km}{80\ km/h} = \frac{1}{80}h = 45s$$

$$\therefore\ \underline{\underline{I_b = (600)(45) = 27\,000\ N\cdot s}}$$

c.) The impulse for 1 s is by Eq. (a)

$$\underline{\underline{I_c = 600\ N\cdot s}}$$

d.) To obtain impulse per second, divide the impulses by the times.

(continued)

16.14 (cont.)

Thus,

$$\frac{I_a}{4.5} = \frac{2700}{4.5} = 600 \ N\cdot s/s$$

$$\frac{I_b}{45} = \frac{27000}{45} = 600 \ N\cdot s/s$$

$$\frac{I_c}{1} = \frac{600}{1} = 600 \ N\cdot s/s$$

The impulses per second are equal in the three cases, since $F = 600N$ is constant [see Eq. (a)].

16.15

GIVEN: Pressure in a gun barrel (Fig. a). The gun fires a shell that weighs 140N and is 64 mm in diameter.

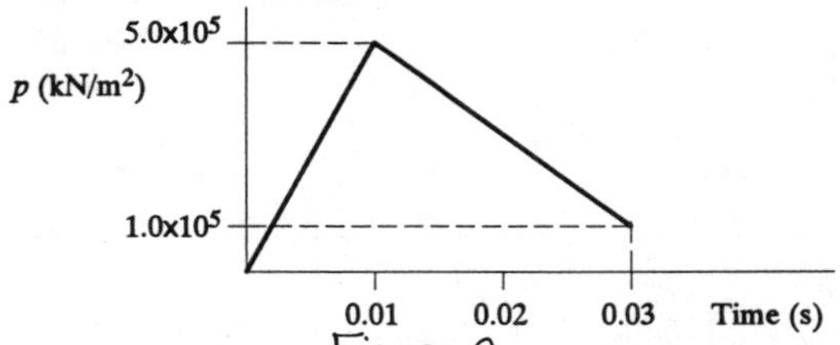

Figure a

FIND:

a) the muzzle velocity of the gun. Ignore the momentum of gases

b) Determine the efficiency of the gun knowing that 40 MJ of energy is expended chemically by the charge.

SOLUTION:

a) By Fig. a, the average pressure for 0.03 seconds is

$$P_{avg} = \frac{\frac{(5\times10^5)(0.01)}{2} + \frac{(4\times10^5)(0.02)}{2} + (1\times10^5)(0.02)}{0.03}$$

$$P_{avg} = 2.833\times10^5 \ kN/m^2$$

Hence, the average force exerted on the shell is

$$F_{avg} = P_{avg}A = (2.833\times10^5)(\pi)(0.032)^2$$

or $F_{avg} = 911.5 \ kN$

Hence, by the principle of impulse-momentum,

$$F_{avg}t = mv$$

or $(911.5\times10^3)(0.03) = \frac{140}{9.81}v$

$$\underline{\underline{v = 1916 \ m/s}}$$

b.) The kinetic energy of the shell at exit is

$$T = \tfrac{1}{2}mv^2 = \tfrac{1}{2}\left(\frac{140}{9.81}\right)(1916)^2$$

$$T = 26.20 \ MJ$$

$$\eta = \text{Efficiency} = \left(\frac{40-26.20}{40}\right)\times100$$

or $\underline{\underline{\eta = 34.5\%}}$

16.16

GIVEN: A body falls freely in a gravitational field with an initial speed $v_i = 10$ m/s. Neglect air resistance and use the law of momentum.

FIND:

a.) The time and distance traveled required to attain a speed of 30 m/s.

b.) The time and distance traveled to increase the speed from 30 m/s to 50 m/s.

c.) Explain why the distances traveled in parts a and b are different, even though the times of travel are the same.

SOLUTION:

a.) Consider the free-body diagram of Fig. a.

Initially, the velocity v of the body at time $t_i = 0$, is $v_i = 10$ m/s.

At time t_f, $v_f = 30$ m/s. Therefore, by impulse-momentum, since $F = mg$ = constant

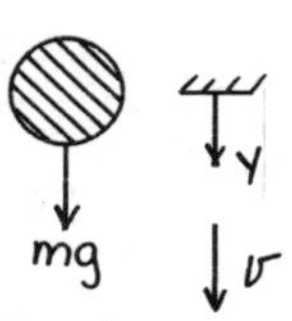

Figure a

$$F\Delta t = m(v_f - v_i) \qquad (a)$$

or the time required to attain a speed of 30 m/s is

$$\underline{\underline{\Delta t_1 = t_f = \frac{m(30-10)}{mg} = \frac{20}{g}}} \qquad (b)$$

By kinematics, since acceleration $a = g$ = Constant, the distance s traveled during this time is

$$s_1 = \tfrac{1}{2}gt_f^2 + v_it_f = \frac{200}{g} + \frac{200}{g}$$

$$\underline{\underline{s_1 = \frac{400}{g}}} \qquad (c)$$

b.) For the interval from $v_i = 30$ m/s to $v_f = 50$ m/s, by Eq. (a), the time required is

$$\underline{\underline{\Delta t_2 = t_f = \frac{m(50-30)}{mg} = \frac{20}{g}}} \qquad (d)$$

(continued)

16.16 (cont.)

This is the same time required to go from 10 m/s to 30 m/s. However, the distance traveled in going from 30 m/s to 50 m/s is

$$s_2 = \tfrac{1}{2} g t_f^2 + v_i t_f = \frac{200}{g} + \frac{600}{g} = \frac{800}{g} \quad (e)$$

Note, by Eqs. (c) and (d), that $s_2 = 2s_1$.

The results of parts a and b are valid for any constant gravitational field. They are true on the moon, on Mars, or on Earth (neglecting air resistance).

For $g = 9.81\ m/s^2$ (on Earth)

$\Delta t_1 = \Delta t_2 = 2.039\ s$, $s_1 = 40.77\ m$, and $s_2 = 81.55\ m$.

c) The difference between $s_1 < s_2$ is due to the fact that in part a $v_i = 10$ m/s and in part b, $v_i = 30$ m/s.

16.17

GIVEN: At standard atmospheric temperature and pressure (STP), the number of molecules is 7.660×10^{23} molecules/ft³, the mass of a molecule of $H_2 = 3.324 \times 10^{-24}$ grams, and the mean molecular speed is $v = 5{,}558$ ft/s.

Hydrogen at STP is contained in a closed vessel with plane walls.

FIND: the pressure (lb/in²) that the hydrogen would exert on a wall if all molecules were to move perpendicular to the wall.

SOLUTION:

Consider 1 in² of wall for 1 s.

How many molecules (therefore, how much mass) will hit 1 in² in 1 s?

Since the molecules travel 5,558 ft/s, all molecules within 5,558 ft will strike the wall. If there are 7.660×10^{23} molecules/ft³, then the number of molecules striking 1 in² in 1 s essentially be a rectangular column of molecules 1 in² × 5,558 ft long. Therefore, (Fig. a),

$$(7.660 \times 10^{23}\ \tfrac{\text{molecules}}{ft^3})(5{,}558\ \tfrac{ft}{s})(\tfrac{1\ ft^2}{144\ in^2})(1 s)$$

$= 2.96 \times 10^{25}$ molecules strike the wall in 1 s.

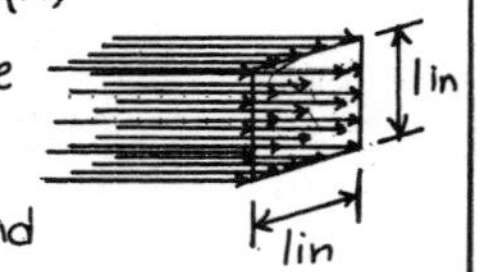

Figure a

By the principle of impulse and momentum,

$$F \Delta t = \Delta p = m v_2 - m v_1 \quad (a)$$

The change in momentum for each molecule is

$$\Delta p = m v_2 - m v_1$$

Assume each molecule rebounds with the same speed with which it strikes (Fig. b)

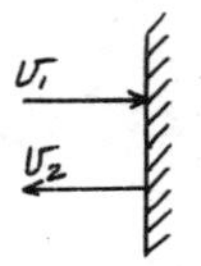

Figure b

$\therefore\ v_2 = -v_1$

$\therefore\ \Delta p = m[v_2 - (-v_2)]$

or $\Delta p = m(2v) \quad (b)$

The mass of each molecule in U.S. Customary units is

$$3.324 \times 10^{-24}\ gm \times 6.852 \times 10^{-5}\ \frac{slug}{gm} = 2.278 \times 10^{-28}\ slugs$$

By Eqs. (a) and (b),

$$F(1 s) = (2.96 \times 10^{25}\ \text{molecules})(2.278 \times 10^{-28}\ \tfrac{slug}{\text{molecule}})(2 \times 5558\ ft/s)$$

or $F = 74.86$ lb

$$\therefore\ \text{Pressure} = \frac{F}{A} = \frac{74.86\ lb}{1\ in^2} = 74.86\ lb/in^2$$

16.18

GIVEN: A solid cone is driven into water at a constant velocity v by a force F that includes the weight of the cone (Fig. a).

FIND: Show that

$F = \pi \rho (v^2 x^2 + \tfrac{1}{3} g x^3)$ where x is the radius of volume of the hemisphere circumscribing the cone. Assume that the volume of water that is driven with the cone is the volume of the hemisphere of radius x minus the volume of the submerged cone.

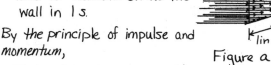

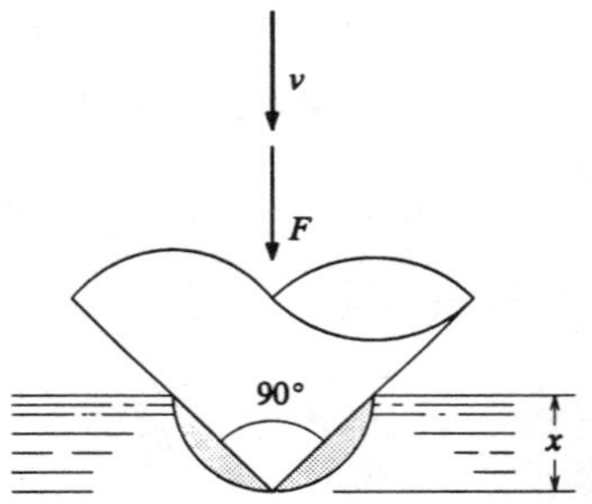

Figure a

SOLUTION:

When the cone is submerged a distance x, the total downward force on the system is

$$F_{total} = F + W_w - F_{pressure} \quad (a)$$

where W_w is the weight of the water driven by the cone and $F_{pressure}$ is the net vertical force due to the pressure acting on the hemisphere surface of radius x. The mass of the water that moves with the cone is

$$m_w = \rho\,(\text{Volume of hemisphere} - \text{Volume of Submerged Cone})$$

(continued)

16.18 (cont.)

$$m_w = \rho\left(\tfrac{2}{3}\pi x^3 - \tfrac{1}{3}\pi x^3\right) = \tfrac{1}{3}\rho\pi x^3 \quad (b)$$

Hence,

$$W_w = m_w g = \tfrac{1}{3}\rho g \pi x^3 \quad (c)$$

The force $F_{pressure}$ is obtained by integration (Fig. b).

At depth ζ, the pressure is, by Fig. b,

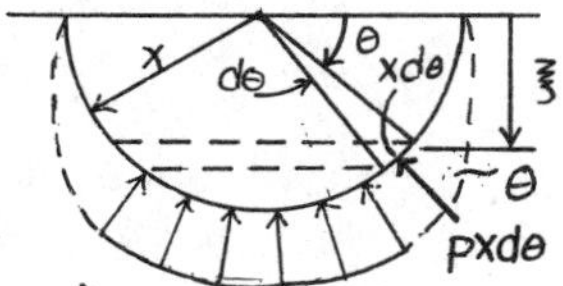

$$p = \rho g \zeta = \rho g x \sin\theta$$

$$\therefore F_{pressure} = \int_0^{\pi/2} (2\pi x\cos\theta)(px \sin\theta \, d\theta)$$

Figure b

$$= 2\pi\rho g x^3 \int_0^{\pi/2} \sin^2\theta \cos\theta \, d\theta$$

or $$F_{pressure} = \tfrac{2}{3}\pi\rho g x^3 \quad (d)$$

Therefore, by Eqs. (a), (b), (c), and (d),

$$F_{total} = F + \tfrac{1}{3}\pi\rho g x^3 - \tfrac{2}{3}\pi\rho g x^3$$

$$F_{total} = F - \tfrac{1}{3}\pi\rho g x^3 \quad (e)$$

The linear momentum of the system in the x direction is

$$p_x = m_{cone}v + m_w v$$

or $$p_x = m_{cone}v + \tfrac{1}{3}\pi\rho x^3 v \quad (f)$$

By the principle of linear momentum, since v is constant, with Eqs. (e) and (f),

$$F - \tfrac{1}{3}\pi\rho g x^3 = \frac{dp_x}{dt} = \pi\rho x^2 v \frac{dx}{dt} = \pi\rho x^2 v^2$$

or $$\underline{F = \pi\rho\left(v^2x^2 + \tfrac{1}{3}gx^3\right)}$$

16.19

GIVEN: A tugboat pulls a barge at speed v. The propellor thrust minus drag produces a net thrust that varies linearly with speed (Fig. a). The tugboat and barge weigh 360 kN.

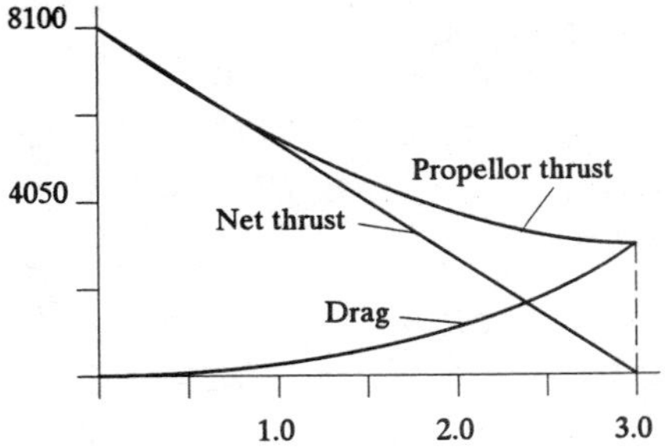

Figure a

FIND:

a.) the time required to increase the speed from v_1 to v_2 in terms of v_1 and v_2.

b.) the distance traveled during this time in terms of v_1 and v_2.

c.) For $v_1 = 1.0$ m/s, $v_2 = 2.5$ m/s, calculate the times and distances for parts a and b respectively.

SOLUTION:

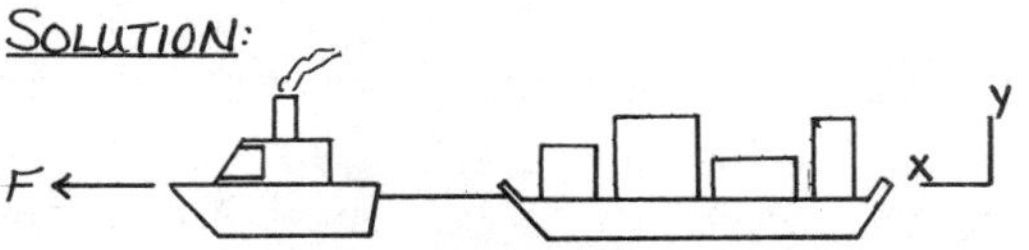

Figure b

a.) By Fig. b and Newton's second law, in terms of momentum with m = constant,

$$F = \frac{d(mv)}{dt} = m\frac{dv}{dt} \quad (a)$$

By Fig. a,

$$F = 8100 - 2700v, \text{ or}$$

$$F = 2700(3 - v) \quad (b)$$

By Eqs. (a) and (b),

$$\frac{2700}{m}dt = \frac{dv}{3-v}$$

Let $t_1 = 0$ when $v = v_1$. Then,

$$\int_{t_1=0}^{t_2} \frac{2700}{m}dt = \int_{v_1}^{v_2}\frac{dv}{3-v}$$

Integration yields

$$\frac{2700}{m}t_2 = -\left[\ln(3-v_2) - \ln(3-v_1)\right]$$

where $m = W/g = \dfrac{360 \text{ kN}}{9.81 \text{ m/s}^2} = 3.670 \times 10^4$ kg

Hence,

$$\underline{t_2 = -13.59\left[\ln\left(\frac{3-v_2}{3-v_1}\right)\right]} \quad (c)$$

b.) By Eq. (a) and the chain rule

$$\frac{dv}{dx}\cdot\frac{dx}{dt} = \frac{dv}{dt}$$

$$F = mv\frac{dv}{dx} \quad (d)$$

By Eqs. (b) and (d):

$$2700(3-v) = mv\frac{dv}{dx}$$

Therefore,

$$\int_0^x \frac{2700}{m}dx = \int_{v_1}^{v_2}\frac{v}{(3-v)}dv$$

Integration yields

$$x = \frac{-m}{2700}\left[v + 3\ln(3-v)\right]_{v_1}^{v_2}$$

or $$\underline{x = -13.59\left[v_2 - v_1 + 3\ln\left(\frac{3-v_2}{3-v_1}\right)\right]} \quad (e)$$

(continued)

16.19 (cont.)

c) For $v_1 = 1.0$ m/s and $v_2 = 2.5$ m/s, Eqs. (c) and (e) yield

$$t = -13.59\left[\ln\left(\frac{3-2.5}{3-1}\right)\right]; \quad \underline{t = 18.84 \text{ s}}$$

$$x = -13.59\left[(2.5-1) + 3\ln\left(\frac{3-2.5}{3-1}\right)\right]; \quad \underline{x = 36.1 \text{ m}}$$

16.20

GIVEN: A particle enters a horizontal repulsion field as shown in Fig. a. It weighs $W_p = 0.05$ N and its speed is $v_p = 10 \text{ m/s}\ \vec{i}$. The force is given by (t in seconds)

$$F = F_x = \frac{0.40}{(t+1)} \text{ (N)} \qquad (a)$$

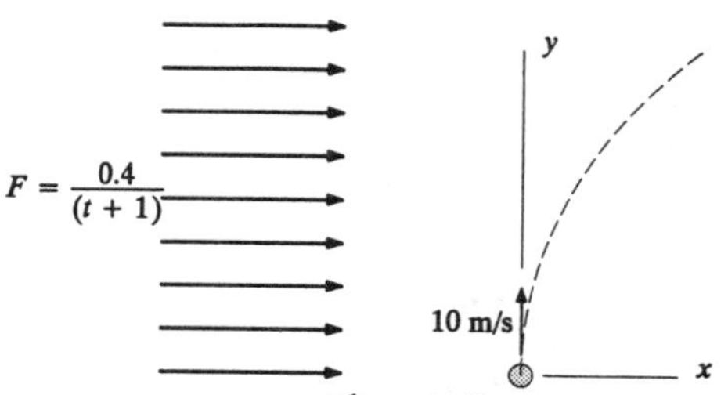

Figure a

FIND:
a) The speed of the particle at time $t = 0.4$ s.
b) The plot of the speed of the particle as a function of time t for the range $0 \le t \le 0.4$ s.
c) The distance the particle travels in the interval $0 \le t \le 0.4$ s.

SOLUTION:
Use the principle of impulse and momentum.

a) Momentum is conserved in the y direction.

$$\therefore \ v_{y_i} = v_{y_f} = 10 \text{ m/s}$$

For the x direction,

$$\int_{t_1}^{t_2} F_x\,dt = m v_{x_f} - m v_{x_i} \qquad (b)$$

Let $t_1 = 0$, $t_2 = 0.4$ s, $v_{x_i} = 0$. Then by Eqs. (a) and (b),

$$\int_0^{0.4} \frac{0.4}{(t+1)}\,dt = m v_{x_f} - 0$$

Integration yields

$$0.4\ln(t+1)\Big|_0^{0.4} = \left(\frac{0.05}{9.81}\right) v_{x_f}$$

or $v_{x_f} = 26.4$ m/s

Hence, $\vec{v}_f = 26.4\vec{i} + 10\vec{j}$ (m/s)

or $\underline{\text{Speed} = |\vec{v}_f| = 28.2 \text{ m/s}}$

b) By $F_x = \dfrac{0.40}{(t+1)} = m\dfrac{dv_x}{dt}$,
where $m = 0.05/9.81 = 0.005097$ kg, we have

$$dv_x = \frac{78.48}{(1+t)}\,dt$$

or $\displaystyle\int_0^{v_x} dv_x = 78.48\int_0^t \frac{dt}{1+t}$

Integration yields

$$v_x = 78.48\ln(1+t) \text{ (m/s)} \qquad (c)$$

Hence, since $v_y = 10$ m/s, the speed of the particle as a function of t is

$$v = \sqrt{v_x^2 + v_y^2} = \sqrt{[78.48\ln(1+t)]^2 + 100} \qquad (d)$$

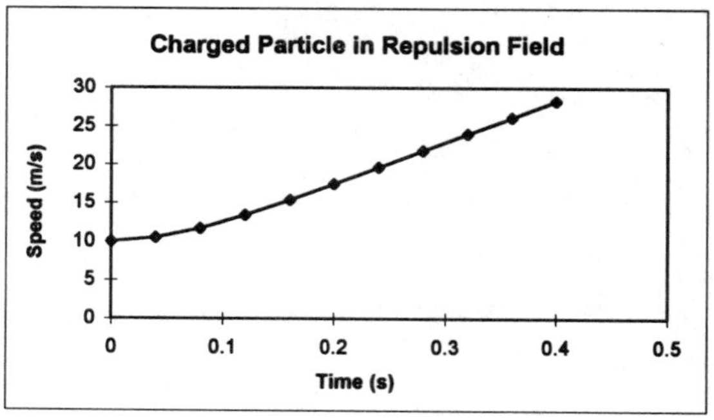

c) Now $v = ds/dt$, and therefore,

$$\int_0^s ds = \int_0^{0.4} v\,dt$$

or $\displaystyle s = \int_0^{0.4} \sqrt{[78.48\ln(1+t)]^2 + 100}\ dt$

By the trapezoidal rule (Appendix E),

$$s = \int_0^{0.40} v\,dt = \frac{0.40}{n}\sum_{i=1}^{n} \frac{(v_{i-1} + v_i)}{2} \qquad (e)$$

where $v_i = \sqrt{[78.48\ln(1+t_i)]^2 + 100}$

Let us take $n = 20$. Then, by Eq. (e),

$$s = \frac{0.020}{2}\sum_{i=1}^{20}(v_{i-1} + v_i) = 0.01\sum_{i=1}^{20}(v_{i-1} + v_i) \qquad (f)$$

where

$v_{1-1} = v_0 = 10$ m/s at $t_0 = 0$

$v_1 = \sqrt{[78.48\ln(1+.02)]^2 + 100} = 10.120$ m/s at $t = 0.02$s

⋮

$v_{10} = \sqrt{[78.48\ln(1+0.2)]^2 + 100} = 17.457$ m/s (g) at $t = 0.20$ s

⋮

$v_{20} = \sqrt{[78.48\ln(1+0.4)]^2 + 100} = 28.236$ m/s at $t_{20} = 0.40$ s

Hence, by Eqs. (f) and (g),

$$\underline{s = 7.16 \text{ m}}$$

This result can also be obtained by calculating the area under the plot of speed vs. time (See plot).

16.21

GIVEN: Example 16.4, where

$k = 1.2\ ft^{-3}$ for $5° \le \alpha \le 15°$

$\rho(\text{sea water}) = 2.0\ lb\cdot s^2/ft^4$

FIND: Chart giving F/L (lb/ft) as a function of x (ft) for $\alpha = 5°, 10°$, and $15°$, for $v = 20$ ft/s and a second chart for $v = 40$ ft/s.

SOLUTION:

From 16.4 (b), $F - B = 2k\rho L x v^2 \tan\alpha$

where $B = \rho g L x^2 \tan\alpha$.

Therefore,

$$F/L = \rho \tan\alpha \left[2kxv^2 + gx^2\right]$$

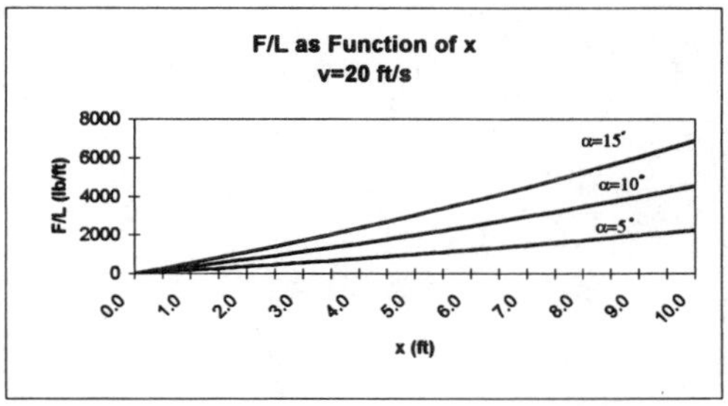

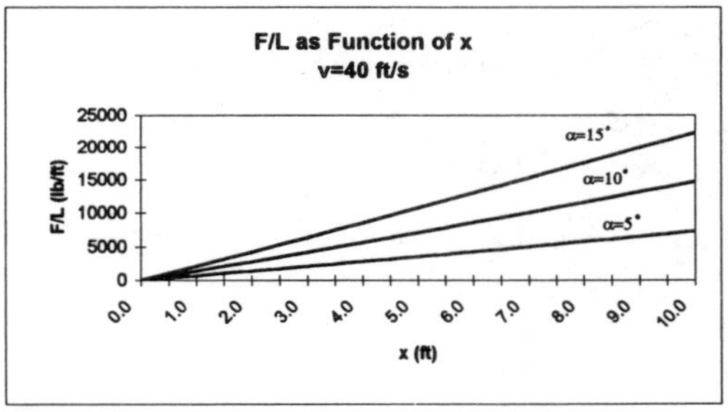

16.22

GIVEN: A pellet with mass $m = 0.50$ grams and a diameter of 4.5 mm is fired from an air rifle. The pellet exits from the muzzle in 0.01 s. Air pressure in the barrel varies initially from 420 kPa to 200 kPa at 0.01 s.

FIND:

a.) The speed v in terms of time t and plot it for $0 \le t \le 0.01$ s.

b.) Determine the exit speed of the pellet.

c.) From the plot, estimate the acceleration of the pellet at $t = 0$ and $t = 0.01$ s. Check the values by your formula for v.

SOLUTION:

a) By Fig. a,

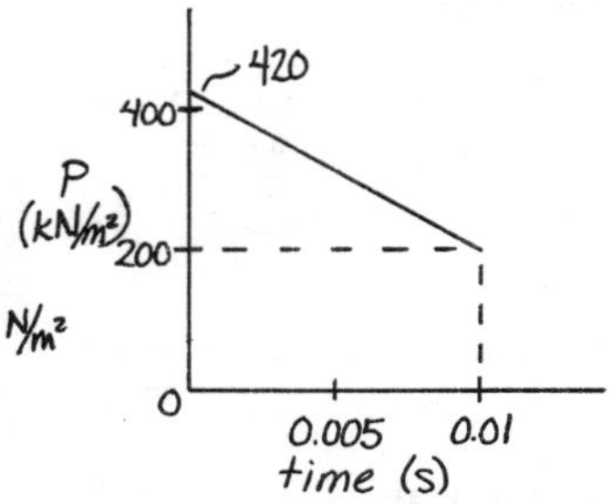

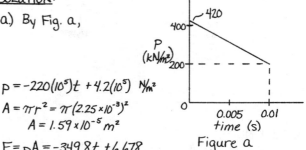

Figure a

$p = -220(10^5)t + 4.2(10^5)\ N/m^2$

$A = \pi r^2 = \pi(2.25\times10^{-3})^2$

$A = 1.59\times10^{-5}\ m^2$

$F = pA = -349.8t + 6.678$

$\int_0^{0.01} F\,dt = mv - 0;\quad m = 5\times10^{-4}$ kg

or $\int_0^{0.01}(-349.8t + 6.678)\,dt = 5(10^{-4})v$

$\therefore\ \underline{\underline{v = -349\,800t^2 + 13\,356t}}$ (m/s)

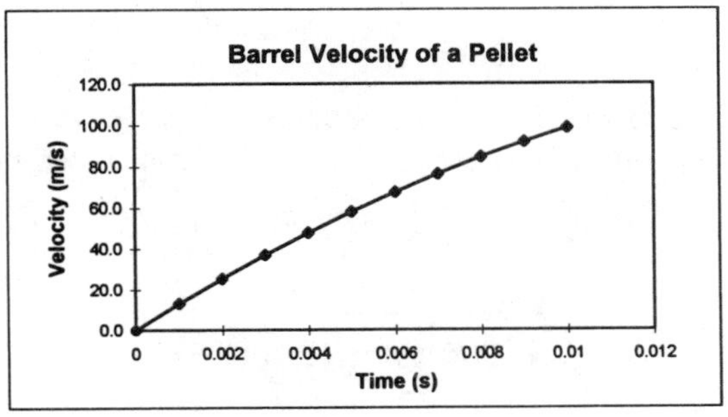

b.) $v_{exit} = (0.04929)/(0.0005)$

$\underline{\underline{v_{exit} = 98.58\ m/s \text{ at } t = 0.01\ s}}$

c.) By the plot, the slopes of the v curve at $t = 0$ and $t = 0.01$ s are $a \approx 13{,}000\ m/s^2$ and $a \approx 6000\ m/s^2$ respectively.

By Eq. (a),

$a(t) = -6\,996\,000t + 13\,356$

$\therefore\ \underline{\underline{a(0) = 13\,356\ m/s^2}}$ and

$\underline{\underline{a(0.01) = 6360\ m/s^2}}$

16.23

GIVEN: Two masses (6 kg and 12 kg) rest on a frictionless table and are connected by an elastic spring. The spring is compressed by moving the masses toward each other. When the spring is released, the 6 kg mass has a velocity of 10 m/s to the left.

FIND: The velocity of the 12 kg. mass.

(continued)

16.23 (cont.)

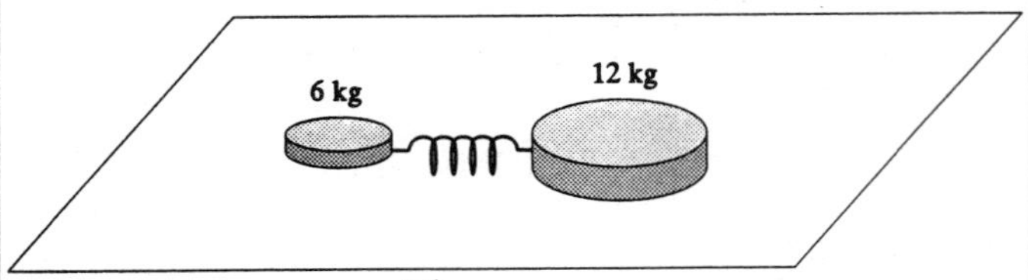

Figure a

SOLUTION:

Since the masses are connected, we can consider them as a system. The only horizontal force acting on the system (the force of the spring) is internal (Fig. b)

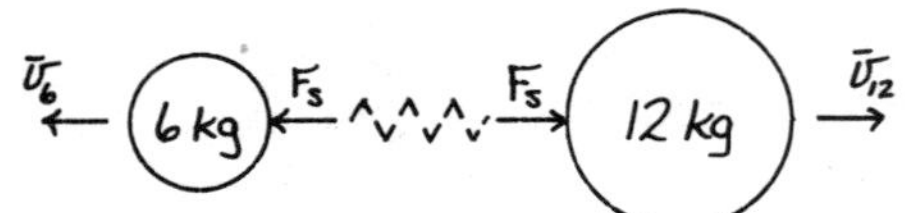

Figure b

Therefore, linear momentum is conserved. Taking velocity positive to the right, we have

$$m_6 v_{6i}^{\,0} + m_{12} v_{12i}^{\,0} = m_6 v_{6f} + m_{12} v_{12f}$$

$$-m_6 v_{6f} = m_{12} v_{12f}$$

$$\therefore\ -(6\ kg)(-10\ m/s) = (12\ kg)(v_{12f})$$

$$\underline{\underline{v_{12f} = 5\ m/s}}$$

16.24

GIVEN: A motionless body subjected to thermal stresses suddenly breaks into two pieces. The pieces fly apart with a separation speed of 32 m/s. The pieces weigh 15 and 25 N.

FIND: Calculate the absolute speed of each piece for the instant that they separate.

SOLUTION:

Schematic:

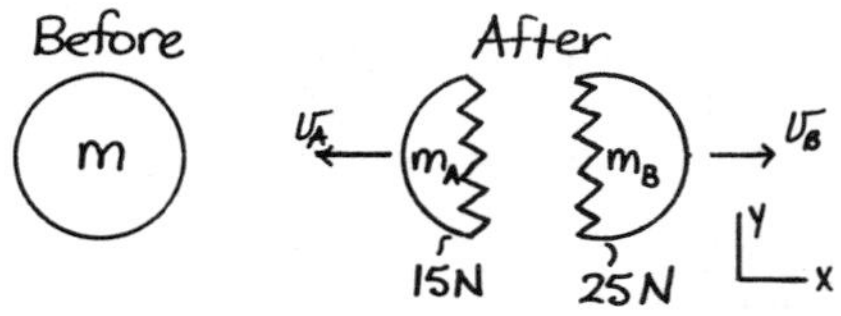

Only internal forces act on the body, so linear momentum is conserved. Thus,

$$(m_A + m_B) v_i^{\,0} = m_A v_A + m_B v_B$$

$$v_A = -\frac{m_B v_B}{m_A} \qquad (a)$$

The separation speed is the relative velocity of the pieces. Hence,

$$v_{B/A} = v_B - v_A \qquad (b)$$

By Eqs. (a) and (b),

$$32 = v_B - \left(\frac{-m_B v_B}{m_A}\right) = v_B\left(1 + \frac{25/9.81}{15/9.81}\right)$$

$$\therefore\ v_B = 12\ m/s$$

By Eq. (b),

$$v_A = v_B - v_{A/B} = 12 - 32 = -20\ m/s$$

Therefore,

$$\underline{\underline{|v_A| = 20\ m/s}} \qquad \underline{\underline{|v_B| = 12\ m/s}}$$

16.25

GIVEN: A motionless block of wood that weighs 22 N rests on a frictionless surface. A revolver bullet that weighs 0.14 N is shot into the side of the block. After the bullet strikes the block, the bullet and block have a speed of 3 m/s (see Fig. a).

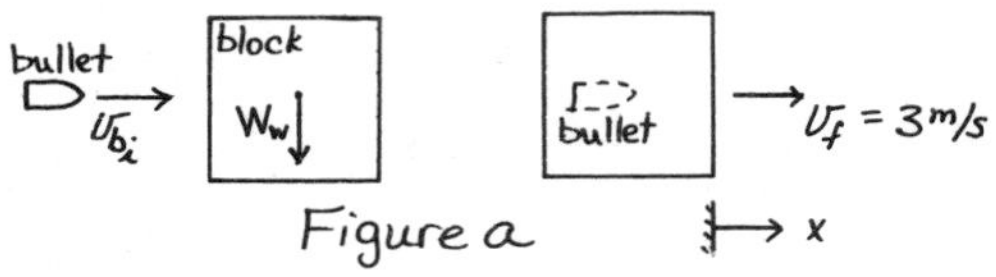

Figure a

FIND:

a.) the speed of the bullet before impact

b.) the initial kinetic energy of the bullet and the final kinetic energy of the block and bullet.

c.) Explain the loss of kinetic energy.

SOLUTION:

a.) Linear momentum is conserved in the x direction; $p_{x_i} = p_{x_f}$. Let b designate the bullet and w designate the wood. Then,

$$m_b v_{b_i} + m_w v_{w_i}^{\,0} = (m_b + m_w) v_f$$

where

$$m_b = \frac{0.14}{9.81} = 0.01427\ kg\ ;\ m_w = \frac{22}{9.81} = 2.24261\ kg$$

Therefore,

$$v_{b_i} = \frac{(m_b + m_w) v_f}{m_b} = \frac{(0.01427 + 2.24261)3}{0.01427},$$

or $\underline{\underline{v_{b_i} = 474.5\ m/s}}$

b.) $T_1 = \frac{1}{2} m_b v_{b_i}^2 = \frac{1}{2}(0.01427)(474.5)^2$

$$\underline{T_1 = 1606.2\ J}$$

$$T_2 = \frac{1}{2}(m_b + m_w) v_f^2 = \frac{1}{2}(0.01427 + 2.24261)(3)^2$$

$$\underline{T_2 = 10.16\ J}$$

c.) The kinetic energy lost is

$$T_1 - T_2 = 1596\ J$$

This energy is lost due to the deformation of the block and the bullet, as well as the heat generated by friction between the bullet and block and sound waves (vibrations)

16.26

GIVEN: A box that weighs 900N is dropped on a railroad freight wagon that weighs 4500N and that rolls along a straight, horizontal track at 1.5 m/s. Neglect friction and air resistance.

FIND:

a) The speed of the wagon after the box has ceased to slide.

b) The initial and final kinetic energy of the system.

c) Explain the loss of kinetic energy.

SOLUTION:

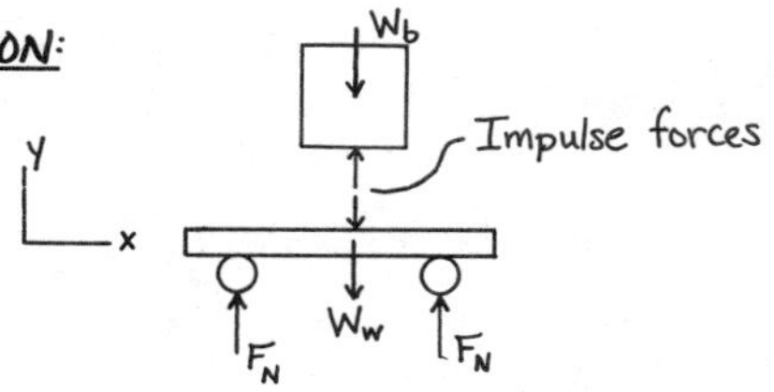

Figure a

a) Since friction and air resistance are negligible, no forces act in the x direction. Therefore, linear momentum in the x direction is conserved. Thus,

$$p_{xi} = p_{xf} \quad (a)$$

where $p_{xi} = m_w v_{wi}$ and $p_{xf} = (m_w + m_b) v_{wf}$

By Eq. (a),

$$m_w v_{wi} = (m_w + m_b) v_{wf}$$

$$\therefore v_{wf} = \frac{m_w v_{wi}}{(m_w + m_b)} \; ; \quad m_w = \frac{4500}{9.81} = 459 \text{ kg}$$

$$m_b = \frac{900}{9.81} = 91.7 \text{ kg}$$

Hence,

$$\underline{\underline{v_{wf} = \frac{(459)(1.5)}{(459 + 91.7)} = 1.250 \text{ m/s}}}$$

b) The kinetic energy initially is

$$T_1 = \tfrac{1}{2} m_w v_{wi}^2 = \tfrac{1}{2}(459)(1.5)^2 = \underline{516 \text{ J}}$$

$$T_2 = \tfrac{1}{2}(m_w + m_b) v_{wf}^2 = \tfrac{1}{2}(459 + 91.7)(1.25)^2 = \underline{430 \text{ J}}$$

c) The loss of KE is $T_1 - T_2 = 86$ J. This loss is due to the box sliding on the wagon and the resulting negative work of friction to bring the box to rest relative to the wagon.

16.27

GIVEN: A submarine that weighs 250 kN including torpedoes, fires two torpedoes from its bow tubes. Each torpedo weighs 1 kN. The submarine's initial speed is 10 m/s, and the speed of the torpedoes upon exit is 20 m/s.

FIND: the momentary reduction of the speed of the submarine.

SOLUTION:

Linear momentum of the system (submarine and torpedoes) is conserved in the firing direction. Therefore,

$$m_s v_{si} = (m_s - 2m_T) v_{sf} + 2 m_T v_T \quad (a)$$

where

$$m_s = \frac{250\,000 \text{ N}}{9.81 \text{ m/s}^2} = 25\,484 \text{ kg}$$

$$m_T = \frac{1000 \text{ N}}{9.81 \text{ m/s}^2} = 101.94 \text{ kg}$$

$v_{si} = 10$ m/s is the initial speed of the sub

$v_T = 20$ m/s is the speed of the torpedoes

v_{sf} is the final speed (at the instant after firing) of the submarine.

Therefore, Eq. (a) yields

$$v_{sf} = \frac{(25\,484)(10) - 2(101.94)(20)}{25\,484 - 2(101.94)} = 9.919 \text{ m/s}$$

Hence, the momentary reduction of the submarine's speed is

$$\Delta v_s = v_{si} - v_{sf} = 10 - 9.919$$

or $\underline{\underline{\Delta v_s = 0.081 \text{ m/s}}}$

16.28

GIVEN: A motionless asteroid splits into two parts of masses m_A and m_B that move apart with velocities of v_A and v_B respectively.

FIND:

a) The speed v_B in terms of m_A, m_B, and v_A

b) Show that the rate of separation of the masses is $\underline{\dfrac{(m_A + m_B) v_A}{m_B}}$

c) Show that the ratio of the kinetic energy T_A of mass m_A to T_B of mass m_B is

$$\frac{T_A}{T_B} = \frac{m_B}{m_A}$$

SOLUTION:

a) In outer space, the asteroid is an isolated system. Therefore, the asteroid experiences only internal forces, and linear momentum is conserved. Therefore, since the asteroid was initially at rest,

$$(m_A + m_B) \cancelto{0}{v_i} = m_A v_A + m_B v_B$$

Hence,

$$\underline{\underline{v_B = -\frac{m_A}{m_B} v_A}}$$

b) The rate of separation is the velocity $v_{A/B}$ of A relative to B (or $v_{B/A}$).

$$\therefore v_{A/B} = v_A - v_B \quad (b)$$

(continued)

16.28 (cont.)

Equations (a) and (b) yield

$$\underline{\underline{v_{A/B} = \frac{(m_B + m_A)}{m_B} v_A}}$$

c) By definition, the kinetic energies of masses A and B are [with Eq. (a)]

$$T_A = \tfrac{1}{2} m_A v_A^2$$

$$T_B = \tfrac{1}{2} m_B v_B^2 = \tfrac{1}{2} \frac{m_A^2 v_A^2}{m_B}$$

Therefore,

$$\underline{\underline{\frac{T_A}{T_B} = \frac{m_B}{m_A}}}$$

16.29

GIVEN: A bullet of mass 0.0087 kg is shot horizontally with a speed of 975 m/s from a hunting rifle of mass 6.80 kg.

FIND:

a) the recoil speed of the rifle at the instant the bullet exits from the muzzle.

b) the average force exerted on the hunter's shoulder if the rifle is stopped in a distance of 50 mm.

c) If the hunter's shoulder is prevented from moving backward by the tree, would the force that the butt exerts on the shoulder be greater, less than, or the same as in part b? Explain.

SOLUTION:

a) The impulsive forces that act on the system (gun and bullet) are internal and cancel, so linear momentum of the system is conserved.

$$\therefore\ (m_B + m_R) v_i^{\,0} = m_B v_B + m_R v_R$$

or

$$m_R v_R = -m_B v_B$$

$$\therefore\ \underline{\underline{v_R = -\frac{(0.0087)(975)}{6.80} = -1.247 \text{ m/s}}}$$

b) To determine the average force exerted on the hunter's shoulder, use $F = ma$, where $a = v\frac{dv}{ds}$ or $a\,ds = v\,dv$.

Therefore,

$$\int_0^{0.050} a\,ds = \int_{1.247}^{0} v\,dv \qquad (a)$$

or $0.050a = -\tfrac{1}{2}(1.247)^2$

Therefore,

$$a = 15.55 \text{ m/s}^2$$

and $\underline{\underline{F_{Av} = (6.80\text{ kg})(15.55\text{ m/s}^2) = 105.7 \text{ N}}}$

c) The force would be much greater, and the acceleration (or deceleration) would be very large as the rifle would be brought to rest in a very short distance s [see Eq. (a), with $s << 0.050$ m]. Thus, by $F = ma$, the force would be greater.

16.30

GIVEN: Two hockey players, A and B, are practicing on ice. Player A weighs $W_A = 780$ N and player B weighs $W_B = 850$ N. Player B is at rest when he is struck by player A traveling at 10 m/s. After the collision, player A has a velocity of 8 m/s at an angle of 35° from his initial direction.

FIND:

a) the velocity of Player B.

b) the loss of kinetic energy as a result of the collision.

SOLUTION:

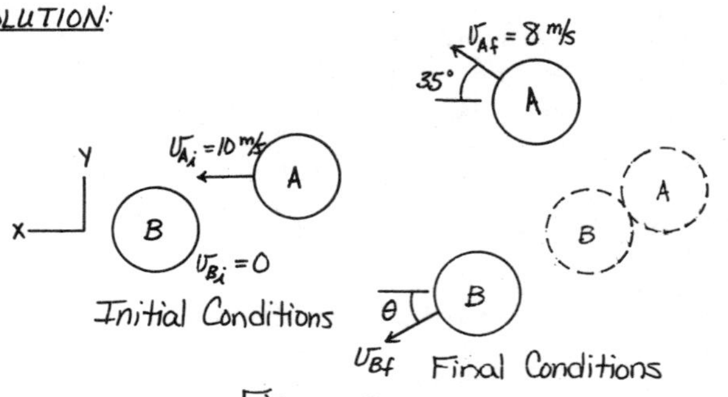

Figure a

a) The initial and final conditions are shown in Fig. a. By the law of conservation of linear momentum

$$m_A \vec{v}_{Ai} + m_B \vec{v}_{Bi} = m_A \vec{v}_{Af} + m_B \vec{v}_{Bf} \qquad (a)$$

where

$$m_A = \frac{780}{9.81} = 79.51 \text{ kg} \quad \text{and} \quad m_B = \frac{850}{9.81} = 86.6 \text{ kg}.$$

Using x, y projections, by Eq. (a) and Fig. a, we have for the x projections

$$m_A v_{Ai} = m_A (v_{Af} \cos 35°) + m_B (v_{Bf} \cos\theta)$$

$$79.51(10) = (79.51)(8)(0.8192) + (86.65)(v_{Bf} \cos\theta)$$

$$\therefore\ v_{Bf} \cos\theta = 3.162 \qquad (b)$$

and for the y projections

$$0 = m_{Af}(v_{Af})(\sin 35°) + m_B (v_{Bf} \sin\theta)$$

$$0 = (79.51)(8)(0.5736) + 86.65 v_{Bf} \sin\theta$$

$$v_{Bf} \sin\theta = -4.227 \qquad (c)$$

Dividing (b) and (c):

$$\frac{v_{Bf} \sin\theta}{v_{Bf} \cos\theta} = \tan\theta = \frac{-4.227}{3.162}$$

(continued)

16.30 (cont.)

or $\theta = 53.20°$ ↺ (See Fig. a)

By Eqs. (b) or (c), $v_{Bf} = 5.28$ m/s. Therefore, after the collision, the velocity of B is

B = 5.28 m/s at a direction of

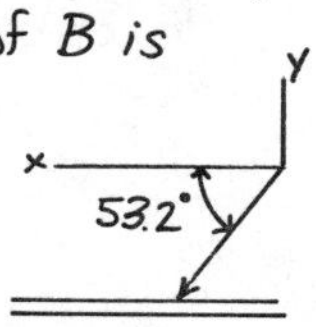

b) Also by Fig. a, with the given data, the kinetic energies T_1 and T_2, before and after the collision, are

$$T_1 = \tfrac{1}{2} m_A v_{Ai}^2 + \tfrac{1}{2} m_B v_{Bi}^2 = \tfrac{1}{2}(79.51)(10)^2 = 3975.5 \text{ J}$$

$$T_2 = \tfrac{1}{2} m_A v_{Af}^2 + \tfrac{1}{2} m_B v_{Bf}^2 = \tfrac{1}{2}(79.51)(8)^2 + \tfrac{1}{2}(86.65)(5.279)^2$$

or $T_2 = 3751.7$ J

Loss of KE $= T_1 - T_2 = 3975.5 - 3751.7$

Loss of KE = 233.8 J

16.31

GIVEN: Figure a. The two hockey players push each other apart on smooth ice with an initial rate of separation of 1.6 m/s. The player on the left weighs $W_L = 800$ N, and the player on the right weighs $W_R = 880$ N.

Figure a

FIND:

a) the absolute speeds of the players.

b) the momentum of each player as they initially separate.

c) the kinetic energy of each player as they initially separate.

d) Explain why the kinetic energies of the players are not the same even though their momenta are equal in magnitude.

SOLUTION:

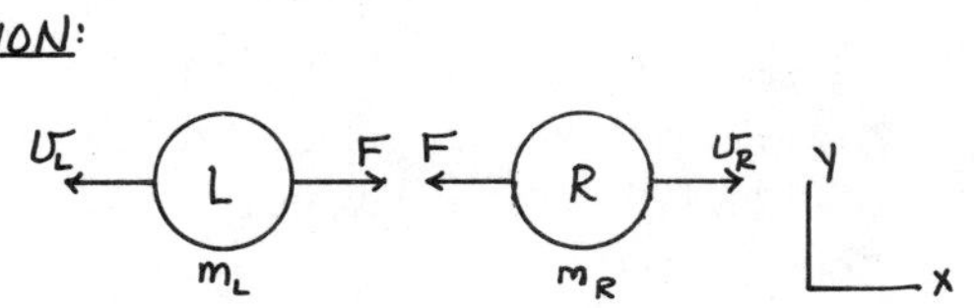

Figure b

a) Figure b is a simplified schematic of the players as they push each other apart.

The initial rate of their separation is the speed of one of the players relative to the other; that is $v_{R/L}$ or $v_{L/R}$. Consider

$$v_{R/L} = v_R + v_L = 1.6 \text{ m/s} \quad \text{(a)}$$

where v_L is the speed of the left player and v_R is the speed of the right player.

The principle of conservation of momentum requires that (see Fig. b) the initial linear momentum equals the final momentum; or by Fig. b, summing momenta in the x direction

$$m_L v_{Li} + m_R v_{Ri} = -m_L v_{Lf} + m_R v_{Rf} \quad \text{(b)}$$

where subscripts i and f denote initial and final values, respectively.

$$m_L = \frac{800 \text{ N}}{9.81 \text{ m/s}^2} = 81.55 \text{ kg}$$

$$m_R = \frac{880 \text{ N}}{9.81 \text{ m/s}^2} = 89.70 \text{ kg}$$

$$v_{Li} = v_{Ri} = 0$$

Therefore, Eq. (b) yields

$$-81.55 v_{Lf} + 89.70 v_{Rf} = 0 \quad \text{(c)}$$

The solution of Eqs. (a) and (c) is

$$v_{Lf} = 0.8381 \text{ m/s}, \quad v_{Rf} = 0.7619 \text{ m/s} \quad \text{(d)}$$

b) The momentum of each player is (relative to the x direction)

$$P_L = -m_L v_{Lf}, \quad P_R = m_R v_{Rf} \quad \text{(e)}$$

Hence, by Eqs. (d) and (e),

$$P_L = -(81.55)(0.8381) = -68.34 \text{ N·s}$$

$$P_R = (89.70)(0.7619) = 68.34 \text{ N·s} \quad \text{(f)}$$

c) The kinetic energies, T_L and T_R, of the players as they separate initially are

$$T_L = \tfrac{1}{2} m_L v_{Lf}^2 = \tfrac{1}{2}(81.55)(0.8381)^2 = 28.64 \text{ J}$$

$$T_R = \tfrac{1}{2} m_R v_{Rf}^2 = \tfrac{1}{2}(89.70)(0.7619)^2 = 26.03 \text{ J}$$

d) Since the momentum is initially zero and linear momentum is conserved, the magnitude of the momenta of the players are equal [see Eqs. (b) and (f)]. However, since then $m_R v_{Rf} = m_L v_{Lf}$ [see Eq. (b)], and therefore $v_{Lf} = \frac{m_R}{m_L} v_{Rf}$. Then $T_L = \tfrac{1}{2} m_L (v_{Lf})^2 = \tfrac{1}{2} \frac{m_R^2}{m_L} (v_{Rf})^2$, or $T_L = \tfrac{1}{2} m_R v_{Rf}^2 \left(\frac{m_R}{m_L}\right) = T_R \left(\frac{m_R}{m_L}\right)$.

Since $m_R > m_L$, $T_L > T_R$.

(continued)

16.31 (cont.)

Alternatively, consider that equal forces F (see Fig. b) act on m_L and m_R. By F=ma, the left player ($m_L < m_R$) is given a larger acceleration and hence a larger velocity at separation.

16.32

GIVEN: Figure a. Residual stresses cause a body to burst into three pieces of masses $m_1 = 2.00$ slugs, $m_2 = 3.00$ slugs, and $m_3 = 5.00$ slugs.

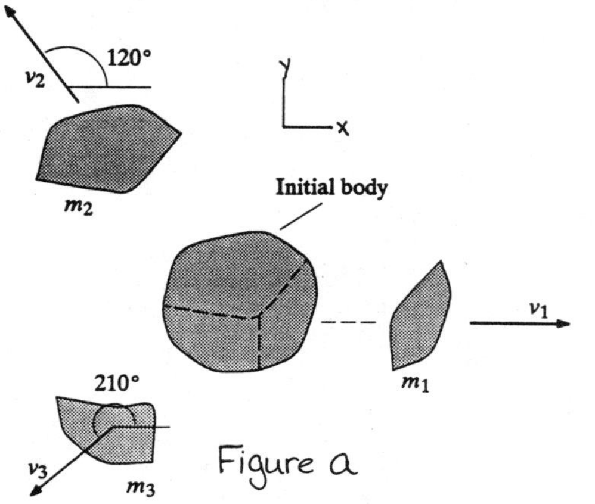

Figure a

The pieces travel in a straight line in the x-y plane with initial speeds v_1, v_2, and v_3, related by the equation

$$2v_1^2 + 3v_2^2 + 5v_3^2 = 540 \text{ ft·lb} \quad (a)$$

FIND: the speeds v_1, v_2, and v_3.

SOLUTION:

Considering the system of three pieces, we see that the forces are internal to the system. Hence, linear momentum is conserved. So, since initially the body is at rest, the x and y projections of linear momentum (Fig. a) are

$$m_1 v_1 + m_2 v_2 \cos 120° + m_3 v_3 \cos 210° = 0$$

$$m_2 v_2 \sin 120° + m_3 v_3 \sin 210° = 0$$

or, with the given values of m_1, m_2, and m_3,

$$2.00v_1 - 1.50v_2 - 4.33v_3 = 0 \quad (b)$$

$$2.60v_2 - 2.50v_3 = 0 \quad (c)$$

The solution of Eqs. (a), (b), and (c) is

$v_1 = 13.57$ ft/s, $v_2 = 4.52$ ft/s, $v_3 = 4.70$ ft/s

16.33

GIVEN: Figure a: A swimmer who weighs 540 N stands in a motionless boat that weighs 1800 N. She dives horizontally out of the boat, and the boat moves away from her. After she dives, her speed relative to the boat is 6 m/s.

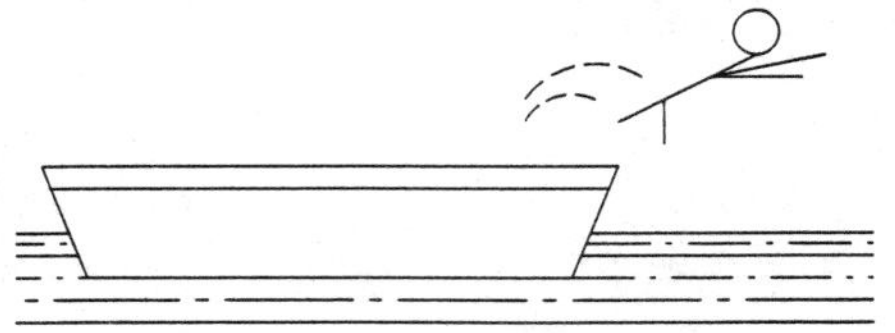

FIND:

a) the speed with which the boat starts to move. Neglect water resistance.

b) whether the speed with which the boat starts to move is higher or lower, taking water resistance into account.

SOLUTION:

a) If water resistance is neglected, linear momentum of the system (boat and swimmer) is conserved. Therefore, since initially the boat and swimmer are at rest, we have (see Fig. b)

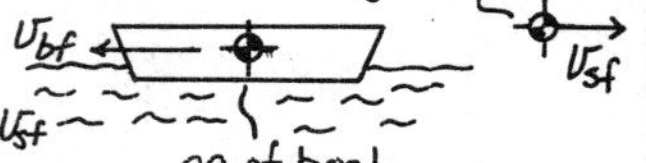

$$m_b v_{bi} + m_s v_{si} = m_b v_{bf} - m_s v_{sf} \quad (v_{bi} = v_{si} = 0)$$

$$m_s v_{sf} - m_b v_{bf} = 0$$

Figure b

where subscripts s and b denote swimmer and boat and subscripts i and f denote initial and final states, respectively. Hence, with $m_s = 540\text{ N}/9.81\text{ m/s}^2$ and $m_B = 1800\text{ N}/9.81\text{ m/s}^2$, we have

$$55.05 v_{sf} - 183.49 v_{bf} = 0 \quad (a)$$

By kinematics, the relative speed $v_{s/b} = 6$ m/s, in terms of v_{sf} and v_{bf}, is (see Fig. b)

$$v_{sf} + v_{bf} = 6 \quad (b)$$

Solving Eqs. (a) and (b), we obtain

$$v_{sf} = 4.615 \text{ m/s} \longrightarrow$$

$$v_{bf} = 1.385 \text{ m/s} \longleftarrow$$

b) If water resistance is included, the speed of the boat would be less, since water resistance (friction) would oppose the boat's motion.

16.34

GIVEN: Two swimmers of mass m_1 and m_2 stand on a motionless raft of mass m_r. When they dive off the raft, it moves away from them. Each swimmer can dive off the raft with a speed V relative to the raft. Neglect water resistance.

FIND:

a) If $m_1 < m_2$ and the swimmers dive off in sequence, which swimmer should dive off first to impart the greater final speed of the raft?

b) Determine whether the swimmers should dive off together or in sequence to impart maximum speed to the raft.

SOLUTION:

Assume the swimmer of mass m_1 dives off first. Since initially the raft and swimmers are motionless, conservation of linear momentum in the horizontal direction yields (see Fig. a), where V_i, the initial speed of the raft is zero,

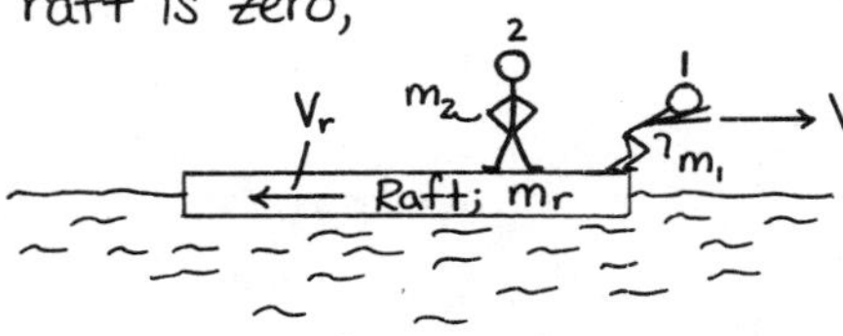

Figure a

$$(m_1 + m_2 + m_r)V_i = 0 = -(m_r + m_2)V_{r_1}^{12} + m_1 V_s \qquad (a)$$

where $V_{r_1}^{12}$ denotes the absolute speed of the raft after swimmer 1 dives (in the sequence m_1 and m_2) and V_s is the absolute speed of diver 1. By kinematics, the speed V is related to V_s and V_i (the initial speed of the raft) by the relation

$$V = V_{s/r} = V_s + V_i = V_s + 0$$

or $\quad V_s = V \qquad (b)$

Substituting Eq. (b) into Eq. (a) and solving for $V_{r_1}^{12}$, we obtain

$$V_{r_1}^{12} = \frac{m_1 V}{m_r + m_2} \qquad (c)$$

Next consider the conditions when swimmer 2 dives. The initial speed of the swimmer 2 and the raft (to the left) is given by Eq. (c). Hence, conservative of linear momentum in the horizontal direction yields

$$-(m_r + m_2)\left(\frac{m_1 V}{m_r + m_2}\right) = -m_r V_{r_2}^{12} + m_2 V_s \qquad (d)$$

where $V_{r_2}^{12}$ denotes the speed of the raft after swimmer 2 dives and where now [with Eq. (c)]

$$V_s = V - V_{r_1}^{12} = \frac{m_r + m_2 - m_1}{m_r + m_2} V \qquad (e)$$

Substitution of Eq. (e) into Eq. (d) yields

$$V_{r_2}^{12} = \left(\frac{m_1}{m_r + m_2} + \frac{m_2}{m_r}\right)V \qquad (f)$$

To obtain the speed of the raft for the sequence m_2, m_1, we need to merely interchange m_1 and m_2 in Eq. (f). Thus,

$$V_{r_2}^{21} = \left(\frac{m_2}{m_r + m_1} + \frac{m_1}{m_r}\right)V \qquad (g)$$

Since $m_2 > m_1$, by Eqs. (f) and (g)

$$\frac{m_1}{m_r + m_2} + \frac{m_2}{m_r} > \frac{m_2}{m_r + m_1} + \frac{m_1}{m_r}$$

Hence,

$$V_{r_2}^{12} > V_{r_2}^{21}$$

or the greater final speed of the raft [given by Eq. (f)] is obtained when the swimmer of mass $m_1 < m_2$ dives off first.

b) If the swimmers dive together from the motionless raft, conservation of linear momentum in the horizontal direction yields, since initially $V_r = V_i = 0$,

$$(m_1 + m_2 + m_r)V_i = 0 = -m_r V_r + (m_1 + m_2)V_s \qquad (h)$$

where V_r is the absolute speed of the raft after the swimmers dive off together with absolute speed V_s. Now by kinematics,

$$V = V_{s/r} = V_s + V_i = V_s + 0$$

or $\quad V_s = V \qquad (i)$

Substitution of Eq. (i) into Eq. (h) and solution for V_r yields

$$V_r = \frac{(m_1 + m_2)V}{m_r} = \left(\frac{m_1}{m_r} + \frac{m_2}{m_r}\right)V \qquad (j)$$

Now, by Eqs. (f) and (g)

$$\frac{m_1}{m_r} + \frac{m_2}{m_r} > \frac{m_1}{m_r + m_2} + \frac{m_2}{m_r}$$

Therefore, the maximum speed of the raft is attained if the two swimmers dive off together.

16.35

GIVEN: A towel weighs ½ lb. In snapping the towel, a boy exerts a 5-lb horizontal force at one corner of the towel.

FIND: Neglecting air resistance, calculate the horizontal acceleration of the center of mass of the towel.

SOLUTION:

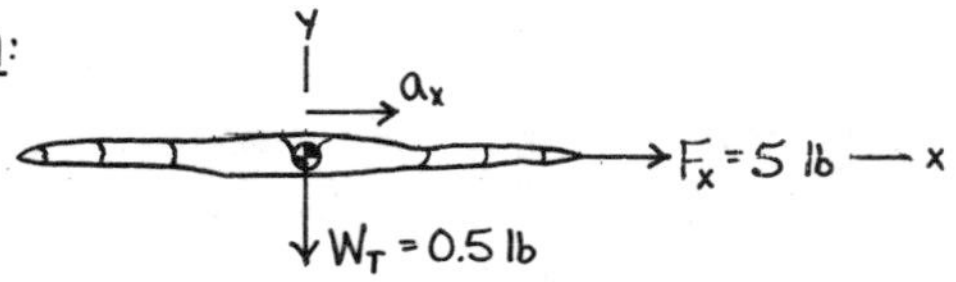

Consider the free-body diagram of the towel (Fig. a). By Fig. a,

$$F_x = m\frac{dv_x}{dt} = ma_x$$

or $5 = \frac{0.5}{32.2}a_x$

$\therefore$ $\underline{\underline{a_x = 322 \text{ ft/s}^2}}$

16.36

GIVEN: A high jumper who weighs 720 N exerts a 2250 N downward force on the ground when he jumps.

FIND: Determine the vertical acceleration of his center of mass.

SOLUTION:

Consider the free-body diagram of the jumper (Fig. a).

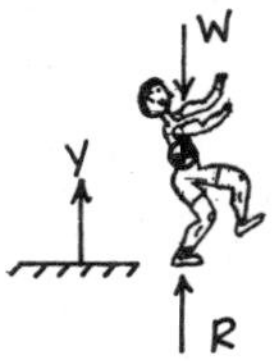

Figure a

By Fig. a, where R denotes the reaction of the ground and W is the weight of the jumper,

$$\Sigma F_y = R - W = ma_y$$

or $2250 - 720 = \frac{720}{9.81}a_y$

so, $\underline{\underline{a_y = 20.85 \text{ m/s}^2}}$

16.37

GIVEN: A car that weighs 14 kN accelerates horizontally at 2 m/s².

FIND:

a) the net horizontal force on the car

b) Draw a free-body diagram showing the horizontal forces that act on the car.

SOLUTION:

a) $\Sigma F_x = F_{net} = m\frac{dv_x}{dt} = ma_x$

$\therefore$ $\underline{\underline{F_{net} = \frac{14(10^3)}{9.81}(2) = 2854 \text{ N}}}$

b) Assume a front-wheel drive (Fig. a)

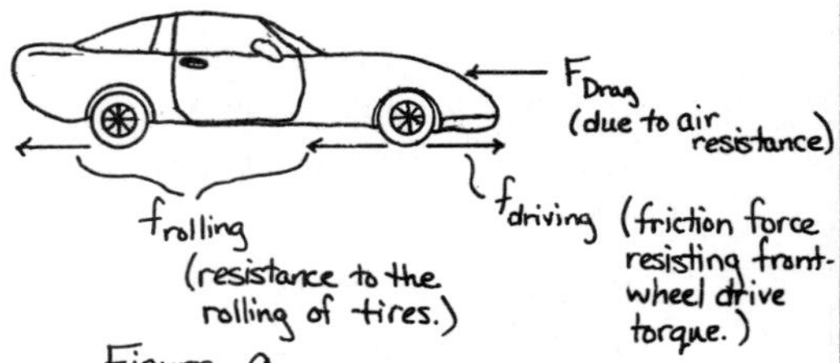

Figure a

By Fig. a,

$$F_{net} = F_{drag} + f_{driving} - f_{rolling}$$

16.38

GIVEN: Figure a, with the sign ($W_s = 1440$ N) subjected to forces shown as it is being hung.

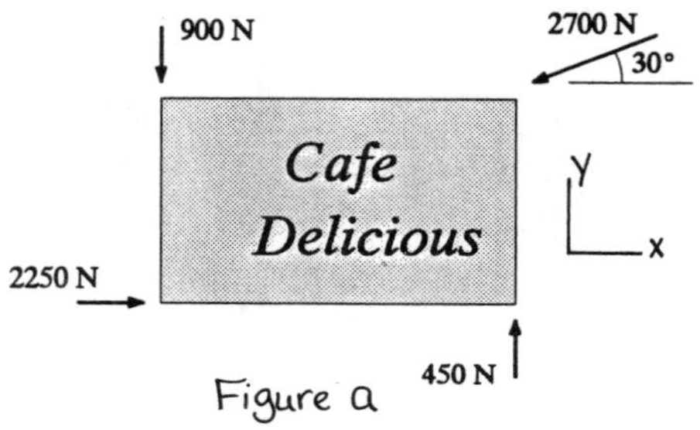

Figure a

FIND: the x and y components of acceleration of the center of mass of the sign.

SOLUTION:

Consider the free-body diagram of the sign (Fig. b)

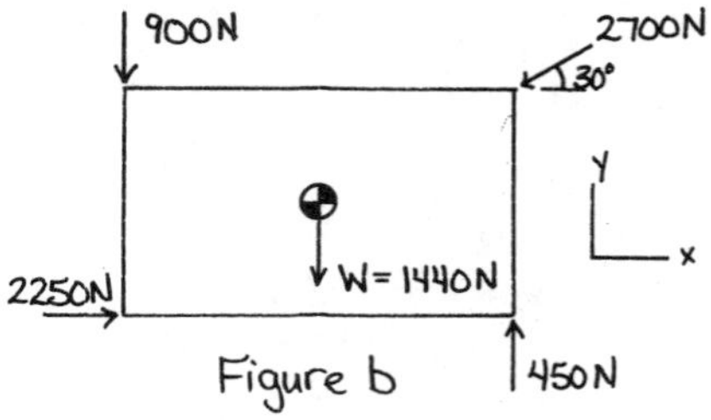

Figure b

$$m_s = \frac{1440}{9.81} = 146.8 \text{ kg}$$

By Fig. b,

$$F_x = m\frac{dv_x}{dt} = ma_x \quad ; \quad F_y = m\frac{dv_y}{dt} = ma_y$$

Therefore,

$$\Sigma F_x = 2250 - 2700\cos 30° = 146.79a_x$$

$$\Sigma F_y = -900 + 450 - 2700\sin 30 - 1440 = 146.79a_y$$

$\underline{\underline{a_x = -0.601 \text{ m/s}^2}}$

$\underline{\underline{a_y = -22.1 \text{ m/s}^2}}$

16.39

GIVEN: A bucket that weighs $W_B = 40$ lb is subjected to two additional forces as shown in Fig. a.

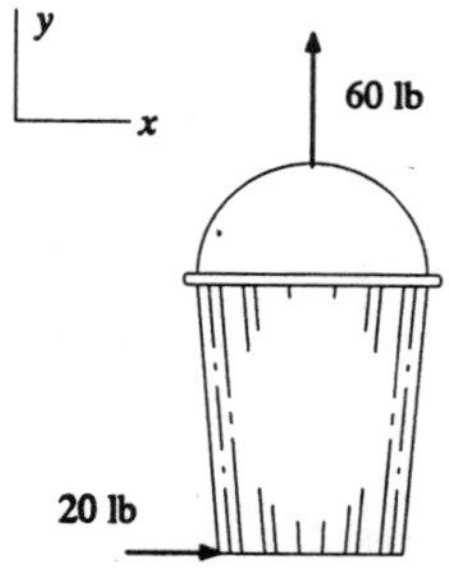

Figure a

FIND: the horizontal and vertical components of acceleration of the center of mass of the system.

SOLUTION:
The free-body diagram of the bucket is shown in Fig. b.

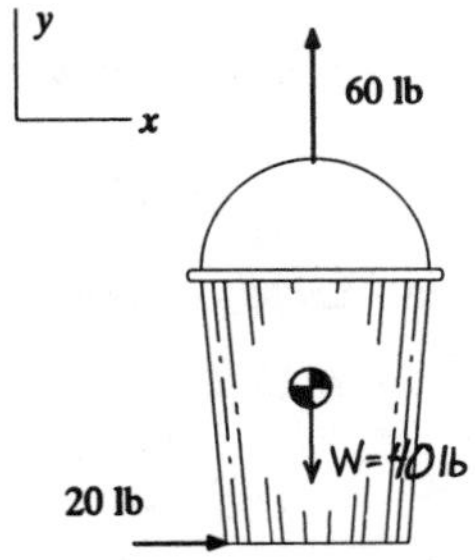

Figure b

By Fig. b and Newton's second law,

$$\Sigma F_x = m\frac{dv_x}{dt} = ma_x$$

$$\Sigma F_y = m\frac{dv_y}{dt} = ma_y$$

where $m = W_B/g = 40\text{lb}/32.2\text{ ft/s}^2$. Hence, by Fig. b,

$$\Sigma F_x = 20 = \left(\frac{40}{32.2}\right)a_x$$

or $\underline{\underline{a_x = 16.10 \text{ ft/s}^2}} \rightarrow$

$$\Sigma F_y = 60 - 40 = \left(\frac{40}{32.2}\right)a_y$$

or $\underline{\underline{a_y = 16.10 \text{ ft/s}^2}} \uparrow$

16.40

GIVEN: A rod that weighs 18N falls freely in a vacuum and strikes one end obliquely against a frictionless surface (see Fig. a)

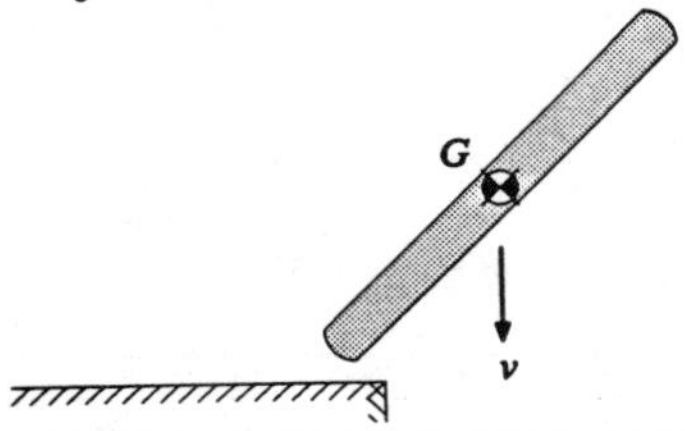

Figure a

The impact reduces the acceleration of the center of mass of the rod to 2.4 m/s².

FIND: the average vertical force that the surface exerts on the rod during impact.

SOLUTION:
The free-body diagram of the rod during impact is shown in Fig. b.

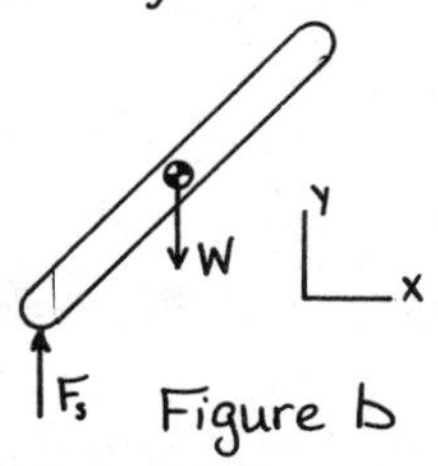

Figure b

By Newton's second law,

$$\Sigma F_y = m\frac{dv_y}{dt} = ma_y \qquad \text{(a)}$$

By Eq. (a) and Fig. b,

$$\Sigma F_y = F_s - W = ma_y \qquad \text{(b)}$$

where $m = 18\text{ N}/9.81\text{ m/s}^2 = 1.835$ kg, $W = 18$N, and $a_y = -2.4$ m/s².

Therefore, by Eq. (b),

$$\underline{\underline{F_s = 18 - (1.835)(2.4) = 13.60 \text{ N}}}$$

16.41

GIVEN: Three particles have the masses (kg) and (x,y) coordinates (meters) shown in Fig. a.

FIND: the coordinates of the center of mass of the system.

Figure a

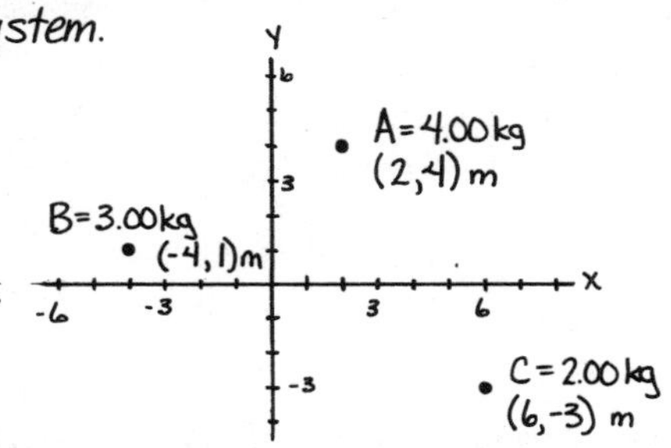

SOLUTION:
By definition, the center of mass coordinates are

$$\bar{x} = \frac{1}{m}\Sigma m_i x_i = \frac{m_A x_A + m_B x_B + m_C x_C}{m}$$

$$\bar{y} = \frac{1}{m}\Sigma m_i y_i = \frac{m_A y_A + m_B y_B + m_C y_C}{m} \qquad \text{(a)}$$

where $m = m_A + m_B + m_C$.

Therefore, by Eq. (a) and the given data (Fig. a)

$$\bar{x} = \frac{4(2) + 3(-4) + 2(6)}{9}$$

$$\bar{y} = \frac{4(4) + 3(1) + 2(-3)}{9}$$

or $\underline{\underline{\bar{x} = 0.889 \text{ m}}}$, $\underline{\underline{\bar{y} = 1.444 \text{ m}}}$

16.42

GIVEN: Three particles have the following masses and (x,y) components of velocity:

A: 4.00 kg (+6.00 m/s, −3.00 m/s);

B: 3.00 kg (+2.00 m/s, +4.00 m/s);

C: 2.00 kg (−4.00 m/s, +1.00 m/s).

FIND: the velocity of the center of mass of the system.

SOLUTION:

By definition,

$$m\bar{v}_x = \sum m_i v_{xi} \; ; \quad m\bar{v}_y = \sum m_i v_{yi}$$

Therefore,

$$m\bar{v}_x = m_A v_{Ax} + m_B v_{Bx} + m_C v_{Cx} \qquad (a)$$
$$m\bar{v}_y = m_A v_{Ay} + m_B v_{By} + m_C v_{Cy}$$

With Eqs. (a) and the given data,

$$\bar{v}_x = \frac{4(6) + 3(2) + 2(-4)}{9.00}$$

$$\bar{v}_y = \frac{4(-3) + 3(4) + 2(1)}{9.00}$$

or $\bar{v}_x = 2.44 \text{ m/s}$, $\bar{v}_y = 0.222 \text{ m/s}$

16.43

GIVEN: A 1200-kg car moves along a straight highway at a speed of 72 km/h. Another car, of mass 1500-kg and speed 84.96 km/h, is 36 m ahead of the first car (Fig. a).

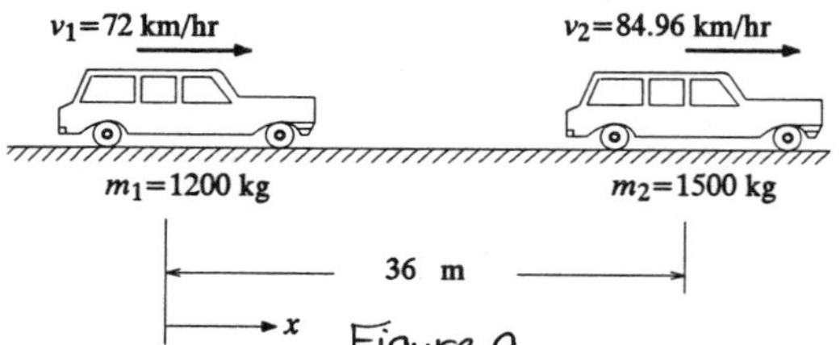

Figure a

FIND:

a) the center of mass of the two cars as a function of time t measured in meters from the initial position of the 1200-kg car.

b) the center of mass of the two cars at $t = 2s$, measured in meters from the initial position of the 1200-kg car.

c) the total momentum of the two cars.

d) the speed v_G of the center of mass of the two cars.

e) the total momentum of the two cars in terms of v_G, and verify the result of part c.

SOLUTION:

A schematic diagram of the initial positions and the speeds of the cars is shown in Fig. b.

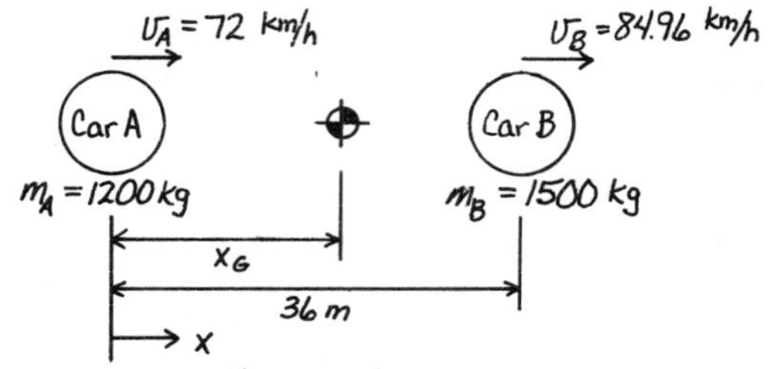

Figure b

In terms of meters per second,

$$v_A = 72 \text{ km/h} = 20 \text{ m/s} \quad \text{and} \quad v_B = 84.96 \text{ km/h} = 23.6 \text{ m/s}$$

a) By definition, the center of mass of the two cars is

$$x_G = \frac{m_A x_A + m_B x_B}{m} \qquad (a)$$

where $m = m_A + m_B$.

Since distance is measured from car A, as a function of time t,

$$x_A = 20t \text{ (m)} , \quad x_B = 36 + 23.6t \text{ (m)} \qquad (b)$$

With the given data and Eqs. (a) and (b),

$$x_G = \frac{(1200)(20t) + 1500(36 + 23.6t)}{2700}$$

or $x_G = 20 + 22t$ (c)

b) By Eq. (c), $\bar{x}_G$ at $t = 2s$ is

$$x_G = 20 + 22(2) = 64 \text{ m}$$

c) By definition, the total momentum of the two cars is

$$Q = m_A v_A + m_B v_B \qquad (d)$$

By Eq. (d), with the given data,

$$Q = (1200)(20) + (1500)(23.6)$$

or $Q = 59.4 \text{ kN·s}$

d) By Eq. (c),

$$v_G = \frac{dx_G}{dt} = 22 \text{ m/s} \qquad (e)$$

e) By definition,

$$m v_G = m_A v_A + m_B v_B \qquad (f)$$

and by Eqs. (d), (e), and (f),

$$Q = m v_G = 2700(22)$$

or $Q = 59.4 \text{ kN·s}$

This result agrees with part c.

16.44

GIVEN: Figure a, in which Rod OC rotates about point O. The sliders A and B generate the coordinates of an ellipse. For the configuration shown, $\overline{OC} = L$, $\overline{AB} = 2L$, and the masses of sliders A and B are equal. That is, $m_A = m_B = m_S$. Assume that the masses of the rods are negligible.

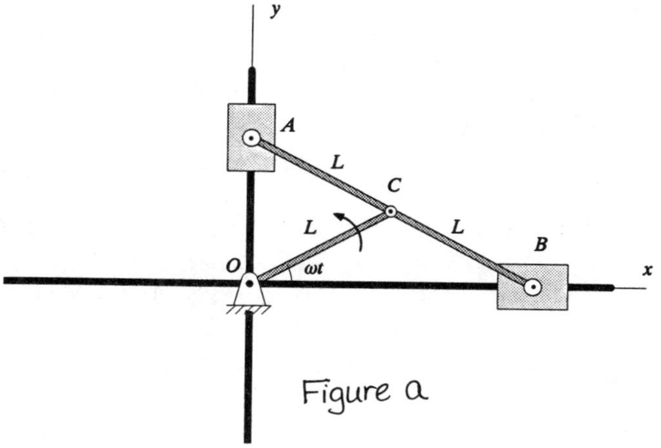

Figure a

FIND: the path of the center of mass of the sliders A and B.

SOLUTION:

Since the masses of A and B are equal, the center of mass is located midway between them, at point C.

By definition, the position vector of the mass center, relative to point O, is

$$\vec{r} = \frac{m_A \vec{r}_A + m_B \vec{r}_B}{m_A + m_B} = \frac{\vec{r}_A + \vec{r}_B}{2} \qquad (a)$$

where, by Fig. a,

$$\vec{r}_A = (2L \sin \omega t)\,\hat{j} \qquad (b)$$
$$\vec{r}_B = (2L \cos \omega t)\,\hat{i}$$

Hence, by Eqs. (a) and (b),

$$\vec{r} = L\left[(\cos \omega t)\hat{i} + (\sin \omega t)\hat{j}\right] \quad (c)$$

Equation (c) is the equation of a circle with origin O and radius L; that is,

$$x = L\cos \omega t\,, \quad y = L \sin \omega t$$
$$r = \sqrt{x^2 + y^2} = L\sqrt{\cos^2 \omega t + \sin^2 \omega t} = L$$

Alternatively, since C is the end of rod OC (Fig. a), which has length L and rotates about O, the path of C (the center of mass) is a circle of radius L and origin O.

16.45

GIVEN: In problem 16.44, let the mass of the uniform rod OC be m_r and that of the uniform rod AB be $2m_r$.

FIND: the path of the center of mass of the ellipsograph in terms of m_r, m_s, L, and ωt.

SOLUTION:

By definition, the position vector of the mass center, relative to point O, is (with Fig. a of Problem 16.44 Solution)

$$\bar{r} = \frac{m_A \bar{r}_A + m_B \bar{r}_B + m_{OC}\bar{r}_{BC} + m_{AB}\bar{r}_{AB}}{m_A + m_B + m_{OC} + m_{AB}} \qquad (a)$$

where,

$$m_A = m_B = m_s\,,$$
$$m_{OC} = m_r\,, \quad m_{AB} = 2m_r\,,$$
$$\bar{r}_A = (2L)(\sin \omega t)\,\hat{j}$$
$$\bar{r}_B = (2L)(\cos \omega t)\,\hat{i}$$
$$\bar{r}_{OC} = \tfrac{1}{2} L\left[(\cos \omega t)\hat{i} + (\sin \omega t)\hat{j}\right]$$
$$\bar{r}_{AB} = L\left[(\cos \omega t)\hat{i} + (\sin \omega t)\hat{j}\right]$$

Hence, by Eqs. (a) and (b),

$$\bar{r} = \left(\frac{4m_s + 5m_r}{4m_s + 6m_r}\right) L\left[(\cos \omega t)\hat{i} + (\sin \omega t)\hat{j}\right]$$

Note that for $m_r = 0$, this result reduces to that of Problem 16.44.

16.46

GIVEN: Two asteroids of masses m_A and m_B move with velocities $\bar{v}_A$ and $\bar{v}_B$, respectively, with respect to an astronaut in a space ship fixed in a Newtonian reference frame.

FIND: Show that the combined kinetic energies of the asteroids, as measured by the astronaut is

$$T = \tfrac{1}{2} m \bar{v}_G^{\,2} + \tfrac{1}{2} m_A \bar{v}_{A/G}^{\,2} + \tfrac{1}{2} m_B \bar{v}_{B/G}^{\,2}$$

where $m = m_A + m_B$, $\bar{v}_G$ is the velocity of the mass center of the asteroids, and $\bar{v}_{A/G}$, $\bar{v}_{B/G}$ are the velocities of masses m_A and m_B, respectively, relative to the center of mass of the asteroids.

SOLUTION:

By definition, the velocity of the center of mass of the asteroids is

$$\bar{v}_G = \frac{m_A \bar{v}_A + m_B \bar{v}_B}{m_A + m_B} \qquad (a)$$

Select center of mass axes. To transfer Eq. (a) to the center of mass axes, we must subtract $\bar{v}_G$ from all the velocities in Eq. (a).

(continued)

16.46 (cont.)

Thus,

$$\bar{v}_G' - \bar{v}_G = \frac{m_A(\bar{v}_A - \bar{v}_G) + m_B(\bar{v}_B - \bar{v}_G)}{m_A + m_B}$$

or $\quad m_A \bar{v}_{A/G} + m_B \bar{v}_{B/G} = 0 \qquad$ (b)

where

$$\bar{v}_{A/G} = \bar{v}_A - \bar{v}_G\,, \quad \bar{v}_{B/G} = \bar{v}_B - \bar{v}_G \qquad (c)$$

By Eq. (c), we have

$$\bar{v}_A = \bar{v}_{A/G} + \bar{v}_G\,, \quad \bar{v}_B = \bar{v}_{B/G} + \bar{v}_G \qquad (d)$$

Now the combined kinetic energies as measured by the astronaut is given by, with Eqs. (d),

$$T = \tfrac{1}{2} m_A \bar{v}_A^2 + \tfrac{1}{2} m_B \bar{v}_B^2 \qquad (e)$$

$$= \tfrac{1}{2} m_A (\bar{v}_{A/G} + \bar{v}_G)\cdot(\bar{v}_{A/G} + \bar{v}_G) + \tfrac{1}{2} m_B (\bar{v}_{B/G} + \bar{v}_G)\cdot(\bar{v}_{B/G} + \bar{v}_G)$$

or expanding

$$T = \tfrac{1}{2}(m_A + m_B)\bar{v}_G^2 + \tfrac{1}{2} m_A \bar{v}_{A/G}^2 + \tfrac{1}{2} m_B \bar{v}_{B/G}^2 + (m_A \bar{v}_{A/G} + m_B \bar{v}_{B/G})\bar{v}_G \qquad (f)$$

However, the last term is zero, by Eq. (b), and $m_A + m_B = m$.

Hence, Eq. (f) reduces to

$$T = \tfrac{1}{2} m \bar{v}_G^2 + \tfrac{1}{2} m_A \bar{v}_{A/G}^2 + \tfrac{1}{2} m_B \bar{v}_{B/G}^2$$

as required.

16.47

GIVEN: Figure a. The boom AB is rotated through an angle θ.

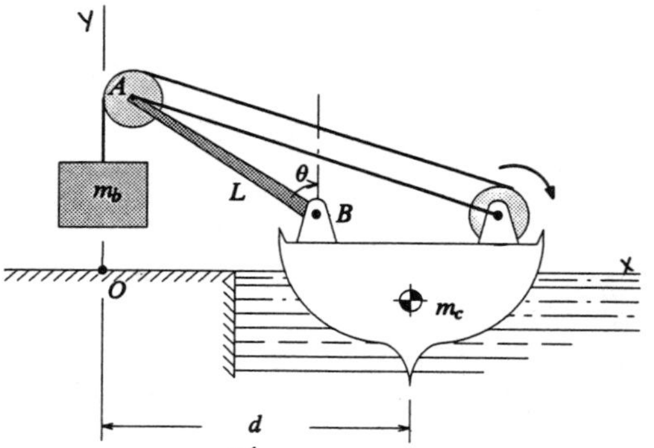

Figure a

FIND: a formula for the distance the floating crane moves in terms of m_B, m_c, L, and θ.

SOLUTION:

Initially, the center of mass of the system is located at

$$\bar{x} = \frac{(m_b)(0) + (m_c)(d)}{m_c + m_b}$$

or $\quad \bar{x} = \dfrac{m_c(d)}{m_c + m_b} \qquad$ (a)

Since no horizontal force acts on the system, the mass center of the system does not move as the boom rotates through θ. Therefore, the crane moves to the left and the load moves to the right.

$$\therefore\ \bar{x} = \frac{m_b(L\sin\theta - x_c) + m_c(d - x_c)}{m_c + m_b} \qquad (b)$$

Equating Eqs. (a) and (b), we obtain

$$x_c = \frac{m_b L \sin\theta}{m_b + m_c} \qquad (c)$$

where x_c is the distance that the crane moves to the left.

16.48

GIVEN: For the system shown in Fig. P16.47, let $m_b = 62$ slugs, $m_c = 620$ slugs and $L = 24$ ft. The boom rotates 35° and ends up 5° to the right of vertical.

FIND: the distance that the floating crane moves.

SOLUTION:

The distance that the box moves to the right relative to the crane is $L(\sin 30° + \sin 5°)$, while the crane moves a distance x_c to the left.

Initially, the center of mass is located at $\bar{x}$ given by

$$(m_b + m_c)\bar{x} = m_b(0) + m_c(d)$$

or $\quad (m_b + m_c)\bar{x} = m_c(d) \qquad$ (a)

Finally, it is located again at $\bar{x}$ since there are no horizontal forces acting. Therefore,

$$(m_b + m_c)\bar{x} = m_b\left[L(\sin 30° + \sin 5°) - x_c\right] + (m_c)(d - x_c) \qquad (b)$$

Equating Eqs. (a) and (b), we find

$$x_c = \frac{(m_b)\left[L(\sin 30° + \sin 5°)\right]}{m_b + m_c}$$

or $\quad x_c = \dfrac{(62)\left[24(\sin 30° + \sin 5°)\right]}{62 + 620} = 1.28 \text{ ft}$

16.49

GIVEN: Figure a, where the mass m_1 moves down the plane so that its center of mass is lowered a distance h.

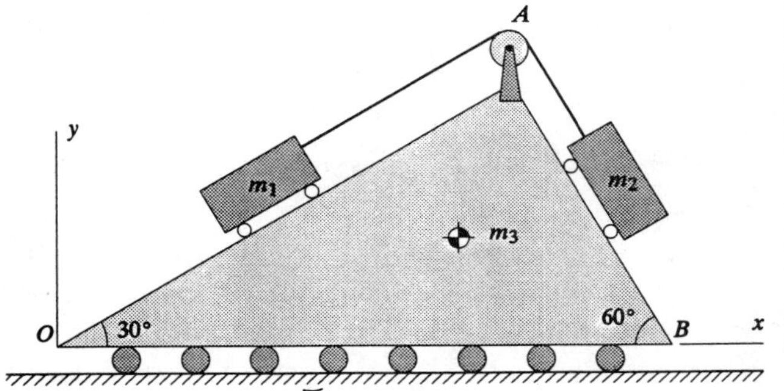

Figure a

FIND: a formula for the horizontal distance the prism moves.

SOLUTION:

Since no horizontal forces exist, the mass center of the system does not move. Also, the mass m_1 moves down the plane a distance $\delta = h/(\sin 30°)$. Therefore, relative to the prism it moves horizontally a distance $\delta(\cos 30°) = h/(\tan 30°)$. At the same time, the prism moves horizontally to the right a distance Δx_3. Therefore, the absolute horizontal displacement of m_1 is

$$\Delta x_1 = \frac{h}{\tan 30°} - \Delta x_3 \qquad (a)$$

Similarly, mass m_2 moves up the prism a distance δ or a horizontal distance to the left relative to the prism $\delta \cos 60° = \frac{h}{\sin 30}(\cos 60) = h$; or an absolute distance to the left of

$$\Delta x_2 = h - \Delta x_3 \qquad (b)$$

Since the mass center of the system does not move $\Delta\bar{x} = 0$; therefore,

$$(m_1 + m_2 + m_3)\Delta\bar{x} = -m_1(\Delta x_1) - m_2(\Delta x_2) + m_3(\Delta x_3) = 0 \quad (c)$$

$\therefore$ By Eqs. (a), (b), and (c),

$$m_3(\Delta x_3) = m_1(\Delta x_1) + m_2(\Delta x_2)$$
$$= m_1\left[\frac{h}{\tan 30°} - \Delta x_3\right] + m_2[h - \Delta x_3]$$

or

$$\Delta x_3 = \frac{(m_1 \cot 30° + m_2)h}{m_1 + m_2 + m_3}$$

16.50

GIVEN: Figure a, with $h = 1.5$ m, $m_3 = 4m_1 = 8m_2$, and m_1 moves down the plane so that its mass center is lowered a distance h.

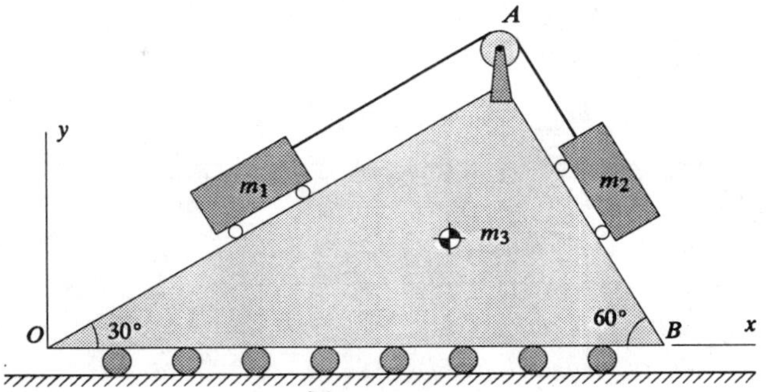

Figure a

FIND: the horizontal displacement of the prism.

SOLUTION:

The formula for the horizontal displacement of the prism is derived in Problem 16.49. It is

$$\Delta x_3 = \frac{(m_1 \cot 30° + m_2)h}{m_1 + m_2 + m_3} \qquad (a)$$

With the given data, Eq. (a) yields

$$\Delta x_3 = \frac{m_3\left[\frac{\cot 30°}{4} + \frac{1}{8}\right](1.5)}{m_3(1/4 + 1/8 + 1)}$$

or

$$\Delta x_3 = 0.609 \text{ m}$$

16.51

GIVEN: An elastic sphere of mass m travels along a straight line in a horizontal plane (x, y), with an initial speed v_i (Fig. a). It strikes a large smooth wall at an initial angle θ_i.

FIND: the final speed v_f of the sphere and its rebound angle θ_f.

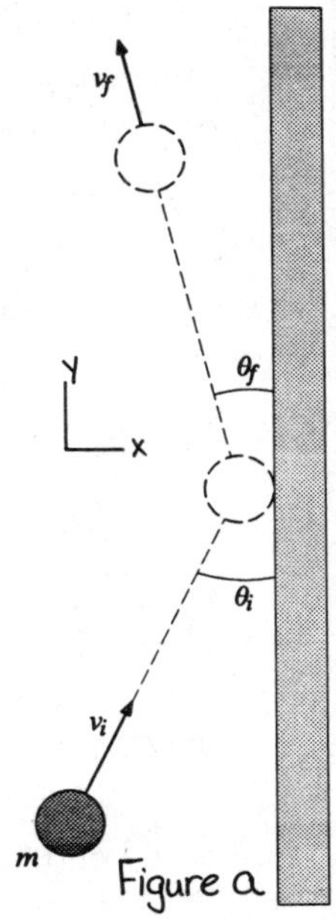

Figure a

SOLUTION:

Since the collision is elastic, the kinetic energy of the sphere is conserved. Thus,

$$\tfrac{1}{2} m v_i^2 = \tfrac{1}{2} m v_f^2$$

or $\quad v_i = v_f \qquad (a)$

Since the wall is smooth, there is no force exerted on the sphere in the y direction (Fig. a). Hence, linear momentum is conserved in the y direction.

(continued)

16.51 (cont.)

Therefore, by Fig. a,

$$mv_{yi} = mv_{yf}$$

or $\quad mv_i \cos\theta_i = mv_f \cos\theta_f \qquad$ (b)

By Eqs. (a) and (b),

$$\underline{\underline{v_f = v_i}} \quad \text{and} \quad \underline{\underline{\theta_f = \theta_i}}$$

16.52

GIVEN: Two identical, elastic balls of mass m move toward each other along a direct-collision course. One ball moves with a velocity v_1 to the right and the other ball moves with a velocity v_2 to the left.

FIND: the velocities of the balls after they collide.

SOLUTION:

Assume that the balls travel along the x axis (Fig. a).

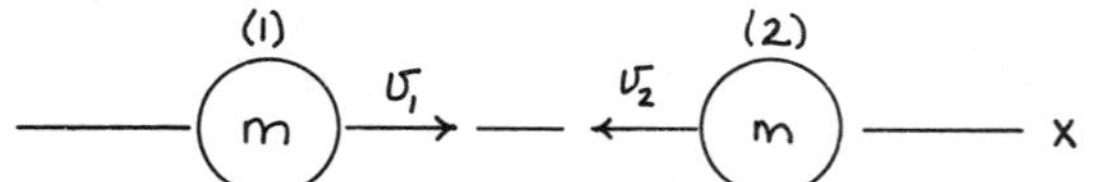

Figure a

Then, the initial velocities of balls 1 and 2 are

$$v_{1i} = +v_1 \ , \quad v_{2i} = -v_2 \qquad (a)$$

Since the balls are elastic, both linear momentum and kinetic energy are conserved in the impact. By conservation of linear momentum,

$$mv_{1i} + mv_{2i} = mv_{1f} + mv_{2f} \qquad (b)$$

where v_{1f} and v_{2f} are the final velocities of balls 1 and 2. By Eqs. (a) and (b),

$$v_1 - v_2 = v_{1f} + v_{2f} \qquad (c)$$

By conservation of kinetic energy [see Eq. (16.40)], with Eq. (a)

$$v_1 + v_2 = v_{2f} - v_{1f} \qquad (d)$$

Solving Eqs. (c) and (d), we obtain

$$\underline{\underline{v_{1f} = -v_2}} \ , \quad \underline{\underline{v_{2f} = v_1}}$$

In words, ball 1 rebounds to the left with speed v_2 and ball 2 rebounds to the right with speed v_1 (Fig. b). That is, the balls interchange velocities.

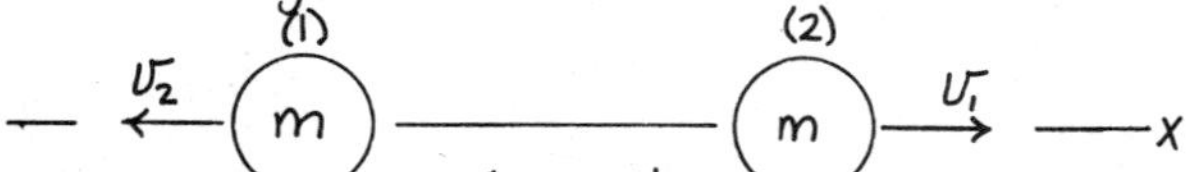

Figure b

16.53

GIVEN: Two elastic spheres A and B of masses m_A and m_B travel toward each other with equal speeds. They undergo a direct-central impact, after which sphere A is at rest.

FIND:

a) the mass ratio m_A/m_B of the spheres.

b) the final velocity of sphere B.

SOLUTION:

The initial conditions are shown in Fig. a, and the final conditions in Fig. b.

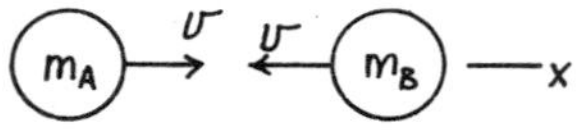

Figure a

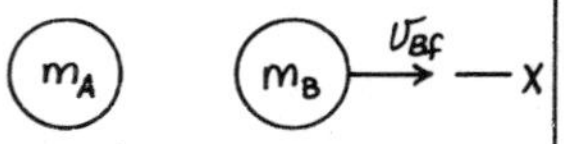

Figure b

a) By conservation of linear momentum,

$$m_A v - m_B v = m_B v_{Bf}$$

or $\quad \dfrac{m_A}{m_B} = 1 + \dfrac{v_{Bf}}{v} \qquad$ (a)

Since the collision is elastic, kinetic energy is conserved. Hence, with $v_{Ai} = v$, $v_{Bi} = -v$, and $v_{Af} = 0$, Eq. (16.40) yields

$$v - (-v) = v_{Bf} \ , \quad \text{or} \quad v_{Bf} = 2v \qquad (b)$$

By Eqs. (a) and (b),

$$\underline{\underline{\frac{m_A}{m_B} = 3}}$$

b) The final velocity of sphere B is v_{Bf}. Therefore, by Eq. (b),

$$\underline{\underline{v_{Bf} = 2v}}$$

16.54

GIVEN: Figure a. Two identical elastic spheres A and B with initial speeds v_i collide with indirect-central impact.

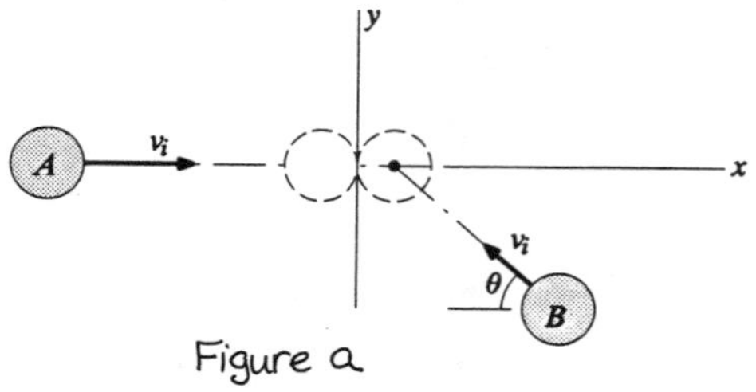

Figure a

FIND: their velocities after impact.

SOLUTION:

By Fig. a,

$$p_{ix} = mv_i - mv_i \cos\theta$$
$$p_{iy} = mv_i \sin\theta \qquad (a)$$

(continued)

16.54 (cont.)

Also, since sphere A will rebound along the x-axis,

$$P_{fx} = m(v_{Af} + v_{Bf_x})$$
$$P_{fy} = m_{Bfy} \qquad (b)$$

By conservation of momentum and Eqs. (a) and (b),

$$P_{ix} = P_{fx}; \quad v_i(1-\cos\theta) = v_{Af} + v_{Bf_x} \qquad (c)$$
$$P_{iy} = P_{fy}; \quad \underline{v_i \sin\theta = v_{Bf_y}} \qquad (d)$$

By conservation of energy [see Eq. (16.40), with

$v_{1ix} = v_{Aix} = v_i$, $v_{2ix} = v_{Bix} = -v_i \cos\theta$

$v_{2fx} = v_{Bfx}$, $v_{1fx} = v_{Af}$], we have

$$v_i + v_i\cos\theta = v_{Bfx} - v_{Af} \qquad (e)$$

By Eqs. (c) and (e),

$$v_i(1-\cos\theta) = v_{Bf_x} + v_{Af}$$
$$v_i(1+\cos\theta) = v_{Bfx} + v_{Af} \qquad (f)$$

Adding Eqs. (f), we find

$$\underline{v_{Bf_x} = v_i} \qquad (g)$$

Subtracting Eqs. (f), we obtain

$$\underline{v_{Af} = -v_i \cos\theta}$$

Thus, after impact the spheres move as shown in Fig. b, where,

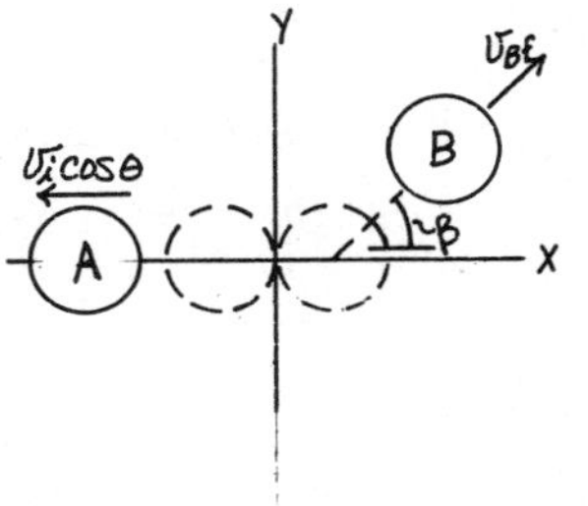

$$\underline{v_{Bf} = v_i\sqrt{1+\sin^2\theta}}$$
$$\underline{\beta = \tan^{-1}(\sin\theta)}$$
$$\underline{v_{Af} = v_i\cos\theta \text{ to the left}}$$

16.55

GIVEN: Figure a. The masses of elastic spheres A and B are $m_A = 0.6\,kg$ and $m_B = 0.8\,kg$. Initially, before impact, sphere A has speed 4 m/s directed along the positive x-axis and sphere B is at rest. After impact, the spheres move off as shown in Fig. a.

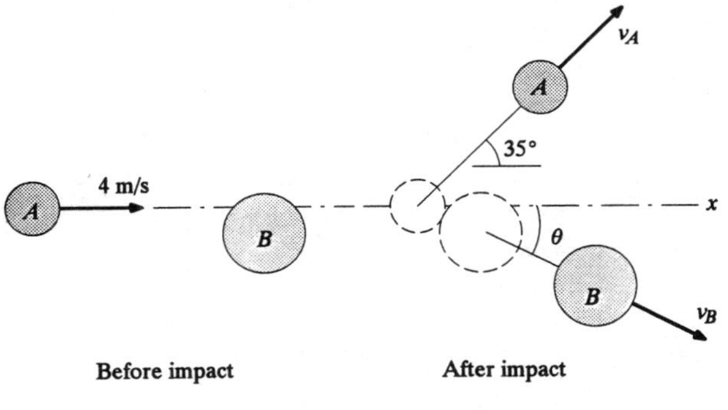

Figure a

FIND: the speeds v_A and v_B, and the angle θ.

SOLUTION:

By the conservation of linear momentum, (see Fig. a)

x direction:

$$m_A v_{Ai} = m_A v_{Af}(\cos 35°) + m_B v_{Bf}(\cos\theta)$$

or $0.4915 v_A + 0.8 v_B \cos\theta = 2.4$ (a)

y direction:

$$0 = m_A v_{Af}(\sin 35°) - m_B v_{Bf}\sin\theta$$

or $0.3441 v_A - 0.8 v_B \sin\theta = 0$ (b)

Since the collision is elastic, kinetic energy is conserved. Hence,

$$\tfrac{1}{2} m_A v_{Ai}^2 + \tfrac{1}{2} m_B v_{Bi}^2 = \tfrac{1}{2} m_A v_{Af}^2 + \tfrac{1}{2} m_B v_{Bf}^2$$

or $0.6 v_A^2 + 0.8 v_B^2 = 9.6$ (c)

Solving Eqs. (a), (b), and (c), we obtain

$$\underline{v_A = 3.468\text{ m/s}}, \quad \underline{v_B = 1.727\text{ m/s}}, \quad \underline{\theta = 1.043\text{ rad} = 59.76°}$$

16.56

GIVEN: Figure a (Newton's pendulum, with 5 balls of masses m_A, m_B, m_C, m_D, and m_E).

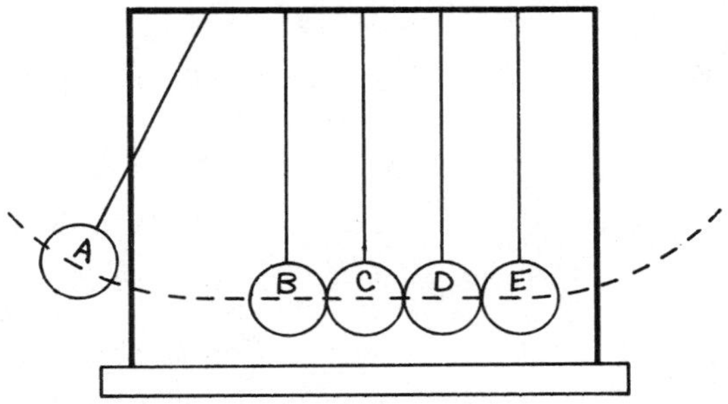

Figure a

FIND: Design the pendulum masses so that when ball A is raised and released, it strikes ball B with velocity v_A, and ball E exits with a speed of $\tfrac{1}{2} v_A$.

SOLUTION:

Initially, ball B is at rest when ball A strikes it with speed v_A. Hence, by the second of Eqs. (16.41), the speed of ball B after impact is

$$v_{Bf} = \frac{2m_A}{m_A + m_B} v_A \qquad (a)$$

Similarly, initially, ball C is at rest when ball B strikes it with speed v_{Bf}. Thus, by the second of Eqs. (16.41), the speed of ball C after impact is [with Eq. (a)]

$$v_{Cf} = \frac{2m_B}{m_B + m_C} v_{Bf} = \left(\frac{2m_A}{m_A+m_B}\right)\left(\frac{2m_B}{m_B+m_C}\right) v_A \qquad (b)$$

(continued)

16.56 (cont.)

Continuing to calculate the speeds of balls D and E in a similar manner, we obtain

$$v_{Df} = \left(\frac{2m_A}{m_A+m_B}\right)\left(\frac{2m_B}{m_B+m_C}\right)\left(\frac{2m_C}{m_C+m_D}\right) v_A \qquad \text{(c)}$$

$$v_{Ef} = \left(\frac{2m_A}{m_A+m_B}\right)\left(\frac{2m_B}{m_B+m_C}\right)\left(\frac{2m_C}{m_C+m_D}\right)\left(\frac{2m_D}{m_D+m_E}\right) v_A \qquad \text{(d)}$$

Design 1:

Let $m_A = m_B = m_C = m_D = m$ and $v_{Ef} = \frac{1}{2} v_A$.
Then, by Eq. (d),

$$\tfrac{1}{2} v_A = \frac{2m}{m+m_E} v_A \qquad \text{(e)}$$

Then by Eq. (e), we find

$$\underline{\underline{m_E = 3m}} \qquad \text{(f)}$$

Design 2:

Let $m_B = m_C = m_D = m_E = m$, and $v_{Ef} = \frac{1}{2} v_A$
Then, by Eq. (d)

$$\tfrac{1}{2} v_A = \frac{2m_A}{m_A+m} v_A \qquad \text{(g)}$$

By Eq. (g), we obtain

$$\underline{\underline{m_A = \frac{m}{3}}} \qquad \text{(h)}$$

Either Eq. (f) or Eq. (h) satisfy the design requirements.

Many other combinations of masses would also work.

Information Item: Note that if the balls are identical, each of balls B, C, D, and E have velocities v_A after they are first struck. Since ball A strikes ball B (at rest) with speed v_A, ball A is stopped [see the first of Eqs. (16.41), with $m_1 = m_2 = m_A = m_B$, $v_{1i} = v_A$, and $v_{2i} = v_{Bi} = 0$] Ball B is then stopped when it strikes ball C, and so on, until ball D is stopped when it strikes E. Then, ball E exits with speed v_A. That is, the speed v_A is transmitted through the balls to ball E.

16.57

GIVEN: A Newton pendulum (see Problem 16.56) with ten balls of masses $m_A, m_B, m_C, \ldots, m_J$ (Fig. a).

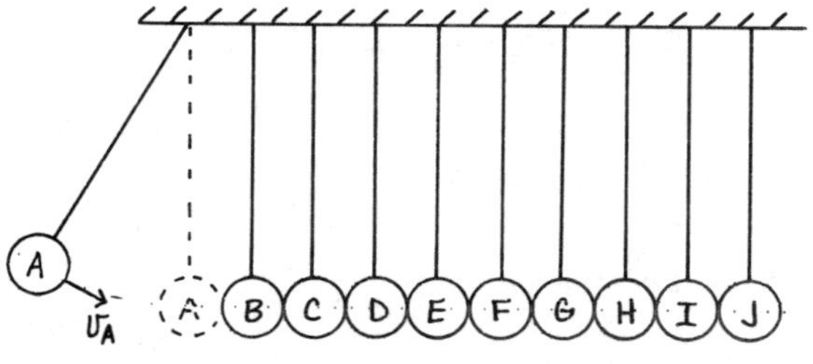

Figure a

FIND: Two combinations of masses $m_A, m_B, \ldots, m_J$ that will satisfy the requirement that ball J leaves with a speed of one-half the speed v_A with which ball A strikes ball B.

SOLUTION:

Equation (d) of the solution for Problem 16.56 may be extended to ten balls in the same manner. Thus,

$$v_{Jf} = \left(\frac{2m_A}{m_A+m_B}\right)\left(\frac{2m_B}{m_B+m_C}\right)\left(\frac{2m_C}{m_C+m_D}\right)\left(\frac{2m_D}{m_D+m_E}\right)\left(\frac{2m_E}{m_E+m_F}\right) \cdot$$
$$\cdot \left(\frac{2m_F}{m_F+m_G}\right)\left(\frac{2m_G}{m_G+m_H}\right)\left(\frac{2m_H}{m_H+m_I}\right)\left(\frac{2m_I}{m_I+m_J}\right) v_A \qquad \text{(a)}$$

Design 1: As in Problem 16.56, we might take $m_A = m_B = \ldots = m_I = m$. Then, for $v_{Jf} = \frac{1}{2} v_A$, Eq. (a) yields

$$\tfrac{1}{2} v_A = \left(\frac{2m}{m+m_J}\right) v_A$$

or $\underline{\underline{m_J = 3m}}$

Design 2: As in Problem 16.56, take $m_B = m_C = \ldots = m_J = m$. Then, for $v_{Jf} = \frac{1}{2} v_A$, Equation (a) yields

$$\tfrac{1}{2} v_A = \left(\frac{2m_A}{m_A+m}\right) v_A$$

or $\underline{\underline{m_A = \tfrac{1}{3} m}}$

Many other combinations satisfy the requirement that $v_{Jf} = \frac{1}{2} v_A$. Each combination will give a unique motion to the balls. As another example, if $m_A = m_B = m_C = m_D = m_E = m$ and $m_F = m_G = m_H = m_I = 2m$, then

$$\underline{\underline{m_J = \tfrac{10}{3} m}}.$$

16.58

GIVEN: A steel ball of mass m collides directly with another ball of mass 3m. The relative speed of approach of the balls is 12 m/s, and the coefficient of restitution of impact is 0.96.

FIND: the speed of separation of the balls.

SOLUTION:

Figures a and b illustrate the conditions before impact and after impact.

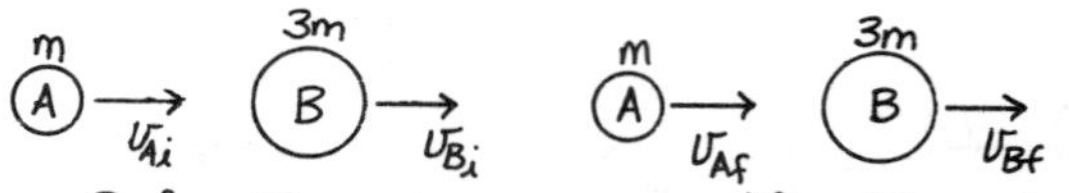

Before Impact
Figure a

After Impact
Figure b

By Eq. (16.45) and Figs. a and b,

$$e = \frac{v_{Bf} - v_{Af}}{v_{Ai} - v_{Bi}} \qquad \text{(a)}$$

(continued)

16.58 (cont.)

The initial rate of approach is, by Fig. a,

$$v_{A/B}^{i} = v_{Ai} - v_{Bi} = 12 \text{ m/s} \qquad (b)$$

The final rate of separation is, by Fig. b,

$$v_{B/A}^{f} = v_{Bf} - v_{Af} \qquad (c)$$

Therefore, by Eqs. (a), (b), and (c), with $e = 0.96$, the rate of separation is

$$\underline{\underline{v_{Bf} - v_{Af} = (0.96)(12) = 11.52 \text{ m/s}}}$$

16.59

GIVEN: A steel ball is dropped onto a horizontal steel plate from a height h and rebounds to 81% of h.

FIND: the coefficient of restitution.

SOLUTION:

The initial and final positions of the ball are shown in Fig. a.

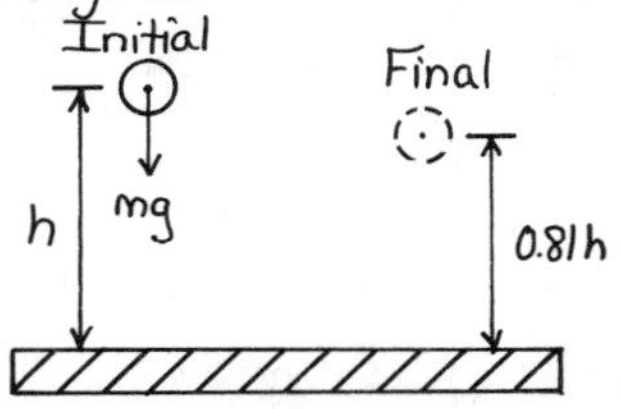

Figure a

To determine the speed with which the ball strikes the plate, we use the energy method. Thus,

$$U = \tfrac{1}{2} m v_i^2$$

or $\quad mgh = \tfrac{1}{2} m v_i^2 \; ; \quad v_i = \sqrt{2gh} \qquad (a)$

The speed with which it leaves the plate is given by

$$U = \tfrac{1}{2} m v_f^2$$

or $\quad mg(0.81h) = \tfrac{1}{2} m v_f^2 \; ; \quad v_f = \sqrt{2(0.81)gh} \qquad (b)$

Since the plate is massive, its speed due to the impact is negligible. Hence, by Eq. (16.45), with $v_{1f} = -v_f = -\sqrt{1.62gh}$, $v_{1i} = \sqrt{2gh}$, and $v_{2i} = v_{2f} = 0$ (for the plate),

$$e = \frac{0 - (-\sqrt{1.62gh})}{\sqrt{2gh} - 0}$$

or $\quad \underline{\underline{e = \sqrt{0.81} = 0.9}}$

16.60

GIVEN: A golf ball that is dropped from a height of 2m above a massive horizontal pavement rebounds to a height of 1.5m.

FIND: the coefficient of restitution.

SOLUTION:

By the energy method, the speed with which the ball strikes the pavement is determined by

$$U = \tfrac{1}{2} m v_i^2$$

or $\quad mg(2) = \tfrac{1}{2} m v_i^2 \; ; \quad v_i = 2\sqrt{g} \qquad (a)$

The speed with which the ball rebounds is determined by

$$U = \tfrac{1}{2} m v_f^2$$

or $\quad mg(1.5) = \tfrac{1}{2} m v_f^2 \; ; \quad v_f = \sqrt{3g} \qquad (b)$

Since the pavement is massive, its initial and final speed is zero. Hence, by Eq. (16.45)

$$e = \frac{v_{2f} - v_{1f}}{v_{1i} - v_{2i}} = \frac{0 - (-\sqrt{3g})}{2\sqrt{g} - 0}$$

or $\quad \underline{\underline{e = \frac{\sqrt{3}}{2} = 0.866}}$

16.61

GIVEN: A locomotive that is traveling at 8 km/h and weighs 540 kN backs into and couples with a 90-kN box car that is at rest.

FIND:

a) the speed v_f of the locomotive and car for the instant after the impact.

b) the kinetic energies before and after the impact.

c) whether the impact is elastic or inelastic.

SOLUTION:

a) By conservation of linear momentum, we have

$$m_L v_{Li} = (m_L + m_c) v_f \qquad (a)$$

where

$$m_L = \frac{540 \text{ kN}}{9.81 \text{ m/s}^2} = 55046 \text{ kg}$$

$$m_c = \frac{90 \text{ kN}}{9.81 \text{ m/s}^2} = 9174 \text{ kg}$$

$$v_{Li} = 8 \text{ km/h} = 2.222 \text{ m/s}$$

Hence, by Eq. (a)

$$\underline{\underline{v_f = 1.905 \text{ m/s} = 6.858 \text{ km/h}}}$$

b) The kinetic energy before the impact is

$$T_i = \tfrac{1}{2} m_L v_{Li}^2 = \tfrac{1}{2}(55046)(2.222)^2$$

or $\quad \underline{\underline{T_i = 135.9 \text{ kN}\cdot\text{m}}}$

(continued)

16.61 (cont.)

The kinetic energy after the impact is

$$T_f = \tfrac{1}{2}(m_L + m_c)v_f^2 = \tfrac{1}{2}(55046 + 9174)(1.905)^2$$

or $\underline{T_f = 116.5 \text{ kN·m}}$

c) Since $T_i > T_f$ (see part b), energy is lost in the impact. Hence, the impact is inelastic.

16.62

GIVEN: A small particle of mass m travels with an initial speed v_i. It collides directly with a massive body of mass M that is traveling with an initial speed kv_i (k = constant) in the same direction as the particle. The coefficient of restitution is e.

FIND: a formula for the speed v_f of the particle with which it rebounds from the body in terms of m, M, k, e, and v_i.

SOLUTION: Since the body is massive and the particle is small, the change in the speed of the body due to the impact is negligible. Therefore, by Eq. (16.45), we have

$$e = \frac{v_{2f} - v_{1f}}{v_{1i} - v_{2i}} = \frac{kv_i - v_f}{v_i - kv_i} \qquad (a)$$

Solving Eq. (a) for v_f, we obtain

$$\underline{v_f = [k(1+e) - e]v_i} \qquad (b)$$

The masses m and M do not enter into the formula for v_f.

Alternatively, by the first of eqs. (16.44) with $M \gg m$,

$$v_f = \frac{(m - eM)v_i + (1+e)Mkv_i}{m + M}$$

$$= \frac{(\frac{m}{M} - e)v_i + (1+e)kv_i}{1 + \frac{m}{M}}$$

or $\underline{v_f \approx [k(1+e) - e]v_i}$

16.63

GIVEN: A motionless bowling ball B is subjected to a direct-central impact by an identical ball A. After the collision, the speed of ball A is 5% of its speed before the collision.

FIND: the coefficient of restitution.

SOLUTION:

Let the mass of ball A be m_1 and that of B be m_2. Let the initial speed of ball A be v_{1i} and that of B be $v_{2i}(=0)$. Then, by the first of Eqs. (16.44), the speed v_{1f} of ball A after the collision is given as

$$v_{1f} = \frac{(m_1 - em_2)v_{1i} + (1+e)m_2v_{2i}}{m_1 + m_2} \qquad (a)$$

Now, since $m_1 = m_2$ and $v_{2i} = 0$, Eq. (a) yields

$$v_{1f} = \tfrac{1}{2}(1-e)v_{1i}$$

or $\frac{v_{1f}}{v_{1i}} = 0.05 = \tfrac{1}{2}(1-e)$

Hence, $\underline{e = 0.90}$

16.64

GIVEN: A freight car that weighs 40 tons rolls along a straight track at 10 mi/h. It strikes a stationary freight car that weighs 30 tons. The two cars latch together.

FIND:
a) the speed v_f of the cars after the collision.
b) the kinetic energies before and after the collision.
c) whether the collision is elastic or inelastic

SOLUTION:

a) Let $m_A = \frac{40 \text{ tons}(2000 \text{ lb/ton})}{32.2 \text{ ft/s}^2} = 2484.5$ slugs,

$$m_B = \frac{30 \text{ tons}(2000 \text{ lb/ton})}{32.2 \text{ ft/s}^2} = 1863.4 \text{ slugs},$$

$$v_{Ai} = 10 \text{ mi/h} = 14.667 \text{ ft/s}, \quad v_{Bi} = 0$$

By conservation of linear momentum,

$$m_A v_{Ai} + m_B v_{Bi} = (m_A + m_B)v_f \qquad (a)$$

or $(2484.5)(14.667) + 0 = (2484.5 + 1863.4)v_f$

Hence, $\underline{v_f = 8.381 \text{ ft/s} = 5.71 \text{ mi/h}}$

b) The kinetic energy before the impact is

$$T_i = \tfrac{1}{2}m_A v_{Ai}^2 = \tfrac{1}{2}(2484.5)(14.667)^2$$

or $\underline{T_i = 2.672 \times 10^5 \text{ ft·lb}}$

The kinetic energy after the cars are latched together is

$$T_f = \tfrac{1}{2}(m_A + m_B)v_f^2 = \tfrac{1}{2}(2484.5 + 1863.4)(8.381)^2$$

or $\underline{T_f = 1.527 \times 10^5 \text{ ft·lb}}$

c) Since $T_i > T_f$, energy is lost in the collision. Hence, the collision is inelastic.

16.65

GIVEN: A body that weighs 200 N travels to the right. It collides centrally with a body that weighs 120 N and travels to the left. The initial speed of each body is 18 km/h, and the coefficient of restitution is 0.40.

FIND:
a) the velocity of each body for the instant that the bodies separate.
b) the velocities of the bodies for the instant that the deformation period ends and the rebound period begins.

SOLUTION:
a) The velocity of each body can be found by either the basic impulse-momentum equations [Eqs. (e), 16.5] and the equation for e [Eq. (16.45)] or directly by Eqs. (16.44).

Let $m_1 = \frac{200\,N}{9.81\,m/s^2} = 20.39\,kg$

$$m_2 = \frac{120\,N}{9.81\,m/s^2} = 12.23\,kg$$

$$v_{1i} = -v_{2i} = 18\,km/h = 5\,m/s$$

v_{1f}, v_{2f} be the final velocities of the bodies at separation.

Then by the first two of Eqs. (16.44),

$$v_{1f} = \frac{(m_1 - em_2)v_{1i} + (1+e)m_2 v_{2i}}{m_1 + m_2}$$

$$v_{1f} = \frac{[20.39 - (0.40)(12.23)]5 + (1+0.40)(12.23)(-5)}{20.39 + 12.23}$$

or $\underline{v_{1f} = -0.249\,m/s} = -0.896\,km/h$

and

$$v_{2f} = \frac{(1+e)m_1 v_{1i} + (m_2 - em_1)v_{2i}}{m_1 + m_2}$$

$$v_{2f} = \frac{(1+0.40)(20.39)(5) + [12.23 - (0.40)(20.39)](-5)}{20.39 + 12.23}$$

or $\underline{v_{2f} = 3.751\,m/s} = 13.50\,km/h$

b) The instant that the deformation period ends both bodies have the same velocity v_{cr} given by the third of Eqs. (16.44). Hence,

$$v_{cr} = \frac{m_1 v_{1i} + m_2 v_{2i}}{m_1 + m_2}$$

$$v_{cr} = \frac{(20.39)(5) + (12.23)(-5)}{20.39 + 12.23}$$

or $\underline{v_{cr} = 1.251\,m/s = 4.50\,km/hr}$

16.66

GIVEN: An automobile collides with a truck at an icy intersection, and the two vehicles move together after the collision. Initially, the car is traveling south and the truck is traveling west. After the collision, they travel in the direction 70° from the south. The truck weighs 1.3 times as much as the car. Neglect friction of the ice.

FIND:
a) the ratio of their speeds before the collision.
b) Which vehicle was traveling faster?

SOLUTION:
a) Let m_T denote the mass of the truck, and m_A the mass of the automobile, v_T the initial speed of the truck, and v_A the initial speed of the automobile (Fig. a)

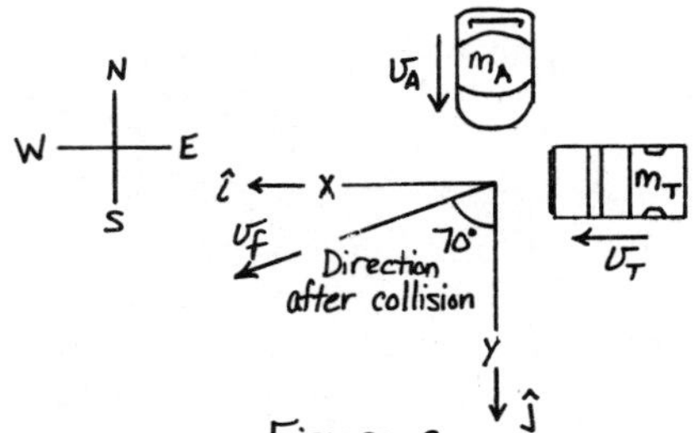

Figure a

Since the vehicles move together after the collision, by conservation of linear momentum and Fig. a, we have

$$m_A \vec{v}_A + m_T \vec{v}_T = (m_A + m_T)\vec{v}_f$$

or $m_A v_A \hat{j} + 1.3 m_A v_T \hat{i} = 2.3 m_A (v_f)[(\sin 70°)\hat{i} + (\cos 70°)\hat{j}]$

Equating $\hat{i}$ and $\hat{j}$ components, we obtain

$\hat{i}$: $1.3 m_A v_T = 2.3 m_A v_f \sin 70°$

or $v_T = 1.6625 v_f$ (a)

$\hat{j}$: $m_A v_A = 2.3 m_A v_f \cos 70°$

or $v_A = 0.7866 v_f$ (b)

Then, by Eqs. (a) and (b),

$$\underline{\frac{v_T}{v_A} = 2.113} \quad (c)$$

b) By Eq. (c),

$$v_T = 2.113 v_A$$

Hence,

<u>the truck was traveling faster.</u>

16.67

GIVEN: A tennis ball is dropped from a height of 10 ft. onto a concrete pavement, and on the third rebound reaches a height of 3 ft.

FIND: the coefficient of restitution.

SOLUTION:

Since the concrete pavement is massive compared to a tennis ball, its initial and final speeds are zero. Hence, for each impact and rebound of the ball, Eq. (16.45) yields

$$e = \frac{v_{2f}^{\,0} - v_{1f}}{v_{1i} - v_{2i}^{\,0}} = -\frac{v_{1f}}{v_{1i}} \qquad (a)$$

where subscript 1 refers to the ball and subscript 2 refers to the pavement (Fig. a)

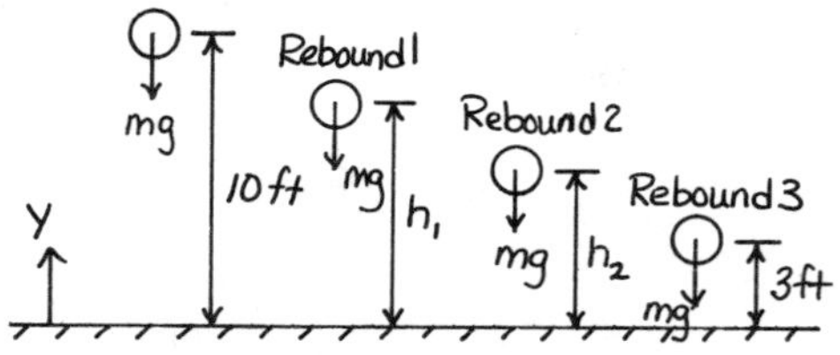

Figure a

To determine the speed of the ball when it strikes the pavement when dropped from a height of 10 ft, we may use the energy method. Thus,

$$U = \tfrac{1}{2} m v^2$$

or $mgh = mg(10) = \tfrac{1}{2} m v^2$

$$v = \pm\sqrt{(20)(32.2)} = \pm 25.377 \text{ ft/s}$$

Since +y is taken upward (Fig. a),

$$v_{1i} = -25.377 \text{ ft/s}$$

Therefore, for the first impact, by Eq. (a), the rebound velocity is

$$v_{1f} = -e v_{1i} = -e(25.377) = 25.377e.$$

Neglecting air resistance, on the second impact the ball strikes the pavement with velocity $v_{2i} = -25.377e$.

Hence, by Eq. (a), the second rebound velocity is

$$v_{2f} = -e v_{2i} = -e(-25.377e) = 25.377e^2$$

Similarly, for the third impact, the third rebound velocity is

$$v_{3f} = -e v_{3i} = -e(-25.377e^2) = 25.377e^3$$

Also, by the energy method, the speed v_{3f} is given by

$$mg(3) = \tfrac{1}{2} m v_{3f}^2$$

or

$$v_{3f} = \sqrt{6g} = \sqrt{6(32.2)} = 13.900 \text{ ft/s}$$

and therefore,

$$25.377e^3 = 13.900$$

Hence, $\underline{e = 0.818}$

16.68

GIVEN: Two asteroids of masses m_A and m_B move with velocities $\mathbf{v}_A$ and $\mathbf{v}_B$ with respect to an astronaut in a spaceship fixed in a Newtonian reference frame. The asteroids collide.

FIND: the minimum possible kinetic energy they can have after the collision, as measured by the astronaut.

SOLUTION:

As shown in P16.46, the kinetic energy as measured by the astronaut is given by

$$T = \tfrac{1}{2} m v_G^2 + \tfrac{1}{2} m_A v_{A/G}^2 + \tfrac{1}{2} m_B v_{B/G}^2$$

where $m = m_A + m_B$.

In an isolated system, the velocity of the center of mass, v_G remains constant. Therefore, the $\tfrac{1}{2} m v_G^2$ component of T is a constant. Thus, T is minimum when both $v_{A/G}$ and $v_{B/G}$ are zero.

This occurs when $v_A = v_B = v_G$; that is, when the collision is perfectly plastic and both asteroids move together with speed v_G. Hence, the minimum possible kinetic energy after the collision is

$$\underline{\underline{T = \tfrac{1}{2} m v_G^2}}$$

16.69

GIVEN: A hard block that weighs 32.2 lb. rests on a fixed table. The block is struck centrally by a ball that is moving horizontally and that weighs ½ lb. An accelerometer shows that the maximum acceleration of the block is 24.12 ft/s². $\mu_k = 0.40$.

FIND: the maximum deceleration (negative acceleration) of the center of mass of the ball.

SOLUTION:

The free-body diagram of the block during the impact is shown in Fig. a, where F is the force exerted by the ball.

(continued)

16.69 (cont.)

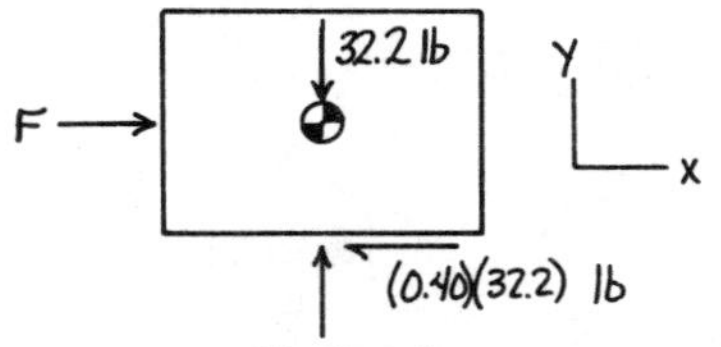

Figure a

By Fig. a,

$$\Sigma F_x = F - (0.40)(32.2) = \frac{32.2}{g} a$$

or with $g = 32.2$ ft/s² and $a = a_{max} = 24.12$ ft/s²

$$F = F_{max} = 37.0 \text{ lb.}$$

The free-body diagram of the ball during the impact is shown in Fig. b, for $F = F_{max}$.

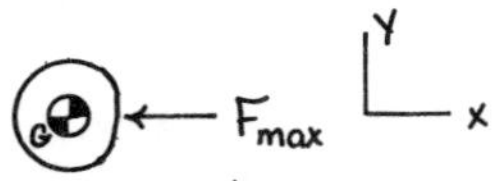

Figure b

By Fig. b,

$$\Sigma F_x = -F_{max} = \frac{0.5}{32.2} a_{G(max)}$$

or $$a_{G(max)} = -\frac{(37)(32.2)}{0.5} = -2383 \text{ ft/s}^2$$

Therefore, the maximum deceleration of the center of mass of the ball is

$$(\text{deceleration})_{max} = -a_{G(max)} = 2383 \text{ ft/s}^2$$

16.70

GIVEN: Two identical balls, A and B, move toward each other along a direct-central collision course. Ball B has an initial velocity v_{Bi}. After the impact, ball A is stopped.

FIND:

a) a formula for the initial velocity v_{Ai} of ball A in terms of v_{Bi} and the coefficient of restitution e.

b) the ratio v_{Ai}/v_{Bi} for the three cases e=0, 0.50, and 1.00, and explain the motion of the balls after the collision for each case.

SOLUTION:

a) The conditions before and after the impact are shown in Fig. a.

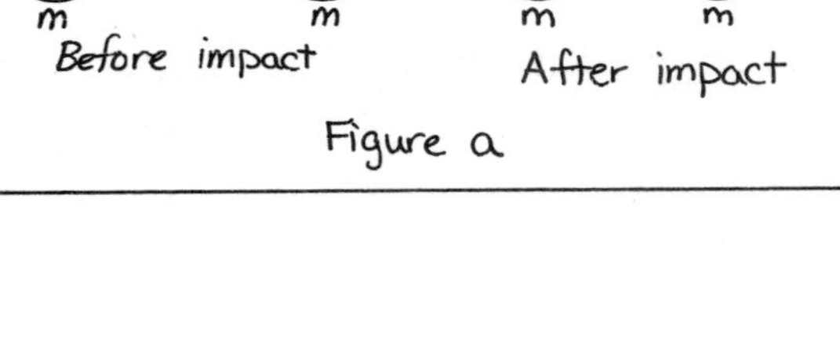

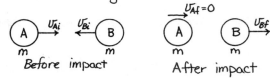

Figure a

By Fig. a and the conservation of linear momentum

$$m v_{Ai} - m v_{Bi} = m v_{Bf} \qquad (a)$$

By Eq. (16.45) with $v_{2f} = v_{Bf}$, $v_{1f} = 0$, $v_{1i} = v_{Ai}$, and $v_{2i} = -v_{Bi}$, we have

$$e = \frac{v_{2f} - v_{1f}}{v_{1i} - v_{2i}} = \frac{v_{Bf}}{v_{Ai} + v_{Bi}} \qquad (b)$$

or, rearranging Eq. (b),

$$v_{Bf} = e(v_{Ai} + v_{Bi}) \qquad (c)$$

Substitution of Eq. (c) into Eq. (a) yields

$$\frac{v_{Ai}}{v_{Bi}} = \frac{1+e}{1-e} \qquad (d)$$

b) By Eq. (d),

For e=0: $$\frac{v_{Ai}}{v_{Bi}} = 1 \qquad (e)$$

For e=0.50: $$\frac{v_{Ai}}{v_{Bi}} = 3 \qquad (f)$$

For e=1.00: $$\frac{v_{Ai}}{v_{Bi}} = \infty \qquad (g)$$

For the case e=0, the collision is perfectly plastic. Then, $v_{Af} = 0$ and $v_{Bf} = 0$ [see Eq. (c)]

For the case e=0.50, $v_{Ai} = 3v_{Bi}$, and by Eq. (c), $v_{Bf} = 2v_{Bi}$; that is, ball A is stopped and ball B rebounds with twice its initial speed.

For the case e=1.00, the result given in Eq. (g) is surprising and needs further explanation. Note that in the derivation of Eq. (d), we divided by (1-e) which is zero when e=1. To revisit the derivation of Eq. (d), substitute Eq. (c) into Eq. (a) to obtain directly

$$v_{Ai} - v_{Bi} = e(v_{Ai} + v_{Bi})$$

or $$(1-e)v_{Ai} = (1+e)v_{Bi} \qquad (h)$$

Then, for e=1, $v_{Bi} = 0$, and by Eqs. (b) or (c), $v_{Bf} = v_{Ai}$. Thus, after the impact (ball B initially at rest; $v_{Bi} = 0$), ball A is stopped and ball B moves to the right with speed v_{Ai} (Fig. a).

Hence, balls A and B interchange their initial speeds. Again,

$$\frac{v_{Ai}}{v_{Bi}} = \frac{v_{Ai}}{0} = \infty$$

16.71

GIVEN: Let $e=0$ for the collision between the disks in Example 16.20 (see Fig. a).

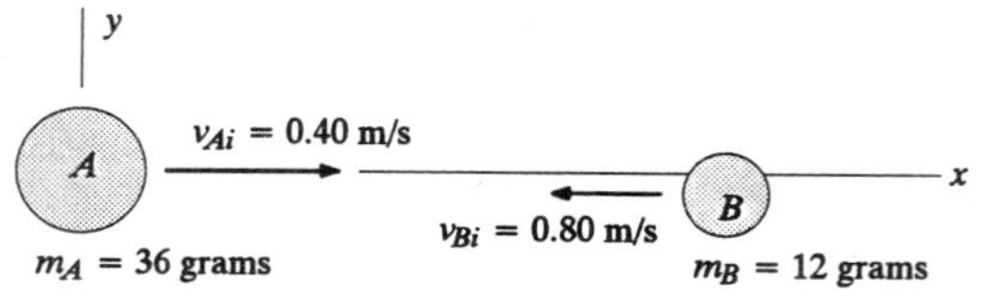

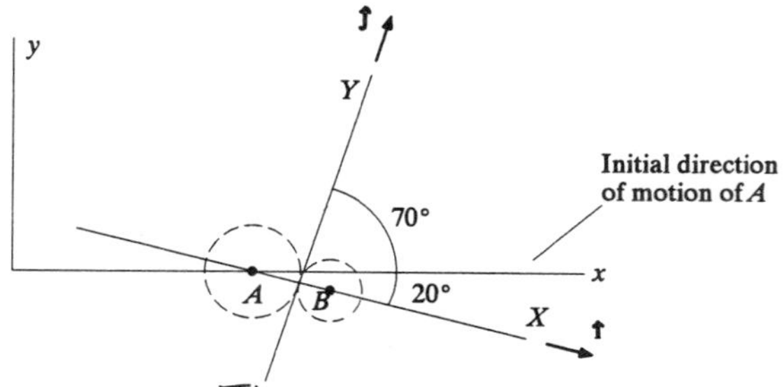

Figure a

FIND:

a) the final velocities of the disks.

b) the percentage of kinetic energy lost in the collision.

SOLUTION:

a) Since the disks are smooth (frictionless), the velocity of each disk in the Y direction (see Fig. a) does not change; that is,

$$v_{AYf} = v_{AYi} = v_{Ai}\sin 20^\circ = (0.40)\sin 20^\circ = 0.1368 \text{ m/s}$$

$$v_{BYf} = v_{BYi} = -v_{Bi}\sin 20^\circ = -(0.80)\sin 20^\circ = -0.2736 \text{ m/s}$$

Since $e=0$, there is no rebound of the bodies in the X direction. In other words, by Eq. (16.46), $v_{BXf} = v_{AXf} = v_{Xf}$. Thus, by conservation of linear momentum in the X direction,

$$m_A v_{Ai}\cos 20^\circ - m_B v_{Bi}\cos 20^\circ = (m_A + m_B)v_{Xf}$$

$$[(0.036)(0.40) - (0.012)(0.8)]\cos 20^\circ = (0.048)v_{Xf}$$

or $v_{Xf} = 0.0940$ m/s.

Hence, with $\hat{\imath}, \hat{\jmath}$ unit vectors in the X, Y directions,

$$\underline{\vec{v}_{Af} = (0.0940\hat{\imath} + 0.1368\hat{\jmath})}\,; \quad v_{Af} = 0.1660 \text{ m/s}$$

$$\underline{\vec{v}_{Bf} = (0.0940\hat{\imath} - 0.2736\hat{\jmath})}\,; \quad v_{Bf} = 0.2893 \text{ m/s}$$

b) The initial kinetic energy is

$$T_i = \tfrac{1}{2}m_A v_{Ai}^2 + \tfrac{1}{2}m_B v_{Bi}^2$$

$$T_i = \tfrac{1}{2}(0.036)(0.40)^2 + \tfrac{1}{2}(0.012)(0.80)^2$$

or $T_i = 6.72\times 10^{-3}$ J

The final kinetic energy is

$$T_f = \tfrac{1}{2}m_A v_{Af}^2 + \tfrac{1}{2}m_B v_{Bf}^2$$

$$T_f = \tfrac{1}{2}(0.036)(0.1660)^2 + \tfrac{1}{2}(0.012)(0.2893)^2$$

or $T_f = 9.98\times 10^{-4}$ J

Therefore, the percent lost of kinetic energy is

$$\underline{\left[\frac{T_i - T_f}{T_i}\right]\times 100 = \left[\frac{6.72 - 0.998}{6.72}\right](100) = 85.1\%}$$

16.72

GIVEN: The disks in Example 16.20 stick together after the collision (Also see Problem 16.71).

FIND:

a) the velocity of the disks for the instant after the collision.

b) Determine the percent loss of kinetic energy in the collision.

SOLUTION:

a) By Fig. E16.20 and conservation of linear momentum in the x and y directions, since the disks stick together,

x direction:

$$m_A v_{Ai} - m_B v_{Bi} = (m_A + m_B)v_{xf}$$

$$(0.036)(0.40) - (0.012)(0.80) = (0.048)v_{xf}$$

or $v_{xf} = 0.10$ m/s

and

y direction: $0 = (m_A + m_B)v_{yf}$

or $v_{yf} = 0$

Therefore,

$$\underline{\vec{v}_{Af} = \vec{v}_{Bf} = 0.10\hat{\imath}} \text{ (m/s)}$$

where $\hat{\imath}$ is a unit vector in the x direction.

b) The initial kinetic energy is

$$T_i = \tfrac{1}{2}m_A v_{Ai}^2 + \tfrac{1}{2}m_B v_{Bi}^2$$

$$T_i = \tfrac{1}{2}(0.036)(0.40)^2 + \tfrac{1}{2}(0.012)(0.8)^2 = 0.00672 \text{ J}$$

The final kinetic energy is

$$T_f = \tfrac{1}{2}(m_A + m_B)v_f^2 = \tfrac{1}{2}(0.048)(0.1)^2$$

$$T_f = 0.00024 \text{ J}$$

Therefore, the percent energy lost is

$$\left[\frac{T_i - T_f}{T_i}\right]\times 100 = \left(\frac{0.00672 - 0.00024}{0.00672}\right)(100)$$

or $\underline{\text{Percent loss} = 96.4\%}$

Note this loss is greater than that of Problem 16.71.

16.73

GIVEN: Equations (16.44)

FIND: Show that Eqs. (16.44) are consistent with the requirement that a direct-central collision does not alter the motion of the center of mass of two colliding bodies.

SOLUTION:

Let the masses of the bodies be m_1 and m_2, their velocities v_1 and v_2, the initial motion of the mass center be v_{Gi}, and the final motion of the mass center be v_{Gf}.

The initial motion of the mass center is given by (with initial velocities v_{1i}, v_{2i})

$$v_{Gi} = \frac{m_1 v_{1i} + m_2 v_{2i}}{m_1 + m_2} \qquad (a)$$

The final motion of the mass center is given by

$$v_{Gf} = \frac{m_1 v_{1f} + m_2 v_{2f}}{m_1 + m_2} \qquad (b)$$

where v_{1f} and v_{2f} are given by the first two of Eqs. (16.44). Hence, by Eqs. (b) and (16.44), we obtain

$$v_{Gf} = \left(\frac{1}{m_1+m_2}\right)^2 \left[m_1(m_1 - em_2)v_{1i} + m_1(1+e)m_2 v_{2i}\right]$$
$$+ \left(\frac{1}{m_1+m_2}\right)^2 \left[m_2(1+e)m_1 v_{1i} + m_2(m_2 - em_1)v_{2i}\right]$$

$$v_{Gf} = \left(\frac{1}{m_1+m_2}\right)^2 \left[m_1(m_1+m_2)v_{1i} + m_2(m_1+m_2)v_{2i}\right]$$

or
$$v_{Gf} = \frac{m_1 v_{1i} + m_2 v_{2i}}{m_1+m_2} \qquad (c)$$

Hence, by Eqs. (a), (b), and (c),

$$\underline{v_{Gf} = v_{Gi}}$$

In words, the motion of the mass center of two colliding bodies is not altered by a direct-central collision.

16.74

GIVEN: Two cars collide at an intersection (Fig. a). Car A weighs 9.0 kN and Car B weighs 10.8 kN. Car A has an initial speed of 32 km/h and Car B has an initial speed of 40 km/h. Car A is struck in the side by car B. Neglect friction.

FIND: the velocity of each car for the instant after the collision for the following cases:

a) The car shells are made of a super-elastic material (e=1) designed to prevent damage to the interior of the car.

b) The car shells are made of a material that absorbs energy and therefore, deforms considerably.

c) In which car would you prefer to be, for part a? For part b?

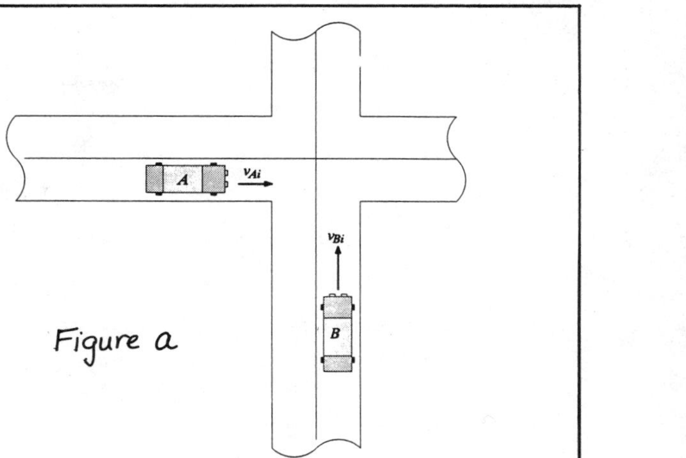

Figure a

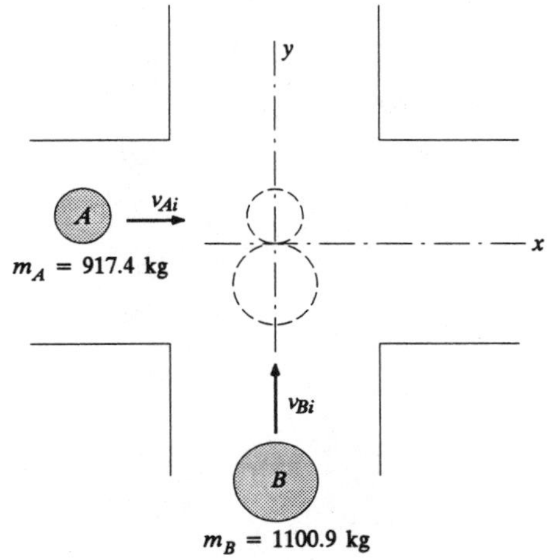

Figure b

SOLUTION:

See Figure b. The collision is indirect-central.

For car A, $m_A = \frac{9000}{9.81} = 917.4$ kg, $v_{Ai} = 32$ km/h $= 8.889$ m/s

For car B, $m_B = \frac{10800}{9.81} = 1100.9$ kg, $v_{Bi} = 40$ km/h $= 11.111$ m/s

a) For e=1, the collision is elastic. By conservation of linear momentum in the x and y directions (see Fig. b),

$$m_A \vec{v}_{Ai} + m_B \vec{v}_{Bi} = m_A \vec{v}_{Af} + m_B \vec{v}_{Bf}$$

Since friction is neglected, the velocities of the bodies in the x direction are unchanged by the collision [see Eqs. (h), Sec. 16.5]. Hence,

$$v_{Afx} = v_{Aix} = 8.889 \text{ m/s} \qquad (a)$$
$$v_{Bfx} = v_{Bix} = 0$$

By conservation of linear momentum in the y direction (see Fig. b)

$$0 + m_B v_{Bi} = m_A v_{Afy} + m_B v_{Bfy}$$

or
$$(1100.9)(11.111) = 917.4 v_{Afy} + 1100.9 v_{Bfy} \qquad (b)$$

Also for the y direction, by Eq. (16.46)

$$e = \frac{v_{Bfy} - v_{Afy}}{v_{Aiy} - v_{Biy}} = \frac{v_{Bfy} - v_{Afy}}{0 - 11.11}$$

(continued)

16.74 (cont.)

or $v_{Bfy} - v_{Afy} = -11.111$ (c)

Solving Eqs. (b) and (c), we find

$v_{Afy} = 12.121$ m/s , $v_{Bfy} = 1.011$ m/s

Hence,

$\vec{v}_{Af} = 8.889\hat{\imath} + 12.121\hat{\jmath} = 15.03$ m/s ∠53.7°

$\vec{v}_{Bf} = 1.011\hat{\jmath} = 1.011$ m/s ∠90°

b) In this case, linear momentum is conserved in both the x and y directions. Also, since the cars stick together after the collision,

$$\vec{v}_{Af} = \vec{v}_{Bf} = \vec{v}_f \quad (d)$$

By conservation of momentum (see Fig. b)

$$m_A v_{Ai}\,\hat{\imath} + m_B v_{Bi}\,\hat{\jmath} = (m_A + m_B)(v_{fx}\hat{\imath} + v_{fy}\hat{\jmath}) \quad (e)$$

Therefore, equating $\hat{\imath}$ and $\hat{\jmath}$ components, we obtain

$$v_{fx} = \frac{m_A v_{Ai}}{m_A + m_B} = \frac{(917.4)(8.889)}{917.4 + 1100.9}$$

$$v_{fy} = \frac{m_B v_{Bi}}{m_A + m_B} = \frac{(1100.9)(11.111)}{917.4 + 1100.9}$$

or $v_{fx} = 4.040$ m/s , $v_{fy} = 6.061$ m/s

Hence,

$\vec{v}_f = 4.040\hat{\imath} + 6.061\hat{\jmath} = 7.284$ m/s ∠56.3°

c) In part a, car A receives a sudden sideways increase of 12.12 m/s (in the y direction) and car B's speed is reduced suddenly by 11.111 − 1.011 = 10.1 m/s. The passenger in car A, therefore, is subjected to a very high sideways acceleration, and ordinarily is not protected by an air bag. Still the passenger in car B is also subjected to a very high acceleration. If the car has a passenger side air bag, it would probably prevent major injuries. Thus, in general, one would rather be in car B.

In part b, the net speed of car A is reduced by 8.889 − 7.284 = 1.605 m/s and the speed of car B is reduced by 11.111 − 7.284 = 3.827 m/s, so that the passenger in car B is subjected to a greater acceleration. However, the passenger in car A may be impacted more severely by the collapse and deformation of the material in the door. In general, again, one would rather be in car B.

CONCLUSION:

Drive carefully and avoid such accidents.

16.75

GIVEN: Figure a, in which a steel pile of mass m_p is driven into the ground by successive impacts of a hammer (driver) of mass m_h that falls freely through a distance h, before striking the pile. Under a single impact, the pile is driven into the ground a distance d.

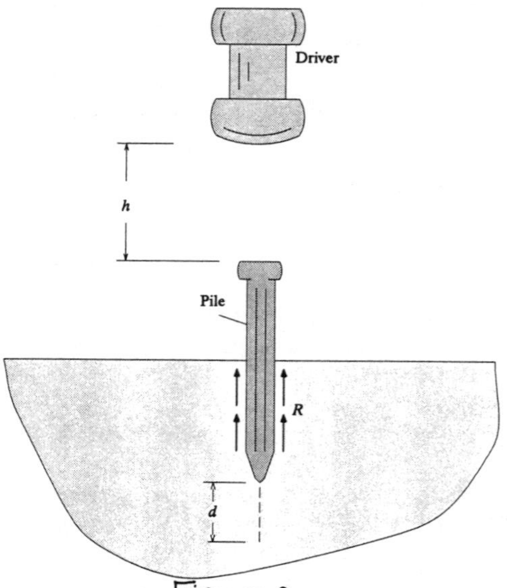

Figure a

FIND:

a) Derive a formula for the total resistive force R exerted on the pile by the soil during the penetration d in terms of m_p, m_h, d, and h. Assume that R is constant, and the impact is perfectly plastic (that is, the hammer and pile move together after impact).

b) Derive a formula for the percent of energy loss in the impact.

c) Based upon the results of part b, what recommendation might you give the operator of the hammer to make the loss of energy small.

SOLUTION:

a) By conservation of linear momentum,

$$m_h v_{hi} = (m_h + m_p) v_f \quad (a)$$

where v_{hi} is the speed of the hammer at initial impact and v_f is the speed of the hammer and pile after impact.

By the work-energy principle, the speed v_{hi} is found from

$$U = m_h g h = \tfrac{1}{2} m_h v_{hi}^2$$

or $v_{hi} = \sqrt{2gh}$ (b)

(continued)

16.75 (cont.)

By Eqs. (a) and (b),

$$v_f = \frac{m_h v_{hi}}{m_h + m_p} = \left(\frac{m_h}{m_h+m_p}\right)\sqrt{2gh} \qquad (c)$$

The resistive force R may also be found by energy principles or by Newton's second law.

By the energy method,

$$U = \Delta T = T_2 - T_1$$

$$(m_h+m_p)gd - Rd = 0 - \tfrac{1}{2}(m_h+m_p)v_f^2$$

or, with Eq. (c),

$$R = m_h g\left[\frac{m_h+m_p}{m_h} + \left(\frac{m_h}{m_h+m_p}\right)\frac{h}{d}\right] \qquad (d)$$

As a check on Eq. (d), by Newton's second law and Fig. b,

$$\Sigma F_y = (m_h+m_p)g - R = (m_h+m_p)a \qquad (e)$$

where $a = \frac{dv}{dt}\frac{ds}{ds} = v\frac{dv}{ds} = \text{constant}$

Hence,

$$\int_0^d a\,ds = ad = \int_{v_f}^0 v\,dv = -\tfrac{1}{2}v_f^2$$

or, with Eq. (c),

$$a = -(1/2d)\,v_f^2 = -\left(\frac{m_h}{m_h+m_p}\right)^2\frac{gh}{d} \qquad (f)$$

Hence, by Eqs. (e) and (f),

$$R = m_h g\left[\frac{m_h+m_p}{m_h} + \left(\frac{m_h}{m_h+m_p}\right)\left(\frac{h}{d}\right)\right] \qquad (g)$$

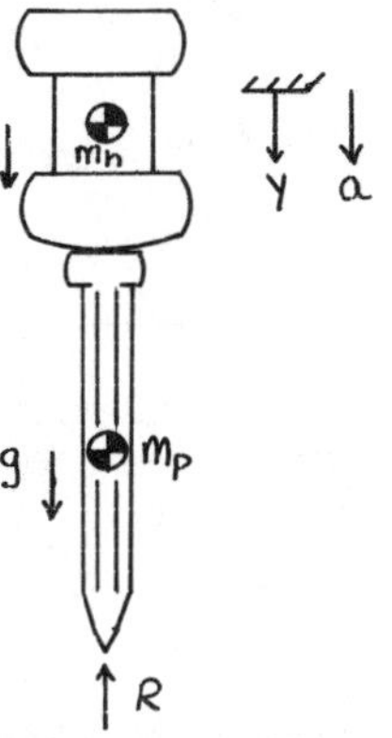

Equation (g) is identical to Eq. (d) Figure b

b) The initial kinetic energy of the system is, with Eq. (b),

$$T_i = \tfrac{1}{2}m_h v_{hi}^2 = \tfrac{1}{2}m_h(2gh) = m_h gh$$

The final kinetic energy after the impact is, with Eq. (c),

$$T_f = \tfrac{1}{2}(m_h+m_p)v_f^2 = m_h gh\left(\frac{m_h}{m_h+m_p}\right)$$

Hence, the percent loss of kinetic energy in the impact is

$$\%\ \text{loss} = \left(\frac{T_i - T_f}{T_i}\right)\times 100 = \left[\frac{m_h gh - m_h gh\left(\frac{m_h}{m_h+m_p}\right)}{m_h gh}\right]\times 100$$

or

$$\%\ \text{loss} = \frac{100(m_p/m_h)}{(1 + m_p/m_h)} \qquad (h)$$

c) By Eq. (h), we see that to make the loss of kinetic energy small per impact, m_p/m_h must be small. Hence, you should recommend that the operator use a hammer of large mass compared to the mass of the pile. Note also that the height h of the hammer drop does not enter into the percent loss of kinetic energy. However, h does affect the penetration d of each impact [see Eq. (e)].

16.76

GIVEN: The impact between the hammer and the pile in Problem 16.75 is perfectly elastic (e=1). Rework the problem for this case.

FIND:

a) To determine the resistive force R of the soil, we must calculate v_{pf}, the speed of the pile for the instant after impact. Let v_{pi} (=0) be the speed of the pile before impact, v_{pf} be the speed of the pile after impact, v_{hi} the speed of the hammer for the instant before impact, and v_{hf} the speed after impact. Then, by the law of conservation of linear momentum,

$$m_h v_{hi} + m_p v_{pi}^{\ 0} = m_h v_{hf} + m_p v_{pf} \qquad (a)$$

Also by Eq. (16.45), with e=1,

$$v_{hi} = v_{pf} - v_{hf} \qquad (b)$$

By the work-energy principle, as the hammer falls from rest,

$$U = T_2 - T_1$$

or

$$m_h gh = \tfrac{1}{2}m_h v_{hi}^2 - 0$$

Therefore,

$$v_{hi} = \sqrt{2gh} \qquad (c)$$

Substitution of Eq. (c) into Eqs. (a) and (b), and the solutions for v_{hf} and v_{pf} yields

$$v_{hf} = \left(\frac{m_h - m_p}{m_h+m_p}\right)\sqrt{2gh} \qquad (d)$$

$$v_{pf} = \left(\frac{2m_h}{m_h+m_p}\right)\sqrt{2gh} \qquad (e)$$

Now the resistive force R of the soil can be determined by Newton's second law and Fig. a. Thus,

$$\Sigma F_y = m_p g - R = m_p a \qquad (f)$$

where $a = -\frac{v_{pf}^2}{2d}$ as "found" in Problem 16.75.

Therefore, by Eqs. (e) and (f),

$$R = m_p g + m_p\frac{v_{pf}^2}{2d}$$

or

$$R = m_p g\left[1 + 4\left(\frac{m_h}{m_h+m_p}\right)^2\left(\frac{h}{d}\right)\right]$$

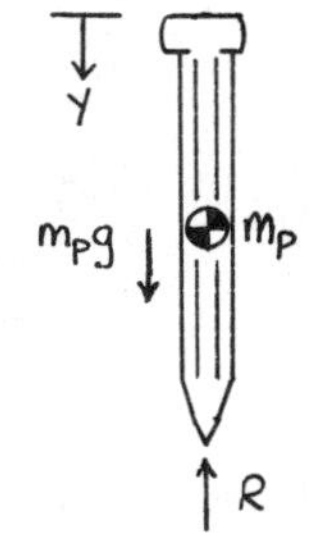

Figure a

b) Since the impact is elastic, kinetic energy is conserved in the impact.

Therefore, there is no energy loss in the impact. This fact can also be verified by calculating the initial and final kinetic energies of the system. The initial kinetic energy is

$$T_i = \tfrac{1}{2}m_h v_{hi}^2 = m_h gh$$

The final kinetic energy is, with Eqs. (d) and (e),

$$T_f = \tfrac{1}{2}m_h v_{hf}^2 + \tfrac{1}{2}m_p v_{pf}^2 = m_h gh \text{ or } T_i - T_f = 0.$$

(continued)

16.76 (cont.)

c) For the case of elastic impact between the hammer and pile, there is no loss of energy. Therefore, you don't need to give the operator any recommendation.

16.77

GIVEN: The pile driver in Problem 16.75 weighs 1000 lb, and is dropped from 20 ft above the pile that weighs 500 lb.

FIND: the depth d of penetration of the pile for an average soil resistance of 30,000 lb. Assume a perfectly plastic impact.

SOLUTION:

To solve this problem you must develop the solution for R in terms of m_h, m_p, d, and h [see Eq. (d) or (g) of the solution of Problem 16.75], namely

$$R = m_h g\left[\frac{m_h + m_p}{m_h} + \left(\frac{m_h}{m_h + m_p}\right)\left(\frac{h}{d}\right)\right] \qquad (a)$$

or solving for d from Eq. (a), we find

$$d = \frac{m_h^2 g h}{(m_h + m_p)[R - (m_h + m_p)g]} \qquad (b)$$

Therefore,

$$d = \frac{(1000)^2(20)}{1500[30{,}000 - 1500]}$$

or

$$\underline{d = 0.4678 \text{ ft}}$$

16.78

GIVEN: The data of Problem 16.77.

FIND: the number of impacts for a penetration of at least 1.5 ft.

SOLUTION:

To solve this problem, you need to derive a formula for d (the penetration per impact) as a function of m_h, m_p, R, and h; see Problems 16.75 and 16.77. Thus, the penetration per impact is

$$d = \left(\frac{m_h}{m_h + m_p}\right)\left\{\frac{m_h g h}{[R - (m_h + m_p)g]}\right\} \qquad (a)$$

For the first impact,

$$d_1 = \left(\frac{1000}{1500}\right)\left\{\frac{(1000)(20)}{[30{,}000 - 1500]}\right\}$$

or $\quad d_1 = 0.4678 \text{ ft} \qquad (b)$

For the second impact, since the hammer is dropped from the same elevation, $h = 20.4678$ ft for the second impact.

Hence, by Eqs. (a) and (b)

$$d_2 = (0.4678)\left(\frac{20.4678}{20}\right) = 0.4788 \text{ ft}$$

Hence,

$$d_1 + d_2 = 0.9466 \text{ ft}$$

Similarly for the third impact,

$$d_3 = (0.4788)\left(\frac{20.9466}{20.4678}\right) = 0.4900$$

Hence, $\quad d_1 + d_2 + d_3 = 1.4366 \text{ ft} < 1.5 \text{ ft}$

One more impact is needed to exceed a penetration of 1.5 ft. If the hammer is dropped from the same elevation again, the penetration due to the fourth impact will be

$$d_4 = (0.4900)\left(\frac{21.4366}{20.9466}\right) = 0.5015 \text{ ft}$$

and $\quad d_1 + d_2 + d_3 + d_4 = 1.9381 \text{ ft} > 1.5 \text{ ft}$

<u>Therefore, four impacts are required for at least 1.5 ft penetration.</u>

16.79

GIVEN: For the pile and hammer of Problem 16.75, $m_h = 900$ kg, $m_p = 450$ kg, $h = 6$ m, $e = 0$, and $R = 270$ kN.

FIND: the distance that the pile penetrates the soil after two impacts of the hammer. Assume that the hammer is dropped from the same elevation for each impact.

SOLUTION:

To solve this problem, you need to derive a formula for d (the penetration per impact) as a function of m_h, m_p, R, and h; see the solutions of Problems 16.75 and 16.77. Thus, the penetration per impact is

$$d = \left(\frac{m_h}{m_h + m_p}\right)\left\{\frac{m_h g h}{[R - (m_h + m_p)g]}\right\} \qquad (a)$$

Then, for the first impact

$$d_1 = \left(\frac{900}{1350}\right)\left\{\frac{(900)(9.81)(6)}{[270{,}000 - 1350(9.81)]}\right\}$$

or $\quad d_1 = 0.1375 \text{ m}. \qquad (b)$

Hence, by Eqs. (a) and (b), for the second impact,

$$d_2 = (0.1375)\left(\frac{6.1375}{6.0}\right) = 0.1407 \text{ m}$$

So the total penetration is

$$\underline{d_1 + d_2 = 0.2782 \text{ m}}$$

16.80

GIVEN: Problem 16.75

FIND:

a) Let the coefficient of restitution be e. Rework the problem for this condition.

b) With the results of part a, let $e = 0.20$, $m_h = 900$ kg, $m_p = 450$ kg, $h = 6$ m, and $R = 270$ kN. Calculate the penetration d for a single impact.

SOLUTION:

a) By the conservation of momentum,

$$m_h v_{hi} = m_h v_{hf} + m_p v_{pf} \quad (a)$$

where v_{hi} is the impact velocity of the hammer, v_{hf} is the velocity of the hammer after the impact, and v_{pf} is the velocity of the pile after the impact. The initial velocity of the pile is $v_{pi} = 0$.

The initial velocity of the hammer is, by the work-energy method,

$$U = \tfrac{1}{2} m_h v_{hi}^2$$

or $$m_h g h = \tfrac{1}{2} m_h v_{hi}^2 ; \quad v_{hi} = \sqrt{2gh} \downarrow \quad (b)$$

By Eq. (16.45),

$$e = \frac{v_{2f} - v_{1f}}{v_{1i} - v_{2i}} ; \quad e = \frac{v_{pf} - v_{hf}}{v_{hi}}$$

or $$v_{pf} - v_{hf} = e v_{hi} \quad (c)$$

By Eqs. (a), (b), and (c),

$$v_{hf} = \left(\frac{m_h - e m_p}{m_h + m_p}\right)\sqrt{2gh} \downarrow \quad (d)$$

$$v_{pf} = \left(\frac{(1+e) m_h}{m_h + m_p}\right)\sqrt{2gh} \downarrow \quad (e)$$

Now the resistive force R is, by Newton's second law and Fig. (a),

$$\Sigma F_y = m_p g - R = m_p a \quad (f)$$

where

$$a = -\frac{v_{pf}^2}{2d}$$

(see Problem 16.75)

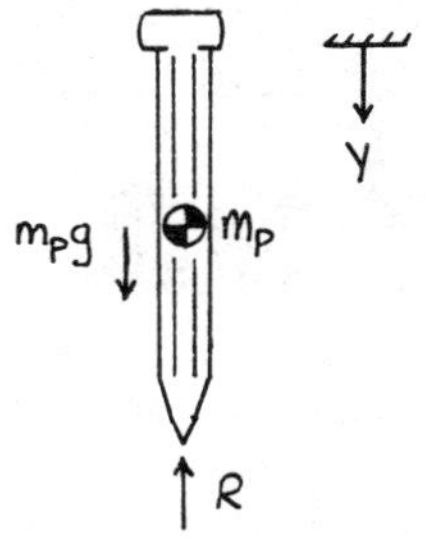

Figure a

Therefore, by Eqs. (e) and (f),

$$R = m_p g + m_p \frac{v_{pf}^2}{2d}$$

or $$R = m_p g\left[1 + (1+e)^2\left(\frac{m_h}{m_h + m_p}\right)^2\left(\frac{h}{d}\right)\right] \quad (g)$$

The initial kinetic energy is

$$T_i = \tfrac{1}{2} m_h v_{hi}^2 = m_h g h$$

The final kinetic energy is (after impact)

$$T_f = \tfrac{1}{2} m_h v_{hf}^2 + \tfrac{1}{2} m_p v_{pf}^2$$

$$T_f = \tfrac{1}{2}\left[m_h\left(\frac{m_h - e m_p}{m_h + m_p}\right)^2 + m_p\left(\frac{(1+e) m_h}{m_h + m_p}\right)^2\right] 2gh$$

or $$T_f = m_h g h\left[1 - \frac{(1-e^2) m_p}{m_h + m_p}\right]$$

Therefore, the loss of kinetic energy is

$$\%\ loss = \left(\frac{T_1 - T_2}{T_1}\right) \times 100 = \left[\frac{(1-e^2)\, m_p/m_h}{1 + m_p/m_h}\right] \times 100 \quad (h)$$

Hence, for m_p/m_h smaller than 1, the energy loss is small. Note that if $e = 1$ (elastic impact), Eq. (h) yields the result that % loss = 0 (See Problem 16.76).

b) For $e = 0.20$, $m_h = 900$ kg, $m_p = 450$ kg, $h = 6$ m, and $R = 270$ kN, by Eq. (g),

$$d = \left[\frac{m_h m_p (1+e)^2}{(m_h + m_p)^2}\right]\left(\frac{m_h g h}{R - m_p g}\right)$$

Hence,

$$d = \frac{(900)(450)(1.20)^2}{(900+450)^2} \cdot \frac{(900)(9.81)(6)}{270{,}000 - (450)(9.81)}$$

or

$$\underline{d = 0.064\ m}$$

Note the strong effect of e; see the solution of Problem 16.79, with $e = 0$, where $d_1 = 0.1375$ m.

16.81

GIVEN: A redesigned fast-acting repetitive machine (see problem statement in text); see Fig. P16.81.

FIND:

a) Determine why the new design does not work properly.

b) What do you recommend to correct the problem, still ensuring that energy is saved.

SOLUTION:

By the suggested method,

1. The problem is one of impact.
2. To simplify the problem, assume that the hammer and rod are elastic. This may not be valid, but it is a starting point. If we assume inelastic behavior, we need to obtain values for the coefficient of restitution.
3. Then momentum and energy are conserved in the impact. Therefore, let

m_h = mass of hammer
m_r = mass of rod
v_{1h} = initial speed of hammer
v_{2h} = final speed of hammer
$v_{1r} = 0$ = initial speed of rod
v_{2r} = final speed of rod

(continued)

16.81 (cont.)

Conservation of momentum yields

$$m_h v_{1h} = m_h v_{2h} + m_r v_{2r} \qquad (a)$$

Conservation of kinetic energy yields

$$\tfrac{1}{2} m_h v_{1h}^2 = \tfrac{1}{2} m_h v_{2h}^2 + \tfrac{1}{2} m_r v_{2r}^2 \qquad (b)$$

The solution of Eqs. (a) and (b) for v_{2h} and v_{2r} is, by Eq. (16.40),

$$v_{2h} = \frac{1-m}{1+m} v_{1h}$$
$$v_{2r} = \frac{2}{1+m} v_{1h} \qquad (c)$$

Equation (c) is valid provided $m = \frac{m_r}{m_h}$ and v_{2r} are not zero, since effectively they were used as divisors in the derivation of Eq. (c).

4. The push-rod and hammer system worked fine before the mass of the hammer was reduced. Therefore, examine the effect of ratio $m = m_r/m_h$ on Eq. (c).

   For $m_r/m_h = m$ small, $m_h >> m_r$. Then, $v_{2h} \approx v_{1h}$, $v_{2r} \approx 2v_{1h}$. In this case, the hammer will continue its motion after the impact.

   For $m_r/m_h = m$ large, $m_h << m_r$. Then, $v_{2h} \approx -v_{1h}$, $v_{2r} \approx 0$. In this case, the hammer rebounds from the heavier rod and the rod remains fixed.

   For $m_r/m_h = m \approx 1$, $m_r \approx m_h$,

   $$v_{2h} \approx 0, \quad v_{2r} \approx v_{1h}.$$

   In this case, the hammer stops.

5. Therefore, it appears that either the mass of the hammer must be made larger than the rod or the rod's mass must be made smaller than that of the hammer for the hammer to follow the rod. Therefore, to save energy, the masses of the follower and the hammer should be reduced, with the mass of the rod smaller than the mass of the hammer; i.e., $m_h > m_r$.

16.82

GIVEN: The moment about a fixed point O of the resultant force that acts on a system of particles as a function of time t is

$$\bar{M}_0 = 6t\hat{\imath} - 3t^2\hat{\jmath} + 5t^4\hat{k}$$

At time $t = 0$,

$$\bar{A}_0 = 2\hat{\imath} + 3\hat{\jmath} + 4\hat{k}$$

FIND: the moment of momentum A_0 of the system about point O.

SOLUTION:

$$\bar{M}_0 = \frac{d\bar{A}_0}{dt}$$

$$\int_{t_1}^{t_2} \bar{M}_0\, dt = \int_{A_{01}}^{A_{02}} d\bar{A}_0 = \bar{A}_{0_2} - \bar{A}_{0_1}$$

or

$$\int_0^t (6t\hat{\imath} - 3t^2\hat{\jmath} + 5t^4\hat{k})\,dt = \bar{A}_{0_2} - (2\hat{\imath} + 3\hat{\jmath} + 4\hat{k})$$

Integration yields

$$3t^2\hat{\imath} - t^3\hat{\jmath} + t^5\hat{k} = \bar{A}_{0_2} - (2\hat{\imath} + 3\hat{\jmath} + 4\hat{k})$$

Therefore,

$$\underline{\bar{A}_{0_2} = (3t^2+2)\hat{\imath} - (t^3-3)\hat{\jmath} + (t^5+4)\hat{k}}$$

16.83

GIVEN: At a particular instant, a particle of mass $m = 3$ grams is at (x, y, z) coordinates (50, 20, 70) in mm, and has (x, y, z) projections of velocity $\bar{v}$ (120, 250, 400) in mm/s.

FIND: the moment of momentum (in N·m·s) of the particle with respect to the (x, y, z) axes.

SOLUTION:

The moment of momentum is $\bar{A} = \bar{r} \times m\bar{v}$ (a)

where $\bar{r} = x\hat{\imath} + y\hat{\jmath} + z\hat{k}$ (Fig. a)

In meters,

$$\bar{r} = 0.050\hat{\imath} + 0.020\hat{\jmath} + 0.070\hat{k} \text{ (m)}$$

and

$$\bar{v} = 0.12\hat{\imath} + 0.25\hat{\jmath} + 0.40\hat{k} \text{ (m/s)}$$

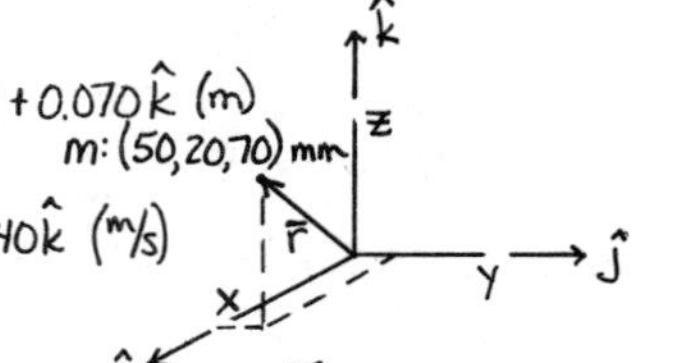

Figure a

Hence, by Eq. (a),

$$A = \begin{vmatrix} \hat{\imath} & \hat{\jmath} & \hat{k} \\ 0.05 & 0.02 & 0.07 \\ 0.12 & 0.25 & 0.40 \end{vmatrix} (3 \times 10^{-3}) \text{ N·m·s}$$

or

$$\underline{\bar{A} = (-2.85\hat{\imath} - 3.48\hat{\jmath} + 3.03\hat{k}) \times 10^{-5} \text{ N·m·s}}$$

16.84

GIVEN: Figure a: Two monkeys, A and B, initially at rest, hold on to free ends of a rope (of negligible mass) that passes over a frictionless pulley also of negligible mass. Monkey A climbs up the rope with speed v_r relative to the rope. See Fig. P16.84.

FIND: the absolute velocities of monkeys A and B in terms of v_r.

SOLUTION:

Consider Fig. a, a schematic of Fig. P16.84. By the law of moment of momentum,

$$A_{0_1} + \int M_0\, dt = A_{0_2} \qquad (a)$$

(continued)

16.84 (cont.)

Since initially the monkeys are at rest, $v_A = v_B = 0$.

Therefore, $A_{O_1} = 0$. Also, by Fig. a,

$$M_O = Wr - Wr = 0$$

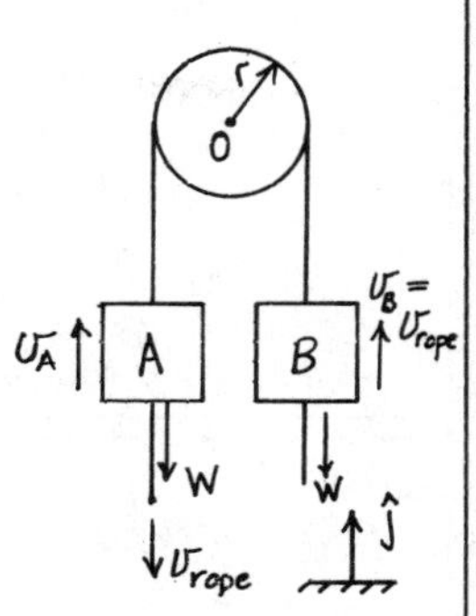

Figure a

Hence, Eq. (a) yields

$$A_{O_2} = 0 \qquad (b)$$

By Fig. a, with m = the mass of each monkey,

$$A_{O_2} = m v_A r - m v_B r = 0$$

or $$v_A = v_B \qquad (c)$$

By kinematics (see Fig. b),

$$\vec{v}_r = v_r \hat{j} = \vec{v}_{A/rope} = \vec{v}_A - \vec{v}_{rope} = v_A \hat{j} - (-v_{rope} \hat{j})$$

Hence,

$$v_r = v_A + v_{rope} = v_A + v_B \qquad (d)$$

Equations (c) and (d) yield,

$$\underline{\underline{v_A = v_B = 1/2\, v_r}}$$

or $$\underline{\underline{\vec{v}_A = \vec{v}_B = 1/2\, v_r \hat{j}}}$$

16.85

GIVEN: The moment of momentum about a fixed point O is

$$A_O = 2t^4 \hat{i} - 4t^2 \hat{j} + 6t \hat{k} \; [\text{N·m·s}] \qquad (a)$$

FIND: the moment about point O of the resultant force that acts on the system.

SOLUTION:

The moment about point O by a resultant force, denoted by $\bar{M}_O$ is given by the time derivative of $\bar{A}_O$. Thus,

$$\bar{M}_O = \frac{d\bar{A}_O}{dt} \qquad (b)$$

Equations (a) and (b) yield

$$\bar{M}_O = d(2t^4 \hat{i} - 4t^2 \hat{j} + 6t \hat{k})/dt$$

or $$\underline{\underline{\bar{M}_O = 8t^3 \hat{i} - 8t \hat{j} + 6\hat{k} \quad \text{N·m}}}$$

16.86

GIVEN: Two heavy counterweights of masses m and 2m are pinned to a rod of negligible mass (Fig. a). The rod rotates about the z-axis, and the masses are a distance L from the axis of rotation. Initially, the angular velocity is ω_i. Pins holding the masses at the distance L are pulled and the masses move to the ends of the rod (Fig. b). Treat the masses as particles, and neglect friction.

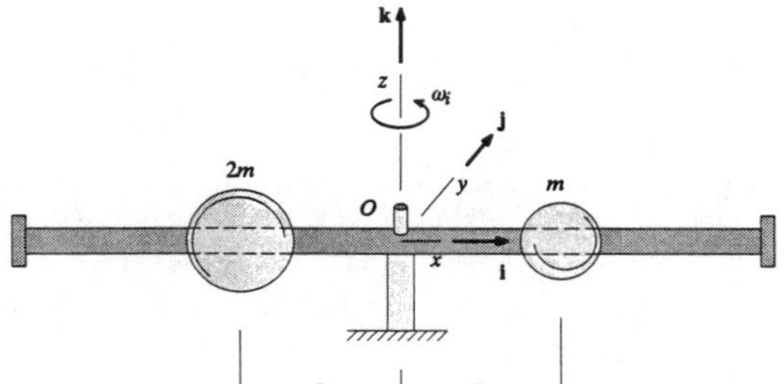

Figure a

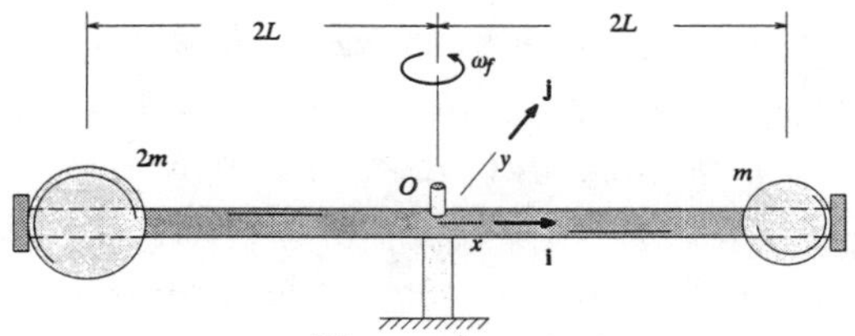

Figure b

FIND: the final angular velocity ω_f of the system.

SOLUTION:

Since friction is neglected, there is no external moment about the z-axis. Hence, angular momentum is conserved. That is,

$$\bar{A}_{O_1} = \bar{A}_{O_2} \qquad (a)$$

where by Figs. a and b,

$$\bar{A}_{O_1} = (L)\hat{i} \times [(mL\omega_i)\hat{j}] + (-L\hat{i}) \times [2m(-L\omega_i)\hat{j}]$$

$$\bar{A}_{O_2} = (2L)\hat{i} \times [m(2L\omega_f)\hat{j}] + (-2L\hat{i}) \times [2m(-2L\omega_f)\hat{j}]$$

or $$\bar{A}_{O_1} = 3mL^2\omega_i$$
$$\bar{A}_{O_2} = 12mL^2\omega_f \qquad (b)$$

So, Eqs. (a) and (b) yield

$$\underline{\underline{\omega_f = \tfrac{1}{4}\omega_i}}$$

16.87

GIVEN: A car that weighs 9.81 kN starts from rest at point O (Fig. a) and accelerates in a straight line. The position vector from O to its center of mass G is

$$\bar{r}(t) = 6t^2\,\hat{\imath}\ \text{(m)} \qquad \text{(a)}$$

where t denotes time in seconds and $\hat{\imath}$ is a unit vector along the x-axis.

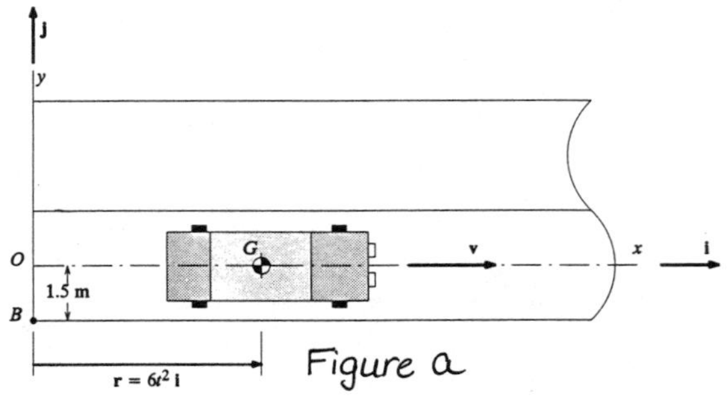

Figure a

FIND:

a) the moment of momentum of G relative to point B as a function of time t.

b) the moment about B of the net force that acts on the car, and hence, the net force.

SOLUTION:

a) The moment of momentum with respect to point B is

$$\bar{A}_B = \bar{r}_{BG} \times m\bar{v}_G \qquad \text{(b)}$$

where $\bar{r}_{BG} = 1.5\hat{\jmath} + \bar{r}(t)$

or $\bar{r}_{BG} = 6t^2\hat{\imath} + 1.5\hat{\jmath} \qquad \text{(c)}$

and $\bar{v}_G = \dfrac{d\bar{r}}{dt} = 12t\,\hat{\imath}\ (\text{m/s}) \qquad \text{(d)}$

By Eqs. (b), (c), and (d), with $m = 9810\ \text{N}/9.81 = 1000\ \text{kg}$

$$\bar{A}_B = \begin{vmatrix} \hat{\imath} & \hat{\jmath} & \hat{k} \\ 6t^2 & 1.5 & 0 \\ 12t & 0 & 0 \end{vmatrix} \times 1000 = -18t\,\hat{k}$$

$$\underline{\bar{A}_B = -18t\,\hat{k}\ (\text{kN}\cdot\text{m}\cdot\text{s})} \qquad \text{(e)}$$

b) The net moment about B is given by

$$\bar{M}_B = \frac{d\bar{A}_B}{dt}$$

Therefore, with Eq. (e)

$$\underline{\bar{M}_B = -18\,\hat{k}\ \ (\text{kN}\cdot\text{m})}$$

Also, $\bar{M}_B = \bar{r}_{BG} \times \bar{F}$

or

$$-18\,\hat{k} = \begin{vmatrix} \hat{\imath} & \hat{\jmath} & \hat{k} \\ 6t^2 & 1.5 & 0 \\ F_x & 0 & F_z \end{vmatrix} = 1.5F_z\hat{\imath} - 6t^2F_z\hat{\jmath} - 1.5F_x\hat{k}$$

Therefore,

$$\underline{F_x = 12\ \text{kN}}$$

$$F_y = F_z = 0$$

Alternatively, by Newton's second law,

$$\bar{F} = m\bar{a} = m\frac{d^2\bar{r}}{dt^2} = (1000\ \text{kg})(12\ \text{m/s}^2)\,\hat{\imath}$$

or

$$\bar{F} = 12\,\hat{\imath}\ (\text{kN})$$

Then, $\bar{M}_B = \bar{r}_{BG} \times \bar{F} = \begin{vmatrix} \hat{\imath} & \hat{\jmath} & \hat{k} \\ 6t^2 & 1.5 & 0 \\ 12 & 0 & 0 \end{vmatrix} = -(18\ \text{kN}\cdot\text{m})\,\hat{k}$

16.88

GIVEN: Three particles with masses m_1, m_2, and m_3 move in the (x,y) plane under the action of mutual attractions or repulsions. No external forces act. The coordinates and velocity projections of the particles are (x_1, y_1), (x_2, y_2), (x_3, y_3) and (u_1, v_1), (u_2, v_2), (u_3, v_3), respectively.

FIND: the equations that express the laws of conservation of momentum and conservation of moment of momentum for the system.

SOLUTION:

The momentum of the system is

$$\bar{Q} = m_1\bar{v}_1 + m_2\bar{v}_2 + m_3\bar{v}_3$$
$$= m_1(u_1\hat{\imath} + v_1\hat{\jmath}) + m_2(u_2\hat{\imath} + v_2\hat{\jmath}) + m_3(u_3\hat{\imath} + v_3\hat{\jmath})$$
$$= (m_1u_1 + m_2u_2 + m_3u_3)\hat{\imath} + (m_1v_1 + m_2v_2 + m_3v_3)\hat{\jmath} = \text{Constant}$$

Hence, for conservation of momentum,

$$\underline{m_1u_1 + m_2u_2 + m_3u_3 = \text{Constant}}$$

$$\underline{m_1v_1 + m_2v_2 + m_3v_3 = \text{Constant}}$$

The moment of momentum is

$$\bar{A}_0 = \sum_{i=1}^{3} r_i \times m_i\bar{v}_i = \bar{r}_1 \times m_1\bar{v}_1 + \bar{r}_2 \times m_2\bar{v}_2 + \bar{r}_3 \times m_3\bar{v}_3$$

or

$$\bar{A}_0 = m_1\begin{vmatrix} \hat{\imath} & \hat{\jmath} & \hat{k} \\ x_1 & y_1 & 0 \\ u_1 & v_1 & 0 \end{vmatrix} + m_2\begin{vmatrix} \hat{\imath} & \hat{\jmath} & \hat{k} \\ x_2 & y_2 & 0 \\ u_2 & v_2 & 0 \end{vmatrix} + m_3\begin{vmatrix} \hat{\imath} & \hat{\jmath} & \hat{k} \\ x_3 & y_3 & 0 \\ u_3 & v_3 & 0 \end{vmatrix}$$

$$= \left[m_1(x_1v_1 - y_1u_1) + m_2(x_2v_2 - y_2u_2) + m_3(x_3v_3 - y_3u_3)\right]\hat{k}$$

Hence, for conservation of moment of momentum,

$$\underline{m_1(x_1v_1 - y_1u_1) + m_2(x_2v_2 - y_2u_2) + m_3(x_3v_3 - y_3u_3) = \text{Constant}}$$

16.89

GIVEN: A flywheel (Fig. a) that consists of a thin rim 1.2m in diameter, of mass = 3600/9.81 = 366.97 kg, and spokes and hub of negligible mass.

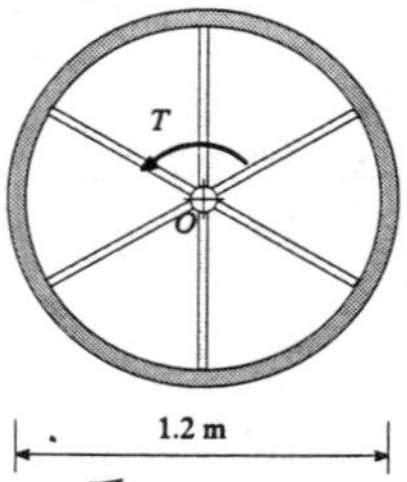

Figure a

FIND: the required torque to increase the angular speed uniformly from 40 to 50 rpm in one revolution. Neglect friction.

(continued)

16.89 (cont.)

SOLUTION:

By definition, the angular acceleration α is

$$\alpha = \frac{d\omega}{dt} = \frac{d\omega}{dt}\frac{d\theta}{d\theta} = \omega\frac{d\omega}{d\theta}$$

since α is constant, integration yields

$$\tfrac{1}{2}\omega^2 = \alpha\theta + \tfrac{1}{2}\omega_0^2 \quad (a)$$

where $\omega = \omega_0$ for $\theta = 0$.

For one revolution, $\theta = 2\pi$ rad.

Hence, by Eq. (a),

$$\alpha = \frac{\omega^2 - \omega_0^2}{4\pi} \quad (b)$$

Also, $40 \text{ rpm} = 40\frac{\text{rev}}{\text{min}} \cdot 2\pi\frac{\text{rad}}{\text{min}} \cdot \frac{1\text{min}}{60\text{s}} = 4.189$ rad/s and similarly, 50 rpm = 5.236 rad/s.

Hence, $\omega = 5.236$ rad/s and $\omega_0 = 4.189$ rad/s and, by Eq. (b),

$$\alpha = \frac{5.236^2 - 4.189^2}{4\pi} = 0.7854 \text{ rad/s}^2 \quad (c)$$

The angular momentum of the rim about O is

$$A_0 = r(mv) = r(mr\omega)$$

and the torque is determined by

$$M_0 = \frac{dA_0}{dt} = mr^2\frac{d\omega}{dt} = mr^2\alpha \quad (d)$$

With $m = 366.97$ kg, $r = 0.6$ m, and $\alpha = 0.7854 \frac{\text{rad}}{\text{s}^2}$, Eq. (d) yields

Torque = M_0 = 103.76 N·m

16.90

GIVEN: The simple pendulum in Fig. a consists of a heavy particle suspended by a massless rod of length L. The pendulum swings in the vertical plane.

FIND: Show by the principle of moment of momentum that the motion of the pendulum is governed by the differential equation

$$\ddot{\theta} + \left(\frac{g}{L}\right)\sin\theta = 0 \quad (a)$$

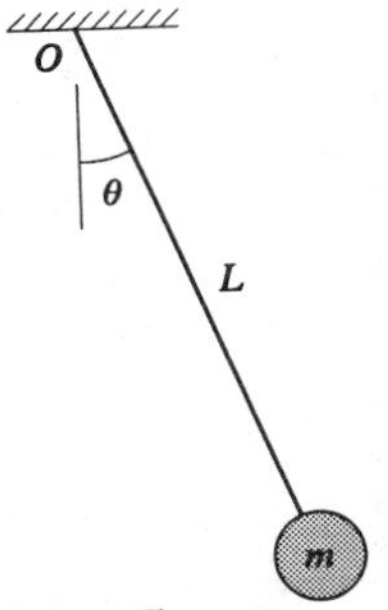

Figure a

SOLUTION:

The moment about the hinge O (see Fig. b) is

$$M_0 = -mgL\sin\theta \quad (b)$$

The moment of momentum about O is

$$A_0 = (L)(mL\omega) = L(mL\dot{\theta}) \quad (c)$$

By the principle of angular momentum,

$$M_0 = \frac{dA_0}{dt} \quad (d)$$

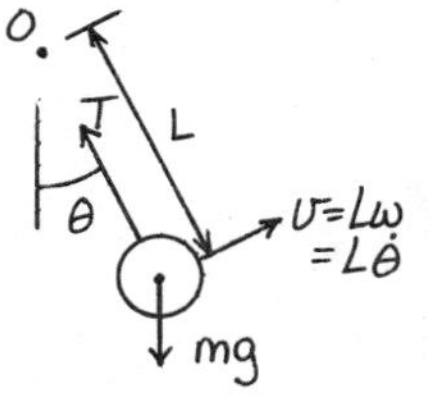

Figure b

Hence, by Eqs. (b), (c), and (d)

$$-mgL\sin\theta = mL^2\ddot{\theta}$$

or

$$\ddot{\theta} + \left(\frac{g}{L}\right)\sin\theta = 0 \quad (e)$$

Equations (a) and (e) are identical.

16.91

GIVEN: Figure a in which the boy's center of mass is initially 6 ft above the ground when he is at rest in a crouched position at A. He swings forward and at the bottom of the swing (point B), instantaneously stands up, raising his center of mass 2 ft.

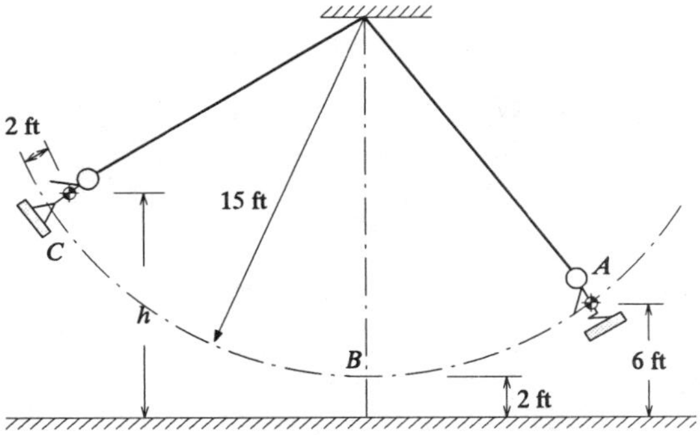

Figure a

FIND: the height h above ground of the boy's center of mass at point C (Fig. a)

SOLUTION:

Since the boy stands up at the bottom of the swing (point B), the moment of his weight mg about the pivot O is not affected (Fig. b), since mg is directed through O. Hence, the moment of momentum A_0 is not affected (not changed), or the moment of momentum is conserved. So,

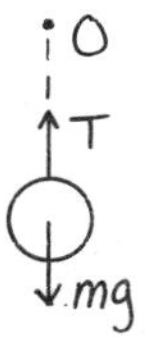

Figure b

$$A_{0_1} = A_{0_2} \quad \text{or}$$

$$(15)mv_1 = (13)mv_2 \quad (a)$$

where v_1 and v_2 are the speeds before and after the boy stands up.

By the work-energy principle, from A to B,

$$U = mg(4) = \tfrac{1}{2}mv_1^2$$

or $v_1 = \sqrt{8g} = \sqrt{(8)(32.2)} = 16.05$ ft/s (b)

So, Eqs. (a) and (b) yield

$$v_2 = 18.52 \text{ ft/s} \quad (c)$$

Now, by the work-energy principle, from B to C,

$$U = T_2 - T_1 \quad \text{or} \quad -mg(h-4) = 0 - \tfrac{1}{2}mv_2^2 \quad (d)$$

By Eqs. (c) and (d), we find h = 9.325 ft

16.92

GIVEN: A rigid pendulum consists of two identical bars of negligible mass carrying equal masses at their ends (Fig. a). The system swings in the plane of the figure about the frictionless hinge O.

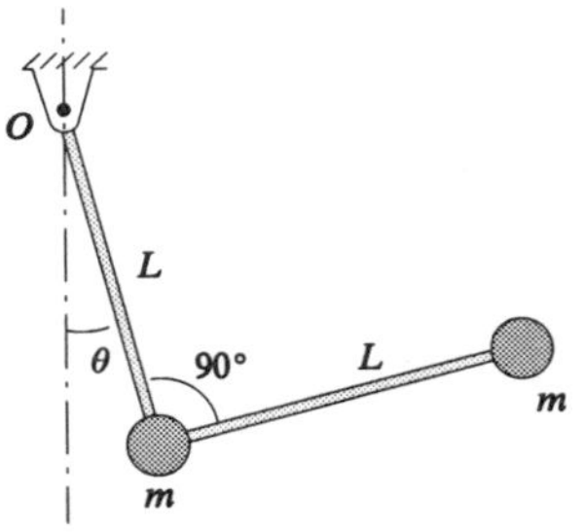

Figure a

FIND: Show that $\ddot{\theta} + \left(\frac{g}{3L}\right)(\cos\theta + 2\sin\theta) = 0$. Use the principle of angular momentum and note that the velocity of a particle that rotates on a circle of radius r is $v = r\dot{\theta}$ in the θ direction.

SOLUTION:

The moment of the weights about O is (see Fig. b)

$$M_o = -mgL\sin\theta - mgL(\sin\theta + \cos\theta)$$

or

$$M_o = -mgL(2\sin\theta + \cos\theta) \quad \text{(a)}$$

The angular momentum about O is (Fig. b)

$$A_o = (L)(mL\dot{\theta}) + (\sqrt{2}L)(m\sqrt{2}L\dot{\theta})$$

or

$$A_o = 3mL^2\dot{\theta} \quad \text{(b)}$$

Hence, by the principle of angular momentum

$$M_o = \frac{dA_o}{dt} \quad \text{(c)}$$

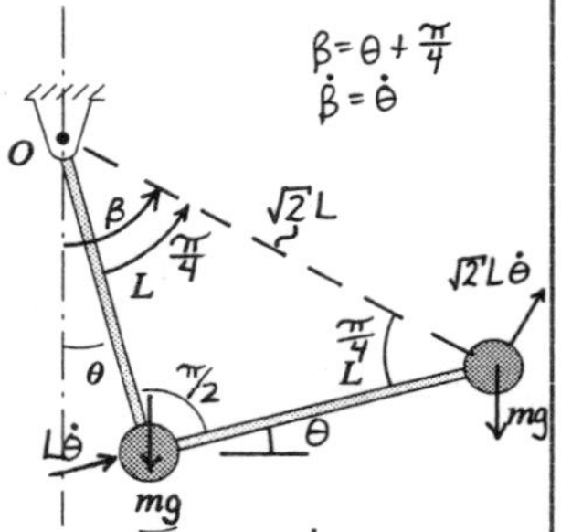

Figure b

So, by Eqs. (a), (b), and (c),

$$-mgL(2\sin\theta + \cos\theta) = 3mL^2\ddot{\theta}$$

or

$$\ddot{\theta} + \left(\frac{g}{3L}\right)(\cos\theta + 2\sin\theta) = 0$$

16.93

GIVEN: Masses m_a and m_b are connected by a light cord that passes over a pulley of radius r and of negligible mass. The pulley rotates about a frictionless axle O (Fig. a). Mass m_a slides on a smooth surface.

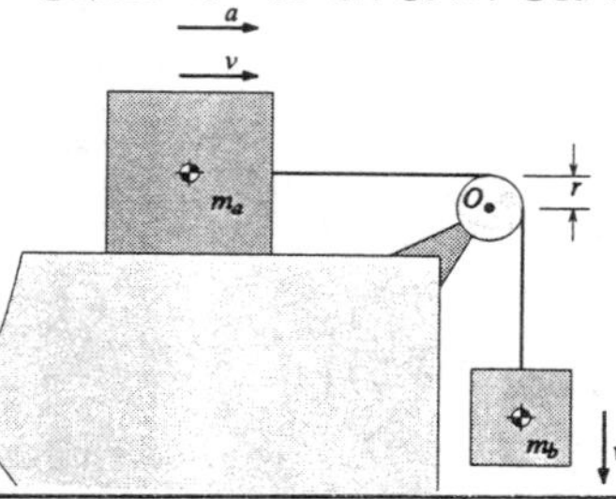

Figure a

FIND:

a) Derive a formula in terms of m_a and m_b for the acceleration of the masses. Use the law of angular momentum relative to O.

b) Determine the magnitude of the acceleration of the masses for $m_a = m_b$.

c) Verify the results by using Newton's second law $\bar{F} = m\bar{a}$ to solve part a.

SOLUTION:

a) In theorem 16.11, let point Q coincide with the fixed point O (Fig. a). Then the summation of the external forces about O yields (see Fig. b)

$$\Sigma M_o = m_b g r \quad \text{(a)}$$

Also by Fig. b, the moment of momentum is

$$A_o = m_a v r + m_b v r \quad \text{(b)}$$

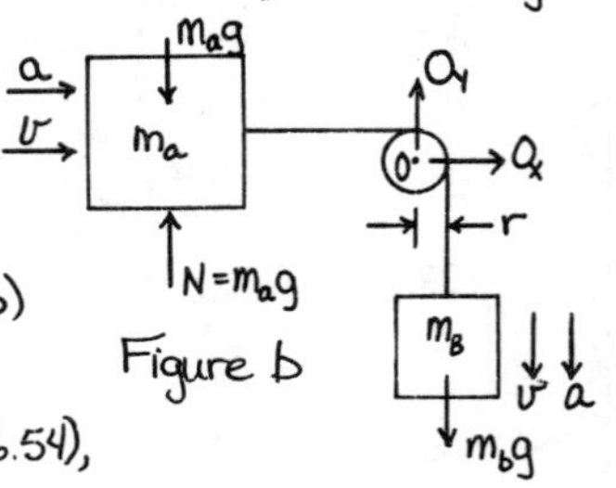

Figure b

Since point O is fixed, by Theorem 16.11, Eq. (16.54),

$$\Sigma M_o = \frac{dA_o}{dt} \quad \text{(c)}$$

Hence, Eqs. (a), (b), and (c) yield

$$m_b g r = (m_a + m_b) r a \quad \text{(d)}$$

where

$$a = \frac{dv}{dt}$$

or by Eq. (d),

$$a = \frac{m_b g}{m_a + m_b} \quad \text{(e)}$$

b) For $m_a = m_b = m$, Eq. (e) yields

$$a = \tfrac{1}{2} g$$

c) To use Newton's second law, consider the free-body diagram of masses m_a and m_b (Fig. c). By the free-body diagram of m_a,

$$\Sigma F_x = T = m_a a \quad \text{(f)}$$

and by the free-body diagram of m_b,

$$\Sigma F_y = T - m_b g = -m_b a$$

or

$$T = m_b g - m_b a \quad \text{(g)}$$

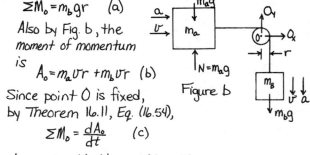

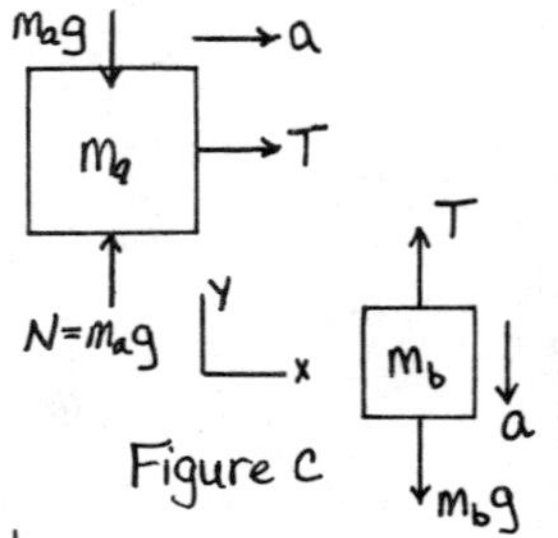

Figure c

Equating Eqs. (f) and (g) and solving for a, we find

$$a = \frac{m_b g}{m_a + m_b} \quad \text{(h)}$$

as in Eq. (e).

16.94

GIVEN: Figure a: The liquid is released from rest at angle θ

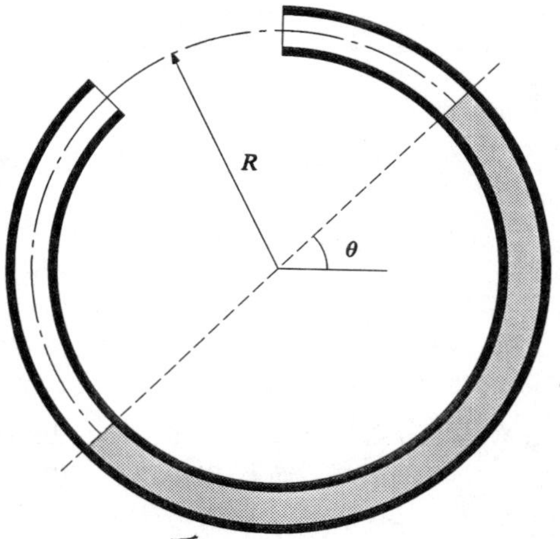

Figure a

FIND: Show by means of the principle of angular momentum that

$$\ddot{\theta} + \frac{2g}{\pi R}\sin\theta = 0 \quad \text{(a)}$$

SOLUTION:

Consider a mass element dm of the liquid at angle ϕ (Fig. b), where

$$dm = \rho R\, d\phi$$

ρ = mass per unit length.

The increment dM_o of the moment about O due to dm is

$$dM_o = (R\cos\phi) g\rho R\, d\phi$$

Hence, the total moment about O is

$$M_o = \rho g R^2 \int_{\theta}^{\pi+\theta} \cos\phi\, d\phi$$

or $\quad M_o = -2\rho g R^2 \sin\theta \quad$ (b)

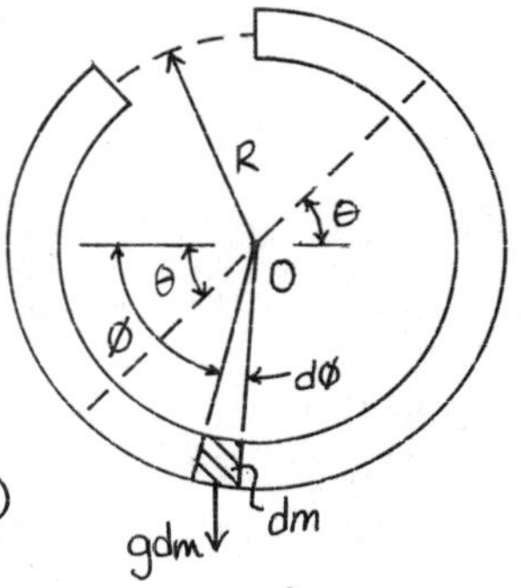

Figure b

The angular momentum relative to O is (Fig. b)

$$A_o = (R)(mv) = R(mR\dot{\theta})$$

where

$$m = \pi\rho R$$

or $\quad A_o = \pi\rho R^3 \dot{\theta} \quad$ (c)

So, by the law of angular momentum

$$M_o = \frac{dA_o}{dt}$$

and Eqs. (b) and (c), we have

$$-2\rho g R^2 \sin\theta = \pi\rho R^3 \ddot{\theta}$$

or $\quad \ddot{\theta} + \frac{2g}{\pi R}\sin\theta = 0 \quad$ (d)

Equation (d) agrees with Eq. (a).

16.95

GIVEN: A particle of mass m is acted on by a force parallel to the x-axis of the Newtonian frame (x, y); see Fig. a

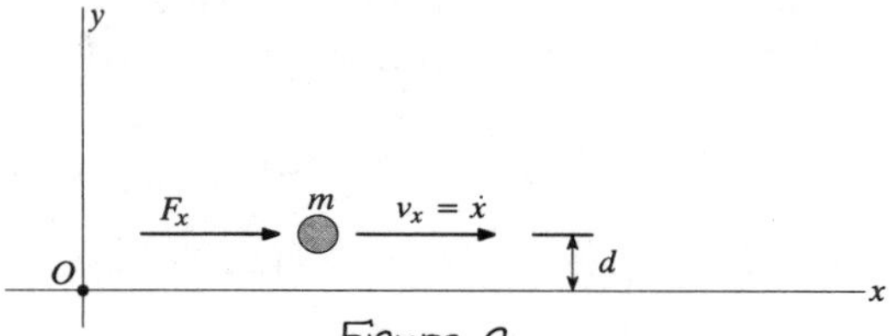

Figure a

FIND: Write the equation for the law of angular momentum relative to the origin O and show that it may be reduced to $F_x = m\ddot{x}$.

SOLUTION:

By Fig. a, the moment of F_x about O is

$$(+)\ M_o = (d)F_x \quad \text{(a)}$$

Also, the angular moment of the mass m is

$$(+)\ A_o = (d)(m\dot{x}) \quad \text{(b)}$$

By the law of angular momentum,

$$M_o = \frac{dA_o}{dt} \quad \text{(c)}$$

Hence, by Eqs. (a), (b), and (c), we obtain

$$(d)F_x = (d)m\ddot{x}$$

or $\quad \underline{\underline{F_x = m\ddot{x}}}$

16.96

GIVEN: Figure a, with $m_a = m_b = m$. Mass m_a slides on the frictionless horizontal surface. The pulley is frictionless and of negligible mass.

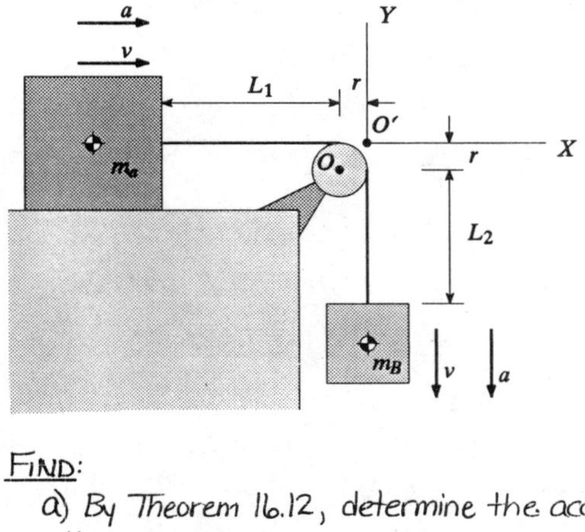

Figure a

FIND:

a) By Theorem 16.12, determine the acceleration of the masses.

b) Verify the result of part a by Newton's law $\bar{F} = m\bar{a}$.

(continued)

16.96 (cont.)

SOLUTION:

a) Let point Q in Theorem 16.12 coincide with G the center of mass of the system. By Fig. a, the center of mass coordinates relative to axes (X,Y) are (since $m_a = m_b = m$)

$$X_G = -\frac{m_a(L_1+r)}{m_a+m_b} = -\tfrac{1}{2}(L_1+r) \qquad \text{(a)}$$

$$Y_G = -\frac{m_b(L_2+r)}{m_a+m_b} = -\tfrac{1}{2}(L_2+r)$$

The free-body diagram of the system is shown in Fig. b.

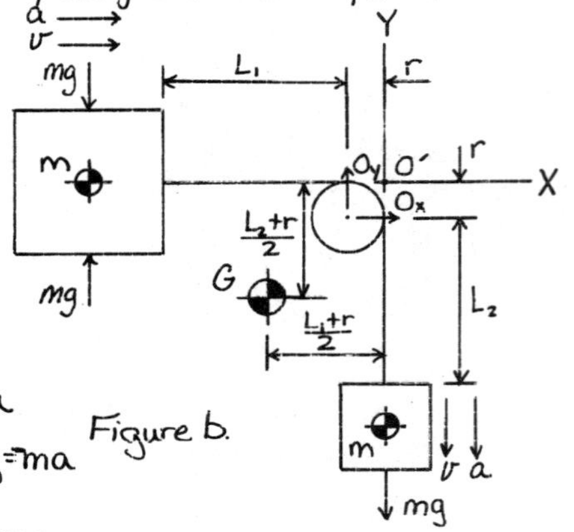

Figure b.

By Fig. b,

$\Sigma F_x = O_x = ma$

$\Sigma F_y = O_y - mg = ma$

Hence,

$$O_x = ma, \quad O_y = mg - ma \qquad \text{(b)}$$

Also by Fig. b,

$$\circlearrowleft\!+\ \Sigma M_G = -\tfrac{1}{2} mg(L_1+r) - O_x(\tfrac{1}{2}(L_2+r)-r) + O_y[\tfrac{1}{2}(L_1+r)-r] \qquad \text{(c)}$$

So, by Eqs. (b) and (c), the total moment about G is

$$M_G = \Sigma M_G = -\tfrac{1}{2} ma(L_1+L_2-2r) - mgr \qquad \text{(d)}$$

The angular momentum with respect to G is, by Fig. b,

$$\circlearrowleft\!+\ A_G = -\tfrac{1}{2}(L_2+r)m\upsilon - \tfrac{1}{2}(L_1+r)m\upsilon$$

or $$A_G = -\tfrac{1}{2}(L_1+L_2+2r)m\upsilon \qquad \text{(e)}$$

By Theorem 16.12, with $Q = G$,

$$M_G = \frac{dA_G}{dt} \qquad \text{(f)}$$

Therefore, by Eqs. (d), (e), and (f),

$$-\tfrac{1}{2} ma(L_1+L_2-2r) - mgr = -\tfrac{1}{2} ma(L_1+L_2+2r) \qquad \text{(g)}$$

where $a = \frac{d\upsilon}{dt}$. Equation (g) yields

$$\underline{\underline{a = \tfrac{1}{2} g}} \qquad \text{(h)}$$

b) To solve the problem by $F = ma$, consider the free-body diagrams of m_a and m_b (Fig. c).

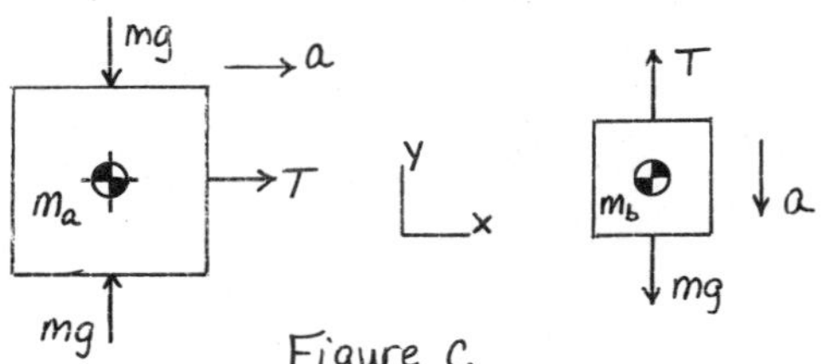

Figure C

For m_a: $\Sigma F_x = T = ma \qquad$ (i)

For m_b: $\Sigma F_y = T - mg = ma \qquad$ (j)

Solving Eqs. (i) and (j) for a, we obtain

$$\underline{\underline{a = \tfrac{1}{2} g}} \qquad \text{(k)}$$

which agrees with Eq. (h).

It is obvious that for this problem, the use of $F = ma$ is much more efficient than the principle of angular momentum.

16.97

GIVEN: Figure a: The equal weight bodies ①, ②, and ③ are released from rest. The fixed disks and body ② are smooth (frictionless).

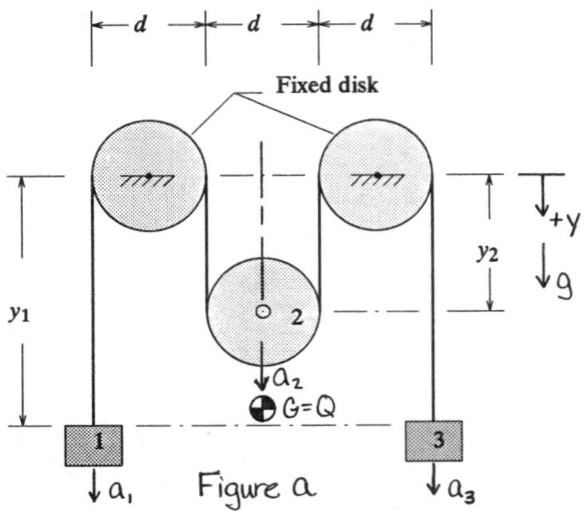

Figure a

FIND:

a) acceleration of the bodies, by the use of free-body diagrams and $F = ma$.

b) With the aid of Theorem 16.12, show that $a_1 = a_3$.

SOLUTION:

a) By kinematics and Fig. a,

$$2y_1 + 2y_2 = \text{Constant}$$

$$\therefore\ 2\upsilon_1 + 2\upsilon_2 = 0 \longrightarrow \upsilon_2 = -\upsilon_1$$

$$2a_1 + 2a_2 = 0 \longrightarrow a_2 = -a_1 \qquad \text{(a)}$$

By inspection, $a_3 = a_1$, since displacement of bodies ① and ③ are the same.

Next consider the free-body diagrams of bodies ① and ② (Fig. b).

(continued)

16.97 (cont.)

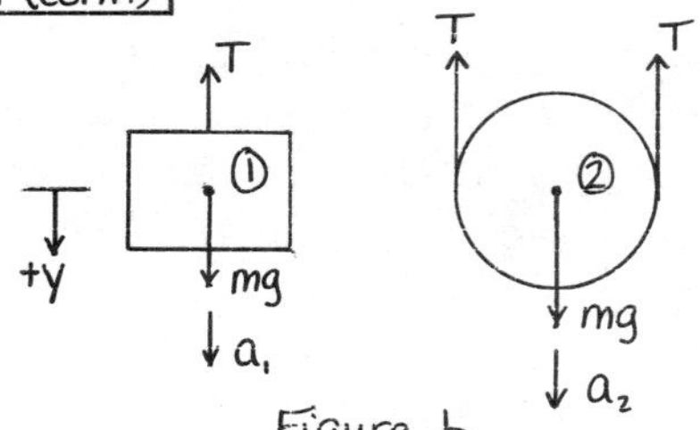

Figure b

By the free-body diagram of body ①,

$$\Sigma F_y = mg - T = ma_1 \quad (b)$$

By the free-body diagram of body ②,

$$\Sigma F_y = mg - 2T = ma_2 \quad (c)$$

or $T = m(g - a_1) = \frac{1}{2}m(g - a_2) = \frac{1}{2}m(g + a_1)$

Solving for a_1, we get $a_1 = \frac{1}{3}g$. Hence,

$$\underline{a_1 = a_3 = \tfrac{1}{3}g\ ;\ a_2 = -\tfrac{1}{3}g}$$

b) Let Q coincide with G (Fig. a). Then, by Theorem 16.12, $M_Q = \frac{dA_Q}{dt}$.

By the free-body diagram of the system (Fig. c),

$M_Q = \Sigma M_G = 0$

and

$A_Q = mv_1\left(\frac{3d}{2}\right) - mv_3\left(\frac{3d}{2}\right)$

$\therefore \frac{dA_Q}{dt} = m\left(\frac{3d}{2}\right)(a_1 - a_3) = 0$

or $\underline{a_1 = a_3}$ (d)

Then, Eqs. (b) and (c) yield

$a_1 = a_3 = \frac{1}{3}g\ ;\ a_2 = -\frac{1}{3}g$

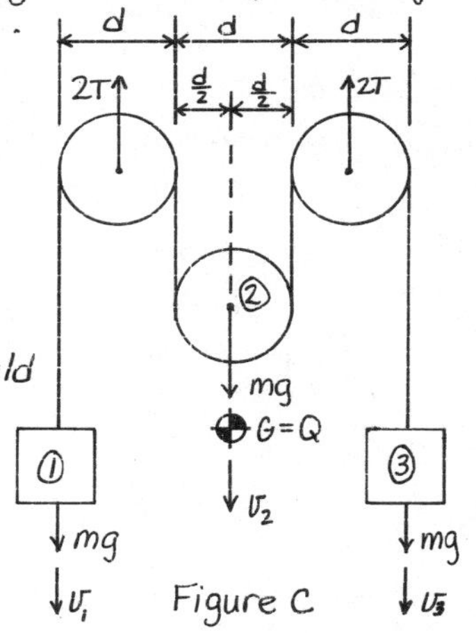

Figure c

16.98

GIVEN: A pendulum whose point support Q moves horizontally such that (Fig. a)

$$x = A \sin \Omega t \quad (a)$$

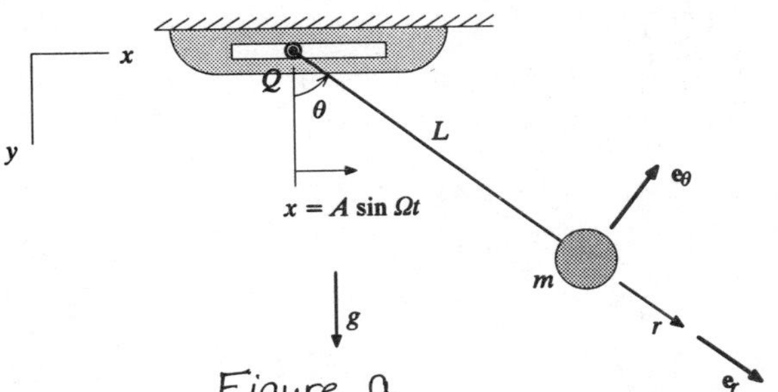

Figure a

FIND: Derive the differential equation of motion of mass m, by means of Eq. (16.59) in terms of θ, Ω, A, L, and t.

SOLUTION:

Equation (16.59) is

$$\vec{M}'_Q = \frac{d\vec{A}'_Q}{dt} + m\vec{r}_{G/Q} \times \vec{a}_Q \quad (b)$$

By Eq. (a) (see Fig. b),

$$\vec{a}_Q = \ddot{x}\,\hat{\imath} = -\Omega^2 A(\sin \Omega t)\hat{\imath} \quad (c)$$

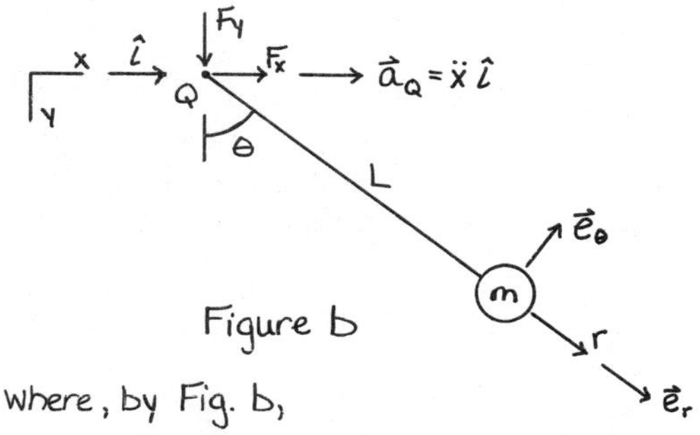

Figure b

where, by Fig. b,

$$\hat{\imath} = \vec{e}_r \sin\theta + \vec{e}_\theta \cos\theta \quad (d)$$

Therefore, by Eqs. (c) and (d),

$$\vec{a}_Q = -\Omega A^2(\sin \Omega t)(\vec{e}_r \sin\theta + \vec{e}_\theta \cos\theta) \quad (e)$$

Also, the position vector of m relative to Q is

$$\vec{r}_{G/Q} = L\,\vec{e}_r \quad (f)$$

Hence, by Eqs. (e) and (f),

$$m\vec{r}_{G/Q} \times \vec{a}_Q = -mL\Omega^2 A(\sin \Omega t)(\cos\theta)\,\vec{e}_z \quad (g)$$

where $\vec{e}_z = \vec{e}_r \times \vec{e}_\theta$ is a unit vector in the z direction.

The moment of momentum about point Q is (Fig. b)

$$\vec{A}'_Q = (mL\dot{\theta})L\,\vec{e}_z = mL^2\dot{\theta}\,\vec{e}_z \quad (h)$$

and the moment about Q is

$$\vec{M}'_Q = -(mg)(L\sin\theta)\,\vec{e}_z \quad (i)$$

Substitution of Eqs. (g), (h), and (i) into Eq. (b) yields, after simplification,

$$\underline{\underline{\ddot{\theta} + \frac{g}{L}\sin\theta - \frac{\Omega^2 A}{L}(\sin \Omega t)(\cos\theta) = 0}}$$

16.99

GIVEN: A pendulum whose point support Q moves vertically such that (Fig. a)

$$y = A\sin\Omega t \quad (a)$$

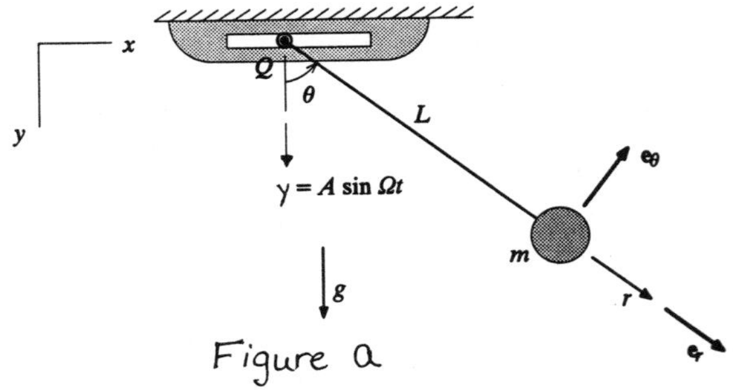

Figure a

FIND: Derive the differential equation of motion of m in terms of $\theta, \Omega, A, L,$ and t.

SOLUTION:

Equation (16.56) is

$$\vec{M}_Q = \frac{d\vec{A}_Q}{dt} + m\vec{r}_{G/Q} \times \vec{a}_Q \quad (b)$$

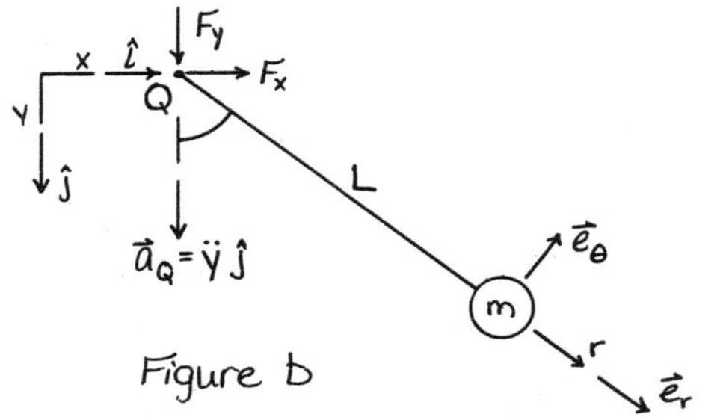

Figure b

By Eq. (a), see Fig. b,

$$\vec{a}_Q = \ddot{y}\,\hat{\jmath} = -\Omega^2 A(\sin\Omega t)\,\hat{\jmath} \quad (c)$$

where

$$\hat{\jmath} = \vec{e}_r\cos\theta - \vec{e}_\theta\sin\theta \quad (d)$$

$$\therefore \quad \vec{a}_Q = -\Omega^2 A(\sin\Omega t)(\vec{e}_r\cos\theta - \vec{e}_\theta\sin\theta) \quad (e)$$

Also, the mass m position vector is

$$\vec{r}_{G/Q} = L\vec{e}_r \quad (f)$$

Hence,

$$\vec{r}_{G/Q} \times \vec{a}_Q = L\Omega^2 A(\sin\Omega t)(\sin\theta)\,\vec{e}_z \quad (g)$$

where $\vec{e}_z = \vec{e}_r \times \vec{e}_\theta$ is a unit vector in the z direction.

The moment of momentum about point Q is (Fig. b)

$$\vec{A}_Q = (mL\dot{\theta})L\,\vec{e}_z = mL^2\dot{\theta}\,\vec{e}_z \quad (h)$$

and the moment about Q is

$$\vec{M}_Q = -(mg)(L\sin\theta)\,\vec{e}_z \quad (i)$$

Substitution of Eqs. (g), (h), and (i) into Eq. (b) yields, after simplification,

$$\underline{\underline{\ddot{\theta} + \left(\frac{g}{L} + \frac{\Omega^2 A}{L}\sin\Omega t\right)\sin\theta = 0}}$$

16.100

GIVEN: Example 16.26 in which two bodies of mass m_1 and m_2 are supported by a mechanism of two light cords and two identical frictionless pulleys of negligible mass (Fig. a). Let $m_1 = 3m_2$.

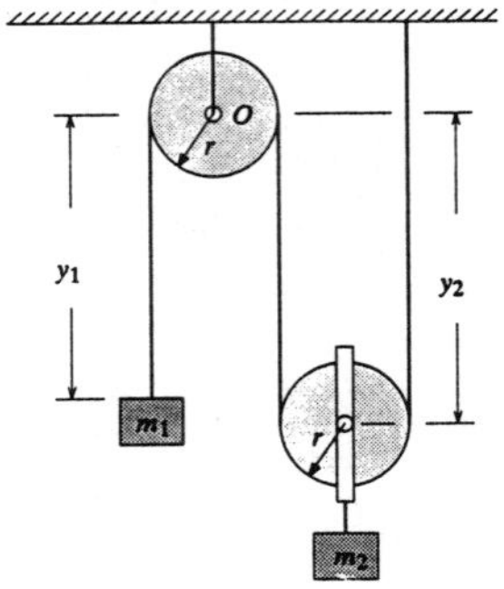

Figure a

FIND:

(a) the accelerations of the bodies by means of free-body diagrams and Newton's second law $F = ma$.

(b) the accelerations of the bodies by means of Theorem 16.12, taking point Q as G, the center of mass of the system.

SOLUTION:

(a) The free-body diagrams of m_1 and m_2 are shown in Fig. b.

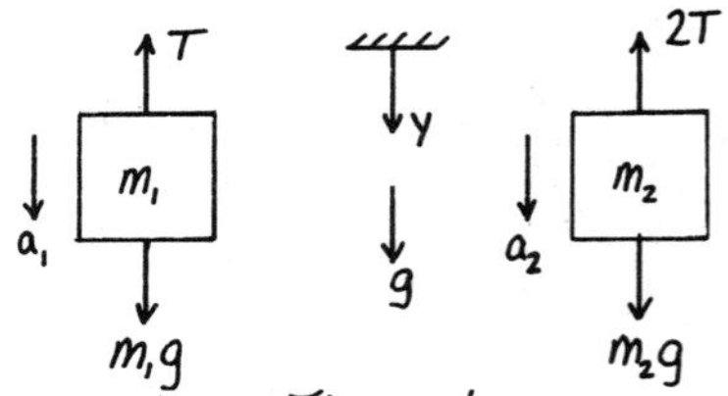

Figure b

From the kinematical constraint of the cord (Fig. a),

$$y_1 + 2y_2 = \text{constant}$$

$$v_1 + 2v_2 = 0; \quad v_1 = -2v_2 \quad (a)$$

$$a_1 + 2a_2 = 0; \quad a_1 = -2a_2 \quad (b)$$

By Fig. b, we have

$$\Sigma F_{y_1} = m_1 g - T = m_1 a_1$$

$$\Sigma F_{y_2} = m_2 g - 2T = m_2 a_2$$

or, since $m_1 = 3m_2$, and $a_1 = -2a_2$

$$T = m_1(g - a_1) = 3m_2(g + 2a_2) \quad (c)$$

$$T = \tfrac{1}{2}m_2(g - a_2) \quad (d)$$

Solving these two equations for a_2, we obtain

$$\underline{\underline{a_2 = -\tfrac{5}{13}g = \tfrac{5}{13}g\uparrow}} \quad (e)$$

and by Eqs. (b) and (e),

$$\underline{\underline{a_1 = \tfrac{10}{13}g\downarrow}}$$

(continued)

16.100 (cont.)

(b) The free-body diagram of the entire system (with the horizontal location of G indicated) is shown in Fig. C.

Figure C

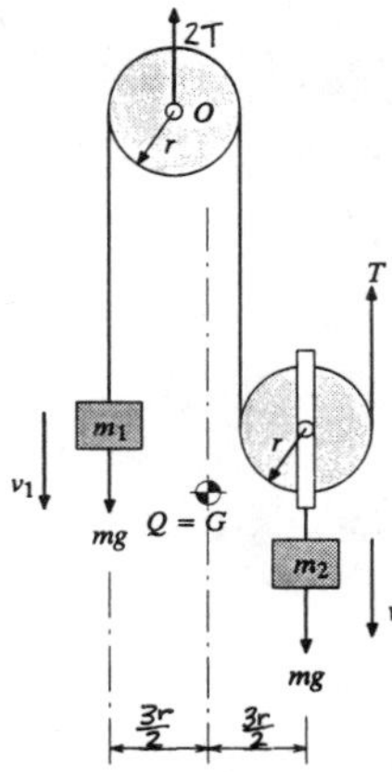

Let point Q = G. Then, by Theorem 16.12,

$$M'_Q = \frac{dA'_Q}{dt} \qquad (g)$$

where, by Fig. C,

$$\circlearrowleft M'_Q = 3m_2 g(\tfrac{3}{4}r) + 2T(\tfrac{1}{4}r) + T(\tfrac{13}{4}r) - m_2 g(\tfrac{9}{4}r)$$

or with Eq. (c),

$$M'_Q = m_2(g + 2a_2)(\tfrac{45}{4}r) \qquad (h)$$

Also, by Fig. C,

$$\circlearrowleft A'_Q = (\tfrac{3}{4}r)m_1 v_1 - (\tfrac{9}{4}r)m_2 v_2$$

or with $m_1 = 3m_2$ and $v_1 = -2v_2$ [see Eq. (a)],

$$A'_Q = -\tfrac{27}{4} m_2 r v_2 \qquad (i)$$

So, by Eqs. (g), (h), and (i),

$$m_2(g + 2a_2)(\tfrac{45}{4}r) = -\tfrac{27}{4} m_2 r a_2$$

or

$$\underline{\underline{a_2 = -\tfrac{5}{13}g = 5/13\, g \uparrow}}$$

and by Eq. (b), again,

$$\underline{\underline{a_1 = \tfrac{10}{13} g \downarrow}}$$

as in part a.

16.101

GIVEN: A stationary rocket engine develops 13.5 kN thrust when the gases are discharged at 1.8 km/s.

FIND: the total weight W of products of combustion that are discharged during 5 minutes of operation.

SOLUTION:

By Eq. (16.62),

$$F_T = m\frac{dv}{dt} = v_{ex}\frac{dm}{dt}$$

Therefore,

$$13.5(10^3)\text{ N} = 1.8(10^3)\text{ m/s}\left(\frac{\Delta m}{300\text{ s}}\right)$$

or,

$$\Delta m = 2250 \text{ kg}$$

Hence,

$$\underline{\underline{W = (\Delta m)g = (2250)(9.81) = 22.07 \text{ kN}}}$$

16.102

GIVEN: A rocket in outer space is initially at rest. Its engine is started and it discharges burned fuel at 3.00 mi/s, relative to the rocket. The initial mass of fuel is four times the mass of the empty rocket.

FIND: the speed that the rocket has acquired after the fuel is exhausted.

SOLUTION:

By Eq. 16.61,

$$v_f = v_i + v_{ex}\ln\left(\frac{m_i}{m_f}\right)$$

where v_f is the final speed of the rocket, $v_{ex} = 3.00$ mi/s, and the rocket is initially at rest ($v_i = 0$).

Thus,

$$v_f = 0 + (3.00\text{ mi/s})\ln\left(\frac{m_{fuel} + m_{rocket}}{m_{rocket}}\right)$$

From the given information,

$$m_{fuel} = 4m_{rocket}.$$

Therefore,

$$\underline{\underline{v_f = (3.00\text{ mi/s})\ln(5) = 4.828\text{ mi/s}}}$$

16.103

GIVEN: A rocket is initially at rest ($v_i = 0$) and has a total mass $(M_{fuel} + M_{rocket}) = M_i$. It launches vertically from the surface of the earth. The fuel is 90% of the total mass ($M_{fuel} = 0.9M_i$). Neglect air resistance and gravity. The exhaust gas speed v_{ex}, relative to the rocket.

FIND:

(a). the maximum speed v_f attainable by the rocket in terms of v_{ex}.

(b). Of what signifigance is the burn time in your answer?

SOLUTION:

By Eq. (16.61)

$$v_f = v_i + v_{ex}\ln\left(\frac{M_i}{M_f}\right) \qquad (a)$$

(continued)

16.103 (cont.)

a.) The maximum speed attainable occurs at the instant before burnout. At this time, the fuel (or 90% of the mass) has burned away. Thus, at burnout $M_f = 0.10 M_i$.

Hence, since $v_i = 0$ at burnout, by Eq. (a),

$$v_f = v_{ex} \ln(10) = 2.30\, v_{ex}$$

b.) The burnout time does not enter into the calculation for v_f, since air resistance and gravity are neglected. However, see Problem 16.105.

16.104

GIVEN: In problem 16.103, assume that the rocket burns fuel at the rate $R = 0.01 M_i$ per second, and exhausts the burned gases at a speed $v_{ex} = 2500$ m/s.

FIND:

(a) the burnout time.

(b) the time at which the absolute velocity of the exhaust gases is zero.

(c) the maximum absolute velocity of the gases.

SOLUTION:

(a). From Problem 16.103, the mass of fuel is $M_{fuel} = 0.90 M_i$. Since the burn rate is $R = 0.01 M_i$ per second, the burnout time is given by

$$t = \frac{0.90 M_i}{0.01 M_i} = 90 \text{ s}$$

(b). The absolute velocity of the exhaust gases is $v^* = v - v_{ex}$

where v is the velocity of the rocket at time t. Hence,

$$v^* = v_{ex} \ln\left(\frac{M_i}{M_i - 0.01 M_i t}\right) - v_{ex} \qquad (a)$$

when $v^* = 0$, Eq. (a) yields,

$$\ln\left(\frac{M_i}{M_i - 0.01 M_i t}\right) = \ln\left(\frac{1}{1 - 0.01t}\right) = 1$$

or $$e = \frac{1}{1 - 0.01t}$$

Therefore,

$$t = 63.2 \text{ s}$$

(c) The maximum absolute velocity of the gases occurs at burnout. Then, $v = v_f$ and $M_i - 0.01 M_i t = 0.10 M_i$.

Hence,

$$v_f = v_{ex} \ln(10) = 2.303\, v_{ex}$$

and, by Eq. (a),

$$v^* = v_f - v_{ex} = 1.303\, v_{ex}$$

So,

$$v^*_{max} = (1.303)(2500) = 3256 \text{ m/s}$$

in the same direction as the rocket which is traveling at $v_f = (2.303)(2500) = 5758$ m/s.

16.105

GIVEN: In Problem 16.103, take the acceleration of gravity to be constant $g = 9.81$ m/s². The fuel burns at a constant rate, the burnout is 20 seconds, and the exhaust speed is 2500 m/s.

FIND:

(a) the maximum speed attainable by the rocket.

(b) the maximum speed if burnout is 40 s.

(c) the effect of burnout on the maximum speed attainable in liftoff from earth.

SOLUTION:

(a) By Eq. (16.64),

$$v(t) = v_i + v_{ex} \ln\left(\frac{M_i}{M_i - Rt}\right) - gt \qquad (a)$$

From Problem 16.103, $0.90 M_i$ is fuel.
The maximum attainable speed v_{max} of the rocket occurs at burnout, that is, when $t = 20$ s.
Also, at burnout, the final mass is

$$M_f = M_i - Rt = 0.10 M_i \qquad (b)$$

Therefore, by Eqs. (a) and (b), with $t = 20$ s, since $v_i = 0$,

$$v(20\text{s}) = v_{max} = 2500 \ln(10) - (9.81)(20) = 5{,}560 \text{ m/s}$$

(b) If the burnout is 40 s, the rate of burning fuel is slower, but at burnout, the final mass is $M_f = 0.10 M_i$ as in part a [Eq. (b)]. By Eq. (a), with $t = 40$ s and $v_i = 0$,

$$v(40\text{s}) = v_{max} = 2500 \ln(10) - (9.81)(40) = 5{,}364 \text{ m/s}$$

(c) By the results of parts a and b, we see that the longer the burnout time in a gravity field, the lower the maximum attainable rocket speed at burnout.

16.106

GIVEN: A spaceship enters the earth's gravitational field in reverse, and travels on a straight line perpendicular to the earth's surface (Fig. a). On entry, its mass is $M = 2250$ kg and its downward speed is $v_i = 16\,200$ km/h.

FIND: the rate R (kg/s) at which fuel must be burned to reduce the speed to 11 880 km/h in 50 s, with $v_{ex} = 1.2$ km/s and the average value of $g = 9$ m/s².

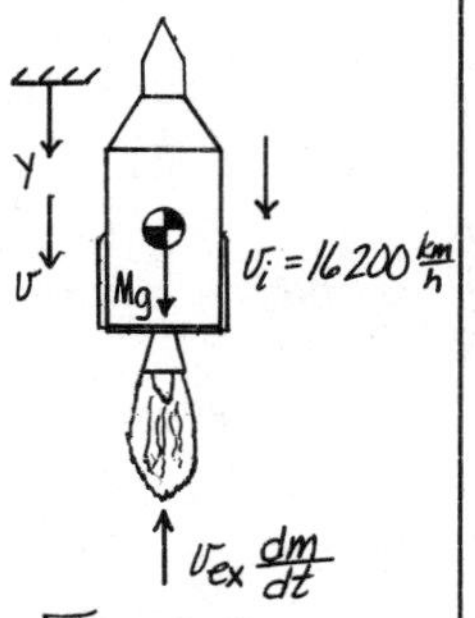

Figure a

SOLUTION:

By Fig. a,

$$\Sigma F_y = Mg - v_{ex}\frac{dm}{dt} = M\frac{dv}{dt}$$

or $dv = g\,dt - \frac{v_{ex}}{M}dm$

But $dm = -dM$. Therefore,

$$\int_{v_i}^{v_f} dv = \int_{t_i}^{t_f} g\,dt + v_{ex}\int_{M_i}^{M_f}\frac{dM}{M} \qquad \text{(a)}$$

Integration of Eq. (a) yields

$$v_f - v_i = g(t_f - t_i) + v_{ex}\ln\left(\frac{M_f}{M_i}\right) \qquad \text{(b)}$$

With, $v_i = 16\,200$ km/h $= 4500$ m/s, $v_f = 11\,800$ km/h $= 3278$ m/s, $g = 9$ m/s², $t_f = 50$ s, $t_i = 0$, $v_{ex} = 1.2$ km/s $= 1200$ m/s, $M_i = 2250$ kg and $M_f = 2250 - 50R$, Eq. (b) yields

$$-1.3933 = \ln\left(\frac{2250 - 50R}{2250}\right)$$

or $$\frac{2250 - 50R}{2250} = e^{-1.3933} \qquad \text{(c)}$$

Solving Eq. (c) for R, we obtain

$$\underline{R = 33.83 \text{ kg/s}}$$

16.107

GIVEN: A rocket that weighs 350 kips (including fuel) is launched vertically with $v_{ex} = 10\,000$ ft/s.

FIND:

(a) the mass rate R (slugs/s) and the corresponding thrust F_T to equal the gravitational force on the rocket.

(b) the values of F_T and v_{ex} to give the rocket an acceleration of $2g$ at takeoff with $R = 35$ slug/s.

SOLUTION:

(a) By Eq. (16.63), with $\frac{dm}{dt} = R$

$$a = \frac{v_{ex}}{M}\frac{dm}{dt} - g = \frac{v_{ex}R}{M} - g \qquad \text{(a)}$$

To overcome gravity, a must be positive. To just equal gravity $a = 0$, and Eq. (a) yields

$$R = \frac{Mg}{v_{ex}} = \frac{350\,000}{10{,}000}$$

or $\underline{R = 35 \text{ slugs/s}}$

and the thrust is, by Eq. (16.62),

$$\underline{\underline{F_T = v_{ex}\frac{dm}{dt} = v_{ex}R = 350 \text{ kips}}}$$

This result is obvious, since F_T must equal the weight $W = 350$ kips of the rocket.

(b) By Eq. (a), with $a = 2g$, $R = 35$ slugs/s, and $M = M_i = \frac{350\,000}{32.2} = 10{,}870$ slugs, we find

$$v_{ex} = \frac{3Mg}{R} = \frac{3(10{,}870)(32.2)}{35}$$

or $\underline{v_{ex} = 30{,}000 \text{ ft/s}}$

Hence, the required thrust is

$$F_T = v_{ex}R = (30{,}000)(35)$$

or $\underline{F_T = 1050 \text{ kips}}$

Alternatively, F_T in part b could have been calculated as

$$\Sigma F = F_T - Mg = 2Mg$$

or $F_T = 3Mg = 3(350 \text{ kips}) = 1050$ kips.

16.108

GIVEN: A rocket initially weighs 300 kN (of which 240 kN is fuel). It is launched vertically with burn mass rate of fuel $R = \frac{2000}{9.81} = 203.9 \frac{kg}{s}$ and exhaust velocity $v_{ex} = 2000$ m/s. From takeoff to burnout, the average acceleration of gravity is $g_{ave} = 9.6$ m/s².

FIND:

(a) the average thrust F_T of the rocket to burnout.

(b) the speed at burnout.

(c) Assume $R = \frac{4000}{9.81} = 407.7$ kg/s, and $v_{ex} = 2200$ m/s. Repeat parts (a) and (b) and explain the results, compared to parts (a) and (b).

SOLUTION:

(a) By Eq. (16.62),

$$F_T = v_{ex}\frac{dm}{dt} = v_{ex}R.$$

$$\therefore\ \underline{F_T = (2000)(203.9) = 407.8 \text{ kN}}$$

(b) By Eq. (16.64),

$$v(t) = v_i + v_{ex}\ln\left(\frac{M_i}{M_f}\right) - g_{ave}t \qquad \text{(a)}$$

Initially, $v_i = 0$, $M_i = 300000/9.81 \doteq 30\,581$ kg.

$\therefore$ $M_f = M_i - Rt = 30\,581 - (203.9)t$. At burnout $M_f = 60000/9.81 = 6116.2$ kg. Therefore, at burnout, $t = t_b = (30581 - 6116.2)/203.9 = 120$ s

(continued)

16.108 (cont.)

Hence, by Eq. (a), at burnout, the velocity is

$$v_f = v(t_b) = 0 + 2000 \ln(5) - (9.6)(120)$$

or $\underline{v_f = 2.067 \text{ km/s}}$

(c) With $R = 4000/9.81 = 407.7$ kg/s and $v_{ex} = 2200$ m/s,

$$\underline{F_T = v_{ex} R = (2200)(407.7) = 897 \text{ kN}}$$

At burnout,

$$t_b = (30\,581 - 6116.2)/(407.7) = 60 \text{ s}$$

Hence, at burnout, by Eq. (a),

$$v_f = v(t_b) = 0 + 2200 \ln(5) - 9.6(60)$$

or $\underline{v_f = 2.965 \text{ km/s}}$

If the gravity effect is ignored, Eq. (a) predicts that in part b, $v_f = 2000 \ln(5) = 3.219$ km/s and in part c, $v_f = 2200 \ln(5) = 3.541$ km/s. In part b, gravity acts for 120s until burnout and reduces v_f to 2.067 km/s. In part c, burnout occurs at 60s, and even though v_{ex} is increased, v_f is reduced to 2.965 km/s. Therefore, in a gravity field, the shorter the burn time, the higher the velocity v_f at burnout.

16.109

GIVEN: In Problem 16.108, assume that $g = 9.81$ m/s² at takeoff. Also, $R = 2$ kN/s, $W_{rocket} = 60$ kN, $v_{ex} = 2$ km/s, $W_{fuel} = 240$ kN, and $t_b = \frac{240}{2} = 120$ s.

FIND:

(a) the initial acceleration of the rocket at takeoff for part a of Problem 16.108.

(b) the acceleration of the rocket, in part a, for the instant before burnout. For the instant after burnout.

SOLUTION:

(a) By Eq. (16.63), and the data of Problem 16.108,

$$a = \frac{v_{ex}}{M}\frac{dm}{dt} - g = \left(\frac{2000}{300000}\right)(2000) - 9.81$$

or $\underline{a = 3.523 \text{ m/s}^2}$

(b) By Eq. (16.64),

$$v(t) = v_{ex} \ln\left(\frac{M_i}{M_i - Rt}\right) - gt \qquad \text{(a)}$$

Differentiation of Eq. (a) with respect to time t yields

$$a(t) = \frac{dv(t)}{dt} = v_{ex}\left(\frac{R}{M_i - Rt}\right) - g \qquad \text{(b)}$$

At burnout, $t_b = 120$ s. At the instant before burnout $t \approx t_b$ and the exhaust velocity is still $v_{ex} = 2000$ m/s.

Also, $M_i = (W_{rocket} + W_{fuel})/g = \frac{300000}{9.81} = 30\,581$ kg and $R = \frac{2000}{9.81} = 203.9$ kg/s since the weights are measured on earth where $g = 9.81$ m/s². However, at burnout the value of g_b is unknown. But since the average value of g from takeoff to burnout is 9.6 m/s² and initially, $g = 9.81$ m/s², we might guess that $g_b \approx 9.4$ m/s². Hence, with the above data, Eq. (b) yields

$$a = (2000)\left(\frac{203.9}{30581 - (203.9)(120)}\right) - 9.4$$

or $\underline{a = 57.3 \text{ m/s}^2}$

Alternatively, by Eq. (16.63), with $M = M_f = \frac{60000}{9.81} = 6116.2$ kg,

$$a = \frac{v_{ex}}{M_f}\frac{dm}{dt} - g = \left(\frac{2000}{6116.2}\right)(203.9) - 9.4$$

or $\underline{a = 57.3 \text{ m/s}^2}$

16.110

GIVEN: A rocket hovers motionless above the earth.

FIND:

(a) Show that $Mg = v_{ex}\frac{dm}{dt}$.

(b) Show that if v_{ex} is constant, the maximum time that the rocket can hover is

$$t_{max} = \left(\frac{v_{ex}}{g}\right)\ln\left(\frac{M_i}{M_f}\right).$$

(c) If $t_{max} = 10$ min, $g = 32.2$ ft/s², $v_{ex} = 3$ mi/s, what is the required ratio M_i/M_f?

SOLUTION:

(a) If the rocket is to hover, the gravity force must be balanced by the thrust force on the rocket. The momentum of gases that exhaust per second is $v_{ex}\frac{dm}{dt}$. This is equal to the thrust on the rocket, and hence must be equal to the weight Mg. Therefore,

$$\underline{Mg = v_{ex}\frac{dm}{dt}} \qquad \text{(a)}$$

Alternatively, by Eq. (16.62),

$$F_T = v_{ex}\frac{dm}{dt}$$

By Fig. a,

$$\Sigma F_y = Mg - v_{ex}\frac{dm}{dt} = 0$$

or $\underline{Mg = v_{ex}\frac{dm}{dt}}$

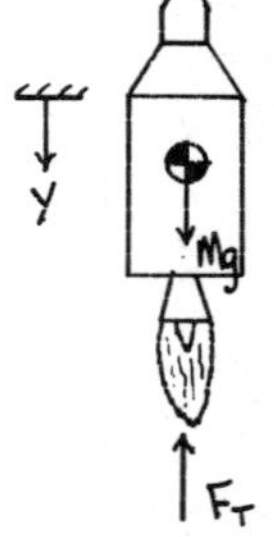

Figure a

(b) Since mass is conserved,

$$\frac{dM}{dt} = -\frac{dm}{dt}$$

(continued)

16.110 (cont.)

Therefore, by Eq. (a), $\frac{dM}{dt} = -\frac{(Mg)}{v_{ex}}$,

or $v_{ex}\frac{dM}{M} = -g\,dt$ (b)

Integration yields

$$v_{ex}\ln(M) = -gt + \text{constant}$$

For $t=0$, $M=M_i$. Therefore, constant $= v_{ex}\ln M_i$. Therefore,

$$v_{ex}\ln\left(\frac{M_i}{M}\right) = gt$$

For $t = t_{max}$, $M = M_f$. Therefore,

$$t_{max} = \left(\frac{v_{ex}}{g}\right)\ln\left(\frac{M_i}{M_f}\right) \quad (c)$$

Alternatively, by Eq. (16.64), for hovering with $t = t_{max}$ and $v(t)=0$, $v_i=0$,

$$v(t) = v_i + v_{ex}\ln\left(\frac{M_i}{M_f}\right) - gt$$

or $0 = 0 + v_{ex}\ln\left(\frac{M_i}{M_f}\right) - gt_{max}$

or $t_{max} = \left(\frac{v_{ex}}{g}\right)\ln\left(\frac{M_i}{M_f}\right)$

(c) By Eq. (c), with $t_{max} = 10$ min $= 600$s, and $v_{ex} = 3$ mi/s $= 15,840$ ft/s,

$$15840\ln\left(\frac{M_i}{M_f}\right) = (32.2)(600)$$

or $\frac{M_i}{M_f} = 3.386$

16.111

GIVEN: A rocket must be designed to boost, from the earth's surface, a payload of 3000 kg mass (3 metric tons) to attain a speed of 8000 m/s.

FIND: the amount of fuel required to deliver the pay for the following designs (Ignore air resistance and gravity):

(a) A design of the rocket engine and fuel that produces an exhaust velocity of $v_{ex} = 2500$ m/s.

(b) A second engine-fuel design that produces an exhaust velocity of $v_{ex} = 5000$ m/s.

(c) What is the effect of doubling the exhaust velocity on fuel consumption?

(d) Which design would you recommend? What are the pros and cons of your recommendation?

SOLUTION:

By Eq. (16.61), with $v_i = 0$

$$v_f = 0 + v_{ex}\ln\left(\frac{M_i}{M_i - Rt}\right) \quad (a)$$

or in terms of the mass of the payload $M_{payload}$ and the mass of the fuel M_{fuel}, Eq. (a) yields, with $v_f = 8000$ m/s and $v_{ex} = 2500$ m/s,

$$8000 = 2500\ln\left(\frac{M_{payload} + M_{fuel}}{M_{payload}}\right)$$

or $8000 = 2500\ln\left(1 + \frac{M_{fuel}}{M_{payload}}\right)$

Hence,

$$1 + \frac{M_{fuel}}{M_{payload}} = e^{8000/2500} = 24.5325$$

or $M_{fuel} = 23.5325\,M_{payload}$

$= 23.5325(3)$

Thus, $M_{fuel} = 70.6$ metric tons $= 70.6 \times 10^3$ kg

(b) Similarly, with $v_f = 8000$ m/s and $v_{ex} = 5000$ m/s,

$$1 + \frac{M_{fuel}}{M_{payload}} = e^{8000/5000} = 4.953$$

or $M_{fuel} = 11.86$ metric ton $= 11.86 \times 10^3$ kg

(c) By designing an engine and fuel (Part b) that doubles the exhaust velocity of part a, the fuel consumption has been reduced approximately 6 fold $(70.6/11.86 = 5.93)$ or $(70.6 - 11.86)(100)/70.6 = 83.2\%$.

(d) On the basis of the fuel savings noted in part c, it appears that the second design should be recommended. However, it is rare that such a revolutionary design that saves such amounts of fuel is achieved. I would be skeptical of the new design, and question the claims. For example,

- What kind of fuel is used?
- What new design features have been made in the engine?
- How expensive are the new fuel and the new engine?
- Are the exhaust velocities of the new design repeatable?
- Is the new fuel readily available?

16.112

GIVEN: A rocket sled burns fuel at a constant rate $R = 120$ lb/s. The weight of the sled is 1800 lb (includes 360 lb of fuel). Neglect friction of the track and drag.

FIND:
(a) Derive a formula for the acceleration a of the sled in terms of the time t and the exhaust velocity v_{ex}. Plot the ratio a/v_{ex} versus time t for the range $0 \le t \le 4$ s, and check the slope of the graph for $t=0$ and $t=4$ s by the formula for a.
(b) Determine the ratio of velocity v_f of the sled at burnout to v_{ex}.

SOLUTION:
(a) The initial mass is $M_i = 1800/32.2 = 55.90$ slugs, and the burn rate is $dm/dt = R = \frac{120 \text{ lb}}{32.2 \text{ ft/s}^2} = 3.727 \frac{\text{slug}}{\text{s}}$

By Eq. (16.63), with $M = M_i - Rt$,

$$a = \frac{v_{ex}}{M}\frac{dm}{dt} = v_{ex}\left(\frac{3.727}{55.90 - 3.727t}\right)$$

or
$$a = v_{ex}\left(\frac{1}{15-t}\right) \qquad \text{(a)}$$

Equation (a) yields
$$\frac{a}{v_{ex}} = \frac{1}{15-t} \qquad \text{(b)}$$

See the plot of Eq. (b).

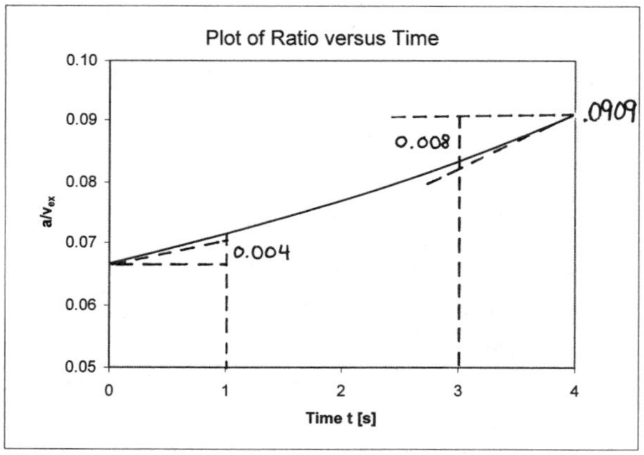

The slope of the graph is, by Eq. (b),

$$\frac{d}{dt}\left(\frac{a}{v_{ex}}\right) = \frac{1}{(15-t)^2} \qquad \text{(c)}$$

By the plot:
For $t=0$, slope $\approx$ 0.004
For $t=4$ s, slope $\approx$ 0.008

By Eq. (c):
For $t=0$, slope $= \frac{1}{225} = 0.00444$
For $t=4$ s, slope $= \frac{1}{121} = 0.00826$

(b) By Eq. (16.61),

$$v_f = v_i + v_{ex} \ln\left(\frac{M_i}{M_b}\right);$$

where $M_b = M_i - \frac{360}{32.2} = 44.72$ slugs.

Therefore,
$$v_f = 0 + v_{ex} \ln\left(\frac{55.90}{44.72}\right)$$

or
$$\frac{v_f}{v_{ex}} = 0.223$$

16.113

GIVEN: A toy rocket contains water that is ejected at a speed of $v_{ex} = 40$ ft/s, by compressed air. The diameter of the jet of water is $d = 0.25$ in., and the mass density of water is $\rho = 1.938$ slugs/ft³.

FIND: the thrust exerted on the rocket, when it is held motionless.

SOLUTION:
The mass (in slugs) of water ejected per second is [see Eq. (16.65)],

$$\frac{dm}{dt} = \rho v_{ex}\left(\frac{\pi d^2}{4}\right) = (1.938)(40)\left(\frac{\pi}{4}\right)\left(\frac{0.25}{12}\right)^2 = 0.0264 \frac{\text{slugs}}{\text{s}}$$

Hence, the linear momentum of the ejected water per second is

$$v_{ex}\frac{dm}{dt} = (40 \text{ ft/s})(0.0264 \text{ slugs/s}) = 1.057 \text{ lb}$$

This force is equal to thrust that acts on the rocket; that is [see also Eq. (16.62)]

$$F_T = v_{ex}\frac{dm}{dt} = 1.057 \text{ lb.}$$

16.114

GIVEN: Air is injected into a turbojet engine at 500 N/s with a speed of 300 m/s (Fig. a). Fuel is injected at 50 N/s normal to the flow of air, and burned fuel is exhausted at a speed of 750 m/s.

FIND: the thrust of the engine.

SOLUTION:
The thrust of the engine is equal to the net change in momentum convected out of the engine. That is, in the direction of air flow,

$$F_T = \left[\left(\frac{dM}{dt}\right)_{fuel} + \left(\frac{dM}{dt}\right)_{air}\right]v_{ex} - \left[\frac{dM}{dt}\right]_{air} v_{air}$$

$$= \left[\frac{50 \text{ N/s}}{9.81 \text{ m/s}^2} + \frac{500 \text{ N/s}}{9.81 \text{ m/s}^2}\right](750 \text{ m/s}) - \left(\frac{500 \text{ N/s}}{9.81 \text{ m/s}^2}\right)(300 \text{ m/s})$$

or
$$F_T = 26\,758 \text{ N} = 26.8 \text{ kN}$$

16.115

GIVEN: Each nozzle of the lawn sprinkler (Fig. a) discharges water at a rate Q (m^3/s). The water weighs w (N/m^3) and has a speed of v (m/s).

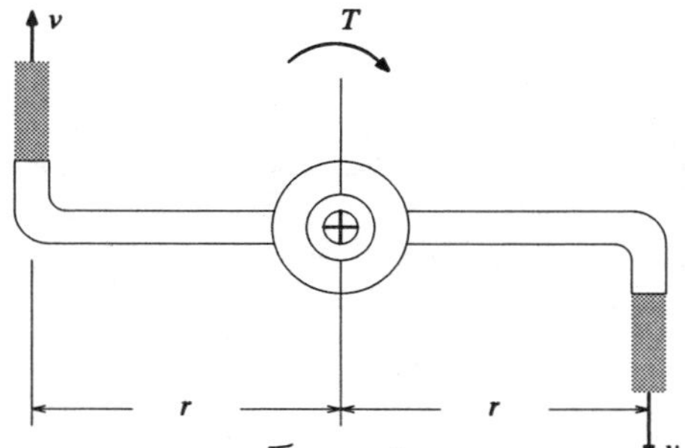

Figure a

FIND: the torque T (N·m), in terms of $Q, r, w,$ and v, required to keep the sprinkler from rotating.

SOLUTION:

The force exerted on the sprinkler by the water that exits from each nozzle is (Fig. b)

$$F_T = v\frac{dm}{dt} = v\frac{Qw}{g}. \quad (a)$$

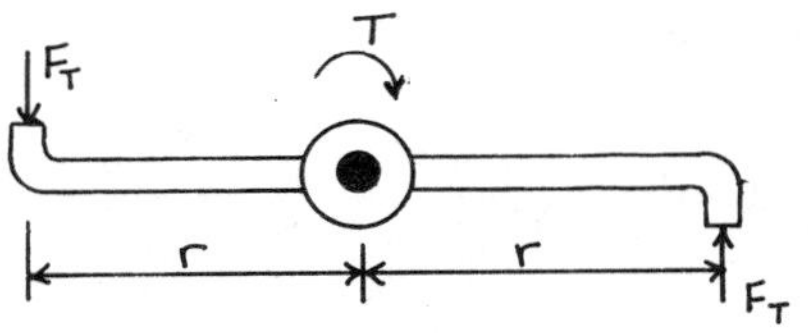

Figure b

Hence, by equilibrium of moments about the axis of the sprinkler,

$$\Sigma M = T - 2rF_T = 0 \quad (b)$$

Equations (a) and (b) yield (with $g = 9.81\ m/s^2$)

$$T = \frac{2rvQw}{g} \quad (N\cdot m)$$

16.116

GIVEN: Figure a. Water ($\rho = 1.938$ slugs/ft^3) enters the spherical vane at a constant rate of 2.50 ft^3/s with a speed of 80 ft/s and leaves with a speed of 20 ft/s.

FIND: the force that the water exerts on the vane.

Figure a

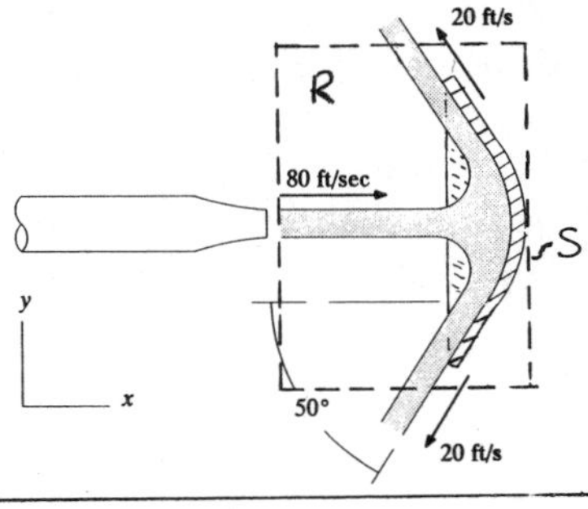

SOLUTION:

By a process similar to Example 16.30, select surface S enclosing region R (Fig. a).

By Theorem 16.13, the force exerted on the water is

$$\vec{F}_w = \vec{V}_{out}\frac{dm}{dt} - \vec{V}_{in}\frac{dm}{dt} \quad (a)$$

For steady flow, $\frac{dm}{dt}$ is constant.

$$\therefore \frac{dm}{dt} = (2.50\ ft^3/s)(1.938\ slugs/ft^3) = 4.845\ slugs/s$$

By (x,y) projections, the (x,y) forces exerted on the water are [with Eq. (a)]

$$F_x = -(20)(\cos 50^\circ)\frac{dm}{dt} - 80\frac{dm}{dt}$$

or $F_x = -450$ lb $= 450$ lb ←

Similarly,

$$F_y = (20)(\sin 50^\circ)\frac{dm}{dt} - 20(\sin 50^\circ)\frac{dm}{dt} = 0$$

Therefore, the force exerted on the vane by the water is

$$F_{vane} = 450 \text{ lb} \rightarrow$$

16.117

GIVEN: Figure a. Water ($\rho = 1$ g/cm^3) enters the vane at a constant rate of 0.09 m^3/s, with a speed of 18 m/s and leaves the vane with the same speed. Two-thirds of the water is deflected upward.

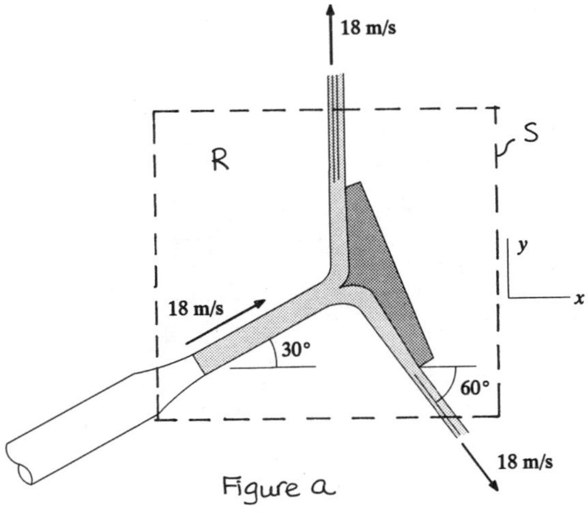

Figure a

FIND: the (x,y) projections of force water exerts on the vane.

SOLUTION:

By a process similar to Example 16.30, select surface S enclosing region R (Fig. a).

By Theorem 16.13, the force exerted on the water is $\vec{F}_{water} = \vec{V}_{out}\frac{dm}{dt} - \vec{V}_{in}\frac{dm}{dt}$ (a) (continued)

16.117 (cont.)

For steady flow $\frac{dm}{dt}$ is constant.

$$\therefore \frac{dm}{dt} = (0.09\ m^3/s)(1\ g/cm^3)\left(\frac{100\ cm}{1m}\right)^3\left(\frac{kg}{1000 g}\right) = 90\ kg/s$$

By Fig. (a) and Eq. (a), the forces exerted on the water are

$$F_x = (18\cos 60°)(\tfrac{1}{3}\tfrac{dm}{dt}) - (18\cos 30°)\frac{dm}{dt}$$

or $F_x = 270 - 1402.9 = -1132.9\ N$

$$F_y = \left[(18)(\tfrac{2}{3}\tfrac{dm}{dt}) - (18\sin 60°)(\tfrac{1}{3}\tfrac{dm}{dt})\right] - (18\sin 30°)\frac{dm}{dt}$$

or $F_y = 1080 - 467.6 - 810 = -197.6\ N$

Hence, the forces that the water exerts on the vane are

$$\underline{(F_x)_{vane} = 1132.9\ N \longrightarrow}$$

$$\underline{(F_y)_{vane} = 197.6\ N \uparrow}$$

16.118

GIVEN: An airplane with weight W and total wing span b flies horizontally with constant speed v (Fig. a). Adopt the airplane as a reference frame. Assume that a cylinder of air with diameter b is deflected downward by the wings through an angle θ (Fig. a).

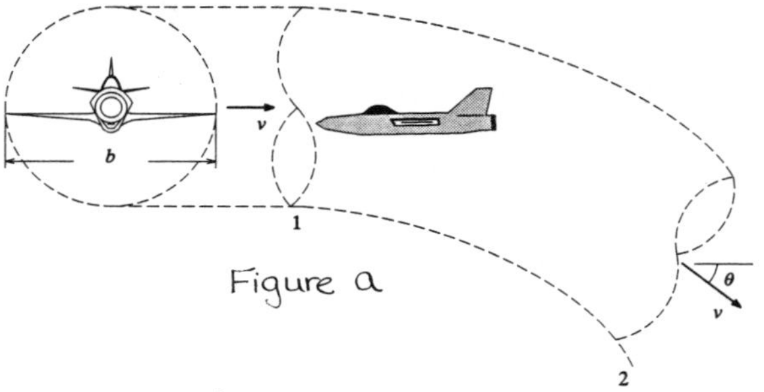

Figure a

FIND: Show that

$$\sin\theta = \frac{4W}{\pi b^2 \rho v^2} \qquad (a)$$

where ρ is the mass density of air.

SOLUTION:

Since the plane flies at a constant speed, the flow is steady. With the airplane as a reference frame (Fig. b)

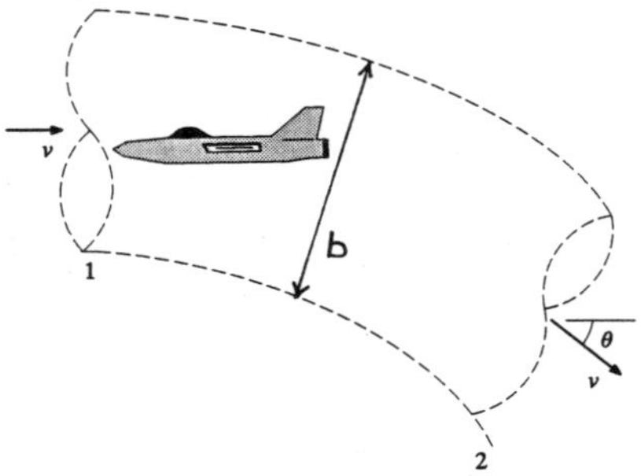

Figure b

The downward momentum leaving at section 2 per second is

$$P_{down} = (v\sin\theta)\left(\frac{dm}{dt}\right)$$

where $\frac{dm}{dt} = \rho v A = \rho v(\frac{\pi}{4}b^2)$

or $\quad P_{down} = \frac{\pi}{4}b^2\rho v^2 \sin\theta$

The reaction to this downward force on the air is the lift acting on the plane, and the magnitude of the lift is equal to the weight W of the plane. Hence,

$$W = \frac{\pi}{4}b^2\rho v^2 \sin\theta$$

or

$$\underline{\sin\theta = \frac{4W}{\pi b^2 \rho v^2}}$$

16.119

GIVEN: A jet of water 100 mm in diameter, moving at 60 m/s, strikes a vane of an impulse wheel moving at 27 m/s (Fig. a). Water exits the vane at 30 m/s, relative to the vane. The distance from the vane to the center of the wheel is 0.75m

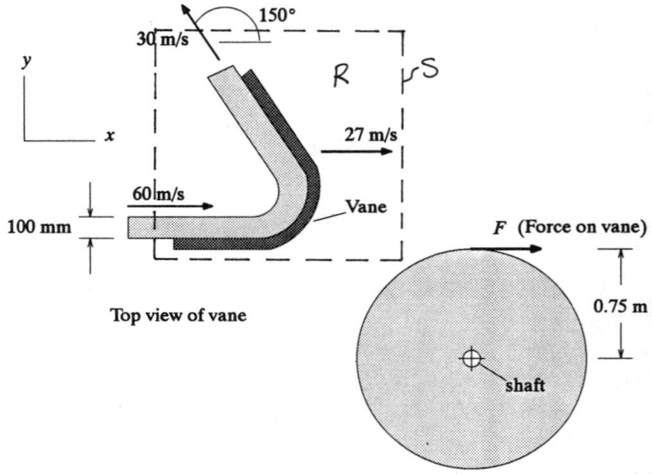

Figure a

Sideview of impulse wheel

FIND: the torque that the wheel transmits to its shaft.

SOLUTION:

By a process similar to Example 16.30, select a surface S enclosing region R (Fig. a). By Theorem 16.13, the force exerted on the water is

$$\vec{F}_{water} = \vec{V}_{out}\frac{dm}{dt} - \vec{V}_{in}\frac{dm}{dt} \qquad (a)$$

For steady flow, $\frac{dm}{dt}$ is constant. Therefore, the rate at which mass enters region R is

$$\frac{dm}{dt} = \rho V_{in} A = (1000\ \tfrac{kg}{m^3})(60\ m/s)(\pi \times 0.05^2\ m^2)$$

or $\frac{dm}{dt} = 471.2\ kg/s$

(continued)

16.119 (cont.)

Therefore, the horizontal momentum of water entering the region per second is

$$V_{in}\frac{dm}{dt} = (60\ \text{m/s})(471.2\ \text{kg/s}) = 28.27\ \text{kN} \quad \text{(b)}$$

The horizontal speed of the water out of region R is (Fig. a)

$$(V_{out})_x = 27\ \text{m/s} - (30\ \text{m/s})\cos 30°$$

or $$(V_{out})_x = 1.019\ \text{m/s}$$

Hence, the horizontal momentum of water exiting Region R per second is

$$(V_{out})_x\frac{dm}{dt} = (1.019\ \text{m/s})(471.2\ \tfrac{\text{kg}}{\text{s}}) = 0.48\ \text{kN} \quad \text{(c)}$$

Therefore, by Eqs. (a), (b), and (c), the horizontal force exerted on the water by the vane is

$$(F_{water})_x = 0.48 - 28.27 = -27.79\ \text{kN}$$

So, the force exerted by the water on the vane is (Fig. P16.118)

$$(F_{vane})_x = 27.79$$

This is the force that determines the torque, since the y component of force that acts on the vane is parellel to the shaft (Fig. a).

Hence, the torque that the wheel transmits to its shaft is

$$\underline{T = (27.79\ \text{kN})(0.75\ \text{m}) = 20.84\ \text{kN}\cdot\text{m}}$$

16.120

GIVEN: Figure a, which represents a thrust augmenter for an airplane. Air ($\rho_1 = 0.00238\ \frac{\text{lbs}^2}{\text{ft}^4}$) enters section 1 with a speed of $v_1 = 240$ ft/s. The cross-sectional areas of sections 1 and 2 are $A_1 = A_2 = 2.25\ \text{ft}^2$. The mass density of the products of combustion leaving section 2 is $\rho_2 = 0.000950\ \frac{\text{lb}\cdot\text{s}^2}{\text{ft}^4}$. The fuel enters at right angles to the airstream and its mass is 3% of the mass of air that passes through the augmenter.

Figure a

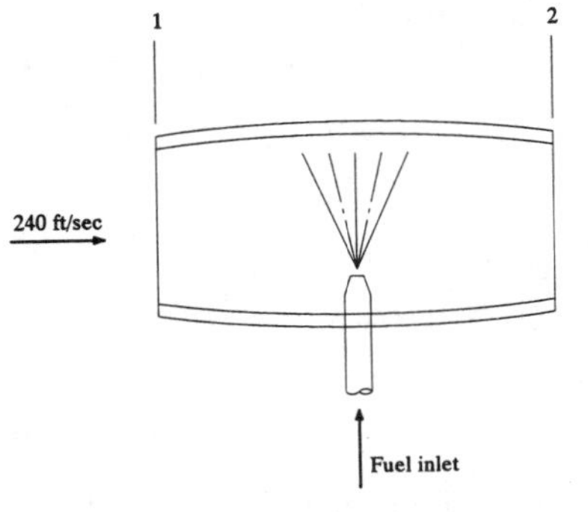

FIND: the thrust that the augmenter exerts on the plane.

SOLUTION:

By Theorem 16.13, since the flow is steady, the net rate at which momentum is convected out of the region equals the net force that acts on the fluid (or the thrust of the augmenter). Thus, the thrust is

$$F_T = v_2\left(\frac{dm}{dt}\right)_2 - v_1\left(\frac{dm}{dt}\right)_1 \quad \text{(a)}$$

where

$$\left(\frac{dm}{dt}\right)_1 = \rho_1 A_1 v_1 = (0.00238)(2.25)(240) = 1.285\ \frac{\text{slugs}}{\text{s}} \quad \text{(b)}$$

Since the mass of the fuel is 3% of the mass of the air that enters at section 1, the mass of the products of combustion that exit at section 2 is

$$(1.03)\left(\frac{dm}{dt}\right)_1 = \left(\frac{dm}{dt}\right)_2 = \rho_2 A_2 v_2 \quad \text{(c)}$$

or $$(1.03)(1.285) = 1.324 = (0.000950)(2.25)v_2$$

Therefore, $$v_2 = 619.2\ \text{ft/s} \quad \text{(d)}$$

Hence, Eqs. (a), (b), (c), and (d) yield

$$F_T = (619.2)(1.324) - (240)(1.285)$$

or $$\underline{F_T = 511.4\ \text{lb.}}$$

16.121

GIVEN: A liquid of mass density ρ flows through a sudden expansion in a pipe (Fig. a).

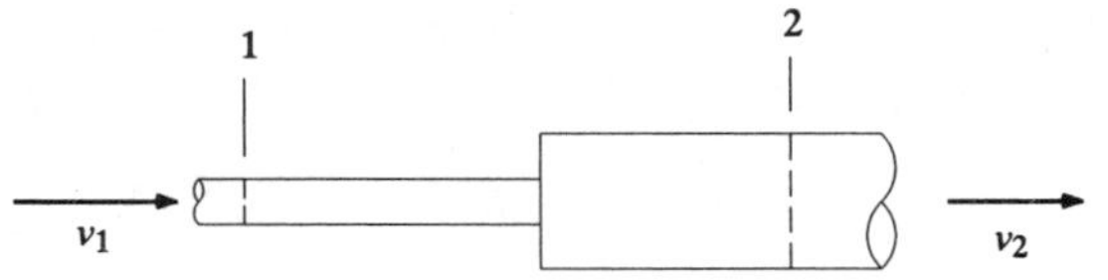

Figure a

FIND: Show that the force that the fluid exerts on the step in the pipe is given by

$$F = (p_2 + \rho V_2^2)A_2 - (p_1 + \rho V_1^2)A_1 \quad \text{(a)}$$

where subscripts 1 and 2 refer to cross sections short distances upstream and downstream from the step, p is pressure, ρ is mass density, and A is area.

SOLUTION:

Consider the free-body diagram of the fluid in the pipe between sections 1 and 2 (Fig. b).

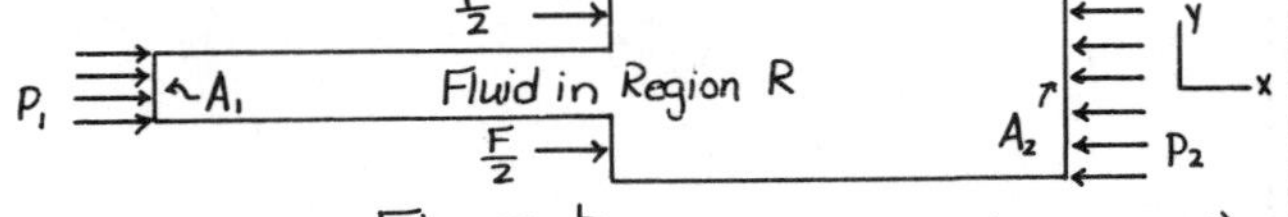

Figure b

(continued)

16.121 (cont.)

By Fig. b, the net force acting on the fluid is

$$\Sigma F_x = p_1 A_1 - p_2 A_2 + F \qquad \text{(b)}$$

where F is the net force that the step exerts on the fluid.

The net momentum per second convected out of the region is

$$V_2\left(\frac{dm}{dt}\right)_2 - V_1\left(\frac{dm}{dt}\right)_1 \qquad \text{(c)}$$

where

$$dm_1 = \rho_1 A_1 V_1$$
$$dm_2 = \rho_2 A_2 V_2 \qquad \text{(d)}$$

By Theorem 16.13, the net force acting on the fluid is equal to the net momentum per second convected out of the region R. Hence, by Eqs. (b), (c), and (d), we have

$$p_1 A_1 - p_2 A_2 + F = \rho_2 A_2 V_2^2 - \rho_1 A_1 V_1^2$$

or $$\underline{\underline{F = (p_2 + \rho_2 A_2)V_2^2 - (p_1 + \rho_1 A_1)V_1^2}}$$

This result verifies Eq. (a).

16.122

GIVEN: A gas flows uniformily with supersonic speed V_1 in a straight uniform pipe. At a certain cross section there is a plane shock at which a sudden transition in the gas occurs. The pressure and mass density upstream and downstream from the shock are p_1, ρ_1 and p_2, ρ_2. Neglect wall friction.

FIND: Show by means of the momentum principle that

$$p_1 + \rho_1 V_1^2 = p_2 + \rho_2 V_2^2. \qquad \text{(a)}$$

Solution:

Consider the free-body diagram of the gas between sections 1 and 2 (Fig. a)

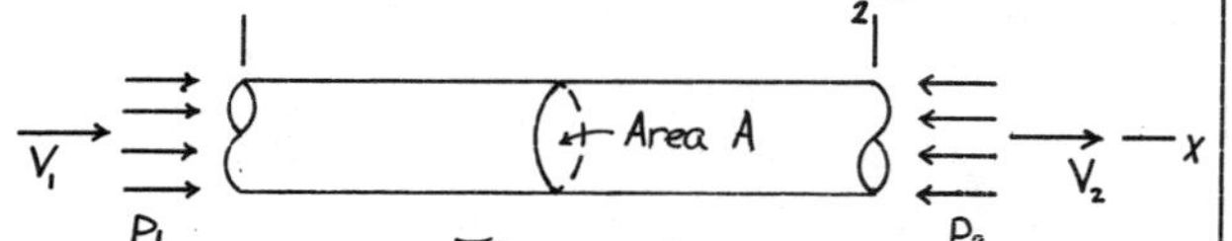

Figure a

Summing forces in the x-direction, we find

$$\Sigma F_x = p_1 A_1 - p_2 A_2 = (p_1 - p_2)A \qquad \text{(b)}$$

since $A_1 = A_2 = A$.

The momentum per second convected out of region 1-2 is

$$V_2\left(\frac{dm}{dt}\right)_2 - V_1\left(\frac{dm}{dt}\right)_1 \qquad \text{(c)}$$

where, since $A_1 = A_2 = A$,

$$\left(\frac{dm}{dt}\right)_1 = \rho_1 A_1 V_1 = \rho_1 A V_1 \qquad \text{(d)}$$
$$\left(\frac{dm}{dt}\right)_2 = \rho_2 A_2 V_2 = \rho_2 A V_2$$

By Theorem 16.13 and Eqs. (b), (c), and (d), we have

$$(p_1 - p_2)A = (\rho_2 V_2^2 - \rho_1 V_1^2)A$$

or $$\underline{\underline{p_1 + \rho_1 V_1 = p_2 + \rho_2 V_2}}$$

which verifies Eq. (a).

16.123

GIVEN: Figure a. A hydraulic jump occurs in a horizontal open channel of rectangular cross-section. Neglect shearing stress on the walls and bottom of the channel and variations in velocity on a cross section.

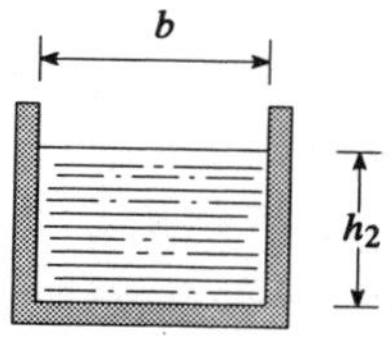

Figure a

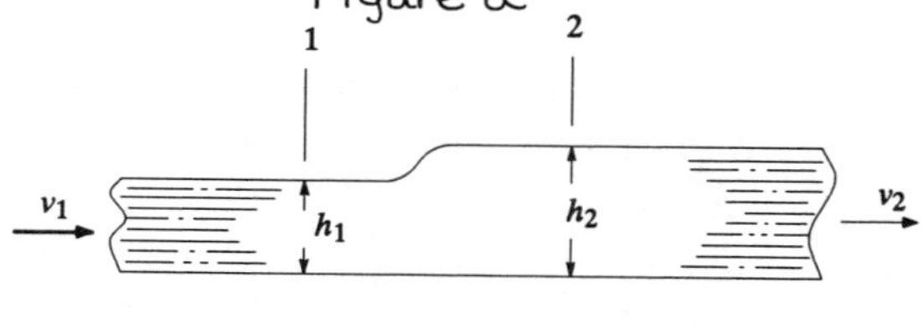

FIND: Show that

$$h_1[v_1^2 + \tfrac{1}{2}(gh_1)] = h_2[v_2^2 + \tfrac{1}{2}(gh_2)] \qquad \text{(a)}$$

SOLUTION:

Consider the free-body diagram of the fluid between sections 1 and 2 (Fig. b), where F_1 and F_2 are the resultant forces due to the pressures that act at sections 1 and 2.

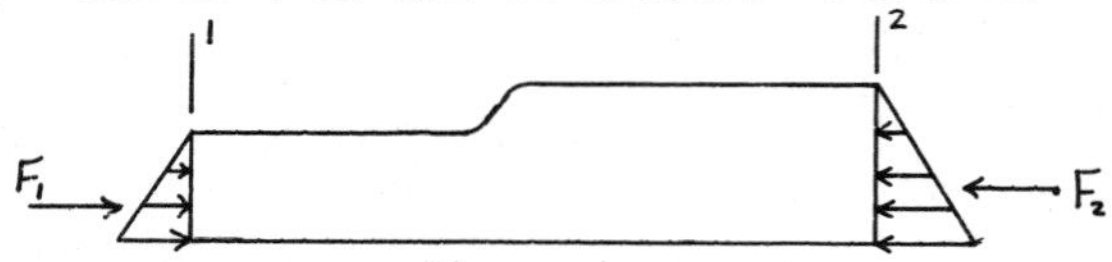

Figure b

These forces are equal to the pressures at the centroids of the areas multiplied by the areas of the cross sections. Hence,

$$F_1 = (\tfrac{1}{2}\rho g h_1)(h_1 b) = \tfrac{1}{2}\rho g h_1^2 b$$
$$F_2 = (\tfrac{1}{2}\rho g h_2)(h_2 b) = \tfrac{1}{2}\rho g h_2^2 b \qquad \text{(b)}$$

The momentum per second convected out of the region is

$$v_2\left(\frac{dm}{dt}\right)_2 - v_1\left(\frac{dm}{dt}\right)_1 = v_2(\rho h_2 b v_2) - v_1(\rho h_1 b v_1) \qquad \text{(c)}$$

Hence, by Theorem 16.13 and Eqs. (b) and (c)

$$F_1 - F_2 = \tfrac{1}{2}\rho g b(h_1^2 - h_2^2) = \rho b[v_2^2 h_2 - v_1^2 h_1]$$

or $$\underline{\underline{h_1[v_1^2 + \tfrac{1}{2}(gh_1)] = h_2[v_2^2 + \tfrac{1}{2}(gh_2)]}}$$

This verifies Eq. (a).

16.124

GIVEN: Figure a. Water ($\rho = 1.0\ \frac{kN s^2}{m^4}$) is discharged from the nozzle at the rate of $0.06\ m^3/s$. Gage pressure upstream from the nozzle is given by $p = \frac{1}{2}\rho(V_2^2 - V_1^2)$.
Six flange bolts fasten the nozzle to the pipe.

FIND: the tension in a bolt, ignoring the tension due to the tightening of the nut.

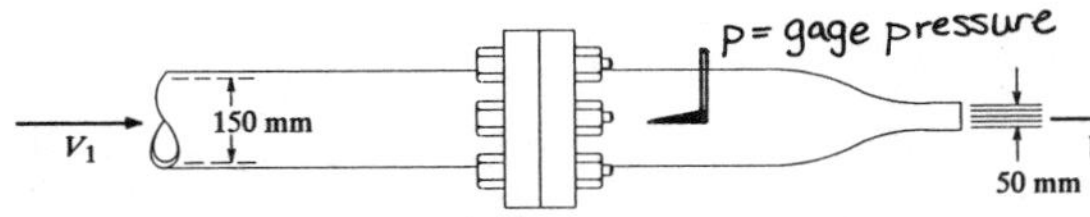

Figure a

SOLUTION:
Consider Fig. b, where section 1-1 is a section upstream from the nozzle and section 2-2 is at the end of the nozzle. Let Q be the volume of water that passes through the pipe per second. Then,

$$Q = A_1V_1 = A_2V_2 = 0.06\ m^3/s \qquad (a)$$

where $A_1 = \pi(0.075)^2 = 0.01767 m^2$ is the cross-sectional area of water at section 1-1 and $A_2 = \pi(0.025)^2 = 0.001963 m^2$. Therefore, by Eq. (a),

$$V_2 = \frac{A_1}{A_2}V_1 = 9V_1 \qquad (b)$$

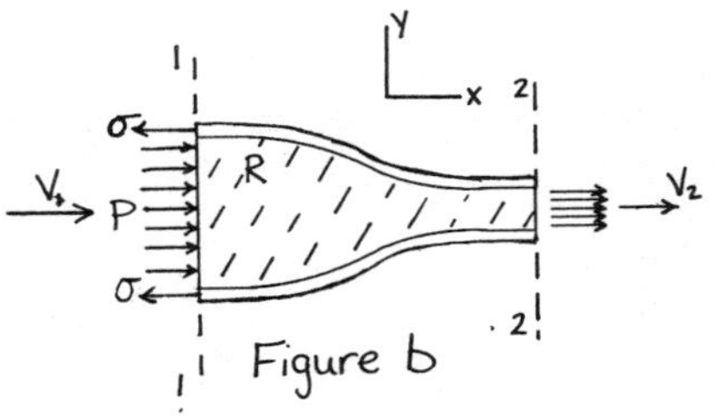

Figure b

Equations (a) and (b) yield

$$V_1 = 3.395\ m/s,\ V_2 = 30.56\ m/s \qquad (c)$$

Therefore, the pressure at section 1-1 is

$$p = \frac{1}{2}\rho(V_2^2 - V_1^2) = 461.2\ kN/m^2 \qquad (d)$$

By the momentum principle (Theorem 16.13) and Fig. b,

$$\Sigma F_x = pA_1 - \sigma A_p = V_2\frac{dm}{dt} - V_1\frac{dm}{dt} \qquad (e)$$

where σ is the stress (Force/Area) in the pipe wall, A_p is the cross-sectional area of the pipe wall and

$$\frac{dm}{dt} = \rho A_1V_1 = (1)(0.01767)(3.395)$$

or

$$\frac{dm}{dt} = 0.06\ kg/s$$

Let $\sigma A_p = F_p$. Then, Eq. (e) yields

$$(461.1)(0.01767) - F_p = (30.56 - 3.395)(0.06)$$

or $F_p = 6.518\ kN$

This is the force that the six bolts must support. Hence, each bolt must carry a tension load of

$$\underline{F_{bolt} = 1.086\ kN}$$

16.125

GIVEN: Figure a, which represents steady flow of a fluid passing between fixed vanes. The cross-sectional areas, pressures, mass densities, and velocities at the inlet 1 and outlet 2 are, respectively, A_1, A_2; p_1, p_2; ρ_1, ρ_2; and v_1, v_2. The velocity and pressure distributions on the inlet and outlet cross-sections are uniform.

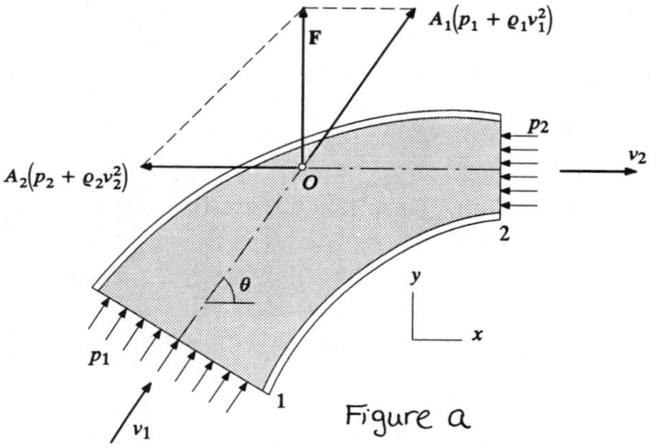

Figure a

FIND:
(a) Show by means of the momentum principle that the force $\vec{F}$ that the fluid exerts on the vanes is determined by the parellelogram construction shown in Fig. a (see Example 16.31).
(b) Show by means of the principle of moment of momentum that the line of action of force $\vec{F}$ passes through point O, the intersection of the center lines through sections 1 and 2 (See Theorem 16.14).

SOLUTION:
(a) The x momentum flowing in at section 1 is

$$P_{x_1} = \rho_1 A_1 v_1^2 \cos\theta \qquad (a)$$

and the x momentum flowing out at section 2 is

$$P_{x_2} = \rho_2 A_2 v_2^2 \qquad (b)$$

(continued)

16.125 (cont.)

Hence, by the momentum principle, the net x component of the force that acts on the fluid between the vanes is, with Eqs. (a) and (b),

$$P_x = A_2 \rho_2 v_2^2 - A_1 \rho_1 V_1^2 \cos\theta \qquad (c)$$

Similarly, the net y component of force is

$$P_y = -A_1 \rho_1 v_1^2 \sin\theta \qquad (d)$$

The forces that act on the fluid are due to the force that the vanes exert on the fluid, namely $(-F_{x_1}, -F_y)$, since $\vec{F}$ is the force exerted by the fluid on the vanes, and the force due to the uniform pressures p_1 and p_2 on the end cross sections, namely, $(p_1 A_1 \cos\theta - p_2 A_2,\ p_1 A_1 \sin\theta)$. Hence,

$$\begin{aligned} P_x &= -F_x + p_1 A_1 \cos\theta - p_2 A_2 \\ P_y &= -F_y + p_1 A_1 \sin\theta \end{aligned} \qquad (e)$$

Consequently, Eqs. (c), (d), and (e) yield

$$\begin{aligned} F_x &= A_1(p_1 + \rho_1 v_1^2)\cos\theta - A_2(p_2 + \rho_2 v_2^2) \\ F_y &= A_1(p_1 + \rho_1 v_1^2)\sin\theta \end{aligned} \qquad (f)$$

where $(F_x, F_y) = \vec{F}$ is the force that the fluid exerts on the vanes.

Equation (f) is also obtained directly from Figure a, if $\vec{F}$ is resolved into x and y components. Hence, the construction in Fig. a is verified except for the location of point O (see the solution of part b, that follows).

(b) Since the velocities are distributed uniformily over the end cross sections, the resultant momentum vectors act along the center line of the cross sections (and perpendicular to the cross sections). Hence, they pass through point O. So, the net moment of momentum of fluid passing out of the region is zero. Therefore, by the principle of moment of momentum for fluids (Theorem 16.14), the forces that act on the fluid between the vanes exert no moment about point O.

The lines of action of the pressure forces $p_1 A_1$ and $p_2 A_2$ pass through point O. Hence, these forces produce no moments about point O. Therefore, the force (F_x, F_y) cannot have a moment about O, and its resultant axis must pass through O, as shown in Fig. a.

16.126

GIVEN: Figure a. The relative speed of water leaving the vanes is 80% of the relative speed of water approaching the vanes (Fig. a). The relative velocity vector striking the vane is tangent to the vanes. Use the data given in Fig. a.

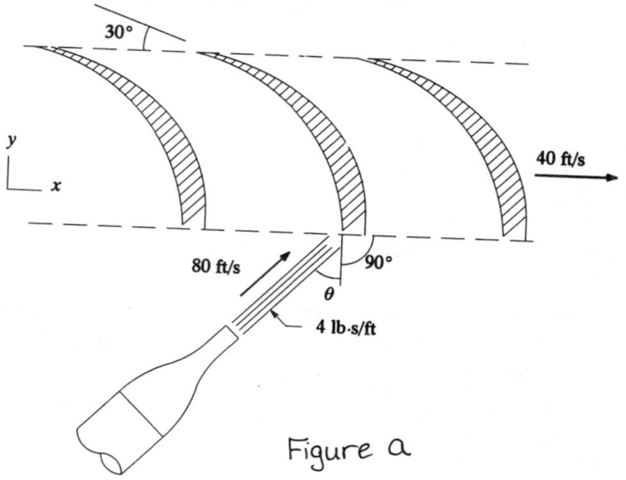

Figure a

FIND:

(a) the angle θ.

(b) the force F_x exerted by the water on the vanes.

SOLUTION:

(a) Consider first the water approaching the vanes. The relative velocity of approach of the water (it is tangent to the vane) is

$$\vec{v}_{W/V}^{(a)} = \vec{v}_{Water}^{(a)} - \vec{v}_{Vane}$$

or $v_{W/V}^{(a)}\,\hat{\jmath} = 80(\sin\theta)\hat{\imath} + 80(\cos\theta)\hat{\jmath} - 40\hat{\imath}$

Hence, equating $\hat{\imath}$ and $\hat{\jmath}$ components

$$80\sin\theta - 40 = 0 \qquad (a)$$

$$v_{W/V}^{a} = 80\cos\theta \qquad (b)$$

Solving Eqs. (a) and (b), we find

$$\underline{\theta = 30°} \qquad (c)$$

$$v_{W/V}^{(a)} = 69.282 \text{ ft/s} \qquad (d)$$

(b) The relative speed of the water at exit is $v_{W/V}^{(e)} = 0.80\, v_{W/V}^{(a)}$, or with Eq. (d)

$$v_{W/V}^{(e)} = (0.80)(69.282) = 55.426 \text{ ft/s}.$$

Hence, the absolute velocity of water at exit is (see Fig. a)

$$\vec{v}_{Water}^{(e)} = \vec{v}_{Vane} + \vec{v}_{W/V}^{(e)}$$

or $\vec{v}_{Water}^{(e)} = 40\hat{\imath} + 55.426[-(\cos 30°)\hat{\imath} + (\sin 30°)\hat{\jmath}]$

or $\vec{v}_{Water}^{(e)} = -8\hat{\imath} + 27.713\hat{\jmath}$ (ft/s) (e)

(continued)

16.126 (cont.)

Now by Fig. a, the x component of momentum per second entering is

$$(4)(80 \sin 30°) = 160 \text{ lb.} \qquad (f)$$

and with Eq. (e), the x component of momentum per second exiting is

$$(4)(-8) = -32 \text{ lb} \qquad (g)$$

Hence, by the momentum principle and Eqs. (f) and (g),

$$\underline{\underline{F_x = 160 - (-32) = 192 \text{ lb}}}$$

16.127

GIVEN: Water flowing at a constant rate in a horizontal flume of triangular cross section, with 90° between walls (Fig. a). At a section, a hydraulic jump occurs (the depth of water increases from 0.6 m to 0.75 m). The velocity is constant over a cross section.

FIND: the rate of flow Q (m^3/s). Neglect friction of the walls.

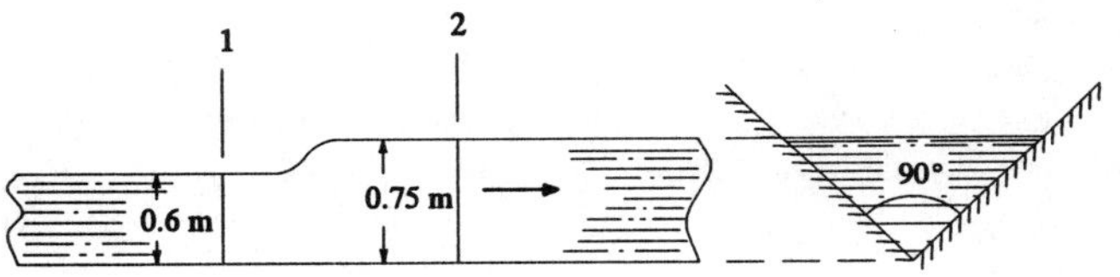

Figure a

SOLUTION:

Consider Fig. b, where F_1 and F_2 are forces

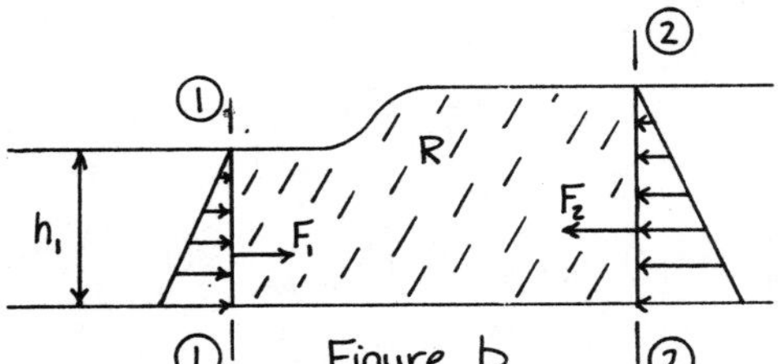

Figure b

that act on the region R of water, due to pressures at sections 1-1 and 2-2, respectively. These forces are equal to the pressures at the centroids of the cross-sectional areas, multiplied by the areas of the cross sections.

Hence, with Fig. c,

$$F_1 = p_c A = (\rho g \tfrac{h_1}{3}) h_1^2$$

or $F_1 = \frac{1}{3} \rho g h_1^3 \qquad (a)$

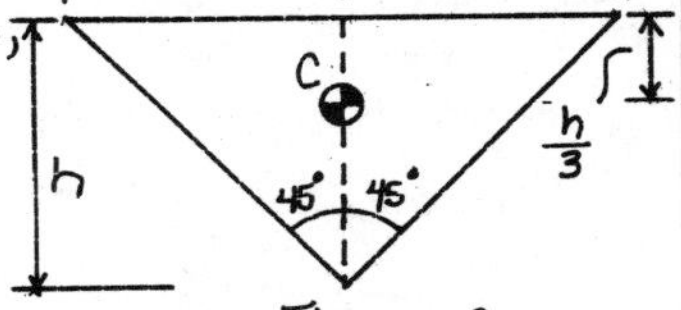

Figure c

and

$$F_2 = \tfrac{1}{3} \rho g h_2^3 \qquad (a)$$

Therefore, the rate of flow is

$$Q = A_1 V_1 = A_2 V_2 = h_1^2 V_1 = h_2^2 V_2 \qquad (b)$$

The momentum flowing into the region R at section 1-1 per second is $\rho Q V_1$, and the momentum flowing out of region R at section 2-2 per second is $\rho Q V_2$. Hence, by Theorem 16.13 and Fig. b,

$$F_1 - F_2 = \rho Q V_2 - \rho Q V_1 \qquad (c)$$

Equations (a), (b), and (c) yield after simplification,

$$\tfrac{1}{3} \rho g (h_1^3 - h_2^3) = \rho Q^2 \left(\frac{1}{h_2^2} - \frac{1}{h_1^2}\right)$$

$$\therefore \quad Q^2 = \tfrac{1}{3} g \frac{(h_2^3 - h_1^3) h_1^2 h_2^2}{h_2^2 - h_1^2} \qquad (d)$$

Since $h_1 = 0.6 m$ and $h_2 = 0.75 m$, with $g = 9.81 \text{ m/s}^2$, Eq. (d) yields

$$\underline{\underline{Q = 0.820 \text{ m}^3/\text{s}}}$$

16.128

GIVEN: Figure a shows the mean velocity distributions at the inlet (section 1) and at a downstream section (section 2) of a pipe in which turbulent flow exists. At section 1, the velocity distribution is uniform ($v = c$), and at section 2, the velocity distribution is given by

$$v = 1.235 c \left(1 - \frac{r}{a}\right)^{1/7} \qquad (a)$$

where c is a constant, a is the inner radius of the pipe and r is the radial coordinate from the centerline of the pipe.

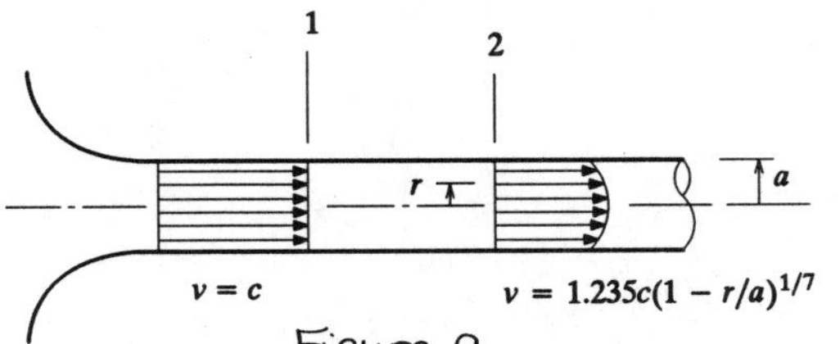

Figure a

FIND: Show that the frictional force of the liquid on the wall of the pipe is

$$F = \pi a^2 (p_1 - p_2 - 0.038 \rho c^2) \qquad (b)$$

where p_1, p_2 are the pressures at sections 1 and 2, and ρ is the mass density of the liquid.

SOLUTION:

The momentum that flows out of section 2 per second is

$$P_{out} = \int v \, dm = \int v (\rho v \, dA) = \int_0^a \rho v^2 (2\pi r) dr \qquad \text{(continued)}$$

16.128 (cont.)

or with Eq. (a),

$$P_{out} = 2\pi\rho(1.235)^2c^2\int_0^a r\left(1-\frac{r}{a}\right)^{2/7}dr \qquad (c)$$

Equation (c) may be integrated by tables, computer software, or substitution. Let's use the last method. Therefore, let

$$1-\frac{r}{a} = x,\ r = a(1-x),\ dr = -dx \qquad (d)$$

Substitution of Eq. (d) into Eq. (c) yields

$$P_{out} = 2\pi\rho(1.235)^2c^2a^2\int_0^1 (x^{2/7} - x^{9/7})dx$$

or

$$P_{out} = 2\pi\rho(1.235)^2c^2a^2\left[\frac{x^{9/7}}{9/7} - \frac{x^{16/7}}{16/7}\right]_0^1$$

Therefore,

$$P_{out} = 2\pi\rho(1.235)^2c^2a^2\left(\frac{49}{144}\right) \qquad (e)$$

Now the momentum that flows in section 1 per second is

$$P_{in} = \rho c^2 A = \rho c^2 \pi a^2 \qquad (f)$$

Hence, the net momentum that flows out of the region 1-2 is, with Eqs. (e) and (f)

$$P_{net} = P_{out} - P_{in} = \pi\rho a^2c^2\left[2(1.235)^2\left(\frac{49}{144}\right) - 1\right]$$

or $$P_{net} = 0.03800\pi\rho a^2c^2 \qquad (g)$$

Consider now the free-body diagram of the fluid in between sections 1 and 2 (Fig. b), where p_1, p_2 are the pressures at sections 1 and 2, and F is the net frictional force exerted by the pipe wall on the fluid.

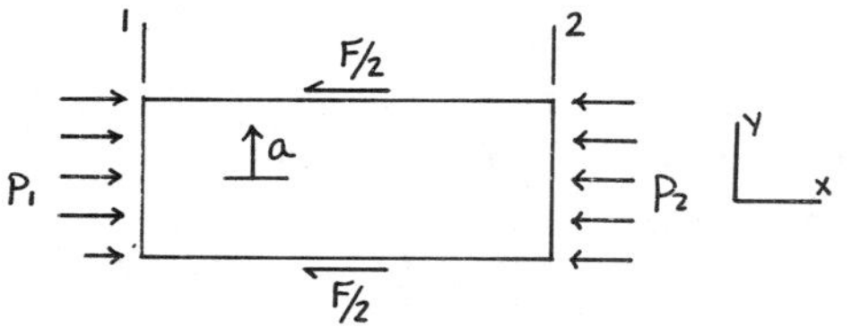

Figure b

By Fig. b,

$$\Sigma F_x = \pi a^2(p_1 - p_2) - F \qquad (h)$$

Hence, by the momentum principle and Eqs. (g) and (h),

$$\pi a^2(p_1 - p_2) - F = 0.03800\pi\rho a^2c^2$$

or $$\underline{F = \pi a^2(p_1 - p_2 - 0.038\rho c^2)}$$

This result verifies Eq. (b).

16.133

GIVEN: Two putty balls of masses 2m, m and speed 21 ft/s, 30 ft/s, respectively, travel along the x axis in the positive sense. The faster ball strikes the slower ball from behind, and the balls stick together.

FIND: the speed of the balls for the instant after the collision.

SOLUTION:

The initial momentum of the balls is

$$p_i = 2m(21) + m(30)$$

or $$p_i = 72m \quad (\text{lb·s}) \qquad (a)$$

The final momentum is

$$p_f = (2m + m)v = 3mv \qquad (b)$$

Momentum is conserved. Hence, by Eqs. (a) and (b),

$$p_i = p_f$$

$$72m = 3mv$$

or $$\underline{v = 24 \text{ ft/s}}$$

16.135

GIVEN: Fluid at a flow rate of 45 kg/s enters a machine with speed 9 m/s and exits with speed 3 m/s. The velocities at the entrance and exit are parallel and in the same sense.

FIND: the net force exerted on the fluid by the machine.

SOLUTION:

By Theorem 16.13,

$$F = v_2\frac{dm}{dt} - v_1\frac{dm}{dt}$$

$$= (3)(45) - 9(45)$$

$$= -270 \text{ lb.}$$

or $$\underline{F = 270 \text{ lb, opposite to the flow}}$$

16.136

GIVEN: A nozzle discharges 4 slugs/s of liquid at a speed of 80 ft/s. The liquid jet strikes normally a fixed flat plate.

FIND: the force the liquid exerts on the plate.

SOLUTION:

Consider Fig. a, in which F is the force exerted by the plate on the liquid.

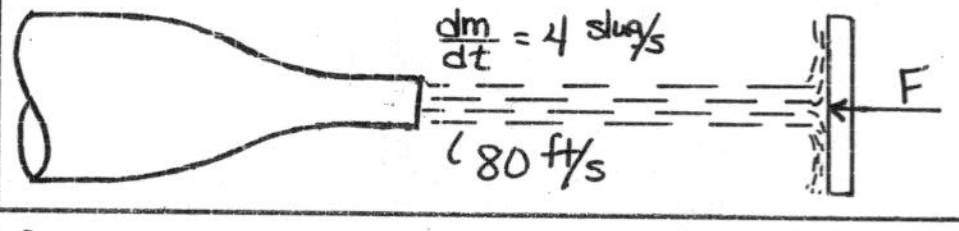

Figure a

(continued)

16.136 (cont.)

By the momentum principle,

$$-F = v_2 \frac{dm}{dt} - v_1 \frac{dm}{dt}$$

$$-F = 0 - (80)(4) = -320 \text{ lb.}$$

$$F = 320 \text{ lb} \leftarrow$$

Therefore, the force exerted by the liquid on the plate is

$$F_L = -F = 320 \text{ lb.} \rightarrow$$

16.137

GIVEN: A fixed vane curves through 120°. A jet of liquid (s.w. = 10 kN/m³) strikes the vane tangentially at a speed of 30 m/s. The cross-sectional area of the jet is 0.0009 m². Neglect friction.

FIND:
(a) the force exerted on the vane in the direction of the jet.
(b) the force exerted on the vane in the direction perpendicular to the jet.

SOLUTION:

(a) The mass density ρ is given by

$$\rho = (\text{s.w.})/(g) = (10 \text{ kN/m}^3)/(9.81 \text{ m/s}^2) = 1019.4 \frac{\text{kg}}{\text{m}^3}$$

and

$$\frac{dm}{dt} = \rho V A = (1019.4 \text{ kg/m}^3)(30 \text{ m/s})(0.0009 \text{ m}^2)$$

or $\frac{dm}{dt} = 27.52$ kg/s

Now, by the momentum principle and Fig. a,

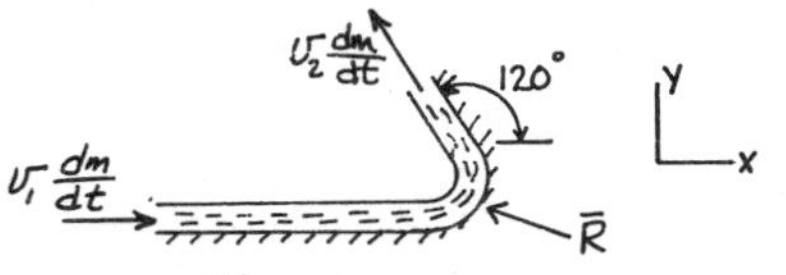

Figure a

where R is the force exerted on the liquid by the vane,

$$\Sigma F_x = R_x = v_2 \frac{dm}{dt} \cos 120° - v_1 \frac{dm}{dt}.$$

Since friction between the fluid and the vane is neglected, $v_1 = v_2$. Then

$$R_x = (30)(27.52)(-0.5 - 1) = -1238 \text{ N}$$

The x component of the force that the liquid exerts on the vane is

$$F_x = -R_x = 1238 \text{ N} \rightarrow$$

(b) Similarly,

$$R_y = v_2 \frac{dm}{dt} \sin 120° = (30)(27.52)(0.866) = 715.0 \text{ N} \uparrow$$

and $F_y = -R_y = -715.0 \text{ N} = 715.0 \text{ N} \downarrow$

16.143

GIVEN: The center of the sun is 93 million miles from the center of the earth, and the sun is 330,000 times more massive than the earth.

FIND:
(a) the center of mass of the sun-earth system.
(b) Compare the distance of the center of mass from the center of the sun to the radius of the sun.

SOLUTION:

By the sketch in Fig. a and the theory of

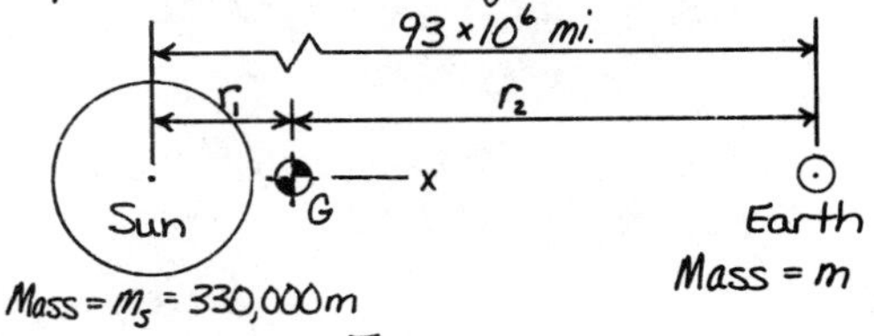

Figure a

mass center, choosing the mass center G as the origin of the x axis, we have

$$r_1 + r_2 = 93 \times 10^6 \qquad \text{(a)}$$

$$(330{,}000m)r_1 - mr_2 = 0$$

The solution of Eqs. (a) is

$$r_1 = 281.8 \text{ mi}, \quad r_2 = 92{,}999{,}718 \text{ mi} \qquad \text{(b)}$$

where r_1 is the distance from the center of the sun to the center of mass G.

(b) The radius of the sun is $r = 864{,}000$ mi. Hence, with r_1 from Eq. (b),

$$\frac{r_1}{r} = 0.000326$$

and the center of mass of the sun-earth system is, relatively, very near the center of the sun.

16.144

GIVEN: A rigid gun that weighs 9 kN rests on frictionless horizontal guides. The gun fires a projectile that weighs 27 N horizontally with an absolute speed of 450 m/s.

FIND: the speed of recoil of the gun.

SOLUTION:

Linear momentum is conserved. Hence,

$$m_g v_{gi} + m_p v_{pi} = m_g v_{gf} + m_p v_{pf} \qquad \text{(a)}$$

where subscripts g and p denote gun and projectile, and subscripts i and f denote

(continued)

16.144 (cont.)

initial and final conditions. Thus, with the given data and Eq. (a),

$$0 + 0 = (9000)v_{gf} + (27)(450)$$

or

$$v_{gf} = -1.35 \text{ m/s} = 1.35 \text{ m/s} \leftarrow$$

That is, the recoil velocity of the gun is 1.35 m/s in the sense opposite to that of the projectile, here assumed $v_{pf} = 450 \text{ m/s} \rightarrow$.

16.146

GIVEN: A particle of mass 3 $\frac{lb \cdot s^2}{ft}$ lies at the point (1, 2, 3) ft. Its (x, y, z) velocity projections are (20, 30, 40) ft/s.

FIND: the moment of momentum of the particle with respect to the z axis.

SOLUTION:

Refer to Figure a.

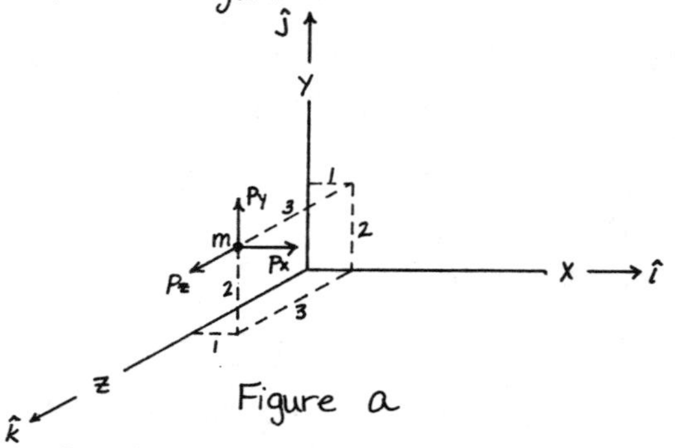

Figure a

By Fig. a and the given data,

$$p_x = mv_x = (3)(20) = 60 \text{ lb·s}$$
$$p_y = mv_y = (3)(30) = 90 \text{ lb·s} \qquad (a)$$
$$p_z = mv_z = (3)(40) = 120 \text{ lb·s}$$

The moment of momentum about axis z is by Fig. a and Eqs. (a)

$$A_z = (p_y)(1) - (p_x)(2)$$

$$A_z = (90)(1) - (60)(2) = -30 \text{ lb·ft·s}$$

Alternatively, by Eqs. (a), the momentum is

$$\bar{p} = p_x\hat{i} + p_y\hat{j} + p_z\hat{k} = 60\hat{i} + 90\hat{j} + 120\hat{k} \quad (b)$$

and the radius vector from O to m (see Fig. a) is $\bar{r} = \hat{i} + 2\hat{j} + 3\hat{k}$ (c)

Then, the moment of momentum with respect to O is

$$\bar{A}_o = \bar{r} \times \bar{p} = \begin{vmatrix} \hat{i} & \hat{j} & \hat{k} \\ 1 & 2 & 3 \\ 60 & 90 & 120 \end{vmatrix}$$

or

$$\bar{A}_o = -30\hat{i} + 60\hat{j} - 30\hat{k} \text{ (lb·ft·s)} \qquad (d)$$

The projection of $\bar{A}_o$ on the z axis gives the moment of momentum about the z axis. Thus, by Eq. (d),

$$A_z = -30 \text{ lb·ft·s}$$

17.1

GIVEN: A particle P moves along a path in the xy plane such that

$$y = 4x^2 - 36 \; ; \quad x, y \text{ in feet.} \qquad (a)$$

FIND: Determine the displacement $\bar{q}_P$ of the particle as it moves from $x = 3$ ft to $x = 6$ ft.

SOLUTION:

At $x=3$, by Eq. (a), $y=0$. At $x=6$ ft, $y=108$ ft. Hence, at $x=0$, the position vector of the particle relative to the origin (0,0) of (x,y) coordinates is

$$\bar{q}_{(x=3)} = 3\hat{\imath} + 0\hat{\jmath} \qquad (b)$$

At $x=6$, the position vector is

$$\bar{q}_{(x=6)} = 6\hat{\imath} + 108\hat{\jmath} \qquad (c)$$

Therefore, the displacement $\bar{q}_P$ of particle P is, by Eqs. (b) and (c),

$$\underline{\underline{\bar{q}_P = \bar{q}_{(x=6)} - \bar{q}_{(x=3)} = 3\hat{\imath} + 108\hat{\jmath}}}$$

17.2

GIVEN: Body AB undergoes plane motion in the xy plane from position 1 to position 2 (Fig. a).

FIND: By graphical construction, the point P about which it can be rotated to move it from position 1 to position 2.

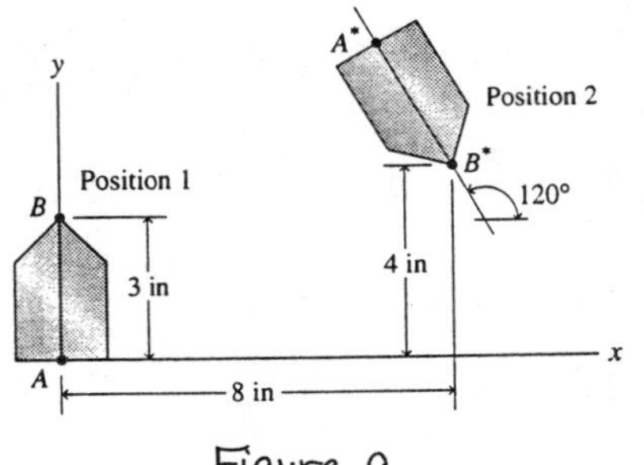

Figure a

SOLUTION:

Construct the perpendicular bisectors of lines AA* and BB*. The bisectors intersect at P (Fig. b). Measure, to scale, the location of P.

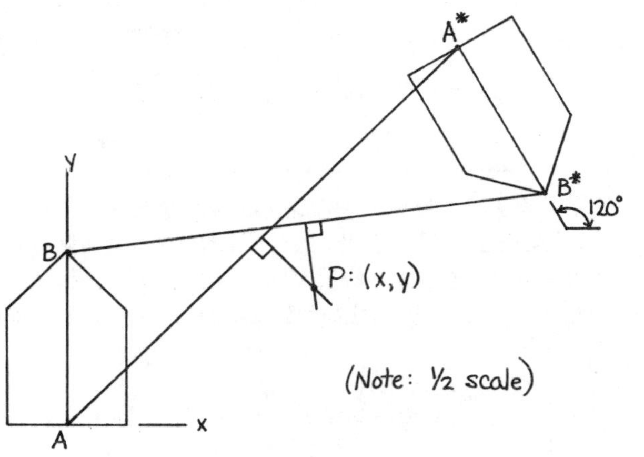

(Note: ½ scale)

$\underline{P(x,y) \simeq (4.1, 2.4) \text{ in.}}$

17.3

GIVEN: A particle P moves from C to D on rod CD as the rod OC rotates from $\theta = 30°$ to $\theta = 150°$ (Fig. a).

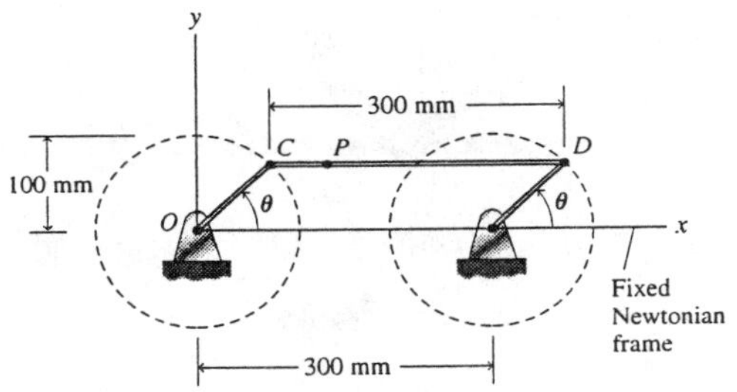

Figure a

FIND:

(a) Determine the absolute displacement of P.

(b) Determine the apparent displacement of P to an observer fixed on rod CD.

SOLUTION:

By Eq. (17.1), the absolute displacement $\bar{q}_{P/A}$ of P is

$$\bar{q}_{P/A} = \bar{q}_{C/A} + \bar{q}_{P/C} \qquad (a)$$

where $q_{C/A}$ is the absolute displacement of C and $q_{P/C}$ is the displacement of P with respect to C.

By Fig. a, for θ initially 30° and finally 150°,

$$\bar{q}_{C/A} = 100(\cos 150° - \cos 30°)\hat{\imath} + \overline{OC}(\sin 150° - \sin 30°)\hat{\jmath} \text{ (mm)} \qquad (b)$$

and

$$\bar{q}_{P/C} = 300\hat{\imath} \text{ (mm)} \qquad (c)$$

(continued)

17.3 (cont.)

where $\hat{\imath}, \hat{\jmath}$ are unit vectors along (x,y) axes, respectively.

By Eqs. (a), (b), and (c),

$$\bar{q}_{P/A} = 126.80\,\hat{\imath}\ (mm)$$

(b) The apparent displacement of P to an observer fixed on rod CD is the same as the displacement of P relative to point C. Hence, by Eq. (c), the apparent displacement of P is

$$\bar{q}_{P/C} = 300\,\hat{\imath}\ (mm)$$

17.4

GIVEN: The wheel in Fig. a rolls 0.90m to the right.

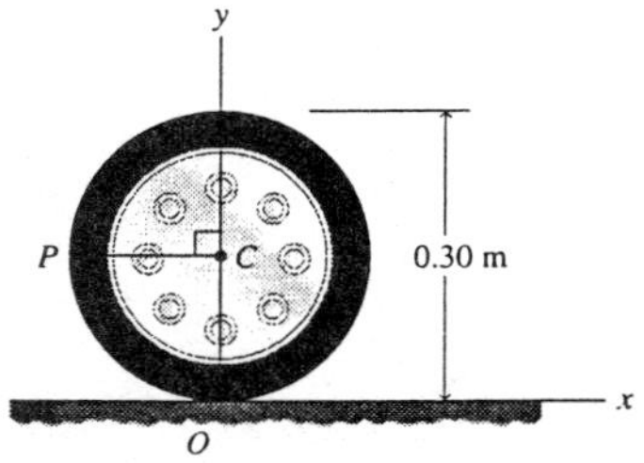

Figure a

FIND: Determine the displacement $\bar{q}_P$ of point P.

SOLUTION:

Since the wheel rolls 0.90m, it rotates through an angle θ given by $0.90 = r\theta = 0.15\theta$.

or $\theta = 6\ rad = 343.77^\circ$ (a)

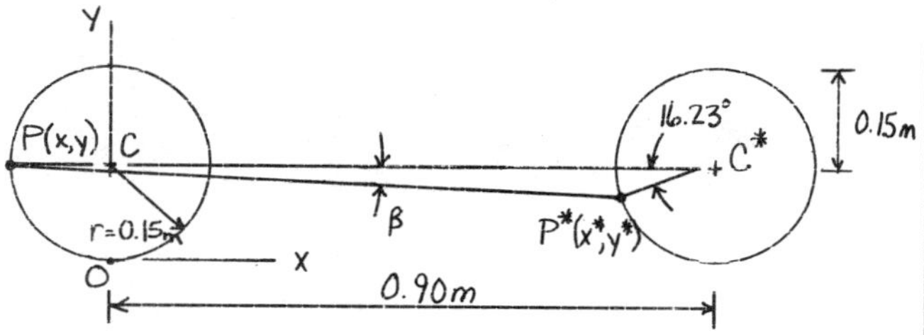

Hence, the radius P^*C^* forms the angle $360^\circ - 343.77^\circ = 16.23^\circ$, with respect to the horizontal line CC^* (Fig. b), where P^* and C^* are the final positions of point P and the center C of the wheel.

By Fig. b,

$$\bar{P}(x,y) = -0.15\,\hat{\imath} + 0.15\,\hat{\jmath}$$

$$\bar{P}^*(x^*, y^*) = (0.90 - 0.15\cos 16.23^\circ)\hat{\imath} + (0.15 - 0.15\sin 16.23^\circ)\hat{\jmath}$$

$$= 0.7560\,\hat{\imath} + 0.10808\,\hat{\jmath}$$

$$\bar{q}_P = \bar{P}^* - \bar{P} = 0.9060\,\hat{\imath} - 0.0419\,\hat{\jmath}\quad (m)$$

or

$$q_P = \sqrt{0.9060^2 + (-0.0419)^2} = 0.9070\ m$$

$$\beta = \tan^{-1}\left(\frac{0.0419}{0.9060}\right) = 2.648^\circ$$

17.5

GIVEN: The body shown in Fig. a undergoes plane motion in the xy plane.

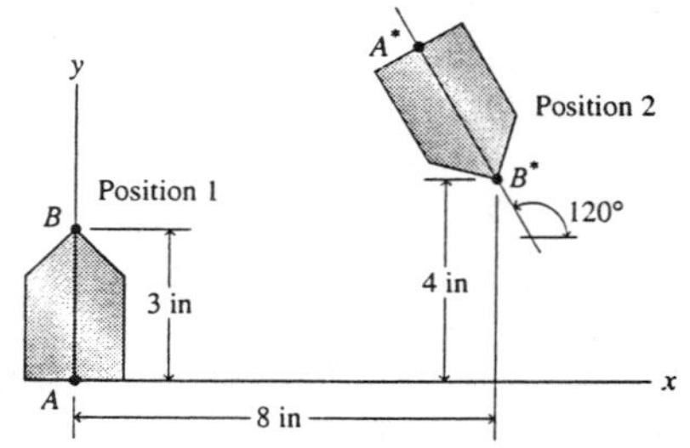

Figure a

FIND: the point P(x,y) about which the body may be rotated to take it from position 1 to position 2, by deriving the equations of the perpendicular bisectors of lines AA^* and BB^* and locating P(x,y) their intersection point analytically.

SOLUTION:

The equation of a straight line is

$$y = mx + b \qquad (a)$$

where m is the line's slope and b is the y axis intercept. By Fig. a, for line AA^*

$$m_{AA^*} = \frac{(y_{A^*} - y_A)}{(x_{A^*} - x_A)} = [(4 + 3\cos 30^\circ) - 0]/(8 - 3\sin 30^\circ)$$

$$m_{AA^*} = 1.015$$

$$b_{AA^*} = 0.$$

Hence,

$$y_{AA^*} = 1.015x \qquad (b)$$

Similarly for line BB^*, $m_{BB^*} = 0.125$ and $b_{BB^*} = 3$.

Hence,

$$y_{BB^*} = 0.125x + 3 \qquad (c)$$

The midpoint of line AA^* is located at

$$\bar{x}_{AA^*} = (8 - 3\sin 30^\circ)/2 = 3.250\ in$$
$$\bar{y}_{AA^*} = (4 + 3\cos 30^\circ)/2 = 3.299\ in \qquad (d)$$

and the midpoint of line BB^* is located at

$$\bar{x}_{BB^*} = 4\ in$$
$$\bar{y}_{BB^*} = 3.5\ in \qquad (e)$$

(continued)

17.5 (cont.)

Hence, the equation of the perpendicular bisector of line AA^* is given by

$$\frac{(y_A - \bar{y}_{AA^*})}{(x_A - \bar{x}_{AA^*})} = -\frac{1}{m_{AA^*}}$$

or, with Eqs. (b) and (d),

$$y_A = -0.9852x_A + 6.501 \text{ (in)} \qquad (f)$$

Similarly, for line BB^*, by Eqs. (c) and (e),

$$\frac{(y_B - \bar{y}_{BB^*})}{(x_B - \bar{x}_{BB^*})} = \frac{-1}{m_{BB^*}}$$

or

$$y_B = -8x_B + 35.5 \text{ (in)} \qquad (g)$$

Equating Equations (f) and (g), we have for the intersection of the perpendicular bisectors

$$-0.9852x + 6.501 = -8x + 35.5$$

or $\quad x = 4.134$ in $\qquad (h)$

By Eq. (h) and either Eq. (f) or (g),

$$y = 2.428 \text{ in} \qquad (i)$$

Hence, by Eqs. (h) and (i),

$$\underline{P(x,y) = (4.134, 2.428) \text{ (in)}}$$

17.6

GIVEN: A wheel of radius r rolls in the xy plane in the positive sense of the x axis. It starts from rest with its center on the y axis and again comes to rest after one-half revolution.

FIND: Derive formulas for the coordinates (x_P, y_P) of point P about which the wheel could have been rotated to cause this displacement and show P on a diagram.

SOLUTION:

Consider Fig. a which shows the initial position 1 and the final position 2 of the wheel.

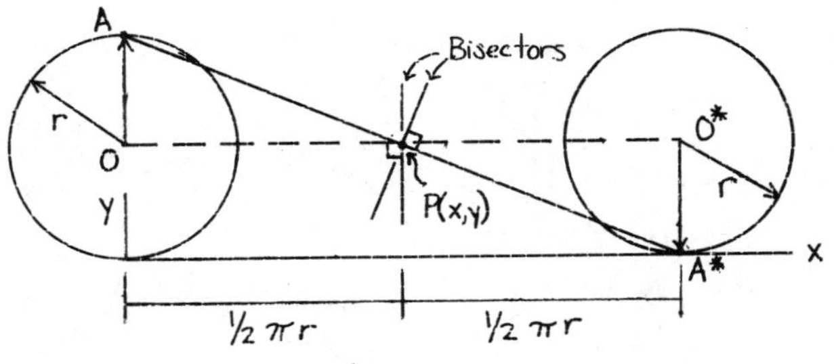

Figure a

By Fig. a, $x_P = \frac{1}{2}\pi r$, $y_P = r$

or $\quad \underline{P(x,y) = \left(\frac{\pi r}{2}, r\right)}$

17.7

GIVEN: Crank OP rotates counterclockwise in the xy plane (Fig. a). The cylindrical tube is pinned to the crank at P, and the rod OD^* is free to slide through the tube.

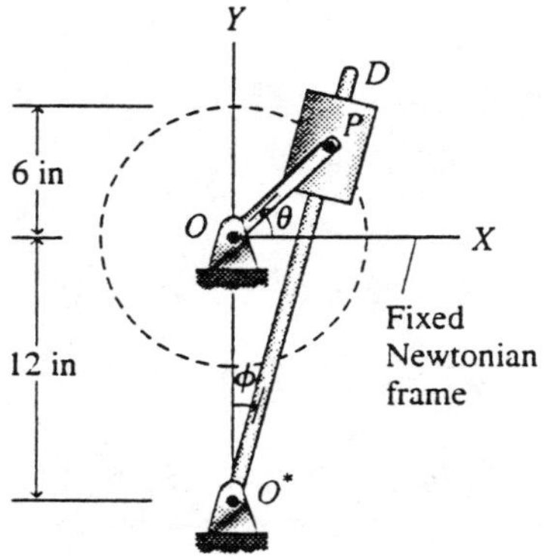

Figure a

FIND:

a) Determine the apparent displacement of P relative to an observer fixed on O^*D, as OP rotates from $\theta = 30°$ to $\theta = 180°$.

b) Determine the absolute displacement of P as OP rotates from $\theta = 30°$ to $\theta = 180°$.

SOLUTION:

Consider Fig. b which shows the crank OP in the initial position 1 and the final position 2.

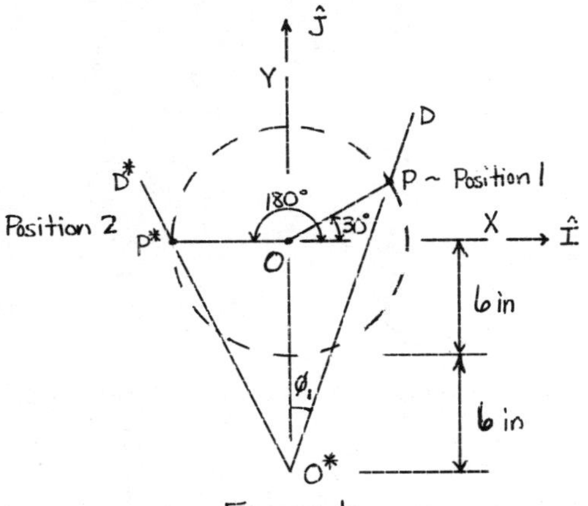

Figure b

At $\theta = 30°$, $O^*P = \sqrt{(12 + 6\sin 30°)^2 + (6\cos 30°)^2}$

$$O^*P = 15.874 \text{ in.}$$

At $\theta = 180°$; $O^*P^* = \sqrt{6^2 + 12^2} = 13.416$ in

(continued)

17.7 (cont.)

Hence, the apparent displacement of P relative to an observer fixed on rod O^*D is

$$q_{P/O^*D} = 15.874 - 13.416 = 2.458 \text{ in}$$

directed toward O^*

b) By Fig. b, the initial coordinates of P are

$$X_P = 6\cos 30^\circ = 5.196 \text{ in}$$
$$Y_P = 6\sin 30^\circ = 3.00 \text{ in.} \qquad (a)$$

The final coordinates of P are

$$X_{P^*} = -6.00 \text{ in}$$
$$Y_{P^*} = 0 \qquad (b)$$

Hence, the absolute displacement of P is

$$\bar{q}_{P/O} = (-6\hat{I} + 0\hat{J}) - (5.916\hat{I} + 3.00\hat{J})$$

or

$$\bar{q}_{P/O} = -11.196\hat{I} - 3.00\hat{J} \quad (\text{in})$$

17.8

GIVEN: A radial line OP on a grinding wheel is located by $\theta = 3t^2 + t$; θ in radians and t in seconds. (Fig. a)

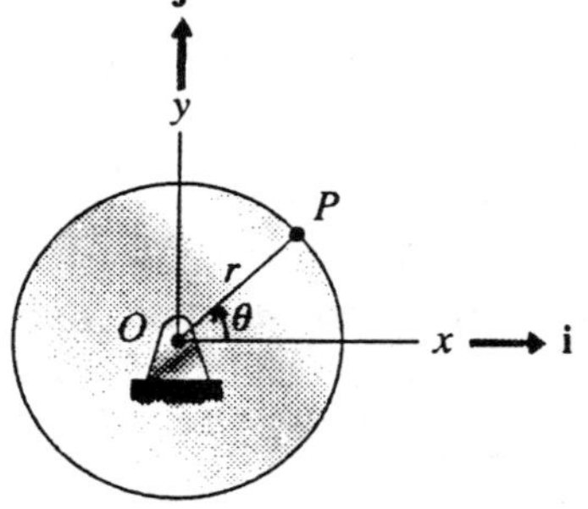

Figure a

FIND:

a) Determine the displacement of P in the time interval $0 \le t \le 4s$, in terms of r and unit vectors $\hat{\imath}$ and $\hat{\jmath}$.

b) For r = 125 mm, calculate the magnitude of P.

SOLUTION:

(a) At t = 0, the position vector of P is

$$\bar{r}_1 = r\hat{\imath} \qquad (a)$$

At t = 4s, $\theta = 52$ rad $\simeq 8.276$ revolutions. Hence, the line OP is located at

$(0.276 \text{ rev})\left(\frac{360^\circ}{\text{rev}}\right) = 99.381^\circ$ counterclockwise (See Fig. b).

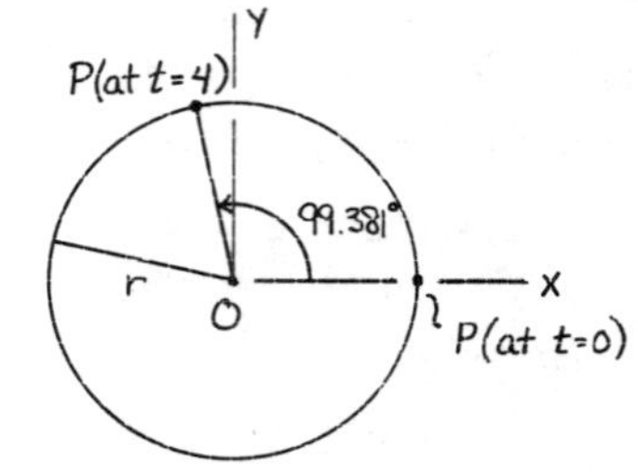

Figure b

Therefore, at t = 4s, the position vector of P is

$$\bar{r}_2 = -r(\sin 9.381^\circ)\hat{\imath} + r(\cos 9.381)\hat{\jmath}$$

or

$$\bar{r}_2 = -(0.1630r)\hat{\imath} + (0.9866r)\hat{\jmath} \qquad (b)$$

Thus, by Eqs. (a) and (b), the displacement of P is

$$\bar{q}_{P/O} = \bar{r}_2 - \bar{r}_1 = -(1.1630r)\hat{\imath} + (0.9866r)\hat{\jmath} \qquad (c)$$

b) For r = 125 mm, Eq. (c) yields

$$\bar{q}_{P/O} = -145.38\hat{\imath} + 123.33\hat{\jmath} \quad [\text{mm}]$$

or

$$q_{P/O} = \sqrt{145.38^2 + 123.33^2} = 190.6 \text{ mm}$$

17.9

GIVEN: A turntable is driven by a friction-driver wheel (Fig. a) which is given an angular acceleration $\alpha = 2$ rad/s².

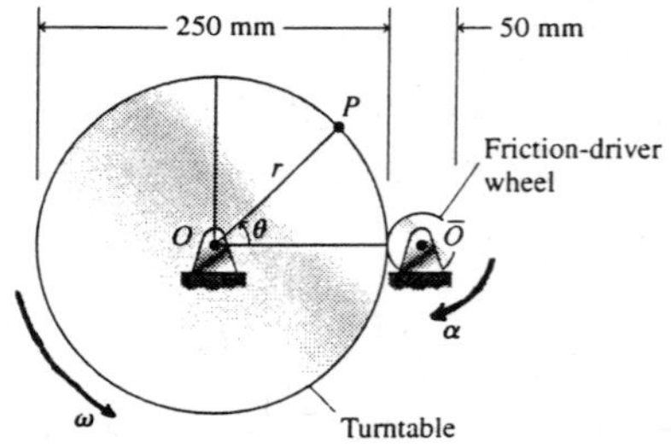

Figure a

FIND:

(a) Determine the time required for the turntable to reach an angular speed ω = 33 rpm, if it starts from rest.

(b) Locate the position of P for the instant ω = 33 rpm, if initially θ = 0.

(c) Determine the displacement of P from θ = 0 to the time that the turntable first rotates at ω = 33 rpm.

(d) Determine the distance between the initial position of P and its position P* for the instant when ω = 33 rpm for the first time.

(continued)

17.9 (cont.)

SOLUTION:

(a) The acceleration of the contact point between the turntable and drive is

$$a = \alpha r = (2)(25) = 50 \text{ mm/s}^2 \uparrow \qquad (a)$$

Hence, the angular acceleration of the turntable is

$$\alpha_T = \frac{50}{125} = 0.4 \text{ rad/s}^2 \qquad (b)$$

Therefore, the angular speed of the turntable is, at time t,

$$\omega_T = \alpha_T t = 0.4t \qquad (c)$$

When $\omega_T = \left(33 \frac{\text{rev}}{\text{min}}\right)\cdot\left(\frac{2\pi \text{ rad}}{\text{rev}}\right)\left(\frac{1 \text{ min}}{60 \text{ s}}\right) = 3.456 \text{ rad/s}$,

Eq. (c) yields

$$t = \frac{3.456 \text{ rad/s}}{0.4 \text{ rad/s}^2} = 8.639 \text{ s}$$

(b) By definition

$$\omega_T = \frac{d\theta_T}{dt} \qquad (d)$$

where θ_T is the angular displacement of the turntable. So, by Eqs. (c) and (d),

$$\theta_T = \int_0^{8.639} \omega_T dt = (0.4)\frac{t^2}{2}\Big|_0^{8.639} = 14.928 \text{ rad.}$$

Hence, point P is located at (when $\omega_T = 33$ rpm)

$$x_P = (125)\cos(14.928 \text{ rad}) = -88.867 \text{ mm}$$
$$y_P = (125)\sin(14.928 \text{ rad}) = 87.907 \text{ mm} \qquad (e)$$

Therefore, P is located at (Fig. b)

$$\bar{r}_{P^*} = -88.867\hat{i} + 87.907\hat{j} \text{ (mm)}$$

or

$$r_{P^*} = 125 \text{ mm @ } \theta = \tan^{-1}\left(\frac{87.907}{-88.867}\right) = 135.3^\circ$$

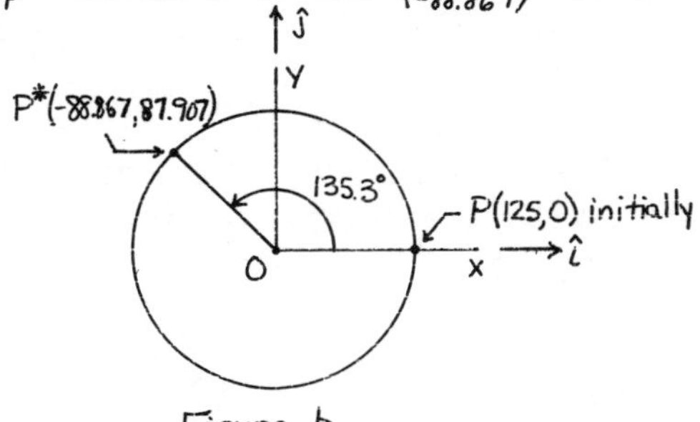

Figure b

(c) By Fig. b, the displacement of P is

$$\bar{q}_P = \bar{P}^* - \bar{P} = -213.87\hat{i} + 87.91\hat{j} \qquad (f)$$

(d) By Eq. (f), the distance between P* and P is

$$q_P = \sqrt{(-213.87)^2 + (87.91)^2} = 231.23 \text{ mm}$$

17.10

GIVEN: An electron moves in the xy plane on the path (x, y in inches, t in seconds)

$$x = 10t\cos(\pi t)$$
$$y = 10t\sin(\pi t) \qquad (a)$$

FIND:

(a) Plot the path of the electron.

(b) Describe the path geometrically.

SOLUTION:

(a) By Eqs. (a), the plot of the path may be constructed (Fig. a).

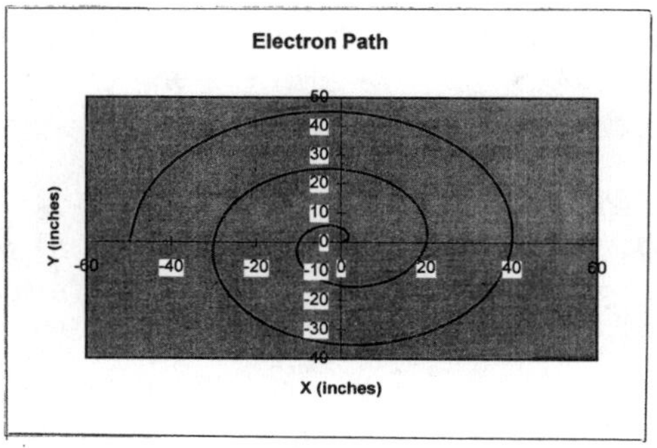

Figure a

(b) By Eqs. (a),

$$x^2 + y^2 = r^2 = (10t)^2$$

or

$$r = 10t \qquad (b)$$

Equation (b) is the equation of a geometric spiral.

17.11

GIVEN: Problem 17.10, where

$$x = 10t\cos(\pi t)$$
$$y = 10t\sin(\pi t) \qquad (a)$$

FIND: Determine the displacement of the electron for the interval $2 \le t \le 3.5$ s.

SOLUTION:

For $t = 2$ s, Eq. (a) yields

$$x = 20 \text{ in}$$
$$y = 0$$

So, its position vector is, for $t = 2$ s,

$$\bar{r}_{(2s)} = 20\hat{i} \text{ (in)} \qquad (b)$$

For $t = 3.5$ s, Eq. (a) yields

$$x = 0$$
$$y = -35 \text{ in.}$$

Hence, its position vector is, for $t = 3.5$ s,

$$\bar{r}_{(3.5s)} = -35\hat{j} \text{ (in)} \qquad (c)$$

(continued)

17.11 (cont.)

Therefore, by Eqs. (b) and (c), the displacement of the electron, in the time interval $t = 2s$ to $t = 3.5 s$, is

$$\underline{\bar{q} = \bar{r}_{(3.5)} - \bar{r}_{(2)} = -20\hat{\imath} - 35\hat{\jmath} \text{ (in)}}$$

17.12

GIVEN: A wheel of radius r rolls in the xy plane in the positive sense of the x axis. It starts from rest with its center on the y axis and comes to rest again after one-quarter revolution.

FIND: Show that the displacement could have been performed by a rotation of the wheel about the point $x = r/4$, $y = r(1 - \pi/4)$.

SOLUTION:

Consider Fig. a which shows the initial and final positions of the wheel. The rotation point is located by

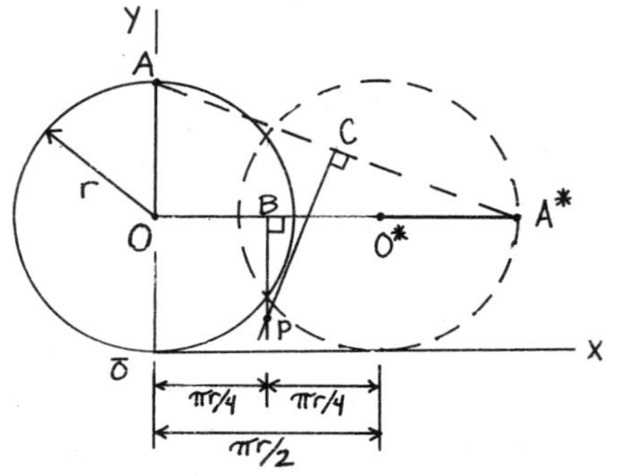

Figure a

The intersection of lines PB and PC, the perpendicular bisectors of lines OO^* and AA^*, respectively. By Fig. a, line PB is $x = \pi r/4$ and line PC is determined by

$$\frac{y - (r + \frac{r}{2})}{x - \frac{1}{2}(\frac{\pi r}{2} + r)} = -\frac{1}{\text{slope of } AA^*} = -\left[\frac{1}{\frac{-r}{\frac{\pi r}{2} + r}}\right]$$

or

$$\frac{y - (\frac{3r}{2})}{x - (\frac{\pi r}{4} + \frac{r}{2})} = 1 + \frac{\pi}{2} \qquad (a)$$

The intersection of lines PB and PC occurs at $x = \pi r/4$. Hence, substitution of this value of x into Eq. (a), the equation of line PC, yields $y = r(1 - \pi/4)$. Thus, the displacement could have been obtained by rotating the wheel as a rigid body about the point

$$\underline{P(x,y) = P\left[\frac{\pi r}{4}, r(1 - \frac{\pi}{4})\right]}$$

17.13

GIVEN: In example 17.1, the book was rotated to go from position 1 [A (0,0), B(0, 9 in)] to position 2 [A^*(13.5 in, 21.79 in), B^*(18 in, 14 in)]. (Fig. a)

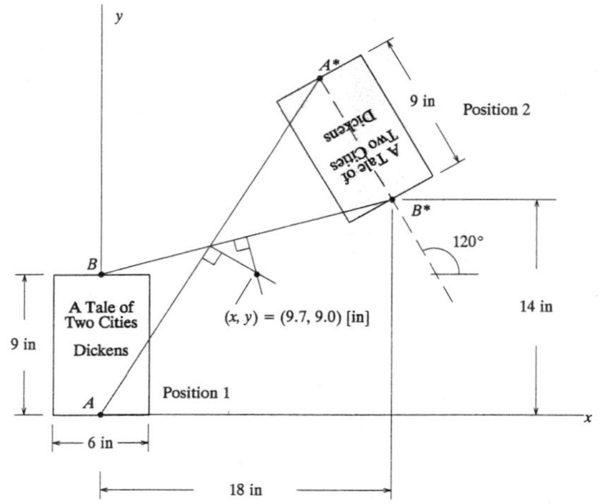

Figure a

FIND: Determine the point (x,y) about which the book may be rotated to go from position 1 to position 2 by deriving the equations for the perpendicular bisectors of lines AA^* and BB^* and locating their intersection point (x, y) analytically.

SOLUTION:

By Fig. a, the slope of line AA^* is

$$m_A = \frac{21.79}{13.5} = 1.614 \qquad (a)$$

The general form of the perpendicular bisector of AA^* is

$$y_{\perp A} = -\frac{1}{m_A} x + b \qquad (b)$$

Since the bisector must pass through the midpoint (6.75 in, 10.895 in) of AA^*, Eqs. (a) and (b) yield

$$y_{\perp A} = -0.6196x + 15.08 \qquad (c)$$

Similarly, the slope of BB^* is

$$m_B = \frac{5}{18} = 0.277\bar{7} \qquad (d)$$

and the general form of the perpendicular bisector of BB^* is

$$y_{\perp B} = -\frac{1}{m_B} x + b \qquad (e)$$

Since the bisector of BB^* must pass through its midpoint (9 in, 11.5 in), Eqs. (d) and (e) yield

$$y_{\perp B} = -3.6x + 43.90 \qquad (f)$$

Equating Eqs. (c) and (f), we find $x = 9.670$ in. Then, with $x = 9.760$ in, we obtain from either Eqs. (c) or (e), $y = 9.088$ in. (continued)

17.13 (cont.)

Hence, the point about which the book may be rotated is

$$\underline{(x,y) = (9.670 \text{ in}, 9.088 \text{ in})} \quad (g)$$

Note: The values estimated in Example 17.1 are $x = 9.7$ in, $y = 9.0$ in.

17.14

GIVEN: The wheel in Problem 17.4 (see Fig. a) rolls 0.90m to the right of the position shown.

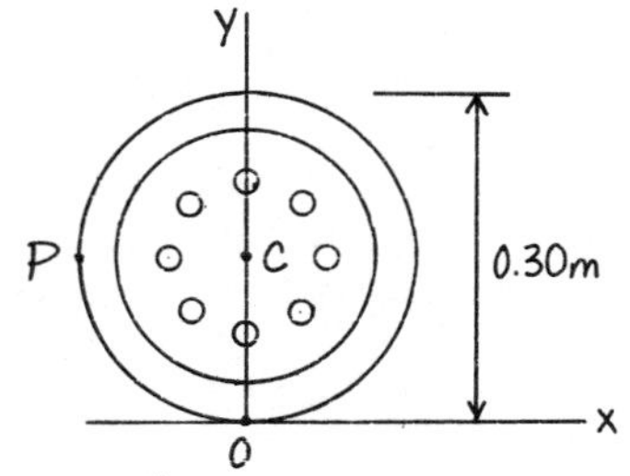

Figure a

FIND: Determine the point Q(x,y) about which the wheel could have been rotated to take it from its initial position to its final position.

SOLUTION:

Consider the particles at points C and P on the wheel (Fig. a). Since the wheel rolls 0.90m to the right, it rotates through the angle

$$\theta = \frac{0.90}{0.15} = 6 \text{ rad} = 343.775^\circ \quad (a)$$

The initial and final positions of C and P are shown in Fig. b.

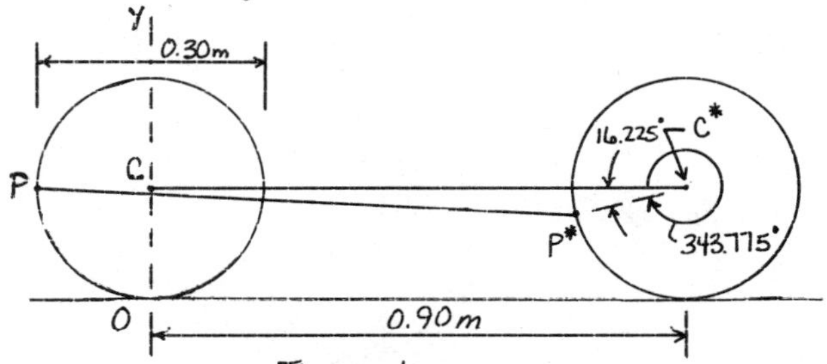

Figure b

The coordinates of P^* are, by Fig. b,

$$x = 0.90 - (0.15)\cos(16.225^\circ) = 0.7560 \text{ m}$$
$$y = 0.15 - (0.15)\sin(16.225^\circ) = 0.1081 \text{ m} \quad (b)$$

The coordinates of the midpoint of PP^* are

$$x_{mp} = (0.7560 - 0.15)/2 = 0.3030 \text{ m}$$
$$y_{mp} = (0.15 + 0.1081)/2 = 0.1290 \text{ m} \quad (c)$$

The coordinates of the midpoint of line CC^* are

$$x_{mc} = 0.450 \text{ m}, \quad y_{mc} = 0.150 \text{ m} \quad (d)$$

The slope of line PP^* is

$$m = \frac{0.1081 - 0.150}{0.7560 + 0.150} = -0.04625 \quad (e)$$

Therefore, the equation of the perpendicular bisector of PP^* is of the form

$$y_P = -\frac{1}{m}x + C \quad (f)$$

By Eqs. (e) and (f),

$$y_P = 21.623x + C \quad (g)$$

Since the bisector must pass through the midpoint of line PP^*, Eqs. (c) and (g) yield $C = -6.423$. Hence, Eq. (g) becomes

$$y_P = 21.623x - 6.423 \quad (h)$$

The perpendicular bisector of line CC^* (see Fig. b) is

$$x_c = 0.450 \text{ [m]} \quad (i)$$

Substitution of Eq. (i) into Eq. (h) yields

$$y_P = 3.308 \text{ m} \quad (j)$$

Hence, the point Q(x,y) about which the wheel could have been rotated is given by Eqs. (i) and (j). That is,

$$Q(x,y) = Q(0.450\text{m}, 3.308\text{m})$$

17.15

GIVEN: The path of a particle is given as

$$x = 3t^2 + 4t \text{ (mm)}$$
$$y = 2t^3 \text{ (mm)} \quad (a)$$

with t in seconds.

FIND:

(a) Determine the particle's displacement from $t = 1$s to $t = 6$s.

(b) Calculate the magnitude of the displacement of part (a).

(c) Plot the path of the particle and show the displacement of the particle in the diagram for $t = 1$s to $t = 6$s.

SOLUTION:

(a) At $t = 1$s, Eqs. (a) yield

$$x_1 = 7\text{mm}, \quad y_1 = 2\text{mm}$$

At $t = 6$s, $x_6 = 132$ mm, $y_6 = 432$ mm

Hence, the displacement is

$$\bar{q} = (x_6 - x_1)\hat{i} + (y_6 - y_1)\hat{j}$$

or $$\underline{\bar{q} = 125\hat{i} + 430\hat{j}} \quad (b)$$

where $\hat{i}, \hat{j}$ are unit vectors along the x, y axis respectively.

(Continued)

17.15 (cont.)

(b) The magnitude of $\bar{q}$ is

$$\underline{\underline{\bar{q} = \sqrt{125^2 + 430^2} = 447.8 \text{ mm}}}$$

(c) The plot of the path and the displacement are shown in Figure a. for $1 \le t \le 6$ s.

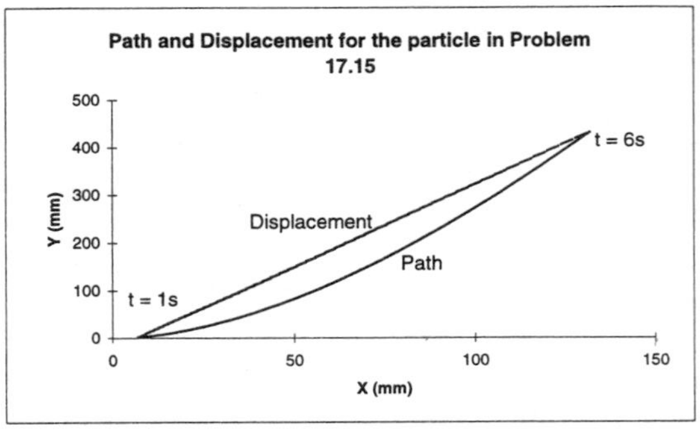

Figure a

17.16

GIVEN: The axle OD of the roller of a rolling mill rotates about the fixed vertical axis OZ (Fig. a). As the axle OD and the rod BC undergo a rotation θ (radians) about axis OZ, the roller of radius r rolls on the rolling surface on a circle of radius R.

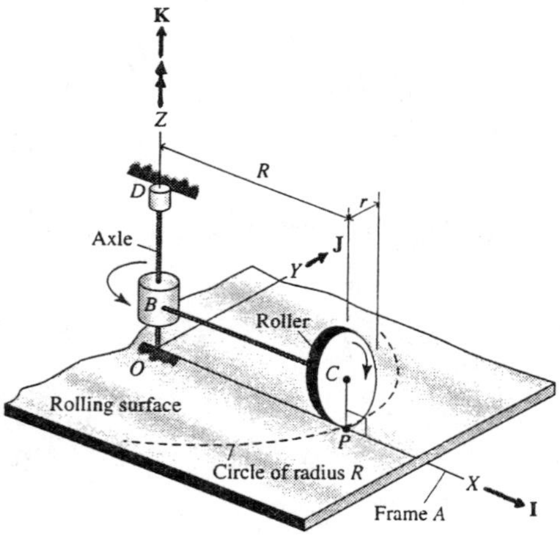

Figure a

FIND:

a) Derive a formula for the absolute displacement $\bar{q}_{P/A}$ of point P on the rim of the roller in terms of R, r, and θ.

b) With the formula derived in part a, calculate the displacement of P for

$R = 3r = 12$ in. and $\theta = 90°$.

c) Verify the result in part b by graphical construction.

SOLUTION:

By kinematics (Fig. a),

$$R\theta = r\phi; \text{ or } \phi = \frac{R}{r}\theta \qquad (a)$$

Now consider a front view of the roller (Fig. b) and a top view of the mill (Fig. c) that show the initial positions of P and C ($\theta = 0$) and the final positions P^* and C^* ($\theta > 0$).

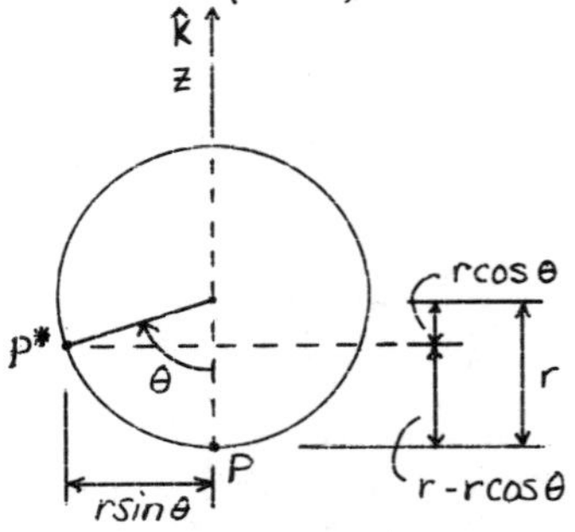

Figure b: Front view of roller

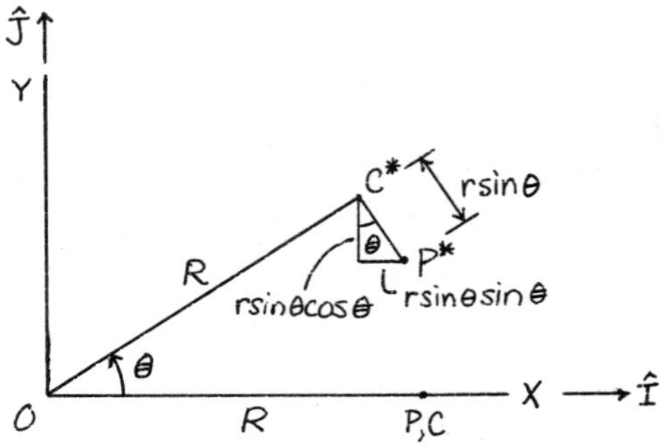

Figure c: Top view of mill

a) By Figs. b and c,

$$\bar{P}^* = (R\cos\theta + r\sin\phi\sin\theta)\hat{I} + (R\sin\theta - r\sin\phi\cos\theta)\hat{J} + (r - r\cos\phi)\hat{K} \qquad (b)$$

$$\bar{P} = R\hat{I}$$

Hence, by Eqs. (a) and (b),

$$\underline{\underline{\bar{q}_{P/A} = \bar{P}^* - \bar{P} = \left[R\cos\theta - R + r\sin\left(\tfrac{R}{r}\theta\right)\sin\theta\right]\hat{I} + \left[R\sin\theta - r\sin\left(\tfrac{R}{r}\theta\right)\cos\theta\right]\hat{J} + \left[r - r\cos\left(\tfrac{R}{r}\theta\right)\right]\hat{K}}} \qquad (c)$$

b) For $R = 12$ in, $r = 4$ in, and $\theta = 90°$, Eq. (c) yields

$$\underline{\underline{\bar{q}_{P/A} = -16\hat{I} + 12\hat{J} + 4\hat{K}}}$$

or $\underline{\underline{q_{P/A} = 20.396 \text{ in.}}}$

c) Since $\phi = \frac{R}{r}\theta = (3)(90°) = 270°$, P^* lies on the horizontal axis of the roller (Fig. d)

(continued)

17.16 (cont.)

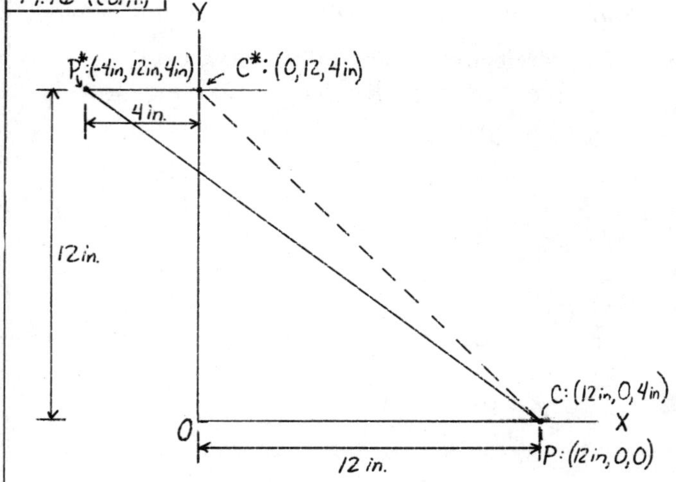

Top view of mill

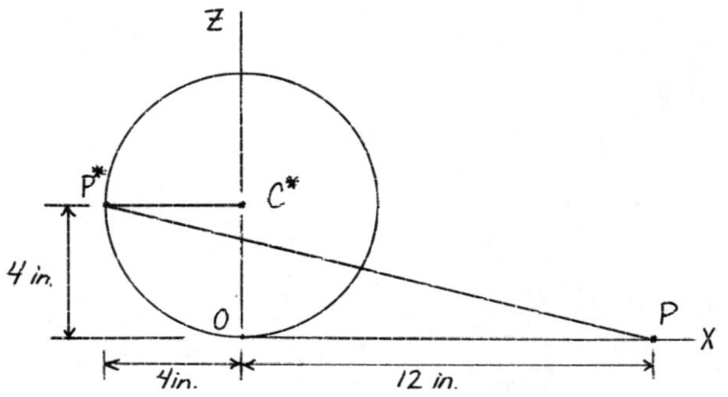

Figure d

By measurement, or by inspection, of Fig. d,

$$\bar{q}_{P/A} = \bar{P}^* - \bar{P} = -16\hat{I} + 12\hat{J} + 4\hat{K} \ \text{(in)}$$

This result verifies the result of part b.

17.17

GIVEN: A wheel of radius r rolls along the positive x axis (Fig. a). It starts from rest with its center on the y axis and again comes to rest after a rotation through angle θ.

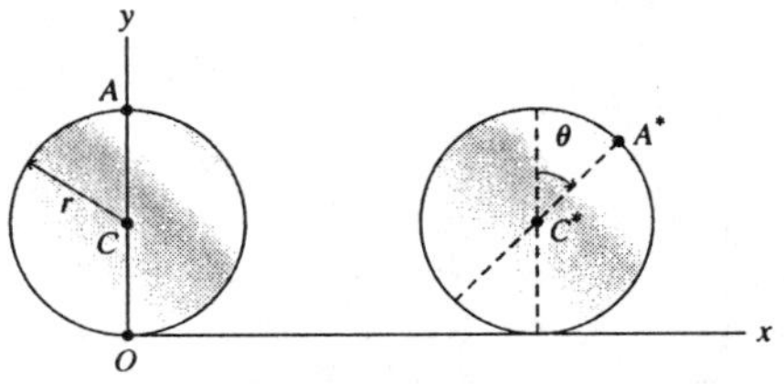

Figure a

FIND:

a) Show that the displacement could have been performed by rotating the w eel about point P with coordinates

$$X_P = \tfrac{1}{2} r\theta, \quad Y_P = r\left[1 - \frac{\theta \sin\theta}{2(1-\cos\theta)}\right].$$

b) Plot the path of point P in the xy plane for $0 \le \theta \le 720°$ and r = 1 ft.

SOLUTION:

Consider Fig. b, in which the perpendicular bisectors PC′ and PA′ of lines CC* and AA* are shown.

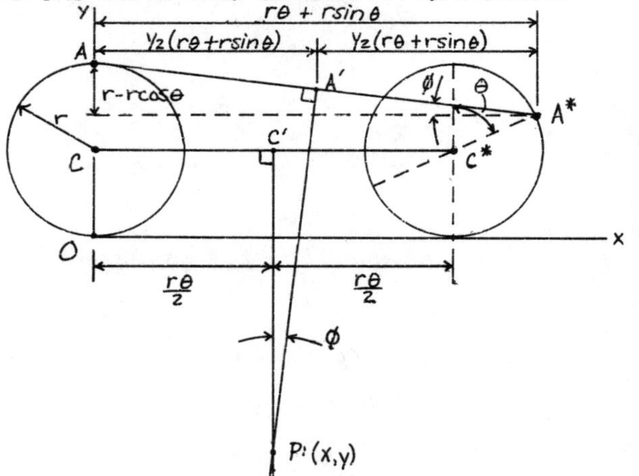

Figure b

By Fig. b, the equation of line PC′ is

$$x = \tfrac{1}{2} r\theta \qquad \text{(a)}$$

Also by Fig. b,

$$\tan\phi = \frac{r - r\cos\theta}{r\theta + r\sin\theta} = \frac{1-\cos\theta}{\theta + \sin\theta} \qquad \text{(b)}$$

Hence, by Eq. (b) and Fig. b, the equation of line PA′ is given by

$$\tan\phi = \frac{1-\cos\theta}{\theta + \sin\theta} = \frac{x - \tfrac{1}{2}(r\theta + r\sin\theta)}{y - [r + r\cos\theta + \tfrac{1}{2}(r - r\cos\theta)]} \qquad \text{(c)}$$

At the intersection P of lines PC′ and PA′, $x = \tfrac{1}{2} r\theta$. [see Fig. b and Eq. (b)]. So with $x = \tfrac{1}{2} r\theta$, Eq. (c) yields

$$y = r\left[1 - \frac{\theta \sin\theta}{2(1-\cos\theta)}\right]$$

Hence, the coordinates of point P are

$$x_P = \tfrac{1}{2} r\theta, \quad y_P = r\left[1 - \frac{\theta \sin\theta}{2(1-\cos\theta)}\right] \qquad \text{(d)}$$

b) With Eq. (d), the path of point P is

$$P(x,y) = \left[\tfrac{1}{2} r\theta,\ r\left(1 - \frac{\theta\sin\theta}{2 - 2\cos\theta}\right)\right]$$

The path is plotted in Fig. c.

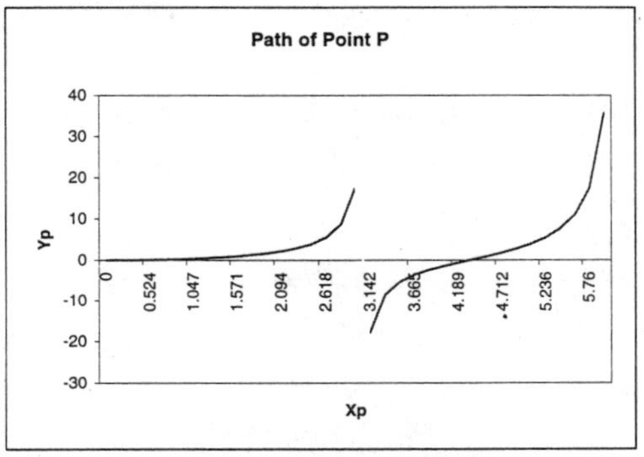

17.18

GIVEN: A rigid body executes plane motion parallel to the xy plane as shown in Fig. a.

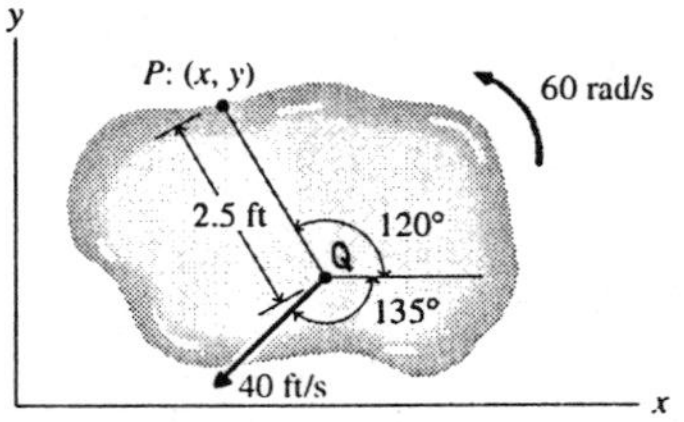

Figure a.

FIND: Calculate the velocity $(\dot{x}, \dot{y})$ of point P.

SOLUTION:

The velocity $\bar{v}_P$ of P is given by

$$\bar{v}_P = \bar{v}_Q + \bar{v}_{P/Q} \qquad (a)$$

By Fig. a, $\bar{v}_{P/Q} = (2.5)(60)\swarrow_{60°} = 150\text{ ft/s}\swarrow_{60°}$

and $\bar{v}_Q = 40\text{ ft/s}\swarrow_{45°}$

Therefore,

$$\bar{v}_P = (-150\sin 60° - 40\sin 45°)\hat{\imath} + (-150\cos 60° - 40\cos 45°)\hat{\jmath}$$

$$= -158.91\hat{\imath} - 103.28\hat{\jmath} \quad (\text{ft/s})$$

where $\hat{\imath}, \hat{\jmath}$ are unit vectors along x and y axes, respectively.

Hence, $\dot{x} = -158.19$ ft/s ; $\dot{y} = -103.28$ ft/s

17.19

GIVEN: A disk of radius 600 mm moves in the xy plane along the x axis. The motion of the center of the disk is given by $x = 600t^3 + 300t$ (mm), with t in seconds. The angular clockwise rotation of the disk is given by $\theta = 3t^2 + 1$ (rad)

FIND:

a) Determine the velocity of the center of the disk for $t = 1$ s.

b) Determine the velocity of the highest point of the disk for $t = 1$ s.

c) Show that skidding occurs if the disk moves on a track below it.

SOLUTION:

a) The velocity of the center of the disk is

$$\bar{v} = \dot{x}\hat{\imath} = (1800t^2 + 300)\hat{\imath} \quad (\text{mm/s}) \qquad (a)$$

For $t = 1$ s, Eq. (a) yields

$$\bar{v} = 2100\hat{\imath} \quad (\text{mm/s})$$

where $\hat{\imath}$ is a unit vector in the positive x sense.

b) Let P = topmost point of the disk and C = the center of the disk. Then the velocity of P is given by (See Fig. a)

$$\bar{v}_P = \bar{v}_C + \bar{v}_{P/C} \qquad (c)$$

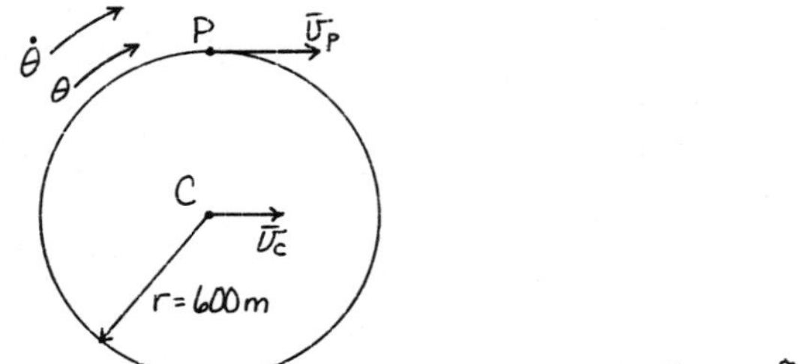

Figure a

By Eq. (b),

$$\bar{v}_C = 2100\hat{\imath} \quad (\text{mm/s}) \qquad (d)$$

Also, by Fig. a,

$$\bar{v}_{P/C} = r\dot{\theta}\hat{\imath} = 600(6t)\hat{\imath} \quad (\text{mm/s})$$

For $t = 1$ s,

$$\bar{v}_{P/C} = 3600\hat{\imath} \quad (\text{mm/s}) \qquad (e)$$

Hence, by Eqs. (c), (d), and (e),

$$\bar{v}_P = 5700\hat{\imath} \quad (\text{mm/s}) \qquad (f)$$

c) If the disk is rolling,

$$\bar{v}_P = 2r\dot{\theta}\hat{\imath} = 2(600)(6t)\hat{\imath}$$

or for $t = 1$ s,

$$\bar{v}_P = 7200\,i \quad (\text{mm/s})$$

Since 7200 mm/s ≠ 5700 mm/s, the disk is skidding.

17.20

GIVEN: One corner of the homogeneous block shown in Fig. a slides along the x axis while the angle θ increases.

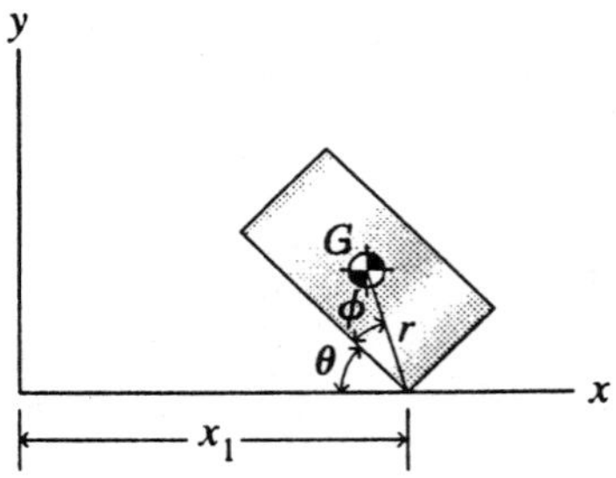

Figure a

FIND: Express the x and y components of velocity $\dot{x}_G, \dot{y}_G$ of the mass center G in terms of $x_1, \theta, \dot{\theta}$, r, and φ. (continued)

17.20 (cont.)

SOLUTION:

By Fig. a,

$$x_G = x_1 - r\cos(\theta+\phi)$$
$$y_G = r\sin(\theta+\phi) \qquad (a)$$

Differentiation of Eq. (a) yields

$$\dot{x}_G = \dot{x}_1 + [r\sin(\theta+\phi)]\dot{\theta}$$
$$\dot{y}_G = [r\cos(\theta+\phi)]\dot{\theta}$$

or

$$\dot{x}_G = \dot{x}_1 + r\dot{\theta}\sin(\theta+\phi)$$
$$\underline{\dot{y}_G = r\dot{\theta}\cos(\theta+\phi)}$$

17.21

GIVEN: The slider-crank mechanism shown in Fig. a.

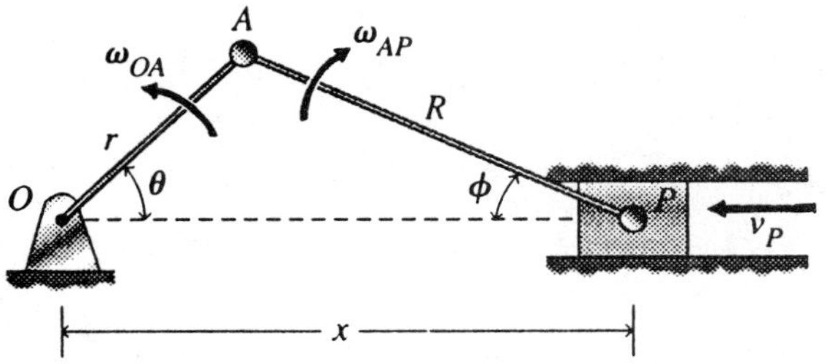

Figure a

FIND: For $r = 75$ mm, $x = 100$ mm, $\theta = 90°$, $\omega_{OA} = 100$ rpm, determine the velocity of the slider P and the angular speed ω_{AP} of the connecting rod by means of Theorem 17.2.

SOLUTION:

Note that $\omega_{OA} = 100$ rpm $= \frac{10\pi}{3}$ rad. Hence, by Figure b, where $r = 75$ mm, $x = 100$ mm, and $\theta = 90°$,

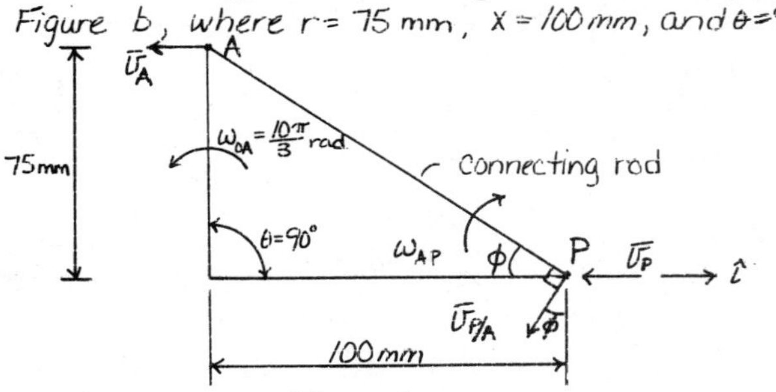

Figure b

$$\bar{v}_A = -(75)\left(\frac{10\pi}{3}\right)\hat{\imath} = -250\pi\,\hat{\imath}\ \text{(mm/s)} \qquad (a)$$

By Theorem 17.2,

$$\bar{v}_P = \bar{v}_A + \bar{v}_{P/A} \qquad (b)$$

By Fig. b and Eqs. (a) and (b),

$$-v_P\,\hat{\imath} = -250\pi\,\hat{\imath} +$$

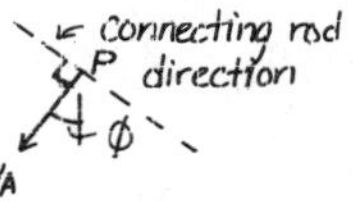

or

$$-v_P\,\hat{\imath} = -250\pi\,\hat{\imath} - v_{P/A}(\sin\phi)\hat{\imath} - v_{P/A}(\cos\phi)\,\hat{\jmath} \qquad (c)$$

Since $\cos\phi \neq 0$ $\quad \underline{v_{P/A} = 0}$

Hence, $\underline{\omega_{AP} = 0}$

and by Eq. (C),

$$\underline{v_P = 250\pi = 785.40 \text{ mm/s}}$$

17.22

GIVEN: A rigid machine link (Fig. a) is constrained to move along the vertical guide C and a horizontal guide D, so that for the position shown the velocity of B is 40 ft/s to the left.

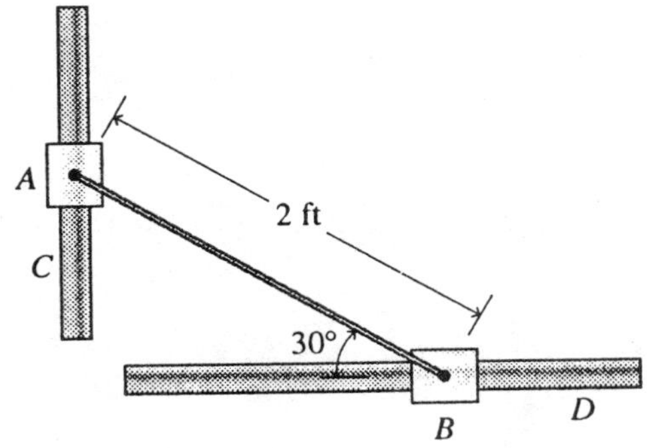

Figure a

FIND:

a) Calculate the velocity of A and the angular velocity of link AB for the position shown. Use Theorem 17.2.

b) Verify the results of part a by differentiating the geometrical displacement equations.

SOLUTION:

a) By Theorem 17.2 and Fig. b,

$$\bar{v}_B = \bar{v}_A + \bar{v}_{B/A} \qquad (a)$$

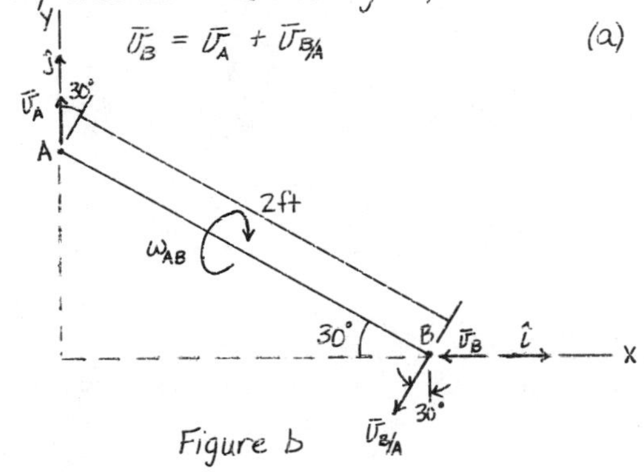

Figure b

By Fig. b and Eq. (a),

$$-40\hat{\imath} = v_A\,\hat{\jmath} - v_{B/A}(\cos 30°)\hat{\jmath} - v_{B/A}(\sin 30°)\hat{\imath}$$

Equating $\hat{\imath}$ and $\hat{\jmath}$ components, we find

$$v_{B/A} = 80 \text{ ft/s}$$
$$v_A = v_{B/A}\cos 30° = 69.28 \text{ ft/s}$$

$$\therefore\ \underline{\omega_{AB} = (v_{B/A})/2 = 40 \text{ rad/s}} \ (\text{clockwise})$$

$$\underline{\bar{v}_A = 69.28\,\hat{\jmath}\ \text{(ft/s)}}$$

(Continued)

17.22 (cont.)

b) Let link AB form the angle θ with the x axis (Fig. c).

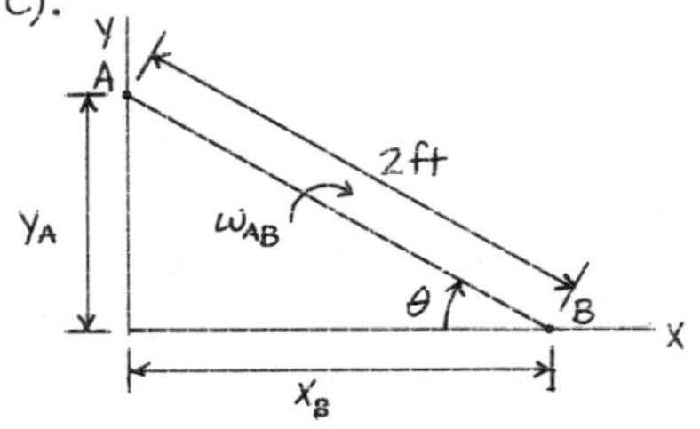

Figure C

By Fig. C,

$$x_B = 2\cos\theta \quad (ft)$$
$$y_A = 2\sin\theta \quad (ft) \qquad (b)$$

Differentiation of Eqs. (b) yields

$$v_B = \dot{x}_B = -2(\sin\theta)\dot{\theta} = -2\,\omega_{AB}\sin\theta$$
$$v_A = \dot{y}_A = 2(\cos\theta)\dot{\theta} = 2\,\omega_{AB}\cos\theta \qquad (c)$$

For $\theta = 30°$, and $v_B = -40$ ft/s, Eqs. (c) yield

$$\omega_{AB} = 40 \text{ rad/s} \circlearrowright$$

$$v_A = 69.28 \text{ ft/s}$$

These results agree with part a.

17.23

GIVEN: The slider B travels along path M-M (Fig. a), and the crank OA rotates at 300 rad/s, clockwise.

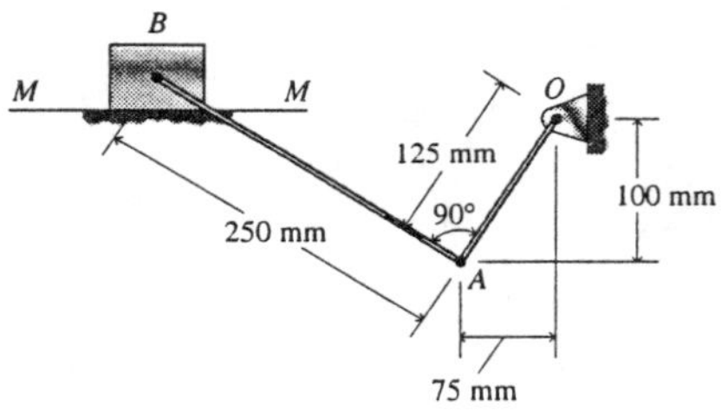

Figure a

FIND: Using Theorem 17.2, determine the velocity of B and the angular speed ω of the connecting rod AB for the configuration shown.

SOLUTION: By Theorem 17.2 and Fig. b,

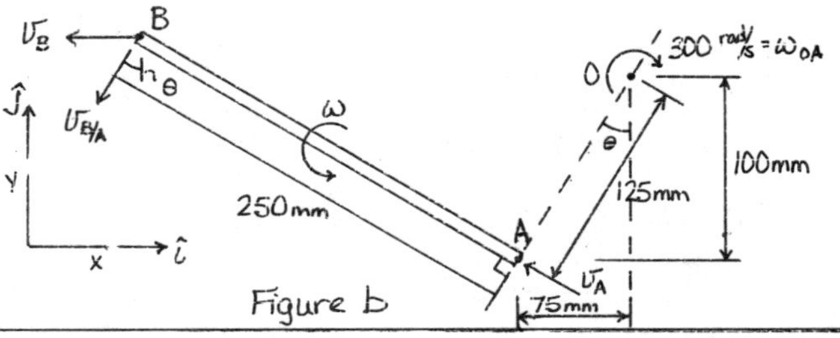

Figure b

$$\vec{v}_B = \vec{v}_A + \vec{v}_{B/A}$$

or

$$-v_B\,\hat{i} = -v_A(\cos\theta)\,\hat{i} + v_A(\sin\theta)\,\hat{j} - v_{B/A}(\sin\theta)\,\hat{i} - v_{B/A}(\cos\theta)\,\hat{j}$$

Hence,

$$v_B = v_A\cos\theta + v_{B/A}\sin\theta$$
$$v_A\sin\theta = v_{B/A}\cos\theta \qquad (a)$$

Also, by Fig. b,

$$v_A = (125)(300) = 37\,500 \text{ mm/s} = 37.50 \text{ m/s} \qquad (b)$$

and

$$\sin\theta = 0.60, \quad \cos\theta = 0.80 \qquad (c)$$

Equations (a), (b), and (c) yield

$$v_{B/A} = 28.125 \text{ m/s}$$
$$v_B = 46.875 \text{ m/s}$$

Then,

$$\omega = \frac{v_{B/A}}{0.25} = 112.5 \text{ rad/s}$$

and

$$\vec{v}_B = -v_B\,\hat{i} = -46.875\,\hat{i} \quad (\text{m/s})$$

17.24

In Fig. a, let $\theta = 30°$

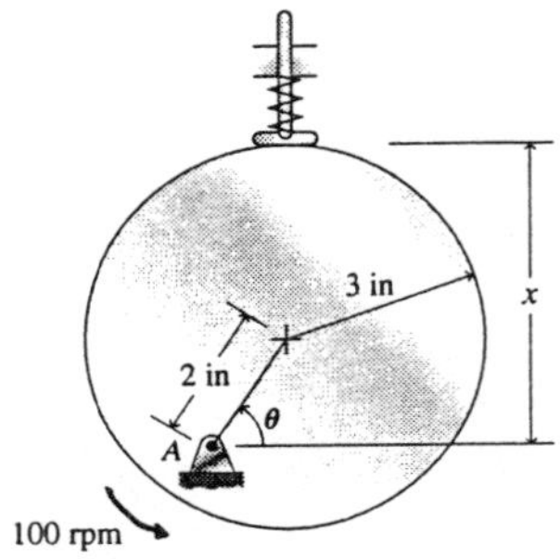

Figure a

FIND: Calculate the velocity of valve B by the theory (Sec. 17.2) for bodies that move in contact with each other.

SOLUTION: Note that 100 rpm $= 10\pi/3$ rad/s. Let P be the point of the cam in contact with the valve B (Fig. b). By Fig. b and the law of cosines,

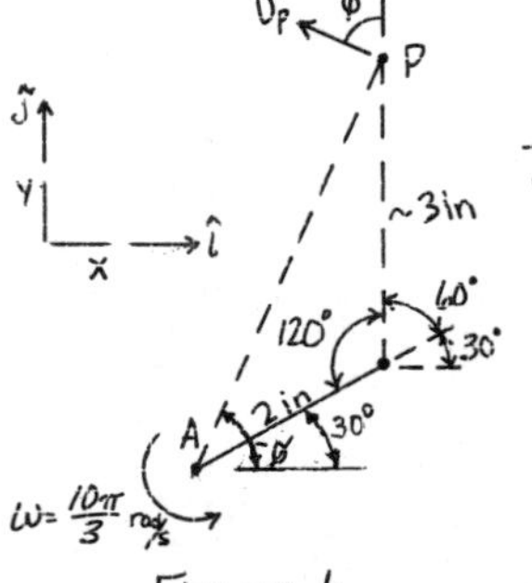

Figure b

$$AP = \sqrt{2^2 + 3^2 - 2(2)(3)\cos 120°}$$
$$AP = \sqrt{19} \quad (\text{in}) \qquad (a)$$

Therefore, by Fig. b and Eq. (a),

$$v_P = (AP)\omega = \tfrac{10\pi}{3}\sqrt{19} = 45.646 \text{ in/s} \qquad (b)$$

and by the theory of motion of bodies in contact, the velocity v_B of the valve is

$$v_B\,\hat{j} = (v_P\cos\phi)\,\hat{j}. \qquad (c)$$

(continued)

17.24 (cont.)

By Fig. b,

$$(AP)\cos\phi = 2\cos 30^\circ$$

or $\cos\phi = 0.3973$ (d)

Hence, Eqs. (b), (c), and (d) yield

$$v_B = 18.14 \text{ in/s}$$

17.25

GIVEN: The crank OA of the mechanism (Fig. a) rotates clockwise at a constant angular speed of 180 rad/s. Link AB is pinned at A and its other end is constrained to move along guide CD.

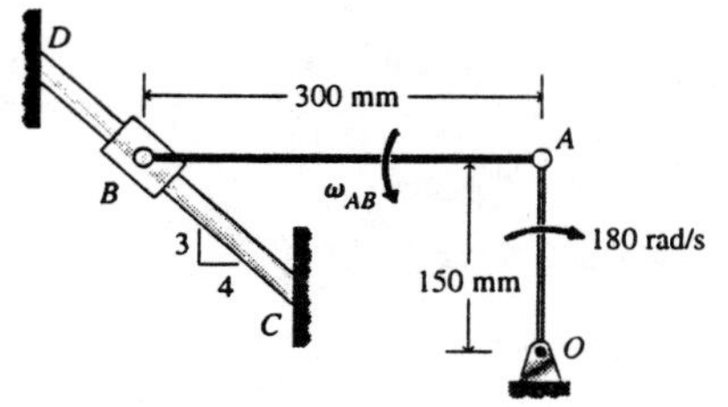

Figure a

FIND:

a) By Theorem 17.2, calculate the velocity of B for the position in which link AB is horizontal.

b) Then, determine the angular speed of link AB.

SOLUTION:

a) Consider the diagram of link AB showing the velocities of points A and B (Fig. b).

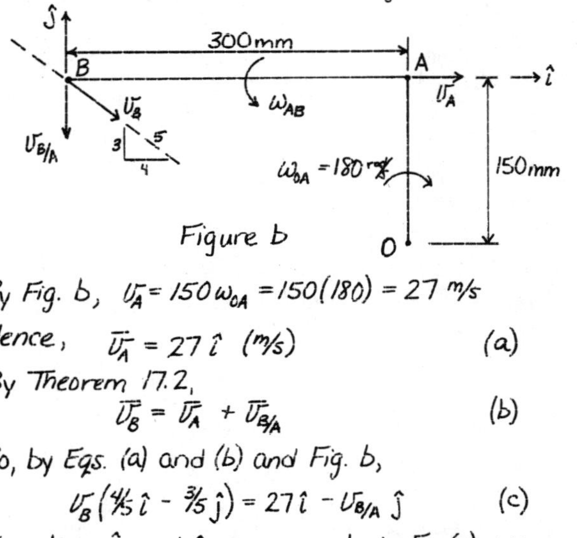

Figure b

By Fig. b, $v_A = 150\,\omega_{OA} = 150(180) = 27$ m/s

Hence, $\bar{v}_A = 27\,\hat{i}$ (m/s) (a)

By Theorem 17.2,

$$\bar{v}_B = \bar{v}_A + \bar{v}_{B/A} \quad (b)$$

So, by Eqs. (a) and (b) and Fig. b,

$$v_B\left(\tfrac{4}{5}\hat{i} - \tfrac{3}{5}\hat{j}\right) = 27\hat{i} - v_{B/A}\,\hat{j} \quad (c)$$

Equating $\hat{i}$ and $\hat{j}$ components in Eq. (c), we obtain

$$\tfrac{4}{5}v_B = 27$$
$$\tfrac{3}{5}v_B = v_{B/A} \quad (d)$$

The solution of Eqs. (d) is

$$v_B = 33.75 \text{ m/s}, \quad v_{B/A} = 20.25 \text{ m/s} \quad (e)$$

Hence, the velocity of B is

$$\bar{v}_B = v_B\left(\tfrac{4}{5}\hat{i} - \tfrac{3}{5}\hat{j}\right) = 27\hat{i} - 20.25\hat{j} \text{ (m/s)}$$

b) By Fig. b,

$$v_{B/A} = 300\,\omega_{AB} \text{ (mm/s)} = 0.30\,\omega_{AB} \text{ (m/s)}$$

Therefore, the angular speed of link AB is

$$\omega_{AB} = (v_{B/A})/0.30 = \frac{20.25}{0.30} = 67.5 \text{ rad/s}$$

17.26

GIVEN: The rigid bar in Fig. a moves in the xy plane. For the position shown, $\dot{x}_1 = 30$ ft/s, $\dot{y}_1 = 18$ ft/s, $\dot{x}_2 = 42$ ft/s. (a)

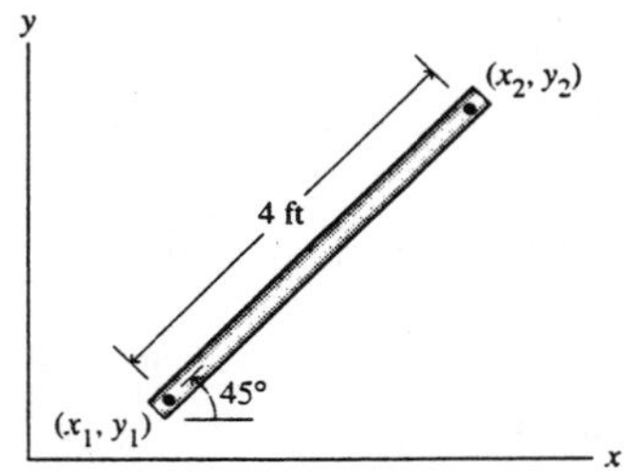

Figure a

FIND:

a) By Theorem 17.2, determine $\dot{y}_2$.

b) Then, determine the x, y projections of the velocity of the center of the bar.

SOLUTION:

a) Consider Fig. b, in which the velocities of the ends of the bar are shown.

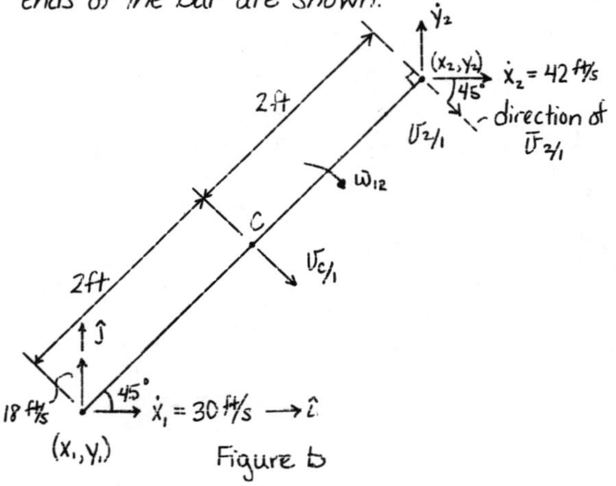

Figure b

By Theorem 17.2,

$$\bar{v}_2 = \bar{v}_1 + \bar{v}_{2/1} \quad (b)$$

or by Fig. b and Eq. (b),

$$\bar{v}_2 - \bar{v}_1 = (\dot{x}_2 - \dot{x}_1)\hat{i} + (\dot{y}_2 - \dot{y}_1)\hat{j} = \bar{v}_{2/1} \quad (c)$$

By Eqs. (a) and (c),

$$12\hat{i} + (\dot{y}_2 - 18)\hat{j} = v_{2/1}\left[(\cos 45^\circ)\hat{i} - (\sin 45^\circ)\hat{j}\right]$$

(continued)

17.26 (cont.)

Equating $\hat{\imath}$ and $\hat{\jmath}$ components, we obtain

$$v_{2/1} \cos 45° = 12$$
$$v_{2/1} \sin 45° + \dot{y}_2 = 18 \quad (d)$$

The solution of Eqs. (d) is

$$v_{2/1} = 16.97 \text{ ft/s}$$
$$\dot{y}_2 = 6 \text{ ft/s} \quad (e)$$

b) To determine the velocity of the center C of the bar, first determine the angular velocity of the bar. By Eq. (e) and Fig. b.

$$\omega_{12} = (v_{2/1})/4 = 4.2426 \text{ rad/s}$$

Hence, the speed of C relative to 1 is (See Fig. b)

$$v_{c/1} = 2\omega_{12} = 8.485 \text{ ft/s}$$

Hence,
$$\bar{v}_{c/1} = 8.485[(\cos 45°)\hat{\imath} - (\sin 45°)\hat{\jmath}]$$
$$= 6\hat{\imath} - 6\hat{\jmath} \quad (f)$$

Then, by Theorem 17.2,

$$\bar{v}_c = \bar{v}_1 + \bar{v}_{c/1}$$

So,
$$\bar{v}_c = 30\hat{\imath} + 18\hat{\jmath} + 6\hat{\imath} - 6\hat{\jmath}$$
$$= 36\hat{\imath} + 12\hat{\jmath} \text{ (ft/s)}$$

and the x,y projections of $\bar{v}_c$ are

$$\dot{x}_c = 36 \text{ ft/s}$$
$$\dot{y}_c = 12 \text{ ft/s}$$

17.27

GIVEN: The gear with center A and the crank AB rotate about an axle at A (Fig. a). The gear and crank are not connected. The gear with center B is riveted to rod BC and pinned to crank AB at B. Crank AB rotates counterclockwise at 40 rpm.

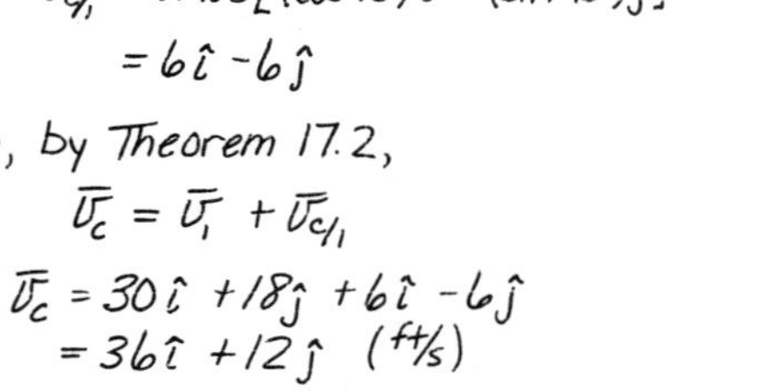

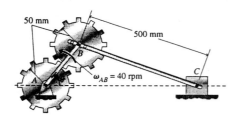

Figure a

FIND: Determine the angular velocity of the gear with center at A for $\theta = 0°, 90°, 180°$, and $270°$.

SOLUTION:

First, consider the gears for $\theta = 0°$ (Fig. b). Let P be the contact point between the gears with centers at A and B.

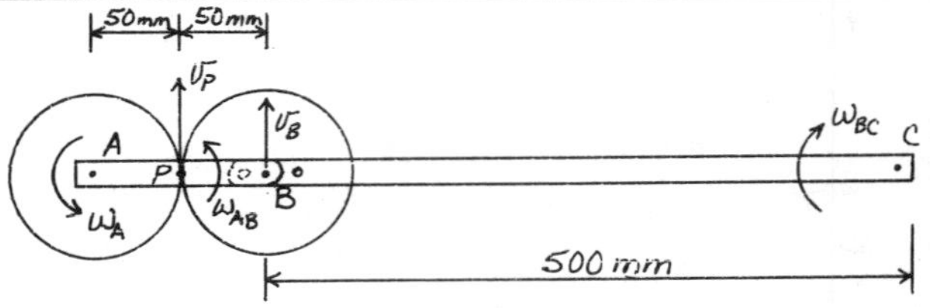

Figure b

By Fig. b,

$$v_B = (0.10)\,\omega_{AB} = (0.50)\,\omega_{BC} \quad (\text{m/s})$$

or
$$\omega_{BC} = \tfrac{1}{5}\,\omega_{AB} \quad (a)$$

Also, by Fig. b and Eq. (a),

$$v_P = (0.55)\omega_{BC} = (0.11)\omega_{AB} = (0.05)\omega_A \quad (\text{m/s})$$

Hence,

$$\omega_A = 2.2\,\omega_{AB} = 2.2(40) = 88 \text{ rpm}\ ; \quad \theta = 0°$$

Similarly, for $\theta = 180°$,

$$v_B = (0.10)\,\omega_{AB} = (0.50)\omega_{BC} \quad (\text{m/s})$$
$$\omega_{BC} = \tfrac{1}{5}\,\omega_{AB} \quad (b)$$
$$v_P = (0.450)\omega_{BC} = (0.09)\omega_{AB} = (0.05)\omega_A \quad (\text{m/s})$$

or
$$\omega_A = 1.8\,\omega_{AB} = 1.8(40) = 72 \text{ rpm}\ ; \quad \theta = 180°$$

Consider the gears for $\theta = 90°$ (Fig. c). Let P be the contact point between the gears with centers at A and B.

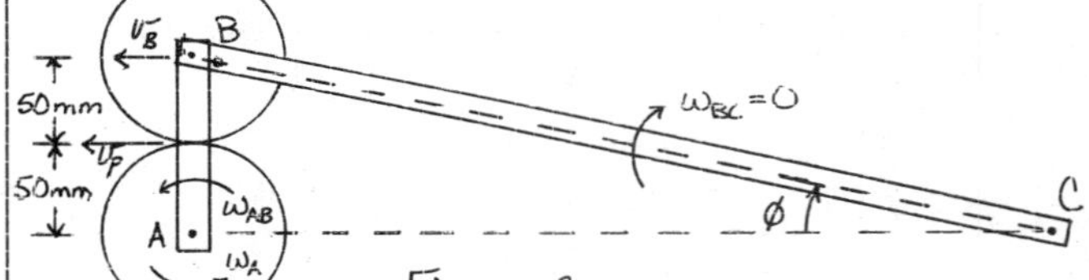

Figure c

For $\theta = 90°$, the rod BC has reached its maximum inclination $\phi = \phi_{max}$. Hence, $\dot{\phi} = \omega_{BC} = 0$. Hence, rod BC and the gear with center at B translates to the left with speed v_B. Hence, the speed of the contact point P is also equal to v_B, that is,

$$v_B = v_P = (0.10)\,\omega_{AB} \quad (c)$$

Also,
$$v_P = (0.05)\,\omega_A \quad (d)$$

Hence, by Eqs. (c) and (d),

$$\omega_A = 2\omega_{AB} = 2(40) = 80 \text{ rpm}\ ; \quad \theta = 90°$$

Similarly for $\theta = 270°$, $\omega_{BC} = 0$, and

$$\omega_A = 2\,\omega_{AB} = 2(40) = 80 \text{ rpm}\ ; \quad \theta = 270°$$

17.28

GIVEN: The arm OP of the quick-return mechanism (Fig. a) rotates counterclockwise at 240 rpm. The cylindrical slider tube is pinned to the arm at P, and rod O^*D is free to slide through the tube.

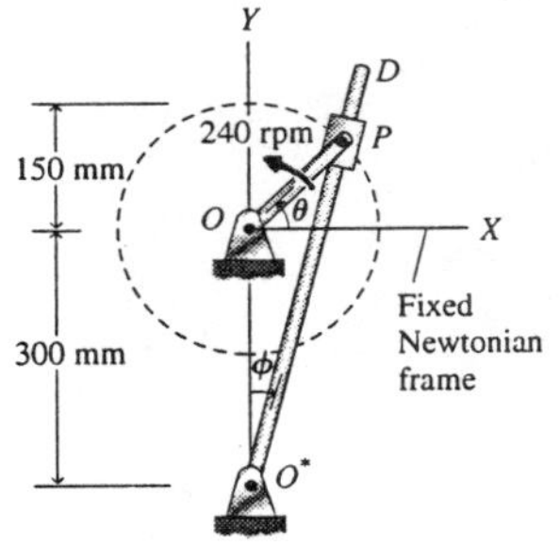

Figure a

FIND: Determine the angular velocity of rod O^*D for $\theta = 30°$.

SOLUTION:

Note that 240 rpm = 25.133 rad/s = ω_{OP}. By Fig. a, geometrical relations yield

$$x_P = (0.15)\cos\theta = \overline{O^*P}\sin\phi \quad (m) \qquad (a)$$
$$y_P = (0.15)\sin\theta = \overline{O^*P}\cos\phi - 0.30 \quad (m)$$

Differentiation of Eqs. (a) yields

$$\dot{x}_P = (0.15)(-\sin\theta)\dot{\theta} = \overline{O^*P}(\cos\phi)\dot{\phi} + (\dot{\overline{O^*P}})\sin\phi \qquad (b)$$
$$\dot{y}_P = (0.15)(\cos\theta)\dot{\theta} = \overline{O^*P}(-\sin\phi)\dot{\phi} + (\dot{\overline{O^*P}})\cos\phi$$

Squaring each of Eqs. (b) and adding, we find

$$(0.15)^2(\dot{\theta})^2 = (\overline{O^*P})^2\dot{\phi}^2 + (\dot{\overline{O^*P}})^2 \qquad (c)$$

By the velocity diagram (Fig. b) and the law of cosines,

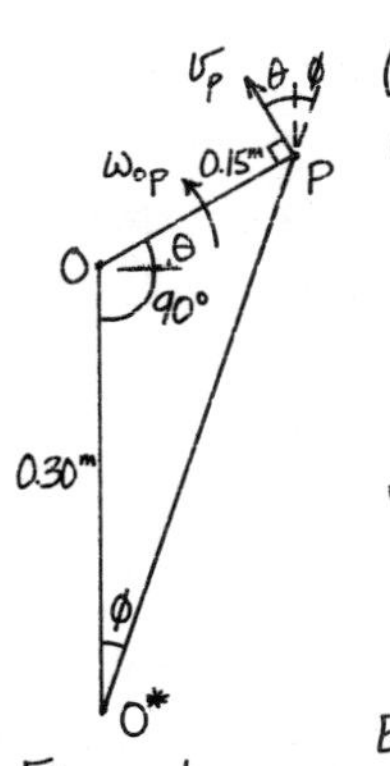

Figure b

$$(\overline{O^*P})^2 = (\overline{OP})^2 + (\overline{OO^*})^2 - 2(\overline{OP})(\overline{OO^*})\cos(\theta + 90°)$$

or for $\theta = 30°$,

$$(\overline{O^*P})^2 = 0.1575 \quad (m^2) \qquad (d)$$

Also, by Fig. b,

$$(\dot{\overline{O^*P}}) = v_P\cos(\theta + \phi) \qquad (e)$$

where by the law of sines,

$$\frac{\sin\phi}{(\overline{OP})} = \frac{\sin(120°)}{(\overline{O^*P})} \qquad (f)$$

Equations (d) and (f) yield

$$\sin\phi = 0.3273; \text{ or } \phi = 19.106° \qquad (g)$$

Then, by Eqs. (e) and (g), with $v_P = (0.15)(\omega_{OP}) = 3.770$ m/s,

$$(\dot{\overline{O^*P}})^2 = 6.090$$

Hence, by Eqs. (c) and (d) with $\dot{\theta} = \omega_{OP} = 25.133$ rad/s,

$$(\dot{\phi})^2 = 51.5716; \quad \dot{\phi} = 7.18 \text{ rad/s}$$

Therefore,

$$\underline{\omega_{O^*P} = \dot{\phi} = 7.18 \text{ rad/s} = 68.57 \text{ rpm}}$$

Alternatively, by Fig. b, the velocity of P perpendicular to $\overline{O^*P}$ is

$$v_{\perp P} = v_P\sin(\theta + \phi) = 3.770\sin(49.106°)$$

or $v_{\perp P} = 2.850$ m/s

Hence,

$$\omega_{O^*D} = \frac{v_{\perp P}}{(\overline{O^*P})} = \frac{2.850}{\sqrt{0.1575}} = 7.18 \text{ rad/s}$$

or $\underline{\omega_{O^*D} = 68.57 \text{ rpm}}$

17.29

GIVEN: Figure a, with $a = 150$ mm, $b = 250$ mm, $OQ = 400$ mm, and $\omega = 80$ rpm.

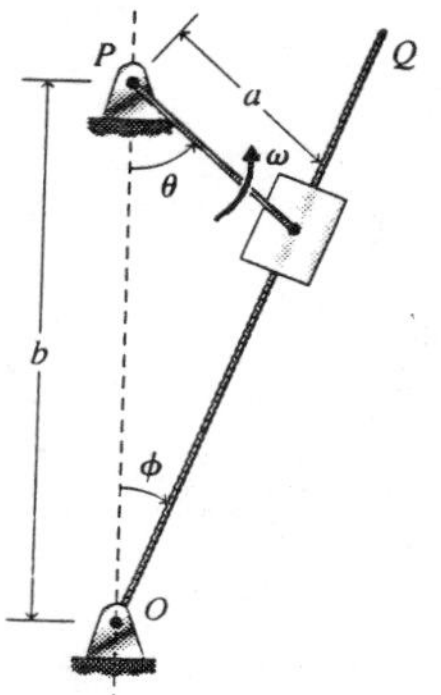

Figure a

FIND:

(a) For $\theta = 30°$, calculate the angular velocity of arm OQ by means of Theorem 17.2.

(b) Determine the velocity of the free end of OQ.

(c) Check your answers by means of Example 13.16.

SOLUTION:

(a) Note that $\omega = 80$ rpm $= 8\pi/3$ rad/s.

Consider Fig. b, in which dimensions and angles of Fig. a are shown.

Figure b:

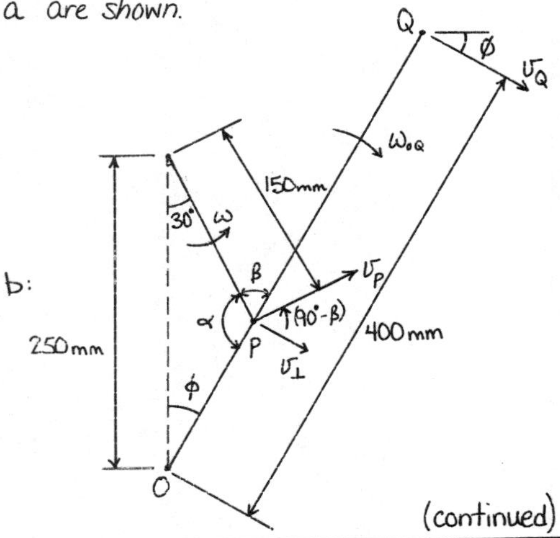

(continued)

17.29 (cont.)

By Fig. b and the law of cosines,

$$\overline{OP}^2 = 0.15^2 + 0.25^2 - 2(0.15)(0.25)\cos 30^\circ$$

or $\overline{OP} = 0.1416\ m$

Also by Fig. b and the law of sines,

$$\frac{\overline{OP}}{\sin 30^\circ} = \frac{0.15}{\sin \phi}\ ; \quad \sin\phi = \frac{0.15 \sin 30^\circ}{0.1416}$$

or $\phi = 31.98^\circ$

Hence,

$$\alpha = 180^\circ - \phi - 30^\circ = 118.02^\circ$$
$$\beta = 180^\circ - \alpha = 61.98^\circ$$
$$90^\circ - \beta = 28.02^\circ$$

Therefore, the component of v_P perpendicular to the arm OQ is (Fig b)

$$v_\perp = v_P \sin 28.02^\circ = (0.15)(8\pi/3)(0.4698)$$

or $v_\perp = 0.5904\ m/s$ ($\phi = 31.98^\circ$)

So, $$\omega_{OQ} = \frac{v_\perp}{\overline{OP}} = \frac{0.5904}{0.1416} = 4.17\ rad/s \circlearrowleft$$

(b) The velocity of the free end of OQ is (Fig. b)

$$v_Q = (0.4)(\omega_{OQ}) = (0.4)(4.17) = 1.67\ m/s$$

(c) By example 13.16,

$$\dot\phi = \omega_{OQ} = -\frac{a\omega(a - b\cos\theta)}{a^2 + b^2 - 2ab\cos\theta} \qquad (a)$$

With $a = 150\ mm$, $b = 250\ mm$, $\omega = \frac{8\pi}{3}\ rad/s$, and $\theta = 30^\circ$, Eq. (a) yields

$$\omega_{OQ} = -\frac{(0.15)(8\pi/3)(0.15 - 0.25\cos 30^\circ)}{(0.15)^2 + (0.25)^2 - 2(0.15)(0.25)(0.866)}$$

or $\omega_{OQ} = 4.17\ rad/s \circlearrowleft$

Then, $v_Q = (0.4)(4.17) = 1.67\ m/s$

These results agree with part b.

17.30

GIVEN: Figure a represents the mechanism of a hydraulic truck lift. The piston moves to the left at 300 mm/s.

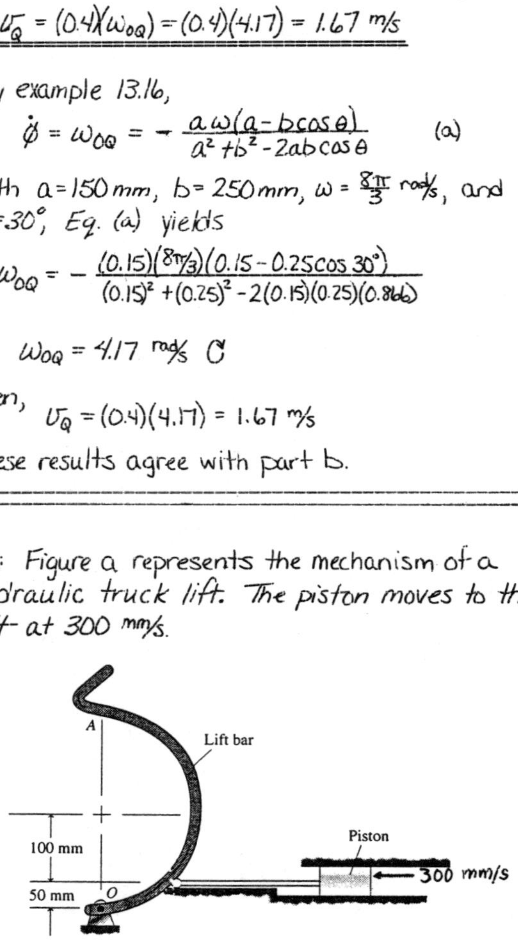

Figure a

FIND: Determine the angular velocity ω of the circular lift bar for the position shown.

SOLUTION:

Let C be the contact point between the lift bar and the piston rod (Fig. b), and v_P be the speed of the piston rod.

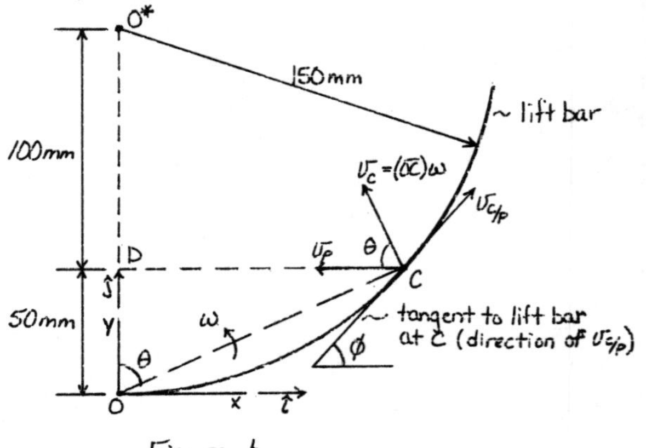

Figure b

By Fig. b,

$$\overline{CD}^2 = 0.15^2 - 0.10^2 = 0.0125$$

or $\overline{CD} = 0.1118\ m$

so $$\tan\theta = \frac{0.1118}{0.05} = 2.236\ ; \quad \theta = 65.904^\circ \qquad (a)$$

By Theorem 17.2 and Fig. b,

$$\bar v_C = \bar v_P + \bar v_{C/P} \qquad (b)$$

where

$$\left.\begin{aligned} \bar v_C &= (\overline{OC})(\omega)[-(\cos\theta)\hat i + (\sin\theta)\hat j] \\ \overline{OC} &= \sqrt{\overline{OD}^2 + \overline{CD}^2} = 0.1225\ m \\ \cos\theta &= 0.4083 \\ \sin\theta &= 0.9129 \end{aligned}\right\} \qquad (c)$$

$$\bar v_P = -0.30\hat i \qquad (d)$$

$$\bar v_{C/P} = v_{C/P}[(\cos\phi)\hat i + (\sin\phi)\hat j] \qquad (e)$$

To determine the angle ϕ, note the equation of a circle of radius 0.15 m with its center at O* is

$$x^2 + (y - 0.15)^2 = 0.15^2 \qquad (f)$$

Differentiation of Eq (f) yields

$$2x\,dx + 2(y - 0.15)\,dy = 0$$

or $$\frac{dy}{dx} = \tan\phi = -\frac{x}{y - 0.15} \qquad (g)$$

So, for $x = 0.1118$, $y = 0.05$ at point C on the circle, Eq. (g) yields

$$\tan\phi = \frac{0.1118}{0.10}\ ; \quad \phi = 48.189^\circ \qquad (h)$$

Hence, by Eqs. (b), (c), (d), (e), and (h),

$$\omega(-0.05\hat i + 0.1118\hat j) = -0.30\hat i + v_{C/P}(0.6667\hat i + 0.7453\hat j)$$

Hence, equating $\hat i$ and $\hat j$ components,

$$0.05\omega + 0.6667 v_{C/P} = 0.30\ ; \quad 0.1118\omega = 0.7453 v_{C/P} \qquad (i)$$

Solving Eq. (i) for ω, we obtain $\omega = 2\ rad/s \circlearrowleft$

17.31

GIVEN: A rigid rod AB 25 in. long moves along a parabolic track $y = 4x^2/45$ [in.] (Fig. a). The end A at the origin moves at a speed of 41 in/s to the right.

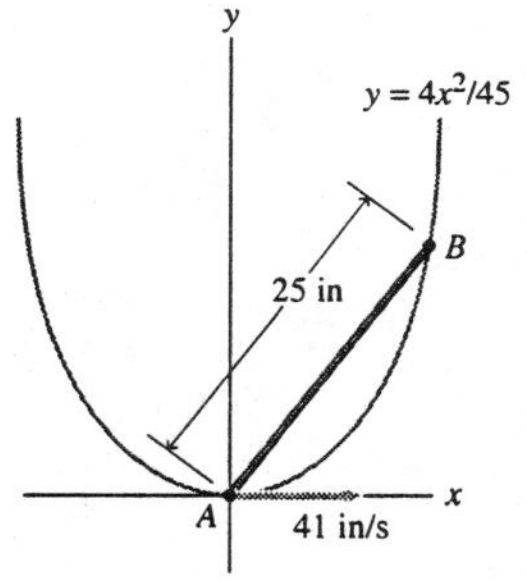

Figure a

FIND: The angular velocity (rad/s) of the rod.

SOLUTION:

Let (x_1, y_1) be the coordinates of point B, when A is at the origin. Therefore,

$$x_1^2 + y_1^2 = 25^2 \qquad \text{(a)}$$

Also the path of point B is

$$y = \frac{4x^2}{45} \qquad \text{(b)}$$

Hence, by Eqs. (a) and (b),

$$x_1^2 + \frac{16x_1^4}{2025} = 625 \qquad \text{(c)}$$

The solution of Eq. (c) is

$$x_1 = 15 \text{ in.} \qquad \text{(d)}$$

So, the slope of the path at point B is

$$\tan\theta = \frac{dy}{dx} = \frac{8x_1}{45} = 2.66\bar{6}, \text{ or } \theta = 69.444° \qquad \text{(e)}$$

Thus, consider the kinematics of the rod (Fig. b).

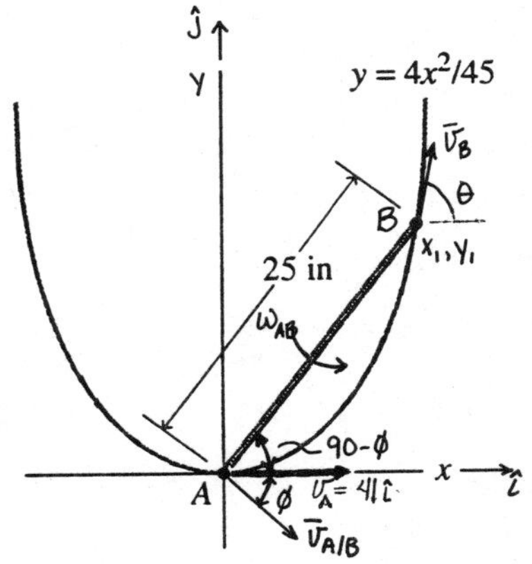

Figure b

By Fig. b,

$$\cos(90 - \phi) = \sin\phi = \frac{x_1}{25} = \frac{15}{25} = 0.60$$

or $\phi = 36.87°$ (f)

Then, by Theorem 17.2 and Eq. (17.6),

$$\bar{v}_A = \bar{v}_B + \bar{v}_{A/B}$$

$$41\hat{\imath} = v_B(\cos\theta)\hat{\imath} + v_B(\sin\theta)\hat{\jmath} + v_{A/B}(\cos\phi)\hat{\imath} - v_{A/B}(\sin\phi)\hat{\jmath} \qquad \text{(g)}$$

Since $v_{A/B} = 25\,\omega_{AB}$, Eqs. (e), (f), and (g) yield

$$41 = 0.3511\,v_B + (25\,\omega_{AB})(0.8)$$
$$0 = 0.9363\,v_B - (25\,\omega_{AB})(0.6) \qquad \text{(h)}$$

The solution of Eq. (h) is

$v_B = 25.63$ in/s ; $\omega_{AB} = 1.60$ rad/s ↺

17.32

GIVEN: The crank OP rotates with angular velocity of 10 rad/s for the position shown in Fig. a. The circular guide AB has a radius of 125 mm with its center at C.

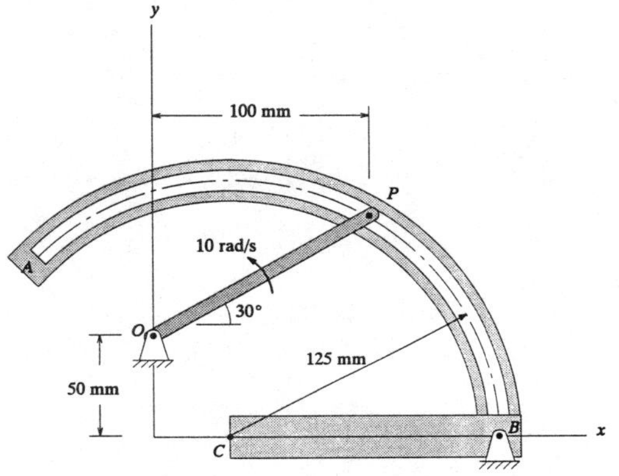

Figure a

FIND: Determine the angular velocity of the guide AB.

SOLUTION:

The velocity of point Q on guide AB that is coincident with point P is $\bar{v}_Q$ (Fig. b).

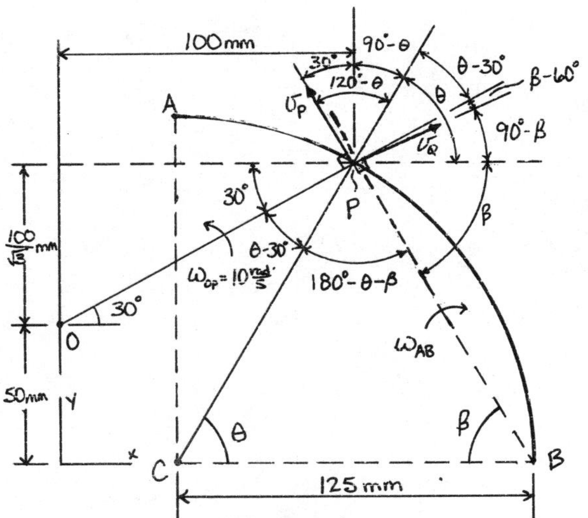

Figure b

(continued)

17.32 (cont.)

Hence, by the theory of Sec. 17.2,

$$v_P \cos(120° - \theta) = v_Q \cos[(\theta+\beta) - 90°] = v_Q \sin(\theta+\beta)$$

or

$$v_Q = v_P \frac{\cos(120°-\theta)}{\sin(\theta+\beta)} \quad (a)$$

By Fig. a, $CP = 125$ mm and $\overline{OP}\cos 30° = 100$ mm.

Hence, $\overline{OP} = 115.473$ mm and

$$v_P = (\overline{OP})(\omega_{op}) = 115.473(10) = 1154.73 \text{ mm/s} \quad (b)$$

Also, by Fig. b,

$$125 \sin\theta = 50 + \frac{100}{\sqrt{3}} = 107.735$$

$$\theta = 59.528° \quad (c)$$

Therefore,

$$120° - \theta = 60.472° \quad (d)$$

Again, by Fig. b and the law of sines,

$$\frac{\sin\beta}{125} = \frac{\sin\theta}{PB}$$

or

$$PB \sin\beta = 125 \sin\theta \quad (e)$$

and

$$PB \cos\beta = 125 - 125\cos\theta \quad (f)$$

Dividing Eq. (e) by Eq (f), we obtain with Eq. (c)

$$\tan\beta = \frac{\sin\theta}{1-\cos\theta} = 1.7486$$

Therefore,

$$\beta = 60.236° \quad (g)$$

and

$$(\theta+\beta) = 119.764°$$

So,

$$\sin(\theta+\beta) = 0.86808 \quad (h)$$
$$\cos(120°-\theta) = 0.49285$$

Hence, by Eqs. (a), (b), and (h), we find

$$v_Q = \frac{(1154.73)(0.49285)}{0.86808} = 655.59 \text{ mm/s} \quad (i)$$

Also,

$$\omega_{AB} = \frac{v_Q}{PB} \quad (j)$$

where by Eqs. (c), (e), and (g)

$$PB = \frac{125 \sin\theta}{\sin\beta} = \frac{125(0.86188)}{0.86808} = 124.11 \text{ mm} \quad (k)$$

Finally, by Eqs. (i), (j), and (k),

$$\omega_{AB} = 5.282 \text{ rad/s} \circlearrowright$$

17.33

GIVEN: The milling-machine mechanism shown in Figure a. Crank BC drives slider C along rod AB and rotates at a constant angular speed of 45 rpm, counterclockwise. Tool E oscillates on the machine bed.

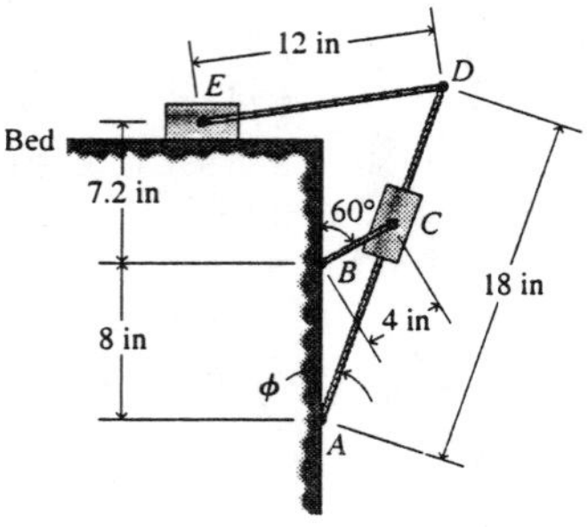

Figure a

FIND: For the position shown in Fig. a, determine the velocity of tool E by graphical construction.

SOLUTION:

Before graphically constructing the solution, we note that the velocity of point C is perpendicular to BC and has magnitude

$$v_C = BC(\omega) = 4\left(45 \times \frac{2\pi}{60}\right) = 6\pi \text{ in/s}.$$

or $v_C = 18.85$ in/s

The mechanism constructed to scale is shown in Fig. b. By measurement of Fig. b, $\phi \approx 19.1°$, $\theta \approx 8.5°$. By construction, since $\bar{v}_D$ is perpendicular to AD, $v_D \approx 24.2$ in/s.

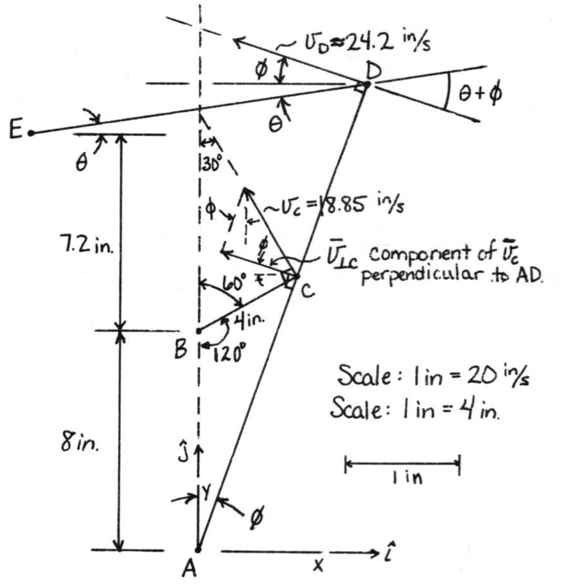

Figure b.

By Figs. b and c and the equation

$$\bar{v}_E = \bar{v}_D + \bar{v}_{E/D}$$

we may construct the vector diagram of Fig. d.

(continued)

17.33 (cont.)

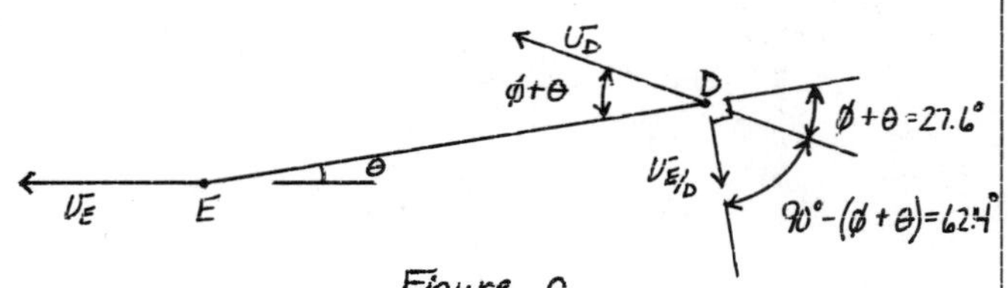

Figure c

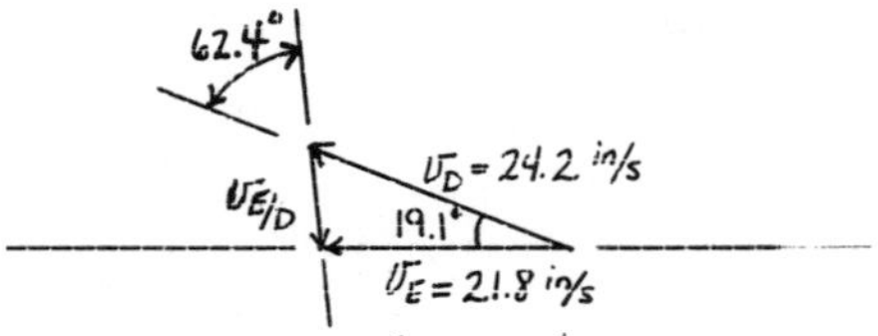

Figure d

Hence, by measurement of Fig. d, we find

$v_E \approx 22$ in/s

17.34

GIVEN: The milling-machine mechanism of Fig. a of the solution of Problem 17.33.

FIND: Determine the angular velocity of Bar AD.

SOLUTION:

One must solve analytically or graphically for the velocity $v_{\perp c}$ (the velocity component of v_c perpendicular to AD) or the velocity v_D. Then,

$$\omega_{AD} = \frac{v_{\perp c}}{AC} = \frac{v_D}{18} \qquad (a)$$

By the graphical solution of Problem 17.33, $v_D \approx 24.2$ in/s. Hence, by Eq. (a),

$$\omega_{AD} \approx \frac{24.2}{18} = 1.34 \text{ rad/s}$$

Alternatively, by Fig. b of the solution of Problem 17.33, the velocity of point C is

$\bar{v}_c = v_c[(-\sin 30°)\hat{i} + (\cos 30°)\hat{j}]$ and since

$$v_c = (BC)\omega = (4)\left(45 \times \frac{2\pi}{60}\right) = 18.85 \text{ in/s}$$

$$\bar{v}_c = -9.425\,\hat{i} + 16.324\,\hat{j} \text{ (in/s)} \qquad (b)$$

Also, by Figs. a and b of the solution of Problem 17.33, by the law of cosines,

$$(AC)^2 = (BC)^2 + (AB)^2 - 2(BC)(AB)\cos 120°$$

or $(AC)^2 = (4)^2 + (8)^2 - 2(4)(8)(-0.50)$

or $AC = 10.583$ in. $\qquad$ (c)

Then, by the law of sines and Figs. a or b,

$$\frac{AC}{\sin 120°} = \frac{BC}{\sin\phi}$$

or $\sin\phi = \frac{BC}{AC}\sin 120° = \frac{4}{10.583}(0.866) = 0.3273$

$\phi = 19.11°$

So, by Eq. (b) and Fig. (b) of Problem 17.33 solution,

$$v_{\perp c} = (9.425)\cos 19.11° + (16.324)\sin 19.11°$$

or $v_{\perp c} = 14.250$ in/s $\qquad$ (d)

Hence, by Eqs (a), (c), and (d),

$$\omega_{AD} = \frac{v_{\perp c}}{AC} = \frac{14.250}{10.583} = 1.346 \text{ rad/s}$$

17.35

GIVEN: The equations (see Example 17.3)

$$v_x = 112R + 112r\cos\theta$$
$$v_y = -112r\sin\theta \qquad (a)$$

FIND:

(a) By eqs. (a), plot $\frac{v_x}{r}$ and $\frac{v_y}{r}$ as functions of θ for $0 \le \theta \le 360°$ and ratios $\frac{R}{r} = 1, \sqrt{2}, \infty$.

(b) From the plots of part a, verify the results given in Eq. (e) of Example 17.3.

SOLUTION:

(a): By Eqs (a),

$$\frac{v_x}{r} = 112\left(\frac{R}{r}\right) + 112\cos\theta$$
$$\frac{v_y}{r} = -112\sin\theta \qquad (b)$$

The plots of Eqs. (b) for $R/r = 1, \sqrt{2}$, and ∞ are shown in Fig. a. Note that for $R/r = \infty$, $v_x/r = \infty$. Also, v_y/r is independent of R/r.

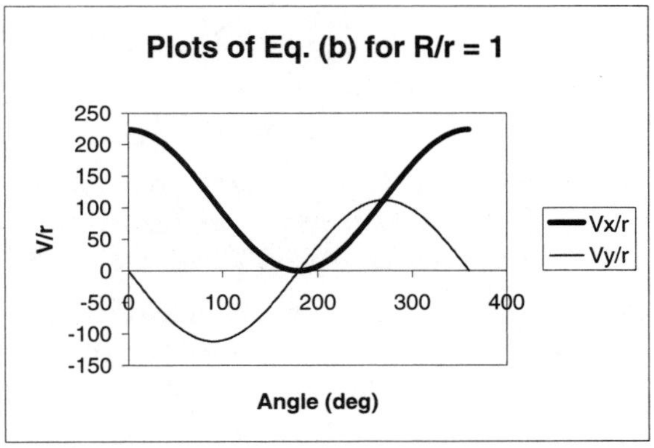

Figure a.1

(continued)

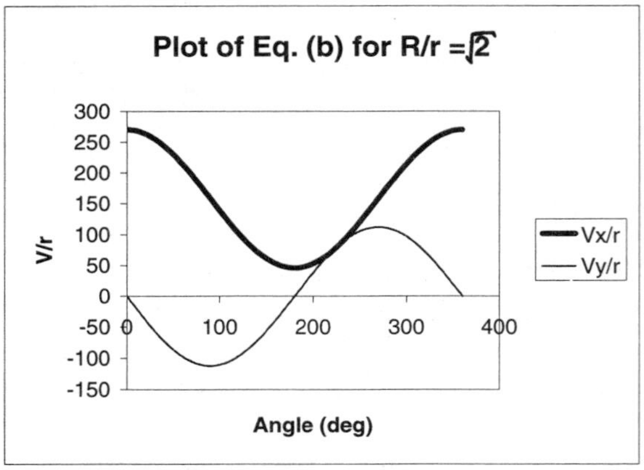

Figure a.2

17.36

GIVEN: The crank OP of a quick-return mechanism rotates counterclockwise in the xy plane (Fig. a). A cylindrical tube is pinned to the crank at P and the rod O^*D is free to slide through the tube.

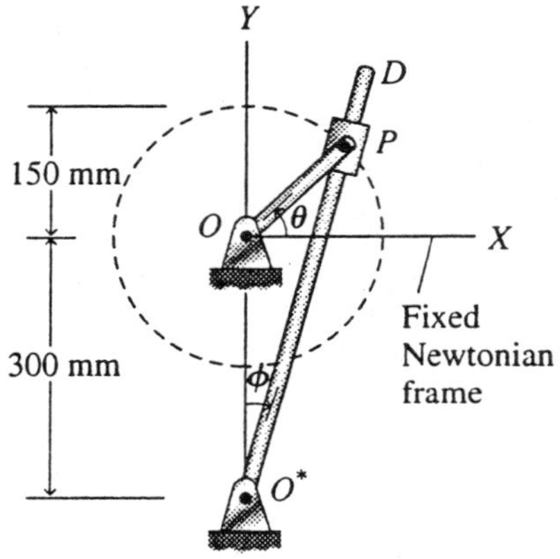

Figure a

FIND:

(a) Plot the magnitude of the displacement of P as a function of θ, measured relative to an observer fixed on rod O^*D for $0° \le \theta \le 360°$.

(b) Plot the magnitude of the absolute displacement of P as a function of θ for $0 \le \theta \le 360°$.

(c) Check the plots of parts (a) and (b) for the special cases $\theta = 90°, 180°, 270°$, and $360°$.

SOLUTION:

(a) Initially for $\theta = 0°$, by Fig. (a),

$$O^*P = O^*P_1 = \sqrt{150^2 + 300^2} = 150\sqrt{5}\ \text{mm} \qquad \text{(a)}$$

At any angle θ, by Fig. a,

$$O^*P = O^*P_2 = \sqrt{(150\cos\theta)^2 + (300 + 150\sin\theta)^2}$$

$$\text{or}\quad O^*P_2 = 150\sqrt{5}\sqrt{1 + 0.8\sin\theta} \qquad \text{(b)}$$

Hence, the magnitude of the displacement relative to an observer fixed on rod O^*D is, with Eqs. (a) and (b),

$$d = |O^*P_2 - O^*P_1|$$

or

$$d = 150\sqrt{5}\left|\sqrt{1 + 0.8\sin\theta} - 1\right| \qquad \text{(c)}$$

The plot of Eq. (c) is shown in Fig. b.

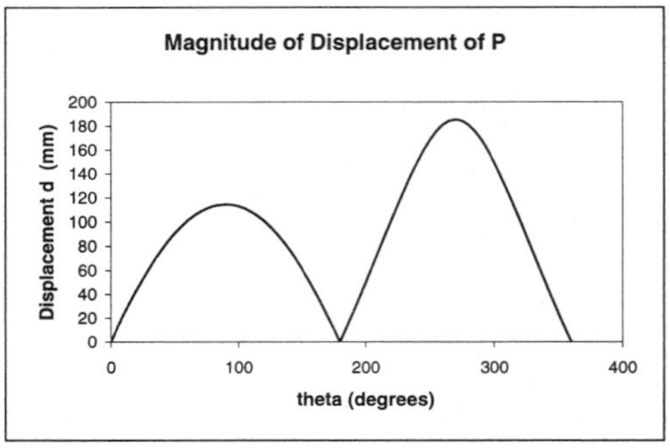

Figure b

(b) The absolute displacement of P when crank OP has rotated through angle θ is (see Fig. c)

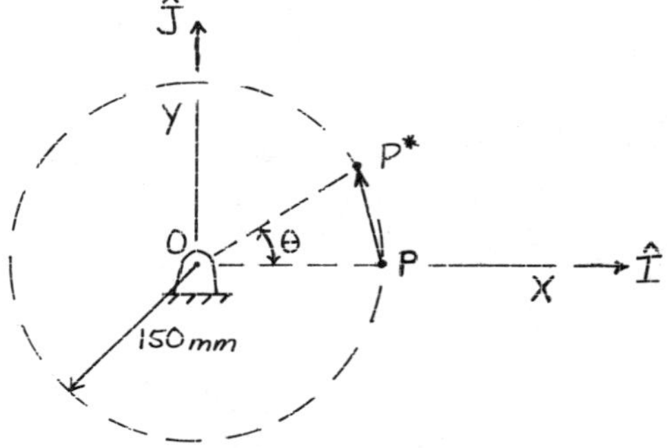

Figure c

$$\overline{P^*P} = (150\cos\theta - 150)\,\hat{I} + (150\sin\theta)\,\hat{J}$$

Hence, the magnitude of $\overline{P^*P}$ is

$$d = 150\sqrt{2 - 2\cos\theta} \qquad \text{(d)}$$

The plot of Eq. (d) is shown in Fig. d.

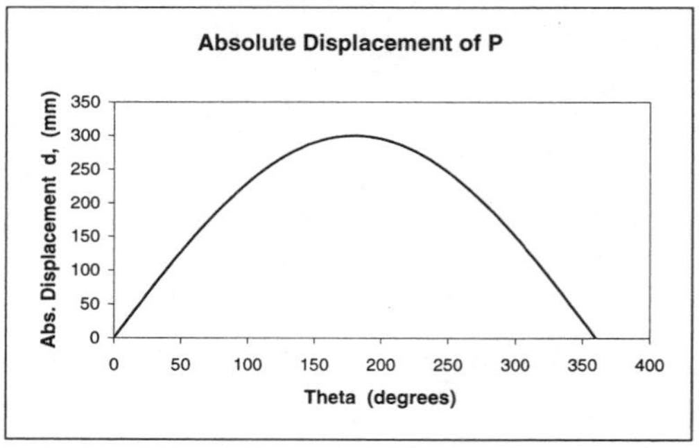

Figure d

(continued)

(c) For $\theta = 90°, 180°, 270°$, and $360°$, Fig. b yields

$d = 114.6\text{ mm}, 0, 185.4\text{ mm}, 0$

These results agree with Fig. a.

For $\theta = 90°, 180°, 270°$, and $360°$, Fig. d yields

$d = 212.1\text{ mm}, 300\text{ mm}, 212.1\text{ mm}, 0$

These results agree with Fig. c.

17.37

GIVEN: A rigid machine link moves along guides C and D (Fig. a), with the speed of B being 40 ft/s to the left for the position shown.

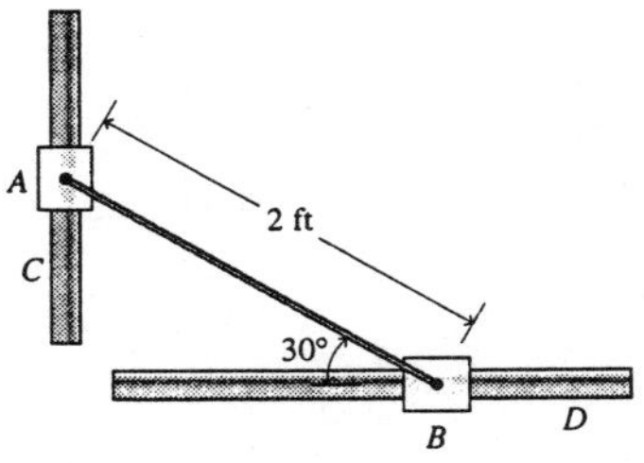

Figure a

FIND:

(a) Determine the instantaneous center of velocity of rod AB.

(b) Determine the angular velocity of AB and the velocity of A.

SOLUTION:

(a) Consider the sketch shown in Fig. b.

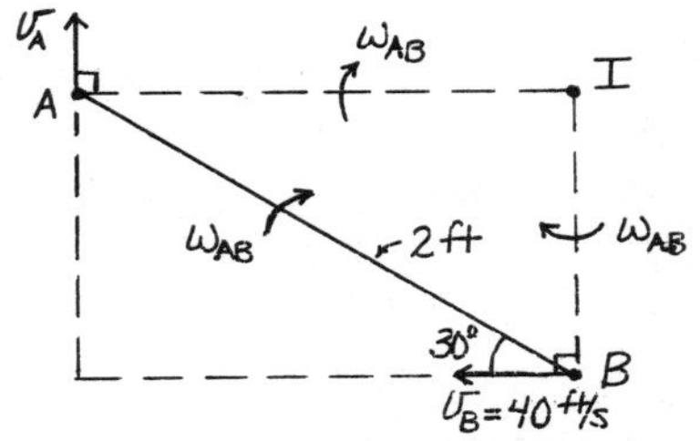

Figure b

Perpendicular lines to v_A and v_B at A and B, respectively, intersect at the instantaneous center I, where by Fig. b,

$$BI = 2\sin 30° = 1\text{ ft}$$
$$AI = 2\cos 30° = 1.732\text{ ft} \qquad \text{(a)}$$

(b) By Fig. b,

$$v_B = \omega_{AB}(BI)$$

or
$$\omega_{AB} = \frac{v_B}{(BI)} = \frac{40\text{ ft/s}}{1\text{ ft}} = 40\text{ rad/s} \qquad \text{(b)}$$

Also, by Fig. b,

$$v_A = \omega_{AB}(AI) \qquad \text{(c)}$$

So by Eqs. (a), (b), and (c),

$$v_A = (40)(1.732) = 69.28\text{ ft/s}$$

17.38

GIVEN: In the mechanism in Fig. a, crank OA rotates with constant speed 180 rad/s. Link AB is pinned at A, and end B is constrained to move along guide CD.

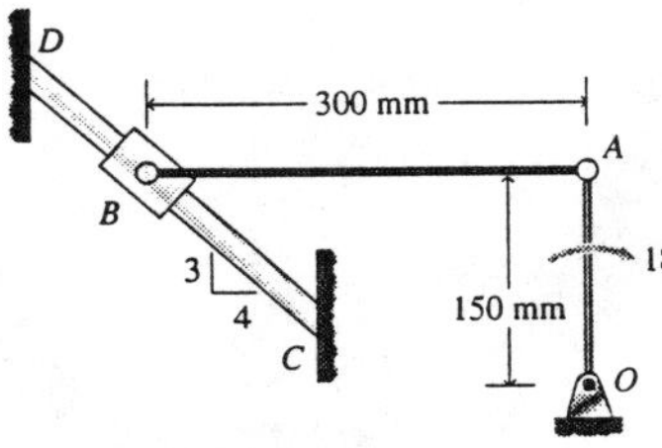

Figure a

FIND:

(a) Determine the instantaneous center of velocity of rod AB for the position shown.

(b) Determine the angular velocity of rod AB and the velocity of B.

SOLUTION:

(a) Consider the sketch shown in Fig. b.

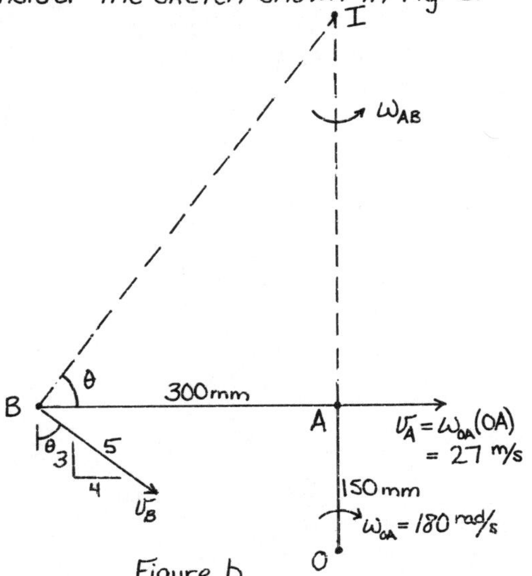

Figure b

Perpendicular lines to v_A and v_B at A and B, respectively, intersect at the instantaneous center I. By the figure, the instantaneous center is located by

$$\tan\theta = \frac{4}{3} = \frac{(AI)}{(AB)}; \qquad \theta = 53.13°$$

and
$$(AI) = \tfrac{4}{3}(AB) = 0.40\text{ m}$$

(b) Hence, by Fig. b,

$$\omega_{AB} = \frac{v_A}{AI} = \frac{27}{0.40} = 67.5\text{ rad/s}$$

Also by Fig. b, $BI = 0.50$ m. Therefore,

$$v_B = \omega_{AB}(BI) = (67.5)(0.50) = 33.75\text{ m/s}$$

17.39

GIVEN: In the mechanism shown in Fig. a, arm OA rotates about O with angular speed ω_O, and wheel B rolls on the circular track with angular speed ω_B.

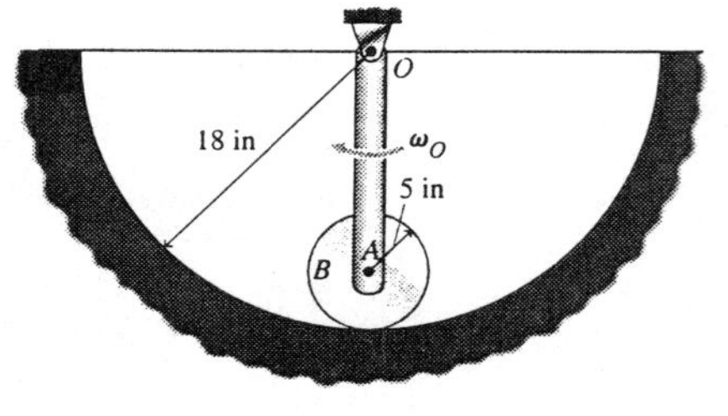

Figure a

FIND: By use of the instantaneous center, determine the ratio ω_B/ω_o and the sense of rotation of the wheel.

SOLUTION:

Consider the sketch shown in Fig. b. The instantaneous center I_w of the velocity of the wheel is the contact point between the wheel and track, and for rod OA it is at O.

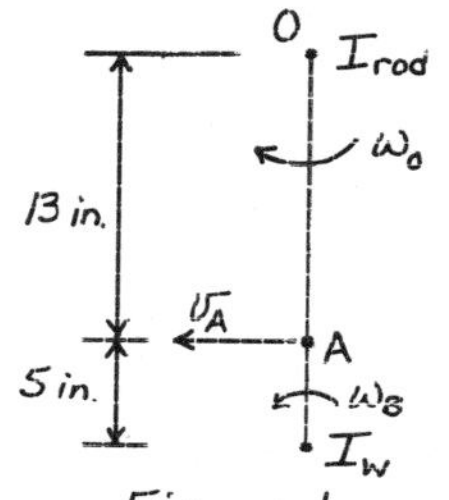

Figure b

So, by Fig. b,

$$v_A = \omega_o(13) = \omega_B(5)$$

or $$\frac{\omega_B}{\omega_o} = \frac{13}{5} = 2.6$$

Since v_A is directed to the left in Fig. b, the angular velocity of the wheel is counterclockwise.

17.40

GIVEN: Crank AB of the four-bar linkage in Figure a rotates at 50 rad/s.

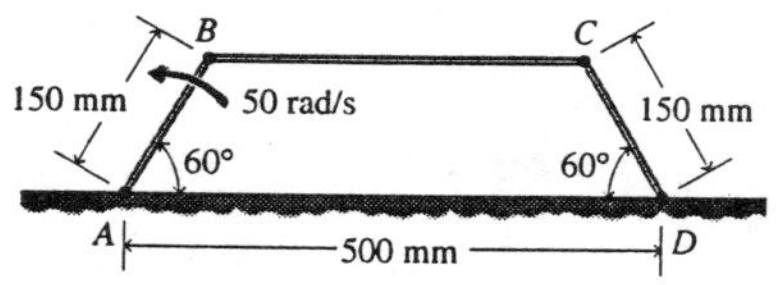

Figure a

FIND:

(a) Locate the instantaneous center of velocity of link BC.

(b) Determine the angular velocity of link BC.

SOLUTION:

Consider the sketch shown in Fig. b.

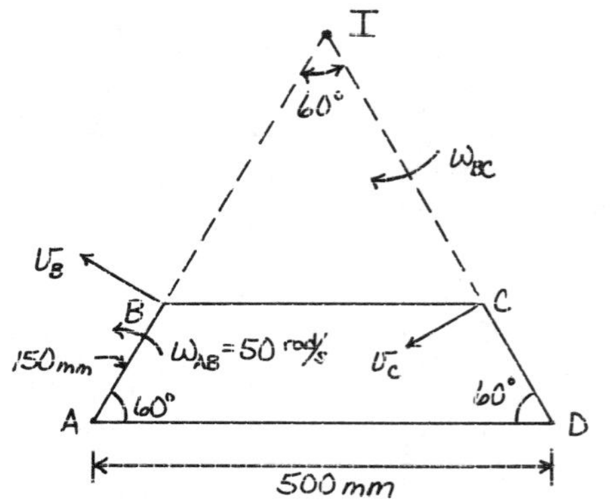

Figure b

(a) By Fig. b, AID is an equilateral triangle. Hence, AI = DI = 500 mm. Therefore, I is located so that

$$BI = 350 \text{ mm} = CI \qquad \text{(a)}$$

(b) So, by Fig. b and Eq. (a),

$$v_B = (AB)\omega_{AB} = 7.5 \text{ m/s} = (BI)\omega_{BC}$$

or $$\omega_{BC} = 21.43 \text{ rad/s clockwise}$$

17.41

GIVEN: In the mechanism shown in Fig. a, the slider B travels along the path M-M. The crank OA rotates at a constant rate of 300 rad/s clockwise.

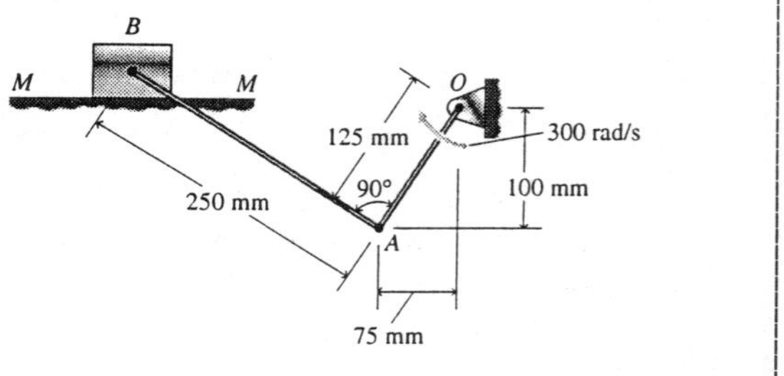

Figure a

FIND:

(a) Determine the instantaneous center of velocity of rod AB for the position shown in Fig. a.

(b) Determine the angular velocity of the connecting rod AB and the velocity of slider B for the position shown in Fig. a. (continued)

17.41 (cont.)

SOLUTION:

(a) Consider the velocity diagram of rod AB (Fig. b).

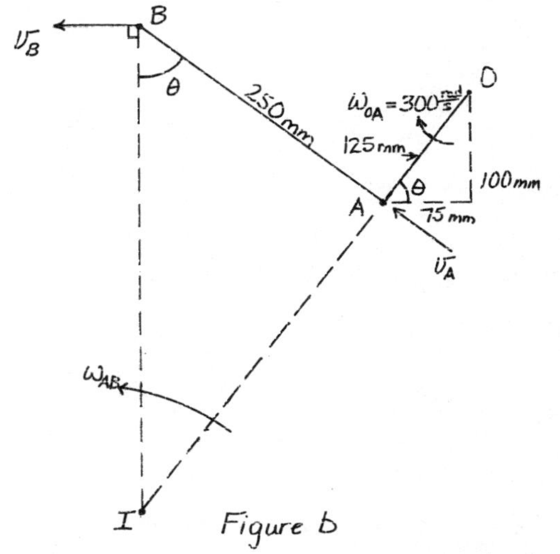

Figure b

By Fig. b, the instantaneous center of velocity of rod AB is located at I, the intersection point of lines $\overline{AI}$ and $\overline{BI}$.

(b) By Fig. b,

$$v_A = (\overline{OA})(\omega_{OA}) = (0.125)(300) = 37.5 \text{ m/s} \qquad (a)$$

Also by Fig. b,

$$v_A = (\overline{AI})\,\omega_{AB} \qquad (b)$$

By Eqs. (a) and (b),

$$\omega_{AB} = \frac{37.5}{\overline{AI}} \qquad (c)$$

By geometry (Fig. b),

$$\cos\theta = \frac{75}{125} = 0.60\ ; \quad \sin\theta = \frac{100}{125} = 0.80 \qquad (d)$$

So, by Eqs. (d) and Fig. b,

$$(\overline{BI})\cos\theta = 0.250\ ; \quad \overline{BI} = \frac{0.250}{0.60} = 0.416\bar{6} \text{ m} \qquad (e)$$

and

$$\overline{AI} = (\overline{BI})\sin\theta = (0.416\bar{6})(0.80) = 0.333\bar{3} \text{ m} \qquad (f)$$

Eqs. (c) and (f) yield

$$\omega_{AB} = \frac{37.5}{0.333\bar{3}} = 112.5 \frac{rad}{s} \qquad (g)$$

Hence, by Eqs. (e) and (g)

$$v_B = (\overline{BI})\,\omega_{AB} = (0.416\bar{6})(112.5) = 46.87 \text{ m/s}$$

17.42

GIVEN: A disc drive is driven by a friction driver (Fig. a). Initially, the angular velocity of the disc is ω_2 [rad/s] clockwise and that of the driver is ω_1 [rad/s] counterclockwise.

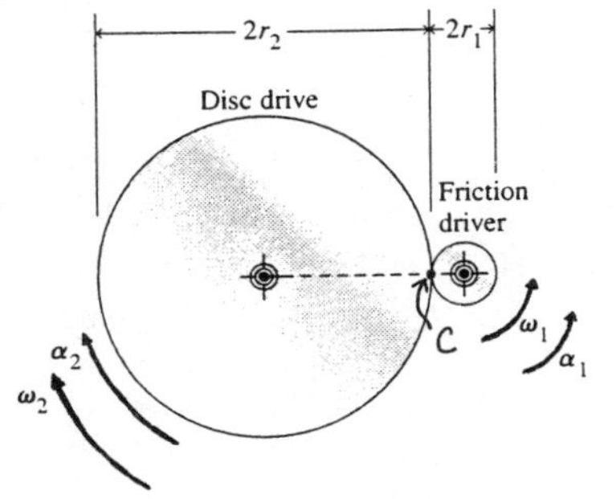

Figure a

FIND: Derive a formula for the constant angular acceleration α_1 [rad/s²] of the driver that is required to increase the disc's angular speed from ω_2 to ω_3 in time t [s]. Express the result in terms of $r_1, r_2, \omega_2, \omega_3$, and t.

SOLUTION:

By Fig. a, where point C is the contact point between the disc and driver, the velocity v_C and the acceleration a_C of point C are

$$v_C = r_1\omega_1 = r_2\omega_2$$

$$a_C = r_1\alpha_1 = r_2\alpha_2$$

or

$$\omega_2 = \frac{r_1\omega_1}{r_2} \qquad (a)$$

$$\alpha_2 = \frac{r_1\alpha_1}{r_2}$$

By kinematics and the second of Eqs. (a),

$$\frac{d\omega_{disc}}{dt} = \alpha_2 = \frac{r_1}{r_2}\alpha_1$$

Hence,

$$\int_{\omega_2}^{\omega_3} d\omega_{disc} = \frac{r_1\alpha_1}{r_2}\int_0^t dt$$

or

$$\omega_3 - \omega_2 = \frac{r_1\alpha_1}{r_2}\,t \qquad (b)$$

By the first of Eqs. (a) and Eq. (b), we obtain

$$\omega_3 = \frac{r_1}{r_2}(\omega_1 + \alpha_1 t)$$

So

$$\alpha_1 = \left(\frac{r_2}{r_1}\omega_3 - \omega_1\right)\frac{1}{t}$$

or

$$\alpha_1 = \frac{r_2\omega_3 - r_1\omega_1}{r_1 t}$$

17.43

GIVEN: The disc drive - driver mechanism of Problem 17.42 has radii $r_1 = 25$ mm and $r_2 = 125$ mm.

FIND: Determine the angular acceleration α_1 required to increase the rate of rotation from 33 rpm to 78 rpm.

SOLUTION:

To solve this problem, α_1 must be expressed in terms of the initial and final rates of rotation and time t. This expression was derived in the solution of problem 17.42. It is

$$\alpha_1 = \frac{r_2\omega_3 - r_1\omega_1}{r_1 t} \quad (a)$$

where here

$$r_1 = 25mm$$
$$r_2 = 125mm$$
$$\omega_3 = 78 \text{ rpm} = 8.168 \text{ rad/s} \quad (b)$$
$$\omega_1 = 33 \text{ rpm} = 3.456 \text{ rad/s}$$
$$t = 2s$$

Substitution of Eqs. (b) into Eq. (a) yields

$$\alpha_1 = 18.69 \text{ rad/s}^2$$

17.44

GIVEN: The angular velocity of crank OB is 100 rpm counterclockwise (Fig. a). Gear C is fixed.

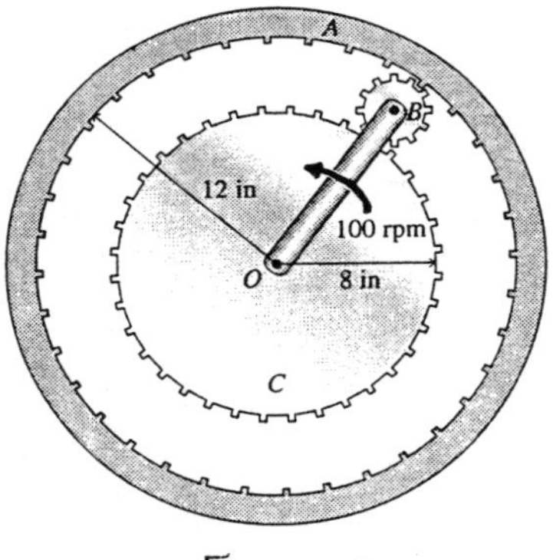

Figure a

FIND:

(a) Determine the angular velocity of the planetary gear B.

(b) Determine the angular velocity of the ring gear A.

SOLUTION:

(a) The instantaneous center of Gear B is I (Fig. b).

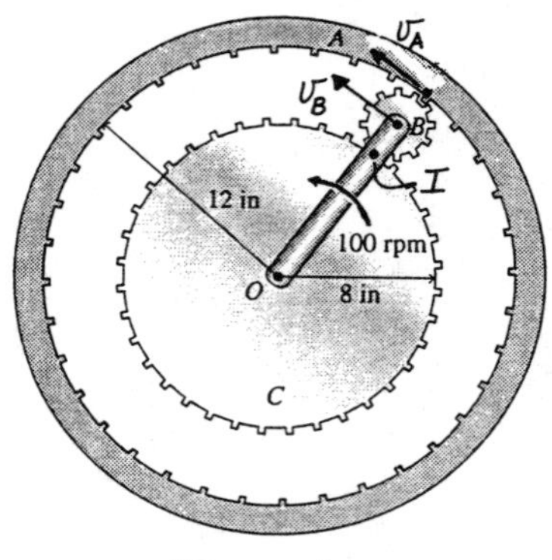

Figure b

In terms of rad/s,

$$\omega_{OB} = 100 \frac{\text{rev}}{\text{min}} \cdot \frac{2\pi}{60} = \frac{10\pi}{3} \frac{\text{rad}}{\text{s}}$$

Therefore,

$$v_B = 10\,\omega_{OB} = \frac{100\pi}{3} \frac{\text{in}}{\text{s}} \quad (a)$$

Also, by the concept of instantaneous center,

$$v_B = (\overline{BI})\,\omega_B = 2\omega_B \quad (b)$$

Equations (a) and (b) yield

$$\omega_B = \frac{50\pi}{3} = 52.36 \text{ rad/s counterclockwise}$$

(b) Hence, by the concept of instantaneous center,

$$v_A = 4\,\omega_B = \frac{200\pi}{3} \text{ in/s} \quad (c)$$

Also,

$$v_A = 12\,\omega_A \quad (d)$$

Equations (c) and (d) yield

$$\omega_A = \frac{50\pi}{9} = 17.45 \text{ rad/s counterclockwise}$$

17.45

GIVEN: In Problem 17.44, the ring gear A is held fixed and the inner gear C can rotate about shaft O. By the data of Problem 17.44, the angular speed of rod OB is

$$\omega_{OB} = 100 \text{ rpm} = \frac{10\pi}{3} \text{ rad/s}.$$

FIND: Determine the angular velocity of gear C.

SOLUTION:

Consider Fig. a. The instantaneous center of gear B is point I.

(continued)

17.45 (cont.)

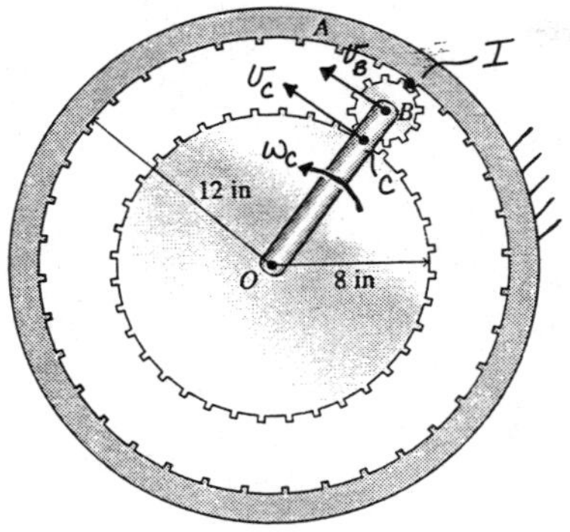

Figure a

The speed of point B is

$$v_B = 10\,\omega_{OB} = \frac{100\pi}{3} \text{ in/s}.$$

Also, by the concept of instantaneous center,

$$v_B = (\overline{IB})\,\omega_B = 2\,\omega_B$$

or

$$\omega_B = \frac{v_B}{2} = \frac{50\pi}{3} \text{ rad/s} \qquad (a)$$

Again by the concept of instantaneous center,

$$v_C = (\overline{IC})\,\omega_B = (4)\left(\frac{50\pi}{3}\right) = \frac{200\pi}{3} \qquad (b)$$

Also, by Fig. a, gear C rotates about O with angular speed ω_C. Hence,

$$v_C = (\overline{OC})\,\omega_C = 8\,\omega_C \qquad (c)$$

Equating Eqs. (b) and (c), we have

$$8\omega_C = \frac{200\pi}{3}$$

$$\omega_C = \frac{25\pi}{3} = 26.18 \text{ rad/s counterclockwise}$$

or

$$\omega_C = \left(\frac{25\pi}{3}\right)\left(\frac{30}{\pi}\right) = 250 \text{ rpm counterclockwise}$$

17.46

GIVEN: The crank OA in Fig. a rotates at a constant rate of 80 rpm.

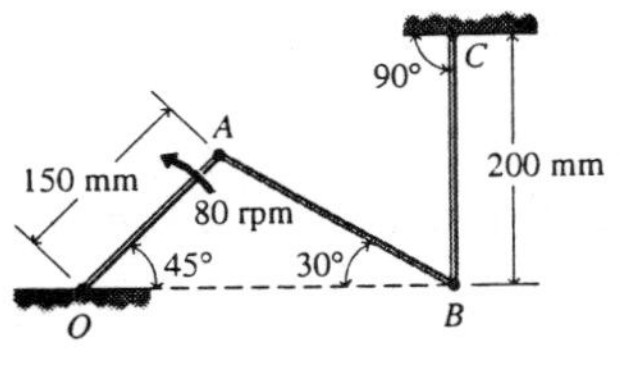

Figure a

FIND:
Using the concept of instantaneous center, determine the angular velocities of rods AB and BC.

SOLUTION:
Consider the velocities of A and B.

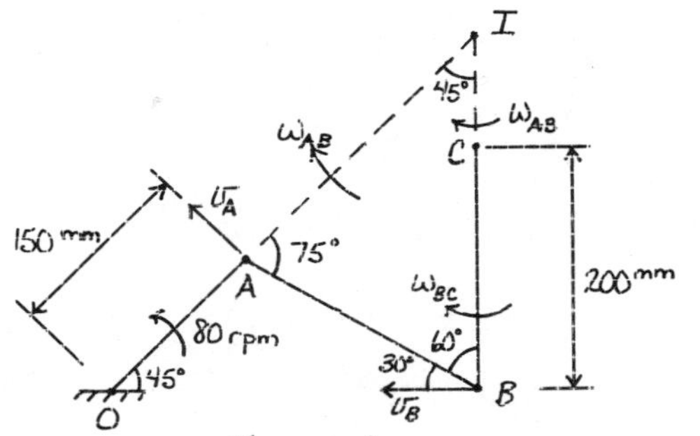

Figure b

By Fig. b, the extensions of line OA and BC intersect at the instantaneous center I of rod AB. Also, by Fig. b,

$$v_A = (150)\,\omega_{OA} = (150)(80)\left(\frac{\pi}{30}\right) = 400\pi \text{ mm/s} \qquad (a)$$

By the concept of instantaneous center,

$$v_A = (\overline{IA})\,\omega_{AB} \quad \text{or} \quad \omega_{AB} = \frac{v_A}{(\overline{IA})} \qquad (b)$$

By geometry and Fig. b,

$$AB \sin 30^\circ = 150 \sin 45^\circ$$

or

$$AB = \frac{150 \sin 45^\circ}{\sin 30^\circ} = 212.13 \text{ mm}$$

Also, by the law of sines

$$\frac{\sin 45^\circ}{AB} = \frac{\sin 60^\circ}{IA}$$

Therefore,

$$IA = AB\,\frac{\sin 60^\circ}{\sin 45^\circ} = 259.8 \text{ mm} \qquad (c)$$

Hence, Eqs. (b) and (c) yield

$$\omega_{AB} = \frac{v_A}{IA} = \frac{400\pi}{259.8} = 4.84 \text{ rad/s}$$

Also, by Fig. b,

$$v_B = \omega_{AB}(\overline{IB}) = \omega_{BC}(\overline{BC})$$

Therefore,

$$\omega_{BC} = \frac{\omega_{AB}(\overline{IB})}{(\overline{BC})} \qquad (d)$$

But,

$$\overline{IB} = (\overline{OI}) \cos 45^\circ = (150 + 259.8) \cos 45^\circ = 289.8 \text{ mm}$$

Hence, by Eq. (d),

$$\omega_{BC} = \frac{(4.84)(289.8)}{200} = 7.01 \text{ rad/s}$$

17.47

GIVEN: For the double pendulum shown in Fig. a, $\theta = \phi$ at all times.

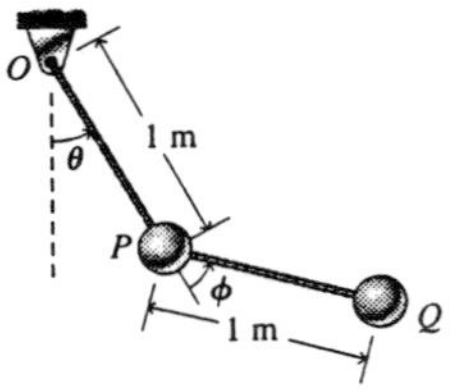

Figure a

FIND: Determine the location of the instantaneous center of link PQ.

SOLUTION:

Consider Fig. b, with $\phi = \theta$.

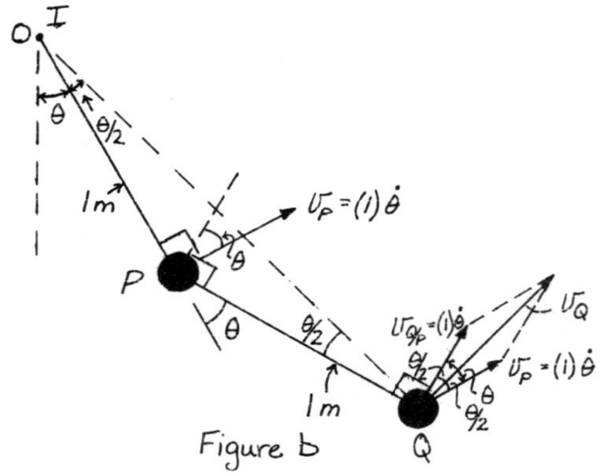

Figure b

By Fig. b, the angle subtended by OP and PQ is $180° - \theta$, and since IPQI is an isosceles triangle, the angle between OP and IQ is $\theta/2$ and between PQ and IQ is also $\theta/2$. By Fig. b,

$$v_P = (OP)\dot{\theta} = 1(\dot{\theta}), \qquad (a)$$

and $$v_{Q/P} = (PQ)\dot{\theta} = 1(\dot{\theta}) \qquad (b)$$

Also,
$$\bar{v}_Q = \bar{v}_P + \bar{v}_{Q/P}$$

where by Fig. b, the angle subtended by $\bar{v}_P$ and $\bar{v}_{Q/P}$ is θ. Since $v_P = v_{Q/P}$, the angle between $\bar{v}_Q$ and $\bar{v}_P$ and between $\bar{v}_Q$ and $\bar{v}_{Q/P}$ is $\theta/2$.

Hence, $\bar{v}_Q$ is perpendicular to line OQ.

Therefore, the instantaneous center I of link PQ coincides with point O.

17.48

GIVEN: The crank OA rotates clockwise at the constant rate $\omega = 100$ rad/s (Fig. a).

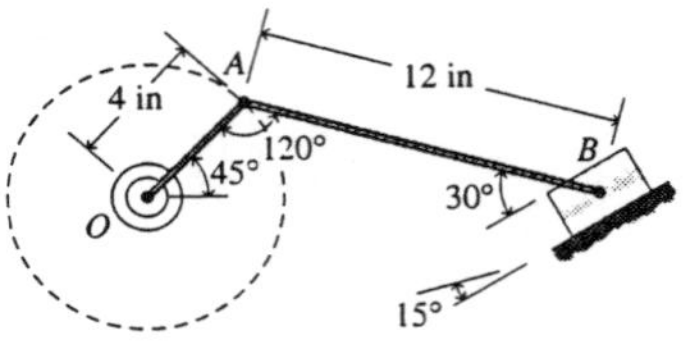

Figure a

FIND:

(a) Using the concept of instantaneous center of velocity, calculate the angular velocity of the connecting rod AB for the position shown.

(b) Determine the velocity of slider B.

SOLUTION:

(a) The instantaneous center is I (see Fig. b)

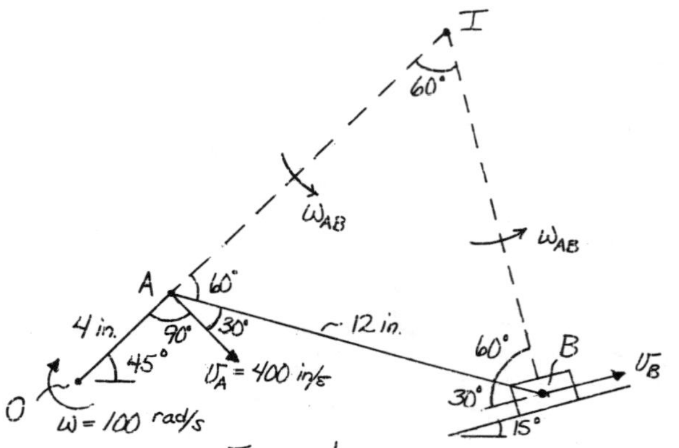

Figure b

By Fig. b, since AIB is an equilateral triangle,

$$AI = BI = AB = 12 \text{ in} \qquad (a)$$

Hence, by the concept of instantaneous center,

$$v_A = (AI)\omega_{AB} \text{ and } v_B = (BI)\omega_{AB} \qquad (b)$$

So, since $AI = BI$,

$$v_A = v_B = 400 \text{ in/s} \qquad (c)$$

Therefore, by Eqs. (a), (b), and (c),

$$\omega_{AB} = \frac{400}{12} = 33.3\bar{3} \text{ rad/s}$$

(b) $$v_B = (BI)\,\omega_{AB} = (12)(33.3\bar{3})$$

$$v_B = 400 \text{ in/s}$$

17.49

GIVEN: The wheel in Example 17.5 rolls and skids in the xy plane (Fig. a), where $x_0 = 12t^3$ [ft], $y_0 = 2$ ft, and $\theta = 8t^4$, where t is in seconds.

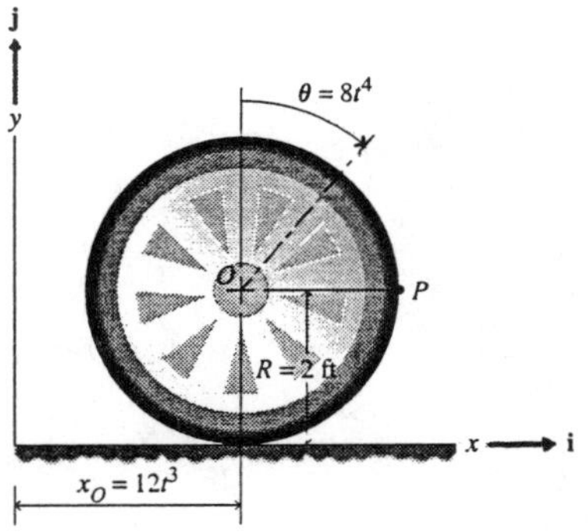

Figure a

FIND: Express the coordinate (x_I, y_I) of the instantaneous center of velocity of the wheel as a function of time t [s].

SOLUTION:

Since the displacement of the center O of the wheel is given by $x_0 = 12t^3$, its velocity is $\bar{v}_0 = \frac{dx_0}{dt}\hat{\imath} = 36t^2$ (a)

Hence, the instantaneous center is located on the vertical line through O. To locate the point on the vertical line that is the instantaneous center, we must know the velocity of a point in the wheel not on the vertical line. A convenient point is point P. The velocity $\bar{v}_P$ of point P may be determined by the equation

$$\bar{v}_P = \bar{v}_0 + \bar{v}_{P/O} \qquad \text{(b)}$$

where $\bar{v}_{P/O}$ is the velocity of P relative to O.

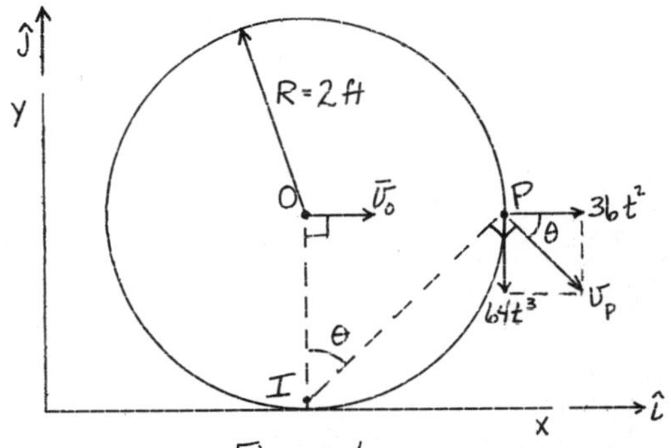

Figure b

By Fig. b,

$$\bar{v}_{P/O} = -R\dot{\theta}\,\hat{\jmath} = -2(32t^3)\hat{\jmath}$$

or $\bar{v}_{P/O} = -64t^3\,\hat{\jmath}$ (c)

So, by Eqs. (a), (b), and (c)

$$\bar{v}_P = (36t^2)\hat{\imath} - (64t^3)\hat{\jmath} \qquad \text{(d)}$$

Thus, $\bar{v}_P$ is directed at the angle

$$\theta = \tan^{-1}\frac{64t^3}{36t^2} = \tan^{-1}\frac{16t}{9} \qquad \text{(e)}$$

Hence, the instantaneous center is point I.

By Fig. b,

$$x_I = x_0 = 12t^3$$

$$y_I = 2 - OI$$

$$OI = \frac{2}{\tan\theta} = \frac{9}{8t}$$

Hence, the x, y coordinates of I are

$$\underline{\underline{x_I = 12t^3,\ y_I = 2 - 9/8t}}$$

17.50

GIVEN: A cylindrical culvert is loaded on a truck (Fig. a) that travels at a constant absolute speed v_A. The stop S is jarred loose and the culvert rolls so that the absolute speed of its center C is v_C.

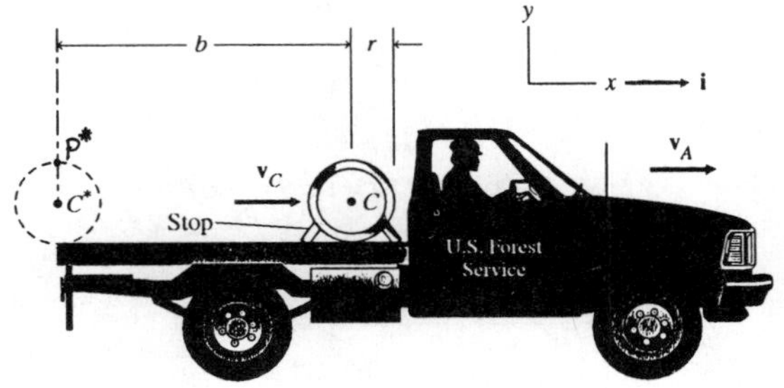

Figure a

FIND:

(a) Derive a formula for the angular velocity $\bar{\omega}_C$ of the culvert in terms of v_A, v_C, and r.

(b) Derive a formula for the time t_f at which the culvert is on the verge of rolling off the truck bed, in terms of v_A, v_C, and b. Take $t = 0$ for the instant that the culvert begins to roll.

(c) Determine the absolute speed v_{P*} of the point P^* of the culvert in terms of v_A and v_C, for the time t_f.

(d) Let $v_A = 12$ m/s, $v_C = 11$ m/s, $r = 450$ mm, and $b = 2$ m. Calculate the values of ω_c, t_f, and v_{P*}.

(continued)

17.50 (cont.)

SOLUTION:

(a) The velocity of the center C of the culvert relative to the truck is

$$\bar{v}_{C/A} = v_{C/A}\,\hat{\imath} = \bar{v}_C - \bar{v}_A = (v_C - v_A)\,\hat{\imath}$$

Since $v_A > v_C$, the culvert rolls to the left relative to the truck; that is, the speed of C relative to the truck is $v_A - v_C$ to the left. Hence, the angular velocity of the culvert is

$$\omega_c = \frac{v_A - v_C}{r} \quad \text{counterclockwise} \qquad (a)$$

(b) The distance b is given by

$$b = (v_A - v_C)\,t_f$$

or

$$t_f = \frac{b}{v_A - v_C} \qquad (b)$$

(c) At the instant that the culvert rolls off the truck, the velocity of point P* relative to the truck is

$$\bar{v}_{P^*/A} = -2r\omega_c\,\hat{\imath} \qquad (c)$$

The absolute velocity of P* is

$$\bar{v}_{P^*} = \bar{v}_A + \bar{v}_{P^*/A} \qquad (d)$$

By Eqs. (a), (c), and (d), we obtain

$$\bar{v}_{P^*} = v_A\,\hat{\imath} - 2(v_A - v_C)\,\hat{\imath}$$

or $\bar{v}_{P^*} = (2v_C - v_A)\,\hat{\imath}$

Hence, the absolute speed of point P* at the instant the culvert rolls off the truck is

$$v_{P^*} = 2v_C - v_A \qquad (e)$$

(d) For $v_A = 12$ m/s, $v_C = 11$ m/s, $r = 450$ mm, and $b = 2$ m, Eqs. (a), (b), and (e) yield

$$\omega_c = \frac{12-11}{0.45} = 2.2\bar{2} \text{ rad/s counterclockwise}$$

$$t_f = \frac{2}{12-11} = 2\text{ s}$$

$$v_{P^*} = 2(11) - 12 = 10 \text{ m/s}$$

17.51

GIVEN: Point O is the center of the circular arc C-C (Fig. a). The rigid link PQ connects the slider and the center Q of the roller.

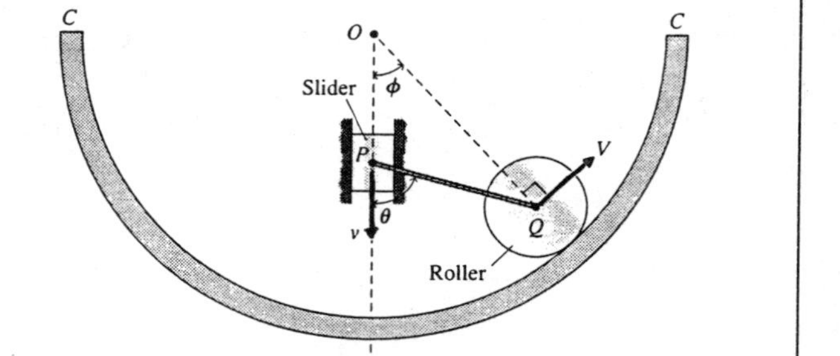

Figure a

FIND:

(a) Locate the instantaneous center of velocity of link PQ.

(b) Show that $v/V = [\sin(\theta - \phi)]/\cos\theta$ where V is the speed of point Q and v is the speed of slider P.

SOLUTION:

(a) By Fig. b, we see that I is the instantaneous center.

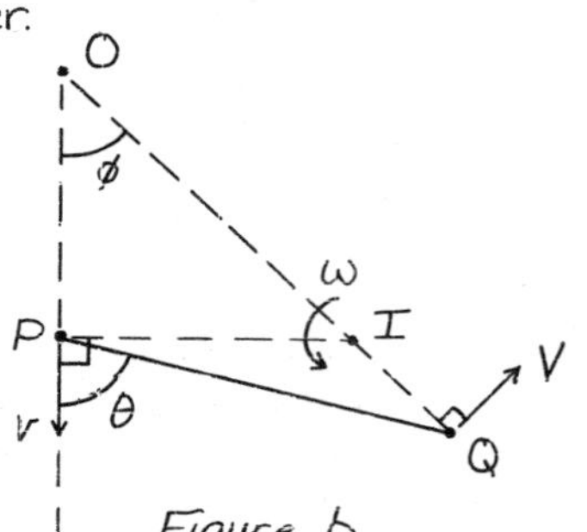

Figure b

(b) By Fig. b,

$$v = (IP)\omega = (OP\tan\phi)\omega \qquad (a)$$

$$V = (IQ)\omega = [(OQ) - (OI)]\omega = \left[(OQ - \frac{OP}{\cos\phi}\right]\omega \qquad (b)$$

Also by Fig. b,

$$(OP) = (OQ)\cos\phi - (PQ)\cos\theta$$
$$(PQ)\sin\theta = (OQ)\sin\phi \qquad (c)$$

Solving Eqs. (c) for (PQ) and (OP), we find

$$(PQ) = (OQ)\frac{\sin\phi}{\sin\theta}$$
$$(OP) = (OQ)\left(\cos\phi - \frac{\sin\phi}{\tan\theta}\right) \qquad (d)$$

Hence, by Eqs. (a), (b), and (d), and the trig. identity $\sin\alpha\cos\beta \pm \sin\beta\cos\alpha = \sin(\alpha \pm \beta)$,

$$v = (OQ)\left[\frac{\tan\phi}{\cos\theta}\,\frac{\sin(\theta-\phi)}{\tan\theta}\right]\omega \qquad (e)$$

$$V = (OQ)\frac{\tan\phi}{\tan\theta}\,\omega \qquad (f)$$

Then, by Eqs. (e) and (f)

$$\frac{v}{V} = \frac{\sin(\theta-\phi)}{\cos\theta}$$

17.52

GIVEN: In the mechanism of Fig. a, PQ = 200 mm and OQ = 300 mm. By part b of Problem 17.51,

$$\frac{v}{V} = \frac{\sin(\theta - \phi)}{\cos\theta} \qquad \text{(a)}$$

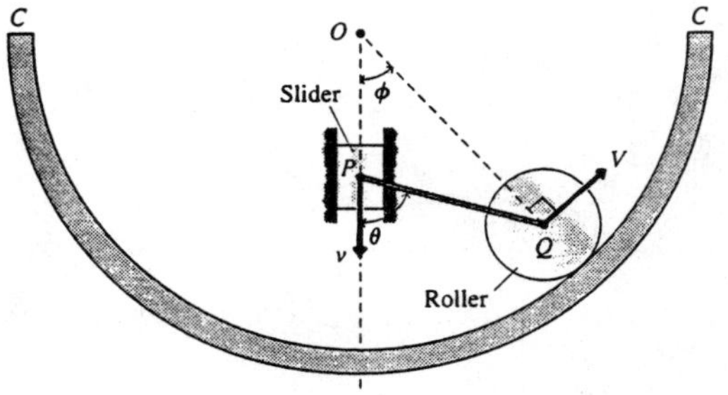

Figure a

FIND:

(a) Plot $\frac{v}{V}$ given by Eq. (a) as a function of ϕ, using an appropriate range $-\phi_o \le \phi \le \phi_o$.

(b) In terms of relative values of v and V, explain the behavior of the plot in the neighborhoods of $\phi = 0$ and $\phi = 41.81°$.

SOLUTION:

(a) To express v/V as a function of ϕ, by Fig. a, we see that

$$OQ \sin\phi = PQ \sin\theta$$

or with OQ = 300 mm and PQ = 200 mm,

$$\sin\theta = 1.5 \sin\phi \qquad \text{(b)}$$

So,

$$\theta = \sin^{-1}(1.5 \sin\phi) \qquad \text{(c)}$$

Substitution of Eq. (c) into Eq. (a) yields

$$\frac{v}{V} = \frac{\sin[\sin^{-1}(1.5\sin\phi) - \phi]}{\cos[\sin^{-1}(1.5\sin\phi)]} \qquad \text{(d)}$$

Note that for $\theta = \pm\frac{\pi}{2}$ $\cos\theta = 0$ and by Eq. (a), $\frac{v}{V} \to \infty$. Hence, let's limit θ to the range $-\frac{\pi}{2} < \theta < \frac{\pi}{2}$

Then, by Eq. (b), ϕ lies in the range

$$-41.81° < \phi < 41.81°$$

See the plot in Fig. b.

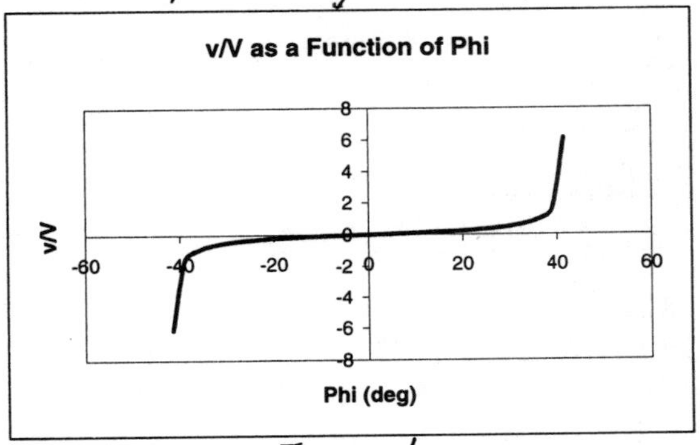

Figure b

(b) For $\phi = 0$, by Fig. a, $v = 0$ and V attains its maximum value. For $\phi = 41.81°$ ($\theta = \pi/2$ rad $= 90°$), bar PQ is horizontal. For $\theta = 90°$, $V = 0$ (See Fig. a) and v attains its maximum value. [See also Eqs. (e) and (f) of the solution of Problem 17.51.]

17.53

GIVEN: In the mechanism of Fig. a, PQ = 10 in. and OQ = 15 in. By Problem 17.51, part b,

$$\frac{v}{V} = \frac{\sin(\theta - \phi)}{\cos\theta} \qquad \text{(a)}$$

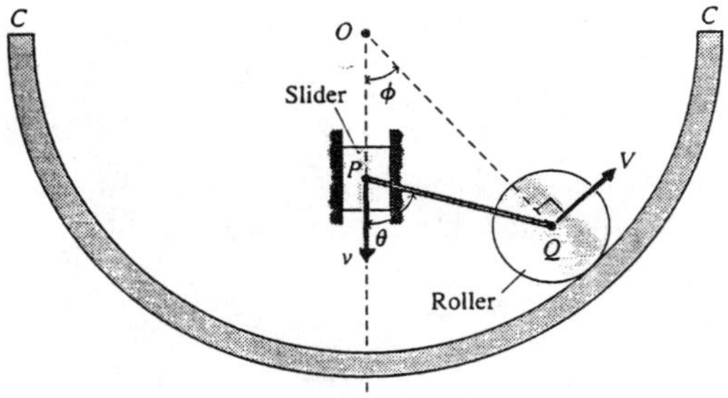

Figure a

FIND:

(a) Plot $\frac{v}{V}$ as a function of ϕ using an appropriate range $-\phi_o < \phi < \phi_o$.

(b) In terms of relative values of v and V, explain the behavior of the plot in the neighborhoods of $\phi = 0$ and $\theta = \pm\frac{\pi}{2}$.

SOLUTION:

(a) To express v/V as a function of ϕ, by Fig. a, we see that

$$OQ \sin\phi = PQ \sin\theta$$

or with OQ = 15 in. and PQ = 10 in,

$$\sin\theta = 1.5 \sin\phi \qquad \text{(b)}$$

So,

$$\theta = \sin^{-1}(1.5 \sin\phi) \qquad \text{(c)}$$

Substitution of Eq. (c) into Eq. (a) yields

$$\frac{v}{V} = \frac{\sin[\sin^{-1}(1.5\sin\phi) - \phi]}{\cos[\sin^{-1}(1.5\sin\phi)]} \qquad \text{(d)}$$

Note by Eq. (b) that for $\theta = 0$, $\phi = 0$ and for $\theta = \pm\frac{\pi}{2}$, $\phi = \pm 0.730$ rad. Hence, an appropriate range for ϕ is

$$-0.730 < \phi < 0.730 \text{ rad}$$

since at $\theta = \pm\pi/2$ ($\phi = \pm 0.730$ rad), $v/V \to \infty$ [See Eq. (a)].

See the plot in Figure b.

(continued)

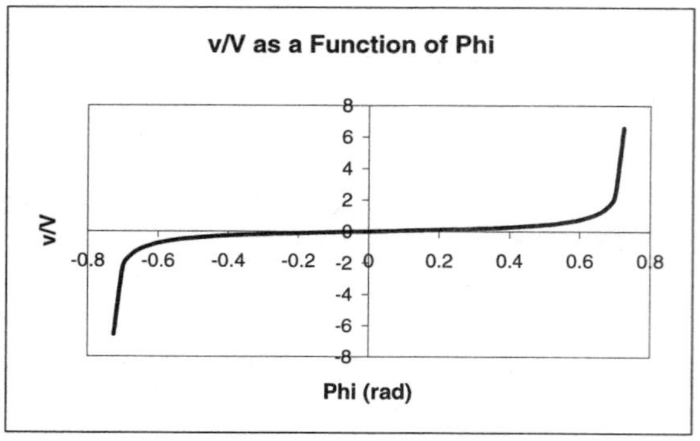

Figure b

(b) For $\theta = 0$, by Fig. a, $v = 0$ and V attains its maximum value. For $\theta \approx \pm \pi/2$, $|\phi| < 0.732$, bar PQ is horizontal. For $\theta = \pi/2$, $V = 0$ (see Fig. a) and v attains its maximum value. [See also Eqs. (e) and (f) of the solution of Problem 17.51.]

17.54

GIVEN: Ball bearings of radius r roll between straight bearing rails that move in opposite directions with constant speeds v_1 and v_2 directed as shown in Fig. a. Assume that $v_1 > v_2$.

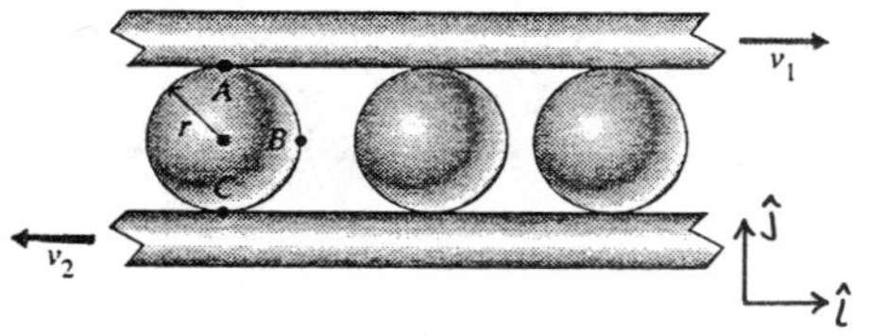

Figure a

FIND:

(a) Derive formulas for the velocities of points A and C and the angular velocity of a typical ball.

(b) Derive formulas for the velocities of points O and B.

(c) Calculate the magnitudes of the velocities of points A, B, C, and O and the angular velocity of a typical ball for the case $v_1 = 6$ m/s, $v_2 = 2$ m/s, $r = 20$ mm.

(d) With the results of part c, determine the instantaneous center of velocity of a ball.

SOLUTION:

(a) Consider Fig. b, with $v_1 > v_2$. Since points A and C are in contact with the upper and lower rails, respectively,

$$v_{A/N} = v_1 \quad \text{and} \quad v_{C/N} = v_2.$$

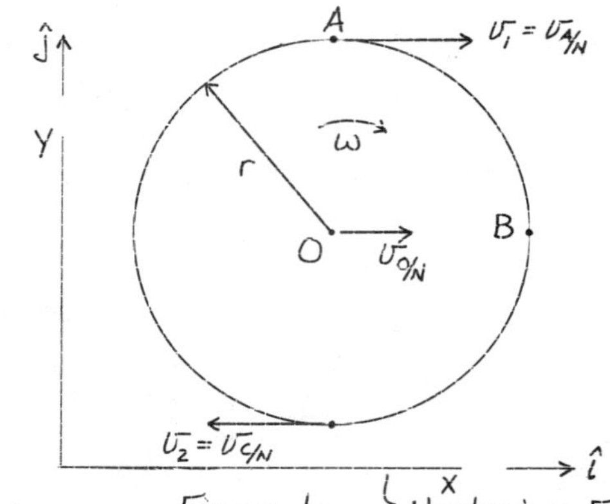

Figure b

Hence,

$$\bar{v}_{A/N} = v_1 \hat{i} \qquad \text{(a)}$$

$$\bar{v}_{C/N} = -v_2 \hat{i} \qquad \text{(b)}$$

Also, by Fig. b and the concept of relative motion,

$$\bar{v}_{O/N} = \bar{v}_{C/N} + \bar{v}_{O/C} = -v_2 \hat{i} + r\omega \hat{i} \qquad \text{(c)}$$

Likewise,

$$\bar{v}_{O/N} = \bar{v}_{A/N} + \bar{v}_{O/A} = v_1 \hat{i} - r\omega \hat{i} \qquad \text{(d)}$$

Equating Eqs. (c) and (d) and solving for ω, we find the angular velocity

$$\omega = \frac{v_1 + v_2}{2r} \quad \text{clockwise} \qquad \text{(e)}$$

(b) By Eqs. (e) and (c) [or (d)], we obtain

$$\bar{v}_{O/N} = \frac{v_1 - v_2}{2} \hat{i} \qquad \text{(f)}$$

By the concept of relative motion and Fig. b,

$$\bar{v}_{B/N} = \bar{v}_{O/N} + \bar{v}_{B/O}$$

or

$$\bar{v}_{B/N} = \tfrac{1}{2}(v_1 - v_2)\hat{i} - \omega r \hat{j} \qquad \text{(g)}$$

Equations (e) and (g) yield

$$\bar{v}_{B/N} = \tfrac{1}{2}(v_1 - v_2)\hat{i} - \tfrac{1}{2}(v_1 + v_2)\hat{j} \qquad \text{(h)}$$

(c) For $v_1 = 6$ m/s, $v_2 = 2$ m/s, $r = 0.02$ m, Eqs. (a), (b), (e), (f), and (h) yield

$$v_{A/N} = 6 \text{ m/s}$$

$$\bar{v}_{B/N} = 2\hat{i} - 4\hat{j}; \text{ or } v_{B/N} = \sqrt{2^2 + 4^2} = 4.472 \text{ m/s}$$

$$v_{C/N} = 2 \text{ m/s}$$

$$v_{O/N} = 2 \text{ m/s}$$

$$\omega = 200 \text{ rad/s}$$

(d) To determine the instantaneous center of velocity of a ball, we may use velocities $\bar{v}_{O/N}$ and $\bar{v}_{B/N}$ (Fig. c). The instantaneous center is point I.

(continued)

17.54 (cont.)

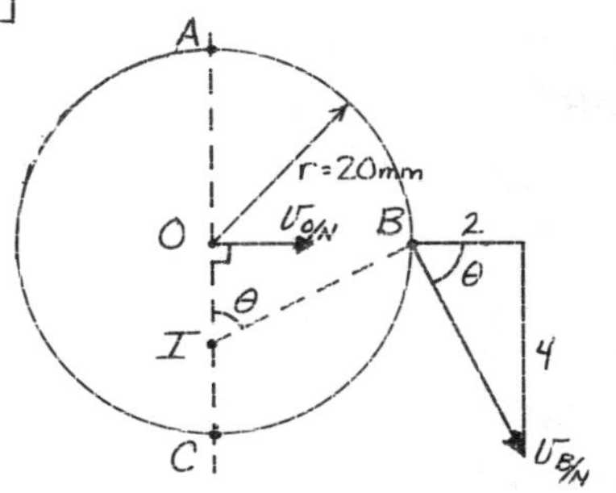

Figure c

By Fig. c,

$$\tan\theta = 2.$$

Hence,

$$OI = \frac{20}{\tan\theta} = 10 \text{ mm}$$

17.55

GIVEN: In Problem 17.54, the lower bearing rail velocity is reversed, and the rail moves to the right with absolute speed v_2 (Fig. a), with $v_1 > v_2$.

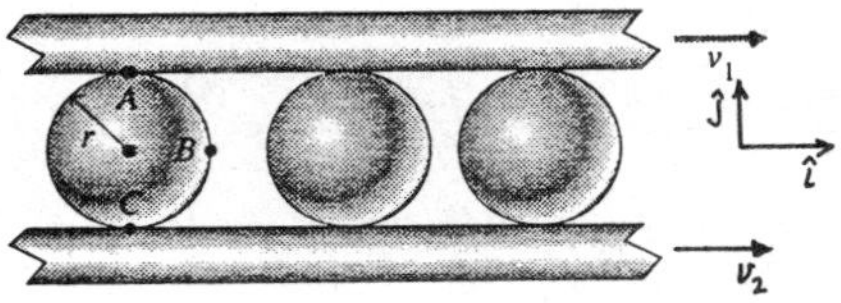

Figure a

FIND:

(a) Derive formulas for the velocities of points A and C and the angular velocity of a typical ball.

(b) Derive formulas for the velocities of points O and B.

(c) Calculate the magnitudes of the velocities of points A, B, C, and O, and the angular velocity of a typical ball for the case $v_1 = 6$ m/s, $v_2 = 2$ m/s, and r = 20 mm.

(d) With the results of part c, determine the instantaneous center of velocity of a ball.

SOLUTION:

(a) Consider Fig. b, with $v_1 > v_2$. Since points A and C are in contact with the upper and lower rails, respectively, $v_{A/N} = v_1$ and $v_{C/N} = v_2$. Hence,

$$\bar{v}_{A/N} = v_1 \hat{i} \quad (a)$$

$$\bar{v}_{C/N} = v_2 \hat{i} \quad (b)$$

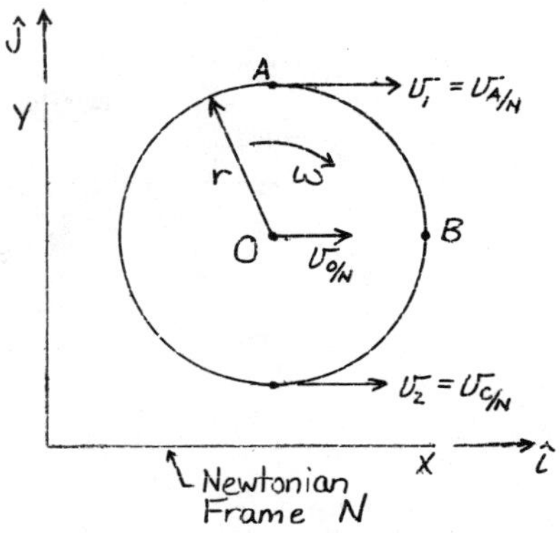

Figure b

Also by Fig. b and the concept of relative motion,

$$\bar{v}_{O/N} = \bar{v}_{C/N} + \bar{v}_{O/C} = v_2 \hat{i} + r\omega \hat{i} \quad (c)$$

Likewise,

$$\bar{v}_{O/N} = \bar{v}_{A/N} + \bar{v}_{O/A} = v_1 \hat{i} - r\omega \hat{i} \quad (d)$$

Equating Eqs. (c) and (d) and solving for ω, we find the angular velocity

$$\omega = \frac{v_1 - v_2}{2r} \text{ clockwise} \quad (e)$$

(b) By Eqs. (e) and (c) [or (d)], we obtain

$$\bar{v}_{O/N} = \frac{v_1 + v_2}{2} \hat{i} \quad (f)$$

By the concept of relative motion and Fig. b,

$$\bar{v}_{B/N} = \bar{v}_{O/N} + \bar{v}_{B/O} = \tfrac{1}{2}(v_1 + v_2)\hat{i} - \omega r \hat{j} \quad (g)$$

Equations (e) and (g) yield

$$\bar{v}_{B/N} = \tfrac{1}{2}(v_1 + v_2)\hat{i} - \tfrac{1}{2}(v_1 - v_2)\hat{j} \quad (h)$$

(c) For $v_1 = 6$ m/s, $v_2 = 2$ m/s, r = 0.02 m, Eqs. (a), (b), (e), (f), and (h) yield

$v_{A/N} = 6$ m/s

$\bar{v}_{B/N} = 4\hat{i} - 2\hat{j}$; or $v_{B/N} = \sqrt{4^2 + 2^2} = 4.472$ m/s

$v_{C/N} = 2$ m/s

$v_{O/N} = 4$ m/s

$\omega = 100$ rad/s

(d) To determine the instantaneous center of velocity of a ball, we may use velocities $\bar{v}_{O/N}$ and $\bar{v}_{B/N}$ (Fig. c). The instantaneous center is point I.

(continued)

17.55 (cont.)

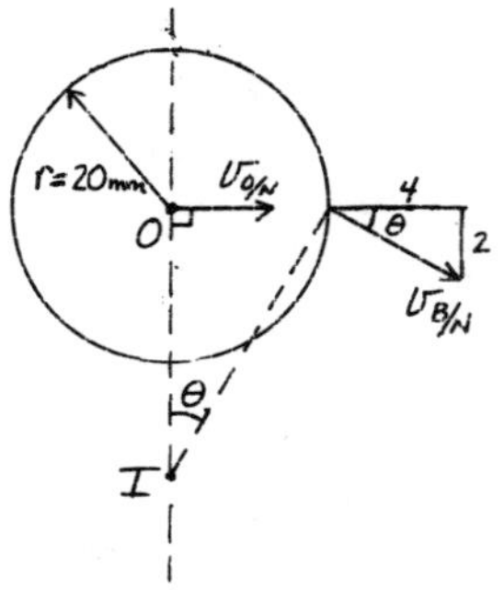

Figure C

By Fig. C, $\tan\theta = 0.5$. Hence,

$$\overline{OI} = \frac{20}{\tan\theta} = 40\ mm$$

17.56

GIVEN: The crank of the slider-crank mechanism shown in Fig. a rotates with constant angular velocity ω_{OA}.

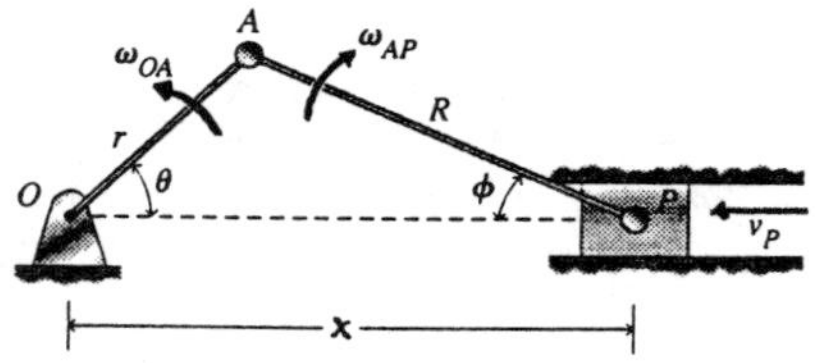

FIND:

(a) Show that the speed v_P of the slider P is given by

$$v_P = r\omega_{OA}\sin\theta\left[1 + \frac{\cos\theta}{\sqrt{\frac{R^2}{r^2} - \sin^2\theta}}\right] \quad (a)$$

(b) On one diagram, plot $v_P/(r\omega_{OA})$ as a function of θ for $0 \le \theta \le 360°$, and for ratios $R/r = 2, 3, 4, 5, 6$, and ∞.

(c) A design requirement for the slider-crank mechanism specifies that $v_P/(r\omega_{OA})$ must lie in the range $1.0 \le |v_P/(r\omega_{OA})| \le 1.5$. Select a range of values of R/r that will satisfy this requirement, and discuss the practicality of the values of R/r in this range.

SOLUTION:

(a) Consider Fig. b.

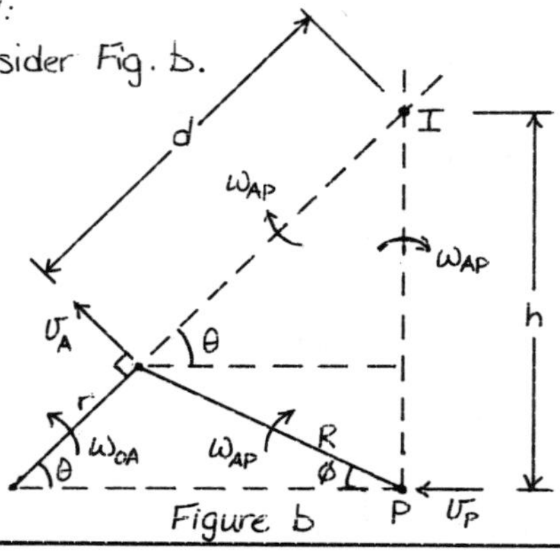

Figure b

By the geometry of fig. b,

$$d\cos\theta = R\cos\phi \quad (a)$$

$$r\sin\theta = R\sin\phi \quad (b)$$

By Eq. (b),

$$\sin\phi = \frac{r}{R}\sin\theta$$

or

$$\cos\phi = \sqrt{1-\sin^2\phi} = \sqrt{1 - \frac{r^2}{R^2}\sin^2\theta} \quad (c)$$

By kinematics,

$$v_A = r\omega_{OA} = (d)(\omega_{AP})$$

or

$$\omega_{AP} = \frac{r\omega_{OA}}{d} \quad (d)$$

where by Eqs. (a) and (c),

$$d = \frac{R\cos\phi}{\cos\theta} = \frac{r}{\cos\theta}\sqrt{\frac{R^2}{r^2} - \sin^2\theta} \quad (e)$$

Then, by Eqs. (d) and (e)

$$\omega_{AP} = \frac{\omega_{OA}\cos\theta}{\sqrt{\left(\frac{R}{r}\right)^2 - \sin^2\theta}} \quad (f)$$

Also by Fig. b,

$$v_P = h\,\omega_{AP} \quad (g)$$

where $h = r\sin\theta + d\sin\theta$

or with Eq. (e),

$$h = r\sin\theta\left[1 + \frac{1}{\cos\theta}\sqrt{\frac{R^2}{r^2} - \sin^2\theta}\right] \quad (h)$$

So, by Eqs. (g) and (h)

$$v_P = r\omega_{OA}\sin\theta\left[1 + \frac{\cos\theta}{\sqrt{\frac{R^2}{r^2} - \sin^2\theta}}\right] \quad (i)$$

(b) See the plot of $v_P/(r\omega_{OA})$ in Fig. C, for $0 \le \theta \le 360°$ and $R/r = 2, 3, 4, 5, 6$, and ∞.

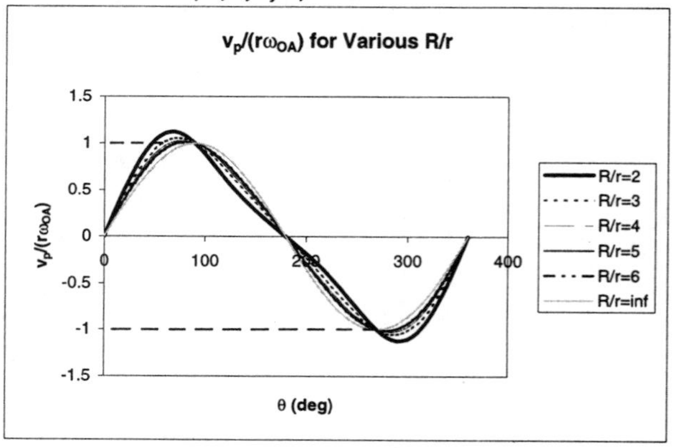

Figure C

(c) Note by Eq. (i) that as $R/r \to \infty$, the ratio $v_P/(r\omega_{OA}) \to \sin\theta$; that is, the curve is a sine wave for the length of the connecting rod is much larger than the length of the crank arm. Since the maximum value of the sine is 1, a very large R/r ratio would meet the design requirement of $1.0 \le |v_P/(r\omega_{OA})| \le 1.5$.

(continued)

17.56 (cont.)

However, a very large R/r is impractical.

By the plot of Fig. c, we see that for $R/r = 2$, the maximum value of $|v_P/(r\omega_{OA})|$ exceeds 1.0 slightly. Hence, the ratios $R/r = 2, 3, 4, 5, 6$ will all meet the design requirements. The particular ratio used will depend on space limitations and limits on the displacement of the slider. Note that all ratios $R/r \geq 2$ will meet the specification $1.0 \leq |v_P/(r\omega_{OA})|$.

17.57

GIVEN: The center O of a disk moves in the positive sense along the x axis with an acceleration of 16 ft/s² (Fig. a), and with angular acceleration and angular velocity, $\alpha = 32$ rad/s² and $\omega = \sqrt{32}$ rad/s.

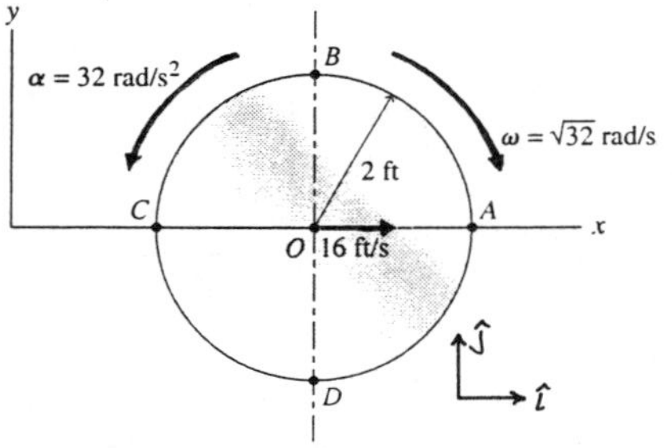

Figure a

FIND: Determine the x and y components of acceleration of points A, B, C, and D of the disk.

SOLUTION:

By Fig. a and the concept of relative acceleration,

$$\bar{a}_A = \bar{a}_O + \bar{a}_{A/O}$$
$$= \bar{a}_O - \omega^2 r\,\hat{i} + \alpha r\,\hat{j}$$
$$= 16\hat{i} - (32)(2)\,\hat{i} + (32)(2)\,\hat{j}$$

or $\bar{a}_A = -48\hat{i} + 64\hat{j}$; $\underline{\bar{a}_{A_x} = -48\hat{i}\ \text{ft/s}^2,\ \bar{a}_{A_y} = 64\hat{j}\ \text{ft/s}^2}$

$$\bar{a}_B = \bar{a}_O + \bar{a}_{B/O}$$
$$= \bar{a}_O - \alpha r\,\hat{i} - \omega^2 r\,\hat{j}$$
$$= 16\hat{i} - 64\hat{i} - 64\hat{j}$$

or $\bar{a}_B = -48\hat{i} - 64\hat{j}$; $\underline{\bar{a}_{B_x} = -48\hat{i}\ \text{ft/s}^2,\ \bar{a}_{B_y} = -64\hat{j}\ \text{ft/s}^2}$

$$\bar{a}_C = \bar{a}_O + \bar{a}_{C/O}$$
$$= \bar{a}_O + \omega^2 r\,\hat{i} - \alpha r\,\hat{j}$$
$$= 16\hat{i} + 64\hat{i} - 64\hat{j}$$

or $\bar{a}_C = 80\hat{i} - 64\hat{j}$; $\underline{\bar{a}_{C_x} = 80\hat{i}\ \text{ft/s}^2,\ a_{C_y} = -64\hat{j}\ \text{ft/s}^2}$

$$\bar{a}_D = \bar{a}_O + \bar{a}_{D/O}$$
$$= \bar{a}_O + \alpha r\,\hat{i} + \omega^2 r\,\hat{j}$$
$$= 16\hat{i} + 64\hat{i} + 64\hat{j}$$

or $\bar{a}_D = 80\hat{i} + 64\hat{j}$; $\underline{\bar{a}_{D_x} = 80\hat{i}\ \text{ft/s}^2 ;\ \bar{a}_{D_y} = 64\hat{j}\ \text{ft/s}^2}$

17.58

GIVEN: The wheel in Fig. a is rolling to the right. The magnitudes of the velocity and acceleration of point O are 1 m/s and 1.5 m/s², respectively.

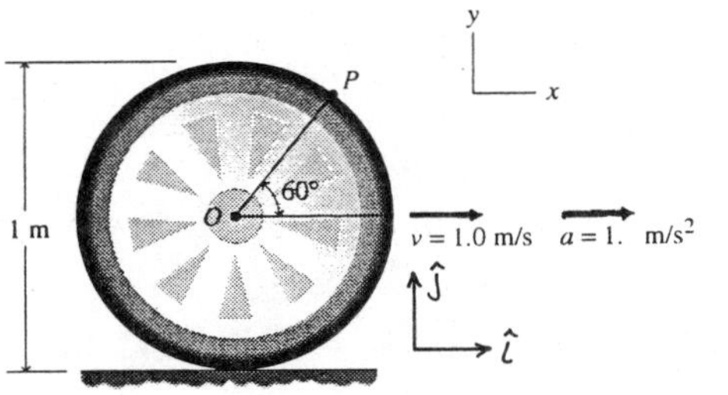

Figure a

FIND: Calculate the x and y projections of the acceleration of point P.

SOLUTION:

By kinematics, $\omega = \frac{v_O}{r} = \frac{1.0}{0.5} = 2$ rad/s ↻ and $\alpha = \frac{a_O}{r} = \frac{1.5}{0.5} = 3$ rad/s² ↻.

Hence, by Fig. a and the concept of relative acceleration,

$$\bar{a}_P = \bar{a}_O + \bar{a}_{P/O}$$
$$= \bar{a}_O + \alpha r \searrow_{30°} + \omega^2 r \swarrow_{60°}$$
$$= 1.5\hat{i} + (3)(0.5) \searrow_{30°} + (2)^2(0.5) \swarrow_{60°}$$
$$= 1.5\hat{i} + (1.5)(\cos 30°)\hat{i} - (1.5)(\sin 30°)\hat{j} - (2)(\cos 60°)\hat{i} - (2)(\sin 60°)\hat{j}$$

or $\bar{a}_P = 1.80\hat{i} - 2.48\hat{j}$ (m/s²)

Hence,

$$\underline{a_{P_x} = 1.80\ \text{m/s}^2}$$
$$\underline{a_{P_y} = -2.48\ \text{m/s}^2}$$

17.59

GIVEN: The rigid body in Fig. a executes plane motion parallel to the xy plane.

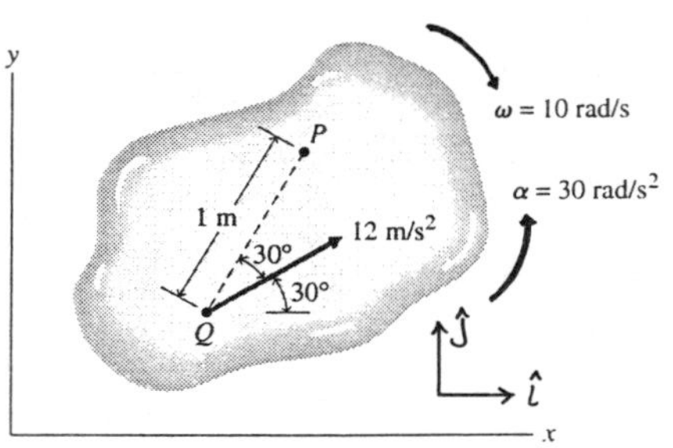

Figure a

FIND: Determine the x and y components of acceleration of point P.

SOLUTION:

By Fig. a and the concept of relative acceleration, with $\alpha = 30$ rad/s², $\omega = 10$ rad/s, and $QP = 1$ m,

$$\bar{a}_P = \bar{a}_Q + \bar{a}_{P/Q}$$

$$\bar{a}_P = a_Q \angle 30° + \alpha r_{QP} \angle 30° + \omega^2 r_{QP} \angle 60°$$

$$\bar{a}_P = (12)\cos 30°\,\hat{i} + (12)\sin 30°\,\hat{j} - (30)(1)\cos 30°\,\hat{i} + (30)(1)\sin 30°\,\hat{j} - (10^2)(1)\cos 60°\,\hat{i} - (10^2)(1)\sin 60°\,\hat{j}$$

or $\bar{a}_P = -65.6\,\hat{i} - 65.6\,\hat{j}$ [m/s²]

Therefore,

$\bar{a}_{P_x} = -65.6$ m/s²

$\bar{a}_{P_y} = -65.6$ m/s²

17.60

GIVEN: A disk 4 ft in diameter rolls in the xy plane in the positive sense of the x axis. (Fig. a) The velocity of its center is

$$v_o = t^2 + 2t + 1 \ [\text{ft/s}] \qquad \text{(a)}$$

where t is in seconds.

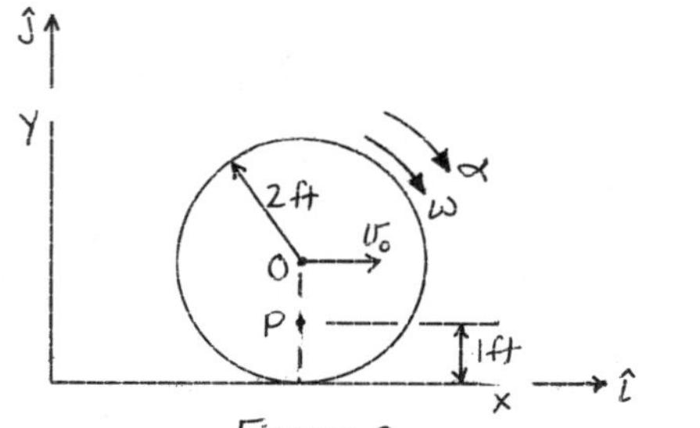

Figure a

FIND: For $t = 1$ s, determine the acceleration of point P directly below the center and 1 ft above the x axis.

SOLUTION:

By kinematics and Eq. (a)

$$a_o = \dot{v}_o = 2t + 2 \qquad \text{(b)}$$

Also,

$$v_o = \omega r = 2\omega \qquad \text{(c)}$$

$$a_o = \alpha r = 2\alpha \qquad \text{(d)}$$

For $t = 1$ s, Eqs. (a) and (b) yield

$$v_o = 4 \text{ ft/s} \qquad \text{(e)}$$

$$a_o = 4 \text{ ft/s}^2 \qquad \text{(f)}$$

Hence, Eqs. (c), (d), (e), and (f) yield

$$\omega = 2 \text{ rad/s} \circlearrowright \qquad \text{(g)}$$

$$\alpha = 2 \text{ rad/s}^2 \circlearrowright \qquad \text{(h)}$$

By kinematics and Fig. a, with Eqs. (d), (f), (g), and (h),

$$\bar{a}_P = \bar{a}_o + \bar{a}_{P/o}$$

$$= \bar{a}_o + \alpha(OP) \leftarrow + \omega^2(OP) \uparrow$$

or $\bar{a}_P = 4\hat{i} - 2\hat{i} + 4\hat{j}$

So, $\bar{a}_P = 2\hat{i} + 4\hat{j}$ [ft/s²]

17.61

GIVEN: The pin A is fixed to bar AB and slides in the slot in bar CD (Fig. a). The magnitude of the acceleration of A is 75 ft/s². The acceleration vector of A is tangent to the circle with center C, and is directed down and to the left.

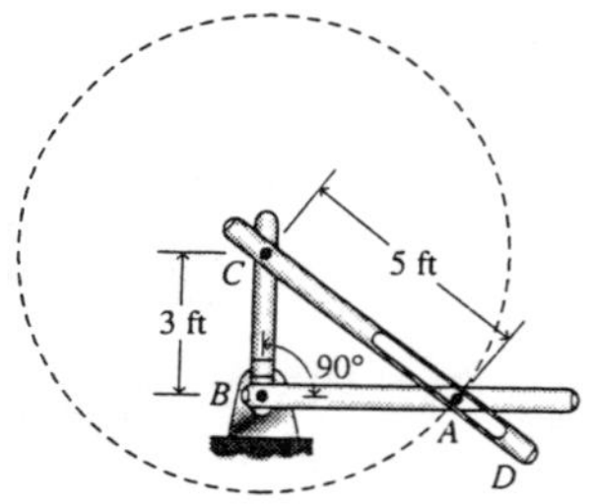

Figure a

FIND:

(a) Determine the angular acceleration of bar AB.

(b) Determine the angular speed of bar AB.

SOLUTION:

(a) Consider the rod in the position shown (Fig. b).

(continued)

17.61 (cont.)

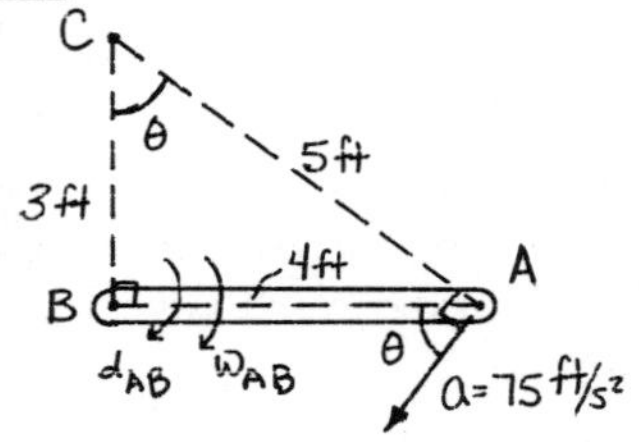

Figure b

By Fig. b and kinematics,

$$a \sin\theta = \alpha_{AB}(AB) \quad (a)$$

Also by Fig. b,

$$\sin\theta = \frac{4}{5}\ ;\ AB = 4\ ft \quad (b)$$

Hence, by Eqs. (a) and (b),

$$\underline{\underline{\alpha_{AB} = (\tfrac{1}{4})(75)(\tfrac{4}{5}) = 15\ rad/s^2}}$$

(b) By Fig. b and kinematics,

$$a\cos\theta = \omega_{AB}^2(AB)$$

where $\cos\theta = 3/5$ and $AB = 4\ ft$.

Hence,

$$\omega_{AB}^2 = (\tfrac{1}{4})(75)(\tfrac{3}{5}) = 11.25$$

or $\underline{\underline{\omega_{AB} = 3.354\ rad/s}}$

17.62

GIVEN: The cylinder (Fig. a) rolls down the inclined plane under the action of forces. At a certain instant, $\omega = 4$ rad/s and $\alpha = 2$ rad/s².

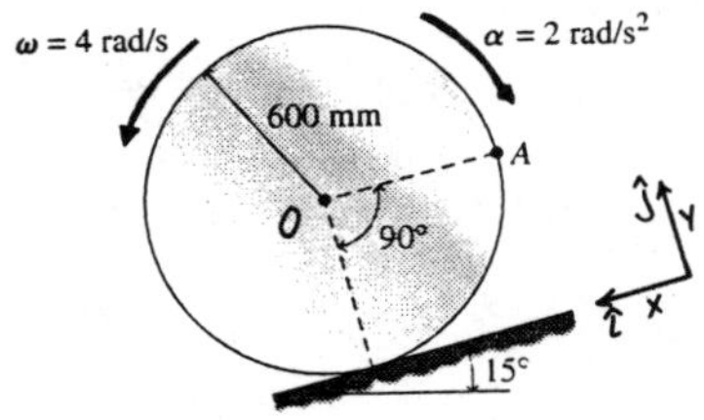

Figure a

FIND: Determine the velocity and acceleration of point A for this instant, and show them in a sketch.

SOLUTION:

Use the xy axes shown in Fig. a. By Fig. a and kinematics,

$$\bar{v}_A = \bar{v}_O + \bar{v}_{A/O} \quad (a)$$

where

$$\bar{v}_O = (\omega r)\hat{\imath} = (4)(0.600)\hat{\imath} = 2.40\hat{\imath}\ m/s \quad (b)$$

$$\bar{v}_{A/O} = (\omega r)\hat{\jmath} = 2.40\hat{\jmath}\ m/s \quad (c)$$

Therefore, by Eqs. (a), (b), and (c),

$$\underline{\underline{\bar{v}_A = 2.4\hat{\imath} + 2.4\hat{\jmath}\ [m/s]}} = 3.39\ m/s\ @\ 45°$$

Likewise, the acceleration of A is

$$\bar{a}_A = \bar{a}_O + \bar{a}_{A/O} \quad (d)$$

where

$$\bar{a}_O = -\alpha r\,\hat{\imath} = -(2)(0.600)\hat{\imath} = -1.2\hat{\imath}\ m/s^2 \quad (e)$$

$$\bar{a}_{A/O} = (\omega^2 r)\hat{\imath} - \alpha r\,\hat{\jmath} = (4^2)(0.600) - (2)(0.600)\hat{\jmath}$$

or $\bar{a}_{A/O} = 9.6\hat{\imath} - 1.2\hat{\jmath}\ m/s^2 \quad (f)$

Then, by Eqs. (d), (e), and (f),

$$\underline{\bar{a}_A = 8.4\hat{\imath} - 1.2\hat{\jmath}\ m/s^2} = 8.485\ m/s^2\ @\ -8.13°$$

The velocity and acceleration of point A are shown in Fig. b.

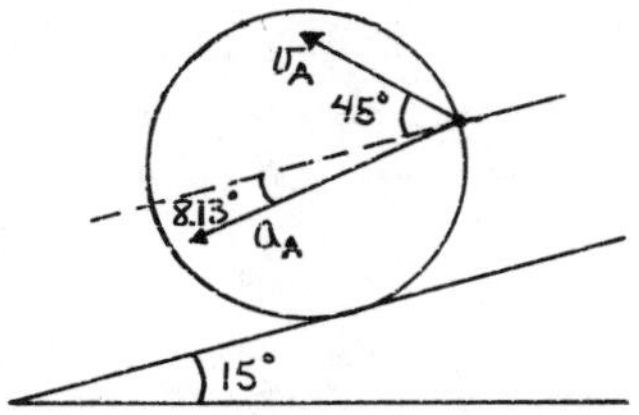

Figure b

17.63

GIVEN: A disk of radius r = 600 mm rotates in the xy plane, and its center C moves parallel to the x axis (Fig. a).

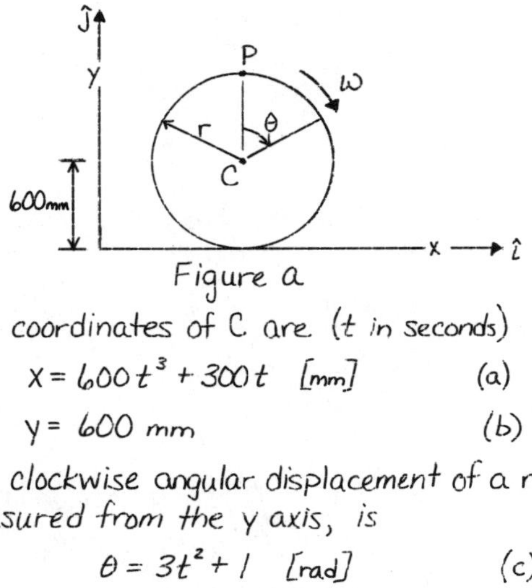

Figure a

The coordinates of C are (t in seconds)

$$x = 600t^3 + 300t\ \ [mm] \quad (a)$$

$$y = 600\ mm \quad (b)$$

The clockwise angular displacement of a radius, measured from the y axis, is

$$\theta = 3t^2 + 1\ \ [rad] \quad (c)$$

FIND: Determine the accelerations of C and the highest point on the disk at t = 1 s.

SOLUTION:

By Eqs. (a), (b), and (c),

$$\dot{x} = 1800t^2 + 300,\quad \ddot{x} = 3600t \quad (d)$$

$$\dot{y} = \ddot{y} = 0 \quad (e)$$

$$\dot{\theta} = 6t\ rad/s,\quad \ddot{\theta} = 6\ rad/s^2 \quad (f)$$

(continued)

17.63 (cont.)

By Fig. a and kinematics,

$$\bar{a}_c = \ddot{x}\,\hat{\imath} + \ddot{y}\,\hat{\jmath} = 3600t\,\hat{\imath} + 0\hat{\jmath} = 3600t\,\hat{\imath}\ \tfrac{mm}{s^2} \quad (g)$$

and

$$\bar{a}_P = \bar{a}_C + \bar{a}_{P/C} \quad (h)$$

where,

$$\bar{a}_{P/C} = \ddot{\theta} r\,\hat{\imath} - (\dot{\theta})^2 r\,\hat{\jmath}$$
$$= 3.6\,\hat{\imath} - (36t^2)(0.600)\,\hat{\jmath}$$

or

$$\bar{a}_{P/C} = 3.6\,\hat{\imath} - 21.6t^2\,\hat{\jmath}\quad m/s^2 \quad (i)$$

For $t = 1$ s, by Eq. (g)

$$\bar{a}_C = 3600\ mm/s^2 \rightarrow = 3.6\ m/s^2\,\hat{\imath}$$

and for the highest point on the disk, Eqs. (g), (h), and (i) yield

$$\bar{a}_P = 3.6\,\hat{\imath} + 3.6\,\hat{\imath} - 21.6\,\hat{\jmath}\ [m/s^2]$$

or

$$\bar{a}_P = 7.2\,\hat{\imath} - 21.6\,\hat{\jmath}\ [m/s^2]$$

17.64

GIVEN: The angular velocity of the body shown in Fig. a is constant. The acceleration of point Q is 100 ft/s² in the direction shown.

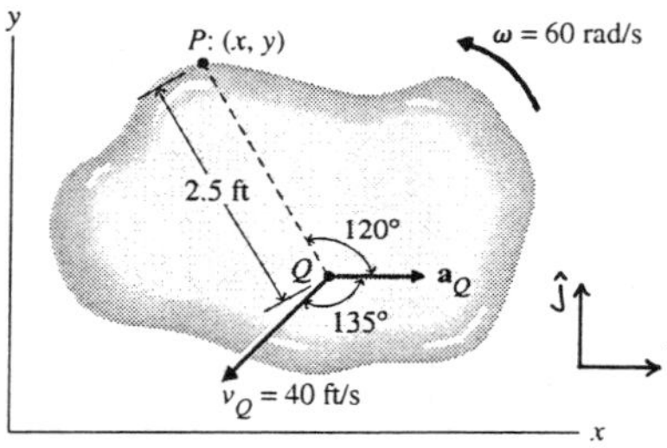

FIND: Determine the acceleration of point P.

SOLUTION:

By Fig. a and kinematics,

$$\bar{a}_P = \bar{a}_Q + \bar{a}_{P/Q} \quad (a)$$

where

$$\bar{a}_Q = 100\,\hat{\imath}\ [ft/s^2] \quad (b)$$

and

$$\bar{a}_{P/Q} = (\omega^2 r \cos 60^\circ)\,\hat{\imath} - (\omega^2 r \sin 60^\circ)\,\hat{\jmath}$$
$$= (3600)(2.5)[0.500\,\hat{\imath} - 0.866\,\hat{\jmath}]$$
$$= 4500\,\hat{\imath} - 7794\,\hat{\jmath} \quad (c)$$

By Eqs. (a), (b), (c),

$$\bar{a}_P = 4600\,\hat{\imath} - 7794\,\hat{\jmath}\ [ft/s^2]$$

17.65

GIVEN: In a mechanism (Fig. a), slider B travels along the path M-M. Crank OA rotates at $\omega_{OA} = 300$ rad/s clockwise. For the position shown, the angular velocity of rod AB is $\omega_{AB} = 112.5$ rad/s counterclockwise.

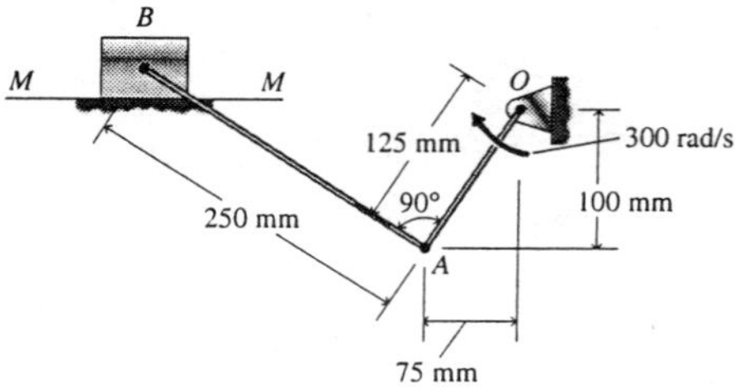

Figure a

FIND: Determine the angular acceleration of rod AB and the acceleration of slider B for the position shown.

SOLUTION:

Consider Fig. a.

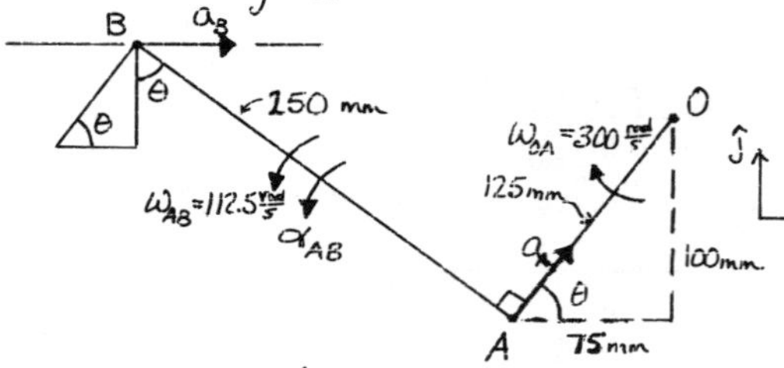

Figure b

By Fig. b and kinematics,

$$\bar{a}_B = \bar{a}_A + \bar{a}_{B/A} \quad (a)$$

Also by Fig. b,

$$\cos\theta = \frac{75}{125} = 0.600, \quad \sin\theta = \frac{100}{125} = 0.800 \quad (b)$$

Then,

$$\bar{a}_A = (\omega_{OA}^2)(OA)[0.600\,\hat{\imath} + 0.800\,\hat{\jmath}]$$
$$= 11250[0.600\,\hat{\imath} + 0.800\,\hat{\jmath}]\ [m/s^2]$$

or

$$\bar{a}_A = 6750\,\hat{\imath} + 9000\,\hat{\jmath}\ [m/s^2] \quad (c)$$

Also,

$$\bar{a}_{B/A} = (\alpha_{AB})(AB)[-(\cos\theta)\,\hat{\imath} - (\sin\theta)\,\hat{\jmath}]$$
$$+ (\omega_{AB}^2)(AB)[(\sin\theta)\,\hat{\imath} - (\cos\theta)\,\hat{\jmath}]$$
$$= (\alpha_{AB})(0.250)[-0.600\,\hat{\imath} - 0.800\,\hat{\jmath}]$$
$$+ (112.5)^2(0.250)[0.800\,\hat{\imath} - 0.600\,\hat{\jmath}]\ [m/s^2]$$

or

$$\bar{a}_{B/A} = \alpha_{AB}[-0.15\,\hat{\imath} - 0.20\,\hat{\jmath}]$$
$$+ 2531.25\,\hat{\imath} - 1898.44\,\hat{\jmath}\ [m/s^2] \quad (d)$$

Hence, by Eqs. (a), (c), and (d)

$$a_A\,\hat{\imath} = -0.15\alpha_{AB}\,\hat{\imath} - 0.20\alpha_{AB}\,\hat{\jmath}$$
$$+ 9281.25\,\hat{\imath} + 7101.56\,\hat{\jmath} \quad (e)$$

(continued)

17.65 (cont.)

Equating the $\hat{\imath}$ and $\hat{\jmath}$ components in Eq. (e), we have

$$7101.56 - 0.20\alpha_{AB} = 0 \qquad (f)$$
$$9281.25 - 0.15\alpha_{AB} = a_A$$

The solution of Eqs. (f) is

$\alpha_{AB} = 35{,}508 \text{ rad/s}^2$ counterclockwise

$\bar{a}_A = 3955\,\hat{\imath}\ [\text{m/s}^2]$

17.66

GIVEN: A rigid body executes plane motion parallel to the xy plane (Fig. a). The velocity and acceleration vectors of Q lie along line L. The velocity and acceleration of point P are shown in Fig. a.

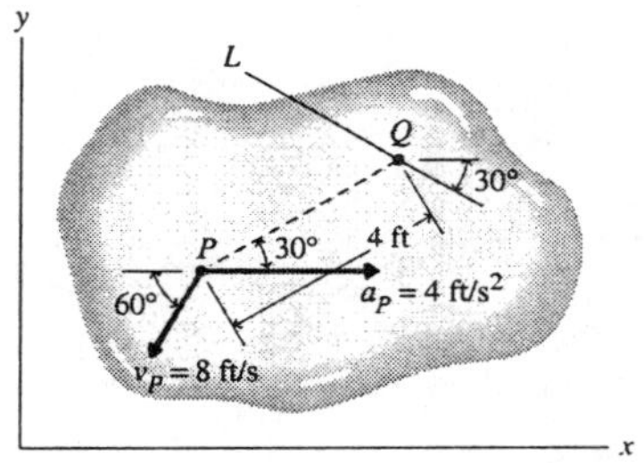

Figure a

FIND: Determine the x and y components of acceleration of point Q.

SOLUTION:

Consider the diagram shown in Fig. b.

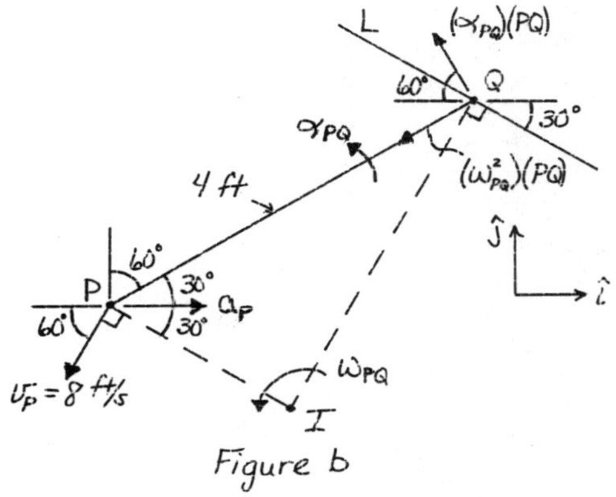

Figure b

To determine ω_{PQ}, we can locate the instantaneous center I, noting that the velocity vector of Q lies along line L.

By Fig. b,

$$PI = 4\cos 60° = 2 \text{ ft}.$$

Therefore,

$$\omega_{PQ} = \frac{v_P}{PI} = \frac{8}{2} = 4 \text{ rad/s}$$

Then, by Fig. a and kinematics,

$$\bar{a}_Q = \bar{a}_P + \bar{a}_{Q/P} \qquad (a)$$

where by Fig. b,

$$\bar{a}_P = 4\,\hat{\imath} \qquad (b)$$

$$\bar{a}_{Q/P} = (\omega_{PQ}^2)(PQ)[-(\cos 30°)\hat{\imath} - (\sin 30°)\hat{\jmath}] + (\alpha_{PQ})(PQ)[-(\cos 60°)\hat{\imath} + (\sin 60°)\hat{\jmath}]$$

or

$$\bar{a}_{Q/P} = -(55.424 + 2\alpha_{PQ})\hat{\imath} - (32 - 3.464\alpha_{PQ})\hat{\jmath} \qquad (c)$$

Assuming that $\bar{a}_Q$ acts up and to the left along line L, we have

$$\bar{a}_Q = a_Q(-\cos 30°)\hat{\imath} + a_Q(\sin 30°)\hat{\jmath}$$

or

$$\bar{a}_Q = -0.866a_Q\,\hat{\imath} + 0.500a_Q\,\hat{\jmath} \qquad (d)$$

By Eqs. (a), (b), (c), and (d), we have

$$-0.866a_Q\,\hat{\imath} + 0.500a_Q\,\hat{\jmath} = -(51.424 + 2\alpha_{PQ})\hat{\imath} - (32 - 3.464\alpha_{PQ})\hat{\jmath}$$

Equating $\hat{\imath}$ and $\hat{\jmath}$ terms on the left and right, we obtain

$$0.866a_Q - 2\alpha_{PQ} = 51.424 \qquad (e)$$
$$0.500a_Q - 3.464\alpha_{PQ} = -32$$

Then, by Cramer's Rule (Appendix A) and Eqs. (e),

$$a_Q = \frac{\begin{vmatrix} 51.424 & -2 \\ -32 & -3.464 \end{vmatrix}}{\begin{vmatrix} 0.866 & -2 \\ 0.500 & -3.464 \end{vmatrix}} = \frac{-242.13}{-2} = 121.1$$

Hence,

$$\bar{a}_Q = (a_{Qx}, a_{Qy}) = 121.1(-0.866\hat{\imath} + 0.500\hat{\jmath})$$

or $a_{Qx} = -104.88 \text{ ft/s}^2$; $a_{Qy} = 60.55 \text{ ft/s}^2$

17.67

GIVEN: Figure a, with $r = 75$ mm, $x = 100$ mm, $\theta = 90°$, $\dot{\theta} = 4$ rad/s, and $\ddot{\theta} = 12$ rad/s².

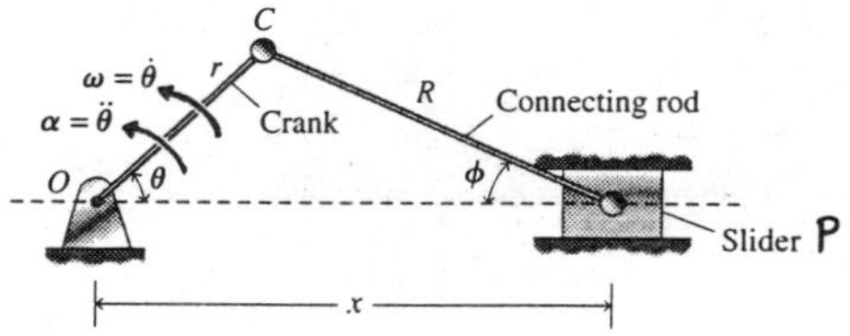

Figure a

FIND:

(a) the velocity and acceleration of the slider.

(b) the angular velocity and angular acceleration of the connecting rod.

(continued)

17.67 (cont.)

SOLUTION:

(a) Consider the diagram of the crank-slider mechanism for $\theta = 90^\circ$ (Fig. b)

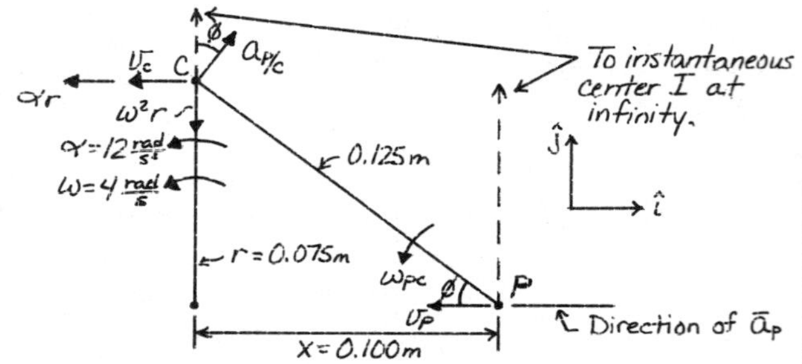

Figure b

Since I is infinitely distant (Fig. b), $\omega_{PC} = 0$, and $\bar{v}_{P/C}$ is zero. Hence, by Fig. b and kinematics,

$$\bar{v}_P = \bar{v}_C + \bar{v}_{P/C} = \bar{v}_C = -r\omega\,\hat{i}$$

or $\bar{v}_P = -0.3\ m/s$

Also, by Fig. b and kinematics

$$\bar{a}_P = \bar{a}_C + \bar{a}_{P/C}$$

$$\bar{a}_P = a_P(\pm)\,\hat{i} \qquad \text{(a)}$$
$$\bar{a}_C = -\alpha r\,\hat{i} - \omega^2 r\,\hat{j}$$
$$\bar{a}_{P/C} = a_{P/C}(\sin\phi\,\hat{i} + \cos\phi\,\hat{j})$$
$$\sin\phi = 0.60, \quad \cos\phi = 0.80$$

Hence, by Eqs. (a),

$$\pm a_P\,\hat{i} = -0.9\hat{i} - 1.2\hat{j} + 0.60 a_{P/C}\,\hat{i} + 0.8 a_{P/C}\,\hat{j} \qquad \text{(b)}$$

So,

$$\pm a_P\hat{i} = -0.90\hat{i} + 0.60\,a_{P/C}\,\hat{i}$$
$$0 = -1.2\hat{j} + 0.80\,a_{P/C}\,\hat{j} \qquad \text{(c)}$$

Equations (c) yield

$$a_{P/C} = 1.5\ m/s^2 \qquad \text{(d)}$$
$$a_P = 0 \quad \text{or} \quad \bar{a}_P = 0 \qquad \text{(e)}$$

(b) As shown in part a, $\omega_{PC} = 0.$

By Eq. (d) and Fig. b,

$$a_{P/C} = \alpha_{P/C}(\overline{PC})$$

or $\alpha_{PC} = \dfrac{a_{P/C}}{\overline{PC}} = \dfrac{1.5}{0.125} = 12\ rad/s^2$

17.68

GIVEN: Figure a, with r = 3 in., R = 12 in., and $\omega = 10\ rad/s$. Each second increases uniformly by 10 rad/s.

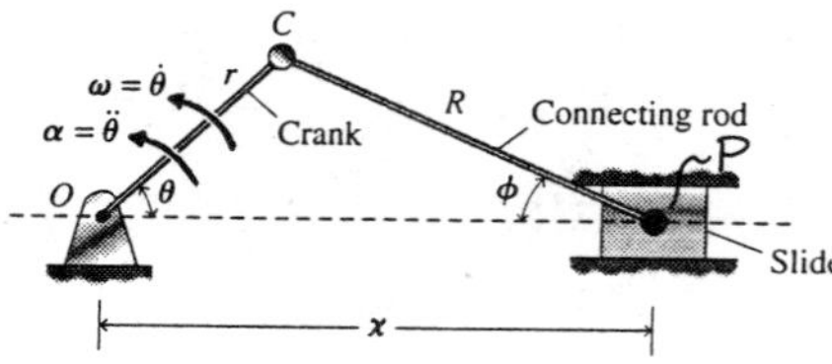

Figure a

FIND:

(a) the velocity and acceleration of the slider, for $\theta = \phi = 0$.

(b) the angular velocity and angular acceleration of the connecting rod, for $\theta = \phi = 0$ (Fig. b).

Solution:

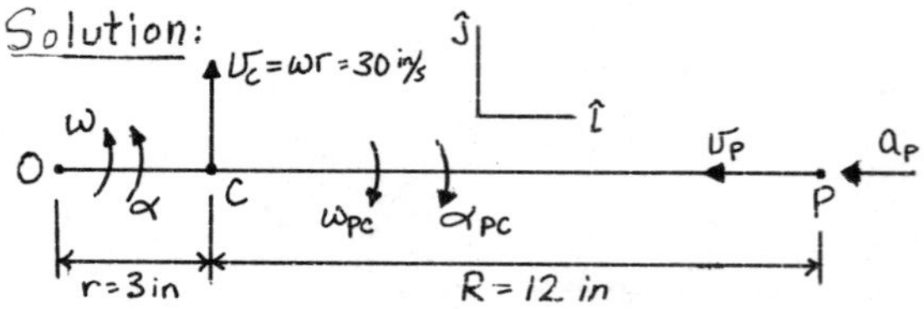

Figure b

(a) By Fig. b and kinematics,

$$\bar{v}_P = \bar{v}_C + \bar{v}_{P/C} \qquad \text{(a)}$$

where

$$\bar{v}_C = 30\,\hat{j} \quad [in/s]$$
$$\bar{v}_{P/C} = -12\,\omega_{PC}\,\hat{j} \qquad \text{(b)}$$
$$\bar{v}_P = -v_P\,\hat{i}$$

Equations (a) and (b) yield

$$\omega_{PC} = 2.5\ rad/s, \ \text{cw} \qquad \text{(c)}$$
$$\bar{v}_P = 0 \qquad \text{(d)}$$

Also, by Fig. b and kinematics,

$$\bar{a}_P = \bar{a}_C + \bar{a}_{P/C} \qquad \text{(e)}$$

where,

$$\left.\begin{aligned} &\bar{a}_P = -a_P\,\hat{i} \\ &\bar{a}_C = -\omega^2 r\,\hat{i} + \alpha r\,\hat{j}; \\ &\alpha = \dot{\omega} = 10\ rad/s^2 \\ &\bar{a}_{P/C} = -\omega_{PC}^2 R\,\hat{i} - \alpha_{PC} R\,\hat{j} \end{aligned}\right\} \qquad \text{(f)}$$

Equations (c), (e), and (f), with r = 3 in, R = 12 in, and $\omega = 10\ rad/s$, yield

$$-a_P\,\hat{i} = -300\hat{i} + 30\hat{j} - 75\hat{i} - 12\alpha_{PC}\,\hat{j}$$

or $a_P = 375\ in/s^2$; $\bar{a}_P = -375\,\hat{i}\ [in/s^2]$ (g)

$$\alpha_{PC} = \frac{30}{12} = 2.5\ rad/s^2\ \text{cw} \qquad \text{(h)}$$

(b) By Eqs. (c) and (h) of part a,

$\omega_{PC} = 2.5\ rad/s$, cw ; $\alpha_{PC} = 2.5\ rad/s^2$, cw

17.69

GIVEN: An automatic control mechanism consisting of gears A and B (Fig. a). Gear A rotates counterclockwise about a fixed shaft with constant angular speed 100 rad/s. Gear B rotates clockwise with constant absolute angular speed of 200 rad/s about a pin fixed to the rim of A.

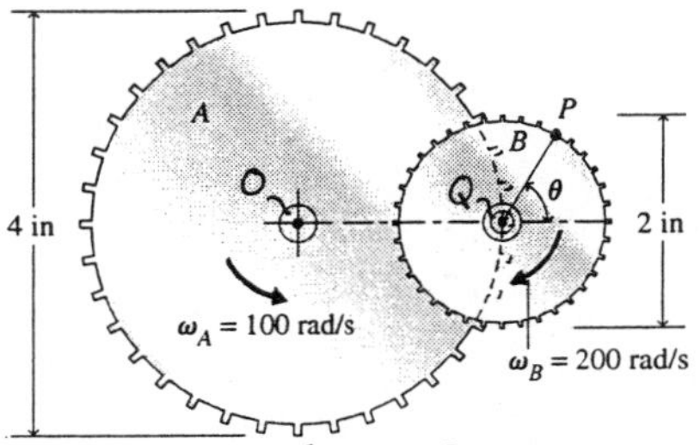

Figure a

FIND:

(a) the angle θ for which the speed of particle P is a maximum.

(b) the angle θ for which the speed of particle P is a minimum.

(c) the angle θ for which the magnitude of the acceleration of particle P is a maximum.

(d) the angle θ for which the magnitude of the acceleration of particle P is a minimum.

(e) the values of the speed in parts a and b, and the magnitudes of the acceleration in parts c and d.

SOLUTION:

Consider the diagram of the components of velocity of point P (Fig. b), for the general angle θ. By Figs. (a) and (b), and kinematics,

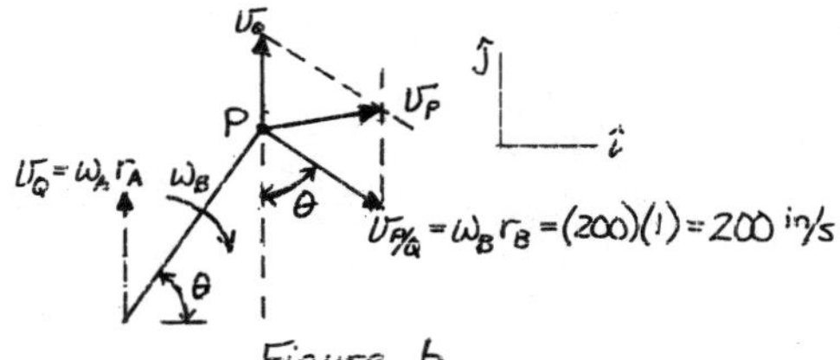

Figure b

$$\bar{v}_P = \bar{v}_Q + \bar{v}_{P/Q} \qquad (a)$$

where

$$\bar{v}_Q = \omega_A r_A \hat{j} = 200\hat{j} \ [\text{rad/s}]$$

$$\bar{v}_{P/Q} = \omega_B r_B [(\sin\theta)\hat{i} - (\cos\theta)\hat{j}] \qquad (b)$$
$$= 200[(\sin\theta)\hat{i} - (\cos\theta)\hat{j}]$$

Equations (a) and (b) yield

$$\bar{v}_P = 200[(\sin\theta)\hat{i} + (1-\cos\theta)\hat{j}]$$

or

$$v_P = 200\sqrt{2(1-\cos\theta)} \qquad (c)$$

(a) By Eq. (c), the angle θ for which v_P is a maximum is (see also Fig. b)

$\underline{\underline{\theta = 180^\circ}}$

(b) By Eq. (c), the angle θ for which v_P is a minimum is (see also Fig. b)

$\underline{\underline{\theta = 0^\circ}}$

Consider next the diagram of the components of acceleration of point P for the general angle θ (Fig. c).

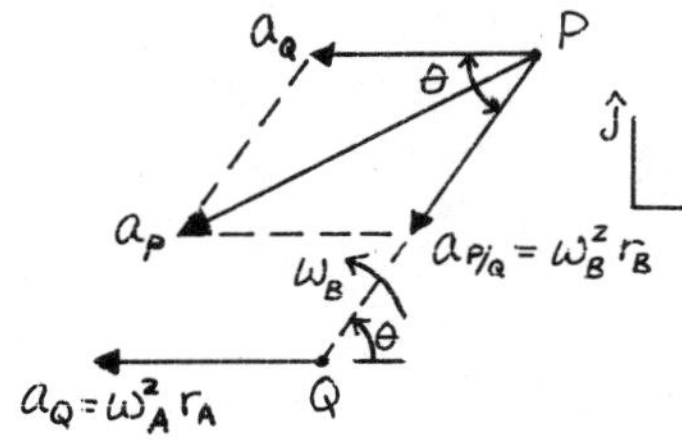

Figure c

By Fig. c and kinematics,

$$\bar{a}_P = \bar{a}_Q + \bar{a}_{P/Q} \qquad (d)$$

where

$$\bar{a}_Q = -\omega_A^2 r_A \hat{i} = -20{,}000\hat{i} \ [\text{in/s}^2]$$

$$\bar{a}_{P/Q} = \omega_B^2 r_B [-(\cos\theta)\hat{i} - (\sin\theta)\hat{j}] \qquad (e)$$
$$= 40{,}000[-(\cos\theta)\hat{i} - (\sin\theta)\hat{j}]$$

By Eqs. (d) and (e),

$$\bar{a}_P = -20{,}000[(1+2\cos\theta)\hat{i} + 2(\sin\theta)\hat{j}]$$

or

$$a_P = 20{,}000\sqrt{5 + 4\cos\theta} \qquad (f)$$

(c) By Eq. (f), the angle for which a_P is a maximum is (see also fig. c)

$\underline{\underline{\theta = 0^\circ}}$

(d) By Eq. (f), the angle for which a_P is a minimum is (see also Fig. c)

$\underline{\underline{\theta = 180^\circ}}$

(e) By Eqs. (c) and (f), or Figs. b and c,

$\theta = 180^\circ$: $\underline{\underline{v_P = v_{P_{max}} = 400 \text{ in/s}}}$

$\theta = 0^\circ$: $\underline{\underline{v_P = v_{P_{min}} = 0}}$

$\theta = 0^\circ$: $\underline{\underline{a_P = a_{P_{max}} = 60{,}000 \text{ in/s}^2}}$

$\theta = 180^\circ$: $\underline{\underline{a_P = a_{P_{min}} = 20{,}000 \text{ in/s}^2}}$

17.70

GIVEN: Figure a, in which rod AB is hinged to the rolling wheel at A and point B maintains contact with surface S.

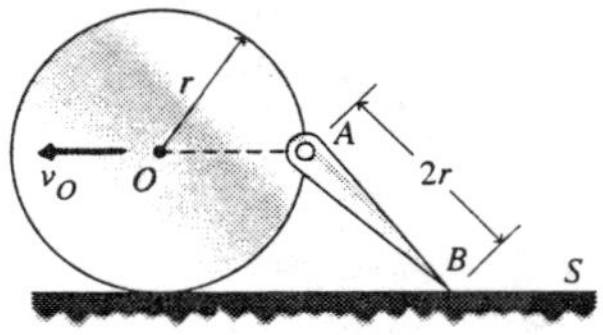

Figure a

FIND: Determine the velocity $\bar{v}_B$ and the acceleration $\bar{a}_B$ of B, for the position shown, in terms of the radius r, the magnitude a_o of the acceleration and the speed v_o of the center O of the wheel.

SOLUTION:

Consider the diagram of velocities (Fig. b).

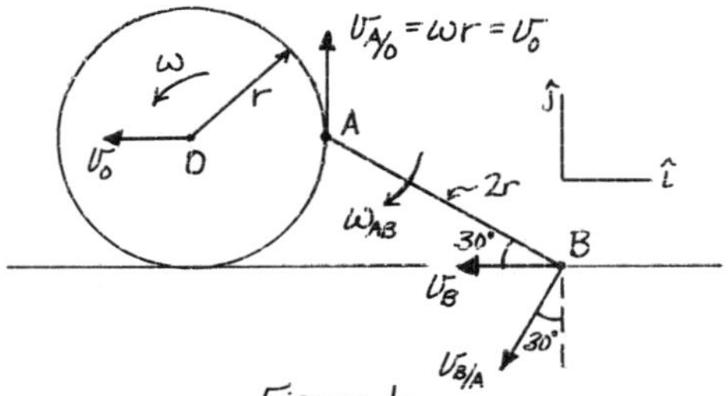

Figure b

By kinematics, $v_o = \omega r$. Therefore, by Fig. b,

$$v_{A/O} = \omega r = v_o\,, \quad v_{B/A} = \omega_{AB}(2r) \qquad (a)$$

By Fig. b and kinematics,

$$\bar{v}_B = \bar{v}_A + \bar{v}_{B/A} \qquad (b)$$

where

$$\left.\begin{aligned} \bar{v}_A &= \bar{v}_o + \bar{v}_{A/O} = -v_o\,\hat{\imath} + v_o\,\hat{\jmath} \\ \bar{v}_{B/A} &= \omega_{AB}(2r)\left(-\tfrac{1}{2}\hat{\imath} - \tfrac{\sqrt{3}}{2}\hat{\jmath}\right) \\ \bar{v}_B &= -v_B\,\hat{\imath} \end{aligned}\right\} \qquad (c)$$

Equations (b) and (c) yield

$$\omega_{AB} = \frac{v_o}{\sqrt{3}\,r} \qquad (d)$$

$$\underline{\underline{v_B = \left(1 + \tfrac{1}{\sqrt{3}}\right)v_o = 1.5774\,v_o}}\,;\ \bar{v}_B = -v_B\,\hat{\imath}$$

The diagram of accelerations is shown in Fig. c. By kinematics, $a_o = \alpha r$. Therefore, by Fig. c, with the first of Eqs. (a) and (d),

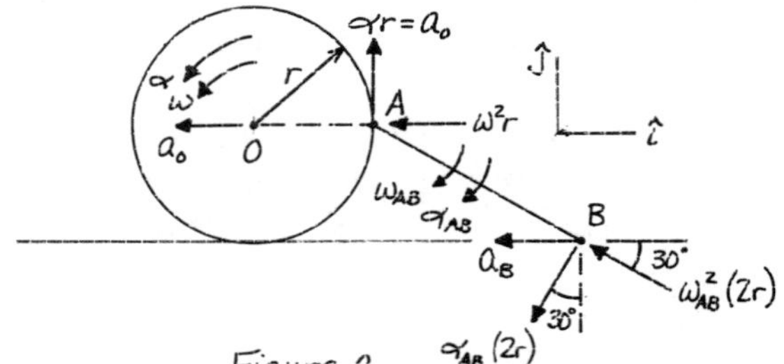

Figure c

$$\left.\begin{aligned} \bar{a}_{A/O} &= -\omega^2 r\,\hat{\imath} + \alpha r\,\hat{\jmath} = -\frac{v_o^2}{r}\hat{\imath} + a_o\,\hat{\jmath} \\ \bar{a}_{B/A} &= \omega_{AB}^2(2r)\left(-\tfrac{\sqrt{3}}{2}\hat{\imath} + \tfrac{1}{2}\hat{\jmath}\right) \\ &\quad + \alpha_{AB}(2r)\left(-\tfrac{1}{2}\hat{\imath} - \tfrac{\sqrt{3}}{2}\hat{\jmath}\right) \\ &= \frac{v_o^2}{r}\left(-\tfrac{\sqrt{3}}{3}\hat{\imath} + \tfrac{1}{3}\hat{\jmath}\right) - \alpha_{AB}r\left(\hat{\imath} + \sqrt{3}\,\hat{\jmath}\right) \end{aligned}\right\} \qquad (e)$$

Also, by kinematics,

$$\bar{a}_B = \bar{a}_A + \bar{a}_{B/A} \qquad (f)$$

where

$$\bar{a}_A = \bar{a}_o + \bar{a}_{A/O} \qquad (g)$$

Hence, Eqs. (e), (f), and (g) yield, with Fig. c,

$$\begin{aligned} -a_B\,\hat{\imath} &= -\left[a_o + \frac{v_o^2}{r}\left(1 + \frac{\sqrt{3}}{3}\right) + r\alpha_{AB}\right]\hat{\imath} \\ &\quad + \left(a_o + \frac{1}{3}\frac{v_o^2}{r} - \sqrt{3}\,r\alpha_{AB}\right)\hat{\jmath} \end{aligned}$$

or

$$r\alpha_{AB} = \frac{1}{\sqrt{3}}\left(a_o + \frac{v_o^2}{3r}\right)$$

$$\begin{aligned} a_B &= \left(1 + \tfrac{1}{\sqrt{3}}\right)a_o + \left(1 + \tfrac{4\sqrt{3}}{9}\right)\frac{v_o^2}{r} \\ &= 1.5774\,a_o + 1.7698\,\frac{v_o^2}{r} \end{aligned} \qquad (h)$$

So, $\underline{\underline{\bar{a}_B = -\left(1.5774\,a_o + 1.7698\,\frac{v_o^2}{r}\right)\hat{\imath}}}$

17.71

GIVEN: The equations

$$\begin{aligned} \ddot{x} &= \ddot{x}_o - r\ddot{\theta}\sin\theta - r(\dot{\theta})^2\cos\theta \\ \ddot{y} &= \ddot{y}_o + r\ddot{\theta}\cos\theta - r(\dot{\theta})^2\sin\theta \end{aligned} \qquad (a)$$

FIND: Show that, by setting $\ddot{x} = \ddot{y} = 0$, the polar coordinates (r, θ) of the instantaneous center of acceleration of Fig. a are determined by the equations

$$r = \sqrt{\frac{\ddot{x}_o^2 + \ddot{y}_o^2}{\alpha^2 + \omega^2}}\,; \quad \tan\theta = \frac{\alpha\ddot{x}_o + \omega^2\ddot{y}_o}{\omega^2\ddot{x}_o - \alpha\ddot{y}_o} \qquad (b)$$

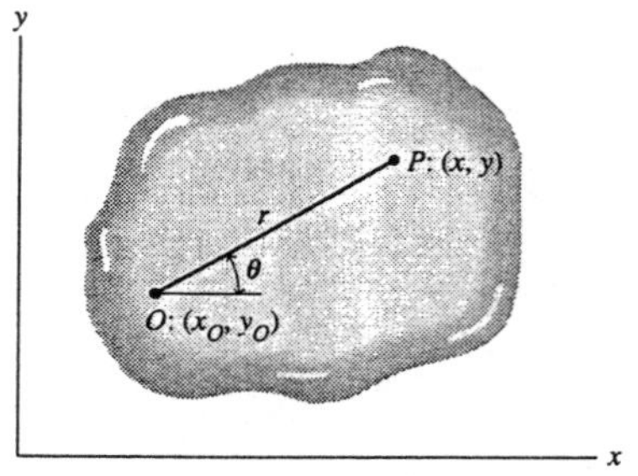

Figure a

(continued)

17.71 (cont.)

SOLUTION:

Since (x,y) are the coordinates of point P (Fig. a), by setting $\ddot{x}=\ddot{y}=0$, we establish point P as the instantaneous center of acceleration. Then, Eqs. (a) may be written as

$$\ddot{x}_0 = r\ddot{\theta}\sin\theta + r(\dot{\theta})^2\cos\theta$$
$$\ddot{y}_0 = -r\ddot{\theta}\cos\theta + r(\dot{\theta})^2\sin\theta \qquad \text{(c)}$$

Multiply the first of Eqs. (c) by $\cos\theta$ and the second by $\sin\theta$ and add to get (with $\dot{\theta}=\omega$)

$$\ddot{x}_0\cos\theta + \ddot{y}_0\sin\theta = r\omega^2 \qquad \text{(d)}$$

Next, multiply the first of Eqs (c) by $\sin\theta$ and the second by $-\cos\theta$ and add to get (with $\alpha=\ddot{\theta}$)

$$\ddot{x}_0\sin\theta - \ddot{y}_0\cos\theta = r\alpha \qquad \text{(e)}$$

Squaring Eqs. (d) and (e), adding and solving for r, we obtain

$$r = \sqrt{\frac{(\ddot{x}_0)^2+(\ddot{y}_0)^2}{\alpha^2+\omega^2}} \qquad \text{(f)}$$

Dividing Eqs. (d) and (e) by $\cos\theta$, dividing the first of the resulting equations by the second and solving for $\tan\theta = \sin\theta/\cos\theta$, we find

$$\tan\theta = \frac{\alpha\ddot{x}_0 + \omega^2\ddot{y}_0}{\omega^2\ddot{x}_0 - \alpha\ddot{y}_0} \qquad \text{(g)}$$

Equations (f) and (g) correspond to Eqs. (b). They determine the location of the instantaneous center (point P), (Fig. b), in terms of polar coordinates (r,θ) relative to point O.

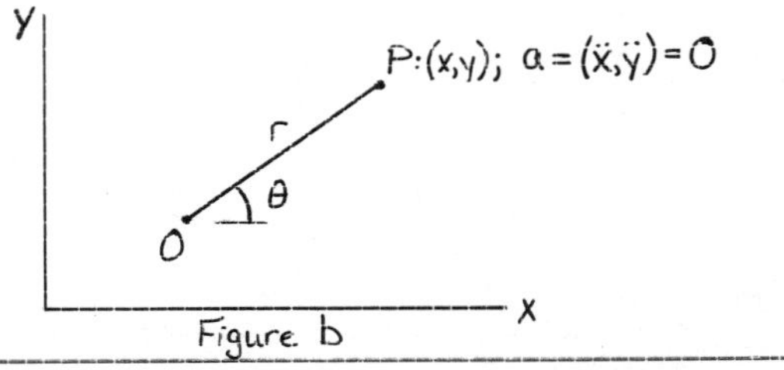

Figure b

17.72

GIVEN: In the mechanism shown in Fig. a, the crank OQ rotates counterclockwise with a constant angular speed ω.

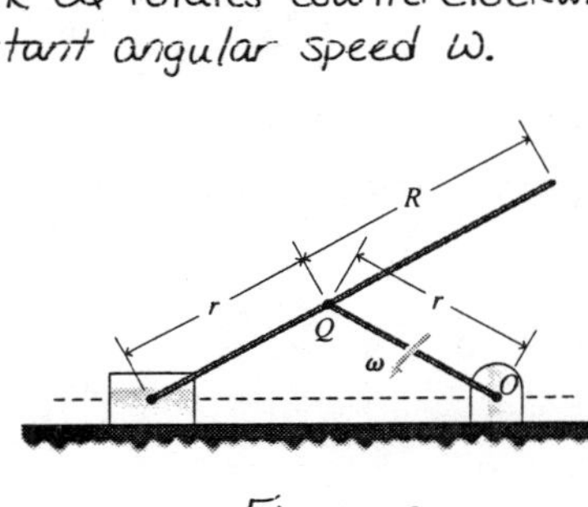

Figure a

FIND:

(a) Represent the acceleration of P by two components, one along line PQ and the other parallel to line OQ.

(b) Show these components with their correct senses on a diagram, and determine their magnitudes in terms of r, R, and ω.

(c) Show that the acceleration vector of P lies on the line OP.

SOLUTION:

(a) Consider the diagram shown in Fig. b.

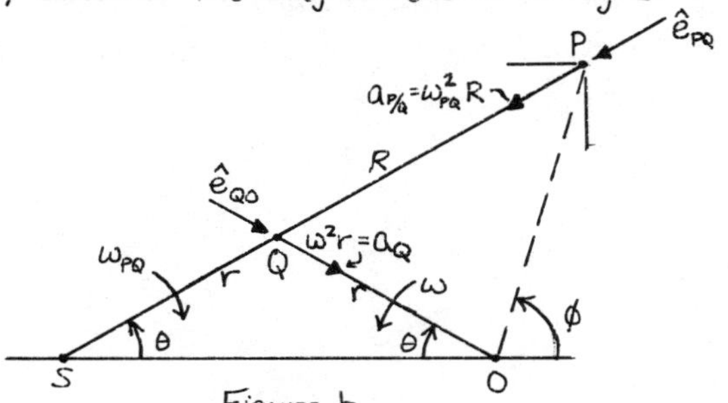

Figure b

By Fig. b and kinematics,

$$\omega = -\dot{\theta} = \omega_{PQ} \qquad \text{(a)}$$

Therefore, by Fig. b, Eq. (a), and kinematics

$$\bar{a}_P = \bar{a}_Q + \bar{a}_{P/Q} \qquad \text{(b)}$$

where since ω is constant,

$$\bar{a}_Q = \omega^2 r\,\hat{e}_{QO}$$
$$\bar{a}_{P/Q} = \omega^2 R\,\hat{e}_{PQ} \qquad \text{(c)}$$

and $\hat{e}_{QO}$, $\hat{e}_{PQ}$ are unit vectors along lines OQ and PQ, respectively. So, by Eqs. (b) and (c),

$$\underline{\bar{a}_P = \omega^2 r\,\hat{e}_{QO} + \omega^2 R\,\hat{e}_{PQ}} \qquad \text{(d)}$$

where the component $\omega^2 r\,\hat{e}_{QO}$ is parallel to line OQ and the component $\omega^2 R\,\hat{e}_{PQ}$ is parallel to line PQ.

(b) The components of $\bar{a}_P$ are shown in Fig. c, with their correct senses and magnitudes.

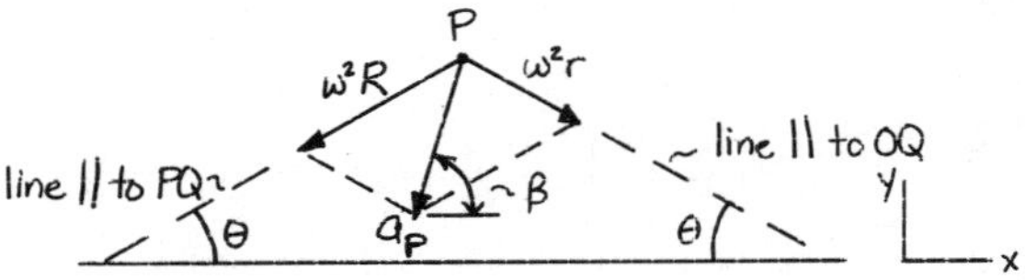

Figure c

(c) By Fig. c,

$$a_x = -\omega^2 R\cos\theta + \omega^2 r\cos\theta$$
$$a_y = -\omega^2 R\sin\theta - \omega^2 r\sin\theta$$

(continued)

17.72 (cont.)

$$\tan\beta = \frac{a_y}{a_x} = \frac{R+r}{R-r}\tan\theta \qquad (e)$$

Also, by Fig. b,

$$\tan\phi = \frac{(R+r)\sin\theta}{(R+r)\cos\theta - 2r\cos\theta} = \frac{R+r}{R-r}\tan\theta \quad (f)$$

Hence, by Eqs. (e) and (f), $\phi = \beta$.

In words, the acceleration vector $\bar{a}_P$ lies on the line OP (Fig. b).

17.73

GIVEN: The Scott-Russell mechanism (Fig. a) is designed to produce straight-line motion of point A.

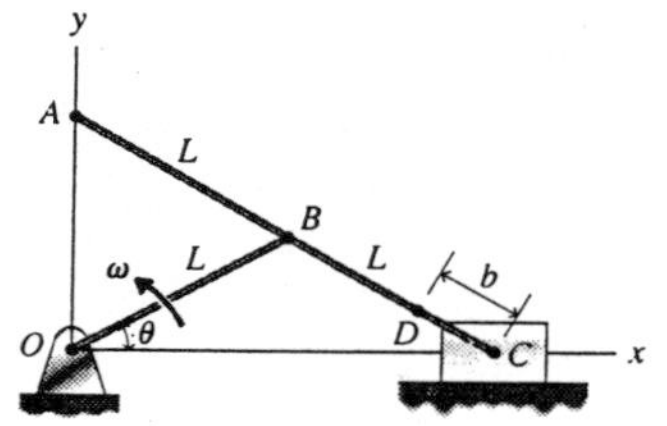

Figure a

FIND:

(a) Show that point A undergoes straight-line motion as crank OB undergoes angular velocity ω.

(b) Determine the angular velocity of rod ABC.

(c) Derive formulas for the xy coordinates of point D in terms of L, b, and θ. Plot the path of D in the xy plane as a function of θ for $L = 150\ mm$, $b = 75\ mm$ and for $L = 150\ mm$, $b = 225\ mm$.

(d) Determine L and ω that meet design specifications that the maximum acceleration of any point in the mechanism is $< 2\ m/s^2$ and the maximum velocity is $< 2\ m/s$ and that L and ω lie in the ranges $75\ mm \le L \le 150\ mm$, $\pi\ rad/s \le \omega \le 1.5\pi\ rad/s$.

SOLUTION:

(a) By Fig. b and geometry,

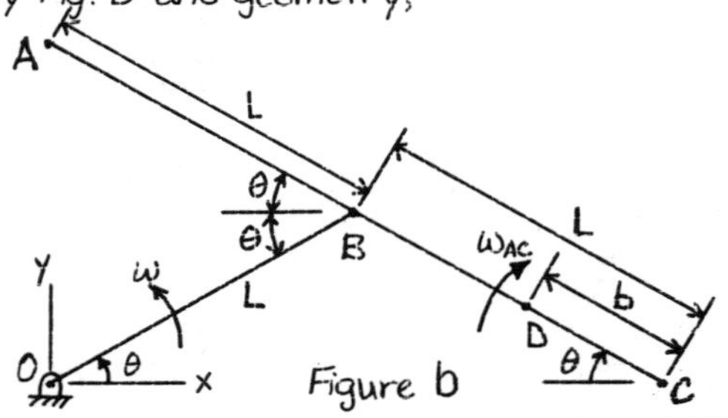

Figure b

$$x_A = \overline{OB}\cos\theta - \overline{BA}\cos\theta$$

$$\therefore\ x_A = L\cos\theta - L\cos\theta = 0 \qquad (a)$$

$$y_A = L\sin\theta + L\sin\theta = 2L\sin\theta$$

Hence, point A moves in a straight line along the y axis and oscillates in the range $-2L \le y \le 2L$.

(b) By Fig. b and geometry,

$$\omega = \dot{\theta} = \omega_{AC}$$

So, $\quad \omega_{AC} = \omega \qquad (b)$

(c) By Fig. b and geometry the xy coordinates of point D are

$$x_D = 2L\cos\theta - b\cos\theta = (2L-b)\cos\theta$$

$$y_D = b\sin\theta \qquad (c)$$

See the plots in Fig. c for ($L = 150\ mm$, $b = 75\ mm$) and ($L = 150\ mm$, $b = 225\ mm$).

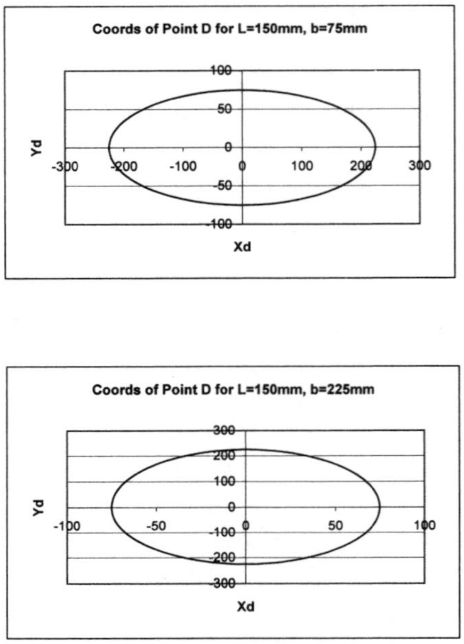

Figure c

(d) The maximum velocity and maximum acceleration occur at point A (or at point C); see Fig. b. Hence, by Eq. (a),

$$v_A = \dot{y}_A = 2L\dot{\theta}\cos\theta = 2L\omega\cos\theta$$

$$a_A = \ddot{y}_A = -2L(\dot{\theta})^2\sin\theta = -2L\omega^2\sin\theta$$

Therefore,

$$v_{max} = 2L\omega$$

$$a_{max} = 2L\omega^2 \qquad (d)$$

(continued)

17.73 (cont.)

With the design requirements, Eqs. (d) yield

$$2L\omega < 2 \text{ m/s} \qquad (e)$$
$$2L\omega^2 < 2 \text{ m/s}^2 \qquad (f)$$

where

$$0.075\text{ m} < L < 0.15\text{ m}$$
$$\pi \text{ rad/s} < \omega < 1.5\pi \text{ rad/s} \qquad (g)$$

By Eqs. (e) and (f),

$$L\omega < 1 \qquad (h)$$
$$L\omega^2 < 1 \qquad (i)$$

Equation (i) is the limiting case. By Eq. (i)

$$\omega^2 < \frac{1}{L} \qquad (j)$$

For $L = 0.075$ m, Eq. (j) yields ω

$$\omega < 3.65 \text{ rad/s} \qquad (k)$$

Since $\pi \le 3.65 \le 1.5\pi$, this value of ω [Eq. (k)], with $L = 0.075$ is acceptable. Also, for $L = 0.150$ m, Eq. (j) yields $\omega < 2.58$ rad/s, which lies outside the acceptable range for ω.

Note also for an intermediate value of L (say $L = 0.10$ m), Eq. (j) yields

$$\omega < 3.16 \text{ rad/s} \qquad (\ell)$$

where $\pi < 3.16 < 1.5\pi$. So, lets take $L = 0.09$ m, as a length that yields $\omega < 3.33$ rad/s. This choice places L and ω safely within their acceptable range.

17.74

GIVEN: The crank OA (Fig. a) rotates with constant angular velocity ω_{OA}.

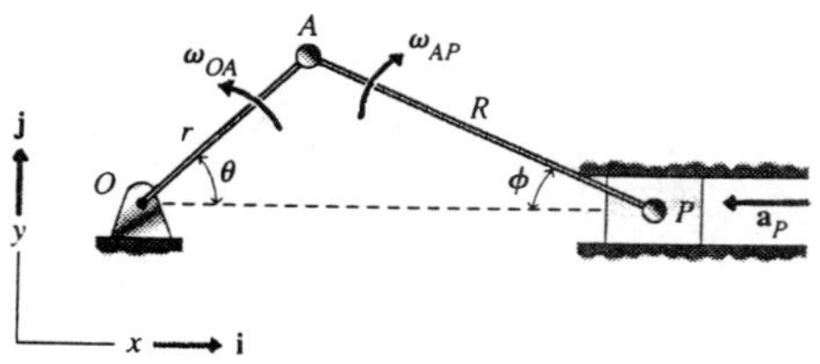

Figure a

FIND:

(a) Show that the acceleration a_P of slider P is

$$\bar{a}_P = -r\omega_{OA}^2\left[\cos\theta + \frac{\frac{R^2}{r^2}(\cos^2\theta - \sin^2\theta) + \sin^4\theta}{\left(\frac{R^2}{r^2} - \sin^2\theta\right)^{3/2}}\right]\hat{\imath}$$

(b) On one diagram, plot the ratio $a_P/r\omega_{OA}^2$ as a function of θ, for $0 \le \theta \le 360°$, for $R/r = 2, 3, 4, 5$, and 6.

(c) Select a range of values of R/r that satisfies the requirement $1.0 \le a_P/r\omega_{OA}^2 \le 1.25$, and discuss the practicality of the values of R/r in the range.

SOLUTION:

(a) Consider the velocities shown in Fig. b and the accelerations shown in Fig. c.

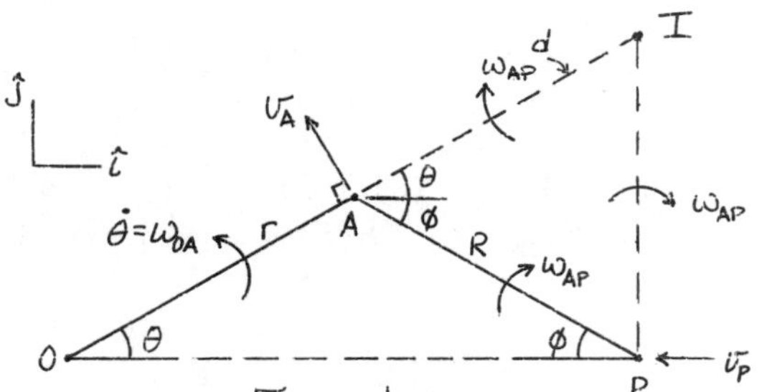

Figure b

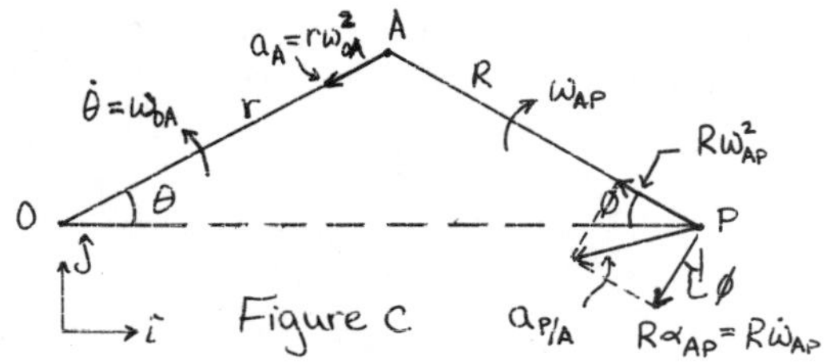

Figure c

By geometry and Fig. b,

$$d\cos\theta = R\cos\phi \qquad (a)$$
$$r\sin\theta = R\sin\phi \qquad (b)$$

By Eq. (b),

$$\sin\phi = \frac{r}{R}\sin\theta \qquad (c)$$

Therefore,

$$\cos\phi = \sqrt{1 - \sin^2\phi} = \sqrt{1 - \frac{r^2}{R^2}\sin^2\theta} \qquad (d)$$

By Fig. b and kinematics, where I is the instantaneous center of velocity,

$$v_A = r\omega_{OA} = d\,\omega_{AP}$$

or

$$\omega_{AP} = \frac{r\omega_{OA}}{d} \qquad (e)$$

where by Eqs. (a) and (d),

$$d = \frac{R\cos\phi}{\cos\theta} = \frac{r}{\cos\theta}\sqrt{\frac{R^2}{r^2} - \sin^2\theta} \qquad (f)$$

Then, by Eqs. (e) and (f),

$$\omega_{AP} = \omega_{OA}\left[\frac{\cos\theta}{\sqrt{\frac{R^2}{r^2} - \sin^2\theta}}\right] \qquad (g)$$

The angular acceleration of the connecting rod AP is $\alpha_{AP} = \dot{\omega}_{AP}$. Hence, By Eq. (g),

$$\alpha_{AP} = -\omega_{OA}^2\left[\frac{\sin\theta}{\sqrt{\frac{R^2}{r^2} - \sin^2\theta}} - \frac{\sin\theta\cos^2\theta}{\left(\frac{R^2}{r^2} - \sin^2\theta\right)^{3/2}}\right]$$

or

$$\alpha_{AP} = -\omega_{OA}^2\left[\frac{(\sin\theta)\left(\frac{R^2}{r^2} - \sin^2\theta\right) - \sin\theta\cos^2\theta}{\left(\frac{R^2}{r^2} - \sin^2\theta\right)^{3/2}}\right] \qquad (h)$$

where $\omega_{OA} = \dot{\theta} =$ constant.

(continued)

17.74 (cont.)

Now by Fig. c and kinematics,

$$\bar{a}_P = \bar{a}_A + \bar{a}_{P/A} \qquad (i)$$

where

$$\bar{a}_A = -r\omega_{OA}^2\left[(\cos\theta)\hat{\imath} + (\sin\theta)\hat{\jmath}\right] \qquad (j)$$

$$\bar{a}_{P/A} = R\omega_{AP}^2\left[-(\cos\phi)\hat{\imath} + (\sin\phi)\hat{\jmath}\right] - R\alpha_{AP}\left[(\sin\phi)\hat{\imath} + (\cos\phi)\hat{\jmath}\right] \qquad (k)$$

With Eqs. (c) and (d), Eq. (k) is expressed in terms of θ. Then, Eqs. (i), (j), and (k) yield after some algebra,

$$\bar{a}_P = -r\omega_{OA}^2\left[\cos\theta + \frac{\frac{R^2}{r^2}(\cos^2\theta - \sin^2\theta + \sin^4\theta)}{\left(\frac{R^2}{r^2} - \sin^2\theta\right)^{3/2}}\right]\hat{\imath} \qquad (\ell)$$

Equation (ℓ) can be used to plot (see Fig. d),

$$\frac{a_P}{r\omega_{OA}^2} = \cos\theta + \frac{\frac{R^2}{r^2}(\cos^2\theta - \sin^2\theta) + \sin^4\theta}{\left(\frac{R^2}{r^2} - \sin^2\theta\right)^{3/2}} \qquad (m)$$

for $R/r = 2, 3, 4, 5, 6$.

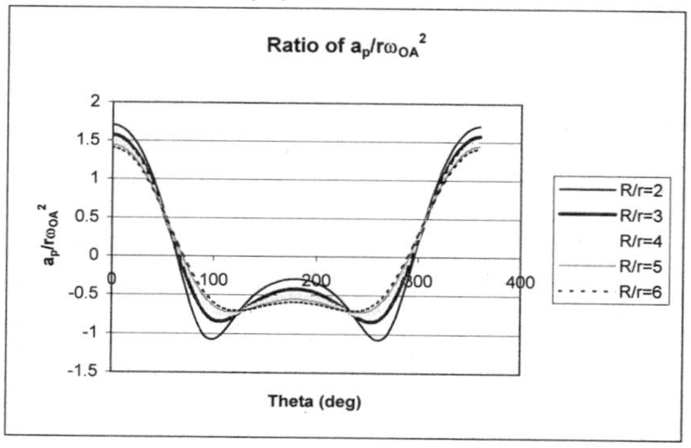

Figure d

(c) By Eq. (m), $[a_P/r\omega_{OA}^2]_{max}$ occurs when $d[a_P/r\omega_{OA}^2]/d\theta = 0$. Then $\sin\theta = 0$ or $\theta = 0$.

Therefore, with $\sin\theta = 0$, Eq. (m) yields

$$\left[\frac{a_P}{r\omega_{OA}^2}\right]_{max} = 1 + \frac{r}{R} \qquad (n)$$

For $1.0 \le a_P/r\omega_{OA}^2 \le 1.25$, Eq. (n) yields (see also Fig. d)

$$1.0 \le 1 + \frac{r}{R} \le 1.25$$

or

$$R \ge 4r$$

or

$$4 \le \frac{R}{r} \le \infty \qquad (o)$$

Based upon space limitations, one must consider the appropriate length of r, and hence, of R. Then, by Eq. (o), it is apparent that the ratio R/r should be somewhat larger than 4, but not so large that space limitations are violated.

17.75

GIVEN: A particle of a deformable body undergoes the rotation $\theta_x = 1°$, $\theta_y = 2°$, and $\theta_z = -0.5°$, where θ_x, θ_y, and θ_z are the (x, y, z) projections of the vector representing the angular displacement.

FIND:
(a) the magnitude of the angular displacement (°)
(b) the direction angles of the axis about which the particle rotates.

SOLUTION:

(a) Since the angular displacements about concurrent axes xyz are very small, they combine approximately by vector addition. Hence, the magnitude of the angular displacement is

$$\theta = \sqrt{\theta_x^2 + \theta_y^2 + \theta_z^2} = \sqrt{1^2 + 2^2 + (.5)^2} = \frac{\sqrt{21}}{2} \text{ degrees}$$

or $\underline{\theta = 2.291°}$

(b) Therefore, the direction cosines of the rotations are

$$\cos\alpha = \frac{\theta_x}{\theta} = \frac{1}{2.291} = 0.4364$$

$$\cos\beta = \frac{\theta_y}{\theta} = \frac{2}{2.291} = 0.8729$$

$$\cos\gamma = \frac{\theta_z}{\theta} = -\frac{0.5}{2.291} = -0.2182$$

Hence, $\underline{\alpha = 64.1°,\ \beta = 29.2°,\ \gamma = 102.6°}$

17.76

GIVEN: In Example 17.18, the workman decided to try pulling the cable from the top of the spool; see Fig. a.

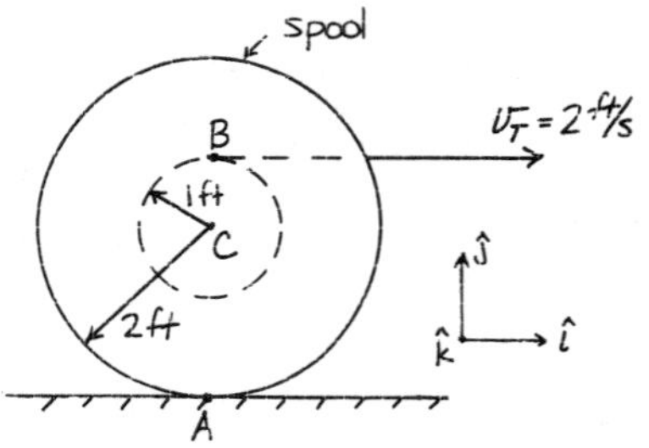

Figure a

FIND: Assuming the spool rolls without slipping, show that the cable is unwound under these conditions.

SOLUTION:

Since the spool rolls, the velocity of Point A is zero (A is the instantaneous center), and the velocity of point P is (Fig. a)

$$\bar{v}_B = 2\hat{\imath} \quad [\text{ft/s}] \qquad (a)$$

(continued)

17.76 (cont.)

Also, by kinematics (Fig. b)

$$\bar{v}_B = \bar{\omega} \times \bar{r}_B \qquad \text{(b)}$$

Where

$$\bar{\omega} = -\omega \hat{k}$$
$$\bar{r}_B = 3\hat{j} \qquad \text{(c)}$$

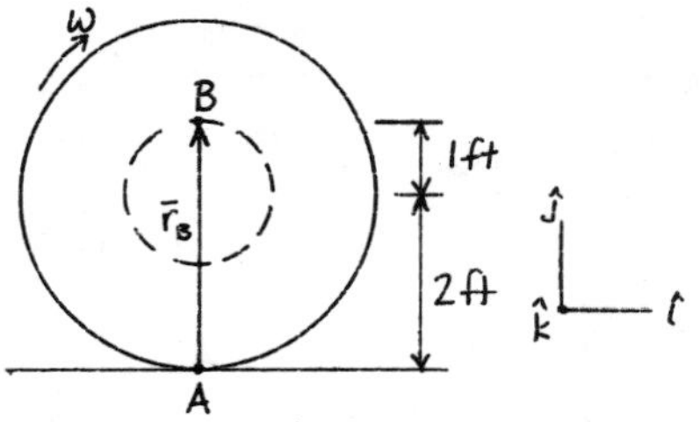

Figure b

By Eqs. (a), (b), and (c),

$$2\hat{i} = -3\omega\, \hat{k}\times\hat{j} = 3\omega\, \hat{i}$$

or

$$\omega = 2/3 \text{ rad/s} \qquad \text{(d)}$$

Similarly, the velocity of point C (Fig. a) is

$$\bar{v}_C = \bar{\omega}\times\bar{r}_C = -\omega\,\hat{k}\times 2\hat{j} = 4/3\ \hat{i}\ [\text{ft/s}] \qquad \text{(e)}$$

By Eqs. (a) and (e)

$$v_B > v_C$$

or

$$2 > 1.3\bar{3}\ [\text{ft/s}]$$

In words, the spool rolls to the right at a speed slower than the truck. Consequently, <u>the cable will unwind</u>.

17.77

GIVEN: A rigid body is subjected to a small angular displacement θ about axis A with the direction of a unit vector $\hat{n}$. Let vector $\hat{r}$ be a vector from any point O on axis A to a particle P of the body.

FIND:

(a) Show that the displacement vector $\bar{q}$ of particle P is approximated by the vector equation $\bar{q} = \bar{\theta}\times\bar{r}$; $\bar{\theta} = \theta\hat{n}$

(b) Explain why this approximation is invalid if θ is large.

SOLUTION:

(a) Consider Fig. a in which axis A, and vectors $\bar{q}$, $\hat{n}$, and $\bar{r}$ are shown. By Fig. a,

$$|\bar{q}| = 2r(\sin\phi)(\sin\theta/2).$$

If θ is small, $\sin\frac{\theta}{2} \approx \frac{\theta}{2}$. So, for small θ,

$$|\bar{q}| \approx \theta r \sin\phi = |\bar{\theta}\times\bar{r}|;\ \bar{\theta} = \theta\hat{n} \qquad \text{(a)}$$

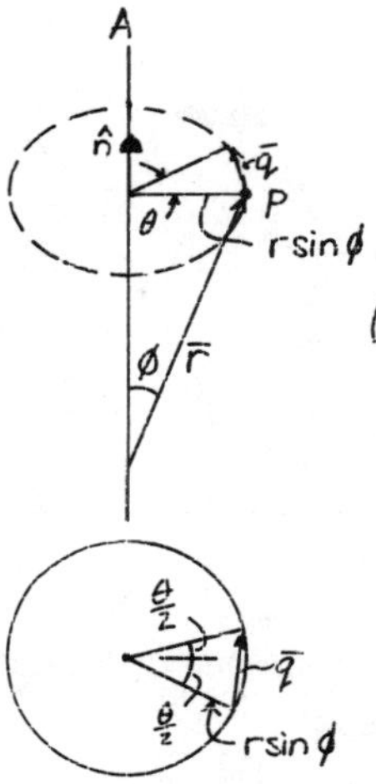

Figure a

Hence, by Fig. a and Eq. (a),

$$\bar{q} = \bar{\theta}\times\bar{r} \qquad \text{(b)}$$

(b) If θ is large, Eq. (a) is not valid. Hence, Eq. (b) is not valid.

17.78

GIVEN: A robotic manipulator consists of three members A, B, and C with axes of rotation a-a, b-b, and c-c (Fig. a). For the position shown, the members lie parallel to the xy plane and the speeds of rotation are $\omega_A = 150$ rad/s, $\omega_B = -50$ rad/s, and $\omega_c = -90$ rad/s.

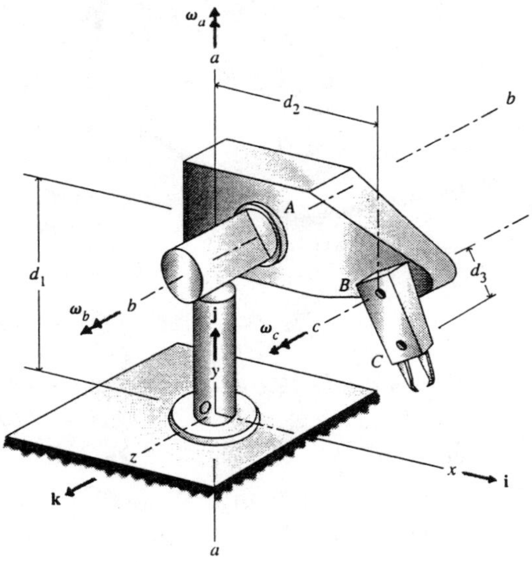

Figure a

FIND: the absolute angular velocity of arm BC.

SOLUTION:

By Fig. a, the angular velocities about axes a-a, b-b, and c-c are [rad/s]

$$\bar{\omega}_A = 150\hat{j},\ \bar{\omega}_B - 50\hat{k},\ \bar{\omega}_c = -90\hat{k} \qquad \text{(a)}$$

Therefore, by Eqs. (a)

$$\underline{\underline{\bar{\omega}_{BC} = \bar{\omega}_A + \bar{\omega}_B + \bar{\omega}_c = 150\hat{j} - 140\hat{k}}}$$

(continued)

17.78 (cont.)

Alternatively,

$$\omega_{BC} = \sqrt{150^2 + (-140)^2} = 205.18 \text{ rad/s}$$

$$\cos\theta_x = \frac{0}{205.18} = 0 \qquad \theta_x = 90°$$

$$\cos\theta_y = \frac{150}{205.18} = 0.73106; \quad \theta_y = 43.03°$$

$$\cos\theta_z = \frac{-140}{205.18} = -0.68232; \quad \theta_z = 133.03°$$

17.79

GIVEN: A spatial robotic manipulator has five rotational axes, with corresponding angular velocities (Fig. a). For the position shown, frame AB is parallel to the xy plane.

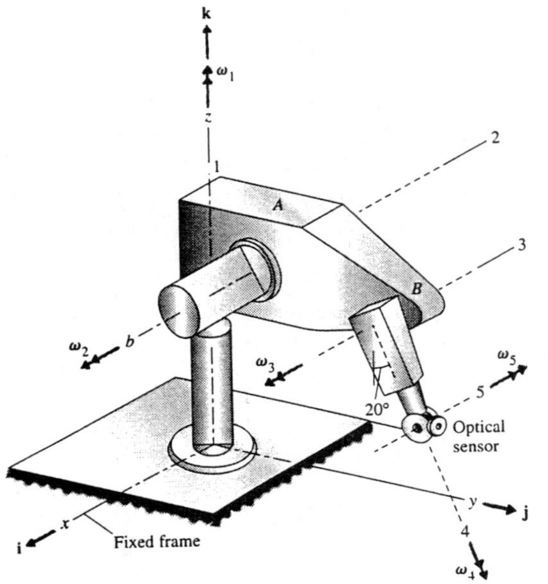

Figure a

FIND:

(a) Determine the absolute angular velocity $\bar{\omega}$ of the optical sensor in terms of the angular speeds ω_1, ω_2, ω_3, ω_4, and ω_5.

(b) For $\omega_1 = 10$, $\omega_2 = 15$, $\omega_3 = 10$, $\omega_4 = 5$, and $\omega_5 = 20 \frac{rad}{s}$, determine the absolute angular velocity and the angular speed of the optical sensor.

SOLUTION:

(a) By Fig. (a), the angular velocities of the components are

$$\bar{\omega}_1 = \omega_1\hat{k}, \quad \bar{\omega}_2 = \omega_2\hat{\imath}, \quad \bar{\omega}_3 = \omega_3\hat{\imath}$$

$$\bar{\omega}_4 = \omega_4[(\sin 20°)\hat{\jmath} - (\cos 20°)\hat{k}] \qquad (a)$$

$$\bar{\omega}_5 = -\omega_5\hat{\imath}$$

Hence, by Eqs. (a), the angular velocity of the sensor is

$$\bar{\omega}_{sensor} = (\omega_2 + \omega_3 - \omega_5)\hat{\imath} + \omega_4(\sin 20°)\hat{\jmath} + [\omega_1 - \omega_4(\cos 20°)]\hat{k}$$

and the angular speed of the sensor is

$$\omega_{sensor} = \sqrt{(\omega_2+\omega_3-\omega_5)^2 + (\omega_4\sin 20°)^2 + (\omega_1 - \omega_4\cos 20°)^2}$$

or

$$\omega_{sensor} = \sqrt{(\omega_2+\omega_3-\omega_5)^2 + \omega_1^2 + \omega_4^2 - 1.87938\,\omega_1\omega_4}$$

(b) For $\omega_1 = 10$, $\omega_2 = 15$, $\omega_3 = 10$, $\omega_4 = 5$, and $\omega_5 = 20$ [rad/s],

$$\bar{\omega}_{sensor} = 5\hat{\imath} + 1.7101\hat{\jmath} + 5.3015\hat{k} \quad [\text{rad/s}]$$

$$\omega_{sensor} = 7.4854 \text{ rad/s}$$

17.80

GIVEN: For the instant shown in Fig. a, the cable is pulled horizontally to the right with a constant speed $v_t = 1.0$ ft/s, causing the spool to roll without slipping. The radii are $r_1 = 12$ in. and $r_2 = 24$ in.

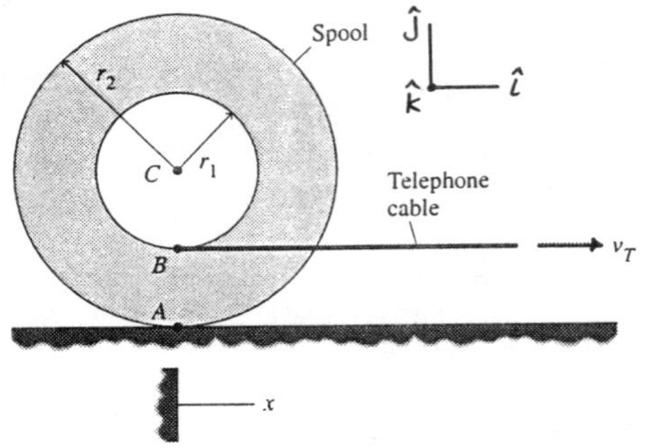

Figure a.

FIND:

(a) Determine the angular velocity of the spool.

(b) Determine the velocity of C, the center of the spool.

(c) Determine the rate of winding (or unwinding) of the cable [ft/s] on (off) the spool.

SOLUTION:

(a) Since the spool rolls without slipping, point A is the instantaneous center ($v_A = 0$). Assume that the disk rolls to the right (Fig. a). Then, by Fig. a and kinematics, the velocity of point B is

$$\bar{v}_B = \bar{v}_A + \bar{v}_{B/A}$$

or

$$\bar{v}_B = 0 + \bar{\omega} \times \bar{r}_B \qquad (a)$$

where

$$\bar{\omega} = -\omega\hat{k} \quad (\text{rad/s})$$

$$\bar{r}_B = (r_2 - r_1)\hat{\jmath} = 1.0\hat{\jmath} \quad (\text{ft}) \qquad (b)$$

Also,

$$\bar{v}_B = 1.0\hat{\imath} \quad [\text{ft/s}] \qquad (c)$$

Equations (a), (b), and (c) yield

$$1.0\hat{\imath} = -\omega(\hat{k}\times\hat{\jmath}) = \omega\hat{\imath}$$

or

$$\omega = 1 \text{ rad/s}$$

Hence,

$$\bar{\omega} = -\hat{k} \quad [\text{rad/s}]$$

(continued)

17.80 (cont.)

(b) By Fig. a and kinematics,

$$\bar{v}_C = \bar{v}_A + \bar{\omega} \times \bar{r}_C$$
$$= 0 + (-\hat{k})(2\hat{j})$$

or $\bar{v}_C = 2\hat{i}$ [ft/s] (d)

(c) Also, by Fig. a and kinematics,

$$\bar{v}_B = \bar{v}_C + \bar{v}_{B/C} \quad (e)$$

By Eqs. (c), (d), and (e),

$$\hat{i} = 2\hat{i} + \bar{v}_{B/C}$$

or $\bar{v}_{B/C} = -\hat{i}$ [ft/s] (f)

Equation (f) means that the cable is wound onto the spool at a rate of 1.0 ft/s.

17.81

GIVEN: The cable in Fig. a is pulled horizontally to the right with a velocity $v_T = 0.6$ m/s and an acceleration $a_T = 0.24$ m/s², causing the spool to roll without slipping.

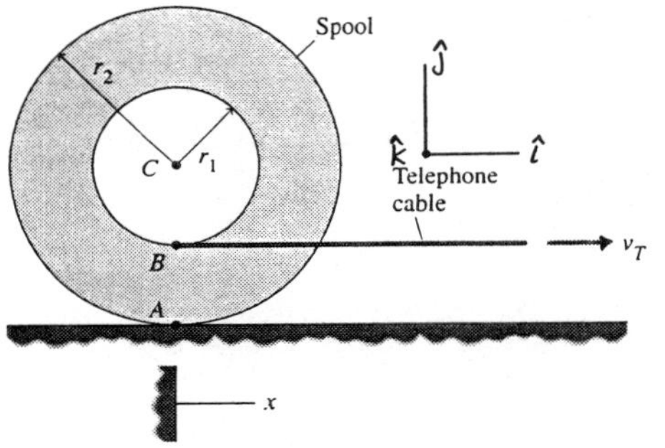

FIND: Determine the acceleration [m/s²] of points A, B, and C of the spool for $r_1 = 300$ mm and $r_2 = 600$ mm.

SOLUTION:

Since the spool rolls, the instantaneous center is point A ($v_A = 0$). Therefore, by kinematics, assuming the spool rolls to the right,

$$\bar{v}_B = 0.6\hat{i} = \bar{\omega} \times \bar{r}_B = (-\omega)\hat{k} \times (0.30)\hat{j}$$

or $0.6\hat{i} = (0.3\omega)\hat{i}$

Therefore, $\omega = 2$ rad/s

$$\bar{\omega} = -2\hat{k} \text{ [rad/s]} \quad (a)$$

Since the spool rolls,

$$\bar{a}_A = \omega^2 r_2 \hat{j} = (2)^2(0.6)\hat{j} = 2.4\hat{j}$$

Also, since we assume the spool rolls to the right, with Fig. a and Eq. (b),

$$\bar{a}_B = \bar{a}_A + \bar{a}_{B/A} \quad (c)$$

where

$$\bar{a}_B = 0.24\hat{i} + a_{By}\hat{j}$$
$$\bar{a}_A = 2.4\hat{j}$$
$$\bar{a}_{B/A} = (0.3\alpha)\hat{i} - \omega^2(0.3)\hat{j}$$
$$= (0.3\alpha)\hat{i} - 1.2\hat{j} \quad (d)$$

Equations (c) and (d) yield

$$0.24\hat{i} + a_{By}\hat{j} = 1.2\hat{j} + (0.3\alpha)\hat{i}$$

Hence,

$$a_{By} = 1.2 \text{ m/s}^2$$
$$\alpha = 0.8 \text{ rad/s}^2$$

So, $\bar{a}_B = 0.24\hat{i} + 1.2\hat{j}$ [m/s²]

and $\bar{a}_C = (0.6\alpha)\hat{i} = 0.48\hat{i}$ [m/s²]

17.82

GIVEN: In the bevel-gear box shown in Fig. a, gear B is fixed, gear A rotates about N-N with angular velocity $\bar{\omega}_A$. Gear C is attached at O by a ball joint to the pedestal fixed to B, and it rotates about the line OD.

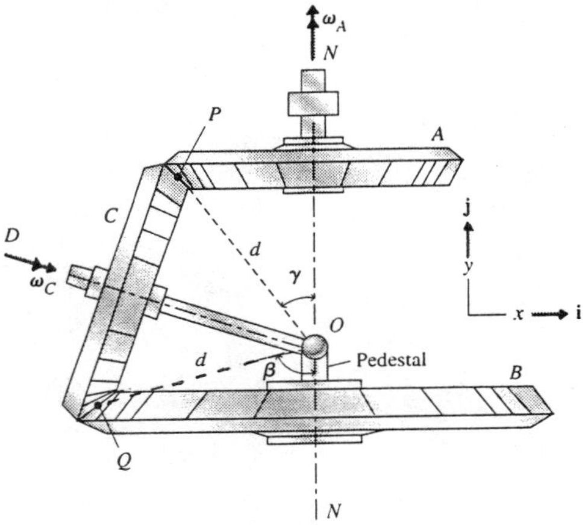

Figure a

FIND: Derive a formula for the angular speed ω_C of gear C in terms of angles γ and β and ω_A.

SOLUTION:

By Fig. a,

$$\bar{v}_P = \bar{\omega}_A \times \overline{OP} = \bar{\omega}_C \times \overline{OP} \quad (a)$$

(continued)

17.82 (cont.)

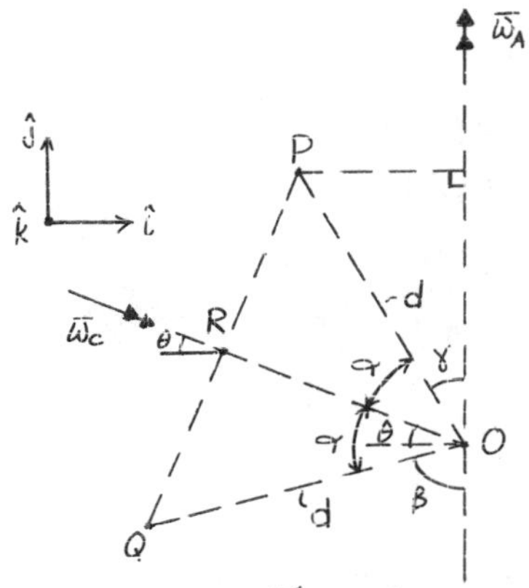

Figure b

By Fig. b,

$$\gamma + 2\alpha + \beta = 180 \;;\quad \alpha = 90° - \tfrac{1}{2}(\beta+\gamma)$$

Also, by Fig. b,

$$\gamma + \alpha + \theta = 90° \;;\quad \theta = 90° - \gamma - \alpha$$

or $\quad \theta = \tfrac{1}{2}(\beta - \gamma)$

Therefore, $\bar{\omega}_c = \omega_c\{\cos[\tfrac{1}{2}(\beta-\gamma)]\hat{i} - \sin[\tfrac{1}{2}(\beta-\gamma)]\hat{j}\}$

$$\overline{OP} = OP[(-\sin\gamma)\hat{i} + (\cos\gamma)\hat{j}] \qquad (b)$$

$$\bar{\omega}_A = \omega_A\hat{j}$$

$$OP = d$$

Equations (a) and (b) yield

$$\bar{v}_P = (\omega_A)(d)(-\sin\gamma)\hat{j}\times\hat{i} = \omega_A d\sin\gamma\,\hat{k} \qquad (c)$$

and

$$\bar{v}_P = (\omega_c)(d)\begin{vmatrix}\hat{i} & \hat{j} & \hat{k}\\ \cos(\frac{\beta-\gamma}{2}) & -\sin(\frac{\beta-\gamma}{2}) & 0\\ -\sin\gamma & \cos\gamma & 0\end{vmatrix}$$

$$= (\omega_c)(d)[(\cos\gamma)\cos(\tfrac{\beta-\gamma}{2}) - (\sin\gamma)\sin(\tfrac{\beta-\gamma}{2})]\hat{k}$$

$$= (\omega_c)(d)[\cos(\gamma + \tfrac{1}{2}(\beta-\gamma))]$$

or

$$\bar{v}_P = (\omega_c)(d)[\cos(\tfrac{\beta+\gamma}{2})]\hat{k} \qquad (d)$$

Equating Eqs. (c) and (d) and solving for ω_c, we have

$$\omega_c = \omega_A\left[\frac{\sin\gamma}{\cos(\frac{\beta+\gamma}{2})}\right] \qquad (e)$$

Alternative Solution:

Directly for Fig. b,

$$v_P = \omega_A d\sin\gamma \qquad (f)$$

and

$$v_P = \omega_c d\sin[90° - \tfrac{1}{2}(\beta+\gamma)]$$

or

$$v_P = \omega_c d\cos(\tfrac{\beta+\gamma}{2}) \qquad (g)$$

Hence, equating Eqs. (f) and (g), we have

$$\omega_c d\cos(\tfrac{\beta+\gamma}{2}) = \omega_A d\sin\gamma$$

or

$$\omega_c = \omega_A\left[\frac{\sin\gamma}{\cos(\frac{\beta+\gamma}{2})}\right]$$

This alternative solution illustrates the fact that sometimes the formal vector approach, although certainly correct, may require more algebra, relative to the direct physical approach used in this alternative solution.

17.83

GIVEN: In the bevel-gear box of Problem 17.82 (Fig. a below), $\omega_A = 30$ rad/s, $\gamma = 45°$, and $\beta = 30°$.

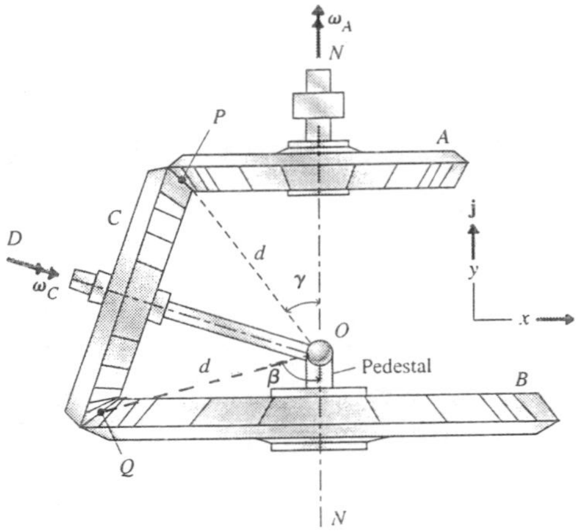

Figure a

FIND: Determine the angular speed ω_c of gear C.

SOLUTION:

If Problem 17.82 has not been assigned, the formula for ω_c will have to be derived here, as in Problem 17.82. By the solution in Problem 17.82,

$$\omega_c = \omega_A\frac{\sin\gamma}{\cos(\frac{\beta+\gamma}{2})} \qquad (a)$$

For $\omega_A = 30$ rad/s, $\gamma = 45°$, and $\beta = 30°$, Eq. (a) yields

$$\omega_c = (30)\frac{\sin 45°}{\cos 37.5°} = 26.74 \text{ rad/s}$$

17.84

GIVEN: A body translates with velocity 9 m/s in the sense and direction of the positive z axis of the xyz axes.

FIND:

(a) By means of a figure, represent this motion as simultaneous rotations about the lines z=0, x=1 m, and z=0, x=2.5m.

(b) What is the angular speed [rad/s] about either line?

(continued)

17.84 (cont.)

SOLUTION:

(a) Refer to Fig. a.

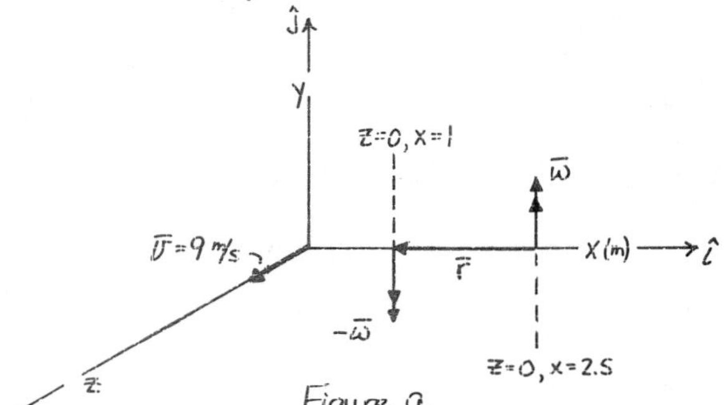

Figure a.

The velocity may be represented by the rotational couple shown in Fig. a.

(b) To determine the angular speed ω (rad/s) about either line, we note that

$$\bar{v} = \bar{\omega} \times \bar{r} \qquad (a)$$

where, by Fig. a,

$$\bar{\omega} = \omega \hat{j} \quad [\text{rad/s}]$$
$$\bar{r} = -1.5\hat{i} \quad [\text{m}]$$
$$\bar{v} = 9\hat{k} \quad [\text{m/s}]$$

By Eqs. (a) and (b)

$$9\hat{k} = (\omega\hat{j}) \times (-1.5\hat{i}) = 1.5\omega\hat{k}$$

or $\omega = \frac{9}{1.5} = 6$ rad/s

17.85

GIVEN: A rigid body is subjected to the angular velocities shown in Fig. a.

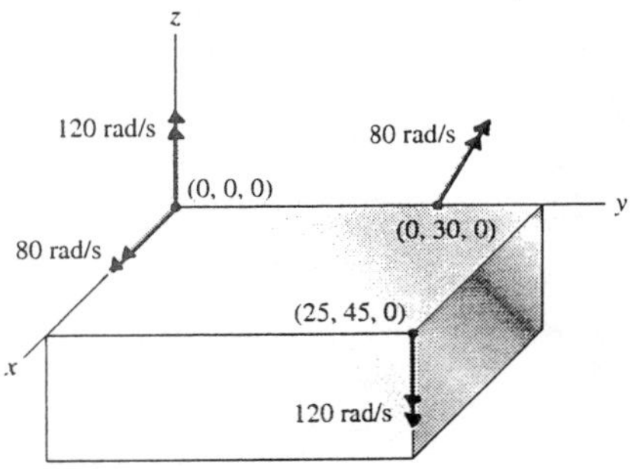

Figure a

FIND: Determine the resultant velocity of translation [in/s].

SOLUTION:

Refer to Fig. b. By the theory of rotational couples, the translational velocity is

$$\bar{v} = \bar{\omega}_1 \times \bar{r}_1 + \bar{\omega}_2 \times \bar{r}_2 \qquad (a)$$

where $\bar{\omega}_1 = 80\hat{i}$, $\bar{r} = 30\hat{j}$, $\bar{\omega}_2 = 120\hat{k}$, $\bar{r}_2 = 25\hat{i} + 45\hat{j}$ (b)

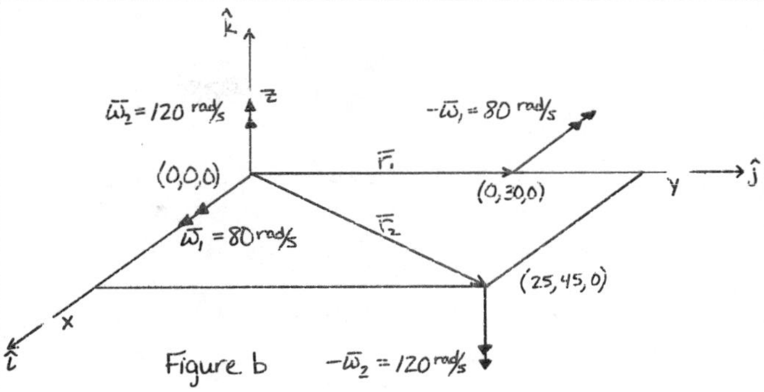

Figure b

By Eqs. (a) and (b)

$$\bar{v} = \begin{vmatrix} \hat{i} & \hat{j} & \hat{k} \\ 80 & 0 & 0 \\ 0 & 30 & 0 \end{vmatrix} + \begin{vmatrix} \hat{i} & \hat{j} & \hat{k} \\ 0 & 0 & 120 \\ 25 & 45 & 0 \end{vmatrix}$$

$$= 2400\hat{k} + (-5400\hat{i} + 3000\hat{j})$$

or $\bar{v} = -5400\hat{i} + 3000\hat{j} + 2400\hat{k}$ [in/s]

17.86

GIVEN: A rotational couple that acts on a rigid body consists of the angular velocity $\bar{\omega} = 100\hat{i} + 300\hat{j} - 500\hat{k}$ [rad/s] about an axis through the point (-125, 25, -50) [mm] and an angular velocity $-\bar{\omega}$ about an axis through the point (50, 225, 175) [mm].

FIND: Determine the velocity $\bar{v}$ of the body.

SOLUTION:

By the theory of rotational couples, the velocity of the body is

$$\bar{v} = \bar{\omega} \times \bar{r} \qquad (a)$$

Where

$$\bar{\omega} = 100\hat{i} + 300\hat{j} - 500\hat{k}$$
$$\bar{r} = [50-(-125)]\hat{i} + (225-25)\hat{j} + [175-(-50)]\hat{k}$$
$$= 175\hat{i} + 200\hat{j} + 225\hat{k} \qquad (b)$$

Equations (a) and (b) yield

$$\bar{v} = \begin{vmatrix} \hat{i} & \hat{j} & \hat{k} \\ 100 & 300 & -500 \\ 175 & 200 & 225 \end{vmatrix}$$

$$\bar{v} = (300 \times 225 + 200 \times 500)\hat{i} + (-500 \times 175 - 100 \times 225)\hat{j} + (100 \times 200 - 175 \times 300)\hat{k}$$

$\bar{v} = 167\,500\hat{i} - 110\,000\hat{j} - 32\,500\hat{k}$ [mm/s]

or $\bar{v} = 167.5\hat{i} - 110.0\hat{j} - 32.5\hat{k}$ [m/s]

17.87

GIVEN: A body rotates with angular speed of 50 rad/s about an axis with direction cosines (0.36, −0.48, 0.80), through the origin of the xyz axes. It is subjected to an equal counter-rotation about a parallel axis through the point (30 in, 28 in, −10 in).

FIND: Calculate the resultant velocity [in/s] of translation of the body.

SOLUTION:

By the theory of rotational couples, the translational velocity is

$$\bar{v} = \bar{\omega} \times \bar{r} \qquad (a)$$

where

$$\bar{\omega} = 50(0.36\hat{\imath} - 0.48\hat{\jmath} + 0.80\hat{k})$$
$$= 18\hat{\imath} - 24\hat{\jmath} + 40\hat{k} \; [\text{rad/s}] \qquad (b)$$
$$\bar{r} = 30\hat{\imath} + 28\hat{\jmath} - 10\hat{k} \; [\text{in}]$$

Equations (a) and (b) yield

$$\bar{v} = \begin{vmatrix} \hat{\imath} & \hat{\jmath} & \hat{k} \\ 18 & -24 & 40 \\ 30 & 28 & -10 \end{vmatrix}$$

$$\bar{v} = (24\cdot 10 - 28\cdot 40)\hat{\imath} + (30\cdot 40 - (18)(-10))\hat{\jmath} + (18\cdot 28 - (30)(-24))\hat{k}$$

or $\underline{\bar{v} = -880\hat{\imath} + 1380\hat{\jmath} + 1224\hat{k} \; [\text{in/s}]}$

17.88

GIVEN: Three angular velocities [rad/s] of a rigid body defined relative to xyz axes are

$$\bar{\omega}_1 = 7\hat{\imath} - 2\hat{\jmath} + 4\hat{k}$$
$$\bar{\omega}_2 = -8\hat{\imath} + 3\hat{\jmath} + 10\hat{k} \qquad (a)$$
$$\bar{\omega}_3 = -10\hat{\imath} + 4\hat{\jmath} - 6\hat{k}$$

about axes through points (3,1,4), (5,8,2), and (1,−4,−3) [ft]. Also, $-\bar{\omega}_1$, $-\bar{\omega}_2$, and $-\bar{\omega}_3$ act about axes through points (6,1,−7), (8,−4,−10), and (7,0,1) [ft].

FIND: Determine the resultant velocity of translation.

SOLUTION:

By the theory of rotational couples,

$$\bar{v} = \bar{\omega}_1 \times \bar{r}_1 + \bar{\omega}_2 \times \bar{r}_2 + \bar{\omega}_3 \times \bar{r}_3 \qquad (b)$$

where

$$\bar{r}_1 = (6-3)\hat{\imath} + (1-1)\hat{\jmath} + (-7-4)\hat{k}$$
$$= 3\hat{\imath} - 11\hat{k} \; [\text{ft}]$$
$$\bar{r}_2 = (8-5)\hat{\imath} + (-4-8)\hat{\jmath} + (-10-2)\hat{k}$$
$$= 3\hat{\imath} - 12\hat{\jmath} - 12\hat{k} \; [\text{ft}] \qquad (c)$$
$$\bar{r}_3 = (7-1)\hat{\imath} + (0-(-4))\hat{\jmath} + (1-(-3))\hat{k}$$
$$= 6\hat{\imath} + 4\hat{\jmath} + 4\hat{k} \; [\text{ft}]$$

Equations (a), (b), and (c) yield

$$\bar{v} = \begin{vmatrix} \hat{\imath} & \hat{\jmath} & \hat{k} \\ 7 & -2 & 4 \\ 3 & 0 & -11 \end{vmatrix} + \begin{vmatrix} \hat{\imath} & \hat{\jmath} & \hat{k} \\ -8 & 3 & 10 \\ 3 & -12 & -12 \end{vmatrix} + \begin{vmatrix} \hat{\imath} & \hat{\jmath} & \hat{k} \\ -10 & 4 & -6 \\ 6 & 4 & 4 \end{vmatrix}$$

$$\bar{v} = 22\hat{\imath} + 89\hat{\jmath} + 6\hat{k} + 84\hat{\imath} - 66\hat{\jmath} + 87\hat{k} + 40\hat{\imath} + 4\hat{\jmath} - 64\hat{k}$$

or $\underline{\bar{v} = 146\hat{\imath} + 27\hat{\jmath} + 29\hat{k} \; [\text{ft/s}]}$

17.89

GIVEN: The L bar in Figure a is subjected to the two angular velocities shown.

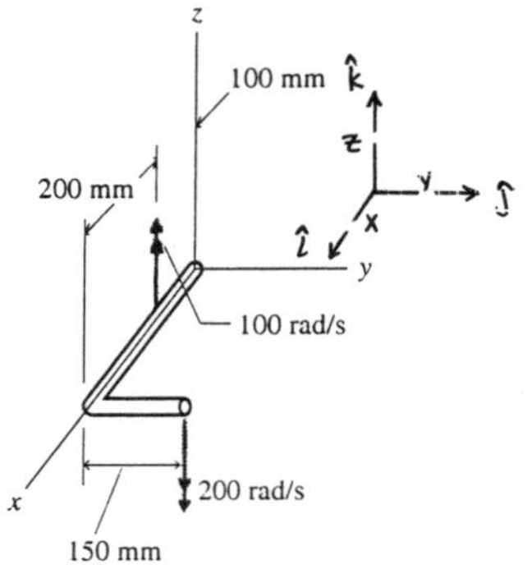

Figure a

FIND: Replace the angular velocities by a single angular velocity [rad/s] about the z axis and a velocity of translation [mm/s].

SOLUTION:

The resultant angular velocity is

$$\bar{\omega} = \bar{\omega}_1 + \bar{\omega}_2 = 100\hat{k} - 200\hat{k}$$

or $\underline{\bar{\omega} = -100\hat{k} \; [\text{rad/s}]} \qquad (a)$

The resultant velocity is given by

$$\bar{v} = \bar{\omega}_1 \times \bar{r}_1 + \bar{\omega}_2 \times \bar{r}_2 \qquad (b)$$

where

$$\bar{\omega}_1 = 100\hat{k}, \quad \bar{r}_1 = -100\hat{\imath}$$
$$\bar{\omega}_2 = -200\hat{k}, \quad \bar{r}_2 = -150\hat{\jmath} - 300\hat{\imath} \qquad (c)$$

Equations (b) and (c) yield

$$\bar{v} = \begin{vmatrix} \hat{\imath} & \hat{\jmath} & \hat{k} \\ 0 & 0 & 100 \\ -100 & 0 & 0 \end{vmatrix} + \begin{vmatrix} \hat{\imath} & \hat{\jmath} & \hat{k} \\ 0 & 0 & -200 \\ -300 & -150 & 0 \end{vmatrix}$$

$$\bar{v} = -10\,000\hat{\jmath} - 30\,000\hat{\imath} + 60\,000\hat{\jmath}$$

$\underline{\bar{v} = -30000\hat{\imath} + 50\,000\hat{\jmath} \; [\text{mm/s}]}$

17.90

GIVEN: In Fig. a, the angular velocities act on the body as shown.

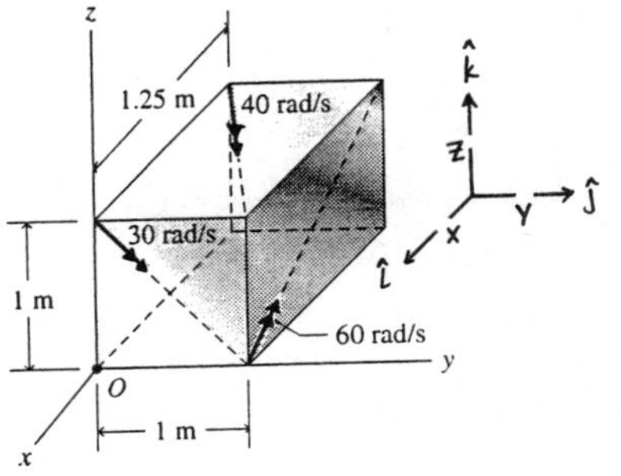

Figure a

FIND: Replace the angular velocities by a single angular velocity through the origin O and a velocity of translation.

SOLUTION:

By Fig. a, define the angular velocities as

$$\bar{\omega}_1 = \frac{40}{\sqrt{2.5625}}(1.25\hat{i} + \hat{j}) = 31.235\hat{i} + 24.988\hat{j}$$

$$\bar{\omega}_2 = \frac{30}{\sqrt{2}}(\hat{j} - \hat{k}) = 21.213\hat{j} - 21.213\hat{k} \qquad (a)$$

$$\bar{\omega}_3 = \frac{60}{\sqrt{2.5625}}(-1.25\hat{i} + \hat{k}) = -46.852\hat{i} + 37.482\hat{k}$$

Hence, $\bar{\omega}$, the single angular velocity at O is [rad/s]

$$\bar{\omega} = \bar{\omega}_1 + \bar{\omega}_2 + \bar{\omega}_3 = -15.617\hat{i} + 46.201\hat{j} + 16.269\hat{k}$$

The velocity of translation is

$$\bar{v} = \bar{\omega}_1 \times \bar{r}_1 + \bar{\omega}_2 \times \bar{r}_2 + \bar{\omega}_3 \times \bar{r}_3 \qquad (b)$$

where, by Fig. a,

$$\bar{r}_1 = 1.25\hat{i} - \hat{k}$$
$$\bar{r}_2 = -\hat{k} \qquad (c)$$
$$\bar{r}_3 = -\hat{j}$$

Equations (a), (b), and (c) yield

$$\bar{v} = \begin{vmatrix} \hat{i} & \hat{j} & \hat{k} \\ 31.235 & 24.988 & 0 \\ 1.25 & 0 & -1 \end{vmatrix} + \begin{vmatrix} \hat{i} & \hat{j} & \hat{k} \\ 0 & 21.213 & 21.213 \\ 0 & 0 & -1 \end{vmatrix} + \begin{vmatrix} \hat{i} & \hat{j} & \hat{k} \\ -46.852 & 0 & 37.482 \\ 0 & -1 & 0 \end{vmatrix}$$

$$\bar{v} = -24.988\hat{i} + 31.235\hat{j} - 31.235\hat{k} - 21.213\hat{i} + 37.482\hat{i} + 46.852\hat{k}$$

or

$$\bar{v} = -8.719\hat{i} + 31.235\hat{j} + 15.617\hat{k} \ \text{[m/s]}$$

17.91

GIVEN: A rigid body is subjected to the designated angular velocities shown in Fig. a.

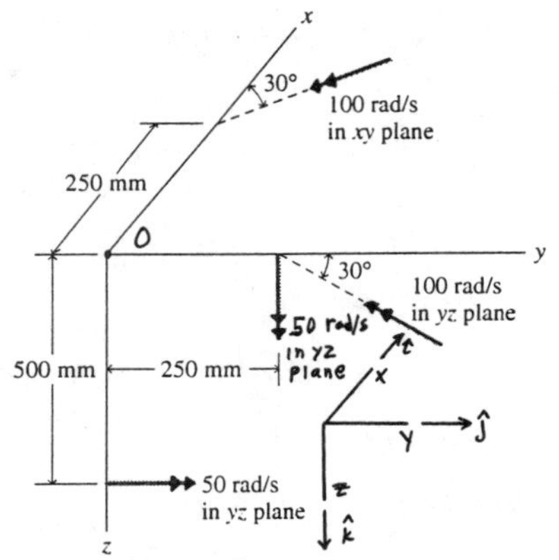

Figure a

FIND: Reduce the motion to a single angular velocity [rad/s] about an axis through the origin O and a velocity of translation [m/s].

SOLUTION:

Designate the angular velocities ω_i and associate radius vectors r_i as follows

$$\bar{\omega}_1 = 100[-(\cos 30°)\hat{i} - (\sin 30°)\hat{j}] = -86.602\hat{i} - 50\hat{j}$$
$$\bar{r}_1 = -0.25\hat{i} \ [m]$$
$$\bar{\omega}_2 = 100[-(\cos 30°)\hat{j} - (\sin 30°)\hat{k}] = -86.602\hat{j} - 50\hat{k}$$
$$\bar{r}_2 = -0.250\hat{j} \ [m] \qquad (a)$$
$$\bar{\omega}_3 = 50\hat{j}$$
$$\bar{r}_3 = -0.500\hat{k} \ [m]$$
$$\bar{\omega}_4 = 50\hat{k}$$
$$\bar{r}_4 = -0.250\hat{j} \ [m]$$

Hence, by Eqs. (a),

$$\bar{\omega} = \bar{\omega}_1 + \bar{\omega}_2 + \bar{\omega}_3 + \bar{\omega}_4 = -86.602\hat{i} - 86.602\hat{j}$$

$$\bar{v} = \bar{\omega}_1 \times \bar{r}_1 + \bar{\omega}_2 \times \bar{r}_2 + \bar{\omega}_3 \times \bar{r}_3 + \bar{\omega}_4 \times \bar{r}_4$$
$$= -12.5\hat{k} - 12.5\hat{i} - 25\hat{i} + 12.5\hat{i}$$

or

$$\bar{v} = -25\hat{i} - 12.5\hat{k}$$

17.92

GIVEN: A rigid body is subjected to the angular velocities [rad/s]

$$\bar{\omega}_1 = 60\hat{i} - 35\hat{j} + 9\hat{k} \text{ about an axis through } (3,8,21)$$
$$\bar{\omega}_2 = -100\hat{i} + 26\hat{j} - 72\hat{k} \text{ about an axis through } (7,-9,-40) \qquad (a)$$
$$\bar{\omega}_3 = 30\hat{i} - 95\hat{j} + 85\hat{k} \text{ about an axis through } (60,-3,25)$$

FIND: Reduce the motion to an angular velocity $\bar{\omega}$ about an axis through (5, 9, -35) and a velocity of translation $\bar{v}$ [in/s].

(continued)

17.92 (cont.)

SOLUTION:

The radial vectors associate with $\bar{\omega}_1, \bar{\omega}_2$, and $\bar{\omega}_3$, respectively, are

$$\bar{r}_1 = (5-3)\hat{\imath} + (9-8)\hat{\jmath} + (-35-21)\hat{k} = 2\hat{\imath} + \hat{\jmath} - 56\hat{k}$$

$$\bar{r}_2 = (5-7)\hat{\imath} + [9-(-9)]\hat{\jmath} + [-35-(-40)]\hat{k} = -2\hat{\imath} + 18\hat{\jmath} + 5\hat{k} \qquad \text{(b)}$$

$$\bar{r}_3 = (5-60)\hat{\imath} + [9-(-3)]\hat{\jmath} + (-35-25)\hat{k} = -55\hat{\imath} + 12\hat{\jmath} - 60\hat{k}$$

By Eqs. (a),

$$\bar{\omega} = \bar{\omega}_1 + \bar{\omega}_2 + \bar{\omega}_3 = -10\hat{\imath} - 104\hat{\jmath} + 22\hat{k} \quad [\text{rad/s}]$$

By Eqs. (a) and (b),

$$\bar{v} = \bar{\omega}_1 \times \bar{r}_1 + \bar{\omega}_2 \times \bar{r}_2 + \bar{\omega}_3 \times \bar{r}_3 \qquad \text{(c)}$$

where

$$\bar{\omega}_1 \times \bar{r}_1 = \begin{vmatrix} \hat{\imath} & \hat{\jmath} & \hat{k} \\ 60 & -35 & 9 \\ 2 & 1 & -56 \end{vmatrix} = 1951\hat{\imath} + 3378\hat{\jmath} + 130\hat{k}$$

$$\bar{\omega}_2 \times \bar{r}_2 = \begin{vmatrix} \hat{\imath} & \hat{\jmath} & \hat{k} \\ -100 & 26 & -72 \\ -2 & 18 & 5 \end{vmatrix} = 1426\hat{\imath} + 644\hat{\jmath} - 1748\hat{k} \qquad \text{(d)}$$

$$\bar{\omega}_3 \times \bar{r}_3 = \begin{vmatrix} \hat{\imath} & \hat{\jmath} & \hat{k} \\ 30 & -95 & 85 \\ -55 & 12 & -60 \end{vmatrix} = 4680\hat{\imath} - 2875\hat{\jmath} - 4865\hat{k}$$

Equations (c) and (d) yield

$$\bar{v} = 8057\hat{\imath} + 1147\hat{\jmath} - 6483\hat{k} \quad [\text{in/s}]$$

17.93

GIVEN: A rigid body is subjected to angular velocities $\bar{\omega}_1, \bar{\omega}_2, \bar{\omega}_3$ [rad/s] and to velocities of translation $\bar{v}_1$ and $\bar{v}_2$ [cm/s] defined as follows:

$$\bar{\omega}_1 = 50\hat{\imath} - 34\hat{\jmath} - 65\hat{k} \text{ about an axis through } (9,7,23)$$

$$\bar{\omega}_2 = -36\hat{\imath} + 55\hat{\jmath} + 72\hat{k} \text{ about an axis through } (30,8,-22) \qquad \text{(a)}$$

$$\bar{\omega}_3 = 86\hat{\imath} - 2.5\hat{\jmath} - 15\hat{k} \text{ about an axis through } (-60,45,10)$$

$$\bar{v}_1 = 105\hat{\imath} - 64\hat{\jmath} + 136\hat{k}$$

$$\bar{v}_2 = 106\hat{\imath} + 95\hat{\jmath}$$

FIND: Replace this system by an equivalent system consisting of an angular velocity $\bar{\omega}$ about an axis through (0,0,20) and a velocity $\bar{v}$ of translation.

SOLUTION:

By the first three of Eqs. (a),

$$\bar{\omega} = \bar{\omega}_1 + \bar{\omega}_2 + \bar{\omega}_3 = 100\hat{\imath} + 18.5\hat{\jmath} - 8\hat{k}$$

By Eqs. (a) and kinematics,

$$\bar{v} = \bar{v}_1 + \bar{v}_2 + \bar{\omega}_1 \times \bar{r}_1 + \bar{\omega}_2 \times \bar{r}_2 + \bar{\omega}_3 \times \bar{r}_3 \qquad \text{(b)}$$

where

$$\bar{r}_1 = -9\hat{\imath} - 7\hat{\jmath} - 3\hat{k}$$

$$\bar{r}_2 = -30\hat{\imath} - 8\hat{\jmath} + 42\hat{k} \qquad \text{(c)}$$

$$\bar{r}_3 = 60\hat{\imath} - 45\hat{\jmath} + 10\hat{k}$$

Therefore,

$$\bar{\omega}_1 \times \bar{r}_1 = \begin{vmatrix} \hat{\imath} & \hat{\jmath} & \hat{k} \\ 50 & -34 & -65 \\ -9 & -7 & -3 \end{vmatrix} = -353\hat{\imath} + 735\hat{\jmath} - 656\hat{k}$$

$$\bar{\omega}_2 \times \bar{r}_2 = \begin{vmatrix} \hat{\imath} & \hat{\jmath} & \hat{k} \\ -36 & 55 & 72 \\ -30 & -8 & 42 \end{vmatrix} = 2886\hat{\imath} - 648\hat{\jmath} + 1938\hat{k} \qquad \text{(d)}$$

$$\bar{\omega}_3 \times \bar{r}_3 = \begin{vmatrix} \hat{\imath} & \hat{\jmath} & \hat{k} \\ 86 & -2.5 & -15 \\ 60 & -45 & 10 \end{vmatrix} = -700\hat{\imath} - 1760\hat{\jmath} - 3720\hat{k}$$

Hence, by the last two of Eqs. (a), Eq. (b), and Eq. (d), we obtain

$$\bar{v} = 2098\hat{\imath} - 1642\hat{\jmath} - 2302\hat{k} \quad [\text{cm/s}]$$

17.94

GIVEN: A rigid body rotates with angular velocity $\bar{\omega} = 10\hat{\imath} - 5\hat{\jmath} + 8\hat{k}$ [rad/s] about an axis through (75, 50, −125) [mm]. Also, the body has a velocity of translation $\bar{v}_0 = 4\hat{\imath} + 2\hat{\jmath} - 3\hat{k}$ [m/s].

FIND:

(a) Replace the system by an angular velocity $\bar{\omega}$ about an axis through (a, b, 0) and a velocity of translation $\bar{v} = v_x\hat{\imath} + v_y\hat{\jmath} + v_z\hat{k}$.

(b) Write the conditions of parallelism ($\bar{\omega} \times \bar{v} = 0$) of vectors $\bar{\omega}$ and $\bar{v}$ in determinant form.

(c) Expand the determinant of part b, put it in scalar form, and solve any two of the resulting equations for a and b.

(d) Show that the third equation is then satisfied automatically.

(e) With the values of a and b, determine v explicitly.

SOLUTION:

(a) By kinematics,

$$\bar{v} = \bar{v}_0 + \bar{\omega} \times \bar{r} \qquad \text{(a)}$$

where

$$\bar{r} = (a - 0.075)\hat{\imath} + (b - 0.050)\hat{\jmath} + 0.125\hat{k} \ [\text{m}] \qquad \text{(b)}$$

Then,

$$\bar{\omega} \times \bar{r} = \begin{vmatrix} \hat{\imath} & \hat{\jmath} & \hat{k} \\ 10 & -5 & 8 \\ a-.075 & b-.05 & .125 \end{vmatrix}, \quad \bar{\omega} = 10\hat{\imath} - 5\hat{\jmath} + 8\hat{k}$$

or

$$\bar{\omega} \times \bar{r} = -(8b + 0.225)\hat{\imath} + (8a - 1.85)\hat{\jmath} + (5a + 10b - 0.875)\hat{k} \qquad \text{(c)}$$

(continued)

17.94 (cont.)

and by Eqs. (a) and (c), with $\bar{v}_0 = 4\hat{\imath} + 2\hat{\jmath} - 3\hat{k}$,

$$\underline{\bar{v} = (-8b + 3.775)\hat{\imath} + (8a + 0.15)\hat{\jmath} + (5a + 10b - 3.875)\hat{k}} \quad (d)$$

(b) By Eq. (d), with $\bar{\omega} = 10\hat{\imath} - 5\hat{\jmath} + 8\hat{k}$, the condition of parallelism of $\bar{\omega}$ and $\bar{v}$ is, in determinant form,

$$\bar{\omega} \times \bar{v} = \begin{vmatrix} \hat{\imath} & \hat{\jmath} & \hat{k} \\ 10 & -5 & 8 \\ (-8b+3.775) & (8a+0.15) & (5a+10b-3.875) \end{vmatrix} = 0 \quad (e)$$

(c) Expanding Eq. (e), we obtain

$$(-89a - 50b + 18.175)\hat{\imath} + (-50a - 164b + 68.95)\hat{\jmath} + (80a - 40b + 20.375)\hat{k} = 0$$

or

$$\begin{aligned} 89a + 50b &= 18.175 \\ 50a + 164b &= 68.95 \\ 80a - 40b &= -20.375 \end{aligned} \quad (f)$$

The first two of Eqs. (f) yield

$$\underline{a = -0.0386\ \text{m}}, \quad \underline{b = 0.432\ \text{m}} \quad (g)$$

(d) Substitution of Eqs. (g) into the last of Eqs. (f) yields, to two decimal places

$$-20.37 = -20.37$$

(e) With the values of a and b [Eqs. (g)], Eq. (d) yields

$$\underline{\bar{v} = 0.319\hat{\imath} - 0.159\hat{\jmath} + 0.252\hat{k}\ [\text{m/s}]}$$

17.95

GIVEN: The crank OC of the grasshopper mechanism (Fig. a) rotates with constant angular speed ω. The xy coordinate axes rotate with the crank.

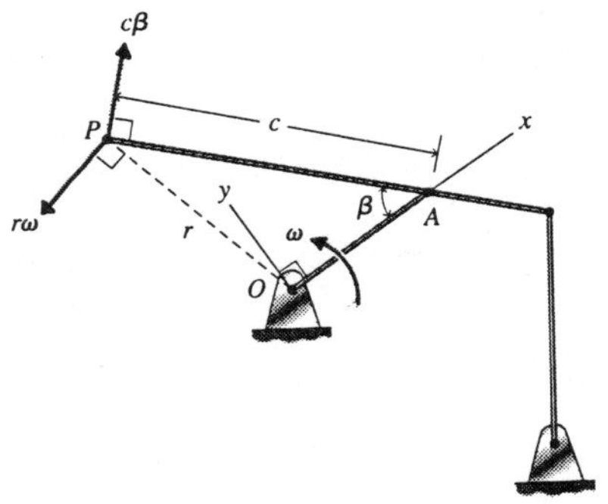

Figure a

FIND: Show by interpretation of the terms in Eq. (17-20) that the velocity of point P is the sum of the two vectors shown in Fig. a.

SOLUTION:

Let capitals A and B denote reference frames. By Eq. (17.20),

$$\bar{v}_{P/A} = \bar{v}_{P/B} + \bar{v}_{O/A} + \bar{\omega} \times \bar{r} \quad (a)$$

Let the xy coordinate system attached to the crank OC be frame B, with origin at O. Let frame A be the Newtonian frame fixed at O (Fig. b).

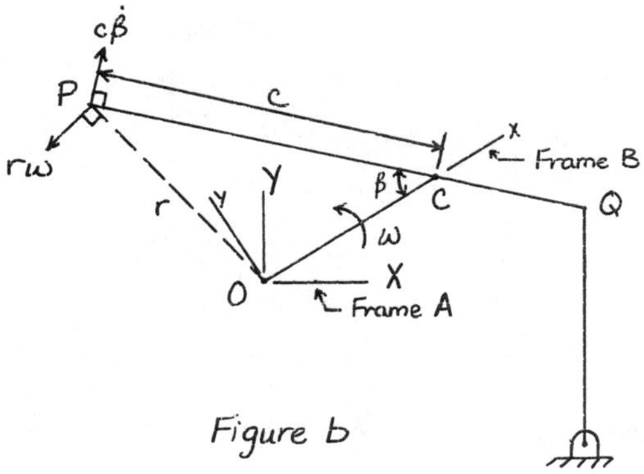

Figure b

By Fig. b, since O is fixed in frame A and line PQ rotates with angular speed $\dot{\beta}$ relative to frame B,

$$\bar{v}_{O/A} = 0 \quad (b)$$

$$v_{P/B} = c\dot{\beta}$$

Therefore,

$$\bar{v}_{P/B} = c\bar{\dot{\beta}} \quad (c)$$

is perpendicular to line PQ at P (Fig. b). Also, since line OP rotates with angular speed ω (Fig. c),

$$|\bar{\omega} \times \bar{r}| = \omega r$$

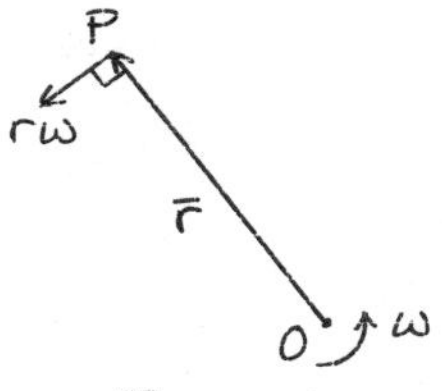

Figure c

Therefore, by Eqs. (a), (b), (c), and (d),

$$\bar{v}_{P/A} = c\bar{\dot{\beta}} + \bar{\omega} \times \bar{r}$$

and this vector equation is represented by Fig. a.

17.96

GIVEN: The crank OC of the grasshopper mechanism (Fig. a) rotates with constant angular speed ω. The xy coordinate axes rotate with the crank.

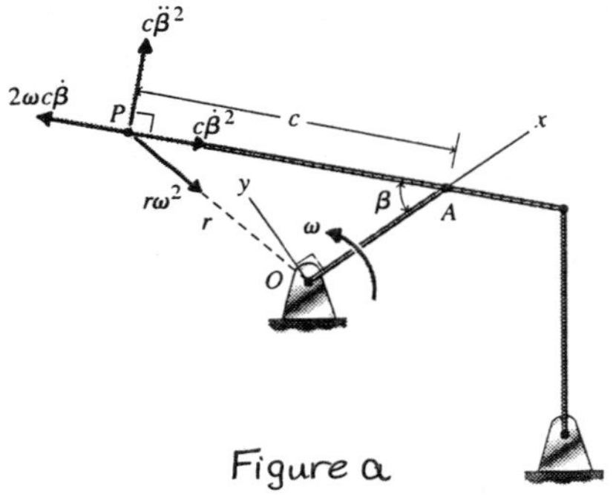

Figure a

FIND: Show by interpretation of the terms in Eq. (17.23) that the acceleration of point P is the sum of the four vectors shown in Figure a.

SOLUTION:

Let capitals A and B denote reference frames. By Eq. (17.23),

$$\bar{a}_{P/A} = \bar{a}_{P/B} + \bar{a}_{O/A} + \bar{\alpha} \times \bar{r} + \bar{\omega} \times (\bar{\omega} \times \bar{r}) + 2\bar{\omega} \times \bar{v}_{P/B} \qquad (a)$$

Let the xy coordinate system attached to the crank OC be frame B, with origin at O. Let frame A be the Newtonian Frame fixed at O (Fig. b).

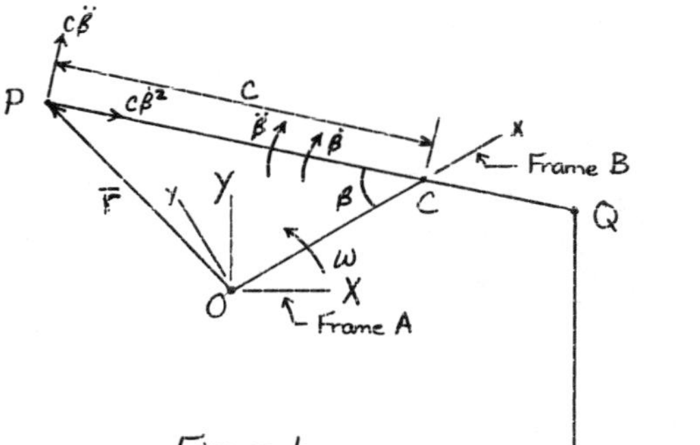

Figure b

In Fig. b, $\dot{\beta}$ and $\ddot{\beta}$ are the angular velocity and angular acceleration of rod PQ, respectively, relative to frame B.

By Fig. b, since O is fixed, in Eq. (a)

$$\bar{a}_{O/A} = 0 \qquad (b)$$

and the term $\bar{a}_{P/B}$ is represented by the two components $c\dot{\beta}^2$ and $c\ddot{\beta}$, or

$$\bar{a}_{P/B} = (c\dot{\beta}^2,\ c\ddot{\beta}) \qquad (c)$$

Also noting that $\bar{\alpha} = \dot{\bar{\omega}} = 0$, since ω is constant, we have

$$\bar{\alpha} \times \bar{r} = 0 \qquad (d)$$

Next to interpret the term $\bar{\omega} \times (\bar{\omega} \times \bar{r})$, we refer to Fig. c. By Fig. c, $\bar{\omega} \times (\bar{\omega} \times \bar{r})$ is a

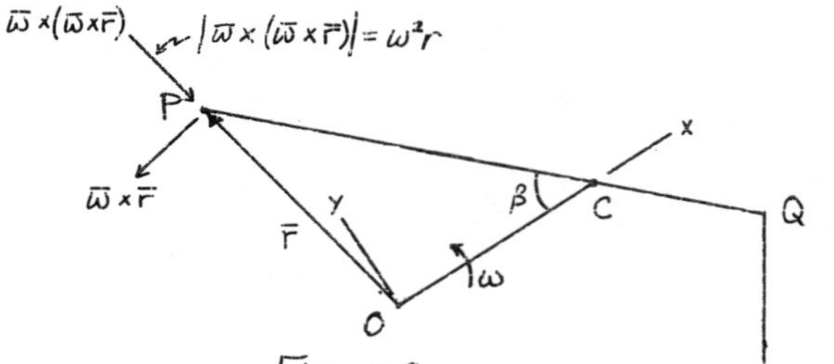

Figure c

vector of magnitude $\omega^2 r$ and is directed toward O, along line PO, as shown.

Finally, to interpret the term $2\bar{\omega} \times \bar{v}_{P/B}$ in Eq. (a), we refer to Fig. d. By Fig. d,

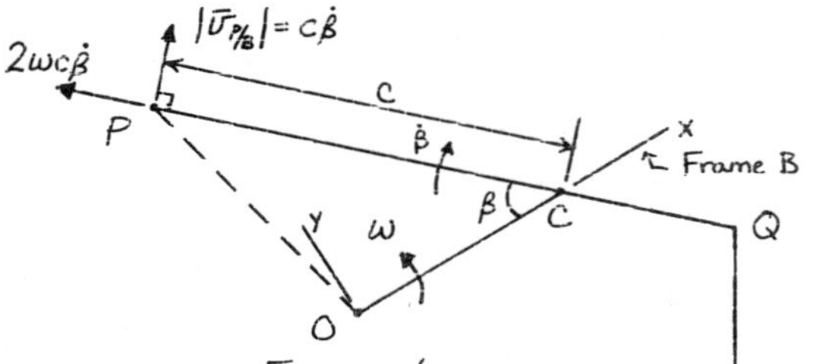

Figure d

$v_{P/B} = |\bar{v}_{P/B}| = c\dot{\beta}$ relative to frame B and the vector $\bar{v}_{P/B}$ is directed as shown. Hence, $\bar{\omega} \times \bar{v}_{P/B}$ has magnitude $\omega c\dot{\beta}$ and it is directed along line PQ, with sense from Q to P. Therefore, $|2\bar{\omega} \times \bar{v}_{P/B}| = 2\omega c\dot{\beta}$ as shown in Fig. d.

By Eqs. (a), (b), (c), (d), and with $\bar{\omega} \times (\bar{\omega} \times \bar{r})$ and $2\bar{\omega} \times \bar{v}_{P/B}$ as defined above,

$$\bar{a}_{P/A} = \bar{a}_{P/B} + \bar{\omega} \times (\bar{\omega} \times \bar{r}) + 2\bar{\omega} \times \bar{v}_{P/B} \qquad (e)$$

Equation (e) is represented graphically by the superposition of Figs. (b), (c), and (d). See also Fig. (a).

17.97

GIVEN: The crank of the slider-crank mechanism of Example 17.29 (see Fig. a below) experiences an angular acceleration of magnitude α.

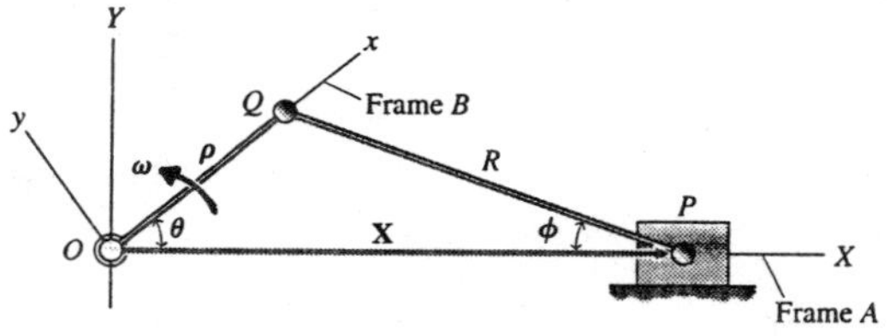

Figure a

FIND:

(a) Show by interpretation of the terms in Eq. (17.23), [see Eq. (a) below] that the acceleration $\bar{a}$ of the slider is represented by the vector sum formed in Fig. b.

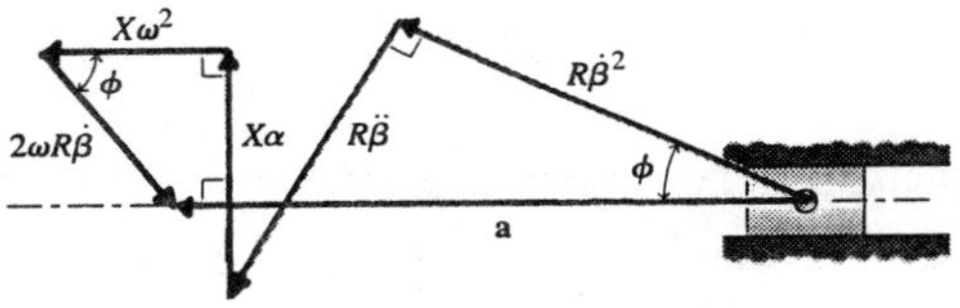

Figure b

(b) Show that this diagram yields the first of Eqs. (f) of Example 17.29. Note that $\bar{r}$ of Eq. (17.23) [See Eq. (a) below] is represented by $\bar{X}$ in Fig. a.

SOLUTION:

(a) Equation (17.23) is repeated here as Eq. (a).

$$\bar{a}_{P/A} = \bar{a}_{P/B} + \bar{a}_{O/A} + \bar{\omega} \times (\bar{\omega} \times \bar{r}) + 2\bar{\omega} \times \bar{v}_{P/B} \qquad (a)$$

where A is a fixed reference (Newtonian) frame and B is a frame that moves in any way with respect to frame A.

Let XYZ axes be attached to frame A and xyz axes be attached to frame B that rotates with crank OQ (Fig. c).

Vectors $\bar{\omega}$ and $\bar{\alpha}$ are in the direction of the positive Z (or z) axis.

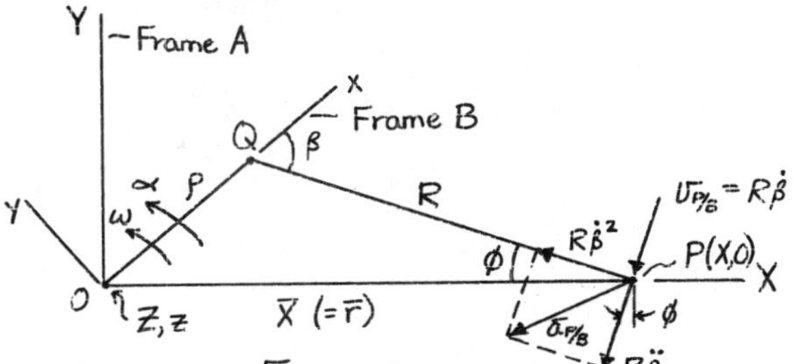

Figure c

By Fig. c,

$$\bar{a}_{O/A} = 0 \qquad (b)$$

$$\bar{a}_{P/B} = \quad R\dot{\beta}^2 \nwarrow,\ \phi,\ P,\ R\ddot{\beta} \swarrow \qquad (c)$$

$$\bar{\alpha} \times \bar{r} = \alpha X \uparrow \qquad (d)$$

$$\bar{\omega} \times (\bar{\omega} \times \bar{r}) = \omega^2 X \leftarrow \qquad (e)$$

$$2\bar{\omega} \times \bar{V}_{P/B} = \quad \phi,\ \searrow 2\omega R\dot{\beta} \qquad (f)$$

Equations (a)-(f) yield Fig. b.

(b) By Fig. b [see also Eqs. (c)-(f)] and (since $a_{py} = 0$), with $\alpha = 0$

$$a_{PY} = R(\dot{\beta})^2 \sin\phi - R\ddot{\beta}\cos\phi - 2\omega R\dot{\beta}\sin\phi = 0$$

or

$$R\ddot{\beta}\cos\phi = [R(\dot{\beta})^2 - 2\omega R\dot{\beta}]\sin\phi \qquad (g)$$

Equation (g) is identical to Eq. (f) of Example 17.29.

17.98

GIVEN: Figure P17.97 (see Fig. a below).

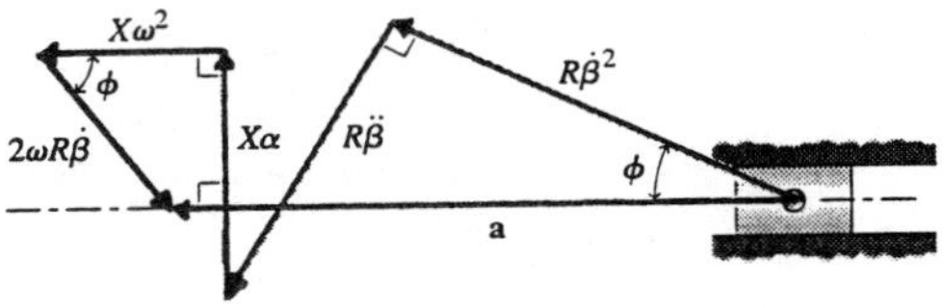

Figure a

FIND: Generalize the second of Eqs. (f) of Example 17.29 [see Eq. (a) below] to include the effect of the magnitude α of the angular acceleration of the crank OQ.

$$R\ddot{\beta}\cos\phi = [R(\dot{\beta})^2 - 2\omega R\dot{\beta}]\sin\phi \qquad (a)$$

SOLUTION:

By Fig. a, since $a_{py} = 0$,

$$a_{py} = R(\dot{\beta})^2 \sin\phi - R\ddot{\beta}\cos\phi + X\alpha - 2\omega R\dot{\beta}\sin\phi = 0$$

or

$$\underline{\underline{R\ddot{\beta}\cos\phi = X\alpha + [R(\dot{\beta})^2 - 2\omega R\dot{\beta}]\sin\phi}}$$

17.99

GIVEN: The linkage shown in Fig. a.

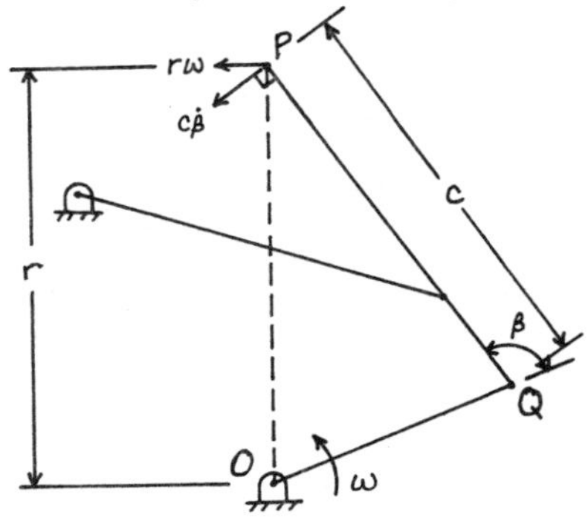

Figure a

FIND: Show by interpretation of the terms in Eq. (17.20) [see Eq. (a) below] that the velocity of point P is the sum of the two vectors shown in Fig. a.

$$\bar{v}_{P/A} = \bar{v}_{P/B} + \bar{v}_{O/A} + \bar{\omega} \times \bar{r} \qquad (a)$$

A = Frame A – a fixed Newtonian frame.
B = Frame B – a frame that moves in any way relative to frame A.

SOLUTION:

Let frame A be a Newtonian frame fixed at O in Fig. b. Let frame B be a frame with xy axes in the plane of the figure and attached to bar OQ, which rotates with angular velocity ω about O. Then, by Fig. b [see Eq. (a)],

$$\bar{v}_{O/A} = 0 \qquad (b)$$

To determine $\bar{v}_{P/B}$, consider Fig. b.

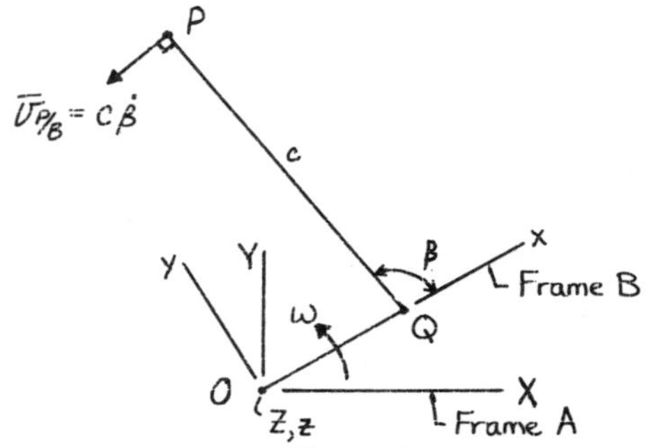

Figure b

By Fig. b, the vector $c\dot{\beta}$ is the velocity $\bar{v}_{P/B}$. Next, by Fig. c, the vector $r\bar{\omega}$ is the velocity $\bar{\omega} \times \bar{r}$. Hence, superposition of Figs. b and c, shows that the velocity of point P is the sum of the two vectors shown in Fig. a. Also, Eq. (a) reduces to

$$\underline{\bar{v}_{P/A} = \bar{v}_{P/B} + \bar{\omega} \times \bar{r}}$$

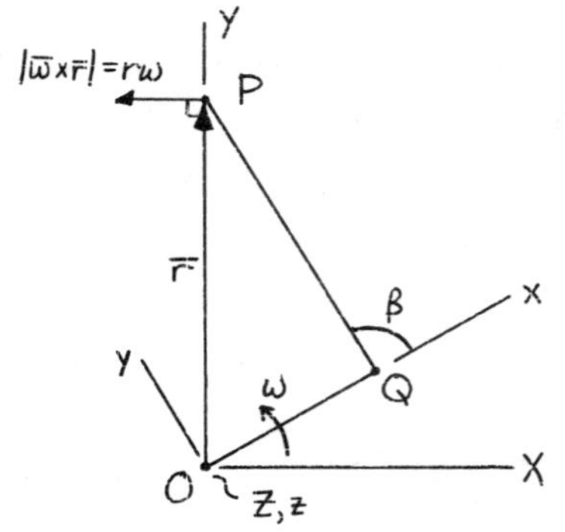

Figure c

17.100

GIVEN: The linkage shown in Fig. a.

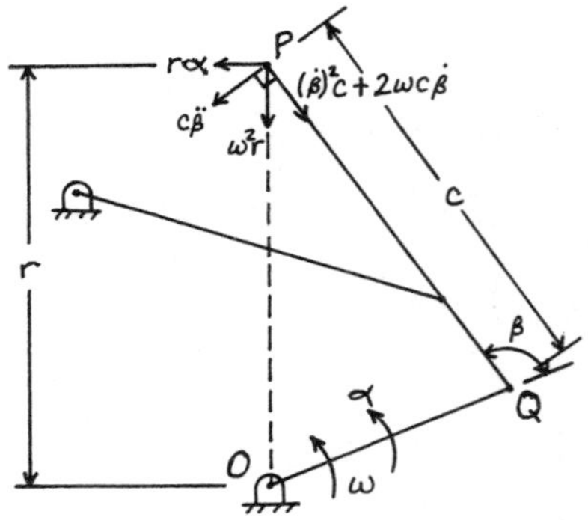

Figure a

FIND: Show by interpretation of the terms in Eq. (17.23) [see Eq. (a) below] that the acceleration of point P is the sum of the four vectors shown in Fig. a.

$$\bar{a}_{P/A} = \bar{a}_{P/B} + \bar{a}_{O/A} + \bar{\alpha} \times \bar{r} + \bar{\omega} \times (\bar{\omega} \times \bar{r}) + 2\bar{\omega} \times \bar{v}_{P/B} \qquad (a)$$

where A is a fixed reference (Newtonian) frame and B is a frame that moves in any way with respect to frame A.

SOLUTION:

Let frame A be a Newtonian frame with XY axes in the plane of the figure. Let frame B be attached to bar OQ, which rotates with angular velocity ω and angular acceleration α about O.

Then, by Fig. b [see Eq. (a)],

$$\bar{a}_{O/A} = 0 \qquad (b)$$

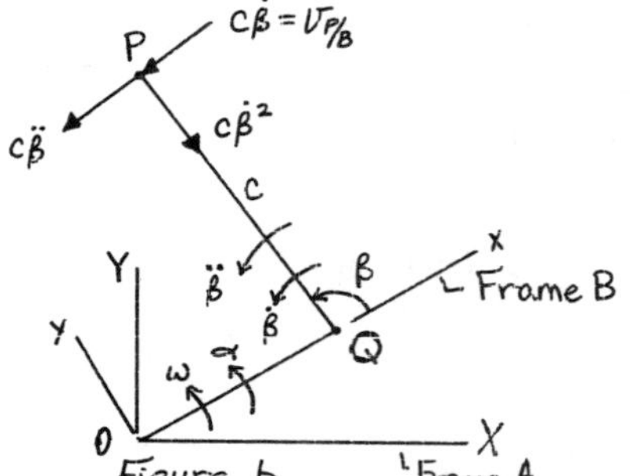

Figure b

(continued)

17.100 (cont.)

Also, by Fig. b, the vectors $\overline{c\dot{\beta}}$ and $\overline{c\dot{\beta}^2}$ are the components of vectors $\bar{a}_{P/B}$ [see Eq. (a)], and the vector $\overline{c\dot{\beta}}$ is the vector $\bar{v}_{P/B}$.

Next, by Fig. c, the vectors $\overline{r\alpha}$ and $\overline{r\omega^2}$ are the vectors $\bar{\alpha}\times\bar{r}$ and $\bar{\omega}\times(\bar{\omega}\times\bar{r})$, respectively.

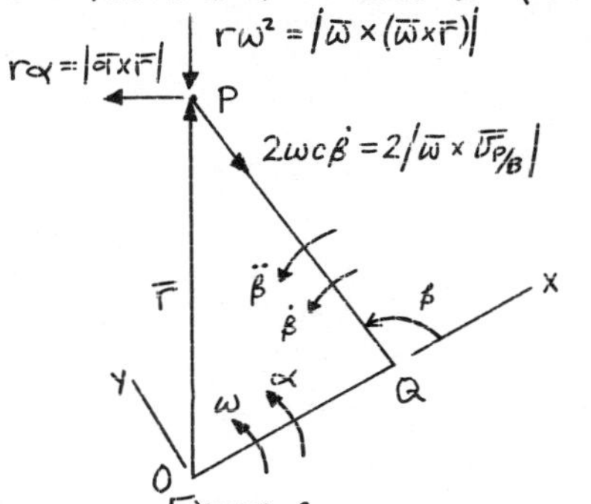

Figure c

By Fig. c, the vector $\overline{2\omega c\dot{\beta}}$ is the vector $2\bar{\omega}\times\bar{v}_{P/B}$. Hence, with the equation

$$\bar{a}_{P/A} = \bar{a}_{P/B} + \bar{\alpha}\times\bar{r} + \bar{\omega}\times(\bar{\omega}\times\bar{r}) + 2\bar{\omega}\times\bar{v}_{P/B},$$

superposition of Figs. b and c yields the four vectors that represent $\bar{a}_{P/A}$ (see Fig. a).

17.101

GIVEN: The crank OC of the four-bar linkage shown in Fig. a rotates at a constant rate of 30 rad/s.

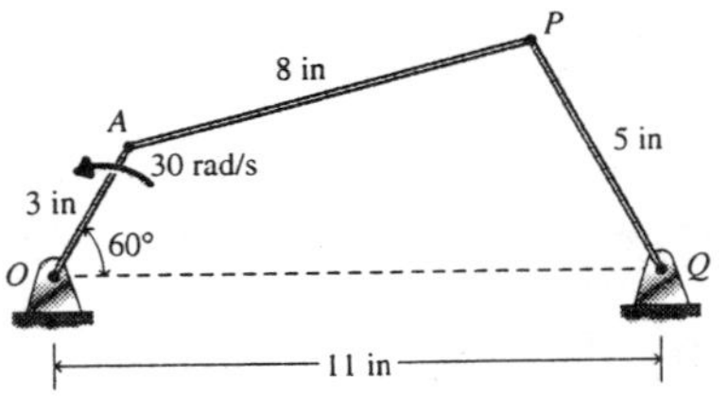

Figure a

FIND:

(a) Adopting a rotating coordinate system that moves with the crank OC, construct to scale a vector diagram for the velocity of point P by means of Eq. (17.20); see Eq. (a) below.

$$\bar{v}_{P/A} = \bar{v}_{P/B} + \bar{v}_{O/A} + \bar{\omega}\times\bar{r} \qquad (a)$$

where A is a fixed reference (Newtonian) frame and B is a frame that rotates with the crank OA.

(b) Determine the speed of point P by measuring the diagram, utilizing the fact that the velocity of point P is perpendicular to the line PQ.

SOLUTION:

(a) By Fig. b and Eq. (a), since point O is fixed,

$$\bar{v}_{O/A} = 0.$$

Therefore, Eq. (a) reduces to

$$\bar{v}_{P/A} = \bar{v}_{P/B} + \bar{\omega}\times\bar{r} \qquad (b)$$

By Eq. (b) and the parallelogram construction, we obtain the vector diagram for $\bar{v}_{P/A}$ (Fig. b).

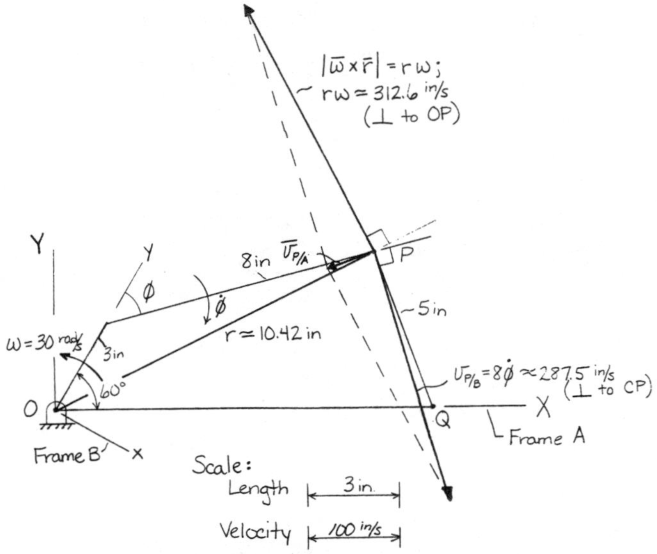

Figure b

(b) $\underline{\bar{v}_P \approx 62.5 \text{ in/s by measurement of Fig. b.}}$

Also note that $8\dot{\phi} \approx 287.5$ in/s, or $\dot{\phi} \approx 35.94$ rad/s

17.102

GIVEN: The crank OC of the four-bar linkage shown in Fig. a rotates at a constant rate of 30 rad/s.

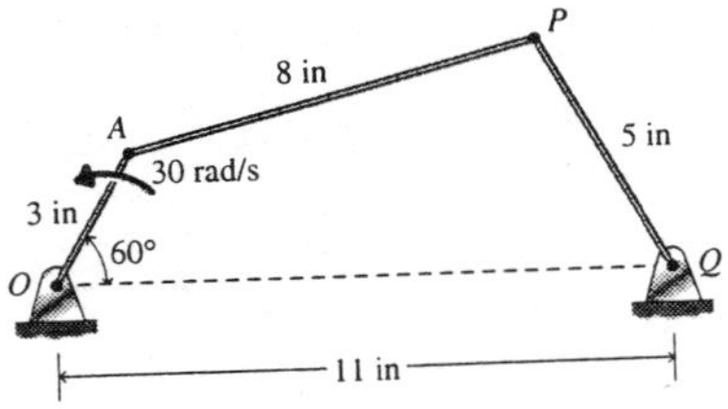

Figure a

FIND:

(a) Adopt a rotating coordinate system that moves with the crank OC, and construct to scale a vector diagram showing the components of the acceleration of point P by means of Eq. (17.23); see Eq. (a) below.

$$\bar{a}_{P/A} = \bar{a}_{P/B} + \bar{a}_{O/A} + \bar{\alpha}\times\bar{r} + \bar{\omega}\times(\bar{\omega}\times\bar{r}) + 2\bar{\omega}\times\bar{v}_{P/B} \quad (a)$$

(continued)

17.102 (cont.)

where A is a fixed reference (Newtonian) frame and B is a frame that rotates with crank OC. Assume that $\bar{v}_{P/A} = v_P = 62.5$ in/s (This value was obtained in the solution of Problem 17.101.)

(b) Construct a polygon of the acceleration components that are to be added to obtain $\bar{a}_{P/A}$.

(c) Utilizing the diagrams of parts a and b, and measuring the polygon, determine the magnitude and direction relative to XY axes with Origin at O of $\bar{a}_{P/A}$. Note that the component of acceleration of point P along line PQ is $v_P^2/5$ [in/s²], directed from P to Q.

SOLUTION:

(a) Since point O is fixed and ω is constant, $\bar{a}_{O/A} = 0$ and $\bar{\alpha} = \dot{\bar{\omega}} = 0$. Hence, Eq. (a) reduces to

$$\bar{a}_{P/A} = \bar{a}_{P/B} + \bar{\omega}\times(\bar{\omega}\times\bar{r}) + 2\bar{\omega}\times\bar{v}_{P/B} \qquad (b)$$

By Fig. b and kinematics, the components of $\bar{a}_{P/B}$ are

$$\bar{a}_{P/B} = (8\dot{\phi}^2, 8\ddot{\phi}) \qquad (c)$$

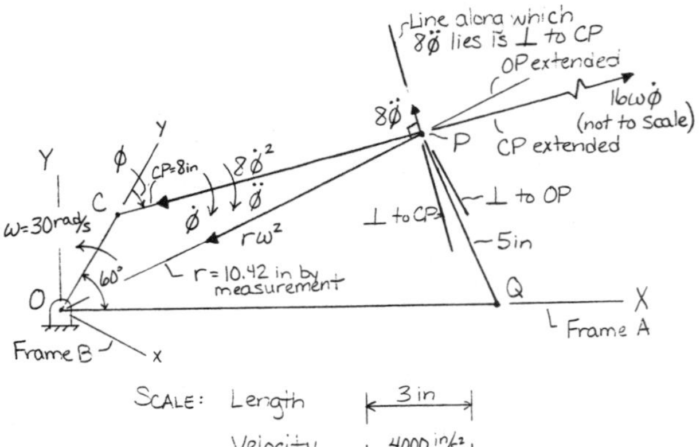

Figure b

Also, the components of $\bar{\omega}\times(\bar{\omega}\times\bar{r})$ and $2\bar{\omega}\times\bar{v}_{P/B}$ are

$$\bar{\omega}\times(\bar{\omega}\times\bar{r}) = r\omega^2 \text{ directed from P to O} \qquad (d)$$

and

$$2\bar{\omega}\times\bar{v}_{P/B} = 16\omega\dot{\phi} \text{ directed from C to P.} \qquad (e)$$

since $\bar{v}_{P/B}$ is perpendicular to CP and directed down and to the right.

To draw the components of $\bar{a}_{P/A}$, we need to know the magnitudes of $\dot{\phi}$ and $\ddot{\phi}$. By the solution of Problem 17.101, $\dot{\phi} \approx 35.94$ rad/s. If Problem 17.101 has not been previously assigned, the value of $\dot{\phi}$ must be determined here. To determine $\ddot{\phi}$, we know that the component of acceleration along PQ is $v_P^2/5$ or $(62.5)^2/5 = 781.25$ in/s². Hence, the projections of $\bar{a}_{P/B}$, $\bar{\omega}\times(\bar{\omega}\times\bar{r})$, and $2\bar{\omega}\times\bar{v}_{P/B}$ [see Eqs. (c), (d), and (e)] along PQ determine $\ddot{\phi}$.

To calculate these projections, we need to determine the angles between the components of $\bar{a}_{P/A}$ and the line PQ.

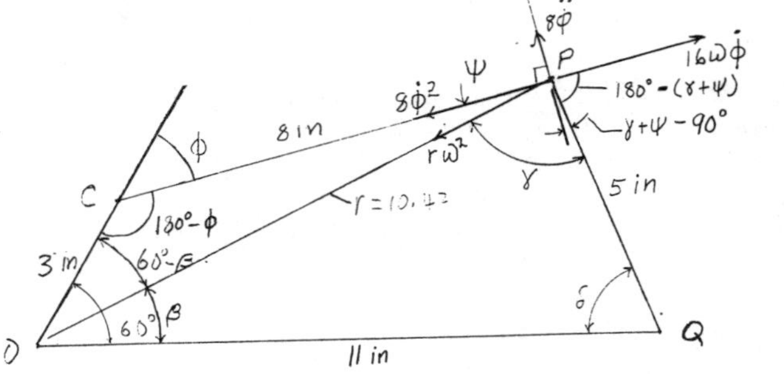

Figure c

By Fig. c, we have

$$\frac{\sin\beta}{5} = \frac{\sin\gamma}{11} = \frac{\sin\delta}{10.42} \qquad (f)$$

$$r^2 = 10.42^2 = 11^2 + 5^2 - 2(5)(11)\cos\delta$$

The solution of Eqs. (f) is (see Fig. c)

$$\beta = 26.82^\circ$$
$$\gamma = 83.07^\circ \qquad (g)$$
$$\delta = 70.11^\circ$$

Also, by Fig. c,

$$\frac{\sin(60^\circ - 26.82^\circ)}{8} = \frac{\sin(180^\circ - \phi)}{10.42} = \frac{\sin\psi}{3} \qquad (h)$$

The solution of Eq. (h) is

$$\phi = 45.46^\circ$$
$$\psi = 11.84^\circ \qquad (i)$$

Hence, by Eqs. (g) and (i) and Fig. c, we have

$$\frac{v_P^2}{5} = 781.25 = -8\ddot{\phi}\cos 4.91^\circ + 8\dot{\phi}^2\cos 94.91^\circ + r\omega^2\cos 83.07^\circ + 16\omega\dot{\phi}\cos 85.09^\circ \qquad (j)$$

where

$$8\dot{\phi}^2 = 8(35.94)^2 = 10\,333.5 \text{ in/s}^2$$
$$r\omega^2 = (10.42)(30)^2 = 9\,378 \text{ in/s}^2 \qquad (k)$$
$$16\omega\dot{\phi} = 16(30)(35.94) = 17\,251.2 \text{ in/s}^2$$

By Eqs. (j) and (k),

$$\ddot{\phi} = 118.23 \text{ rad/s}^2$$

So,

$$8\ddot{\phi} = 945.84 \text{ in/s}^2 \qquad (\ell)$$

Then, by Eqs. (k) and (ℓ), the components of $\bar{a}_{P/A}$ are drawn to scale in Fig. b.

(b) With Fig. b, the polygon of the components of the acceleration is constructed in Fig. d, along with the vector $\bar{a}_{P/A}$.

(continued)

17.102 (cont.)

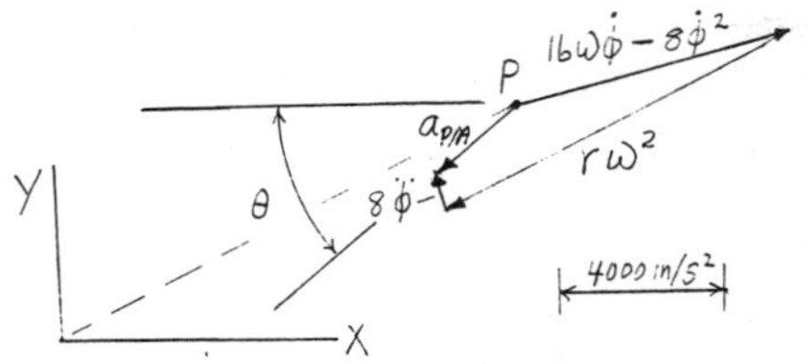

Figure d

(c) By measurement of Fig. d,

$$a_{P/A} = 2750 \text{ in/s}^2$$
$$\theta = 41°$$

17.103

GIVEN: "The Scrambler" is a carnival ride consisting of three main arms that rotate at an angular speed of $\omega_1 = \dot{\theta}_1$ about a center pivot shaft $\bar{O}$ plus three sets of four secondary arms that rotate at an absolute angular speed $\omega_2 = \dot{\theta}_2$ about a moving pivot point O at the end of each main arm (Fig. a). Each secondary arm carries a bench that can hold up to three riders. The initial configuration is $\theta_1 = \theta_2 = 0$, when time $t = 0$.

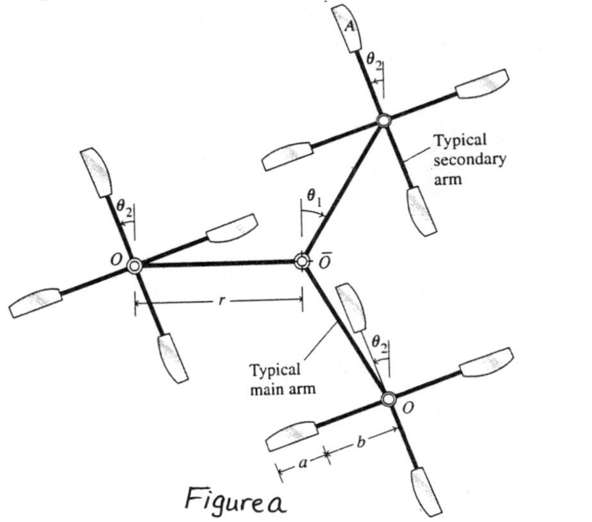

Figure a

FIND:

(a) Assume that the rider at A is on the extreme outside edge of the bench. Derive a formula for the velocity of the rider at A as a function of ω_1, ω_2, r, a, and b.

(b) Determine the time the maximum speed is first attained in terms of ω_1 and ω_2.

(c) The Scrambler operates at constant angular speeds $\omega_1 = 12$ rpm and $\omega_2 = 15$ rpm, with r = 13 ft, a = 4 ft, and b = 6 ft. Determine the maximum speed experienced by the rider and the associated configuration.

SOLUTION:

(a) Let axes xy be attached to O and translate with O. Also, let XY be Newtonian axes with origin at $\bar{O}$. (Fig. b)

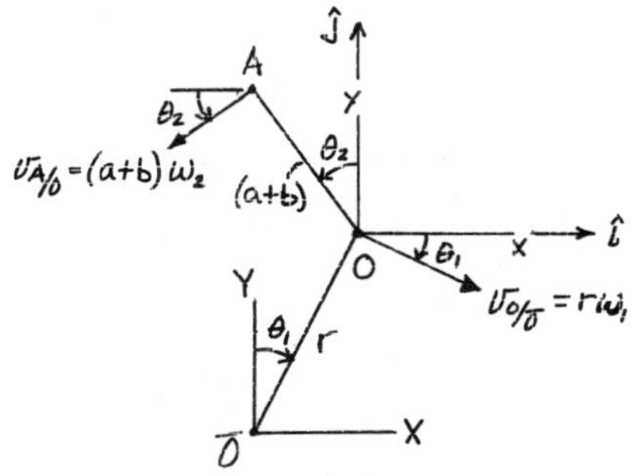

Figure b

The velocity of A relative to axes xy is $\bar{v}_{A/O}$ and the velocity of point O relative to $\bar{O}$ is $\bar{v}_{O/\bar{O}}$. Hence, by kinematics, the velocity of A relative to $\bar{O}$ is

$$\bar{v}_{A/\bar{O}} = \bar{v}_{A/O} + \bar{v}_{O/\bar{O}} \qquad (a)$$

By Fig. b,

$$\bar{v}_{A/O} = (a+b)\omega_2[-(\cos\theta_2)\hat{\imath} - (\sin\theta_2)\hat{\jmath}]$$
$$\bar{v}_{O/\bar{O}} = r\omega_1[(\cos\theta_1)\hat{\imath} - (\sin\theta_1)\hat{\jmath}] \qquad (b)$$

Since $\theta_1 = \theta_2 = 0$ when $t = 0$, $\theta_1 = \omega_1 t$ and $\theta_2 = \omega_2 t$. Then, by Eqs. (a) and (b),

$$\bar{v}_{A/\bar{O}} = [r\omega_1\cos(\omega_1 t) - (a+b)\omega_2\cos(\omega_2 t)]\hat{\imath} - [r\omega_1\sin(\omega_1 t) + (a+b)\omega_2\sin(\omega_2 t)]\hat{\jmath} \qquad (c)$$

(b) By Eq. (c),

$$\bar{v}_{A/\bar{O}}^2 = r^2\omega_1^2 + (a+b)^2\omega_2^2 - 2r(a+b)\omega_1\omega_2\cos[(\omega_1+\omega_2)t] \qquad (d)$$

For maximum (or minimum) $\bar{v}_{A/\bar{O}}$, $d\bar{v}_{A/\bar{O}}/dt = 0$, or by Eq. (d),

$$\sin(\omega_1+\omega_2)t = 0$$

So, $(\omega_1+\omega_2)t = \pi$ or 0. For maximum $\bar{v}_{A/\bar{O}}$ [Eq. (d)],

$$t = \frac{\pi}{\omega_1+\omega_2} \qquad (e)$$

(c) In terms of rad/s,

$$\omega_1 = 12\,\frac{\text{rev}}{\text{min}}\left(\frac{2\pi\text{ rad}}{\text{rev}}\right)\left(\frac{1\text{ min}}{60\text{ sec}}\right) = 0.4\pi \text{ rad/s}$$
$$\omega_2 = 15\,\frac{\text{rev}}{\text{min}}\left(\frac{2\pi\text{ rad}}{\text{rev}}\right)\left(\frac{1\text{ min}}{60\text{ sec}}\right) = 0.5\pi \text{ rad/s} \qquad (f)$$

and by Eqs. (e) and (f)

$$t = \frac{1}{0.9} = 1.\overline{1} \text{ sec} \qquad (g)$$

Then, by Eqs. (d), (f), and (g), with r = 13 ft, a = 4 ft, and b = 6 ft,

$$(\bar{v}_{A/\bar{O}})_{max} = 32.04 \text{ ft/s}$$

(continued)

17.103 (cont.)

Since $t = 1.11\bar{1}$ s, when $v_{A/\bar{O}}$ is maximum,

$$\theta_1 = \omega_1 t = (0.4\pi)(1.11\bar{1}) = 1.3963 \text{ rad} = 80^\circ$$

$$\theta_2 = \omega_2 t = (0.5\pi)(1.11\bar{1}) = 1.7453 \text{ rad} = 100^\circ$$

Hence, the configuration of the Scrambler is shown in Fig. C. Note by Fig. C that

$$(v_{A/\bar{O}})_{max} = 15.708 + 16.336 = 32.04 \text{ ft/s}.$$

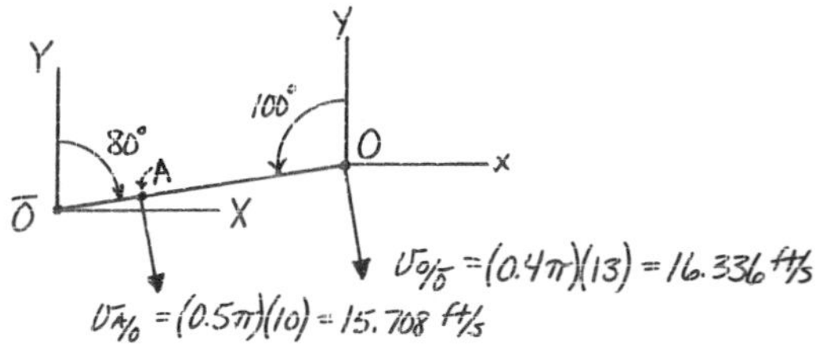

Figure C

17.104

GIVEN: Refer to the description of "The Scrambler" and the problem statement of Problem 17.103. See Fig. a below.

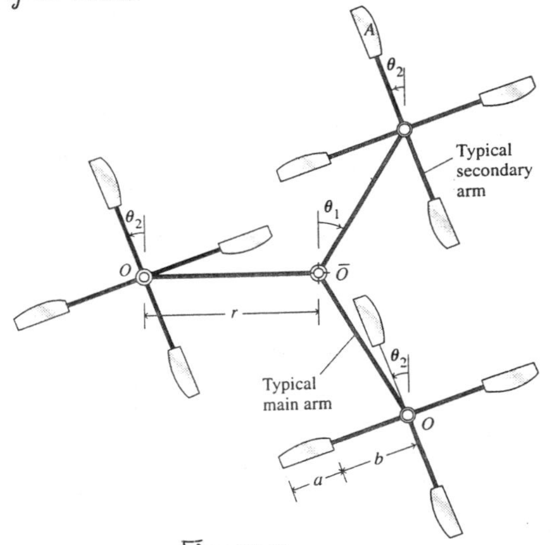

Figure a

FIND:

(a) Derive a formula for the acceleration $\bar{a}_{P/\bar{O}}$ of the rider A as a function of ω_1, ω_2, r, a, and b, where as noted in Problem 17.103, since $t = 0$ when $\theta_1 = \theta_2 = 0$, and $\omega_1 t = \theta_1$; $\omega_2 t = \theta_2$.

(b) Determine the maximum acceleration $(a_{P/\bar{O}})_{max}$ experienced by the rider.

SOLUTION:

Let axes xy be attached to point O and translate with O, and axes XY be Newtonian axes with origin at $\bar{O}$ (Fig. b).

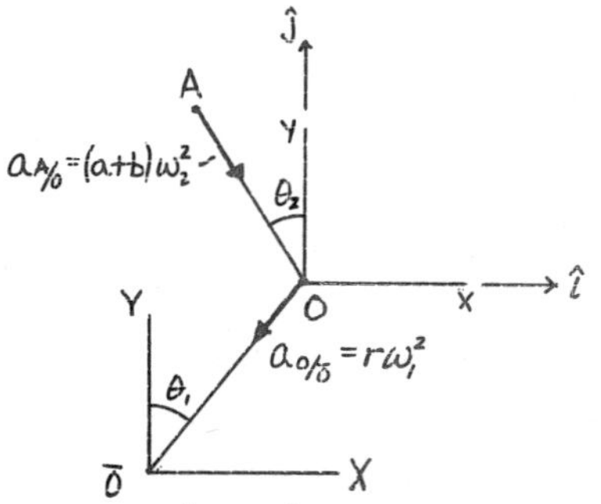

Figure b

Since ω_1 and ω_2 are constants, the accelerations $\bar{a}_{A/O}$ and $\bar{a}_{O/\bar{O}}$ are directed as shown in Fig. b. Hence,

$$\bar{a}_{A/O} = (a+b)\omega_2^2[(\sin\theta_2)\hat{i} - (\cos\theta_2)\hat{j}]$$

$$\bar{a}_{O/\bar{O}} = r\omega_1^2[-(\sin\theta_1)\hat{i} - (\cos\theta_1)\hat{j}]$$

or since $\omega_1 t = \theta_1$ and $\omega_2 t = \theta_2$,

$$\bar{a}_{A/O} = (a+b)\omega_2^2[(\sin\omega_2 t)\hat{i} - (\cos\omega_2 t)\hat{j}]$$

$$\bar{a}_{O/\bar{O}} = r\omega_1^2[-(\sin\omega_1 t)\hat{i} - (\cos\omega_1 t)\hat{j}] \qquad \text{(a)}$$

By Kinematics,

$$\bar{a}_{A/\bar{O}} = \bar{a}_{A/O} + \bar{a}_{O/\bar{O}} \qquad \text{(b)}$$

By Eqs. (a) and (b),

$$\bar{a}_{A/\bar{O}} = [(a+b)\omega_2^2 \sin\omega_2 t - r\omega_1^2 \sin\omega_1 t]\hat{i} - [(a+b)\omega_2^2\cos\omega_2 t + r\omega_1^2\cos\omega_1 t]\hat{j} \qquad \text{(c)}$$

(b) By Eq. (c),

$$(a_{A/\bar{O}})^2 = (a+b)^2\omega_2^4 + r^2\omega_1^4 + 2(a+b)r\omega_1^2\omega_2^2\cos(\omega_1+\omega_2)t \qquad \text{(d)}$$

For $a_{A/\bar{O}}$ to be a maximum (or minimum), $\frac{da_{A/\bar{O}}}{dt} = 0$. Hence, by Eq. (d)

$$\sin(\omega_1+\omega_2)t = 0$$

$$t = \frac{\pi}{\omega_1+\omega_2} \quad \text{or} \quad t = 0 \qquad \text{(e)}$$

With $\omega_1 = 12$ rpm $= 0.4\pi$ rad/s and $\omega_2 = 15$ rpm $= 0.5\pi$ rad/s, Eq. (e) yields $t = 1/0.9 = 1.11\bar{1}$ sec or $t = 0$. Then, by Eq. (d), with $r = 13$ ft, $a = 4$ ft, and $b = 6$ ft (See Problem 17.103)

$$(a_{P/\bar{O}})_{max} = 45.20 \text{ ft/s}^2 \text{ for } t = 0.$$

$$(a_{P/\bar{O}})_{min} = 4.14 \text{ ft/s}^2 \text{ for } t = 1.11\bar{1} \text{ sec}.$$

Note that 45.20 ft/s² = 1.4g; or the maximum acceleration of the passenger is 1.4 times the acceleration of gravity.

17.105

GIVEN: Refer to the detailed description given in Problem 17.105 of the text and to Fig. a below.

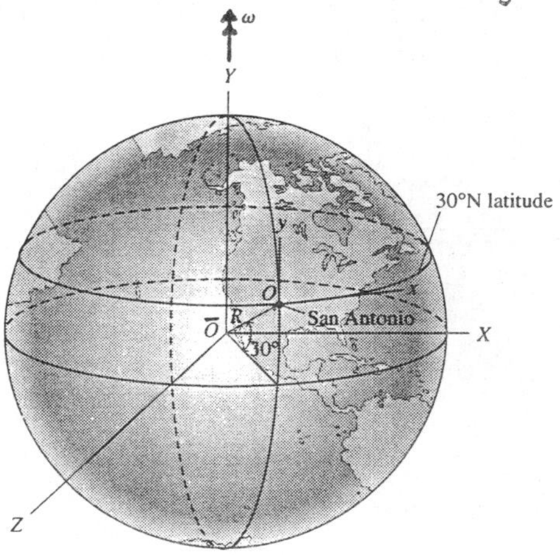

Figure a

FIND: Design a track for experimental studies of high speed missiles, with restrictions outlined in the problem statement of Problem 17.105 in the text.

SOLUTION:

Consider fixed axes (X, Y, Z) with origin $\bar{O}$ at the center of the earth and axes (x, y, z) with origin O at the intersection of a longitude and the 30°N Latitude (Figs. a and b).

Case (a) -- Try first laying the track along the 30°N latitude. Let the missile travel from West to East. Use Eq. (17.23), namely

$$\bar{a}_{P/A} = \bar{a}_{P/B} + \bar{a}_{O/A} + \bar{\alpha}\times\bar{r} + \bar{\omega}\times(\bar{\omega}\times\bar{r}) + 2\bar{\omega}\times\bar{v}_{P/B} \qquad (a)$$

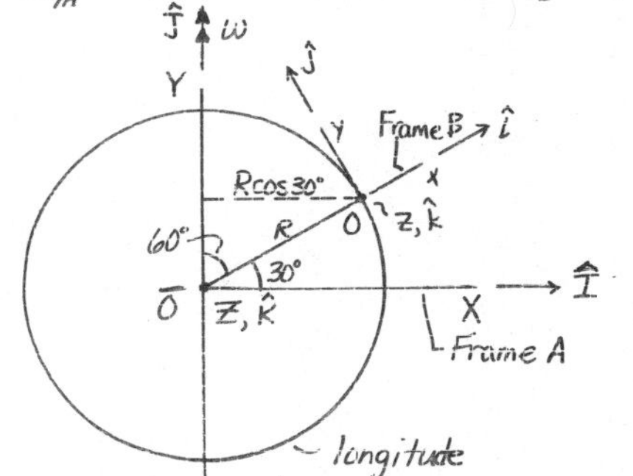

Figure b

Let the missile coincide with O. Then, $\bar{r} = 0$ (see Fig. 17.25). The velocity of the missile relative to Frame B is therefore

$$\bar{v}_{P/B} = -600\,\hat{k} \qquad (b)$$

with $R = 6.437\times10^6$ m, $R\cos 30° = 5.5746\times10^6$ m. Then, the acceleration

$$\bar{a}_{O/A} = \omega^2 R\cos 30°(-\cos 30°\,\hat{\imath} + \sin 30\,\hat{\jmath}) \qquad (c)$$

since axes (x, y, z) rotate with angular speed of the earth's spin relative to frame A.

Thus,

$$\omega = \frac{2\pi}{(24)(60)(60)} = 7.2722\times10^{-5}\ \text{rad/s} = \text{constant};\ \alpha = 0.$$

and $\omega^2 = 5.2885\times10^{-9}$ rad²/s²

and $\bar{a}_{O/A} = -0.025530\,\hat{\imath} + 0.01474\,\hat{\jmath}$ (m/s²) (d)

Since the missile travels a circular path of radius $R\cos 30° = 5.5746\times10^6$ m about the Y axis, with speed $v_{P/B} = 600$ m/s,

$$\bar{a}_{P/B} = \frac{v_{P/B}^2}{R\cos 30°}(-\cos 30°\,\hat{\imath} + \sin 30°\,\hat{\jmath})$$

$$= -0.0559267\,\hat{\imath} + 0.032289\,\hat{\jmath} \qquad (e)$$

Since $\bar{r} = 0$, $\bar{\omega}\times(\bar{\omega}\times\bar{r}) = 0$ (see Fig. 17.25).

Also,

$$2\bar{\omega}\times\bar{v}_{P/B} = 2(7.2722\times10^{-5})[0.50\,\hat{\imath} + 0.866\,\hat{\jmath}]\times(-600\,\hat{k})$$

$$= -0.07557\,\hat{\imath} + 0.04363\,\hat{\jmath} \qquad (f)$$

Then, by Eq. (a), with $\bar{\alpha}\times\bar{r} = 0$,

$$\bar{a}_{P/A} = -0.157031\,\hat{\imath} + 0.090666\,\hat{\jmath}\ (\text{m/s}^2) \qquad (g)$$

The horizontal acceleration perpendicular to the track is

$a_\perp = 0.09066$ m/s² (See Fig. b)

Since $a_\perp = 0.09066 > 0.06$ (m/s²) (h)

this design is not satisfactory.

Case (b) -- Let the track lie along the 30°N latitude, but let the missile travel for East to West. For this case, $\bar{a}_{O/A}$, $\bar{a}_{P/B}$, $\bar{\omega}\times(\bar{\omega}\times\bar{r})$ remain the same as in case a. However, $\bar{v}_{P/B} = 600\hat{k}$.

Now $2\bar{\omega}\times\bar{v}_{P/B} = 0.0755\,\hat{\imath} - 0.04363\,\hat{\jmath}$ (m/s²)

and hence,

$$\bar{a}_{P/A} = -0.009227\,\hat{\imath} - 0.00341\,\hat{\jmath}\ (\text{m/s}^2) \qquad (i)$$

and $a_\perp = 0.00341$ m/s² < 0.06 m/s².

This design meets the acceleration requirement.

Case (c) -- Let the track be laid along a longitude and the missile travel South to North. Again, $a_{O/A}$ and $\bar{\omega}\times(\bar{\omega}\times\bar{r})$ are the same as in cases (a) and (b). However, now $\bar{v}_{P/B} = 600\,\hat{\jmath}$;

$$\bar{a}_{P/B} = \frac{-v_{P/B}^2}{R}\,\hat{\imath} = 0.0559267\,\hat{\imath}$$

$$2\bar{\omega}\times\bar{v}_{P/B} = 2(7.2722\times10^{-5})(0.5\,\hat{\imath} + 0.866\,\hat{\jmath})\times(600\,\hat{\jmath})$$

$$= 0.04363\,\hat{k}$$

and

$$\bar{a}_{P/A} = -0.08146\,\hat{\imath} + 0.01474\,\hat{\jmath} + 0.04363\,\hat{k}$$

The horizontal acceleration perpendicular to the track is

$a_\perp = 0.04363$ m/s² < 0.06 m/s²

This design is also satisfactory.

Case (d) -- Again let the track be along a longitude, but with missile traveling North to South. Then, $\bar{v}_{P/B} = -600\,\hat{\jmath}$.

The terms $\bar{a}_{O/A}$, $\bar{a}_{P/B}$, and $\bar{\omega}\times(\bar{\omega}\times\bar{r})$ are the same as in case (c); also $\bar{\alpha}\times\bar{r} = 0$. (continued)

17.105 (cont.)

However, now

$$2\bar{\omega} \times \bar{v}_{P/B} = -0.04363\,\hat{k}$$

and again the horizontal acceleration perpendicular to the track is

$$a_\perp = 0.04363 \text{ m/s}^2 < 0.06 \text{ m/s}^2.$$

This design is also satisfactory.

Recommendation: Depending on the availability of sites, cases (b), (c), and (d) meet the design criteria on acceleration. Case (b) is the most conservative ($0.00341 \ll 0.06$ m/s²), but the missile must travel from East to West. Cases (c) and (d) show that the missile may travel either from South to North, or from North to South (with $0.04363 < 0.06$ m/s²). If this flexibility is desirable, the track should be laid along a longitude. If this condition is not necessary, Case (b) is the best design since it is most conservative ($0.00341 \ll 0.06$ m/s²)

There may be other directions (non-longitudinal or non-latitudinal) that may meet the design requirements, but the analysis for such directions is more complex. See also Problem 13.153.

17.111

GIVEN: A body rotates with angular speed ω about the z axis (Fig. a).

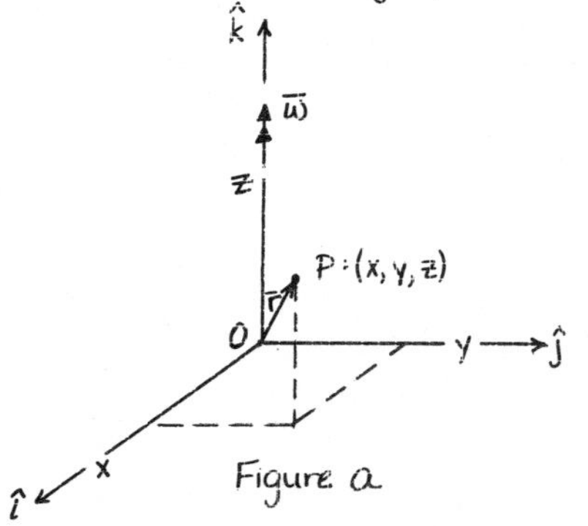

Figure a

FIND: Prove that the (x, y, z) projections of the velocity of the particle at the point (x, y, z) are

$$v_x = -y\omega\,, \quad v_y = x\omega\,, \quad v_z = 0 \qquad \text{(a)}$$

SOLUTION:

By Fig. a and kinematics,

$$\bar{v}_P = \bar{\omega} \times \bar{r} \qquad \text{(b)}$$

where

$$\bar{\omega} = \omega\hat{k}$$

$$\bar{r} = x\hat{\imath} + y\hat{\jmath} + z\hat{k} \qquad \text{(c)}$$

Equations (b) and (c) yield

$$v_P = \begin{vmatrix} \hat{\imath} & \hat{\jmath} & \hat{k} \\ 0 & 0 & \omega \\ x & y & z \end{vmatrix} = -y\omega\hat{\imath} + x\omega\hat{\jmath} + 0\hat{k}$$

Hence, (see Eq. (a))

$$v_x = -y\omega\,, \quad v_y = x\omega\,, \quad v_z = 0$$

18.1

Given: A package that weighs 45 N is gently lowered on to a conveyor belt that moves horizontally at 5.4 m/s. The coefficient of sliding friction is $\mu_k = 0.30$.

Calculate the time required for the package to stop sliding on the belt.

Solution: Consider the free-body diagram of the package (Fig. a).

By Fig. a,

$$\Sigma F_x = \mu_k N = ma$$
$$\Sigma F_y = N - mg = 0 \qquad (a)$$

Solving Eqs. (a) for the acceleration a, we find (with $\mu_k = 0.30$)

$$a = \mu_k g = (0.30)(9.81) = 2.94 \frac{m}{s^2}$$

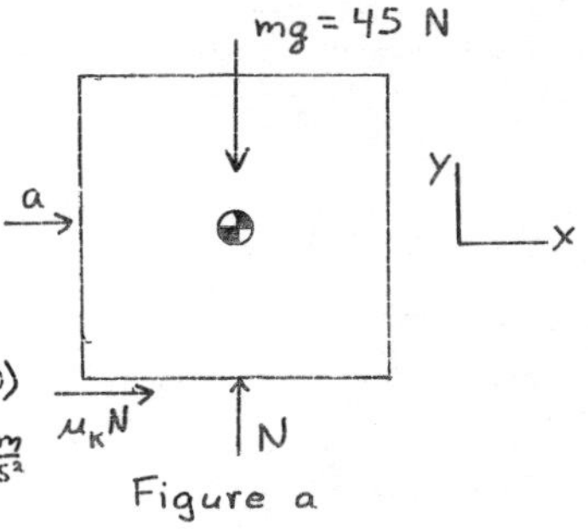

Figure a

Now $a = dv/dt$. So, by integration,

$$\int_0^t a\,dt = \int_0^{5.4} dv \quad \text{or} \quad 2.94t = 5.4$$

Hence, $\underline{t = 1.835\,s}$

18.2

Given: A uniform block weighs 180 N. It is subjected to the forces shown in Fig. a and slides on the horizontal surface. The coefficient of kinetic friction is $\mu_k = 0.20$.

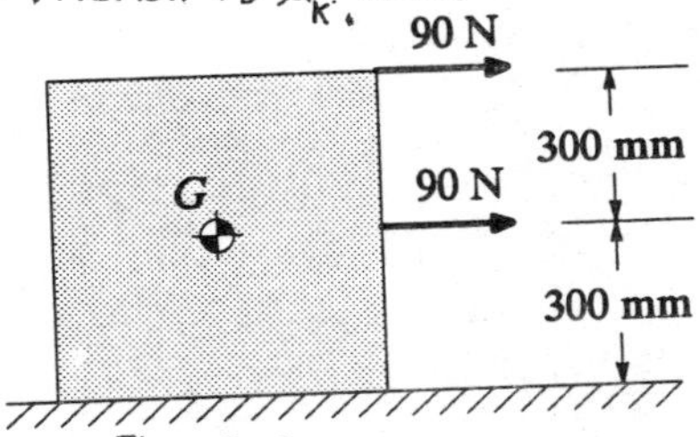

Figure a

Determine the location of the resultant normal force that the surface exerts on the block.

Solution: The free-body diagram of the block is shown in Fig. b.

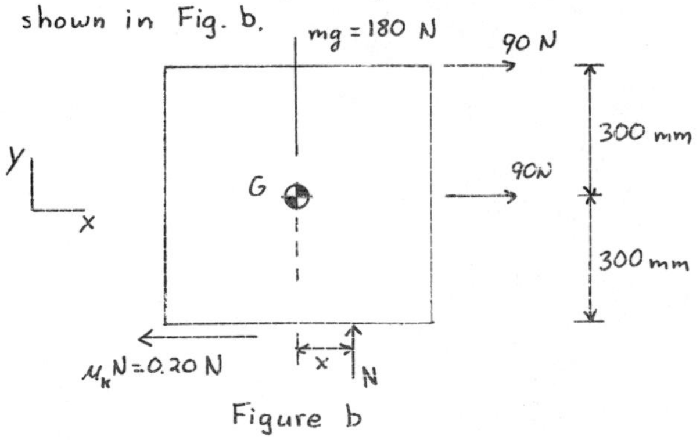

Figure b

By Fig. b,

$$\Sigma F_y = N - mg = 0; \quad N = mg = 180 \text{ N}$$
$$\circlearrowleft \Sigma M_G = x(180) - 90(0.30) - (0.2)(180)(0.30) \qquad (a)$$

Solving Eq. (a), we obtain

$$\underline{x = 0.21\,m}$$

18.3

Given: A homogeneous cubical crate rests on a horizontal floor. A workman pushes the crate at the middle of the top edge. The coefficient of kinetic friction is μ_K

Derive a formula in terms of μ_K and g for the maximum acceleration that the crate can have, if it does not start to tip over.

Solution: Consider the free-body diagram of the crate (Fig. a), at impending tipping.

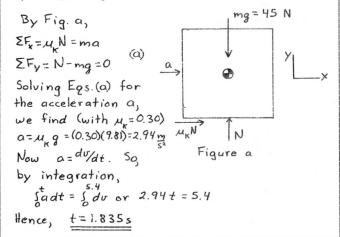

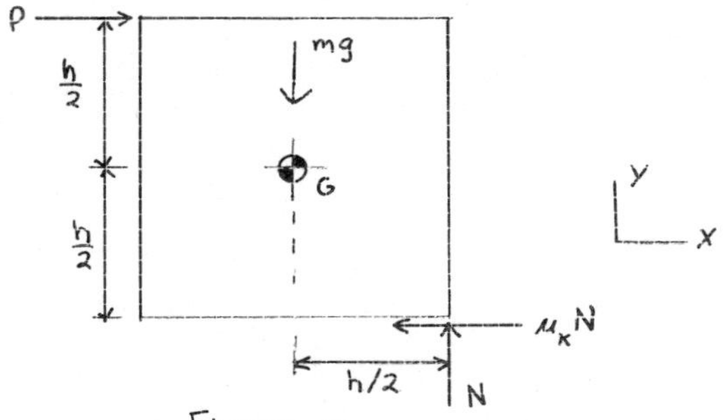

Figure a

By Fig. a and kinetics,

$$\Sigma F_x = P - \mu_K N = m\,a_{max}$$
$$\Sigma F_y = N - mg = 0 \qquad (a)$$
$$\circlearrowleft \Sigma M_G = N(\tfrac{h}{2}) - \mu_K N(\tfrac{h}{2}) - P(\tfrac{h}{2}) = 0$$

The solution of Eq. (a) is

$$N = mg$$
$$P = mg(1 - \mu_K)$$
$$\underline{a_{max} = g(1 - 2\mu_K)}$$

18.4

Given: A homogeneous triangular prismatic test model of a race car weighs 20 lb (Fig. a). It is propelled by a jet that exerts a horizontal force of 10 lb, and it rolls on frictionless wheels of negligible mass.

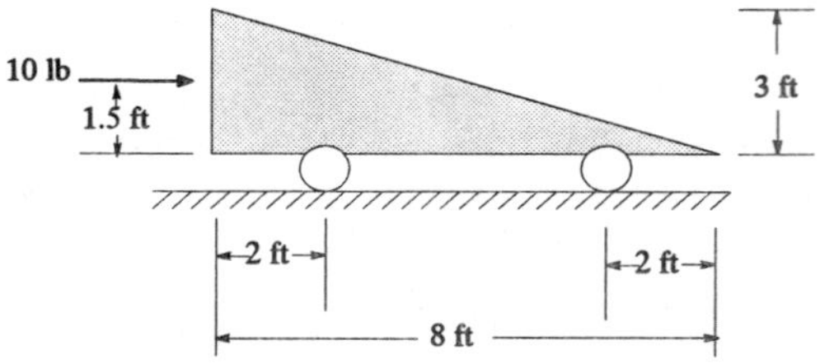

Figure a

Determine the acceleration of the model and the reaction of the horizontal surface on the wheels.

Solution: The free-body diagram of the model is shown in Fig. b.

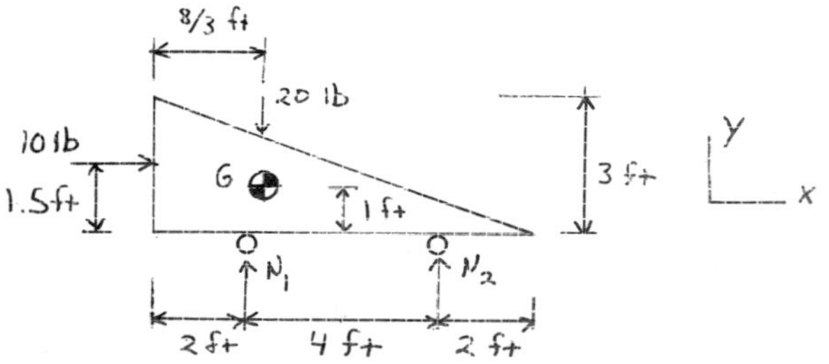

Figure b

By Fig. b and kinetics,

$\Sigma F_x = 10 = ma = \frac{20}{g} a$

$\Sigma F_y = N_1 + N_2 - 20 = 0$ (a)

$\circlearrowleft \Sigma M_G = 3.33\bar{3}\, N_2 - 0.66\bar{6}\, N_1 - (0.5)(10) = 0$

The solution of Eqs. (a) is

$a = \frac{10g}{20} = 16.1 \text{ ft/s}^2$

$N_1 = 15.416\bar{6}$ lb

$N_2 = 4.583\bar{3}$ lb

18.5

Given: An automobile has rear-wheel drive. The coefficient of friction between the tires and the pavement is 0.70. The distance between the front and rear axles is 12 ft. The center of mass is 8 ft forward of the rear axes and 2 ft above the pavement.

Calculate the maximum possible acceleration of the car on a horizontal pavement.

Solution: The free-body diagram of the car is shown in Fig. a. Since no information is given on drag due to air, assume it is negligible

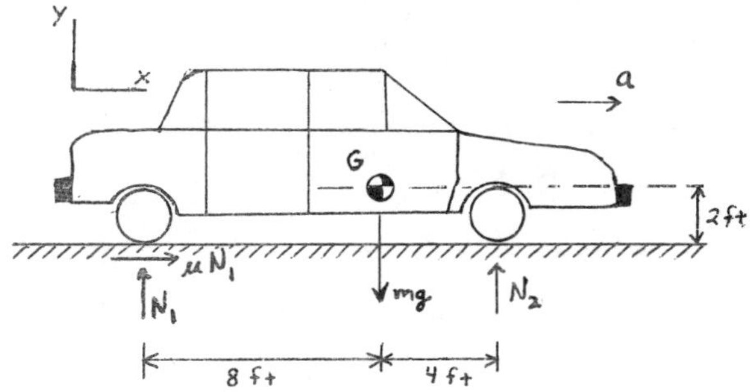

Figure a

By Fig. a and kinetics,

$\Sigma F_x = \mu N_1 = m a_{max}$

$\Sigma F_y = N_1 + N_2 - mg = 0$ (a)

$\circlearrowleft \Sigma M_G = 4N_2 + \mu N_1 (2) - 8 N_1 = 0$

The maximum acceleration will occur at impending slipping of the rear tires. For impending slipping $\mu = 0.70$. Then, by Eqs. (a)

$N_1 = 0.3774\ mg$

$N_2 = 0.6226\ mg$

and

$a_{max} = 8.51 \text{ ft/s}^2$

18.6

Given: A 90-N cylinder translates, without rolling, up a plane with an acceleration of 3 m/s² (Fig. a).

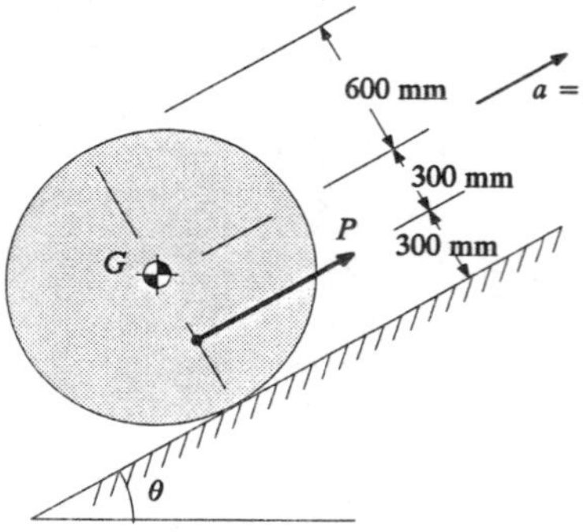

Figure a

Determine the force P and the coefficient of sliding friction in terms of θ.

Continued

18.6 Cont.

Solution: Figure b is the free-body diagram of the cylinder. By Fig. b,

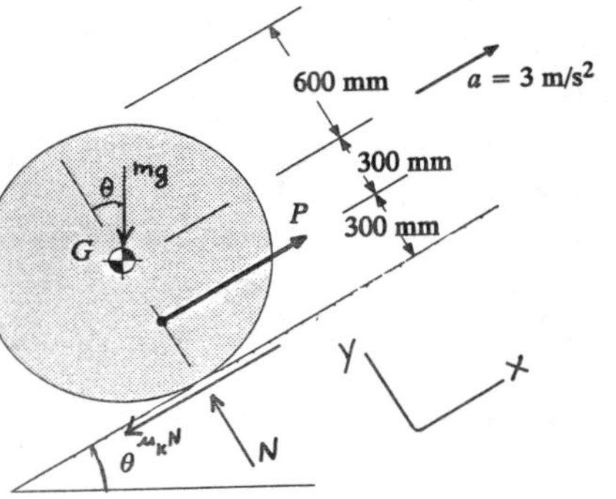

Figure b

By Fig. b

$$\left.\begin{aligned} \Sigma F_x &= P - \mu_K N - mg\sin\theta = ma \\ \Sigma F_y &= N - mg\cos\theta = 0 \\ \circlearrowleft \Sigma M_G &= (0.30)P - (0.6)\mu_K N = 0 \\ mg &= 90\ \text{N} \ ; \quad a = 3\ \text{m/s}^2 \end{aligned}\right\} \quad (a)$$

The solution of Eqs. (a) is

$N = mg\cos\theta = 90\cos\theta$ [N]

$\underline{P = 180\sin\theta + 55.05}$

$\underline{\mu_K = \tan\theta + 0.3058\sec\theta}$

18.7

Given: Part E of a mechanism weighs 64.4 lb (Fig. a). Arm OA rotates with angular velocity of 10 rad/s and angular acceleration of 4 rad/s², both clockwise. Arms OA and BC are constrained to remain parallel by tie rod DD. Pin C moves in a frictionless horizontal guide.

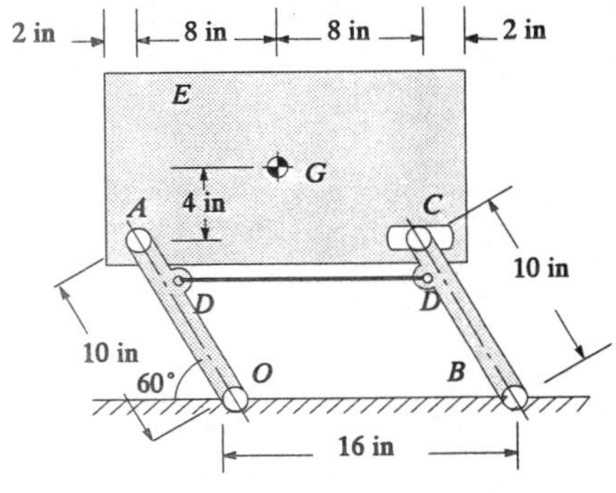

Figure a

For the position shown in Fig. a, determine the horizontal and vertical forces that act on Part E at point A and C

Solution:

The radial and tangential components of acceleration of point A are (see Fig. b)

$$a_r = r\omega^2 = (10)(10)^2 = 1000 \ \text{in/s}^2$$

$$a_t = r\alpha = (10)(4) = 40 \ \text{in/s}^2$$

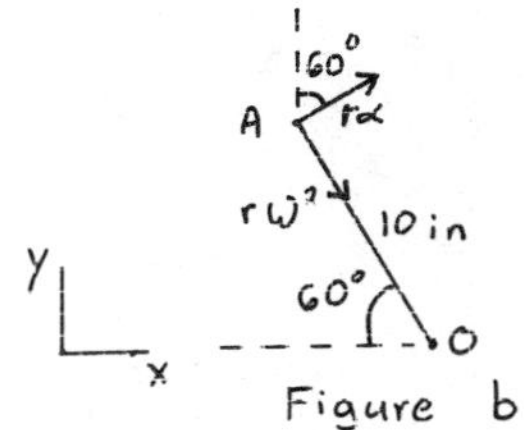

Figure b

All points in part E have the same acceleration (a_x, a_y), since the motion is a curvilinear translation. Hence, by Fig. b,

$$a_x = r\omega^2\cos 60° + r\alpha\sin 60°$$

$$a_y = -r\omega^2\sin 60° + r\alpha\cos 60°$$

or

$$\left.\begin{aligned} a_x &= 534.64\ \text{in/s}^2 = 44.55\ \text{ft/s}^2 \\ a_y &= -846.0\ \text{in/s}^2 = -70.50\ \text{ft/s}^2 \end{aligned}\right\} \quad (a)$$

The free-body diagram of part E is shown in Fig. c. By Fig. c,

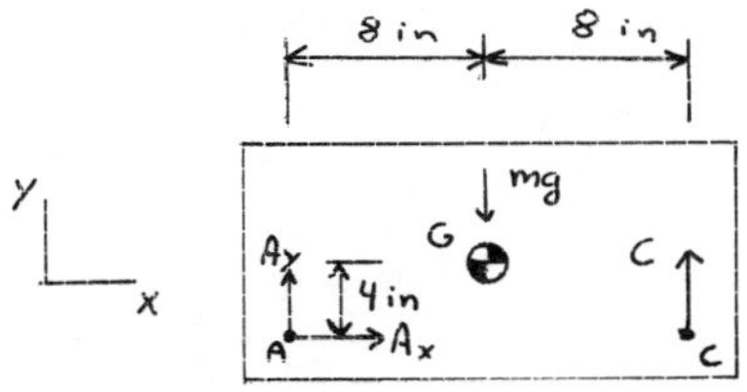

Figure c

$$\left.\begin{aligned} \Sigma F_x &= A_x = m a_x \\ \Sigma F_y &= A_y + C - mg = m a_y \\ \circlearrowleft \Sigma M_G &= 4A_x + 8C - 8A_y = 0 \end{aligned}\right\} \quad (b)$$

$$mg = 64.4\ \text{lb}$$

By Eqs. (a) and (b)

$\underline{A_x = 89.10\ \text{lb}}$

and

$$\left.\begin{aligned} A_y + C &= -76.6 \\ A_y - C &= 44.55 \end{aligned}\right\} \quad (c)$$

The solution of Eqs. (c) is

$\underline{A_y = -16.03\ \text{lb}}$

$\underline{C = -60.58\ \text{lb}}$

18.8

Given: A large rectangular crate with center of mass at the geometric center is dragged by a rope (Fig. a). The coefficient of sliding friction is 0.10 and $b/h = 1/2$.

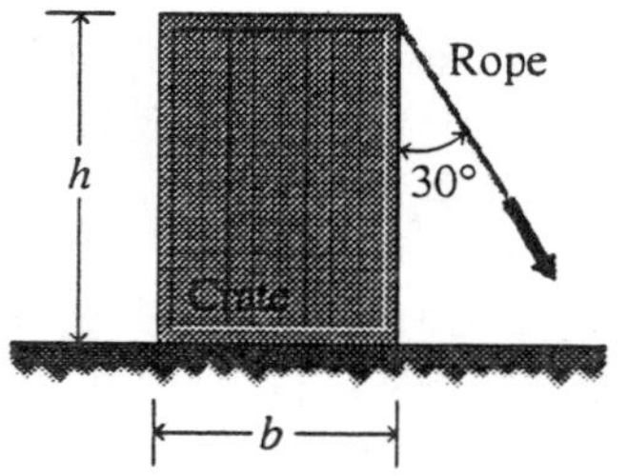

Figure a

Determine the acceleration at which the crate will start to tip.

Solution:

The free-body diagram of the crate on the verge of tipping is shown in Fig. b. By Fig. b,

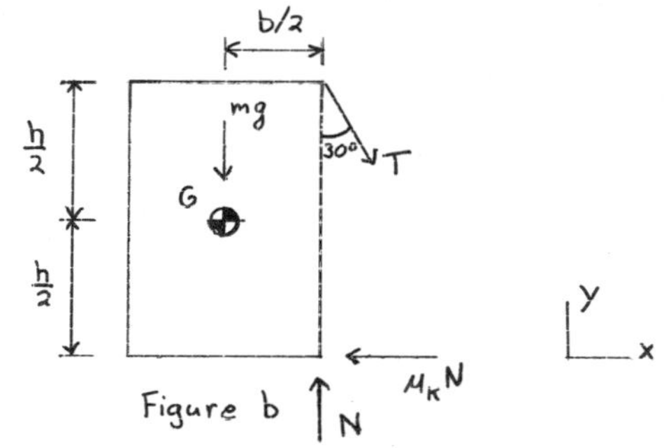

Figure b

$$\Sigma F_x = T\sin 30° - \mu_k N = m a_x \qquad (a)$$

$$\Sigma F_y = N - T\cos 30° - mg = 0 \qquad (b)$$

$$\circlearrowleft \Sigma M_G = N\left(\tfrac{b}{2}\right) - T(\cos 30°)\left(\tfrac{b}{2}\right) - T(\sin 30°)\left(\tfrac{h}{2}\right) - \mu_k N\left(\tfrac{h}{2}\right) = 0 \qquad (c)$$

Since $b/h = 1/2$ and $\mu_k = 0.10$, Eq. (c) yields

$$0.40N - 0.933T = 0$$

or

$$N = 2.333T \qquad (d)$$

Equations (b) and (d) yield

$$mg = 1.466T \qquad (e)$$

Equations (a), and (d) yield

$$ma_x = 0.2667T \qquad (f)$$

Dividing Eq. (f), by Eq. (e), we obtain

$$a_x = 0.1819g \qquad (g)$$

In SI units, Eq. (g) yields

$$a_x = 0.1819(9.81) = 1.78 \text{ m/s}^2$$

and in U.S. Customary units

$$a_x = 0.1819(32.2) = 5.86 \text{ ft/s}^2$$

18.9

Given: A homogeneous triangular prismatic plate weighs 18 N. It is suspended by cords AC and BD (Fig. a). For the position shown, the angular velocity of the cords is 10 rad/s counter clockwise.

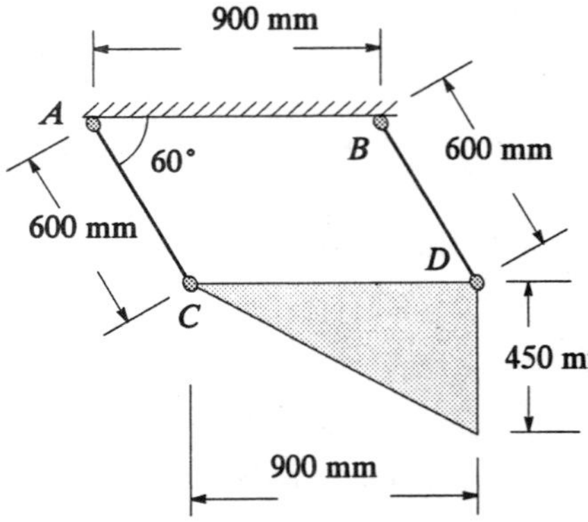

Figure a

Determine the tensions in the cords AC and BD and the angular acceleration of the cords.

Solution:

The normal and tangential components of the mass center are shown in Fig. b

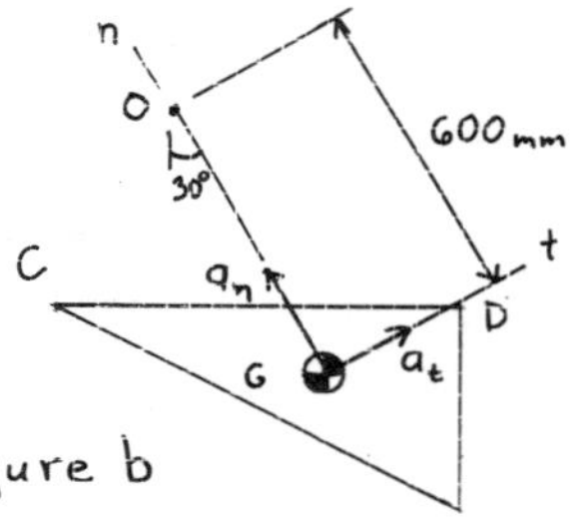

Figure b

The free-body diagram of the plate is shown in Fig. C

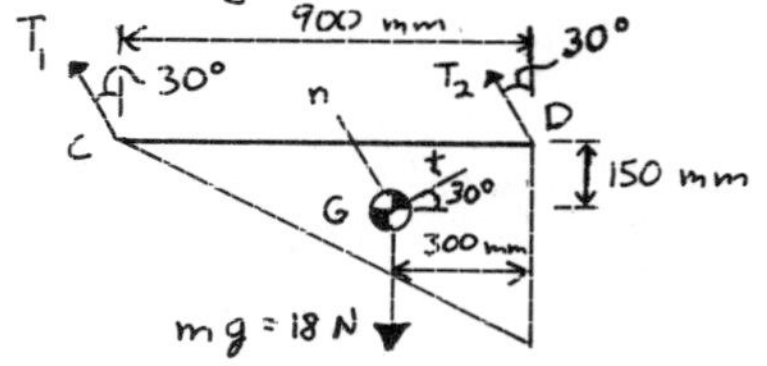

Figure C

Continued

18.9 Cont.

By Fig. b, since each point in the plate moves on a circular arc of radius $r = 0.60$ m,

$$a_t = r\alpha = 0.60\alpha \quad [m/s^2]$$
$$a_n = r\omega^2 = (0.60)(10)^2 = 60\ m/s^2 \qquad (a)$$

By Fig. c,

$$\Sigma F_t = mg(\cos 120°) = m a_t \qquad (b)$$
$$\Sigma F_n = T_1 + T_2 - mg(\cos 30°) = m a_n \qquad (c)$$
$$\circlearrowleft + \Sigma M_G = T_2(\cos 30°)(0.30) + T_2(\sin 30°)(0.15) - T_1(\cos 30°)(0.60) + T_1(\sin 30°)(0.15) = 0 \qquad (d)$$

By Eqs. (a) and (b)

$$a_t = 0.6\alpha = -0.5g$$

or $\underline{\alpha = -(0.5)(9.81)/0.6 = -8.175\ rad/s^2}$

By Eq. (c),

$$T_1 + T_2 = 125.68 \qquad (e)$$

By Eq. (d),

$$T_1 - 0.753T_2 = 0 \qquad (f)$$

Equations (e) and (f) yield

$\underline{T_2 = 71.69\ N}$

$\underline{T_1 = 53.99\ N}$

18.10

Given: When the triangular plate of Problem 18.9 reaches its lowest position (Fig. a), the speed of the center of mass is 2.4 m/s.

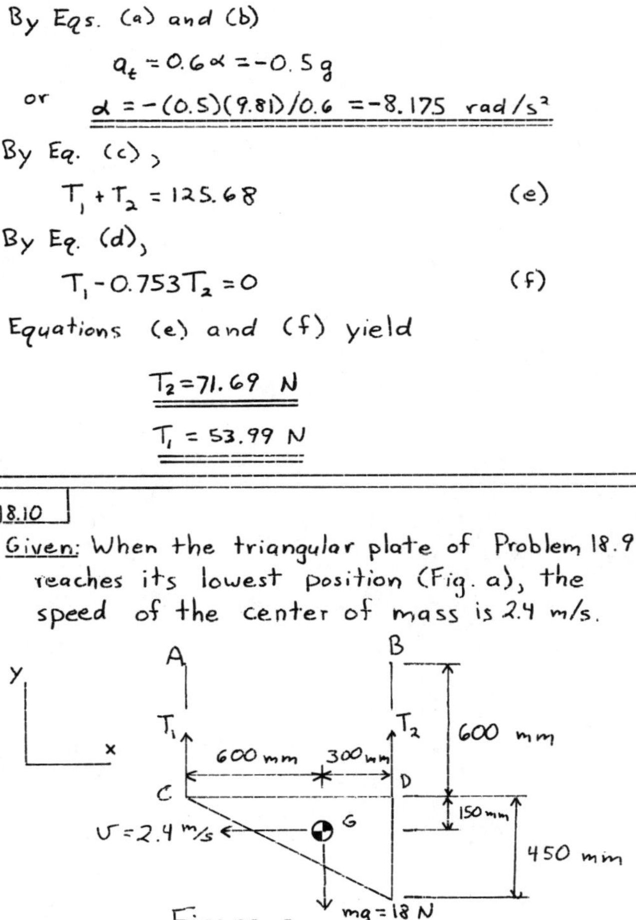

Figure a

Determine the tensions in the cords AC and BD for this position

Solution: By Fig. a,

$$\Sigma F_y = T_1 + T_2 - mg = m\frac{v^2}{r}$$
$$\circlearrowleft + \Sigma M_G = 0.30T_2 - 0.60T_1 = 0$$

or
$$\left.\begin{aligned} T_1 + T_2 &= 18 + \frac{18}{9.81}\frac{(2.4)^2}{0.6} = 35.61 \\ T_2 &= 2T_1 \end{aligned}\right\} \qquad (a)$$

The solution of Eqs. (a) is

$\underline{T_1 = 11.87\ N}$

$\underline{T_2 = 23.74\ N}$

18.11

Given: A frictionless mechanism is acted on by a force P and translates up the incline as shown in Fig. a. The uniform bar ACD and the base AB are connected by a pin at A and by the cord CB. The bar and the base each weigh 9.81 N.

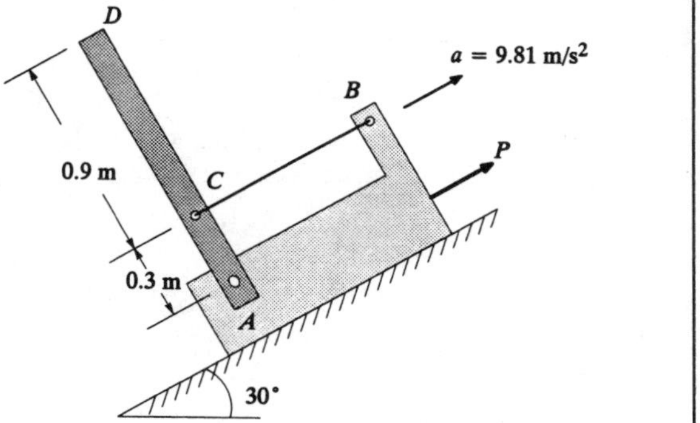

Figure a

Determine the force P and the tension in cord CB.

Solution: The free-body diagram of bar ACD is shown in Fig. b.

By Fig. b,

$$\Sigma F_x = T + A_x - mg\sin 30° = ma \qquad (a)$$
$$\Sigma F_y = A_y - mg\cos 30° = 0 \qquad (b)$$
$$\circlearrowleft \Sigma M_G = 0.6A_x + 0.3T = 0 \qquad (c)$$

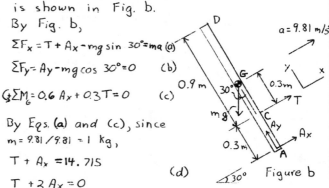

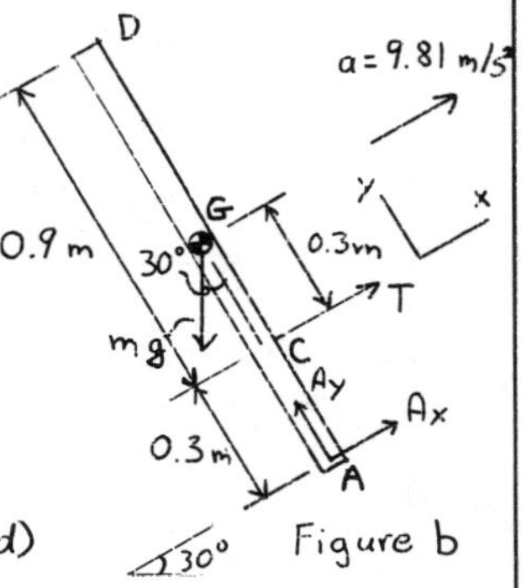

Figure b

By Eqs. (a) and (c), since $m = 9.81/9.81 = 1$ kg,

$$\left.\begin{aligned} T + A_x &= 14.715 \\ T + 2A_x &= 0 \end{aligned}\right. \qquad (d)$$

The solution of Eqs. (d) yields

$\underline{T = 29.43\ N}$

Summing forces up the plane for the entire system (see Figs. a and b), we have

$$\Sigma F_x = P - 2mg\sin 30° = 2ma$$

or since $m = 1$ kg and $a = 9.81$ m/s

$\underline{P = 29.43\ N}$

8.12

<u>Given:</u> An iron pipe 1.2 m in diameter is laid laterally across the bed of a truck, and it is held in position by two blocks of height h (Fig. a). The truck has only rear-wheel brakes, and 60% of the weight of the truck is on the rear wheels when the truck is braked and skids to a stop. The coefficient of kinetic friction between tires and pavement is $\mu_k = 0.60$.

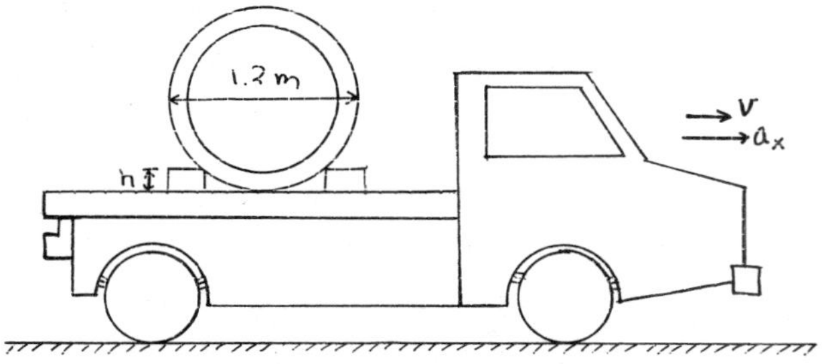

Figure a

Determine the minimum height h [mm], if the pipe is not to roll over a block when the truck is braked and skids to a stop.

<u>Solution:</u> First we determine the acceleration of the truck. By the free-body diagram of the truck (Fig. b),

$$\Sigma F_x = -\mu_k N_R = m a_x \qquad (a)$$

where

$$N_R = 0.60\,W = 0.60\,mg \qquad (b)$$

and $\mu_K = 0.60$

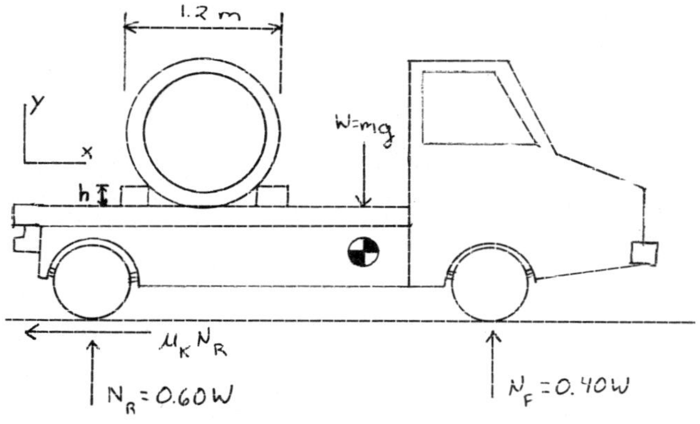

Figure b

By Eqs. (a) and (b),

$$a_x = -3.532 \text{ m/s}^2 \qquad (c)$$

Next by the free-body diagram of the pipe for the instant that it lifts off the bed of the truck (Fig. c),

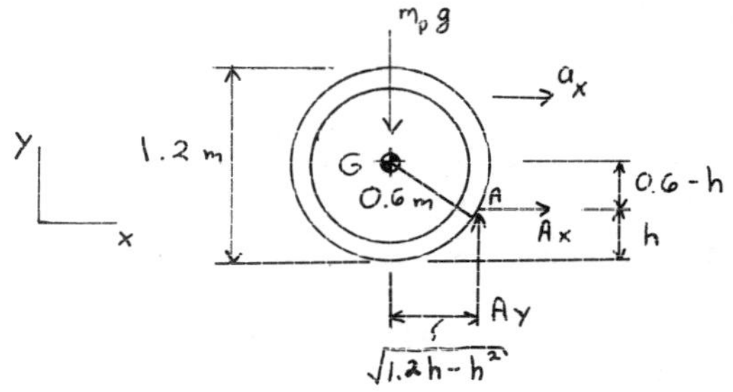

Figure c

$$\Sigma F_x = A_x = m_p a_x \qquad (d)$$

$$\Sigma F_y = A_y - m_p g = 0 \qquad (e)$$

$$\circlearrowleft \Sigma M_G = (0.6 - h)A_x + (\sqrt{1.2h - h^2})A_y = 0 \qquad (f)$$

where m_p is the mass of the pipe.

By Eqs. (c), (d), and (e), we find,

$$A_x = -0.36\,A_y \qquad (g)$$

and by Eqs. (f) and (g),

$$0.36(0.6 - h) = \sqrt{1.2h - h^2}$$

Squaring both sides of this equation and simplifying,

We get,

$$h^2 - 1.2h + 0.04130 = 0 \qquad (h)$$

The meaningful solution of Eq. (h) is

$$\underline{\underline{h = 0.0355 \text{ m} = 35.5 \text{ mm}}}$$

8.13

<u>Given:</u> The uniform crate shown in Fig. a weighs 16.1 lb. It is subjected to a horizontal force $P = t^3$ (P in pounds and t in seconds). The coefficient of static and kinetic friction are equal at 0.50

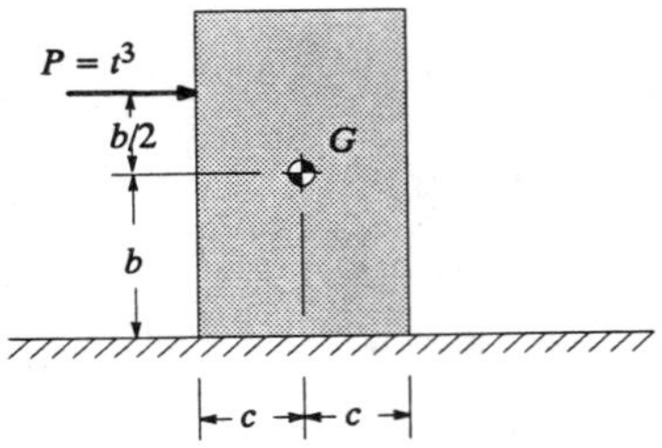

Figure a

a) Calculate the acceleration of the crate for t = 1 s. For t = 3 s. Assume the crate does not tip for $t \le 3$ s.

(Continued)

18.13 Cont.

b) In terms of b and c, derive a formula the time t at which the crate is on the verge of tipping, supposing that c is sufficiently large so that it slides before it tips.

c) Plot the acceleration of the crate for the time interval $0 \le t \le 6$ s, assuming that the crate does not tip.

d) Plot the time t at which the crate is on the verge of tipping as a function of c/b for the range $1.5 \le c/b \le 7.5$.

e) On the basis of the plots of parts c and d, discuss the assumption used in part c.

Solution:

a) The free-body diagram of the crate is shown in Fig. b.

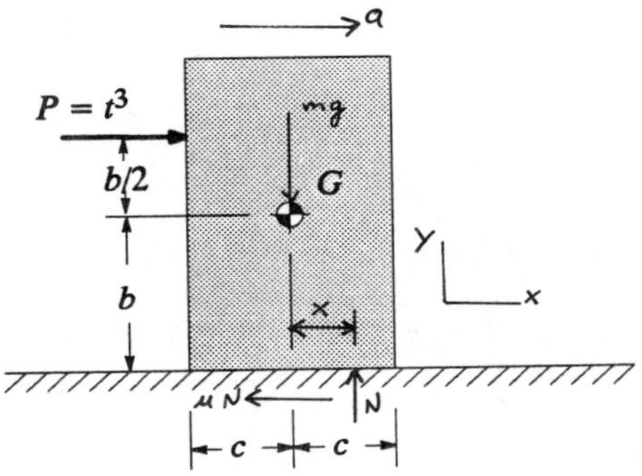

Figure b

By Fig. b,

$\Sigma F_x = t^3 - \mu N = ma$ (a)

$\Sigma F_y = N - mg = 0$ (b)

$\Sigma M_G = Nx - \mu N b - t^3\left(\frac{b}{2}\right) = 0$ (c)

By Eqs. (a) and (b), with $\mu = 0.50$,

$N = mg = 16.1$ lb (d)

$t^3 - 8.05 = 0.5a$ (e)

By Eq. (e), the crate does not begin to move (accelerate) until $t^3 > 8.05$ or until

$t > 2.004$ s

Hence, for $t = 1$s : $a = 0$

and, by Eq. (e),

for $t = 3$s : $a = 37.9$ ft/s²

b) When $x = c$ (Fig. b), the crate is on the verge of tipping. Then, Eqs. (c) and (d) yield

$t = \left[16.1\left(2\frac{c}{b} - 1\right)\right]^{1/3}$ (f)

c) By Eq. (e),

$a = 2t^3 - 16.1$ [ft/s²] (g)

For $0 \le t \le 6$ s, a is plotted in Fig. c

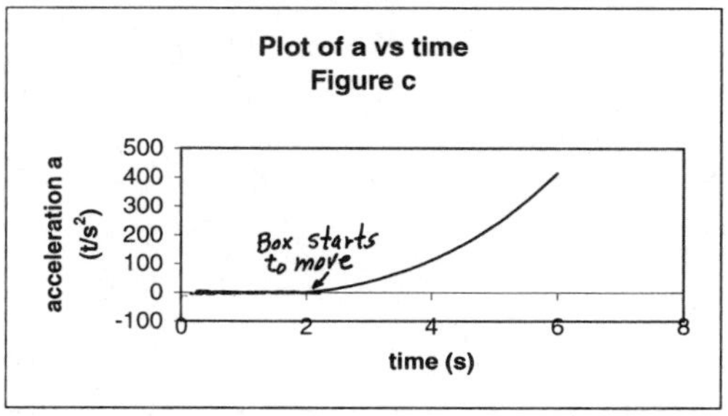

d) By Eq. (f), the plot of t, at tipping, as a function of c/b, for $1.5 \le c/b \le 7.5$, is shown in Fig. d

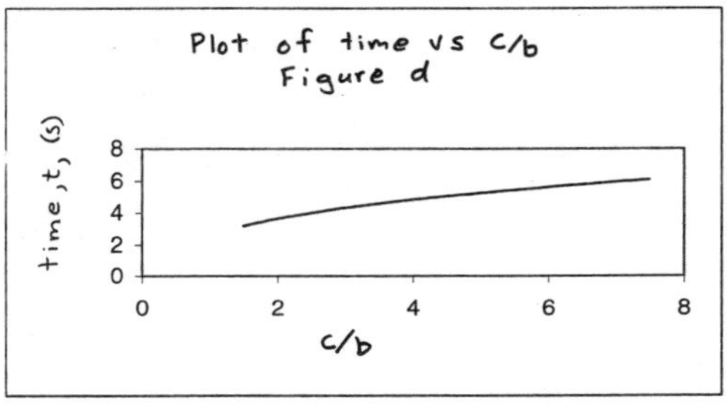

e) For the crate not to tip in the time interval $0 \le t \le 6$ s, the ratio c/b must be greater than 7.208.

18.14

Given: Part E of the mechanism shown in Fig. a weighs 64.4 lb. Arm OA rotates with angular velocity 10 rad/s and angular acceleration 4 rad/s², both clockwise. Arms

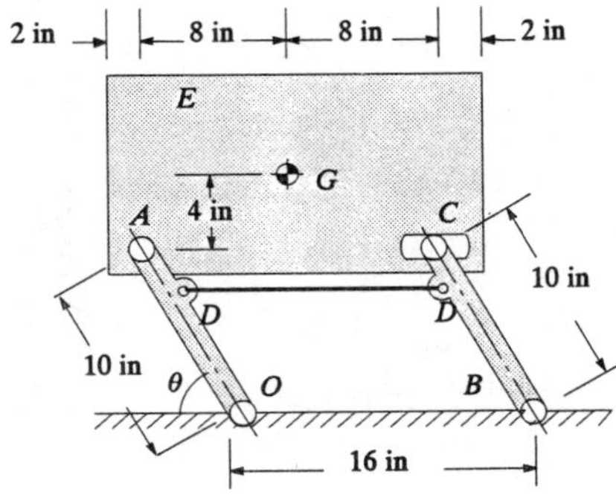

Figure a

OA and BC remain parallel, and pin C moves in a frictionless horizontal guide. (Continued)

18.14 Cont.

a) Determine the horizontal and vertical forces A_x, A_y, and C that act on part E at points A and C, as functions of θ.

b) Plot A_x, A_y, and C as functions of θ for $0 \le \theta \le 180°$.

c) By the plots of part b, verify that for $\theta = 60°$, $A_x = 89.1$ lb, $A_y = -16.0$ lb, and $C = -60.6$ lb (see Problem 18.7)

Solution:

a) Since the part E undergoes curvilinear translations, its acceleration is the same as the acceleration of point A on arm OA (Fig. b)

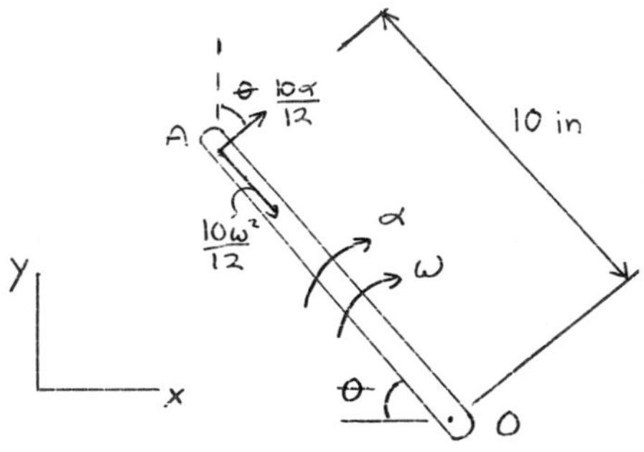

Figure b

The free-body diagram of block E is shown in Fig. c

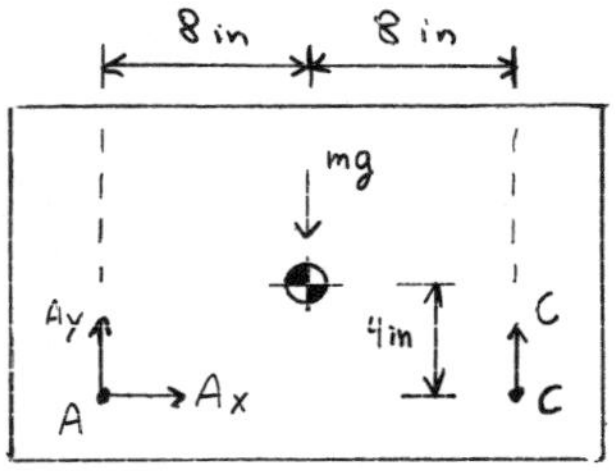

Figure c

By Figs. b and c,

$$\Sigma F_x = A_x = m a_x = \left(\frac{64.4}{32.2}\right)\left(\frac{10}{12}\omega^2 \cos\theta + \frac{10}{12}\alpha \sin\theta\right) \quad (a)$$

$$\Sigma F_y = A_y + C - mg = m a_y = \left(\frac{64.4}{32.2}\right)\left(\frac{10}{12}\alpha \cos\theta - \frac{10}{12}\omega^2 \sin\theta\right) \quad (b)$$

$$\circlearrowleft \Sigma M_G = 8C + 4A_x - 8A_y = 0 \quad (c)$$

The solution of Eqs. (a), (b), (c) is, with $m = 2$ slugs, $\omega = 10$ rad/s, and $\alpha = 4$ rad/s^2;

$$A_x = 6.6\bar{6} \sin\theta + 166.6\bar{6} \cos\theta \quad [\text{lb}] \quad (d)$$

$$A_y = 32.2 - 81.6\bar{6} \sin\theta + 45 \cos\theta \quad [\text{lb}] \quad (e)$$

$$C = 32.2 - 85.0 \sin\theta - 38.3\bar{3} \cos\theta \quad [\text{lb}] \quad (f)$$

b) The plots of A_x, A_y, and C [Eqs. (d) - (f)] are shown in Fig. d.

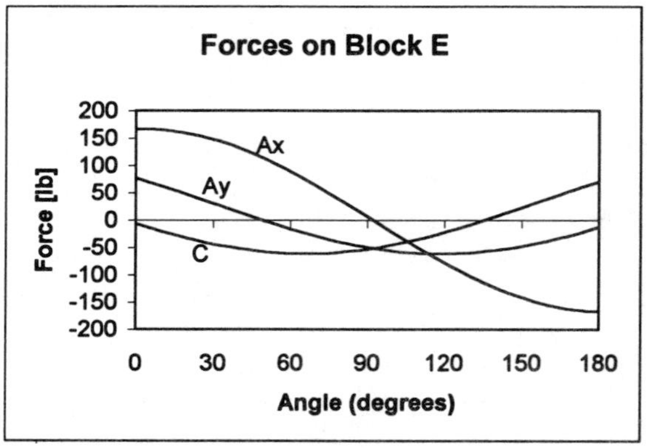

Figure d

c) By Fig. d, $A_x \approx 89.1$ lb, $A_y \approx -16.0$ lb, and $C \approx -60.6$ lb, for $\theta = 60°$.

18.15

Given: The homogeneous triangular plate shown in Fig. a weighs 18 N. It is suspended by light cords AC and BD. The system is released from rest when $\theta = \theta_0$

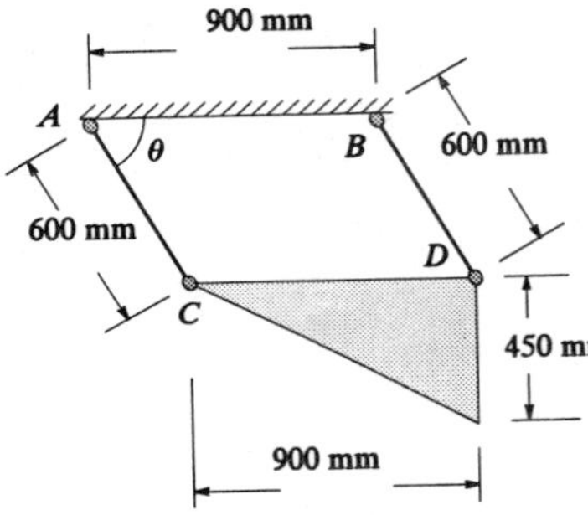

Figure a

a) In terms of θ and θ_0, derive formulas for the angular velocity $\omega = \dot{\theta}$, the angular acceleration $\alpha = \ddot{\theta}$, and the tensions T_{AC} and T_{BD} in the cords.

b) Plot ω, α, T_{AC}, and T_{BD} as functions of θ for $\theta_0 = 25°$, for the range $25° \le \theta \le 155°$.

c) What is the smallest angle θ_0 for which neither of the cords will become slack? In view of this angle, discuss the significance of the plots in part b.

Solution:

a) Consider Fig. a. Since the plate undergoes curvilinear translation, every point in the plate has the same velocity and acceleration. Hence, the velocity of the mass center is (Fig. b).

(Continued)

18.15 Cont.

$$v_G = r\omega \qquad (a)$$

where $r = 0.6$ m and $\omega = \dot{\theta}$, since each particle in the plate swings on an arc of radius r. So, $OG = r = 0.6$ m. To determine ω in terms of θ and θ_o, consider the positions θ and θ_o (Fig. b). By the work-energy principle,

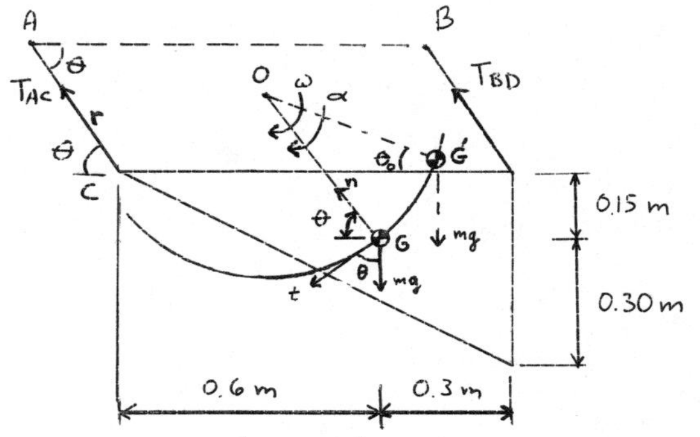

Figure b

$$U_{1\text{-}2} = \Delta T = T_2 - T_1 \qquad (b)$$

where $U_{1\text{-}2}$ is the work done on the plate from θ_o to θ and ΔT is the corresponding change in kinetic energy of the plate.

By Fig. b, the work done on the plate is

$$U_{1\text{-}2} = mgr(\sin\theta - \sin\theta_o) \qquad (c)$$

Since the plate is released from rest,

$$T_1 = 0$$

$$T_2 = \frac{1}{2} m v_G^2 = \frac{1}{2} m r^2 \omega^2 \qquad (d)$$

Equations (b), (c), and (d) yield

$$\omega^2 = \frac{2g}{r}(\sin\theta - \sin\theta_o)$$

or

$$\omega = \dot{\theta} = \left[\frac{2g}{r}(\sin\theta - \sin\theta_o)\right]^{1/2} \qquad (e)$$

Differentiation of Eq. (e) yields

$$\ddot{\theta} = \alpha = \frac{g}{r}\cos\theta \qquad (f)$$

To determine the tensions T_{AC} and T_{BD}, we sum forces in the n and t directions (Fig. b). Thus,

$$\Sigma F_n = T_{AC} + T_{BD} - mg\sin\theta = m a_n = mr\omega^2 = mr(\dot{\theta})^2 \qquad (g)$$

$$\Sigma F_t = mg\cos\theta = m a_t = mr\alpha = mr\ddot{\theta} \qquad (h)$$

Also, summation of moments about G yields (Fig. c)

$$\circlearrowleft \Sigma M_G = T_{BD}(\sin\theta)(0.30) + T_{BD}(\cos\theta)(0.15) - T_{AC}(\sin\theta)(0.60) + T_{AC}(\cos\theta)(0.15) = 0$$

or,

$$T_{BD} = T_{AC}\left[\frac{0.6\sin\theta - 0.15\cos\theta}{0.3\sin\theta + 0.15\cos\theta}\right] \qquad (i)$$

Note by Eq. (h),

$$\alpha = \ddot{\theta} = \frac{g}{r}\cos\theta$$

which agrees with Eq. (f). By Eqs. (e), (f), (g), and (i), we find

$$T_{AC} = mg\left[\frac{(3\sin\theta - 2\sin\theta_o)(0.3\sin\theta + 0.15\cos\theta)}{0.9\sin\theta}\right]$$

$$T_{BD} = mg\left[\frac{(3\sin\theta - 2\sin\theta_o)(0.6\sin\theta - 0.15\cos\theta)}{0.9\sin\theta}\right] \qquad (j)$$

For $mg = 18$ N and $\theta_o = 25°$, Eqs. (j) yield

$$T_{AC} = (3.0)\left[\frac{(3\sin\theta - 0.8452)(2\sin\theta + \cos\theta)}{\sin\theta}\right]$$

$$T_{BD} = (3.0)\left[\frac{(3\sin\theta - 0.8452)(4\sin\theta - \cos\theta)}{\sin\theta}\right] \qquad (k)$$

b) By Eqs. (e) and (f), plots of ω and α are shown in Fig. c, and by Eqs. (k), plots of T_{AC} and T_{BD} are shown in Fig. d.

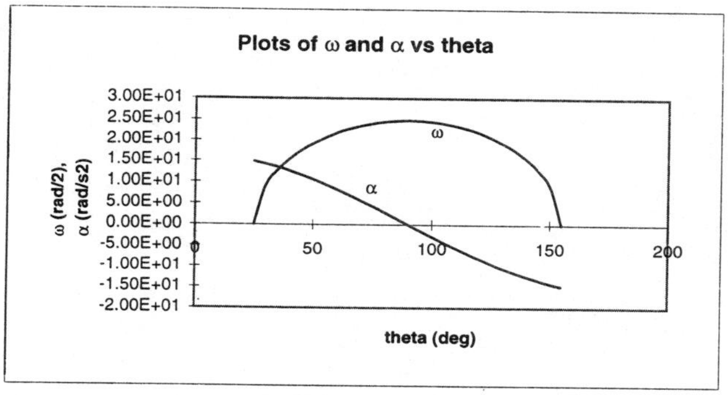

Figure c

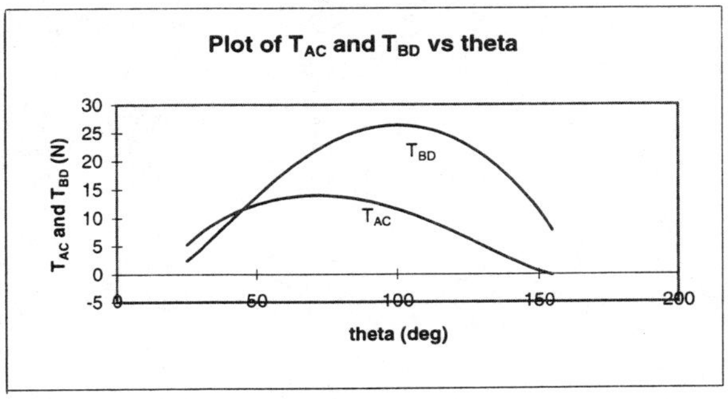

Figure d

c) In general, $\theta = \theta_{max} = 180° - \theta_o$ is the value of θ for which a cord goes slack. Therefore, at the slack condition

$$\sin\theta = \sin(180° - \theta_o) = \sin\theta_o$$

$$\cos\theta = \cos(180° - \theta_o) = -\cos\theta_o \qquad (l)$$

(Continued)

18.15 CONT.

By Eqs. (j) and (ℓ), with $mg = 18$ N,

$$T_{AC} = 20(0.3 \sin\theta_0 - 0.15 \cos\theta_0) \quad (m)$$
$$T_{BD} = 20(0.6 \sin\theta_0 + 0.15 \cos\theta_0) \quad (n)$$

By Eq. (m), $T_{AC} = 0$ (goes slack) when $\theta_0 = 26.565°$ and for $T_{BD} = 0$, by Eq. (n), $\theta_0 = -14.036°$. Hence, T_{AC} goes slack first for $\theta_0 = 26.565°$. This is the smallest value of θ_0 allowable for which the cord AC goes slack. Thus, $\theta_{max} = 180° - \theta_0 = 153.435°$. The plots of Figs. C and d must be redone, with $\theta_0 \geq 26.565°$ to be valid.

18.16

Given: To save time in unloading trucks, it is proposed that two identical crates be slid down a chute, one crate on top of the other (Fig. a).

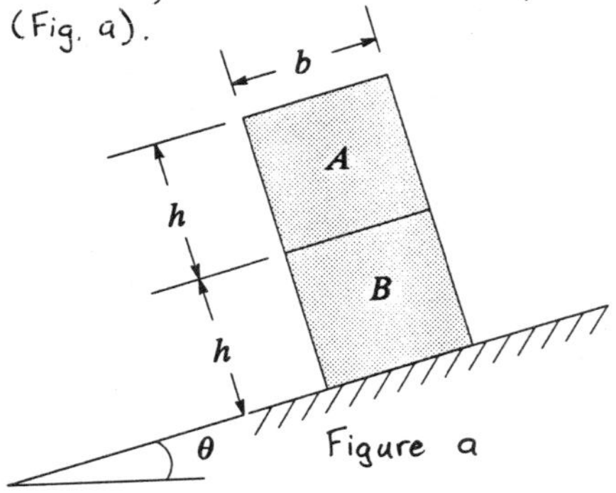

Figure a

Discuss this proposal on the basis of mechanics. Assume that the coefficients μ of static and kinetic frictions are equal. Between the crates, $\mu = \mu_b$ and between the lower crate and the chute $\mu = \mu_c$.

Solution:

Case 1: Consider first a single crate (Fig. b)

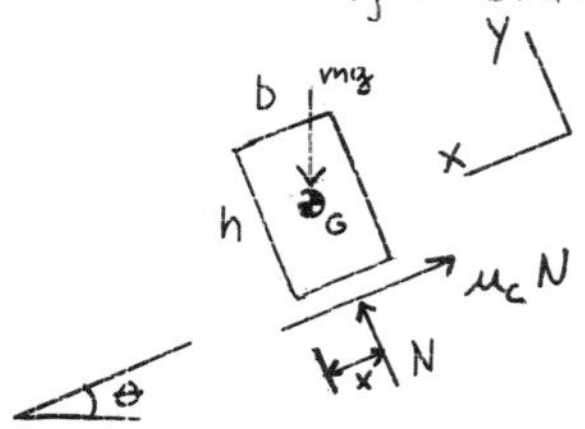

By Fig. b, for sliding of the crate,

$$\Sigma F_x = mg \sin\theta - \mu_c N = ma$$
$$\Sigma F_y = N - mg \cos\theta = 0\ ; \quad N = mg \cos\theta$$

These equations yield

$$a = g(\sin\theta - \mu_c \cos\theta) \quad (a)$$

So, a is independent of m, b, and h, if the crate does not tip.

For impending tipping, $x = 0$ (Fig. b). Then,

$$\circlearrowleft \Sigma M_G = \mu_c N\left(\tfrac{h}{2}\right) - N\left(\tfrac{b}{2}\right) = 0$$

or

$$\mu_c = \frac{b}{h}$$

Therefore, to prevent tipping of a single crate,

$$\mu_c < \frac{b}{h} \quad (b)$$

Likewise to prevent tipping of the upper crate on the lower crate (Fig. a),

$$\mu_b < \frac{b}{h} \quad (c)$$

Case 2: Consider next two crates (Fig. c)

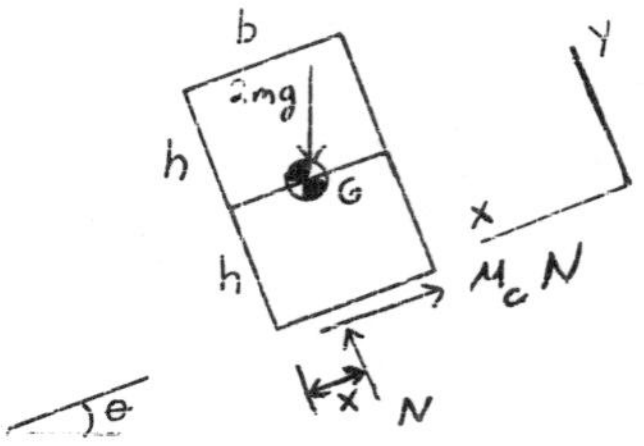

Figure c

By Fig. c, for sliding of the crates, as for a single crate,

$$\Sigma F_x = 2mg \sin\theta - \mu_c N = 2ma$$
$$\Sigma F_y = N - 2mg\cos\theta = 0\ ; \quad N = 2mg \cos\theta$$
$$a = g(\sin\theta - \mu_c \cos\theta)$$

For impending tipping of the crates as a unit, $x = 0$ (Fig. c) Then,

$$\circlearrowleft \Sigma M_G = \mu_c N h - N \frac{b}{2} = 0$$

or

$$2\mu_c = \frac{b}{h}$$

Thus, to prevent tipping of the crates as a unit,

$$2\mu_c < \frac{b}{h} \quad (d)$$

Consequently, by Eqs. (b) and (d), if

$$\mu_c < \frac{b}{h} < 2\mu_c \quad (e)$$

a single crate will not tip but the double crates will.

Case 3: Consider the possibility of the upper crate sliding on the lower crate (Fig. d)

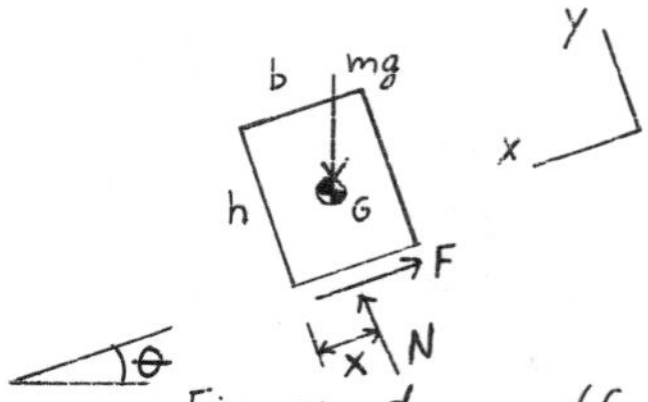

Figure d

(Continued)

18.16 Cont.

By Fig. d and Eq. (a),

$$\Sigma F_x = mg\sin\theta - F = ma = mg(\sin\theta - \mu_c\cos\theta)$$

$$\Sigma F_y = N - mg\cos\theta = 0\ ;\ N = mg\cos\theta$$

Solving these equations for F, we obtain

$$F = mg\mu_c\cos\theta \qquad (f)$$

To prevent sliding of the upper crate on the lower, where μ_b is the coefficient between crates

$$F < \mu_b N = \mu_b mg\cos\theta \qquad (g)$$

By Eqs. (f) and (g), to prevent sliding of the upper crate,

$$\mu_c < \mu_b$$

In summary, the proposal is workable provided,

$2\mu_c < \frac{b}{h}$ (to prevent tipping of the two crates as a unit)

$\mu_b < \frac{b}{h}$ (to prevent tipping of the upper crate)

$\mu_c < \mu_b$ (to prevent sliding of the upper crate)

Thus, to insure these requirements simultaneous

$$\mu_c < \mu_b < \frac{b}{2h}$$

Remark: The above analysis assumes that the lower crate can support the weight of the upper crate without being crushed and that the chute can safely support the weight of two crates.

18.17

Given: Part E of the mechanism (Fig. a) wheighs 64.4 lb. Arm OA rotates with constant angular velocity ω [rad/s] clockwise. Arms OA and BC remain parallel. Pin C moves in a frictionless horizontal guide.

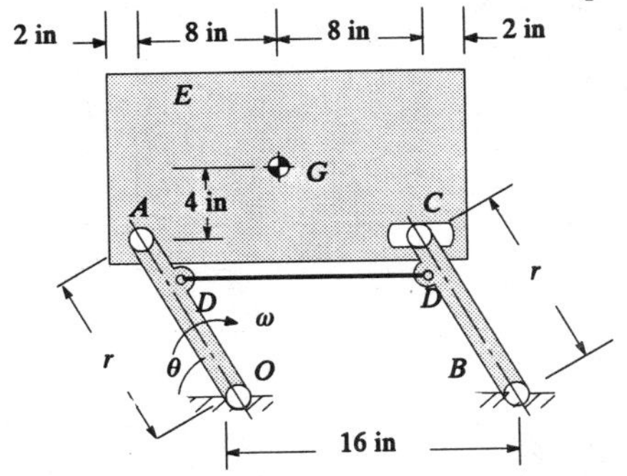

Figure a

a) Derive formulas for the horizontal and vertical forces A_x, A_y, and C that are exerted on part E at A and C as functions of r, ω, and θ. Check the formulas for 60° and θ = 300° by considering free-body diagrams in these positions.

b) With the formulas derived in part a, determine the angles θ at which A_x, A_y, and C attain maximum values, and the corresponding formulas for A_x, A_y, and C in terms of r and ω.

c) Design specifications for the mechanism require that r lies in the range 8 in ≤ r ≤ 12 in and that ω lies in the range 32 rad/s ≤ ω ≤ 38 rad/s. Also required are $A_x \le 2000$ lb, $A_y \le 2000$ lb, and $C \le 1050$ lb. Determine values of r and ω that meet these requirement.

Solution:

a) Since arm OA rotates at ω rad/s, and since the part E undergoes curvilinear translation, the acceleration of point A (and every other point in part E) is $a_n = r\omega^2$ directed parallel to arm OA.

Figure b is the free-body diagram of part E

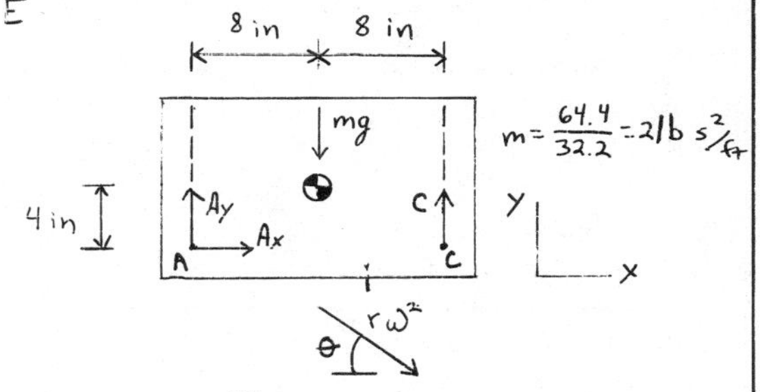

Figure b

By Fig. b,

$$\Sigma F_x = A_x = ma_x = mr\omega^2\cos\theta \qquad (a)$$

$$\Sigma F_y = A_y + C - mg = ma_y = -mr\omega^2\sin\theta \qquad (b)$$

$$\circlearrowleft \Sigma M_G = 4A_x - 8A_y + 8C = 0 \qquad (c)$$

Solving Eqs. (a), (b), and (c), we find with $m = 2\ \text{lb}\cdot\text{s}^2/\text{ft}$,

$$A_x = 2r\omega^2\cos\theta \qquad (d)$$

$$A_y = 32.2 - r\omega^2\sin\theta + \tfrac{1}{2}r\omega^2\cos\theta \qquad (e)$$

$$C = 32.2 - r\omega^2\sin\theta - \tfrac{1}{2}r\omega^2\cos\theta \qquad (f)$$

b) By Eq. (d),

$\theta = 0°$ for $A_{x\,max}$

By Eq. (e), for $A_{y\,max}$, $\frac{dA_y}{d\theta} = 0$, or

$$\frac{dA_y}{d\theta} = -r\omega^2\cos\theta - \tfrac{1}{2}r\omega^2\sin\theta = 0$$

or for $A_{y\,max}$ (or $A_{y\,min}$)

$\tan\theta = -2.0$; $\theta = -63.435°$ (116.565°)

(Continued)

18.17 Cont.

By Eq. (f) for C_{max}, $\frac{dC}{d\theta} = 0$, or

$$\frac{dC}{d\theta} = -r\omega^2 \cos\theta + \frac{1}{2} r\omega^2 \sin\theta = 0$$

so, for C_{max}, (or C_{min})

$$\tan\theta = 2.0 \; ; \; \theta = (63.435^\circ), 243.435^\circ$$

Hence, by Eqs. (d), (e), and (f), with corresponding values of θ,

For $\theta = 0$; $A_{x\,max} = 2r\omega^2$ [lb]

For $\theta = -63.435^\circ$; $A_{y\,max} = 32.2 + 1.1180 r\omega^2$ [lb]

For $\theta = 243.435^\circ$; $C_{max} = 32.2 + 1.1180 r\omega^2$ [lb]

c) For $8\text{ in} \le r \le 12\text{ in}$, $32\text{ rad/s} \le \omega \le 38\text{ rad/s}$, $A_x \le 2000$ lb, $A_y \le 2000$ lb and $C \le 1050$ lb, by the results of b,

$A_{x\,max} = 2r\omega^2 \le 2000$ lb ; or $r\omega^2 \le 1000$

$A_{y\,max} = 32.2 + 1.118 r\omega^2 \le 2000$ lb; or $r\omega^2 \le 1760.1$

$C_{max} = 32.2 + 1.118 r\omega^2 \le 1050$ lb; or $r\omega^2 \le 910.38$

Therefore, if $r\omega^2 \le 910.38$, where $\frac{8}{12}\text{ ft} \le r \le 1\text{ ft}$ and $32\text{ rad/s} \le \omega \le 38\text{ rad/s}$, the specifications of part C are met. As seen in Table a, various combinations of r and ω are satisfactory.

Table a : $r\omega^2 = 910.38$

r [ft]	ω^2	ω [rad/s]
8/12	1365.57	36.95
9/12	1213.84	34.84
10/12	1092.46	33.05
11/12	993.14	31.51
12/12	910.38	30.17

} These values of r and ω meet specifications (rows 8/12 – 10/12)

18.18

Given: A package that weighs 45 N is gently lowered on to a conveyor belt that moves horizontally at 5.4 m/s. The coefficient of sliding friction is $\mu_K = 0.30$.

By the inertial-force method, calculate the time required for the package to stop sliding on the belt.

Solution:

The free-body diagram of the package, including the inertial force, is shown in Fig. a.

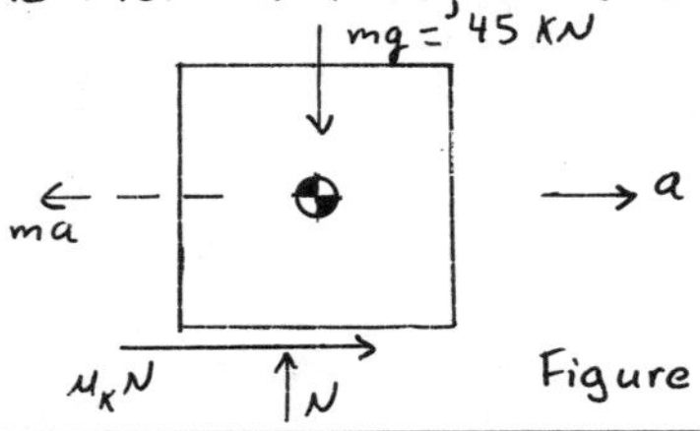

Figure a

By Fig. a,

$\Sigma F_x = 0 : \mu_K N - ma = 0$ (a)

$\Sigma F_y = 0 : N - mg = 0$

with $\mu_K = 0.30$, Eqs (a) yield

$$a = \mu_K g = (0.30)(9.81) = 2.943 \text{ m/s}^2$$

By kinematics,

$$a = dv/dt \quad (b)$$

Integration of Eq. (b) yields

$$\int_0^t a\,dt = \int_0^{5.4} dv$$

or $2.943\, t = 5.4$

Hence, $t = 1.835$ s

18.19

Given: A uniform block weighs 180 N. It is subjected to the forces shown in Fig. a and slides on the horizontal surface. The coefficient of kinetic friction is $\mu_K = 0.20$.

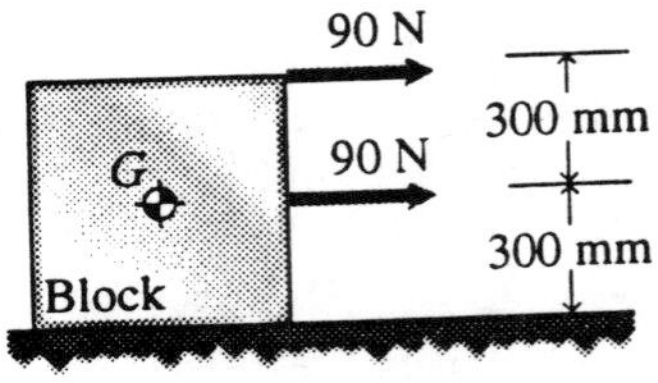

Figure a

Using the inertial-force method, determine the location of the resultant normal force that the surface exerts on the block.

Solution

The free-body diagram of the block, including the inertial force, is shown in Fig. b.

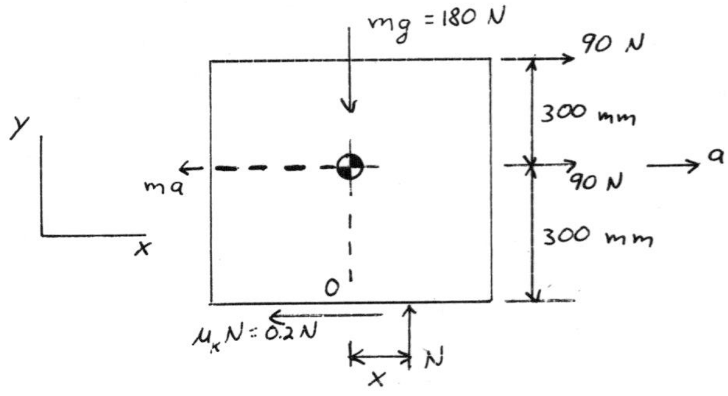

Figure b

$\Sigma F_x = 90 + 90 - 0.2N - ma = 0$; $ma = 180 - 0.2N$

$\Sigma F_y = N - 180 = 0$; $N = 180$ N (a)

$\therefore ma = 144$ N (b)

Then,

$\circlearrowleft + \Sigma M_O = xN - (0.3)(90) - (0.6)(90) + (0.3)ma = 0$ (c)

By Eqs. (a), (b), and (c), $x = 0.21$ m

18.20

Given: A homogeneous cubical crate rests on a horizontal floor. A workman pushes the crate horizontally at the middle of a top edge. The coefficient of kinetic friction is μ_K.

By the inertial-force method, derive a formula for the maximum acceleration in terms of μ_K and g that the crate can have if it does not start to tip over.

Solution: The free-body diagram of the crate, including the inertial force, is shown in Fig. a, at impending tipping.

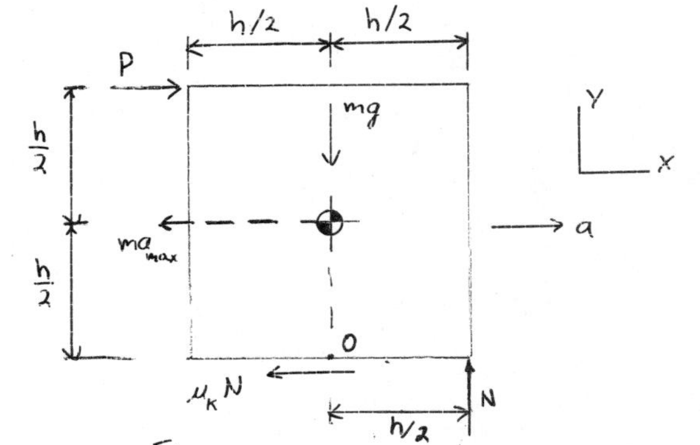

Figure a

By Fig. a,

$$\Sigma F_x = P - \mu_K N - m a_{max} = 0 \qquad (a)$$

$$\Sigma F_y = N - mg = 0 \qquad (b)$$

$$\circlearrowleft \Sigma M_O = \tfrac{h}{2} N + \tfrac{h}{2} m a_{max} - hP = 0 \qquad (c)$$

Solving Eqs. (a), (b), and (c), we obtain

$$N = mg$$

$$P = mg(1 - \mu_K)$$

$$\underline{\underline{a_{max} = g(1 - 2\mu_K)}}$$

18.21

Given: A homogeneous triangular prismatic test model of a race car weighs 20 lb (Fig. a). It is propelled by a jet that exerts a horizontal force of 10 lb, and it rolls on frictionless wheels of negligible mass

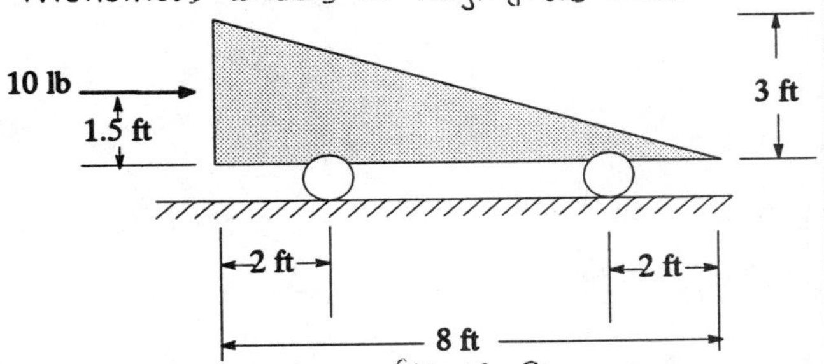

figure a

Using the inertial-force method, determine the acceleration of the model and the reactions of the horizontal surface on the wheels.

Solution: The free-body diagram of the model, including the inertial force, is shown in Fig. b.

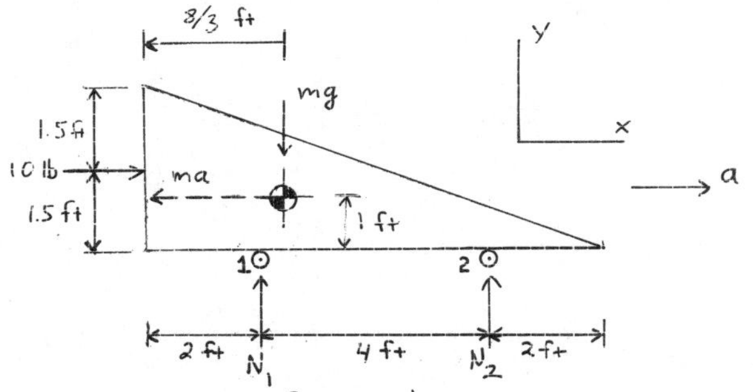

Figure b

By Fig. b,

$$\Sigma F_x = 10 - ma = 0 \qquad (a)$$

$$\Sigma F_y = N_1 + N_2 - mg = 0 \qquad (b)$$

$$\circlearrowleft \Sigma M_1 = (4)N_2 - (\tfrac{2}{3})mg + (1)ma - (1.5)(10) = 0 \qquad (c)$$

By Eq. (a), with $m = 20/g$,

$$a = \frac{10g}{20} = 16.1 \text{ ft/s}^2 \qquad (d)$$

Eqs. (b), (c), and (d) yield

$$\underline{\underline{N_1 = 15.416\bar{6} \text{ lb}}}$$

$$\underline{\underline{N_2 = 4.583\bar{3} \text{ lb}}}$$

18.22

Given: An automobile with rear-wheel drive has its center of mass 8 ft forward of the rear axle and 2 ft above the pavement. The distance between its front and rear axles is 12 ft, and the coefficient of friction between the tires and the pavement is 0.70.

Using the inertial-force method, calculate the maximum possible acceleration of the car on a horizontal pavement

Solution:

The free-body diagram of the car, including the inertial-force, is shown in Fig. a,

(Continued)

18.22 Cont.

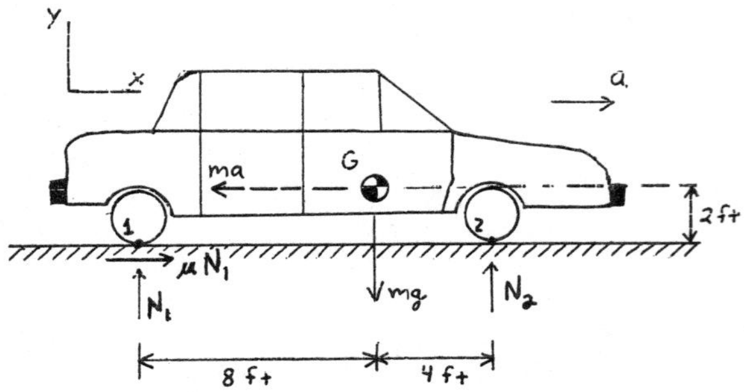

Figure a

By Fig. a,

$\Sigma F_x = \mu N_1 - ma = 0$; $ma = \mu N_1$ (a)

$\Sigma F_y = N_1 + N_2 - mg = 0$ (b)

$\circlearrowleft \Sigma M_1 = 12 N_2 + 2ma - 8mg = 0$ (c)

By Eqs. (a), (b), and (C), with $\mu = 0.70$,

$$\left.\begin{aligned} N_1 + N_2 &= mg \\ 0.70 N_1 + 6 N_2 &= 4\,mg \end{aligned}\right\} \quad (d)$$

The solution of Eqs. (d) is

$$\begin{aligned} N_1 &= 0.3774\ mg \\ N_2 &= 0.6226\ mg \end{aligned} \quad (e)$$

By Eqs. (a) and (e),

$a_{max} = 0.2642\,g = 8.51\ ft/s^2$

18.23

Given: A cylinder that weighs 90 N translates, without rolling, up a plane with acceleration $3\ m/s^2$ (Fig. a)

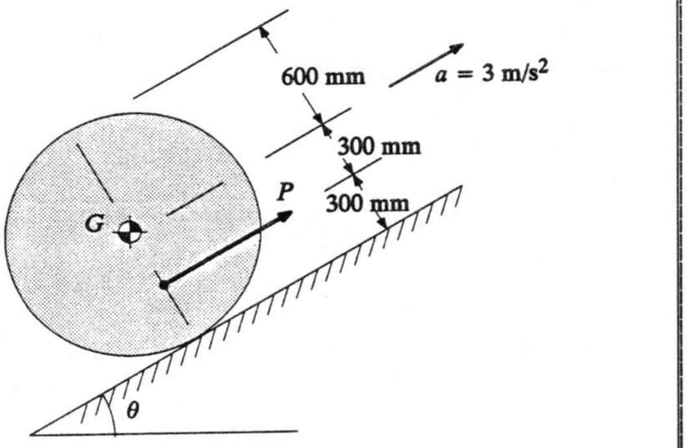

Figure a

Using the inertial-force method, determine the force P and the coefficient of sliding friction in terms of θ.

Solution: Figure b is the free-body diagram of the cylinder, including the inertial force.

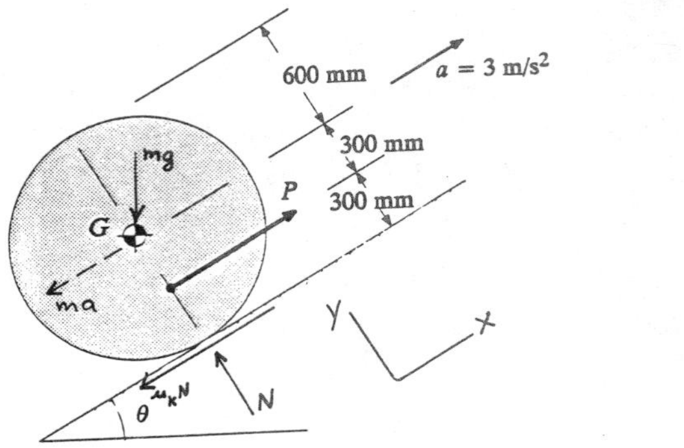

Figure b

By Fig. b,

$\Sigma F_x = P - \mu_k N - ma - mg \sin\theta = 0$ (a)

$\Sigma F_y = N - mg\cos\theta = 0$; $N = mg\cos\theta$ (b)

$\circlearrowleft \Sigma M_0 = (0.60)ma + (0.60)mg\sin\theta - (0.30)P = 0$ (c)

By Eq. (c), with $a = 3\ m/s^2$, $mg = 90\ N$ and $m = 90/9.81 = 9.174\ kg$,

We obtain,

$P = 180\sin\theta + 55.05\ [N]$ (d)

Then, Eqs. (a), (b), and (d) yield

$\mu_k = \tan\theta + 0.306\sec\theta$

18.24

Given: Part E of a mechanism weighs 64.4 lb (Fig. a). Arm OA rotates with angular velocity 10 rad/s and angular acceleration 1 rad/s², both clockwise. Arms CA and BC are constrained to remain parallel by tie rod DD. Pin C moves in a frictionless horizontal guide.

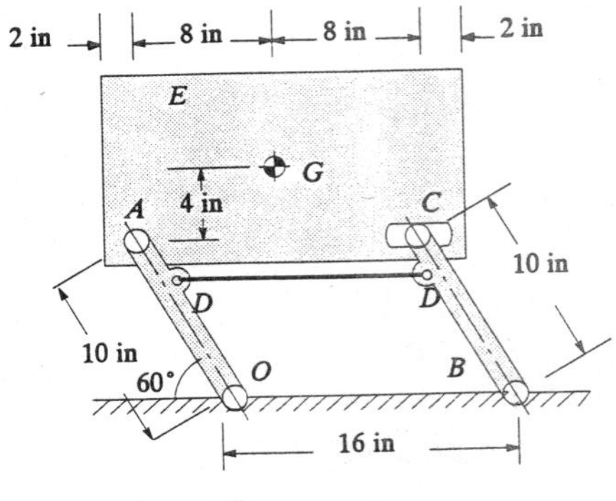

Figure a

For the position shown in Fig. a, using the inertial-force method, determine the horizontal and vertical forces that act on part E at points A and C.

(Continued)

18.24 Cont.

Solution: The radial and tangential components of acceleration of point A are (see Fig. b)

$$a_r = r\omega^2 = (10)(10)^2 = 1000 \text{ in/s}^2$$

$$a_t = r\alpha = (10)(4) = 40 \text{ in/s}^2$$

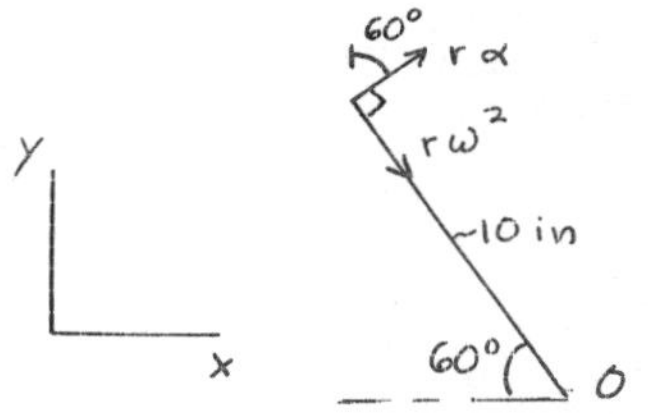

Figure b

All points in part E have the same acceleration (a_x, a_y), since the motion is a curvilinear translation. Hence, by Fig. b,

$$a_x = r\omega^2 \cos 60° + r\alpha \sin 60°$$
$$a_y = -r\omega^2 \sin 60° + r\alpha \cos 60°$$

or

$$\left.\begin{aligned} a_x &= 534.6 \text{ in/s}^2 = 44.55 \text{ ft/s}^2 \\ a_y &= -846.0 \text{ in/s}^2 = -70.50 \text{ ft/s}^2 \end{aligned}\right\} \quad (a)$$

The free-body diagram of part E, including inertial forces, is shown in Fig. c.

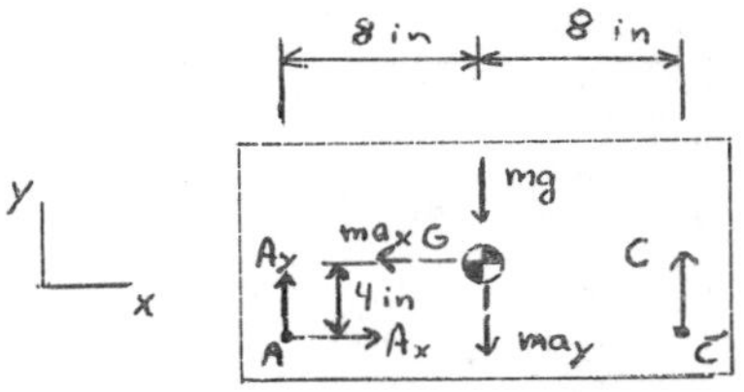

Figure c

By Fig. c,

$$\Sigma F_x = A_x - ma_x = 0 \quad (b)$$
$$\Sigma F_y = A_y + C - mg - ma_y = 0 \quad (c)$$
$$\circlearrowleft \Sigma M_A = 16C + 4ma_x - 8mg - 8ma_y = 0 \quad (d)$$

With Eqs. (a), (b), and (d), we find

$$\underline{\underline{A_x = 89.10 \text{ lb}}}$$
$$\underline{\underline{C = -60.58 \text{ lb}}}$$

Then, Eq. (c) yields

$$\underline{\underline{A_y = -16.03 \text{ lb}}}$$

18.25

Given: A large rectangular crate, with center of mass at the geometric center, is dragged by a rope (Fig. a). The coefficient of sliding friction is 0.10 and $b/h = 1/2$.

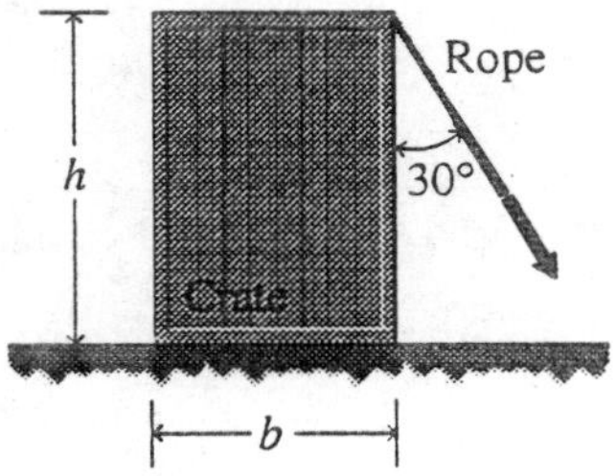

Figure a

Determine the acceleration at which the crate will start to tip, by the inertial-force method.

Solution: The free-body diagram of the crate on the verge of tipping, including the inertial force, is shown in Fig. b.

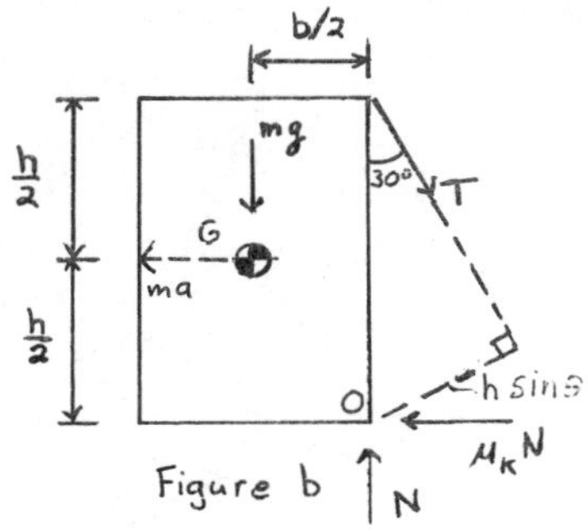

Figure b

By Fig. b,

$$\Sigma F_x = T \sin 30° - ma - \mu_k N = 0 \quad (a)$$
$$\Sigma F_y = N - mg - T\cos 30° = 0 \quad (b)$$
$$\circlearrowleft \Sigma M_O = \left(\tfrac{h}{2}\right)ma + \left(\tfrac{b}{2}\right)mg - (h \sin 30)T = 0 \quad (c)$$

By Eqs. (a) and (b), since $h/b = 1/2$ and $\mu_k = 0.10$,

$$T = 2.41896\,ma + 0.241896\,mg \quad (d)$$

Equations (c) and (d) yield

$$a = 0.1819\,g \quad (e)$$

In SI units, Eq. (e) yields

$$\underline{\underline{a = (0.1819)(9.81) = 1.78 \text{ m/s}^2}}$$

and in U.S. Customary units,

$$\underline{\underline{a = (0.1819)(32.2) = 5.86 \text{ ft/s}^2}}$$

18.26

Given: A homogeneous triangular prismatic plate weighs 18 N. It is suspended by cords AC and BD (Fig. a). For the position shown, the angular velocity of the cords is 10 rad/s counterclockwise.

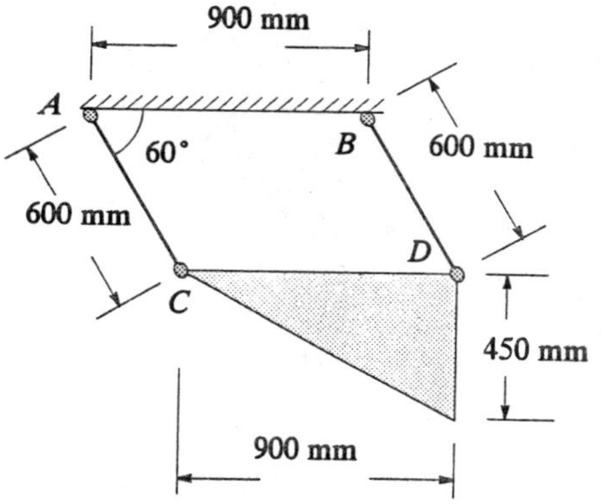

Figure a

Determine the tensions in the cords AC and BD, and the angular acceleration of the cords, by the inertial-force method.

Solution: The normal and tangential components of acceleration are shown in Fig. b,

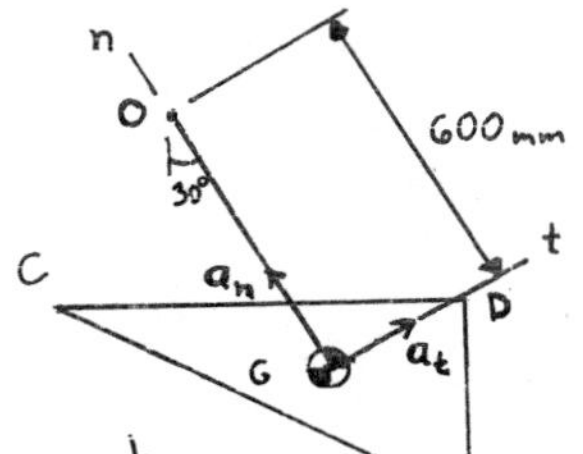

Figure b

The free-body diagram of the plate, including the inertial forces, is shown in Fig. c.

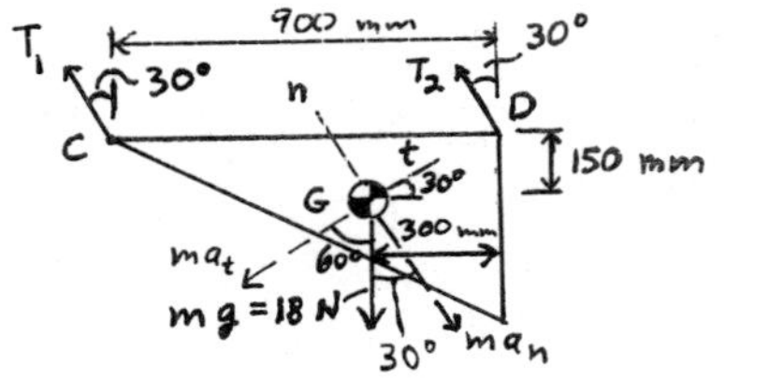

Figure c

By Fig. b, since each point in the plate moves on a circular arc of radius $r = 0.60\,m$,

$$a_t = r\alpha = 0.60\alpha \quad [m/s^2] \qquad (a)$$

$$a_n = r\omega^2 = (0.60)(10^2) = 60\,m/s^2 \qquad (b)$$

By Fig. c,

$$\Sigma F_t = -mg\sin 30° - ma_t = 0 \qquad (c)$$

$$\Sigma F_n = T_1 + T_2 - mg\cos 30° - ma_n = 0 \qquad (d)$$

$$\circlearrowleft\!+\ \Sigma M_C = (0.90\cos 30°)T_2 - (0.60)mg - (0.15)(ma_t\sin 60°) - (0.60)(ma_t\cos 60°) - (0.60)(ma_n\cos 30°) + (0.15)(ma_n\sin 30°) = 0 \qquad (e)$$

By Eqs. (a) and (c),

$$a_t = 0.60\alpha = -9.81\sin 30°$$

or,

$$\underline{\alpha = -8.175 \ \text{rad/s}^2} \qquad (f)$$

By Eqs. (a), (b), (e), and (f),

$$\underline{T_2 = 71.69 \ N} \qquad (g)$$

Equations (b), (d), and (g) yield

$$\underline{T_1 = 53.99 \ N}$$

18.27

Given: When the triangular plate of Problem 18.9 reaches its lowest point (see the free-body diagram in Fig. a), the speed of the center of mass is 2.4 m/s.

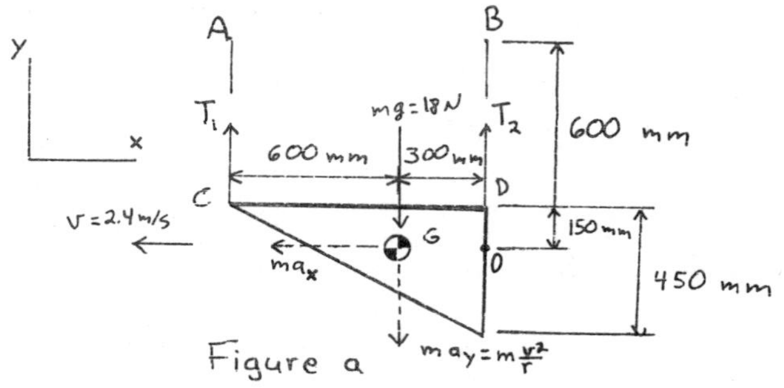

Figure a

Determine the tensions in the cords AC and BD for the position shown, by the inertial-force method.

Solution:

Since each point in the plate moves on a circular arc of radius $r = 0.60\,m$, by Fig. a,

$$\Sigma F_y = T_1 + T_2 - mg - m\frac{v^2}{r} = 0$$

or

$$T_1 + T_2 = 35.615 \quad [N] \qquad (a)$$

$$\circlearrowleft\!+\ \Sigma M_O = (0.30)mg + (0.30)\left(m\frac{v^2}{r}\right) - (0.90)T_1 = 0 \qquad (b)$$

By Eq. (b),

$$\underline{T_1 = 11.87 \ N} \qquad (c)$$

By Eqs. (a) and (c),

$$\underline{T_2 = 23.74 \ N}$$

18.28

Given: A frictionless mechanism is acted on by a force P and translates up an inclined plane as shown in Fig. a. The uniform bar ACD and the base AB are connected by the cord CB. The bar and the base each weigh 9.81 N.

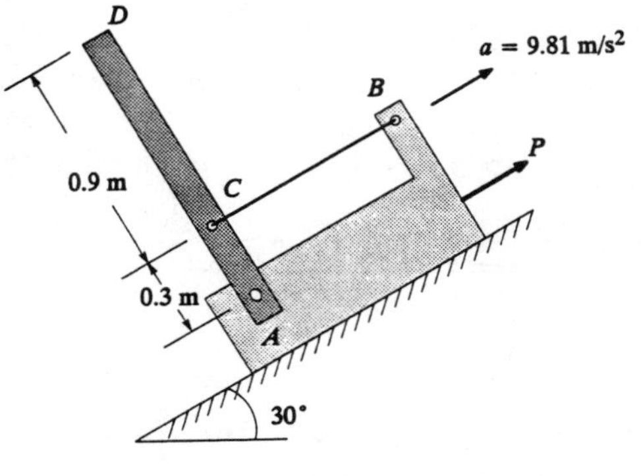

Figure a

Determine the force P and the tension in cord CB, by the inertial-force method

Solution: The free-body diagram of bar ACD, including the inertial-force, is shown in Fig. b,

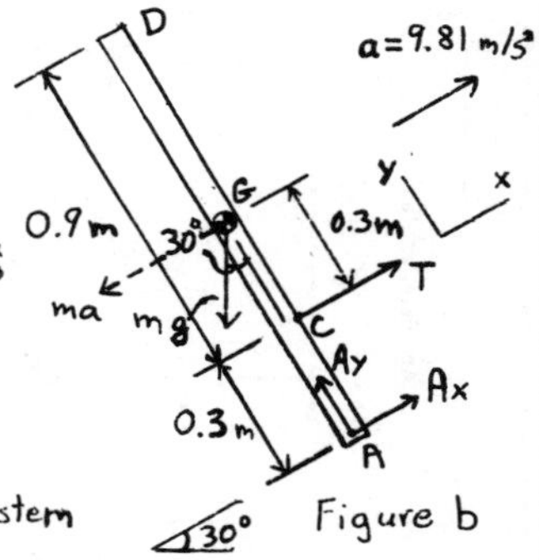

Figure b

By Fig. b,

$\circlearrowleft \Sigma M_A = (0.6)ma + (0.6)mg\sin 30° - (0.3)T = 0$ (a)

By Eq. (a), since $a = 9.81\ m/s^2$, $mg = 9.81\ N$, and $m = 1\ N\cdot s^2/m$,

$\underline{T = 29.43\ N}$

Summing forces up the plane for the entire system (see Figs. a and b), we have, since the inertial-force 2ma acts down the plane,

$\Sigma F = P - 2mg\sin 30° - 2ma = 0$

Hence, with $m = 1\ kg$, $mg = 9.81\ N$, and $a = 9.81\ N/s^2$,

$\underline{\underline{P = 29.43\ N}}$

18.29

Given: An iron pipe 1.2 m in diameter is laid laterally across the bed of a truck, and it is held in position by two blocks of height h (Fig. a). The truck has only rear-wheel brakes, and 60% of the weight of the truck is on the rear wheels, when the truck is braked and skids to a stop. The coefficient of kinetic friction between tires and pavement is $\mu_k = 0.60$

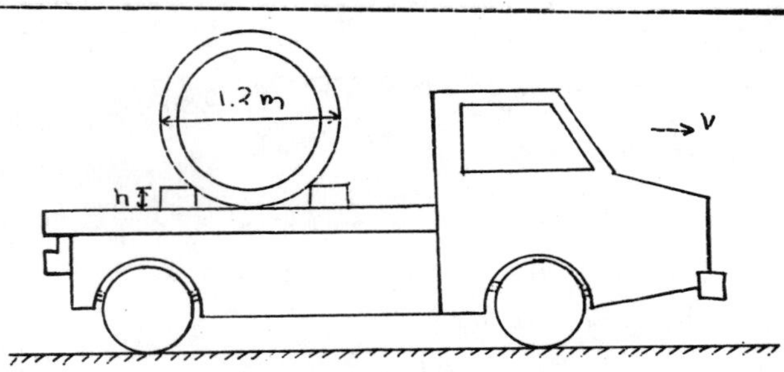

Figure a

Using the inertial-force method, determine the minimum height h [mm], if the pipe is not to roll over a block when the truck is braked and skids to a stop.

Solution:

First we determine the acceleration of the truck. By the free-body diagram of the truck, including the inertial force (Fig. b),

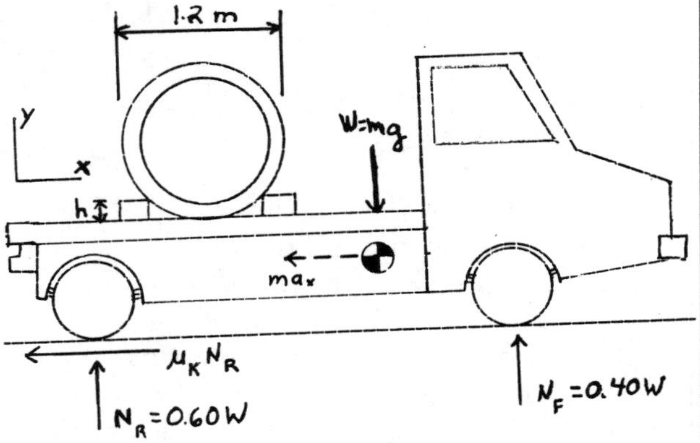

Figure b

We find,

$\Sigma F_x = -\mu_K N_R - ma_x = 0$

where,

$N_R = 0.60mg$

$\mu_K = 0.60$ (b)

By Eqs. (a) and (b), we obtain

$a_x = -3.532\ m/s^2$ (c)

Next, by the free-body diagram of the pipe, including the inertial force, for the instant that the pipe lifts off the bed of the truck (Fig. c)

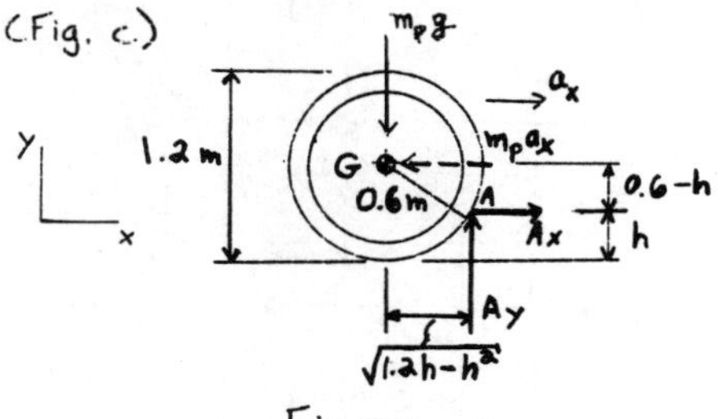

Figure c

18.29 Cont.

$$\circlearrowleft \Sigma M_A = (0.6-h)\, m_p a_x + \left(\sqrt{1.2h-h^2}\right) m_p g = 0 \qquad (d)$$

Equations (c) and (d) yield

$$\sqrt{1.2h-h^2} = 0.36(0.6-h) \qquad (e)$$

Squaring both sides of Eq. (e) and simplifying, we get

$$h^2 - 1.2h + 0.04130 = 0 \qquad (f)$$

or

$$\underline{\underline{h = 0.0355\ \text{m} = 35.5\ \text{mm}}}$$

18.30

Given: The uniform crate shown in Fig. a weighs 16.1 lb. It is subjected to a horizontal force $P = t^3$ (P in pounds and t in seconds). The coefficient of static and kinetic friction are equal at 0.50.

a) Using the inertial-force method, calculate the acceleration of the crate for $t = 1$ s. For $t = 3$ s. Assume the crate does not tip for $t \le 3$ s.

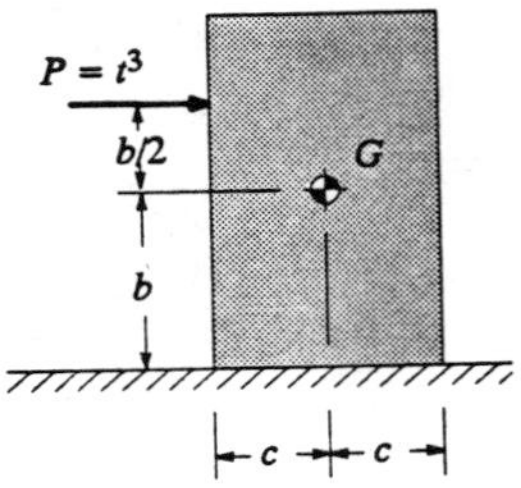

Figure a

b) In terms of b and c, derive a formula for the time t at which the crate is on the verge of tipping, supposing that c is sufficiently large so that the crate slides before it tips.

c) Plot the acceleration of the crate for the time interval $0 \le t \le 6$ s, assuming that the crate does not tip.

d) Plot the time t at which the crate is on the verge of tipping as a function of c/b for the range $1.5 \le c/b \le 7.5$.

e) On the basis of the plots of parts c and d, discuss the assumption used in part a.

Solution:

a) The free-body diagram of the crate, including the inertial force, is shown in Fig. b.

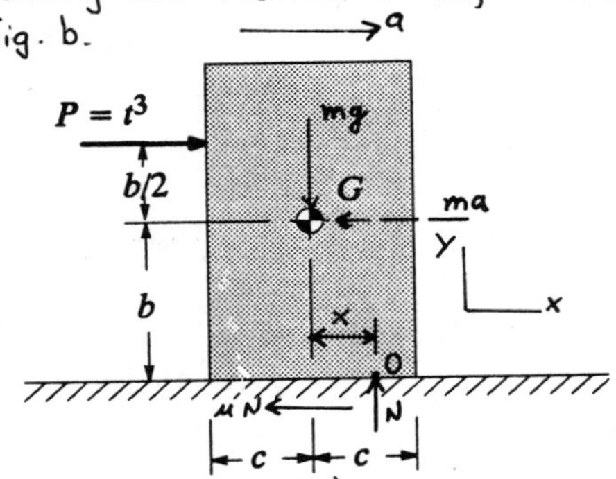

Figure b

By Fig. b,

$$\Sigma F_x = t^3 - \mu N - ma = 0 \qquad (a)$$

$$\Sigma F_y = N - mg = 0;\ N = mg \qquad (b)$$

$$\circlearrowleft \Sigma M_O = (x)(mg) + (b)(ma) - \tfrac{3}{2} b t^3 = 0 \qquad (c)$$

By Eqs. (a) and (b), with $\mu = 0.50$, $m = \frac{1}{2}\ \frac{\text{lb}\cdot\text{s}^2}{\text{ft}}$,

$$N = mg = 16.1\ \text{lb} \qquad (d)$$

$$t - 8.05 = 0.5\,a \qquad (e)$$

By Eq. (e), the crate does not begin to move (accelerate) until $t^3 > 8.05$, or until $t > 2.004$ s.

Hence, $\underline{\underline{\text{for } t = 1\text{ s},\ a = 0}}$

and, by Eq. (e),

$$\underline{\underline{\text{for } t = 3\text{ s},\ a = 37.9\ \text{ft/s}^2}}$$

b) When $x = c$ (Fig. b), the crate is on the verge of tipping. Then, Eqs. (c) and (d) yield

$$t = \left[16.1\left(2\frac{c}{b} - 1\right)\right]^{1/3} \qquad (f)$$

c) By Eq. (e)

$$a = 2t^3 - 16.1 \qquad (g)$$

For $0 \le t \le 6$ s, a is plotted in Fig. c.

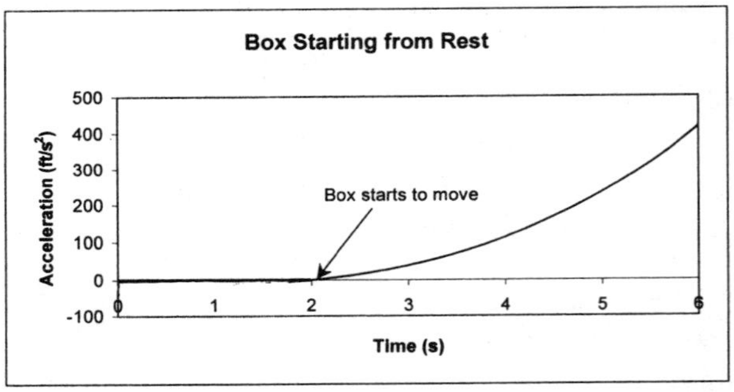

Figure c

d) By Eq. (f), the plot of t, at tipping, as a function of c/b, for $1.5 \le c/b \le 7.5$, is shown in Fig. d.

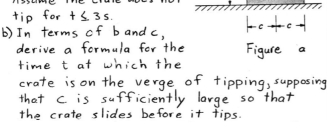

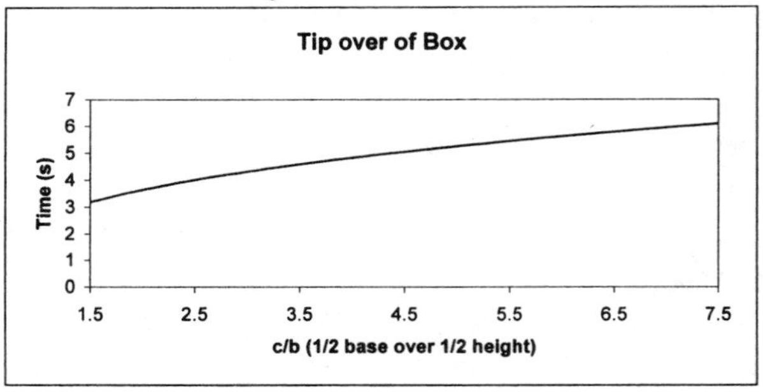

Figure d

e) For the crate not to tip in the time interval $0 \le t \le 6$ s, the ratio c/b must be greater than 7.208.

18.31

<u>Given:</u> An automobile is traveling at 70 ft/s. Its center of mass is 2 ft above the pavement and 4 ft behind the front axle. The distance between front and rear axles is 10 ft, and the coefficient of sliding friction between tires and pavement is 0.75. The brakes are applied, causing the tires to skid.

By the inertial-force method, determine the distance in which the car stops if
a) only the rear wheels are braked,
b) only the front wheels are braked,
c) all four wheels are braked.

<u>Solution:</u>

a) Figure a is the free-body diagram of the car, including the inertial force, if only the rear wheels are braked.

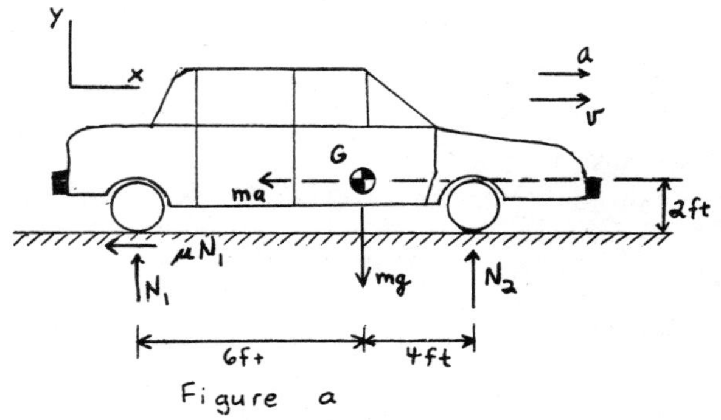

Figure a

By Fig. a,

$$\Sigma F_x = -\mu N_1 - ma = 0 \qquad (a)$$
$$\circlearrowleft + \Sigma M_{N_2} = 10N_1 - 4mg - 2ma = 0$$

with $\mu = 0.75$, Eqs. (a) yield

$$N_1 = \frac{8mg}{23} \ [\text{lb}] \qquad (b)$$
$$a = -\frac{6g}{23} \ [\text{ft/s}^2] \qquad (c)$$

By kinematics and Eq. (c),

$$a = v\frac{dv}{dx} = -\frac{6g}{23} \qquad (d)$$

Integration of Eq. (d) yields

$$\frac{v^2}{2} = -\frac{6g}{23}x + \frac{v_0^2}{2} \qquad (e)$$

where $v_0 = 70$ ft/s. Hence, when the car comes to rest, $v = 0$, and Eq. (e) yields,

$$x = \frac{23}{12(32.2)}(70)^2 = 291.7 \text{ ft}$$

(b) If only the front wheels are braked, the frictional force μN_2 acts on the back wheels (there is no frictional force on the back wheels). Then, (see Fig. a)

$$\Sigma F_x = -\mu N_2 - ma = 0 \qquad (f)$$
$$\circlearrowleft + \Sigma M_{N_1} = 10N_2 - 6mg + 2ma = 0$$

With $\mu = 0.75$, Eqs. (f) yield

$$N_2 = \frac{12}{17}mg \ [\text{lb}] \qquad (g)$$
$$a = -\frac{9}{17}g \ [\text{ft/s}^2] \qquad (h)$$

By kinematics and Eq. (h),

$$a = v\frac{dv}{dx} = -\frac{9}{17}g \qquad (i)$$

Integration of Eq. (i) yields

$$\frac{v^2}{2} = -\frac{9}{17}gx + \frac{v_0^2}{2}$$

Since $v = 0$ when the car comes to rest

$$x = \frac{17(70)^2}{(18)(32.2)} = 143.7 \text{ ft}$$

c) If all wheels are braked, frictional forces μN_1 and μN_2 act on the rear and front wheels, respectively. Then (see Fig. a),

$$\Sigma F_x = -\mu N_1 - \mu N_2 - ma = 0$$
$$\Sigma F_y = N_1 + N_2 - mg = 0$$

or

$$a = -0.75g = v\frac{dv}{dx} \qquad (j)$$

Integration of Eq. (j) yields

$$\frac{v^2}{2} = -0.75gx + \frac{v_0^2}{2}$$

or, since $v = 0$ when the car is stopped,

$$x = \frac{(70)^2}{(1.5)(32.2)} = 101.4 \text{ ft}$$

18.32

<u>Given:</u> A hollow cylinder of mass m has inner radius a and outer radius b (Fig. a)

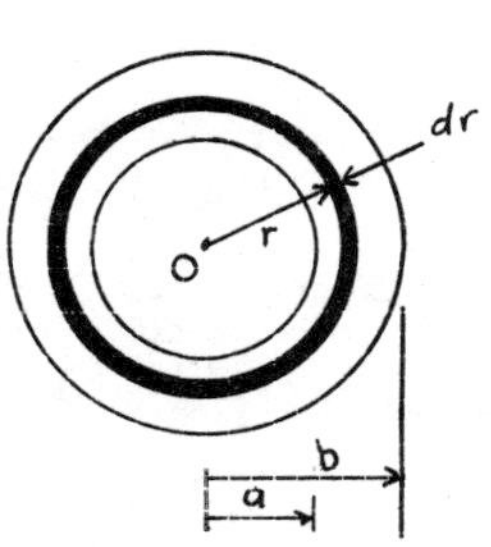

Show that the moment of inertia of the cylinder with respect to its geometric axis O is

$$I_o = \frac{1}{2}m(a^2 + b^2) \qquad (a)$$

(Continued)

18.32 Cont

Solution: Consider a circular element of radius r and thickness dr (Fig. a). The mass dm of this element is

$$dm = \rho L 2\pi r\, dr \qquad (b)$$

where ρ is the mass density and L is the length of the element perpendicular to the plane of Fig. a. By definition, the moment of inertia of the cylinder with respect to axis O is

$$I_o = \int r^2 dm \qquad (c)$$

By Eqs. (b) and (c),

$$I_o = 2\pi\rho L \int_a^b r^3 dr = \frac{2\pi\rho L}{4}(b^4 - a^4) \qquad (d)$$

Also, the mass of the hollow cylinder is

$$m = \pi(b^2 - a^2)L\rho \qquad (e)$$

or

$$\pi\rho L = \frac{m}{b^2 - a^2}$$

Equations (d) and (e) yield

$$I_o = \frac{1}{2}m(a^2 + b^2) \qquad (f)$$

Equation (f) is the same as Eq. (a)

18.33

Given: A homogeneous prism whose cross section is a circular sector has mass m (Fig. a)

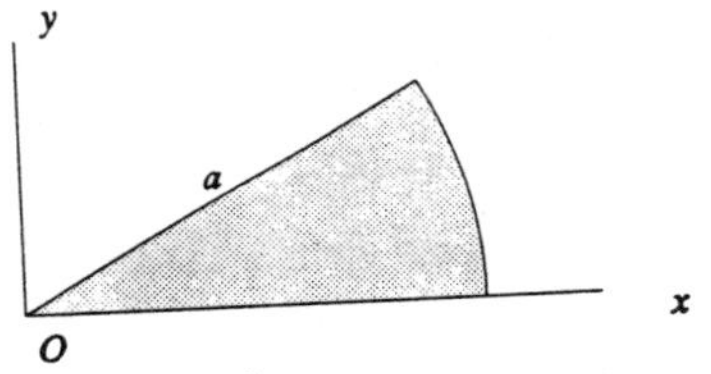

Figure a

Show that its mass moment of inertia relative to axis O is $I_o = \frac{1}{2}ma^2$.

Solution: For a circular cylinder of radius a and mass m_c (see Appendix F, Table F 1)

$$I_o = \frac{1}{2}m_c a^2$$

Hence, for a circular sector of mass m and radius a,

$$I_o = \frac{1}{2}m_c a^2\left(\frac{m}{m_c}\right) = \frac{1}{2}ma^2$$

Alternatively, consider a volume element (Fig. b)

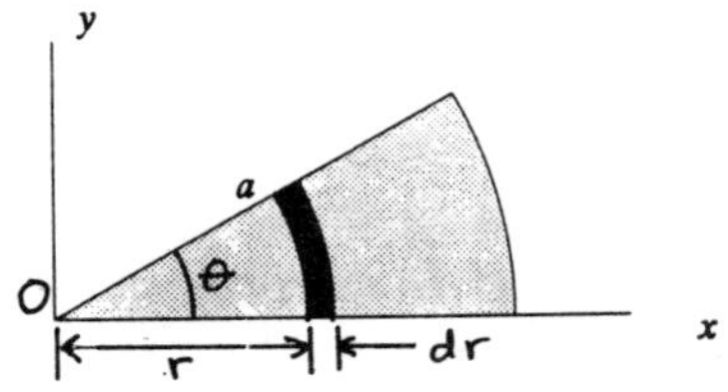

Figure b

$dV = r\theta L dr$, where L is the length of the prism.

The associated elemental mass is $dm = \rho r\theta L dr$. Hence,

$$I_o = \int r^2 dm = \rho\theta L \int_0^a r^3 dr = \frac{1}{4}\rho\theta L a^4 \quad (a)$$

and

$$m = \int dm = \rho\theta L \int_0^a r\, dr = \frac{1}{2}\rho\theta L a^2 \quad (b)$$

Equations (a) and (b) yield

$$I_o = \frac{1}{2}ma^2$$

18.34

Given: A thin ($h \ll r$) circular disk, shown in Fig. a, has radius r, thickness h and mass density ρ.

Derive a formula for the mass moment of inertia relative to axis a-a.

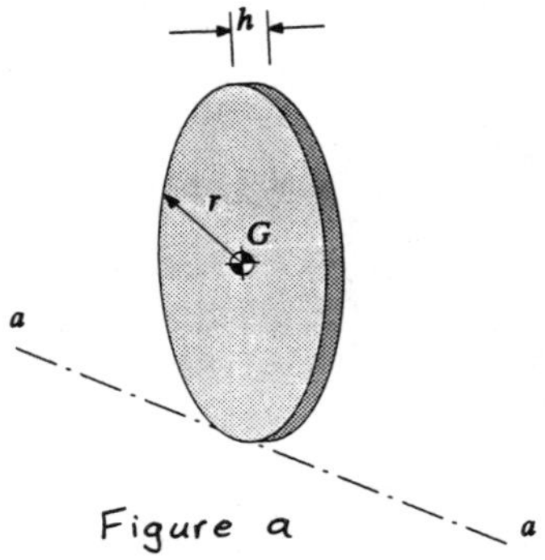

Figure a

Solution:

Refer to Fig. b, a front view of the disk.

By Appendix F, Table F1

$$I_{o\text{-}o} = \frac{1}{4}mr^2 = \frac{1}{4}(\pi r^2 h\rho)r^2$$

or $$I_{o\text{-}o} = \frac{\pi}{4}\rho h r^4$$

Then, by the parallel axes theorem (steiner's theorem)

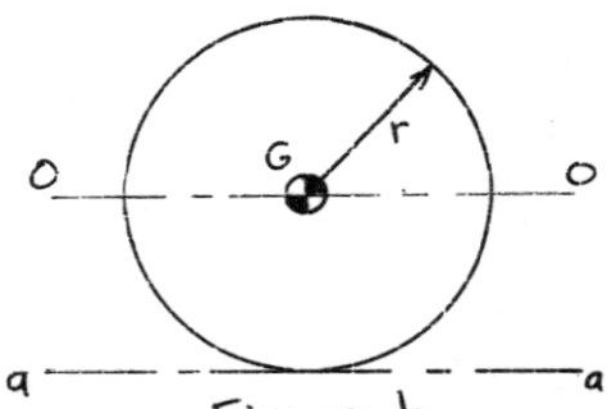

Figure b

$$I_{a\text{-}a} = I_{oo} + mr^2 = \frac{\pi}{4}\rho h r^4 + (\pi r^2 h\rho)r^2$$

or

$$I_{a\text{-}a} = \frac{5\pi}{4}\rho h r^4$$

18.35

Given: A homogeneous cube has a mass m and dimension a of an edge.

Derive formulas for the mass moment of inertia and the radius of gyration with respect to an axis that coincides with one of its edges

Solution: Refer to Fig. a, a view of one of its faces.

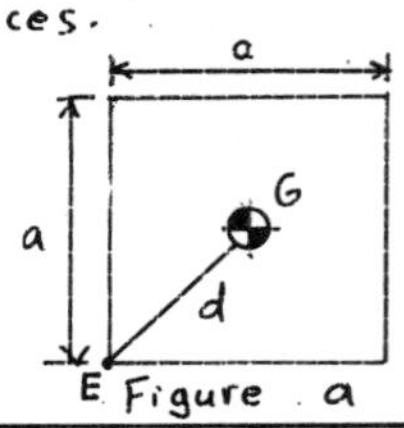

Figure a

(Continued)

18.35 Cont.

The mass moment of inertia of the cube relative to an axis through G, perpendicular to its face, is (see Appendix F for a rectangular parallelepiped with a = b)

$$I_G = \frac{1}{6} m a^2$$

Hence, by the parallel axes theorem (Steiner's theorem) the mass moment of inertia with respect to the axis E along an edge is

$$\underline{\underline{I_E = I_G + md^2 = \frac{1}{6} ma^2 + m\left(\frac{a^2}{4} + \frac{a^2}{4}\right) = \frac{2}{3} ma^2}}$$

Then, the radius of gyration is

$$\underline{\underline{k_E = \sqrt{I_E/m} = \sqrt{\tfrac{2}{3}}\, a}}$$

18.36

Given: A hollow thin right-circular cone has mass m and constant thickness t << r (Fig. a)

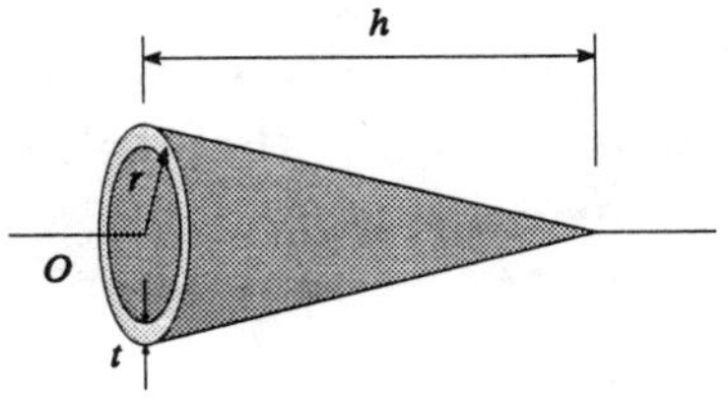

Figure a

Show that the mass moment of inertia relative to its geometric axis is approximately $\frac{1}{2} mr^2$

Solution:

Consider a longitudinal cross section of the cone (Fig. b).

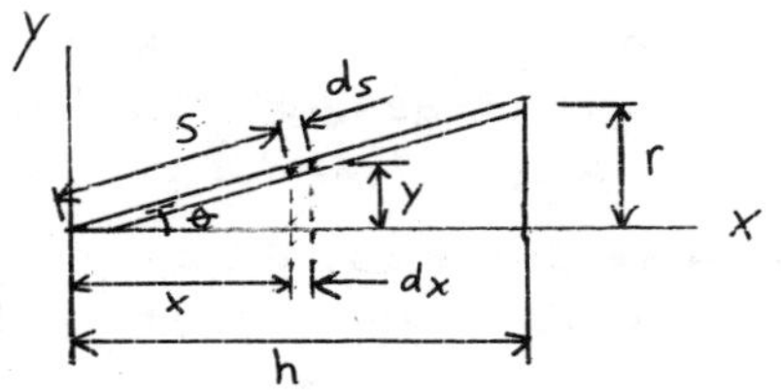

Figure b

By Fig. b,

$$y = \frac{r}{h} x \qquad (a)$$

and the elemental mass dm at s is

$$dm \approx \rho(2\pi y\, ds) \qquad (b)$$

where ρ is the mass per unit area

By the definition, the mass moment of inertia with respect to the cone's geometric axis is

$$I_x = \int y^2 dm \qquad (c)$$

By Eqs. (a), (b), and (c), and Fig. b,

$$I_x = 2\pi\rho \frac{r^3}{h^3} \int x^3 ds \;;\; ds = dx/\cos\theta. \;(d)$$

Then, by Eqs. (b) + (d), integration from x = 0 to x = h, yields

$$I_x = \frac{\pi}{2}\rho r^3 h \sec\theta ;\; m \approx \pi\rho r h \sec\theta \qquad (e)$$

By Eq. (e),

$$\underline{\underline{I_x \approx \frac{1}{2} mr^2}}$$

18.37

Given: A homogeneous rectangular parallelepiped has a specific weight of 68 kN/m³ (Fig. a)

Determine the mass moment of inertia [N·m·s²] of the parallelepiped with respect to axis A-A.

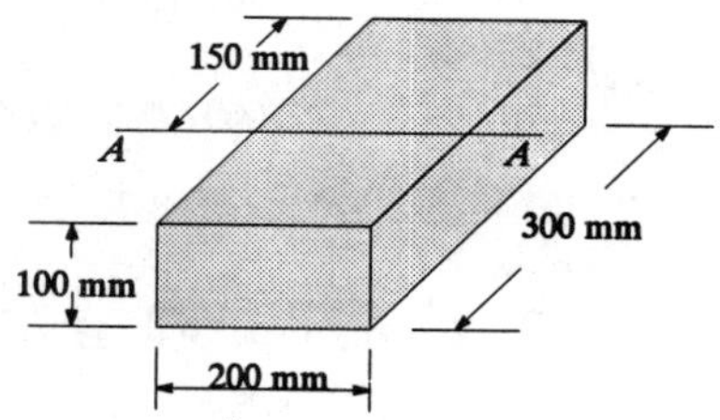

Figure a

Solution:

Consider a view of the right face of the body (Fig. b)

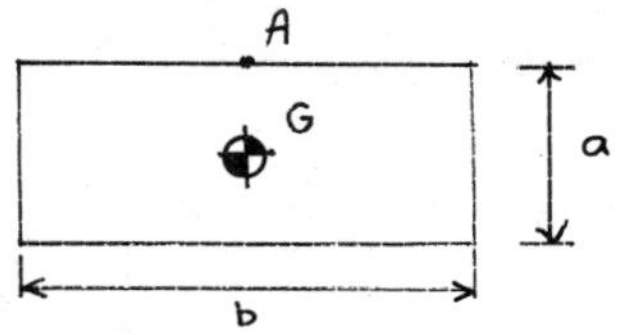

Figure b

By Appendix F, Table F1 and Figs. a and b,

$$I_G = \frac{1}{12} m(a^2 + b^2)$$

$$m = (0.10)(0.20)(0.30)(68000)/9.81 = 41.59 \text{ kg}$$

$$I_G = \frac{1}{12}(41.59)(0.10^2 + 0.30^2) = 0.3466 \text{ N·m·s}^2$$

Hence, by the parallel axes theorem (Steiner's Theorem),

$$I_A = I_G + m\left(\frac{a}{2}\right)^2 = 0.3466 + (41.59)\left(\frac{0.10}{2}\right)^2 =$$

$$\underline{\underline{I_A = 0.451 \text{ N·m·s}^2}}$$

18.38

Given: A rectangular prism has a circular cutout centered on its longitudinal geometric axis (Fig. a). The prism is 450 mm long and has a mass density ρ = 8.46 kN·s²/m⁴.

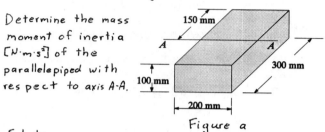

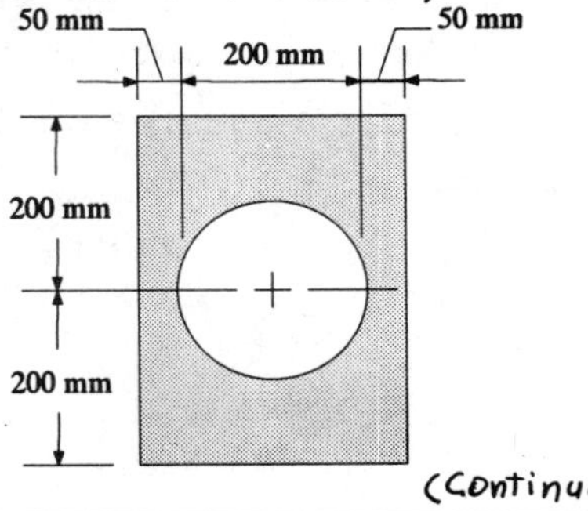

Figure a

(Continued)

18.38 Cont.

Calculate the mass moment of inertia [N·m·s²] of the prism relative to its longitudinal geometric axis.

Solution:

Consider the prism with cut out to be a composite body as a prism of mass m_p without cutout and a cylinder with negative mass $-m_c$, Thus

$$m_p = (8460)(0.40)(0.30)(0.450) = 456.84 \text{ kg}$$

and

$$-m_c = -(8460)(\pi)(0.10^2)(0.450) = -119.60 \text{ kg}$$

For the prism without a hole (see Appendix F, Table F1),

$$I_P = \frac{1}{12} m_p (a^2 + b^2) = \frac{1}{12}(456.84)(0.40^2 + 0.30^2)$$

$$I_P = 9.518 \text{ N·m·s}$$

Similarly, for the cylinder,

$$I_C = \frac{1}{2} m_c r^2 = -\frac{1}{2}(119.60)(0.10^2) = -0.598 \text{ N·m·s}^2$$

Therefore, $\underline{I_O = I_P + I_C = 8.92 \text{ N·m·s}^2}$

18.39

Given: A half-cylinder (Fig. a) has mass m.

Derive a formula for its moment of inertia about an axis through point P, making use of the formula of a complete cylinder about its geometric axis.

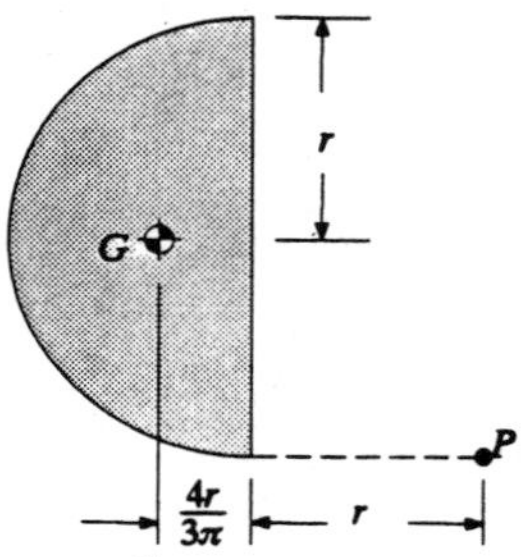

Figure a

Solution:

First we determine the moment of inertia about O (Fig. a). For a complete cylinder of radius r and mass 2m (see Appendix F, Table F 1),

$$I_O^c = \frac{1}{2}(2m)r^2 = mr^2$$

Thus, for the half-cylinder,

$$I_O = \frac{1}{2} I_O^c = \frac{1}{2} mr^2$$

Next, by the parallel axis theorem (Steiner's theorem) we find the moment of inertia about the mass center G, thus,

$$I_G = I_O - m\left(\frac{4r}{3\pi}\right)^2 = mr^2\left(\frac{1}{2} - \frac{16}{9\pi^2}\right) \quad \text{(a)}$$

Then, by the parallel axis theorem,

$$I_P = I_G + m\left[r^2 + \left(r + \frac{4r}{3\pi}\right)^2\right]$$

$$= mr^2\left[\frac{1}{2} - \frac{16}{9\pi^2} + 1 + \left(1 + \frac{8}{3\pi} + \frac{16}{9\pi^2}\right)\right]$$

$$= mr^2\left(\frac{5}{2} + \frac{8}{3\pi}\right)$$

or $\underline{\underline{I_P = 3.349\, mr^2}}$

18.40

Given: The area bounded by the curve $y^3 = 8x$, the line $x = 4$, and the x axis is rotated about the x axis to generate a homogeneous solid of mass density ρ.

Determine the mass moment of inertia of the solid with respect to the x axis.

Solution:

The bounded area is shown in Fig. a.

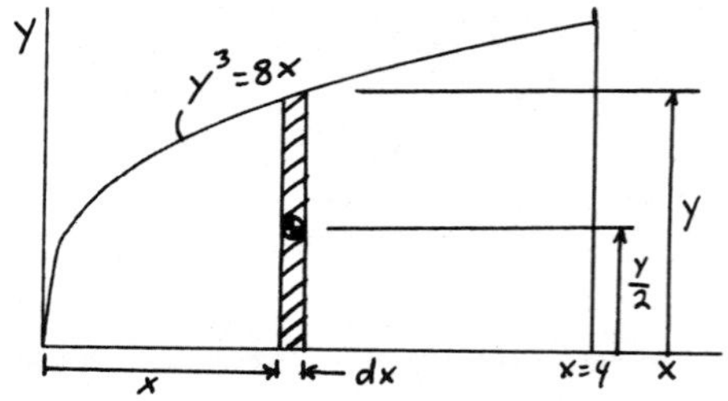

Figure a

By Fig. a, the differential mass element of a circular disk of the solid is

$$dm = \rho \pi y^2 dx$$

and the corresponding element of mass moment of inertia is, relative to the x axis,

$$dI_x = \frac{1}{2} y^2 dm = \frac{\pi}{2} \rho y^4 dm = 8\pi\rho x^{4/3} dx$$

Hence, $I_x = \int_0^4 8\pi\rho x^{4/3} dx = 8\pi\rho \left(\frac{3}{7}\right) x^{7/3}\Big|_0^4$

or $\underline{\underline{I_x = 273.6\,\rho}}$

18.41

Given: The area bounded by the curve $y^3 = ax^2$ and the line $y = a$ is rotated about the x axis to generate a homogeneous solid of mass density ρ.

Determine the mass moment of inertia of the solid with respect to the x axis

Solution: The bounded area is shown in Fig. a.

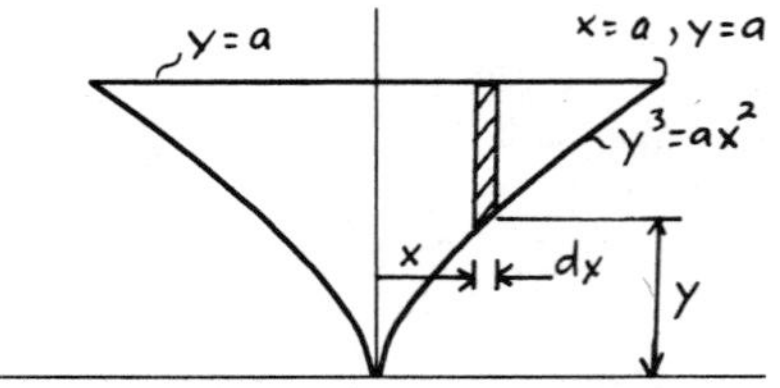

Figure a

By Fig. a, the differential mass moment of inertia of the circular disk of radii $r = y$ and $r = a$ and thickness dx is, relative to the x axis,

(Continued)

18.41 Cont.

$$dI_x = \frac{\pi}{2}(a^4 - y^4)\rho\,dx = \frac{\pi\rho}{2}(a^4 - a^{4/3}x^{8/3})\,dx$$

and the corresponding total mass moment of inertia is, relative to the x axis,

$$I_x = 2\int_0^a dI_x = \pi\rho\int_0^a (a^4 - a^{4/3}x^{8/3})\,dx \quad (a)$$

Integration of Eq. (a) yields

$$I_x = \pi\rho\left[a^4x - \frac{3}{11}a^{4/3}x^{11/3}\right]\Big|_0^a$$

or

$$I_x = \frac{8}{11}\pi\rho a^5$$

Alternatively, taking an element parallel to the x axis (Fig. b),

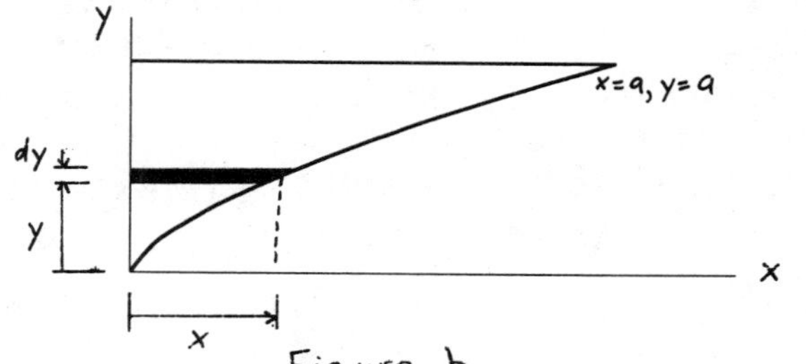

Figure b

$$dm = \rho 2\pi y x\,dy = \frac{2\pi\rho}{\sqrt{a}}\,y^{5/2}\,dy$$

Hence, $I_x = 2\int y^2\,dm = \frac{4\pi\rho}{\sqrt{a}}\int_0^a y^{9/2}\,dy = \frac{4\pi\rho}{\sqrt{a}}\left(\frac{2}{11}\right)y^{11/2}\Big|_0^a$

or

$$I_x = \frac{8}{11}\pi\rho a^5$$

18.42

Given: The area bounded by the ellipse $(x/a)^2 + (y/b)^2 = 1$ is rotated about the x axis to generate a homogeneous solid of mass density ρ.

Determine the mass moment of inertia of the solid with respect to the x axis.

Solution: The bounded region is shown in Fig. a.

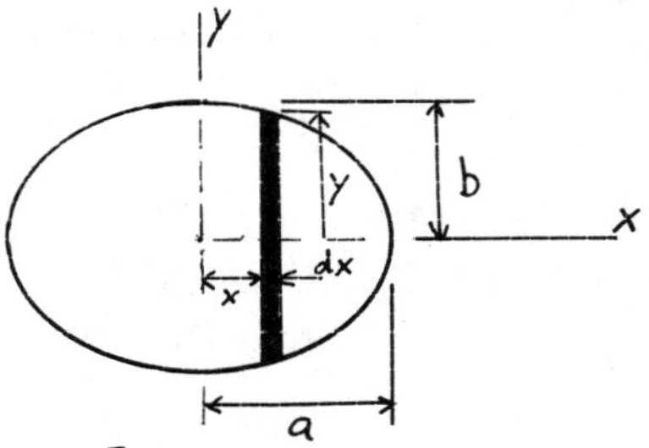

Figure a

By Fig. a, the mass of the disk element of radius y and thickness dx is

$$dm = \rho\pi y^2\,dx$$

Hence, for the disk, the mass moment of inertia is,

$$dI_x = \frac{1}{2}y^2\,dm = \frac{\pi\rho}{2}y^4\,dx = \frac{\pi\rho b^4}{2a^4}(a^2 - x^2)^2\,dx$$

So, the total mass moment of inertia of the solid is, integrating from $x=0$ to $x=a$,

$$I_x = 2\int dI_x = \frac{\pi\rho b^4}{a^4}\int_0^a (a^4 - 2a^2x^2 + x^4)\,dx$$

or

$$I_x = \frac{8\pi}{15}\rho a b^4$$

18.43

Given: The area bounded by the curve $8y = x^3$, the line $y = 4$, and the y axis ($x>0$) is rotated about the x axis to generate a homogeneous solid of mass density ρ.

Determine the mass moment of inertia of the solid with respect to the x axis.

Solution: The bounded area is shown in Fig. a.

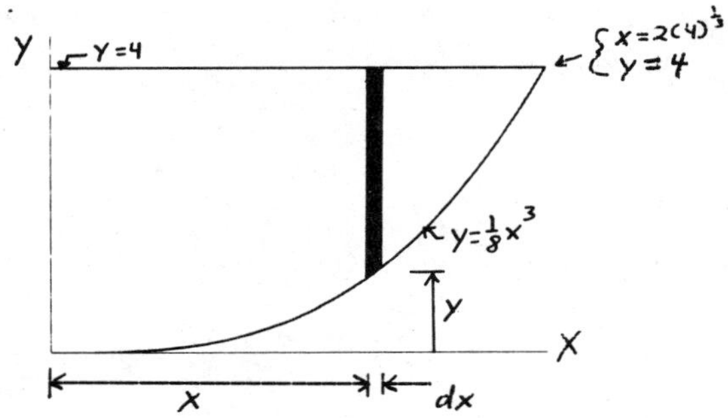

Figure a

By Fig. a, for the hollow disk of thickness dx, the element of mass moment of inertia relative to the x axis is

$$dI_x = \frac{\pi\rho}{2}(4^4 - y^4)\,dx = \frac{\pi\rho}{2}\left(256 - \frac{x^{12}}{8^4}\right)dx$$

or

$$I_x = \frac{\pi\rho}{2}\int_0^{2(4)^{1/3}}\left(256 - \frac{x^{12}}{8^4}\right)dx$$

Integration yields

$$I_x = 1178\rho$$

18.44

Given: The area bounded by the lines $y = x+2$, $y = -x+2$, and $y = 4$ is rotated about the x axis to generate a homogeneous solid of mass density ρ.

Determine the mass moment of inertia of the solid with respect to the x axis

Solution: The bounded area is shown in Fig. a.

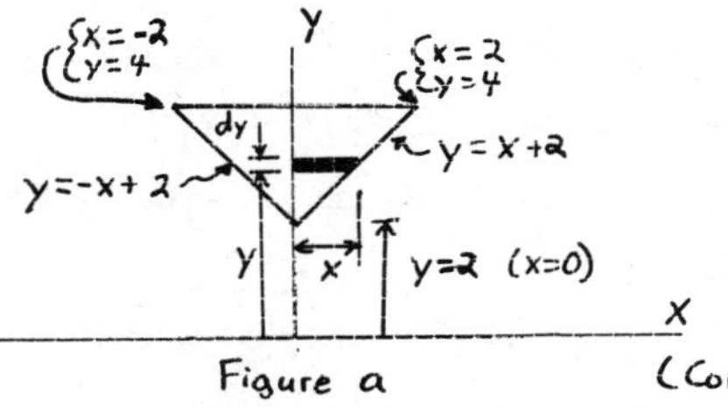

Figure a

(Continued)

18.44 Cont.

For the element $x\,dy$ shown in Fig. a, the mass element (after rotation about the x axis) is

$$dm = \rho\,(2\pi y)\,x\,dy$$

and the corresponding mass moment of inertia is,

$$dI_x = y^2 dm = 2\pi\rho y^3 (y-2)dy$$

Hence, the total mass moment of inertia of the solid is

$$I_x = 2\int dI_x = 4\pi\rho\int_2^4 (y^4 - 2y^3)\,dy \quad (a)$$

so,

$$\underline{\underline{I_x = 985.2\,\rho}}$$

18.45

Given: A thick hoop of mass m is supported on a fulcrum (Fig. a).

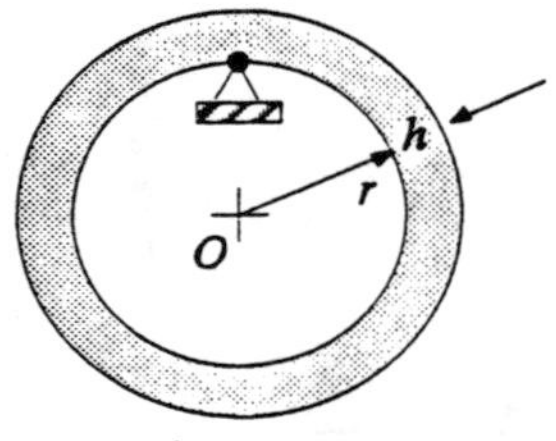

Figure a

In terms of the radius r, the thickness h, and the mass m, derive an expression for the moment of inertia of the hoop about the fulcrum.

Solution: Consider a circular element of the hoop of thickness dR (Fig. b)

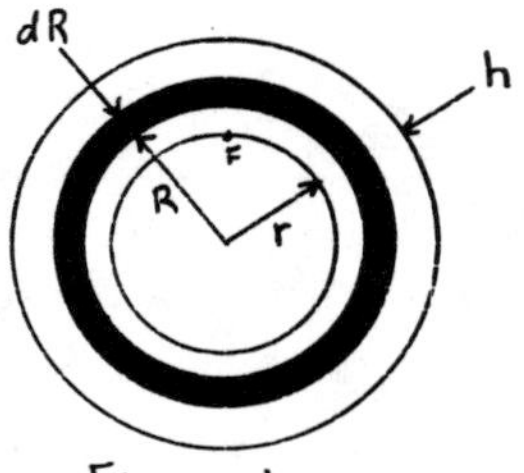

Figure b

The differential mass of the element is

$$dm = \rho\,2\pi R\,dR \quad (a)$$

where ρ is the mass per unit area.

The corresponding mass of the hoop is, by Eq. (a),

$$m = \int dm = \int_r^{r+h} 2\pi\rho R\,dR = \pi\rho[(r+h)^2 - r^2] \quad (b)$$

The mass moment of inertia relative to O is, with Eq. (a),

$$I_o = \int R^2 dm = 2\pi\rho\int_r^{r+h} R^3 dR \quad (c)$$

Integrating Eq. (c), we get

$$I_o = \tfrac{1}{2}\pi\rho[(r+h)^4 - r^4] = \tfrac{1}{2}\pi\rho[(r+h)^2 - r^2][(r+h)^2 + r^2]$$

with Eq. (b), the above equation becomes (see also Appendix F, Table F1)

$$I_o = \tfrac{1}{2}m[(r+h)^2 + r^2]$$

Then, by Steiner's theorem (parallel axis theorem), the mass moment of inertia relative to the fulcrum is,

$$\underline{\underline{I_F = I_o + mr^2 = \tfrac{1}{2}m[3r^2 + (r+h)^2]}}$$

18.46

Given: A hollow concrete cylinder is reinforced by eight longitudinal steel rebars (Fig. a). The length of the cylinder is 36 in. The mass density of the concrete is $\rho_c = 0.000245\ \mathrm{lb\cdot s^2/in^4}$ and that of the steel is $\rho_s = 0.000733\ \mathrm{lb\cdot s^2/in^4}$.

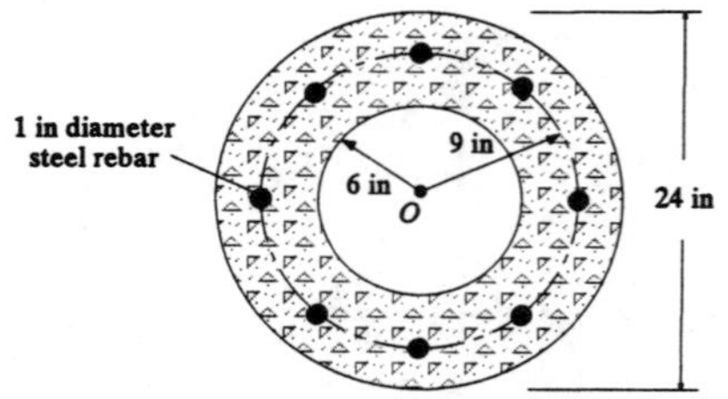

Figure a

Calculate the mass moment of inertia [$\mathrm{lb\cdot ft\cdot s^2}$] of the cylinder with respect to its longitudinal axis.

Solution:

By Appendix F, Table F1, for the solid concrete cylinder (as if the steel bars were concrete)

$$I'_{oc} = \tfrac{1}{2}m'_c(r_1^2 + r_2^2) \quad (a)$$

where,

$$m'_c = \pi\rho_c(r_2^2 - r_1^2)L$$
$$r_1 = 6\text{ in} = 0.5\text{ ft}$$
$$r_2 = 12\text{ in} = 1.0\text{ ft} \quad (b)$$
$$\rho_c = 0.000245\ \frac{\mathrm{lb\cdot s^2}}{\mathrm{in^4}} = 5.080\ \mathrm{lb\cdot s^2/ft^4}$$
$$L = 36\text{ in} = 3\text{ ft}$$

By Eqs. (a) and (b),

$$I'_{oc} = \frac{\pi}{2}(5.080)(3)(1^2 - 0.5^2)(1^2 + 0.5^2) =$$

$$I'_{oc} = 22.443\ \mathrm{lb\cdot ft\cdot s^2} \quad (c)$$

The mass moment of inertia of the concrete replaced by the steel bars is

$$I''_{oc} = \tfrac{1}{2}m''_c r_s^2 + m''_c d^2$$

where

$$m''_c = 8\rho_c\pi r_s^2 L$$
$$r_s = 0.5\text{ in} = 0.04166\text{ ft}$$
$$d = 9\text{ in} = 0.75\text{ ft}$$

(Continued)

18.46 Cont.

Hence,
$$I''_{oc} = (8)(5.080)(\pi)(0.04166)^2(3)[\tfrac{1}{2}(0.04166)^2 + 0.75^2]$$
$$I''_{oc} = (0.66497)(0.000868 + 0.5625)$$
or
$$I''_{oc} = 0.3746 \quad lb\cdot ft\cdot s^2 \qquad (d)$$

Likewise, the mass moment of inertia of the steel bars is
$$I_{os} = \tfrac{1}{2} m_s r_s^2 + m_s d^2$$
where,
$$m_s = 8\rho_s \pi r_s^2 L$$
$$\rho_s = 0.000733 \ \frac{lb\cdot s^2}{in^4} = 15.199 \ \frac{lb\cdot s^2}{ft^4}$$

Hence,
$$I_{os} = (8)(15.199)\pi(0.04166)^2(3)[\tfrac{1}{2}(0.04166)^2 + 0.75^2]$$
or
$$I_{os} = 1.1208 \quad lb\cdot ft\cdot s^2 \qquad (e)$$

So the total mass moment of inertia is, by Eqs. (c), (d), and (e)
$$I_o = I'_{oc} - I''_{oc} + I_{os}$$
$$= 22.443 - 0.3746 + 1.1208$$
or
$$\underline{\underline{I_o = 23.19 \quad lb\cdot ft\cdot s^2}}$$

18.47

Given: A wooden block is 8 in deep, 10 in wide, and 24 in long. A 2 in diameter cylindrical hole is drilled through the block, with the axis of the hole parallel to the longest centroidal axis and 3 in from that centroidal axis.
The wood weighs 40 lb/ft³.

Calculate the mass moment of inertia of the final shape with respect to the longest centroidal axis of the undrilled block.

Solution: Figure a is a cross section sketch of the block perpendicular to the longest centroidal axis.

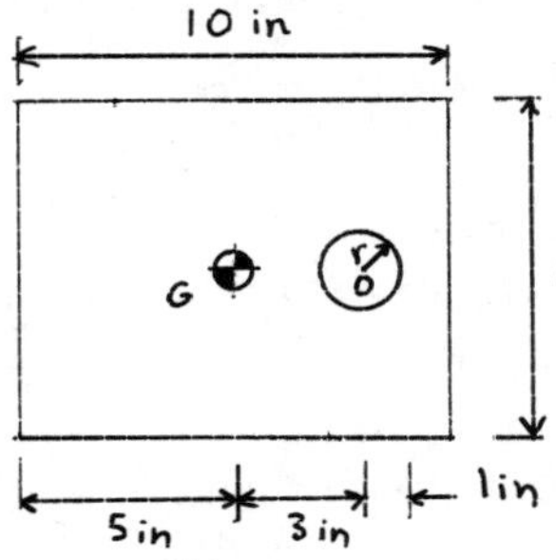

Figure a

The mass of the undrilled block is
$$m' = \rho abL = (40)(\tfrac{8}{12})(\tfrac{10}{12})(\tfrac{24}{12})(\tfrac{1}{32.2})$$
or
$$m' = 1.380 \quad lb\cdot s^2/ft$$

By Appendix F, Table F1, the mass moment of inertia of the undrilled block is
$$I'_G = \tfrac{1}{12} m'(a^2+b^2) = \tfrac{1}{12}(1.380)[(\tfrac{8}{12})^2 + (\tfrac{10}{12})^2]$$
or
$$I'_G = 0.1310 \quad lb\cdot ft\cdot s^2 \qquad (a)$$

For the circular cylinder of wood removed, the mass is
$$m'' = \frac{\rho \pi r^2 L}{g} = \pi(40)(\tfrac{1}{12})^2(\tfrac{24}{12})(\tfrac{1}{32.2})$$
or
$$m'' = 0.0542 \quad lb\cdot s^2/ft$$

By Appendix F, Table F1 and Steiner's theorem, the mass moment of inertia of the removed cylinder is (since r = 1 in and d = 3 in)
$$I''_G = \tfrac{1}{2} m'' r^2 + m'' d^2$$
$$= \tfrac{1}{2}(0.0542)(\tfrac{1}{12})^2 + (0.0542)(\tfrac{3}{12})^2$$
or
$$I''_G = 0.00358 \quad lb\cdot ft\cdot s^2$$

So, the mass moment of inertia of the drilled block is
$$\underline{\underline{I_G = I'_G - I''_G = 0.1310 - 0.00358 = 0.1274 \quad lb\cdot ft\cdot s^2}}$$

18.48

Given: A solid right-circular cone has altitude h and base radius r (Fig. a)

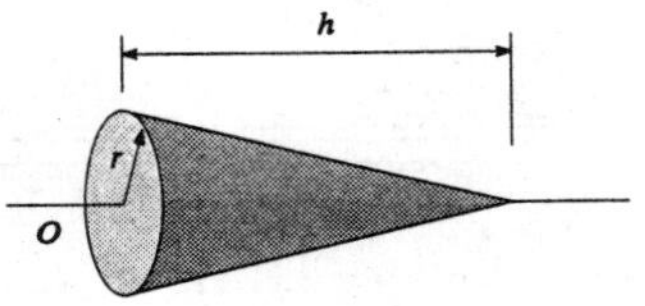

Figure a

Determine the mass moment of inertia and the radius of gyration of the cone with respect to the geometric axis of the cone.

Solution: Consider the centroidal section of the cone shown in Fig. b.

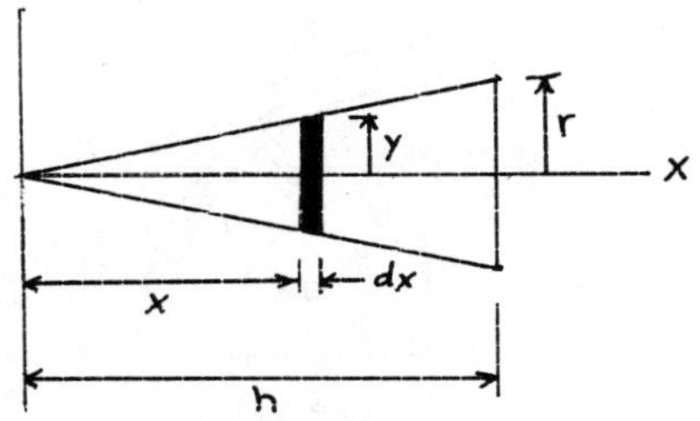

Figure b

By Fig. b,
$$y = \frac{r}{h} x \qquad (a)$$

(Continued)

18.48 Cont.

The element dm of mass of the disk of thickness dx and radius y is

$$dm = \rho \pi y^2\, dx \qquad (b)$$

where ρ is the mass density of the disk.

Hence, the mass moment of inertia of the disk is, with respect to the centroid axis x,

$$dI_x = \frac{1}{2} y^2 dm = \frac{1}{2}\pi\rho y^4 dx \qquad (c)$$

Equations (a) and (c) yield

$$dI_x = \frac{1}{2}\pi\rho \frac{r^4}{h^4} x^4\, dx$$

So the moment of inertia of the cone is

$$I_x = \int dI_x = \frac{1}{2}\pi\rho\frac{r^4}{h^4}\int_0^h x^4 dx$$

or by integration

$$I_x = \frac{1}{10}\pi\rho h r^4 \qquad (d)$$

Also, by Eqs. (a) and (b), the mass of the cone is

$$m = \int dm = \pi\rho\frac{r^2}{h^2}\int_0^h x^2 dx = \frac{1}{3}\pi\rho h r^2 \qquad (e)$$

Hence, Eqs. (d) and (e) yield

$$I_x = \frac{3}{10} m r^2$$

and the radius of gyration of the cone is

$$k_x = \sqrt{I_x/m} = r\sqrt{3/10}$$

18.49

Given: A uniform, hollow, thin homogeneous sphere has a mass m and a mean radius r.

Show that the mass moment of inertia of the sphere with respect to a diametral axis is $I = \frac{2}{3}mr^2$

Solution: Consider the cross section of a quadrant of the sphere (Fig. a)

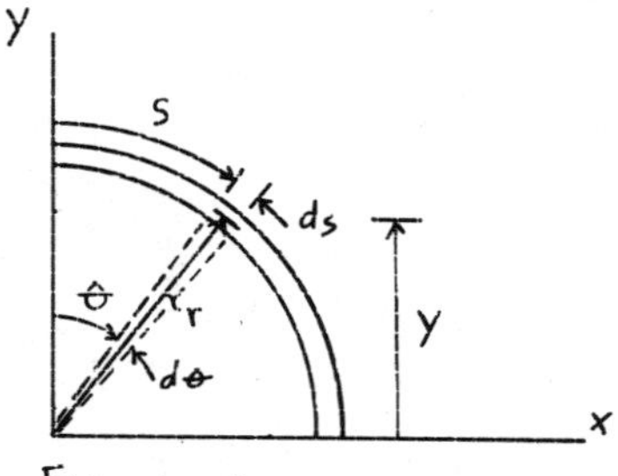

Figure a

Let ρ be the mass per unit area. The surface area generated by the rotation of the surface element ds about the x axis is $dA = 2\pi y\, ds$. Hence, the mass of the element is

$$dm = \rho\, dA = 2\pi\rho y\, ds \qquad (a)$$

and the corresponding mass moment of inertia is, relative to the diametral axis x,

$$dI_x = y^2 dm = 2\pi\rho y^3 ds \qquad (b)$$

By Fig. a,

$$y = r\cos\theta\,, \qquad ds = r\,d\theta \qquad (c)$$

Equations (b) and (c) yield

$$dI_x = 2\pi\rho r^4 \cos^3\theta\, d\theta$$

Hence, the mass moment of inertia of the sphere is

$$I_x = \int dI_x = 2\pi\rho r^4 \int_{-\frac{\pi}{2}}^{\pi/2} \cos^3 d\theta$$

Integration yields

$$I_x = 2\pi\rho r^4\left[\tfrac{1}{3}(\sin\theta)(\cos^2\theta + 2)\right]_{-\frac{\pi}{2}}^{\pi/2}$$

or $$I_x = \frac{8}{3}\pi\rho r^4 \qquad (d)$$

Also, by Eqs. (a) and (c),

$$m = \int dm = \int 2\pi\rho y\, ds = 2\pi\rho r^2 \int_{-\pi/2}^{\pi/2} \cos\theta\, d\theta$$

or

$$m = 4\pi\rho r^2 \qquad (e)$$

Equations (d) and (e) yield

$$I_x = \frac{2}{3} m r^2$$

18.50

Given: A slotted steel electric rotor has the dimensions shown in Fig. a.

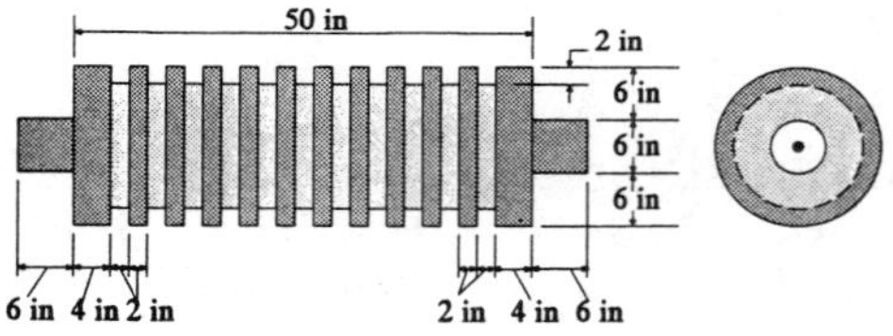

Figure a

Determine its radius of gyration with respect to its geometric axis.

Solution: The electric rotor is equivalent to

1) a solid cylinder of length 50 in and a radius of 7 in
2) a cylindrical shell of length 28 in, an inner radius of 7 in and an outer radius of 9 in
3) a solid cylinder of length 12 in and a radius of 3 in

Hence, by Appendix F, Table F1.

$$I_1 = \frac{1}{2}mr^2 = \frac{1}{2}(\pi\rho r_1^2 L_1) r_1^2$$

$$I_3 = \frac{1}{2}mr^2 = \frac{1}{2}(\pi\rho r_3^2 L_3) r_3^2 \qquad (a)$$

$$I_2 = \frac{1}{2}m(R_2^2 + R_1^2) = \frac{1}{2}\left[\pi\rho L_2 (R_2^2 - R_1^2)\right](R_2^2 + R_1^2)$$

(Continued)

18.50 Cont.

where, $r_1 = 7$ in , $L_1 = 50$ in
$r_3 = 3$ in , $L_3 = 12$ in (b)
$R_2 = 9$ in , $R_1 = 7$ in , $L_2 = 28$ in

Hence, by Eqs. (a) and (b),

$$I = I_1 + I_2 + I_3 = \frac{1}{2}\pi\rho[(50)(7)^4 + (28)(9^4 - 7^4) + (12)(3)^4]$$

or

$$I = 118,751\pi\rho \qquad (c)$$

also,

$$m_1 = \pi\rho r_1^2 L_1 = 2450\pi\rho$$
$$m_2 = \pi\rho(R_2^2 - R_1^2)L_2 = 896\pi\rho$$
$$m_3 = \pi\rho r_3^2 L_3 = 108\pi\rho$$

or

$$m = m_1 + m_2 + m_3 = 3454\pi\rho \qquad (d)$$

By Eqs. (c) and (d), we obtain the radius of gyration as

$$k = \sqrt{\frac{I}{m}} = \sqrt{34.38} = 5.86 \text{ in}$$

18.51

Given: The area bounded by the curve $y = x^2 + 2$ and the line $y = 4$ generates a homogeneous solid when revolved about the x-axis.

In terms of the mass density ρ, determine the mass moment of inertia of the solid with respect to the x axis.

Solution: Figure a represents the bounded area. Consider the area element $x\,dy$

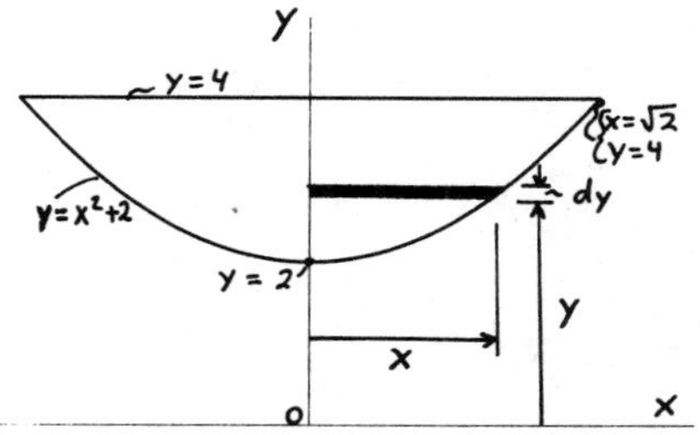

Figure a

When this element is rotated about the x axis, it generates a volume $dV = 2\pi y x\,dy$. The corresponding element of mass is

$$dm = \rho\, 2\pi y x\, dy$$

where, $y = x^2 + 2$ and $dy = 2x\,dx$

Hence, $dm = \rho 4\pi (x^2+2)x^2\,dx$

Therefore, the element of mass moment of inertia is

$$dI_x = y^2 dm = 4\pi\rho (x^2+2)^3 x^2 dx$$

So the total moment of inertia of the solid is $I_x = 2\int dI_x = 8\pi\rho \int_0^{\sqrt{2}} x^2(x^2+2)^3\,dx$

or $I_x = 8\pi\rho \int_0^{\sqrt{2}} (x^8 + 6x^6 + 12x^4 + 8x^2)\,dx$

so,

$$I_x = 837.7\rho$$

18.52

Given: The mass density ρ of a plastic pipe manufactured by a centrifugal casting device varies linearly through the thickness. At the inner surface $\rho = \rho_0$ and at the outer surface $\rho = 1.10\rho_0$. The inner radius of the pipe is r_0. Its thickness is $0.20r_0$, and its length is L.

In terms of r_0, ρ_0, and L derive an expression for the mass moment of inertia about the pipe's longitudinal axis.

Solution:

Since ρ varies linearly through the thickness

$$\rho = a + br \qquad (a)$$

where, a and b are constants and r is the radial coordinate (Fig. a).

Figure a

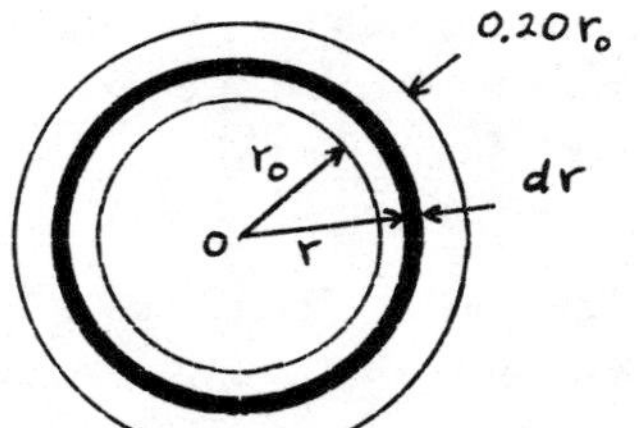

By Eq. (a),

for $r = r_0$, $\rho_0 = a + br_0$

for $r = r_0 + 0.20r_0$, $1.10\rho_0 = a + b(r_0 + 0.20r_0)$

Hence,

$$a = \frac{1}{2}\rho_0$$
$$b = \frac{1}{2}\rho_0/r_0 \qquad (b)$$

Therefore, by Eqs. (a), and (b),

$$\rho = \frac{1}{2}\rho_0(1 + r/r_0) \qquad (c)$$

So, the differential mass dm of a solid element of length L and thickness dr (see Fig. a) is, with Eq. (c),

$$dm = \rho(2\pi r L)dr = \pi\rho_0\left(r + \frac{r^2}{r_0}\right)L\,dr$$

and the corresponding mass moment of inertia is

$$dI_0 = r^2 dm = \pi\rho_0\left(r^3 + \frac{r^4}{r_0}\right)L\,dr$$

The total mass moment of inertia is

$$I_0 = \pi\rho_0 L \int_{r_0}^{1.2r_0}\left(r^3 + \frac{r^4}{r_0}\right)dr$$

(Continued)

18.52 Cont.

Integration yields

$$I_o = \pi \rho_o L \left[\frac{r^4}{4} + \frac{r^5}{5 r_o} \right] \Big|_{r_o}^{1.2 r_o}$$

or

$$\underline{I_o = 1.778\, \rho_o L r_o^4}$$

18.53

<u>Given:</u> A wheel, consisting of a rim, a hub and 8 spokes, has dimensions shown in Fig. a. The mass density of the material is $\rho = 0.00070$ lb·s²/in⁴.

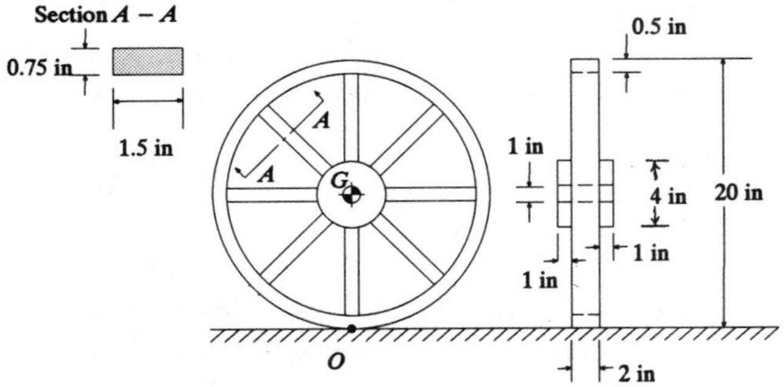

Figure a

Calculate the mass moment of inertia [lb·ft·s²] of the wheel relative to the axis passing through point O and perpendicular to the plane of the wheel.

<u>Solution:</u>

First, determine the mass moments of inertia of the various parts of the wheel relative to the center of mass axis perpendicular to the plane of the wheel. Then, use the parallel axis theorem (Steiner's theorem) to transfer the mass moment of inertia to the axis at O.

For the rim:

$$dm_{rim} = \rho (2\pi r\, dr) L_{rim}$$

$$I_{(rim)G} = \rho 2\pi L_{rim} \int_{9.5}^{10} r^3 dr = \rho 2\pi L_{rim} \frac{r^4}{4} \Big|_{9.5}^{10}$$

or $I_{(rim)G} = (0.00070)(2\pi)(2)\left(\frac{10^4}{4} - \frac{9.5^4}{4}\right)$

$$I_{(rim)G} = 4.079 \text{ lb·in·s}^2 \qquad (a)$$

Alternatively, since the rim is thin,

$$I_{(rim)G} \approx m_{rim} r_{mean}^2$$

where, $r_{mean} = 9.75$ in

$$m_{rim} \approx \rho 2\pi r_{mean} L_{rim} t = (0.0007)(2\pi)(9.75)(2)(0.5)$$

or

$$m_{rim} \approx 0.04288 \text{ lb·s}^2/\text{in}$$

So, $I_{(rim)_G} \approx (0.04288)(9.75)^2 = 4.076$ lb·in·s²

For the hub:

By Appendix F, Table F1, and Fig. a,

$$I_{(hub)G} = \frac{1}{2} m_{hub} (r_1^2 + r_2^2)$$

where

$$m_{hub} = \rho \pi (r_2^2 - r_1^2) L_{hub}$$

$r_1 = 2$ in

$r_2 = 0.5$ in

$L_{hub} = 4$ in

Hence,

$$m_{hub} = (0.0007)(\pi)(2^2 - 0.5^2)(4) = 0.03299 \ \frac{\text{lb·s}^2}{\text{in}}$$

$$I_{(hub)_G} = \frac{(0.03299)}{2}(2^2 + 0.5^2) = 0.0701 \text{ lb·in·s}^2 \qquad (b)$$

For the spokes:

The mass moment of inertia of one spoke about the center of mass of the wheel is (see Fig. a)

$$I_{(spoke)G} = \rho A_{spoke} \int_2^{9.5} r^2 dr$$

where (see Fig. a)

$$A_{spoke} = (0.75)(1.5) = 1.125 \text{ in}^2$$

Hence,

$$I_{spoke} = (0.0007)(1.125) \frac{r^3}{3} \Big|_2^{9.5}$$

or $I_{spoke} = 0.2230$ lb·in·s²

For 8 spokes

$$I_{(spokes)G} = 8(0.2230) = 1.7840 \text{ lb·in·s}^2 \qquad (c)$$

Hence, the mass moment of inertia of the wheel about its mass center is by Eqs. (a), (b), and (c)

$$I_{(wheel)_G} = I_{(rim)_G} + I_{(hub)_G} + I_{(spokes)_G}$$

$$= 4.079 + 0.0701 + 1.7840$$

or

$$I_{(wheel)_G} = 5.933 \text{ lb·in·s}^2 \qquad (d)$$

Then, the mass moment of inertia about the axis through O is

$$I_{(wheel)O} = I_{(wheel)G} + m_{wheel} d^2 \qquad (e)$$

where (see Fig. a),

$$d = 10 \text{ in} \qquad (f)$$

$$m_{wheel} = m_{rim} + m_{hub} + m_{spokes}$$

$$m_{rim} = \rho 2\pi L_{rim} \int_{9.5}^{10} r\, dr = (0.0007)(2\pi)(2) \frac{r^2}{2} \Big|_{9.5}^{10}$$

$= 0.04288$ lb·s²/in

$m_{hub} = 0.03299$ lb·s²/in

$$m_{spokes} = \rho (A_{spokes})(L_{spokes})$$

$= (0.0007)(8)(0.75)(1.5)(7.5) = 0.04725$ lb·s²/in

Hence,

$$m_{wheel} = 0.1231 \text{ lb·s}^2/\text{in} \qquad (g)$$

and, by Eqs. (d), (e), (f), and (g),

$$I_{(wheel)_O} = 5.933 + (0.1231)(10)^2$$

or $\underline{I_{(wheel)_O} = 18.245 \text{ lb·in·s}^2 = 1.52 \text{ lb·ft·s}^2}$

18.54

Given: The area bounded by the curve $x^2+(y-4)^2=4$ generates a homogeneous solid when revolved about the x-axis.

In terms of the mass density ρ, determine the mass moment of inertia of the solid with respect to the x axis.

Solution: The bounded area is shown in Fig. a. Consider the strip $(y_2-y_1)\,dx$

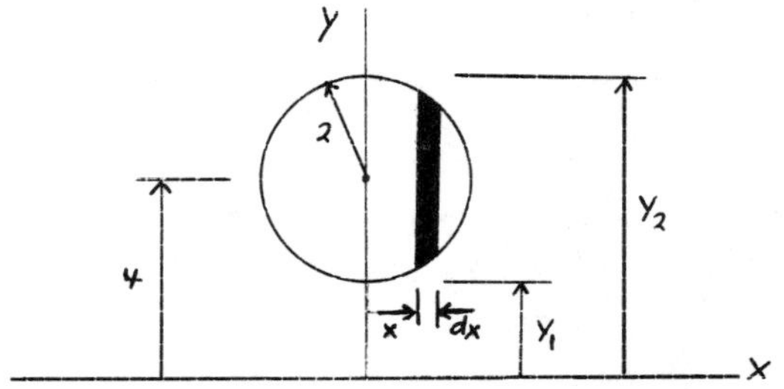

Figure a

when this strip is revolved about the x axis, the generated solid has the mass moment of inertia

$$dI_x = \frac{1}{2}\pi\rho\,(y_2^4 - y_1^4) \qquad (a)$$

By $x^2+(y-4)^2=4$, we have

$$y-4 = \pm\sqrt{4-x^2}$$

Hence,

$$y_1 = 4-\sqrt{4-x^2}$$
$$y_2 = 4+\sqrt{4-x^2}$$

Let $u=\sqrt{4-x^2}$, Then,

$$y_1 = 4-u$$
$$y_2 = 4+u$$

and

$$y_2^4 - y_1^4 = 32\,(16u+u^3) \qquad (b)$$

By Eqs. (a) and (b), the mass moment of inertia of the generated solid is

$$I_x = 2\left(\frac{\pi\rho}{2}\right)(32)\int_0^2 (16u+u^3)\,dx$$

$$I_x = 32\pi\rho\int_0^2\left[16\sqrt{4-x^2}+(4-x^2)\sqrt{4-x^2}\right]dx \quad (c)$$

By integration tables or computer software

$$\int_0^2\sqrt{4-x^2}\,dx = \frac{1}{2}\left[x\sqrt{4-x^2}+4\sin^{-1}\left(\frac{x}{2}\right)\right]\Big|_0^2 \quad (d)$$
$$=\pi$$

$$\int x^2\sqrt{4-x^2}\,dx = \left\{-\frac{x}{4}\sqrt{(4-x^2)^3}+\frac{1}{2}\left[x\sqrt{4-x^2}+4\sin^{-1}\left(\frac{x}{2}\right)\right]\right\}\Big|_0^2 \quad (e)$$
$$=\pi$$

Hence, by Eqs. (c), (d), and (e),

$$\underline{\underline{I = 32\pi\rho\,(20\pi-\pi) = 608\pi^2\rho = 6001\rho}}$$

18.55

Given: The height of a solid right-circular cone is h and the radius of the base is r.

a) Derive a formula for the mass moment of inertia of the cone about an axis through its tip and parallel to its base.

b) Derive a formula for the mass moment of inertia of the cone about a diameter of its base.

Solution:

a) Consider the cone with xyz axes shown in Fig. a.

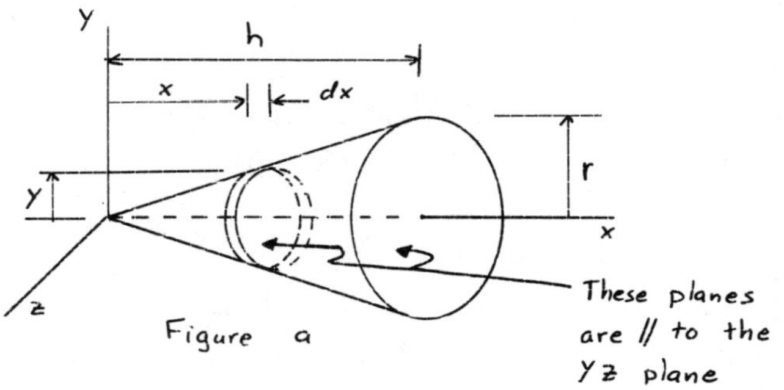

Figure a

The mass moment of inertia of the cone about the z-axis is required. By Appendix F, Table F1, the mass moment of inertia of the cross sectional disk of thickness dx about a diametral axis is

$$dI = \frac{1}{4}\,y^2\,dm \qquad (a)$$

The mass moment of inertia of the disk about the z axis is, with Eq. (a) and the parallel axis theorem,

$$dI_z = dI + dm\,(x^2) \qquad (b)$$

Also, $dm = \rho\pi y^2\,dx \qquad (c)$

Therefore, by Eqs. (a), (b), and (c), the mass moment of inertia of the cone about the z axis is

$$I_z = \int dI_z = \pi\rho\int_0^h\left(x^2y^2+\frac{1}{4}y^4\right)dx \quad (d)$$

By Fig. a,

$$y = \frac{r}{h}\,x \qquad (e)$$

Equations (d) and (e) yield

$$I_z = \pi\rho\int_0^h\left(\frac{r^2}{h^2}+\frac{r^4}{4h^4}\right)x^4\,dx$$

Integration yields

$$I_z = \pi\rho\left[\frac{1}{5}r^2h^3+\frac{1}{20}r^4h\right] \qquad (f)$$

Also, by Eqs. (c) and (e),

$$m = \pi\rho\int_0^h\frac{r^2}{h^2}x^2\,dx = \frac{1}{3}\pi\rho r^2h \qquad (g)$$

(Continued)

18.55 Cont.

So, by Eqs. (f) and (g),

$$I_z = \frac{3}{5} m \left(h^2 + \frac{r^2}{4} \right) \qquad (h)$$

b) The center of mass of the cone is located at $x_1 = \frac{3}{4} h$ (see Fig. b, and Appendix D_1 Table D. 3).

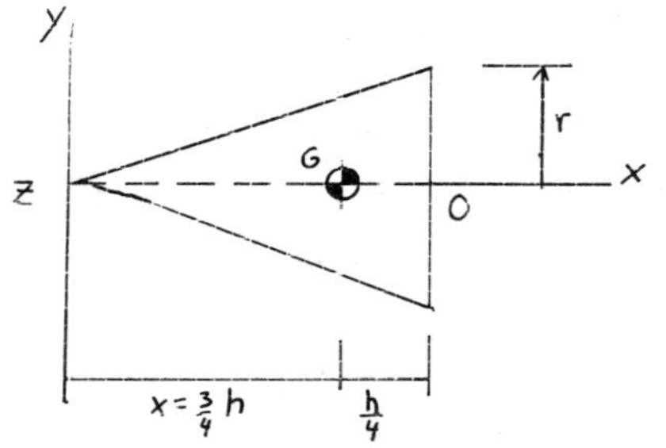

Figure b

Let I_G be the moment of inertia about a transverse axis through the center of mass G. Then, by Steiner's theorem,

$$I_z = I_G + m \left(\tfrac{3}{4} h \right)^2$$
$$I_o = I_G + m \left(\tfrac{1}{4} h \right)^2 \qquad (i)$$

where I_o is the moment of inertia about a diameter of the base (Fig. b).

Therefore, by Eqs. (h) and (i)

$$I_o = I_z - \frac{1}{2} m h^2 = \frac{3}{5} m \left(h^2 + \frac{r^2}{4} \right) - \frac{1}{2} m h^2$$

or

$$I_o = \frac{1}{10} m \left(h^2 + \frac{3}{2} r^2 \right)$$

18.56

Given: The prism shown in Fig. a is to be used in a rotating mechanism. By P18.33, $I_o = \frac{1}{2} m a^2$. However, a design specification requires that additional mass of the same material be added so that $I_o = m a^2$ where m is the mass of the original prism. Additionally, the length L (perpendicular to the cross section of the prism) must not be changed, and the volume of the modified prism must not be greater than twice the initial volume.

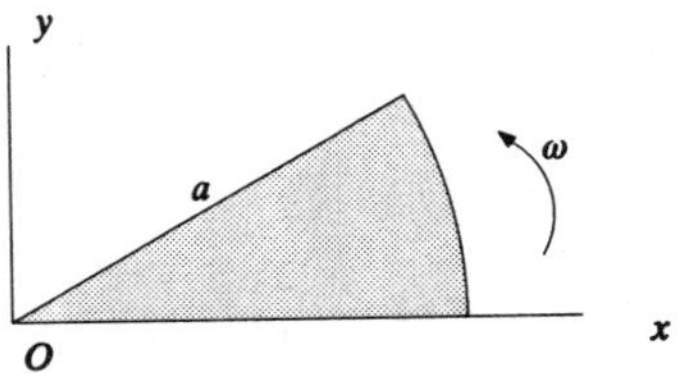

Figure a

a) Determine at least two configurations (of additional mass and its distribution) that meet the specifications.

b) List some reasons for choosing one of the configurations choosen in part a over the othe.

Solution:

a) Two obvious ways of adding mass are to add radial mass m_1 (Fig. b) or angular mass m_2 (Fig. c).

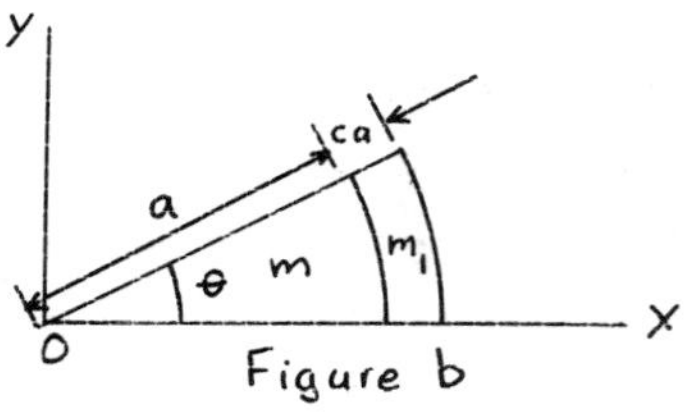

Figure b

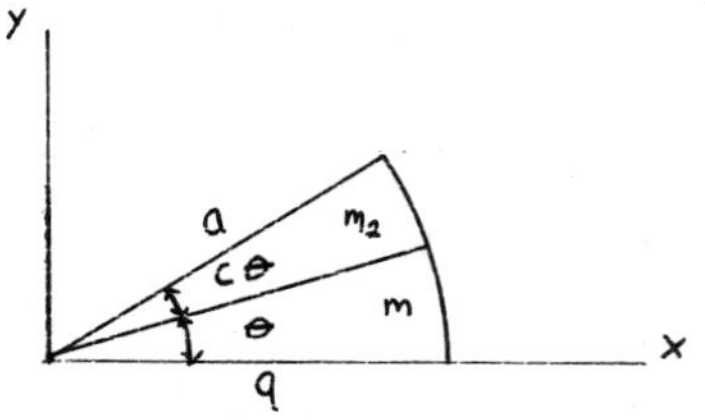

Figure c

Consider the case shown in Fig. b. By Fig. b,

$$m = \rho \pi a^2 L \left(\frac{\theta}{2\pi} \right) = \frac{1}{2} \rho a^2 L \theta \qquad (a)$$

or

$$\rho = \frac{2m}{a^2 L \theta} \qquad (b)$$

also,

$$m_1 = \rho \pi \left[(a + ca)^2 - a^2 \right] L \left(\frac{\theta}{2\pi} \right) \qquad (c)$$

By Eqs. (a), (b), and (c),

$$m_1 = m \left(c^2 + 2c \right) \qquad (d)$$

Hence, with Eq. (d), we have

$$I_o = \frac{1}{2} (m + m_1)(a + ca)^2 = m a^2$$

or

$$\frac{1}{2} m a^2 (1 + 2c + c^2)(1 + c)^2 = m a^2 \qquad (e)$$

Therefore

$$(1 + c)^4 = 2 \qquad (f)$$

The solution of this equation is

$$c = 2^{1/4} - 1 = 0.189207 \qquad (g)$$

also, by Eq. (f),

$$c^2 + 2c = \sqrt{2} - 1$$

and therefore, by Eq. (d),

$$m_1 = (\sqrt{2} - 1) m = 0.41421 m$$

Substitution of Eq. (g) into Eq. (e) yields

$$m a^2 = m a^2$$

(Continued)

18.56 Cont.

Now initially the volume of the prism is $V_1 = \pi a^2 L\left[\frac{\theta}{2\pi}\right] = \frac{1}{2} a^2 L \theta$

The volume of the modified prism is (Fig. b)

$$V_2 = \pi (a+ca)^2 L\left(\frac{\theta}{2\pi}\right) = \frac{1}{2}(a+ca)^2 L\theta$$

or with $c = 0.189207$ [Eq. (g)]

$$V_2 = \frac{1}{2}(1.4142)a^2 L\theta = 1.4142\, V_1$$

Hence,

$$V_2 < 2V_1$$

and the specifications are met.

Next consider Fig. c. For this case,

$$m_2 = \rho \pi a^2 L \frac{c\theta}{2\pi} = \frac{1}{2}\rho a^2 L c\theta \qquad (h)$$

By Eqs. (a) and (h),

$$m_2 = cm \qquad (i)$$

Then, with Eq. (i) and Fig. c,

$$I_o = \frac{1}{2}(m+m_2)a^2 = ma^2$$

or,

$$\frac{1}{2}ma^2(1+c) = ma^2$$

Therefore, $c = 1$

with $c = 1$, the volume of the modified prism is $V_2 = \pi a^2 L\left(\frac{2\theta}{2\pi}\right) = a^2 L\theta = 2V_1$

So adding a second prism equal in dimensions to the original prism will meet specifications so that

$$I_o = ma^2$$

$$V_2 \le 2V_1$$

b) The modification of Fig. b requires an additional radially placed mass $m_1 = 0.41421m$. It also requires an additional radial space of length $ca = 0.189207a$.

The modification of Fig. c requires additional angular mass $m_2 = m$, thus, more mass than that of Fig. b. However, it requires no additional radial space, Consequently, available space may dictate the choice.

18.57

Given: An airplane lands at a speed of 36 m/s. The outside diameter of a tire is 1.2 m. The radius of gyration of a wheel (with the tire) is 450 mm. The mass of the wheel is 150 kg. When the airplane lands, the wheel skids until it gains enough angular velocity to roll. The frictional force on the skidding wheel is 4050 N.

Determine how far the air plane moves while the wheel is skidding. Neglect deceleration of the plane and flattening of the wheel.

Solution: By the principle of angular momentum, with respect to the axle of the wheel,

$$M = I\frac{d\omega}{dt} \qquad (a)$$

where M is the moment due to the frictional force and $I = mk^2$ is the mass moment of inertia of the wheel. Since M is constant, Eq. (a) yields,

$$Mt = I\omega = mk^2\omega \qquad (b)$$

for the skidding wheel.

Let rolling begin when $t = t_1$. For $t = t_1$ the speed v of the plane is

$$v = r\omega \qquad (c)$$

where $r = 0.6$ m is the outside radius of the tire. Since deceleration of the plane is neglected,

$v = 36$ m/s. Hence, by Eq. (c),

$$\omega = 36/(0.6) = 60 \text{ rad/s} \qquad (d)$$

Therefore, by Eqs. (b) and (d), with $t = t_1$,

$$Mt_1 = mk^2(60)$$

Also,

$$M = (4050)(0.60) = 2430 \text{ N}\cdot\text{m}$$

Hence,

$$2430\, t_1 = (150)(0.450)^2(60)$$

or

$$t_1 = 0.75 \text{ s}$$

and the distance the plane travels in time t_1 is

$$s = vt_1 = (36)(0.75) = 27 \text{ m}$$

18.58

Given: A top spins on a horizontal surface. Its initial angular speed is 1500 rpm. The radius of gyration is 1.50 in. The mass of the top is 0.02 slug. The top stops spinning in 4 min.

Determine the resisting torque [lb·in], assuming that it is constant. Neglect wobbling of the top as it slows down.

Solution: By the principle of angular momentum with respect to the axis of the top,

$$M = I\frac{d\omega}{dt} = mk^2\frac{d\omega}{dt}$$

Since M is constant, integration yields

$$Mt = mk^2\omega + \text{constant}$$

For $t = 4$ min $= 240$ s, $\omega = 0$. Therefore,

$$\text{constant} = 240\, M$$

(Continued)

18.58 Cont.

and

$$Mt = mk^2\omega + 240M$$

or

$$M = \frac{mk^2\omega}{t-240}$$

Hence, for $t=0$,

$$M = -\frac{mk^2\omega_0}{240} \qquad (a)$$

where $\omega_0 = (1500)(2\pi/60) = 50\pi$ rad/s, and $mk^2 = (0.02)(1.5/12)^2 = 0.0003125$ slug·ft²

Therefore, by Eq. (a),

$$M = -\frac{(0.0003125)(50\pi)}{240} = -0.0002045 \text{ lb·ft}$$

So the resisting torque is 0.002454 lb·in

18.59

Given: A homogeneous circular cylinder of mass 45 kg and radius 150 mm is attached at its ends to frictionless axles that coincides with the geometric axis of the cylinder (Fig. a). A block of mass 15 kg is suspended by a cord that is wrapped around the cylinder. The system is released from rest.

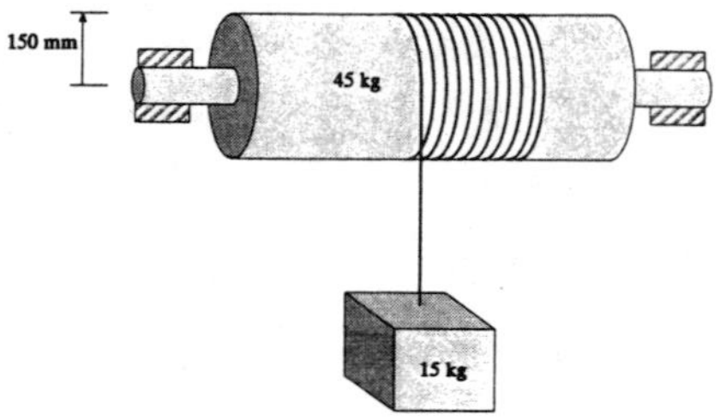

Figure a

What is the speed of the block 2 s after it is released? Neglect the masses of the axles.

Solution: An end view of the cylinderblock system is shown in Fig. b.

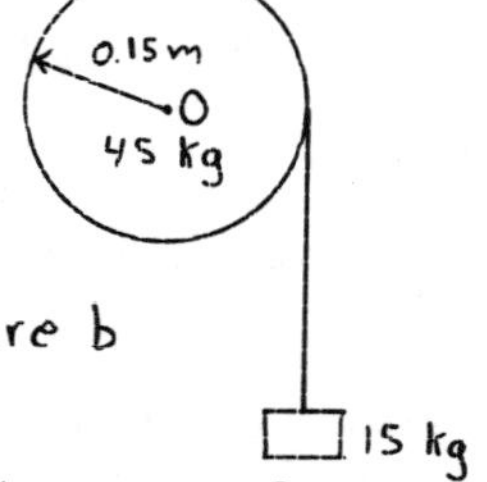

Figure b

Free-body diagrams of the cylinder and the block are shown in Figs. c and d, where T is the tension in the cord.

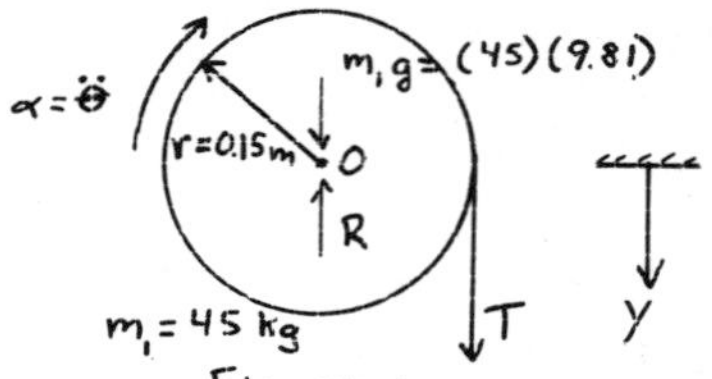

Figure c

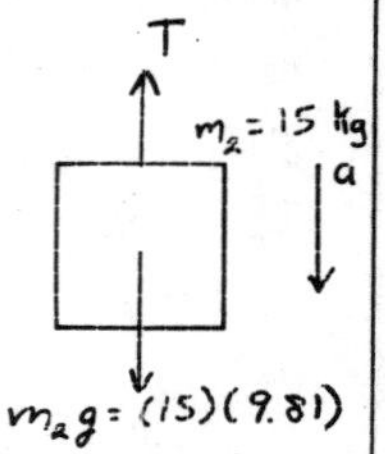

Figure d

By Fig. c,

$$\Sigma M_O = Tr = I_O\ddot{\theta} = \tfrac{1}{2}m_1 r^2\alpha$$

or

$$T = \tfrac{1}{2}m_1 r\alpha \qquad (a)$$

Next, by Fig. d,

$$\Sigma F_y = m_2 g - T = m_2 a \qquad (b)$$

By kinematics,

$$a = r\alpha \qquad (c)$$

Then, by Eqs. (a), (b), and (c), we find

$$a = \frac{m_2 g}{m_2 + 0.5m_1} = \frac{15}{37.5}g = 0.40g \qquad (d)$$

Then, since a is constant,

$$v = at \qquad (e)$$

For $t = 2$ s, Eqs. (d) and (e) yield

$$v = (0.40)(9.81)(2) = 7.85 \text{ m/s}$$

18.60

Given: A homogeneous cylinder (Fig. a) rotates about a fixed axes O with initial angular velocity of 144 rad/s. A light, rigid braking bar AB is held by a frictionless pin at A. The coefficient of kinetic friction between the bar and cylinder is $\mu_k = 0.40$. The mass of the cylinder is 5 lb·s²/ft. The braking force at B is 100 lb.

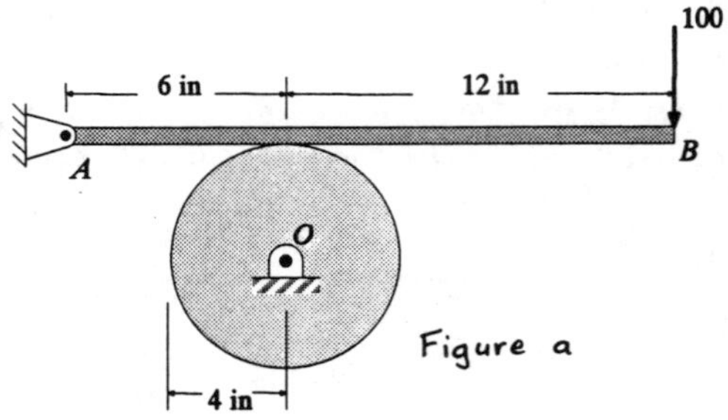

Figure a

Determine the time required to stop the cylinder.

(Continued)

18.60 Cont.

Solution: The free-body diagrams of the bar and the cylinder are shown in Figs. b and c.

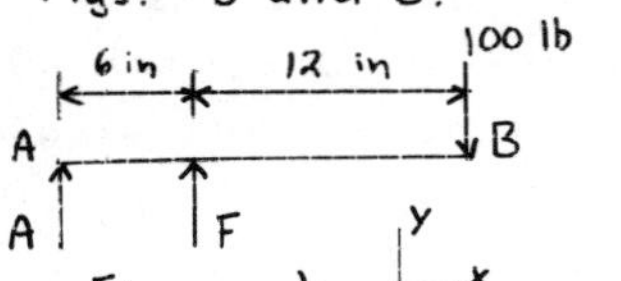

Figure b

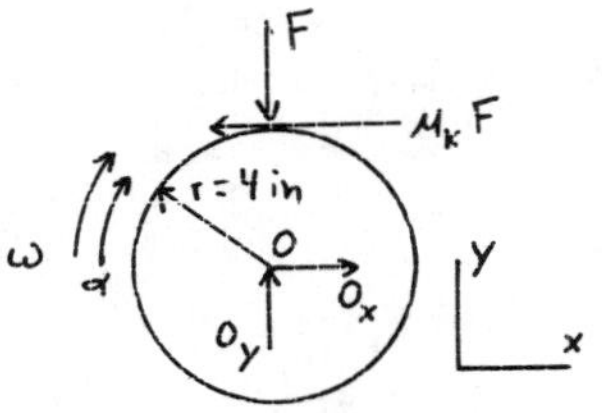

Figure c

Since the bar is light, assume that the mass of the rod is negligible. Then, by Fig. b and statics,

$$\circlearrowleft \Sigma M_A = 6F - (18)(100) = 0 \; ; \quad \text{or} \quad F = 300 \text{ lb}$$

There fore,

$$\mu_k F = (0.40)(300) = 120 \text{ lb}$$

Consequently, by Fig. c,

$$\circlearrowleft + \Sigma M_o = -(\mu_k F)(r) = I_o \alpha = \tfrac{1}{2} m r^2 \alpha$$

Hence, $\alpha = -\dfrac{2\mu_k F}{mr} = -\dfrac{2(120)}{(5)(\frac{4}{12})} = -144 \text{ rad/s}^2$

By kinematics,

$$\alpha = \frac{d\omega}{dt}$$

and since $\alpha = -144 \text{ rad/s}^2$ is constant, integration yields

$$\omega = -144t + \omega_o = -144t + 144$$

when the cylinder stops $\omega = 0$ so,

$$-144t + 144 = 0$$

or the time required for the cylinder to stop is

$$\underline{\underline{t = 1 \text{ s}}}$$

18.61

Given: Two masses (15 kg and 30 kg) are suspended from opposite ends of a light cord. The cord is wrapped around a homogeneous solid cylinder that is attached at its ends to frictionless axles that coincide with the geometric axis of the cylinder (Fig. a). The mass of the cylinder is 60 kg and its radius is 150 mm. The system is released from rest.

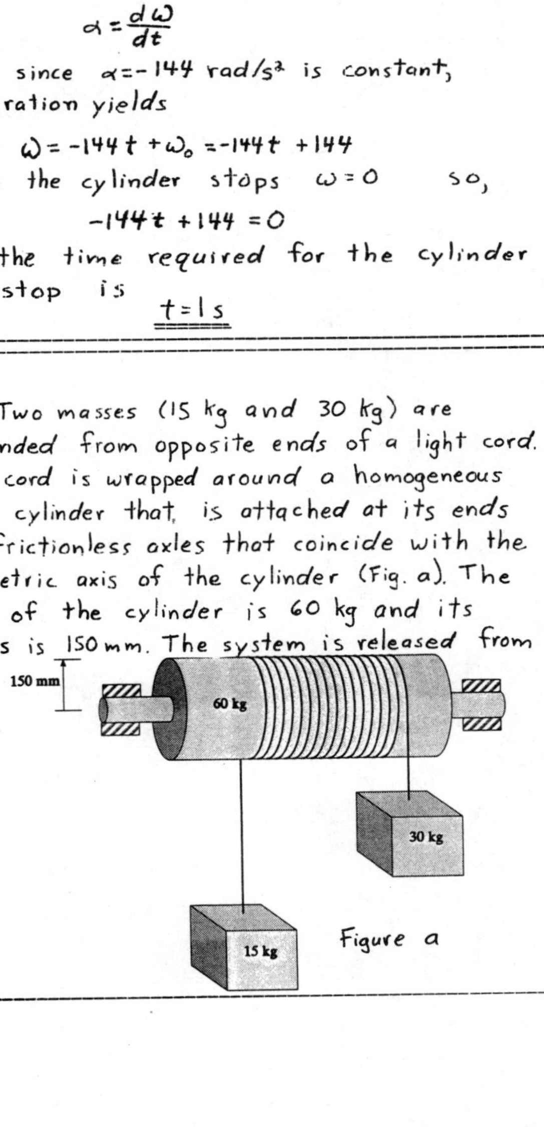

Figure a

Determine the time [s] required for the larger mass to descend 1.2 m. Neglect the masses of the axles.

Solution: An end view of the system is shown in Fig. b.

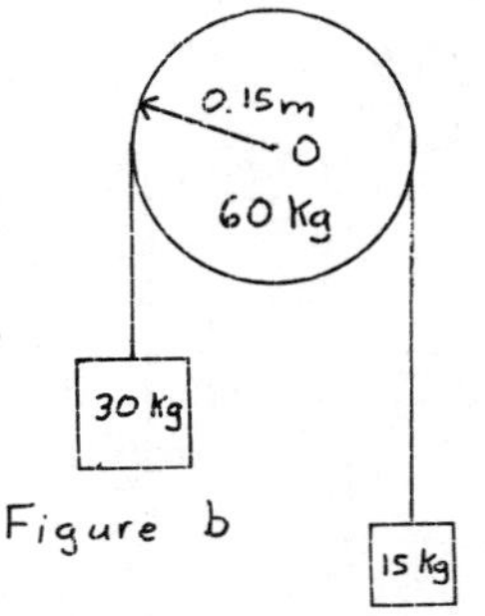

Figure b

Free-body diagrams of the cylinder and masses are shown in Figs. c, d, and e, where T_1 and T_2 are tensions in the cord.

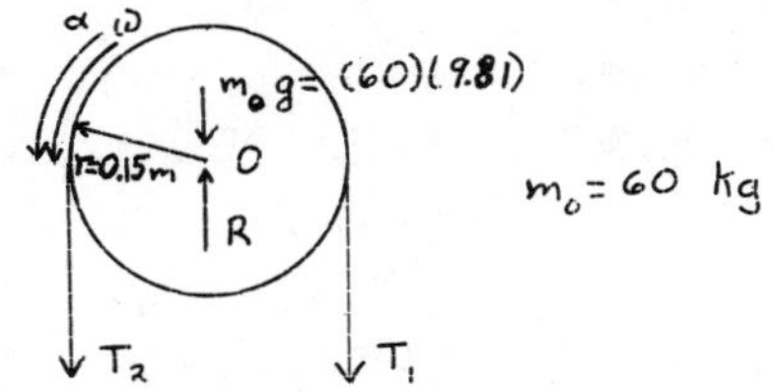

Figure c

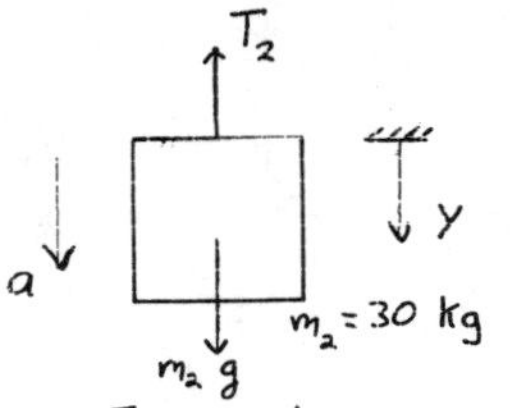

Figure d

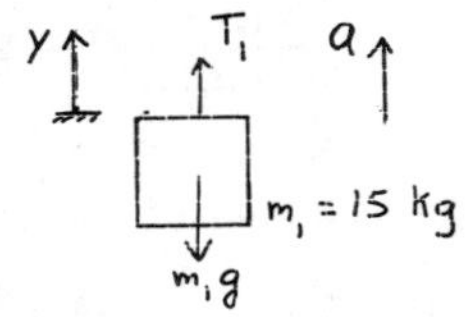

Figure e

By Fig. c,

$$\circlearrowleft \Sigma M_o = T_2 r - T_1 r = I_o \alpha = \tfrac{1}{2} m_o r^2 \alpha \qquad (a)$$

By Fig. e,

$$\Sigma F_y = -m_1 g + T_1 = m_1 a \qquad (b)$$

and by Fig. d,

$$\Sigma F_y = m_2 g - T_2 = m_2 a \qquad (c)$$

Also, by kinematics,

$$a = \alpha r \qquad (d)$$

Solving Eqs. (a), (b), (c), and (d) for a, we get

$$a = \frac{2(m_2 - m_1)g}{m_o + 2m_1 + 2m_2}$$

Hence, with $m_o = 60$ kg, $m_1 = 15$ kg, and $m_2 = 30$ kg,

$$a = \tfrac{1}{5} g = \tfrac{1}{5}(9.81) = 1.962$$

Then, since a is constant, by kinematics,

$$s = \tfrac{1}{2} a t^2 \quad \text{or} \quad t = \sqrt{2s/a}$$

when $s = 1.2$ m, $\underline{\underline{t = \sqrt{\frac{2.4}{1.962}} = 1.106 \text{ s}}}$

18.62

Given: The modified Atwood machine (Fig. a) is on a planet where $g = 53\ ft/s^2$. The bearing is frictionless, and the cord does not slip on the pulley. The apparatus is released from rest.

Determine the angle [rad] through which the system turns in 3 s.

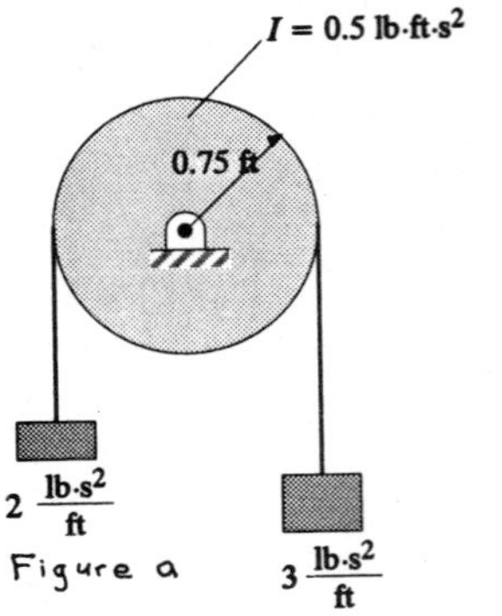

Figure a

Solution:
The free-body diagrams of the wheel and the masses are shown in Figs. b, c, and d.

By Fig. b,

$$\circlearrowleft + \Sigma M_o = (T_2 - T_1) r = I_o \alpha \qquad (a)$$

By Fig. c,

$$\Sigma F_y = T_1 - m_1 g = m_1 a \qquad (b)$$

By Fig. d,

$$\Sigma F_y = m_2 g - T_2 = m_2 a \qquad (c)$$

Also, by Kinematics,

$$a = r\alpha \qquad (d)$$

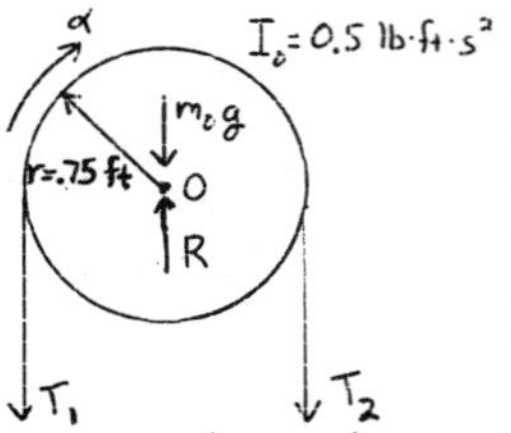

Figure b

Equations (a), (b), (c), and (d) yield,

$$a = \frac{(m_2 - m_1) g}{m_1 + m_2 + I_o/r^2}$$

or with $m_1 = 2\ lb \cdot s^2/ft$, $m_2 = 3\ lb \cdot s^2/ft$, $I_o = 0.5\ lb \cdot ft \cdot s^2$, and $r = 0.75\ ft$,

$$a = \frac{9}{53} g = \frac{9}{53}(53) = 9\ ft/s^2 \qquad (e)$$

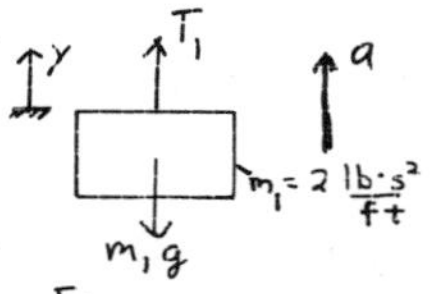

Figure c

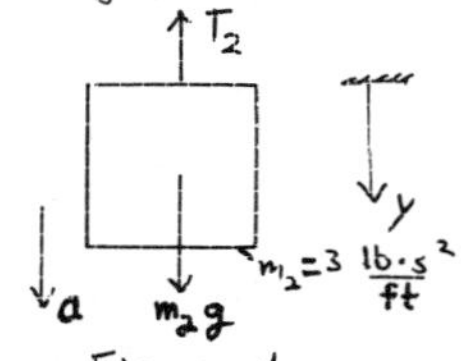

Figure d

Since a is constant, by kinematics,

$$s = \frac{1}{2} a t^2$$

Therefore, when $t = 3\ s$,

$$s = \frac{1}{2}(9)(3)^2 = 40.5\ ft \qquad (f)$$

Also, by kinematics, $s = r\theta$. Hence,

$$\theta = \frac{s}{r} = \frac{40.5}{0.75} = 54\ rad$$

18.63

Given: A thin circular disk with a radius of 3 in rests flat on a table. It is set spinning about its vertical axis of symmetry with an angular velocity of 10 rad/s. The coefficient of kinetic friction is 0.10.

a) Assuming that the pressure between the face of the disk and the table is uniform, calculate how long it will take the disk to stop spinning.

b) Calculate how many revolutions it performs before it stops.

Solution:

a) Since the pressure between the disk and table is uniform, the force acting on the face of the disk per unit area is

$$p = \frac{mg}{\pi a^2} \qquad (a)$$

Hence, the frictional force acting on an elemental area $dA = r\,dr\,d\theta$ is (Fig. a)

$$dF = \mu_K\, p\, dA = \mu_K\, p\, r\, dr\, d\theta$$

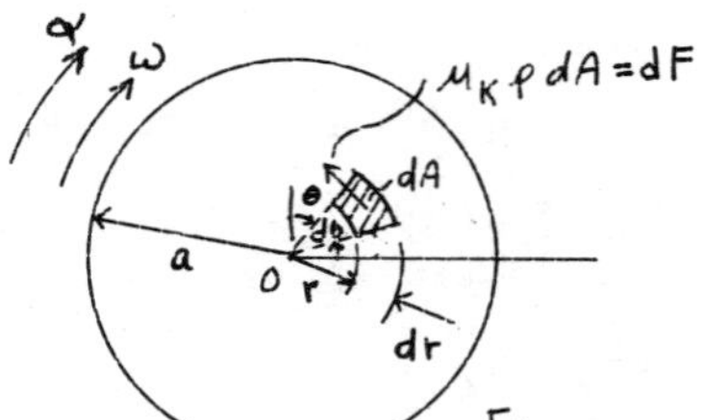

Figure a

The frictional force dF on dA exerts the resisting moment about O

$$dM = r\,dF = \mu_K\, p r^2\, dr\, d\theta \qquad (b)$$

By integration of Eq. (b), with Eq. (a), we have

$$M = \frac{0.10\, mg}{\pi a^2} \int_0^a r^2 dr \int_0^{2\pi} d\theta$$

or

$$M = \frac{mga}{15} \qquad (c)$$

Then, by Fig. a,

$$\circlearrowleft + \Sigma M_o = -M = I_o \alpha = \frac{1}{2} m a^2 \alpha \qquad (d)$$

Equations (c) and (d) yield

$$\alpha = -\frac{2g}{15a} \qquad (e)$$

By kinematics, $\alpha = d\omega/dt$. Hence

$$\frac{d\omega}{dt} = -\frac{2g}{15a}$$

Integration of this equation yields

$$\omega = \frac{-2gt}{15a} + \omega_o \qquad (f)$$

Let $t = t_1$, when the disk stops. Then, $\omega = 0$, and with $\omega_o = 10\ rad/s$, Eq. (f) yields

$$t_1 = \frac{15 a \omega_o}{2g} = \frac{(15)(3)(10)}{(2)(386)} = 0.5835$$

b) Note that $\frac{d\omega}{dt} = \frac{d\omega}{d\theta}\frac{d\theta}{dt} = \omega \frac{d\omega}{d\theta}$. Then by Kinematics,

$$\omega \frac{d\omega}{d\theta} = \alpha = -\frac{2g}{15a}$$

Integration yields

$$\frac{\omega^2}{2} = \frac{-2g}{15a}\theta + \frac{\omega_o^2}{2}$$

When the disk stops $\omega = 0$. Hence

$$\theta = \frac{15 a \omega_o^2}{4g} = \frac{(15)(3)(10)^2}{(4)(386)} = 2.914\ rad$$

or, $\theta = 0.464$ revolutions

18.64

Given: A ring hangs on a shaft S, which is initially motionless (Fig. a). The stop B is frictionless. The shaft S is given a sudden constant angular velocity ω.

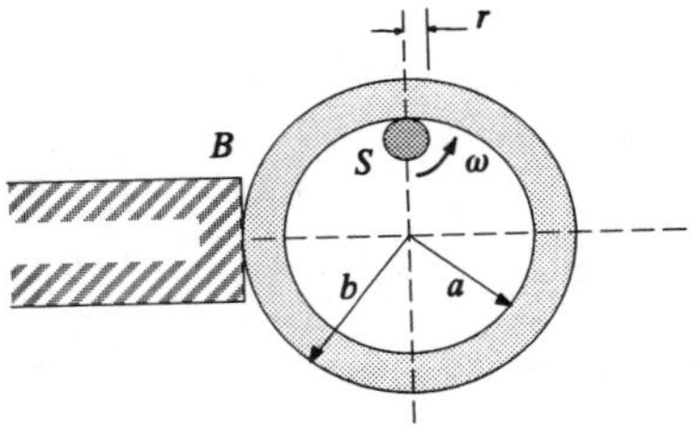

Figure a

Show that the time t required for the ring to stop slipping on the shaft is

$$t = \frac{r\omega}{2\mu_K g}\left(1+\frac{b^2}{a^2}\right)$$

where μ_K is the coefficient of kinetic friction between the shaft and the ring.

Solution:

By Appendix F, Table F1, the mass moment of inertia of the ring about its centroidal axis O is

$$I_o = \frac{1}{2} m (a^2+b^2) \qquad (a)$$

The free-body diagram of the ring is shown in Fig. b.

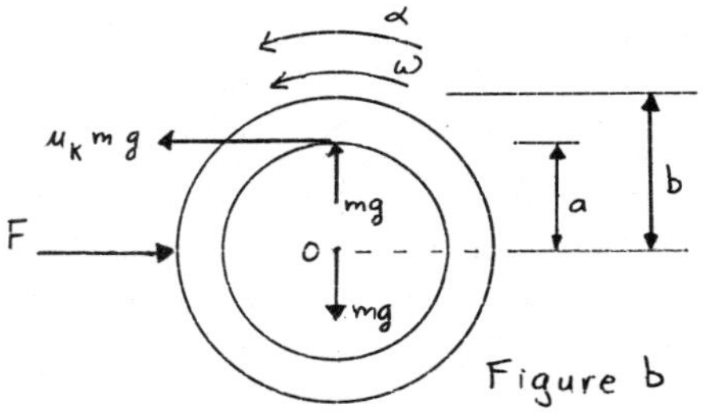

Figure b

By Fig. b, and Eq. a,

$$\circlearrowleft\!\Sigma M_o = \mu_K m g a = I_o \alpha = \frac{1}{2} m (a^2+b^2)\alpha \qquad (b)$$

Solving Eq. (b) for α, we obtain

$$\alpha = \frac{2\mu_K g a}{a^2+b^2} \qquad (c)$$

and by kinematics, where Ω is the angular velocity of the ring,

$$\alpha = \frac{d\Omega}{dt} = \frac{2\mu_K g a}{a^2+b^2} \qquad (d)$$

Integration of Eq. (d) yields, since initially Ω=0,

$$\Omega = \frac{2\mu_K g a}{a^2+b^2}\, t \qquad (e)$$

when sliding stops, since the shaft's angular velocity is ω

$$a\Omega = r\omega \quad \text{or} \quad \Omega = \frac{r}{a}\omega \qquad (f)$$

and $t = t_1$. Hence, Eqs. (e) and (f) yield

$$t_1 = \frac{r\omega}{2\mu_K g}\cdot\left(1+\frac{b^2}{a^2}\right)$$

18.65

Given: The circular cam of mass m_c (Fig. a) has its camshaft hole offcenter a distance e. The flat-face follower has mass m_f and is constrained by a preloaded spring with constant k to maintain contact with the cam. The cam rotates with constant angular velocity ω, so that θ=ωt is the angular rotation of the cam and t denotes time. The displacement of the follower is x; when θ=0, x=0. Use units s, kg, m, and rad.

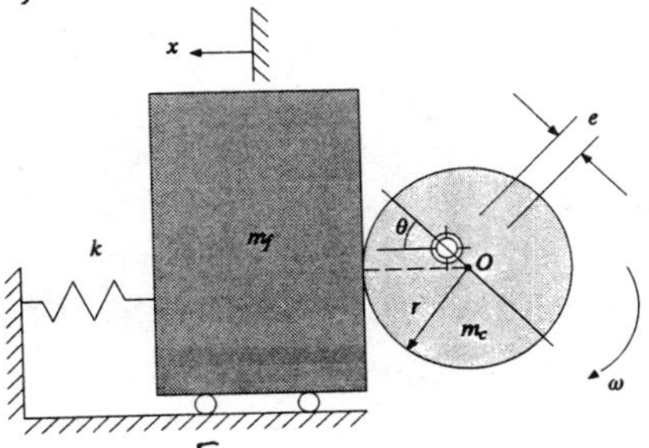

Figure a

a) Show that $x = e - e\cos\omega t$

b) Plot x, $\dot{x}$, and $\ddot{x}$, the displacement, velocity, and acceleration, respectively, of the follower, as functions of θ=ωt, for $0 \le \theta \le 2\pi$. Let e=ω=1.

c) The compressive force in the spring is $P = k\delta$, where δ is the total distance the spring is compressed from its unstretched length. At x=0, $\delta=\delta_o$. Determine a formula for the follower force F exerted on the cam in terms of k, e, δ_o, m_f, and ω, and the condition to ensure that F>0.

d) Derive a formula for the torque T applied to the cam by the shaft.

e) Let k=20 N/m, e=0.05 m, δ_o=0.30 m, m_f=0.25 kg, m_c=0.30 kg, and ω=π rad/s. Plot F and T as functions of θ=ωt, for $0 \le \theta \le 2\pi$. Is contact between the cam and the follower maintained? Is T a simple harmonic function?

(Continued)

18.65 Cont.

Solution:

a) Consider the configuration of the system at $\theta = 0$ (Fig. b).

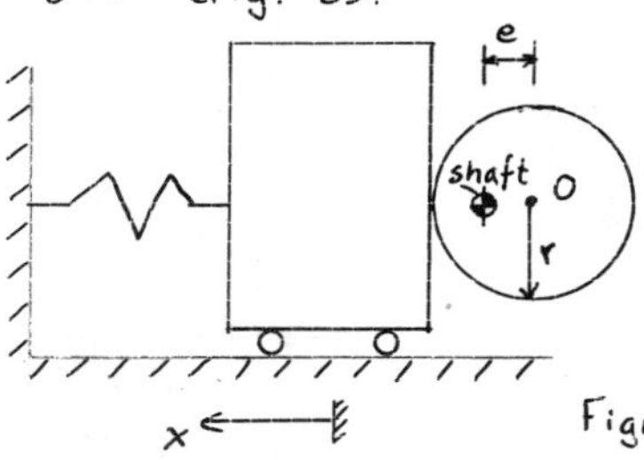

Figure b

In this position, the follower is a distance r−e to the left of the shaft, and the center O of the cam is a distance e to the right of the shaft.

Next consider the configuration of the system after a rotation θ (Fig. a). Now the center O of the cam is a horizontal distance $e\cos\theta$ to the right of the shaft, and the follower is a horizontal distance $r - e\cos\theta$ to the left of the shaft. Hence, under a rotation θ of the cam, the follower has moved left a distance $x = (r - e\cos\theta) - (r-e)$ or $x = e - e\cos\theta$.

Hence, since $\theta = \omega t$,

$$x = e - e\cos\omega t \qquad (a)$$

b) Differentiation of Eq. (a) with respect to time t yields

$$\dot{x} = e\omega \sin\omega t$$
$$\ddot{x} = e\omega^2 \cos\omega t \qquad (b)$$

Then for $e = \omega = 1$, and $\theta = \omega t$,

$$x = 1 - \cos\theta$$
$$\dot{x} = \sin\theta \qquad (c)$$
$$\ddot{x} = \cos\theta$$

The plots of $x, \dot{x}$, and $\ddot{x}$ [using Eq. (c)], for $0 \le \theta \le 2\pi$, are shown in Fig. c.

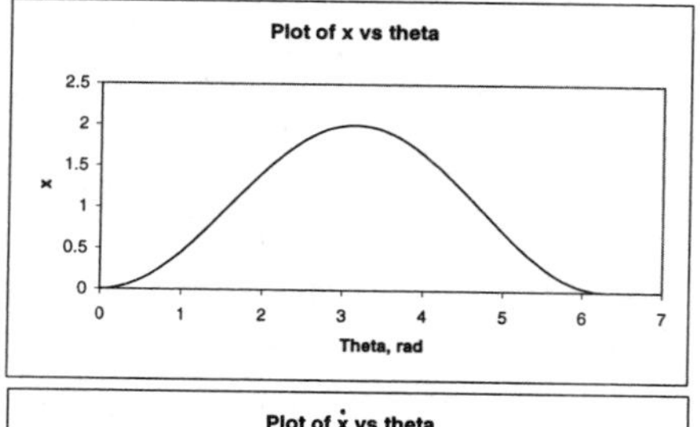

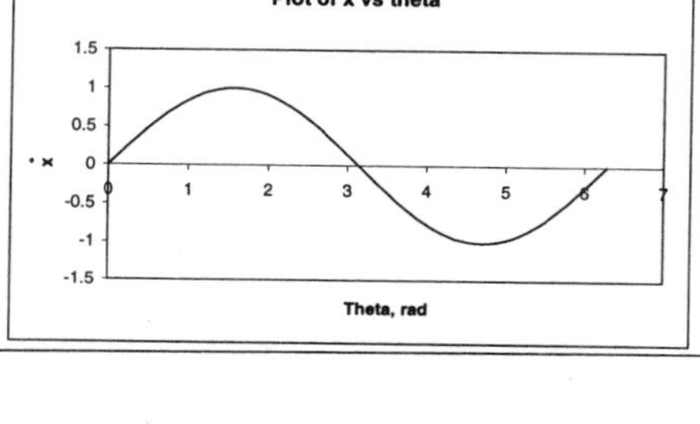

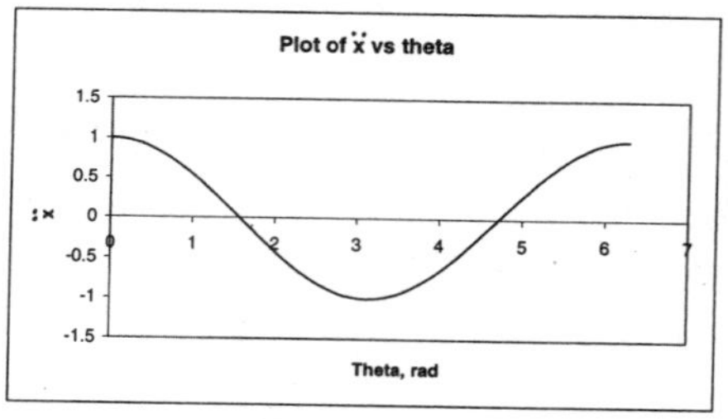

Figure c

c) To derive a formula for the force F that the follower exerts on the cam, consider the free-body diagram of the follower in the position $x = e - e\cos\theta$ (Fig. d).

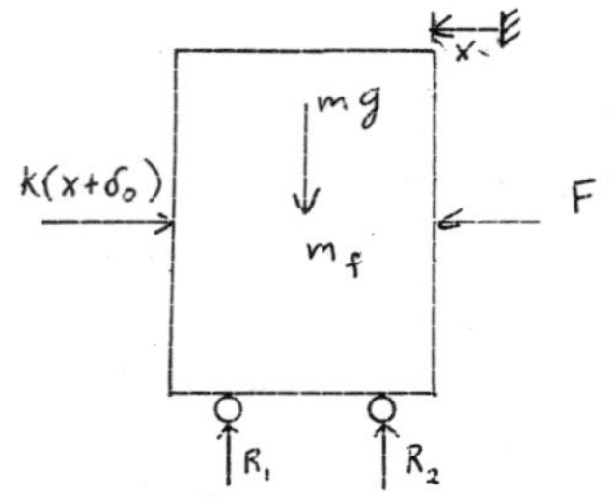

Figure d

By Fig. d

$$\Sigma F_x = F - k(x + \delta_0) = m_f \ddot{x}$$

Substituting for x and $\ddot{x}$ from Eqs. (a) and (b), we find, with $\theta = \omega t$

$$F = k(e + \delta_0) + (m_f\omega^2 - k)e\cos\omega t \qquad (d)$$

To ensure that $F > 0$, consider the extreme values of F (maximum and minimum values of F). They are determined by the condition $dF/d\theta = 0$. Hence, by Eq. (d), with $\theta = \omega t$,

$$\frac{dF}{d\theta} = -e\omega(m_f\omega^2 - k)\sin\theta = 0$$

or

$$\theta = 0, \pi.$$

For $\theta = 0$, Eq. (d) yields,

$$F = k\delta_0 + m_f e\omega^2$$

Therefore, for $F > 0$, $k > -\frac{m_f e\omega^2}{\delta_0}$. This inequality is satified automatically, since k is a positive constant. For $\theta = \pi$, Eq. (d) yields

$$F = k(2e + \delta) - m_f e\omega^2$$

Therefore, for $F > 0$, the required restriction on k is

$$k > \frac{m_f e\omega^2}{2e + \delta}$$

d) To derive a formula for the torque T applied to the cam by the shaft, consider the free-body diagram of the cam for the position θ (Fig. e).

(Continued)

18.65 Cont.

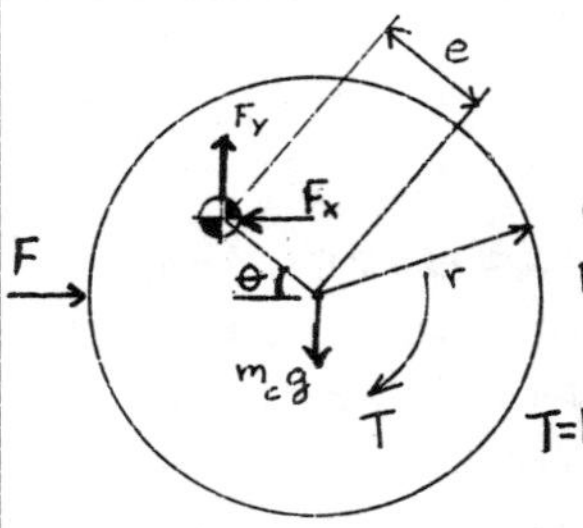

Figure e

By Fig. e, since ω = constant $(\alpha = \frac{d\omega}{dt} = 0)$

$\circlearrowleft \Sigma M_s = F_e \sin\theta - m_c g e \cos\theta - T = I\alpha = 0$ (e)

By Eqs. (d) and (e), we obtain with $\omega t = \theta$

$$T = ke(e+\delta_0)\sin\omega t + \frac{1}{2}e^2(m_f\omega^2 - k)\sin 2\omega t - m_c g e \cos\omega t \quad (f)$$

e) For $k = 20$ N/m, $e = 0.05$ m, $\delta_0 = 0.30$ m, $m_f = 0.25$ kg, $m_c = 0.30$ kg, and $\omega = \pi$ rad/s, Eqs. (d) and (e) become

$$F = 7 - 0.8766\cos\theta$$
$$T = 0.35\sin\theta - 0.02192\sin 2\theta - 0.1472\cos\theta \quad (g)$$

with Eqs. (g), plots of F and T for $0 \le \theta \le 2\pi$ may be generated. (see Figs. (f) and (g))

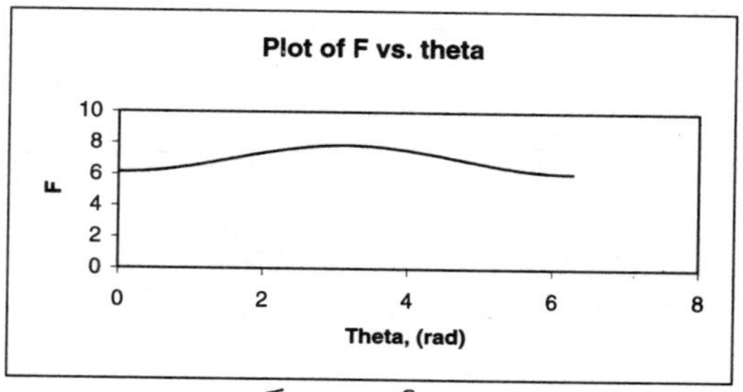

Figure f

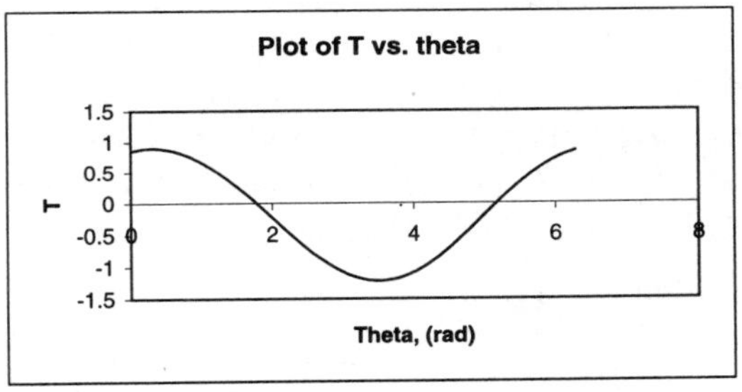

Figure g

Note that T is composed of the functions $\sin\theta$, $\sin 2\theta$, and $\cos\theta$. Since $\theta = \omega t$, the angular frequency of $\sin\theta$ (also $\cos\theta$) is ω. But the angular frequency of $\sin 2\theta$ is 2ω. Hence, T cannot be reduced to a single harmonic function (see Section 13.5).

18.66

Given: A nonuniform circular cylinder has mass $m_c = 3$ slugs and radius $r = 6$ in. It is mounted on a frictionless axle that coincides with its geometric axis O (Fig. a). The center of mass is located 2.25 in from O, and the radius of gyration relative to O is $k_O = 3.75$ in. A cord wrapped around the cylinder passes over a smooth roller at A and is attached to a block B of mass $m_B = 1$ slug. The system is released from rest when $\theta = 0$.

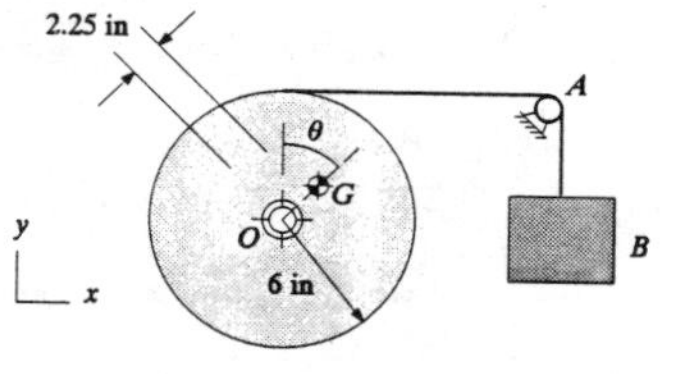

Figure a

a) Derive formulas for the velocity v and the acceleration a of block B and the tension T in the cord as functions of θ.

b) Plot v, a, and T versus θ, for $0 \le \theta \le 360°$

c) From the plots of part b, determine the maximum magnitudes of v, a, and T, and verify these magnitudes using the formulas derived in part a.

Solution: a) The free-body diagrams of the cylinder and the block are shown in Figs. b and c. By Fig. b,

$$\circlearrowleft + \Sigma M_O = \frac{6}{12}T + m_c g\left(\frac{2.25}{12}\sin\theta\right) = m_c k^2\alpha$$

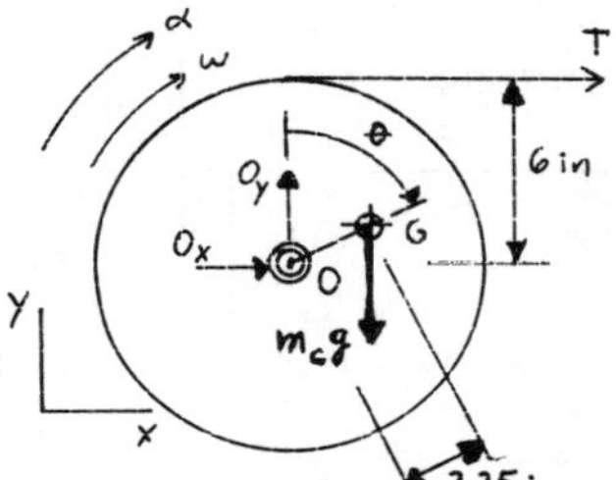

Figure b

Hence, with $m_c = 3$ slugs and $k = 3.75/12 = 0.3125$ ft, we find

$$T = 0.58594\alpha - 36.225\sin\theta \quad (a)$$

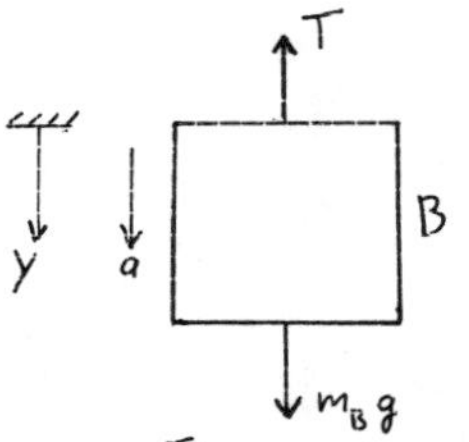

Figure c

By Fig. c and the relationship $a = (6/12)\alpha$,

$$\Sigma F_y = m_B g - T = m_B a = \left(\tfrac{6}{12}\right)m_B\alpha \quad (b)$$

By Eqs. (a) and (b), we find, with $m_B = 1$ slug,

$$\alpha = 29.653 + 33.358\sin\theta \quad (c)$$

Hence, $a = \left(\frac{6}{12}\right)\alpha = 14.826 + 16.679\sin\theta$ [ft/s²] (d)

and by Eqs. (a) and (c),

$$T = 17.375 - 16.679\sin\theta \text{ [lb]} \quad (e)$$

Now by kinematics $\alpha = \frac{d\omega}{dt} = \left(\frac{d\omega}{d\theta}\right)\left(\frac{d\theta}{dt}\right) = \omega\frac{d\omega}{d\theta}$

So, by Eq. (c)

$$\omega\frac{d\omega}{d\theta} = 29.653 + 33.358\sin\theta$$

or

$$\int_0^\omega \omega\, d\omega = \int_0^\theta (29.653 + 33.358\sin\theta)\, d\theta$$

Integration yields

$$\frac{1}{2}\omega^2 = 29.653\theta + 33.358(1 - \cos\theta)$$

(Continued)

18.66 Cont.

Hence,

$$\omega = \sqrt{59.306\,\theta + 66.716(1-\cos\theta)}$$

and

$$v = r\omega = (0.5)\omega = 0.5\sqrt{59.306\theta + 66.716(1-\cos\theta)}\ [\tfrac{ft}{s}] \quad (f)$$

b) By Eqs. (f), (d), and (e), v, a, and T may be plotted for $0 \le \theta \le 360°$ (see Figs. d, e, and f.)

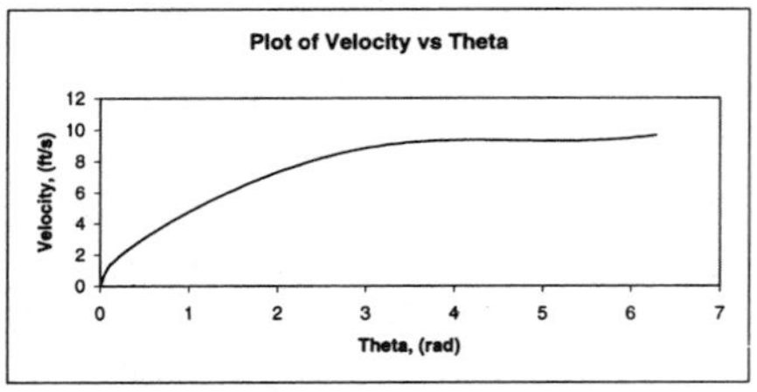

Figure d

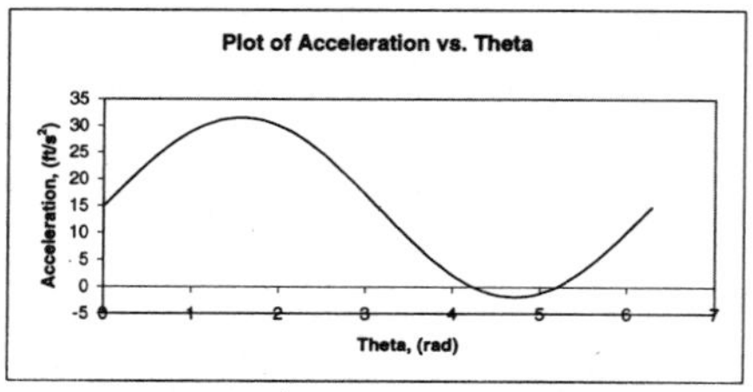

Figure e

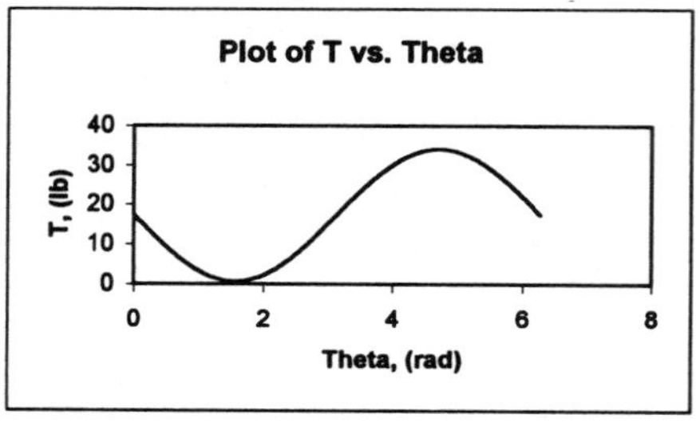

Figure f

c) By Figs. d, e, and f,

$V_{max} = 9.6$ ft/s

$a_{max} = 31.5$ ft/s²

$T_{max} = 34.05$ lb

By the formulas derived in Part a, the velocity v is a maximum (or minimum) when $dv/d\theta = 0$. Hence, by Eq. (f),

$$\frac{dv}{d\theta} = \frac{(0.5)(\frac{1}{2})[59.306 + 66.716\sin\theta]}{[59.306\theta + 66.716(1-\cos\theta)]^{1/2}} = 0$$

or

$$\sin\theta = -0.88893$$

Hence,

$\theta = -1.0950$ rad or in the range $0 \le \theta \le 2\pi$,

$\theta = 4.2366$ rad or 5.1882 rad

Hence, by Eq. (f),

For $\theta = 4.2366$ rad,

$$v = 0.5\sqrt{(59.306)(4.2366) + 66.716(1.4580)}$$

or $v = 9.334$ ft/s

Similarily for $\theta = 5.1882$ rad,

$v = 9.2716$ ft/s

For $\theta = 2\pi$ (360°),

$v = 9.6518$ ft/s

Therefore, in the range $0 \le \theta \le 2\pi$, the velocity v reaches a maximum of 9.334 ft/s for $\theta = 4.2366$ rad, then decreases until $\theta = 5.1882$ rad (where v is a minimum of 9.2716 ft/s). Then v begins to increase again until at $\theta = 360°$, $v = 9.6518$ ft/s

Hence, in the range $0 \le \theta \le 360°$, the maximum value of v is

$$V_{max} = 9.6518 \text{ ft/s}$$

By Eqs. (d) and (e), we have by inspection

$a = a_{max}$ for $\theta = \pi/2$ rad

$T = T_{max}$ for $\theta = 3\pi/2$ rad

Hence,

$$a_{max} = 31.51 \text{ ft/s}^2$$

$$T_{max} = 34.05 \text{ lb}$$

18.67

Given: The system of Problem 18.66 (Fig. a below) has a cylinder of mass $m_c = 3$ slugs and radius $r = 6$ in. It is mounted on a frictionless axis at its geometric center O, with a radius of gyration $k_o = 3.75$ in. The center of mass is located 2.25 in from O. Block B of mass $m_B = 1$ slug is suspended from a cord that passes over a smooth roller at A and wrapped around the cylinder. The system is released from rest when $\theta = 0$.

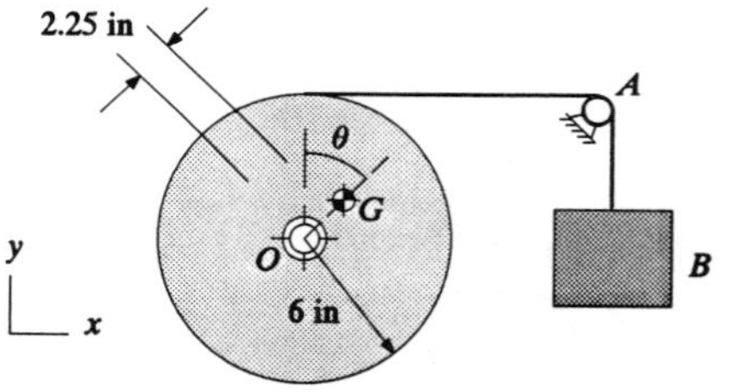

Figure a

a) As functions of θ, derive formulas for O_x and O_y, the forces exerted on the cylinder by the axle.

b) Plot O_x and O_y versus θ for $0 \le \theta \le 360°$.

c) From the plots of part b, determine the maximum magnitudes of O_x and O_y for $0 \le \theta \le 360°$, and verify these magnitudes using the formulas derived in part a.

(Continued)

18.67 Cont.

Solution: The free-body diagrams of the cylinder and the block are shown in Figs. b and c.

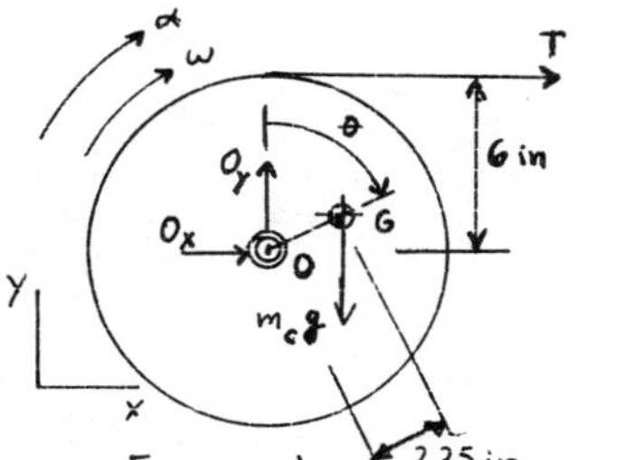

Figure b

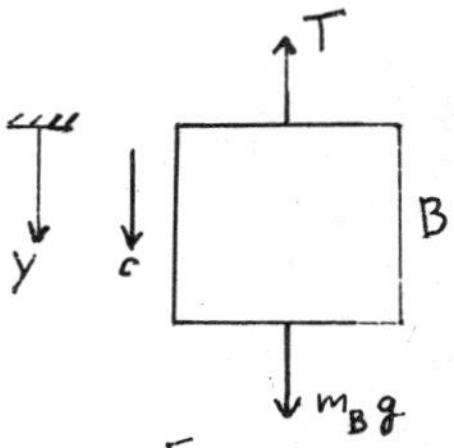

Figure c

By Fig. b,

$$\circlearrowright + \Sigma M_o = \frac{6}{12}T + m_c g\left(\frac{2.25}{12}\right)\sin\theta = m_c k^2 \alpha$$

or, with $m_c = 3$ slugs, and $k = 3.75/12 = 0.3125$ ft,

$$T = 0.58594\alpha - 36.225\sin\theta \qquad (a)$$

Also by Fig. b,

$$\Sigma F_x = O_x + T = m_c\left(\frac{2.25}{12}\right)(\alpha\cos\theta - \omega^2\sin\theta) \qquad (b)$$

$$\Sigma F_y = O_y - m_c g = m_c\left(\frac{2.25}{12}\right)(-\alpha\sin\theta - \omega^2\cos\theta) \qquad (c)$$

By Fig. c, and the kinematic relation $a = \left(\frac{6}{12}\right)\alpha$,

$$\Sigma F_y = m_B g - T = m_B a = \left(\frac{6}{12}\right) m_B \alpha \qquad (d)$$

By Eqs. (a) and (d), with $m_B = 1$ slug, we find

$$\alpha = 29.653 + 33.358\sin\theta \qquad (e)$$

$$T = 17.374 - 16.679\sin\theta \qquad (f)$$

Next, by kinematics, $\alpha = d\omega/dt = \frac{d\omega}{d\theta}\frac{d\theta}{dt} = \omega\frac{d\omega}{d\theta}$

So, by Eq. (e),

$$\omega\frac{d\omega}{d\theta} = 29.653 + 33.358\sin\theta$$

or

$$\int_0^\omega \omega\, d\omega = \int_0^\theta (29.653 + 33.358\sin\theta)\, d\theta$$

Integration yields

$$\omega^2 = 59.306\theta + 66.716(1 - \cos\theta) \qquad (g)$$

By Eqs. (b), (c), (e), (f), and (g), we obtain

$$O_x = -17.374 - 20.849\sin\theta + 16.680\cos\theta - 33.360\theta\sin\theta + 28.146\sin 2\theta \qquad (h)$$

$$O_y = 96.6 - 16.680\sin\theta - 37.528\cos\theta - 33.360\theta\cos\theta - 18.764\sin^2\theta + 37.528\cos^2\theta \qquad (i)$$

b) By Eqs. (h) and (i), O_x and O_y may be plotted versus θ for $0 \le \theta \le 360°$. See Figs. d and e.

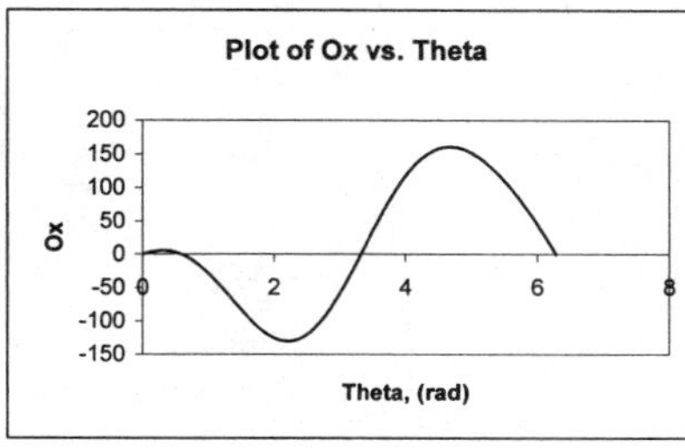

Figure d

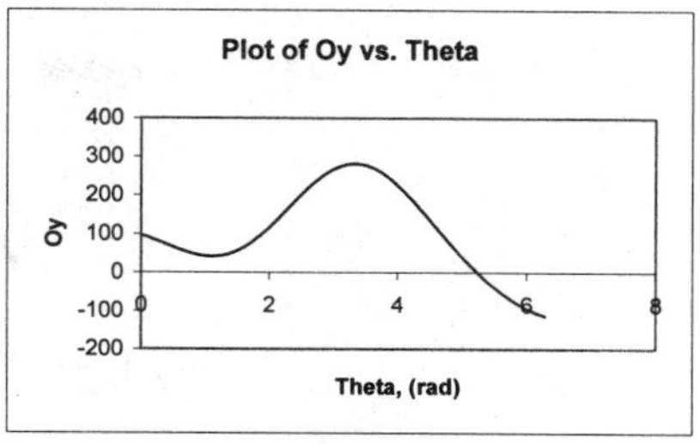

Figure e

c) By Figs. d and e,

$(O_x)_{max} = 160$ lb

$(O_y)_{max} = 281$ lb

The maximum (or minimum) values of O_x and O_y occur when $dO_x/d\theta = 0$ and $dO_y/d\theta = 0$. So, by Eq. h,

$$\frac{dO_x}{d\theta} = -20.849\cos\theta - 50.040\sin\theta - 33.360\theta\cos\theta + 56.292\cos(2\theta) = 0$$

The roots are,

$\theta = 0.31623$ rad $= 18.119°$

$\theta = 2.20986$ rad $= 126.616°$

$\theta = 4.6776$ rad $= 268.007°$

By the graph the maximum O_x occurs at $\theta = 268.007°$.

Hence, by Eq. (h),

$$(O_x)_{max} = 160.79 \text{ lb}$$

Similarly, by Eq. (i), $\frac{dO_y}{d\theta} = 0$ yields the maximum value of O_y is at $\theta = 3.3323$ rad or $\theta = 190.93°$ and then by Eq. (i),

$$(O_y)_{max} = 281.3 \text{ lb}$$

18.68

Given: A disk hangs from a frictionless horizontal axle O (Fig. a). A 100 lb horizontal force acting in the plane of the disk is applied suddenly to the lowest point of the disk.

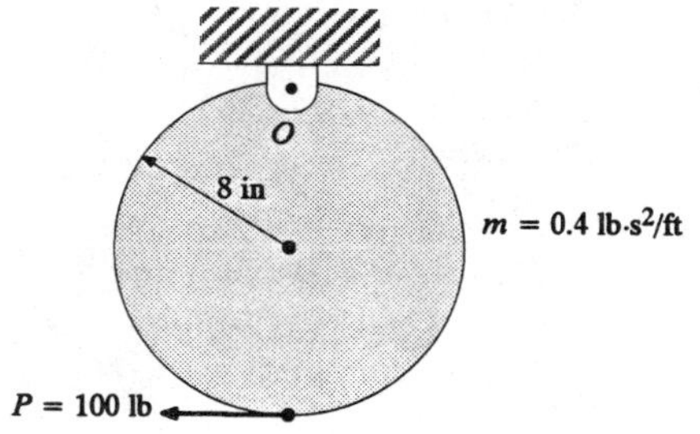

Figure a

For the instant that the force is applied, determine the angular acceleration of the disk and the horizontal and vertical forces exerted on the disk by the axle. (Continued)

18.68 Cont.

Solution: The free-body diagram of the disk is shown in Fig. b.

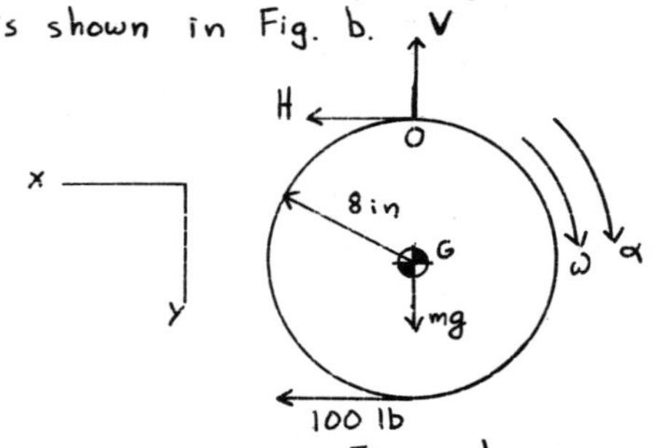

Figure b

The mass moment of inertia relative to the axle O is

$$I_O = \frac{1}{2}mr^2 + mr^2 = \frac{3}{2}mr^2$$

where $m = 0.4\ \text{lb}\cdot\text{s}^2/\text{ft}$ and $r = 8/12 = 2/3$ ft.
Therefore,

$$I_O = \left(\frac{3}{2}\right)(0.4)\left(\frac{4}{9}\right) = \frac{2.4}{9}\ \text{lb}\cdot\text{ft}\cdot\text{s}^2 \qquad (a)$$

By Fig. b,

$$\circlearrowright + \Sigma M_O = (100)\left(\frac{16}{12}\right) = I_O\alpha \qquad (b)$$

Equations (a) and (b) yield

$$\underline{\underline{\alpha = 500\ \text{rad/s}^2}} \qquad (c)$$

Also by Fig. b,

$$\Sigma F_x = H + 100 = m\,a_G = (0.4)(500)\left(\frac{8}{12}\right)$$

or $\underline{\underline{H = 33.3\bar{3}\ \text{lb}}}$

and, since initially $\omega = 0$,

$$\Sigma F_y = mg - V = 0$$

or $\underline{\underline{V = mg = (0.4)(32.2) = 12.88\ \text{lb}}}$

18.69

Given: A disk hangs from a frictionless horizontal axle O (Fig. a).

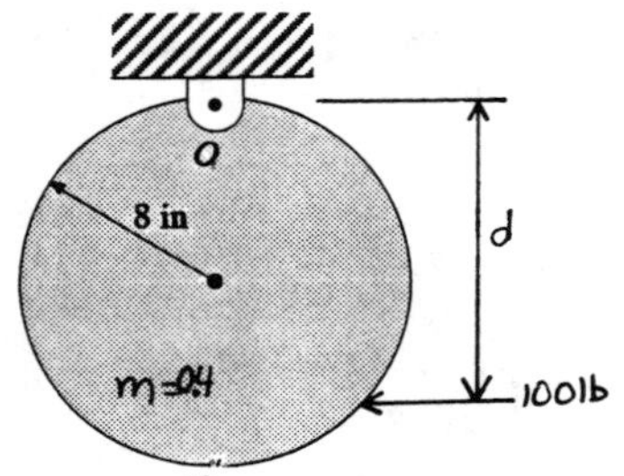

Figure a

Determine the distance d below the axle at which the 100 lb horizontal force must be applied so that there is no initial horizontal reaction of the axle.

Solution: Figure b is a free-body diagram of the disk, assuming that the horizontal force at O is zero.

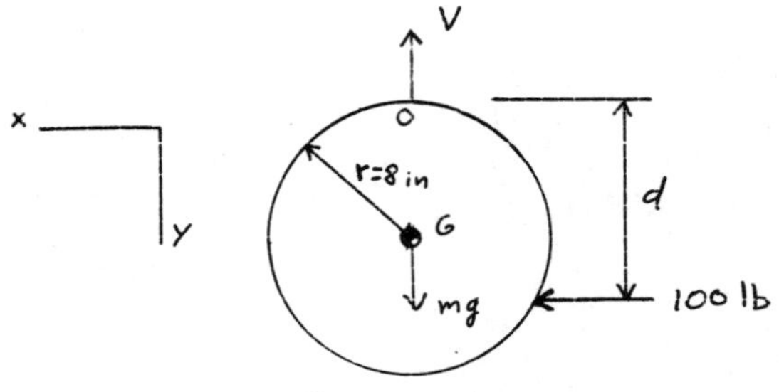

Figure b

The mass moment of inertia of the disk relative to O is

$$I_O = \frac{1}{2}mr^2 + mr^2 = \frac{3}{2}mr^2 = \left(\frac{3}{2}\right)(0.4)\left(\frac{8}{12}\right)^2$$

or

$$I_O = 0.266\bar{6}\ \text{lb}\cdot\text{ft}\cdot\text{s}^2 \qquad (a)$$

By Fig. b and Eq. (a),

$$\circlearrowright + \Sigma M_O = 100d = I_O\alpha = (0.266\bar{6})\alpha$$

or

$$d = 0.00266\bar{6}\,\alpha \qquad (b)$$

$$\Sigma F_x = 100 = m\,a_G = mr\alpha = (0.40)\left(\frac{8}{12}\right)\alpha$$

or

$$\alpha = \frac{1200}{3.2}\ \text{rad/s}^2 \qquad (c)$$

Equations (b) and (c) yield

$$\underline{\underline{d = 1.0\ \text{ft}}}$$

Alternatively, by Eq. (18.22),

$$d = k_O^2/r \qquad (d)$$

where

$$k_O = \sqrt{I_O/m} = \sqrt{0.266\bar{6}/0.40} = 0.8165 \qquad (e)$$

Then, by Eqs. (d) and (e),

$$\underline{\underline{d = \frac{(0.8165)^2}{(8/12)} = 1.0\ \text{ft}}}$$

18.70

Given: A circular disk rotates in the plane of the page about a fixed horizontal pin through point O (Fig. a). The disk is subjected to the force of gravity and the forces exerted by the pin. For the position shown, the disk rotates counterclockwise with angular velocity $\omega = 40$ rad/s.

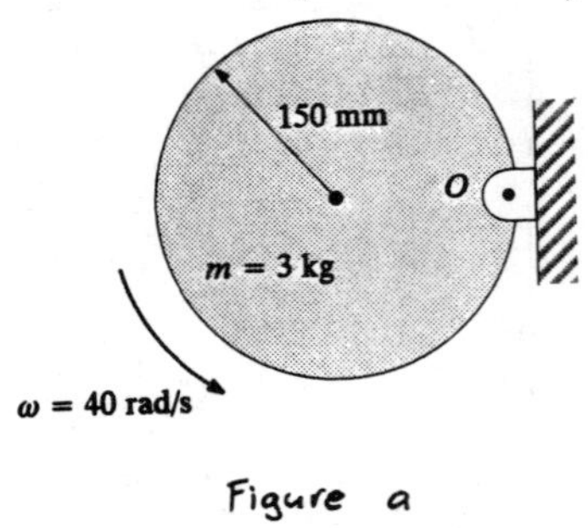

Figure a

18.70 Cont.

a) Calculate the angular acceleration of the disk.
b) Determine the horizontal and vertical forces exerted on the disk by the pin.

Solution:

a) Figure b is the free-body diagram of the disk. By Fig. b,

$\circlearrowleft + \sum M_o = I_o \alpha$ or, $mgr = \frac{3}{2} mr^2 \alpha$

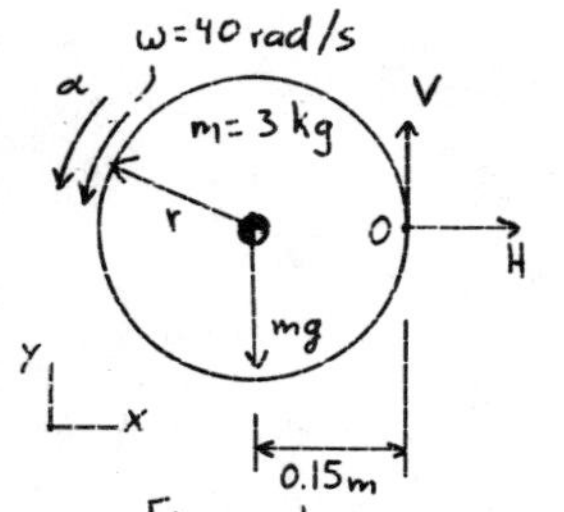

Figure b

Hence, with $r = 0.15$ m, and $g = 9.81$ m/s^2,

$$\alpha = \frac{2g}{3r} = \frac{2(9.81)}{3(0.15)} = 43.6 \ \frac{\text{rad}}{\text{s}^2} \quad (a)$$

b) By Fig. b,

$$\sum F_x = H = m a_x = m r \omega^2$$
$$\sum F_y = V - mg = m a_y = -mr\alpha \quad (b)$$

Therefore, with Eq. (a), Eq. (b) yields

$$H = (3)(0.15)(40)^2 = 720 \text{ N}$$

$$V = (3)[9.81 - (0.15)(43.6)] = 9.81 \text{ N}$$

18.71

Given: A wedge-shaped pendulum of constant thickness (perpindicular to the plane of the page) is released from rest from the horizontal position (Fig. a). The pendulum weighs 36 N, and it is 900 mm long. The support Q exerts a constant frictional resisting couple equal in magnitude to 2.7 N·m.

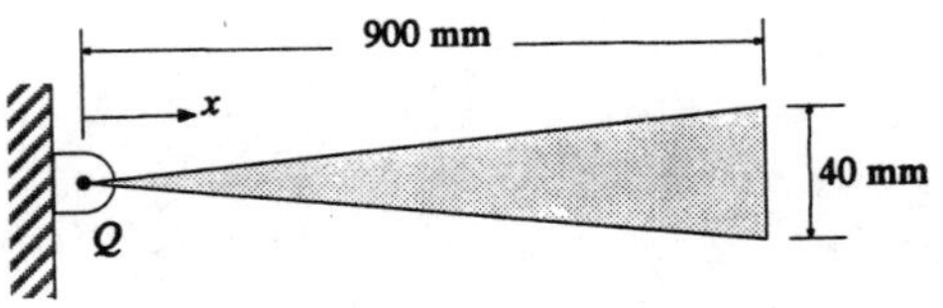

Figure a

Determine the direction and the magnitude of the force the pendulum exerts on the support the instant after it is released.

Solution: The center of mass of the pendulum is located 600 mm (0.60 m) from the support Q (see Fig. b, the free-body diagram of the pendulum).

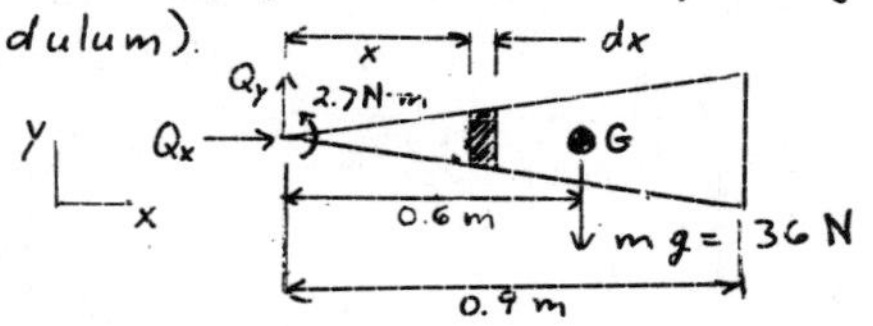

Figure b

By Fig. b,

$$\circlearrowleft + \sum M_Q = [(36)(0.6) - 2.7] \text{ N}\cdot\text{m} = I_Q \alpha \quad (a)$$

To determine I_Q, let

$$\rho = cx = \text{mass per unit length}$$

Then, by Fig. b,

$$dm = cx\,dx$$

or

$$\int dm = m = \int_0^L cx\,dx = \frac{1}{2} c L^2 \quad (b)$$

where $L = 0.90$ m the length of the pendulum. With $m = 36/9.81$ kg, Eq. (b) yields,

$$c = (2)\left(\frac{36}{9.81}\right)/(0.90)^2 = 9.061$$

So, $dm = 9.061\,x\,dx$

or $I_Q = \frac{9.061}{4} x^4 \Big|_0^{0.90} = 1.486 \text{ N}\cdot\text{m}\cdot\text{s}^2 \quad (c)$

Equations (a) and (c) yield

$$\alpha = 12.719 \text{ rad/s}^2 \circlearrowright \quad (d)$$

By Fig. b, since initially $\omega = 0$,

$$\sum F_x = Q_x = m r \omega^2 = 0 \ ; \quad Q_x = 0$$

$$\sum F_y = Q_y - 36 = -mr\alpha = \left(\frac{-36}{9.81}\right)(0.6)(12.719)$$

$$Q_y = 7.995 \text{ N}$$

18.72

Given: A non uniform bar of weight W is pivoted at its upper end A (Fig. a), and its center of mass G is a distance r from the pivot at A. Its mass moment of inertia is I. The bar is released from rest at the 45° angle. Neglect friction.

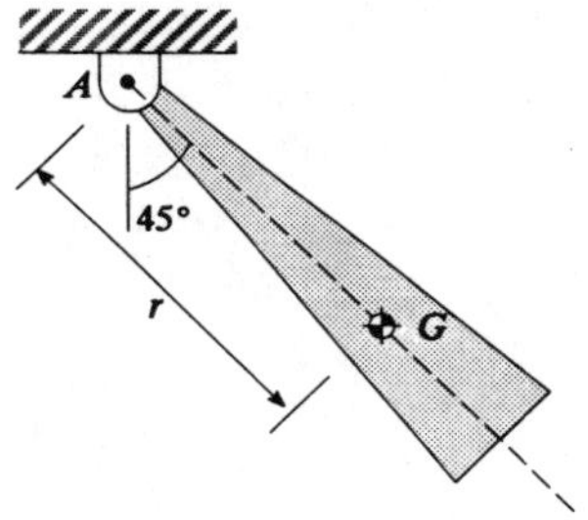

Figure a

a) Derive a formula in terms of I, W, r, and g for the radially and tangentially directed forces P_t and P_n exerted on the bar by the pivot A for the instant after the bar is released.

b) Derive a formula in terms of I, W, r, and g for the angular velocity of the bar for the instant it passes through the vertical position.

(Continued)

18.72 Cont.

Solution:

a) The forces P_n and P_t (Fig. b) are determined by Eqs. (18·20), repeated below as Eqs. (a), where A replaces O.

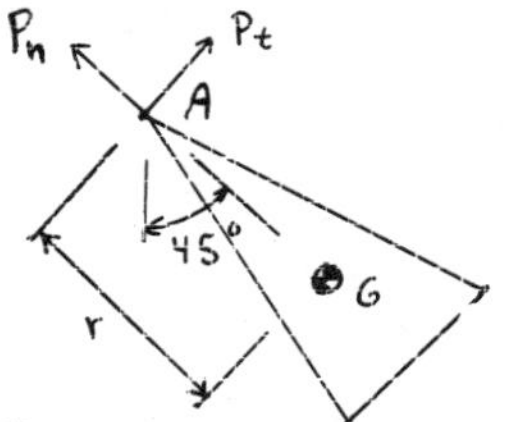

Figure b

$$P_n = mg\left[\frac{2r^2}{k_A^2}(\cos\theta - \cos\theta_o) + \cos\theta\right]$$

$$P_t = mg\left(1 - \frac{r^2}{k_A^2}\right)\sin\theta \qquad (a)$$

For $\theta = \theta_o = 45°$, Eqs. (a) yield

$$P_n = mg\cos 45° = 0.7071\ mg$$

$$P_t = mg\left(1 - \frac{mr^2}{I}\right)\sin 45° = 0.7071 mg\left(1 - \frac{mr^2}{I}\right)$$

where $I = mk_A^2$

b) The angular velocity of the bar is determined by Eq. (18.17), repeated here as Eq. (b)

$$\omega = \frac{1}{k_A}\sqrt{2gr(\cos\theta - \cos\theta_o)} \qquad (b)$$

For $\theta_o = 45°$ and $\theta = 0$, Eq. (b) yields

$$\omega = \sqrt{\frac{2gr}{k_A^2}}(1 - 0.7071) = 0.765\sqrt{\frac{mgr}{I}}$$

18.73

Given: An irregular body of mass 12 lb·s²/ft is suspended by cords AB and CD (Fig. a). Its radius of gyration with respect to the axis through G and perpendicular to the plane of the cords is $k_G = 2.50$ ft.

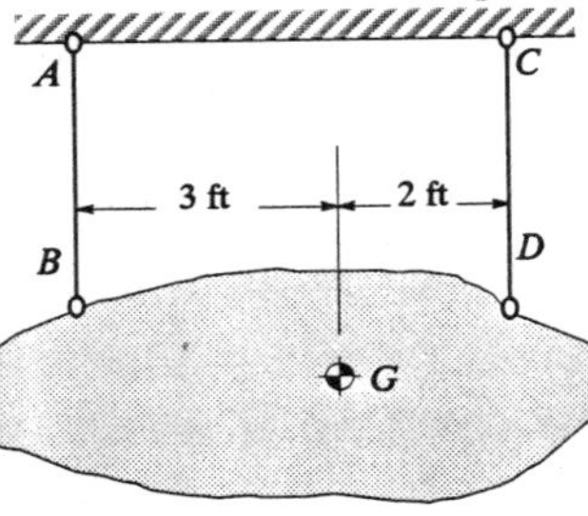

Figure a

Determine the increase of tension in cord AB for the instant after cord CD is cut

Solution:

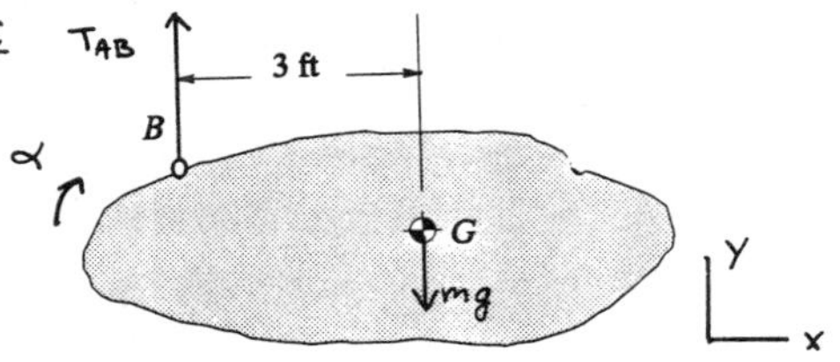

Figure b

The free-body diagram of the body, the instant after the cord CD is cut, is shown in Fig. b.

By Fig. b,

$$\circlearrowleft \Sigma M_G = 3T_{AB} = I_G\alpha$$
$$\Sigma F_y = mg - T_{AB} = m a_G \qquad (a)$$

Also, by kinematics

$$a_G = 3\alpha \qquad (b)$$

and

$$I_G = mk_G^2 \qquad (c)$$

with $m = 12$ lb·s²/ft, $g = 32.2$ ft/s² and $k_G = 2.5$ ft. Equations (a), (b), and (c) yield,

$$\alpha = 6.334\ \text{rad/s}^2$$
$$T_{AB} = 158.36\ \text{lb} \qquad (d)$$

The free-body diagram of the body before cord CD is cut is shown in Fig. c.

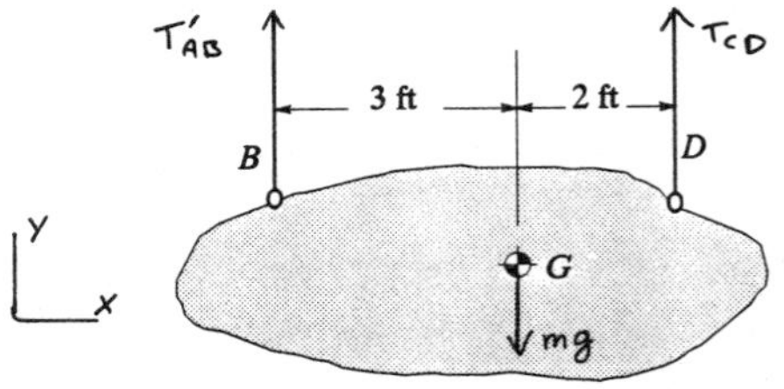

Figure c

By Fig. c and statics,

$$\circlearrowleft \Sigma M_D = 5T'_{AB} - 2mg = 0$$

or

$$T'_{AB} = \frac{2}{5}mg = \frac{2}{5}(12)(32.2) = 154.56\ \text{lb} \qquad (e)$$

Hence, by Eqs. (d) and (e), the change in T_{AB} is

$$\Delta T_{AB} = T_{AB} - T'_{AB} = 158.36 - 154.56 = 3.80\ \text{lb}$$

18.74

Given: The linkage (Fig. a) is connected by frictionless pins at A, B, C, and D, and is held in equilibrium by force P. Links AB and CD each weigh 10 lb, and link BD weighs 20 lb. The force P is removed suddenly.

a) Determine the initial angular acceleration of link AB.

b) Determine the horizontal and vertical forces that act on link AB at point B, for the instant after P is removed

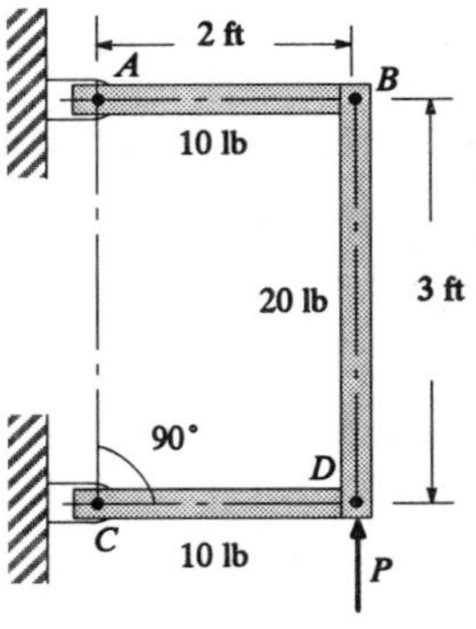

Figure a

(Continued)

18.74 Cont.

Solution: The free-body diagrams of links BD, AB, and CD are shown in Figs. b, c, and d.

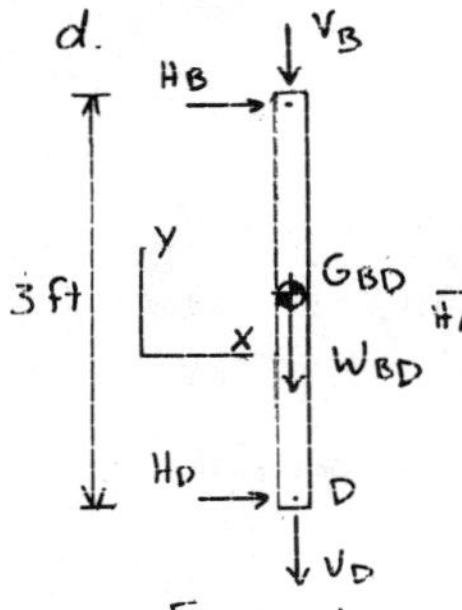

Figure b

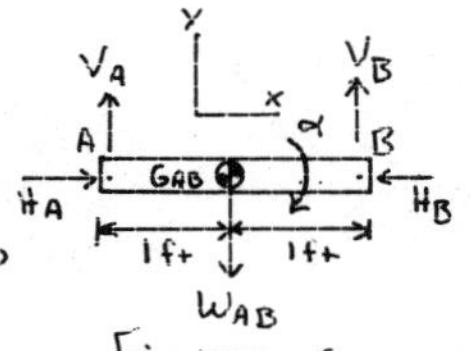

Figure c

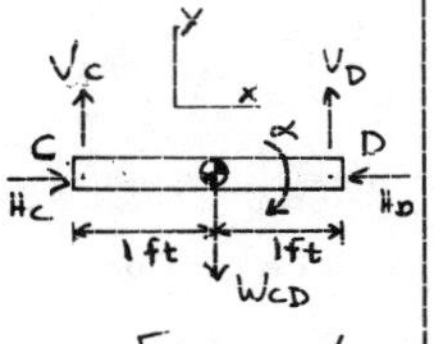

Figure d

By Figs. (a) and (b), the motion of link BD is a translation. Hence, by Fig. b,

$$\circlearrowleft \Sigma M_{G_{BD}} = \tfrac{3}{2} H_B - \tfrac{3}{2} H_D = 0 \quad \text{or} \quad H_B = H_D \qquad (a)$$

$$\Sigma F_x = H_B + H_D = m_{BD}(a_{BD})_x = m(-2\omega^2) = 0 \qquad (b)$$

since $\omega = 0$ initially for the horizontal links. Therefore, by Eqs. (a) and (b)

$$H_B = H_D = 0 \qquad (c)$$

By Fig. b,

$$\Sigma F_y = -V_B - V_D - W_{BD} = m_{BD}(a_{BD})_y = m_{BD}(-2\alpha)$$

or

$$V_B + V_D + 20 = \frac{40}{g}\alpha \qquad (d)$$

where α is the initial angular acceleration of either link AB or link CD.

Next, by Fig. c,

$$\circlearrowleft + \Sigma M_A = -(2)V_B + (1)W_{AB} = I_A \alpha = \tfrac{1}{3} m_{AB}(2)^2 \alpha$$

or

$$-2V_B + 10 = \frac{40}{3g}\alpha \qquad (e)$$

Likewise, by Fig. d,

$$-2V_D + 10 = \frac{40}{3g}\alpha \qquad (f)$$

Equations (e) and (f) yield $V_B = V_D$. Hence, by Eq. (d),

$$2V_B + 20 = \frac{40}{g}\alpha \qquad (g)$$

Eqs. (e) and (g), we find with $g = 32.2\ \text{ft/s}^2$,

$$\alpha = \frac{9}{16} g = 18.11\ \text{rad/s}^2 \qquad (h)$$

b) By Eqs. (e) and (h) [or by Eqs. (g) and (h)], we obtain $V_B = 1.25$ lb. Hence, the horizontal and vertical forces acting on link AB at B are

$$H_B = 0, \quad V_B = 1.25\ \text{lb}$$

18.75

Given: For the simple pendulum shown in Fig. a, $k = r = L$.

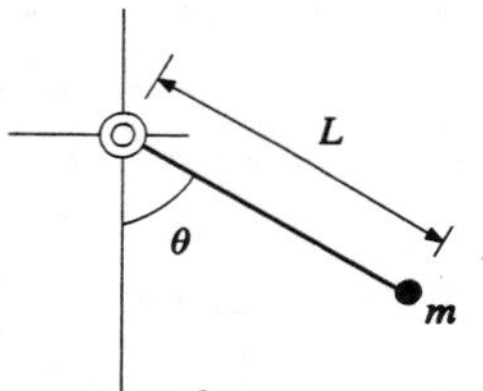

Figure a

a) Use the small-angle approximation $\cos\theta = 1 - \theta^2/2$ to show that [see Eqs. (18.20)]

$$P_n = mg\left(1 - \tfrac{3}{2}\theta^2 + \theta_0^2\right)$$
$$P_t = 0 \qquad (a)$$

b) With Eq. (18.20), derive a formula for the error E in approximating P_n by Eq. (a).

c) Plot the formula derived in Part b as a function of θ, for $0 \le \theta \le \theta_0$, with $\theta_0 = \pi/3, \pi/6, \pi/10$, and $\pi/30$.

d) By the plot of part c, estimate the maximum value of θ_0 for which $|E| < 5\%$ for all values of θ, where $|E|$ denotes the absolute value of E. Verify your estimate by the formula of part b.

Solution: a) By Eq. (18.20)

$$P_n = mg\left[\frac{2r^2}{k_0^2}(\cos\theta - \cos\theta_0) + \cos\theta\right]$$
$$P_t = mg\left(1 - \frac{r^2}{k_0^2}\right)(\sin\theta) \qquad (b)$$

with $k = r = L$, Eqs. (b) yield,

$$P_n = mg(3\cos\theta - 2\cos\theta_0)$$
$$P_t = 0 \qquad (c)$$

Then, with $\cos\theta = 1 - \theta^2/2$ and $\cos\theta_0 = 1 - \theta_0^2/2$, Eqs. (c) yield

$$P_{n(\text{approx})} = mg\left(1 - \tfrac{3}{2}\theta^2 + \theta_0^2\right) \qquad (d)$$

$$P_t = 0 \qquad (e)$$

b) By Eqs. (b) and (d), the error in approximating P_n by Eq. (d) is

$$E = \left[\frac{P_n - P_{n(\text{approx})}}{P_n}\right](100\%)$$

or

$$E = \left[1 - \frac{(1 - \tfrac{3}{2}\theta^2 + \theta_0^2)}{(3\cos\theta + 2\cos\theta_0)}\right](100\%) \qquad (f)$$

c) Using Eq. (f), we obtain the plots of E for $0 \le \theta \le \theta_0$, for $\theta_0 = \pi/3, \pi/6, \pi/10$, and $\pi/30$ (see Fig. b)

(Continued)

18.75 Cont.

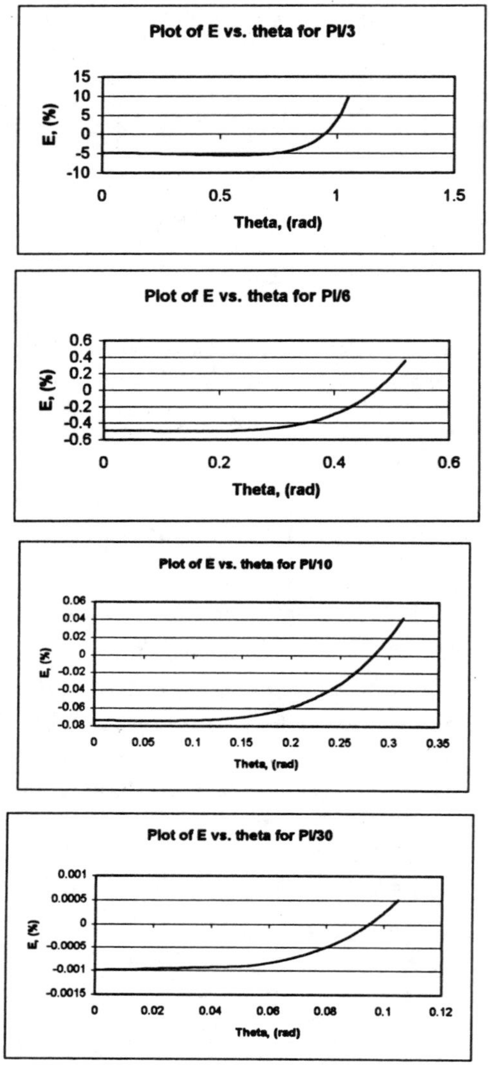

Figure b

d) By Fig. b, it can be seen that for $\theta_0 = \pi/3$ rad, the minimum value of E is approximately -5.5 % for $\theta \approx 0.56$ rad, and for $\theta = \theta_0 = \pi/3$ rad, by Eq. (f), E = 9.66 %. Accordingly, for $\theta_0 = \pi/3$ rad and $0 \le \theta \le \theta_0$,

$$-5.5\% \le E \le 9.66\%$$

Hence, for $|E| < 5\%$, θ_0 must be less than $\pi/3$ rad.

Also, by Fig. (b), for $\theta_0 = \pi/6$, $\pi/10$, and $\pi/30$, we see that $-5\% \le E \le 5\%$ for $0 \le \theta \le \theta_0$. Thus, for these values of θ_0, $|E| < 5\%$ for all values of θ in the range $0 \le \theta \le \theta_0$. In particular, for $|E| < 5\%$, the maximum allowable value of θ_0 must lie in the range $\pi/6 < \theta_0 < \pi/3$.

We also can see by Fig. b, that the maximum positive value of E occurs at $\theta = \theta_0$. Hence, by Eq. (f) for E = +5 %, at $\theta = \theta_0$, we have,

$$E = \left[1 - \frac{(1 - \frac{1}{2}\theta_0^2)}{\cos\theta_0}\right] \times 100\% = 5\%$$

or after simplification

$$95\cos\theta_0 + 50\theta_0^2 = 100 \qquad (g)$$

The solution of Eq. (g) is $\theta_0 = 0.92766$ rad. Thus, for E = 5 %, $\theta_0 = 0.92766$ rad.

To determine the largest minimum value of E for $\theta_0 = 0.92766$ rad, we determine the corresponding value of θ by the condition that $dE/d\theta = 0$. Thus, by Eq. (f), after simplification,

$$\frac{dE}{d\theta} = 3\theta(3\cos\theta - 2\cos\theta_0) - (1 + \theta_0^2 - 1.5\theta^2)(3\sin\theta) = 0$$

or with $\theta_0 = 0.92766$ rad,

$$3\theta(3\cos\theta - 1.19942) - (5.58166 - 4.5\theta^2)\sin\theta = 0$$

The solution of this equation is $\theta = 0.46244$ rad. Hence, with $\theta_0 = 0.92766$ rad and $\theta = 0.46244$ rad, Eq. (f) yields,

$$E = \left[1 - \frac{1 - 1.5(0.46244)^2 + 0.92766^2}{2.68490 - 1.19942}\right] \times 100$$

or

$$E = -3.655\ \%$$

Hence, for $\theta_0 \le 0.92766$ rad $\le 53.15°$, $|E| \le 5\%$

For $|E| < 5°$, $\theta_0 < 0.92766$ rad = 53.15°

The plot of E for $\theta_0 = 0.92766$ rad is shown in Fig. c.

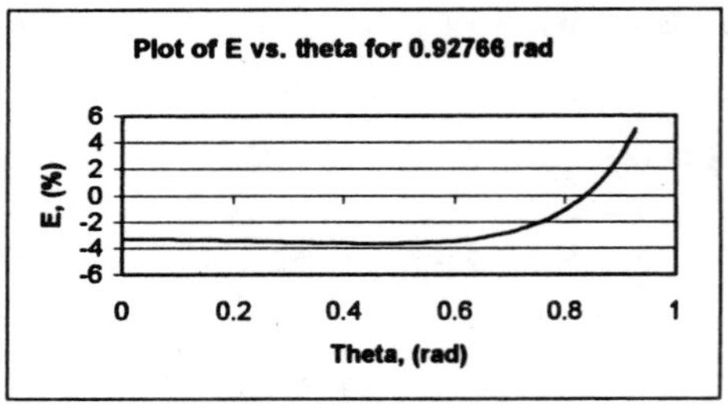

Figure c

18.76

Given: For the simple pendulum, Eq. (18.17) can be written

$$dt = \frac{d\theta}{\sqrt{\frac{2g}{L}(\cos\theta - \cos\theta_0)}} \qquad (a)$$

and the half-period of the pendulum is

$$\frac{T}{2} = \sqrt{\frac{L}{g}} \int_{-\theta_0}^{\theta_0} \frac{d\theta}{\sqrt{2(\cos\theta - \cos\theta_0)}} \qquad (b)$$

(Continued)

18.76 Cont.

a) Determine T by integration of Eq. (b) using Simpson's rule or an alternative computer-based method, for $\theta_0 = \pi/2$, $\pi/4$, $\pi/10$, and $\pi/20$ radians.

b) Determine the percent error E in the period T calculated using Eq. 18.21, compared to T calculated with Eq. (b).

Solution:

a) By the Simpson's rule, with $n = 20$, let

$$f(\theta) = \frac{1}{\sqrt{2(\cos\theta - \cos\theta_0)}} \qquad (c)$$

Then, with θ_0 replaced by $0.999837\theta_0$ in the limits

$$\int_{-0.999837\theta_0}^{0.999837\theta_0} f(\theta)d\theta = \frac{1}{3}h\left[f_0 + 4(f_1 + f_3 + \cdots + f_{19}) + 2(f_2 + f_4 + \cdots + f_{18}) + f_{20}\right] \quad (d)$$

where $h = \frac{\theta_0}{20}$ (e)

and by Eq. (c),

$f_i = f(\theta)$ at θ_i (f)

The limits $-0.999837\theta_0$ and $0.999837\theta_0$ were selected by trial and error so that for $\theta_0 = \pi/20 = 9°$, $T \approx 2\pi\sqrt{L/g}$.

The periods for $\theta_0 = \pi/2$, $\pi/4$, $\pi/10$, and $\pi/20$, as determined by Simpson's rule are listed in Table a. These values were obtained using the spreadsheet software program EXCEL.

Table a

θ_0	h	$T_{SR}/\sqrt{L/g}$
$\frac{\pi}{2}$	0.078539	7.6302965
$\frac{\pi}{4}$	0.0392699	6.5646586
$\frac{\pi}{10}$	0.015708	6.3156875
$\frac{\pi}{20}$	0.007854	6.2815516

b) By small-displacement theory [see Eq. (18.21)]

$T_{SD} = 2\pi\sqrt{L/g} \approx 6.2831853\sqrt{L/g}$.

Hence, the percent error is

$$E = \left[\frac{T_{SR}' - T_{SD}}{T_{SR}}\right] \times 100 \qquad (g)$$

By Table a and Eq. (g), the percent errors for $\theta_0 = \pi/2$, $\pi/4$, $\pi/10$, and $\pi/20$ are listed in Table b.

Table b

θ_0	E
$\frac{\pi}{2}$	17.655
$\frac{\pi}{4}$	4.288
$\frac{\pi}{10}$	0.515
$\frac{\pi}{20}$	0.026

Alternatively, by the software program DERIVE, using the approximate solution option, we obtain the periods and the percent errors listed in Table c.

Table c

θ_0	$T_D/\sqrt{L/g}$	E
$\frac{\pi}{2}$	7.38193	14.884
$\frac{\pi}{4}$	6.48885	3.169
$\frac{\pi}{10}$	6.30889	0.407
$\frac{\pi}{20}$	6.24861	0.553

The values of $f_0, \cdots f_{20}$ for the Simpson rule are listed in Table d.

f_i	$\frac{\pi}{2}$	$\frac{\pi}{4}$	$\frac{\pi}{10}$	$\frac{\pi}{20}$
f_0	44.191	74.322	177.76	353.33
f_1	1.7878	3.0628	7.3573	14.632
f_2	1.2720	2.2150	5.3412	10.628
f_3	1.0494	1.8535	4.4847	8.9281
f_4	0.9223	1.6489	4.0012	7.9689
f_5	0.8409	1.5187	3.6945	7.3605
f_6	0.7862	1.4316	3.4897	6.9544
f_7	0.7491	1.3729	3.3518	6.6811
f_8	0.7251	1.3349	3.2627	6.5044
f_9	0.7115	1.3135	3.2125	6.4049
f_{10}	0.7071	1.3066	3.1962	6.3727
f_{11}	0.7115	1.3135	3.2125	6.4049
f_{12}	0.7251	1.3349	3.2627	6.5044
f_{13}	0.7491	1.3729	3.3518	6.6811
f_{14}	0.7862	1.4316	3.4897	6.9544
f_{15}	0.8409	1.5187	3.6945	7.3605
f_{16}	0.9223	1.6489	4.0012	7.9689
f_{17}	1.0494	1.8535	4.4847	8.9281
f_{18}	1.2720	2.2150	5.3412	10.628
f_{19}	1.7878	3.0628	7.3573	14.632
f_{20}	44.191	74.322	177.76	353.33

18.77

Given: A NASCAR driver drives his car around a horizontal unbanked curve, so that it is just on the verge of tipping. The radius of the circle described by the mass center of the car is 100 ft. The width of the car between wheels is 5 ft. The center of mass is 2 ft above the pavement and midway between the wheels.

Determine the speed of the car.

(Continued)

18.77 Cont.

Solution: A front view free-body diagram of the car is shown in Fig. a.

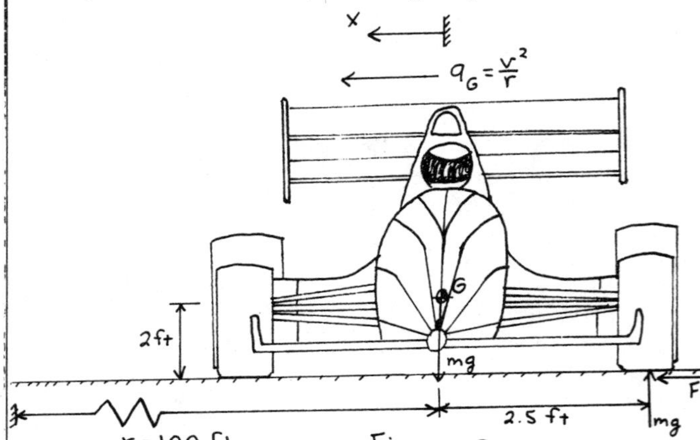

Figure a

By Fig. a,

$$\Sigma F_x = F = m a_G = m \frac{v^2}{r} \qquad (a)$$

$$\circlearrowleft^{+} \Sigma M_G = mg(2.5) - F(2) = 0 \qquad (b)$$

Equations (a) and (b) yield

$$2.5\, mg = 2 \frac{m v^2}{100}$$

or

$$\underline{v = 63.44 \text{ ft/s}}$$

18.78

Given: A homogeneous circular cylinder is suspended like a yo-yo from a cord that is wrapped around it. It is allowed to fall so that its axis remains horizontal and the cord unwinds evenly.

Determine the acceleration of the center of mass of the cylinder.

Solution: Figure a is a free-body diagram of the cylinder.

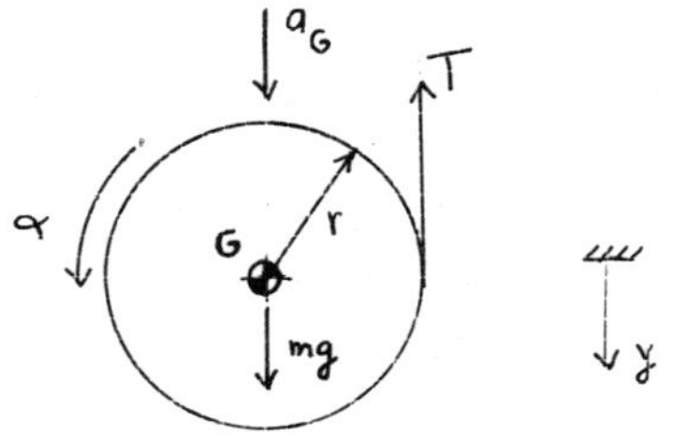

Figure a

By Fig. a,

$$\Sigma F_y = mg - T = m a_G \qquad (a)$$

$$\circlearrowleft^{+} \Sigma M_G = Tr = I_G \alpha = \tfrac{1}{2} m r^2 \alpha \qquad (b)$$

By kinematics,

$$a_G = r\alpha \qquad (c)$$

By Eqs. (b) and (c),

$$T = \tfrac{1}{2} m a_G \qquad (d)$$

and by Eqs. (a) and (d),

$$\underline{\underline{a_G = (2/3)\, g}}$$

Note: Since symbols were used in this solution, the results are valid for any consistent set of units.

18.79

Given: A homogeneous circular disk 1.22 m in diameter rotates and skids with backspin on a horizontal surface. The initial angular velocity of the disk is 30 rad/s. After 3 s, the angular velocity is 6 rad/s. Determine the coefficient of kinetic friction between the disk and surface.

Solution: The free-body diagram of the disk is shown in Fig. a.

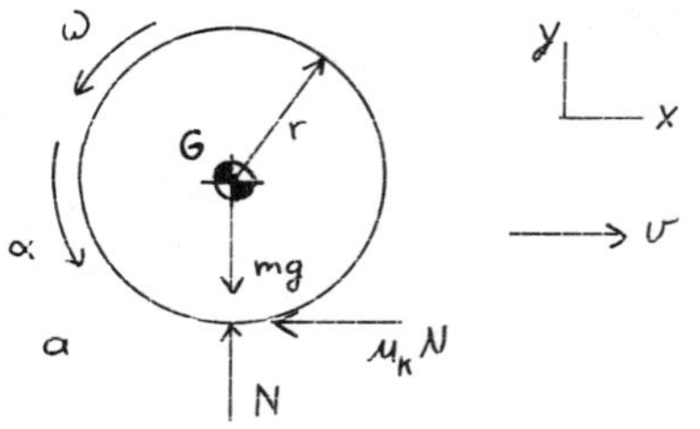

Figure a

By Fig. a,

$$\Sigma F_y = N - mg = 0 ; \quad N = mg \qquad (a)$$

$$\circlearrowleft^{+} \Sigma M_G = -\mu_K N r = I_G \alpha = \tfrac{1}{2} m r^2 \alpha \qquad (b)$$

By Eqs. (a) and (b),

$$\alpha = -\frac{2\mu_K g}{r} \qquad (c)$$

By kinematics and Eq. (c),

$$\alpha = \frac{d\omega}{dt} = -\frac{2\mu_K g}{r}$$

Integration yields

$$\omega = -\frac{2\mu_K g}{r} t + \omega_0 \qquad (d)$$

where $\omega = \omega_0 = 30$ rad/s when $t = 0$. For $t = 3$ s, Eq. (d) yields, since $\omega = 6$ rad/s,

$$6 = -\frac{6\mu_K (9.81)}{0.61} + 30$$

or

$$\underline{\underline{\mu_K = 0.249}}$$

18.80

Given: A bowling ball is pushed by a horizontal bar that slides between guides (Fig. a). The horizontal force that the bar exerts on the ball is P. The coefficient of sliding friction between the end of the bar and the ball is $\mu_K = 0.20$. The ball rolls on the horizontal surface.

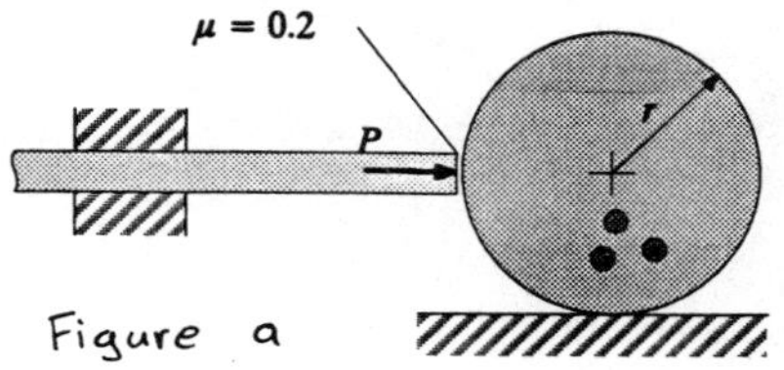

Figure a

Calculate the numerical value of P/ma, where m is the mass of the ball and a is the acceleration of the bar.

Solution: The free-body diagram of the ball is shown in Fig. b.

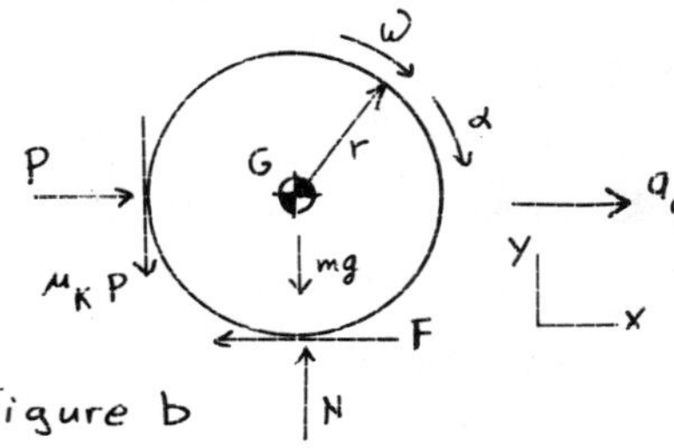

Figure b

By Fig. b,

$$\Sigma F_x = P - F = m a_G \qquad (a)$$

$$\Sigma F_y = N - mg - \mu_K P = 0 \qquad (b)$$

$$\circlearrowleft\!+ \Sigma M_G = Fr - \mu_K Pr = I\alpha = \tfrac{2}{5} mr^2\alpha \qquad (c)$$

By kinematics,

$$a_G = r\alpha \qquad (d)$$

Eqs. (c) and (d) yield, with $\mu_K = 0.20$,

$$F - 0.20P = \tfrac{2}{5} m a_G \qquad (e)$$

dding Eqs. (a) and (e), we obtain

$$0.80P = \tfrac{7}{5} m a_G$$

Hence, for any consistent set of units,

$$\frac{P}{m a_G} = \frac{7}{4} = 1.75$$

Note: Equation (b) was not needed to obtain this result.

18.81

Given: A motionless solid ball hanging by a cord is suddenly subjected to a horizontal force F at its lowest point.

a) Derive formulas for the initial angular acceleration of the ball and the initial acceleration of its center.

b) Determine the distance from the center of the ball to the point P on the vertical axis that has no acceleration.

Solution: a) Figure a is the free-body diagram of the ball.

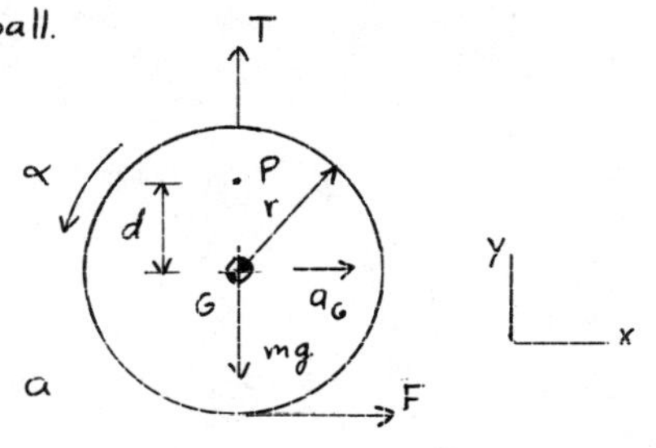

Figure a

By Fig. a,

$$\Sigma F_x = F = m a_G \qquad (a)$$

$$\circlearrowleft\!+ \Sigma M_G = Fr = I_G\alpha = \tfrac{2}{5} mr^2\alpha$$

or

$$F = \tfrac{2}{5} mr\alpha \qquad (b)$$

$$\alpha = \frac{5}{2}\frac{F}{mr} \qquad (c)$$

and by Eq. (a),

$$a_G = \frac{F}{m} \qquad (d)$$

b) By kinematics, the acceleration of a point P (Fig. a) is,

$$a_P = a_G - \alpha d \qquad (e)$$

For $a_P = 0$, Eq. (e) yields, with Eqs. (c) and (d)

$$d = \frac{a_G}{\alpha} = \tfrac{2}{5} r$$

18.82

Given: The system shown in Fig. a is released from rest. The homogeneous disk A rolls on the horizontal surface. The pulley B is frictionless and has negligible mass.

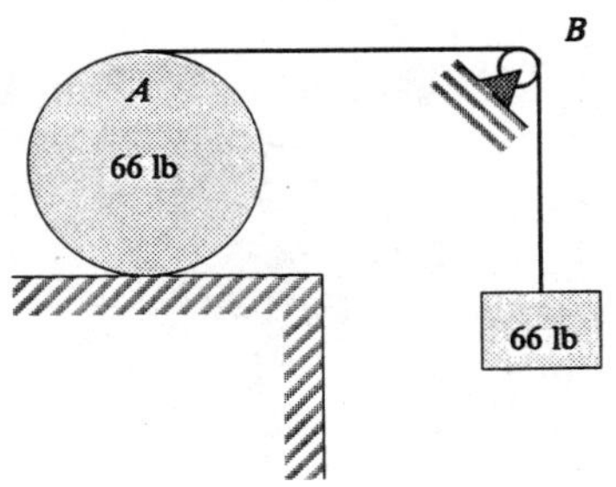

Figure a

(Continued)

18.82 Cont.

Determine the acceleration of the center of the disk and the tension in the cord.

Solution: Free-body diagrams of the 66 lb disk and the 66 lb weight are shown in Figs. b and c.

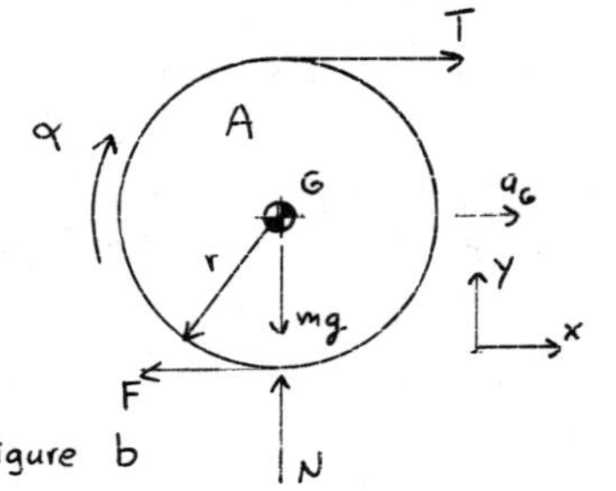

Figure b

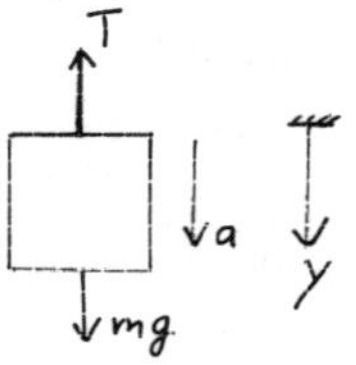

Figure c

By Fig. b,

$$\Sigma F_x = T - F = m a_G \qquad (a)$$

$$\circlearrowleft \Sigma M_G = Tr + Fr = I_G \alpha = \tfrac{1}{2} m r^2 \alpha \qquad (b)$$

By Kinematics,

$$a_G = r\alpha \qquad (c)$$

$$a = 2 a_G \qquad (d)$$

By Fig. c,

$$\Sigma F_y = mg - T = ma \qquad (e)$$

Equations (b) and (c) yield,

$$T + F = \tfrac{1}{2} m a_G \qquad (f)$$

Then Eqs. (a) and (f) yield

$$T = \tfrac{3}{4} m a_G \qquad (g)$$

and Eqs. (d), (e), and (g) yield

$$a_G = \tfrac{4}{11} g = \tfrac{4}{11}(32.2) = 11.71 \text{ ft/s}^2 \qquad (h)$$

Finally, by Eqs. (g) and (h)

$$T = \tfrac{3}{11} mg = \tfrac{3}{11}(66) = 18 \text{ lb}$$

18.83

Given: A homogeneous circular disk rotates in a horizontal plane about a fixed vertical axle O (Fig. a). The axle is 2 ft from the mass center G of the disk. The mass of the disk is 10 lb·s²/ft. A clockwise couple, $M = 10t$ is applied to the disk in its plane; M is in pound-feet and t is in seconds. Initially, the disk is at rest.

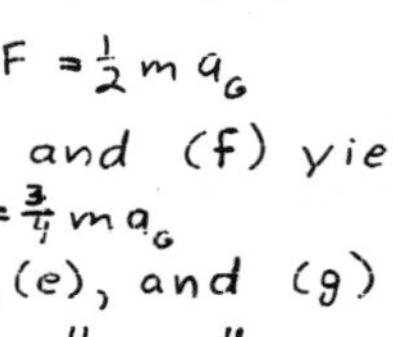

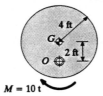

Figure a

For $t = 4$ s, determine the perpendicular and parallel components, to the line OG, of the force exerted on the disk by the axle.

Solution: The free-body diagram of the disk is shown in Fig. b, where O_x, O_y are components parallel and perpendicular, respectively, to OG.

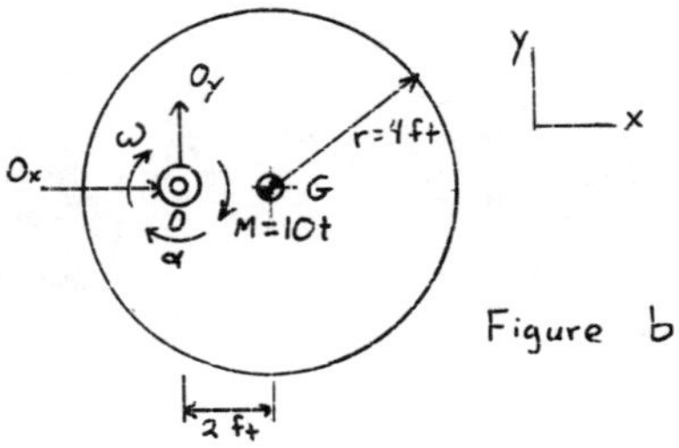

Figure b

By Fig. b, the mass moments of inertia about the mass center G of the disk is

$$I_G = \tfrac{1}{2} m r^2 = \tfrac{1}{2}(10)(4)^2 = 80 \text{ lb·ft·s}^2 \qquad (a)$$

By the parallel axis theorem and Eq. (a), the mass moment of inertia about the axle O is,

$$I_O = I_G + md^2 = 80 + (10)(2)^2 = 120 \text{ lb·ft·s}^2 \qquad (b)$$

Then, by Fig. b,

$$\circlearrowleft \Sigma M_O = M = I_O \alpha$$

or

$$10t = 120\alpha = 120 \frac{d\omega}{dt} \qquad (c)$$

Integration of Eq. (c) yields

$$12\omega = t^2/2 \qquad (d)$$

Hence, for $t = 4$ s,

$$\omega = \tfrac{2}{3} \text{ rad/s} \qquad (e)$$

The acceleration of the mass center is, with Eq. (c) and Fig. b,

$$a_{Gx} = -2\omega^2 = -\tfrac{8}{9} \text{ ft/s}^2$$

$$a_{Gy} = -2\alpha = -2\left(\tfrac{1}{3}\right) = -\tfrac{2}{3} \text{ ft/s}^2 \qquad (f)$$

since by Eq. (c), $\alpha = \frac{1}{3}$ rad/s² for $t = 4$s.

Now by Fig. b,

$$\Sigma F_x = O_x = m a_{Gx}$$

$$\Sigma F_y = O_y = m a_{Gy} \qquad (g)$$

and by Eqs. (f) and (g),

$$O_x = (10)\left(-\tfrac{8}{9}\right) = -\tfrac{80}{9} = -8.889 \text{ lb}$$

$$O_y = (10)\left(-\tfrac{2}{3}\right) = \tfrac{-20}{3} = -6.667 \text{ lb}$$

18.84

Given: A ball B and a cylinder C, with equal masses m and equal radii, rest on a horizontal plank (Fig. a). The plank is given an acceleration a, directed to the right as shown. Assume that the bodies do not slip on the plank.

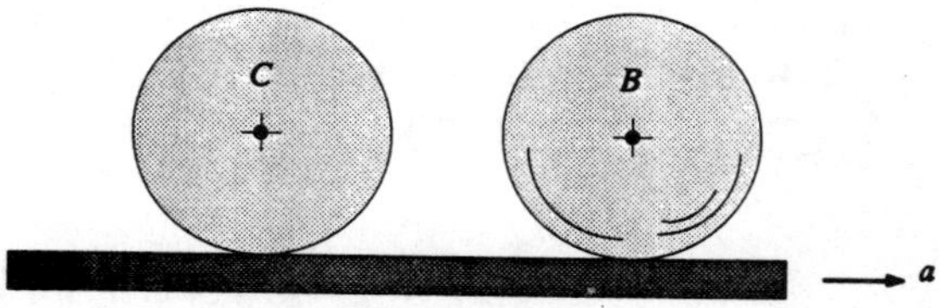

Figure a

Determine whether or not they approach each other.

Solution: The free-body diagram of either body is shown in Fig. b.

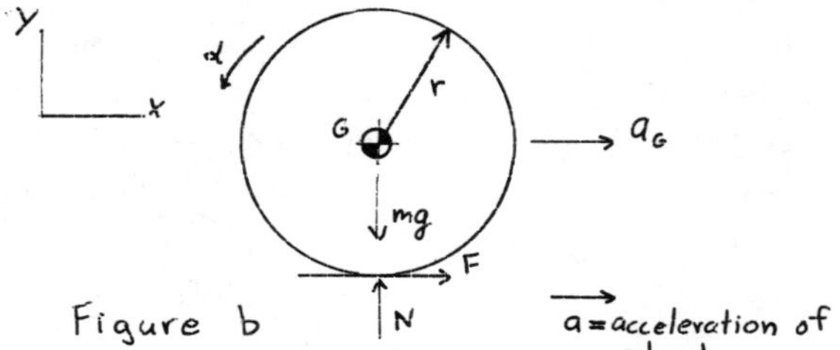

Figure b

By Fig. b, for either body,

$$\Sigma F_x = F = m a_G \qquad (a)$$

$$\Sigma F_y = N - mg = 0\ ;\ N = mg \qquad (b)$$

$$\circlearrowleft \Sigma M_G = Fr = I_G \alpha \qquad (c)$$

By kinematics,

$$a_G = a - r\alpha \qquad (d)$$

Therefore, by Eqs. (a), (c), and (d)

$$F = m(a - r\alpha) = I_G \alpha / r \qquad (e)$$

Equation (e) yields

$$\alpha = \frac{mar}{I_G + mr^2} \qquad (f)$$

For ball B, $I_G = \frac{2}{5}mr^2$. Therefore, for ball B,

$$\alpha_B = \frac{5}{7}\frac{a}{r}, \text{ counterclockwise}$$

Similarly for cylinder C, $I_G = \frac{1}{2}mr^2$ and

$$\alpha_C = \frac{2}{3}\frac{a}{r}; \text{ counterclockwise}$$

Therefore, $\alpha_B > \alpha_C$, and by Eq. (d), the acceleration a_{GC} of cylinder C is greater than the acceleration a_{GB} of ball B. Hence, they approach each other.

18.85

Given: A homogeneous cylinder rolls in the clockwise sense on a horizontal surface (Fig. a). The mass of the cylinder is m. It is subjected to a clockwise couple $M = a\theta^2 + b\theta$ that acts in a cross-sectional plane of the cylinder, where a and b are constants and θ is the angular displacement in radians.

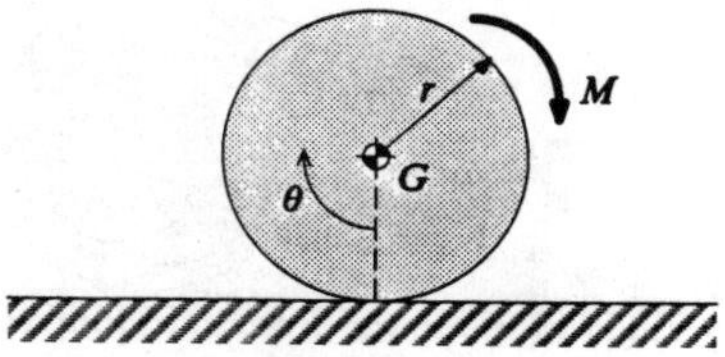

Figure a

Express the speed of the center of the cylinder as a function of a, b, r, θ, m, and ω_0, where ω_0 is the angular velocity for $\theta = 0$.

Solution: The free-body diagram of the cylinder is shown in Fig. b.

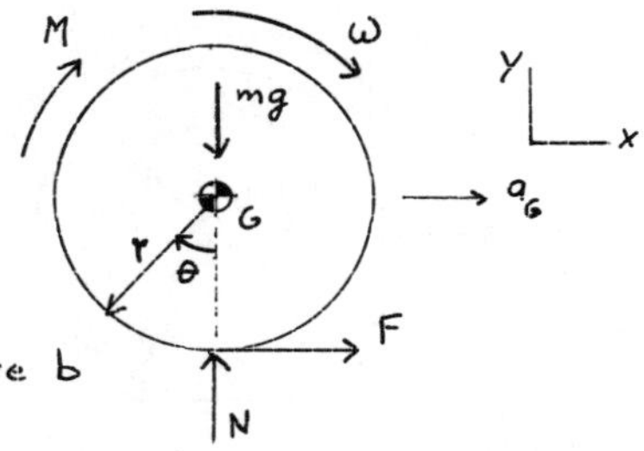

Figure b

By Fig. b,

$$\Sigma F_x = F = m a_G \qquad (a)$$

where, by kinematics,

$$a_G = r\alpha = r\ddot{\theta} \qquad (b)$$

$$\circlearrowright \Sigma M_G = M - Fr = I_G \alpha = \tfrac{1}{2} m r^2 \ddot{\theta} \qquad (c)$$

with $M = a\theta^2 + b\theta$, Eqs. (a), (b), and (c) yield

$$M = a\theta^2 + b\theta = \tfrac{3}{2} m r^2 \ddot{\theta} \qquad (d)$$

By kinematics, $\ddot{\theta} = \frac{d\omega}{dt} = \frac{d\omega}{d\theta}\frac{d\theta}{dt} = \omega \frac{d\omega}{d\theta}$. Then by Eq. (d),

$$\tfrac{3}{2} m r^2 \omega\, d\omega = (a\theta^2 + b\theta)\, d\theta \qquad (e)$$

Integration of Eq. (e) yields

$$\tfrac{3}{4} m r^2 (\omega^2 - \omega_0^2) = \tfrac{1}{3} a\theta^3 + \tfrac{1}{2} b\theta^2$$

where $\omega = \omega_0$ when $\theta = 0$. Hence,

$$\omega^2 = \omega_0^2 + \frac{4}{3mr^2}\left(\frac{a}{3}\theta^3 + \frac{b}{2}\theta^2\right) \qquad (f)$$

By kinematics and Eq. (f), we have

$$v = r\omega = r\sqrt{\omega_0^2 + \frac{4}{3mr^2}\left(\frac{a}{3}\theta^3 + \frac{b}{2}\theta^2\right)}$$

18.86

Given: A solid homogeneous spherical body wheighs 64 lb. It rests on a horizontal table and is pushed by a moving block that slides on the table (Fig. a). The normal force between the sphere and the block at the contact point is 20 lb. The coefficient of friction between the sliding surfaces is 0.25.

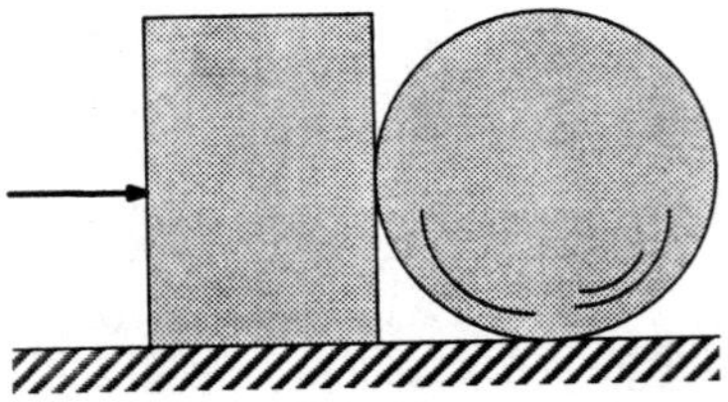

Figure a

a) Assuming that the sphere rolls on the table, determine the acceleration of its center.

b) Determine the frictional force that the table exerts on the sphere, and verify that the sphere does not slide on the table.

Solution: a) The free-body diagram of the sphere is shown in Fig. b.

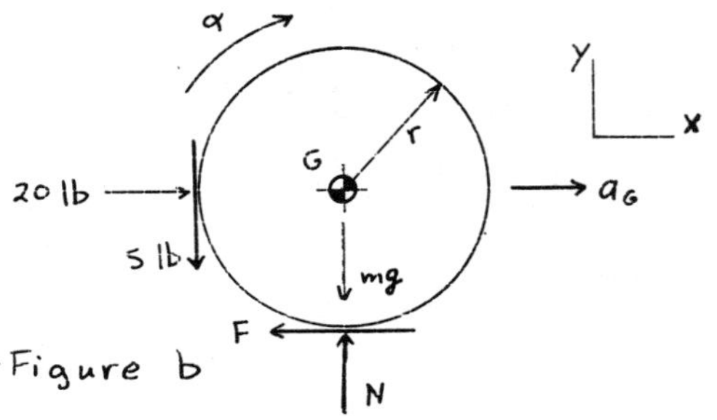

Figure b

By Fig. b,

$$\Sigma F_x = 20 - F = m a_G \quad \text{(a)}$$

$$\Sigma F_y = N - mg - 5 = 0 \quad \text{(b)}$$

$$\circlearrowright \Sigma M_G = Fr - 5r = I_G \alpha = \tfrac{2}{5} m r^2 \alpha \quad \text{(c)}$$

By kinematics,

$$a_G = r\alpha \quad \text{(d)}$$

Therefore, by Eqs. (c) and (d),

$$F - 5 = \tfrac{2}{5} m a_G \quad \text{(e)}$$

Adding Eqs. (a) and (e), we obtain

$$15 = \tfrac{7}{5} m a_G$$

or

$$a_G = \frac{75}{7m} = \frac{75(32.2)}{(7)(64)} = 5.391 \text{ ft/s}^2 \quad \text{(f)}$$

b) By Eqs. (e) and (f), we find

$$F = 5 + \tfrac{2}{5} m \left(\frac{75}{7m}\right) = \frac{65}{7} = 9.286 \text{ lb}$$

By Eq. (b),

$$N = 5 + mg = 5 + 64 = 69 \text{ lb}$$

Hence, the frictional force at impending slipping is

$$F_{max} = \mu N = (0.25)(69) = 17.25 \text{ lb}$$

Since $F < F_{max}$, the sphere does not slide.

18.87

Given: A mechanical system consists of a drum A (mass of $m_A = 150$ kg and radius of gyration of 0.5 m) and a solid homogeneous cylinder (mass $m_B = 75$ kg) connected by a flexible inextensible cord that is wrapped around A, passed over a frictionless pulley and tied to a light rod pinned to B (Fig. a).

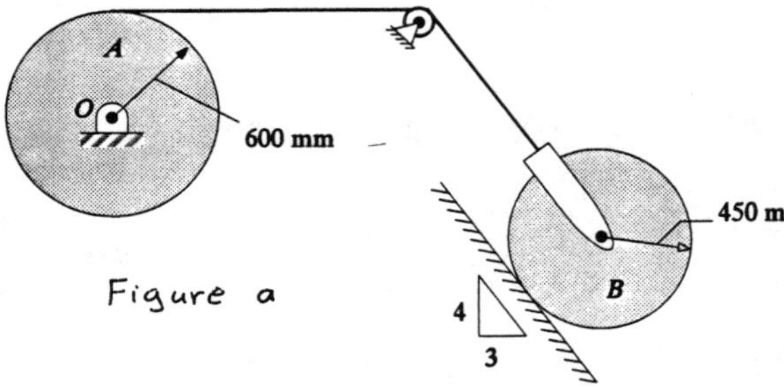

Figure a

Neglecting friction in the bearing O, determine the acceleration of the center of roller B.

Solution: The free-body diagrams of the drum and the cylinder are shown in Figs. b and c.

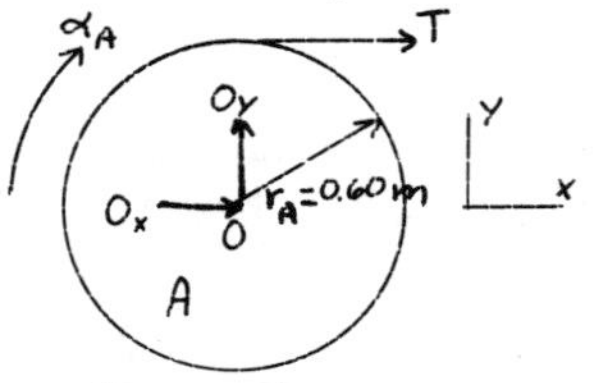

Figure b

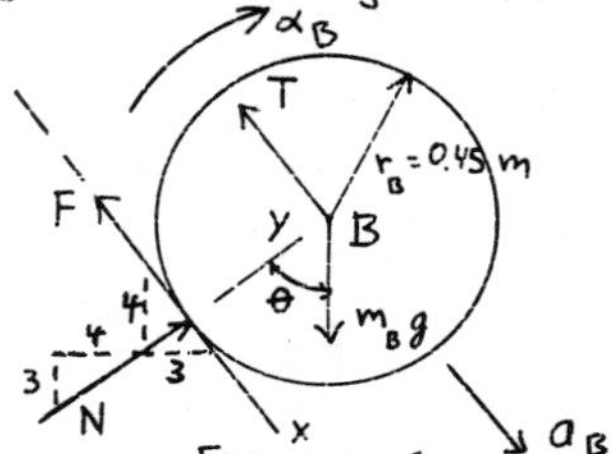

Figure c

By Fig. b,

$$\circlearrowleft \Sigma M_O = T(0.60) = I_O \alpha_A = (150)(0.5)^2 \alpha_A = 37.5\alpha_A$$

or

$$T = 62.5\,\alpha_A \quad \text{(a)}$$

By Fig. c,

$$\Sigma F_x = m_B g \sin\theta - T - F = m_B a_B \quad \text{(b)}$$

where $\sin\theta = 4/5 = 0.80$.

By kinematics,

$$(0.45)\alpha_B = (0.60)\alpha_A = a_B \quad \text{(c)}$$

also by Fig. b,

$$\circlearrowright \Sigma M_B = F r_B = I_B \alpha_B$$

where, $I_B = \frac{1}{2} m_B r_B^2 = \frac{1}{2}(75)(0.45)^2 = 7.594 \text{ N·m·s}^2$

Therefore,

$$F = 16.875\,\alpha_B \text{ [N]} \quad \text{(d)}$$

By Eqs. (a) and (c),

$$T = 104.17\,a_B \quad \text{(e)}$$

and by Eqs. (c) and (d),

$$F = 37.5\,a_B \quad \text{(f)}$$

Finally, by Eqs. (b), (e), and (f), we have

$$a_B = 60g/216.67 = (60)(9.81)/216.67$$

or

$$a_B = 2.72 \text{ m/s}^2$$

18.88

Given: The system shown in Fig. a is used to lower a drum C to a lower floor. Initially, F = 0, and the system is released from rest.

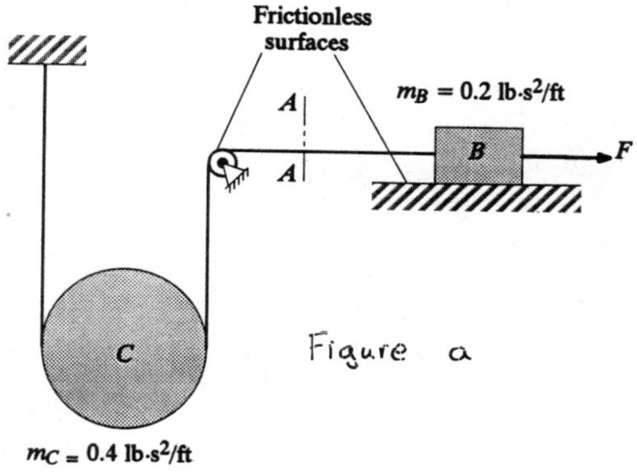

Figure a

a) Assuming that the flexible, inextensible cord does not slip on drum C, determine the tensions in the cord at section A-A.

b) After the drum has dropped for 1 s, the force F is applied, and the drum comes to rest when it has dropped a total distance 10 ft to the lower floor. Determine the force F.

Solution: a) The free-body diagrams of the drum C and the block B are shown in Figs. b and c.

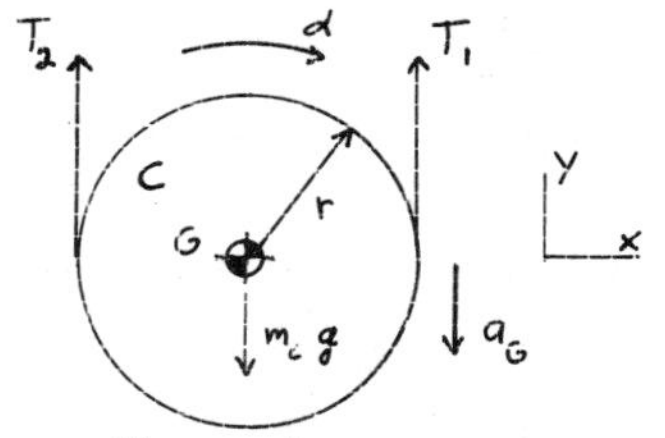

Figure b

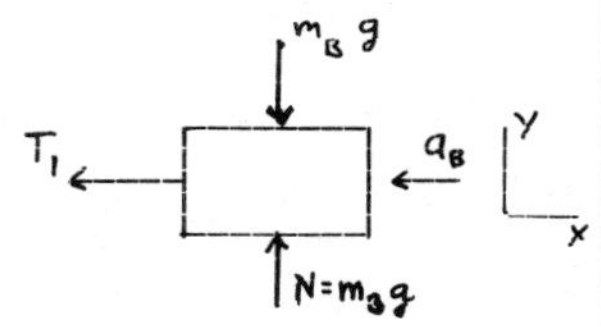

Figure c

By Fig. b,

$$\circlearrowright\Sigma M_G = T_2 r - T_1 r = I_G \alpha \qquad (a)$$

$$\Sigma F_y = T_1 + T_2 - m_c g = -m_c a_G \qquad (b)$$

By kinematics,

$$a_G = \alpha r \qquad (c)$$

By Fig. c,

$$\Sigma F_x = -T_1 = -m_B a_B \qquad (d)$$

By kinematics, if the block moves a distance d to the left, the cylinder descends a distance d/2. Hence,

$$a_G = 1/2\, a_B \qquad (e)$$

Equations (a), (b), (c), and (e) yield

$$2T_1 = m_c g - \tfrac{1}{2} m_c a_B - \tfrac{1}{2}\frac{I_G}{r^2} a_B \qquad (f)$$

Then, Eqs. (d) and (f) yield

$$[2m_B + \tfrac{1}{2}(m_c + \frac{I_G}{r^2})]a_B = m_c g \qquad (g)$$

Since $I_G = \frac{1}{2} m_c r^2$, Eq. (g) yields

$$a_B = \frac{m_c g}{2m_B + \frac{3}{4} m_c} = \frac{0.4 g}{0.4 + 0.3} = \frac{4}{7} g \qquad (h)$$

Equation (d) and (h) yield

$$\underline{\underline{T_1 = m_B a_B = (0.2)(\tfrac{4}{7})(32.2) = 3.68 \text{ lb}}}$$

b) First, we determine the distance that cylinder C descends in 1 s. Since a_G is constant, the distance s that the cylinder drops is, with Eqs. (e) and (h)

$$s = \tfrac{1}{2} a_G t^2 = \tfrac{1}{2}(\tfrac{1}{2} a_B)(1)^2 = \tfrac{1}{4}(\tfrac{4}{7} g)$$

or

$$s = \tfrac{1}{7} g = \tfrac{1}{7}(32.2) = 4.6 \text{ ft}.$$

Thus, at the end of one second, the cylinder is above the floor a distance of

$$s_1 = 10 - 4.6 = 5.4 \text{ ft}$$

and it is traveling at a speed of

$$v = a_G t = \tfrac{1}{2} a_B t = \tfrac{1}{2}(\tfrac{4}{7} g)(1)$$

or

$$v = 9.2 \text{ ft/s}$$

By kinematics,

$$a_G = \frac{dv}{dt} = \frac{dv}{ds}\frac{ds}{dt} = v\frac{dv}{ds}$$

Hence, since $v = 0$, when the cylinder touches the floor, integration yields since a_G=constant

$$a_G \int_0^{5.4} ds = \int_{9.2}^{0} v\,dv$$

$$(a_G)\left[s\Big|_0^{5.4}\right] = \tfrac{1}{2} v^2 \Big|_{9.2}^{0}$$

or

$$a_G = -7.837 \text{ ft/s}^2 \qquad (i)$$

Thus, $a_G = 7.837$ ft/s² up (Fig. b). Then, by Eq. (e), $a_B = 2a_G = 15.674$ ft/s² (j) to the right (Fig. d).

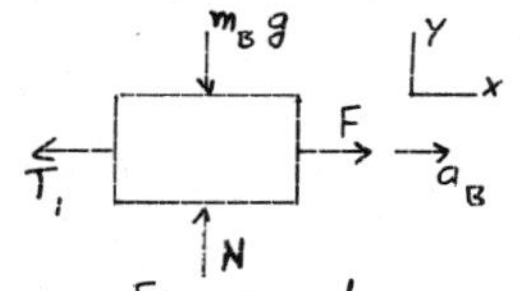

Figure d

By Fig. d and Eq. (j),

$$\Sigma F_x: F - T_1 = m a_B = 3.135 \text{ lb} \quad (k)$$

By Eqs. (f) and (j),

$$T_1 = \tfrac{1}{2}[m_c g - \tfrac{3}{4} m_c a_B]$$

$$T_1 = \tfrac{1}{2}[(0.4)(32.2) - \tfrac{3}{4}(0.4)(15.674)]$$

or

$$T_1 = 4.089 \text{ lb} \qquad (\ell)$$

Equations (k) and (ℓ) yield

$$\underline{F = 4.089 + 3.135 = 7.22 \text{ lb}}$$

18.89

Given: A cord is wrapped around a spool (Fig. a), whose radius of gyration is k. The coefficient of sliding friction between the spool and inclined plane is μ_k. The spool is released from rest.

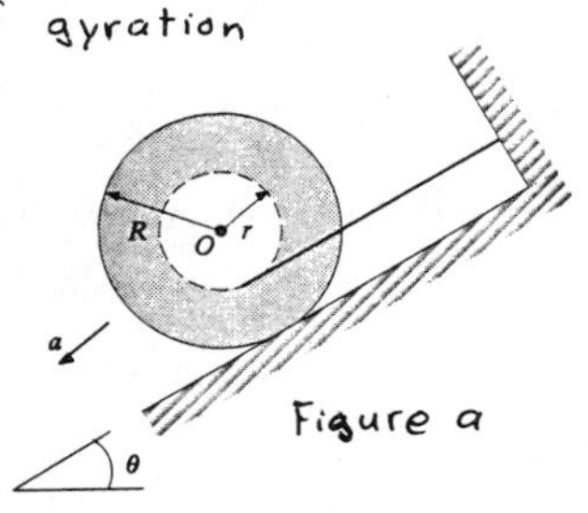

Figure a

(Continued)

18.89 Cont.

a) Show that if the angle θ is large enough to permit motion, the acceleration a of the center of the spool is

$$a = \frac{r^2 g[\sin\theta + \mu_K(1 - R/r)(\cos\theta)]}{r^2 + k^2} \quad (a)$$

b) From Eq. (a), derive an equation for the smallest angle θ at which motion occurs.

Solution: Since $r < R$, the spool must skid and rotate if it moves. Then, as the spool unwinds, the acceleration a of the center of the spool (Fig. a) is, in terms of the angular acceleration α,

$$a = r\alpha \quad (b)$$

The free-body diagram of the spool is shown in Fig. b.

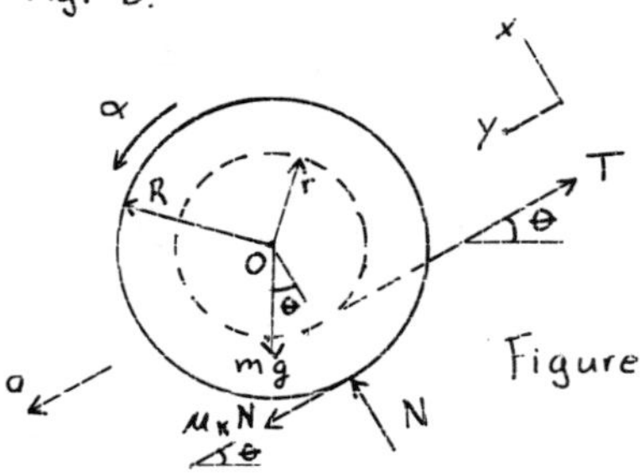

Figure b

By Fig. b,

$$\Sigma F_x = N - mg\cos\theta = 0 \quad (c)$$

$$\Sigma F_y = mg\sin\theta + \mu_K N - T = ma \quad (d)$$

$$\circlearrowleft \Sigma M_o = Tr - \mu_K NR = I_o\alpha = mk^2\alpha \quad (e)$$

Eqs. (b), (c), and (d) yield

$$mg\sin\theta + \mu_K mg\cos\theta - T = mr\alpha \quad (f)$$

Multiplying Eq. (f) by r and adding the resulting equation to Eq. (e), we obtain (since $N = mg\cos\theta$ and $a = r\alpha$)

$$a = \frac{r^2 g[\sin\theta + \mu_K(1 - R/r)\cos\theta]}{r^2 + k^2} \quad (g)$$

Equation (a) is identical to Eq. (g).

b) Equation (g) is valid only if $a \geq 0$, since the spool will not move up the plane. Hence, the smallest value of θ for impending motion (called it θ_o) is given by Eq. (g) with $a = 0$. Hence,

$$\sin\theta_o + \mu_K(1 - R/r)\cos\theta_o = 0$$

or

$$\tan\theta_o = \mu_K(R/r - 1)$$

Thus, motion will occur if

$$\theta > \theta_o = \tan^{-1}[\mu_K(\tfrac{R}{r} - 1)]$$

18.90

Given: The bowling ball shown in Fig. a has backspin as it moves along the horizontal lane. The coefficient of kinetic friction between the ball and the floor is 0.25. The ball's initial speed is 1.0 m/s. After it travels a certain distance to the right, it stops and remains motionless.

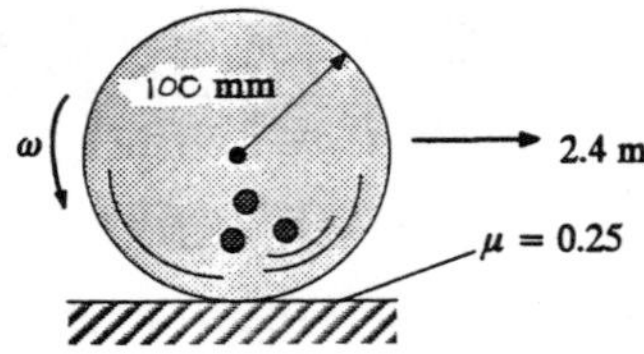

Figure a

a) Determine the distance the ball traveled.

b) Determine the time [s] required for the ball to stop.

c) Determine the initial value of the ball's angular velocity.

Solution:

a) Figure b is the free-body diagram of the ball

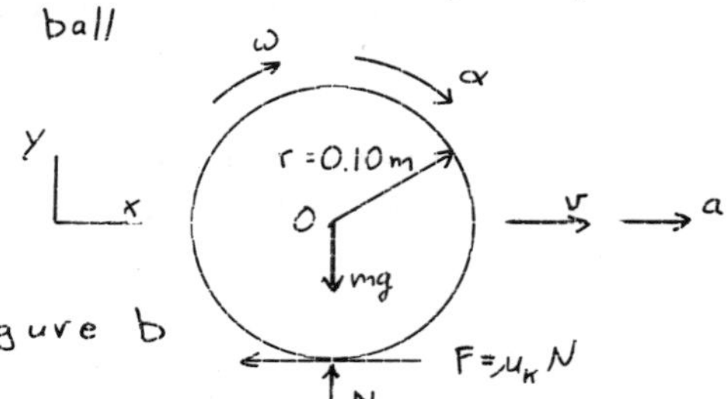

Figure b

By Fig. b,

$$\Sigma F_x = -\mu_K N = ma \quad (a)$$

$$\Sigma F_y = N - mg = 0\ ; \quad N = mg \quad (b)$$

By Eqs. (a) and (b), with $\mu_K = 0.25$, we find

$$a = -\tfrac{1}{4}g \quad (c)$$

By kinematics and the chain rule,

$$a = \frac{dv}{dt} = \frac{dv}{dt}\frac{dx}{dx} = \frac{dx}{dt}\frac{dv}{dx} = v\frac{dv}{dx} \quad (d)$$

and by Eqs. (c) and (d),

$$v\,dv = -\tfrac{1}{4}g\,dx$$

Integration yields

$$\frac{v^2}{2} - \frac{v_o^2}{2} = -\frac{1}{4}gx \quad (e)$$

The ball comes to rest when $v = 0$; hence, by Eq. (e),

$$x_{max} = \frac{2v_o^2}{g} = \frac{2(1.0)^2}{9.81} = 0.2039 \text{ m}$$

(Continued)

18.90 Cont.

b) Since a is constant [see Eq. (c)], by kinematics,

$$v = v_0 + at = v_0 - \frac{1}{4} g t$$

Hence, when the ball comes to rest ($v = 0$),

$$t = \frac{4v_0}{g} = \frac{4(1.0)}{9.81} = 0.4077 \text{ s}$$

c) By Fig. b,

$$\circlearrowleft + \Sigma M_0 = (\mu_k N)(r) = I_0 \alpha = \frac{2}{5} m r^2 \alpha$$

or

$$\mu_k N = \frac{2}{5} m r \alpha \qquad (f)$$

By Eqs. (a), (b), (c), and (f),

$$\mu_k N = \frac{1}{4} mg = \frac{2}{5} m r \alpha$$

Hence, with $r = 0.10$ m, we find

$$\alpha = \frac{5}{8}\frac{g}{r} = \frac{5}{8}\left(\frac{9.81}{0.10}\right) = 61.31 \text{ rad/s}^2$$

(Note that if the ball was rolling without skidding, α would be $\alpha = a/r = -\frac{1}{4} g/r = -24.53$ rad/s²)

Hence, by kinematics,

$$\omega = \omega_0 + \alpha t \qquad (g)$$

When $t = 0.4077$ s, $\omega = 0$. Then, Eq. (g) yields (see Fig. b)

$$\omega_0 = -(61.31)(0.4077) = -25 \text{ rad/s}$$

That is, the initial angular velocity ω_0 is opposite in sense to the angular acceleration α.

18.91

Given: The connecting rod AB of the mechanism shown in Fig. a weighs 16.1 lb. sliders A and B are frictionless and of negligible mass. Initially, the mechanism is at rest as shown, when a 10 lb horizontal force acts on A.

For the instant after the 10 lb force is applied, calculate the acceleration of A and B, the force B exerts on the vertical guides, and the force A exerts on the horizontal surface.

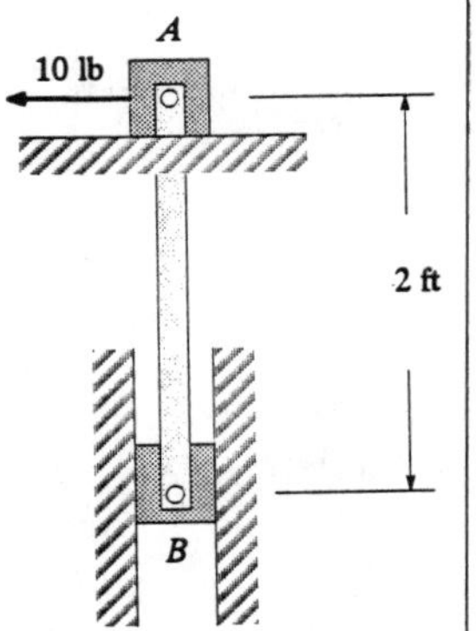

Figure a

Solution: The free-body diagram of rod AB is shown in Fig. b. The accelerations of A, B and G are shown in Fig. c. By Fig. b,

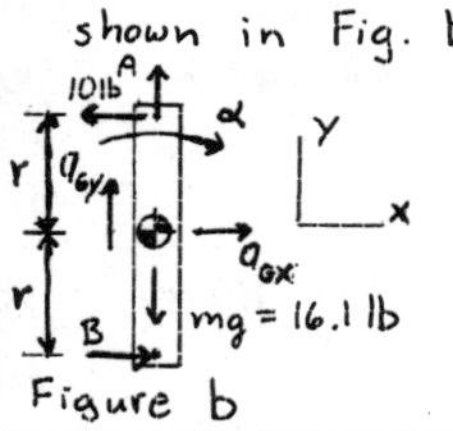

Figure b

$$\Sigma F_x = B - 10 = m a_{Gx} \qquad (a)$$

$$\Sigma F_y = A - mg = m a_{Gy} \qquad (b)$$

$$\circlearrowleft \Sigma M_G = -Br - 10r = I_G \alpha = \frac{1}{12} m (2r)^2 \alpha \qquad (c)$$

where $m = \frac{16.1}{32.2} = 0.5$ lb·s²/ft, $r = 1$ ft.

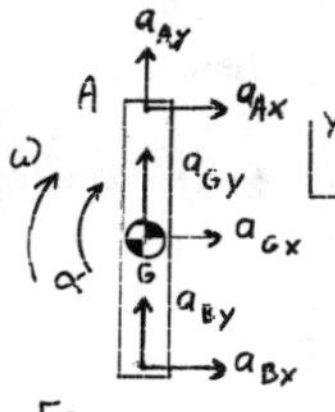

Figure c

By kinematics and Fig. c, which shows accelerations and the angular velocity,

$$a_{Ax} = a_{Gx} + r\alpha \qquad (d)$$

$$a_{Ay} = a_{Gy} - r\omega^2 \qquad (e)$$

$$a_{Bx} = a_{Gx} - r\alpha \qquad (f)$$

$$a_{By} = a_{Gy} + r\omega^2 \qquad (g)$$

Initially, $\omega = 0$. Also because of the constraints $a_{Ay} = 0$ and $a_{Bx} = 0$. Hence, Eqs. (e) and (f) yield

$$a_{Gy} = 0 \qquad (h)$$

$$a_{Gx} = r\alpha \qquad (i)$$

Therefore, Eqs. (b) and (h) yield

$$A = mg = 16.1 \text{ lb}$$

and therefore slider A exerts a downward force of 16.1 lb on the horizontal surface. Equations (a), (c), and (i) yield

$$\left.\begin{aligned} B - 10 &= \tfrac{1}{2}\alpha \\ B + 10 &= -\tfrac{1}{6}\alpha \end{aligned}\right\} \qquad (j)$$

The solution of Eqs. (j) is (see Figs. b and c)

$$B = -5 \text{ lb}$$

$$\alpha = -30 \text{ rad/s}^2 \qquad (k)$$

Therefore, slider B exerts a force of 5 lb on the vertical guide, directed horizontally to the right. So by Eqs. (d), (g), (h), (i), and (k),

$$a_{Ax} = 2r\alpha = -60 \text{ ft/s}^2$$

$$a_{By} = 0$$

Hence, the accelerations of A and B are, respectively (see Fig. c),

$$a_{Ax} = -60 \text{ ft/s}^2, \quad a_{Ay} = 0$$

and

$$a_{Bx} = a_{By} = 0$$

18.92

Given: A point mass M is attached to one end of a uniform bar of mass m and length L. The bar hangs by one end from a light cord, with the mass M at the top (Fig. a)

a) Determine the point at which the bar can be struck laterally without causing an immediate acceleration of M.

b) Repeat part a for the case in which M is at the lower end of the bar.

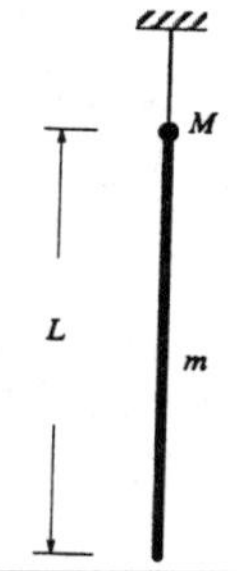

Figure a

(Continued)

18.92 Cont.

Solution: a) The free-body diagram of the bar and mass M is shown in Fig. b in which G denotes the mass center of the system.

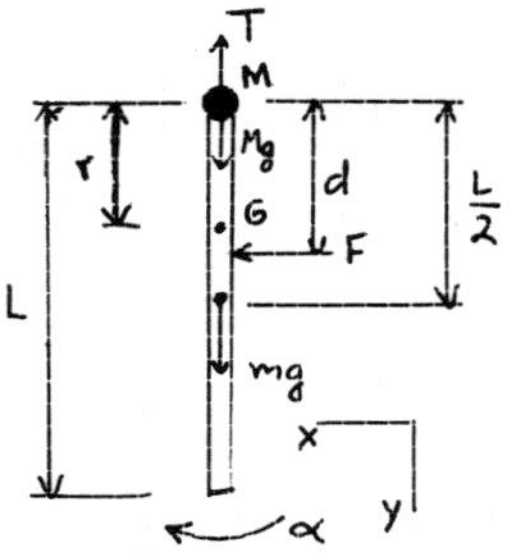

Figure b

The location of mass center is given by

$$(m+M)r = m\frac{L}{2}$$

or

$$r = \frac{1}{2}\left[\frac{mL}{M+m}\right] \qquad (a)$$

By Fig. b,

$$\Sigma F_x = F = (M+m)a_{Gx} \qquad (b)$$

$$\circlearrowleft \Sigma M_G = F(d-r) = I_G\alpha \qquad (c)$$

Let a_M be the acceleration of mass M. Then, by kinematics, since we require $a_M = 0$,

$$a_M = a_{Gx} - r\alpha = 0$$

or

$$a_{Gx} = r\alpha \qquad (d)$$

Equations (b), (c), and (d) yield

$$(M+m)r\alpha = \frac{I_G}{d-r}\alpha \qquad (e)$$

Hence,

$$d = r + \frac{I_G}{(M+m)r} \qquad (f)$$

where, with the parallel-axis theorem, the mass moment of inertia of the system is (Fig. b)

$$I_G = \frac{1}{12}mL^2 + m\left(\frac{L}{2}-r\right)^2 + Mr^2$$

With Eq. (a), this equation yields

$$I_G = \frac{1}{12}mL^2\left(\frac{4M+m}{M+m}\right) \qquad (g)$$

Equations (f) and (g) yield

$$\underline{\underline{d = \frac{2}{3}L}}$$

The above solution was relative to the mass center of the system [Eqs. (b) and (c)].

Alternatively, since we require that the acceleration of mass M be zero, we may solve the problem relative to the Newtonian frame attached to M. Then, by Fig. b,

$$\Sigma F_x = F = ma \qquad (h)$$

$$\circlearrowleft \Sigma M_M = Fd = I_M\alpha \qquad (i)$$

where,

$$I_M = \frac{1}{3}mL^2 + M(0)^2 = \frac{1}{3}mL^2 \qquad (j)$$

and

$$a = \frac{L}{2}\alpha \qquad (k)$$

is the acceleration of the mass center of the rod. Then Eqs. (h), (i), (j), and (k) yield

$$(ma)(d) = \left(\frac{1}{3}mL^2\right)\left(\frac{2a}{L}\right)$$

or

$$\underline{\underline{d = \frac{2}{3}L}}$$

b) For the mass M at the lower end of the bar (Fig. c), let us locate F at a distance d from M. As in part a

$$r = \frac{1}{2}\left(\frac{m}{M+m}\right)L \qquad (a)$$

and

$$I_G = \frac{1}{12}mL^2\left(\frac{4M+m}{M+m}\right) \qquad (g)$$

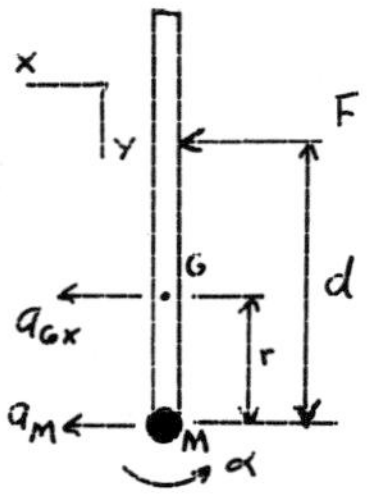

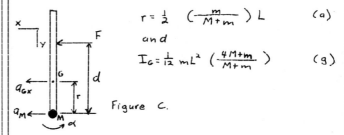

Figure c.

Then, by Fig. c,

$$\Sigma F_x = F = (M+m)a_{Gx} \qquad (l)$$

$$\circlearrowleft \Sigma M_G = F(d-r) = I_G\alpha \qquad (m)$$

Since a_M is required to be zero, by kinematics,

$$a_M = a_{Gx} - r\alpha = 0$$

$$a_{Gx} = r\alpha \qquad (n)$$

Then, Eqs. (a), (g), (l), (m), and (n) yield

$$\underline{\underline{d = r + \frac{I_G\alpha}{(M+m)r\alpha} = \frac{2}{3}L}}$$

Therefore, in both parts (a) and (b), the force must be applied at the distance $d = \frac{2}{3}L$ from mass M.

18.93

Given: In Fig. a, let the weight of the sphere be W and let the horizontal force that the block exerts on the ball be P. Let the coefficient of sliding friction between contacting surface be μ.

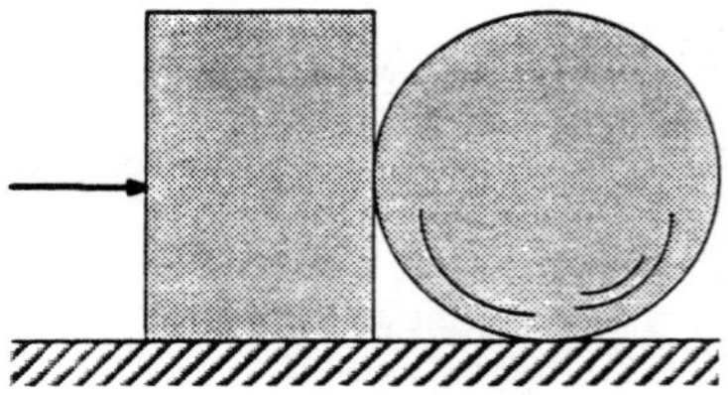

Figure a

Show that the ball will skid on the table, if

$$\frac{P}{W} > \frac{7\mu}{(1-\mu)(2+7\mu)}$$

Solution: Suppose first that the ball rolls (Fig. b).

By Fig. b,

$$\Sigma F_x = P - F = ma_G \qquad (a)$$

$$\Sigma F_y = N - \mu P - mg = 0 \qquad (b)$$

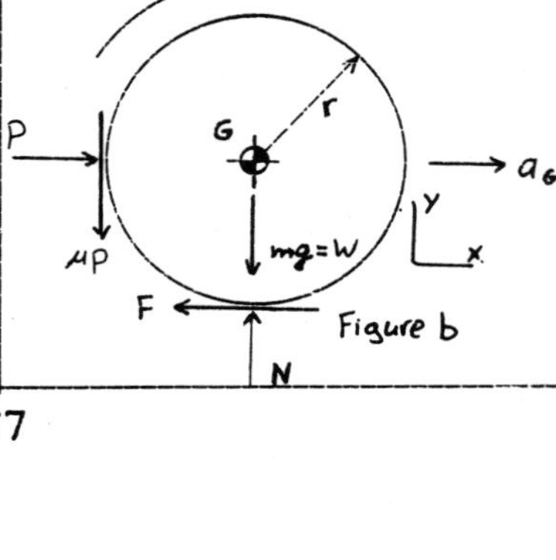

Figure b

(Continued)

18.93 Cont.

$\circlearrowleft + \Sigma M_G = Fr - \mu Pr = I_G \alpha = \frac{2}{5} m r^2 \alpha$ (c)

For rolling,

$a_G = r\alpha$ (d)

Hence, Eqs. (c) and (d) yield

$F - \mu P = \frac{2}{5} m a_G$ (e)

Adding Eqs. (a) and (e), and solving for $m a_G$ we find,

$m a_G = \frac{5}{7}(1-\mu)P$ (f)

Then, Eqs. (e) and (f) yield

$F = \frac{1}{7}(2+5\mu)P$ (g)

also, by Eq. (b),

$N = \mu P + W$ (h)

If the ball does not skid, $F < \mu N$ or by Eq. (h),

$F < \mu^2 P + \mu W$ (i)

Then, by Eqs. (g) and (i)

$\frac{1}{7}(2+5\mu)P < \mu^2 P + \mu W$

So, $(2+5\mu-7\mu^2)P < 7\mu W$

or factoring,

$(1-\mu)(2+7\mu)P < 7\mu W$

Therefore, if the ball does not skid

$$\frac{P}{W} < \frac{7\mu}{(1-\mu)(2+7\mu)}$$

and if the ball skids

$$\frac{P}{W} > \frac{7\mu}{(1-\mu)(2+7\mu)}$$

18.94

Given: The mechanism shown in Fig. a consists of a thin uniform bar A that weighs 200 lb and a part B that weighs 48 lb and that is connected to A by a light flexible inextensionable cord. Bar A rotates about a fixed frictionless horizontal axis O.

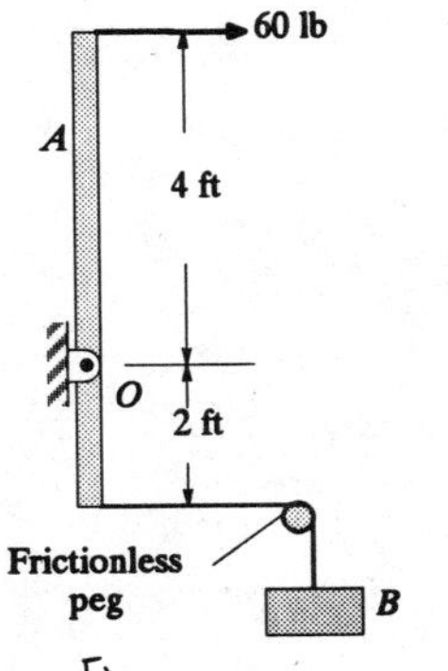

Figure a

When A is vertical, B has a velocity of 6 ft/s downward. For this position, determine the acceleration of B and the vertical and horizontal components of the force exerted on A by the pin O.

Solution: The free-body diagrams of bar A and part B are shown in Figs. b and c.

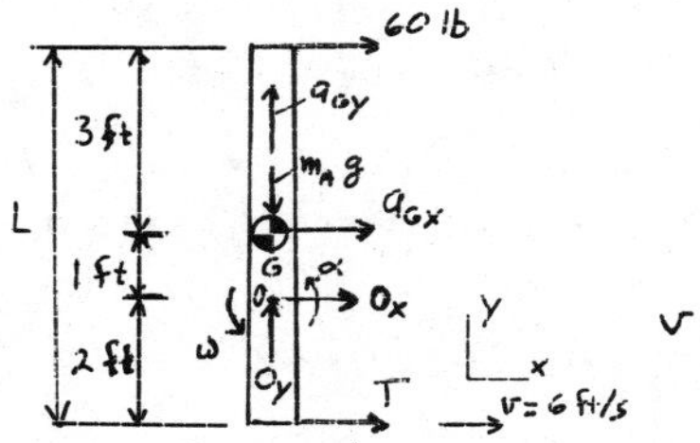

Figure b

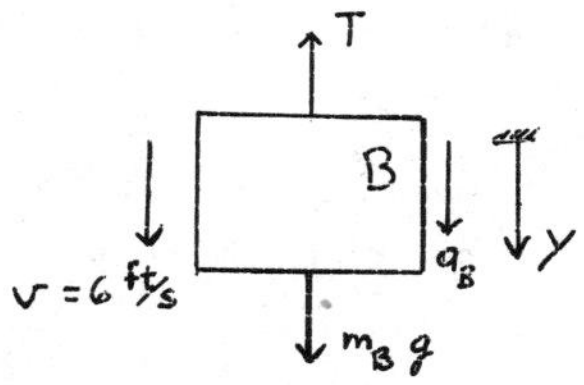

Figure c

By Fig. b,

$\Sigma F_x = O_x + T + 60 = m_A a_{Gx}$ (a)

$\Sigma F_y = O_y - m_A g = m_A a_{Gy}$ (b)

$\circlearrowleft \Sigma M_o = 2T - (4)(60) = I_o \alpha$ (c)

where by the parallel-axis theorem (Steiner's theorem)

$I_o = I_G + m_A d^2$

$= \frac{1}{12} m_A L^2 + m_A d^2$

$= \left(\frac{200}{32.2}\right)\left[\frac{1}{12}(6)^2 + (1)^2\right]$

or

$I_o = 24.845 \ \text{lb}\cdot\text{f}\cdot\text{s}^2$ (d)

For the given position (Fig. b),

$\omega = \frac{v}{2} = \frac{6}{2} = 3 \ \text{rad/s}$

Therefore, $a_{Gy} = -(1)\omega^2 = -9 \ \text{ft/s}^2$ (e)

Then, by Eqs. (b) and (e),

$O_y = 144.1 \ \text{lb}$

Also, by Fig. b,

$a_{Gx} = -(1)(\alpha) = -\alpha \ [\text{ft/s}^2]$ (f)

Equations (a) and (f) yield

$O_x + T + 60 = -(6.211)\alpha$ (g)

and by Eqs. (c) and (d),

$T = 120 + 12.422\alpha$ (h)

Also, by Figs. b and c,

$\Sigma F_y = m_B g - T = m_B a_B$

$a_B = 2\alpha \ [\text{ft/s}^2]$ (i)

or $T = 48 - 2.981\alpha$ (j)

Equations (h) and (i) yield

$\alpha = -4.674 \ \text{rad/s}^2$ (k)

$T = 61.93 \ \text{lb}$ (l)

So, by Eqs. (i) and (k),

$a_B = 2(-4.674) = -9.35 \ \text{ft/s}^2$

and by Eqs. (g), (k), and (l),

$O_x = -92.90 \ \text{lb}$

18.95

Given: A bowling ball of radius r is thrown down a bowling lane with initial speed v_0 and backspin ω_0. The coefficient of sliding friction is μ_K.

a) Show that the distance s through which the ball skids is

$$s = \frac{2}{49\mu_K g}(v_0 + r\omega_0)(6v_0 - r\omega_0)$$

provided the ball does not roll backward.

b) Show that, to prevent rolling backward,

$$v_0 \geq \frac{2}{5} r\omega_0$$

Solution: a) The free-body diagram of the ball is shown in Fig. a.

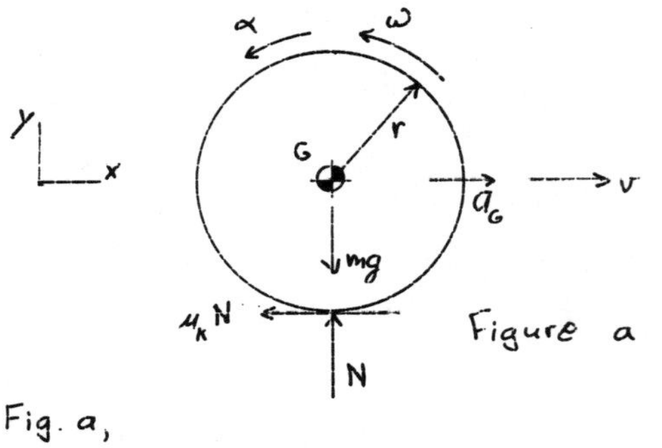

Figure a

By Fig. a,

$$\Sigma F_x = -\mu_K N = m a_G \quad (a)$$

$$\Sigma F_y = N - mg = 0 \;;\; N = mg \quad (b)$$

$$\circlearrowleft \Sigma M_G = -\mu_K N r = I_G \alpha = \frac{2}{5} m r^2 \alpha \quad (c)$$

Equations (a) and (b) yield

$$a_G = -\mu_K g \quad (d)$$

Equations (b) and (c) yield

$$\alpha = -\frac{5}{2}\mu_K \frac{g}{r} \quad (e)$$

By kinematics, $\alpha = d\omega/dt$. Therefore, by Eq. (e),

$$\frac{d\omega}{dt} = -\mu_K \frac{5g}{2r}$$

and integration yields

$$\omega - \omega_0 = -\mu_K \frac{5g}{2r} t \quad (f)$$

where $\omega = \omega_0$ when $t = 0$

Also, by kinematics and Eq. (d),

$$a_G = \frac{dv}{dt} = -\mu_K g$$

so,

$$v - v_0 = -\mu_K g t \quad (g)$$

where $v = v_0$ when $t = 0$

When skidding stops,

$$v = -r\omega \quad (h)$$

and Eqs. (f), (g), and (h) yield

$$v_0 - \mu_K g t_1 = \mu_K \left(\frac{5g}{2}\right) t_1 - r\omega_0$$

where t_1 is the time at which the ball stops skidding. Hence,

$$t_1 = \frac{2}{7\mu_K g}(v_0 + r\omega_0) \quad (i)$$

gain, by kinematics and Eq. (g),

$$v = \frac{ds}{dt} = v_0 - \mu_K g t$$

or by integration

$$s = v_0 t - \frac{1}{2}\mu_K g t^2 \quad (j)$$

When $t = t_1$, Eqs. (i) and (j) yield

$$s = \frac{2v_0}{7\mu_K g}(v_0 + r\omega_0) - \frac{2}{49\mu_K g}(v_0 + r\omega_0)^2$$

or

$$s = \frac{2}{49\mu_K g}(v_0 + r\omega_0)(6v_0 - r\omega_0)$$

where s is the distance that the ball travels while skidding. This result is what was required in part a.

b) The ball will start to roll back if the initial speed v_0 is too small. To ensure that the ball does not roll backward, we must have $v > 0$ for $t < t_1$. Hence, by Eq. (g), $v = v_0 - \mu_K g t > 0$, or since $t < t_1$

$$v_0 > \mu_K g t > \mu_K g t_1$$

so, we must have

$$v_0 > \mu_K g t_1$$

to ensure the ball does not roll backward. Therefore, by Eq. (i)

$$v_0 > \frac{2}{7}(v_0 + r\omega_0$$

or

$$v_0 > \frac{2}{5} r\omega_0$$

18.96

Given: A uniform half-disk rolls freely on a horizontal surface (Fig. a). For $\theta = 0$, the angular velocity is ω_0.

Show that the angular acceleration corresponding to $\theta = 0$ is

$$\alpha_0 = \frac{8}{9\pi}\left(\omega_0^2 + \frac{g}{r}\right)$$

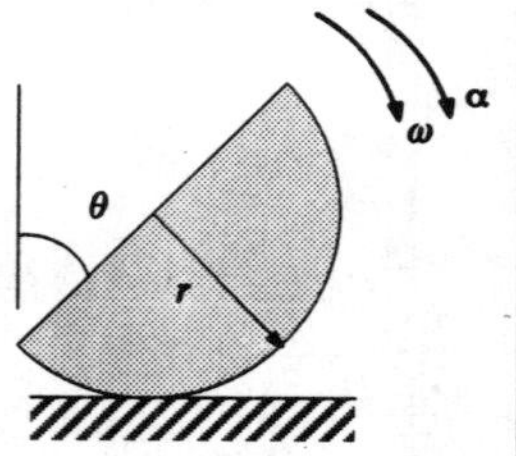

Figure a

(Continued)

18.96

Solution: The free-body diagram of the half-disk is shown in Fig. b at $\theta = 0$, where m is the mass of the half-disk.

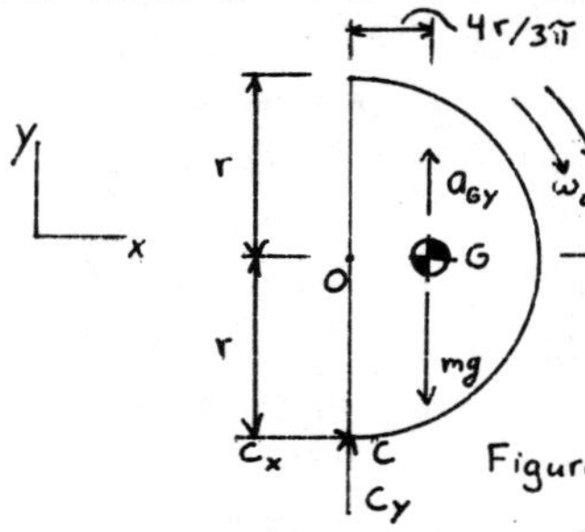

Figure b

By Appendix D, Table D2, the center of mass is located at $4r/3\pi$ from the diameter of the half-disk.

By Fig. b,

$$\Sigma F_x = C_x = m a_{Gx} \qquad (a)$$

$$\Sigma F_y = C_y - mg = m a_{Gy} \qquad (b)$$

$$\circlearrowleft \Sigma M_G = \left(\frac{4r}{3\pi}\right) C_y - r C_x = I_G \alpha \qquad (c)$$

where (a_{Gx}, a_{Gy}) is the acceleration of the mass center G, m is the mass of the half-disk, and I_G is the mass moment of inertia of the half-disk relative to G.

Note that the relation "Torque = $I\alpha$" does not apply for moments about the instantaneous center C of velocity, since the acceleration of point C is not directed along the line CG through the center of mass (see Section 16.7, Theorem 16.12).

The mass moment of inertia of a complete disk about its center O is (Appendix F, Table F1)

$$I_{DO} = \frac{1}{2} m_D r^2$$

where $m_D = 2m$ is the mass of the complete disk. Hence, the mass moment of inertia of the half-disk, about point O is

$$I_O = \frac{1}{2} I_{DO} = \frac{1}{2}\left[\frac{1}{2}(2m)r^2\right] = \frac{1}{2} m r^2 \qquad (d)$$

Then by the parallel-axis theorem (Steiner's theorem) and Eq. (d),

$$I_O = I_G + m\left(\frac{4r}{3\pi}\right)^2$$

or

$$I_G = \frac{1}{2} m r^2 - \frac{16 r^2}{9\pi^2} m = m r^2\left(\frac{1}{2} - \frac{16}{9\pi^2}\right) \qquad (e)$$

By kinematics and Fig. b,

$$a_{Gx} = (a_o)_x + (a_{G/o})_x = \alpha_o r - \frac{4r}{3\pi}\omega_o^2 \qquad (f)$$

$$a_{Gy} = (a_o)_y + (a_{G/o})_y = 0 - \alpha_o \frac{4r}{3\pi} = -\frac{4r}{3\pi}\alpha_o \qquad (g)$$

Hence, by (a), (b), (c), (e), (f), and (g)

$$C_x = m\left(r\alpha_o - \frac{4r}{3\pi}\omega_o^2\right) \qquad (h)$$

$$C_y = m\left(g - \frac{4r}{3\pi}\alpha_o\right) \qquad (i)$$

$$\frac{4r}{3\pi} C_y - r C_x = m r^2\left(\frac{1}{2} - \frac{16}{9\pi^2}\right)\alpha_o \qquad (j)$$

Multiplying Eq. (h) by r and Eq. (i) by $\frac{4r}{3\pi}$, substituting into Eq. (j), and solving for α_o, we obtain, as required,

$$\alpha_o = \frac{8}{9\pi}\left(\omega_o^2 + \frac{g}{r}\right)$$

18.97

Given: A uniform rod OA is welded to a thin hoop at A (Fig. a). The mass of the hoop is one-half the mass of the rod. The coefficient of static friction between the hoop and the horizontal surface is $\mu_s = 0.40$. A horizontal force P in the plane of the hoop is applied to the rod at point O. For the position shown, the hoop is rolling to the right with angular velocity 200 rpm, and is on the verge of skidding.

Determine the acceleration [m/s²] of the center O for this position.

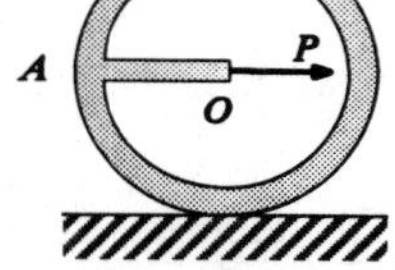

Figure a

Solution: By Fig. b, taking moments about point O, we find the center of mass to be given by (where m is the mass of the rod)

$$\circlearrowleft \Sigma M_o = \left(m + \frac{m}{2}\right) g r_G = mg\frac{r}{2} + \frac{1}{2} mg(0) = \frac{1}{2} mgr$$

or

$$r_G = \frac{1}{3} r$$

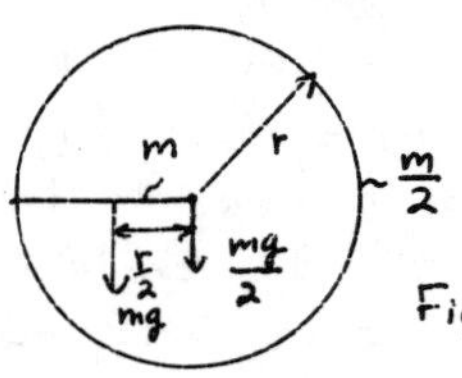

Figure b

The free-body diagram of hoop-rod system is shown in Fig. c.

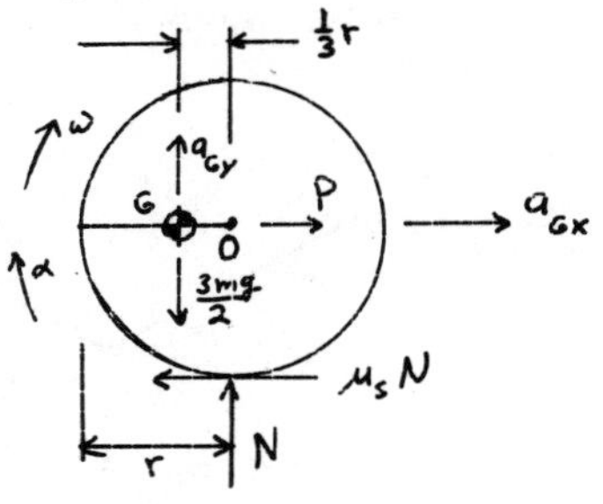

Figure c

(Continued)

18.97 Cont.

By Fig. c,

$$\Sigma F_x = P - \mu_s N = \frac{3}{2} m a_{Gx} \qquad (a)$$

$$\Sigma F_y = N - \frac{3mg}{2} = \frac{3}{2} m a_{Gy} \qquad (b)$$

$$\circlearrowleft \Sigma M_G = \mu_s N(r) - N\left(\frac{r}{3}\right) = I_G \alpha \qquad (c)$$

where, by Fig. c,

$$I_G = \frac{1}{12} m r^2 + m\left(\frac{r}{6}\right)^2 + \left(\frac{m}{2}\right) r^2 + \left(\frac{m}{2}\right)\left(\frac{r}{3}\right)^2$$

or

$$I_G = \frac{2}{3} m r^2 \qquad (d)$$

Also, by kinematics and Fig. c,

$$(a_o)_x = r\alpha, \quad (a_o)_y = 0 \qquad (e)$$

$$a_{Gx} = (a_o)_x + (a_{G/o})_x = r\alpha + \frac{1}{3} r\omega^2 \qquad (f)$$

$$a_{Gy} = (a_o)_y + (a_{G/o})_y = 0 + \frac{r\alpha}{3} = \frac{r\alpha}{3} \qquad (g)$$

Equations (a), (b), (c), (d), (f), and (g) yield, with $\mu_s = 0.40$,

$$P - 0.40 N = \frac{3}{2} m \left(r\alpha + \frac{1}{3} r\omega^2\right) \qquad (h)$$

$$N - \frac{3}{2} mg = \frac{3}{2} m \left(\frac{r}{3}\alpha\right) \qquad (i)$$

$$0.4 N(r) - N\left(\frac{r}{3}\right) = \frac{2}{3} m r^2 \alpha \qquad (j)$$

By Eq. (i),

$$N = \frac{3}{2} m \left(g + \frac{1}{3} r\alpha\right) \qquad (k)$$

Then, by Eqs. (j) and (k),

$$r\alpha = \frac{3}{19} g \qquad (\ell)$$

Therefore, by Eqs. (e) and (ℓ)

$$(a_o)_x = a_o = \frac{3}{19} g = \frac{3}{19}(9.81) = 1.549 \text{ m/s}^2$$

18.98

Given: Figure a represents a simplified model of a milling machine mechanism. Crank OA is rotated at constant angular speed ω by torque M. The mass of the milling block B is m, and the horizontal resisting force the milling bed exerts on B is $F = kN$, k = constant and N = normal (vertical) force exerted on the block by the bed. When the block moves to left, $F = k_1 N$ to the right. When the block moves to the right, $F = k_2 N$ to the left; the constants k_1 and k_2 are different. The weights and masses of the equal length bars are negligible.

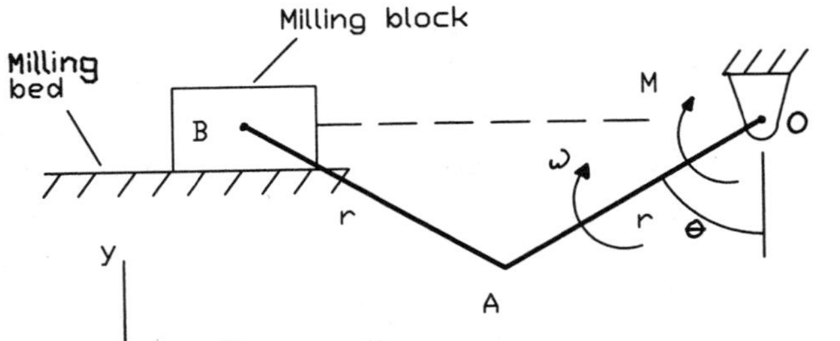

Figure a

a) In terms of k_1, m, r, ω, g, and θ, derive formulas for the force components A_x, A_y, B_x, B_y, and for the moment M required to maintain a constant ω.

b) For m = 6 kg, r = 250 mm, ω = π rad/s, $k_1 = 0.40$, and g = 9.81 m/s², plot A_x, A_y, B_x, B_y, and M as functions of θ, for $0 \le \theta \le 90°$.

c) From the plots of part b, estimate the maximum magnitudes of A_x, A_y, B_x, B_y, and M and the corresponding values of θ.

d) What happens to the equation of part a for the range $90° \le \theta \le 180°$? Use $k_2 = 0.20$.

Solution: First determine the angular velocity and angular acceleration of points A and B of bar AB. By Fig. b, since ω is the constant angular velocity of bar AB,

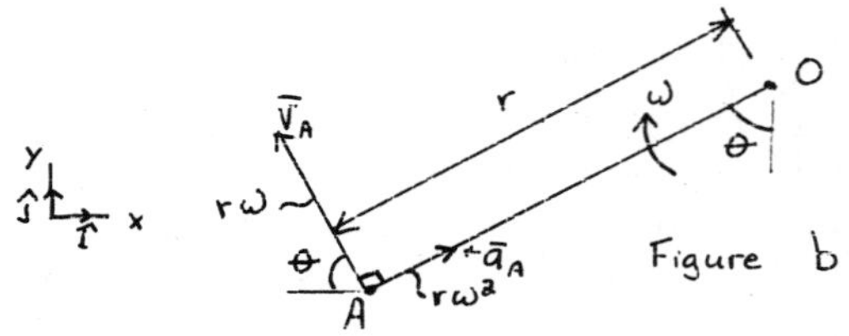

Figure b

$$\bar{v}_A = r\omega[-(\cos\theta)\hat{\imath} + (\sin\theta)\hat{\jmath}] \qquad (a)$$

$$\bar{a}_A = r\omega^2[(\sin\theta)\hat{\imath} + (\cos\theta)\hat{\jmath}] \qquad (b)$$

By kinematics, the velocity and acceleration of point B are

$$\bar{v}_B = \bar{v}_A + \bar{v}_{B/A} \qquad (c)$$

$$\bar{a}_B = \bar{a}_A + \bar{a}_{B/A} \qquad (d)$$

where $\bar{v}_{B/A}$ is the velocity of B relative to A and $\bar{a}_{B/A}$ is the acceleration of B relative to A.

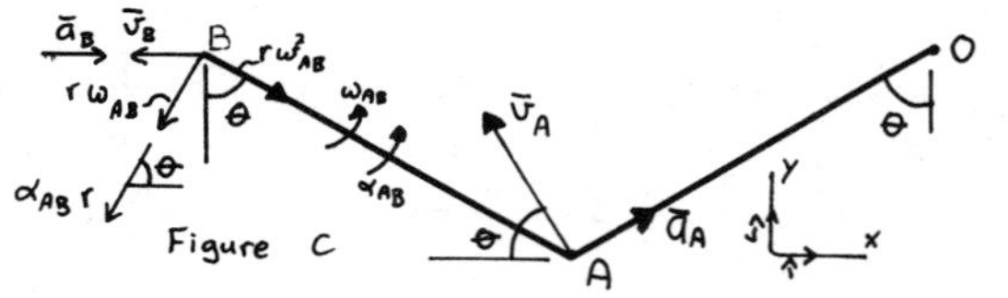

Figure c

By Fig. c,

$$\bar{v}_B = -v_B \hat{\imath} \qquad (e)$$

$$\bar{v}_{B/A} = r\omega_{AB}[-(\cos\theta)\hat{\imath} - (\sin\theta)\hat{\jmath}] \qquad (f)$$

$$\bar{a}_B = a_B \hat{\imath} \qquad (g)$$

$$\bar{a}_{B/A} = r\omega_{AB}^2[(\sin\theta)\hat{\imath} - (\cos\theta)\hat{\jmath}] + r\alpha_{AB}[-(\cos\theta)\hat{\imath} - (\sin\theta)\hat{\jmath}] \qquad (h)$$

By Eqs. (a), (c), (e) and (f), $\bar{v}_B = -v_B\hat{\imath} = -r[(\omega + \omega_{AB})(\cos\theta)\hat{\imath}] + r[(\omega - \omega_{AB})(\sin\theta)\hat{\jmath}]$,

Hence, $\omega = \omega_{AB}$, and

$$v_B = 2r\omega\cos\theta \qquad (i)$$

Thus, rods OA and OB rotate with the same angular velocity.

(Continued)

18.98 Cont.

Also, by Fig. b, and Eqs. (b), (d), (g), and (h), we have

$$\bar{a}_B = a_B\hat{\imath} = r[(2\omega^2 \sin\theta - \alpha_{AB}(\cos\theta))\hat{\imath} - (\alpha_{AB}\sin\theta)\hat{\jmath}]$$

Therefore, $\alpha_{AB} = 0$, and

$$a_B = 2r\omega^2 \sin\theta \qquad (j)$$

The fact that $\alpha_{AB}=0$ could have been observed from the fact that $\omega_{AB} = \omega =$ constant.

Now consider the free-body diagram of bar OA (Fig. d)

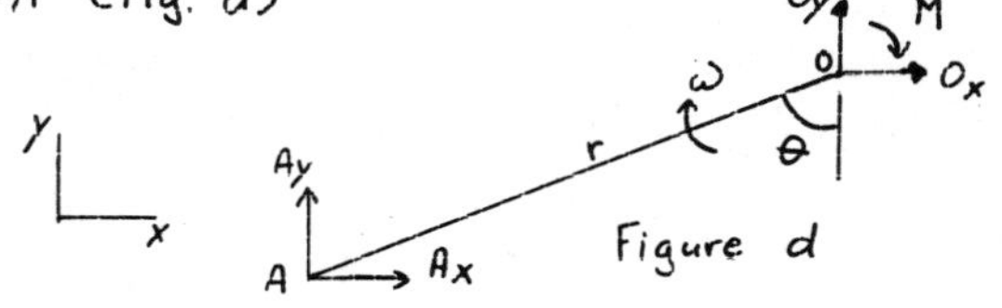

Figure d

By Fig. d, since $\omega =$ constant ($\alpha_{OA} = 0$),

$$\circlearrowleft \Sigma M_O = A_x r\cos\theta - A_y r\sin\theta - M = 0$$

or

$$M = r(A_x \cos\theta - A_y \sin\theta) \qquad (k)$$

Then, by the free-body diagram of bar AB (Fig. e), since the mass of bar AB is negligible,

$$\Sigma F_x = B_x - A_x = 0;\quad B_x = A_x \qquad (l)$$

$$\Sigma F_y = B_y - A_y = 0;\quad B_y = A_y \qquad (m)$$

$$\circlearrowleft \Sigma M_B = A_x r\cos\theta + A_y r\sin\theta = 0$$

or

$$A_x\cos\theta + A_y \sin\theta = 0 \qquad (n)$$

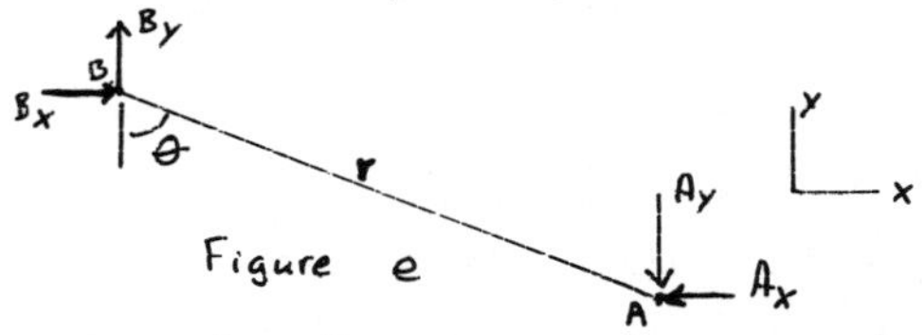

Figure e

Next, consider the free-body diagram of the milling block (Fig. f).

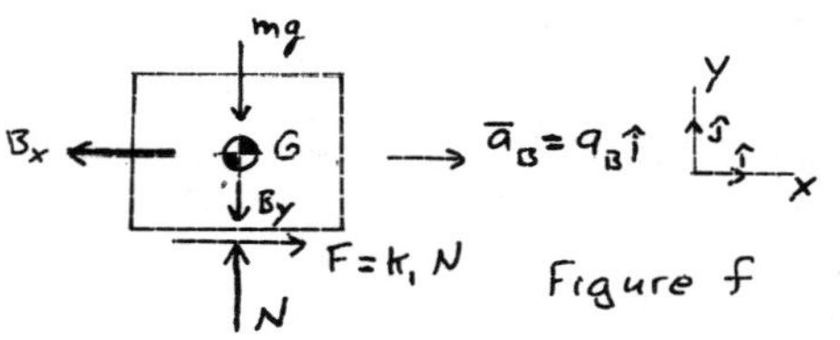

Figure f

By Fig. f,

$$\Sigma F_x = k_1 N - B_x = m a_B$$

$$\Sigma F_y = N - B_y - mg = 0$$

or with Eq. (j),

$$k_1 N - B_x = 2mr\omega^2 \sin\theta$$

$$N = B_y + mg$$

Elimination of N from these equations yields

$$B_x - k_1 B_y = k_1 mg - 2mr\omega^2 \qquad (o)$$

collecting Eqs. (l), (m), (n), and (o), we obtain four equations in A_x, A_y, B_x, and B_y, namely,

$$A_x - B_x = 0$$

$$A_y - B_y = 0$$

$$A_x\cos\theta + A_y\sin\theta = 0$$

$$B_x - k_1 B_y = k_1 mg - 2mr\omega^2 \sin\theta$$

The solutions of these equations are

$$A_x = B_x = \frac{(k_1 mg - 2mr\omega^2\sin\theta)\sin\theta}{\sin\theta + k_1\cos\theta}$$

$$A_y = B_y = \frac{(2mr\omega^2 \sin\theta - k_1 mg)\cos\theta}{\sin\theta + k_1\cos\theta} \qquad (p)$$

By Eqs. (h) and (p),

$$M = \frac{(k_1 mgr - 2mr^2\omega^2\sin\theta)\sin 2\theta}{\sin\theta + k_1\cos\theta} \qquad (q)$$

b) For $m = 6$ kg, $r = 0.250$ m, $\omega = \pi$ rad/s, $k_1 = 0.40$, and $g = 9.81$ m/s², Eqs. (p) and (q) yield,

$$A_x = B_x = \frac{(23.544 - 29.609\sin\theta)\sin\theta}{\sin\theta + 0.40\cos\theta} \quad [N]$$

$$A_y = B_y = \frac{(29.609\sin\theta - 23.544)\cos\theta}{\sin\theta + 0.40\cos\theta} \quad [N]$$

$$M = \frac{(5.886 - 7.402\sin\theta)\sin 2\theta}{\sin\theta + 0.40\cos\theta} \quad [N\cdot m]$$

The plots of A_x, B_x, A_y, B_y, and M are shown in Fig. g.

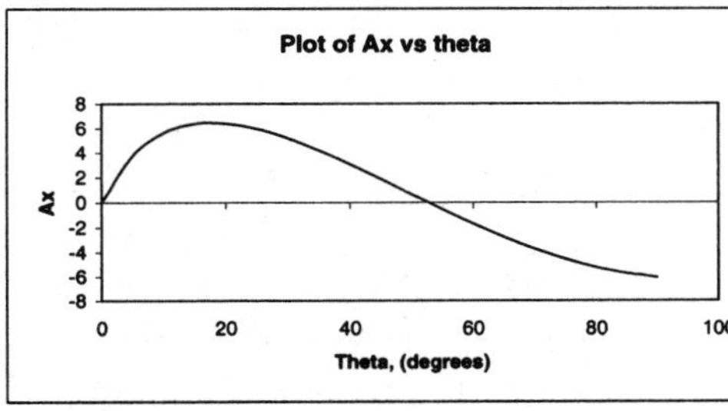

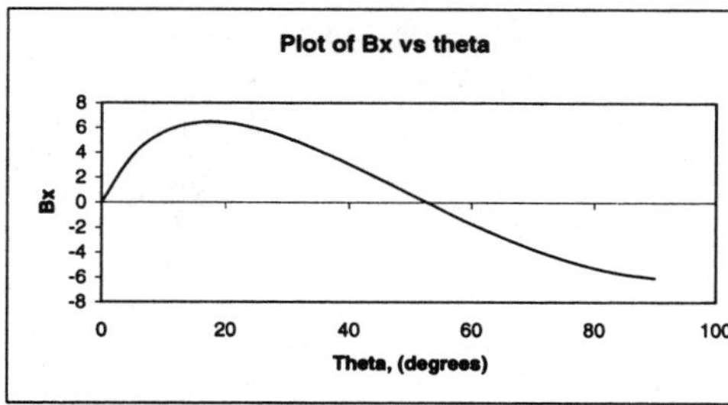

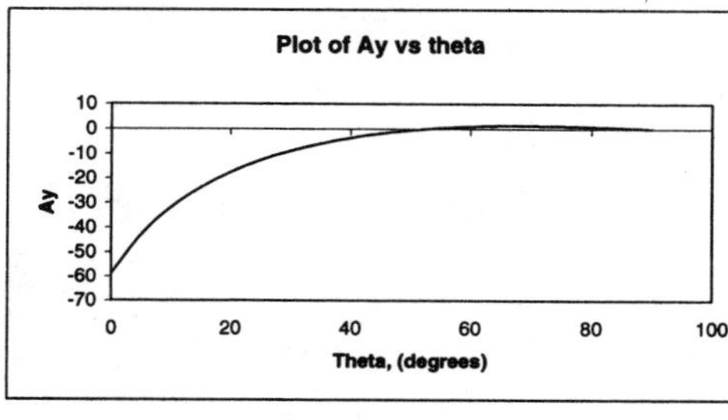

Figure g (continued)

(Continued)

18.98 Cont.

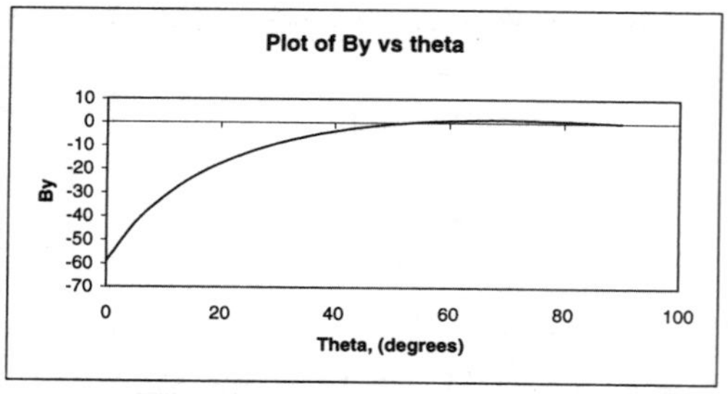

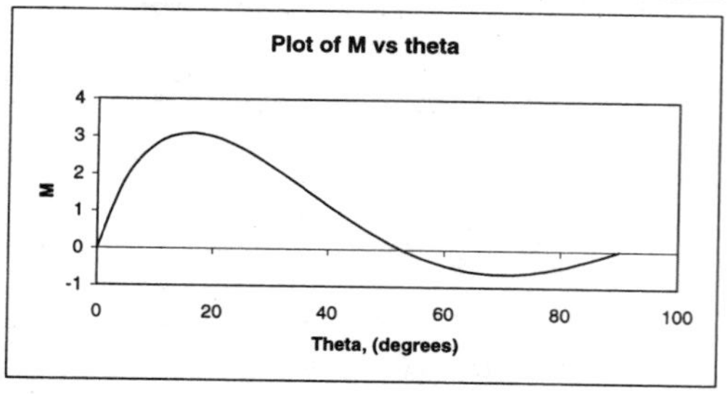

Figure g (Cont.)

c) By Fig. g, the maximum magnitudes are

Maximum $A_x = B_x \approx 6.4$ N, $\theta \approx 19°$

Maximum $A_y = B_y \approx 58.9$ N, $\theta \approx 0°$

Maximum $M \approx 3.1$ Nm, $\theta \approx 15°$

d) The constant k_1 is replaced by $-k_2$ in Eqs. (p) and (q). Then, the solution proceeds as with k_1 for what ever value k_2 may have.

18.99

Given: The mass of each homogeneous bar, OA and AB, in Problem 18.98 is 1 kg and the mass of the block is 6 kg (Fig. a). The weights of the bars OA and AB are small compared to other forces and friction at pins O, A, and B is small.

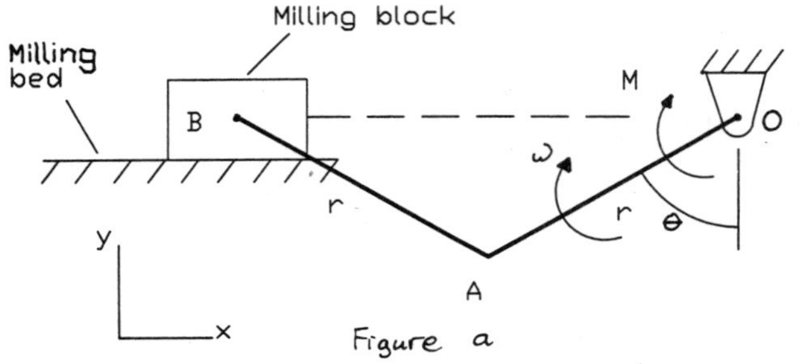

Figure a

a) For $r = 250$ mm, $\omega = \pi$ rad/s, $k_1 = 0.40$, and $g = 9.81$ m/s², derive formulas for A_x, A_y, B_x, B_y, and M in terms of θ, for the range $0 \le \theta \le 90°$

b) Plot A_x, A_y, B_x, B_y, M versus θ, for $0 \le \theta \le 90°$

c) From the plots obtained in part b, estimate the maximum magnitudes of A_x, A_y, B_x, B_y, and M and the corresponding values of θ.

d) What happens to the equations derived in part a as θ increases beyond 90° into the range $90 \le \theta \le 180°$?

Solution:

a) By the results of Problem 18.98,

$$\omega = \omega_{AB}\ ;\ \alpha_{OA} = \alpha_{AB} = 0 \quad (a)$$

$$v_B = 2r\omega\cos\theta\ ;\ \bar{v}_B = -v_B\hat{\imath} = -2r\omega(\cos\theta)\hat{\imath} \quad (b)$$

$$\bar{a}_A = r\omega^2[(\sin\theta)\hat{\imath} + (\cos\theta)\hat{\jmath}] \quad (c)$$

$$\bar{a}_B = -2r\omega^2\sin\theta\ ;\ \bar{a}_B = -a_B\hat{\imath} = 2r\omega^2(\sin\theta)\hat{\imath} \quad (d)$$

$$\bar{a}_{B/A} = r\omega^2[(\sin\theta)\hat{\imath} - (\cos\theta)\hat{\jmath}] \quad (e)$$

In the present case, the effects of the masses of bars OA and AB on the motion are included, but the weights of the bars are assumed to be small compared to other forces acting. Therefore, by the free-body diagram of bar OA (Fig. b), we have, where m_1 is the mass of the bar,

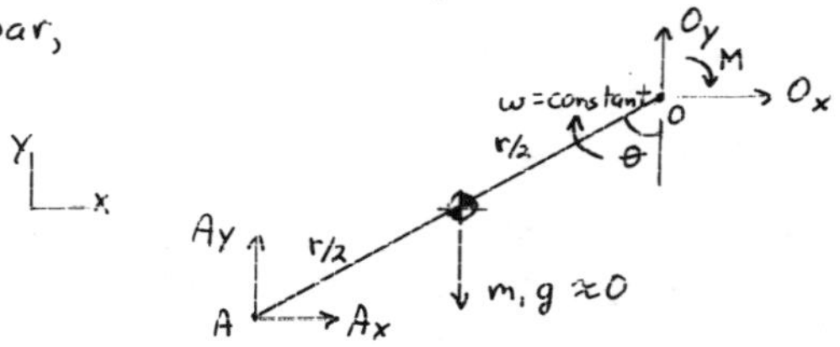

Figure b

and since $\alpha_{OA} = 0$,

$$\circlearrowleft + \Sigma M_0 = A_x(r\cos\theta) - A_y(r\sin\theta) - M = 0$$

or

$$M = r(A_x\cos\theta - A_y\sin\theta) \quad (f)$$

Next, consider the free-body diagram of bar AB (Fig. c)

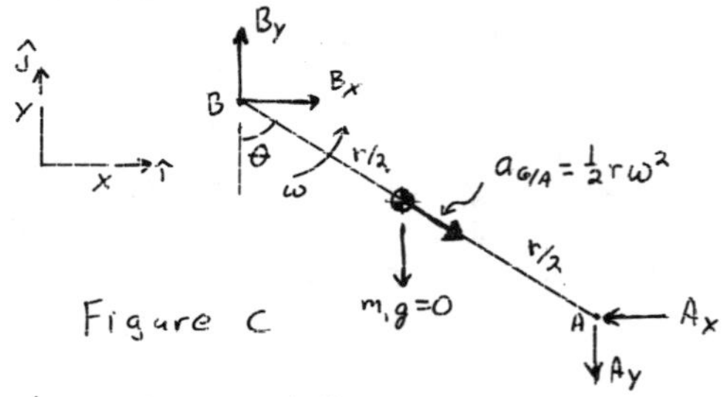

Figure c

By kinematics and Fig. c,

$$\bar{a}_G = \bar{a}_A + \bar{a}_{G/A}$$

$$\bar{a}_{G/A} = \tfrac{1}{2}r\omega^2[(\sin\theta)\hat{\imath} - (\cos\theta)\hat{\jmath}]$$

Therefore, with these equations and Eq. (c), we find,

$$\bar{a}_G = \tfrac{1}{2}r\omega^2[3(\sin\theta)\hat{\imath} + (\cos\theta)\hat{\jmath}] \quad (g)$$

(Continued)

18.99 Cont.

By Fig. c and Eq. (g),

$$\Sigma F_x = B_x - A_x = m_1 a_{Gx} = 3/2\, m_1 r\omega^2 \sin\theta \quad (h)$$

$$\Sigma F_y = B_y - A_y = m_1 a_{Gy} = \tfrac{1}{2} m_1 r\omega^2 \cos\theta \quad (i)$$

Since $\omega_{AB} = \omega = \text{constant}$, $\alpha_{AB} = 0$ [see also Eqs. (a)]. Then, by Fig. c,

$$\circlearrowleft + \Sigma M_G = A_x\left(\tfrac{r}{2}\cos\theta\right) + A_y\left(\tfrac{r}{2}\sin\theta\right) + B_x\left(\tfrac{r}{2}\cos\theta\right) + B_y\left(\tfrac{r}{2}\sin\theta\right) = 0 \quad (j)$$

Next, consider the free-body diagram of the block (Fig. d).

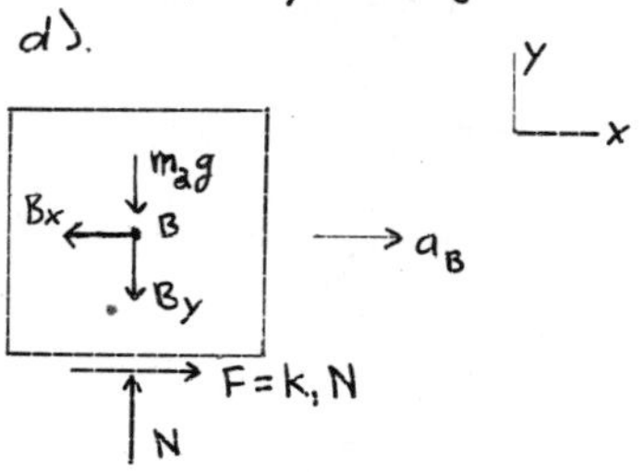

Figure d

By Fig. d and Eq. (d), we have, where m_2 is the mass of the block,

$$\Sigma F_x = k_1 N - B_x = m_2 a_B = 2 m_2 r\omega^2 \sin\theta \quad (k)$$

$$\Sigma F_y = N - B_y - m_2 g = 0$$

or

$$N = B_y + m_2 g \quad (\ell)$$

Equations (k) and (ℓ) yield

$$B_x - k_1 B_y = k_1 m_2 g - 2 m_2 r\omega^2 \sin\theta \quad (m)$$

Collecting Eqs. (h), (i), (j), and (m), we have

$$\left.\begin{aligned} A_x - B_x &= -\tfrac{3}{2} m_1 r\omega^2 \sin\theta \\ A_y - B_y &= -\tfrac{1}{2} m_1 r\omega^2 \cos\theta \\ A_x\cos\theta + A_y\sin\theta &+ B_x\cos\theta + B_y\sin\theta = 0 \\ B_x - k_1 B_y &= k_1 m_2 g - 2 m_2 r\omega^2 \sin\theta \end{aligned}\right\} (n)$$

With $r = 250$ mm $= 0.250$ m, $\omega = \pi$ rad/s, $k_1 = 0.40$, $m_1 = 1$ kg, $m_2 = 6$ kg, and $g = 9.81$ m/s², Eqs. (n) become for $0 \le \theta \le 90°$

$$\left.\begin{aligned} A_x - B_x &= -3.7011 \sin\theta \quad [N] \\ A_y - B_y &= -1.2337 \cos\theta \quad [N] \\ A_x\cos\theta + A_y\sin\theta + B_x\cos\theta + B_y\sin\theta &= 0 \quad [N] \\ B_x - 0.4 B_y &= 23.5440 - 29.6088 \sin\theta \quad [N] \end{aligned}\right\} (o)$$

The solution of Eqs. (o) is, in [N],

$$\left.\begin{aligned} A_x &= -\frac{(66.6198\sin^2\theta + 0.98696\sin\theta\cos\theta - 47.088\sin\theta)}{2\sin\theta + 0.80\cos\theta} \\ A_y &= -\frac{(0.98696\cos^2\theta - 61.685\sin\theta\cos\theta + 47.088\cos\theta)}{2\sin\theta + 0.80\cos\theta} \\ B_x &= -\frac{(59.2176\sin^2\theta - 1.9739\sin\theta\cos\theta - 47.088\sin\theta)}{2\sin\theta + 0.80\cos\theta} \\ B_y &= -\frac{(-64.1524\sin\theta\cos\theta + 47.088\cos\theta)}{2\sin\theta + 0.80\cos\theta} \end{aligned}\right\} (p)$$

Then, by Eq. (f) and the first two of Eqs. (p)

$$M = \frac{(23.544 - 32.0762\sin\theta)(\sin\theta)(\cos\theta)}{2\sin\theta + 0.80\cos\theta} \; [N\cdot m] \quad (q)$$

b) The plots of A_x, A_y, B_x, B_y, and M versus θ, for $0 \le \theta \le 90°$ are shown in Fig. e.

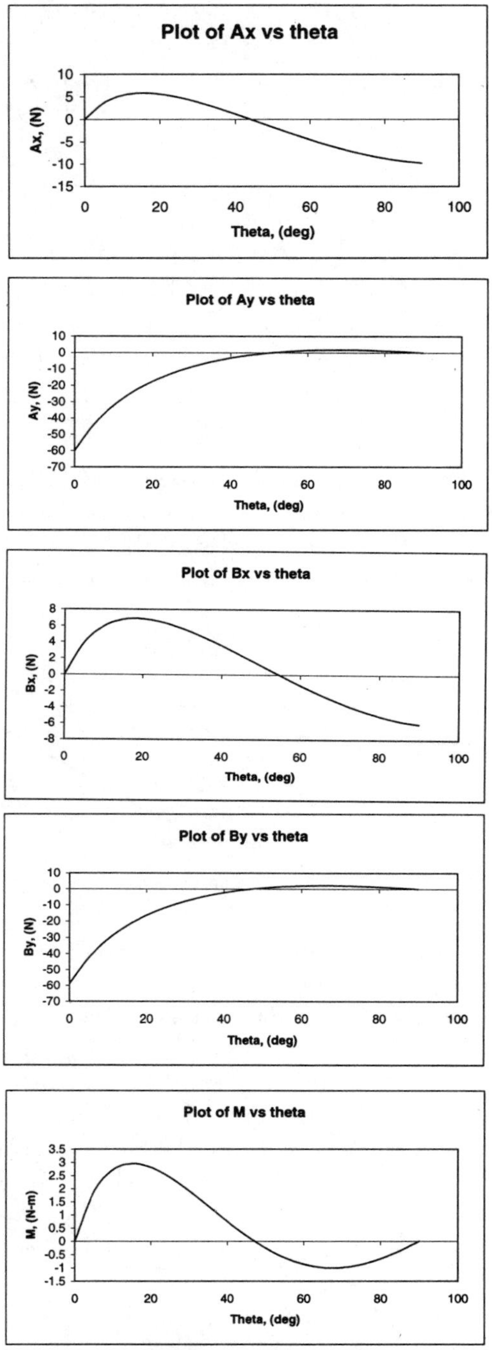

Figure e

(Continued)

18.99 Cont.

c) By Fig. e, the maximum magnitudes of A_x, A_y, B_x, B_y, and M and the corresponding values of θ are listed below:

Maximum Values		Theta
$A_x \approx 5.8$ N	;	$\theta \approx 15°$
$A_y \approx 60$ N	;	$\theta \approx 0°$
$B_x \approx 6.8$ N	;	$\theta \approx 20°$
$B_y \approx 58.9$ N	;	$\theta \approx 0°$
$M \approx 3.0$ N·m	;	$\theta \approx 15°$

d) For θ in the range $90° \le \theta \le 180°$, the constant k_1 in Eq. (n) is replaced by the constant $-k_2$. Then, the solution proceeds as with k_1.

18.100

Given: The mass of each homogeneous bar, OA and AB, in Problem 18.98 (Fig. a below) is m_1, and the mass of the block is m_2. The corresponding weights are $m_1 g$ and $m_2 g$.

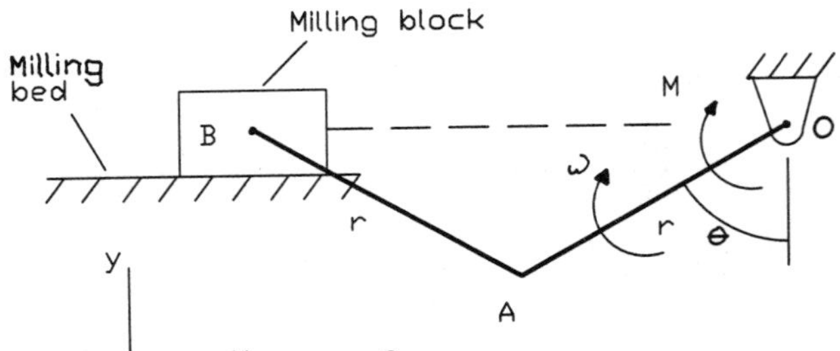

Figure a

a) Derive the five equations that the force components A_x, A_y, B_x, B_y, and the moment M, required to maintain constant ω, must satisfy in terms of k_1, m_1, m_2, r, ω, g, and θ, for the range $0 \le \theta \le 90°$. Include the effects of the weight of each bar.

b) For $m_1 = 1$ kg, $m_2 = 6$ kg, $r = 250$ mm, $\omega = \pi$ rad/s, $k_1 = 0.40$, and $g = 9.81$ m/s², solve the equations derived in Part a for A_x, A_y, B_x, B_y, and M. Plot A_x, A_y, B_x, B_y, and M as functions of θ, for $0 \le \theta \le 90°$.

c) From the plots of part b, estimate the maximum magnitudes of A_x, A_y, B_x, B_y, and M and the corresponding values of θ.

d) What happens to the equations derived in part a as θ increases beyond 90° into the range $90° \le \theta \le 180°$?

Solution: (a) By the results of Problem 18.98,

$$\omega_{AB} = \omega = \text{constant} \; ; \; \alpha_{OA} = \alpha_{AB} = 0 \qquad (a)$$

$$v_B = 2r\omega\cos\theta \; ; \; \bar{v}_B = -v_B\hat{\imath} = -2r\omega(\cos\theta)\hat{\imath} \qquad (b)$$

$$\bar{a}_A = r\omega^2[(\sin\theta)\hat{\imath} + (\cos\theta)\hat{\jmath}] \qquad (c)$$

$$a_B = -2r\omega^2\sin\theta \; ; \; \bar{a}_B = -a_B\hat{\imath} = 2r\omega^2\sin\theta\,\hat{\imath} \qquad (d)$$

$$\bar{a}_{B/A} = r\omega^2[(\sin\theta)\hat{\imath} - (\cos\theta)\hat{\jmath}] \qquad (e)$$

In the present case, the effects of both the mass and weights of bars OA and AB on the motion are to be included in the analysis.

The free-body diagram of bar OA is shown in Fig. b.

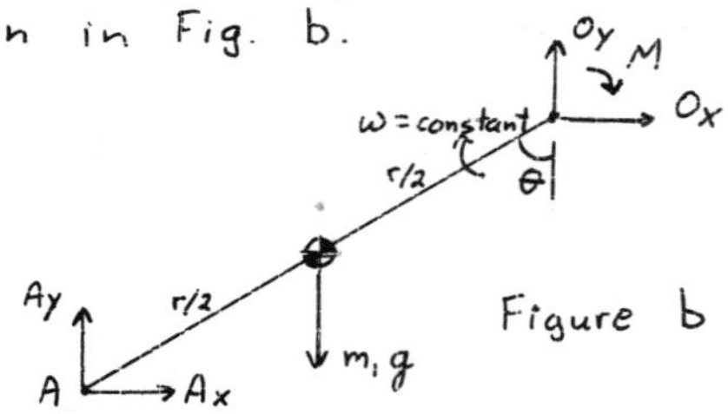

Figure b

By Fig. b, since $\alpha_{OA} = 0$,

$$\circlearrowleft\!+\sum M_O = A_x(r\cos\theta) - A_y(r\sin\theta) + m_1 g(\tfrac{r}{2}\sin\theta) - M = 0$$

or

$$M = r(A_x\cos\theta - A_y\sin\theta + \tfrac{1}{2}m_1 g\sin\theta) \qquad (f)$$

Next, consider the free-body diagram of bar AB (Fig. c)

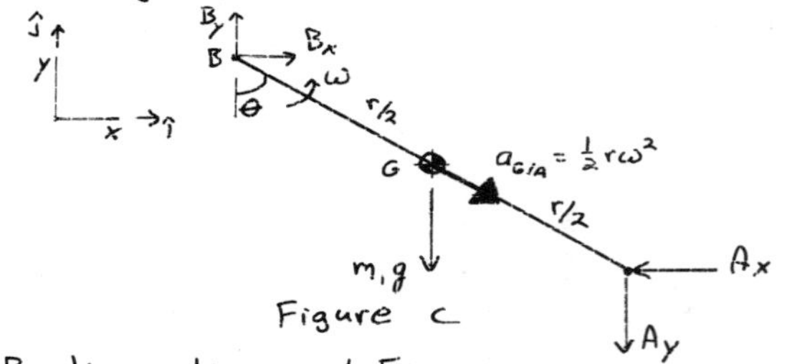

Figure c

By kinematics and Fig. c,

$$\bar{a}_G = \bar{a}_A + \bar{a}_{G/A}$$

$$\bar{a}_{G/A} = \tfrac{1}{2}r\omega^2[(\sin\theta)\hat{\imath} - (\cos\theta)\hat{\jmath}]$$

Therefore, with these equations and Eq. (c), we find,

$$\bar{a}_G = \tfrac{1}{2}r\omega^2[3(\sin\theta)\hat{\imath} + (\cos\theta)\hat{\jmath}] \qquad (g)$$

By Fig. (c) and Eq. (g),

$$\sum F_x = B_x - A_x = m_1 a_{Gx} = \tfrac{3}{2}m_1 r\omega^2\sin\theta \qquad (h)$$

$$\sum F_y = B_y - A_y - m_1 g = m_1 a_{Gy} = \tfrac{1}{2}m_1 r\omega^2\cos\theta \qquad (i)$$

Since $\omega_{AB} = \omega = \text{constant}$, $\alpha_{AB} = 0$ [see also Eqs. (a)]. Then, by Fig. c,

$$\circlearrowleft\!+\sum M_G = A_x(\tfrac{r}{2}\cos\theta) + A_y(\tfrac{r}{2}\sin\theta) + B_x(\tfrac{r}{2}\cos\theta) + B_y(\tfrac{r}{2}\sin\theta) = 0$$

or

$$A_x\cos\theta + A_y\sin\theta + B_x\cos\theta + B_y\sin\theta = 0 \qquad (j)$$

(Continued)

18.100

Next, consider the free-body diagram of the block (Fig. d).

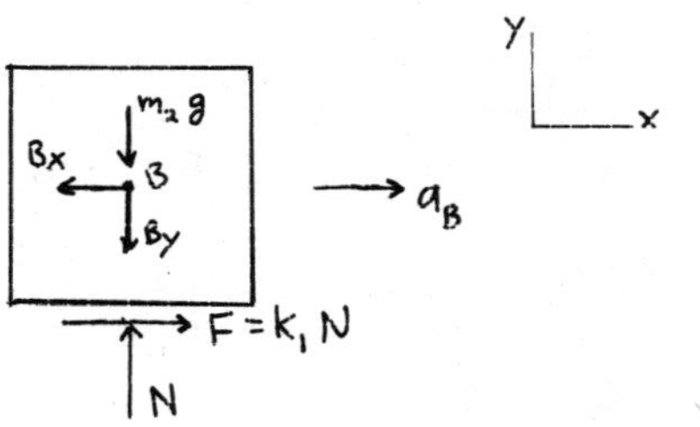

Figure d

By Fig. d and Eq. (d), we have, where m_2 is the mass of the block,

$$\Sigma F_x = k_1 N - B_x = m_2 a_B = 2 m_2 r \omega^2 \sin\theta \qquad (k)$$
$$\Sigma F_y = N - B_y - m_2 g = 0$$
or
$$N = B_y + m_2 g \qquad (\ell)$$

Equations (k) and (ℓ) yield

$$B_x - k_1 B_y = k_1 m_2 g - 2 m_2 r \omega^2 \sin\theta \qquad (m)$$

Collecting Eqs. (h), (i), (j), and (m), we have,

$$\left.\begin{aligned} &A_x - B_x = -\tfrac{3}{2} m_1 r \omega^2 \sin\theta \\ &A_y - B_y = -\tfrac{1}{2} m_1 r \omega^2 \cos\theta - m_1 g \\ &A_x \cos\theta + A_y \sin\theta + B_x \cos\theta + B_y \sin\theta = 0 \\ &B_x - k_1 B_y = k_1 m_2 g - 2 m_2 r \omega^2 \sin\theta \end{aligned}\right\} \quad (n)$$

b) With $r = 250$ mm $= 0.250$ m, $\omega = \pi$ rad/s, $k_1 = 0.40$, $m_1 = 1$ kg, $m_2 = 6$ kg, and $g = 9.81$ m/s^2, Eqs. (n) become for $0 \le \theta \le 90°$

$$\left.\begin{aligned} &A_x - B_x = -3.7011 \sin\theta \quad [N] \\ &A_y - B_y = -9.81 - 1.2337 \cos\theta \quad [N] \\ &A_x \cos\theta + A_y \sin\theta + B_x \cos\theta + B_y \sin\theta = 0 \quad [N] \\ &B_x - 0.4 B_y = 23.5440 - 29.6088 \sin\theta \quad [N] \end{aligned}\right\} \quad (o)$$

The solution of Eqs. (o) is, in newtons,

$$A_x = -\frac{(66.6198 \sin^2\theta + 0.98696 \sin\theta\cos\theta - 51.012 \sin\theta)}{2\sin\theta + 0.8\cos\theta}$$

$$A_y = -\frac{(0.98696 \cos^2\theta - 61.685 \sin\theta\cos\theta + 54.936 \cos\theta + 9.81 \sin\theta)}{2\sin\theta + 0.8\cos\theta}$$

$$B_x = -\frac{(59.2176 \sin^2\theta - 1.9739 \sin\theta\cos\theta - 51.012 \sin\theta)}{2\sin\theta + 0.8\cos\theta} \qquad (p)$$

$$B_y = -\frac{(-64.1524 \sin\theta\cos\theta + 47.088 \cos\theta - 9.81 \sin\theta)}{2\sin\theta + 0.8\cos\theta}$$

Then, by Eq. (f) and the first two of Eqs. (p),

$$M = \frac{(27.468 - 32.076 \sin\theta)\sin\theta\cos\theta + 4.905 \sin^2\theta}{2\sin\theta + 0.8\cos\theta} \quad [N \cdot m] \quad (q)$$

The plots of A_x, A_y, B_x, B_y, and M versus θ, for $0 \le \theta \le 90°$ are shown in Fig. e.

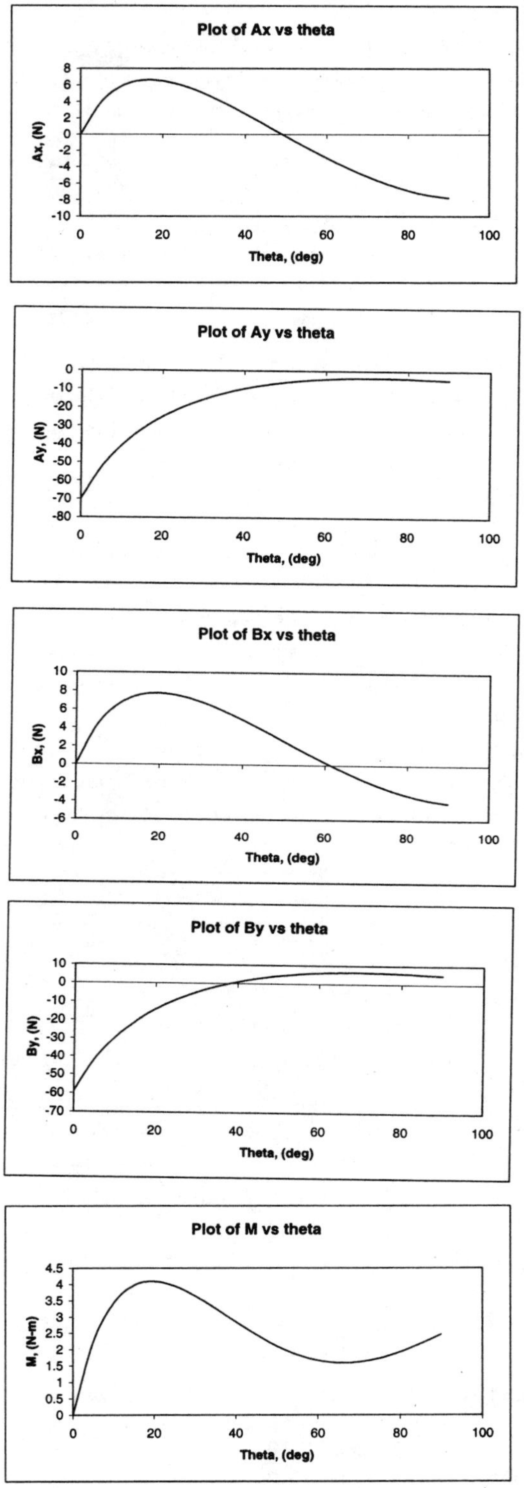

Figure e

(Continued)

18.100 Cont.

c) From the plots of part b (Fig. e), the maximum magnitudes of A_x, A_y, B_x, B_y, and M and the corresponding values of θ are listed below.

Maximum Magnitudes		Theta
$A_x \approx 6.6$ N	;	$\theta \approx 15°$
$A_y \approx 70$ N	;	$\theta \approx 0°$
$B_x \approx 7.8$ N	;	$\theta \approx 20°$
$B_y \approx 58.9$ N	;	$\theta \approx 0$
$M \approx 4.1$ N·m	;	$\theta \approx 20°$

d) For θ in the range $90° \le \theta \le 180°$, the constant k_1 in Eq. (n) is replaced by the constant $-k_2$. Then the solution proceeds as with k_1.

18.101

Given: The problem statement of Problem 18.77. Solve Problem 18.77 by the inertial-force method, that is, determine the speed of the race car as it rounds a horizontal unbanked curve of radius r = 100 ft, and is on the verge of tipping.

Solution: The free-body diagram of the race car of Problem 18.77 is shown in Fig. a, including the inertial force and appropriate dimensions.

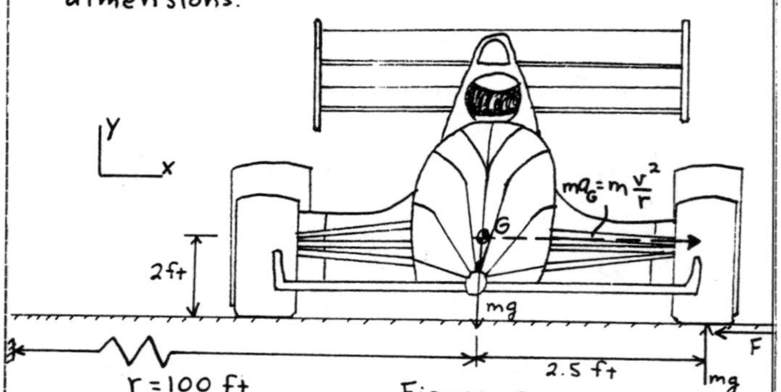

Figure a

Hence, for dynamic equilibrium, by Fig. a,

$$\Sigma F_x = -F + \frac{mv^2}{r} = 0 \qquad (a)$$

$$\Sigma M_G = 2.5\,mg - 2F = 0 \qquad (b)$$

Equations (a) and (b) yield, with r = 100 ft,

$$2.5\,mg = 2m\frac{v^2}{100}$$

or,

$$\underline{v = 63.44 \text{ ft/s}}$$

18.102

Given: With reference to Problem 17.8, a homogeneous circular cylinder is suspended like a yo-yo from a cord that is wrapped around it. It falls so that its axis remains horizontal and the cord unwinds evenly.

By the inertial-force method, determine the acceleration of the center of mass of the cylinder.

Solution:

The free-body diagram of the cylinder, including the inertial force and inertial couple is shown in Fig. a.

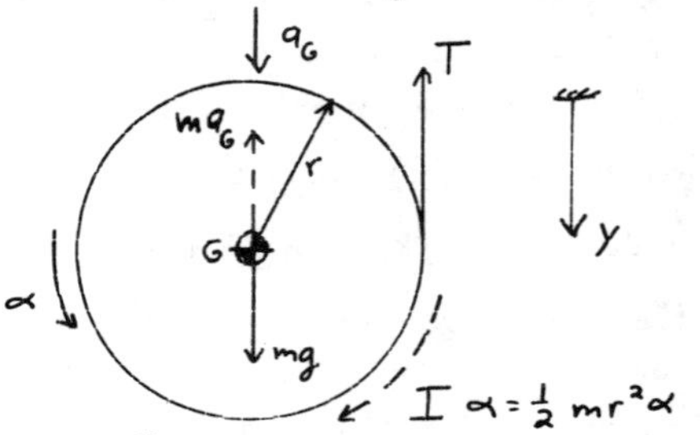

Figure a

By Fig. a,

$$\Sigma F_y = mg - T - ma_G = 0 \qquad (a)$$

$$\circlearrowleft\ \Sigma M_G = rT - \tfrac{1}{2}mr^2\alpha = 0 \qquad (b)$$

where by kinematics,

$$a_G = r\alpha \qquad (c)$$

By Eqs. (b) and (c),

$$T = \tfrac{1}{2}ma_G \qquad (d)$$

Then, by Eqs. (a) and (d),

$$\underline{a_G = \tfrac{2}{3}g}$$

18.103

Given: A homogeneous circular disk 1.22 m in diameter rotates and skids with backspin on a horizontal surface. The initial angular velocity of the disk is 30 rad/s. After 3 s, the angular velocity is 6 rad/s.

Determine the coefficient of kinetic friction between the disk and surface by the inertial-force method.

Solution: The free-body diagram of the disk, including the inertial force and the inertial couple is shown in Fig. a.

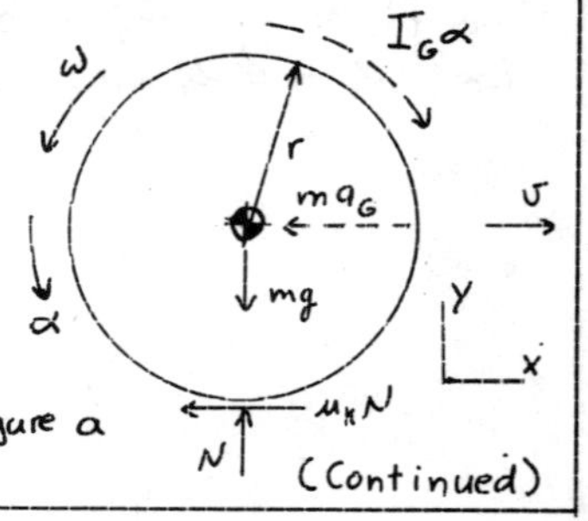

Figure a

(Continued)

18.103 Cont.

By Fig. a,

$$\Sigma F_y = N - mg = 0\,;\quad N = mg \qquad \text{(a)}$$

$$\circlearrowleft\ \Sigma M_G = -\mu_K N r - I_G\alpha = 0 \qquad \text{(b)}$$

By Eqs. (a) and (b), with $I_G = \frac{1}{2}mr^2$,

$$\alpha = -\frac{2\mu_K g}{r} \qquad \text{(c)}$$

By Eq. (c) and kinematics,

$$\alpha = \frac{d\omega}{dt} = -\frac{2\mu_K g}{r}$$

Integration yields

$$\omega = -\frac{2\mu_K g}{r}t + \omega_0 \qquad \text{(d)}$$

where $\omega = \omega_0 = 30$ rad/s when $t = 0$. For $t = 3$s, $\omega = 6$ rad/s. Hence, Eq. (d) yields, with $r = 0.61$ m,

$$6 = -\frac{2\mu_K g}{0.61}(3) + 30$$

or $\underline{\mu_K = 0.249}$

18.104

Given: Refer to Problem 18.80, in which a bowling ball that rolls on a horizontal surface is pushed by a horizontal bar with a force P (Fig. a). The coefficient of sliding friction between the bar and the ball is $\mu_K = 0.20$.

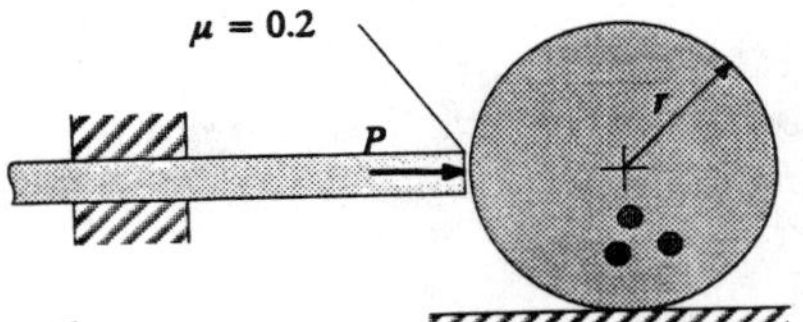

Figure a

Calculate the numerical value of P/ma, where m = mass of the ball and a is the acceleration of the bar by the inertial-force method.

Solution: The free-body diagram of the ball, including the inertial force and inertial couple.

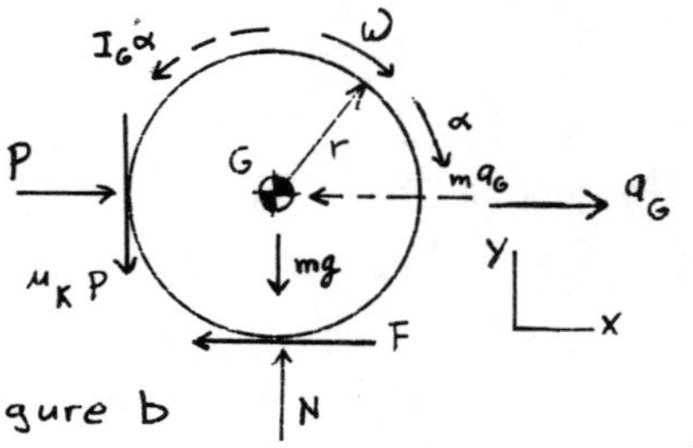

Figure b

By Fig. b,

$$\circlearrowleft\ \Sigma M_0 = I_G\alpha + mar + \mu_K Pr - Pr = 0 \qquad \text{(a)}$$

For a sphere $I_G = \frac{2}{5}mr^2$ and by kinematics $\alpha = a/r$. Hence, Eq. (a) yields,

$$Pr(1 - \mu_K) = 0.80\,Pr = \frac{7}{5}mar$$

or $\underline{\dfrac{P}{ma} = \dfrac{7}{4} = 1.75}$

18.105

Given: Two spheres are rigidly connected by a light rod that rotates about a frictionless ball joint at O with a constant angular velocity of 120 rpm in a horizontal plane on top of a vertical column 10 ft high (Fig. a). The larger sphere weighs 300 lb and the smaller sphere weighs 50 lb.

a) Determine the horizontal force that the rod exerts on the column. Use the inertial-force method.

b) Determine the bending moment in the column at the ground level.

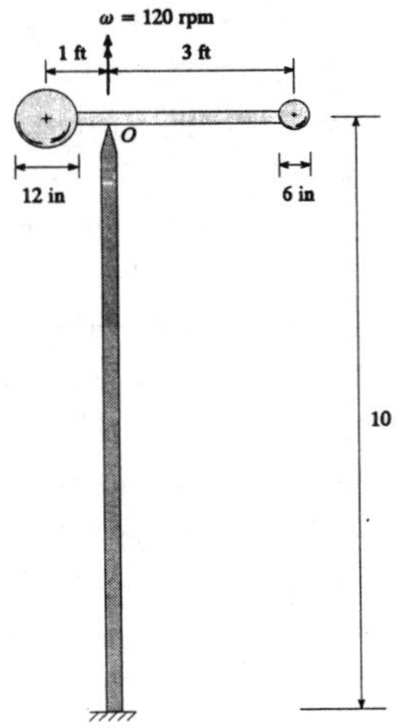

Figure a

Solution:

The horizontal forces including inertial forces, that act on the rod are shown in Fig. b.

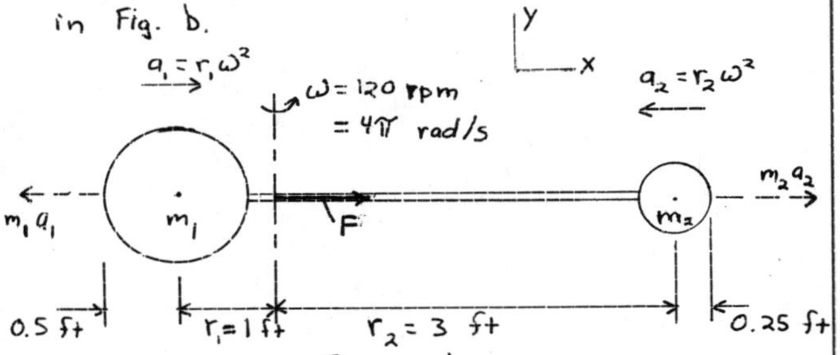

Figure b

By Fig. b,

$$\Sigma F_x = F + m_2a_2 - m_1a_1 = 0$$

or

$$F = m_1a_1 - m_2a_2 = \left[\left(\frac{300}{32.2}\right)(1) - \left(\frac{50}{32.2}\right)(3)\right](4\pi)^2$$

So, $\underline{F = 735.6 \text{ lb}}$

(Continued)

18.105 Cont.

Then, by Fig. c, the moment in the column at ground level is given by,

$\circlearrowleft + \Sigma M_O = M - 10F = 0$

or

$M = 10F = 7356$ lb-ft

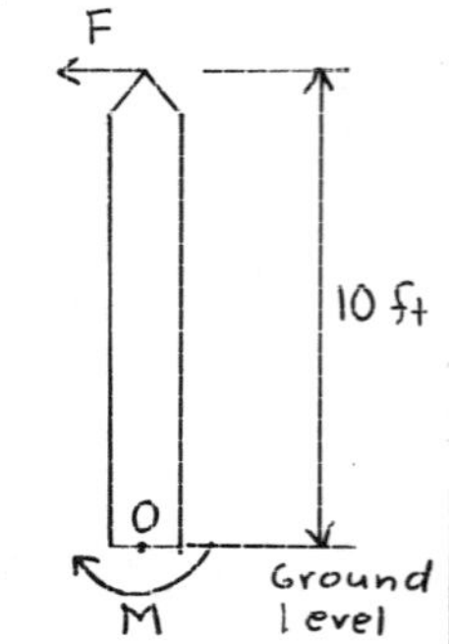

Figure c

18.106

Given: A race car loses a wheel of diameter 750 mm. The wheel rotates with constant angular velocity along a straight path on the horizontal track. Due to uneven tread loss, the center of mass of the wheel lies 100 mm from its geometric center.

a) Determine the maximum angular speed (rpm) with which the wheel can rotate and still maintain continuous contact with the track. Use the inertial-force method.

b) Given that the speed of the car was 240 km/h when the wheel come off, determine whether or not the wheel maintains continuous contact with the track.

Solution:

a) The critical case is when the center of mass is directly above the geometric center of the wheel (Fig. a).

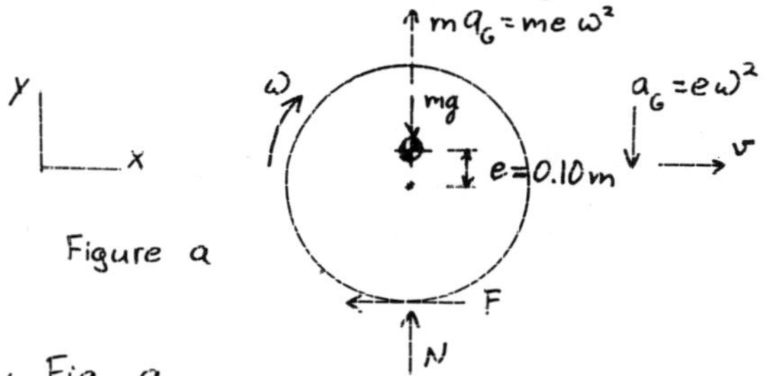

Figure a

By Fig. a,

$\Sigma F_y = N + me\omega^2 - mg = 0$

or, $N = m(g - e\omega^2)$

For the wheel to remain on the track, $N \geq 0$, or $g - e\omega^2 \geq 0$. Hence, we must have,

$\omega \leq \sqrt{g/e} = \sqrt{9.81/0.10} = 9.905$ rad/s

So, for the wheel to remain on the ground, the maximum angular speed (rpm) is

$\omega = (9.905)\left(\frac{60}{2\pi}\right) = 94.58$ rpm

b) Since the speed of the car (and therefore of the geometrical center of the wheel) is $v = 240$ km/h $(= 240{,}000/3600 = 66.66\bar{6}$ m/s$)$, by kinematics, with $r = 0.375$ m,

$\omega = v/r = 66.66\bar{6}/0.375 = 177.8$ rad/s

Since $\omega > 9.905$ rad/s, the wheel will not maintain continuous contact with the track.

18.107

Given: A uniform slender rod AB of mass 9 kg is placed in a box in a horizontal attitude (Fig. a). The box and rod rotate with angular velocity $\omega = 9$ rad/s about the vertical axis through the center O.

Determine the forces exerted on the rod by the three contacting sides of the box. Neglect friction and use the inertial-force method.

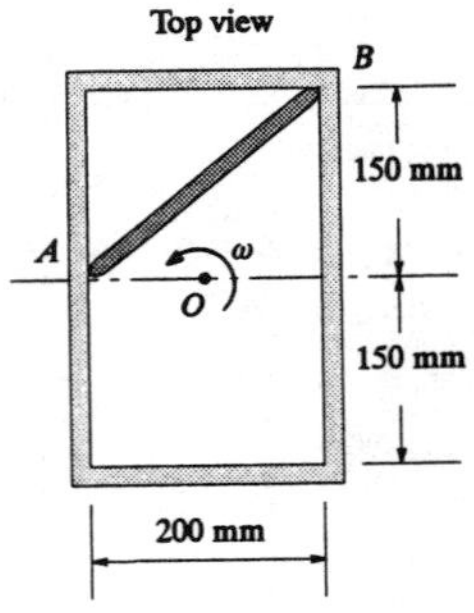

Figure a

Solution:

The free-body diagram of the rod, including the inertial force, is shown in Fig. b for the horizontal plane

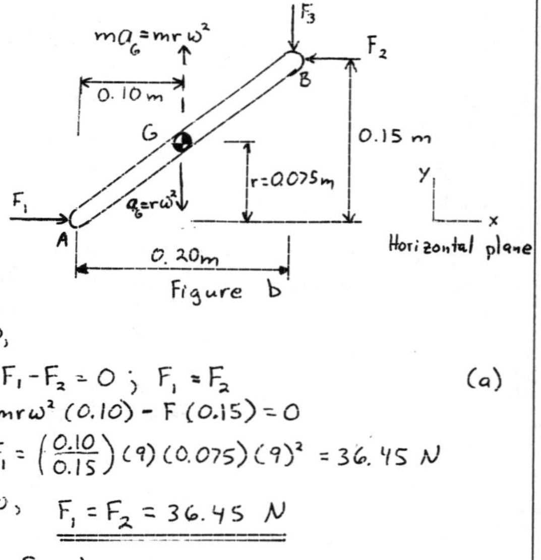

Figure b

By Fig. b,

$\Sigma F_x = F_1 - F_2 = 0; \quad F_1 = F_2$ (a)

$\circlearrowleft + \Sigma M_B = mr\omega^2 (0.10) - F(0.15) = 0$

or $F_1 = \left(\frac{0.10}{0.15}\right)(9)(0.075)(9)^2 = 36.45$ N

so, $F_1 = F_2 = 36.45$ N

Also, by Fig. b,

$\Sigma F_y = mr\omega^2 - F_3 = 0 \quad ; \quad F_3 = mr\omega^2$

So, $F_3 = (9)(0.075)(9)^2 = 54.66$ N

18.108

Given: By Problem 18.81, a motionless solid ball hanging by a cord is subjected to a horizontal force F at its lowest point. By the inertial-force method.

a) Derive formulas for the initial angular acceleration of the ball and the initial acceleration of its center, in terms of F and the mass m and the radius r of the ball.

b) Determine the distance d from the center of the ball to the point P on the vertical axis that has no acceleration.

Solution: The free-body diagram of the ball, including the inertial force and inertial couple, is shown in Fig. a, at the instant that F is applied.

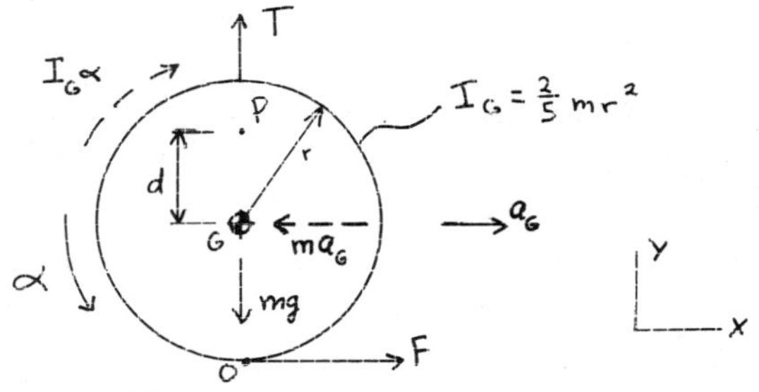

Figure a

By Fig. a,

$$\circlearrowleft \Sigma M_O = I_G\alpha - (ma_G)(r) = 0$$

or

$$a_G = \frac{2}{5} r\alpha \qquad \text{(a)}$$

Also, by Fig. a,

$$\Sigma F_x = F - ma_G = 0$$

or

$$a_G = F/m \qquad \text{(b)}$$

Equations (a) and (b) yield

$$\alpha = \frac{5}{2}\frac{F}{mr} \qquad \text{(c)}$$

b) By kinematics, the acceleration of a point P distance d from G is

$$a_P = a_G - \alpha d \qquad \text{(d)}$$

For $a_P = 0$, Eq. (d) yields, with Eqs. (b) and (c)

$$d = \frac{a_G}{\alpha} = \frac{2}{5} r$$

18.109

Given: By Problem 18.82, the system shown in Fig. a is released from rest. The homogeneous disk A rolls on the horizontal surface. The pulley B is frictionless and has negligible mass.

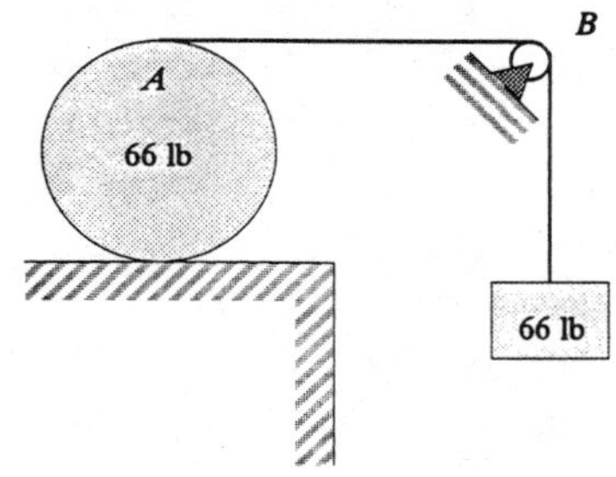

Figure a

Determine the acceleration of the center of the disk and the tension in the cord.

Solution: Free-body diagrams of the 66 lb disk and the 66 lb weight are shown in Figs. b and c, including the inertial force and the inertial couple.

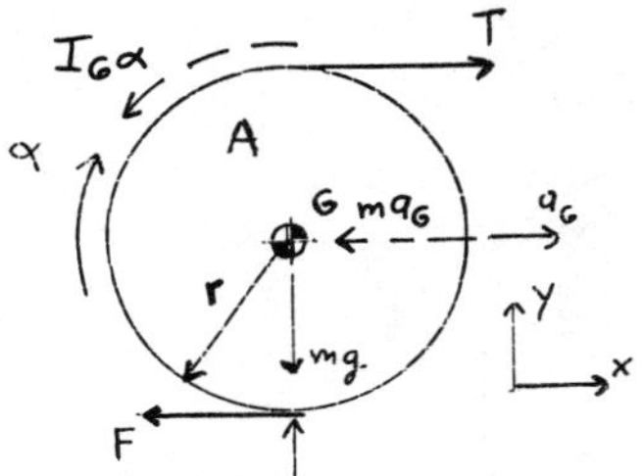

Figure b

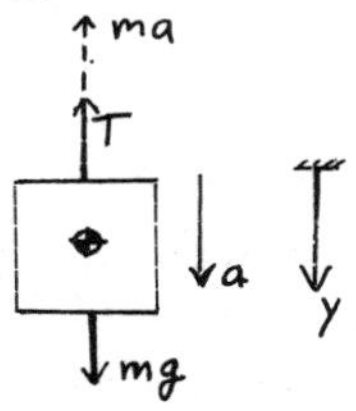

Figure c

By Fig. b,

$$\circlearrowleft \Sigma M_O = 2Tr - I_G\alpha - ma_G r = 0 \qquad \text{(a)}$$

By kinematics,

$$a_G = \alpha r \qquad \text{(b)}$$

$$a = 2a_G \qquad \text{(c)}$$

Equations (a) and (b) yield, with $I_G = \frac{1}{2}mr^2$,

$$T = \frac{3}{4} m a_G \qquad \text{(d)}$$

By Fig. c,

$$\Sigma F_y = mg - T - ma = 0 \qquad \text{(e)}$$

By Eqs. (c) and (e),

$$T = mg - 2m a_G \qquad \text{(f)}$$

Equations (d) and (f) yield

$$a_G = \frac{4}{11} g = \frac{4}{11}(32.2) = 11.709 \text{ ft/s}^2$$

$$T = \frac{3}{4}\left(\frac{66}{g}\right)\left(\frac{4}{11} g\right) = 18 \text{ lb}$$

18.110

Given: By Problem 18.83, a homogeneous circular disk rotates in a horizontal plane about a fixed vertical axle O (Fig. a), two feet from the mass center G of the disk. The mass of the disk is $m = 10$ lb·s²/ft. A clockwise couple $M = 10t$ [lb·ft], with t in seconds, is applied to the disk. Initially, the disk is at rest.

For $t = 4$ s, determine the perpendicular and parallel components to the line OG of the force exerted on the disk by the axle.

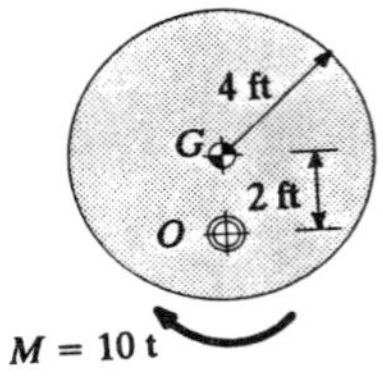

Figure a

Solution:

The free-body diagram including the inertial force and inertial couple, of the disk is shown in Fig. b, where O_x, O_y are components of the axle reaction parallel and perpendicular, respectively, to OG.

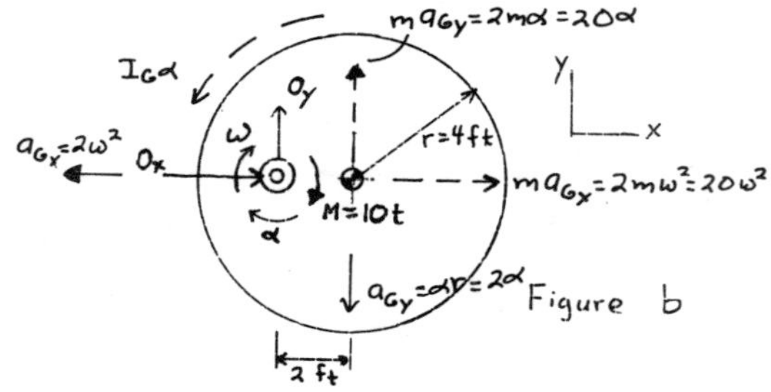

Figure b

By Fig. b, the mass moment of inertia about the mass center G of the disk is,

$$I_G = \tfrac{1}{2}mr^2 = \tfrac{1}{2}(10)(4)^2 = 80 \text{ lb·ft·s}^2 \qquad (a)$$

Then, by Fig. b,

$$\Sigma F_x = m a_{Gx} + O_x = 0 \;;\quad O_x = -20\omega^2 \qquad (b)$$

$$\Sigma F_y = m a_{Gy} + O_y = 0 \;;\quad O_y = -20\alpha \qquad (c)$$

$$\circlearrowright \Sigma M_O = M - I_G\alpha - (m a_{Gy})2 = 0$$

or

$$10t = (80)(\alpha) + (20\alpha)(2) = 120\alpha = 120\frac{d\omega}{dt} \qquad (d)$$

Integration of Eq. (d) yields

$$\omega = \frac{1}{24}t^2 \qquad (e)$$

For $t = 4$ s, Eqs. (e) and (d) yield,

$$\omega = \tfrac{2}{3} \text{ rad/s} \qquad (f)$$

$$\alpha = \tfrac{1}{3} \text{ rad/s}^2 \qquad (g)$$

Equations (b), (c), (f), and (g) yield

$$\underline{O_x = -\frac{80}{9} = -8.89 \text{ lb}}$$

$$\underline{O_y = -\frac{20}{3} = -6.67 \text{ lb}}$$

18.111

Given: By Problem 18.84, a ball B and a cylinder C with equal masses and equal radii rest on a horizontal plank (Fig. a). The plank is given an acceleration a, directed to the right as shown. Assume that the bodies do not slip on the plank.

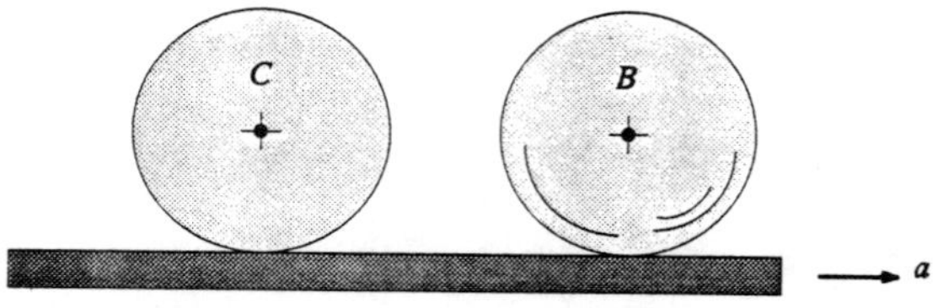

Figure a

Determine whether or not the bodies approach each other.

Solution:

The free-body diagram of either body, including inertial force and inertial couple is shown in Fig. b.

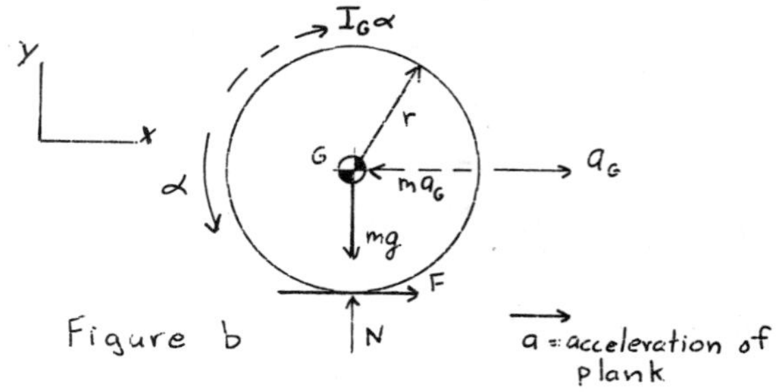

Figure b

By Fig. b, for either body,

$$\circlearrowleft \Sigma M_O = m a_G r - I_G\alpha = 0$$

or

$$a_G = \frac{I_G}{mr}\alpha \qquad (a)$$

By kinematics and Fig. b,

$$a_G = a - \alpha r \qquad (b)$$

Equating Eqs. (a) and (b) and solving for α, we obtain,

$$\alpha = \frac{m a r}{I_G + m r^2} \qquad (c)$$

For ball B, $I_G = \frac{2}{5}mr^2$. Therefore, for ball B, by Eq. (c),

$$\alpha_B = \frac{5}{7}\frac{a}{r}, \text{ counterclockwise}$$

Similarly for cylinder C, $I_G = \frac{1}{2}mr^2$, and

$$\alpha_C = \frac{2}{3}\frac{a}{r}, \text{ counterclockwise}$$

Therefore, $\alpha_B > \alpha_C$, and by Eq. (b), the acceleration a_{GC} of cylinder C is greater than acceleration a_{GB} of ball B. Hence, the bodies approach each other (see Fig. a).

18.112

Given: By Problem 18.85, a homogeneous cylinder of mass m rolls on a horizontal surface when subjected to a clockwise couple $M = a\theta^2 + b\theta$; a and b are constants and θ is the angular displacement of the cylinder in radians (Fig. a).

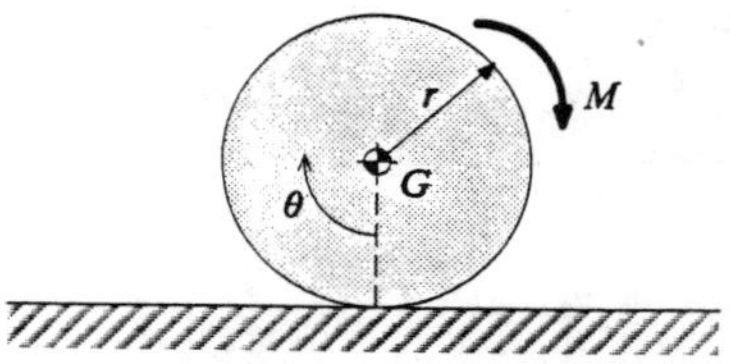

Figure a

Express the speed of the center of the cylinder as a function of a, b, r, θ, m, and ω_0, where ω_0 is the angular velocity when $\theta = 0$.

Solution:

The free-body diagram, including the inertial force and the inertial couple, is shown in Fig. b,

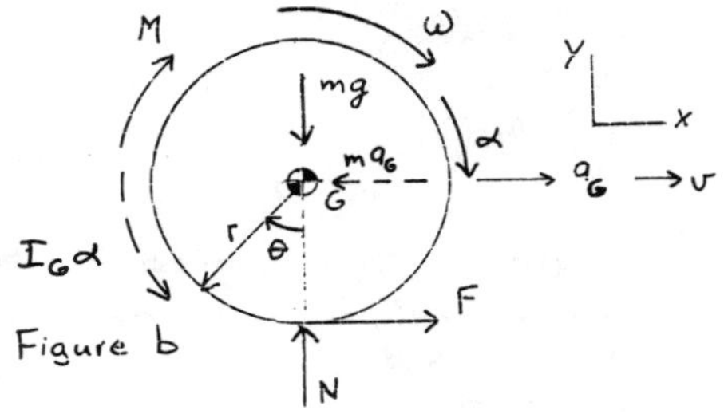

Figure b

By Fig. b,

$$\circlearrowleft \Sigma M_0 = M - I_G\alpha - ma_G r = 0 \qquad (a)$$

where $I_G = \frac{1}{2}mr^2$, and by kinematics,

$$a_G = r\alpha = r\ddot{\theta} \qquad (b)$$

By Eqs. (a) and (b), with $M = a\theta^2 + b\theta$,

$$a\theta^2 + b\theta = \frac{3}{2}mr^2\ddot{\theta} \qquad (c)$$

Also, by kinematics and the chain rule,

$$\ddot{\theta} = \frac{d\omega}{dt} = \frac{d\omega}{d\theta}\frac{d\theta}{dt} = \omega\frac{d\omega}{d\theta}$$

Then, by Eq. (c),

$$\frac{3}{2}mr^2\omega\,d\omega = (a^2\theta + b\theta)\,d\theta \qquad (d)$$

Integration of Eq. (d) yields

$$\frac{3}{4}mr^2(\omega^2 - \omega_0^2) = \frac{1}{3}a\theta^3 + \frac{1}{2}b\theta^2$$

where $\omega = \omega_0$ when $\theta = 0$. Hence,

$$\omega^2 = \omega_0^2 + \frac{4}{3mr^2}\left(\frac{a}{3}\theta^3 + \frac{b}{2}\theta^2\right) \qquad (e)$$

By kinematics and Eq. (e), we obtain

$$v = r\omega = r\sqrt{\omega_0^2 + \frac{4}{3mr^2}\left(\frac{a}{3}\theta^3 + \frac{b}{2}\theta^2\right)}$$

18.113

Given: By Problem 18.86, a solid homogeneous spherical body that weighs 64 lb rests on a horizontal table and is pushed by a moving block (Fig. a). The normal force between the sphere and the block is 20 lb. The coefficient of friction between sliding surfaces is 0.25.

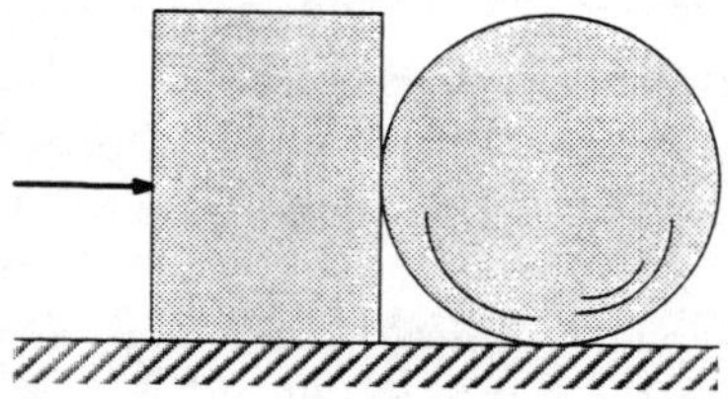

Figure a

a) Assuming that the sphere rolls, determine the acceleration of its center.

b) Determine the frictional force that the table exerts on the sphere, and verify that the sphere does not slide on the table.

Solution:

a) The free-body diagram of the sphere, including the inertial force and the inertial couple, is shown in Fig. b.

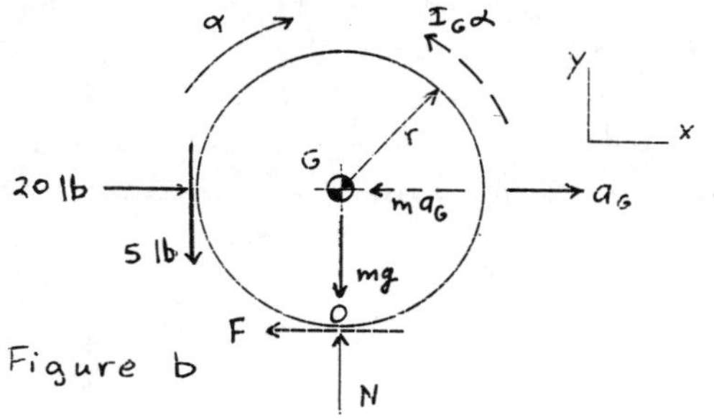

Figure b

By Fig. b,

$$\circlearrowleft + \Sigma M_0 = ma_G r + I_G\alpha + 5r - 20r = 0 \qquad (a)$$

where, $\quad I_G = \frac{1}{2}mr^2$

By kinematics, $\quad a_G = \alpha r \qquad (b)$

Hence by Eqs. (a) and (b), with $m = \frac{64}{32.2}$ lb·s²/ft,

$$\underline{a_G = 5.391 \text{ ft/s}^2}$$

(b) By Fig. b,

$$\Sigma F_x = 20 - F - ma_G = 0 \qquad (d)$$

$$\Sigma F_y = N - mg - 5 = 0 \qquad (e)$$

Equations (c), (d), and (e) yield

$$\underline{F = 9.286 \text{ lb}} \qquad (f)$$

$$N = 69 \text{ lb} \qquad (g)$$

18.113 Cont.

The frictional force at impending slipping is, with Eq. (g),

$$F_{max} = \mu N = (0.25)(69) = 17.25 \text{ lb}$$

Since, $F < F_{max}$, the sphere does not slip.

18.114

Given: A wrecking-ball mechanism consists of a sphere that weighs 236 lb and has a 1 ft diameter, a light rod, and a shaft (Fig. a). Starting from rest, the rod and sphere rotate about the axis of the shaft through an angular displacement $\theta = t^3 + t - 2$ [rad], with t in seconds. At $t = 1$ s, the rod coincides with the x axis (Fig. a)

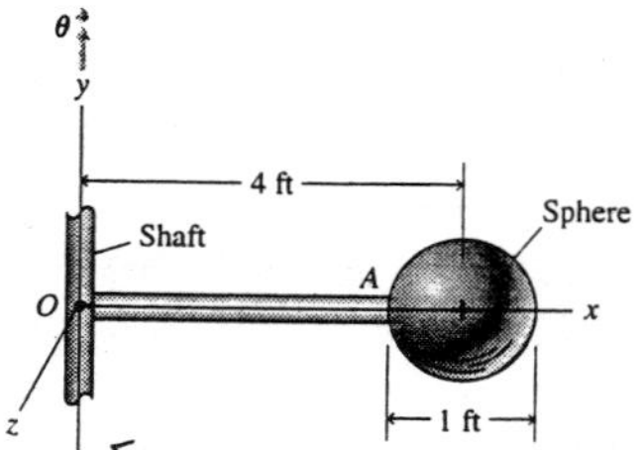

Figure a

For $t = 1$s, determine the force components (F_{x_1}, F_{y_1}, F_z) and the couple components (M_x, M_y, M_z) that the vertical shaft exerts on rod OA.

Solution: The free-body diagram of the rod and sphere, including the inertial force and inertial couple, is shown in Fig. b.

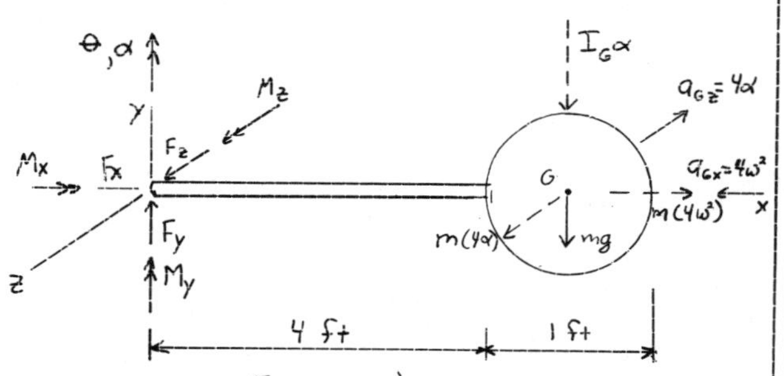

Figure b

By kinematics, at $t = 1$s,

$$\omega = \frac{d\theta}{dt} = 3t^2 + 1 = 4 \text{ rad/s} \qquad (a)$$

$$\alpha = \frac{d\omega}{dt} = 6t = 6 \text{ rad/s}^2 \qquad (b)$$

Also, for a sphere, with $m = 236/32.2 = 7.329 \text{ lb·s}^2/\text{ft}$,

$$I_G = \frac{2}{5} mr^2 = \frac{2}{5}(7.329)(0.5)^2 = 0.7329 \text{ lb·ft·s}^2$$

Hence, $\quad I_G \alpha = 4.398 \text{ lb·ft} \qquad (c)$

Then, by Fig. b and Eqs. (a), (b), and (c), for $t = 1$s,

$\Sigma F_x = F_x + 4m\omega^2 = 0$; $\underline{F_x = -117.3 \text{ lb}}$

$\Sigma F_y = F_y - mg = 0$; $\underline{F_y = 236 \text{ lb}}$

$\Sigma F_z = F_z + 4m\alpha = 0$; $\underline{F_z = 175.9 \text{ lb}}$

$\Sigma M_x = M_x = 0$; $\underline{M_x = 0}$

$\Sigma M_y = M_y - I_G\alpha - (4m\alpha)(4) = 0$; $\underline{M_y = 708 \text{ lb·ft}}$

$\Sigma M_z = M_z - (mg)(4) = 0$; $\underline{M_z = 944 \text{ lb·ft}}$

18.115

Given: A thin uniform bar of mass $m = 0.005 \text{ lb·s}^2/\text{in}$ is welded to a disk at C (Fig. a). The disk rotates about axis O with constant angular velocity $\omega = 20$ rad/s.

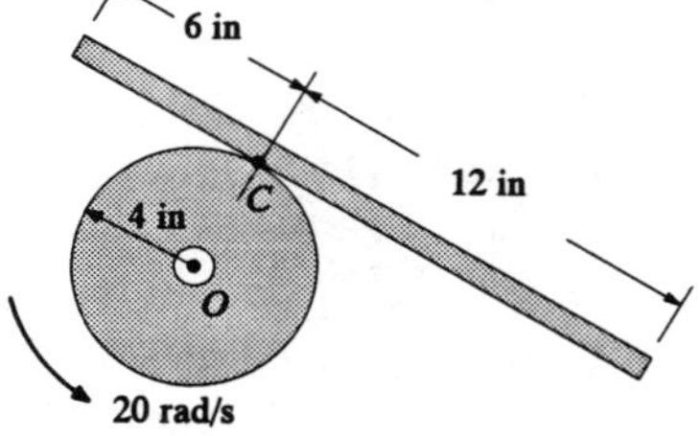

Figure a

Neglecting gravity, calculated the tangential force F, the normal force N, and the moment M exerted on the bar by the weld.

Solution:

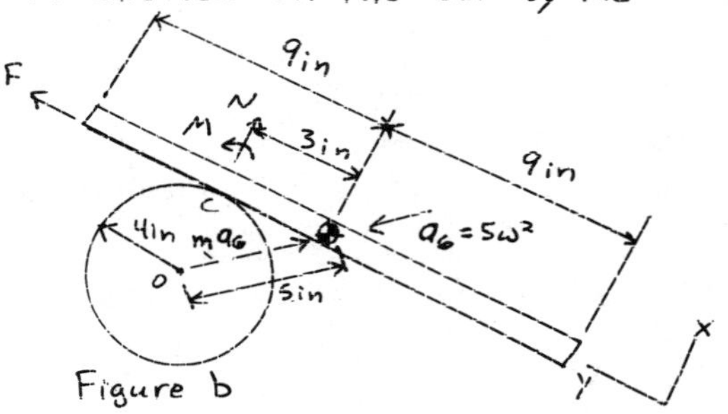

Figure b

Figure b is the free-body diagram of the rod, including the inertial force. The inertial couple $I_G\alpha = 0$, since ω is constant.

By Fig. b,

$$\Sigma F_x = N + \frac{4}{5} m a_G = 0$$

$$\Sigma F_y = F - \frac{3}{5} m a_G = 0$$

$$\Sigma M_C = M + \frac{4}{5} m a_G (3) = 0$$

with $m = 0.005 \text{ lb·s}^2/\text{in}$ and $a_G = 5\omega^2 = 2000 \text{ in/s}^2$; these equations yield,

$\underline{N = -8 \text{ lb}}$

$\underline{F = 6 \text{ lb}}$

$\underline{M = -24 \text{ lb·in} = -2 \text{ lb·ft}}$

18.116

Given: By Problem 18.87, a mechanical system consists of a drum A (mass $m_A = 150$ kg and radius of gyration $k = 0.5$ m) and a solid homogeneous cylinder (mass $m_B = 75$ kg), connected by a flexible inextensible cord that is wrapped around A, passed over a frictionless pulley, and tied to a light rod pinned to B (Fig. a).

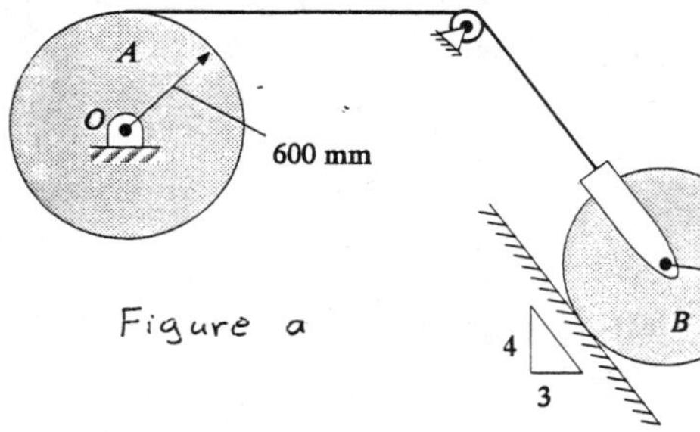

Figure a

Neglecting friction in the bearing O, determine the acceleration of the center of roller B.

Solution: Figure b shows the free-body diagrams of the drum and the cylinder, including inertial forces and inertial couples.

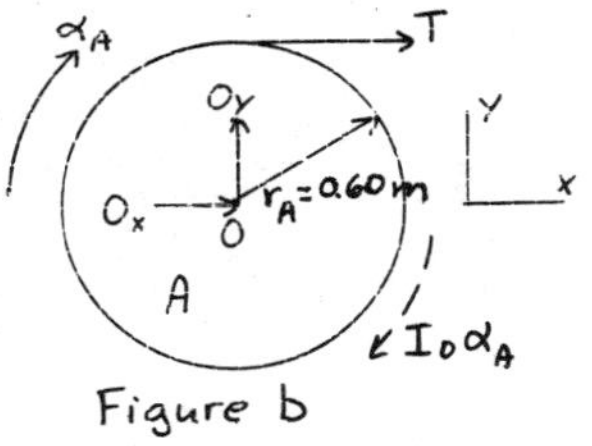

Figure b

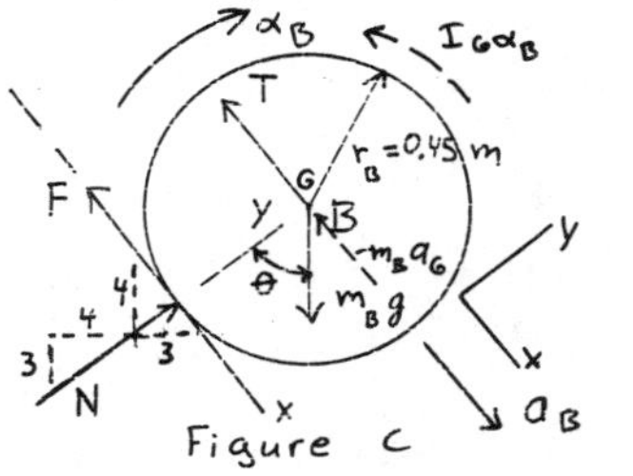

Figure c

By Fig. b,

$$\circlearrowleft + \Sigma M_O = I_O\alpha_A - Tr_A = 0 \qquad (a)$$

where $I_O = m_A k^2 = (150)(0.5)^2 = 37.5\ \text{N}\cdot\text{m}\cdot\text{s}^2$ and $r_A = 0.60$ m

By kinematics and Figs. b and c,

$$r_B\alpha_B = r_A\alpha_A = a_G \qquad (b)$$

By Eqs. (a) and (b), we obtain

$$T = 104.166\, a_G \qquad (c)$$

By Fig. c,

$$\circlearrowleft + \Sigma M_C = I_G\alpha_B + Tr_B + m_B a_G r_B - m_B g(\sin\theta)r_B = 0 \qquad (d)$$

where, $I_G = \frac{1}{2}m_B r_B^2$ and $\sin\theta = 4/5$.

Then, by Eqs. (b) and (d), we find

$$T = 0.8\, m_B g - 1.5\, m_B a_G$$

or with $m_B = 75$ kg, $\quad T = 588.6 - 112.5 a_G \qquad (e)$

Equating Eqs. (c) and (e), we obtain

$$\underline{a_G = 2.72\ \text{m/s}^2}$$

18.117

Given: By Problem 18.88, the system shown in Fig. a is used to lower a drum C to a lower floor. Initially $F = 0$, and the system is released from rest.

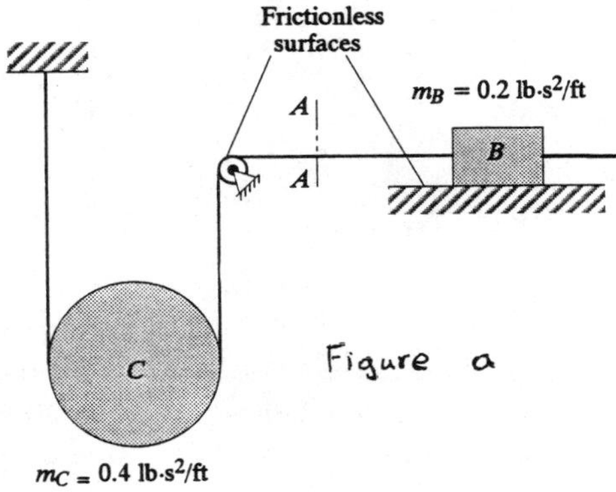

Figure a

a) Assuming that the cord does not slip on drum C, determine the tension in the cord at section A-A.

b) After the drum has dropped 1 s, the force F is applied, and the drum comes to rest when it has dropped a total of 10 ft to the lower floor. Determine the force F.

Solution:

a) Figures b and c are the free-body diagrams of drum C and block B, respectively, including the inertial forces.

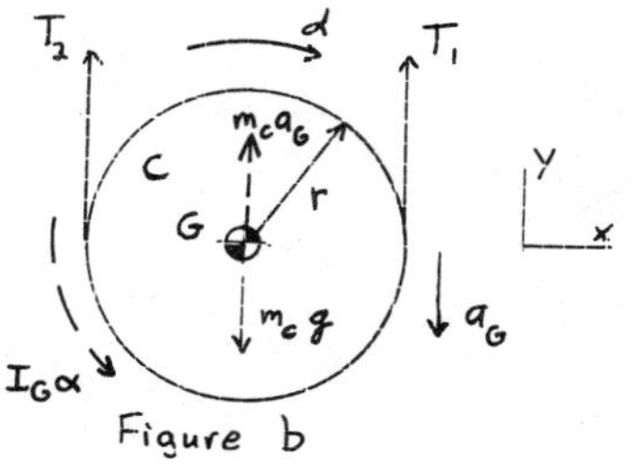

Figure b

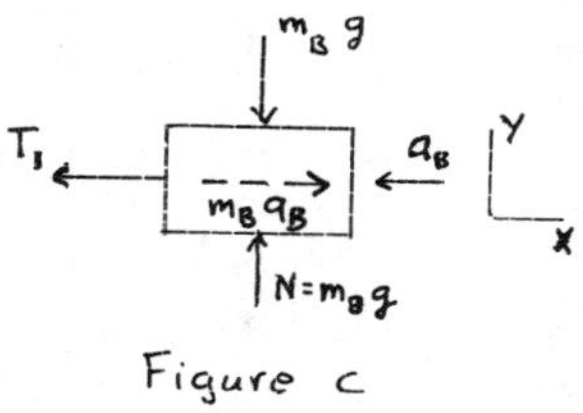

Figure c

By Figs. (a) and (b) and kinematics,

$$a_G = \alpha r \qquad (a)$$

Also by Fig. b,

$$\circlearrowright + \Sigma M_G = (T_2 - T_1)r - I_G\alpha = 0 \qquad (b)$$

$$\Sigma F_y = T_1 + T_2 + m_C a_G - m_C g = 0 \qquad (c)$$

where $I_G = \frac{1}{2}m_C r^2$

By kinematics and Figs. (a), (b), and (c), if the block moves a distance d to the left, the cylinder descends a distance d/2. Hence,

$$a_G = \tfrac{1}{2}a_B \qquad (d)$$

(Continued)

18.117 Cont.

By Fig. c,

$$\Sigma F_x = m_B a_B - T_1 = 0$$

or

$$T_1 = m_B a_B \qquad (e)$$

Equations (a), (b), (c), and (d) yield

$$2T_1 = m_c g - \tfrac{3}{4} m_c a_B \qquad (f)$$

Equations (e) and (f) yield, with $m_B = 0.2\ \text{lb·s}^2/\text{ft}$ and $m_c = 0.4\ \text{lb·s}^2/\text{ft}$,

$$a_B = \frac{4 m_c g}{8 m_B + 3 m_c} = 18.40\ \text{ft/s}^2 \qquad (g)$$

Equations (e) and (g) yield

$$T_1 = m_B a_B = (0.2)(18.40) = 3.68\ \text{lb}$$

b) First, we determine the distance that cylinder C descends in 1s. Since $a_G = \frac{1}{2} a_B$ is constant, the distance that the cylinder drops is,

$$s = \tfrac{1}{2} a_G t^2 = \tfrac{1}{2}(\tfrac{1}{2} a_B)(1)^2 = 4.6\ \text{ft}$$

So, at the end of one second, the cylinder is a distance above the floor of

$$s_1 = 10 - 4.6 = 5.4\ \text{ft}$$

and it is traveling at a speed of,

$$v_1 = a_G t = \tfrac{1}{2} a_B t = \tfrac{1}{2}(18.40)(1) = 9.20\ \text{ft/s}.$$

By kinematics and the chain rule,

$$a_G = \frac{dv}{dt} = \frac{dv}{ds}\frac{ds}{dt} = v\frac{dv}{ds}$$

Hence, since $v = 0$, when the cylinder touches the floor, integration yields since a_G = constant

$$a_G \int_0^{5.4} ds = \int_{9.2}^{0} v\, dv$$

or

$$a_G \left[s \Big|_0^{5.4} \right] = \tfrac{1}{2} v^2 \Big|_{9.2}^{0}$$

so

$$a_G = -7.837\ \text{ft/s}^2 \qquad (h)$$

Thus, $a_G = 7.837\ \text{ft/s}^2$ up (Fig. b), and by Eq. (d),

$$a_B = 2a_G = 15.674\ \text{ft/s}^2 \qquad (i)$$

to the right (Fig. d)

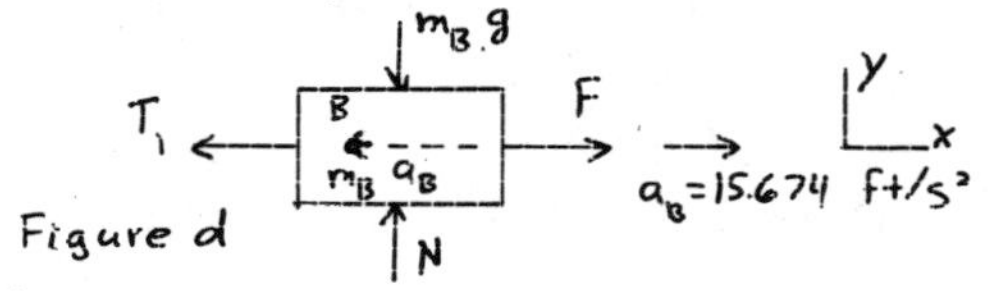

Figure d

By Fig. d and Eq. (i),

$$\Sigma F_x = F - T_1 - m_B a_B = 0$$

or

$$F - T_1 = m_B a_B = (0.2)(15.674)$$

so

$$F - T_1 = 3.135\ \text{lb} \qquad (j)$$

By Eqs. (f) and (i),

$$T_1 = \tfrac{1}{2}(m_c g - \tfrac{3}{4} m_c a_B) = 4.089\ \text{lb} \qquad (k)$$

Equations (i) and (k) yield

$$F = 4.089 + 3.135 = 7.224\ \text{lb}$$

18.118

Given: By Problem 18.89, a cord is wrapped around a spool (Fig. a), whose radius of gyration is k. The coefficient of sliding friction between the spool and inclined plane is μ_K. The spool is released from rest.

a) Show that if the angle θ is large enough to permit motion, the acceleration a of the center of the spool is

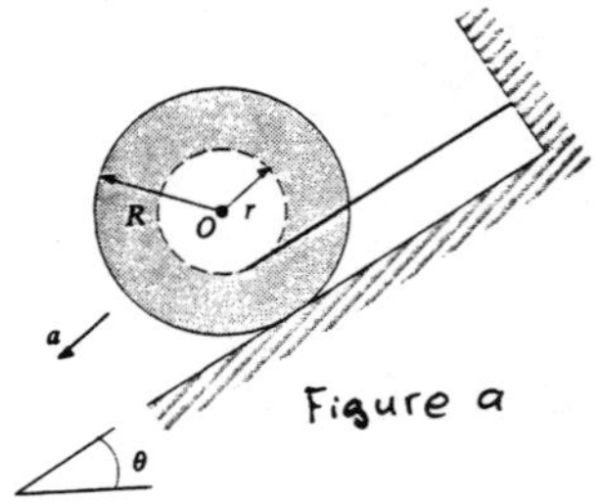

Figure a

$$a = \frac{r^2 g[\sin\theta + \mu_K(1 - \frac{R}{r})(\cos\theta)]}{r^2 + k^2} \qquad (a)$$

b) From Eq. (a), derive an equation for the smallest angle θ at which motion occurs.

Solution:

a) Since $r < R$, the spool must skid and rotate if it moves. Then, as the spool unwinds, the acceleration a of the center of the spool (Fig. a) is, in terms of the angular acceleration α,

$$a = r\alpha \qquad (b)$$

The free-body diagram of the spool, including the inertial force and the inertial couple, is shown in Fig. b.

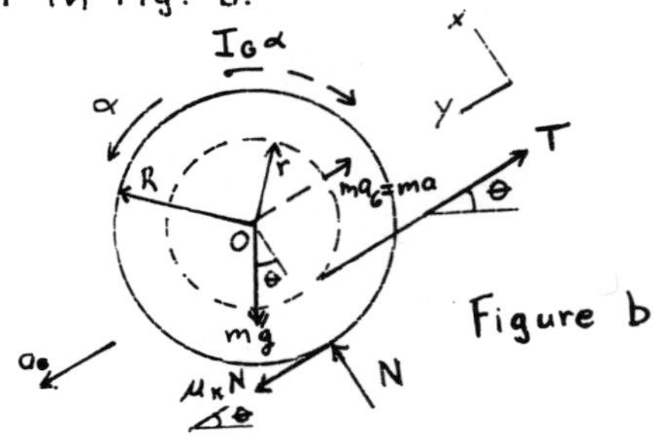

Figure b

By Fig. b,

$$\Sigma F_x = ma - \mu_K N + T - mg\sin\theta = 0 \qquad (c)$$

$$\Sigma F_y = N - mg\cos\theta = 0;\ N = mg\cos\theta \qquad (d)$$

$$\circlearrowleft + \Sigma M_G = I_G\alpha - Tr + \mu_K NR = 0 \qquad (e)$$

where, $I_G = mk^2$ and with Eq. (b)

$$I_G\alpha = \frac{mk^2 a}{r} \qquad (f)$$

(Continued)

18.118 Cont.

By Eqs. (c) and (d)

$$Tr = mgr\sin\theta + \mu_k mgr\cos\theta - mar \qquad (g)$$

Then, by Eqs. (d), (e), (f), and (g), we obtain, solving for a,

$$a = \frac{r^2 g\left[\sin\theta + \mu_k\left(1-\frac{R}{r}\right)\cos\theta\right]}{r^2 + k^2} \qquad (h)$$

Equation (h) agrees with Eq. (a)

b) Equation (h) is valid only if $a \geq 0$, since the spool will not move up the plane. Hence, the smallest value of θ for impending motion (called it θ_0) is given by Eq. (h) with $a=0$. Hence,

$$\sin\theta_0 + \mu_k\left(1 - R/r\right)\cos\theta_0 = 0$$

or

$$\tan\theta_0 = \mu_k\left(\frac{R}{r} - 1\right)$$

Thus, motion will occur if θ is slightly greater than

$$\theta_0 = \tan^{-1}\left[\mu_k\left(\frac{R}{r} - 1\right)\right]$$

18.119

Given: By Problem 18.90, the bowling ball has a backspin as it moves along the horizontal lane (Fig. a). The coefficient of kinetic friction is 0.25. The initial speed of the ball is 1.0 m/s. After it travels a certain distance to the right it stops and remains motionless.

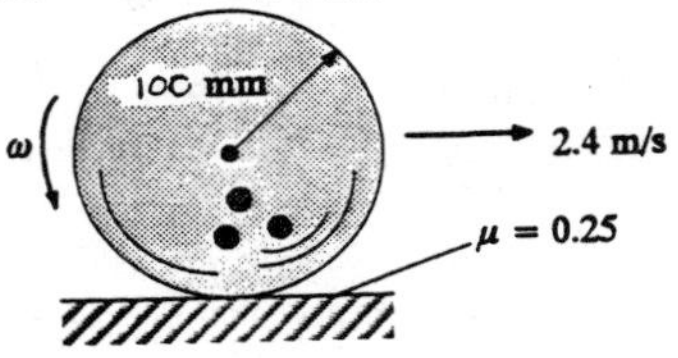

Figure a

a) Determine the distance the ball travels
b) Determine the time required for the ball to stop.
c) Determine the initial value of the ball's angular velocity

Solution:

a) Figure b is the free-body diagram of the ball, including the inertial force and the inertial couple.

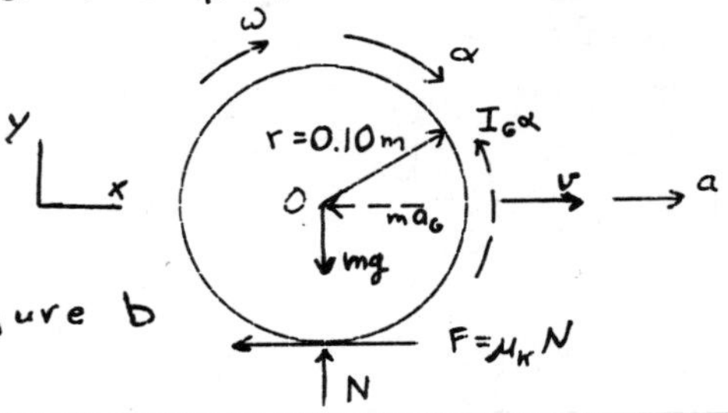

Figure b

By Fig. b,

$$\Sigma F_x = -m a_G - \mu_k N = 0 \qquad (a)$$

$$\Sigma F_y = N - mg = 0;\ N = mg \qquad (b)$$

By Eqs. (a) and (b), with $\mu_k = 0.25$,

$$a_G = -\frac{1}{4}g \qquad (c)$$

By kinematics and the chain rule,

$$a_G = \frac{dv}{dt} = \frac{dv}{dx}\frac{dx}{dt} = v\frac{dv}{dx} \qquad (d)$$

Then, by Eqs. (c) and (d), $v\,dv = -\frac{1}{4}g\,dx$

Integration yields

$$\frac{v^2}{2} - \frac{v_0^2}{2} = -\frac{1}{4}gx \qquad (e)$$

The ball comes to rest when $v=0$; hence, by Eq. (e),

$$x_{max} = \frac{2v_0^2}{g} = \frac{2(1.0)^2}{9.81} = 0.2039 \text{ m}$$

b) Since a_G is constant, by kinematics,

$$v = v_0 + a_G t = v_0 - \frac{1}{4}gt$$

when the ball comes to rest, $v=0$. Hence,

$$t = \frac{4v_0}{g} = \frac{4(1.0)}{9.81} = 0.4077 \text{ s}$$

c) By Fig. b,

$$\circlearrowleft+\ \Sigma M_O = m a_G r + I_G\alpha = 0 \qquad (f)$$

where $I_G = \frac{2}{5}mr^2$. By Eq. (f),

$$\alpha = -\frac{5}{2r}a_G = -\frac{5}{2(0.10)}\left(-\frac{1}{4}g\right)$$

or

$$\alpha = 61.31 \text{ rad/s}^2$$

(Note that if the ball were rolling $\alpha = a_G/r = -\frac{1}{4}\frac{g}{r} = -24.53 \text{ rad/s}^2$)

Since α is constant, by kinematics

$$\omega = \omega_0 + \alpha t \qquad (g)$$

when $t = 0.4077$ s $\omega = 0$. Then, Eq. (g) yields, (see Fig. b)

$$\omega_0 = -(61.31)(0.4077) = -25 \text{ rad/s}$$

That is, the initial angular velocity ω_0 is opposite in sense to the angular acceleration α.

18.120

Given: By Problem 18.91, the connecting rod AB weighs 16.1 lb (Fig. a). Siders A and B are frictionless and of negligible mass. The mechanism is at rest when a 10 lb horizontal force is applied to A. For the instant after the force is applied, calculate the accelerations of A and B, the force B exerts on the vertical guides, and the force A exerts on the horizontal surface.

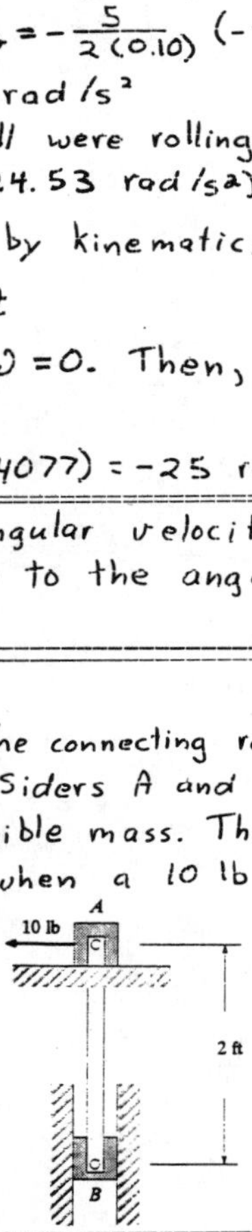

Figure a

(Continued)

18.120

Solution:

Figure b is the free-body diagram of rod AB, including the inertial force and the inertial couple.

By Fig. b,

$\Sigma F_x = B - 10 - m a_{Gx} = 0$ (a)

$\Sigma F_y = A - mg - m a_{Gy} = 0$ (b)

$\circlearrowleft \Sigma M_B = (10)(2r) + I_G \alpha + m a_{Gx}(r) = 0$ (c)

where $I_G = \frac{1}{12} m (2r)^2 = \frac{1}{3} m r^2$

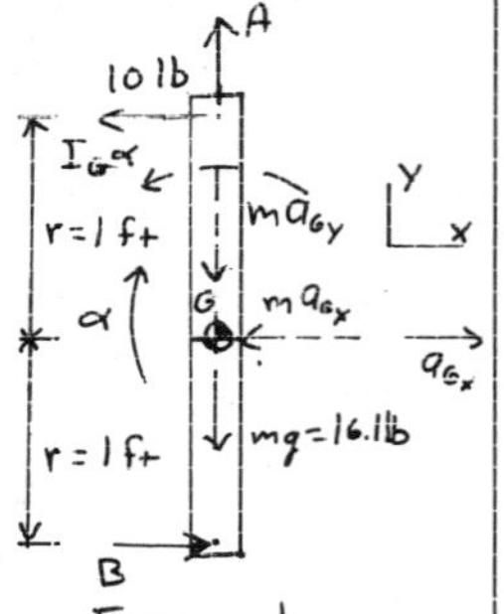

Figure b

The accelerations of points A, B, and G are shown in Fig. c.

By kinematics and Fig. c,

$a_{Ax} = a_{Gx} + r\alpha$ (d)

$a_{Ay} = a_{Gy} - r\omega^2$ (e)

$a_{Bx} = a_{Gx} - r\alpha$ (f)

$a_{By} = a_{Gy} + r\omega^2$ (g)

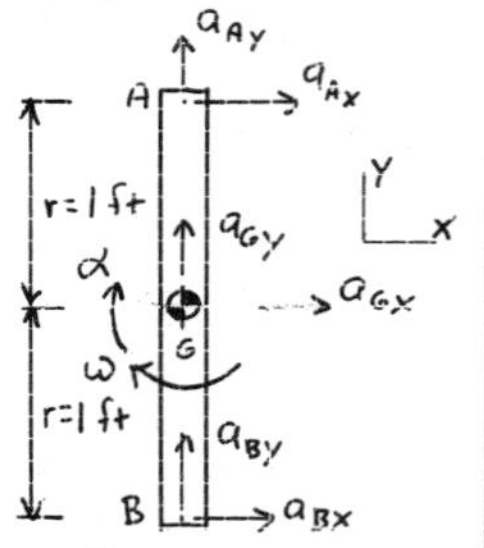

Figure c

Initially, $\omega = 0$, Also, because of the constraints, $a_{Ay} = 0$ and $a_{Bx} = 0$. Then, by Eqs. (e) and (f),

$a_{Gy} = 0$ (h)

$a_{Gx} = r\alpha$ (i)

and Eqs. (b) and (h) yield

$\underline{\underline{A = mg = 16.1 \text{ lb}}}$

Therefore, slide A exerts a downward force of 16.1 lb on the horizontal surface.

By Eqs. (c) and (i), with r = 1 ft and $m = 16.1/32.2 = 0.5 \text{ lb}\cdot\text{s}^2/\text{ft}$,

$\alpha = -30 \text{ rad/s}^2$ (j)

Then, Eqs. (a), (i), and (j) yield

$\underline{\underline{B = -5 \text{ lb}}}$ (k)

Hence, slider B exerts a force of 5 lb on the vertical guides, directed horizontally to the right.

By Eqs. (d), (g), (h), (i), and (j)

$a_{Ax} = 2r\alpha = -60 \text{ ft/s}^2$

$a_{By} = 0$

Hence, the acceleration of A and B are, respectively,

$\underline{\underline{a_{Ax} = -60 \text{ ft/s}^2,\ a_y = 0}}$ and $\underline{\underline{a_{Bx} = a_{By} = 0}}$

18.121

Given: By Problem 18.92, a point mass M is attached to one end of a uniform bar of mass m and length L that hangs by one end from a light cord, with mass M at the top (Fig. a)

a) Determine the point at which the bar can be struck laterally without causing an immediate acceleration of M.

b) Repeat part a for the case in which M is at the lower end of the bar.

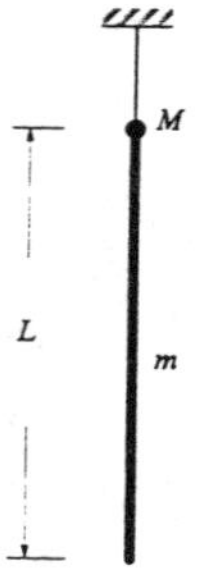

Figure a

Solution:

a) Since we require the acceleration of M to be initially zero, take a Newtonian frame xy with origin O at mass M (Fig. b).

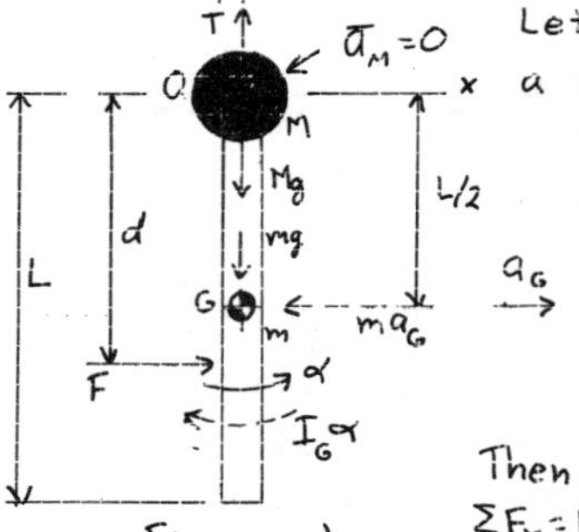

Figure b

Let the force F be applied a distance d from O and G be the mass center of the bar. Then since $\bar{a}_M = 0$, and $\omega = 0$ initially, by Fig. b,

$a_G = \frac{L}{2}\alpha$ (a)

Then, by Fig. b and Eq. (a)

$\Sigma F_x = F - m a_G = 0;\ F = \frac{1}{2} m L \alpha$ (b)

$\circlearrowleft \Sigma M_O = Fd - (m a_G)\frac{L}{2} - I_G \alpha = 0$ (c)

where $I_G = \frac{1}{12} m L^2$. Therfore, by Eqs. (b) and (c)

$(\frac{1}{2} m L \alpha) d = \frac{1}{12} m L^2 \alpha + \frac{1}{4} m L^2 \alpha$

Hence, solving for d, we find

$\underline{\underline{d = \frac{2}{3} L}}$ (d)

b) Similarly for M at the lower end of the rod, we take a Newtonian frame xy with origin O at mass M (Fig. c)

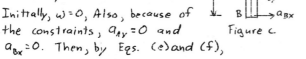

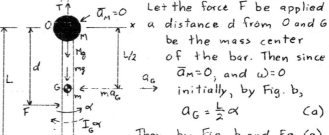

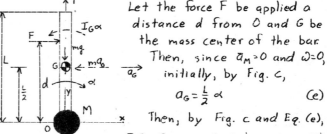

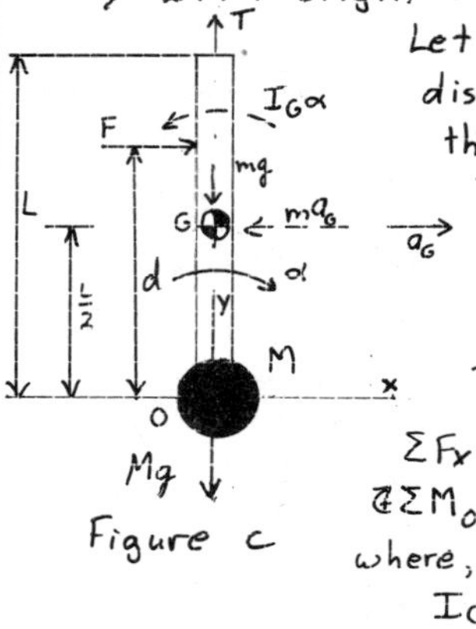

Figure c

Let the force F be applied a distance d from O and G be the mass center of the bar. Then, since $\bar{a}_M = 0$ and $\omega = 0$, initially, by Fig. c,

$a_G = \frac{L}{2}\alpha$ (e)

Then, by Fig. c and Eq. (e),

$\Sigma F_x = F - m a_G = 0;\ F = \frac{1}{2} m L \alpha$ (f)

$\circlearrowleft \Sigma M_O = Fd - (m a_G) L/2 - I_G \alpha$ (g)

where,

$I_G = \frac{1}{12} m L^2$

(Continued)

18.121 Cont.

Therefore, by Eqs. (e), (f), and (g),

$$\left(\tfrac{1}{2}mL\alpha\right)d = \tfrac{1}{12}mL^2\alpha + \tfrac{1}{4}mL^2\alpha$$

Hence, solving for d, we obtain

$$\underline{d = 2/3\,L}$$

[See Eq. (d)]
In both parts a and b, the force must be applied at the distance

$$d = \tfrac{2}{3}L \text{ from mass } M.$$

18.122

Given: By Problems 18.93 and 18.86, let the weight of the sphere be W, the horizontal force exerted on the ball be P, and the coefficient of friction between contacting surfaces be μ.

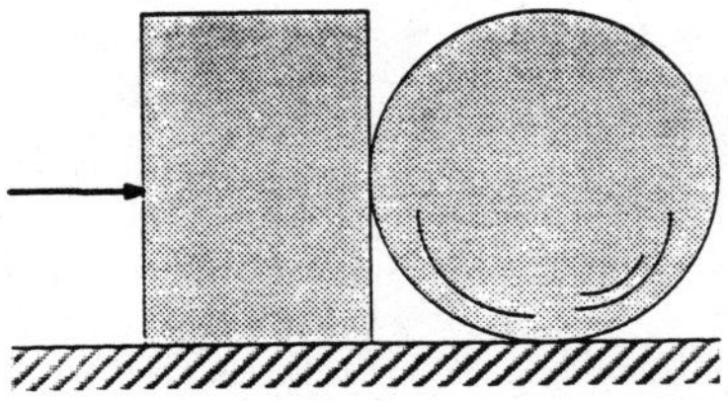

Figure a

Show that the ball will skid on the table if,

$$\frac{P}{W} = \frac{7\mu}{(1-\mu)(2+7\mu)}$$

Solution: Suppose first that the ball rolls (Fig. b).

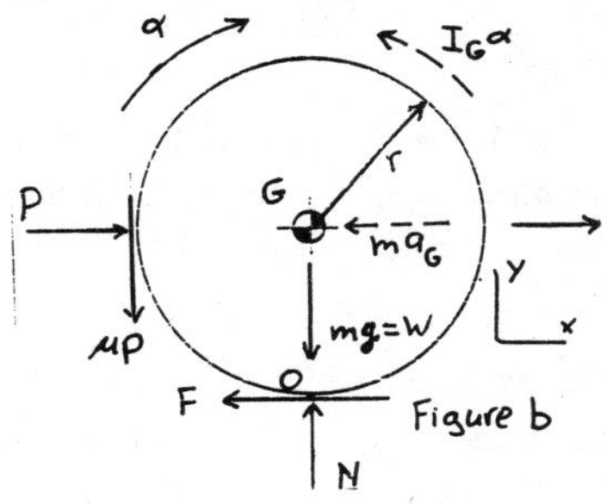

Figure b

For rolling, $a_G = r\alpha$ (a)

By Fig. b and Eq. (a),

$$\Sigma F_x = P - F - ma_G = 0; \quad F = P - ma_G \tag{b}$$

$$\Sigma F_y = N - \mu P - W = 0; \quad N = \mu P + W \tag{c}$$

$$\Sigma M_O = Pr - \mu Pr - I_G\alpha - ma_G r = 0$$

or, since $I_G = \frac{2}{5}mr^2$ and $\alpha = a_G/r$,

$$ma_G = \tfrac{5}{7}P(1-\mu) \tag{d}$$

Then, by Eqs. (b) and (d),

$$F = \tfrac{1}{7}(2+5\mu)P \tag{e}$$

$$\tfrac{1}{7}(2+5\mu)P < \mu^2 P + \mu W$$

or

$$\frac{P}{W} < \frac{7\mu}{(1-\mu)(2+7\mu)}$$

If the ball skids, as was to be shown,

$$\underline{\underline{\frac{P}{W} > \frac{P\mu}{(1-\mu)(2+7\mu)}}}$$

18.123

Given: By Problem 18.94, the mechanism shown in Fig. a consists of a thin uniform bar A that weighs 200 lb and part B that weighs 48 lb and that is connected to A by a cord. Bar A rotates about a fixed frictionless horizontal axes O. When A is vertical, B has a velocity of 6 ft/s downward.

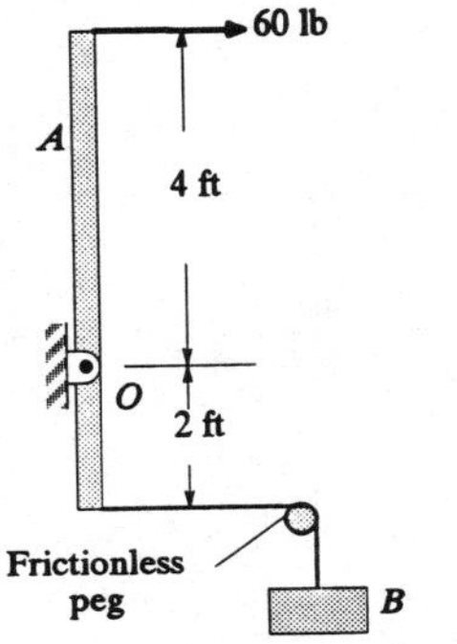

Figure a

For the given position, determine the acceleration of B and the vertical and horizontal components of the force exerted on A by the pin O.

Solution: The free-body diagrams of bar A and part B, including inertial forces and inertial couple are shown in Figs. b and c.

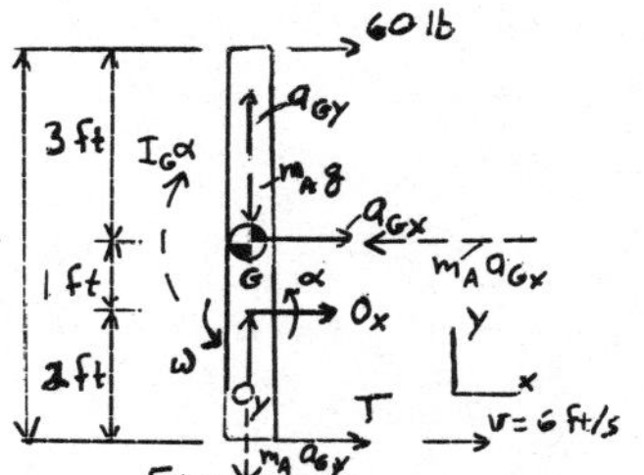

Figure b

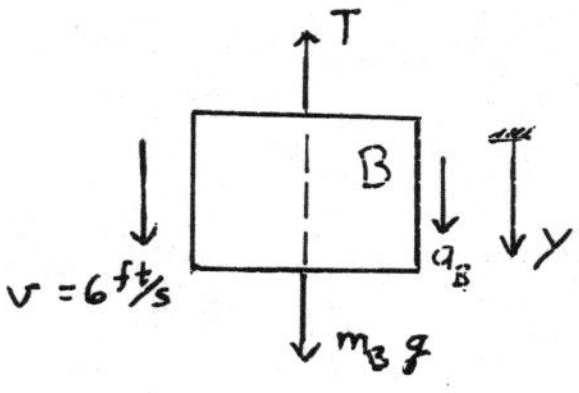

Figure c

By the given data,

$$m_A = 200/32.2 = 6.211 \text{ lb·s}^2/\text{ft} \tag{a}$$

$$m_B = 48/32.2 = 1.491 \text{ lb·s}^2/\text{ft} \tag{b}$$

By Fig. b and kinematics,

$$\omega = \frac{v}{2} = \frac{6}{2} = 3 \text{ rad/s}$$

$$a_{Gy} = -(1)(\omega)^2 = -9 \text{ ft/s}^2 \tag{c}$$

$$a_{Gx} = -(1)(\alpha) = -\alpha \ [\text{ft/s}^2] \tag{d}$$

$$a_B = 2\alpha \tag{e}$$

(Continued)

18.123 (cont.)

$\Sigma F_x = O_x + T + 60 - m_A a_{Gx} = 0$

or $\quad O_x + T = -6.211\alpha - 60 \qquad (f)$

$\Sigma F_y = O_y - m_A g - m_A a_{Gy} = 0$

or, $\quad \underline{\underline{O_y = 200 - 55.899 = 144.1 \text{ lb}}}$

$\circlearrowleft + \Sigma M_G = O_x(1) + T(3) - (60)(3) - I_G\alpha = 0$

where, $I_G = \frac{1}{12} m_A L^2 = \frac{1}{12}(6.211)(6)^2 = 18.633 \text{ lb}\cdot\text{ft}\cdot\text{s}^2$

or $\quad O_x + 3T = 18.633\alpha + 180 \qquad (g)$

By Eqs. (f) and (g),

$T = 12.422\alpha + 120 \qquad (h)$

$O_x = -18.633\alpha - 180 \qquad (i)$

By Fig. c,

$\Sigma F_y = m_B g - T - m_B a_B = 0$

or, with Eqs. (b) and (e)

$T = 48 - 2.982\alpha \qquad (j)$

Equating Eqs. (h) and (j), we find

$\alpha = -4.674 \text{ rad/s}^2 \qquad (k)$

Then, Eqs. (e) and (k) yield

$\underline{\underline{a_B = -9.348 \text{ ft/s}^2}}$

and by Eqs. (i) and (k), we obtain

$\underline{\underline{O_x = -92.91 \text{ lb}}}$

Summarizing, we have

$a_B = -9.348 \text{ ft/s}^2$

$O_x = -92.91 \text{ lb}$

$O_y = 144.1 \text{ lb}$

18.124

Given: A cylindrical vessel partially filled with a liquid rotates about its vertical axis with constant angular velocity ω (Fig. a).

Using the fact that the resultant R of the external force and the inertial force that acts on a particle of water at the free surface is normal to the surface, show that the equation of a cross section of the free surface is (Fig. a),

$$y = \frac{x^2\omega^2}{2g} + \text{constant}$$

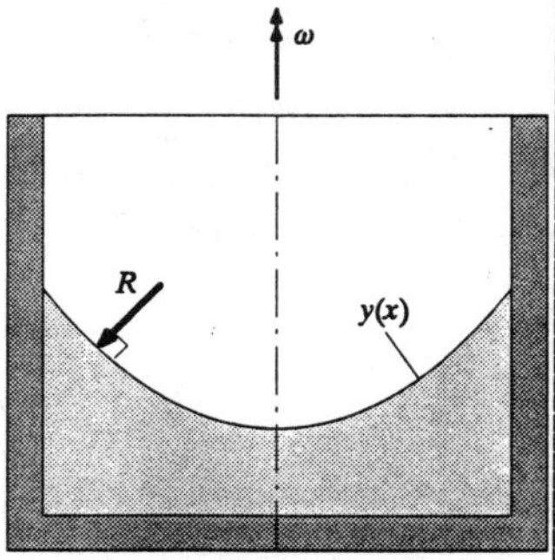

Figure a

Solution:

The free-body diagram of a particle P of water at the free surface is shown in Fig. b.

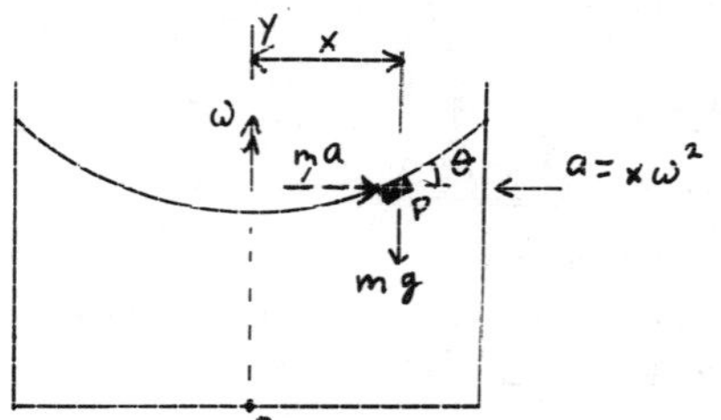

Figure b

The slope of the free surface is

$$\frac{dy}{dx} = \tan\theta \qquad (a)$$

Also, by Fig. c,

$$\tan\theta = \frac{mx\omega^2}{mg} = \frac{x\omega^2}{g} \qquad (b)$$

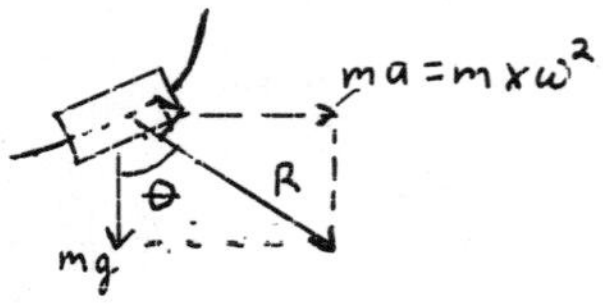

Figure c

By Eqs. (a) and (b),

$$\frac{dy}{dx} = \frac{x\omega^2}{g} \qquad (c)$$

Integration of Eq. (c) yields

$$\underline{\underline{y = \frac{x^2\omega^2}{2g} + \text{constant}}}$$

the required result

18.125

Given: A thin uniform rigid L-shaped bar of mass $m = 4.5$ kg is attached to a turntable by a frictionless pin at A, and it rests against a frictionless peg at B (Fig. a).

At a given instant, the turntable rotates about the vertical axis O with an angular velocity of 6 rad/s and an angular acceleration of 10 rad/s.

Determine the forces exerted on the bar by the pin at A and the peg at B.

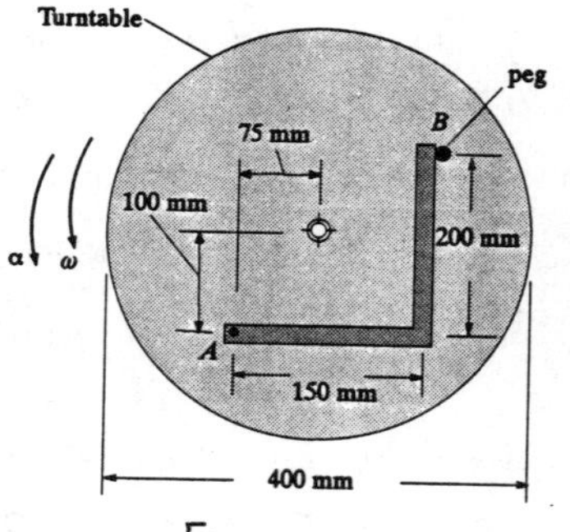

Figure a

(Continued)

18.125 Cont.

Solution:

The in-plane forces acting on the bar, including inertial forces and couples, are shown in Fig. b. In Fig. b, the mass m=4.5 kg has been divided between the segments AC and BC so that,

$$m_1 = \frac{150}{350}(4.5) = 1.929 \text{ kg} \quad (a)$$

$$m_2 = \frac{200}{350}(4.5) = 2.571 \text{ kg} \quad (b)$$

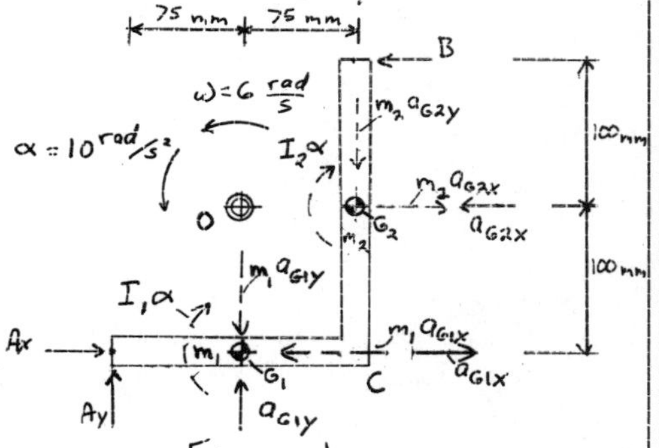

Figure b

By Fig. b and kinematics, the accelerations of the mass centers of bars AC and BC are, respectively,

$$a_{G1x} = r\alpha = (0.100)(10) = 1 \text{ m/s}^2$$
$$a_{G1y} = r\omega^2 = (0.100)(6)^2 = 3.6 \text{ m/s}^2 \quad (c)$$

and

$$a_{G2x} = r\omega^2 = (0.075)(6)^2 = 2.7 \text{ m/s}^2$$
$$a_{G2y} = r\alpha = (0.075)(10) = 0.75 \text{ m/s}^2 \quad (d)$$

The mass moments of inertia of bars AC and BC, relative to their mass centers are

$$I_1 = \tfrac{1}{12} m_1 (AC)^2 = \tfrac{1}{12}(1.929)(0.150)^2 = 0.003617 \text{ N·m·s}^2 \quad (e)$$

$$I_2 = \tfrac{1}{12} m_2 (BC)^2 = \tfrac{1}{12}(2.571)(0.200)^2 = 0.00857 \text{ N·m·s}^2 \quad (f)$$

By Fig. b,

$$\Sigma F_x = A_x - B + m_2 a_{G2x} - m_1 a_{G1x} = 0 \quad (g)$$

$$\Sigma F_y = A_y - m_1 a_{G1y} - m_2 a_{G2y} = 0 \quad (h)$$

$$\circlearrowleft\Sigma M_A = B(0.20) - m_2 a_{G2y}(0.150) - m_2 a_{G2x}(0.100) - m_1 a_{G1y}(0.075) - I_1\alpha - I_2\alpha = 0 \quad (i)$$

Substitution of Eqs. (a), (b), (c), (d), (e), and (f) into Eqs. (g), (h), and (i) yields

$$A_x - B = -5.013 \quad (j)$$

$$\underline{A_y = 8.873 \text{ N}}$$

$$\underline{B = 8.131 \text{ N}} \quad (k)$$

Finally, Eqs. (j) and (k) yield

$$\underline{A_x = 3.118 \text{ N}}$$

18.126

Given: By Problem 18.95, a bowling ball of radius r is thrown down a bowling lane with initial speed v_0 and backspin ω_0. The coefficient of sliding friction is μ_K.

a) Show that the distance s through which the ball skids is

$$s = \frac{2}{49\mu_K g}(v_0 + r\omega_0)(6v_0 - r\omega_0)$$

provided the ball does not roll backward

b) Show that, to prevent rolling backward,

$$v_0 \geq \frac{2}{5} r\omega_0$$

Solution:

a) The free-body diagram of the ball is shown in Fig. a, including the inertial force and the inertial couple.

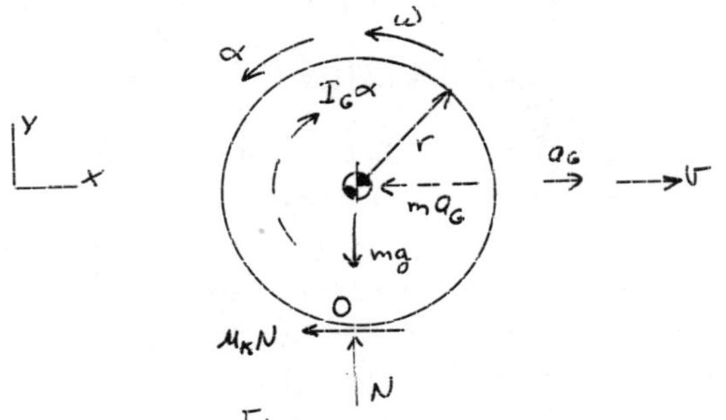

Figure a

By Fig. a,

$$\Sigma F_x = -\mu_K N - m a_G = 0 \quad (a)$$

$$\Sigma F_y = N - mg = 0;\ N = mg \quad (b)$$

Equations (a) and (b) yield

$$a_G = -\mu_K g \quad (c)$$

also, by Fig. a,

$$\circlearrowleft\Sigma M_0 = (m a_G)(r) - I_G\alpha = m a_G r - \tfrac{2}{5} m r^2 \alpha = 0 \quad (d)$$

Equations (c) and (d) yield

$$\alpha = -\tfrac{5}{2}\mu_K \frac{g}{r} \quad (e)$$

By kinematics, $\alpha = d\omega/dt$. Therefore, by Eq. (e),

$$\frac{d\omega}{dt} = -\mu_K \frac{5g}{2r}$$

and integration yields

$$\omega - \omega_0 = -\mu_K \frac{5g}{2r} t \quad (f)$$

where $\omega = \omega_0$ when $t = 0$

Also, by kinematics and Eq. (c)

$$a_G = \frac{dv}{dt} = -\mu_K g$$

So,

$$v - v_0 = -\mu_K g t \quad (g)$$

where $v = v_0$ when $t = 0$

(Continued)

18.126 Cont

when skidding stops

$$v = -r\omega \qquad \text{(h)}$$

Then, Eqs (f), (g), and (h) yield

$$v_0 - \mu_k g t_1 = \mu_k \frac{5g}{2} t_1 - r\omega_0$$

where t_1 is the time at which the ball stops skidding. Hence,

$$t_1 = \frac{2}{7\mu_k g}(v_0 + r\omega_0) \qquad \text{(i)}$$

Again, by kinematics and Eq. (g),

$$v = \frac{ds}{dt} = v_0 - \mu_k g t$$

and integration yields

$$s = v_0 t - \frac{1}{2}\mu_k g t^2 \qquad \text{(j)}$$

When $t = t_1$, Eqs. (i) and (j) yield

$$s = \frac{2v_0}{7\mu_k g}(v_0 + r\omega_0) - \frac{2}{49\mu_k g}(v_0 + r\omega_0)^2$$

or after simplification

$$s = \frac{2}{49\mu_k g}(v_0 + r\omega_0)(6v_0 - r\omega_0)$$

where s is the distance that the ball travels while skidding. This result is what is required in part a.

b) The ball will start to roll back if the initial speed v_0 is too small. To ensure that the ball does not roll backward, we must have $v > 0$ for $t < t_1$. Hence, by Eq. (g), $v = v_0 - \mu_k g t > 0$ for $t < t_1$. Therefore, $v_0 > \mu_k g t > \mu_k g t_1$. So to ensure that the ball does not roll backward, by Eq. (i), we have,

$$v_0 > \frac{2}{7}(v_0 + r\omega_0)$$

or

$$v_0 > \frac{2}{5} r\omega_0$$

This is the result required.

18.127

Given: By Problem 18.96, a uniform half-disk rolls freely on a horizontal surface (Fig. a). For $\theta = 0$, the angular velocity is ω_0.

Show that the angular acceleration corresponding to $\theta = 0$ is

$$\alpha_0 = \frac{8}{9\pi}\left(\omega_0^2 + \frac{g}{r}\right)$$

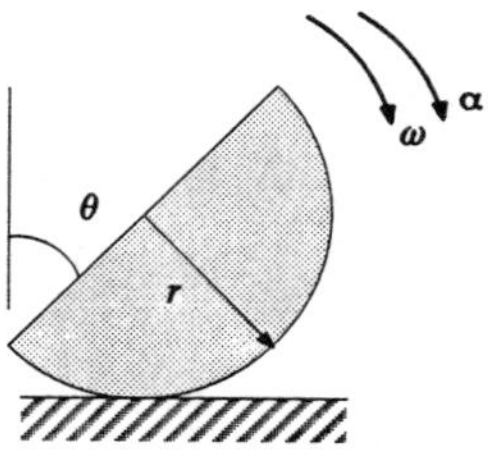

Figure a

Solution:

The free-body diagram of the half-disk, included the inertial forces and the inertial couple, is shown in Fig. b at $\theta = 0$, where m is the mass of the half-disk.

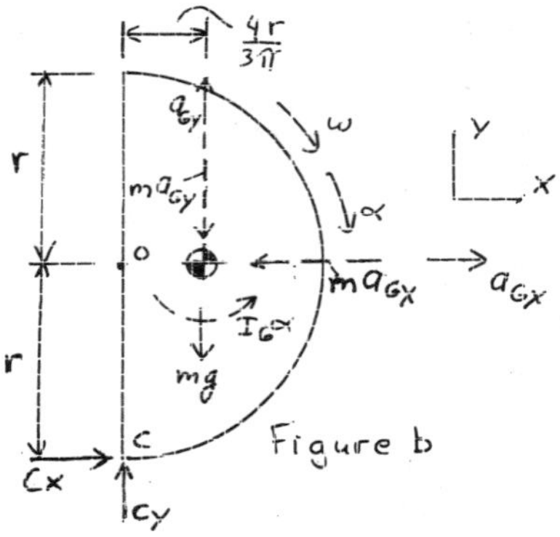

Figure b

By Appendix D, Table D2, the center of mass of a half-disk is located at $4r/3\pi$ from its diameter (Fig. b).

By Fig. b,

$$\sum M_C = I_G\alpha + (ma_{Gx})r - (ma_{Gy})\frac{4r}{3\pi} - (mg)\frac{4r}{3\pi} = 0$$

or

$$I_G\alpha = mr\left[(a_{Gy} + g)\frac{4}{3\pi} - a_{Gx}\right] \qquad \text{(a)}$$

The mass center of inertia of a complete disk about its center O is (Appendix F, Table F1)

$$I_{DO} = \frac{1}{2} m_D r^2$$

where $m_D = 2m$ is the mass of the complete disk. Hence, the mass moment of inertia of the half-disk about O is

$$I_O = \frac{1}{2} I_{DO} = \frac{1}{2}\left[\frac{1}{2}(2m)r^2\right] = \frac{1}{2}mr^2 \qquad \text{(b)}$$

Then, by the parallel-axis theorem (Steiners theorem) and Eq. (b),

$$I_O = I_G + m\left(\frac{4r}{3\pi}\right)^2$$

or

$$I_G = \frac{1}{2}mr^2 - \frac{16r^2}{9\pi^2}m = mr^2\left(\frac{1}{2} - \frac{16}{9\pi^2}\right) \qquad \text{(c)}$$

By kinematics and Fig. b, for $\theta = 0$,

$$a_{Gx} = (a_O)_x + (a_{G/O})_x = \alpha_0 r - \frac{4r}{3\pi}\omega_0^2 \qquad \text{(d)}$$

$$a_{Gy} = (a_O)_y + (a_{G/O})_y = 0 - \alpha_0\frac{4r}{3\pi} = -\frac{4r}{3\pi}\alpha_0 \qquad \text{(e)}$$

Hence, by Eqs. (a), (c), (d), and (e), for $\theta = 0$,

$$mr^2\left(\frac{1}{2} - \frac{16}{9\pi^2}\right)\alpha_0 = mr\left[\left(-\frac{4r}{3\pi}\alpha_0 + g\right)\frac{4}{3\pi} - \alpha_0 r + \frac{4r}{3\pi}\omega_0^2\right]$$

Solving for α_0, we obtain, as required,

$$\alpha_0 = \frac{8}{9\pi}\left(\omega_0^2 + \frac{g}{r}\right)$$

Note that in this inertial-force method, the moment dynamic equilibrium equation [Eq. (a)] and kinematics were sufficient to obtain the solution. Compare the solution obtained in Problem 18.96.

18.128

Given: By Problem 18.97, a uniform rod OA is welded to a thin hoop at A (Fig. a). The mass of the hoop is one-half the mass of the rod. The coefficient of static friction between the hoop and the horizontal surface is $\mu_s = 0.40$. A horizontal force P in the plane of the hoop is applied to the rod at point O. For the position shown the hoop is rolling to the right with angular velocity of 200 rpm and is on the verge of skidding.

Determine the acceleration [m/s²] of the center O for this position.

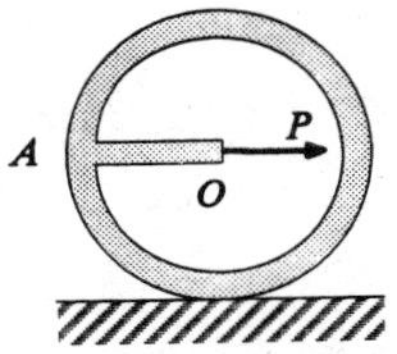

Figure a

Solution:

By Fig. b, taking moments about point O, we find the center of mass to be given by (where m is the mass of the rod)

$$\circlearrowleft + \Sigma M_O = (m + \tfrac{1}{2}m) g r_G = mg(\tfrac{1}{2}r) + \tfrac{1}{2}mg(0) = \tfrac{1}{2}mgr$$

or

$$r_G = \tfrac{1}{3}r$$

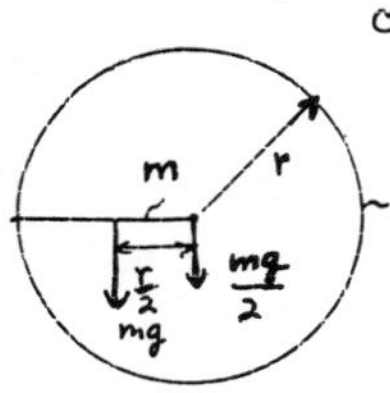

Figure b

The free-body diagram of the hoop-rod system, including the inertial forces and inertial couple, is shown in Fig. c.

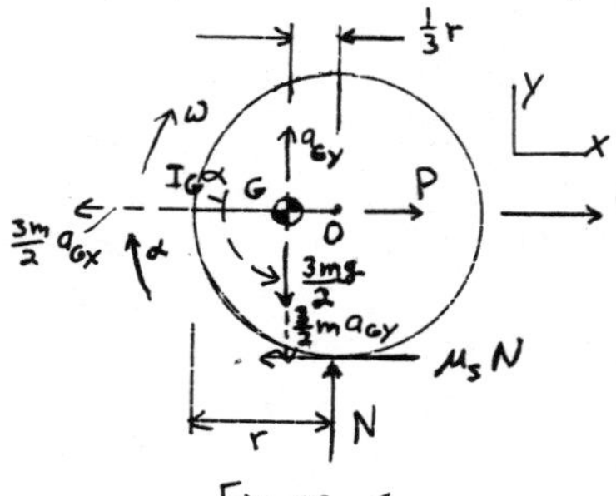

Figure c

By Fig. c,

$$\circlearrowleft + \Sigma M_O = I_G\alpha + (\tfrac{3}{2}mg)(\tfrac{1}{3}r) + (\tfrac{3}{2}m a_{Gy})(\tfrac{1}{3}r) - \mu_s N r = 0 \quad (a)$$

$$\Sigma F_y = N - \tfrac{3}{2}mg - \tfrac{3}{2}m a_{Gy} = 0\,; \quad N = \tfrac{3}{2}m(g + a_{Gy}) \quad (b)$$

By Fig. c and the parallel-axis theorem,

$$I_G = \tfrac{1}{2}mr^2 + m(\tfrac{r}{6})^2 + (\tfrac{m}{2})r^2 + (\tfrac{m}{2})(\tfrac{r}{3})^2$$

or $$I_G = \tfrac{2}{3}mr^2 \quad (c)$$

Also by kinematics and Fig. c,

$$(a_O)_x = r\alpha, \ (a_O)_y = 0\,; \text{ or } a_O = r\alpha \quad (d)$$

$$a_{Gy} = (a_O)_y + (a_{G/O})_y = 0 + \tfrac{1}{3}r\alpha = \tfrac{1}{3}a_O \quad (e)$$

By Eqs. (a), (b), (c), and (e), we have

$$\tfrac{2}{3}mr^2\alpha + \tfrac{1}{2}mgr + (\tfrac{3}{2}m)(\tfrac{1}{3}a_O)(\tfrac{1}{3}r) - \tfrac{3}{2}\mu_s m(g + \tfrac{1}{3}a_O)r = 0$$

or since by Eq. (d), $r\alpha = a_O$, canceling m and r, we find

$$\tfrac{2}{3}a_O + \tfrac{1}{2}g + \tfrac{1}{6}a_O - \tfrac{3}{2}(0.4)g - \tfrac{1}{2}(0.4)a_O = 0$$

or

$$a_O = \frac{0.6}{3.8}g = \frac{(0.6)(9.81)}{3.8} = 1.549 \text{ m/s}^2$$

Compare this solution method to that of Problem 18.97

18.129

Given: A thin homogeneous triangular plate rotates about the x-axis with angular speed ω (Fig. a). The thickness of the plate is constant.

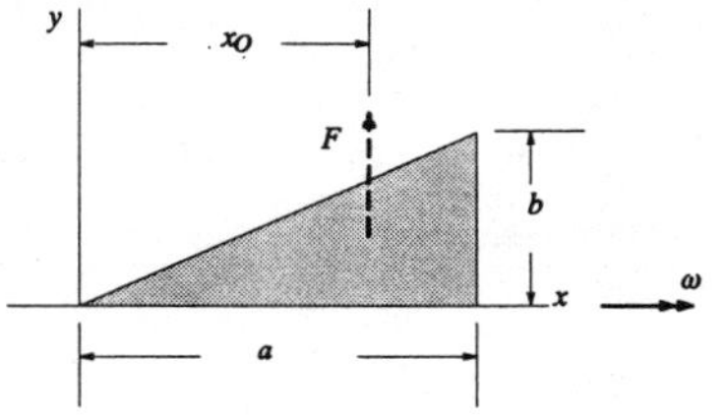

Figure a

a) Show that the magnitude of the resultant centrifugal force on the plate is $F = mb\omega^2/3$, where m is the mass of the plate.

b) Show that the resultant axis of the centrifugal force is located at $x_0 = 3a/4$.

Solution:

a) Consider an elemental strip y dx of mass dm (Fig. b).

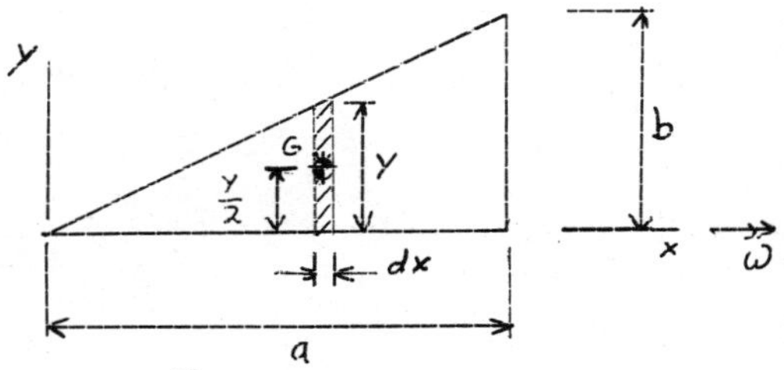

Figure b

Let ρ be the mass per unit area of the plate. Then,

$$dm = \rho y\, dx \quad (a)$$

(Continued)

18.129 Cont

and the element dF of centrifugal force is

$$dF = (dm)(a_G) = (\rho y dx)\left(\frac{y}{2}\omega^2\right) \qquad (b)$$

By Fig. b,

$$y = \frac{b}{a}x \qquad (c)$$

By Eqs. (b) and (c), we have

$$dF = \frac{1}{2}\rho\omega^2 y^2 dx = \frac{1}{2}\rho\omega^2\left(\frac{b^2}{a^2}\right)x^2 dx \qquad (d)$$

Integration yields

$$F = \frac{1}{2}\rho\frac{\omega^2 b^2}{a^2}\int_0^a x^2 dx = \frac{\rho\omega^2 b^2}{2a^2}\left.\frac{x^3}{3}\right|_0^a$$

or

$$F = \frac{\rho\omega^2 a b^2}{6} = \frac{1}{3}mb\omega^2 \qquad (e)$$

since

$$m = \frac{1}{2}\rho a b \qquad (f)$$

b) The centrifugal moment about the z axis (Fig. b) is, with Eq. (d),

$$dM = x\,dF = \frac{\rho\omega^2 b^2}{2a^2}x^3 dx \qquad (g)$$

Integration of Eq. (g) yields, with Eq. (f)

$$M = \frac{\rho\omega^2 b^2}{2a^2}\int_0^a x^3 dx = \frac{1}{8}\rho\omega^2 a^2 b^2 = \frac{1}{4}ab\omega^2 \qquad (h)$$

Hence, by Eqs. (e) and (h)

$$x_0 = \frac{M}{F} = \frac{3}{4}a$$

18.130

Given: A shaft that rotates at 1200 rpm (40π rad/s) passes through the mass center of, and is welded to, an oblique rectangular wobble plate (Fig. a). The plate is 10 in wide (perpendicular to the plane of the page) and has a specific weight $\gamma = 0.280$ lb/in^3

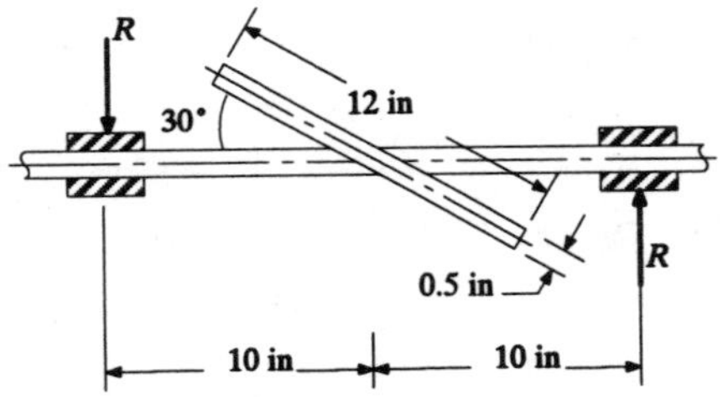

Figure b

Determine the reaction R of the bearings on the shaft.

Solution: Take an element along the width of the plate of thickness ds and depth 0.50 in (Fig. b)

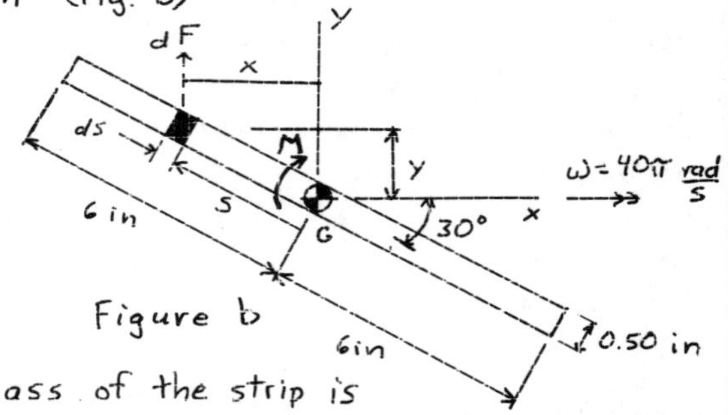

Figure b

The mass of the strip is

$$dm = \left(\frac{0.280}{386.4}\right)(0.50)(10)ds = \frac{1.4}{386.4}ds \qquad (a)$$

Hence, the centrifugal force on the strip is

$$dF = y\omega^2 dm \qquad (b)$$

where by Fig. b,

$$y = s(\sin 30^\circ) = 0.5\,s \qquad (c)$$

Equations (a), (b), and (c) yield

$$dF = \left(\frac{0.70}{386.4}\right)\omega^2 ds \qquad (d)$$

The moment of dF about the z axis is (positive clockwise)

$$dM = -x\,dF \qquad (e)$$

where,

$$x = -s\cos 30^\circ = -0.866\,s \qquad (f)$$

Hence, by Eqs. (d), (e), and (f), with $\omega = 40\pi$ rad/s

$$dM = 24.775\,s^2 ds$$

By integration, we obtain

$$M = 24.775\int_{-6}^{6} s^2 ds = 24.775\left.\frac{s^3}{3}\right|_{-6}^{6} = 3567.6 \text{ lb·in} \quad (g)$$

By Figs. a and b, $M = 20R$. So, with Eq. (g),

$$R = \frac{3567.6}{20} = 178.4 \text{ lb}$$

18.131

Given: A thin uniform rod is bent into a quarter-circle (Fig. a). It rotates about the vertical axis 1-1 with angular speed ω. It is held by a frictionless hinge at O.

Show that $\omega = \sqrt{2g/r}$ if the tangent to the rod at point O is horizontal.

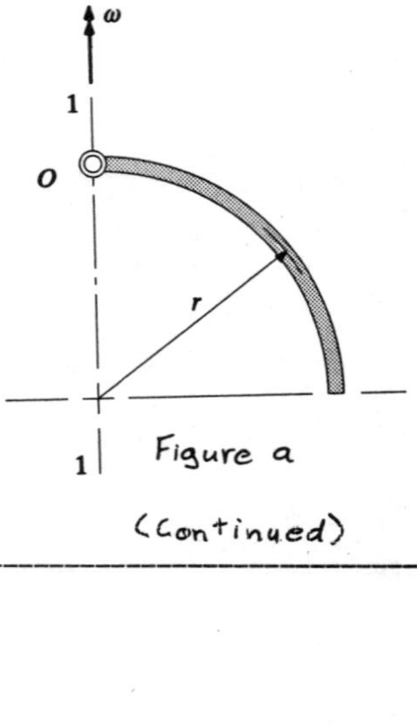

Figure a

(Continued)

18.131 Cont.

Solution:

Consider an element $r\,d\theta$ of the rod (Fig. b)

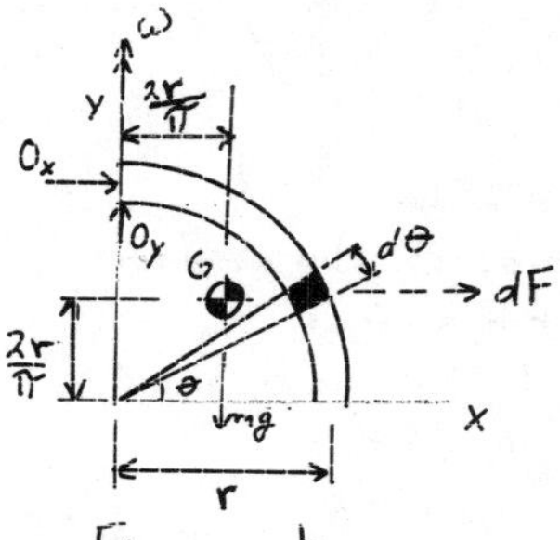

Figure b

The mass of the element is

$$dm = \rho r\, d\theta \qquad (a)$$

where ρ is the mass per unit length. The centrifugal force of the element is

$$dF = (r\cos\theta)\,\omega^2\, dm \qquad (b)$$

By Eqs. (a) and (b),

$$dF = \rho r^2 \omega^2 \cos\theta\, d\theta \qquad (c)$$

and the moment of dF about the hinge O is (Fig. b)

$$\circlearrowleft\!+\; dM_F = (r - r\sin\theta)\, dF$$

or $\circlearrowleft\!+\; dM_F = \rho r^3 \omega^2 \cos\theta\,(1-\sin\theta)\, d\theta$

Integration yields

$$\circlearrowleft\!+\; M_F = \rho r^3 \omega^2 \int_0^{\pi/2} (1-\sin\theta)\cos\theta\, d\theta$$

or

$$\circlearrowleft\!+\; \Sigma M_F = \tfrac{1}{2}\rho r^3 \omega^2 \qquad (d)$$

The moment of the weight mg about O is, where $x_G = 2r/\pi$ (Appendix D, Table D1),

$$\circlearrowleft\!+\; M_W = \frac{2r}{\pi}\, mg$$

where by Eq. (a), $m = \rho r \int_0^{\pi/2} d\theta = \frac{\pi r \rho}{2}$.

Therefore,

$$\circlearrowleft\!+\; M_W = \rho r^2 g \qquad (e)$$

Then, with Eqs. (d) and (e), summation of moments about O yields

$$\circlearrowleft\!+\; \Sigma M_O = \tfrac{1}{2}\rho r^3\omega^2 - \rho r^2 g = 0$$

Hence, $\omega = \sqrt{2g/r}$

18.132

Given: A homogeneous semiconical body rotates about the x axis with angular speed ω.

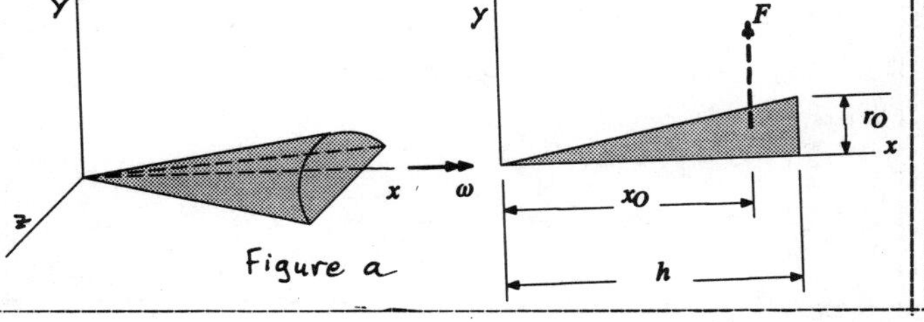

Figure a

a) Show that the magnitude of the resultant centrifugal force is $F = m r_0 \omega^2/\pi$.

b) Show that the resultant axis of the centrifugal force is determined by $x_0 = 4h/5$.

Solution: Consider a semicircular element of radius r and thickness dx (Fig. b).

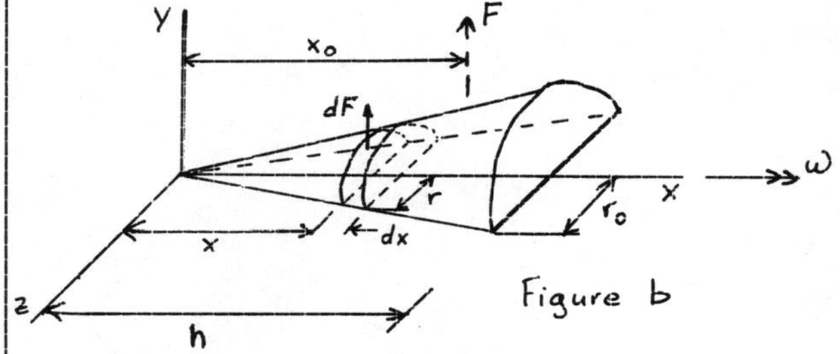

Figure b

The mass of the element is

$$dm = \tfrac{1}{2}\pi r^2 \rho\, dx \qquad (a)$$

where ρ is the mass density. The corresponding centrifugal force is, since the mass center of the element is located at $y_G = \frac{4r}{3\pi}$ (Appendix D, Table D2); see Fig. c.

$$dF = \left(\frac{4r}{3\pi}\omega^2\right) dm = \tfrac{2}{3}\rho\omega^2 r^3\, dx \qquad (b)$$

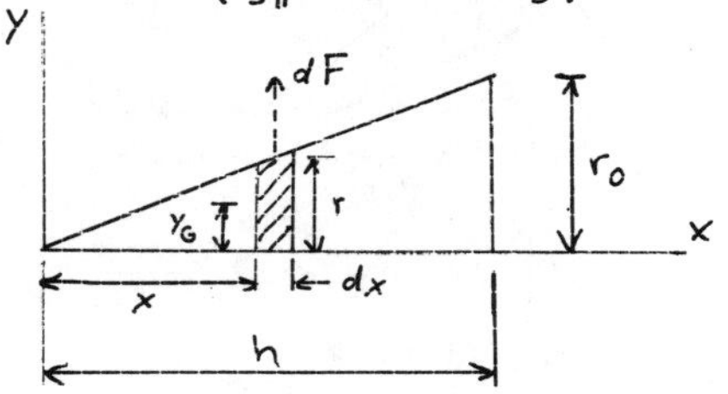

Figure c

By Fig. c,

$$r = \frac{r_0}{h}x \qquad (c)$$

Then, Eqs. (b) and (c) yield

$$dF = \tfrac{2}{3}\rho\omega^2 \frac{r_0^3}{h^3} x^3\, dx \qquad (d)$$

Integration yields

$$F = \tfrac{2}{3}\rho\omega^2 \frac{r_0^3}{h^3}\int_0^h x^3\, dx = \tfrac{1}{6}\rho\omega^2 r_0^3 h \qquad (e)$$

Also, by Eqs. (a) and (c),

$$m = \tfrac{1}{6}\rho\pi r_0^2 h \qquad (f)$$

So, Eqs. (e) and (f) yield

$$F = \frac{1}{\pi} m r_0 \omega^2$$

(b) The moment about the z axis of the centrifugal force dF is (Fig. b), with Eq. (d),

$$dM = x\, dF = \tfrac{2}{3}\rho\omega^2 \frac{r_0^3}{h^3} x^4\, dx$$

and integration yields,

$$M = \tfrac{2}{3}\rho\omega^2 \frac{r_0^3}{h^3}\int_0^h x^4\, dx = \tfrac{2}{15}\rho\omega^2 r_0^3 h^2 \qquad (g)$$

Then, Eqs. (e) and (g) give,

$$x_0 = \frac{M}{F} = \tfrac{4}{5}h$$

18.133

Given: The sector of a circular plate (Fig. a) has a small constant thickness perpendicular to the plane of the page. It lies in a vertical plane and rotates about the vertical axis 1-1 with constant angular speed ω, and is held by a frictionless hinge at O.

Show that, if $\phi \ge \beta$,

$$\cos\phi = \frac{4g\sec\beta}{3a\omega^2}$$

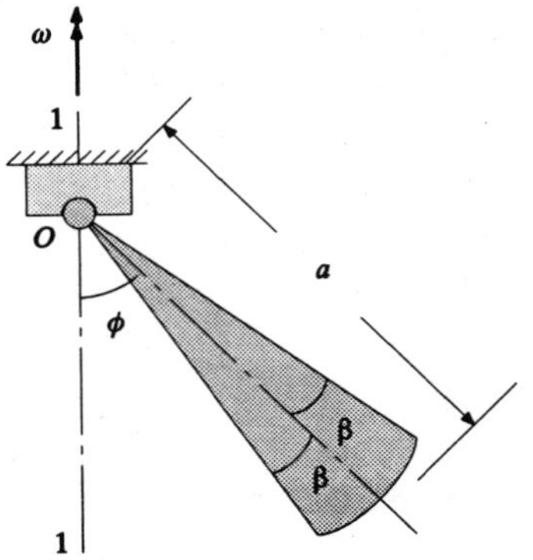

Figure a

Solution: Consider the element of mass (Fig. b)

$$dm = \rho R\,d\theta\,dR \qquad (a)$$

where ρ is the mass per unit area

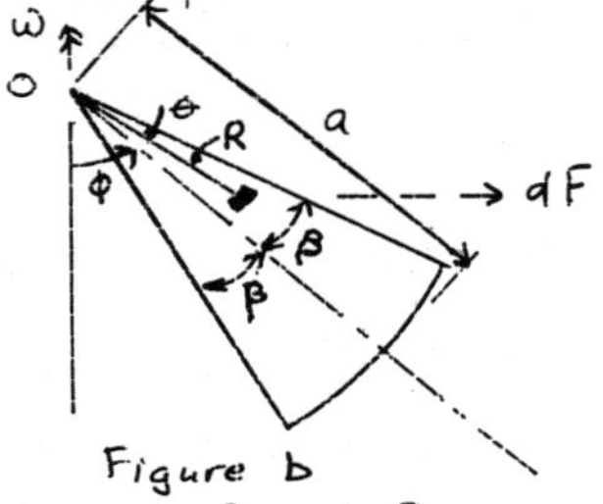

Figure b

The associated centrifugal force is (Fig. b)

$$dF = \rho[R\sin(\theta+\phi)]\omega^2 R\,d\theta\,dR \qquad (b)$$

The moment of the centrifugal force dF about the hinge O is

$$\circlearrowleft dM = R\cos(\theta+\phi)\,dF \qquad (c)$$

or by Eqs. (b) and (c)

$$\circlearrowleft dM = \rho\omega^2 R^3 \sin(\theta+\phi)\cos(\theta+\phi)\,d\theta\,dR$$

Integration of this equation yields the centrifugal moment

$$\circlearrowleft M_F = \rho\omega^2 \int_0^a R^3 dR \int_{-\beta}^{\beta} \frac{\sin[2(\theta+\phi)]}{2}\,d\theta$$

or

$$M_F = \tfrac{1}{8}\rho\omega^2 a^4 \int_{-\beta}^{\beta} \sin[2(\theta+\phi)]\,d\theta$$

$$= -\tfrac{1}{16}\rho\omega^2 a^4 \cos[2(\theta+\phi)]\Big|_{-\beta}^{\beta}$$

Expanding, we obtain

$$\circlearrowleft M_F = \tfrac{1}{2}\rho\omega^2 a^4 \sin\phi\cos\phi\,\sin\beta\cos\beta \qquad (d)$$

The center of mass of a circular sector is located at (Appendix D, Table D2)

$$R_G = \frac{2a\sin\beta}{3\beta} \qquad (e)$$

See Fig. c.

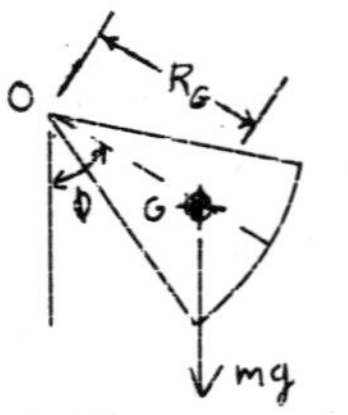

Figure c

By Fig. c, the moment of the weight about the hinge O is

$$\circlearrowleft M_W = mg\,R_G \sin\phi \qquad (f)$$

Also, by Eq. (a)

$$m = \rho\beta a^2 \qquad (g)$$

or by Eqs. (e), (f), and (g) we find

$$\circlearrowleft M_W = \tfrac{2}{3}\rho g a^3 \sin\beta\sin\phi \qquad (h)$$

Hence, by equilibrium of moments about O, Eqs. (d) and (h) yield

$$\circlearrowleft + \Sigma M_O = \tfrac{1}{2}\rho\omega^2 a^4 \sin\phi\cos\phi\sin\beta\cos\beta - \tfrac{2}{3}\rho g a^3 \sin\beta\sin\phi = 0$$

or

$$a\omega^2\cos\phi\cos\beta = \tfrac{4}{3}g$$

Hence,

$$\cos\phi = \frac{4g}{a\omega^2\cos\beta} = \frac{4g\sec\beta}{a\omega^2}$$

18.134

Given: A uniform slender rod lies on a frictionless horizontal surface. It is set spinning clockwise about its center with an angular speed of 120 rpm. One end of the rod is stopped suddenly by impact with a bumper.

Determine the angular speed [rpm] of the rod and the sense of the rod's rotation following the collision.

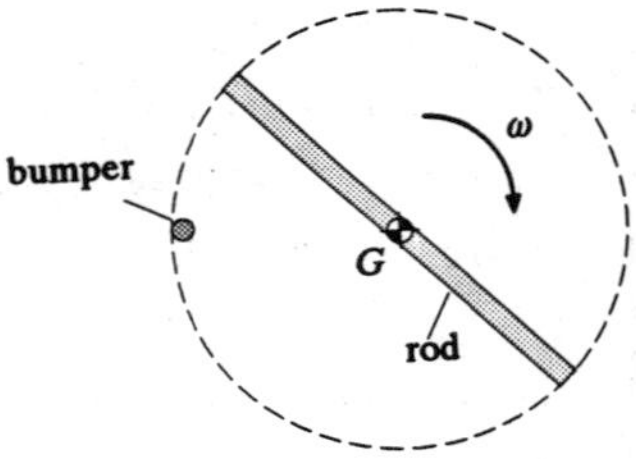

Figure a

Solution:

Figure b is a diagram of the rod, where F is the reaction of the bumper at any instant and v_G is the velocity of the mass center of the rod.

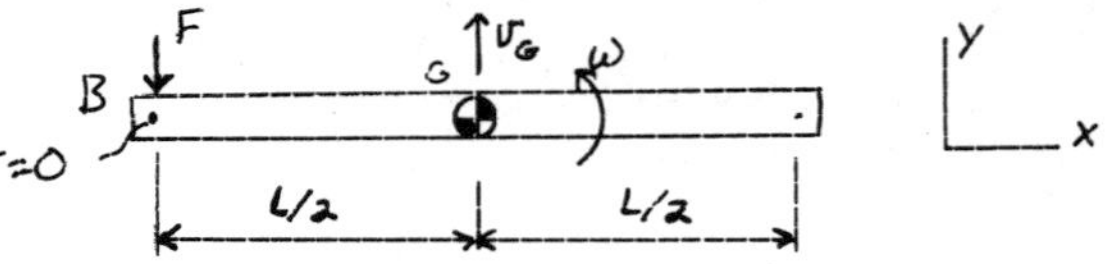

Figure b

By Fig. b,

$$\Sigma F_y = -F = m\frac{dv_G}{dt} \qquad (a)$$

$$\circlearrowleft \Sigma M_G = F\left(\tfrac{L}{2}\right) = I_G\alpha = \tfrac{1}{12}mL^2\alpha \qquad (b)$$

(Continued)

18.134 Cont.

By Eq. (a), the linear impulse acting on the rod is, from $t=0$ to $t=t_s$, the time when the rod is stopped (see Section 16.5),

$$-\int_0^{t_s} F\,dt = m v_G \Big|_0^{t_s} = m v_{GS} \qquad (c)$$

where v_{GS} is the velocity of the mass center when the rod end is stopped. Note initially when the rod was spinning about its mass center $v_G = 0$. Also, by Eq. (b), the linear impulse is

$$\int_0^{t_s} F\,dt = \tfrac{1}{6} mL(\omega_s - \omega_o) \qquad (d)$$

where ω_s is the angular velocity of the rod when the end is stopped and ω_o is the initial angular speed (120 rpm) of the rod.

By Eqs. (c) and (d),

$$v_{GS} = \tfrac{1}{6} L(\omega_o - \omega_s) \qquad (e)$$

Also, by kinematics and Fig. b,

$$v_{GS} = L/2\ \omega_s \qquad (f)$$

Equating Eqs. (e) and (f), we obtain

$$\omega_s = \tfrac{1}{4}\omega_o$$

Now $\omega_o = -120$ rpm (since we have taken positive ω as counter clockwise). Therefore,

$$\omega_s = -30 \text{ rpm}$$

or

$$\underline{\omega_s = 30 \text{ rpm} \quad \text{clockwise}}$$

An alternative, but more subtle approach is to note the initial angular momentum of the rod before impact is

$$\circlearrowleft\!+\ A_i = I\omega_o = \tfrac{1}{12} mL^2 \omega_o \qquad (g)$$

The instant after impact, end B of the rod is stopped and the final angular momentum is

$$\circlearrowleft\!+\ A_f = I\omega_s = \tfrac{1}{3} mL^2 \omega_f \qquad (h)$$

Then, by the principle of angular impulse and angular momentum (Section 16.6) we have, with Eqs. (g) and (h),

$$\int_0^{t_s} M_B\,dt = A_f - A_i = \tfrac{1}{3} mL^2\omega_f - \tfrac{1}{12} mL^2 \omega_o \qquad (i)$$

But M_B, the moment about B due to the impact force F, is zero, since F acts at B. Therefore, Eq. (i) yields, as above,

$$\underline{\underline{\omega_f = \tfrac{1}{4}\omega_o = \tfrac{1}{4}(-120) = -30 \text{ rpm}}}$$

18.135

Given: A rigid uniform rod 1 m long falls, maintaining a horizontal orientation (Fig. a). One end of the rod strikes a ledge. The speed of the rod is 15 m/s when it strikes the ledge. The coefficient of restitution of the impact is $e = 0.75$.

Determine the angular velocity of the rod and the speed of its center of mass for the instant after the collision.

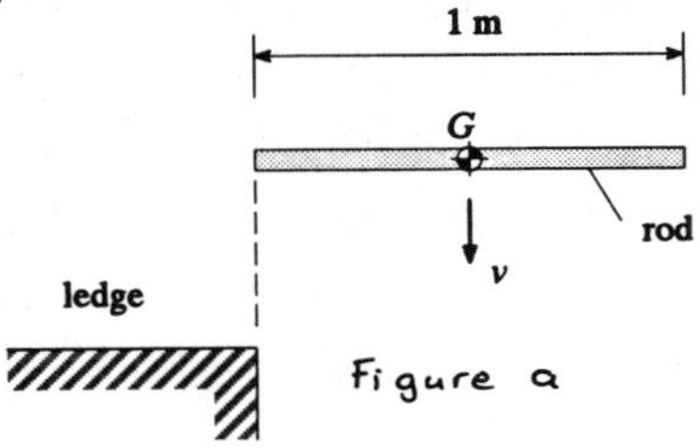

Figure a

Solution:

The free-body diagram of the rod of mass m during the impact is shown in Fig. b.

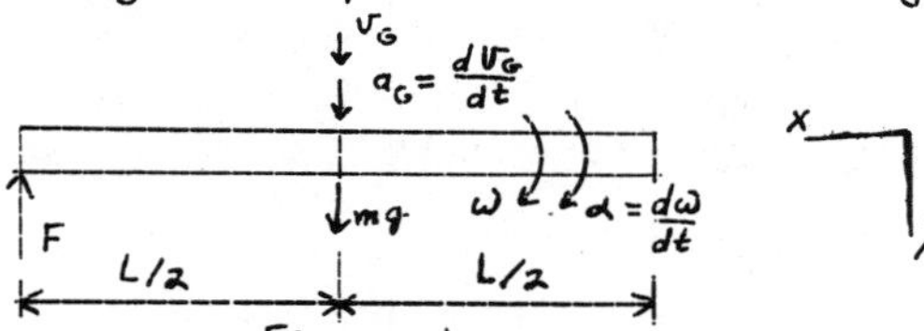

Figure b

By Fig. b, recalling that in impacts the impact force is much larger than ordinary forces such as the weight of a body, that is, $F \gg mg$, we have by Fig. b,

$$\Sigma F_y = -F + mg \approx -F = m\frac{dv_G}{dt} \qquad (a)$$

$$\circlearrowleft\!+\ \Sigma M_G = F\left(\tfrac{L}{2}\right) = I_G\alpha = \tfrac{1}{12} mL^2 \frac{d\omega}{dt}$$

or

$$F = \tfrac{1}{6} mL \frac{d\omega}{dt} \qquad (b)$$

The impulse of F during the compression phase of the impact (see Section 16.5) is

$$I_c = \int_0^{t_{cr}} F\,dt \qquad (c)$$

where t_{cr} denotes the time at which the compression phase ends and the rebound phase begins.

Hence, by Eqs. (a) and (c)

$$I_c = \int_0^{t_{cr}} F\,dt = m(v_{Gi} - v_{Gc}) \qquad (d)$$

where v_{Gc} is the velocity of the center of mass of the rod at the end of the compression phase and v_{Gi} is the initial velocity of the center of mass ($v_{Gi} = 15$ m/s)

The impulse the rod receives during the rebound phase is (refer to Section 16.5)

$$\int_{t_{cr}}^{t_f} F\,dt = eI_c \qquad (e)$$

where e is the coefficient of restitution and t_f denotes the time at the end of the rebound phase. Hence, by Eq. (a),

$$eI_c = \int_{t_{cr}}^{t_f} F\,dt = m(v_{Gc} - v_{Gf}) \qquad (f)$$

(Continued)

18.135 Cont.

where v_{Gf} is the velocity of the center of mass of the rod after rebound

By Eq. (b), we obtain similarily,

$$I_c = \int_0^{t_{cr}} F\,dt = \tfrac{1}{6} mL\omega_c \qquad (g)$$

where ω_c is the angular velocity of the rod at the end of the compression phase (initially, $\omega = 0$)

Also, by Eq. (b),

$$eI_c = \int_{t_{cr}}^{t_f} F\,dt = \tfrac{1}{6} mL(\omega_f - \omega_c) \qquad (h)$$

where ω_f is the angular velocity of the rod at the end of the rebound phase.

Also, by kinematics and Fig. b,

$$v_{Gc} = L/2\,\omega_c \qquad (i)$$

since the end of the rod in contact with the edge of the ledge is motionless at the end of the compression phase, that is, when $t = t_{cr}$.

Equations (d), (f), (g), (h), and (i) yield

$$v_{Gf} = \tfrac{1}{4}(3-e)v_{Gi}$$

$$\omega_f = \tfrac{3}{2L}(1+e)v_{Gi}$$

Then, with $L = 1\text{m}$, $e = 0.75$, and $v_{Gi} = 15$ m/s, we obtain,

$v_{Gf} = 8.44$ m/s

$\omega_f = 39.38$ rad/s

18.139

Given: An open tank of water moves horizontally with an acceleration of 16.1 ft/s^2 (Fig. a). Using the fact that the resultant of the external force and the inertial force that acts on any particle of water is perpendicular to the free surface of the water, determine the slope of the free surface.

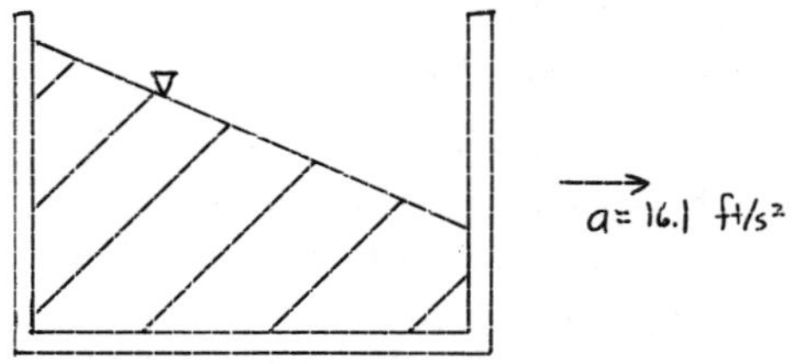

Figure a

Solution:

The free-body diagram of a particle on the free surface is shown in Fig. b, including the inertial force, and the resultant force R.

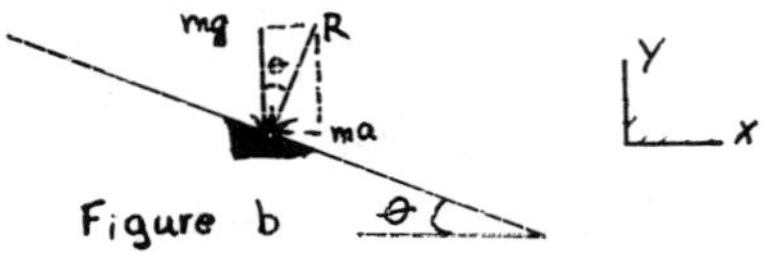

Figure b

By Fig. b,

$$\tan\theta = \frac{ma}{mg} = \frac{a}{g} = \frac{16.1}{32.2} = 0.5$$

Hence, the slope of the free surface relative to the xy axes shown is

slope = −0.50

18.140

Given: A body of mass 2 lb·s²/ft has a mass moment of inertia of 4 lb·ft·s² about an axis 1 ft from the center of mass.

Calculate the mass moment of inertia about a parallel axis 2 ft from the center of mass

Solution:

By the parallel axis theorem, in terms of the mass moment of inertia I_G of the body relative to mass center.

$$I_1 = 4 = I_G + (2)(1)^2$$

Therefore, $I_G = 2$ lb·ft·s²

Hence, by the parallel-axis theorem, the moment of inertia I_2 about an axis 2 ft from the mass center is

$$I_2 = I_G + (2)(2)^2 = 10 \text{ lb·ft·s}^2$$

18.141

Given: A thin hollow cylinder of radius 0.6 m weighs 290 N

Calculate its radius of gyration with respect to its geometric axis.

Solution: The mass moment of inertia of the cylinder is $I = mr^2 = \left(\frac{290}{9.81}\right)(0.6)^2 = 10.642$ N·m·s²

Then, since the radius of gyration is given by $k = \sqrt{\frac{I}{m}} = 0.6$ m

Alternatively,

$$I = mr^2 = mk^2$$

Therefore, $k = r = 0.6$ m

18.142

Given: A transverse axis 1-1 intersects a thin uniform rod of length L, at the distance $L/3$ from one end of the rod.

Derive a formula for the radius of gyration about the axis 1-1.

Solution: The rod and axis are shown in Fig. a

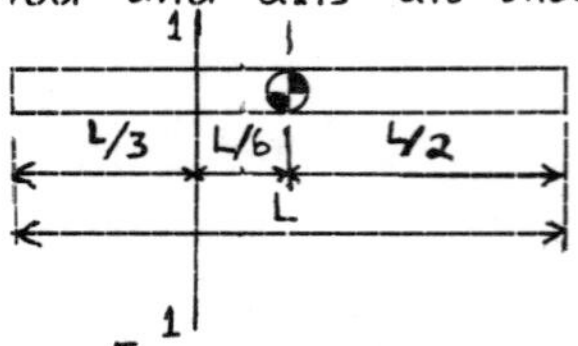

Figure a

(Continued)

18.142 Cont.

Since axis 1-1 is $L/3$ from the end, it is $L/6$ from the parallel axis through the center of mass G. By the parallel-axis theorem,

$$I_{1-1} = I_G + m\left(\frac{L}{6}\right)^2$$

$$= \frac{1}{12}mL^2 + \frac{1}{36}mL^2$$

or,

$$I_{1-1} = \frac{1}{9}mL^2 = mk^2$$

where k is the radius of gyration relative to axis 1-1.

Hence,

$$k = \frac{L}{3}$$

18.143

Given: A T-shaped piece of sheet metal that is 4 in high consists of two rectangles, each 3 in long and 1 in wide. The mass of the piece is $m = 0.001$ lb·s²/ft.

Calculate the mass moment of inertia about a line 1-1 that coincides with the bottom edge of the stem of the tee.

Solution: The piece is shown in Fig. a.

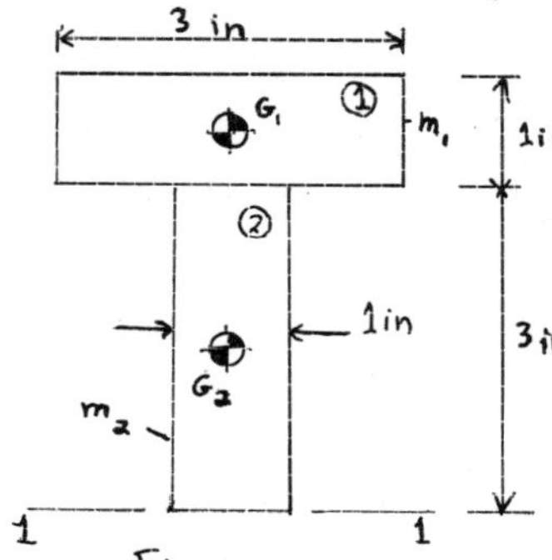

Figure a

By Fig. a, the mass moments of parts 1 and 2 relative to axes parallel to axes 1-1 and through their respective mass centers G_1 and G_2 are (see Appendix F, Table F1), with $m_1 = m_2 = \frac{1}{2}m = 0.0005$ lb·s²/ft,

$$I_{G1} = \frac{1}{12}m_1\left(\frac{1}{12}\right)^2 = \frac{1}{12}(0.0005)\left(\frac{1}{144}\right) = 2.894\times10^{-7}$$

$$I_{G2} = \frac{1}{12}m_2\left(\frac{3}{12}\right)^2 = \frac{1}{12}(0.0005)\left(\frac{1}{16}\right) = 2.604\times10^{-6}$$

Hence, by the parallel-axis theorem, we have

$$I_{1-1} = I_{G1} + m_1d_1^2 + I_{G2} + m_2d_2^2$$

$$= 2.894\times10^{-7} + (0.0005)\left(\frac{3.5}{12}\right)^2 + 2.604\times10^{-6} + (0.0005)\left(\frac{1.5}{12}\right)^2$$

or,

$$I_{1-1} = 5.324\times10^{-5} \text{ lb·ft·s}^2$$

18.148

Given: An idling shaft rotating at 1000 rpm has a fly wheel with mass moment of inertia of 20 lb·ft·s². It is suddenly coupled to a motionless shaft that has a fly wheel with mass moment of inertia of 30 lb·ft·s²

Calculate the angular velocity of the shaft at the instant after they are coupled.

Solution: Since there are no external angular moments applied to the shaft, angular momentum is conserved. Initially, the angular momentum is

$$A_i = I_i\omega_i = (20)\left[\left(1000\frac{\text{rev}}{\text{min}}\right)\left(\frac{2\pi\text{ rad}}{\text{rev}}\right)\left(\frac{1\text{ min}}{60\text{ s}}\right)\right]$$

or $A_i = 2094.4$ lb·ft·s (a)

After the shaft are coupled, the final angular moment is,

$$A_f = I_f\omega_f = (20+30)\omega_f = 50\omega_f \quad \text{(b)}$$

Equating Eqs. (a) and (b), we obtain

$$\omega_f = 41.888 \text{ rad/s} = 400 \text{ rpm}$$

18.153

Given: A block with dimensions 1 ft by 1 ft by 6 ft rests on its square end on a truck bed with its sides parallel to the direction of motion. The coefficient of static friction between the block and the truck is $\mu_s = 0.17$.

a) Determine whether the block tips first or slides first, if the truck stops suddenly.

b) Determine the value of the acceleration that would cause tipping or sliding of the block.

Solution: a) Figure a shows the block on the verge of tipping, including the inertial force.

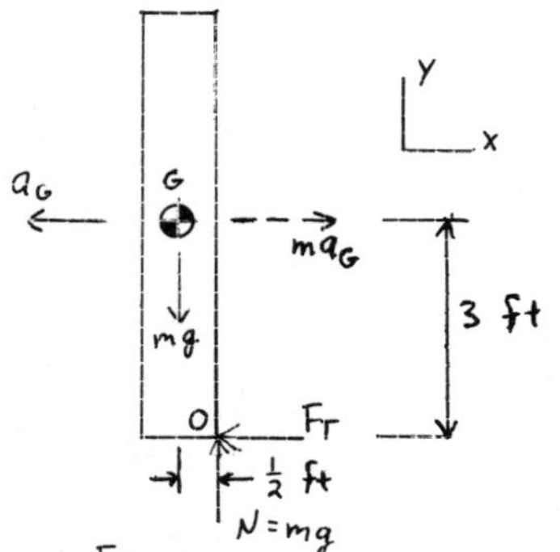

Figure a

By Fig. a,

$$\Sigma F_X = ma_G - F_T = 0 \quad \text{(a)}$$

$$\circlearrowleft \Sigma M_O = mg\left(\tfrac{1}{2}\right) - ma_G(3) = 0 \quad \text{(b)}$$

(Continued)

18.153 Cont.

By Eq. (a), $a_G = \frac{1}{6} g$ (c)

Then, Eqs. (a) and (c) yield for tipping

$$F_T = \frac{1}{6} mg = 0.1666\, mg \quad (d)$$

For sliding,

$$F_S = 0.17\, mg \quad (e)$$

By Eqs. (d) and (e),

$$F_T < F_S$$

Hence, tipping occurs first.

b) Since tipping occurs first, Eq. (c) yields,

$$\underline{\underline{a_G = \frac{1}{6} g = \frac{1}{6}(32.2) = 5.366 \text{ ft/s}^2}}$$

19.1

GIVEN The couple M of magnitude 660 N·m at O, Clockwise 45° in the xy plane from the position shown in Fig. a.

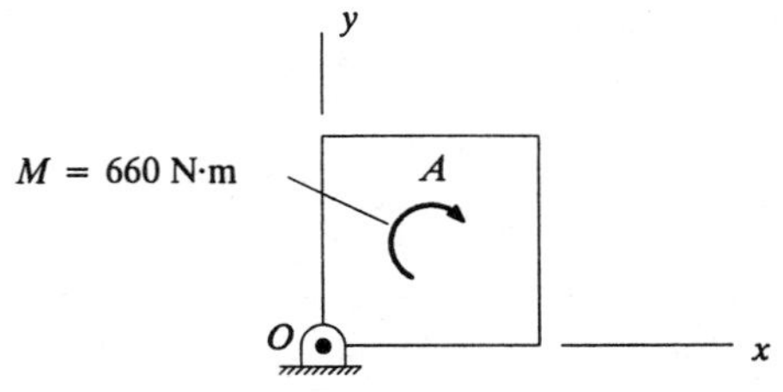

Figure a.

FIND Calculate the work performed by the couple during this displacement

SOLUTION

The work performed is given by

$$U = \int_{\theta_1}^{\theta_2} M\,d\theta = \int_0^{\pi/4} 660\,d\theta = (660)\left(\frac{\pi}{4}\right)$$

$$= 129.59 \text{ N·m}$$

19.2

GIVEN A torque of magnitude $M = a\theta^2 + b\theta$ is applied to a rotating shaft ; θ = angular displacement of the shaft and **a** and **b** are constants. The sense of the torque is the same as the positive sense of θ.

FIND Derive a formula for the work U as a function of θ_0 and θ_1 the initial and final values of θ.

SOLUTION

By definition,

$$U = \int_{\theta_0}^{\theta_1} M\,d\theta = \int_{\theta_0}^{\theta_1} (a\theta^2 + b\theta)\,d\theta$$

$$= \left(\tfrac{1}{3}a\theta^3 + \tfrac{1}{2}b\theta^2\right)\Big|_{\theta_0}^{\theta_1}$$

or

$$U = \tfrac{1}{3}a\left(\theta_1^3 - \theta_0^3\right) + \tfrac{1}{2}b\left(\theta_1^2 - \theta_0^2\right)$$

19.3

GIVEN A machinist applies equal and opposite forces of 5 lb, at 6 in from the center, to the T-handle of a finishing tool to ream a drilled hole (Fig a.). The tool is first rotated through 1080° and then the 5 lb forces are reversed and the tool is rotated through 540°.

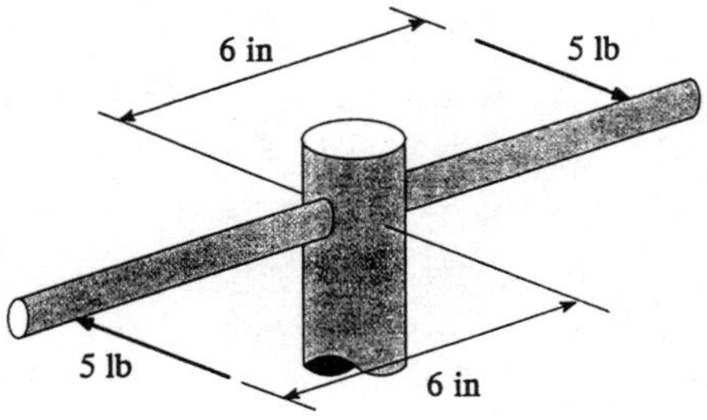

Figure a.

FIND Determine the total work performed by the machinist. The sense of the couple exerted by the machinist is always in the same sense as the rotation θ.
Hence, for the 1080° rotation,

SOLUTION

since 1080° = 18.849 rad

the work done is;

$$U = \int_0^{18.849} M\,d\theta = \int_0^{18.849} (5)(1)\,d\theta = (5)(1)(18.849)$$

$$= 94.248 \text{ lb·ft}$$

and for the 540° rotation,

$$U = \int_0^{9.425} M\,d\theta = \int_0^{9.425} (5)(1)\,d\theta = (5)(1)(9.425)$$

$$= 47.124 \text{ lb·ft}$$

Hence, the total work performed by the machinist is

$$U = U_1 + U_2 = 141.37 \text{ lb·ft}$$

19.4

GIVEN Figure a, with a rotational spring with resisting moment $M = k\theta$ inserted at joint **B**.

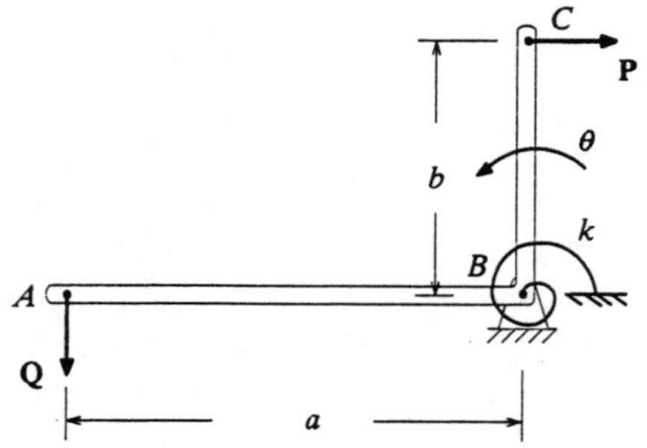

Figure a.

FIND

a.) Derive a formula for **P** in terms of Q, a, b, k and θ, for equilibrium

b.) For Q = 500 N, a = 600 mm, b = 500 mm, and k = 90 N·m, calculate P for $\theta = 0°$; $\theta = +30°$; $\theta = -30°$. Use the formula derive in part a.

SOLUTION

a.) Consider a positive rotation θ and draw the free-body diagram of the bell crank (Fig b).

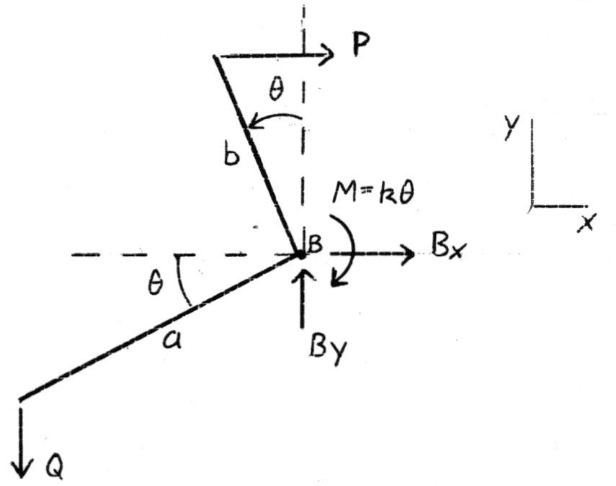

Figure b

By Fig. b,

$$\circlearrowleft\!+ \Sigma M_B = Qa\cos\theta - Pb\cos\theta - k\theta = 0$$

or

$$P = \frac{a}{b}Q - \frac{k\theta}{b\cos\theta} \qquad (a)$$

b.) By Eq. (a), with given data,

$$P = 600 - \frac{90\,\theta}{(0.50)\cos\theta} = 600 - \frac{180\,\theta}{\cos\theta} \qquad (b)$$

By Eq. (b),

For $\theta = 0°$, $P = 600$ N

For $\theta = +30°$, $P = 600 - \frac{(180)(\pi/6)}{\cos 30°} = 491.2$ N

For $\theta = -30°$, $P = 600 - \frac{(180)(-\pi/6)}{\cos(-30°)} = 708.8$ N

19.5

GIVEN A rotor weighs 260 N and has a radius of gyration $k = 0.50$ m. A couple of moment $M = 25t$ [N·m], where t is in seconds, is applied to the rotor shaft.

FIND Determine the work [N·m] performed on the rotor by the couple during the time the rotor undergoes 10 revolutions, starting from rest. Neglect friction

SOLUTION

Since 10 revolutions is equal to $(10)(2\pi) = 20\pi$ radians, the work done by the couple is

$$U = \int_0^{20\pi} (25t)\,d\theta \qquad (a)$$

In order to determine U, we must determine dθ in terms of time t and the time required for the rotor to rotate 10 revolutions. Since friction is neglected,

$$\Sigma M_o = M = I_o\,\alpha = mk^2\frac{d\omega}{dt}$$

where α denotes angular acceleration, ω denotes angular velocity, and I_o is the mass moment of inertia about the axis of the rotor.

(continued)

19.5 cont.

Hence,

$$M = 25t = \left(\frac{260}{9.81}\right)(0.50)^2 \frac{d\omega}{dt}$$

$$= 6.626 \frac{d\omega}{dt} \qquad (b)$$

Multiplying Eq. (b) by dt and integrating, we obtain,

$$12.5\,t^2 = 6.626\,\omega = 6.626 \frac{d\theta}{dt}$$

or

$$d\theta = 1.8865\,t^2\,dt \qquad (c)$$

By Eqs. (a) and (c), we have

$$U = 47.163 \int_0^{t_f} t^3\,dt \qquad (d)$$

where t_f is the time required to make 10 revolutions.

Integration of Eq. (d) yields

$$U = 47.163\left(\frac{t^4}{4}\right)\Big|_0^{t_f} = 11.791\,t_f^{\,4} \qquad (e)$$

To determine t_f, we have by integrating Eq. (c);

$$\int_0^{20\pi} d\theta = 1.8865 \int_0^{t_f} t^2\,dt$$

or

$$20\pi = 1.8865\left(\frac{t^3}{3}\right)\Big|_0^{t_f} = 0.6288\,t_f^{\,3}$$

So

$$t_f = 4.640 \text{ s} \qquad (f)$$

Then, Eqs. (e) and (f) yields

$$U = 5466.8 \quad \text{N}\cdot\text{m}$$

19.6

GIVEN The planetary gear drive in Fig. a lies in a horizontal plane. Arm OD is connected to the axles of the three identical contacting gears and is rotated at angular velocity ω by the torsional moment M_o. Gear A is fixed (it cannot rotate). Gears B and C rotate without slipping.

Moments $M_B = 20$ N·m and $M_C = 15$ N·m act on gears **B** and **C**.

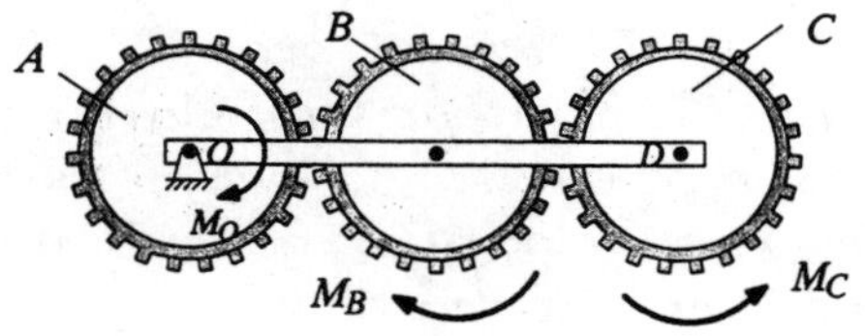

Figure a.

FIND Determine the work performed by M_B and M_C as arm **OD** rotates through 360°

SOLUTION

Let r be the radii of the gears. Then the velocities of the centers of gears B and C are $2r\omega$ and $4r\omega$, respectively (Fig. b)

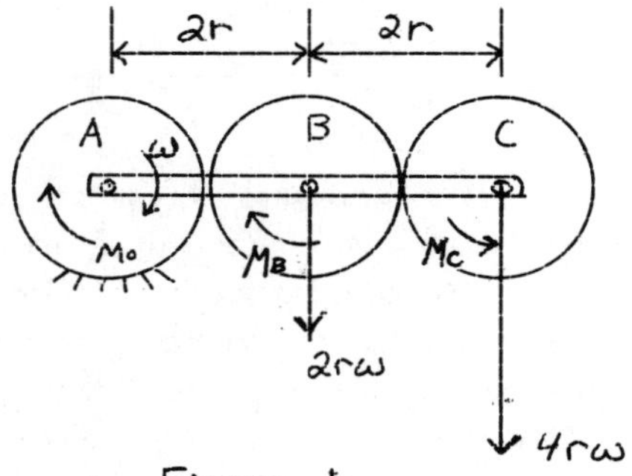

Figure b.

Since gear B does not slip on gear **A**, the contact point between gears A and B is the instantaneous center of velocity of gear **B**. Hence, the velocity of the gear **B** at the contact point between gears **B** and **C** is $4r\omega$ (Fig. c) and gear B rotates with angular velocity

$$\omega_B = 2r\omega/r$$

$$= 2\omega$$

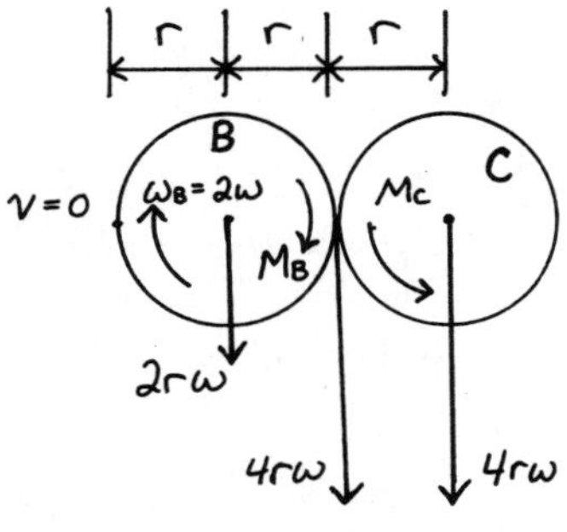

Figure C

(continued)

19.6 cont.

Also, since gear C rotates without slipping, the velocity of the contact point of gear C with gear B is $4r\omega$. The velocity of the center of gear C is also $4r\omega$. Therefore, gear C undergoes curvilinear translation; it does not rotate. However, since the angular velocity of gear B is 2ω, it undergoes a rotational displacement twice that of rod **OD**. Therefore, the work of couples M_B and M_C as the rod **OD** rotates through $360° = 2\pi$ rad is given by

$$W_B = M_B(4\pi) = (20)(4\pi) = 251.3 \text{ N·m}$$

$$W_C = M_C(0) = (10)(0) = 0 \text{ N·m}$$

That is, M_C does no work since gear C does not rotate.

19.7

GIVEN A homogeneous flywheel of uniform thickness is mounted on a shaft and rotates about it's centroidal axis perpendicular to the plane of the flywheel. The flywheel weighs 5 kN and has a radius of 1.5 m. The flywheel, initially at rest, is subjected to a torque $M = 5t^{1/2}$ [kN·m]; t in seconds

FIND
a.) Determine the angular rotation [rad] of the flywheel for $t = 4$s.
b.) Determine the work [J] performed by the torque as a function of t and calculate the work for $t = 4$s.

SOLUTION
a.) By kinetics, $M = I\frac{d\omega}{dt}$, where I is the mass moment of inertia of the flywheel about its centroidal axis and ω is the angular velocity.

Hence,
$$5000\, t^{1/2} = \frac{1}{2}\left(\frac{5000}{9.81}\right)(1.5)^2 \frac{d\omega}{dt}$$

or
$$t^{1/2}dt = 0.11468\, d\omega \qquad (a)$$

Integration of Eq. (a) yields, since $\omega = 0$ at $t = 0$;

$$\int_0^t t^{1/2}\,dt = 0.11468\int_0^\omega d\omega = 0.11468\,\omega$$

or
$$\omega = \frac{1}{0.11468}\left(\frac{2}{3}t^{3/2}\right)$$
$$= 5.813\, t^{3/2} \quad [\text{rad}] \qquad (b)$$

By Eq. (b) and kinematics

$$\omega = \frac{d\theta}{dt} = 5.813\, t^{3/2}$$

or
$$d\theta = 5.813\, t^{3/2}\,dt \qquad (c)$$

Integration of Eq (c), with $\theta = 0$ for $t = 0$, yields;

$$\int_0^\theta d\theta = \int_0^t 5.813\, t^{3/2}\,dt$$

$$\theta = 5.813\left(\frac{2}{5}\right)t^{5/2} = 2.325\, t^{5/2}$$

Therefore, for $t = 4$s,

$$\theta = 74.4 \text{ rad}$$

b.) The work of the couple is given by

$$U = \int M\,d\theta = 5000\int t^{1/2}\,d\theta \qquad (d)$$

By Eqs. (c) and (d), we obtain

$$U = 29{,}065\int_0^4 t^2\,dt = \frac{29065}{3}t^3\Big|_0^4$$

or
$$U = 620 \times 10^3 \text{ [J]}$$

19.8

GIVEN A couple $\vec{M} = 5t\hat{\imath} + 16t^2\hat{\jmath} + 12k$, where t is a parameter, is applied to a rigid body that undergoes a rotation $\vec{\theta} = e^{-t}\hat{\imath} + (1/t)\hat{\jmath} + t^{1/2}\hat{k}$.

FIND Determine the work performed on the body by the couple over the range $0 \le t \le 10$.

(continued)

19.8 cont.

SOLUTION

The work of a couple $M\hat{k}$ under a rotation $d\theta\hat{k}$ is [see Eq. (19.1)]

$$U = \int_{\theta_0}^{\theta_1} (M\hat{k}) \cdot (d\theta\hat{k}) = \int_{\theta_0}^{\theta_1} M d\theta$$

Hence, for a couple $\bar{M} = M_x\hat{i} + M_y\hat{j} + M_z\hat{k}$ that undergoes a rotation $d\bar{\theta} = (d\theta_x)\hat{i} + (d\theta_y)\hat{j} + (d\theta_z)\hat{k}$, the work is

$$U = \int_{\theta_0}^{\theta_1} \bar{M} \cdot d\bar{\theta} = \int_{\theta_0}^{\theta_1} (M_x d\theta_x + M_y d\theta_y + M_z d\theta_z) \qquad (a)$$

where

$$M_x = 5t, \quad M_y = 16t^2, \quad M_z = 12 \qquad (b)$$

and

$$d\bar{\theta} = [(-e^{-t})\hat{i} - (\tfrac{1}{t^2})\hat{j} + (\tfrac{1}{2}t^{-1/2})\hat{k}]dt$$

or

$$d\theta_x = -e^{-t}dt, \quad d\theta_y = -\tfrac{1}{t^2}\,dt$$
$$d\theta_z = \tfrac{1}{2}t^{-1/2}\,dt \qquad (c)$$

Equations (a), (b) and (c) yield, where $t = 0$ for $\theta = \theta_0$ and $t = 10$ s for $\theta = \theta_1$

$$U = \int_0^{10} [-5te^{-t} - 16 + 6t^{-1/2}]\,dt \qquad (d)$$

Integration of Eq. (d) yields

$$U = [5(1+t)e^{-t} - 16t + 12t^{1/2}]\Big|_0^{10}$$

or

$$\underline{\underline{U = -122.1 \quad [FL]}}$$

19.9

GIVEN A homogeneous disk of radius a and uniform thickness (h) is supported by a thin elastic rod (Fig a.) When the disk is rotated through an angle θ from its equilibrium position, the rod exerts a restoring torque $T = k\theta$, where k [FL] is a constant. The disk is rotated through an angle $\theta = \theta_0$ and then released from rest.

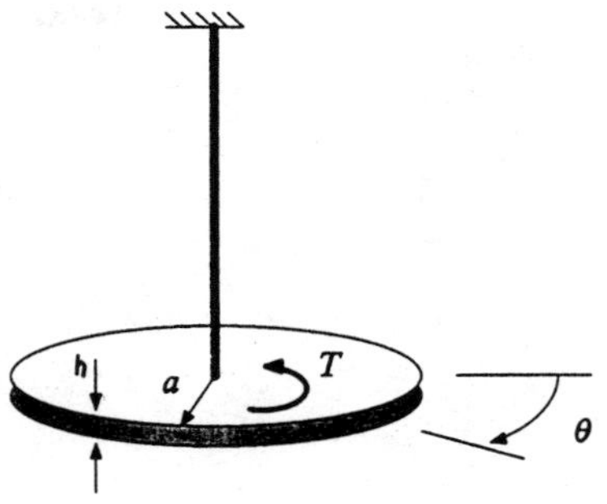

Figure a

FIND

a.) Show that the differential equation of motion of the disk is

$$\ddot{\theta} + \frac{k}{I}\theta = 0 \qquad (a)$$

where I is the mass moment of inertia of the disk about the axis of the rod, and the dots denote derivatives with respect to time t.

b.) Verify that

$$\theta = \theta_0 \cos\left(\sqrt{\frac{k}{I}}\right)t \qquad (b)$$

is the solution of Eq. (a), given the initial conditions $\theta = \theta_0$ and $\dot{\theta} = 0$ for $t = 0$

c.) Let $\theta_0 = \pi/2$. Determine the work performed by the torque $T = k\theta$ from the time the disk is released to the time that $\theta = 0$ for the first time.

SOLUTION

a.) Consider the free-body diagram of the disk (Fig. b)

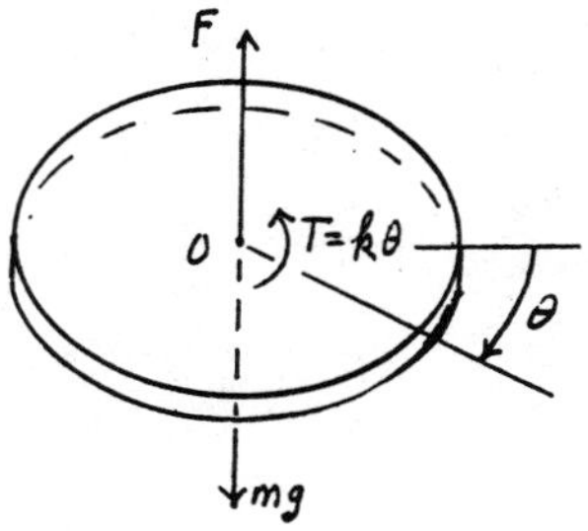

Figure b.

(continued)

19.9 cont.

By Fig. b, and kinetics,

$$\circlearrowleft \Sigma M_o = -k\theta = I\alpha = I\frac{d^2\theta}{dt^2} = I\ddot{\theta}$$

or

$$\ddot{\theta} + \frac{k}{I}\theta = 0$$

This verifies Eq (a).

b.) By Eq. (b), when the disk is released from rest at $t=0$, We obtain

$$\theta = \theta_0$$

By the time derivative of Eq. (b), we find;

$$\dot{\theta} = -\theta_0\sqrt{\frac{k}{I}}\sin\sqrt{\frac{k}{I}}\,t$$

Therefore, for $t=0$, $\dot{\theta}=0$; that is, the disk starts from rest at $t=0$. Finally, substitution of Eq. (b) into Eq. (a) yields

$$-\theta_0\left(\frac{k}{I}\right)\cos\sqrt{\frac{k}{I}}\,t + \theta_0\left(\frac{k}{I}\right)\cos\sqrt{\frac{k}{I}}\,t = 0$$

Thus, Eq. (b) is the solution of Eq. (a) with $\theta=\theta_0$ and $\dot{\theta}=0$ at $t=0$.

c.) For $\theta_0 = \pi/2$, Eq. (b) yields

$$\theta = \pi/2 \cos\sqrt{\frac{k}{I}}\,t \qquad (c)$$

Then, differentiation yields

$$d\theta = -\frac{\pi}{2}\left(\sqrt{\frac{k}{I}}\right)\left(\sin\sqrt{\frac{k}{I}}\,t\right)dt$$

also,

$$T = k\theta = k\left(\frac{\pi}{2}\right)\cos\sqrt{\frac{k}{I}}\,t$$

Consequently, the work performed by the torque from $\theta=\theta_0$ to $\theta=0$ (for the first time) is

$$U = \int_{\theta_0}^{0} T\,d\theta$$

$$= -\frac{\pi^2}{4}k\sqrt{\frac{k}{I}}\int_0^{t_\theta=0}\left(\sin\sqrt{\frac{k}{I}}\,t\right)\left(\cos\sqrt{\frac{k}{I}}\,t\right)dt \qquad (d)$$

To determine $t_{\theta=0}$; by Eq. (c), we have

$$\theta = \frac{\pi}{2}\cos\sqrt{\frac{k}{I}}\,t_{\theta=0} = 0$$

Therefore,

$$\sqrt{\frac{k}{I}}\,t_{\theta=0} = \pi/2$$

or

$$t_{\theta=0} = \frac{\pi}{2}\sqrt{\frac{I}{k}} \qquad (e)$$

With Eq (e), integration of Eq. (d) yields

$$U = -\frac{\pi^2}{4}k\sqrt{\frac{k}{I}}\int_0^{\frac{\pi}{2}\sqrt{\frac{I}{k}}}\left(\frac{\sin 2\sqrt{\frac{k}{I}}\,t}{2}\right)dt$$

$$= -\frac{\pi^2}{8}k\left(\frac{1}{2}\cos 2\sqrt{\frac{k}{I}}\,t\right)\Big|_0^{\frac{\pi}{2}\sqrt{\frac{I}{k}}}$$

or

$$U = -\frac{\pi^2 k}{16}(\cos\pi - \cos 0) = \frac{\pi^2}{8}k$$

19.10

GIVEN An elevator (Fig a.) has mass m and is supported by a cable wrapped around a cylindrical drum of radius r. The drum is connected by a shaft to an electric motor. The drum and its parts have mass m_2 and radius of gyration k. Two counter weights, each of mass m_1 are connected to the elevator by light cables that pass around frictionless pulleys of negligible mass. The motor supplies a constant counterclockwise torque T to the shaft.

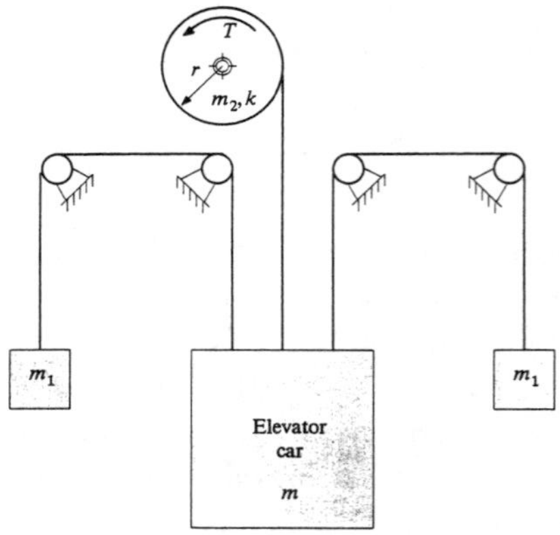

Figure a.

(continued)

FIND

a.) Using $F = ma$ and free-body diagrams, derive the equations that can be used to solve for the tension T_1 in the drum cable, the tension T_2 in the counterweight cables, and the acceleration a of the elevator in terms of the given parameters.

b.) For $T = 40$ lb·ft, $m = 20$ lb·s²/ft, $m_1 = 7$ lb·s²/ft, $m_2 = 2.7$ lb·s²/ft, $r = 1.60$ ft, and $k = 1.3$ ft, evaluate T_1, T_2, and a

c.) The elevator moves 20 ft while the torque T acts. Determine the work performed by T.

SOLUTION

a.) Figures b. and c. are free-body diagrams of the elevator and one of the counterweights, respectively.

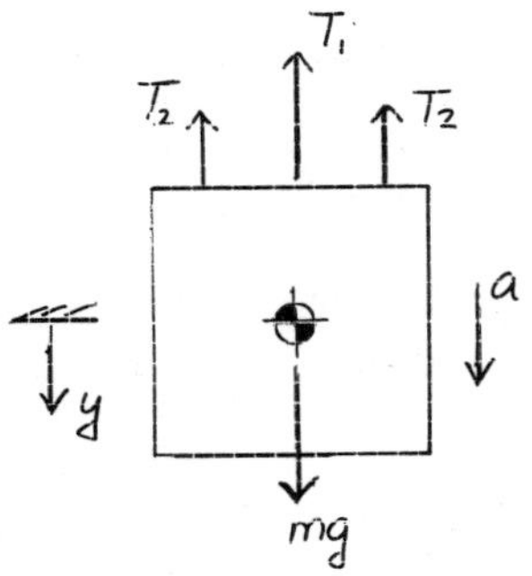

Figure b

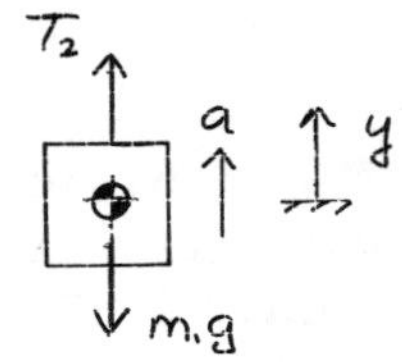

Figure c

By Fig. b,

$$\Sigma F_y = mg - T_1 - 2T_2 = ma$$

or $\quad T_1 + 2T_2 = m(g-a) \qquad$ (a)

By Fig. c,

$$\Sigma F_y = T_2 - m_1 g = m_1 a$$

or $\quad T_2 = m_1(g+a) \qquad$ (b)

Next, consider the free-body diagram of the drum (Fig d).

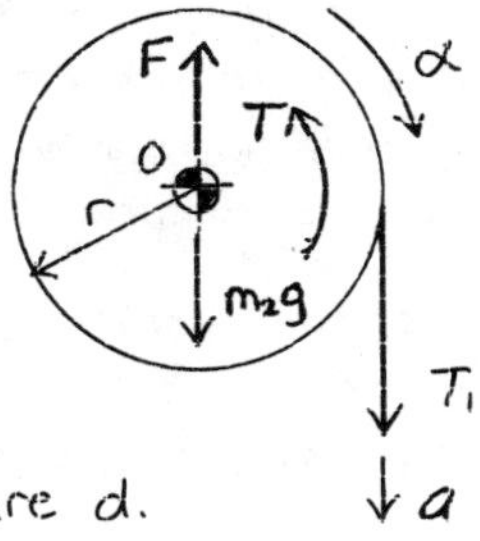

Figure d.

By Fig. d,

$$\circlearrowleft \Sigma M_0 = T_1 r - T = I_0 \alpha = m_2 k^2 \qquad (c)$$

By Eq. (c) and the kinematic relation $a = r\alpha$, we obtain

$$T_1 = \frac{T}{r} + m_2 \left(\frac{k}{r}\right)^2 a \qquad (d)$$

Equations (a), (b), and (d) are three independent equations in terms of T_1, T_2, and a. For given values of the parameters T, m, m_1, m_2, r, and k, they may be solved for T_1, T_2, and a.

b.) For $T = 40$ lb·ft, $m = 20$ lb·s²/ft, $m_1 = 7$ lb·s²/ft, $m_2 = 2.7$ lb·s²/ft, $r = 1.60$ ft and $k = 1.3$ ft, Eqs (a), (b), and (d) yield

$$T_1 - 2T_2 + 20a = 644$$
$$T_2 - 7a = 225.4$$
$$T_1 - 1.782a = 25$$

The solution of these equations is

$$T_1 = 33.38 \text{ lb}$$

$$T_2 = 258.30 \text{ lb}$$

$$a = 4.70 \text{ ft/s}^2$$

Since a is positive, the elevator moves down.

c.) Since the elevator moves down 20 ft, the drum rotates clockwise $\theta = 20/r = 20/1.6 = 12.5$ rad. Hence, the counterclockwise torque $T = 40$ lb·ft performs work

$$U = -T\theta = -(40)(12.5) = -500 \text{ ft·lb}$$

19.11

GIVEN An airplane that has mass of 9810 kg is raised to an altitude of 6 km to a cruising speed of 640 km/h by its engines.

FIND Determine the work that the engines performed. Neglect air resistance and assume gravity is constant.

SOLUTION

The work done is composed of two parts, namely, the work done in overcoming gravity and the work done to increase the speed from zero to 640 km/h. Hence, the total work is, noting that 640 km/h ≈ $177.7\bar{7}$ m/s,

$$U = \int_0^{6000\,m} mg\,dh + \int_0^{177.7\bar{7}} mv\,dv$$

$$= (9810)(9.81)(6000) + \tfrac{1}{2}(9810)(177.7\bar{7})^2$$

or

$$U = 7.324 \times 10^8 \text{ J}$$

19.12

GIVEN A homogeneous concrete block (specific weight $\gamma = 150$ lb/ft³) has the cross-sectional dimensions shown in Fig. a, and it is 10 ft long. The block must be placed with its side BC on the ground.

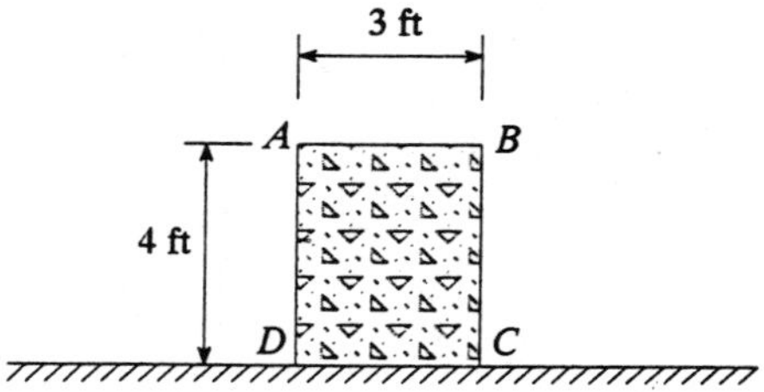

Figure a.

FIND

a.) Determine the work required to slowly rotate the block so that diagonal AC is vertical.

b.) Determine the total work performed on the block as it is slowly rotated from the position shown in Fig. a to the position with side BC on the ground.

SOLUTION

a.) The weight of the block is

$$W = (3)(4)(10)(150) = 18{,}000 \text{ lb}$$

When the block is on edge C, with diagonal AC vertical, its center of gravity is 2.5 ft above the ground (Fig. b)

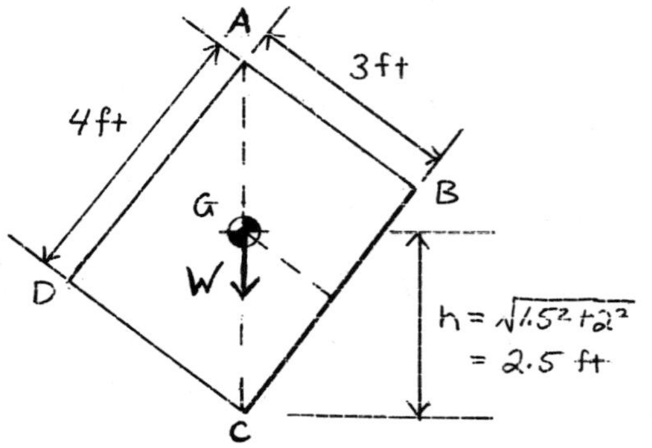

Figure b

Hence, the work performed from its initial position (Fig a.) to the position shown in Fig. b is.

$$U = W(h-2) = 18{,}000\,(2.5-2)$$

or

$$U = 9000 \text{ ft}\cdot\text{lb}$$

b.) When the block is on its side BC (Fig. c), its center of gravity is 1.5 ft above the ground. Hence, the total work performed on the block from its initial position (Fig. a) to its final position (Fig c.) is

$$U = W(h-2) = 18{,}000\,(1.5-2)$$

or

$$U = -9000 \text{ ft}\cdot\text{lb}$$

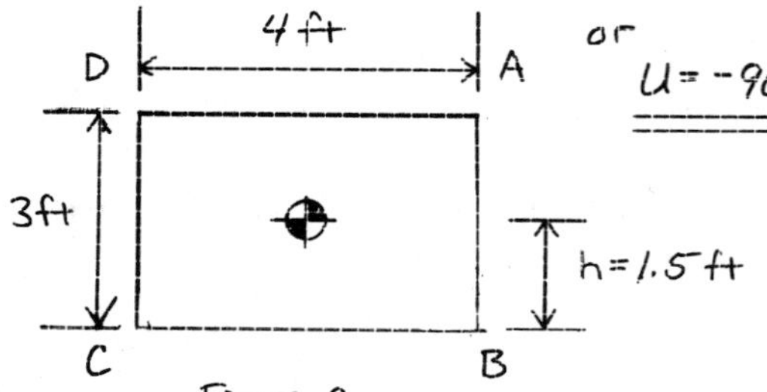

Figure C

19.13

GIVEN The pulley system is released from rest (Fig. a), and the weight W_1 moves down. The masses of the pulleys are negligible as is friction.

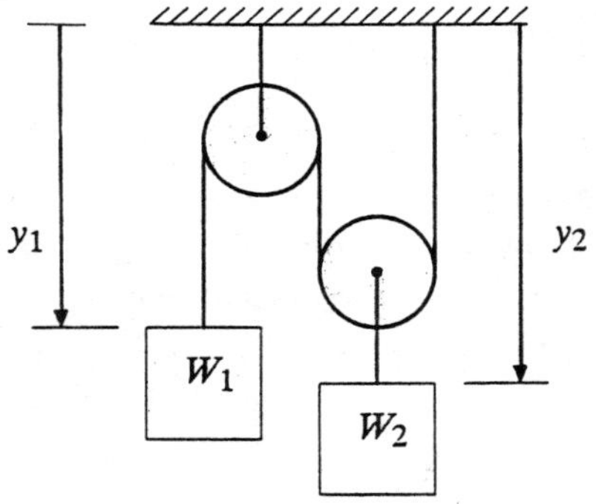

Figure a

FIND

a.) What is the minimum weight W_1 for it to move down?

b.) Determine the work done on the system as W_1 drops a distance h.

SOLUTION

a.) Assume that the system is in equilibrium, and determine the magnitude of W_1 for equilibrium. For W_1 to move down, its weight must be greater than for equilibrium.

The free-body diagrams of W_1 and W_2 are shown in Figs. b and c.

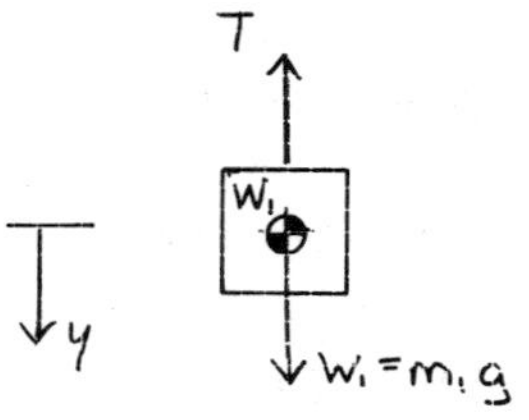

Figure b

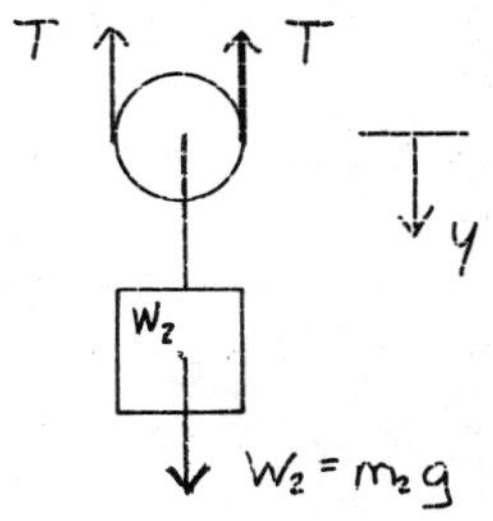

Figure c

By Fig. b, for equilibrium,

$$\Sigma F_y = W_1 - T = 0;$$

$$W_1 = T \qquad (a)$$

By Fig. c, for equilibrium,

$$\Sigma F_y = W_2 - 2T = 0;$$

$$T = \tfrac{1}{2} W_2 \qquad (b)$$

and by Eqs (a) and (b),

$$W_1 = \tfrac{1}{2} W_2 \qquad (c)$$

Therefore, for W_1 to move down

$$W_1 > \tfrac{1}{2} W_2$$

b.) By Fig. a, the geometrical constraint is

$$y_1 + 2y_2 = \text{constant} \qquad (d)$$

So, when W_1 moves down a distance h, W_2 moves a distance Δy_2.

Then, by Eq. (d)

$$y_1 + 2y_2 = (y_1 + h) + 2(y_2 + \Delta y_2)$$

or $$h + 2(\Delta y_2) = 0$$

Therefore, $$\Delta y_2 = -h/2 \qquad (e)$$

Thus, W_2 moves up a distance $h/2$.

Consequently, the work done on the system is.

$$U = W_1 h + W_2 (\Delta y_2) = \left(W_1 - \frac{W_2}{2}\right) h$$

19.14

GIVEN The pulley system shown in Fig. a is released from rest. The masses of the pulleys are negligible, as is frictions.

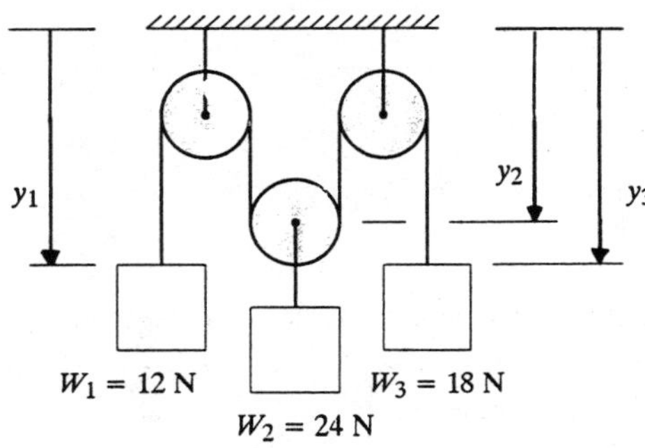

Figure a.

(continued)

19.14 cont.

FIND

a.) Determine how the weights will move. That is, which of the weights moves up and which of the weights moves down?

b.) Using the results of part a, determine the work done on the system as a weight moves down 2m

SOLUTION

a.) Assume that the system is in equilibrium and determine the cord tension required to maintain each weight in equilibrium. To do this, consider the free-body diagram of each weight shown in Figs. b, c, and d.

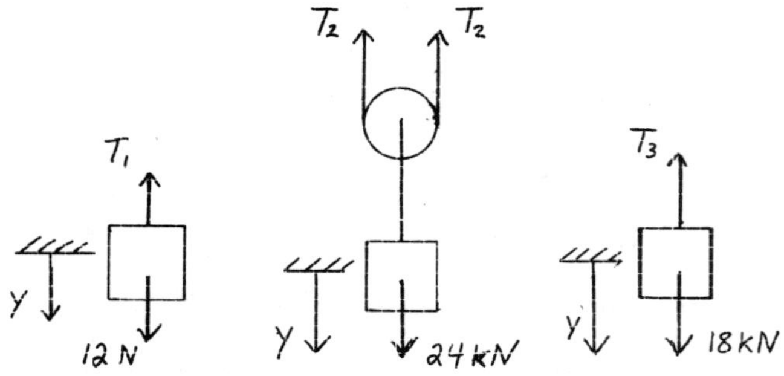

Figure b   Figure c   Figure d

By Fig. b, for equilibrium,

$$\Sigma F_y = 12N - T_1 = 0;$$

$$T_1 = 12N \qquad (a)$$

Similarly, by Eqs. c and d

$$\Sigma F_y = 24N - 2T_2 = 0; \qquad T_2 = 12N \qquad (b)$$

$$\Sigma F_y = 18N - T_3 = 0; \qquad T_3 = 18N \qquad (c)$$

By Eqs. (a), (b), and (c),

$$T_3 > T_2 = T_1$$

Hence, weight W_3 will move down, and with $T_3 > 18N$, W_1 and W_2 will move up.

b.) By Fig. a, the geometrical constraint is

$$y_1 + 2y_2 + y_3 = \text{constant}$$

and by differentiation with respect to time,

$$\ddot{y}_1 + 2\ddot{y}_2 + \ddot{y}_3 = 0 \qquad (d)$$

By Figs. a, b, c, and d, with $T_1 = T_2 = T_3 = T$,

$$\Sigma F_y = T - 12 = \frac{-12}{g}\ddot{y}_1 \qquad (e)$$

$$\Sigma F_y = 2T - 24 = \frac{-24}{g}\ddot{y}_2 \qquad (f)$$

$$\Sigma F_y = T - 18 = -\frac{18}{g}\ddot{w} \qquad (g)$$

Hence, by Eqs. (e), (f), and (g), the accelerations of weights W_1, W_2, and W_3 are, respectively,

$$\ddot{y}_1 = a_1 = g\left(1 - \frac{T}{12}\right) \qquad (h)$$

$$\ddot{y}_2 = a_2 = g\left(1 - \frac{T}{12}\right) \qquad (i)$$

$$\ddot{y}_3 = a_3 = g\left(1 - \frac{T}{18}\right) \qquad (j)$$

Then, Eqs. (d), (h), (i), and (j) yield

$$4 - \frac{11}{36}T = 0$$

or

$$T = \frac{144}{11} \qquad (k)$$

Substitution of Eq. (h) into Eqs. (h), (i), and (j) yields

$$a_1 = -\frac{1}{11}g, \qquad a_2 = -\frac{1}{11}g$$

$$a_3 = \frac{3}{11}g$$

So, the distances traveled by weights W_1, W_2, and W_3 are, respectively, in time t

$$s_1 = \frac{1}{2}a_1 t^2, \qquad s_2 = \frac{1}{2}a_2 t^2$$

$$s_3 = \frac{1}{2}a_3 t^2$$

Hence, since $s_3 = 2\,m$ down, $t^2 = \frac{44}{3g}$. Therefore,

$$s_1 = s_2 = \frac{1}{2}\left(-\frac{g}{11}\right)\left(\frac{44}{3g}\right) = -\frac{2}{3}m = \frac{2}{m}\text{ up}$$

So, the work done on the system is

$$U = W_1 s_1 + W_2 s_2 + W_3 s_3$$

$$= 12(-2/3) + 24(-2/3) + 18(2)$$

or

$$\underline{\underline{U = 12\ N\cdot m}}$$

19.15

GIVEN A homogeneous semi-cylinder of precast concrete (specific weight γ), of radius r and length L, is unloaded at a construction site in the initial position shown in Fig. a. However, the construction crew wants the cylinder placed so that face AB is on the ground.

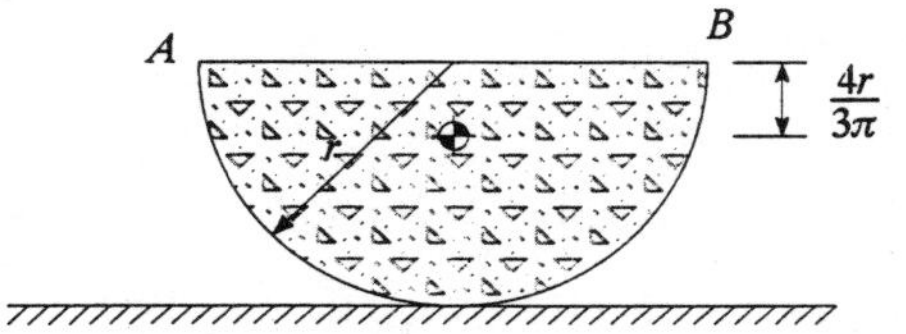

Figure a

FIND

a.) In terms of r, L, and γ, determine the work required to roll the cylinder so that the center of gravity is directly over B.

b.) In terms of r, L, and γ, determine the total work performed on the cylinder to move it from its initial position to the position with force AB on the ground.

c.) For $r = 0.6\text{ m}$, $L = 3\text{ m}$, and $\gamma = 2.2\text{ kN/m}^3$, calculate the work performed in parts a and b.

SOLUTION

a.) Consider the cylinder with the center of gravity directly above B (Fig. b).

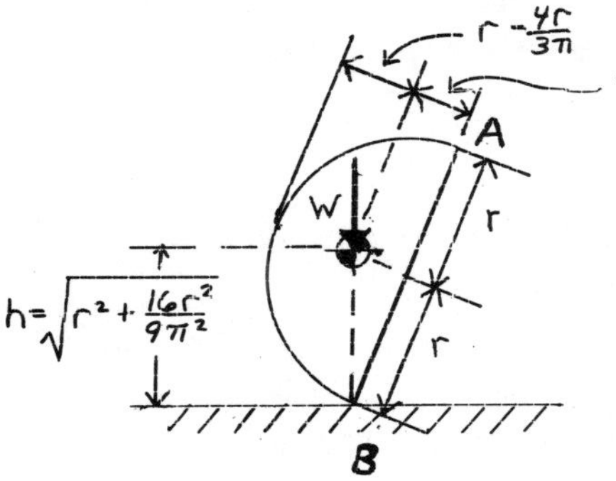

Figure b

The work required to raise the semicylinder to this position is, by Figs. a and b,

$$U = W\left[h - \left(r - \frac{4r}{3\pi}\right)\right] \quad \text{(a)}$$

where

$$W = \tfrac{1}{2}\pi r^2 L\gamma \quad \text{(b)}$$

$$h = \sqrt{r^2 + \frac{16r^2}{9\pi^2}} = \frac{r}{3\pi}\sqrt{9\pi^2 + 16} \quad \text{(c)}$$

Substituting Eqs. (b) and (c) into Eq (a), we obtain

$$\underline{\underline{U = 0.8023\, r^3 L\gamma}} \quad \text{(d)}$$

b.) Consider the semicylinder with its face AB on the ground (Fig. c).

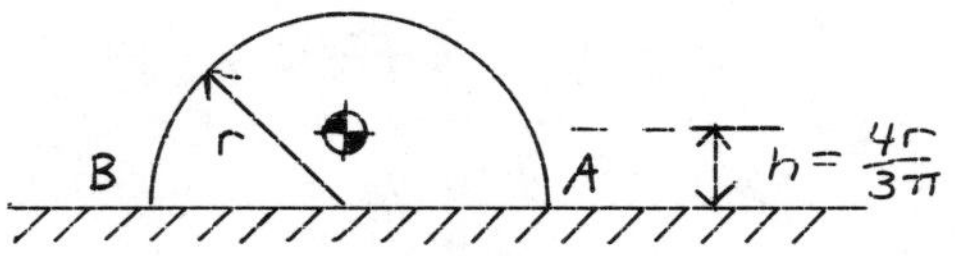

Figure c

The total work performed on the cylinder from its initial position (Fig. a) is, with Eq. (b),

$$U = W\left[h - \left(r - \frac{4r}{3\pi}\right)\right]$$

$$= \frac{1}{2}\pi r^2 L\gamma\left[\frac{4r}{3\pi} - \left(r - \frac{4r}{3\pi}\right)\right]$$

$$= \left(\frac{4}{3} - \frac{\pi}{2}\right) r^3 L\gamma$$

or

$$\underline{\underline{U = -0.2375\, r^3 L\gamma}} \quad \text{(e)}$$

c.) For $r = 0.6$, $L = 3\text{m}$, and $\gamma = 2.2\text{ kN/m}^3$, Eq. (d) yields

$$U = 0.8023\,(0.6)^3(3)(2.2) = 1.144\text{ kN}\cdot\text{m}$$

or

$$\underline{\underline{U = 1144\text{ N}\cdot\text{m}}}$$

and Eq (e) yields

$$U = -0.2375\,(0.6)^3(3)(2.2) = -0.3386\text{ kN}\cdot\text{m}$$

or

$$\underline{\underline{U = -338.6\text{ N}\cdot\text{m}}}$$

19.16

GIVEN A ball of mass m and radius r slides and rotates as it travels down an inclined plane (Fig a.)

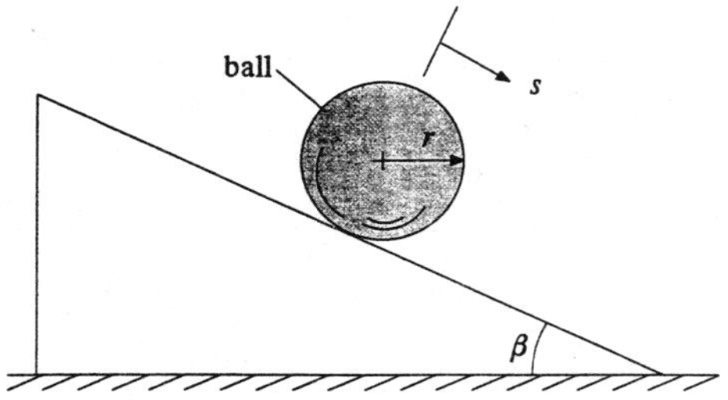

Figure a.

FIND

Prove that the work performed by the frictional force F on the ball is

$$U = -\mu mg\,(s - r\theta)\cos\beta$$

where μ is the coefficient of friction, s is the distance the center of the ball travels, and θ is the angle through which the ball rotates.

SOLUTION

The free-body diagram of the ball is shown in Fig. b

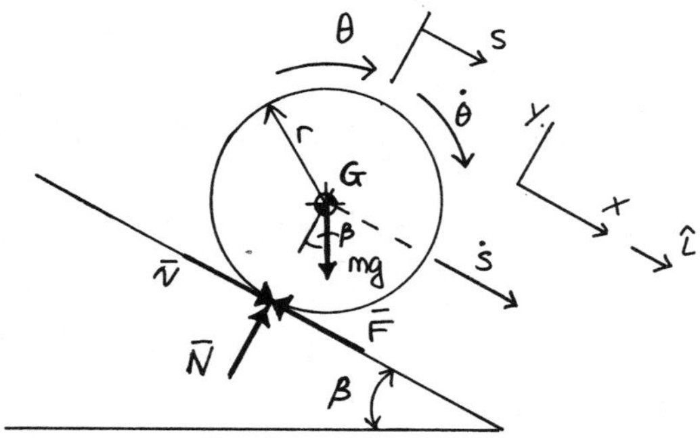

Figure b.

The work performed on the ball by the force of friction is [see Eq (19.2)]

$$U = \int \bar{F}\cdot\bar{v}\,dt \qquad (a)$$

where $\bar{F}$ is the force of friction and $\bar{v}$ is the velocity of the point c of the ball in contact with the inclined plane (Fig. b)

By Fig. b,

$$\Sigma F_y = N - mg\cos\beta = 0$$

or

$$N = mg\cos\beta$$

Since the slides, the friction force magnitude is

$$F = \mu N = \mu mg\cos\beta$$

Hence,

$$\bar{F} = -\mu mg\,(\cos\beta)\,\hat{\imath} \qquad (b)$$

also, by Fig. b and kinematics,

$$\bar{v} = (\dot{s} - r\dot{\theta})\,\hat{\imath} = \left(\frac{ds}{dt} - r\frac{d\theta}{dt}\right)\hat{\imath} \qquad (c)$$

Therefore, by Eqs (a), (b), and (c),

$$U = -\mu mg\cos\beta\left[\int_0^s ds - r\int_0^\theta d\theta\right]$$

or

$$\underline{\underline{U = -\mu mg\,(s - r\theta)\cos\beta}}$$

19.17

GIVEN An airplane lands with a speed of 120 ft/s. The outside diameter of a tire is 4 ft, and the radius of gyration of the tire and wheel is 1.5 ft. The mass of the wheel is 10 lb·s²/ft. Upon landing, the tire skids until it gains enough angular velocity to roll. The friction force of the pavement on the skidding tire is constant at 900 lb. Neglect deceleration of the airplane during skidding.

FIND

a.) Calculate the time required for the wheel to begin to roll.

b.) Calculate the work the force of skidding performs on the wheel during the skidding period.

(continued)

19.17 cont.

SOLUTION

a.) By $\Sigma M = I\, d\omega/dt$, we have

$$(900)(2) = (10)(1.5)^2 \frac{d\omega}{dt}$$

or

$$d\omega = 80\, dt$$

Hence,

$$\omega = 80t \qquad (a)$$

Since the plane's speed is 120 ft/s, rolling begins when

$$120 = \omega r = (80t)(2)$$

or

$$t = \frac{120}{160} = 0.75 \text{ s}$$

b.) The work performed by friction on the wheel may be calculated by

$$U = -\int F v\, dt \qquad (b)$$

where v is the speed of the point on the skidding tire in contact with the pavement. Since the plane's speed is 120 ft/s, say to the right, and the wheel is rotating clockwise about the wheel axle, the speed of skidding is, with Eq (a),

$$v = 120 - \omega r = 120 - 160t \qquad (c)$$

Hence, for the skidding period, $0 \le t \le 0.75$, Eqs. (b) and (c) yield, with F = 900 lb,

$$U = -\int_0^{0.75} 900\,(120 - 160t)\, dt$$

$$= -108{,}000t + 72{,}000t^2 \Big|_0^{0.75}$$

or

$$U = -40{,}500 \text{ ft}\cdot\text{lb}$$

19.18

GIVEN A homogeneous ball of mass m rests on a flat horizontal board that moves horizontally with constant acceleration a (Fig. a).

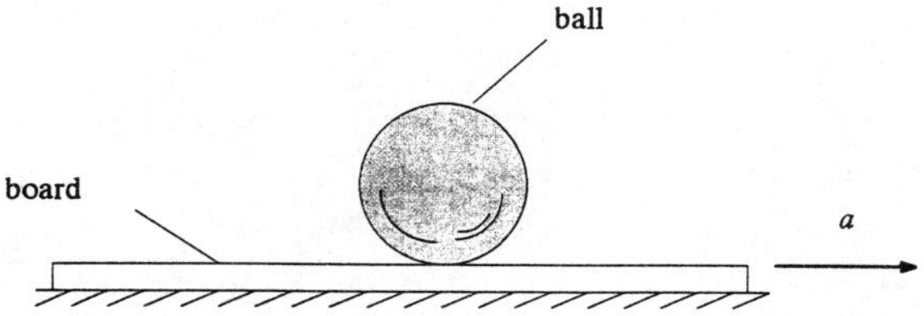

Figure a.

At time t_1 the velocity of the board is v. The ball does not slip on the board.

FIND Show that the work the frictional force performs on the ball in the time interval $t_0 \le t \le t_1$ is:

$$U = \frac{1}{7} ma\left[a(t_1^2 - t_0^2) + 2v_0(t_1 - t_0)\right]$$

SOLUTION

Consider the free-body diagram of the ball (Fig b.)

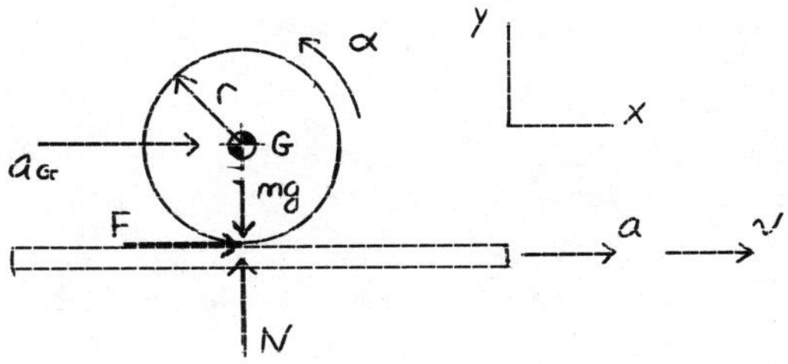

Figure b

By Fig. b,

$$\Sigma F_x = F = m a_G \qquad (a)$$

$$\circlearrowleft \Sigma M_G = Fr = I_G \alpha = \frac{2}{5} mr^2 \alpha \qquad (b)$$

By Fig. b, and kinematics

$$a_G = a - r\alpha \qquad (c)$$

Equation (a), (b), and (c) yield

$$F = \frac{2}{7} ma$$

(continued)

19.18 cont.

At time t_0, the velocity of the board is v_0. Hence, the velocity of the board at time t is

$$v = v_0 + at \qquad (d)$$

Since the ball does not slip on the board, v is also the velocity of the contact point of the ball. Hence, the work done by the frictional force, see Fig. b, is, in the interval $t_0 \le t \le t_1$

$$U = \int_{t_0}^{t_1} F v\, dt = \frac{2}{7} ma \int_{t_0}^{t_1} (v_0 + at)\, dt \qquad (e)$$

Integration of Eq (e) yields

$$U = \frac{1}{7} ma \left[a(t_1^2 - t_0^2) + v_0(t_1 - t_0) \right]$$

19.19

GIVEN A homogeneous solid flywheel with mass moment of inertia I is subjected to a constant couple $M_1 = k_1$. A brake produces a resisting couple $M_2 = k_2\omega$, where ω is the angular velocity [rad/s] of the flywheel and k_2 is a constant. Initially, the flywheel is at rest.

FIND

a.) Derive a formula for the total work performed on the flywheel by the couples M_1 and M_2 as functions of time t [s]. Neglect friction of the axle of the flywheel.

b.) Calculate the total work performed by the couples during the time interval $0 \le t \le 5$ s, for $k_1 = 50$ N·m, $k_2 = 25$ N·m·s, and $I = 50$ N·m·s²

SOLUTION

a.) By the free-body diagram of the flywheel (Fig. a), we have

$$\circlearrowleft \Sigma M_O = M_1 - M_2 = I\alpha = I \frac{d\omega}{dt}$$

or

$$k_1 - k_2\omega = I \frac{d\omega}{dt} \qquad (a)$$

Separating variables of Eq. (a), we find

$$dt = I \frac{d\omega}{k_1 - k_2\omega}$$

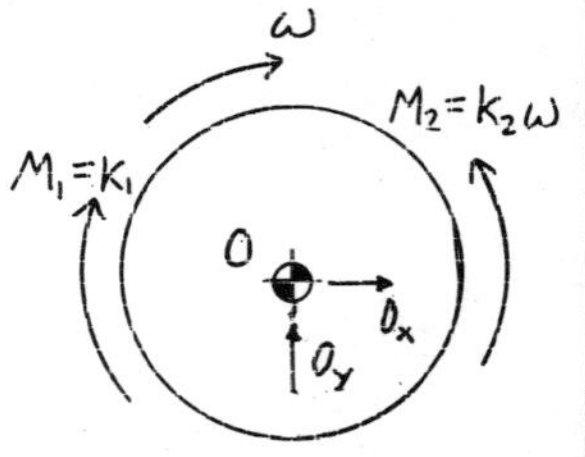

Figure a

Integration yields

$$\int_0^t dt = I \int_0^\omega \frac{d\omega}{k_1 - k_2\omega}$$

$$t = I \left[-\frac{1}{k_2} \ln (k_1 - k_2\omega) \right] \Big|_0^\omega$$

or

$$t = \frac{I}{k_2} \ln \frac{k_1}{k_1 - k_2\omega} \qquad (b)$$

Solving Eq (b) for ω, we obtain

$$\omega = \frac{d\theta}{dt} = \frac{k_1}{k_2} \left(1 - e^{-k_2 t/I} \right) \qquad (c)$$

or

$$d\theta = \frac{k_1}{k_2} \left(1 - e^{-k_2 t/I} \right) dt \qquad (d)$$

The work of the couple is

$$U = \int (M_1 - M_2)\, d\theta \qquad (e)$$

So, with the definitions of M_1 and M_2, Eqs. (c), (d), and (e) yield

$$U = \frac{k_1^2}{k_2} \int_0^t \left(e^{-k_2 t/I} - e^{-2k_2 t/I} \right) dt \qquad (f)$$

Integration yields

$$U = \frac{-k_1^2}{k_2^2} I \left(e^{-k_2 t/I} - \frac{1}{2} e^{-2k_2 t/I} \right) \Big|_0^t$$

Evaluating the limits and simplifying, we find

$$U = \frac{1}{2} \frac{k_1^2}{k_2^2} I \left(1 - e^{-k_2 t/I} \right)^2 \qquad (g)$$

b.) For $t = 5$ s, $k_1 = 50$ N·m, $k_2 = 25$ N·m·s, and $I = 50$ N·m·s²; Eq. (g) yields

$$U = \frac{1}{2} \left(\frac{50}{25} \right)^2 (50) \left(1 - e^{-25(5)/50} \right)^2$$

or

$$U = 84.26 \text{ N·m}$$

19.20

GIVEN A cannon of mass 1500 kg fires a projectile of mass 3 kg horizontally with a muzzle speed of 300 m/s. Friction is negligible.

FIND
a.) Determine the recoil speed of the cannon.
b.) Calculate the work performed on the system by the explosive charge.

SOLUTION
a.) Since the explosive forces are internal to the system (consisting of the cannon and projectile), linear momentum is conserved. Hence,

$$(m_c v_c + m_p v_p)_{initial} = (m_c v_c + m_p v_p)_{final} \qquad (a)$$

where subscript c denotes cannon and subscript p denotes projectile. Initially, the system is at rest; hence, Eq. (a) yields

$$(m_c v_c + m_p v_p)_{final} = 0 \qquad (b)$$

Therefore, with $m_c = 1500$ kg, $m_p = 3$ kg, and $v_p = 300$ m/s, Eq. (a) yields

$$v_c = -\frac{m_p v_p}{m_c} = -\frac{(3)(300)}{1500} = -0.6 \text{ m/s} \qquad (c)$$

That is, the recoil speed of the cannon is 0.6 m/s, with sense opposite to that of the projectile velocity.

b.) By the law of kinetic energy [Eq. (19.6)] the work performed on the system is

$$U_{1-2} = T_2 - T_1 \qquad (d)$$

where the initial knetic energy $T_1 = 0$ and the final kinetic energy is

$$T_2 = \frac{1}{2} m_c v_c^2 + \frac{1}{2} m_p v_c^2$$

or

$$T_2 = \frac{1}{2}(1500)(0.6)^2 + \frac{1}{2}(3)(300)^2$$
$$= 135\,270 \text{ J} \qquad (e)$$

Therefore, by Eqs. (d) and (e),

$$U_{1-2} = 1.3527 \times 10^5 \text{ J}$$

19.21

GIVEN A belt drive consists of two identical pulleys, each with mass moment of inertia I_0 (Fig. a). Pully A rotates with angular velocity ω and the belt (specific weight γ per unit length) does not slip.

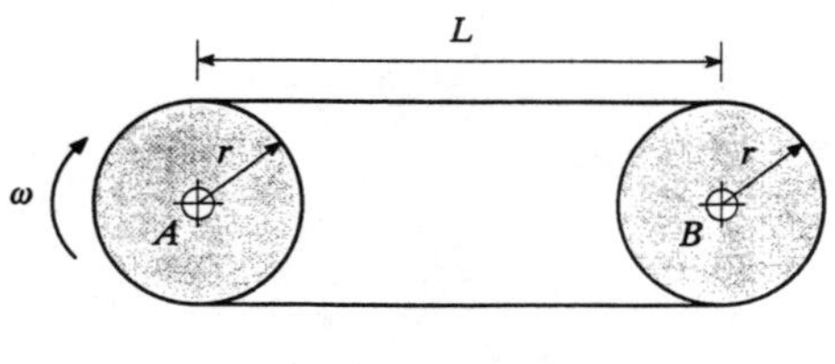

Figure a.

FIND
a.) Derive a formula for the kinetic energy T of the system in terms of I_0, ω, γ, r, and L.
b.) For $r = 150$ mm, $L = 1$ m, $\omega = 10\pi$ rad/s, $I_0 = 0.060$ kg·m², and $\gamma = 4$ N/m, calculate T [N·m].

SOLUTION
a.) By Fig. a, the length of the belt is

$$L_B = 2L + 2\pi r = 2(L + \pi r)$$

Therefore, its mass is

$$m_B = 2(L + \pi r)\gamma/g$$

Since the belt does not slip on the pulleys, its speed is $v_c = r\omega$. Therefore, the kinetic energy of the system is

$$T = (2)\left(\frac{1}{2} I_0 \omega^2\right) + \frac{1}{2}\left[\frac{2(L+\pi r)\gamma}{g}\right](r\omega)^2$$

or

$$T = \left[I_0 + \frac{(L+\pi r)\gamma r^2}{g}\right]\omega^2$$

(continued)

19.21 cont.

b.) For $r = 150$ mm, $L = 1$ m, $\omega = 10\pi$ rad/s, $I_o = 0.060$ kg·m², and $\gamma = 4$ N/m, Eq (a) yields

$$T = \left[0.060 + \frac{(1 + \pi \times 0.150)(4)(.150)^2}{9.81}\right](100\pi^2)$$

or

$$T = 72.54 \text{ N·m}$$

19.22

GIVEN The uniform cube shown in Fig. a is pinned at O and it is rotated clockwise 90° in the horizontal xy plane by the couple $\bar{M}$. The cube measures 150 mm on an edge and weighs 9N.

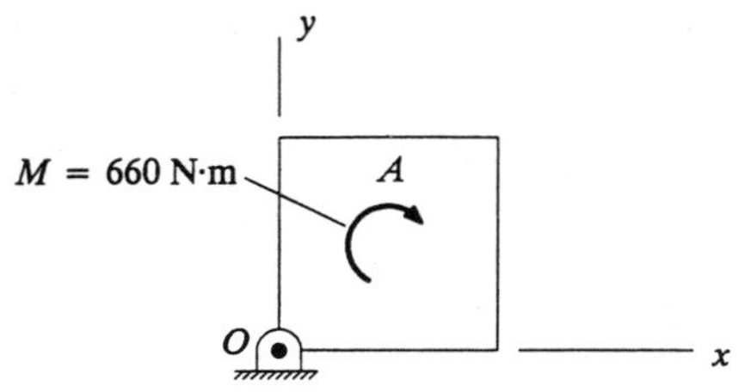

Figure a.

FIND Determine the angular velocity [rad/s] of the cube at the end of the 90° rotation.

SOLUTION

The mass moment of inertia of the cube about its mass center G (Fig. b) is (see Appendix F, Table F1.)

$$I_G = \frac{1}{6} ma^2 \quad [\text{kg·m}^2] \qquad (a)$$

where

$$m = 9/9.81 = 0.9174 \text{ kg} \qquad (b)$$

$$a = 150 \text{ mm} = 0.150 \text{ m}$$

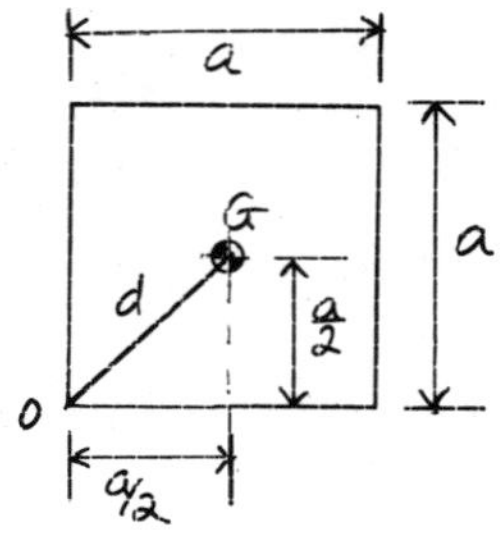

Figure b.

By the parallel-axis theorem and Fig. b, the mass moment of inertia about the axis along edge O of the cube is since $d^2 = a^2/2$

$$I_o = I_G + md^2 = \frac{1}{6} ma^2 + \frac{1}{2} ma^2 = \frac{2}{3} ma^2$$

or, with Eqs. (b),

$$I_o = \frac{2}{3}(0.9174)(0.150)^2 = 0.01376 \text{ kg·m}^2 \qquad (c)$$

Now, by the law of kinetic energy,

$$U_{1-2} = T_2 - T_1 \qquad (d)$$

where

$$U_{1-2} = \int_0^{\pi/2} M d\theta = (600)\frac{\pi}{2} = 942.48 \text{ N·m} \qquad (e)$$

$$T_1 = 0 \qquad (f)$$

$$T_2 = \frac{1}{2} I_o \omega^2 = 0.006881\,\omega^2 \quad [\text{N·m}]$$

Hence, Eqs. (d), (e), and (f) yield

$$942.48 = 0.006881\,\omega^2$$

or

$$\omega = 370 \text{ rad/s}$$

19.23

GIVEN A thin uniform rod is balanced vertically on a table. The table is jarred slightly, and the rod topples over. Its angular velocity when it strikes the table is 5 rad/s. The rod does not slip on the table.

FIND Calculate the length [m] of the rod.

SOLUTION

Let the length of the rod be denoted by L. Then, the work performed by gravity as the rod falls from the vertical to the horizontal position is

$$U_{1-2} = W\frac{L}{2} \qquad (a)$$

where W is the weight of the rod. Assuming that initially the angular speed of the rod is negligible, the initial kinetic energy of the rod is $T_1 = 0$. When the rod strikes the table, its kinetic energy is.

$$T_2 = \frac{1}{2} I \omega^2 \qquad (b)$$

where

$$I = \frac{1}{3} m L^2 = \frac{1}{3}\frac{W}{g}L^2 \qquad (c)$$

$$\omega = 5 \text{ rad/s}$$

Hence, by the law of kinetic energy and Eqs. (a), (b), and (c), with $T_1 = 0$,

$$U_{1-2} = T_2 - T_1$$

$$W\frac{L}{2} = \frac{1}{2}\left(\frac{WL^2}{3g}\right)\omega^2 = \frac{WL}{6g}(5)^2$$

or

$$\underline{L = 1.177 \text{m}}$$

19.24

GIVEN A homogeneous ball that weighs 45 N starts from rest and rolls down an inclined plane.

FIND Determine the kinetic energy [N·m] of the ball when its center is moving at $v = 3$ m/s.

SOLUTION

The kinetic energy T of the ball consists of translation energy and rotational energy. Thus,

$$T = \frac{1}{2} m v^2 + \frac{1}{2} I \omega^2 \qquad (a)$$

where

$$I = \frac{2}{5} m r^2 \qquad (b)$$

and by kinematics,

$$v = r\omega \qquad (c)$$

By Eqs. (a), (b), and (c),

$$T = \frac{1}{2} m v^2 + \frac{1}{5} m v^2 = \frac{7}{10} m v^2$$

or

$$\underline{T = \frac{7}{10}\left(\frac{45}{9.81}\right)(3)^2 = 28.90 \quad \text{N·m}}$$

19.25

GIVEN A cannon of mass 1500 kg rests on an inclined plane with slope 1/10. It fires a projectile of mass 3 kg directly down the plane, with a muzzle speed of 300 m/s. Friction is negligible.

FIND Determine the distance the cannon moves up the plane, assuming it is free to recoil.

SOLUTION

During the internal explosion, linear momentum is conserved. Therefore,

$$(m_c v_c + m_p v_p)_{initial} = (m_c v_c + m_p v_p)_{final} \qquad (a)$$

where subscripts c and p denote cannon and projectile, respectively.

(continued)

19.25 cont.

Since initially, linear momentum is zero, Eq. (a) yields, the relation between final momenta of the cannon and projectiles as

$$m_c v_c = -m_p v_p \qquad (b)$$

Since $m_c = 1500$ kg, $m_p = 3$ kg, and $v_p = 300$ m/s. Eq. (b) yields

$$v_c = 0.60 \text{ m/s} \qquad (c)$$

The free-body diagram of the cannon is shown in Fig. a.

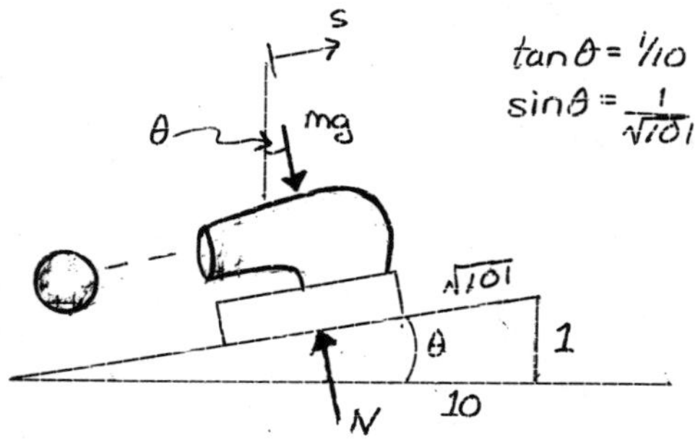

$\tan\theta = 1/10 = 0.100$

$\sin\theta = \frac{1}{\sqrt{101}} = 0.099504$

Figure a

The work done by gravity as the cannon moves up the plane a distance s is (Fig. a)

$$U_{1-2} = -mg(s)(\sin\theta) \qquad (d)$$

The initial kinetic energy of the cannon is

$$T_1 = \frac{1}{2} m v_c^2 \qquad (e)$$

and the final kinetic energy is

$$T_2 = 0 \qquad (f)$$

By Eqs. (d), (e), and (f), and the law of kinetic energy

$$U_{1-2} = T_2 - T_1$$

$$-mg(s)(\sin\theta) = 0 - \tfrac{1}{2} m v_c^2$$

or

$$s = \tfrac{1}{2}\frac{v_c^2}{g(\sin\theta)} = \frac{1}{2}\,\frac{(0.60)^2}{(9.81)(0.099504)}$$

So

$$s = 0.1844 \text{ m}$$

19.26

GIVEN The radius of pulley B of Problem 19.21, Fig. 19.21-a, is increased to R so that its moment of inertia is I_1 (Fig. a). The old belt is replaced by a new belt of weight W.

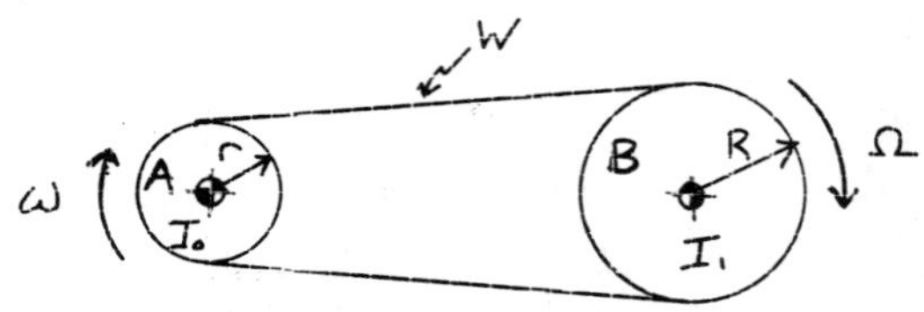

Figure a

FIND

a.) Derive a formula for the kinetic energy T of the system in terms of I_0, I_1, ω, r, R, and W, where I_0 is the mass moment of inertia of pulley A.

b.) For $r = 150$ mm, $R = 300$ mm, $\omega = 10\pi$ rad/s, $I_0 = 0.060$ kg·m², $I_1 = 0.60$ kg·m², and $W = 15$ N, calculate T [N·m].

SOLUTION

a.) By Fig. a and kinematics,

$$r\omega = R\Omega \qquad (a)$$

Therefore,

$$T = \tfrac{1}{2} I_0 \omega^2 + \tfrac{1}{2} I_1 \Omega^2 + \tfrac{1}{2}\frac{W}{g} v^2 \qquad (b)$$

where, by kinematics, the speed of the belt is

$$v = r\omega \qquad (c)$$

Hence, by Eqs. (a), (b), and (c),

$$T = \frac{1}{2}\left[I_0 + \left(\frac{r}{R}\right)^2 I_1 + \left(\frac{W}{g}\right) r^2\right]\omega^2 \qquad (d)$$

b.) For $r = 150$ mm, $R = 300$ mm, $\omega = 10\pi$ rad/s, $I_0 = 0.060$ kg·m², $I_1 = 0.60$ kg·m², and $W = 15$ N, Eq. (d) yields

$$T = \frac{1}{2}\left[0.060 + \left(\frac{0.15}{0.30}\right)^2 (0.60) + \left(\frac{15}{9.81}\right)(0.150)^2\right](10\pi)^2$$

or

$$T = 120.61 \text{ N·m}$$

19.27

GIVEN A cubical concrete block with edges $a = 2\text{m}$ long is initially at rest on one edge on a horizontal floor at 30° (Fig. a). It is released from that position, and does not slip on the floor.

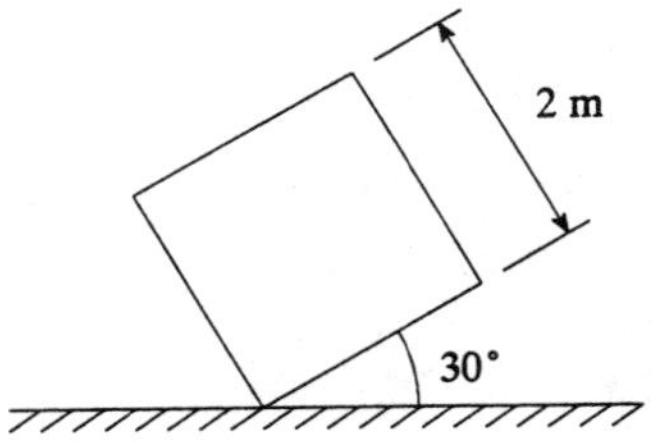

Figure a.

FIND Determine the angular speed ω [rad/s] of the cube when its face strikes the floor.

SOLUTION

Initially, the kinetic energy of the block is

$$T_1 = 0 \qquad (a)$$

When its face strikes the floor, its kinetic energy is

$$T_2 = \frac{1}{2} I_o \omega^2 \qquad (b)$$

where I_o is the mass moment of inertia of the block about the edge in contact with the floor (Fig. a). The mass moment of inertia of the cubical block about the parallel axis through its mass center is (Appendix F, Table F1)

$$I_G = \frac{1}{6} ma^2 = \frac{1}{6} m(2)^2 = \frac{2}{3} m \qquad (c)$$

Hence, by Eq. (c) and the parallel-axis theorem,

$$I_o = I_G + md^2 = \frac{2}{3}m + m(\sqrt{2})^2 = \frac{8}{3}m \qquad (d)$$

Hence, by the law of kinetic energy and Eqs. (a), (b), (c), (d), and (e)

$$U_{1-2} = T_2 - T_1$$

$$0.3660\, mg = \tfrac{1}{2}\left(\tfrac{8}{3} m\right)\omega^2$$

or

$$\omega = \sqrt{0.2745 g} = \sqrt{0.2745(9.81)} = 1.64 \text{ rad/s}$$

19.28

GIVEN The armature of a large electric generator weighs 14,000 lb. Its radius of gyration is $k = 1.10$ ft.

FIND Disregarding friction and electromagnetic effects, use the law of kinetic energy to calculate the constant torque M that must be applied to the armature to increase its angular velocity ω from 4 to 10 rad/s in 1 revolution

SOLUTION

The work performed by the torque in 1 revolution ($= 2\pi$ rad) is

$$U_{1-2} = M(2\pi) \qquad (a)$$

The initial kinetic energy of the armature is

$$T_1 = \frac{1}{2} I\omega^2 = \frac{1}{2}(mk^2)\omega^2$$

$$= \frac{1}{2}\left(\frac{14{,}000}{32.2}\right)(1.10)^2(4)^2$$

or

$$T_1 = 4208.7 \text{ ft}\cdot\text{lb} \qquad (b)$$

The final kinetic energy is

$$T_2 = \frac{1}{2}\left(\frac{14000}{32.2}\right)(1.10)^2(10)^2$$

or

$$T_2 = 26{,}304.3 \text{ ft}\cdot\text{lb} \qquad (c)$$

Then, by the law of kinetic energy and Eqs. (a), (b), and (c),

$$U_{1-2} = T_2 - T_1$$

$$2\pi M = 26{,}304.3 - 4208.7$$

or

$$M = 3516.6 \text{ lb}\cdot\text{ft}$$

19.29

GIVEN A homogeneous cylinder starts from rest and rolls down an inclined plane.

FIND By the law of kinetic energy, derive a formula for the speed v of the cylinder after it has rolled a distance x.

SOLUTION

When the cylinder has rolled a distance x down the plane, the work that gravity has performed (Fig. a) is

$$U_{1-2} = (mg \sin\theta)x \qquad (a)$$

where θ is the angle the plane forms with the horizontal and m is the cylinder's mass.

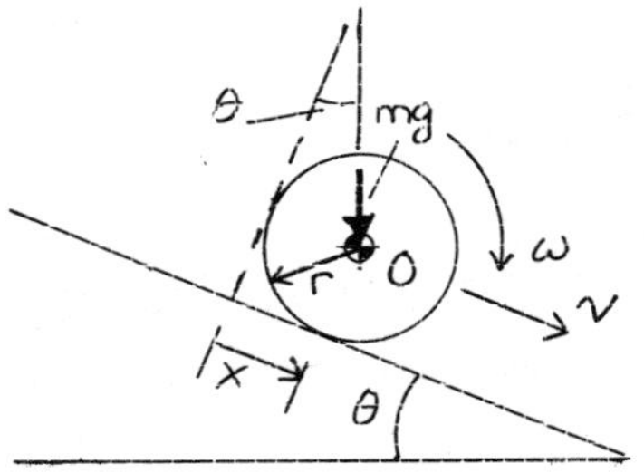

Figure a

The initial kinetic energy of the cylinder is

$$T_1 = 0 \qquad (b)$$

The kinetic energy at x is

$$T_2 = \frac{1}{2}mv^2 + \frac{1}{2}I_o\omega^2 \qquad (c)$$

where by Fig. a,

$$v = r\omega \qquad (d)$$

and for the cylinder,

$$I_o = \frac{1}{2}mr^2 \qquad (e)$$

Hence, by the law of kinetic energy and Eqs (b), (c), (d) and (e),

$$T_1 + U_{1-2} = T_2$$

$$0 + (mg\sin\theta)x = \frac{3}{4}mv^2$$

or

$$v = \sqrt{\frac{4}{3}gx\sin\theta}$$

19.30

GIVEN Each of the four wheels of the wagon shown in Fig. a is a solid disk, 1.2 m in diameter and weighs 1500 N. The wagon starts from rest and rolls down the inclined plane of slope 5/12. Neglect friction.

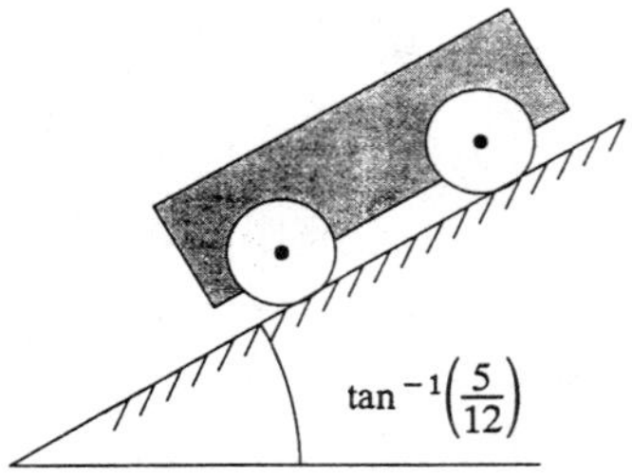

Figure a.

FIND Calculate the speed of the wagon after it has rolled 30 m down the plane.

SOLUTION

The total weight of the wagon is

$$W = 4(300) + 1500 = 2700 \text{ N} \qquad (a)$$

Hence, the mass of the wagon is $m = 2700/9.81$ or

$$m = 275.23 \text{ kg} \qquad (b)$$

For a wheel,

$$I = \frac{1}{2}m_w r^2 = \frac{1}{2}\left(\frac{300}{9.81}\right)(0.6)^2$$

$$I = 5.5046 \text{ kg}\cdot\text{m}^2 \qquad (c)$$

Hence, the kinetic energy of the wagon after it rolls 30 m down the plane is, by Eqs (b) and (c),

(continued)

19.30 cont.

$$T_2 = \frac{1}{2}mv^2 + 4\left(\frac{1}{2}I\omega^2\right)$$

$$= 137.61 v^2 + 11.01 \omega^2 \qquad (d)$$

Since the speed of the wagon is (Fig a.)

$$v = r\omega = 0.6\omega$$

Eq. (d) yields

$$T_2 = 167.19\, v^2 \quad [N \cdot m] \qquad (e)$$

The work performed by gravity after it has rolled 30 m down the plane is

$$U_{1-2} = W(30 \sin\theta) = (2700)(30)\sin\theta$$

$$= 81\,000 \sin\theta \quad [N \cdot m]$$

where, since $\tan\theta = 5/12$, $\sin\theta = 5/13$

Hence,

$$U_{1-2} = 31\,153.8 \ N \cdot m \qquad (f)$$

Therefore, by the law of kinetic energy and Eqs. (e) and (f), with $T_1 = 0$, we obtain

$$T_1 + U_{1-2} = T_2$$

$$0 + 31\,153.8 = 167.19\, v^2$$

or

$$v = 13.65 \ m/s$$

19.31

GIVEN A light cord is unwound from a spool by a constant 10 lb force (Fig. a). The spool weighs 64.4 lb and has a radius of gyration of 1 ft relative to its center. Friction prevents skidding.

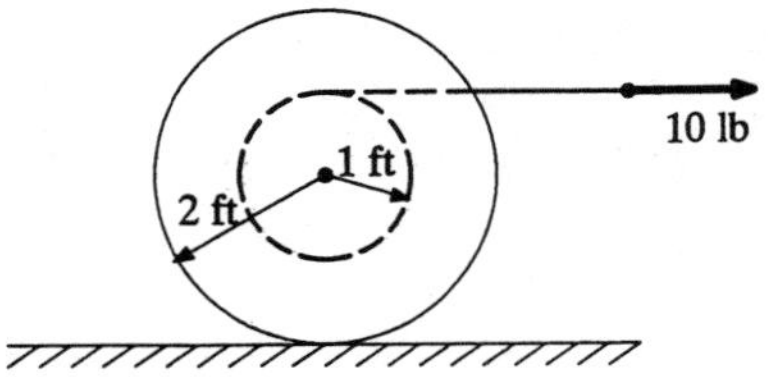

Figure a.

FIND By energy principles, determine the speed of the spool after it has moved 5 ft.

SOLUTION

Since the spool rolls for 5 ft, its angle θ of rotation is given by (Fig. a)

$$r\theta = 5 \text{ ft}$$

or

$$\theta = \frac{5}{2} = 2.5 \text{ rad} \qquad (a)$$

By Fig. a, and Eq (a), the distance through which the 10 ft force moves is given by $3\theta = 7.5$ ft. Therefore, the work performed on the spool is

$$U_{1-2} = (10)(7.5) = 75 \text{ ft} \cdot \text{lb} \qquad (b)$$

Initially, the kinetic energy of the spool is

$$T_1 = 0 \qquad (c)$$

When it has moved 5 ft, the kinetic energy is

$$T_2 = \frac{1}{2}mv^2 + \frac{1}{2}I\omega^2 \qquad (d)$$

where m is the mass of the spool, v is the speed of its center, I is the mass moment of inertia, and ω is the angular velocity. By Fig. a and kinematics

$$v = r\omega = 2\omega \qquad (e)$$

By Eqs. (d) and (e)

$$T_2 = \frac{1}{2}mv^2 + \frac{1}{2}(mk^2)\frac{v^2}{4}$$

$$= \frac{1}{2}\left(\frac{64.4}{32.2}\right)\left(1 + \frac{1}{4}\right)v^2$$

or

$$T_2 = \frac{5}{4}v^2 \qquad (f)$$

Hence, by the law of kinetic energy and Eqs. (b), (c), and (f),

$$U_{1-2} = T_2 - T_1$$

$$75 = \frac{5}{4}v^2$$

$$v = 7.746 \ ft/s$$

19.32

GIVEN A cannon of mass $m_c = 50\ lb\cdot s^2/ft$ fires a projectile of mass $m_p = 0.10\ lb\cdot s^2/ft$ horizontally. The projectile muzzle velocity relative to the cannon is 1200 ft/s.

FIND

a.) Determine the recoil speed of the cannon and the absolute speed of the projectile.

b.) Calculate the work performed on the system from the time that the charge explodes to the time the projectile exits from the muzzle.

SOLUTION

a.) By the concept of relative motion,

$$v_p = v_c + v_{P/c} = v_c + 1200\ [ft/s] \quad (a)$$

where v_p and v_c are the absolute velocities of the projectile and cannon, respectively. Also, since initially the system is at rest, the initial momentum is zero. Therefore, by the principle of conservation of momentum, $m_c v_c + m_p v_p = 0$, or

$$50 v_c + 0.10 v_p = 0 \quad (b)$$

By Eqs. (a) and (b), we find

$$v_c = -2.395\ ft/s$$

$$v_p = 1197.6\ ft/s \quad (c)$$

→ Equations (c) indicate that the senses of v_c and v_p are opposite; that is, if the projectile travels to the right, the cannon travels to the left.

b.) By the law of kinetic energy for the system,

$$U_{1-2} = T_2 - T_1 \quad (d)$$

where U_{1-2} is the work performed on the system by the explosive charge and T_1, T_2 are the initial kinetic energy and the final kinetic energy (at the time the projectile exits from the muzzle), respectively. Since initially the system is at rest,

$$T_1 = 0$$

$$T_2 = \frac{1}{2} m_c v_c^2 + \frac{1}{2} m_p v_p^2 \quad (e)$$

Eqs. (c), (d), and (e) yield

$$U_{1-2} = \frac{1}{2}[(50)(-2.395)^2 + (0.10)(1197.6)^2]$$

or

$$U_{1-2} = 7.186 \times 10^4\ ft\cdot lb$$

19.33

GIVEN A wooden sphere weighs 30 N. It is suspended by a cord from a fixed support, with its center 1 m below the support (Fig. a). A bullet that weighs 0.3 N is fired horizontally into the sphere with a speed of 600 m/s, and is imbedded at the center of the sphere.

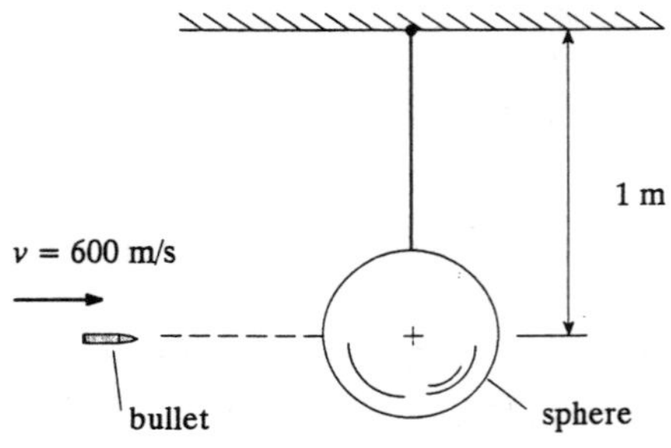

Figure a.

FIND

a.) Determine the height to which the sphere rises above its initial position.

b.) Determine the energy lost in the impact.

SOLUTION

a) The velocity, v, of the sphere and bullet immediately after the bullet is embedded is determined by the momentum equation

(continued)

19.33 cont.

$$m_b v_b + m_s v_s = (m_b + m_s) v \quad (a)$$

where the subscripts b and s denote the bullet and sphere, respectively. Now initially $v_b = 600$ m/s and $v_s = 0$. Therefore, with $m_b = 0.3/9.81$ and $m_s = 30/9.81$, Eq (a) yields

$$v = 5.941 \text{ m/s} \quad (b)$$

Then, by the law of kinetic energy, the height h to which the sphere and bullet rise is given by

$$U_{1-2} = T_2 - T_1 \quad (c)$$

where

$$U_{1-2} = -(m_b + m_s) g h$$
$$= -(0.3 + 30) h = -30.3 h \ [\text{N}\cdot\text{m}] \quad (d)$$

$$T_1 = 1/2 (m_b + m_s) v^2$$
$$= \frac{1}{2}\left(\frac{30.3}{9.81}\right)(5.941)^2 = 54.51 \ [\text{N}\cdot\text{m}] \quad (e)$$
$$T_2 = 0$$

Equations (c), (d), and (e) yield

$$-30.3 h = 0 - 54.51$$

or

$$\underline{h = 1.80 \text{ m}} \quad (f)$$

b.) The energy lost in the impact is

Energy lost = Initial energy − Final energy

The initial energy is

$$\frac{1}{2} m_b v_b^2 = \frac{1}{2}\left(\frac{0.3}{9.81}\right)(600)^2 = 5504.6 \ [\text{N.m}] \quad (g)$$

The final energy is, by Eqs (d) and (f),

$$30.3 h = 30.3(1.80) = 54.54 \ \text{N}\cdot\text{m} \quad (h)$$

Equations (g) and (h) yield

$$\underline{\text{Energy lost} = 5504.6 - 54.54 = 5450 \ \text{N}\cdot\text{m}}$$

19.34

GIVEN Two consecutive extreme positions of a pendulum of weight W is shown in Fig. a. The pivot contains a spring that exerts a resisting couple $k\theta$, where k is a constant, and the pivot also exerts a constant frictional couple C.

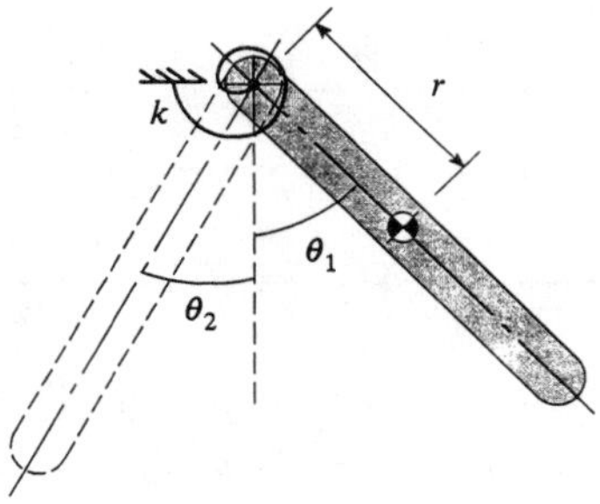

Figure a.

FIND Show by the law of kinetic energy that, for small angles ($\cos\theta \approx 1 - \theta^2/2$)

$$\theta_1 - \theta_2 = \frac{2C}{k + Wr}$$

SOLUTION

Since the angular velocity is zero at the end of each extreme position, the law of kinetic energy yields

$$U_{1-2} = T_2 - T_1 = 0$$

Then, by Fig. a, starting from rest at $\theta = \theta_1$ and returning to rest at $\theta = \theta_2$, we have

$$U_{1-2} = mgr(\cos\theta_2 - \cos\theta_1) - C(\theta_1 + \theta_2) - \frac{1}{2}k(\theta_2^2 - \theta_1^2) = 0 \quad (a)$$

With the small-angle approximations

$$\cos\theta_1 = 1 - \frac{\theta_1^2}{2}, \qquad \cos\theta_2 = 1 - \frac{\theta_2^2}{2},$$

Eq. (a) yields

$$mgr\left(\frac{\theta_1^2}{2} - \frac{\theta_2^2}{2}\right) - C(\theta_1 + \theta_2) + \frac{1}{2}k(\theta_1^2 - \theta_2^2) = 0$$

(continued)

19.34 cont.

or

$$mgr(\theta_1+\theta_2)(\theta_1-\theta_2) - 2C(\theta_1+\theta_2) + k(\theta_1+\theta_2)(\theta_1-\theta_2) = 0$$

Therefore, canceling $(\theta_1+\theta_2)$ and letting $mg = W$, we obtain

$$\theta_1 - \theta_2 = \frac{2C}{k+Wr}$$

19.35

GIVEN Refer to Problem 19.17, where an airplane lands with a speed of 120 ft/s. The outside diameter of a tire is 4 ft, the radius of gyration of the wheel and tire is $k = 1.5$ ft, and the mass of the wheel and tire is 10 lb·s²/ft. Upon landing, the tire skids for a time t_s then rolls. The friction force is constant at $F = 900$ lb. During skidding, the deceleration of the plane is negligible.

FIND Compute the work of all the forces that act on the wheel during the first 0.5 s after landing, and show that this work equals the increase of kinetic energy of the wheel.

SOLUTION

The forces that do work on the wheel are the force exerted by friction and the force exerted by the axle. Since the speed of the plane is constant during skidding these forces are equal in magnitude F but opposite in sense (Fig a.)

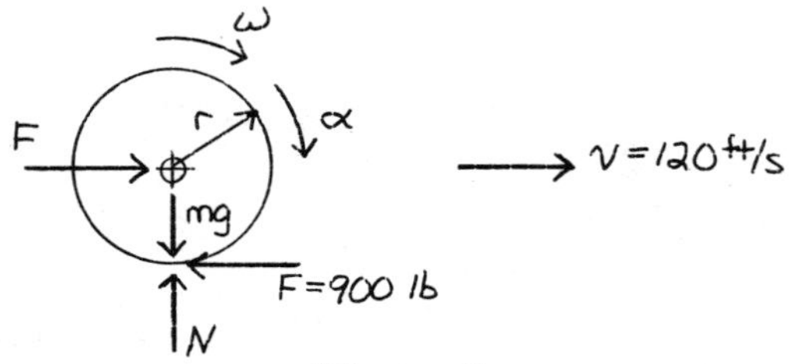

Figure a

By Fig. a and kinematics,

$$\circlearrowleft\!+ \sum M_G = Fr = I_G\alpha = mk^2\frac{d\omega}{dt}$$

or

$$(900)(2) = (10)(1.5)^2\frac{d\omega}{dt}$$

so

$$d\omega = 80\,dt$$

Hence, since $\omega = 0$ for $t = 0$, integration yields

$$\int_0^\omega d\omega = \int_0^t 80\,dt; \qquad \omega = 80t \qquad (a)$$

Thus, by kinematics,

$$\omega = \frac{d\theta}{dt} = 80t$$

or

$$d\theta = 80t\,dt \qquad (b)$$

By Fig. a, we see that the constant couple $M = Fr$ acts on the wheel. Therefore, the work done on the wheel during the first 0.50 s is, with Eq. (b),

$$U = \int M\,d\theta = Fr\int_0^{1/2} 80t\,dt$$

or

$$U = (900)(2)(40)t^2\Big|_0^{1/2} = 18{,}000 \text{ ft·lb} \qquad (c)$$

The increase in kinetic energy of the wheel in the first 0.5 s is, since the translation energy ($\frac{1}{2}mv^2$) is constant,

$$T = \frac{1}{2}I\omega^2 \qquad (d)$$

where, by Eq. (a), the angular speed at $t = 0.5$ s is $\omega = 40$ rad/s. Hence, by Eq. (d)

$$T = \frac{1}{2}(10)(1.5)^2(40)^2 = 18{,}000 \text{ ft·lb} \qquad (e)$$

Hence, by Eqs. (c) and (e) $U = T$, which was to be proved.

Note that skidding ceases when

(continued)

19.35 cont.

$$120 \text{ ft/s} = \omega r = 2\omega = 160t$$

or when $t = 0.75$ s. Hence, the above results remain valid, since the wheel is skidding at $t = 0.5$ s

19.36

GIVEN Two bodies of masses m_1 and m_2 with velocities v_1 and v_2, respectively, collide head on.

FIND

a.) Show by means of the equations for inelastic impact [Eqs. (16.44)] that the loss of kinetic energy ΔT is

$$\Delta T = \frac{m_1 m_2 (1-e^2)(v_1 - v_2)^2}{2(m_1 + m_2)}$$

where e is the coefficient of restitution. Explain how the energy is lost.

b.) Let $m_1 = 1$ kg, $v_1 = 6$ m/s, and $v_2 = 4$ m/s. Plot ΔT as a function of e for $0 \le e \le 1$ for $m_2/m_1 = 0.4$, 0.8, 1.0, and 1.2. Explain the behavior of ΔT for the various ratios of m_2/m_1.

SOLUTION

a.) Before the collision, with $v_1 = v_{1i}$ and $v_2 = v_{2i}$, the initial kinetic energy is

$$T_i = \frac{1}{2} m_1 v_{1i}^2 + \frac{1}{2} m_2 v_{2i}^2 \qquad (a)$$

after the collision, the final kinetic energy is

$$T_f = \frac{1}{2} m_1 v_{1f}^2 + \frac{1}{2} m_2 v_{2f}^2 \qquad (b)$$

where v_{1f}, v_{2f} are the velocities after the collision. Hence, the loss of kinetic energy is, by Eqs. (a) and (b),

$$\Delta T = T_i - T_f \qquad (c)$$

The velocities v_{1f} and v_{2f} are related to v_{1i} and v_{2i} by Eqs. (16.44a) and (16.44b). Thus, where e is the coefficient of restitution,

$$v_{1f} = \frac{(m_1 - e m_2) v_{1i} + (1+e) m_2 v_{2i}}{m_1 + m_2}$$

$$v_{2f} = \frac{(1+e) m_1 v_{1i} + (m_2 - e m_1) v_{2i}}{m_1 + m_2} \qquad (d)$$

Substituting Eqs. (d) into Eq. (b) and simplifying, we get:

$$T_f = \frac{1}{2(m_1 + m_2)} \left[m_1 v_{1i}^2 (m_1 + e^2 m_2) + 2(1-e^2) m_1 m_2 v_{1i} v_{2i} + m_2 v_{2i}^2 (m_2 + e^2 m_1) \right] \qquad (e)$$

Substitution of Eqs. (a) and (e) into Eq. (c) yields

$$\Delta T = \frac{m_1 m_2 (1-e^2)(v_{1i} - v_{2i})^2}{2(m_1 + m_2)} \qquad (f)$$

This loss of kinetic energy results from the fact that energy is used to deform the bodies inelastically.

(b) Dividing numerator and denominator of Eq. (f) by m_1, we obtain

$$\Delta T = \frac{m_2 (1-e^2)(v_{1i} - v_{2i})^2}{2(1 + m_2/m_1)} \qquad (g)$$

Also, let $m_2/m_1 = k$, where k is constant. Then, Eq. (g) may be written as

$$\Delta T = \frac{m_1 k (1-e^2)(v_{1i} - v_{2i})^2}{2(1+k)} \qquad (h)$$

For $m_1 = 1$ kg, $v_{1i} = 6$ m/s, and $v_{2i} = 4$ m/s Eq (h) yields

$$\Delta T = \frac{2k(1-e^2)}{1+k} \quad [FL] \qquad (i)$$

(continued)

19.36 cont.

The plot of ΔT, for $0 \le e \le 1$, is shown in Fig. a, for $k = m_2/m_1 = 0.4$, 0.8, 1.0, and 1.2

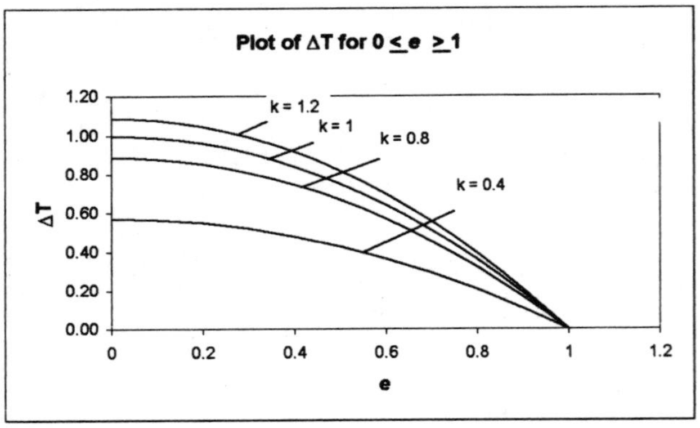

Figure a

By Fig. a, the larger the ratio $k = m_2/m_1$ the greater the loss of kinetic energy. This follows from the factor that the greater the ratio, the larger the initial energy and hence, for a given value of e (except for the elastic case $e = 1$), the larger the deformation of the bodies (the larger the loss of kinetic energy).

19.37

GIVEN A homogeneous ball of mass m and radius r rolls outward on an oscillating seesaw (Fig. a). The moment of inertia of the board about its hinge is I.

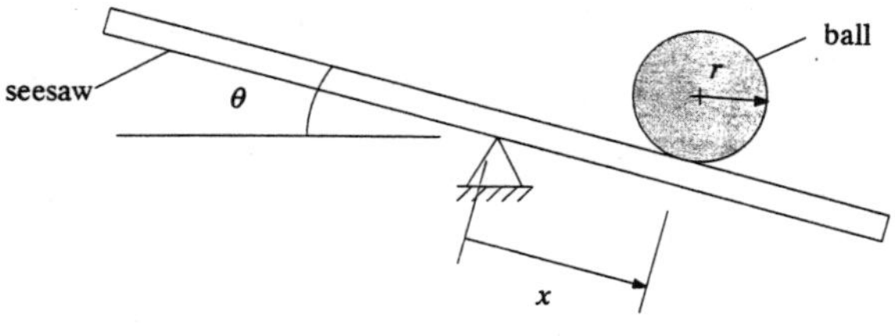

Figure a.

FIND Show that the kinetic energy of the system is

$$T = \frac{7}{10} m(\dot{x} + r\dot{\theta})^2 + \frac{1}{2}(I + mx^2)(\dot{\theta})^2$$

SOLUTION

To compute the kinetic energy of translation of the ball, we need to determine the coordinates of the center of the ball and their time derivatives.

By Fig. b, the coordinates are, relative to Newtonian axes X, Y,

$$X = x\cos\theta + r\sin\theta$$
$$Y = -x\sin\theta + r\cos\theta \qquad (a)$$

The time derivatives of Eqs. (a) are

$$\dot{X} = \dot{x}\cos\theta - x\dot{\theta}\sin\theta + r\dot{\theta}\cos\theta$$
$$\dot{Y} = -\dot{x}\sin\theta - x\dot{\theta}\cos\theta - r\dot{\theta}\sin\theta \qquad (b)$$

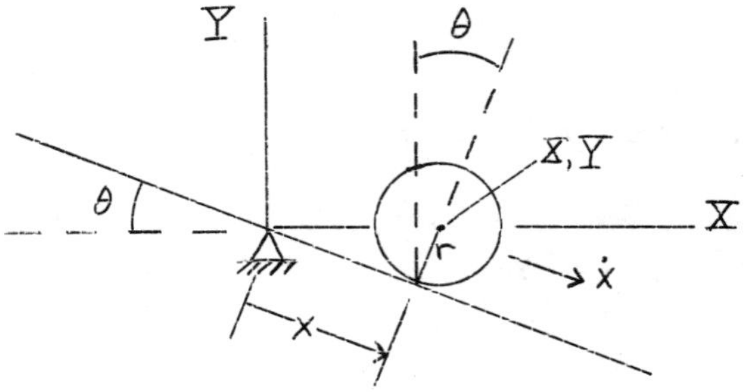

Figure b

The kinetic energy of translation of the ball is

$$T_1 = \frac{1}{2} m(\dot{X}^2 + \dot{Y}^2) \qquad (c)$$

Hence, by Eqs. (b) and (c), we obtain

$$T_1 = \frac{1}{2} m(\dot{x}^2 + x^2\dot{\theta}^2 + r^2\dot{\theta}^2 + 2r\dot{x}\dot{\theta})$$

or

$$T_1 = \frac{1}{2} m[(\dot{x} + r\dot{\theta})^2 + x^2\dot{\theta}^2] \qquad (d)$$

The angular velocity of the ball is (Fig b.)

$$\omega = \dot{\theta} + \dot{x}/r \qquad (e)$$

and the moment of inertia of the ball is

$$I_{ball} = \frac{2}{5} mr^2 \qquad (f)$$

(continued)

19.37 cont.

Hence, the kinetic energy of rotation of the ball is, by Eqs. (e) and (f),

$$T_2 = \frac{1}{2} I_{ball}\, \omega^2 = \frac{1}{5} mr^2 \left(\dot{\theta} + \frac{\dot{x}}{r}\right)^2 \qquad (g)$$

In addition, the kinetic energy of the board is

$$T_3 = \frac{1}{2} I \dot{\theta}^2 \qquad (h)$$

Then, the total kinetic energy of the system is, by Eqs. (d), (g), and (h),

$$T = T_1 + T_2 + T_3$$

or

$$T = \frac{7}{10} m(\dot{x} + r\dot{\theta})^2 + \frac{1}{2}(I + mx^2)\dot{\theta}^2$$

as was to be proved.

19.38

GIVEN A block of mass m, fastened to an inextensible string that wraps around a vertical post of radius 75 mm, moves on a horizontal surface (Fig. a). The string remains taut. The coefficient of sliding friction of the block is $\mu_k = 0.30$. The length s [m] of the path of the mass is $s = L\theta - \frac{1}{2} a\theta^2$, where θ [rad] is the angle through which the string is displaced, a [m] is the radius of the post, and L [m] is the length of unwrapped string for $\theta = 0$. When $r = 1.8$ m, the angular speed of the string is 12 rad/s.

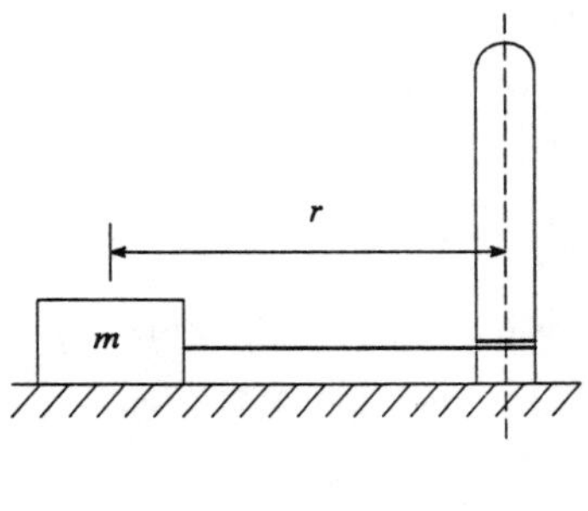

Figure a.

FIND Determine the angular speed ω [rad/s] of the string for $r = 0.9$ m.

SOLUTION

By Fig. b, the initial length of unwrapped string when $r = r_i = 1.8$ m is

$$L_i = \sqrt{(1.8)^2 - (0.075)^2} = 1.7984 \text{ m} \qquad (a)$$

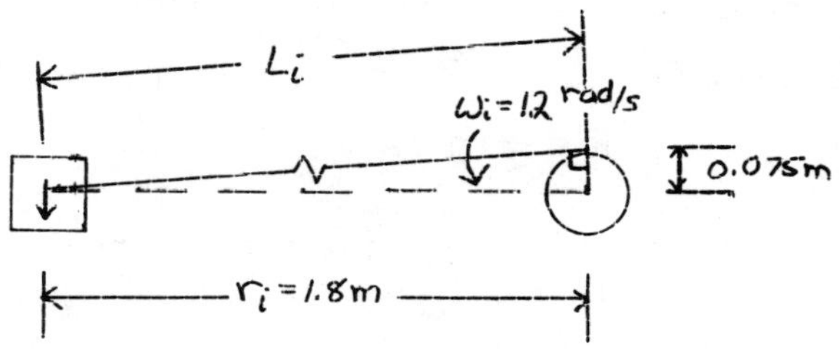

Figure b.

Similarly, when $r = r_f = 0.9$ m, the final length of the string is

$$L_f = \sqrt{(0.9)^2 - (0.075)^2} = 0.8967 \text{ m} \qquad (b)$$

Then, by Eqs. (a) and (b), $L_i - L_f = 0.9015$ m of string is wrapped around the post. Hence, the angle θ through which the string is displaced is given by $a\theta = L_i - L_f$, or

$$\theta = \frac{L_i - L_f}{a} = \frac{0.9015}{0.075} = 12.02 \text{ rad} \qquad (c)$$

Therefore, the distance travelled by the mass m is, with Eqs. (a) and (c),

$$s = L_i\theta - \frac{1}{2} a\theta^2$$
$$= (1.7984)(12.02) - \frac{1}{2}(0.075)(12.02)^2$$

or

$$s = 16.20 \text{ m} \qquad (d)$$

The work performed by friction on the mass is, as the mass moves the distance s:

$$U = -\mu_k mg \qquad (e)$$

(continued)

19.38 cont.

or by Eqs (d) and (e), with $\mu_k = 0.30$,

$$U = -(0.30)(mg)(16.20) = -4.86 \qquad (f)$$

The initial kinetic energy of the mass is

$$T_i = \frac{1}{2} m v_i^2 \qquad (g)$$

where v_i is the initial speed of the mass. Similarly, the final kinetic energy [for $L = L_f$; see Eq. (b)] is

$$T_f = \frac{1}{2} m v_f^2 \qquad (h)$$

Then, by the law of kinetic energy and Eqs (f), (g), and (h),

$$U = T_f - T_i$$

or

$$-4.86\, mg = \frac{1}{2} m (v_f^2 - v_i^2)$$

So,

$$v_f^2 = v_i^2 - 9.72 g \qquad (i)$$

By Fig. b and Eq. (a)

$$v_i = L_i \omega = 1.7984 (12)$$
$$= 21.58 \text{ m/s} \qquad (j)$$

Equations (i) and (j) yield $v_f^2 = (21.58)^2 - (9.72)(9.81)$. Hence, the velocity of the mass, when $L = L_f$ is,

$$v_f = 19.25 \text{ m/s} \qquad (k)$$

Therefore, the angular velocity of the string when $L = L_f$ is, with Eqs (b) and (k),

$$\omega_f = \frac{v_f}{L_f} = \frac{19.25}{0.8967} = 21.5 \text{ rad/s}$$

19.39

GIVEN A homogeneous cylinder A of mass m starts from rest when $\theta = \phi = 0$, and rolls on a fixed cylinder B (Fig. a).

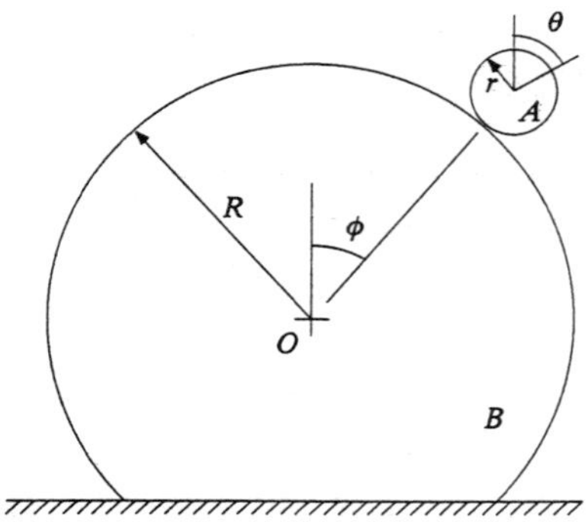

Figure a.

FIND

a.) Draw a free-body diagram of cylinder A, and derive a formula for the normal contact force N cylinder B exerts on cylinder A, in terms of R, r, m, g, ϕ, and $\Omega = \dot{\phi}$

b.) Derive a formula for Ω in terms of R, r, g, and ϕ. Then, derive a formula for the angle ϕ_{max} at which A loses contact with B, using the results of part a.

c.) Show that as $r \to 0$, ϕ_{max} is the same angle found in Problem 14.33, namely, $\phi_{max} \to 48.19°$.

d.) For R = 10 in and R/r = 5, 1, and 0.5, plot Ω as a function of ϕ, for $0 \le \phi \le \phi_{max}$. With the aid of this plot, estimate the angular velocity of cylinder A and the speed of its center of mass as it loses contact with B.

SOLUTION

a.) The kinematical constraint of rolling is (Fig a.)

$$r\theta = R\phi \qquad (a)$$

(continued)

19.39 cont.

Differentiation of Eq (a) yields

$$r\dot{\theta} = r\omega = R\dot{\phi} = R\Omega \qquad (b)$$

$$r\ddot{\theta} = r\alpha = R\ddot{\phi} \qquad (c)$$

By fig. b, the acceleration components of the mass center G are, with Eqs. (b) and (c),

$$(a_G)_n = -(R+r)(\dot{\phi})^2 = -(R+r)\Omega^2 \qquad (d)$$

$$(a_G)_t = R\ddot{\phi} + r\ddot{\theta} = 2R\ddot{\phi} \qquad (e)$$

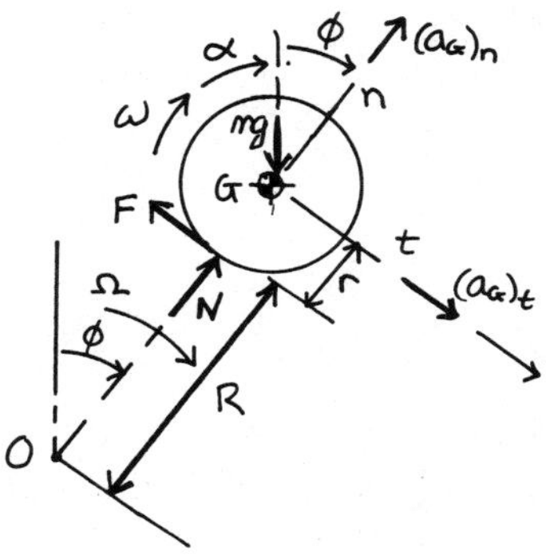

Figure b

By Fig. b and Eqs. (c), (d), and (e),

$$\Sigma F_n = N - mg\cos\phi = m\,(a_G)_n = -m(R+r)\Omega^2 \qquad (f)$$

$$\Sigma F_t = mg\sin\phi - F = m\,(a_G)_t = 2mR\ddot{\phi} \qquad (g)$$

$$\circlearrowleft + \Sigma M_G = Fr = I_G\alpha = \tfrac{1}{2}mr^2\left(\tfrac{R}{r}\right)\ddot{\phi} = \tfrac{1}{2}mrR\ddot{\phi} \qquad (h)$$

By Eq. (f), the normal contact force is

$$N = mg\cos\phi - m\,(R+r)\Omega^2 \qquad (i)$$

b.) By the law of kinetic energy for A at ϕ,

$$U_{1-2} = T_2 - T_1 \qquad (j)$$

where, since the cylinder A starts at rest when $\theta = \phi = 0$,

$$U_{1-2} = mg\,(R+r)(1-\cos\phi)$$

$$T_1 = 0 \qquad (k)$$

$$T_2 = \tfrac{1}{2}m v_G^2 + \tfrac{1}{2}I_G\omega^2$$

Also, by Fig. b,

$$v_G = (R+r)\Omega \qquad (\ell)$$

Hence, Eqs. (j), (k), and (ℓ) yield, with $I_G = \frac{1}{2}mr^2$ and $\omega = R\Omega/r$ [see Eq. (b)],

$$\Omega = 2\sqrt{\frac{g(R+r)(1-\cos\phi)}{2(R+r)^2 + R^2}} \qquad (m)$$

Now the cylinder A loses contact with B when $N = 0$ [see Eq. (i)]. Hence, with $N=0$, Eqs. (i) and (m) yield

$$\cos\phi = \frac{4\,(R+r)^2}{6\,(R+r)^2 + R^2}$$

Hence,

$$\phi_{max} = \cos^{-1}\left[\frac{4\,(R+r)^2}{6\,(R+r)^2 + R^2}\right] \qquad (n)$$

c.) As $r \to 0$, $I_G \to 0$, and Eqs. (i), (j), (k), and (ℓ) yield

$$mgR\,(1-\cos\phi) = \tfrac{1}{2}mR^2\Omega^2$$

$$\Omega^2 = \frac{g\cos\phi}{R}$$

Elimination of Ω^2 from these equations yields

$$\cos\phi = \tfrac{2}{3}$$

or

$$\phi = \cos^{-1}\left(\tfrac{2}{3}\right) = 48.19°$$

as was to be shown.

(continued)

19.39 cont.

d.) For $R = 10$ in and $R/r = 5$, 1, and 0.5, the plot of Ω [see Eq. (m)] is shown in Fig. c for $0 \le \phi \le \phi_{max}$, where by Eq (m)

$$\phi_{max} = \cos^{-1}\left[\frac{4(1+R/r)^2}{6(1+R/r)^2 + (R/r)^2}\right]$$

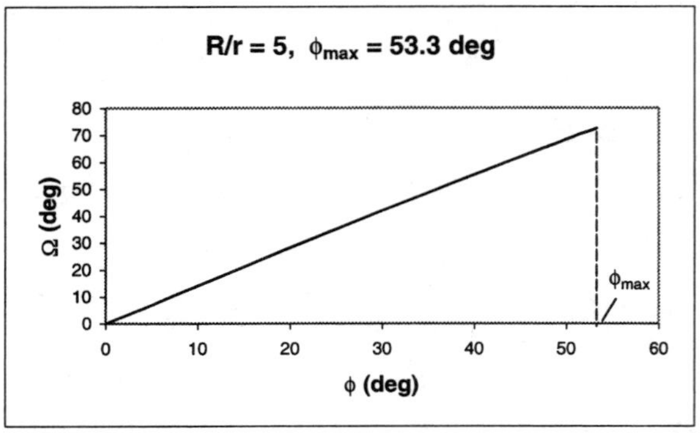

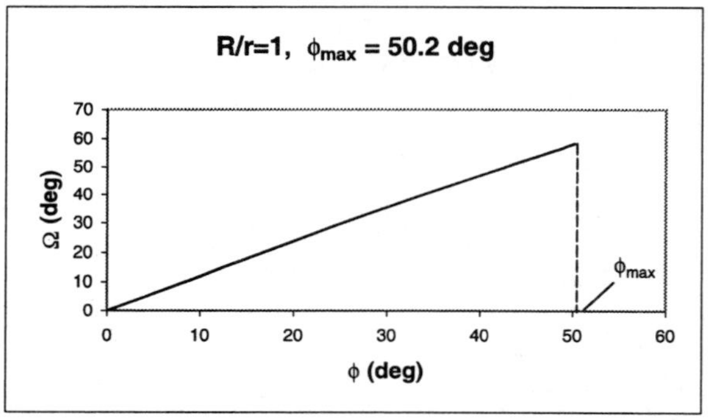

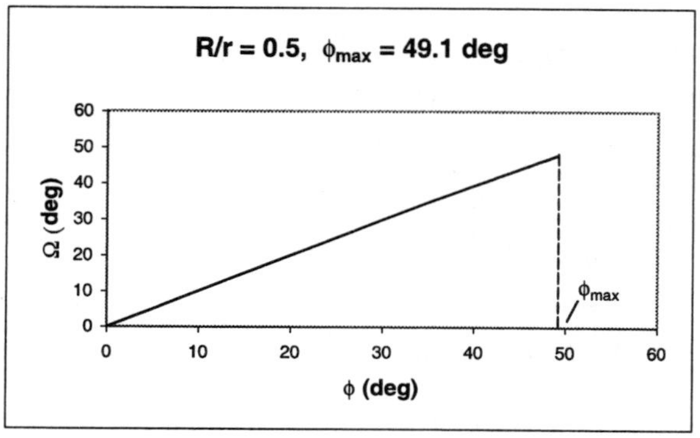

For $R/r = 5$, $\phi_{max} = 53.36°$, then by the plot for $R/r = 5$,

$\omega = \frac{R}{r}\Omega = 5($ and

$v = (R + \frac{R}{5})\Omega = 12($

For $R/r = 1$, $\phi_{max} = 50.21°$.

Then $\omega = R/r\ \Omega = (1)($ and

$v = (R+R)\Omega = (20)$

For $R/r = 0.5$, $\phi_{max} = 49.11°$

Then $\omega = R/r\ \Omega = (0.5)($ and

$v = (R + \frac{R}{0.5})\Omega = (30)($

19.40

GIVEN A thin hoop of mass m and radius R hangs over a fixed horizontal shaft of radius r (Fig. a) The hoop swings in its plane, perpendicular to the axis of the shaft, without slipping on the shaft.

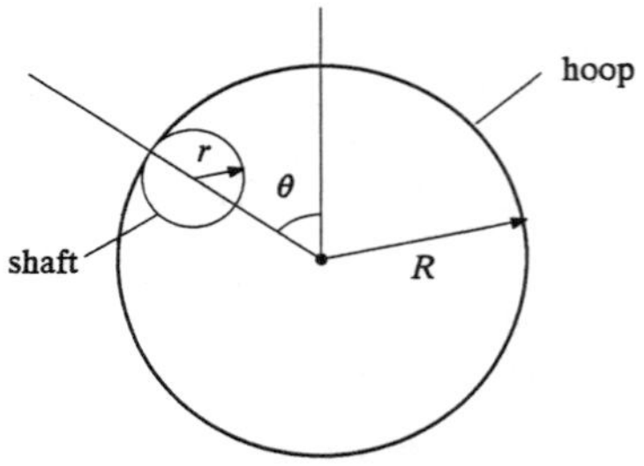

Figure a.

FIND Show that the kinetic energy of the hoop is

$$T = m(R-r)^2(\dot{\theta})^2$$

where θ is the angle the diameter of the hoop, passing through the center of the shaft, forms with the vertical.

(continued)

19.40 cont.

SOLUTION

Consider a particle P fixed on the hoop that coincides with the topmost point of the shaft when $\theta = 0$ (Fig. b)

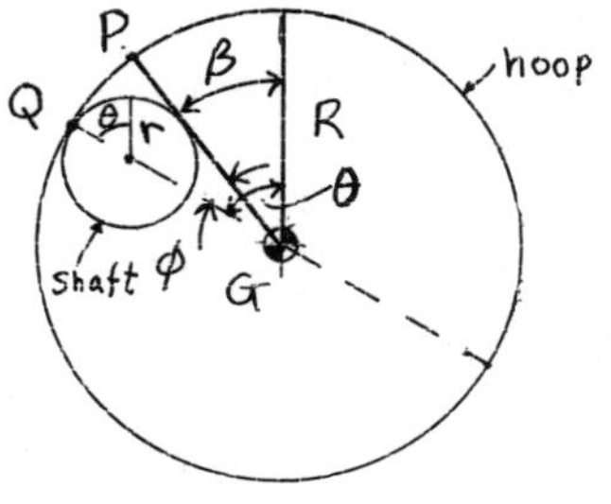

Figure b

By Fig. b, the rolling constraint yields

$$PQ = r\theta = R\phi \ ; \quad \phi = \frac{r}{R}\theta \qquad (a)$$

The angular velocity of the hoop is

$$\omega = \dot{\beta} \qquad (b)$$

where $\beta = \theta - \phi$. Hence, Eqs. (a) and (b) yield

$$\omega = \dot{\theta} - \dot{\phi} = \left(1 - \frac{r}{R}\right)\dot{\theta} \qquad (c)$$

Point Q is the instantaneous center of the hoop. Therefore, the kinetic energy of the hoop is

$$T = \tfrac{1}{2} I_Q \omega^2 \qquad (d)$$

where I_Q is the mass moment of inertia of the hoop relative to Q.

The mass moment of inertia of the hoop relative to the geometrical axis G is [Appendix F; Table F2.]

$$I_G = mR^2 \qquad (e)$$

Hence, by the parallel axis theorem and Eq. (e),

$$I_Q = I_G + mR^2 = 2mR^2 \qquad (f)$$

Then, by Eqs. (c), (d), and (f),

$$T = m(R-r)^2(\dot{\theta})^2$$

19.41

GIVEN The muzzle velocity of a 10 lb projectile as it leaves a 7000 lb gun is 1400 ft/s. The gun is directed horizontally and it recoils compressing an initially unstretched spring 6 in.

FIND Determine the spring constant, k [lb/in]

SOLUTION

By conservation of momentum, since initially the system is at rest,

$$m_g v_g + m_p v_p = 0 \qquad (a)$$

where subscripts g and p denote gun and projectile, respectively, and v_g, v_p are the velocities of the gun and projectile as the projectile leaves the gun. Hence, by Eq. (a),

$$v_g = -\frac{m_p}{m_g} v_p = -\left(\frac{10}{7000}\right)(1400)$$

or the recoil speed of the gun is

$$|v_g| = 2 \text{ ft/s}$$

Then, by conservation of energy of the gun-spring system,

$$T_i + V_i = T_f + V_f \qquad (b)$$

where T_i, V_i are the initial kinetic energy and the initial potential energy of the gun-spring system when the projectile leaves the gun and T_f, V_f the final energies, when the gun stops and the spring is compressed.

Since initially the kinetic energy of the gun is

$$T_i = \frac{1}{2} m_g v_g^2 = \frac{1}{2}\left(\frac{7000}{32.2}\right)(2)^2 = 434.78 \text{ (ft·lb)} \qquad (c)$$

and since initially the spring is unstretched,

$$V_i = 0 \qquad (d)$$

The final energies are (since the gun comes to rest and the spring is compressed $x = 6 \text{ in} = 0.5 \text{ ft}$).

(continued)

19.41 cont.

$T_f = 0$

$V_f = \frac{1}{2} kx^2 = \frac{1}{2} k(0.5)^2$

$= 0.125\, k$ [ft·lb] (e)

Hence, Eqs. (b), (c), (d), and (e) yield

$434.78 = 0.125\, k$

or

$k = 3478$ lb/ft $= 289.9$ lb/in.

19.42

GIVEN A thin, uniform rod of mass m is welded to a hoop of mass $m/2$ (Fig. a). In position A, the hoop rolls with angular velocity $\omega = 4$ rad/s.

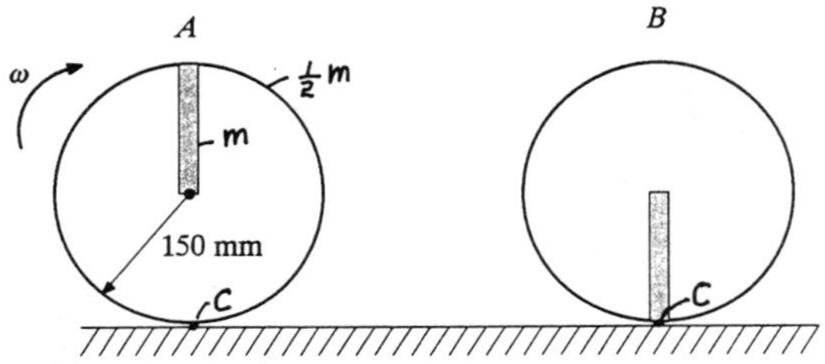

Figure a

FIND What is the angular velocity ω_B of the hoop in position B?

SOLUTION

By conservation of energy,

$$T_A + V_A = T_B + V_B \qquad (a)$$

where subscripts A and B denote positions A and B, respectively. Let the potential energy reference be taken as the horizontal surface on which the hoop rolls (Fig. a). Then,

$$V_A = \left(\tfrac{1}{2} mg\right) R + (mg)\left(\tfrac{3}{2} R\right) = 2mgR$$

$$V_B = \left(\tfrac{1}{2} mg\right) R + (mg)\left(\tfrac{1}{2} R\right) = mgR \qquad (b)$$

where R is the radius of the hoop. Also, by Fig. (a), the kinetic energies in position A and B are, respectively, with respect to the instantaneous center C (the point of contact of the hoop on the horizontal surface on which the hoop rolls.)

$$T_A = \tfrac{1}{2} I_{CA}\, \omega_A^2$$
$$T_B = \tfrac{1}{2} I_{CB}\, \omega_B^2 \qquad (c)$$

where, for position A,

$$I_{CA} = I_{hoop} + I_{rod} \qquad (d)$$

and by the parallel axis theorem,

$$I_{hoop} = \left(\tfrac{1}{2} m\right) R^2 + \left(\tfrac{1}{2} m\right) R^2 = mR^2$$
$$I_{rod} = \tfrac{1}{12} mR^2 + m\left(\tfrac{3}{2} R\right)^2 = \tfrac{7}{3} mR^2 \qquad (e)$$

By Eqs. (d) and (e),

$$I_{CA} = \tfrac{10}{3} mR^2 \qquad (f)$$

Similarly, for position B,

$$I_{CB} = I_{hoop} + I_{rod}$$

$$I_{hoop} = mR^2$$
$$I_{rod} = \tfrac{1}{3} mR^2$$

or

$$I_{CB} = \tfrac{4}{3} mR^2 \qquad (g)$$

Hence, Eqs. (c), (f), and (g) yield

$$T_A = \tfrac{1}{2}\left(\tfrac{10}{3} mR^2\right)\omega_A^2 = \tfrac{5}{3} mR\,\omega_A^2$$
$$T_B = \tfrac{1}{2}\left(\tfrac{4}{3} mR^2\right)\omega_B^2 = \tfrac{2}{3} mR^2 \omega_B^2 \qquad (h)$$

Substitution of Eqs. (b) and (h) into Eq. (a) yields

$$\omega_B^2 = \tfrac{5}{2}\omega_A^2 + \tfrac{3}{2}\tfrac{g}{R} \qquad (i)$$

With $\omega_A = 4$ rad/s, $g = 9.81$ m/s², and $R = 0.150$ m, Eq (i) yields

$\omega_B = 11.75$ rad/s

19.43

GIVEN In the position shown (Fig. a), the elliptic disk rolls on the horizontal surface with initial angular speed $\omega_i = 10$ rad/s.

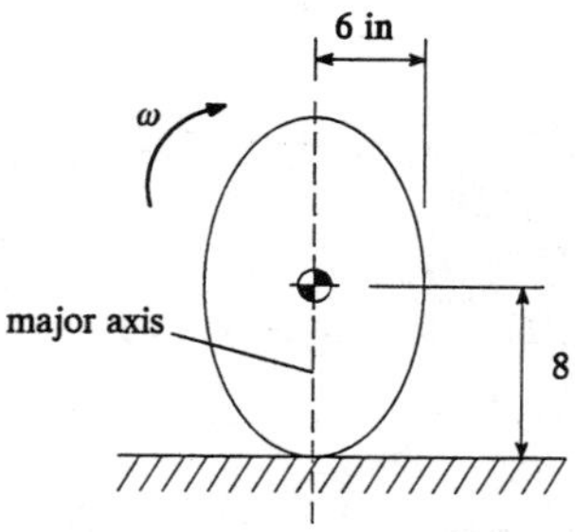

Figure a.

FIND Determine the angular speed ω_f it will have when its major axis is horizontal.

SOLUTION

By conservation of energy,

$$T_i + V_i = T_f + V_f \qquad (a)$$

where i, f denote initial and final energies. Hence, by Fig. a,

$$V_i = mg\,a$$
$$T_i = \tfrac{1}{2} m v_{Gi}^2 + \tfrac{1}{2} I_G \omega_i^2 \qquad (b)$$

Similarly, when the major axis is horizontal,

$$V_f = mg\,b$$
$$T_f = \tfrac{1}{2} m v_{Gf}^2 + \tfrac{1}{2} I_G \omega_f^2 \qquad (c)$$

By Fig. a, initially and finally

$$v_{Gi} = a\,\omega_i$$
$$v_{Gf} = b\,\omega_f \qquad (d)$$

Then, Eqs. (a), (b), (c), and (d) yield, with

$$I_G = \tfrac{1}{4} m (a^2 + b^2)$$

$a = 8$ in, and $b = 6$ in,

$$61\omega_f^2 = 89\omega_i^2 + 4g \qquad (e)$$

With $\omega_i = 10$ rad/s and $g = 386.4$ in/s², Eq. (e) yields

$$\omega_f = 13.086 \text{ rad/s}$$

19.44

GIVEN Each arm in the frame ABCD weighs 40 N (Fig. a). Joints A, B, C, and D are frictionless.

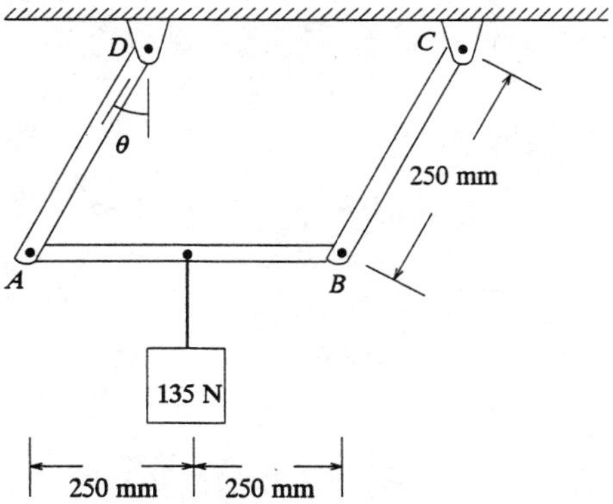

Figure a

FIND

a.) Take $\theta = 0$ as the datum for potential energy. Determine the potential energy of the system as a function of θ.

b.) The frame is released from rest at $\theta = 90°$. Determine the maximum angular speed of the frame.

SOLUTION

a.) By Fig. a,

$$V = 0.250(1-\cos\theta)[135 + 40] + \tfrac{1}{2}(0.250)(1-\cos\theta)(80)$$

or

$$V = 63.75\,(1-\cos\theta)$$

(continued)

19.44 cont.

b.) Let $m_1 = 40/9.81 = 4.077$ kg, be the mass of each of bars AB, AD, and CB, and $m_2 = 135/9.81 = 13.761$ kg be the mass of the block.
Also, let $r = 250$ mm $= 0.250$ m

Initially at $\theta = 90°$,

$$T_i = 0 \qquad (a)$$

$$V_i = 2(m_1 g)(\tfrac{1}{2}r) + m_1 g r + m_2 g r$$

or
$$V_i = 2m_1 g r + m_2 g r \qquad (b)$$

When the system is released from rest, it attains its maximum angular speed when $\theta = 0°$. At $\theta = 0°$,

$$V_f = 0 \qquad (c)$$

$$T_f = 2(\tfrac{1}{2} I_1 \omega_f^2) + \tfrac{1}{2} m_1 v_{AB}^2 + \tfrac{1}{2} m_2 v_{AB}^2$$

where
$$I_1 = \tfrac{1}{3} m_1 r^2$$
$$v_{AB} = r\omega_f$$

Then,
$$T_f = \tfrac{1}{6}(5m_1 + 3m_2) r^2 \omega_f^2 \qquad (d)$$

Note that bar AB undergoes curvilinear translation, so its kinetic energy is $\tfrac{1}{2} m_1 v_{AB}^2$.

By the law of conservation of energy,

$$T_i + V_i = T_f + V_f \qquad (e)$$

Then, by Eqs. (a), (b), (c), (d), and (e)

$$(2m_1 + m_2) g = \tfrac{1}{6}(5m_1 + 3m_2) r \omega_f^2 \qquad (f)$$

Inserting numerical values for m_1, m_2, g, and r, we obtain for ω_f,

$$\omega_f = 9.147 \text{ rad/s}$$

19.45

GIVEN In the frame in Fig. P19.44, rotational springs are inserted at joints D and C. They exert moments $M = -4.2\theta$ [N·m] for any rotation.

FIND Rework 19.44 for this case.

SOLUTION

At $\theta = 90°$ the potential energy stored in the spring is.

$$V_{springs} = 2\int_0^{\pi/2} 4.2\,\theta\, d\theta = 4.2\,\theta^2 \Big|_0^{\pi/2}$$

or
$$V_{springs} = 10.363 \text{ N·m}$$

This additional potential energy must be added to the Initial potential energy of the system in Problem 19.44 [see Eq. (b) in Problem 19.44]. Hence, the total initial potential energy is,

$$V_i = 2m_1 g r + m_2 g r + 10.362 \text{ [N·m]} \qquad (a)$$

The potential energy of the springs is zero, when $\theta = 0$. Hence, the final potential energy at $\theta = 0$ is the same as in Problem 19.44. Also, the kinetic energy terms are the same as in Problem 19.44. Consequently, by Eq. (a) above and Eqs. (a), (c), (d), and (e) of Problem 19.44.

$$2m_1 g r + m_2 g r + 10.362 = \tfrac{1}{6}(5m_1 + 3m_2) r^2 \omega_f^2 \qquad (b)$$

Substituting $m_1 g = 40$ N, $m_2 g = 135$ N, $g = 9.81$ m/s², $r = 0.25$ m, $m_1 = 4.077$ kg, and $m_2 = 13.761$ kg into Eq. (b), we find for ω_f,

$$\omega_f = 9.990 \text{ rad/s}$$

19.46

GIVEN A uniform trap door weighs 1000 lb and can rotate in a vertical plane about a frictionless pin O (Fig. a). It is held in its equilibrium position by a spring of constant $k = 300$ lb/ft, when $\theta = 30°$.

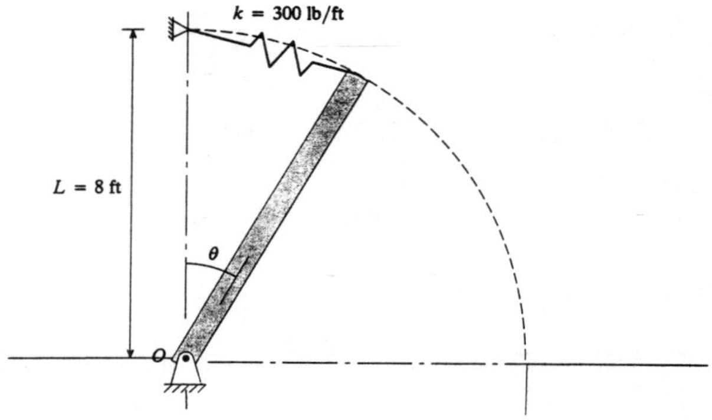

FIND

a.) Determine the work required to pull the door down to its closed position ($\theta = 90°$).

b.) The door is released from rest from its closed position. Determine its angular velocity as it passes through the equilibrium position ($\theta = 30°$)

SOLUTION

a.) For any angle θ, the length of the spring is (Fig. b)

$$l = 2L\sin\frac{\theta}{2} \qquad (a)$$

Hence, for $\theta = 30°$, by Eq. (a), with $L = 8$ ft,

$$l_1 = (2)(8)\sin 15° = 4.1411 \text{ ft} \qquad (b)$$

Similarly for $\theta = 90°$,

$$l_2 = (2)(8)\sin 45° = 11.3137 \text{ ft} \qquad (c)$$

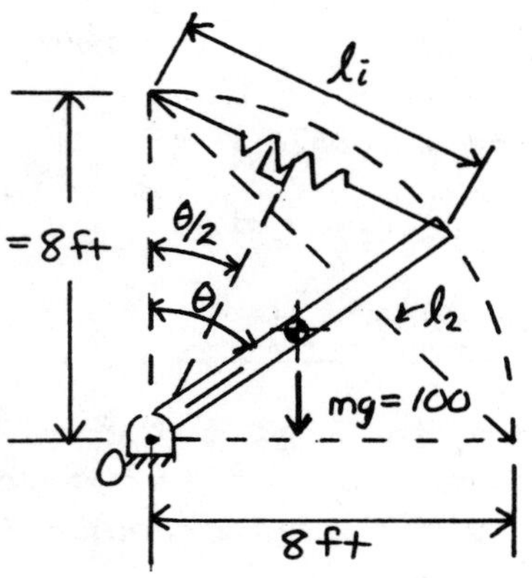

Figure b

By Fig. b, in the equilibrium position ($\theta = 30°$)

$$\Sigma M_O = (ke_1)\left(L\cos\frac{\theta}{2}\right) - (mg)\left(\frac{L}{2}\sin\theta\right) = 0 \qquad (d)$$

where e_1 is the extension of the spring beyond its unstretched length l_0. By Eq. (d), with $k = 300$ lb/ft, $L = 8$ ft, $mg = 1000$ lb, and $\theta = 30°$, we find

$$e_1 = \frac{250}{289.7\overline{7}} = 0.8627\overline{7} \qquad (e)$$

By Eqs. (b) and (e), with Fig. b, we obtain the unstretched length of the spring as

$$l_0 = l_1 - e_1$$
$$= 4.1411 - 0.8627\overline{7} = 3.2783 \text{ ft} \qquad (f)$$

For $\theta = 90°$, the extension of the spring is, by Eqs. (c) and (f); see Fig. b,

$$e_2 = l_2 - l_0$$
$$= 11.3137 - 3.2783 = 8.0354 \text{ ft} \qquad (g)$$

Therefore, the work required to lower the door from $\theta = 30°$ to $\theta = 90°$ is

$$U = V_2 - V_1$$

where

$$V_1 = \frac{1}{2}ke_1^2 = \frac{1}{2}(300)(0.8627\overline{7})^2$$

$$V_2 = -mg\left(\frac{L}{2}\cos 30°\right) + \frac{1}{2}ke_2^2$$
$$= -1000(4\cos 30°) + \frac{1}{2}(300)(8.0354)^2$$

(continued)

19.46 cont.

or
$$V_1 = 111.658$$
$$V_2 = 6221.055$$

Hence, the work required is

$$U = 6109.4 \text{ ft}\cdot\text{lb}$$

b.) By part a, the work of the spring as the door rotates from the closed position $\theta = 90°$ to the equilibrium position $(\theta = 30°)$ is

$$U_{spring} = \frac{1}{2}ke_2^2 - \frac{1}{2}ke_1^2$$
$$= \frac{1}{2}(300)(8.0354^2 - 0.86277^2)$$

or

$$U_{spring} = 9573.5 \text{ ft}\cdot\text{lb}$$

The corresponding work of gravity is

$$U_{gravity} = -(mg)\left(\frac{L}{2}\right)\cos 30°$$
$$= -1000\,(4)(0.8650)$$

or
$$U_{gravity} = -3464.1 \text{ ft}\cdot\text{lb}$$

Hence, the total work is

$$U = U_{spring} + U_{gravity}$$
$$= 9573.5 - 3464.1 = 6109.4 \text{ ft}\cdot\text{lb} \quad \text{(h)}$$

Note: this is the same as the work required to lower the door.

The kinetic energy of the door as it moves through the equilibrium position is

$$T = \frac{1}{2}I_o\omega^2 = \frac{1}{2}\left(\frac{1}{3}mL^2\right)\omega^2$$

or
$$T = \frac{1}{2}\left(\frac{1}{3}\right)\left(\frac{1000}{32.2}\right)(8)^2\omega^2$$
$$= 331.263\,\omega^2 \quad \text{(i)}$$

Then, by the law of kinetic energy, since the kinetic energy at $\theta = 90°$ is zero.

$$U = T$$

or
$$6109.4 = 331.263\,\omega^2$$

So,

$$\omega = 4.29 \text{ rad/s}$$

Alternatively by the law of conservation of energy

$$T_1 + V_1 = T_2 + V_2$$

$$0 + \frac{1}{2}ke_2^2 = \frac{1}{2}I_o\omega^2 + mg\left(\frac{L}{2}\right)\cos 30° + \frac{1}{2}ke_1^2$$

or
$$9685.155 = 331.263\,\omega^2 + 3464.1 + 111.658$$

and
$$\omega = 4.29 \text{ rad/s}$$

19.47

GIVEN Three rigid bodies (a thin ring, a solid cylinder, and a solid sphere) have the same radius r and mass m (Fig. a). They are released from rest at the same elevation and at the same instant. They roll down identical planes for a distance of $x = 1\text{m}$.

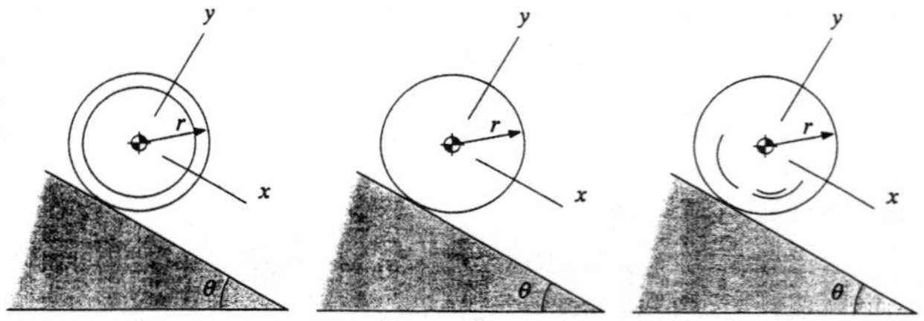

Ring $I_R = mr^2$ | Cylinder $I_C = \frac{1}{2}mr^2$ | Sphere $I_S = \frac{2}{5}mr^2$

Figure a.

FIND

a.) By the law of conservation of energy, determine which body first travels the 1-m distance.

b.) Determine the time required for the fastest of the three bodies to travel 1m.

(continued)

19.47 cont.

SOLUTION

a.) By conservation of energy,

$$T_1 + V_1 = T_2 + V_2 \qquad (a)$$

Consider a general body of circular cross section that rolls down an inclined plane (Fig. b)

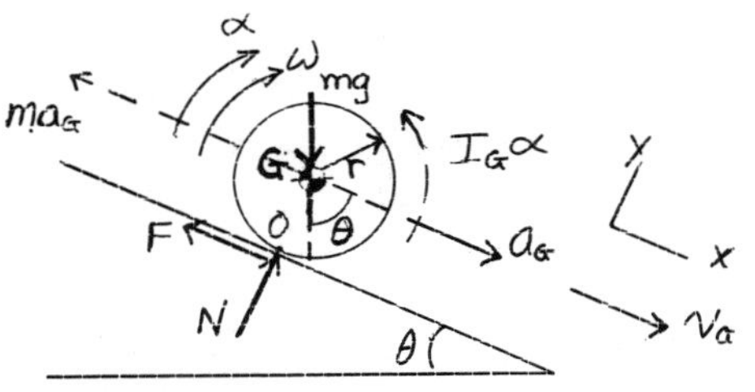

Figure b.

Let the body start from rest at position 1. and roll through position 2 at $x = 1m$. Then, taking the potential energy datum at $x = 1m$, we obtain, by Eq. (a),

$$0 + mgx\sin\theta = [\tfrac{1}{2} m v_G^2 + \tfrac{1}{2} I_G \omega^2] + 0 \qquad (b)$$

By kinematics and Fig. b,

$$v_G = r\omega \qquad (c)$$

Substituting Eq. (c) for ω and solving for v_G^2; we obtain

$$v_G^2 = \frac{2mr^2 g\sin\theta}{mr^2 + I_G} \qquad (d)$$

Since the thin ring, the solid cylinder, and the solid sphere have the same mass, m, and the same radius r, the body with the smallest mass moment of inertia relative to its mass center will have the greatest **speed** (v_G). Now,

For the sphere: $I_G = \frac{2}{5} mr^2$

For the cylinder: $I_G = \frac{1}{2} mr^2$

For the thin ring: $I_G = mr^2$

Then, since $\frac{2}{5}mr^2 < \frac{1}{2}mr^2 < mr^2$ the sphere will travel 1m fastest, followed by the cylinder, and then the ring.

b.) By Fig. b and the inertial-force method, for the sphere,

$$\circlearrowleft \Sigma M_O = mgr\sin\theta - ma_G r - \tfrac{2}{5} mr^2\alpha = 0 \qquad (e)$$

By kinematics and Fig. b,

$$a_G = r\alpha \qquad (f)$$

Eliminating α by Eqs. (e) and (f) and solving for a_G, we obtain

$$a_G = \tfrac{5}{7} g\sin\theta \qquad (g)$$

By kinematics, since a_G is constant and the sphere starts from rest,

$$x = 1 = \tfrac{1}{2} a_G t^2 \qquad (h)$$

with $g = 9.81\ m/s^2$; Eqs. (g) and (h) yield

$$t = \frac{0.5342}{\sqrt{\sin\theta}} \quad [s]$$

19.48

GIVEN A homogeneous yo-yo that weighs 1 N is released from rest (Fig. a). The thickness and the weight of the string is negligible.

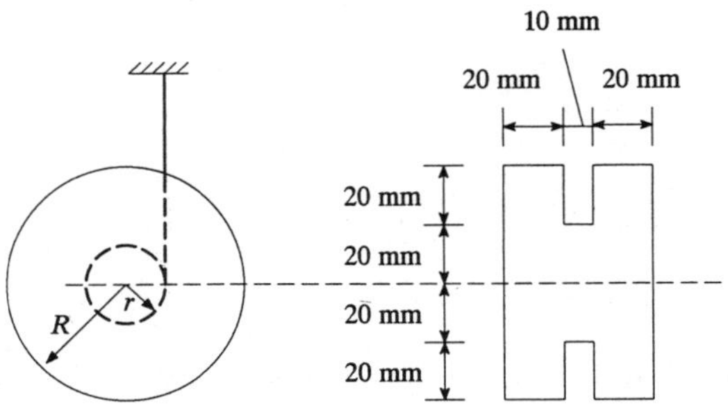

Figure a

(continued)

19.48 cont.

FIND Determine the velocity [m/s] of the mass center of the yo-yo at the instant that it has dropped 1m.

SOLUTION

Since the yo-yo is dropped from rest, the law of conservation of energy yields

$$T_1 + V_1 = T_2 + V_2$$

$$0 + 0 = \frac{1}{2} I_G \omega^2 + \frac{1}{2} m v_G^2 - mg(1) \quad (a)$$

where I_G is the mass momentum of inertia relative to the mass center G, m is the mass of the yo-yo, v_G is the velocity of the mass center, and ω the angular velocity of the yo-yo.

By Fig. a,

$$I_G = I_{core} + I_{sides}$$

$$I_{core} = \frac{1}{2} m_{core} r^2 \quad (b)$$

$$I_{sides} = \frac{1}{2} m_{sides} R^2$$

Also, the total mass of the yo-yo is

$$m = \frac{1\,N}{9.81\ ft/s^2} = 0.10194\ kg \quad (c)$$

The volume of the yo-yo is the sum of the volume of the sides plus the volume of the core. By Fig. (a), the volume of the sides is

$$V_{sides} = 2\pi(0.040)^2(0.02) = 6.4\pi \times 10^{-5}\ m^3 \quad (d)$$

and similarly, the volume of the core is

$$V_{core} = \pi(0.02)^2(0.01) = 4\pi \times 10^{-6}\ m^3 \quad (e)$$

Hence, the total volume is, by Eqs. (d) and (e),

$$V_{total} = V_{side} + V_{core} = 6.8\pi \times 10^{-5}\ m^3 \quad (f)$$

and the mass density, by Eqs. (c) and (f), is

$$\rho = \frac{m}{V_{total}} = 477.184\ kg/m^3 \quad (g)$$

Hence, the masses of the sides and the core are, respectively, by Eqs. (d), (e), and (g)

$$m_{sides} = \rho V_{side} = 0.09594\ kg$$
$$m_{core} = \rho V_{core} = 0.005996\ kg \quad (h)$$

By Eqs. (b) and (h), we obtain

$$I_{core} = \frac{1}{2}(0.005996)(0.02)^2 = 1.1992 \times 10^{-6}\ kg \cdot m^2$$

$$I_{sides} = \frac{1}{2}(0.09594)(0.04)^2 = 7.6752 \times 10^{-5}\ kg \cdot m^2$$

and

$$I_G = 1.1992 \times 10^{-6} + 7.67652 \times 10^{-5} = 7.796 \times 10^{-5}\ kg \cdot m^2 \quad (i)$$

By Kinematics and Fig. a,

$$v_G = r\omega = 0.020\,\omega \quad (j)$$

Therefore, by Eqs. (a), (c), (i), and (j), we have

$$0.14843\, v_G^2 = 1$$

or

$$\underline{\underline{v_G = 2.956\ m/s}}$$

19.49

GIVEN A constant torque M is applied to the homogeneous circular drum (weight W_d and radius r) of the system used to raise crates (weight W_c) from a dock to the floor of a warehouse (Fig. a). Initially, a crate is at rest on the ramp, and the coefficient of kinetic friction between the crate and the ramp is μ_k.

(continued)

19.49 cont.

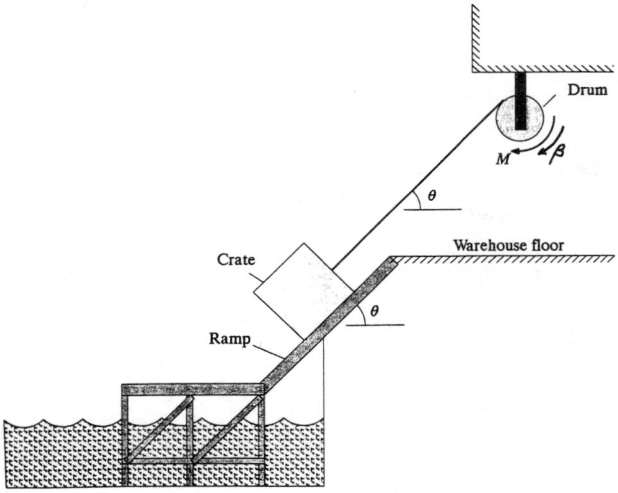

Figure a.

FIND Derive a formula for the speed v of the crate at the instant when the drum has rotated through an angle β, in terms of M, W_d, W_c, r, θ, and β. Then express the angular velocity ω_d of the drum in terms of the same quantities.

SOLUTION

Since no information is given about the cable from the drum to the crate, assume that the mass of the cable is negligible.

Figure b is the free-body diagram of the crate.

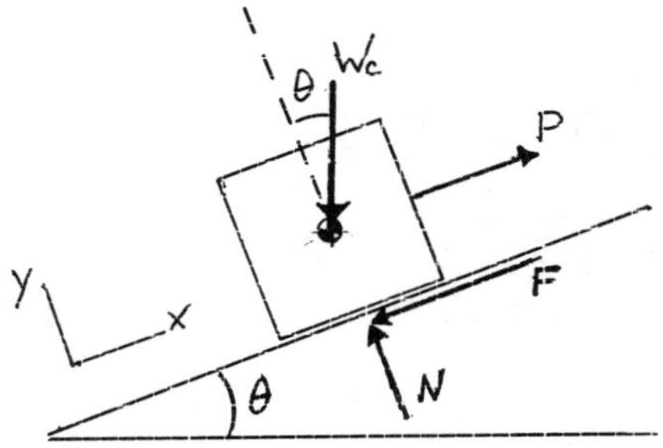

Figure b.

By Fig. b,

$$\Sigma F_y = N - W_c \cos\theta = 0; \quad N = W_c \cos\theta \qquad (a)$$

Then, with Eq. (a), the friction force F is

$$F = \mu_k N = \mu_k W_c \cos\theta \qquad (b)$$

The law of kinetic energy is

$$T_1 + V_1 + U_{1-2} = T_2 + V_2 \qquad (c)$$

and since initially the crate is at rest, $T_1 = 0$. Also, with the datum of potential energy at the rest-position of the crate, $V_1 = 0$. The work U_{1-2} performed on the system under a rotation of the drum is, with Eq. (b.)

$$U_{1-2} = M\beta - (\mu_k W_c \cos\theta) r\beta \qquad (d)$$

where $r\beta$ is the distance that the crate moves up the ramp (Fig. a). [Note that the force P in the cable (Fig. b) does no work since it is internal to the system.]

The kinetic energy T_2 and the potential energy V_2 are given by

$$T_2 = \frac{1}{2}\left(\frac{W_c}{g}\right)v^2 + \frac{1}{2}\left[\frac{1}{2}\left(\frac{W_d}{g}\right)r^2\right]\omega^2$$
$$V_2 = W_c (r\beta)\sin\theta \qquad (e)$$

where by kinematics (Fig. a)

$$v = r\omega; \quad \omega = \frac{v}{r} \qquad (f)$$

Hence, by Eqs. (c), (d), (e), and (f),

$$M\beta - \mu_k W_c r\beta \cos\theta =$$
$$= \frac{1}{4g}(2W_c + W_d)v^2 + W_c r\beta \sin\theta$$

or

$$v = 2\sqrt{\frac{g\,[M\beta - (\mu_k \cos\theta + \sin\theta) W_c r\beta]}{2W_c + W_d}} \qquad (g)$$

(continued)

19.49 cont.

Then by Eqs. (f) and (g),

$$\omega = \frac{1}{r} v$$

$$= \frac{2}{r}\sqrt{\frac{g[M\beta - (\mu_k \cos\theta + \sin\theta) W_c r \beta]}{2W_c + W_d}}$$

19.50

GIVEN The torque applied to the shaft of a car is

$$T = 240 \text{ N·m} \qquad (a)$$

when the shaft is rotating at 4200 rpm.

FIND Determine the power (in kW and in hp) transmitted through the drive shaft of the car.

SOLUTION

$$\text{Power} = E = \bar{F}\cdot\bar{v} = T\omega \qquad (b)$$

where

$$\omega = 4200\left(\frac{\text{rev}}{\text{min}}\right)\left(\frac{2\pi \text{ rad}}{\text{rev}}\right)\left(\frac{1\text{min}}{60\text{ s}}\right)$$

or

$$\omega = 439.82 \text{ rad/s} \qquad (c)$$

Then, by Eqs. (a), (b), and (c)

$$E = (240)(439.82)$$

or

$$E = 105.56 \text{ kW} \quad (141.50 \text{ hp})$$

19.51

GIVEN The drive shaft of a car can transmit 260 hp when it is rotating at 3200 rpm.

FIND Determine the torque T [lb·ft] in the shaft.

SOLUTION

$$\text{Power} = E = T\omega$$

Therefore

$$E = 260 \text{ hp}\left(\frac{550 \text{ ft·lb/s}}{1 \text{ hp}}\right)$$

$$= 143{,}000 \text{ ft·lb/s} \qquad (b)$$

and

$$\omega = \left(3200 \frac{\text{rev}}{\text{min}}\right)\left(\frac{2\pi \text{ rad}}{\text{rev}}\right)\left(\frac{1\text{min}}{60\text{ s}}\right)$$

$$= 335.10 \text{ rad/s} \qquad (c)$$

Hence, by Eqs. (a), (b), and (c),

$$T = 426.7 \text{ lb·ft}$$

19.52

GIVEN The drive shaft of a car can transmit 200 kW when it is rotating at 3000 rpm.

FIND Determine the torque T [N·m] in the shaft

SOLUTION

$$\text{Power} = E = T\omega$$

where

$$E = 200 \text{ kW} = 200\,000 \text{ Watts}$$

$$= 200\,000 \text{ N·m/s} \qquad (b)$$

and

$$\omega = \left(3000 \frac{\text{rev}}{\text{min}}\right)\left(\frac{2\pi \text{ rad}}{\text{rev}}\right)\left(\frac{1\text{ min}}{60\text{ s}}\right)$$

$$= 314.159 \text{ rad/s} \qquad (c)$$

By Eqs. (a), (b), and (c),

$$T = \frac{200\,000}{314.159} = 636.6 \text{ N·m}$$

19.53

GIVEN The shaft of a motor delivers ½ hp at 1800 rpm.

FIND

a.) Determine the moment [lb·ft] of the couple acting on the shaft.

b.) Determine the work [ft·lb] performed by the couple in 10 min.

SOLUTION

a.) The power is given by

$$\text{Power} = E = M\omega \qquad (a)$$

where, by the given data

$$E = \frac{1}{2}\text{ hp} = \left(\frac{1}{2}\right)(550 \text{ ft}\cdot\text{lb/s})$$

$$= 275 \text{ ft}\cdot\text{lb/s} \qquad (b)$$

$$\omega = \left(1800 \frac{\text{rev}}{\text{min}}\right)\left(\frac{2\pi \text{ rad}}{\text{rev}}\right)\left(\frac{1 \text{ min}}{60 \text{ s}}\right)$$

$$= 60\pi \text{ rad/s} \qquad (c)$$

Then, by Eqs. (a), (b), and (c),

$$M = \frac{275}{60\pi} = 1.459 \text{ lb}\cdot\text{ft} \qquad (d)$$

b.) In 10 min, by Eq. (c), the motor rotates through

$$\left(60\pi \frac{\text{rad}}{\text{s}}\right)(10 \times 60 \text{ s}) = 36{,}000\pi \text{ rad}$$

Hence, the work performed by M is, with Eq (d)

$$U = \int_0^{36{,}000\pi} M\,d\theta = M\theta \Big|_0^{36{,}000\pi}$$

$$= (1.459)(36{,}000\pi) = 165{,}000 \text{ ft}\cdot\text{lb}$$

19.54

GIVEN A rotor weighs 64.4 lb, and its radius of gyration is 2 ft. A couple of magnitude 40 lb·ft is required to keep the shaft rotating at 30 rad/s.

FIND

a.) Determine the horsepower required

b.) An additional couple of magnitude 60 lb·ft, in the same sense as the 40 lb·ft couple, is applied to the shaft. At what rate does the angular speed increase?

SOLUTION

a.) The power is given by

$$\text{Power} = E = M\omega \qquad (a)$$

where

$$M = 40 \text{ lb}\cdot\text{ft} \qquad (b)$$

$$\omega = 30 \text{ rad/s} \qquad (c)$$

Therefore, by Eqs. (a), (b), and (c),

$$E = (40)(30) = 1200 \text{ ft}\cdot\text{lb/s}$$

or

$$E = (1200 \text{ ft}\cdot\text{lb/s})\left(\frac{1 \text{ hp}}{550 \text{ ft}\cdot\text{lb/s}}\right)$$

$$= 2.18\overline{18} \text{ hp}$$

b.) The additional couple of magnitude $M = 60$ lb·ft produces the change $d\omega/dt$ of the angular speed ω. Hence, by kinetics of rotation.

$$\Sigma M = M = 60 \text{ lb}\cdot\text{ft}$$

$$= I\alpha = I\frac{d\omega}{dt} \qquad (d)$$

where

$$I = mk^2 = \left(\frac{64.4}{32.2}\right)(2)^2 = 8 \qquad (e)$$

Hence, by Eqs. (d) and (e),

$$\frac{d\omega}{dt} = \frac{M}{I} = \frac{60}{8} = 7.5 \text{ rad/s}^2$$

19.55

GIVEN The torsion constant for the shaft of a dynamometer is 56 N·m per degree of twist of the shaft. The shaft is twisted 6° when it rotates 1800 rpm and delivers power to a machine.

FIND Determine the power [kW] delivered to the machine.

SOLUTION

The moment delivered by the shaft to the machine is

$$M = k\theta = (56\ N\cdot m)\ (6 deg)$$
$$= 336\ N\cdot m \qquad (a)$$

The rotational speed of the shaft is

$$\omega = \left(1800\ \frac{rev}{min}\right)\left(\frac{2\pi\ rad}{rev}\right)\left(\frac{1\ min}{60\ s}\right)$$
$$= 60\pi\ \text{rad/s} \qquad (b)$$

Hence, by the power relation

$$E = M\omega$$

and Eqs. (a) and (b), we have

$$E = (336)(60\pi) = 63\,334.5\ W$$

or

$$E = 63.33\ kW$$

19.56

GIVEN A rotating disk sander with diameter D [ft] is pressed against a flat surface with force F [lb]. The sander rotates at an angular speed ω [rpm]. The coefficient of friction is μ.

FIND Assuming that the pressure on the disk is uniform, show that the power E [hp] required to drive the sander is

$$E = \frac{\pi \mu F D \omega}{49{,}500}$$

SOLUTION

The shear stress [lb/ft²] is (see Fig a)

$$S = \mu p \qquad (a)$$

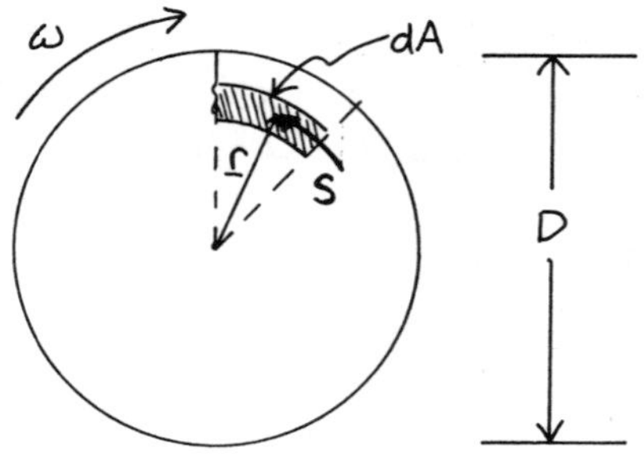

Figure a.

where p is the uniform pressure on the disk. Hence,

$$p = \frac{F}{\frac{\pi}{4}D^2} = \frac{4F}{\pi D^2} \qquad (b)$$

The frictional force on the elemental area $dA = r\,d\theta\,dr$ is (Fig. a)

$$dR = S\,dA \qquad (c)$$

The element of torque on dA is, with Eqs. (a), (b), and (c),

$$dM = r\,dR = \frac{4\mu F}{\pi D^2} r^2 dr\,d\theta \qquad (d)$$

Integration of Eq. (d) yields

$$M = \frac{4\mu F}{\pi D^2}\int_0^{D/2}\int_0^{2\pi} r^2 dr\,d\theta$$

$$= \frac{4\mu F}{\pi D^2}\,\frac{r^3}{3}\Bigg|_0^{D/2}\,\theta\Bigg|_0^{2\pi}$$

or

$$M = \frac{1}{3}\mu F D \qquad (e)$$

Also, the angular speed in rad/s is

$$\Omega = \left(\omega\ \frac{rev}{min}\right)\left(\frac{2\pi\ rad}{1\ rev}\right)\left(\frac{1\ min}{60\ s}\right) = \frac{\pi}{30}\omega\ [\text{rad/s}] \qquad (f)$$

where ω is in rpm (revolutions per minute).

(continued)

19.56 cont.

Then, the required energy in horsepower is, with Eqs. (e) and (f),

$$E = \frac{M\Omega}{550} = \left(\frac{1}{3}\mu F D\right)\left(\frac{\pi}{30}\omega\right) / (550)$$

or

$$E = \frac{\pi \mu F D \omega}{49{,}500} \quad [\text{hp}]$$

where F is in [lb], D [ft], and ω [rpm].

19.57

GIVEN A moving freight car weighs 270 kN. It collides with a stationary boxcar that weighs 180 kN. The cars couple together and roll with a speed 1 m/s.

FIND
a.) Determine the initial speed of the freight car.
b.) Calculate the loss of kinetic energy of the system due to the collision.
c.) Explain the loss of energy.

SOLUTION
a.) By the law of linear momentum,

$$m_1 v_{1i} + m_2 v_{2i} = m_1 v_{1f} + m_2 v_{2f} \qquad (a)$$

where subscripts 1 and 2 denote the freight car and the boxcar, respectively, and the subscripts i and f denote the initial and final velocities.

With the given values of m_1 and m_2, $v_{2i} = 0$, and with $v_{1f} = v_{2f} = 1$ m/s, Eq. (a) yields

$$\frac{270{,}000}{9.81} v_{1i} + 0 = \left(\frac{270{,}000}{9.81} + \frac{180000}{9.81}\right)(1)$$

Therefore,

$$v_{1i} = 1.666 \text{ m/s} \qquad (b)$$

b.) By the law of kinetic energy

$$T_1 + V_1 + U_{1-2} = T_2 + V_2 \qquad (c)$$

where

$$T_1 = \tfrac{1}{2} m_1 v_{1i}^2$$
$$V_1 = V_2 = 0 \quad \left(\begin{array}{c}\text{assuming track}\\ \text{is horizontal}\end{array}\right) \qquad (d)$$
$$T_2 = \tfrac{1}{2}(m_1 + m_2) v^2 = \tfrac{1}{2}(m_1 + m_2)(1)^2$$

and

$$U_{1-2} = U_{1-2}^{c} + U_{1-2}^{nc} = 0 + U_{1-2}^{nc} \qquad (e)$$

Since there are no conservative forces, that produce work, acting, the work U_{1-2}^{c} due to conservative forces is zero. Hence, $U_{1-2} = U_{1-2}^{nc}$, the work done by nonconservative forces that act during the impact.

Hence, by Eqs. (b), (c), (d), and (e)

$$U_{1-2}^{nc} = T_2 - T_1 = \frac{1}{2}\left(\frac{270000 + 180000}{9.81}\right)(1)^2 - \frac{1}{2}\left(\frac{270\,000}{9.81}\right)(1.666)^2$$

or

$$U_{1-2}^{nc} = -15\,290 \text{ N·m}$$

The work done by the nonconservative forces is the negative of the kinetic energy loss. Hence, the energy loss is,

$$-U_{1-2}^{nc} = T_1 - T_2 = 15\,290 \text{ N·m} \qquad (f)$$

c.) The energy loss [see Eq. (f)] that occurs in the collision is due to work done by nonconservative forces that act during the collision. These nonconservative forces produce nonconservative work that produces deformations of the connection between the cars, vibrations, and sound waves at impact.

19.58

GIVEN A system of two masses m_1 and m_2 collide

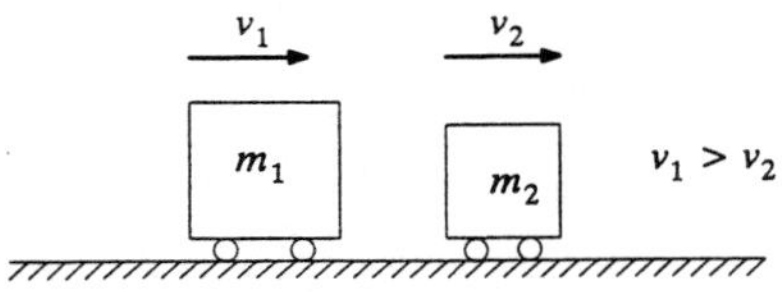

Figure a.

FIND Show that in the absence of external forces, the velocity of the center of mass of the system remains unchanged by the collision.

SOLUTION

By the law of linear momentum

$$m_1 \bar{v}_{1i} + m_2 \bar{v}_{2i} = m_1 \bar{v}_{1f} + m_2 \bar{v}_{2f}$$

or, since the velocities have the same sense,

$$m_1 v_{1i} + m_2 v_{2i} = m_1 v_{1f} + m_2 v_{2f} \qquad (a)$$

where the i and f subscripts denote the initial and final speeds respectively. Hence, by Eq. (a) and the definition of mass center G

$$m_1 v_{1i} + m_2 v_{2i} = (m_1 + m_2) v_{Gi} \qquad (b)$$

$$m_1 v_{1f} + m_2 v_{2f} = (m_1 + m_2) v_{Gf} \qquad (c)$$

So, by Eqs. (a), (b), and (c)

$$(m_1 + m_2) v_{Gi} = (m_1 + m_2) v_{Gf}$$

or

$$v_{Gi} = v_{Gf}$$

19.59

GIVEN Two freight cars A and B of weight 40 tons and 30 tons, respectively, collide (Fig. a). After the collision, car B has a speed of 32 mi/h.

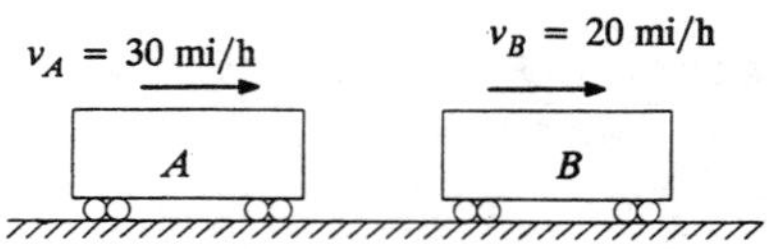

Figure a.

FIND Determine the speed of car A after the collision:

a.) By the law of conservation of linear momentum

b.) By the fact that the velocity of the mass center is unchanged by the collision.

SOLUTION

a.) By linear momentum,

$$m_A v_{Ai} + m_B v_{Bi} = m_A v_{Af} + m_B v_{Bf} \qquad (a)$$

or

$$(40)(30) + (30)(20) = (40) v_{Af} + (30)(32)$$

$$\therefore v_{Af} = 21 \text{ mi/h}$$

b.) The mass center velocity is defined as

$$\bar{v}_G = (m_A \bar{v}_A + m_B \bar{v}_B)/(m_A + m_B)$$

Hence, with $\bar{v}_{Gi} = \bar{v}_{Gf}$, we have

$$\frac{m_A \bar{v}_{Ai} + m_B \bar{v}_{Bi}}{m_A + m_B} = \frac{m_A \bar{v}_{Af} + m_B v_{Bf}}{m_A + m_B}$$

Hence, cancelling $m_A + m_B$, we obtain Eq. (a) or

$$v_{Af} = 21 \text{ mi/h}$$

19.60

GIVEN A force of constant magnitude $F = 16$ lb acts horizontally to the right on a body from time $t=0$ to $t=3$s. At $t=3$s, the force instantaneously reverses its direction and acts horizontally to the left, from time $t=3$s to $t=5$s.

FIND Determine the linear impulse exerted by the force for the time interval $0 \le t \le 5$ s.

SOLUTION

For $0 \le t \le 3$s, the linear impulse is

$$\bar{I}_{L(0-3)} = 3\bar{F} = (3)(16\hat{\imath}) = 48\hat{\imath} \qquad (a)$$

where the unit vector $\hat{\imath}$ is directed to the right.

For $3\text{s} \le t \le 5$s, the linear impulse is

$$\bar{I}_{L(3-5)} = (2)(-\bar{F}) = (2)(-16\hat{\imath}) = -32\hat{\imath} \qquad (b)$$

Hence, the impulse over the interval $0 \le t \le 5$s is, by Eqs. (a) and (b),

$$\bar{I}_{L(0-5)} = \bar{I}_{L(0-3)} + \bar{I}_{L(3-5)} = 48\hat{\imath} - 32\hat{\imath} = 16\hat{\imath}$$

19.61

GIVEN The force $F = 16$ lb, initially directed to the right, acts on a crate of mass 1 slug. At time $t=0$, the crate is at rest on a smooth platform. From time $t=0$ to $t=3$s, the force F acts to the right and from $t=3$s to $t=5$s acts to the left.

FIND Determine the velocity v_G of the center of mass of the crate at $t=5$s.

SOLUTION

The net impulse for the interval $0 \le t \le 5$s is, where unit vector $\hat{\imath}$ is directed to the right,

$$\bar{I}_{L(0-5)} = \int_0^3 \bar{F}\,dt + \int_3^5 (-\bar{F})\,dt = (16\hat{\imath})(3) + (-16\hat{\imath})(5-3)$$

or

$$\bar{I}_{L(0-5)} = 16\hat{\imath} \ [\text{lb}\cdot\text{s}] \qquad (a)$$

The law of linear momentum is

$$m\bar{v}_{G0} + \bar{I}_{L(0-5)} = m\bar{v}_{G5} \qquad (b)$$

Since at $t=0$, the 1-slug crate is at rest, $\bar{v}_{G0} = 0$. Hence, with $m = 1$ slug, Eqs. (a) and (b) yield at $t=5$s,

$$\bar{v}_{G5} = 16\hat{\imath} \ [\text{ft/s}]$$

19.62

GIVEN A railroad boxcar wheel weighs $W = 1$ kN and has a radius $r = 0.5$ m. The boxcar is traveling along a straight track, and the wheel rolls with angular speed of 60 rpm. The boxcar has four wheels on each side, two at the front and two at the back.

FIND Determine the total linear momentum of the eight wheels.

SOLUTION

The angular speed of a wheel in rad/s is given by

$$\omega = \left(60\ \frac{\text{rev}}{\text{min}}\right)\left(\frac{2\pi\ \text{rad}}{1\ \text{rev}}\right)\left(\frac{1\ \text{min}}{60\ \text{s}}\right) = 2\pi\ \text{rad/s} \qquad (a)$$

Therefore, since the wheel has a radius $r = 0.5$m, the speed of the mass center G of the wheel (the geometrical center of the wheel) is

$$v_G = \omega r = (2\pi)(0.5) = \pi\ \text{m/s} \qquad (b)$$

Hence, the linear momentum of the wheel is

$$Q = m v_G = \frac{W}{g} v_G = \left(\frac{1000\ \text{N}}{9.81\ \text{m/s}^2}\right)(\pi\ \text{m/s}) = 101.94\pi\ \text{N}\cdot\text{s}$$

So, the total linear momentum of 8 wheels is,

$$Q_{Total} = 8Q = 8(101.94\pi) = 2562\ \text{N}\cdot\text{s}$$

19.63

GIVEN A body that weighs $W = 16$ N has a mass-center velocity

$$\bar{v}_G = 4t^2\hat{i} - 16t\hat{j} + 4\hat{k} \quad [\text{m/s}] \qquad (a)$$

t is in seconds.

FIND

a.) Determine the linear momentum $\bar{Q}$ [N·s] of the body for $t = 0$

b.) Determine the linear momentum $\bar{Q}$ of the body for $t = 5$ s

SOLUTION

a.) The linear momentum of the body is

$$\bar{Q} = m\bar{v}_G \qquad (b)$$

For $t = 0$, Eq. (a) yields

$$\bar{v}_G = 4\hat{k} \qquad (c)$$

Therefore, by Eqs. (b) and (c), with $m = \frac{W}{g} = \frac{16}{9.81} = 1.631$ kg

$$\bar{Q}_0 = (1.631)(4\hat{k}) = 6.524\hat{k} \quad [\text{N·s}]$$

b.) Similarly, for $t = 5$ s, Eq (a) yields

$$\bar{v}_G = 100\hat{i} - 80\hat{j} + 4\hat{k} \qquad (d)$$

Then, by Eqs. (b) and (d),

$$\bar{Q}_5 = (1.631)(100\hat{i} - 80\hat{j} + 4\hat{k})$$

or

$$\bar{Q}_5 = 163.1\hat{i} - 130.5\hat{j} + 6.52\hat{k} \quad [\text{N·s}]$$

19.64

GIVEN A force $\bar{F}$ acts on a 16-N body to produce a mass-center velocity

$$\bar{v}_G = 4t^2\hat{i} - 16t\hat{j} + 4\hat{k} \quad [\text{m/s}] \qquad (a)$$

where t is in seconds.

FIND Determine the linear impulse $\bar{I}_L$ of the force for the time interval $0 \le t \le 5$ s.

SOLUTION

By the law of linear momentum

$$m\bar{v}_{G(0)} + \int_0^t \bar{F}\,dt = m\bar{v}_{G(t)} \qquad (b)$$

For $t = 0$, by Eq. (a), $\bar{v}_{G(0)} = 4\hat{k}$, and for $t = 5$ s, $\bar{v}_{G(5)} = 100\hat{i} - 80\hat{j} + 4\hat{k}$. Hence, by Eq. (b)

$$\bar{I}_L = \int_0^5 \bar{F}\,dt = m\left[\bar{v}_{G(5)} - \bar{v}_{G(0)}\right]$$

$$= \left(\frac{16}{9.81}\right)\left[100\hat{i} - 80\hat{j} + 4\hat{k} - 4\hat{k}\right]$$

or

$$\bar{I}_L = 163.1\hat{i} - 130.5\hat{j} \quad [\text{N·S}]$$

19.65

GIVEN A force F varies with time t (Fig. a). When F is positive, it acts in the positive sense of the x axis; when negative, it acts in the negative sense of the x axis. The force acts on a body of mass $m = 60$ kg that moves along the x axis with initial velocity (at $t = 0$) of its mass center $v_{Gi} = 8$ m/s in the positive sense of the x axis.

(continued)

19.65 cont.

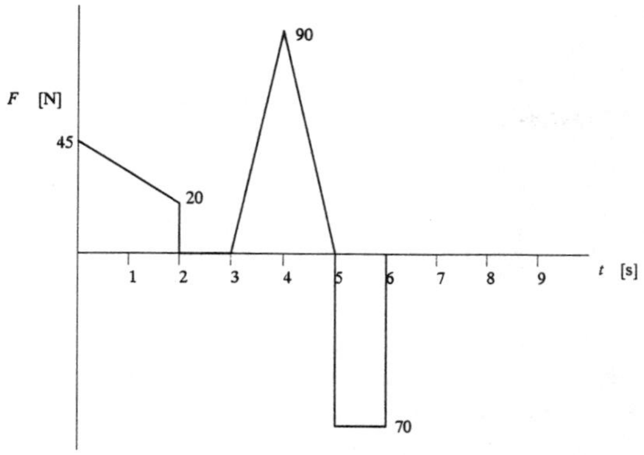

Figure a.

FIND

a.) Determine the average force exerted on the body for the time interval $0 \le t \le 8s$.

b.) Determine the final velocity $\bar{v}_{Gf}$ of the mass center at $t = 8s$.

SOLUTION

a.) By Fig (a), the linear impulse is the area of the force - time diagram. Thus, for $0 \le t \le 8$ s,

$$\bar{I}_L = \int_0^8 \bar{F}\,dt = (F_{ave})(8)\hat{\imath}$$

Therefore,

$$F_{ave} = \frac{1}{8}\left[\frac{1}{2}(45+20)(\;\;) + \frac{1}{2}(90)(2) - (70)(1)\right]$$

or $$F_{ave} = \frac{1}{8}(85) = 10.625 \text{ N}$$

b.) By the law of linear momentum,

$$m\bar{v}_{Gi} + \int_0^8 \bar{F}\,dt = m\bar{v}_{Gf}$$

Therefore,

$$\left(\frac{60}{9.81}\right)8\hat{\imath} + 85\hat{\imath} = \left(\frac{60}{9.81}\right)v_{Gf}\,\hat{\imath}$$

or $$v_{Gf} = 21.90 \text{ m/s}$$

19.66

GIVEN Starting from rest at $t=0$ a 12-N block slides down a 30° inclined plane, with $\mu_k = 0.1$ (Fig/a)

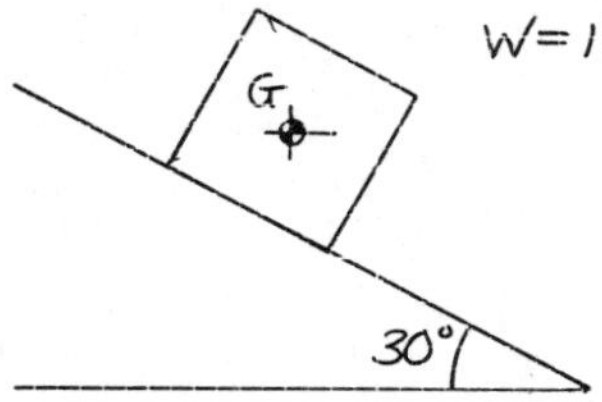

FIND Determine the speed, v_G, of the block at the end of 3s. Use momentum principles.

SOLUTION

Consider the free-body diagram of the block (Fig b.)

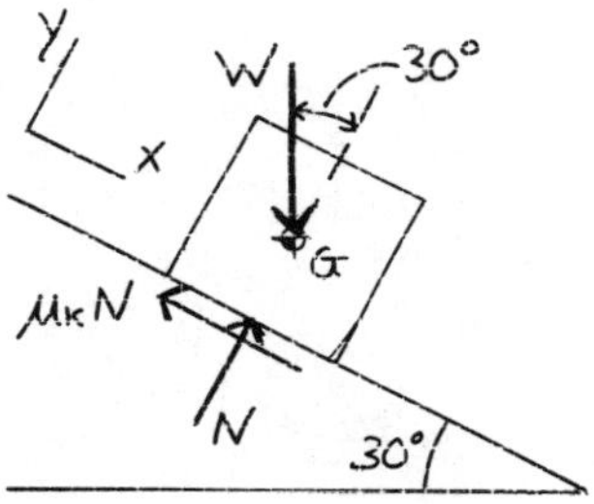

Figure b

By Fig b,

$$\Sigma F_y = N - W\cos 30° = 0$$

or $$N = 12\sqrt{3}/2 = 10.392 \text{ N}$$

Hence,

$$\mu_k N = (0.1)(10.392) = 1.039 \text{ N}$$

Also, by Fig. b,

$$\Sigma F_x = W\sin 30° - \mu_k N$$
$$= (12)\left(\tfrac{1}{2}\right) - 1.039$$

or $$\Sigma F_x = 4.961 \text{ N}$$

(continued)

19.66 cont.

By the law of linear momentum

$$\int_0^3 (\Sigma F_x)\,dt = m v_G$$

or

$$(4.961)(3) = \frac{12}{9.81} v_G$$

So,

$$v_G = 12.17 \text{ m/s}$$

19.67

GIVEN Part of a drop-hammer system consists of masses $m_1 = 0.50$ lb·s²/ft and $m_2 = 1.0$ lb·s²/ft connected by an inextensible cord that passes over a uniform pulley of mass $m_3 = 0.70$ lb·s²/ft (Fig. a). At a given instant, mass m_1 moves downward at 10 ft/s.

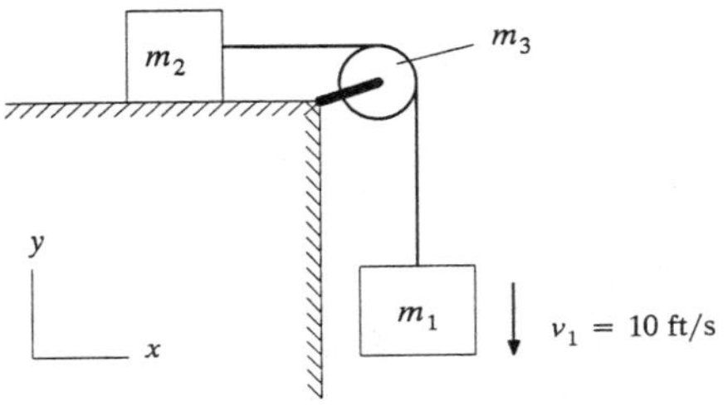

Figure a.

FIND Determine the linear momentum of the system for that instant.

SOLUTION

Since the mass center of the pulley does not move, its linear momentum is zero. The linear momentum of mass m_1 is (Fig. b)

$$\bar{Q}_1 = m_1 \bar{v}_1 = (0.50)(10)(-\hat{j}) = -5.0\hat{j} \text{ [lb·s]} \quad (a)$$

and of mass m_2 is

$$\bar{Q}_2 = m_2 \bar{v}_2 = (1.0)(10)\hat{i} = 10\hat{i} \text{ [lb·s]} \quad (b)$$

since the speed of m_2 is the same as that of m_1.

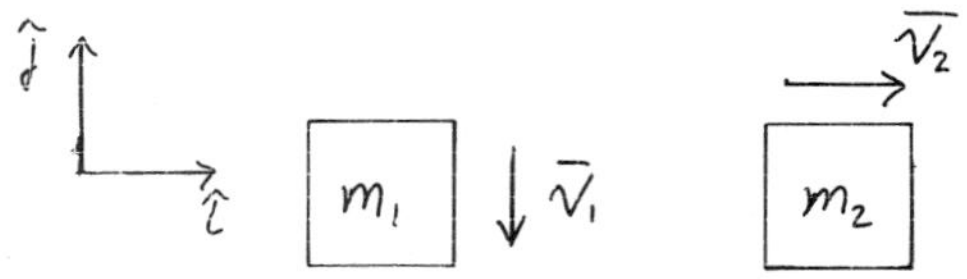

Figure b.

Therefore, by Eqs. (a) and (b), the linear momentum of the system is

$$\bar{Q} = \bar{Q}_1 + \bar{Q}_2 = 10\hat{i} - 5\hat{j} \text{ [lb·s]}$$

or

$$\bar{Q} = 11.18 \text{ lb·s, at } \theta = -26.565°$$

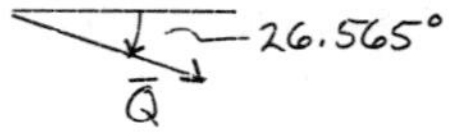

19.68

GIVEN In the drop-hammer mechanism of Problem 19.67, let the mass m_3 of the pulley and the friction between mass m_2 and the support be negligible (Fig. a). Mass $m_1 = 0.50$ lb·s²/ft and mass $m_2 = 1.0$ lb·s²/ft. The system is released from rest.

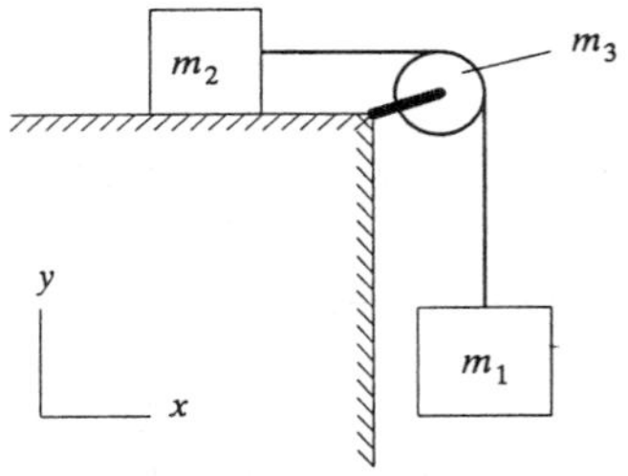

Figure a.

FIND

a.) Determine the linear impulse that acts on the mechanism in the time interval $0 \le t \le 3$s.

b.) Determine the speed of the masses at $t = 3$s. Use the law of linear momentum.

(continued)

19.68 cont.

c.) Check the result of part b, using $\bar{F} = m\bar{a}$ for masses m_1 and m_2.

SOLUTION

a.) The linear impulse that acts on the system is the sum of the linear impulses that act on masses m_1 and m_2, since the linear impulse that acts on the pully is zero.

By the free-body diagrams of masses m_1 and m_2 (Fig. b), the linear impulse of the system is, for $0 \le t \le 3$s,

$$\bar{I}_L = \left(\int_0^3 T dt\right)\hat{\imath} + \left[\int_0^3 (m_1 g - T)dt\right](-\hat{\jmath}) \quad (a)$$

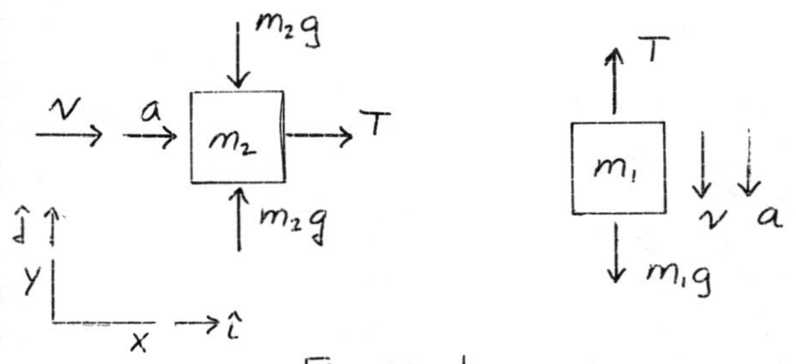

Figure b.

To determine T, we may use Newton's second law for masses m_1 and m_2. Thus, by Fig. b,

$$\left.\begin{aligned} m_1 g - T &= m_1 a \\ T &= m_2 a \end{aligned}\right\} \quad (b)$$

Solving Eqs. (b) for T, we find

$$T = \frac{m_1 m_2 g}{m_1 + m_2} \quad (c)$$

Then, Eqs. (a) and (c) yield

$$\bar{I}_L = \frac{3m_1 g}{m_1 + m_2}(m_2\hat{\imath} - m_1\hat{\jmath}) \quad (d)$$

With $m_1 = 0.50$ lb·s²/ft, $m_2 = 1.0$ lb·s²/ft, and $g = 32.2$ ft/s, Eq. (d) yield

$$\bar{I}_L = 32.2\hat{\imath} - 16.1\hat{\jmath} \text{ [lb·s]} \quad (e)$$

b.) The speed v of the masses may be determined by the law of linear momentum. Thus, for the time interval $0 \le t \le 3$s,

$$m_1 \bar{v}_{1(0)} + m_2 \bar{v}_{2(0)} + \bar{I}_L = $$
$$= m_1 \bar{v}_{1(3)} + m_2 \bar{v}_{2(3)} \quad (f)$$

Since m_1 and m_2 start from rest, their initial speeds (at $t=0$) are zero. That is,

$$\bar{v}_{1(0)} = \bar{v}_{2(0)} = 0 \quad (g)$$

Also, since their speeds are equal, at 3s,

$$v_{1(3)} = v_{2(3)} = v \quad (h)$$

So, by Eqs. (e), (f), (g), and (h),

$$32.2\hat{\imath} - 16.1\hat{\jmath} = v\hat{\imath} - 0.5v\hat{\jmath}$$

Hence,

$$v = 32.2 \text{ ft/s} \quad (i)$$

c.) To check Eq. (i) by $\bar{F} = m\bar{a}$, we have from Eqs. (b),

$$a = \frac{m_1 g}{m_1 + m_2} = \frac{(0.5)(32.2)}{0.5 + 1.0} = \frac{1}{3}(32.2) \text{ ft/s}^2$$

Since the system starts from rest and the acceleration a is constant, at $t = 3$s,

$$v = at = \left[\frac{1}{3}(32.2)\right](3) = 32.2 \text{ ft/s} \quad (j)$$

Equations (i) and (j) provide the desired check.

19.69

GIVEN The belt-pulley system shown in Fig. a consists of a belt of mass m_b and two uniform pulleys A and B of masses m_A and m_B, respectively. Pulley A rotates with angular velocity ω_A.

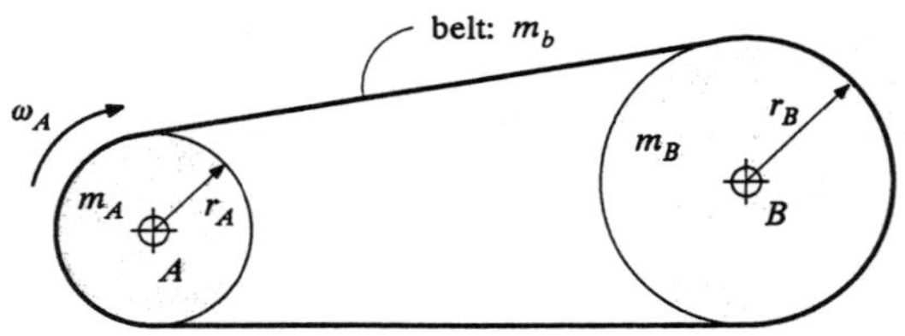

Figure a.

FIND Determine the linear momentum of the system.

SOLUTION

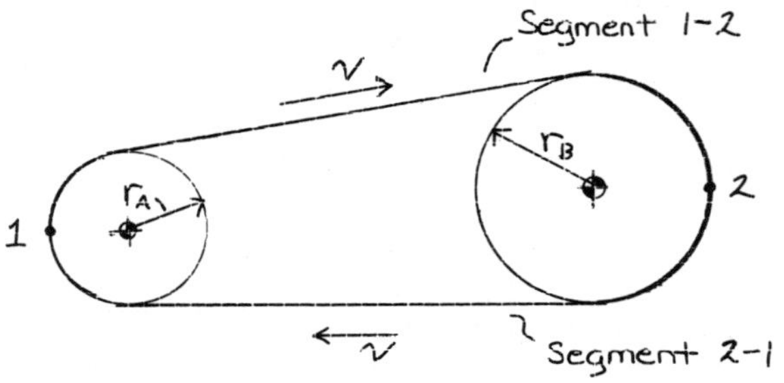

Figure b.

Since the mass centers (the geometrical center) of the uniform pulleys do not move, their linear momentum is zero. Also, since the linear momentum of segment 1-2 of the belt is directed opposite to the linear momentum of segment 2-1 and since the magnitudes of these momenta are equal, the linear momentum of the belt is zero. Hence, the linear momentum of the system is zero.

Alternatively; Since the mass centers of the belt and the pulleys do not move, the linear momentum of the system is zero.

19.70

GIVEN If the resultant force that acts on a body has a short duration – for example, if it is a force due to an impact with another body or an explosive force – it is characterized by a rapid increase in magnitude to a maximum value followed by a rapid decrease. Such a force may be represented as a function of time t by

$$F = a t e^{-bt} \;; \quad t > 0 \qquad (a)$$

where a and b are constants.

FIND

a.) Determine the average value of F over the time interval $0 \le t \le t_1$ in terms of a, b, and t_1

b.) Determine the maximum value of F in terms of a and b.

c.) For F applied horizontally to a body of mass m, initially at rest when $t=0$, on a frictionless horizontal surface, determine the speed v_G of the mass center of the body in terms of t, a, b, and m.

d.) Determine the limiting value of v_G as $t \to \infty$.

SOLUTION

a.) By definition,

$$\int_0^{t_1} F dt = (F_{ave})(t_1) \qquad (b)$$

Substitution of Eq. (a) into Eq (b), we have,

$$F_{AVE} = \frac{1}{t_1}\int_0^{t_1} a t e^{-bt} dt$$

$$= -\frac{a}{b^2}(e^{-bt})(1+bt)\Big|_0^{t_1}$$

or

$$F_{ave} = \frac{a}{t_1 b^2}\left[1 - e^{-bt_1}(1+bt_1)\right] \qquad (c)$$

b.) The maximum and minimum values of F occur when $dF/dt = 0$.

(continued)

19.70 cont.

Hence, by Eq. (a),

$$\frac{dF}{dt} = ae^{-bt} - abte^{-bt} = ae^{-bt}(1-bt) = 0$$

Therefore, $dF/dt = 0$ for either

$$e^{-bt} = 0; \quad t \to \infty$$

or

$$t = \frac{1}{b}$$

For $t \to \infty$, by Eq. (a), $F \to 0$ (the minimum value).

For $t = \frac{1}{b}$, by Eq. (a),

$$F = F_{max} = \frac{a}{eb}$$

c.) By the law of linear momentum,

$$(F_{ave})t = mv_G(t) - mv_G(0) \qquad (d)$$

By Eqs. (c) and (d), with $t_1 = t$ and $v_G(0) = 0$ at $t = 0$, the velocity v_G as a function of t is

$$v_G(t) = \frac{a}{mb^2}\left[1 - e^{-bt}(1+bt)\right] \qquad (e)$$

d.) As $t \to \infty$, Eq. (e) yields

$$v_G \to \frac{a}{mb^2}$$

19.71

GIVEN James Bond (Agent 007) and his jet-propelled escape vehicle weigh 289.8 lb (Fig a.). The jet produces a vertical force $F = 257.6 + 3t^3$ [lb], t in seconds.

FIND Determine the speed [mi/h] of the combined mass center of Bond and his vehicle at $t = 4\text{s}$ after ignition of the jet. Neglect the weight of the burned fuel.

Figure a

SOLUTION

The jet must exert a force of 289.8 lb before it can lift the escape vehicle and Bond. The time at which $F = 289.8$ lb is given by the equation

$$257.6 + 3t^3 = 289.8$$

or, at lift-off,

$$t = 2.206\text{ s}$$

Then, by the law of linear momentum,

$$mv_{G_1} + \int F dt = mv_{G2}$$

we have for $2.206\text{ s} \le t \le 4\text{s}$, with $v_{G1} = 0$,

$$\int_{2.206}^{4} (257.6 + 3t^3)dt = \left(\frac{289.8}{32.2}\right)v_{G2} = 9v_{G2}$$

Integration yields

$$257.6t + \frac{3t^4}{4}\Bigg|_{2.206}^{4} = 9v_{G2}$$

Therefore,

$$v_{G2} = 70.708 \text{ ft/s} = 48.21 \text{ mi/h}$$

19.72

GIVEN The pulley system shown in Fig. a is released from rest at time $t=0$, and the weight W_1 moves downward. The pulleys are frictionless and of negligible weight.

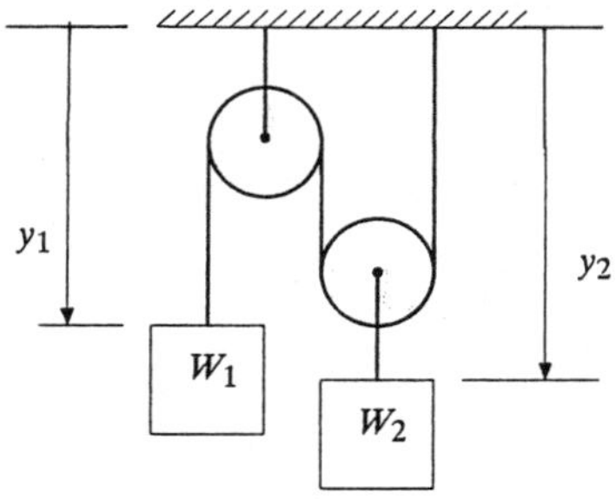

Figure a.

FIND

a.) Using the law of linear momentum, derive formulas for the velocities v_{G1} and v_{G2} of the weights W_1 and W_2, respectively, in terms of W_1, W_2 and t.

b.) For $W_1 = 12$ N, $W_2 = 4$ N, and $t = 2$ s, calculate the magnitudes of v_{G1} and v_{G2}.

SOLUTION

a.) Consider the free-body diagrams of W_1 (Fig. b) and W_2 (Fig. c)

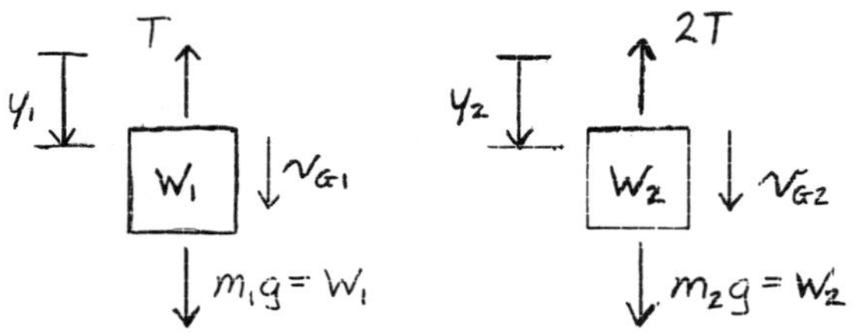

Figure b Figure c

By Figure b, the net-force on W_1 is

$$F_1 = W_1 - T = \text{constant} \tag{a}$$

and by Fig. c, the net force on W_2 is

$$F_2 = W_2 - 2T = \text{constant} \tag{b}$$

By the principle of linear momentum and Eqs. (a) and (b),

$$\int_0^t F_1\,dt = (W_1 - T)t = m_1 v_{G1} = \frac{W_1}{g} v_{G1} \tag{c}$$

and

$$\int_0^t F_2\,dt = (W_2 - 2T)t = m_2 v_{G2} = \frac{W_2}{g} v_{G2} \tag{d}$$

also, by Fig. a, the equation of constraint is $y_1 + 2y_2 = \text{constant}$. Hence, differentiation yields

$$v_{G1} + v_{G2} = 0$$

or

$$v_{G1} = -2v_{G2} \tag{e}$$

Substituting Eq. (e) into Eq. (c) and solving the resulting equation and Eq. (d) for v_{G2}, we obtain

$$v_{G2} = -\frac{(2W_1 - W_2)gt}{4W_1 + W_2} \tag{f}$$

Then, Eqs. (e) and (f) yield

$$v_{G1} = \frac{2(2W_1 - W_2)gt}{4W_1 + W_2} \tag{g}$$

Since W_1 moves downward (Fig. b), $2W_1 - W_2 > 0$. Hence, v_{G2} [Eq. (f)] is negative; W_2 moves up (Fig. c).

b.) For $W_1 = 12$ N, $W_2 = 4$ N, and $t = 2$ s, Eqs. (g) and (f) yield

$$v_{G1} = 15.09 \text{ m/s}$$

$$|v_{G2}| = 7.546 \text{ m/s}$$

19.73

GIVEN Water from a reservoir is ejected downstream through a large pipe. At the end of the pipe, the water passes through a 90° elbow (Fig. a). The water enters the elbow at a speed of 10 m/s. The inside diameter of the cross section of the elbow is d=1.0 m.

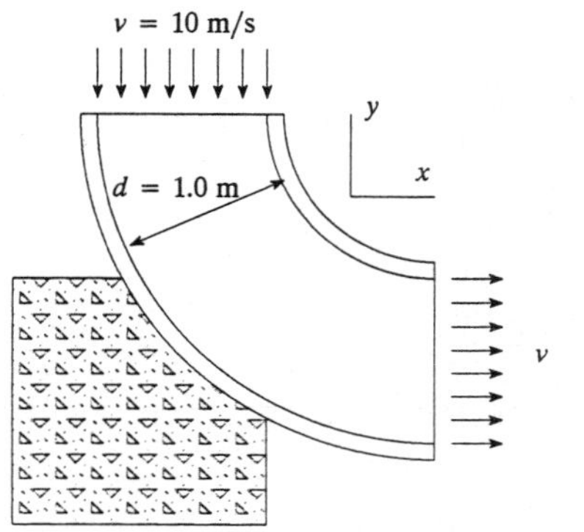

Figure a.

FIND Determine the horizontal and vertical components of the force that the concrete support exerts on the elbow due to the flowing water. Ignore gravity and friction in the elbow.

SOLUTION

The volume of water flowing into the elbow per second is

$$\frac{Vol.}{s} = Av$$

where

$$A = \frac{\pi d^2}{4} = \frac{\pi}{4}(1)^2 = \frac{\pi}{4}\, m^2$$

$$v = 10\ m/s$$

or

$$\frac{Vol}{s} = \left(\frac{\pi}{4}\right)(10) = 2.5\pi\ m^3/s$$

Since friction and gravity are neglected, this is also the volume of water exiting from the elbow. Hence, the mass of water that passes through the elbow per second is,

$$m = \left(\frac{Vol.}{s}\right)(\gamma_{water})\left(\frac{1}{g}\right)$$

$$= \left(2.5\pi\ \frac{m^3}{s}\right)\left(9801\ \frac{N}{m^3}\right)\left(\frac{1}{9.81\ m/s^2}\right)$$

or

$$m = 800.6\ N\cdot s/m = 800.6\ kg/s \qquad (a)$$

The velocity entering the elbow is (Fig. b)

$$\bar{V}_1 = -10\hat{j}\ [m/s] \qquad (b)$$

and the velocity exiting the elbow is

$$\bar{V}_2 = 10\hat{i}\ [m/s] \qquad (c)$$

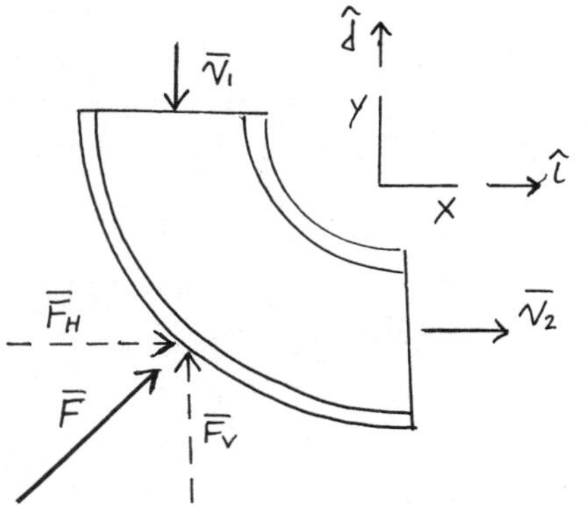

Figure b.

The law of linear momentum is (since the flow is steady and $\bar{F}$ is constant), with m in kg/s and Fig. b;

$$m_1\bar{V}_1 + \bar{F} = m\bar{V}_2 \qquad (d)$$

where $\bar{F}$ is the force exerted on the elbow by the concrete support. Also, since m is kg/s, $\bar{F}$ is the impulse per second, that is $\bar{F} = \bar{I}_L/s\ \left[\frac{F\cdot s}{s}\right]$.

Equations (a), (b), (c), and (d) yield

$$\left(800.6\ \frac{kg}{s}\right)\left(-10\ \frac{m}{s}\hat{j}\right) + \bar{F} = \left(800.6\ \frac{kg}{s}\right)\left(10\ \frac{m}{s}\hat{j}\right)$$

so,

$$\bar{F} = 800.6\,(\hat{i}+\hat{j})\ [kg\cdot m/s^2]$$

or

$$\bar{F} = 800.6\,(\hat{i}+\hat{j})\ [N] \qquad (e)$$

(continued)

19.73 cont.

Hence, the horizontal and vertical components of $\bar{F}$ are, by Eq. (e), see also Fig. b,

$$\bar{F}_H = 800.6\ \hat{\imath}\quad [N]$$

$$\bar{F}_V = 800.6\ \hat{\jmath}\quad [N]$$

19.74

GIVEN A fixed symmetric vane (Fig. a) redirects a horizontal stream of water flowing from a nozzle at velocity v [m/s] and volumetric flow rate R_v [m³/s]. The specific weight of water is γ [N/m³]. The water leaves the vane with velocity v_2 at angle θ.

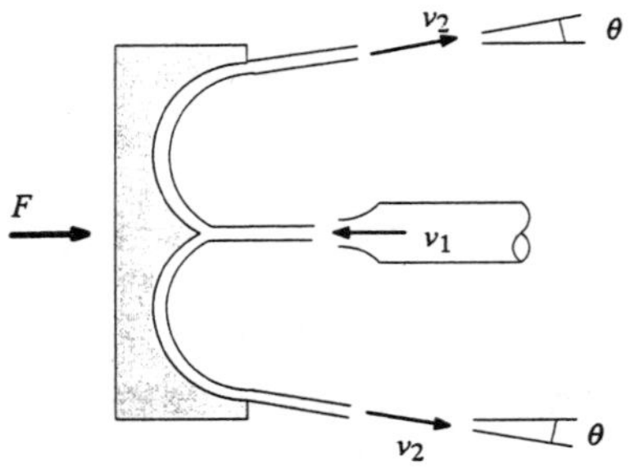

Figure a

FIND Determine the force F exerted on the water by the vane in terms of R_v, γ, v_1, v_2, and θ.

SOLUTION

The weight of water redirected per second is

$$W = \left(R_v \frac{m^3}{s}\right)\left(\gamma \frac{N}{m^3}\right) = R_v\gamma\ \ [N/s] \qquad (a)$$

Therefore, the mass of water redirected per second is, with Eq. (a),

$$m = \frac{W}{g} = \frac{R_v\gamma}{g}\quad \left[\frac{N}{s}\cdot\frac{s^2}{m} = \frac{kg}{s}\right] \qquad (b)$$

By the law of linear momentum per second

$$(m\bar{v}_1)_x + \bar{F} = (m\bar{v}_2)_x \qquad (c)$$

By Eqs. (b) and (c), and Fig. a,

$$-\frac{R_v\gamma}{g}v_1 + F = \frac{R_v\gamma}{g}v_2\cos\theta \qquad (d)$$

Hence, Eq. (d) yields

$$F = \frac{R_v\gamma}{g}(v_1 + v_2\cos\theta)$$

19.75

GIVEN The drop-hammer mechanism shown in Fig. a. When the system is released from rest at time $t=0$, mass m_1 moves down, with speed v at time t. The mass and friction of the pulley are negligible.

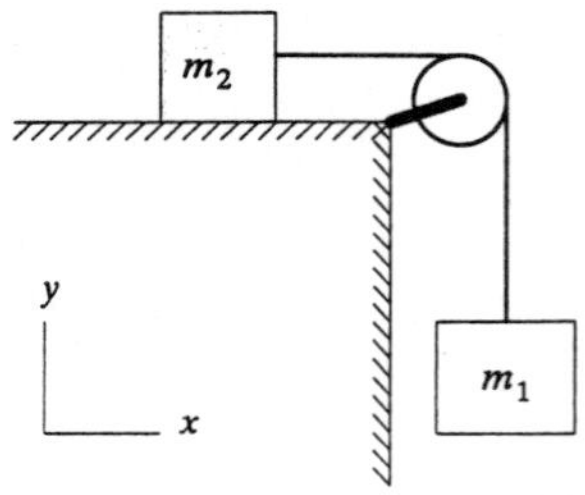

Figure a

FIND

a.) In terms of m_1, m_2, v, and t, derive a formula for μ_K, the coefficient of sliding friction between m_2 and its support. Use the law of linear momentum.

b.) Design the system so that $v/t = 16$ ft/s², $\mu_K \le 0.10$, and T, the tension in the cord, not be greater than 480 lb. Use $g = 32$ ft/s²

(continued)

19.75 cont.

SOLUTION

a.) Consider the free-body diagrams of m_1 and m_2 (Figs. b and c).

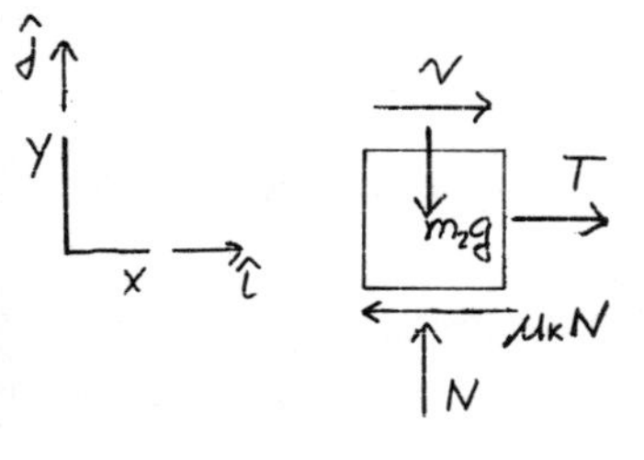

Figure b.

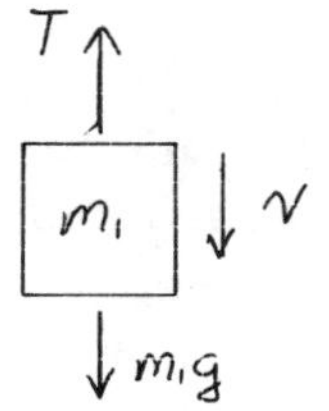

Figure c.

By Figs. (b) and (c) and the law of linear momentum

$$\left.\begin{aligned} \int_0^t (T - \mu_K N)\,dt &= m_2 v \\ \int_0^t (m_1 g - T)\,dt &= m_1 v \end{aligned}\right\} \qquad (a)$$

Also, by Fig. b, $\Sigma F_y = N - m_2 g = 0$ or $N = m_2 g$. Therefore,

$$\mu_K N = \mu_K m_2 g \qquad (b)$$

Equations (a) and (b) yield, after integration,

$$(m_1 g - T)t = m_1 v \qquad (c)$$

$$(T - \mu_K m_2 g)t = m_2 v \qquad (d)$$

Solving Eqs. (c) and (d) for μ_K and T, we find

$$\mu_K = \frac{m_1}{m_2} - \left(1 + \frac{m_1}{m_2}\right)\frac{v}{gt} \qquad (e)$$

$$T = m_1\left(g - \frac{v}{t}\right) \qquad (f)$$

b.) For $v/t = 16$ ft/s², Eq. (e) yields, with $g = 32$ ft/s²,

$$\mu_K = \frac{1}{2}\left(\frac{m_1}{m_2} - 1\right) \qquad (g)$$

If $\mu_K \le 0.10$, Eq. (g) yields

$$\frac{m_1}{m_2} \le 1.20 \qquad (h)$$

By Eq. (f), with $v/t = 16$ ft/s² and $g = 32$ ft/s²,

$$m_1 = \frac{T}{16}\left(\frac{\text{lb s}^2}{\text{ft}}\right) \qquad (i)$$

If $T \le 480$ lb, $m_1 \le 30$ slugs. Hence, with Eq. (h), $m_2 \ge 25$ slugs. Consequently, if $m_1 = 30$ slugs, $m_2 = 25$ slugs, $m_1/m_2 = 1.2$ and $T = 16\,m_1 = 480$ lb. This design would be at the limit of acceptable requirements. Good engineering judgement may require that $m_1/m_2 < 1.20$ and $T < 480$ lb. For example, for $m_1 = 20$ slugs and $m_2 = 25$ slugs, $m_1/m_2 = 0.8$ and $T = 300$ lb. The selection of m_1 may also depend upon the amount of energy required in the forge, the larger m_1 the more energy delivered.

Note also that if $\mu_K < 0.10$, the ratio m_1/m_2 is reduced accordingly.

19.76

GIVEN A cubical crate of mass $m = 50$ kg is pulled up a ramp of a warehouse at $v = 5$ m/s (Fig. a). The center of mass of the crate is located at its geometrical center.

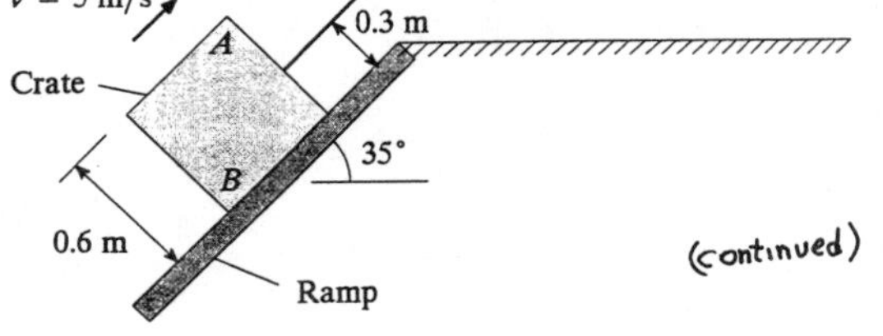

(continued)

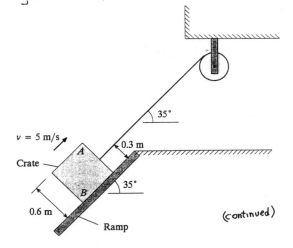

Figure a.

19.76 cont.

FIND Determine the angular moments of the crate about the fixed points in space that coincide with point A and point B of the crate.

SOLUTION

The linear momentum of the crate is (Fig. b)

$$\bar{Q} = m\bar{v}_G = (50)(5)\hat{i} = 250\,\hat{i} \ [\text{N·s}] \qquad (a)$$

where $\hat{i}$ is a unit vector directed up the ramp.

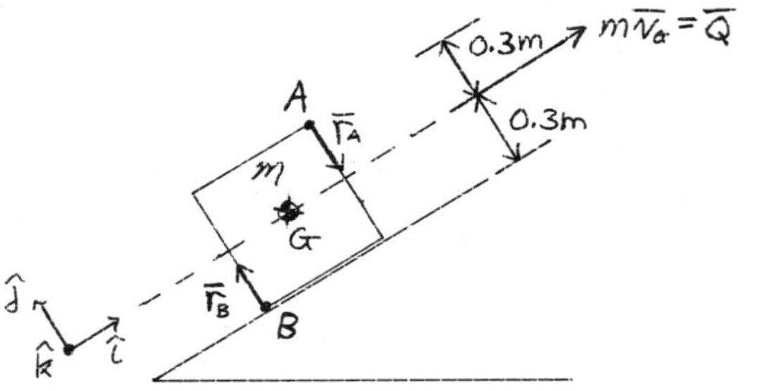

Figure b.

Hence, the angular momentum about the point that coincides with A is

$$\bar{A}_A = \bar{r}_A \times \bar{Q} = (-0.3\hat{j}) \times (250\,\hat{i}) = 75\,\hat{k}$$

where unit vectors $\hat{i}$, $\hat{j}$, and $\hat{k}$ form a right-hand triad, (Fig. b).

Similarly, the angular momentum about the point that coincides with B is

$$\bar{A}_B = \bar{r}_B \times \bar{Q} = (0.3\hat{j})(250\,\hat{i}) = -75\,\hat{k}$$

19.77

GIVEN A uniform cylinder, of weight 216 N and diameter 300 mm, rolls to the right on a horizontal plane, so that the speed of its mass center G is $v_G = 3$ m/s.

FIND

a.) The angular momentum of the cylinder with respect to the longitudinal axis through G.

b.) The magnitude, direction, and location of the linear momentum of the cylinder and show it in a sketch.

SOLUTION

a.) The angular momentum about the axis throug G is

$$A_G = I_G\,\omega = \left(\tfrac{1}{2} m r^2\right)\left(\frac{v_G}{r}\right) = \tfrac{1}{2} m r\, v_G$$

or

$$A_G = \left(\tfrac{1}{2}\right)\left(\tfrac{216}{9.81}\right)(0.150)(3) = 4.954 \ \text{N·m·s} \ \circlearrowright$$

b.) The linear momentum is

$$\bar{Q} = m\bar{v}_G = \left(\frac{216}{9.81}\right)(3)\,\hat{i}$$

or

$$\bar{Q} = 66.055\,\hat{i} \ [\text{N·s}]$$

where $\hat{i}$ is a unit vector directed horizontally to the right. The angular momentum about the axis through G is equal to the moment of the linear momentum about the axis through G. Hence, $A_G = Q\,y_G \circlearrowright$ where y_G is the vertical distance above G. Therefore,

$$y_G = \frac{A_G}{Q} = \frac{4.954}{66.055} = 0.075 \text{ m.}$$

See Fig. a,

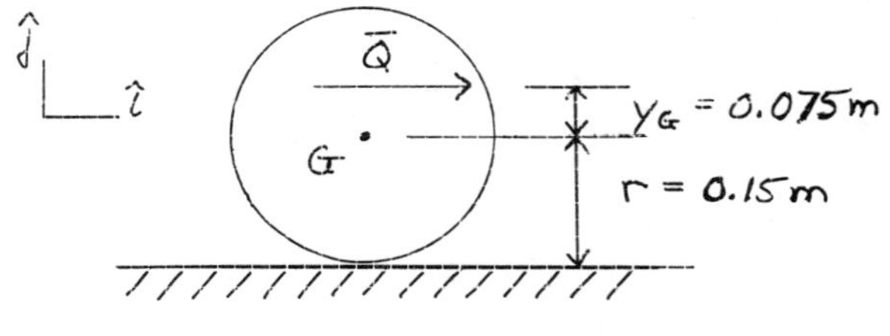

Figure a.

19.78

GIVEN Refer to Problem 19.77, where a uniform cylinder (weight = 216 N, diameter = 300 mm) rolls to the right on a horizontal plane, with speed of its mass center $v_G = 3$ m/s.

FIND Determine the angular momentum of the cylinder with respect to the longitudinal axis through the instantaneous center O of velocity and locate the linear momentum relative to O and show it in a sketch.

SOLUTION

The angular momentum about the axis through O is

$$A_o = I_o \omega$$

where

$$I_o = I_G + mr^2 = \frac{1}{2}mr^2 + mr^2 = \frac{3}{2}mr^2$$

$$\omega = v_G/r$$

or

$$I_o = \frac{3}{2}\left(\frac{216}{9.81}\right)(0.150)^2 = 0.7431 \text{ N·m·s}^2$$

$$\omega = (3)\left(\frac{1}{0.150}\right) = 20 \text{ rad/s}$$

Therefore,

$$A_o = (0.7431)(20) = 14.862 \text{ N·m·s} \circlearrowright$$

The linear momentum is (Fig. a)

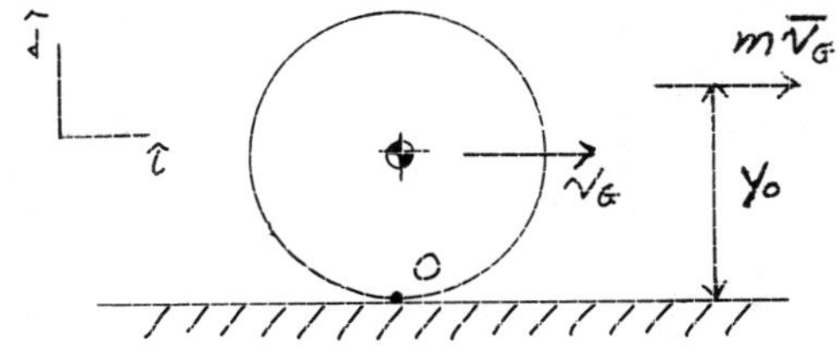

Figure a

$$\bar{Q} = m\bar{v}_G = \left(\frac{216}{9.81}\right)(3)\hat{\imath} = 66.055\,\hat{\imath} \text{ N·s}$$

Now, the angular momentum of the cylinder is the momentum of the linear momentum.

Hence,

$$A_o = (Q_G)\,y_o \circlearrowright$$

or

$$y_o = \frac{A_o}{Q_G} = \frac{14.862}{66.055}$$

$$\therefore y_o = 0.225 \text{ m}$$

Note that $y_o - r = 0.225 - 0.150 = 0.075 = y_G$ as obtained in Problem 19.77

19.79

GIVEN The core of a communication satellite has radius $r = 1.5$ ft, length $L = 4$ ft, and mass density $\rho = 1.5$ slug/ft³ (Fig. a). It spins about the y-axis at a rate of 8 rpm.

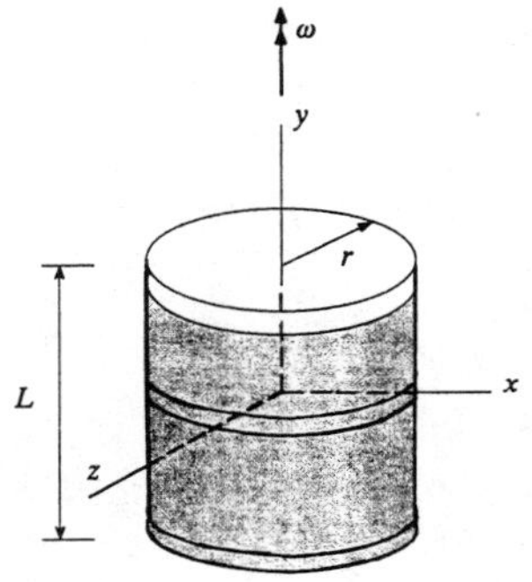

Figure a.

FIND Determine the angular momentum of the core.

SOLUTION

In rad/s, the spin rate is

$$\omega = \left(8\frac{\text{rev}}{\text{min}}\right)\left(2\pi\frac{\text{rad}}{\text{rev}}\right)\left(\frac{1\text{min}}{60\text{s}}\right)$$

$$= 0.8378 \text{ rad/s}$$

The mass moment of inertia about y, the spin axis, is

$$I_y = \frac{1}{2}mr^2$$

(continued)

19.79 cont.

where

$$m = \rho \times Vol = \left(1.5\ \frac{slugs}{ft^3}\right)\left(\pi(1.5)^2\right)(4)$$

$$= 42.412\ slugs$$

Hence,

$$I_y = \left(\frac{1}{2}\right)(42.412)(1.5^2) = 47.713\ slug\cdot ft^2$$

and

$$A_y = I_y\omega = (47.713)(0.8378) = 39.97\ lb\cdot ft\cdot s$$

19.80

GIVEN The spin of the communication satellite core of Problem 19.79 is changed to $\bar{\omega} = \hat{\imath} + 0.5\hat{\jmath}$.

FIND Determine the magnitude and the direction (relative to x, y, z axes) of the angular momentum of the core.

SOLUTION

The angular momentum of the core may be expressed as, since there is no rotation about the z axis,

$$\bar{A} = A_x\hat{\imath} + A_y\hat{\jmath}$$

where

$$A_x = I_x\omega_x$$

$$A_y = I_y\omega_y$$

and

$$\omega_x = 1\ rad/s, \quad \omega_y = 0.5\ rad/s$$

I_x = mass moment of inertia about the x axis

I_y = mass moment of inertia about the y axis

By Appendix F, Table F.1,

$$I_x = \frac{1}{12}m(3r^2 + L^2)$$

$$I_y = \frac{1}{2}mr^2$$

where

$$m = (\rho)(Vol) = \left(1.5\ \frac{slug}{ft^3}\right)(\pi\times 1.5^2)(4)$$

$$= 42.412\ slugs$$

$$r = 1.5 ft$$

Hence,

$$I_x = \frac{1}{12}(42.412)\left[(3)(1.5)^2 + 4^2\right]$$

$$= 80.406\ slug\cdot ft^2$$

$$I_y = \frac{1}{12}(42.412)(1.5)^2 = 47.714\ slug\cdot ft^2$$

Therefore,

$$A_x = (80.406)(1) = 80.406\ lb\cdot ft\cdot s$$

$$A_y = (47.714)(0.5) = 23.857\ lb\cdot ft\cdot s$$

Hence, the angular momentum of the core is

$$\bar{A} = 80.406\hat{\imath} + 23.857\hat{\jmath}\ [lb\cdot ft\cdot s]$$

The magnitude of $\bar{A}$ is

$$A = \sqrt{(80.406)^2 + (23.857)^2}$$

$$= 83.87\ lb\cdot ft\cdot s$$

The direction of $\bar{A}$ is given by

$$\tan\theta_x = \frac{23.857}{80.406} = 0.2967$$

So, $\theta_x = 16.526°$ (see Fig. a)

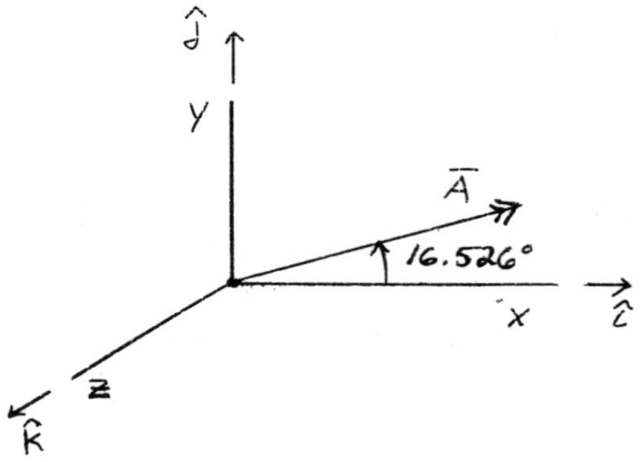

Figure a.

19.81

GIVEN Refer to Example 19.18.

FIND Determine the energy lost during the impact.

SOLUTION

At the instant before impact, the container is translating. Hence, its kinetic energy is

$$T_1 = \frac{1}{2} m v_{G1}^2 \tag{a}$$

where, from Example 19.18

$$m = 2054 \ \text{lb}\cdot\text{s}^2/\text{ft}$$
$$v_{G1} = 25.38 \ \text{ft/s} \tag{b}$$

By Eqs. (a) and (b),

$$T_1 = \frac{1}{2}(2054)(25.38)^2 = 6.615 \times 10^5 \ \text{ft}\cdot\text{lb} \tag{c}$$

At the instant after the impact, the container is translating and rotating. Hence, its kinetic energy is

$$T_2 = \frac{1}{2} m v_{G2}^2 + \frac{1}{2} I_G \omega_2^2 \tag{d}$$

where, from Example 19.18

$$\omega_2 = 0.984 \ \text{rad/s}$$
$$v_{G2} = \sqrt{34}\,\omega_2 = 5.738 \ \text{ft/s} \tag{e}$$
$$I_G = 5153 \ \text{lb}\cdot\text{ft}\cdot\text{s}^2$$

Equations (d) and (e) yield

$$T_2 = \frac{1}{2}(2054)(5.738)^2 + \frac{1}{2}(5153)(0.984)^2$$

or $$T_2 = 3.63 \times 10^4 \ \text{ft}\cdot\text{lb} \tag{f}$$

Therefore, by Eqs. (c) and (f), the energy lost during the impact is

$$T_1 - T_2 = 6.615 \times 10^5 - 3.63 \times 10^4$$
$$= 6.252 \times 10^5 \ \text{ft}\cdot\text{lb}$$

19.82

GIVEN Refer to Example 19.18

FIND Determine the angular velocity ω_3 of the container for the instant before the face ADEH strikes the slab.

SOLUTION

By Example 19.18, (see also Problem 19.81), for the instant after the impact,

$$\omega_2 = 0.984$$
$$v_{G2} = \sqrt{34}\,\omega_2 = 5.738 \ \text{ft/s} \tag{a}$$

Also, by Example 19.18,

$$I_G = 5153 \ \text{lb}\cdot\text{ft}\cdot\text{s}^2$$
$$m = 2054 \ \text{lb}\cdot\text{s}^2/\text{ft} \tag{b}$$

Hence, the kinetic energy at the instant after impact is, with Eqs. (a) and (b),

$$T_2 = \frac{1}{2} m v_{G2}^2 + \frac{1}{2} I_G \omega_2^2$$
$$= 3.630 \times 10^4 \ \text{ft}\cdot\text{lb} \tag{c}$$

At this instant its potential energy relative to the slab is (see Fig. E19.18d of Example 19.18)

$$V_2 = mg L_{AG*} \sin(45° + \theta)$$
$$= (2054)(32.2)(\sqrt{34}) \sin 75.96°$$

or $$V_2 = 3.741 \times 10^5 \ \text{ft}\cdot\text{lb} \tag{d}$$

Similarly for the instant before the container face ADEH strikes the slab, the kinetic energy and potential energy are

$$T_3 = \frac{1}{2} m v_{G3}^2 + \frac{1}{2} I_G \omega^2$$
$$V_3 = mg L_{AG*} \sin\theta$$

where $$v_{G3} = L_{AG*}\,\omega_3 = \sqrt{34}\,\omega_3$$

Hence,

$$T_3 = \frac{1}{2}(2054)(\sqrt{34}\,\omega_3)^2 + \frac{1}{2}(5153)\omega_3^2$$
$$= 37{,}494\,\omega^2 \tag{e}$$

(continued)

19.82 cont.

$$V_3 = (2054)(32.2)\sqrt{34}\sin 30.96^\circ$$
$$= 1.984 \times 10^5 \ \text{ft}\cdot\text{lb} \qquad (f)$$

Then, by the principle of conservation of energy,

$$T_2 + V_2 = T_3 + V_3 \qquad (g)$$

By Eqs. (c), (d), (e), (f), and (g) we obtain

$$4.104\times10^5 = 3.749\times10^4\omega^2 + 1.984\times10^5$$

or

$$\omega = 2.378 \ \text{rad/s}$$

19.83

GIVEN A rotor that weighs 260 N with a radius of gyration of 0.5m is mounted on a shaft with axial moment of inertia of 2.0 $kg\cdot m^2$ (Fig. a). The shaft is at rest when a torque $T = 25t$ [N·m], t=seconds is applied.

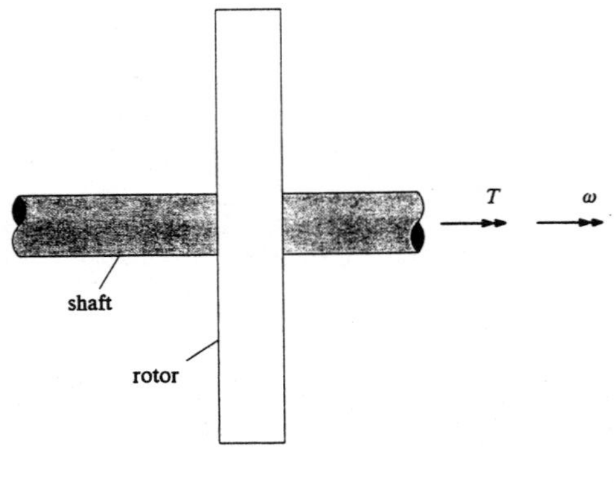

Figure a.

FIND

a.) Determine the angular momentum and the angular velocity of the rotor-shaft system as a function of time t.

b.) For $t = 4s$, how many revolutions has the rotor-shaft system undergone?

SOLUTION

a.) The moment of inertia of the rotor is

$$I_r = mk_o^2 = \left(\frac{260}{9.81}\right)(0.5)^2 = 6.626 \ kg\cdot m^2$$

The total moment of inertia of the system is

$$I = 6.626 + 2.0 = 8.63 \ kg\cdot m^2$$

So, since initially the system is at rest, the law of angular momentum yields

$$A_{o1} + \int_0^t M dt = A_{o2}$$

$$0 + \int_0^t 25t\,dt = I\omega$$

Therefore, the angular momentum of the shaft is,

$$A_{o2} = I\omega = 12.5t^2 \qquad (a)$$

and hence, by Eq (a), the angular velocity of the shaft is

$$\omega = \frac{12.5t^2}{I} = \frac{12.5t^2}{8.63} = 1.45t^2 \qquad (b)$$

b.) Since $\omega = d\theta/dt$, Eq. (b) yields

$$\theta = \int_0^4 \omega dt = \int_0^4 1.45t^2 dt = \left.\frac{1.45t^3}{3}\right|_0^4$$

or

$$\theta = 30.93 \text{ rads} = 4.92 \text{ rev}$$

So, the rotor-shaft system undergoes 4.92 revolutions in the first 4s.

19.84

GIVEN A rigid horizontal uniform slender tube, with mass $m_t = 0.010$ lb·s²/in, rotates freely in a horizontal plane about a vertical axis with frictionless bearings through the center of the tube. A steel ball bearing of mass $m_b = 0.0008$ lb·s²/in moves in the tube without friction. When the ball crosses the axis of rotation, its speed is $v_1 = 10$ ft/s and the angular velocity of the tube is $\omega_1 = 32$ rad/s.

FIND

a.) By the law of conservation of angular momentum, determine the angular velocity, ω_2, of the tube at the instant the ball reaches the end of the tube.

b.) By the law of conservation of energy, determine the speed, v_2, of the ball at the instant it reaches the end of the tube.

SOLUTION

a.) By the principle of conservation of angular momentum,

$$A_1 = A_2$$

or

$$I_{G1}\omega_1 = I_{G2}\omega_2 \qquad (a)$$

where, since the tube is slender (Table F.2, Appendix F)

$$I_{G1} \approx \frac{1}{12} m_t L^2 = \frac{1}{12}(0.01)(36)^2 = 1.080 \text{ lb·in·s}^2$$

$$\omega_1 = 32 \text{ rad/s} \qquad (b)$$

$$I_{G2} = I_{G1} + m_b r^2 = 1.080 + (0.0008)(18)^2 = 1.339 \text{ lb·in·s}^2$$

By Eqs. (a) and (b), we find

$$\omega_2 = \frac{I_{G1}\omega_1}{I_{G2}} = \frac{(1.080)(32)}{1.339} = 25.806 \text{ rad/s} \qquad (c)$$

b.) By the law of conservation of energy,

$$T_1 + V_1 = T_2 + V_2 \qquad (d)$$

where

$$T_1 = \frac{1}{2} m_b v_1^2 + \frac{1}{2} I_{G1}\omega_1^2$$

$$T_2 = \frac{1}{2} m_b v_2^2 + \frac{1}{2} I_{G1}\omega_2^2 \qquad (e)$$

and since the system rotates in a horizontal plane,

$$V_1 = V_2 \qquad (f)$$

By Eqs. (d), (e), and (f), we have

$$v_2^2 = v_1^2 + \frac{I_{G1}}{m_b}(\omega_1^2 - \omega_2^2) \qquad (g)$$

Finally, by Eqs. (b), (c), and (g)

$$v_2^2 = (120)^2 + \left(\frac{1.080}{0.0008}\right)(32^2 - 25.806^2)$$

or

$$v_2 = 705.53 \text{ in/s} = 58.79 \text{ ft/s}$$

19.85

GIVEN A uniform sphere, with a 300-mm diameter and 216-N weight, rolls to the right so that the velocity of its mass center G is $v_G = 3$ m/s.

FIND

a.) The sphere's angular momentum with respect to the horizontal axis that passes through G and that is perpendicular to the plane of motion.

b.) Repeat part a for the axis that passes through O, the instantaneous center of velocity.

c.) The magnitude, direction and line of action of the linear momentum of the sphere.

SOLUTION

a.) The sphere's angular momentum with respect to the axis through G is

$$A_G = I_G\omega = \left(\frac{2}{5}mr^2\right)\left(\frac{v_G}{r}\right) = \frac{2}{5}mrv_G$$

or

$$A_G = \frac{2}{5}\left(\frac{216}{9.81}\right)(0.15)(3) = 3.963 \text{ N·m·s} \circlearrowright$$

(continued)

19.85 cont.

b.) With respect to the axis O, the instantaneous center of velocity (the point of the sphere in contact with the horizontal plane);

$$A_o = I_o \omega$$

where

$$I_o = I_G + mr^2 = \frac{2}{5}mr^2 + mr^2 = \frac{7}{5}mr^2$$

$$\omega = v_G / r$$

Hence,

$$A_o = \frac{7}{5}(mr^2)\left(\frac{v_G}{r}\right) = \frac{7}{5} m r v_G$$

or

$$A_o = \left(\frac{7}{5}\right)\left(\frac{216}{9.81}\right)(0.15)(3) = 13.872 \text{ N·m·s} \circlearrowright$$

c.) The angular momentum with respect to the axis through G is also equal to the linear momentum, mv_G, times the distance from G; that is, the location of mv_G is given by

$$A_G = (mv_G)\, y_G \circlearrowright$$

or

$$y_G = \frac{A_G}{mv_G} = \frac{3.963}{\left(\frac{216}{9.81}\right)(3)} = 0.06 \text{ m above G}$$

Alternatively,

$$A_o = (mv_G)\, y_o \circlearrowright$$

or

$$y_o = \frac{A_o}{mv_G} = \frac{13.872}{\left(\frac{216}{9.81}\right)(3)} = 0.21 \text{ m above O}$$

Note that $y_o = r + y_G = 0.15 + 0.06 = 0.21$ m (see Fig. a)

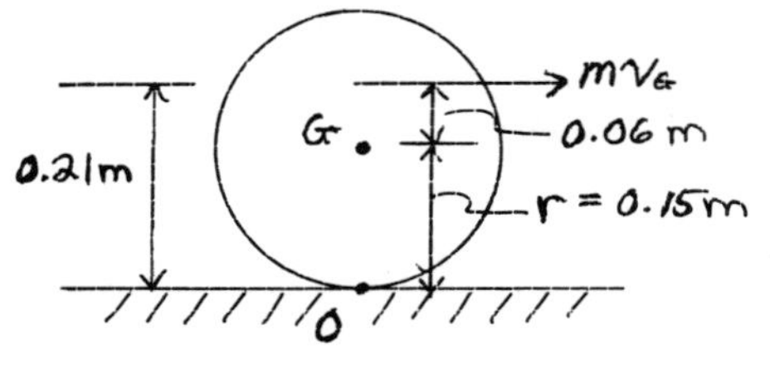

Figure a

19.86

GIVEN Refer to Example 19.18. At the instant before impact, the speed of the mass center of the container is $v_{G1} = 25.38$ ft/s and the counter-clockwise angular velocity of the container is $\omega_1 = 3$ rad/s.

FIND Determine the angular velocity ω_2 of the container for the instant after impact.

SOLUTION

By the principle of angular momentum, relative to the edge AD for the impact time $\Delta t \ll 1$,

$$A_{(AD)1} + \int_0^{\Delta t} (mg)(L_{AG^*}) \cos(45° + \theta)\, dt = A_{(AD)2} \quad \text{(a)}$$

where now (see Figs. a and b),

$$A_{(AD)1} = I_G \omega_1 + (mv_{G1})[(L_{AG^*})\cos(45° + \theta)]$$

$$A_{(AD)2} = I_G \omega_2 + (mv_{G2})(L_{AG^*}) \quad \text{(b)}$$

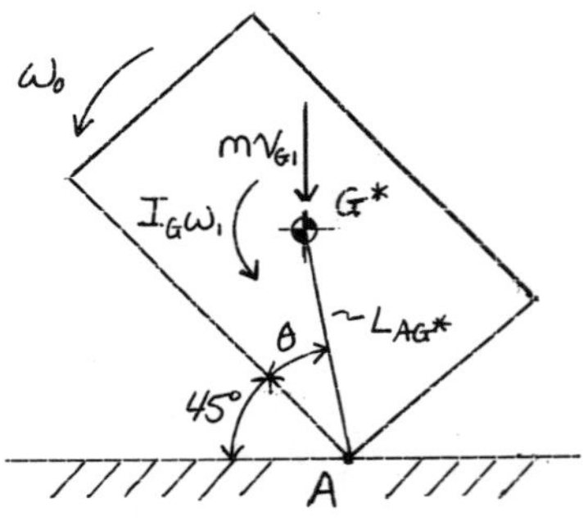

Figure a. Angular momentum at t = 0

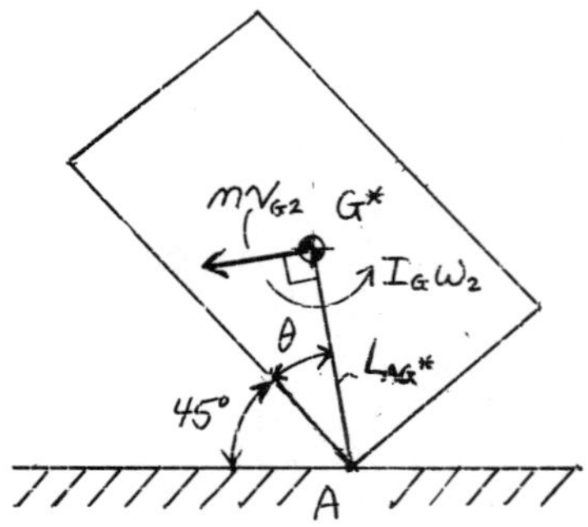

Figure b. Angular momentum at end of impact $t = \Delta t$

(continued)

19.86 cont.

Since mg is of ordinary magnitude (not an impact force) and since $\Delta t << 1$, the integral in Eq. (a) is negligible. Also, by kinematics, since the container rotates about the edge AD, $v_{G2} = (L_{AG*})\omega_2$. Hence, by Eqs. (a) and (b),

$$\omega_2 = \frac{I_G \omega_1 + (m v_{G1})[(L_{AG*})\cos(45° + \theta)]}{I_G + m(L_{AG*})^2} \qquad (c)$$

By the data of Example 19.18,

$$I_G = 5135 \text{ lb}\cdot\text{ft}\cdot\text{s}^2$$
$$m = 2054 \text{ lb}\cdot\text{s}^2/\text{ft} \qquad (d)$$
$$\theta = 30.96°$$
$$L_{AG*} = \sqrt{34} \text{ ft}$$

Hence, by Eqs. (c) and (d), with $\omega_1 = 3$ rad/s and $v_{G0} = 25.38$ ft/s,

$$\omega_2 = \frac{(5135)(3) + (2054)(23.38)(\sqrt{34})\cos 75.96°}{5135 + (2054)(34)}$$

or

$$\omega_2 = 1.189 \text{ rad/s}$$

19.87

GIVEN A large flywheel is subjected to moments $M_1 = k_1 v$ and $M_2 = k_2$, where k_1 and k_2 are constants (Fig. a; also refer to the problem statement for details on M_1 and M_2). The flywheel has radius r, moment of inertia I, and initial angular speed ω_0.

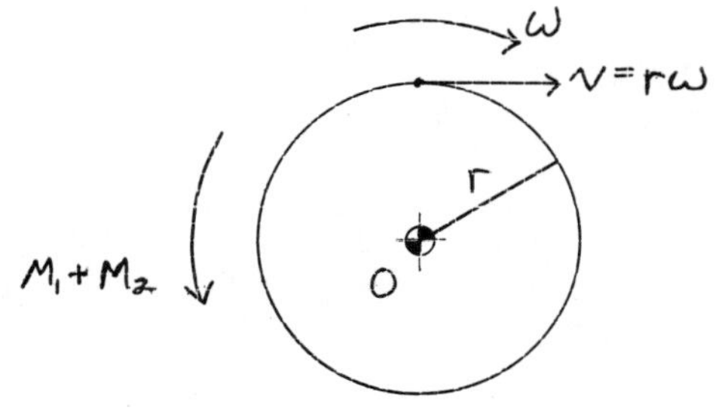

Figure a

FIND

a.) Determine the time required to stop the flywheel in terms of k_1, k_2, r, I, and ω_0.

b.) Determine the total work performed by the moments M_1 and M_2 in bringing the flywheel to a stop.

SOLUTION

a.) By Fig. (a) and the law of angular momentum,

$$\Sigma M_0 = -(k_2 + k_1 r \omega) = I \frac{d\omega}{dt}$$

or

$$I\,d\omega = -(k_2 + k_1 r\omega)\,dt$$

$$\int_0^{t_f} dt = -I \int_{\omega_0}^{0} \frac{d\omega}{k_2 + k_1 r \omega}$$

Integration yields

$$t_f = -\frac{I}{k_1 r}\left[\ln(k_2 + k_1 r\omega)\right]_{\omega_0}^{0}$$

or

$$t_f = \frac{I}{k_1 r} \ln\left(1 + \frac{k_1 r \omega_0}{k_2}\right)$$

b.) By definition of work of a couple,

$$\text{Work} = \int_0^{\theta_{max}} M\,d\theta = -\int_0^{\theta_{max}} (k_2 + k_1 r\omega)\,d\theta \qquad (a)$$

By the angular moment principle,

$$\Sigma M_0 = I\frac{d\omega}{dt} = I\frac{d\omega}{dt}\frac{d\theta}{d\theta} = I\omega\frac{d\omega}{d\theta}$$

or

$$-(k_2 + k_1 r\omega)\,d\theta = I\omega\,d\omega$$

and

$$d\theta = \frac{I\omega\,d\omega}{k_2 + k_1 r\omega} \qquad (b)$$

By Eqs (a) and (b) we obtain

$$\text{Work} = I\int_{\omega_0}^{0} \omega\,d\omega = -\tfrac{1}{2} I\omega_0^2 \qquad (c)$$

Alternatively, by work-energy we have directly,

$$\text{Work} = T_2 - T_1 = 0 - \tfrac{1}{2} I\omega_0^2 = -\tfrac{1}{2} I\omega_0^2$$

19.88

GIVEN A rod spins about its mass center on a frictionless horizontal floor, with initial angular velocity $\omega_0 = 120$ rpm (Fig. a). One end of the rod is stopped suddenly by impact with a bumper, and the rod rotates about point O; that is, the impact is fully plastic.

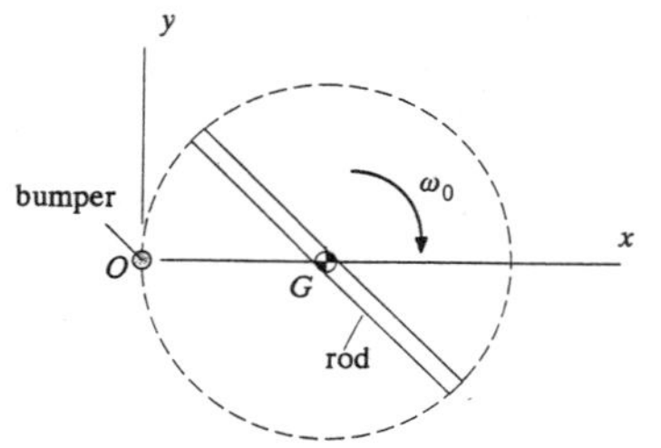

Figure a

FIND Determine the angular velocity ω of the rod following the collision

SOLUTION

Initially, the velocity of the mass center of the rod is zero. Hence, the initial angular momentum of the rod (taken positive clockwise) relative to point O is, by Fig. a and Eq (19.44),

$$A_{01} = m x_G v_{Gy} - m y_G \, v_{Gx} + I_G \omega_0$$
$$= I_G \omega_0$$

or

$$A_{01} = \frac{1}{12} m L^2 \omega_0 \qquad \text{(a)}$$

Since initially $\bar{v}_G = 0$ and hence $v_{Gy} = v_{Gx} = 0$. Upon collision, the mass center attains a velocity $\bar{v}_G$ and angular velocity ω (Fig. b), where F is the force exerted on the rod by the bumper.

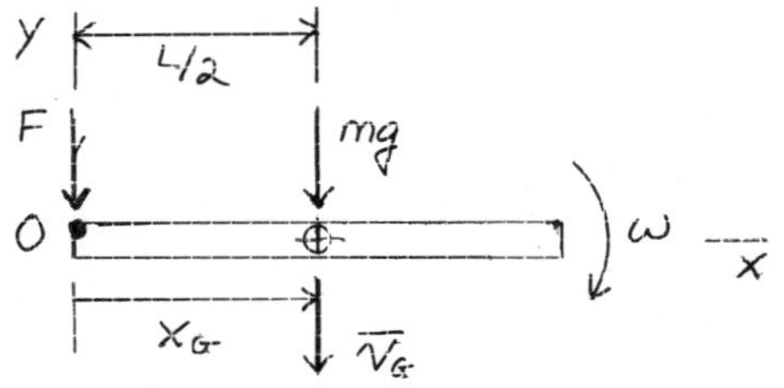

Figure b.

The angular momentum of the rod is now [see Eq. (19.44)], taking clockwise as positive,

$$A_{02} = m x_G v_G + I_G \omega$$
$$= m\left(\frac{L}{2}\right)\left(\frac{L}{2}\omega\right) + \frac{1}{12} m L^2 \omega$$

or

$$A_0 = \frac{1}{3} m L^2 \omega \qquad \text{(b)}$$

since $x_G = \frac{L}{2}$ and $v_G = \left(\frac{L}{2}\right)\omega$, since the rod rotates about point O.

By the law of angular momentum [Eq. (19.37)],

$$A_{01} + \int_0^{\Delta t} M_0 \, dt = A_{02} \qquad \text{(c)}$$

where Δt is the impact time ($\Delta t \ll 1$). By Fig. b, $\int_0^{\Delta t} M_0 \, dt = \int_0^{\Delta t} (mg)\left(\frac{L}{2}\right) dt$ is negligible since $\Delta t \ll 1$ and mg is of ordinary magnitude; that is, mg is not an impulse force. Hence, Eqs. (a), (b), and (c) yield

$$A_{01} = A_{02}$$
$$\frac{1}{12} m L^2 \omega_0 = \frac{1}{3} m L^2 \qquad \text{(d)}$$

or

$$\omega = \frac{1}{4}\omega_0 = \frac{1}{4}(120) = 30 \text{ rev/min}, \ \text{CW}$$

Note: Equation (d) signifies that angular momentum is conserved.

19.89

GIVEN The mass m_1 is released from rest and the cylinder (mass m_2) rolls without slipping (Fig. a).

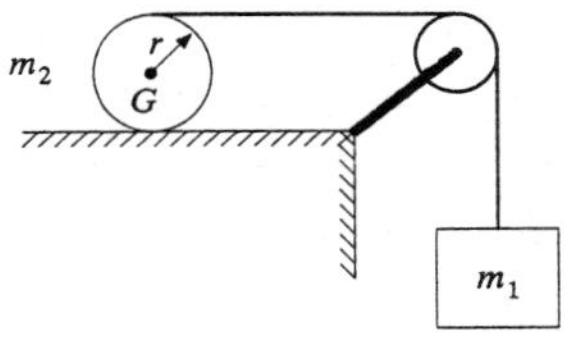

Figure a.

FIND

a.) Determine the speeds v_1 and v_2 as functions of t in terms of m_1, m_2, r and g

b.) Calculate v_1, v_2 for $m_1 g = 200$ lb, $m_2 g = 100$ lb, and $g = 32.2$ ft/s²

SOLUTION

a.) Consider the free-body diagrams of m_1 and m_2 (Fig. b and c.),

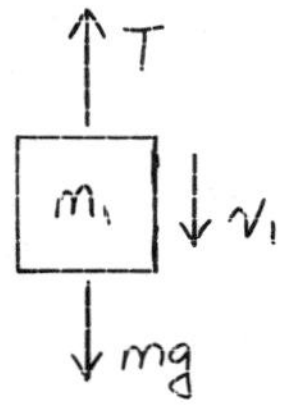

Figure b.

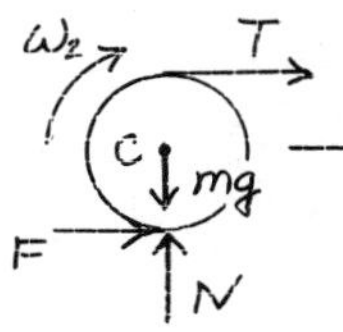

Figure c

By Figure (b), the principle of linear momentum yields

$$\int_0^t (mg - T)\,dt = m_1 v_1$$

or

$$(m_1 g - T)t = m_1 v_1 \qquad (a)$$

By Fig. (c), the principle of linear momentum yields

$$\int_0^t (T+F)\,dt = m_2 v_2$$

or

$$(T+F)t = m_2 v_2 \qquad (b)$$

and similarly the principle of angular momentum yields

$$\int_0^t (T-F)r\,dt = I\omega_2 = \frac{1}{2} m_2 r^2 \omega_2 = \frac{1}{2} m_2 r v_2$$

or

$$(T-F)t = \frac{1}{2} m_2 v_2 \qquad (c)$$

By kinematics (Fig. a), $v_1 = 2v_2$, There fore, Eq. (a) becomes

$$(m_1 g - T)t = 2m_1 v_2 \qquad (d)$$

Equations (b), (c), and (d) may be solved for the quantities T, F, and v_2 in terms of t. Since we are interested in v_2, we solve to obtain v_2. Thus,

$$v_2 = \frac{4m_1 g t}{8m_1 + 3m_2}$$

$$v_1 = 2v_2 = \frac{8m_1 g t}{8m_1 + 3m_2}$$

The radius r does not enter into the formulas.

b.) For $m_1 g = 200$ lb, $m_1 = 200/g$, $m_2 = 100/g$, and $t = 2$s,

$$v_1 = 54.23 \text{ ft/s}; \quad v_2 = 27.12 \text{ ft/s}$$

19.90

GIVEN A block of mass m moves on a frictionless horizontal surface (Fig a).

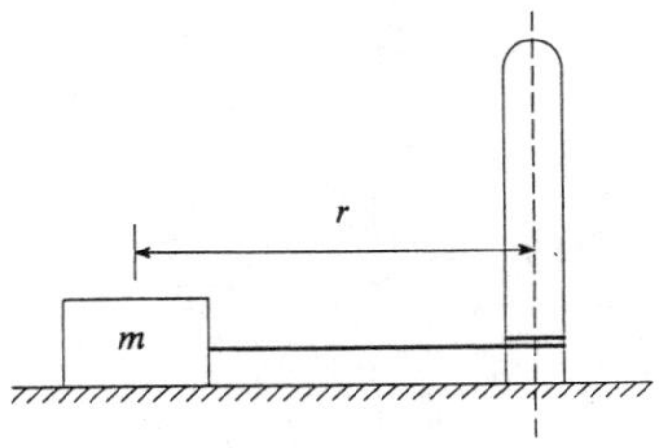

Figure a

(continued)

19.90 cont.

It is fastened to one end of an inextensional string that wraps around a vertical pole. The string remains taut as the mass moves. When $r = 1.8$ m, the angular speed of the string is $\omega = 12$ rad/s

FIND

a.) Determine the angular speed, ω_2, of the string for $r = 0.9$ m. Assume that the radius of the pole is small compared to r.

b.) Determine the angular momenta, with respect to the axis of the pole, of the mass at $r = 1.8$m and at $r = 0.9$m

c.) Explain why angular moment of the mass relative to the axis of the pole is not conserved.

SOLUTION

a.) Since there are no net external forces acting on the system consisting of the mass, the string, and the pole, Kinetic energy is conserved.

$$T_1 = T_2$$

$$\therefore \frac{1}{2} m v_1^2 = \frac{1}{2} m v_2^2 \rightarrow v_1 = v_2$$

Neglecting the radius of the pole, we have

$$r_1 \omega_1 = r_2 \omega_2$$

or

$$(1.8)(12) = (0.9)\omega_2$$

$$\therefore \omega_2 = 24 \text{ rad/s}$$

b.) For $r_1 = 1.8$ m, $\omega_1 = 12$ rad/s. Therefore,

$$A_1 = (m r_1^2)\omega_1 = m(1.8)^2(12) = 38.88 \text{ N·m·s}$$

For $r_2 = 0.9$m, $\omega_2 = 24$ rad/s, Therefore,

$$A_2 = (m r_2^2)\omega_2 = m(0.9)^2(24) = 19.44 \text{ N·m·s}$$

c.) Since $A_1 \neq A_2$, angular momentum is not conserved. The reason that angular momentum is not conserved is that the foundation exerts a torque on the pole. Hence, the moment of the external force about the axis of the pole is not zero. Therefore, the angular momentum of the system consisting of the mass, string, and pole is not conserved.

19.91

GIVEN A rigid, uniform slender rod of length L (m) falls in a horizontal attitude (Fig. a). One end of the rod strikes the edge O of a ledge. The coefficient of restitution is e, and the speed of the rod is v_1 just before it strikes the ledge.

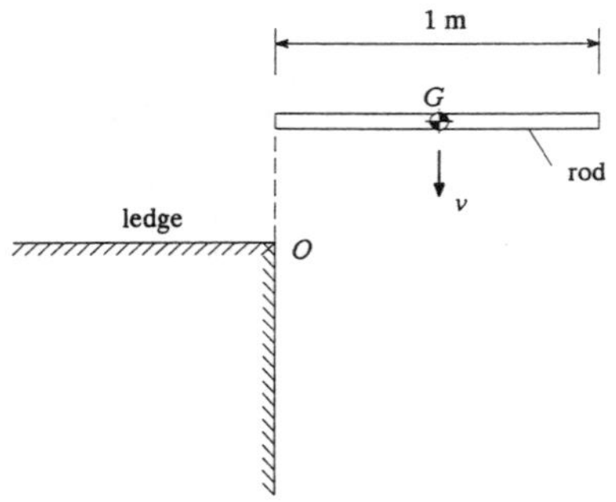

Figure a.

FIND

a.) Derive formulas in terms of v_1, e, and L for the angular speed ω_2 of the rod and the speed v_2 of the mass center G of the rod immediately after the collision.

b.) For $v_1 = 15$ m/s and $L = 1$m, calculate ω_2 and v_2 for $e = 1$, 0.5, and 0.

SOLUTION

Consider the free-body diagram of the rod during the impact (Fig. b).

(continued)

19.91 cont.

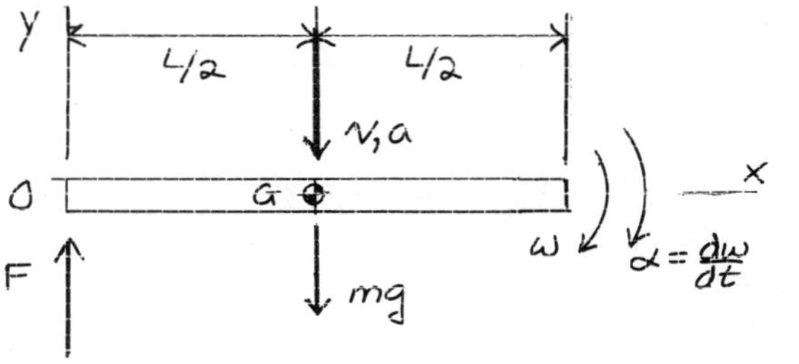

Figure b.

By Fig. b, since $F \gg mg$

$$\Sigma F_y = F = -ma = -m\frac{dv}{dt} \qquad (a)$$

$$\Sigma M_G = F\frac{L}{2} = I_G\alpha = \frac{1}{12}mL^2\frac{d\omega}{dt} \qquad (b)$$

The linear impulse during the time of compression is (see section 16.5 and Fig. 16.15a)

$I_c = \int_0^{t_{cr}} F\,dt$. Hence, by Eq. (a),

$$I_c = \int_0^{t_{cr}} F\,dt = -m\int_{v_1}^{v} dv = m(v_1 - v) \qquad (c)$$

where v is the speed of the center of mass at time t_{cr} (the end of the compression phase).

The linear impulse during the rebound phase is

$I_r = \int_{t_{cr}}^{t_f} F\,dt = e\,I_c$. Hence, by Eq. (a),

$$e\,I_c = \int_{t_{cr}}^{t_f} F\,dt = -\int_{v}^{v_2} m\,dv = m(v - v_2) \qquad (d)$$

where t_f is the time at which the end of the rod leaves the ledge. Similarly, by Eq. (b), we obtain the angular impulse relations during the compression phase.

$$I_{Ac} = \int_0^{t_{cr}} F\frac{L}{2}dt = \frac{1}{12}mL^2\int_0^{\omega} d\omega = \frac{1}{12}mL^2\omega$$

where ω is the angular velocity of the rod at $t = t_{cr}$.

Hence,

$$\int_0^{t_{cr}} F\,dt = \frac{1}{6}mL\omega = I_c \qquad (e)$$

During the rebound phase, we have,

$$I_{Ar} = \int_{t_{cr}}^{t_f} F\frac{L}{2}dt = \frac{1}{12}mL^2\int_{\omega}^{\omega_2} d\omega = \frac{1}{12}mL^2(\omega_2 - \omega)$$

or

$$\int_{t_{cr}}^{t_f} F\,dt = \frac{1}{6}mL(\omega_2 - \omega) = e\,I_c \qquad (f)$$

Also, by kinematics,

$$v = \frac{L}{2}\omega \qquad (g)$$

since the end of the rod in contact with the edge of the ledge is at rest when $t = t_{cr}$ (the end of the compression period).

Equations (c), (d), (e), (f), and (g) yield, after some algebra,

$$v_2 = \frac{1}{4}(3 - e)v_1$$

$$\omega_2 = \frac{3}{2L}(1 + e)v_1 \qquad (h)$$

b.) By Eq. (h), with $v_1 = 15$ m/s and $L = 1$m,

$$v_2 = 11.25 - 3.75e \quad (\text{m/s})$$
$$\omega_2 = 22.5 + 22.5e \quad (\text{m/s}) \qquad (i)$$

Thus, with Eqs. (i),

$v_2 = 7.5$ m/s, $\omega_2 = 45$ rad/s } $e = 1$

$v_2 = 9.375$ m/s, $\omega_2 = 33.75$ rad/s } $e = 0.5$

$v_2 = 11.25$ m/s, $\omega_2 = 22.5$ rad/s } $e = 0$

Note that in the case $e = 0$, the rod rotates about O. Therefore, angular momentum is conserved, and

$$A_{O1} = A_{O2} \qquad (j)$$

(continued)

19.91 cont.

and $A_{O1} = m v_1 \frac{L}{2}$

$$A_{O2} = \frac{1}{3} m L^2 \omega_2 = \frac{1}{3} m L^2 \left(\frac{2v_2}{L}\right) = \frac{2}{3} m L v_2$$

and by Eq. (j), as before,

$$v_2 = \frac{3}{4} v_1 = \frac{3}{4}(15) = 11.25 \text{ m/s}$$

$$\omega_2 = \frac{2v_2}{1} = 22.5 \text{ rad/s}$$

19.92

GIVEN The system shown in Fig. a, with $m_1 = 0.5$ slug, $m_2 = 1.0$ slug, and $m_3 = 0.70$ slug is released from rest at time $t = 0$. Neglect friction.

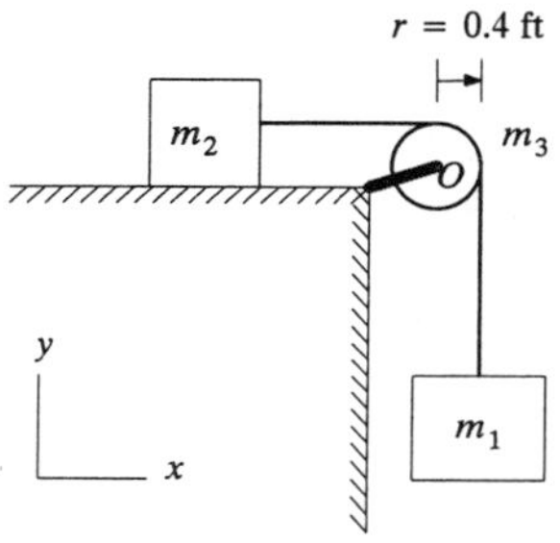

Figure a.

FIND

a.) Using momentum principles, determine the speed of the masses at time $t = 3$ s and the tension T_1 and T_2 exerted by the cord on m_1 and m_2.

b.) Check the result of part a by $\bar{F} = m\bar{a}$ and $\bar{M} = I\bar{\alpha}$ applied to the masses and the pulley

SOLUTION

a.) First, apply the law of linear momentum to the system. The linear momentum of the system is the linear momentum of masses m_1 and m_2, since the pulley's mass center (its geometrical center) is fixed. Thus, at time $t = 3$ s,

for m_1:

$$\bar{Q}_{m_1} = -m_1 v_{(3)} \hat{j} \quad (a)$$

for m_2:

$$\bar{Q}_{m_2} = m_2 v_{(3)} \hat{i} \quad (b)$$

where $v_{(3)}$ is the speed at $t = 3$ s.

To determine the linear impulse of the system, consider the free-body diagrams of m_1 and m_2 (Fig. b).

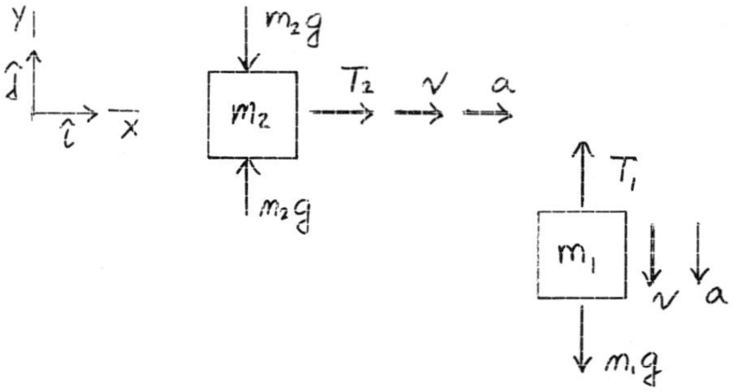

Figure b.

The linear impulse of the system is, by Fig. b,

$$\bar{I}_L = \left(\int_0^3 T_2\,dt\right)\hat{i} - \left(\int_0^3 (m_1 g - T_1)\,dt\right)\hat{j}$$

or

$$\bar{I}_L = 3T_2\hat{i} - 3(m_1 g - T_1)\hat{j} \quad (c)$$

By the law of linear momentum,

$$m_2 v_{(0)}\hat{i} + m_1 v_{(0)}\hat{j} + I_L = m_2 v_{(3)}\hat{i} - m_1 v_{(3)}\hat{j} \quad (d)$$

where $v_{(0)} = 0$ is the speed at $t = 0$. Hence, with $m_1 = 0.5$ slug, $m_2 = 1.0$ slug, Eqs. (c) and (d) yield,

$$3T_2\hat{i} - 3(m_1 g - T_1)\hat{j} = m_2 v_{(3)}\hat{i} - m_1 v_{(3)}\hat{j}$$

or

$$\left.\begin{aligned} 3T_2 &= (1.0)v_{(3)} \\ 3T_1 &= -(0.5)v_{(3)} + 48.3 \end{aligned}\right\} \quad (e)$$

(continued)

19.92 cont.

Equations (e) are two equations in three unknowns [T_1, T_2, and $v_{(3)}$]. To obtain a third equation in these unknowns, consider the principle of angular momentum relative to point O, the center of the pulley (Fig. c).

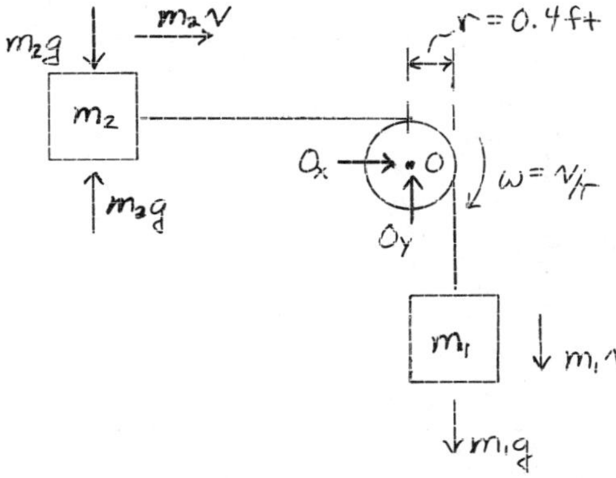

Figure c.

At time $t = 3$ s, the angular momentum about point O is

$$A_{2O} = (m_2 v)r + (m_1 v)r + \frac{1}{2} m_3 r v \qquad (f)$$

and the angular impulse is

$$\int_0^3 M_O dt = \int_0^3 m_1 g r dt = 3 m_1 g r \qquad (g)$$

The law of angular momentum is

$$A_{1O} + \int M_O dt = A_{2O} \qquad (h)$$

But $A_{1O} = 0$, since the system starts from rest at $t = 0$. Hence, Eqs. (f) and (g) yield

$$3 m_1 g = (m_1 + m_2 + \tfrac{1}{2} m_3) v_{(3)}$$

or

$$v_{(3)} = \frac{3(0.5)(32.2)}{0.5 + 1.0 + 0.35} = 26.11 \text{ ft/s}$$

Hence, by Eqs. (e),

$T_1 = 11.75$ lb $\qquad T_2 = 8.70$ lb

b.) By Fig. b,

$$\Sigma F_x = T_2 = m_2 a$$

$$\Sigma F_y = m_1 g - T_1 = m_1 a$$

or

$$T_2 = a$$

$$T_1 = m_1(g - a) = 16.1 - 0.5a \qquad (i)$$

By the free-body diagram of the pulley (Fig. d),

$$\Sigma M_O = (T_1 - T_2) r = \tfrac{1}{2} m_3 r^2 \alpha = \tfrac{1}{2} m_3 r a$$

or

$$T_1 - T_2 = \tfrac{1}{2}(m_3) a = 0.35 a \qquad (j)$$

The solution of Eqs. (i) and (j) is

$T_1 = 11.75$ lb

$T_2 = 8.703$ lb

$a = 8.703$ ft/s²

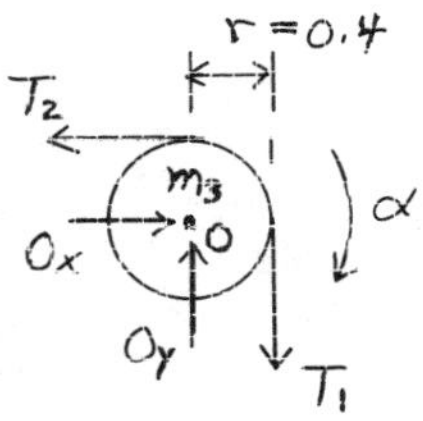

Figure d

Hence,

$$v_{(3)} = at = (8.703)(3) = 26.11 \text{ ft/s}$$

These results agree with the results of part a.

19.93

GIVEN Two monkeys of equal mass m_m hang from a rope that passes over a wheel of radius r and mass m_w (Fig. a). The monkey at A climbs up the rope with speed v_A relative to the rope, while the monkey at B continues to hang on the rope at B. The wheel bearing at O is frictionless, and the rope does not slip on the wheel.

(continued)

19.93 cont.

Figure a

FIND

a.) Using momentum principles, derive a formula for the velocity $\bar{v}_B$ of the monkey at B in terms of m_m, m_w, r, and v_A. Assume that the mass m_w of the wheel is uniformly distributed around the rim and the mass of the spokes is negligible.

b.) Plot the ratio v_B/v_A of the speeds of the monkeys as a function of the ratio m_w/m_m for the range $0 \le m_w/m_m \le 8$, and describe what happens as $m_w/m_m \to \infty$.

SOLUTION

a.) Consider the free-body diagrams of the monkeys and the wheel (Figs. b, c, and d).

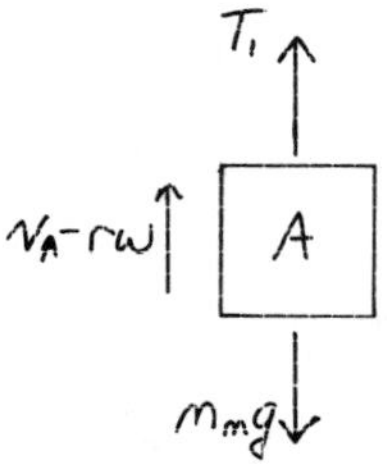

Figure b

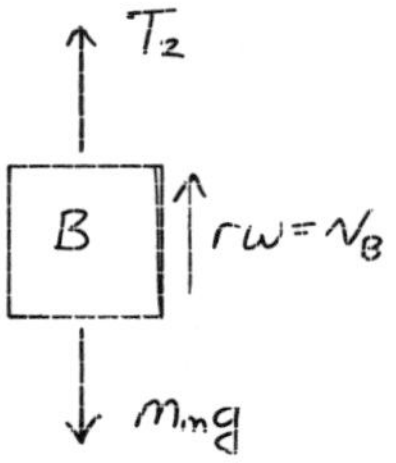

Figure c

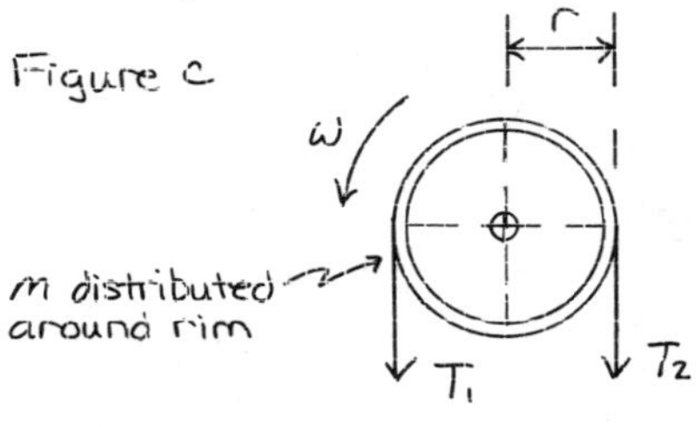

Figure c

By Figs. b. and c, and the law of linear momentum,

$$\int (T_1 - m_m g)\,dt = (m_m)(v_A - r\omega) \qquad (a)$$

$$\int (T_2 - m_m g)\,dt = (m_m)(r\omega) \qquad (b)$$

By Fig. d and the law of angular momentum,

$$\int (T_1 - T_2) r\,dt = (m_w r^2)\omega = m_w r\, v_B \qquad (c)$$

Subtracting Eq. (b) from Eq (a), we find:

$$\int (T_1 - T_2)\,dt = m_m (v_A - 2r\omega)$$
$$= m_m (v_A - 2v_B) \qquad (d)$$

By Eqs. (c) and (d), cancelling r from Eq. (c), we obtain

$$v_B = \frac{v_A}{2 + \frac{m_w}{m_m}}$$

or

$$\frac{v_B}{v_A} = \frac{1}{2 + m_w/m_m} \qquad (d)$$

b.) See Fig. e for the plot of Eq. (d) for $0 \le m_w/m_m \le 8$.

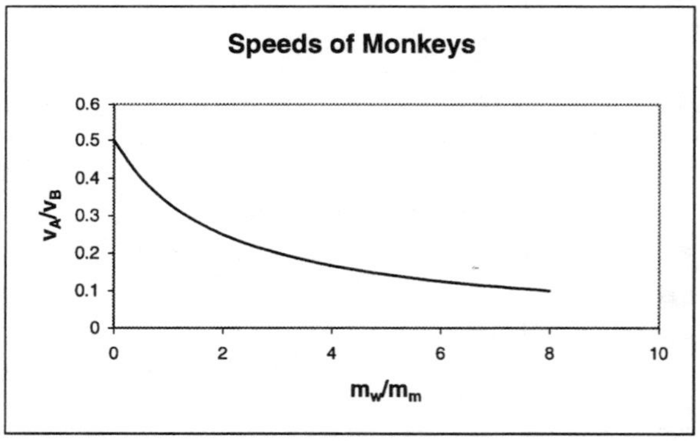

Figure d

As $m_w/m_m \to \infty$, $v_B \to 0$. Hence, $\omega \to 0$, and monkey A has absolute speed v_A up the rope which is at rest, and monkey B does not move, but stays at point B on the stationary rope.

19.94

GIVEN A proposed design for a speed regulator consists of a thin homogeneous rod, of length $2L$ and weight W, that rotates about axis s–s (Fig. a). The two balls, each of mass m, can slide along the rod. Initially, they are held by a cord at equal distances ℓ from the axis of rotation. When the angular velocity ω reaches a critical value ω_{cr}, the cord should break and the balls should slide to the stops A and B, where they will remain, reducing the rotational speed of the system. The system is tested by spinning the rod about axis s–s until the cord breaks.

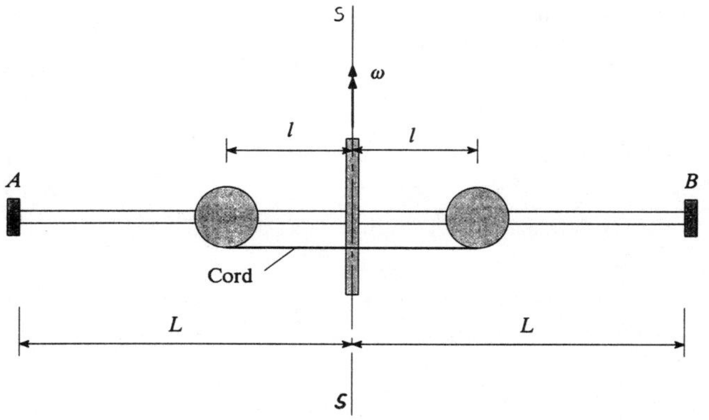

Figure a.

FIND

a.) Derive a formula for the final angular velocity ω of the system in terms of W, m, L, ℓ, and ω_{cr}. Treat the masses as particles and assume that friction and the masses of the cord and stops are negligible.

b.) Derive a formula for ℓ/L in terms of W, m, ω_{cr}, and ω, using the results of part a.

c.) What restrictions must be placed on W, m, and ℓ for the system to work properly, if the final rotational speed is to be reduced by 50% from ω_{cr}.

d.) Determine the required ratio ℓ/L for $W = 20$ N, $mg = 40$ N, and $\omega/\omega_{cr} = 0.50$.

e.) Will the system work properly with the ratio determined in part d for $\omega/\omega_{cr} = 0.50$, if $W = 40$ N and $mg = 20$ N? Explain.

SOLUTION

a.) Angular velocity of the system is conserved. Therefore, initially the angular momentum relative to axis s–s is, when the cord breaks,

$$A_i = (I + 2m\ell^2)\omega_{cr} = \left[\frac{1}{12}\frac{W}{g}(2L)^2 + 2m\ell^2\right]\omega_{cr} \quad (a)$$

The final angular momentum is,

$$A_f = \left[\frac{1}{12}\frac{W}{g}(2L)^2 + 2mL^2\right]\omega \quad (b)$$

Since angular momentum is conserved, $A_i = A_f$, and by Eqs. (a) and (b),

$$\frac{\omega}{\omega_{cr}} = \frac{W + 6mg\left(\frac{\ell}{L}\right)^2}{W + 6mg} \quad (c)$$

b.) By Eq. (c),

$$\frac{\ell}{L} = \sqrt{\left(1 + \frac{W}{6mg}\right)\frac{\omega}{\omega_{cr}} - \frac{W}{6mg}} \quad (d)$$

c.) By Eq. (c),

$$\frac{\omega}{\omega_{cr}} = \frac{\frac{WL^2}{mg} + 6\ell^2}{\frac{WL^2}{mg} + 6L^2} \leq \frac{1}{2}$$

or

$$\frac{W}{mg} + \frac{12\ell^2}{L^2} \leq 6 \quad (e)$$

d.) For $\omega/\omega_{cr} = 0.50$, $W = 20$ N, $mg = 40$ N, Eq. (e) yields

$$\frac{20}{40} + 12\frac{\ell^2}{L^2} = 6$$

or

$$\frac{\ell}{L} = 0.677$$

(continued)

19.94 cont.

e.) For $\omega/\omega_{cr} = 0.50$, $W = 40$ N, and $mg = 20$ N, Eq. (e) yields, for $\omega/\omega_{cr} \leq 0.50$

$$\frac{40}{20} + \frac{12\ell^2}{L^2} \leq 6 \qquad (f)$$

or with $\ell/L = 0.677$

$$2 + 5.5 = 7.5 > 6$$

Hence, the system will not reduce the angular speed by 50%, since it does not satisfy Eq. (f).

19.96

GIVEN A grinder bears on a fixed steel plate. The peripheral speed of the grinder is 100 ft/s, and the frictional force is 10 lb.

FIND

a.) How much work does the grinder perform on the plate per second?

b.) How much work does the plate perform on the grinder?

SOLUTION

The work performed per second by a force is

$$\text{Work/s} = \bar{F} \cdot \bar{v} \qquad (a)$$

where $\bar{v}$ is the velocity of the particle on which the force $\bar{F}$ acts. Note it is not the velocity of the geometrical point of action of $\bar{F}$. Hence,

a.) By Eq. (a), the work the grinder performs on the plate is zero, since the particle of the plate on which the grinder acts is zero.

b.) By Eq. (a), the work the plate performs on the grinder is (Fig. a)

$$\text{Work/s} = (-F)v = (-10)(100) = -1000 \text{ ft·lb/s}$$

since the velocity of the particle on which the plate acts is 100 ft/s and F is directed opposite to v.

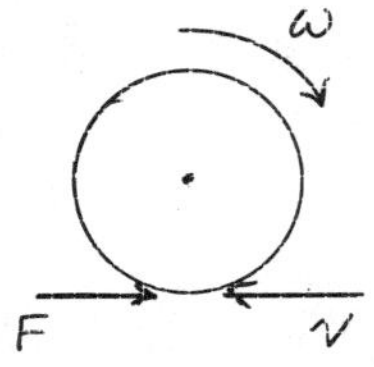

Figure a.

19.100

GIVEN A simple pendulum swings about its pivot with angular velocity ω and angular acceleration α.

FIND Use the principle of conservation of mechanical energy to derive the equation for angular velocity ω. Differentiate this equation to obtain the equation for the angular acceleration α of the pendulum.

SOLUTION

Figure a is a sketch of the pendulum in an initial position θ_0. Let the pendulum be released from rest from this position. Figure b. is a sketch of the pendulum in position θ.

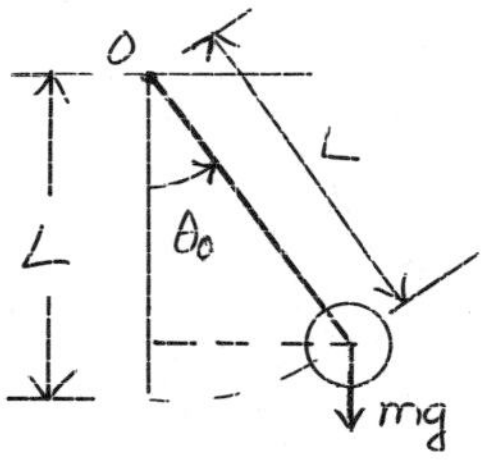

Figure a.

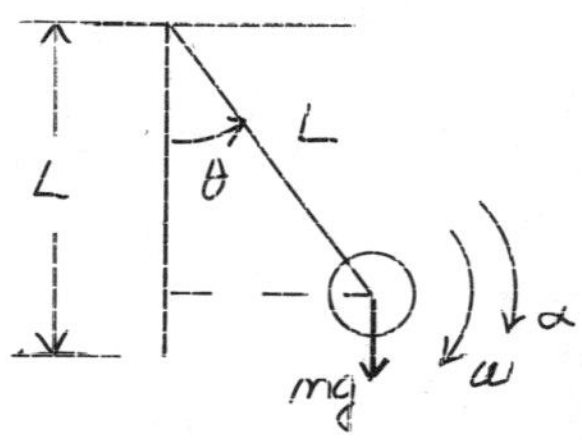

Figure b.

By the principle of conservation of energy,

$$T_1 + V_1 = T_2 + V_2 \qquad (a)$$

where

$$T_1 = 0, \quad V_1 = mgL(1 - \cos\theta_0)$$

$$T_2 = \frac{1}{2}m(L\omega)^2 \qquad (b)$$

$$V_2 = mgL(1 - \cos\theta)$$

By Eqs. (a) and (b)

$$g(1 - \cos\theta_0) = \frac{1}{2}L\omega^2 + g(1 - \cos\theta) \qquad (c)$$

(continued)

19.100 cont.

This equation defines ω; that is

$$\omega^2 = \frac{2g}{L}(\cos\theta - \cos\theta_0) \qquad (d)$$

Since θ_0 is a constant, differentiation of Eq. (d) yields, since $\dot{\theta} = -\omega$ (Fig. b),

$$2\omega \frac{d\omega}{dt} = \frac{2g\dot{\theta}}{L}(-\sin\theta) = \frac{2g\omega}{L}\sin\theta$$

or

$$\alpha = \frac{d\omega}{dt} = 2\frac{g}{L}\sin\theta$$

19.101

GIVEN A belt driver pulley is 0.6 m in diameter. The pulley rotates at 1200 rpm. The tensions in the tight and slack sides of the belt are 2 kN and 800 N respectively.

FIND Determine the horsepower transmitted to the pulley.

SOLUTION

By Fig. a,

$$\Sigma M_0 = (T_1 - T_2) r = M \qquad (a)$$

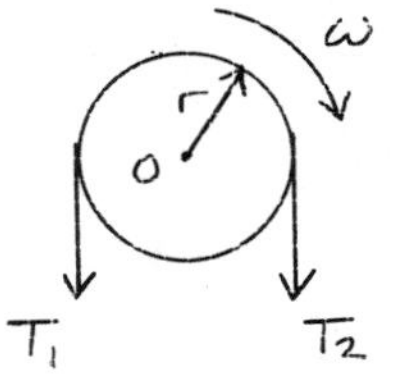

Figure a.

The power transmitted is, with Eq. (a),

$$E = M\omega = (T_1 - T_2) r\omega \qquad (b)$$

With the given data, Eq. (b) yields

$$E = (2000 - 800)(0.3)\left(1200 \cdot \frac{\pi}{30}\right)$$

or

$$E = 45.239 \text{ kW} \qquad (c)$$

Then, since 1 hp = 0.746 kW, by Eq (c)

$$E = \frac{45.239}{0.746} = 60.64 \text{ hp}$$

19.102

GIVEN A baton, consisting of two equal masses connected by a rigid rod of negligible mass, is thrown end over end into the air.

FIND Show that the tension in the rod performs no net work on the masses.

SOLUTION

The tension in the rod exerts forces $\bar{F}$ and $-\bar{F}$ on the masses (Fig. a)

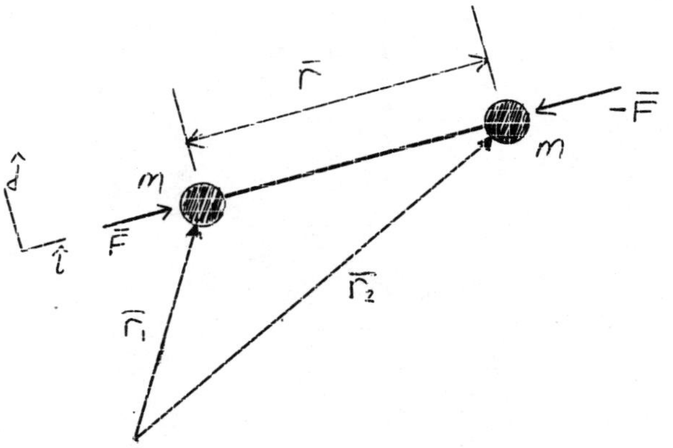

Figure a.

By definition of work of a force, the increment of work that $\bar{F}$ and $-\bar{F}$ perform on the mass is

$$dU_{1-2} = \bar{F}\cdot d\bar{r}_1 - \bar{F}\cdot d\bar{r}_2$$
$$= \bar{F}\cdot(d\bar{r}_1 - d\bar{r}_2) \qquad (a)$$

Now, by Fig. a, $\bar{r}_1 + \bar{r} = \bar{r}_2$ or

$$\bar{r}_1 - \bar{r}_2 = -\bar{r} \qquad (b)$$

Then, by Eqs. (a) and (b)

$$dU_{1-2} = -\bar{F}\cdot d\bar{r} \qquad (c)$$

Also, by Fig. a, since $\bar{F}$ is directed along the rod,

$$\bar{F} = F\frac{\bar{r}}{r} \qquad (d)$$

where r is the length of the rod.

(continued)

19.102 cont.

Therefore, by Eqs. (c) and (d),

$$dU_{1-2} = -\frac{F}{r}\bar{r}\cdot d\bar{r} \qquad (e)$$

and since $\bar{r}\cdot\bar{r} = r^2 =$ constant,

$$d(\bar{r}\cdot\bar{r}) = \bar{r}\cdot dr + d\bar{r}\cdot\bar{r}$$
$$= 2\bar{r}\cdot d\bar{r} = 0 \qquad (f)$$

Hence, by Eqs. (e) and (f),

$$dU_{1-2} = 0$$

or for any finite displacement, the work performed by the tension in the rod is

$$U_{1-2} = \int dU_{1-2} = 0$$

See also Section 19.3, subsection–Work Performed on a Rigid Body by Internal Forces.

20.1

Given: The system shown in Fig. a.

Derive the differential equation of motion for vertical vibrations of the spring/mass system shown in Fig. a. Measure x from the static equilibrium configuration.

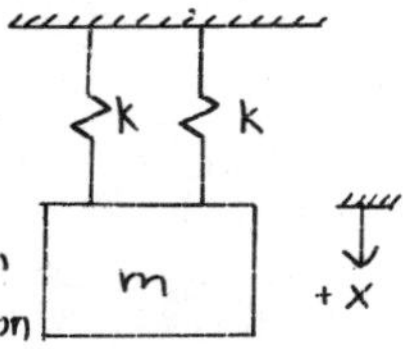

Figure a

Solution:

Consider the free-body diagram (Fig. b) of the mass in the static equilibrium position, where, Δ is the elongation of each spring.

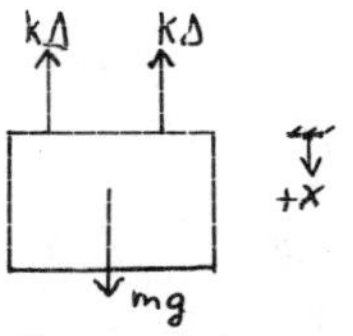

Figure b

For equilibrium: (Fig. b)

$$\Sigma F_x = mg - 2k\Delta = 0 \quad \text{or} \quad k\Delta = \tfrac{1}{2}mg \qquad \text{(a)}$$

Next, Consider the free-body diagram (Fig. c) of the mass displaced an amount x from its static equilibrium position. By Newton's second law: (Fig. c)

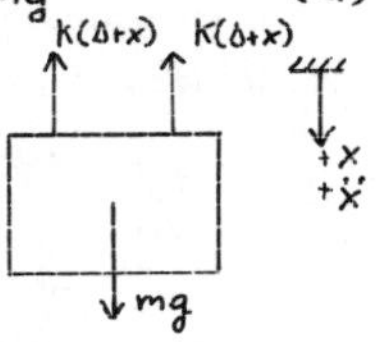

Figure c

$$\Sigma F_x = mg - 2k(\Delta + x) = m\ddot{x}$$

$$mg - 2k\Delta - 2kx = m\ddot{x} \qquad \text{(b)}$$

By Eqs. (a) and (b), we find,

$$mg - mg - 2kx = m\ddot{x}$$

or $\underline{\underline{m\ddot{x} + 2kx = 0}}$

20.2

Given: The system shown in Figure a

Determine the effective spring constant for the system.

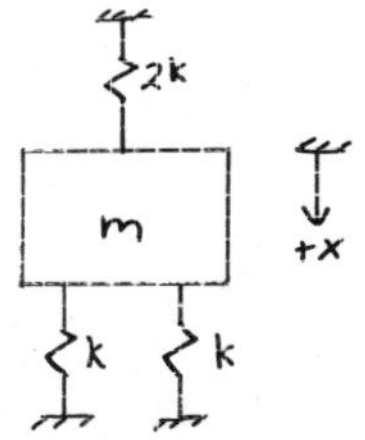

Figure a

Solution:

Consider the free-body diagram (Fig. b) of the mass in the static equilibrium position ($x=0$).

For equilibrium: (Fig. b)

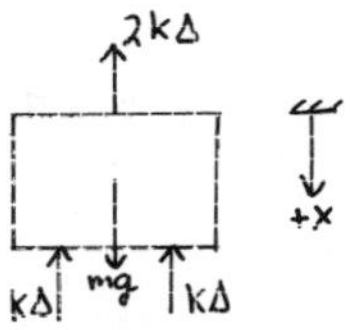

Figure b

$$\Sigma F_x = mg - 4k\Delta = 0$$

$$k\Delta = \tfrac{1}{4}mg \qquad \text{(a)}$$

Here we have assumed that the upper spring is stretched and the lower springs are compressed when the mass is in static equilibrium.

The free-body diagram (Fig. c) shows the mass displaced an amount x from its static equilibrium position. By Newton's second law, (Fig. c)

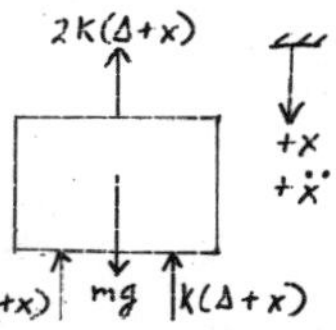

Figure c

$$\Sigma F_x = mg - 4k(\Delta + x) = m\ddot{x} \qquad \text{(b)}$$

By Eqs. (a) and (b), we obtain,

$$m\ddot{x} + 4kx = 0 \qquad \text{(c)}$$

From Eq. (c), the effective stiffness is

$\underline{\underline{k_{eff} = 4k}}$

20.3

Given: The system shown in Fig. a is in the static equilibrium position. The bar is uniform.

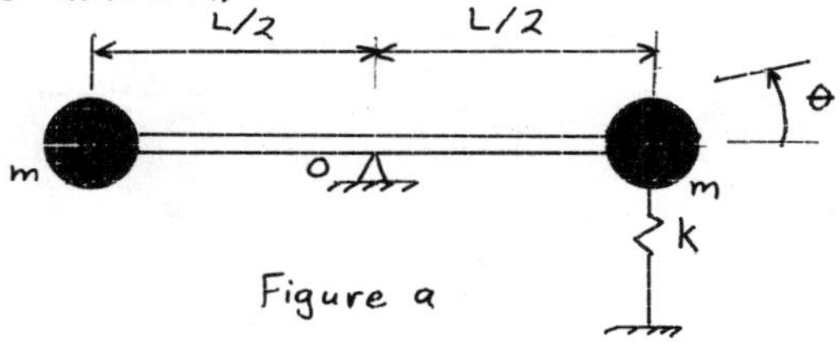

Figure a

Derive the differential equation of motion for the system shown in Fig. a. Neglect friction.

Solution:

Use the rotation angle θ as the independent coordinate. Consider the free-body diagram (Fig. b) of the system in its static equilibrium position.

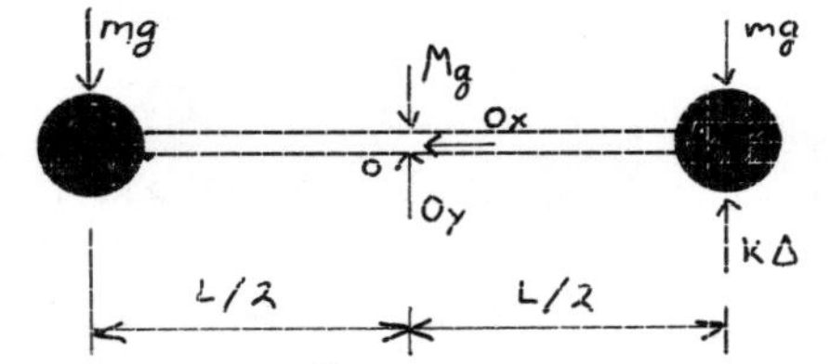

Figure b

For moment equilibrium:

$$\circlearrowleft + \Sigma M_o = mg(L/2) - mg(L/2) + k\Delta(L/2) = 0$$

$$\therefore k\Delta = 0$$

The force in the spring is zero for static equilibrium. Now, displace the system through an angle θ and draw another free-body diagram (Fig. c)

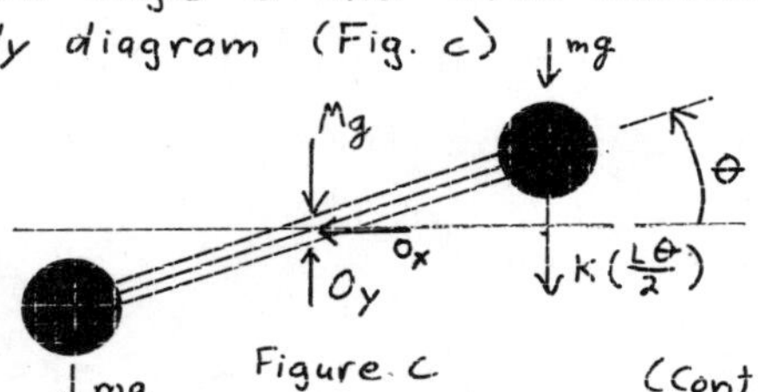

Figure c

(Continued)

20.3 Cont.

The equation of motion is given by

$$\circlearrowleft + \Sigma M_o = -k\left(\frac{L\theta}{2}\right)\left(\frac{L}{2}\right) = I\ddot{\theta} \qquad (a)$$

The moment of inertia of the bar and two balls is

$$I = \tfrac{1}{12} M L^2 + 2m\left(\tfrac{L}{2}\right)^2$$

$$I = \left(\tfrac{1}{12} M + \tfrac{1}{2} m\right) L^2$$

Now, substitute into Eq. (a) for I and simplify:

$$\left(\tfrac{1}{12} M + \tfrac{1}{2} m\right) L^2 \ddot{\theta} + k\left(\tfrac{L^2}{4}\right)\theta = 0$$

$$\left(\tfrac{M}{3} + 2m\right)\ddot{\theta} + k\theta = 0$$

Note: Any multiple of this equation is also a proper solution. For instance, we could also write:

$$(M + 6m)\ddot{\theta} + 3k\theta = 0$$

20.4

Given: The system shown in Fig. a is in the static equilibrium position. The bar is uniform

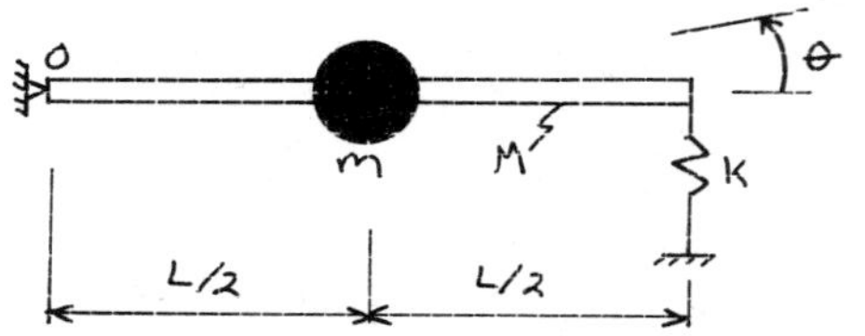

Figure a

Derive the differential equation of motion for the system shown in Fig. a. Neglect friction.

Solution:

Use the rotation angle θ as the independent coordinate. Start with a free-body diagram (Fig. b) of the system in its static equilibrium position.

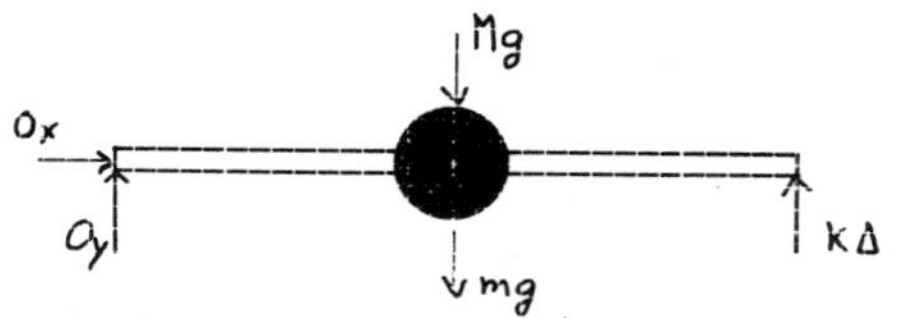

Figure b

For moment equilibrium:

$$\circlearrowleft + \Sigma M_o = k\Delta L - (M+m)g\,(L/2) = 0$$

$$k\Delta = \tfrac{1}{2}(M+m)g \qquad (a)$$

Now displace the system through an angle θ. Draw another free-body diagram (Fig. c).

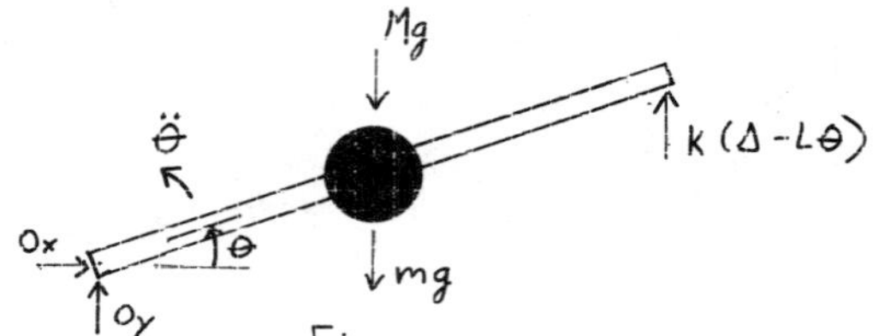

Figure c

The equation of motion is given by

$$\circlearrowleft + \Sigma M_o = -\tfrac{L}{2}(M+m)g + k(\Delta - L\theta)L = I\ddot{\theta} \qquad (b)$$

Substitute for kΔ from Eq. (a) into Eq. (b):

$$-k\theta L^2 = I\ddot{\theta} \qquad (c)$$

The moment of inertia of the bar and ball about the hinge is (see Appendix F):

$$I = \tfrac{1}{3} M L^2 + m\left(\tfrac{L}{2}\right)^2$$

$$I = \left(\tfrac{1}{3} M + \tfrac{1}{4} m\right) L^2$$

Now substitute into Eq. (c) for I and simplify,

$$-k\theta L^2 = \left(\tfrac{1}{3} M + \tfrac{1}{4} m\right) L^2 \ddot{\theta}$$

$$(4M + 3m)\ddot{\theta} + 12k\theta = 0$$

20.5

Given: The system shown in Fig. a is in static equilibrium.

Derive the differential equation of motion for the system shown in Fig. a.

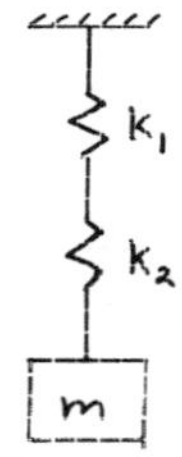

Figure a

Solution:

Begin by replacing the two springs in series by a single spring with effective stiffness k_{eff}; see Fig. b. Due to load F, Both spring systems stretch an amount Δ. For the original pair of springs:

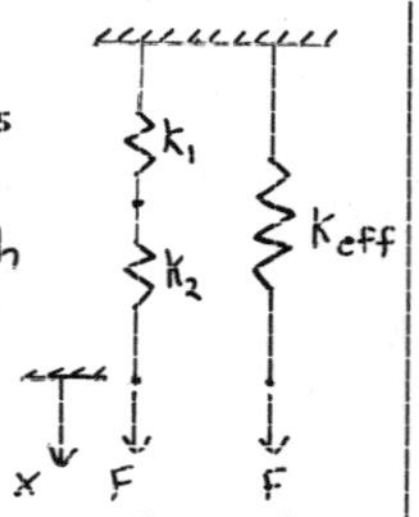

Figure b

$$\Delta = \delta_1 + \delta_2 \quad \text{where}$$

$$\delta_1 = F/k_1 \quad \& \quad \delta_2 = F/k_2$$

For the single spring

$\Delta = F/k_{eff}$ ∴ Equating these elongations gives:

$$F/k_{eff} = F/k_1 + F/k_2$$

$$1/k_{eff} = 1/k_1 + 1/k_2 \qquad \therefore\ k_{eff} = \frac{k_1 k_2}{k_1 + k_2}$$

So the equation of motion becomes

$$m\ddot{x} + \left(\frac{k_1 k_2}{k_1 + k_2}\right)x = 0$$

20.6

Given: The system shown in Fig. a is in the static equilibrium position. The bar is uniform.

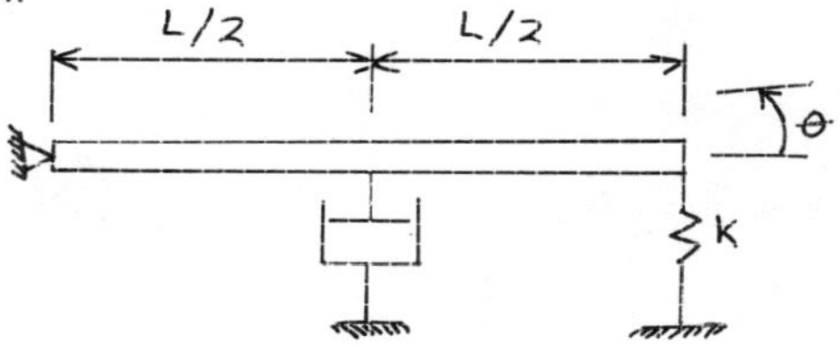

Figure a

Derive the differential equation of motion for the system shown in Fig. a. Neglect friction.

Solution:

Use the rotation angle θ as the independent coordinate. Start with a free-body diagram (Fig. b) of the system in its static equilibrium position.

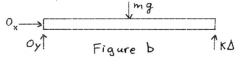

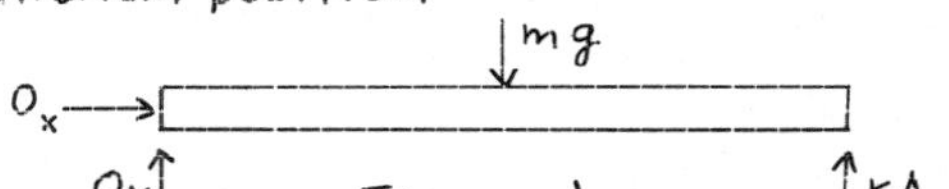

Figure b

For moment equilibrium:

$$\circlearrowleft + \Sigma M_o = k\Delta L - mg\,(L/2) = 0$$

$$k\Delta = \tfrac{1}{2}mg \qquad (a)$$

Now, displace the system through an angle θ. Draw another free-body diagram (Fig. c).

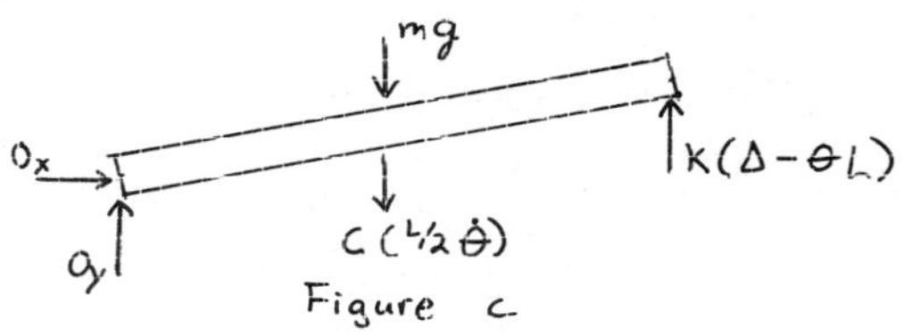

Figure c

The equation of motion is given by

$$\circlearrowleft + \Sigma M_o = k(\Delta - \theta L)L - mg\,(L/2) - c(\tfrac{L}{2}\dot\theta)(L/2) = I\ddot\theta \qquad (b)$$

Substitute for kΔ from Eq. (a) into Eq. (b)

$$-k\theta L^2 - c\dot\theta\,\frac{L^2}{4} = I\ddot\theta \qquad (c)$$

The moment of inertia of the bar about the hinge is (see Appendix F):

$$I = \tfrac{1}{3}mL^2$$

Now substitute into Eq. (c) for I and simplify:

$$k\theta L^2 + c\dot\theta\,\frac{L^2}{4} + \tfrac{1}{3}mL^2\ddot\theta = 0$$

$$\underline{4m\ddot\theta + 3c\dot\theta + 12k\theta = 0}$$

20.7

Given: The system shown in Fig. a is in the static equilibrium position. The bar is uniform.

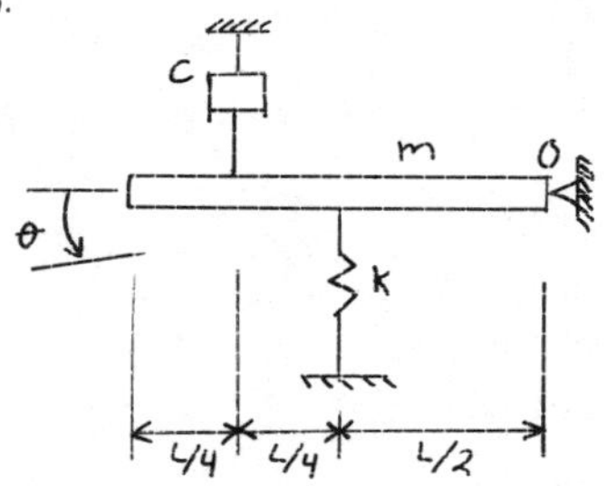

Figure a

Derive the differential equation of motion for the system shown in Fig. a. Neglect Friction.

Solution:

Use the rotation angle θ as the independent coordinate. Start with a free-body diagram (Fig. b) of the system in its static equilibrium position.

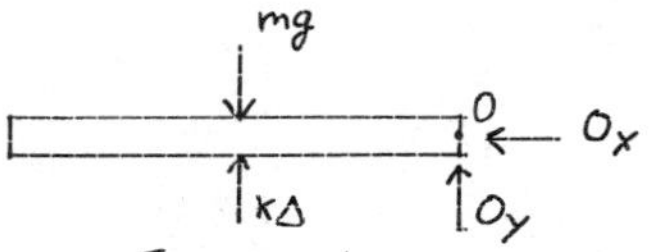

Figure b

For moment equilibrium:

$$\circlearrowleft + \Sigma M_o = L/2\,(mg) - L/2\,(k\Delta) = 0$$

$$k\Delta = mg \qquad (a)$$

Now, displace the system through an angle θ. Draw another free-body diagram (Fig. c).

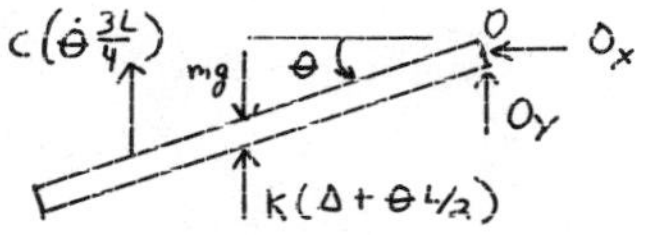

Figure c

The equation of motion is given by

$$\circlearrowleft + \Sigma M_o = mg\,(\tfrac{L}{2}) - c\dot\theta\,(\tfrac{3L}{4})^2 - k(\Delta + \theta\tfrac{L}{2})\tfrac{L}{2} = I\ddot\theta \qquad (b)$$

Substitute for kΔ from Eq. (a) into Eq. (b)

$$-c\dot\theta\,(\tfrac{9L^2}{16}) - k\theta\,\frac{L^2}{4} = I\ddot\theta \qquad (c)$$

The moment of inertia of the rod is (see Appendix): $I = \tfrac{1}{3}mL^2$

Now substitute into Eq. (c) for I and simplify,

$$c\dot\theta\,(\tfrac{9L^2}{16}) + k\theta\,(\tfrac{L^2}{4}) + \tfrac{1}{3}m\ddot\theta L^2 = 0$$

$$\underline{16m\ddot\theta + 27c\dot\theta + 12k\theta = 0}$$

20.8

Given: The system shown in Fig. a which is in the static equilibrium position. The bar is uniform.

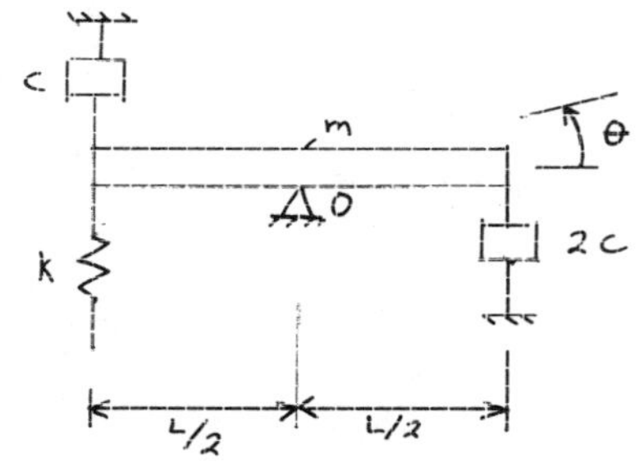

Figure a

Derive the differential equation of motion for the system shown in Fig. a. Neglect friction.

Solution:

Use the rotation angle θ as the independent coordinate. Start with a free-body diagram (Fig. b) of the system in its equilibrium position.

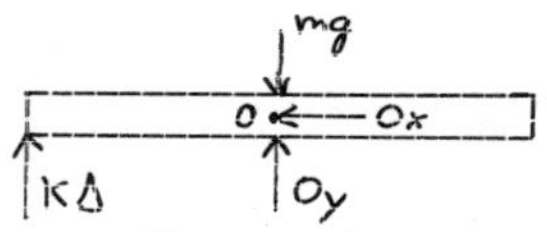

Figure b

For moment equilibrium:

$$\circlearrowleft + \sum M_0 = -k\Delta = 0$$

$\therefore$ The spring force is zero at static equilibrium. Now, draw a free-body diagram (Fig. c) of the mass displaced through an angle θ.

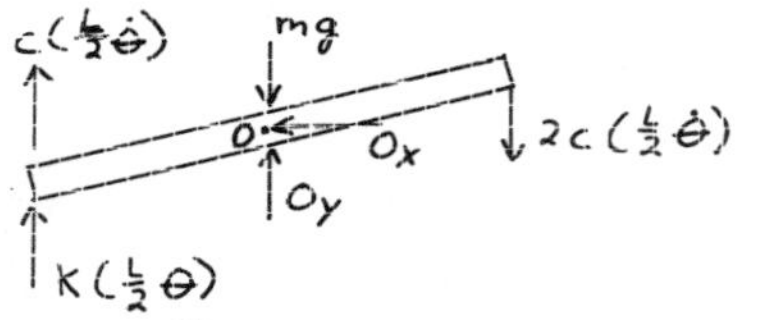

Figure c

The equation of motion is given by

$$\circlearrowleft + \sum M_0 = -3c\left(\tfrac{L}{2}\dot{\theta}\right)\tfrac{L}{2} - k\left(\tfrac{L}{2}\theta\right)\tfrac{L}{2} = I\ddot{\theta} \qquad (a)$$

The moment of inertia of the bar about its center is (see Appendix F):

$$I = \tfrac{1}{12} m L^2$$

Substitute for I in Eq. (a) and simplify:

$$\tfrac{3}{4} c L^2 \dot{\theta} + \tfrac{1}{4} k L^2 \theta + \tfrac{1}{12} m L^2 \ddot{\theta} = 0$$

$$\underline{\underline{m\ddot{\theta} + 9c\dot{\theta} + 3k\theta = 0}}$$

20.9

Given: The system shown in Fig. a which is in the static equilibrium position. The bars are uniform.

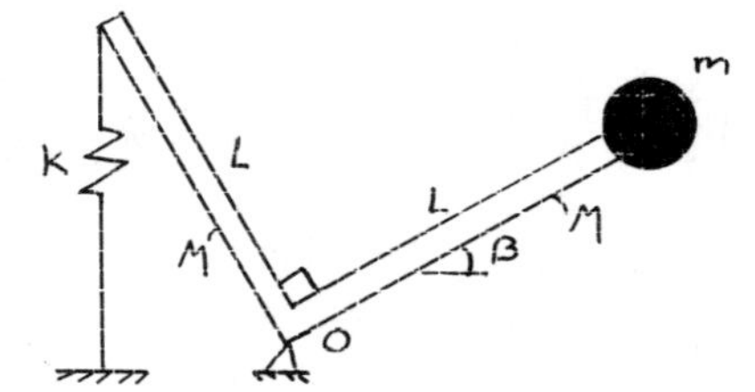

Figure a

Derive the differential equation of motion for the system shown in Fig. a. Neglect friction.

Solution:

Start with a free-body diagram (Fig. b) of the system in its equilibrium position.

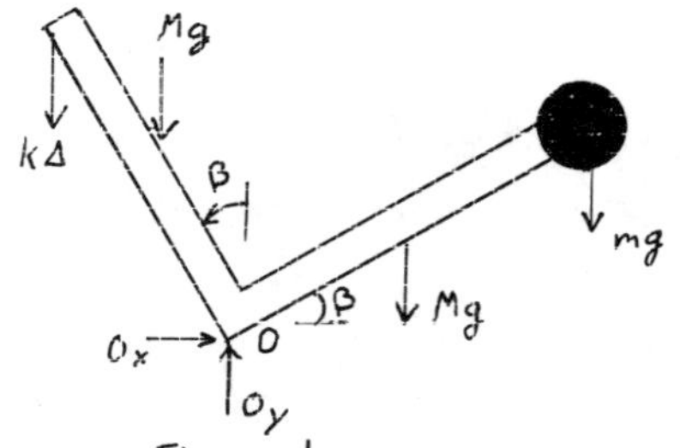

Figure b

For moment equilibrium:

$$\circlearrowleft + \sum M_0 = (k\Delta)L\sin\beta + (Mg)\tfrac{L}{2}\sin\beta - (Mg)\tfrac{L}{2}\cos\beta - mgL\cos\beta = 0$$

$$k\Delta\sin\beta = \tfrac{Mg}{2}(\cos\beta - \sin\beta) + mg\cos\beta \qquad (a)$$

Now, displace the system through an angle θ. Draw another free-body diagram (Fig. c).

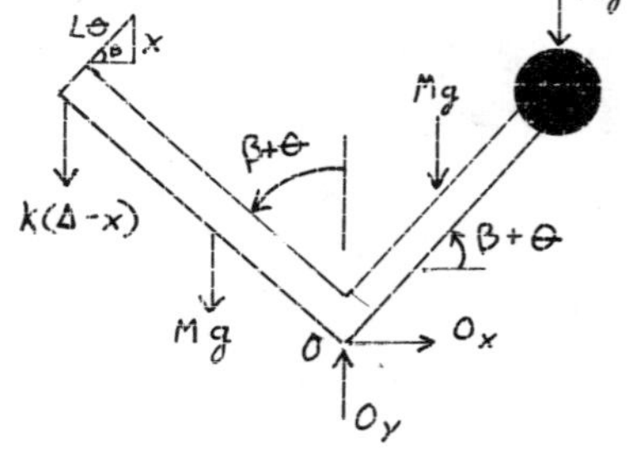

Figure c

The change in length of the spring under this rotation is: $x = L\theta\sin\beta$. The equation of motion is given by

$$\circlearrowleft + \sum M_0 = k(\Delta - x)L\sin\beta + (Mg)\tfrac{L}{2}(\sin\beta - \cos\beta) - mgL\cos\beta = I\ddot{\theta} \qquad (b)$$

(Continued)

20.9 Cont.

Substitute from Eq. (a) for $k\Delta \sin\beta$ into Eq. (b) and simplify:

$$kxL\sin\beta + I\ddot{\theta} = 0$$

Now write x in terms of θ

$$k(L\theta \sin\beta)L\sin\beta + I\ddot{\theta} = 0 \qquad (c)$$

The moment of inertia of the system about the hinge O is:

$$I = 2\left(\tfrac{1}{3}ML^2\right) + mL^2 \qquad (d)$$

Substitute Eq. (d) into Eq. (c) and simplify

$$(k\sin^2\beta)\theta + \left(\tfrac{2}{3}M + m\right)\ddot{\theta} = 0$$

$$\underline{\underline{(2M+3m)\ddot{\theta} + (3k\sin^2\beta)\theta = 0}}$$

20.10

<u>Given:</u> The spring system shown in Fig. a. The springs are unstretched when $x = 0$.

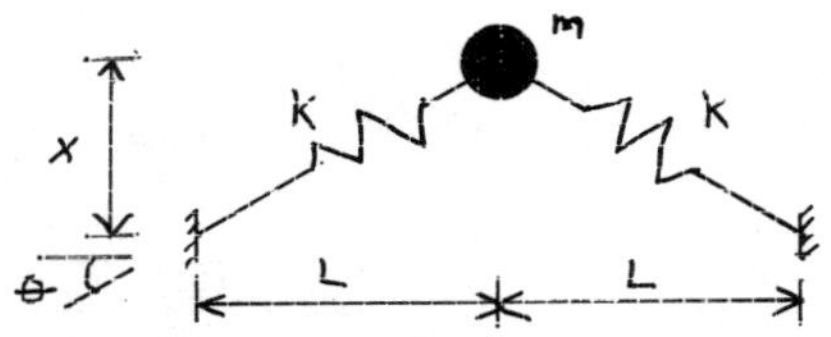

Figure a

Derive the differential equation of motion for large vertical oscillations. Show that, for small oscillations, the restoring force is approximately proportional to x^3. Neglect gravity.

<u>Solution:</u>

The elongation δ of the spring is:

$$\delta = \sqrt{L^2+x^2} - L$$

The spring force is $F = k\delta = k[\sqrt{L^2+x^2} - L]$.

The vertical force on the ball from the two springs is:

$$F_x = 2F\sin\theta = 2k[\sqrt{L^2+x^2} - L]\frac{x}{\sqrt{L^2+x^2}}$$

By Newton's second law, the equation of motion is:

$$\underline{\underline{m\ddot{x} + 2kx\left[1 - \frac{L}{\sqrt{L^2+x^2}}\right] = 0}} \qquad (a)$$

Rewrite Eq. (a) in terms of θ.

$$m\ddot{x} + 2kx(1 - \cos\theta) = 0 \qquad (b)$$

Now write $\cos\theta$ in the form of a series

$$\cos\theta = 1 - \frac{\theta^2}{2!} + \frac{\theta^4}{4!} - \frac{\theta^6}{6!} + \ldots$$

For small vibrations, we can retain only the first two terms. So,

$$\cos\theta = 1 - \frac{\theta^2}{2!}$$

Also, for small vibrations, $\theta \approx \tan\theta = \frac{x}{L}$

$$\therefore \quad \cos\theta \approx 1 - \tfrac{1}{2}\left(\tfrac{x}{L}\right)^2 \qquad (c)$$

Substitute from Eq. (c) for $\cos\theta$ in Eq. (b),

$$m\ddot{x} + 2kx\left(1 - \left[1 - \tfrac{1}{2}\left(\tfrac{x}{L}\right)^2\right]\right) = 0$$

$$m\ddot{x} + \frac{kx^3}{L^2} = 0$$

The restoring force is

$\underline{\underline{F_s = \dfrac{kx^3}{L^2}}}$ which is proportional to x^3.

20.11

<u>Given:</u> The cylindrical plug slides in a hole in a tank (Fig. a). The plug is in static equilibrium with its lower face at section 1-1. If the plug oscillates, the air in the tank is compressed according to $pV^{1.4} = \text{constant}$. p is pressure, V is volume.

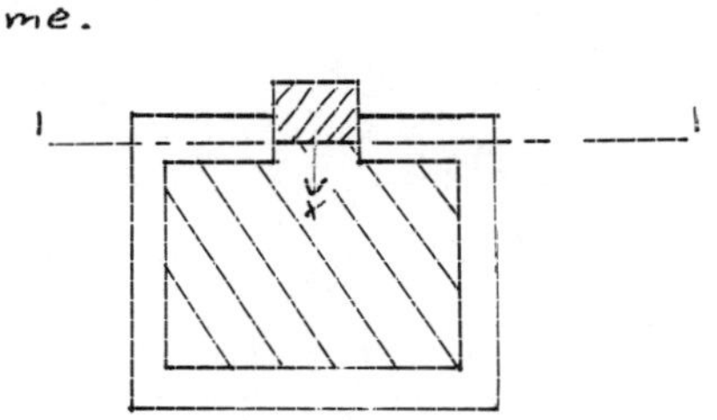

Figure a

Show that for small oscillations, the spring constant is approximately $k = 1.4A^2p_o/V_o$. A = area of the plug. p_o & V_o are static equilibrium values of p & V. Hint: $p_oV_o^{1.4} = pV^{1.4}$ = constant & $V = V_o - Ax$.

<u>Solution:</u>

Write the "spring force" ($F = pA$) on the plug in terms of the displacement of the plug.

$$pV^{1.4} = p_oV_o^{1.4}$$

$$pA(V_o - Ax)^{1.4} = p_oAV_o^{1.4}$$

$$pAV_o^{1.4}\left(1 - \frac{Ax}{V_o}\right)^{1.4} = p_oAV_o^{1.4}$$

$$pA = p_oA\left(1 - \frac{Ax}{V_o}\right)^{-1.4} \qquad (a)$$

Now we must write the spring force pA as a linear function of x in order to find the spring constant k ($F = pA = kx$). By the binomial theorem, we can write

$$\left(1 - \frac{Ax}{V_o}\right)^{-1.4} = 1 + 1.4\frac{Ax}{V_o} \qquad (b)$$

To two terms. Substitute Eq. (b) into Eq. (a) to obtain the spring force.

$$pA = p_oA\left(1 + 1.4\frac{Ax}{V_o}\right)$$

$$pA = p_oA + \frac{1.4p_oA^2x}{V_o} \qquad (c)$$

(Continued)

20.11 Cont.

Now, draw a free-body diagram (Fig. b) of the plug and write the equation of motion from Newton's second law.

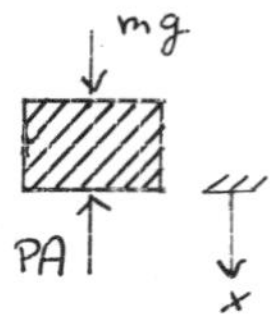

Figure b

$$\Sigma F_x = mg - pA = m\ddot{x} \qquad (d)$$

For static equilibrium

$$\Sigma F_x = mg - p_o A_o = 0 \;;\quad mg = p_o A \qquad (e)$$

Substitute from Eqs. (c) & (e) into Eq. (d) and simplify,

$$p_o A - \left(p_o A + \frac{1.4\, p_o A^2 x}{V_o}\right) = m\ddot{x}$$

$$m\ddot{x} + \left(\frac{1.4\, p_o A^2}{V_o}\right) x = 0$$

Therefore the spring constant is

$$k = \frac{1.4\, p_o A^2}{V_o}$$

20.12

Given: Displacement x is measured from the static equilibrium position shown in Fig. a.

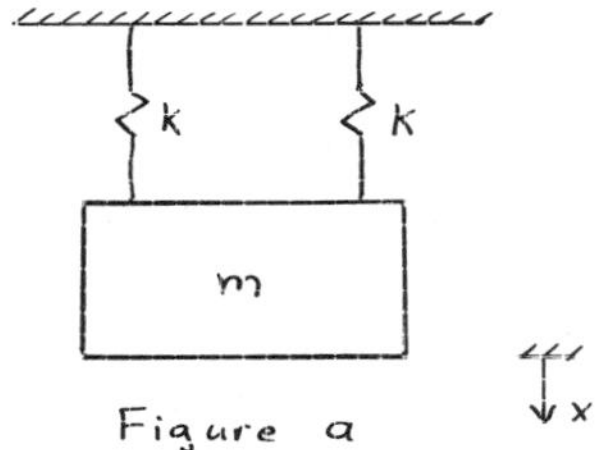

Figure a

Determine the frequency and period of vibration of the spring/mass system shown in Fig. a.

Solution:

First write the equation of motion for the system. Begin with a free-body diagram (Fig. b) of the mass in the static equilibrium position.

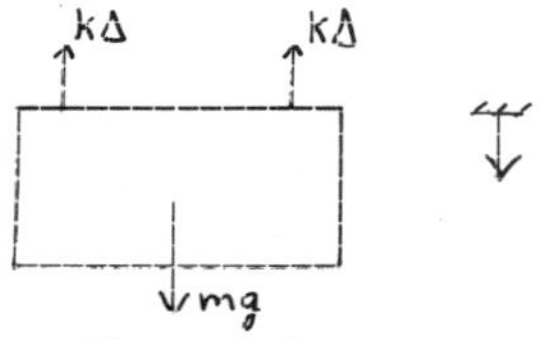

Figure b

Here, Δ is the elongation of each spring. For equilibrium:

$$\Sigma F_x = mg - 2k\Delta = 0 \;;\quad k\Delta = \tfrac{1}{2} mg \qquad (a)$$

Now, draw a free-body diagram (Fig. c) of the mass displaced an amount x from its static equilibrium position.

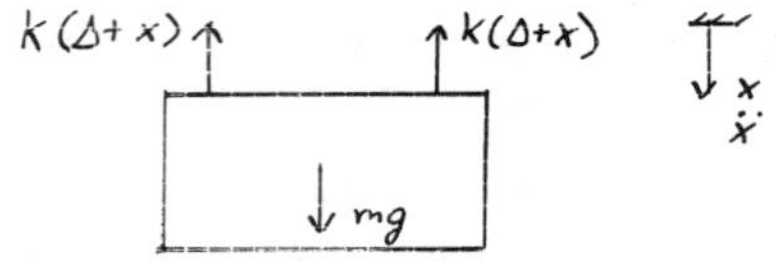

Figure c

By Newton's second law:

$$\Sigma F_x = mg - 2k(\Delta + x) = m\ddot{x}$$

$$mg - 2k\Delta - 2kx = m\ddot{x} \qquad (b)$$

Substitute for $k\Delta$ from Eq. (a) into Eq. (b)

$$mg - mg - 2k\Delta = m\ddot{x}$$

$$m\ddot{x} + 2kx = 0 \qquad (c)$$

Now, from Eq. (c), we see that the effective mass and effective stiffness are,

$$m_{eff} = m \qquad k_{eff} = 2k$$

Hence, from Eq. (20.7), the frequency of vibration is:

$$\omega = \sqrt{2k/m} \quad \text{or} \quad f = \frac{\omega}{2\pi} = \frac{1}{2\pi}\sqrt{\frac{2k}{m}}$$

From, Eq. (20.8), the period is

$$T = \frac{2\pi}{\omega} = 2\pi\sqrt{\frac{m}{2k}}$$

20.13

Given: System shown in Fig. a.

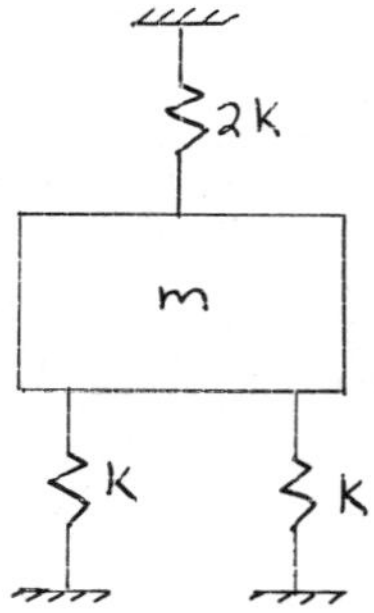

Figure a

Determine the frequency of vibration of the system shown in Fig. a.

Solution:

First write the equation of motion for the system. Start with a free-body diagram (Fig. b) of the mass in the static equilibrium position ($x=0$).

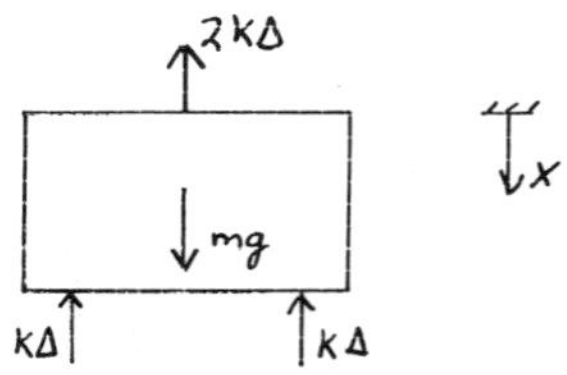

Figure b

(Continued)

20.13 Cont.

For equilibrium:

$$\Sigma F_x = mg - 4k\Delta = 0$$
$$k\Delta = \tfrac{1}{4} mg \qquad (a)$$

Here we have assumed that the upper spring is stretched and the lower springs are compressed when the mass is in static equilibrium. Next, draw a free-body diagram (Fig. c) and write the equation of motion for the mass when it is displaced an amount x from its equilibrium position.

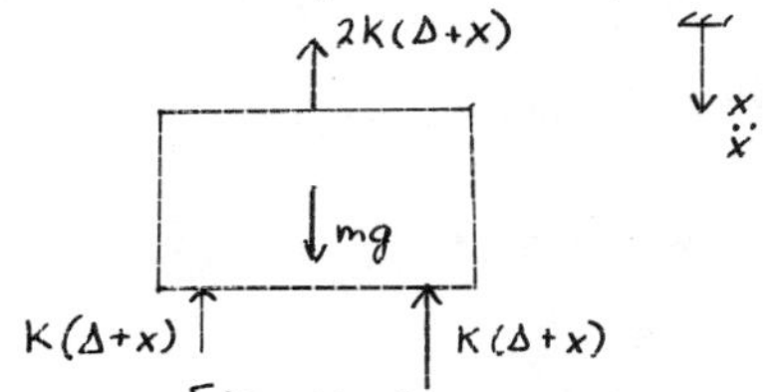

Figure c

$$\Sigma F_x = mg - 4k(\Delta + x) = m\ddot{x} \qquad (b)$$

Substitute for $k\Delta$ from Eq. (a) into Eq. (b) and simplify:

$$m\ddot{x} + 4kx = 0 \qquad (c)$$

Now, from Eq. (c), we see that the effective mass and effective stiffness are,

$$m_{eff} = m \quad , \quad k_{eff} = 4k$$

Hence, from Eq. (20.7), the frequency of vibration is:

$$\omega = 2\sqrt{\frac{k}{m}} \quad \text{or} \quad f = \frac{\omega}{2\pi} = \frac{1}{\pi}\sqrt{\frac{k}{m}}$$

From Eq. (20.8), the period is

$$T = \frac{2\pi}{\omega} = \pi\sqrt{\frac{m}{k}}$$

20.14

Given: The system shown in Fig. a where the bars are uniform.

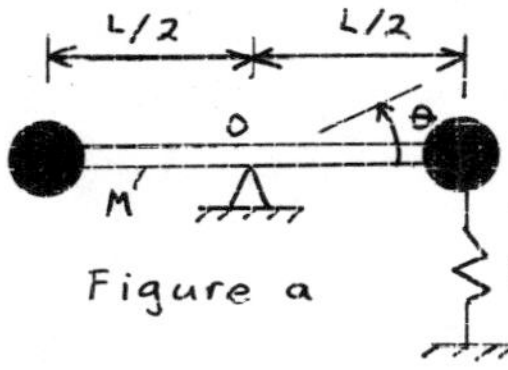

Figure a

Determine the frequency and the period of vibration of the system shown in Fig. a. Neglect friction.

Solution:

First write the equation of motion for the system. Begin with a free-body diagram (Fig. b) of the system in its static equilibrium position: $\theta = 0$, where θ is the independent coordinate.

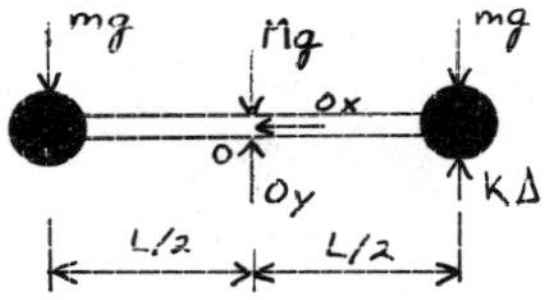

Figure b

For moment equilibrium:

$$\circlearrowleft\!+\Sigma M_o = mg\left(\tfrac{L}{2}\right) - mg\left(\tfrac{L}{2}\right) + k\Delta\left(\tfrac{L}{2}\right) = 0$$
$$\therefore\ k\Delta = 0$$

Therefore, the force in the spring is zero for static equilibrium. Now, displace the system through an angle θ and draw another free-body diagram (Fig. c).

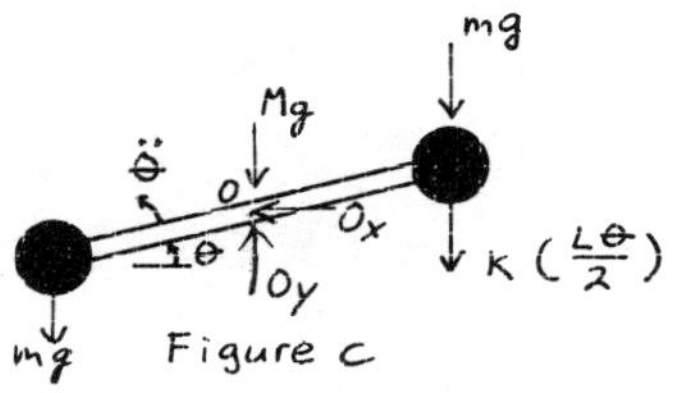

Figure c

The equation of motion is given by

$$\circlearrowleft\!+\Sigma M_o = -k\left(\tfrac{L\theta}{2}\right)\tfrac{L}{2} = I\ddot{\theta} \qquad (a)$$

The moment of inertia of the bar and the two balls is:

$$I = \tfrac{1}{12}ML^2 + 2m\left(\tfrac{L}{2}\right)^2$$
$$I = \left(\tfrac{1}{12}M + \tfrac{1}{2}m\right)L^2 \qquad (b)$$

Substitute into Eq. (a) for I from Eq. (b) and simplify:

$$\left(\tfrac{1}{12}M + \tfrac{1}{2}m\right)L^2\ddot{\theta} + k\left(\tfrac{L^2}{4}\right)\theta = 0$$
$$\left(\tfrac{M}{3} + 2m\right)\ddot{\theta} + k\theta = 0 \qquad (c)$$

So, by Eq. (c), we see that the effective mass and effective stiffness of the system are:

$$m_{eff} = \left(\tfrac{M}{3} + 2m\right) \quad , \quad k_{eff} = k$$

Hence, from Eq. (20.7), the frequency of vibration is:

$$\omega = \sqrt{\frac{k}{\frac{M}{3} + 2m}} \quad \text{or} \quad f = \frac{\omega}{2\pi} = \frac{1}{2\pi}\sqrt{\frac{k}{\frac{M}{3} + 2m}}$$

By Eq. (20.8), the period is

$$T = \frac{2\pi}{\omega} = 2\pi\sqrt{\frac{\frac{M}{3} + 2m}{k}}$$

20.15

Given: The system shown in Fig. a where the bar is uniform.

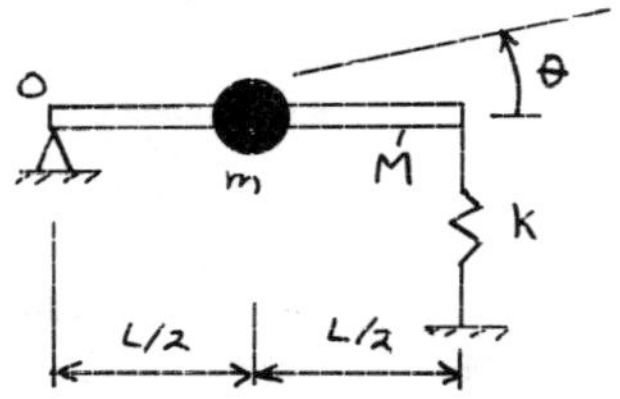

Figure a

Determine the frequency and period of vibration of the system. Neglect friction.

Solution:

First, write the equation of motion for the system. Start with a free-body diagram (Fig. b) of the system in the static equilibrium position: $\theta = 0$, where θ is the independent coordinate.

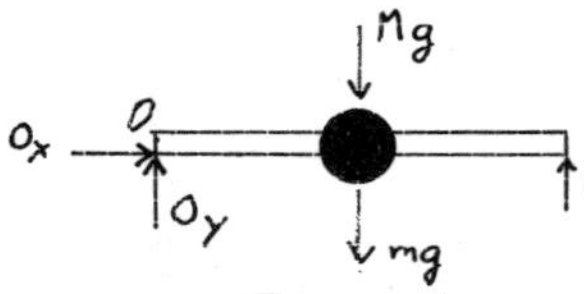

Figure b

For moment equilibrium:

$$\circlearrowleft + \Sigma M_O = k\Delta L - (M+m)g\,(L/2) = 0$$

$$k\Delta = \tfrac{1}{2}(M+m)g \qquad (a)$$

Now, displace the system through an angle θ & draw another free-body diagram (Fig. c).

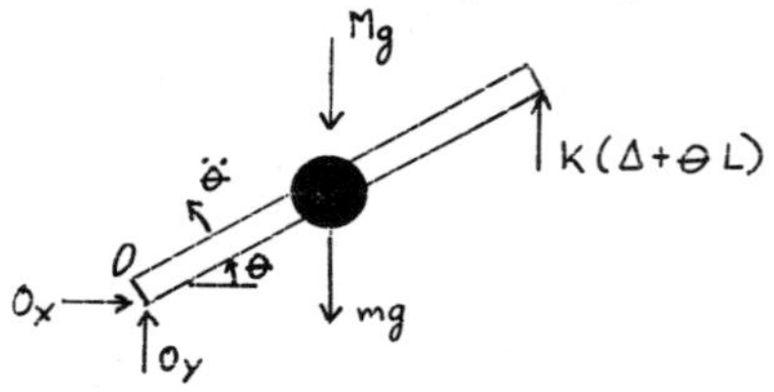

Figure c

The equation of motion is given by

$$\circlearrowleft + \Sigma M_O = -\tfrac{L}{2}(M+m)g + k(\Delta - L\theta)L = I\ddot{\theta} \qquad (b)$$

Substitute for $k\Delta$ from Eq. (a) into Eq. (b):

$$-k\theta L^2 = I\ddot{\theta} \qquad (c)$$

The moment of inertia of the bar and ball about the hinge is (see Appendix F):

$$I = \tfrac{1}{3}ML^2 + m\left(\tfrac{L}{2}\right)^2$$

$$= \left(\tfrac{1}{3}M + \tfrac{1}{4}m\right)L^2 \qquad (d)$$

Now substitute for I from Eq. (d) into Eq. (c) and simplify:

$$-k\theta L^2 = \left(\tfrac{1}{3}M + \tfrac{1}{4}m\right)L^2\ddot{\theta}$$

$$(4M+3m)\ddot{\theta} + (12k)\theta = 0 \qquad (e)$$

From Eq. (e), we see that the effective mass and effective stiffness are

$$m_{eff} = 4M + 3m \quad , \quad k_{eff} = 12k$$

Thus, from Eq. (20.7), the frequency of vibration is:

$$\omega = \sqrt{\frac{12k}{4M+3m}} \qquad \text{or} \qquad f = \frac{\omega}{2\pi} = \frac{1}{\pi}\sqrt{\frac{3k}{4M+3m}}$$

From Eq. (20.8), the period is

$$T = \frac{2\pi}{\omega} = \pi\sqrt{\frac{4M+3m}{3k}}$$

20.16

Given: The system shown in Fig. a

Determine the frequency and period of vibration of the system shown in Fig. a.

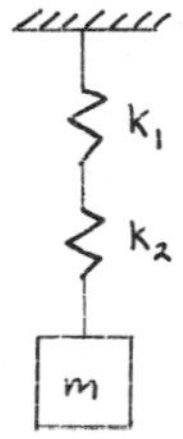

Figure a

Solution:

Begin by replacing the two springs in series by a single spring with effective stiffness k_{eff}; see Fig. b. Due to load F, both spring systems stretch an amount Δ. For the original pair of springs:

$\Delta = \delta_1 + \delta_2$ where $\delta_1 = \frac{F}{k_1}$ & $\delta_2 = \frac{F}{k_2}$

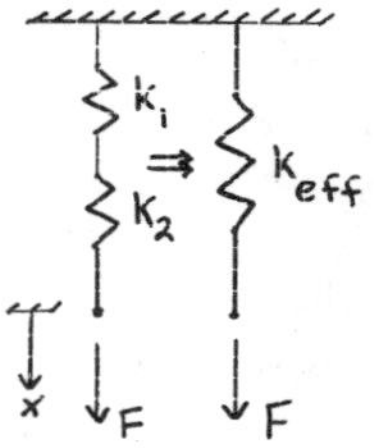

Figure b

For the single spring: $\Delta = F/k_{eff}$

Equating these elongations gives:

$$F/k_{eff} = F\left(1/k_1 + 1/k_2\right) \Rightarrow k_{eff} = \frac{k_1 k_2}{k_1 + k_2} \qquad (a)$$

The effective mass is simply $m_{eff} = m$ (b)

Now, from Eq. (20.7), the frequency of vibration is:

$$\omega = \sqrt{\frac{k_1 k_2}{m(k_1+k_2)}} \qquad \text{or} \qquad f = \frac{\omega}{2\pi}\sqrt{\frac{k_1 k_2}{m(k_1+k_2)}}$$

From Eq. (20.8), the period is

$$T = \frac{3\pi}{\omega} = 2\pi\sqrt{\frac{m(k_1+k_2)}{k_1 k_2}}$$

20.17

Given: An electric motor that weighs 200 lb is mounted on four identical springs, one at each corner of the housing (Fig. a). The springs are eached compressed 1-in by the weight of the motor.

Determine the frequency of vibration of the motor.

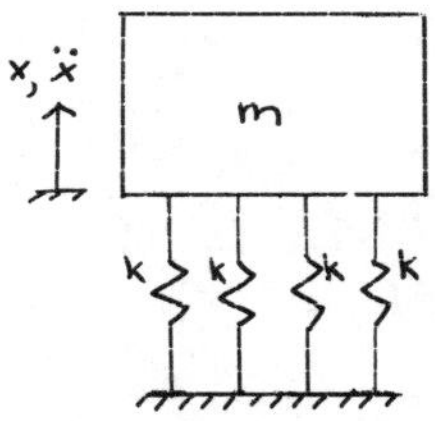

Figure a

Solution:

First find the effective stiffness and mass of the system in Fig. a.

$$m_{eff} = W/g = 200/(32.2)(12) = 0.5176 \frac{\text{lb}\cdot\text{s}^2}{\text{in}}$$

$$k_{eff} = 4k = 200/1 = 200 \text{ lb/in}$$

So, we can write the equation of motion as: $0.5176\ddot{x} + 200x = 0$

From Eq. (20.7), the frequency of vibration is:

$$\omega = \sqrt{\frac{200}{0.5176}} = 19.657 \text{ rad/s}$$

or

$$f = \frac{\omega}{2\pi} = 3.129 \text{ Hz}$$

Alternate Solution:

We can solve the problem in a general form and obtain an interesting result. Then as data to use: $W = 200$ lb, $\delta = 1$ in & $g = 386.4$ in/s².

The effective mass is, $m_{eff} = W/g$

The effective stiffness is, $k_{eff} = W/\delta$

From Eq. (20.7), the frequency is:

$$\omega = \sqrt{\frac{k_{eff}}{m_{eff}}} = \sqrt{\frac{W/\delta}{W/g}} = \sqrt{\frac{g}{\delta}} = 19.657 \text{ rad/s}$$

$$f = \frac{\omega}{2\pi} = \frac{1}{2\pi}\sqrt{\frac{g}{\delta}} = 3.129 \text{ Hz}$$

For the problem as described, the weight and the stiffness of the spring do not appear in the final expression for frequency.

20.18

Given: A 9 N body hangs from a spring with free length of 1 m. Its period of vibration is $T = 0.5$ s.

Find the extension of the spring if the body hangs motionless.

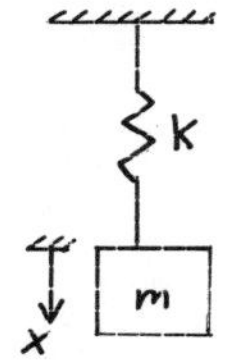

Figure a

Solution:

For the spring-mass system in Fig. a, we can find the mass of the body from its weight. Then the spring stiffness is found from the mass and the period.

$$m = \frac{W}{g} = \frac{9}{9.81} = 0.9174 \text{ kg}$$

By Eq. (20.8):

$$T = 2\pi\sqrt{\frac{m}{k}} \Rightarrow k = m\left(\frac{2\pi}{T}\right)^2$$

$$\therefore k = 0.9174\left(\frac{2\pi}{0.5}\right)^2 = 144.9 \text{ N/m}$$

Now, under a static load of 9 N, the spring elongation δ is:

$$\delta = \frac{W}{k} = \frac{9}{144.9} = 0.0621 \text{ m} = 62.1 \text{ mm}$$

Alternative Solution:

We can solve the problem in a general form and obtain an interesting result. Then, as data, use: $W = 9$ N, $T = 0.5$ s, $g = 9.81$ m/s².

The effective mass is $m_{eff} = W/g$

The effective stiffness is found from Eq. (20.8):

$$T = \frac{2\pi}{\omega} = 2\pi\sqrt{\frac{m_{eff}}{k_{eff}}}$$

$$\therefore k_{eff} = m_{eff}\left(\frac{2\pi}{T}\right)^2 = \frac{4\pi^2 W}{g T^2}$$

The elongation of the spring is:

$$\delta = \frac{W}{k_{eff}} = \frac{gT^2}{4\pi} = 0.0621 \text{ m} = 62.1 \text{ mm}$$

For the problem as described, the weight of the body does not appear in the final expression for spring elongation.

20.19

Given: A spring/mass system is in equilibrium in the position shown (Fig. a). The mass m oscillates on the frictionless rod under the action of gravity and the spring.

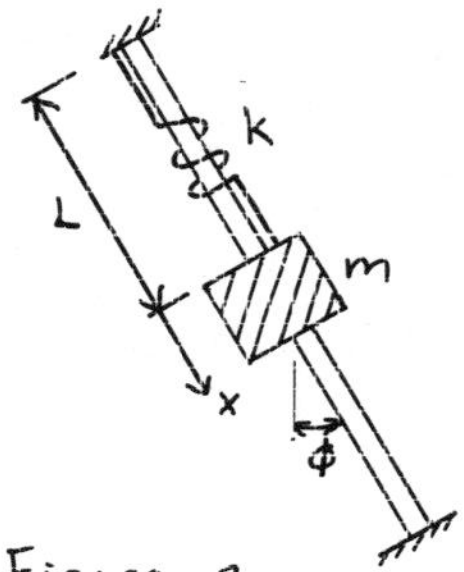

Figure a

Derive the equation of motion.

Solution:

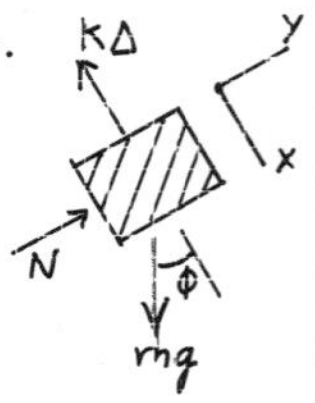

Figure b

Start with a free-body diagram of the mass in the equilibrium position (Fig. b). The static elongation of the spring is Δ.

$\Sigma F_x = mg\cos\phi - k\Delta = 0$

$$k\Delta = mg\cos\phi \qquad (a)$$

$\Sigma F_y = N - mg\sin\phi = 0$

$$N = mg\sin\phi \qquad (b)$$

Now, displace the mass an amount x and draw another free-body diagram (Fig. c). Since the mass is not moving in the Y direction, Eq. (b) still applies here. For motion in the x direction:

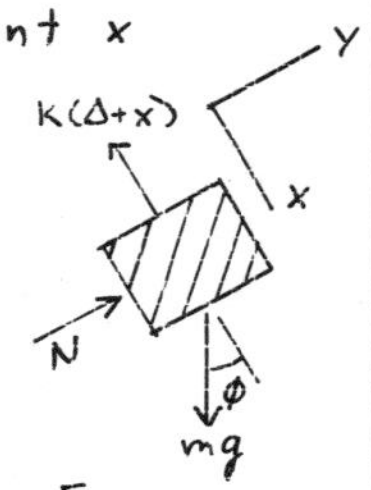

Figure c

$\Sigma F_x = -k(\Delta + x) + mg\cos\phi = m\ddot{x} \qquad (c)$

Substitute from Eq. (a) into Eq. (c) and simplify to obtain the equation of motion:

$$\underline{m\ddot{x} + kx = 0}$$

20.20

Given: A block rests on a horizontal plank. The plank undergoes simple harmonic motion with an Amplitude of 250 mm, $\mu_s = 0.40$.

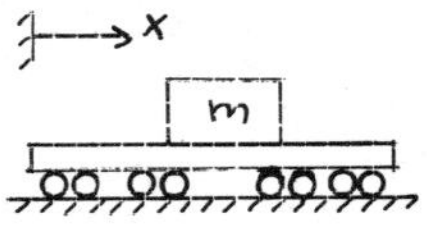

Figure a

Find the period at which the block is on the verge of slipping on the plank.

Solution:

Since the motion is simple harmonic with amplitude 0.25 m, the displacement of the system is, where ω is the angular frequency and t is time,

$$x = 0.25\sin\omega t \qquad (a)$$

Therefore, by differentiation of Eq. (a),

$$\ddot{x} = -0.25\omega^2\sin\omega t \qquad (b)$$

Then, by Fig. b,

$\Sigma F_x = -F = m\ddot{x} \qquad (c)$

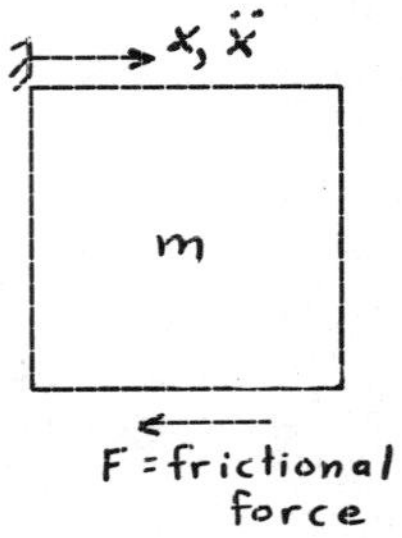

Figure b

The maximum acceleration is,

$$(\ddot{x}_{max}) = 0.25\omega^2$$

Hence, since at impending sliding, $F_{max} = \mu_s mg = 0.4mg$, Eq. (c) yields,

$$0.4mg = m(0.25\omega^2)$$

or

$$\omega^2 = 1.6g = 15.696, \text{ or } \omega = 3.962 \text{ rad/s}$$

So,

$$T = \frac{2\pi}{\omega} = \frac{2\pi}{3.962} = 1.586 \text{ s}$$

20.21

Given: Two identical masses that weigh 675 N each rest on a frictionless plane, (Fig. a).

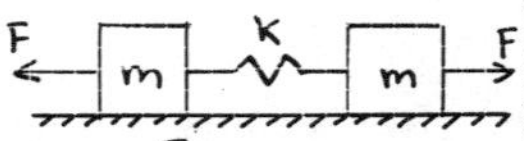

Figure a

They are connected by a spring with stiffness 36 N/mm. Forces F are applied to pull the masses apart. Then the forces are suddenly removed.

Find the period and frequency of vibration.

Solution:

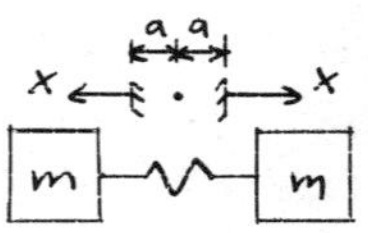

Figure b

The system is symmetric; it vibrates about its center, (Fig. b). For a displacement x of each mass, the spring force is $F = (2x)k$. From a free-body diagram (Fig. c) of one mass, we have

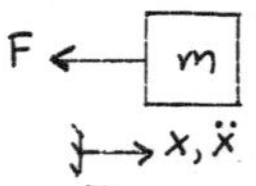

Figure c

$\Sigma F_x = -2kx = m\ddot{x}$ or $m\ddot{x} + 2kx = 0$

(Continued)

20.21 Cont.

By Eq. (20.7), the frequency is:

$$\omega = \sqrt{\frac{2k}{m}} = \sqrt{\frac{2(3600)}{675/9.81}} = 32.35 \text{ rad/s}$$

$$f = \frac{\omega}{2\pi} = 5.148 \text{ Hz}$$

By Eq. (20.8), the period is:

$$T = \frac{2\pi}{\omega} = 0.194 \text{ s}$$

20.22

Given: Mass m is attached to the mid-point of a light, elastic string (Fig. a) and it vibrates horizontally. Neglect gravity and assume the spring tension F is constant.

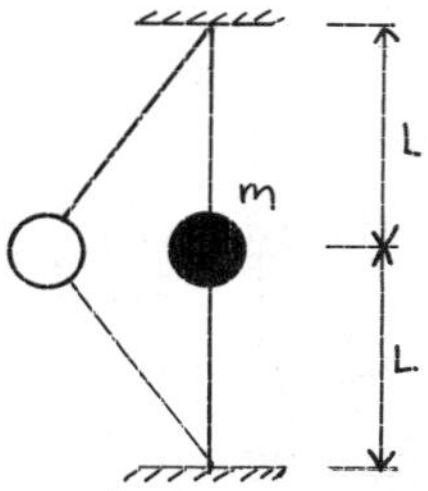

Figure a

Derive a formula for the frequency of small vibrations.

Solution:

Start with a free-body diagram of the ball in the displaced position (Fig. b). The equation of motion for the ball is:

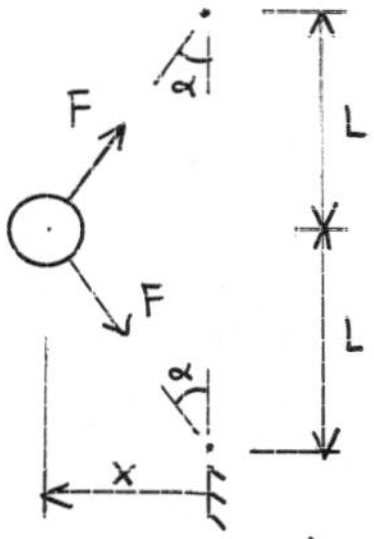

Figure b

$$\Sigma F_x = -2F\sin\alpha = m\ddot{x} \qquad (a)$$

If α is small, $\sin\alpha \approx \alpha$.

Also $\ddot{x} = L\sin\ddot{\alpha} = L\ddot{\alpha}$

Substitution into Eq. (a) yields

$$mL\ddot{\alpha} + 2F\alpha = 0$$

By Eq. (20.7),

$$\omega = \sqrt{\frac{2F}{mL}}, \qquad f = \frac{\omega}{2\pi} = \frac{1}{2\pi}\sqrt{\frac{2F}{mL}}$$

20.23

Given: To find the moment of inertia I of a car wheel about the axis of the axle, engineers suspend it by a wire as shown in Fig. a. The torsional spring constant for the wire is k = 0.0920 lb·in/rad. The period of angular vibration for the wheel is T = 62 s.

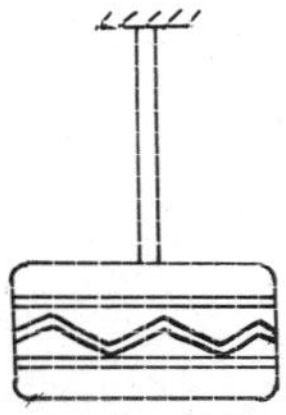
Figure a

Find the moment of inertia I.

Solution:

Draw a free-body diagram (Fig. b) of the wheel, looking down the axis of the wire. Under a rotation θ, the wire applies a restoring force $M = k\theta$ as shown. The equation of motion for the wheel is,

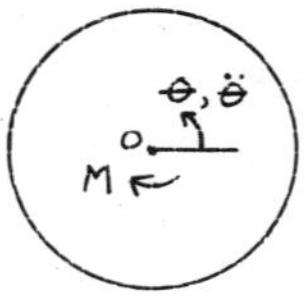

Figure b

$$\circlearrowleft + \Sigma M_o = -k\theta = I\ddot{\theta}$$

$$\therefore\ I\ddot{\theta} + k\theta = 0$$

From Eq. (20.8), $T = \frac{2\pi}{\omega} = 2\pi\sqrt{\frac{I}{k}}$

$$\therefore\ I = k\left(\frac{T}{2\pi}\right)^2$$

with the given data for k and T:

$$I = 8.96 \text{ lb·in·sec}^2$$

20.24

Given: The disk in Fig. a is supported by a hinge at O.

a) Derive the differential equation for large vibration.

b) Use a linearizing approximation to obtain the differential equation for small vibration and the corresponding natural frequency

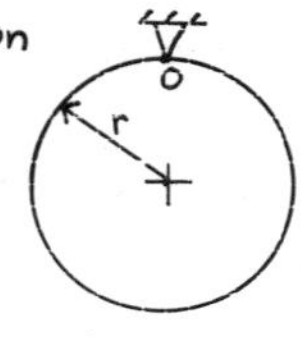

Figure a

Solution:

a) Since the disk rotates about O, we must have its moment of inertia about an axis through O and normal to the plane of the disk. From Appendix F

$$I_G = \tfrac{1}{2} mr^2$$

By Steiner's Theorem (Eq. 18.6)

$$I_o = I_G + mr^2 = \tfrac{3}{2} mr^2$$

Now, with the free-body diagram (Fig. b) of the disk in the displaced configuration, we write the equation of motion as:

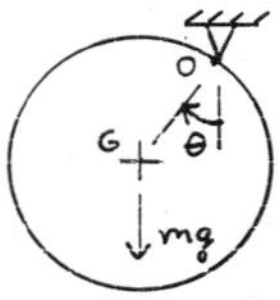

Figure b

$$\circlearrowleft + \Sigma M_o = -mgr\sin\theta = I_o\ddot{\theta}$$

Hence, the equation of motion for large vibration simplifies to:

$$\left(\tfrac{3}{2} r\right)\ddot{\theta} + g\sin\theta = 0 \qquad (a)$$

(Continued)

20.24

b) For small vibrations, $\sin\theta \approx \theta$. Therefore Eq. (a) reduces to

$$\left(\tfrac{3}{2}r\right)\ddot{\theta} + g\theta = 0$$

From Eq. (20.7), the natural frequency is,

$$\omega = \sqrt{\frac{g}{\frac{3}{2}r}} = \sqrt{\frac{2g}{3r}}$$

20.25

Given: The disk in Fig. a rotates about its center O. Its mass is $m = 1.5\ \text{lb·s}^2/\text{in}$ and the constant for each spring is $k = 10\ \text{lb/in}$. The disk is rotated slightly from its equilibrium position & released.

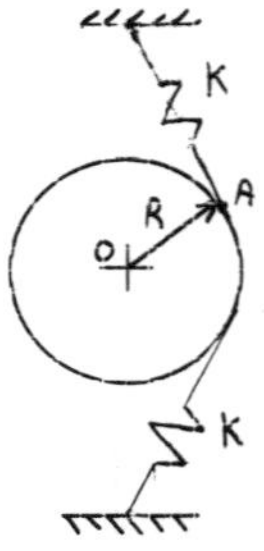

Figure a

Find the frequency of oscillation. Assume the spring forces act tangent to the disk during oscillation.

Solution:

During rotation, point A moves an amount:

$$\delta = R\theta$$

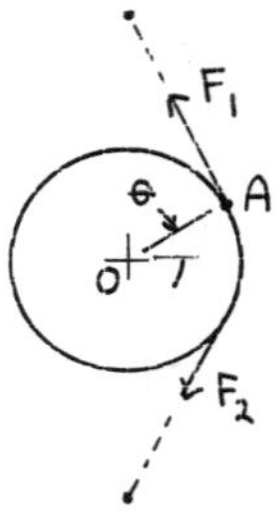

Figure b

The resulting force in the top spring is $F_1 = F - kR\theta$ (a)

Likewise, the resulting force in the bottom spring is

$$F_2 = F + kR\theta \qquad \text{(b)}$$

where F is the initial spring force. The equation for moments about the center of the disk is, by Fig. b,

$$\circlearrowleft + \Sigma M_o = (F_1 - F_2)R = I\ddot{\theta} \qquad \text{(c)}$$

Substitute from Eqs. (a) & (b) into (c):

$$I_o\ddot{\theta} + 2kR\theta = 0 \qquad \text{(d)}$$

From Appendix F, the moment of inertia for the disk is,

$$I = \tfrac{1}{2}mR^2$$

Substitute into Eq. (d) for I and simplify:

$$\tfrac{1}{2}m\ddot{\theta} + 2k\theta = 0$$

From Eq. (20.7) we have,

$$\omega = \sqrt{\frac{2k}{0.5m}} = 2\sqrt{\frac{k}{m}} \quad \text{or} \quad f = \frac{\omega}{2\pi} = \frac{1}{\pi}\sqrt{\frac{k}{m}}$$

with $k = 10\ \text{lb/in}$ & $m = 1.5\ \text{lb·s}^2/\text{in}$, this gives:

$$\underline{\underline{\omega = 5.164\ \text{rad/s}}} \quad \& \quad \underline{\underline{f = 0.822\ \text{Hz}}}$$

20.26

Given: A block of mass hangs from a string that is wrapped around a disk and is attached to a spring with stiffness k, (Fig. a). The disk has moment of inertia I_o.

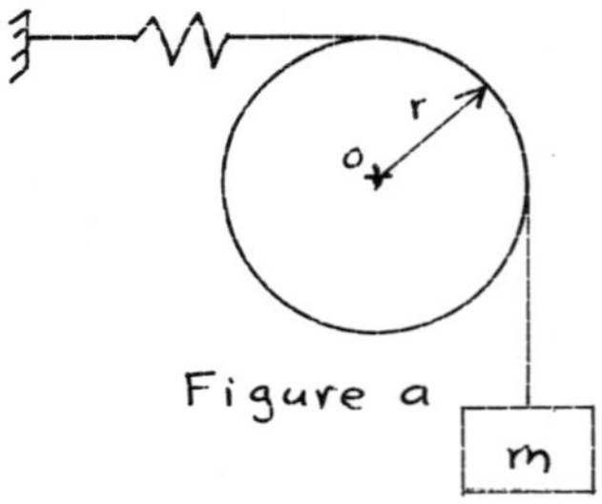

Figure a

Derive an expression for the frequency of vibration of the system. Assume the string does not slip on the disk

Solution:

Consider the free-body diagram of the disk in its static equilibrium position (Fig. b).

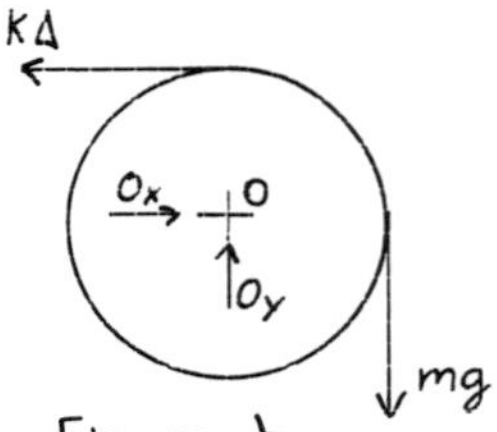

Figure b

By Fig. b,

$$\circlearrowleft + \Sigma M_o = k\Delta r - mgr = 0$$

$$\therefore\ k\Delta = mg \qquad \text{(a)}$$

wher Δ is the static elongation of the spring. Next, consider the free-body diagram of the disk after it has been displaced an angle θ (Fig. c)

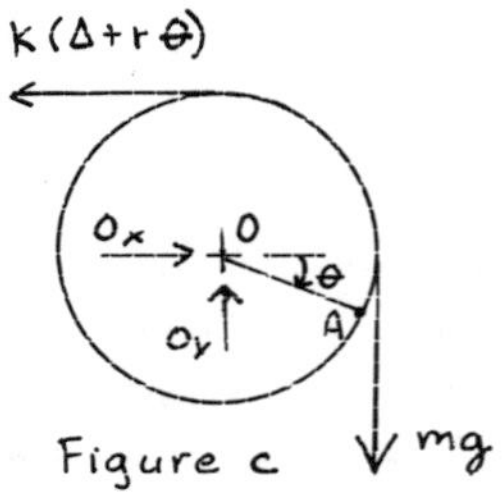

Figure c

By Fig. c,

$$\circlearrowleft + \Sigma M_o = -k(\Delta + r\theta)r + mgr = I\ddot{\theta} \qquad \text{(b)}$$

By Eqs. (a) and (b), we find,

$$I\ddot{\theta} + kr^2\theta = 0 \qquad \text{(c)}$$

(Continued)

20.26 Cont.

Since both the disk and the mass are vibrating about the center point O, the moment of inertia of the system about O is,

$$I = I_o + mr^2$$

Hence, the equation of motion is,

$$(I_o + mr^2)\ddot{\theta} + kr^2\theta = 0$$

By Eq. (20.7), the frequency of vibration is

$$\omega = \sqrt{\frac{kr^2}{I_o + mr^2}} \quad \text{or} \quad f = \frac{1}{2\pi}\sqrt{\frac{kr^2}{I_o+mr^2}}$$

20.27

Given: The mass-spring system shown in Fig. a is in its static equilibrium position. The bars are uniform.

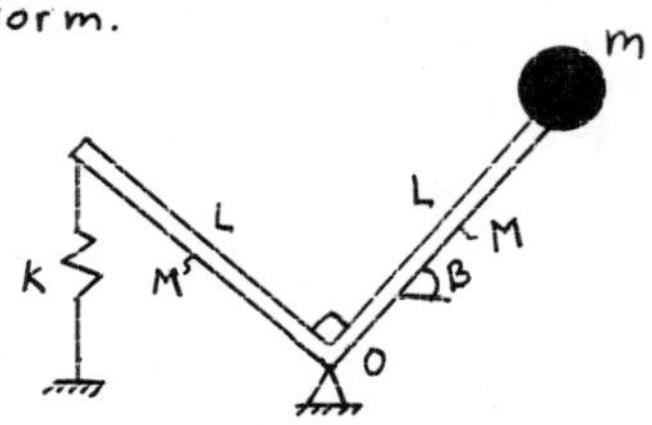

Figure a

Determine the frequency and period of vibration for the system. Neglect friction.

Solution:

Consider the free-body diagram of the system in its static equilibrium position (Fig. b).

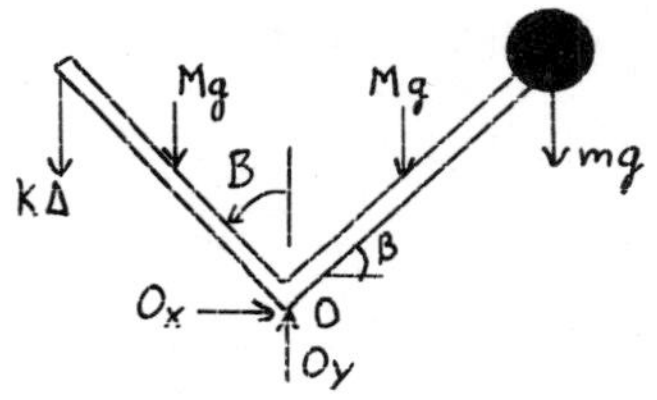

Figure b

By Fig. b,

$$\circlearrowleft + \Sigma M_o = (k\Delta)L\sin\beta + (Mg)\tfrac{L}{2}\sin\beta - (Mg)\tfrac{L}{2}\cos\beta - mgL\cos\beta = 0$$

or, $$k\Delta\sin\beta = \left(\frac{Mg}{2}\right)(\cos\beta - \sin\beta) + mg\cos\beta \qquad (a)$$

Next, consider the free-body diagram of the system displaced an angle θ (Fig. c).

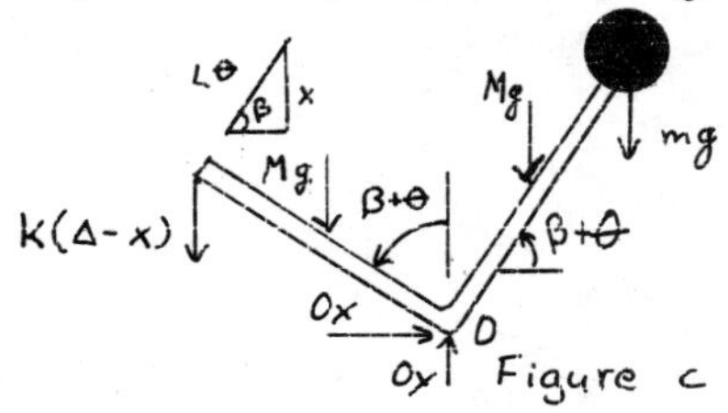

Figure c

The change in length of the spring under this rotation is,

$$x = L\theta\sin\beta$$

By Fig. c,

$$\circlearrowleft + \Sigma M_o = k(\Delta - x)L\sin\beta + (Mg)\tfrac{L}{2}(\sin\beta - \cos\beta) - mgL\cos\beta = I\ddot{\theta} \qquad (b)$$

By Eq. (a) and (b),

$$kxL\sin\beta + I\ddot{\theta} = 0$$

Now, writing x in terms of θ,

$$k(L\theta\sin\beta)L\sin\beta + I\ddot{\theta} = 0 \qquad (c)$$

The moment of inertia of the system about the hinge O is (see Appendix F):

$$I = 2\left(\tfrac{1}{3}ML^2\right) + mL^2 \qquad (d)$$

By Eq (c) and (d), we obtain,

$$(k\sin^2\beta)\theta + \left(\tfrac{2}{3}M + m\right)\ddot{\theta} = 0$$

$$(2M+3m)\ddot{\theta} + (3k\sin^2\beta)\theta = 0 \qquad (e)$$

Now, by Eqs. (e) and (20.7), the frequency of vibration is,

$$\omega = \sin\beta\sqrt{\frac{3k}{(2M+3m)}}, \quad f = \frac{\sin\beta}{2\pi}\sqrt{\frac{3k}{(2M+3m)}}$$

By Eq. (20.8), the period is,

$$T = \frac{2\pi}{\omega} = \frac{2\pi}{\sin\beta}\sqrt{\frac{(2M+3m)}{3k}}$$

20.28

Given: The uniform bar AB in Fig. a weighs 360 N. The mass of body C is 15 kg. $k_1 = 1.5$ kN/m, $k_2 = 3.0$ kN/m.

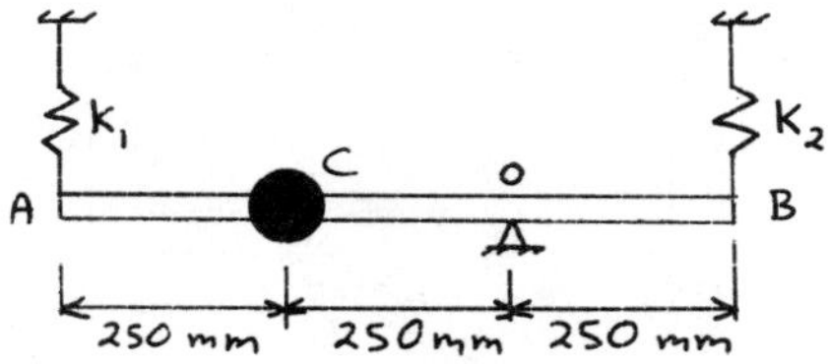

Figure a

Find the frequency of small vibrations

Solution:

As with all previous problems, the system will vibrate about its static equilibrium position. The forces that act in this position are invariant, and hence, they do not influence the vibration. So, we can consider a free-body diagram (Fig. b).

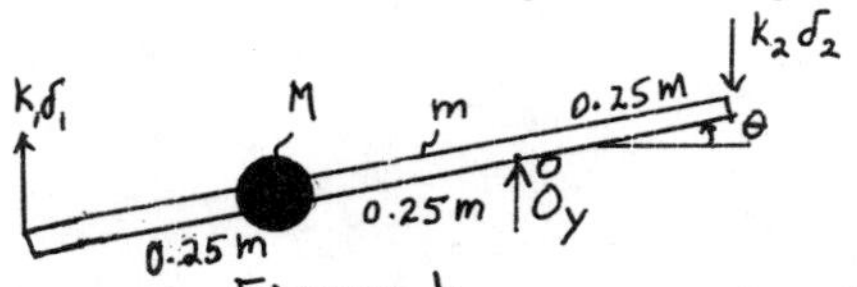

Figure b

(Continued)

20.28 Cont.

The equation of motion is given by (Fig. b)

$$\circlearrowleft + \Sigma M_o = -k_1\delta_1(0.5) - k_2\delta_2(0.25) = I_o\ddot{\theta} \qquad (a)$$

where, $\delta_1 = 0.5\,\theta$ and $\delta_2 = 0.25\,\theta$

The moment of inertia of the system about point O is:

$$I_o = [\tfrac{1}{12}mL^2 + m(0.125)^2] + M(0.25)^2 \qquad (b)$$

where, $m = \frac{360}{9.81} = 36.7$ kg and $M = 15$ kg

By Eqs. (a),(b), δ_1 and δ_2, we obtain,

$$(-0.5^2 k_1 - 0.25^2 k_2)\theta = [(\tfrac{0.75^2}{12} + 0.125^2)m + (0.25^2)M]\ddot{\theta}$$

Now, values for k_1, k_2, M, and m gives,

$$3.23125\ddot{\theta} + 562.5\,\theta = 0$$

Hence, by Eq. (20.7),

$$\omega = \sqrt{\frac{562.5}{3.23125}} = 13.19 \text{ rad/s}, \qquad f = \frac{\omega}{2\pi} = 2.10 \text{ Hz}$$

20.29

Given: A cylindrical body floats upright in a liquid with density ρ. The body has area A and mass m. It is given an initial velocity v_0. Account for the liquid by increasing the cylinder's mass 50%.

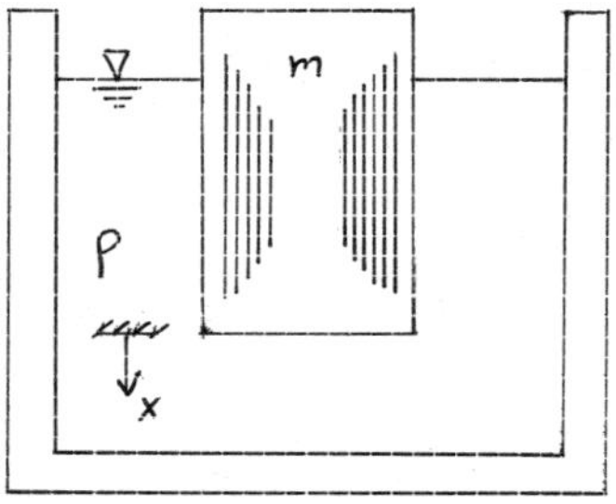

Figure a

Derive the differential equation for free vibration and determine the frequency.

Solution:

The body is shown in its static equilibrium position in Fig. b,

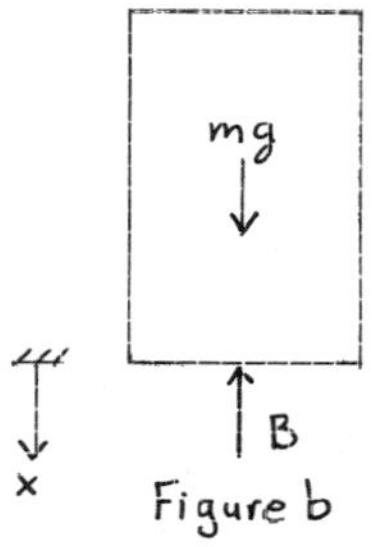

Figure b

From Fig. b, we can see that under static equilibrium,

$$B = mg$$

where B is the buoyant force on the cylinder. So from Fig. c for the vibrating cylinder,

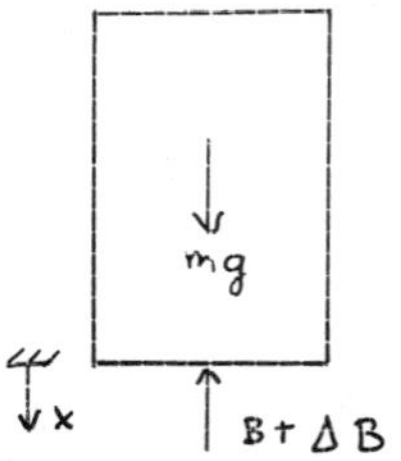

Figure c

The equation of motion is,

$$\Sigma F_x = -\Delta B = (1.5\,m)\ddot{x}$$

where $\Delta B = (\rho g)Ax$ is the weight of the liquid displaced during the vibration. Hence, the equation of motion is,

$$1.5\,m\ddot{x} + \rho g A x = 0$$

From Eq. (20.7), the frequency of vibration is,

$$\omega = \sqrt{\frac{\rho g A}{1.5\,m}} \quad \text{or} \quad f = \frac{1}{2\pi}\sqrt{\frac{\rho g A}{1.5\,m}}$$

20.30

Given: The cylinder, in Fig. a, has mass m and is displaced from its equilibrium position. It rolls on the surface.

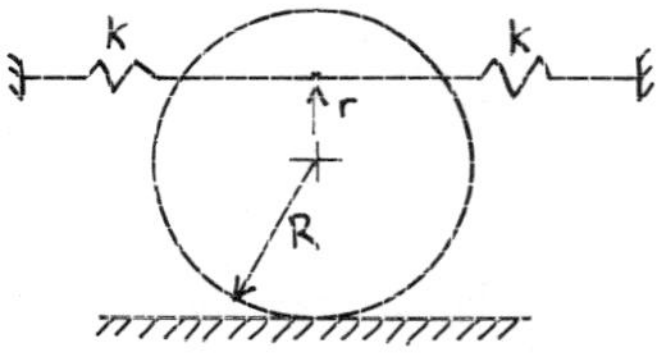

Figure a

a) Derive the differential equation of motion for small vibrations.

b) Derive a formula for the period of the motion.

Solution:

a) Let the initial tension in each spring be F_o. (Fig. b)

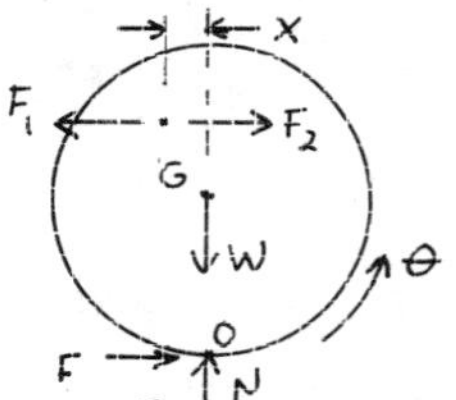

Figure a

(Continued)

20.30 Cont.

After the cylinder is displaced, the spring forces (Fig. b) are,

$$F_1 = F_0 - kx \quad , \quad F_2 = F_0 + kx \qquad (a)$$

where $x = (r+R)\theta$, for small motions. The sum of moments about point O is,

$$\circlearrowleft + \Sigma M_O = (F_1 - F_2)(r+R) = I_O \ddot{\theta} \qquad (b)$$

By Eq. (a) and (b),

$$I_O \ddot{\theta} + 2k(r+R)^2 \theta = 0$$

From Appendix F and Steiner's Theorem (Eq. 18.6), the moment of inertia of the cylinder is,

$$I_O = \frac{1}{2} m R^2 + m R^2 = \frac{3}{2} m R^2$$

So the equation of motion becomes

$$3 m R^2 \ddot{\theta} + 4k(r+R)^2 \theta = 0$$

b) From Eq. (20.8), the period of the motion is

$$T = \frac{\pi R}{(r+R)} \sqrt{\frac{3m}{k}}$$

20.31

Given: The cylinder shown in Fig. a undergoes small vibrations as it rolls on the inclined plane. $k = 15$ lb/in, $m = 6$ lb·s²/ft.

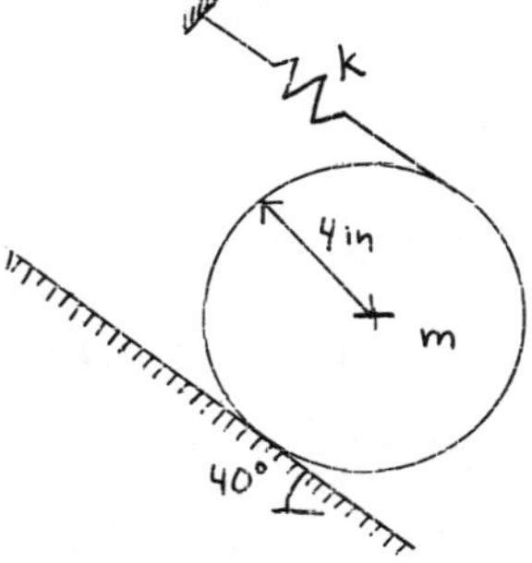

Figure a

Find its period.

Solution:

As with the previous problems, the forces that act on the system in the static equilibrium configuration drop out of the equation of motion. We need consider only the increment in spring force during vibration (Fig. b).

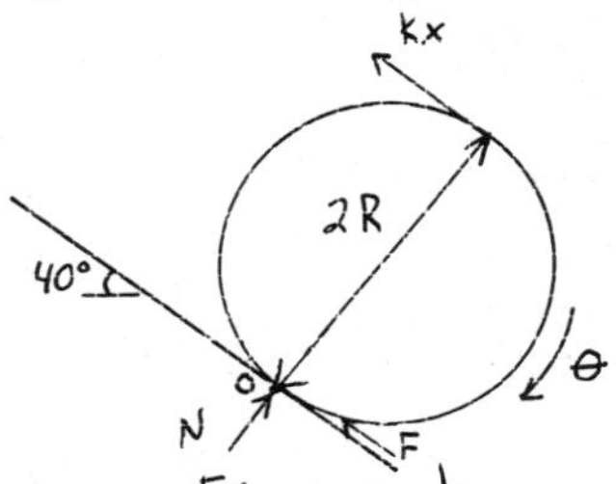

Figure b

For the cylinder in Fig. b, the sum of moments about O gives,

$$\circlearrowleft + \Sigma M_O = -kx(2R) = I_O \ddot{\theta} \qquad (a)$$

where $x = 2R\theta$ and from Appendix F and Steiner's Theorem (Eq. 18.6),

$$I_O = \frac{1}{2} m R^2 + m R^2 = \frac{3}{2} m R^2 \qquad (b)$$

By Eq. (a) and (b),

$$\frac{3}{2} m R^2 \ddot{\theta} + 4kR^2 \theta = 0$$

or

$$3m\ddot{\theta} + 8k\theta = 0$$

With $k = 15$ lb/in and $m = 6$ lb·s²/ft $= 0.5$ lb·s²/in, the period is given by Eq. (20.8) as,

$$T = 2\pi \sqrt{\frac{3m}{8k}} = 0.702 \text{ sec}$$

20.32

Given: A uniform slender bar of length L is suspended in a vertic plane by a frictionless hinge located a distance s above the center of mass of the bar (Fig. a).

a) Derive an equation relating the period T of small vibration of the bar to the distance s.

b) Determine s such that T is minimum. Show that this value of s is equal to the radius of gyration of the bar relative to its center.

c) Plot T/T_{min} vs s/L for $0.1 \le s/L \le 0.5$. What happens at $s/L = 0$?

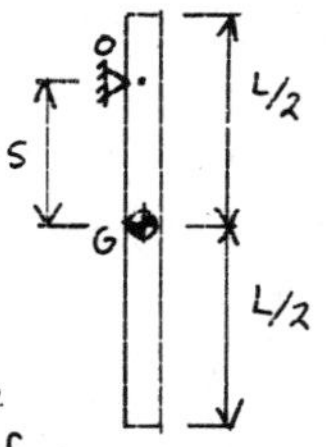

Figure a

Solution:

a) The bar is shown in its displaced position in Fig. b.

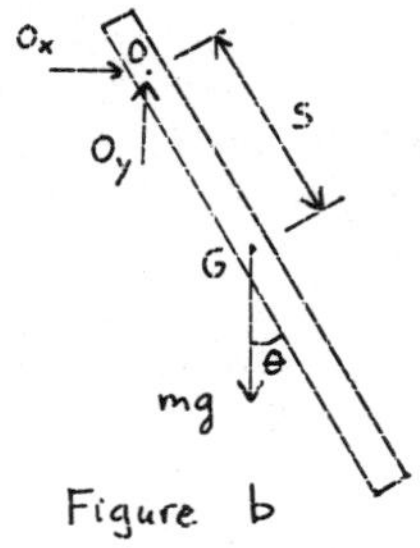

Figure b

By Fig. b,

$$\circlearrowleft + \Sigma M_O = -mgs(\sin\theta) = I_O \ddot{\theta}$$

For small vibration, $\sin\theta \approx \theta$ and we have,

$$I_O \ddot{\theta} + mgs\theta = 0 \qquad (a)$$

(Continued)

20.32 Cont.

From Steiner's Theorem (Eq. 18.6) and Appendix F, the moment of inertia of the bar about the pivot is,

$$I_o = I_G + ms^2 = m(\tfrac{1}{12}L^2 + s^2) \quad (b)$$

By Eq. (a) and (b),

$$m_{eff}\ddot{\theta} + k_{eff}\,\theta = 0 \quad (c)$$

where,

$$m_{eff} = m(\tfrac{1}{12}L^2 + s^2), \quad k_{eff} = mgs$$

Then, by Eqs. (20.7) and (20.8), we have

$$\omega = \sqrt{\frac{12gs}{L^2+12s^2}}, \quad \text{and} \quad T = 2\pi\sqrt{\frac{L^2+12s^2}{12gs}} \quad (d)$$

b) To find the minimum period, we find s such that $dT/ds = 0$. Differentiation of Eq. (d) yields,

$$\frac{dT}{ds} = \pi\left(\frac{12s^2 - L^2}{12gs^2}\right)\Bigg/\sqrt{\frac{L^2+12s^2}{12gs}}$$

For T to be a minimum (T_{min}) we must have $dT/ds = 0$. This occurs when,

$$12s^2 - L^2 = 0 \quad \text{or} \quad s = L/\sqrt{12} = 0.2887L$$

For a slender bar, the moment of inertia about the center is $I_G = mL^2/12$. The radius of gyration is given by,

$$k = \sqrt{\frac{I_G}{m}} = L/\sqrt{12}$$

Therefore, $T = T_{min}$ at $s = k$

c) When $s = L/\sqrt{12}$, $T = T_{min}$. By Eq. (d), we obtain,

$$T_{min} = 2\pi\sqrt{\frac{2L}{g\sqrt{12}}}$$

The plot of T/T_{min} vs. s/L is shown in Fig. c. As $s/L \to 0$, the period $T \to \infty$. When $s/L = 0$, the bar is hinged at its center of gravity and there is no restoring force to cause vibration.

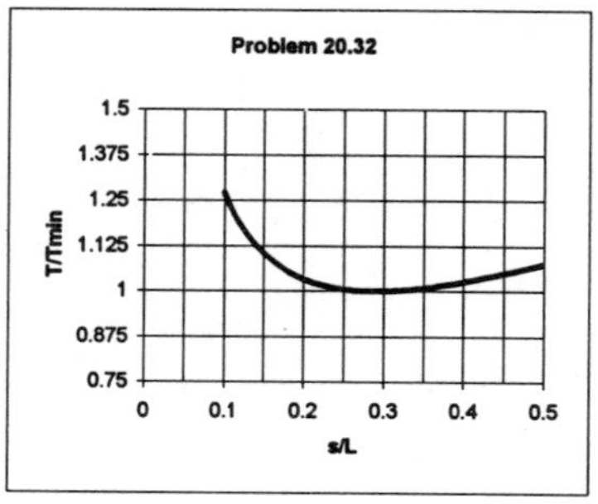

Figure c

20.33

Given: A metronome consists of a uniform bar of mass $m/4$, a mass m, and a torsional spring (Fig. a). The restoring moment from the spring is $k\theta$. When $b = 3$, $f = 1$ Hz.

Find b such that $f = 3$ Hz

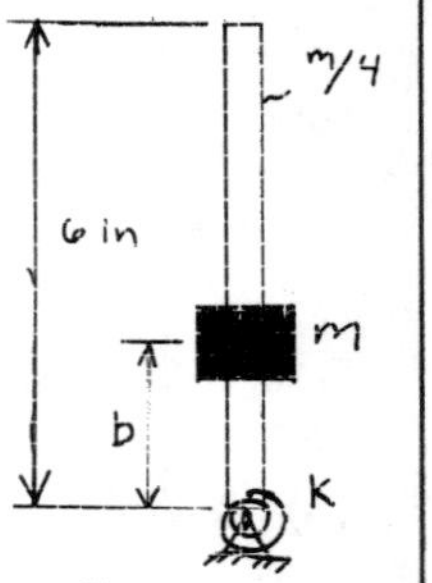

Figure a

Solution:

The general equation of motion for small vibrations of the metronome may be obtained using the free-body diagram in Fig. b. Thus, by Fig. b,

$$\circlearrowleft + \Sigma M_o = (mg)b\sin\theta + \left(\frac{mg}{4}\right)3\sin\theta - k\theta = I_o\ddot{\theta} \quad (a)$$

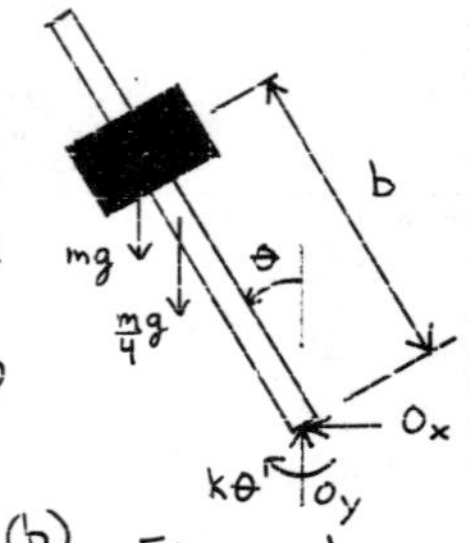

Figure b

For small vibrations, $\sin\theta = \theta$. The moment of inertia of the metronome about point O is (see Appendix F)

$$I_o = mb^2 + \tfrac{1}{3}\left(\tfrac{m}{4}\right)6^2 = m(b^2+3) \quad (b)$$

Substitution of Eq. (b) into Eq. (a) gives,

$$m(b^2+3)\ddot{\theta} + [-mg(b+3/4) + k]\theta = 0 \quad (c)$$

From Eq. (c) we see that,

$$m_{eff} = m(b^2+3), \quad k_{eff} = k - mg[b + 3/4] \quad (d)$$

Hence, by Eqs. (d) and (20.7), the frequency is,

$$f = \frac{1}{2\pi}\sqrt{\frac{k_{eff}}{m_{eff}}} = \frac{1}{2\pi}\sqrt{\frac{k - mg[b+3/4]}{m(b^2+3)}} \quad (c)$$

We must solve this problem with the given data but without knowing k and m explicitly. So, we manipulate Eq. (c) to isolate k and m.

Multiplying by 2π and squaring both sides,

$$(2\pi f)^2 = \frac{k - mg[b+3/4]}{m(b^2+3)}$$

Multiplying by (b^2+3) and simplifying,

$$(2\pi f)^2(b^2+3) = \frac{k}{m} - g[b+3/4]$$

Therefore, $k/m = (2\pi f)^2(b^2+3) + g[b+3/4] \quad (d)$

Now, for $b = 3''$, $f = 1$ Hz (use $g = 386.4$ in/s^2)

$$k/m = (2\pi)^2(3^2+3) + 386.4(3.75) = 1922.74$$

(Continued)

20.33 Cont.

For, $k/m = 1922.74$ and $f = 3$ Hz, find b from Eq. (d),

$$1922.74 = (2\pi(3))^2 (b^2+3) + 386.4(b+0.75)$$

After simplification,

$$b^2 + 1.0875b - 1.596 = 0$$

from which, $\underline{b = 0.8316 \text{ in}}$

20.34

Given: The series circuit shown in Fig. a has a voltage E (t) applied.

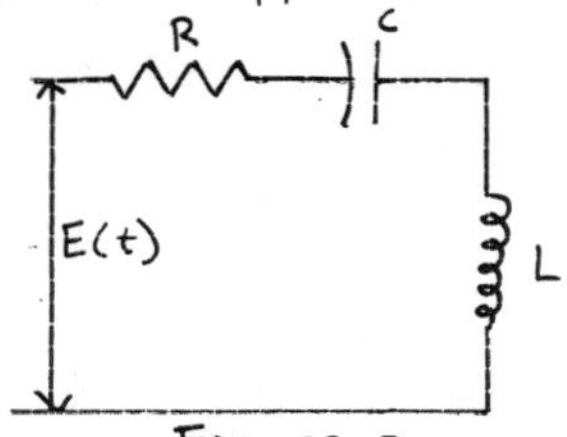

Figure a

Show that the angular frequency of free oscillation of charge Q of the series circuit of Fig. a is,

$$\omega = \sqrt{\frac{1}{LC} - \left(\frac{R}{2L}\right)^2}$$

Solution:

By analogy to a mechanical (spring, mass, damper) system,

$$m \equiv L$$
$$c \equiv R \qquad (a)$$
$$k \equiv \frac{1}{C}$$

For oscillation to occur, the system must be underdamped

$$4km > c^2 \qquad (b)$$

Which by analogy (see Eq. a) requires that

$$\frac{4L}{C} > R^2$$

For a damped spring-mass system, Eq. (20.14b) gives,

$$\omega = \frac{\sqrt{4km - c^2}}{2m} \qquad (c)$$

For the analogous electric circuit,

$$\omega = \frac{\sqrt{\frac{4L}{C} - R^2}}{2L} = \sqrt{\frac{1}{CL} - \left(\frac{R}{2L}\right)^2}$$

20.35

Given: A capacitor with $C = 4(10^{-5})$ F, a resistor with $R = 40\,\Omega$, and an inductor with $L = 0.10$ H are connected in a series circuit, Fig. a. The voltage E is removed suddenly, and simultaneously a switch is closed across the terminals

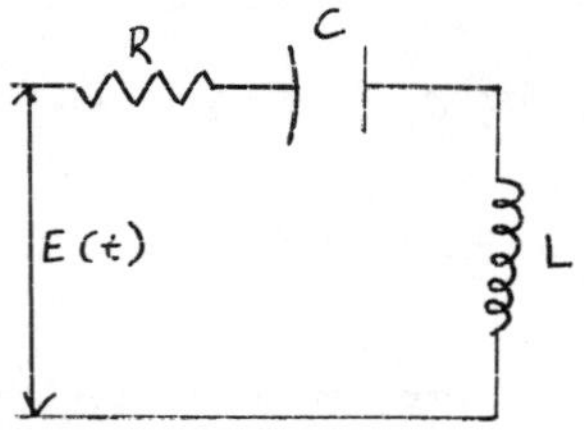

Figure a

a) Determine the period of oscillation of charge Q.

b) Derive the ratio of amplitudes of successive oscillations.

Solution:

a) By analogy to a mechanical (spring, mass, damper) system,

$$m \equiv L$$
$$c \equiv R \qquad (a)$$
$$k \equiv 1/C$$

For oscillation to occur, the system must be underdamped,

$$4km > c^2 \qquad (b)$$

or by analogy (see Eq. a), we must have

$$\frac{4L}{C} > R^2 \qquad (c)$$

For the given data, Eq. (c) gives,

$$\frac{4(0.1)}{4(10^{-5})} > 40^2$$

$$10{,}000 > 1600$$

Therefore, oscillation occurs.
From Eq. (20.14 b), the damped spring-mass system vibrates at,

$$\omega = \frac{\sqrt{4km - c^2}}{2m} \qquad (d)$$

which, by analogy to the electric circuit, gives,

$$\omega = \frac{1}{2L}\sqrt{\frac{4L}{C} - R^2}$$

with the given data,

$$\omega = \frac{1}{2(0.1)}\sqrt{10{,}000 - 1600} = 458 \text{ rad/s}$$

by Eq. (20.8),

$$\underline{T = \frac{2\pi}{\omega} = 0.013711 \text{ s}}$$

(Continued)

20.35

b) For this system, the logarithmic decrement is given by,

$$\delta = \frac{2\pi\lambda}{\omega}$$

where by analogy to Eq. (20.14a)

$$\lambda = \frac{R}{2L}$$

Therefore, $\delta = \frac{2\pi}{\omega}\left(\frac{R}{2L}\right) = 2.742$

so from Eq. (20.24)

$$x_1/x_2 = e^{\delta} = 15.52$$

20.36

Given: The spring-mass-damper system, shown in Fig. a, undergoes small angular vibration about the hinge. It is in the static equilibrium position.

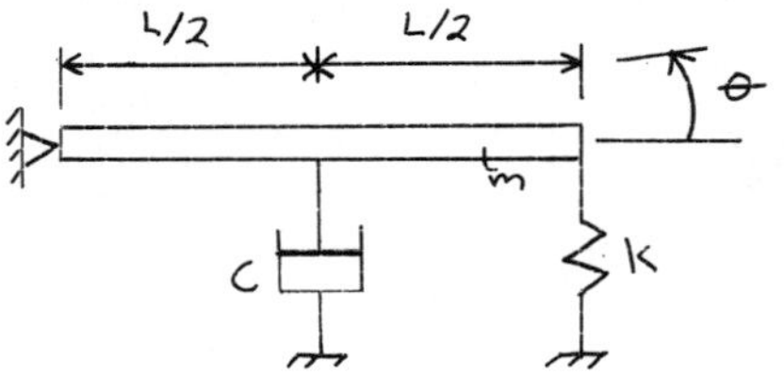

Figure a

Derive a formula for the frequency of the system. Neglect friction of the support.

Solution:

First, we find the effective stiffness, mass, and damping constant for the system. The free-body diagram of the system in its static equilibrium position is shown in Fig. b.

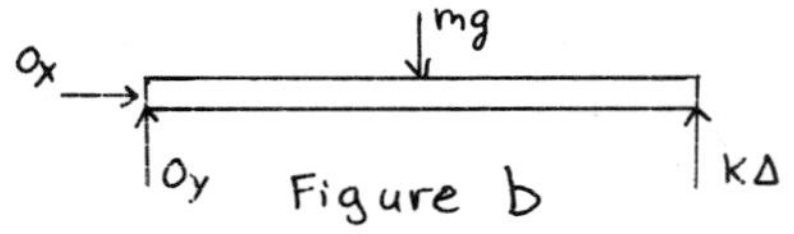

Figure b

By Fig. b,

$$\circlearrowleft + \Sigma M_0 = k\Delta L - mg\left(\frac{L}{2}\right) = 0 \quad \text{or} \quad k\Delta = \frac{1}{2}mg \qquad \text{(a)}$$

Now, consider the free-body diagram of the system displaced through an angle θ (Fig. c).

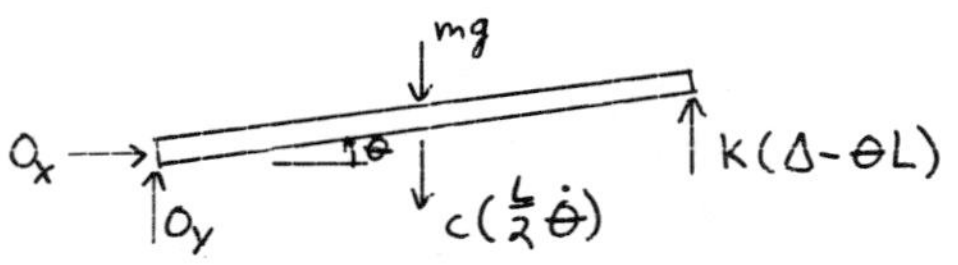

Figure c

By Fig. c,

$$\circlearrowleft + \Sigma M_0 = k(\Delta - \theta L)L - mg\left(\frac{L}{2}\right) - c\left(\frac{L}{2}\dot{\theta}\right)\frac{L}{2} = I\ddot{\theta} \qquad \text{(b)}$$

Then, by Eqs. (a) and (b), we may write,

$$I\ddot{\theta} + c\left(\frac{L}{2}\right)^2\dot{\theta} + kL^2\theta = 0 \qquad \text{(c)}$$

The moment of inertia of the bar about the hinge is (see Appendix F),

$$I = \frac{1}{3}mL^2 \qquad \text{(d)}$$

So, by Eqs. (c) and (d), we obtain the equation of motion in the form,

$$m_{eff}\ddot{\theta} + c_{eff}\dot{\theta} + k_{eff}\theta = 0$$

where,

$$m_{eff} = 4m, \quad c_{eff} = 3c, \quad k_{eff} = 12k$$

Assuming that the system is underdamped, we get the frequency of small vibration from Eq. (20.14b) as,

$$\omega = \frac{\sqrt{4(12k)(4m) - (3c)^2}}{2(4m)} \qquad \text{(e)}$$

Simplification of Eq. (e) yields,

$$\omega = \sqrt{\frac{3k}{m} - \left(\frac{3c}{8m}\right)^2} \quad \text{or} \quad f = \frac{1}{2\pi}\sqrt{\frac{3k}{m} - \left(\frac{3c}{8m}\right)^2}$$

20.37

Given: The spring-mass-damper system shown in Fig. a is in its static equilibrium position.

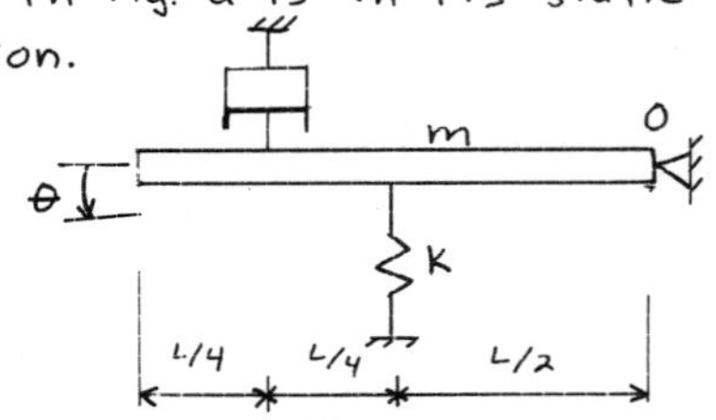

Figure a

Derive a formula for the frequency of small vibration for the system. Neglect friction of the support.

Solution:

First, we find the effective stiffness, mass, and damping constant for the system. The free-body diagram of the system in its equilibrium position is shown in Fig. b.

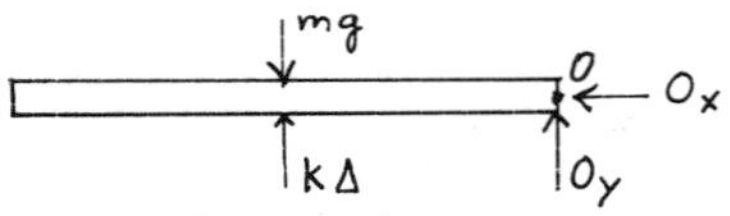

Figure b

By Fig. b,

$$\circlearrowleft + \Sigma M_0 = L/2\,(mg) - L/2\,(k\Delta) = 0$$

or

$$k\Delta = mg \qquad \text{(a)}$$

(Continued)

20.37 Cont.

Next, consider the free-body diagram of the system rotated through an angle θ (Fig. c).

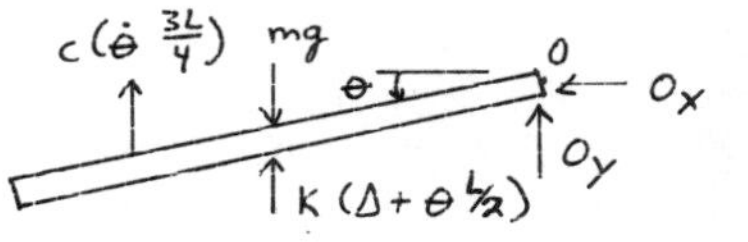

Figure c

By Fig. c,

$$\circlearrowleft + \Sigma M_o = mg\left(\frac{L}{2}\right) - c\,\dot{\theta}\left(\frac{3L}{4}\right)^2 - k\left(\Delta + \theta\frac{L}{2}\right)\frac{L}{2} = I\ddot{\theta} \qquad (b)$$

Then, by Eqs. (a) and (b), we have after simplification,

$$I\ddot{\theta} + \left(\frac{9L^2}{16}\right)c\dot{\theta} + k\left(\frac{L^2}{4}\right)\theta = 0 \qquad (c)$$

Where the moment of inertia of the bar about the hinge is (see Appendix F),

$$I = \frac{1}{3}mL^2 \qquad (d)$$

Then, with Eqs. (c) and (d), we obtain the equation of motion in the form,

$$16m\ddot{\theta} + 27c\dot{\theta} + 12k\theta = 0 \qquad (e)$$

By Eq. (e), see that,

$$m_{eff} = 16m\ , \quad c_{eff} = 27c\ , \quad k_{eff} = 12k$$

Assuming that the system is underdamped, we get the frequency of small vibration from Eq. (20.14 b) as

$$\omega = \frac{\sqrt{4(12k)(16m) - (27c)^2}}{2(16m)} \qquad (f)$$

Simplification of Eq. (f) yields

$$\omega = \sqrt{\frac{3k}{4m} - \left(\frac{27c}{32m}\right)^2} \quad \text{or} \quad f = \frac{1}{2\pi}\sqrt{\frac{3k}{4m} - \left(\frac{27c}{32m}\right)^2}$$

20.38

Given: The spring-mass-damper system shown in Fig. a is in its static equilibrium position.

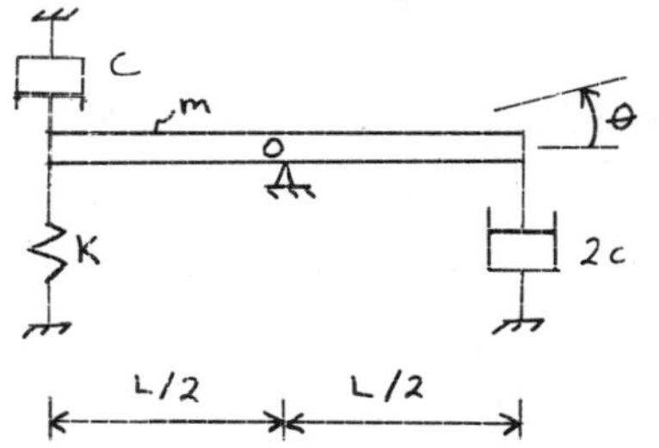

Figure a

Derive a formula for the frequency of small vibration for the system. Neglect friction of the support.

Solution:

First, we find the effective stiffness, mass, and damping constant for the system. The free-body diagram of the system in its static equilibrium position is shown in Fig. b.

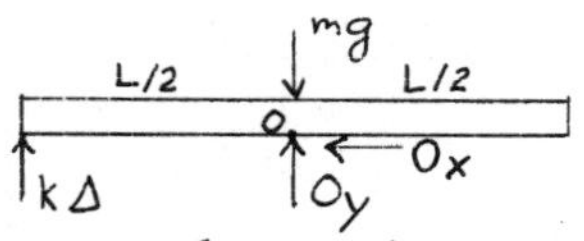

Figure b

By Fig. b,

$$\circlearrowleft + \Sigma M_o = -k\Delta L/2 = 0$$

Hence, the spring elongation Δ is zero in the static equilibrium position. The free-body diagram of the system when it is rotated through the angle θ is shown in Fig. c.

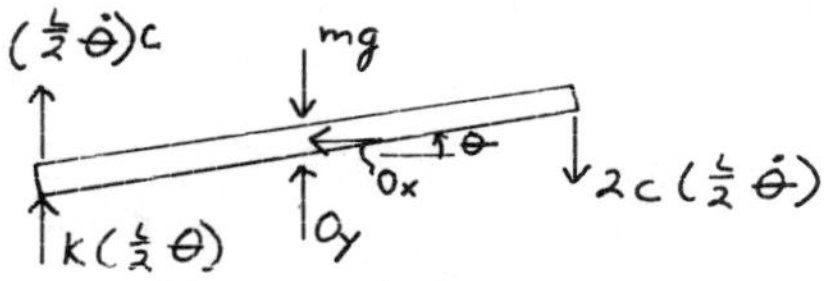

Figure c

By Fig. c,

$$\circlearrowleft + \Sigma M_o = -3c\left(\frac{L}{2}\dot{\theta}\right)\left(\frac{L}{2}\right) - k\left(\frac{L}{2}\theta\right)\frac{L}{2} = I\ddot{\theta} \qquad (a)$$

Where the moment of inertia of the bar about its center is (see Appendix F),

$$I = \frac{1}{12}mL^2 \qquad (b)$$

By Eqs. (a) and (b), we may write the equation of motion in the form,

$$m\ddot{\theta} + 9c\dot{\theta} + 3k = 0 \qquad (c)$$

By Eq. (c), we see that,

$$m_{eff} = m\ , \quad c_{eff} = 9c\ , \quad k_{eff} = 3k$$

Assuming that the system is underdamped, we get the frequency of small vibration from Eq. (20.14) as,

$$\omega = \frac{\sqrt{4(3k)(m) - (9c)^2}}{2m} \qquad (d)$$

Simplification of Eq. (d) yields,

$$\omega = \sqrt{\frac{3k}{m} - \left(\frac{9c}{2m}\right)^2} \quad \text{or} \quad f = \frac{1}{2\pi}\sqrt{\frac{3k}{m} - \left(\frac{9c}{2m}\right)^2}$$

20.39

<u>Given:</u> The bell-crank system, shown in Fig. a, undergoes small oscillations about the equilibrium position.

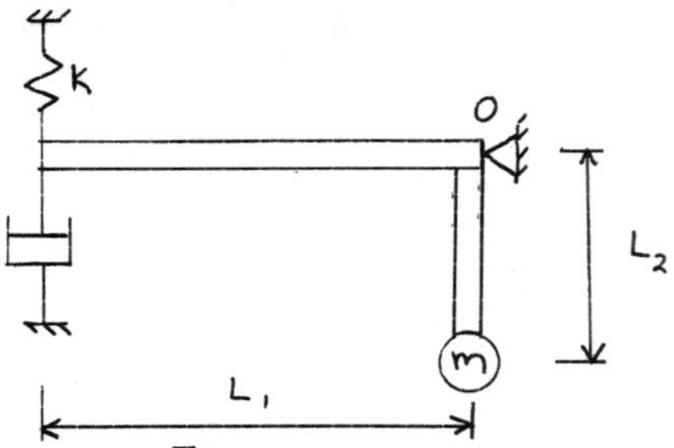

Figure a

a) Derive the equation of motion of the system. Neglect the mass and weight of the arms and the friction of the bearing.

b) Determine the critical damping constant for the system.

<u>Solution:</u>

a) Since the arms are assumed weightless, there is no force in the spring in the equilibrium position. We apply a rotation θ to the crank and draw its free-body diagram (Fig. b).

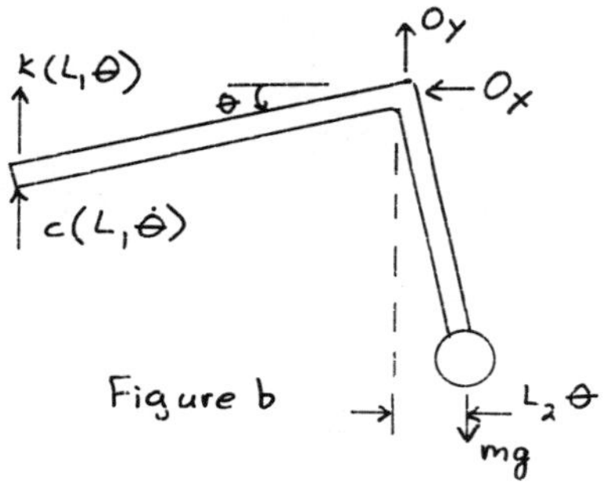

Figure b

By Fig. b,

$$\circlearrowleft + \Sigma M_O = -k(L_1\theta)L_1 - c(L_1\dot{\theta})L_1 - mgL_2\theta = I\ddot{\theta}$$

where, $I = mL_2^2$

Hence, the equation of motion is,

$$mL_2^2\ddot{\theta} + cL_1^2\dot{\theta} + (kL_1^2 + mgL_2)\theta = 0 \qquad (a)$$

b) From Eq. (a), the effective stiffness, damping constant, and mass are,

$$k_{eff} = kL_1^2 + mgL_2, \quad c_{eff} = cL_1^2, \quad m_{eff} = mL_2^2 \qquad (b)$$

Critical damping is given by Eq. (20.19) as,

$$c_{cr} = 2\sqrt{km} \qquad (c)$$

For this problem, Eqs. (b) and (c) yield

$$(cL_1^2)_{cr} = 2\sqrt{(kL_1^2 + mgL_2)(mL_2^2)}$$

Hence, the critical damping ratio is,

$$c_{cr} = \frac{2mL_2}{L_1}\sqrt{\frac{k}{m} + \frac{gL_2}{L_1^2}}$$

20.40

<u>Given:</u> The energy of recoil of a large cannon is absorbed in a recoil spring. At the end of the recoil, a damping device engages to return the gun to the firing position without oscillation. A particular gun weighs 40 tons and recoils 10 ft with an initial speed of 30 ft/s.

a) Find the spring constant.

b) Find the minimum damping constant for no oscillation.

<u>Solution:</u>

a) The mass of the gun is,

$$m = \frac{40\,(2000)}{32.2} = 2484 \ \frac{\text{lb}\cdot\text{s}^2}{\text{ft}}$$

The kinetic energy at the start of recoil is,

$$T = \tfrac{1}{2}mv^2 = \tfrac{1}{2}(2484)(30^2)$$

$$T_1 = 1,118,000 \ \text{ft}\cdot\text{lb} \qquad (a)$$

At the end of recoil, the potential energy of the spring is,

$$V_2 = \tfrac{1}{2}k\Delta^2 = 50\,k \qquad (b)$$

If energy is conserved during recoil

$$T_1 = V_2 \qquad (c)$$

Then, Eqs. (a), (b), and (c) yield

$$50k = 1,118,000$$

or $k = 22,360$ lb/ft

b) The minimum damping constant to prevent oscillation is [by Eq. (20.19)]

$$c_{CR} = 2\sqrt{km}$$

So, for this cannon,

$$c_{min} = c_{CR} = 2\sqrt{(22,360)(2484)}$$

$$c_{min} = 14,905 \ \text{lb}\cdot\text{s/ft}$$

20.41

Given: A body with mass of 7.2 kg is suspended from a spring with a constant of 1.80 kN/m. A damper provides a force of 36 N when the mass has a speed of 600 mm/s.

Determine the frequency [Hz] of free vibrations.

Solution:

The force in the damper is,

$$f_d = cv$$

So, the damping constant is $c = f_d/v$. For the data given,

$$c = 36/0.60 = 60 \text{ N·s/m}$$

The critical damping constant is [by Eq. (20.19)]

$$c_{CR} = 2\sqrt{km} = 2\sqrt{(1800)(7.2)}$$

$$c_{CR} = 228 \text{ N·s/m}$$

Since $c < c_{CR}$, the system is underdamped. The frequency of vibration is given by Eq. (20.14b) with $(c^2 < 4km)$.

So,

$$\omega = \frac{\sqrt{4km - c^2}}{2m} = \frac{\sqrt{4(1800)7.2 - 60^2}}{2(7.2)} = 15.81 \frac{\text{rad}}{\text{s}}$$

or,

$$f = \frac{\omega}{2\pi} = 2.516 \text{ Hz}$$

20.42

Given: A body with mass 7.2 kg is suspended from a spring with a constant of 1.8 kN/m. A damper provides a damping force of 36 N when the speed of the mass is 600 mm/s

a) Determine the logarithmic decrement of the motion.

b) Calculate the ratio of maximum amplitudes of two successive cycles.

c) Determine the ratio of the maximum amplitude of a cycle to that of another that follows it 3 cycles later.

Solution:

a) The force in the damper is $f_d = cv$, so the damping constant is $c = f_d/v$. For the data given,

$$c = 36/0.6 = 60 \text{ N·s/m}$$

The critical damping constant is

$$c_{CR} = 2\sqrt{km}$$

or

$$c_{CR} = 2\sqrt{(1800)(7.2)} = 228 \text{ N·s/m}$$

So, the relative damping ratio [Eq. (20.22)] is

$$\zeta = \frac{c}{c_{CR}} = 0.263$$

The logarithmic decrement is given by Eq. (20.25). So,

$$\delta = \frac{2\pi\zeta}{\sqrt{1-\zeta^2}} = 1.713$$

b) Let x_1 and x_2 be the maximum amplitudes of two successive cycles of vibration. By Eq. (20.24),

$$\ln\left(\frac{x_1}{x_2}\right) = \delta$$

or

$$\frac{x_1}{x_2} = e^{\delta} = 5.54$$

c) For 3 successive cycles, the amplitude ratio is,

$$\frac{x_1}{x_2} = e^{3\delta} = 170.5$$

20.43

Given: The mass of the body in Fig. a is $m = 0.800$ lb·s²/in. The constant of each spring is $k = 20$ lb/in. The shaking force is $F(t) = 0$. The mass has an initial displacement $x_0 = 4$ in. After 25 cycles, the amplitude has decreased to 0.10 in.

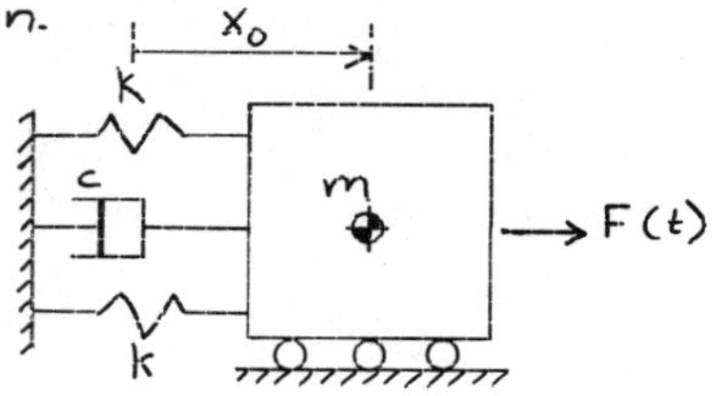

Figure a

What is the damping constant c of the dashpot?

Solution:

For the system shown, $k = 40$ lb/in, $m = 0.80$ lb·s²/in, and the critical damping ratio is [By Eq. (20.19)]

$$c_{CR} = 2\sqrt{km} = 11.31 \text{ lb·s/in}$$

The amplitude ratio for 25 cycles is obtained by Eq. (20.24) as,

$$\frac{x_1}{x_{26}} = e^{25\delta} = \frac{4}{0.10}$$

So, the logarithmic decrement is,

$$\delta = \frac{1}{25}\ln\left(\frac{4}{0.10}\right) = 0.1476$$

The relative damping ratio is found from Eq. (20.25),

$$\delta = \frac{2\pi\zeta}{\sqrt{1-\zeta^2}}$$

(Continued)

20.43 Cont.

which gives $\zeta = 0.0235$ (from a computer-based equation solver).

The damping constant can now be found from Eq. (20.22),

$$c = \zeta c_{CR} = 0.0235(11.31)$$

$$\underline{\underline{c = 0.266 \text{ lb}\cdot\text{s/in}}}$$

20.44

Given: The spring/mass/damper system, in Fig. a, vibrates with a frequency of 6 Hz. After 10 s, the amplitude of the motion is 70% of its initial amplitude.

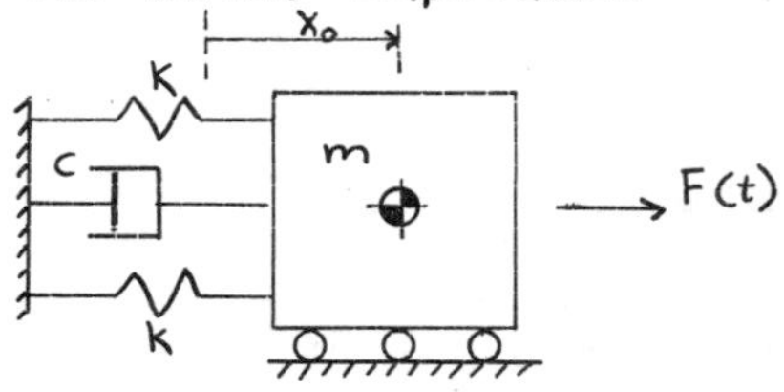

Figure a

a) Determine the value of the logarithmic decrement.

b) Calculate the relative damping ratio ζ.

Solution:

(a) The frequency of vibration is $f = 6$ Hz. So, the period is $T = 1/f = 0.1667$ s. The 10 s interval corresponds to,

$$n = 10/0.1667 = 60 \text{ cycles}$$

The amplitude ratio for 60 cycles is obtained by Eq. (20.24) as

$$\frac{x_1}{x_{61}} = e^{60\delta} = \frac{1}{0.7}$$

So, the logarithmic decrement is

$$\underline{\underline{\delta = \frac{1}{60}\ln\left(\frac{1}{0.7}\right) = 0.005945}}$$

b) The relative damping ratio is defined by Eq. (20.25). With the aid of a computer-based equation solver, we find,

$$\delta = \frac{2\pi\zeta}{\sqrt{1-\zeta}} \quad ; \quad \underline{\underline{\zeta = 0.0009462}}$$

20.45

Given: A seismic pickup used to record vertical vibrations of the earth consists of a mass m supported on two springs. Damping is represented by a dashpot (Fig. a). When the casing of the instrument is displaced a vertical distance z_c, the mass m is displaced a vertical distance z_m from its equilibrium position (Fig. b)

The relative displacement between the mass and the casing is $z_r = z_m - z_c$.

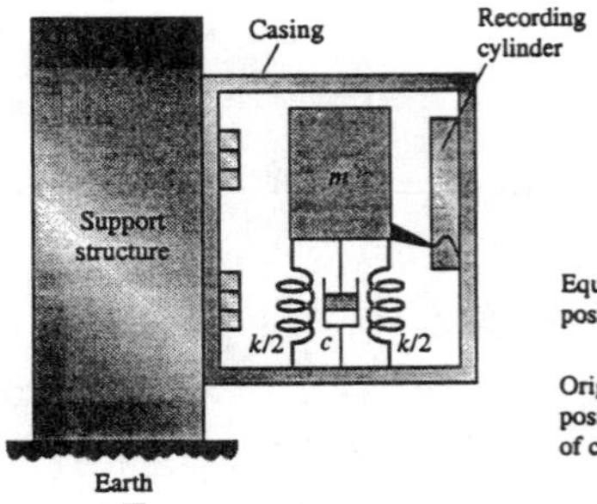

Figure a

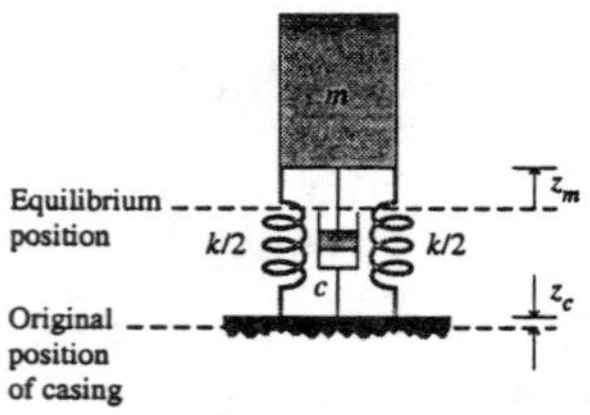

Figure b

Assuming that the vibration of the earth, and hence, the casing, is

$$z_c = Z_c \sin \Omega t \qquad (a)$$

Show that the differential equation for relative displacement of the mass m is,

$$m\ddot{z}_r + c\dot{z}_r + kz_r = Z_c\Omega^2 m \sin\Omega t \qquad (b)$$

Solution:

The free-body diagram of the mass is shown in Fig. c.

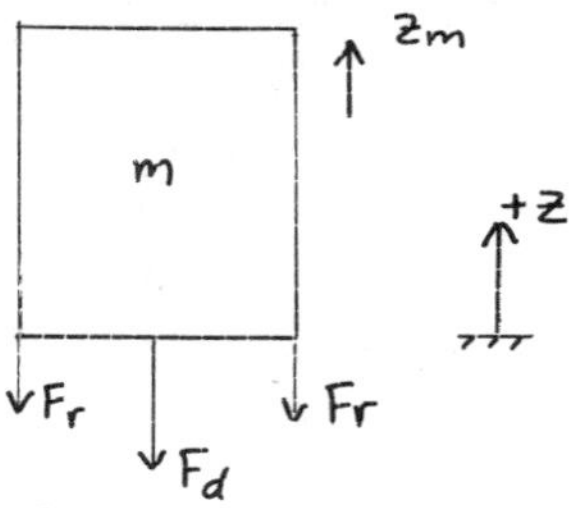

Figure c

The displacement coordinate is the absolute displacement z_m of the mass. The restoring forces F_r and the damping force F_d depends on the relative displacement z_r between the mass and the casing.

$$F_r = k/2\,(z_m - z_c) \quad , \quad F_d = c(\dot{z}_m - \dot{z}_c) \qquad (c)$$

By Fig. c,

$$\Sigma F_z = -2F_r - F_d = m\ddot{z}_m \qquad (d)$$

By Eqs. (c) and (d), we find,

$$k(z_m - z_c) + c(\dot{z}_m - \dot{z}_c) + m\ddot{z}_m = 0 \qquad (e)$$

(Continued)

20.45 Cont.

Now, substituting Eq. (e) for $(z_m - z_c)$ noting that,

$$\ddot{z}_m = \ddot{z}_r + \ddot{z}_c$$

Therefore,

$$k z_r + c\dot{z}_r + m(\ddot{z}_r + \ddot{z}_c) = 0$$

$$k z_r + c\dot{z}_r + m\ddot{z}_r = -m\ddot{z}_c \qquad (f)$$

From Eq. (a), the second derivative of z_c is,

$$\ddot{z}_c = -\Omega^2 z_c \sin\Omega t \qquad (g)$$

Substitution of Eq. (g) into Eq. (f) gives,

$$m\ddot{z}_r + c\dot{z}_r + k z_r = m z_c \Omega^2 \sin\Omega t$$

which is identical to Eq. (b).

20.46

Given: A seismic pickup, used to record vertical vibrations of the earth, consists of a mass m supported on two springs. Damping is represented by a dashpot (Fig. a). When the casing of the instrument is displaced a vertical distance z_c, the mass m is displaced a vertical distance z_m from its equilibrium position (Fig. b). The relative displacement between the mass and the casing is $z_r = z_m - z_c$.

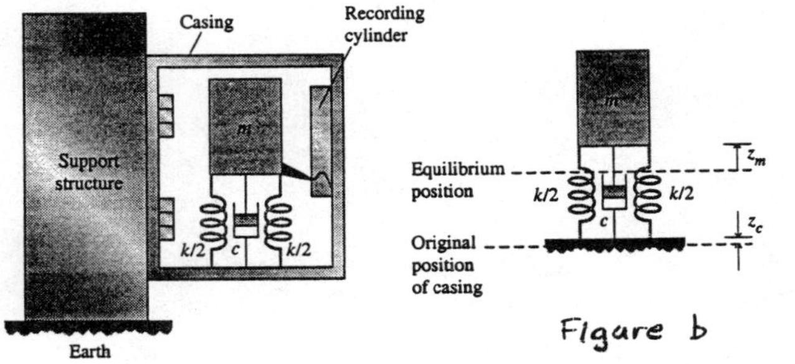

Figure a

Figure b

Assuming that the vibration of the earth, and hence, the casing, is,

$$z_c = z_c \sin\Omega t \qquad (a)$$

The differential equation for relative displacement of the mass m is,

$$m\ddot{z}_r + c\dot{z}_r + k z_r = z_c \Omega^2 m \sin\Omega t \quad (b)$$

a) Show that the amplitude z_r of relative displacement z_r is,

$$z_r = \frac{z_c(\Omega/\omega_0)^2}{\sqrt{[1-(\Omega/\omega_0)^2]^2 + (2\zeta\Omega/\omega_0)^2}}$$

b) Show that for flexible springs ($\omega_0 << \Omega$), z_r is approximately equal to z_c, and the phase angle between z_r and z_c is approximately 180°.

Solution:

a) The steady state solution is given by Eq. (20.31),

$$z_r = \frac{(F_0/k)\sin(\Omega t - \gamma)}{\sqrt{(1-(\Omega/\omega_0)^2)^2 + (2\zeta\Omega/\omega_0)^2}} \qquad (c)$$

where F_0 is the amplitude of the forcing function. In this case, from Eq. (b),

$$F_0 = z_c \Omega^2 m \qquad (d)$$

The amplitude of z_r, from Eq. (c), with Eq. (d), is,

$$z_r = \frac{\frac{z_c \Omega^2 m}{k}}{\sqrt{(1-(\Omega/\omega_0)^2)^2 + (2\zeta\Omega/\omega_0)^2}}$$

and, since $m/k = 1/\omega_0^2$, we have,

$$z_r = \frac{z_c(\Omega/\omega_0)^2}{\sqrt{(1-(\Omega/\omega_0)^2)^2 + (2\zeta\Omega/\omega_0)^2}} \qquad (d)$$

b) For flexible springs $(\Omega/\omega_0) >> 1$

So, from Eq. (d),

$$z_r \approx \frac{z_c(\Omega/\omega_0)^2}{\sqrt{(\Omega/\omega_0)^4 + 4\zeta^2(\Omega/\omega_0)^2}}$$

$(\Omega/\omega_0)^4$ gets large faster than $4\zeta^2(\Omega/\omega_0)^2$, so

$$z_r \approx \frac{z_c(\Omega/\omega_0)^2}{\sqrt{(\Omega/\omega_0)^4}}$$

Therefore,

$$z_r \approx z_c$$

The phase angle γ is found from Eq (20.28),

$$\tan\gamma = \frac{c\Omega}{k - m\Omega^2} = \frac{\frac{c}{k}\Omega}{1 - \frac{m\Omega^2}{k}} = \frac{(c/k)\Omega}{1 - \frac{\Omega^2}{\omega_0^2}}$$

As $\frac{\Omega}{\omega_0}$ gets large $\tan\gamma \to 0$ from the negative side. Therefore,

$$\gamma \to 180°$$

20.47

Given: A seismic pickup used to record vertical vibrations of the earth consists of a mass m supported on two springs. Damping is represented by a dashpot (Fig. a). When the casing of the instrument is displaced a vertical distance z_c, the mass m is displaced a vertical distance z_m from its equilibrium position (Fig. b). The relative displacement between the mass and the casing is $z_r = z_m - z_c$.

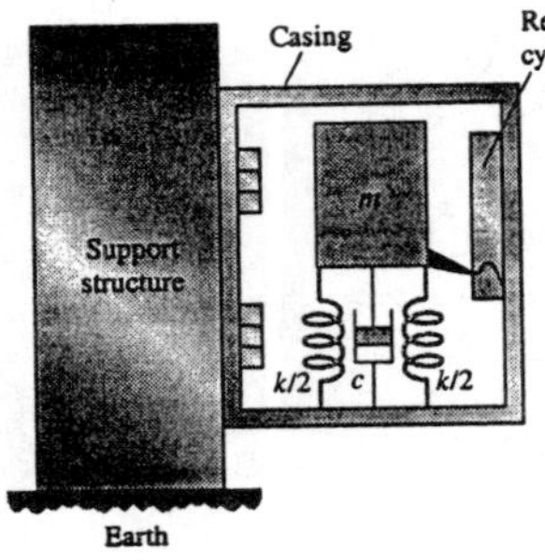

Figure a

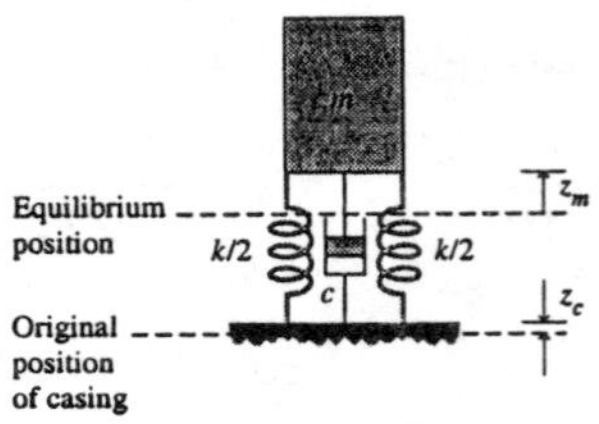

Figure b

Assuming that the vibration of the earth, and hence, the casing, is,

$$z_c = Z_c \sin \Omega t \qquad (a)$$

The differential equation for relative displacement of the mass m is,

$$m\ddot{z}_r + c\dot{z}_r + kz_r = Z_c \Omega^2 m \sin \Omega t \qquad (b)$$

Show that for stiff springs ($\omega_o >> \Omega$), the amplitude Z_r of the relative displacement z_r of the seismic pickup is approximately $Z_c \Omega^2/\omega_o^2$.

Solution:

First, we must find the amplitude Z_r of the vibration z_r. By Eq. (20.31),

$$z_r = \frac{(F_o/k)\sin(\Omega t - \gamma)}{\sqrt{(1-(\Omega/\omega_o)^2)^2 + (2\zeta\Omega/\omega_o)^2}} \qquad (c)$$

where F_o is the amplitude of the forcing function. In this case, from Eq. (b),

$$F_o = Z_c \Omega^2 m \qquad (d)$$

The amplitude of z_r, from Eq. (c), with Eq. (d), is

$$Z_r = \frac{\frac{Z_c \Omega^2 m}{m}}{\sqrt{(1-(\Omega/\omega_o)^2)^2 + (2\zeta\Omega/\omega_o)^2}}$$

since $m/k = 1/\omega_o^2$ we have

$$Z_r = \frac{Z_c(\Omega/\omega_o)^2}{\sqrt{(1-(\Omega/\omega_o)^2)^2 + (2\zeta\Omega/\omega_o)^2}} \qquad (e)$$

For stiff springs (Ω/ω_o) is small and $(\Omega/\omega_o)^2 \to 0$.

So, $\left(\frac{\Omega}{\omega_o}\right)$ drops out of the denominator of Eq. (e) and we have,

$$\underline{\underline{Z_r = Z_c(\Omega/\omega_o)^2}}$$

20.48

Given: A horizontal force $P\sin\Omega t$ is applied to the mass m shown in Fig. a. Neglect gravity and assume that the tension in the string remains constaint.

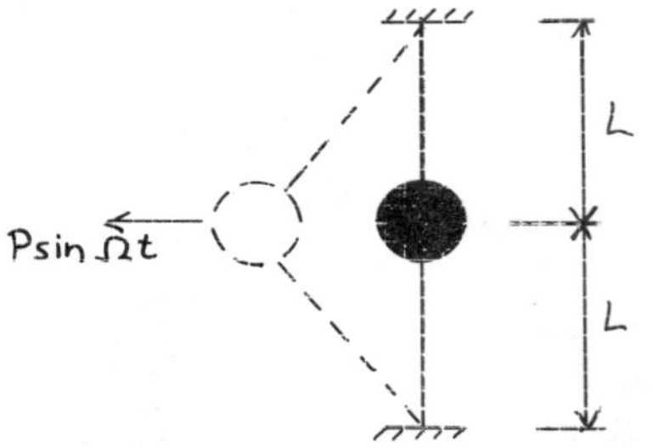

Figure a

a) Derive an equation for the magnification factor as a function of Ω.

b) Determine the ratio of Ω to the undamped angular frequency ω_o such that the magnification factor is 0.10.

Solution:

a) We start by writing the equation of motion for the system and finding the angular frequency ω_o. The free-body diagram of the mass in the displaced position is shown in Fig. b.

(Continued)

20.48 Cont.

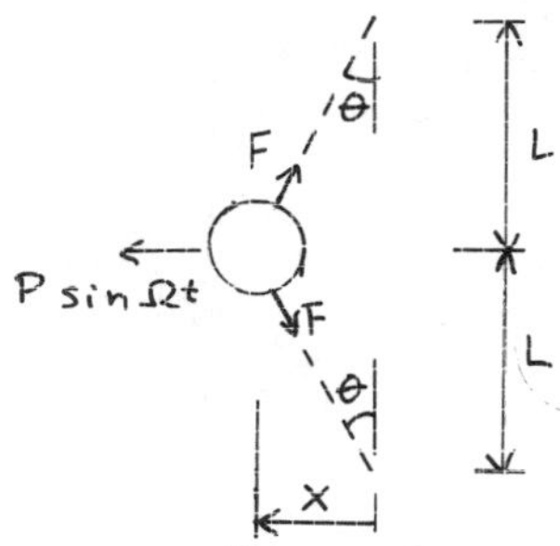

Figure b

The equation of motion is, by Fig. b,

$$\Sigma F_x = P\sin\Omega t - 2F\sin\theta = m\ddot{x} \qquad (a)$$

Since the tension in the string remains constant, θ is small. Then $\sin\theta \approx \theta$.
Also $\ddot{x} = L\sin\ddot{\theta} = L\ddot{\theta}$
Substitution into Eq. (a) yields,

$$m(L\ddot{\theta}) + 2F\theta = P\sin\Omega t \qquad (b)$$

From Eq. (b), we see that,

$$m_{eff} = mL \quad , \quad k_{eff} = 2F$$

Hence,

$$\omega_o = \sqrt{\frac{k}{m}} = \sqrt{\frac{2F}{mL}} \qquad (c)$$

Since the system is undamped, the magnification factor is given by Eq. (20.34) as,

$$\frac{\theta_{max}}{\Delta} = \left| \frac{1}{1-\Omega^2/\omega_o^2} \right| \qquad (d)$$

where θ_{max} is the maximum angular displacement of the vibration and Δ is the static angular displacement due to P.
By Eqs. (c) and (d), we have,

$$\frac{\theta_{max}}{\Delta} \left| \frac{1}{1-\Omega^2/\left(\frac{2F}{mL}\right)} \right|$$

b) From Eq. (d), for $\theta_{max}/\Delta = 0.10$

$$0.10 = \left| \frac{1}{1-\Omega^2/\omega_o^2} \right| \qquad (e)$$

Assume that Ω is less than ω_o, and drop the absolute value bars from Eq. (e),
Then,

$$0.1 = \frac{1}{1-\Omega^2/\omega_o^2}$$

or

$$\frac{\Omega^2}{\omega_o^2} = -9$$

However, this can not be true since $\frac{\Omega^2}{\omega_o^2}$ is >0.

So we must have $\Omega > \omega_o$ Then, by Eq. (e),

$$0.1 = \frac{1}{\frac{\Omega^2}{\omega_o^2} - 1}$$

There fore,

$$\frac{\Omega}{\omega_o} = \sqrt{11} = 3.317$$

Alternatively, we may derive the equation for the magnification factor as follows:

Let $\qquad \theta = \Theta \sin\Omega t \qquad (f)$

Then, by Eqs. (b) and (f), we obtain, solving for Θ,

$$\Theta = \frac{P}{2F}\left(\frac{1}{1-\frac{\Omega^2}{2F/mL}}\right) \qquad (g)$$

Also, by Eq. (d), we see that the square of the undamped angular frequency is given by,

$$\omega_o^2 = \frac{2F}{mL} \qquad (h)$$

So, by Eqs. (g) and (h), the magnification factor may be expressed as,

$$\left| \frac{1}{1-\frac{\Omega^2}{\omega_o^2}} \right| = \frac{2F\Theta}{P} = \frac{\theta_{max}}{\Delta}$$

where $\Delta = P/2F$ is the small static angular displacement of the strings under the action of a constant horizontal force of magnitude P (see Fig. b), acting on the mass m.

20.49

Given: The magnification factor is given by Eq. (20.32).

a) Determine an expression for the frequency ratio Ω/ω_o for which the magnification factor is a maximum.

b) For the frequency ratio determined in part a, show that $\omega_o \geq \Omega$.

(Continued)

20.49 Cont.

Solution:

a) Let MF be the magnification factor given by Eq. (20.32),

$$MF = \frac{1}{\sqrt{(1-\frac{\Omega^2}{\omega_0^2})^2 + 4\zeta^2(\Omega^2/\omega_0^2)}} \quad (a)$$

we want to maximize MF in terms of Ω/ω_0. To simplify notation, let $y=\left(\frac{\Omega}{\omega_0}\right)^2$. Substituting y into Eq. (a) and squaring both sides, gives,

$$MF^2 = \frac{1}{(1-y)^2+4\zeta^2 y} = \frac{1}{1+(4\zeta^2-2)y+y^2} \quad (b)$$

Now, differentiating Eq. (b) with respect to y and setting the derivative equal to zero,

$$\frac{d(MF^2)}{dy} = \frac{4\zeta^2-2+2y}{[1+(4\zeta^2-2)y+y^2]^2} = 0$$

Therefore,

$$4\zeta^2 - 2 + 2y = 0$$

or,

$$y = 1-2\zeta^2$$

So MF is maximum when,

$$y = \left(\frac{\Omega}{\omega_0}\right)^2 = 1-2\zeta^2$$

or

$$\frac{\Omega}{\omega_0} = \sqrt{1-2\zeta^2} \quad (c)$$

b) The relative damping ratio ζ is always a positive number. Also, the expression $\sqrt{1-2\zeta^2}$ is real since Ω and ω_0 are real. Therefore,

$$0 \le \zeta < 0.5$$

hence,

$$\Omega/\omega_0 \le 1.0$$

which means,

$$\omega_0 \ge \Omega$$

20.50

Given: The center of mass of the flywheel of an engine that is mounted on four springs is located at a distance e from the axis of the shaft [see Fig. a]. The mass of the flywheel is m and the mass of the engine, including the fly wheel, is m. A viscous dashpot is attached to the housing of the engine. The flywheel rotates with constant angular velocity Ω.

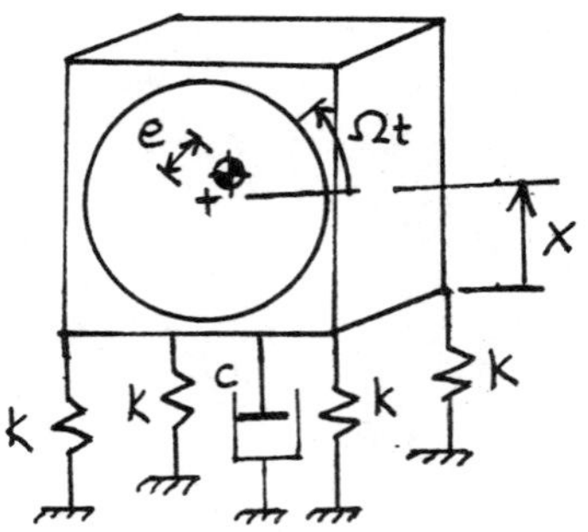

Figure a

a) Assume no horizontal motion of the engine. Show that the differential equation of small vibration is

$$M\ddot{x} + c\dot{x} + 4kx = me\Omega^2 \sin \Omega t$$

Hint: Note that $Mx_G = (M-m)x + m(x + e\sin\Omega t) = Mx + me\sin\Omega t$, where x_G is the coordinate of the mass center of the system.

b) What is the frequency of vertical motion?

Solution:

a) The displacement of the mass (M−m) is x. The displacement of the mass m is $x + e\sin\Omega t$. From the hint, we see that the momentum of the system is,

$$M\dot{x}_G = (M-m)\dot{x} + m(\dot{x} + e\Omega\cos\Omega t) \quad (a)$$

Differentiating Eq. (a) gives,

$$M\ddot{x}_G = (M-m)\ddot{x} + m(\ddot{x} - e\Omega^2 \sin\Omega t) \quad (b)$$

Now, we can write the equation of motion for the engine using a free-body diagram (Fig. b) and Newtons second law.

$$\Sigma F_x = -4F_s - F_d = M\ddot{x}_G \quad (c)$$

For the springs: $F_s = kx$
For the damper: $F_d = c\dot{x}$ (d)

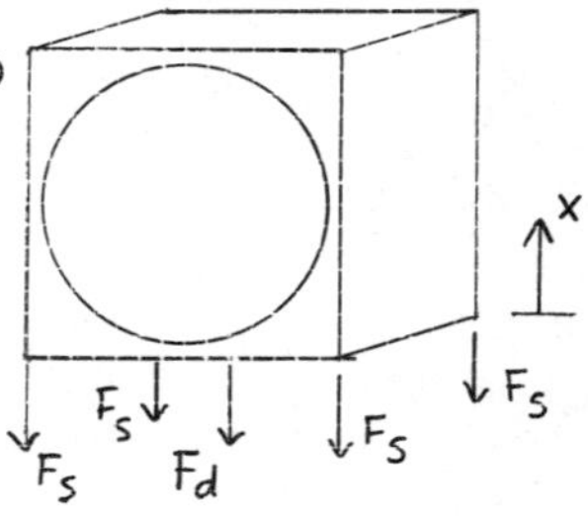

Figure b

By Eqs. (b), (c), and (d)

$$-4kx - c\dot{x} = (M-m)\ddot{x} + m(\ddot{x} - e\Omega^2 \sin\Omega t)$$

or,

$$M\ddot{x} + c\dot{x} + 4kx = me\Omega^2 \sin\Omega t$$

(Continued)

20.50 Cont.

b) Since the system is damped, the transient motion will die out. The frequency of the steady state motion is then the frequency of the internal shaking force.

$$f = \frac{\Omega}{2\pi}$$

20.51

Given: The center of mass of the flywheel of an engine is located at a distance e from the axis of the shaft [see Fig. a]. The mass of the flywheel is m and the mass of the engine, including the flywheel, is M. The flywheel rotates with constant angular velocity $\Omega = 300$ RPM. $M = 1.1658\ \frac{lb\cdot s^2}{in}$, $m = 0.12953\ lb\cdot s^2/in$, $e = 0.10$ in, and $k = 280$ lb/in.

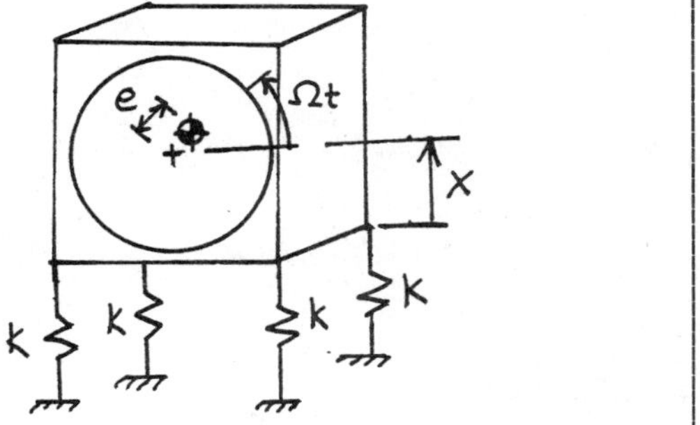

Figure a

Determine the amplitude of the forced vibration. Use $g = 386\ in/s^2$.

Solution:

In Problem 20.50, we are given the equation of motion of the system (after removing damping) as,

$$M\ddot{x} + 4kx = me\Omega^2 \sin \Omega t \qquad (a)$$

where $\Omega = 300$ RPM $= 10\pi$ rad/s.

The static deflection of the spring is,

$$\Delta = \frac{me\Omega^2}{4k} = \frac{(0.12953)(0.10)(10\pi)^2}{4(280)}$$

$$\Delta = 0.01141 \text{ in}$$

The natural angular frequency is,

$$\omega_0 = \sqrt{\frac{4k}{M}} = \sqrt{\frac{4(280)}{1.1658}} = 30.995 \text{ rad/s}$$

By Eq. (20.34), the magnification factor is,

$$\frac{X_{max}}{\Delta} = \left| \frac{1}{1 - (\Omega/\omega_0)^2} \right| = 36.57$$

Hence the amplitude of the forced vibration is,

$$X_{max} = (36.57)(0.01141)$$

$$X_{max} = 0.417 \text{ in}$$

20.52

Given: The center of mass of the flywheel of an engine is located at a distance $e = 0.10$ in from the axis of the shaft [see Fig. a]. The mass of the flywheel is $m = 0.12953\ lb\cdot s^2/in$, and the mass of the engine, including the flywheel, is $M = 1.1658\ lb\cdot s^2/in$. The flywheel rotates at constant angular velocity $\Omega = 300$ rpm and $k = 280$ lb/in. Damping is half of its critical value.

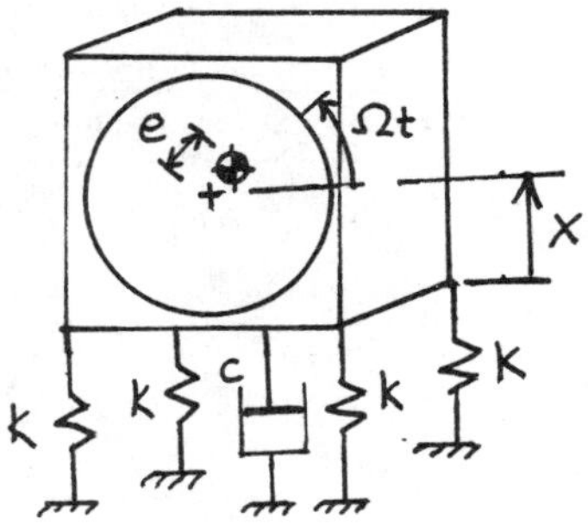

Figure a

Calculate the amplitude of the forced vibration and the phase angle γ of the vibration. The differential equation of motion is,

$$M\ddot{x} + c\dot{x} + 4kx = me\Omega^2 \sin \Omega t$$

Solution:

The frequency of the shaking force is,

$$\Omega = 300 \text{ rpm} = 10\pi \text{ rad/s}$$

The static deflection of the springs is,

$$\Delta = \frac{me\Omega^2}{4k} = \frac{(0.12953)(0.10)(10\pi)^2}{4(280)}$$

$$\Delta = 0.01141 \text{ in}$$

The relative damping ratio is $\zeta = \frac{c}{c_{cr}} = 0.5$

The natural angular frequency is

$$\omega_0 = \sqrt{\frac{4k}{M}} = \sqrt{\frac{4(280)}{1.1658}} = 30.995 \text{ rad/s}$$

(Continued)

20.52 Cont.

The magnification factor from Eq. (20.32) is,

$$\frac{X_{max}}{\Delta} = \frac{1}{\sqrt{\left(1-\left(\frac{\Omega}{\omega_o}\right)^2\right)^2 + 4\,\zeta^2\left(\frac{\Omega}{\omega_o}\right)^2}}$$

where,

$\left(\frac{\Omega}{\omega_o}\right) = 1.01358$ Therefore,

$$\frac{X_{max}}{\Delta} = \frac{1}{\sqrt{(0.0273)^2 + 4(0.5)^2(1.01358)^2}}$$

$$\frac{X_{max}}{\Delta} = 0.98624$$

Hence,

$$X_{max} = 0.98624(0.01141) = 0.01125 \text{ in}$$

b) The phase angle γ is given by Eq. (20.30)

$$\tan\gamma = \frac{2\zeta(\Omega/\omega_o)}{1-(\Omega/\omega_o)^2} = \frac{2(0.5)(1.01358)}{1-1.01358^2}$$

$\tan\gamma = -37.07$, or $\underline{\gamma = 91.55°}$

Note: By Eq. (g) of sec. 20.5, $\sin\gamma$ is always positive.

Therefore, $0 \le \gamma \le 180°$.

20.53

Given: A body vibrates on a straight line. Damping is critical. The body is displaced 50 mm from its equilibrium position by a 36 N static force in a test. In service, the body is subjected to a harmonic shaking force with amplitude of 72 N and a frequency equal to twice the frequency of free vibration of the undamped system.

Determine the amplitude of the forced vibration.

Solution:

For critical damping, $\zeta = 1.0$.

The spring constant is $k = 36/50 = 0.72$ N/m.

The shaking force has amplitude $F_o = 72$ N and frequency $\Omega = 2\omega_o$.

So, with $\Omega/\omega_o = 2.0$ and $\zeta = 1.0$, the magnification factor is (From Eq. 20.32)

$$\frac{X_{max}}{\Delta} = \frac{1}{\sqrt{(1-2^2)^2 + 4(1^2)(2^2)}} = 0.20$$

Under service, $\Delta = F_o/k = 72/0.72 = 100$ mm

$\therefore\ X_{max} = 0.20\Delta = 20$ mm

20.54

Given: The lever shown in Fig. a is rigid and massless. The angular speed of the motor is much larger than the natural angular frequency of the system. The length L is large compared to e so the motion of the slider may be assumed equal to $e \sin\Omega t$.

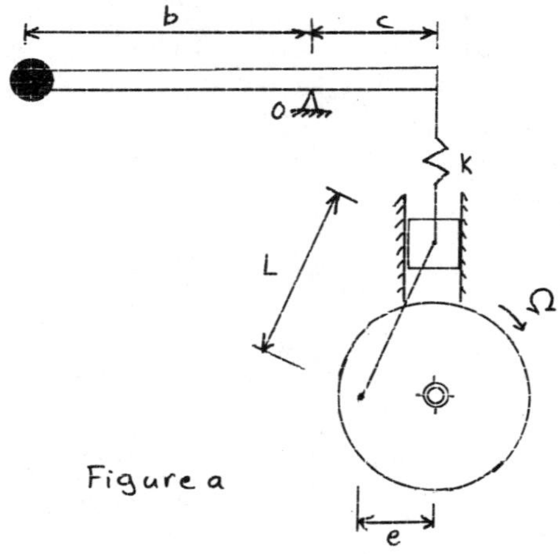

Figure a

a) Derive the differential equation of vibration of the lever.

b) Derive an expression for the magnification factor and show that it is nearly zero.

c) Show that the phase angle γ is approximately 180°.

Solution:

a) The free-body diagram of the lever in the deflected position is shown in Fig. b.

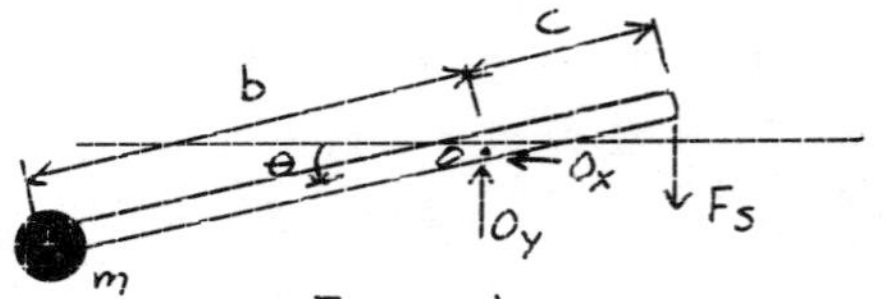

Figure b

The elongation δ of the spring is given by the motion of the right end of the lever and that of the slider.

Thus, $\delta = c\theta - e\sin\Omega t$

The spring force is,

$$F_s = k\delta = k(c\theta - e\sin\Omega t)$$

(Continued)

20.54 Cont.

The equation of motion is obtained by summing moments about the hinge.

$$\circlearrowleft + \sum M_o = -k(c\theta - e\sin\Omega t)c = I\ddot{\theta}$$

where $I = mb^2$ since the lever is massless. Therefore,

$$\underline{mb^2\ddot{\theta} + kc^2\theta = kce\sin\Omega t}$$

b) Since the system is undamped, the magnification factor is given by Eq. (20.34),

$$\frac{X_{max}}{\Delta} = \left|\frac{1}{1-\left(\frac{\Omega}{\omega_o}\right)^2}\right|$$

If $\Omega >> \omega_o$ then $\frac{\Omega}{\omega_o} >> 1$

So,

$$\frac{X_{max}}{\Delta} \approx 0$$

The motion of the lever is very small.

c) From Eq. (20.30), $\tan\gamma = 0$ since $\zeta = 0$. Likewise, from Eq. (g) of section 20.5, $\sin\gamma = 0$. But, Eq. (f) of Sec. 20.5 says that,

$$\cos\gamma = \frac{k - m\Omega^2}{\sqrt{(k-m\Omega^2)^2 + (c\Omega^2)}} \quad (a)$$

Note that $\omega_o^2 = \frac{k}{m}$. We can rearrange Eq. (a) to get,

$$\cos\gamma = \frac{m\left(\frac{k}{m} - \Omega^2\right)}{m\sqrt{\left(\frac{k}{m} - \Omega^2\right)^2 + \left(\frac{c}{m^2}\Omega^2\right)}} \quad (b)$$

Now, substituting into Eq. (b) for k/m and setting $c = 0$,

$$\cos\gamma = \frac{\omega_o^2 - \Omega^2}{\sqrt{(\omega_o^2 - \Omega^2)^2}}$$

Since $\Omega >> \omega$, we have $\cos\gamma \approx -1$

Therefore,

$$\underline{\gamma \approx 180°}$$

20.55

Given: A body with a mass of 4.5 kg is subjected to a harmonic shaking force with an amplitude of 9 N. A spring with a constant equal to the critical value are attached to the body. The amplitude of motion must be less than 5 mm. Determine the range of angular frequency (if any) to which the shaking force must be confined.

Solution:

From the given data, we have,

$m = 4.5$ kg, $k = 3.6$ N/mm, $\zeta = 1.0$,

$X_{max} = 5$ mm, $F = F_o\sin\Omega t$ with $F_o = 9$ N.

First, we find the magnification factor such that $X_{max} = 5$ mm. The static deflection is,

$$\Delta = \frac{F_o}{k} = \frac{9}{3.6} = 2.5 \text{ mm}$$

So, the maximum value MF_{max} of the magnification factor is,

$$MF_{max} = \frac{X_{max}}{\Delta} = \frac{5}{2.5} = 2.0$$

Next, we determine Ω such that $M \le 2.0$.

The angular frequency of the undamped system is,

$$\omega_o = \sqrt{\frac{k}{m}} = \sqrt{\frac{3.6(1000)}{4.5}} = 28.28 \text{ rad/s}$$

To simplify notation, let $y = \left(\frac{\Omega}{\omega_o}\right)^2$. Then, by Eq (20.32) the magnification factor is,

$$MF = \frac{1}{\sqrt{(1-y)^2 + 4\zeta^2 y}} \le MF_{max}$$

Substituting for ζ and MF_{max}, we have,

$$\frac{1}{\sqrt{(1-y)^2 + 4y}} \le 2.0$$

Squaring both sides and solving for y,

$$1/4 \le (1-y)^2 + 4y = (1+y)^2$$

or, $1/2 \le 1 + y$

Therefore,

$$-1/2 \le y$$

Since $y = \left(\frac{\Omega}{\omega_o}\right)^2 > 0$ for any value of Ω there is no restriction on Ω such that

$X_{max} \le 5$ mm; that is, all values of frequency are allowable.

20.56

Given: A sinusoidal voltage $E(t) = 120\cos 120t$ is applied to the terminals of the series circuit shown in Fig. a, and $c = 4(10^{-5})$ Farad, $R = 40\,\Omega$, and $L = 0.10$ Henry.

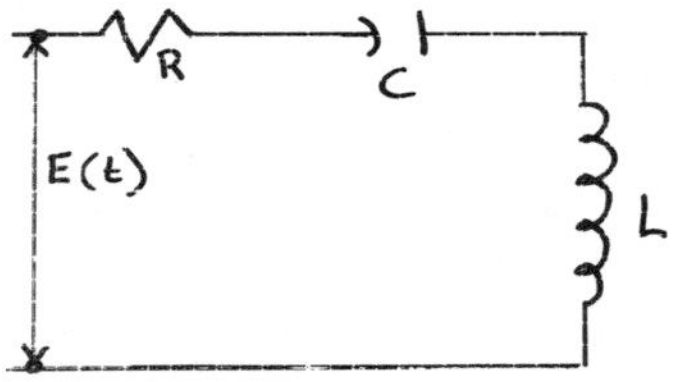

Figure a

Determine the resulting periodic current i of the system. Use the relation $i = \dot{Q}$.

Solution:

By Eq. (20.4), the differential equation for Q is,

$$L\ddot{Q} + R\dot{Q} + \frac{Q}{c} = E_o \cos \Omega t \qquad (a)$$

where $E_o = 120$ volts and $\Omega = 120$ rad/s.

To determine the current, we must first obtain the solution for Q [see Eq. (a)]. Therefore, let

$$Q = A \sin \Omega t + B \cos \Omega t \qquad (b)$$

where A, B are constants. By differentiation of Eq. (b), we have,

$$\dot{Q} = A\Omega \cos \Omega t - B\Omega \sin \Omega t \qquad (c)$$

$$\ddot{Q} = -A\Omega^2 \sin \Omega t - B\Omega^2 \cos \Omega t \qquad (d)$$

Substituting Eqs. (b), (c), and (d) into Eq. (a) and equating the coefficients of sines and cosines, we obtain,

$$\left(\frac{1}{c} - L\Omega^2\right)A - R\Omega B = 0$$
$$R\Omega A + \left(\frac{1}{c} - L\Omega^2\right)B = E_o \qquad (e)$$

The solution of Eqs. (e) is,

$$A = \frac{R\Omega E_o}{(\frac{1}{c} - L\Omega^2)^2 + R^2\Omega^2}$$
$$B = \frac{(\frac{1}{c} - L\Omega^2)E_o}{(\frac{1}{c} - L\Omega^2)^2 + R^2\Omega^2} \qquad (f)$$

Equations (b) and (f) yield,

$$Q = \frac{E_o}{(\frac{1}{c} - L\Omega^2)^2 + R^2\Omega^2}\left[R\Omega \sin \Omega t + (\tfrac{1}{c} - L\Omega^2)\cos \Omega t\right]$$

or

$$Q = \frac{E_o \sin(\Omega t + \beta)}{\sqrt{(\frac{1}{c} - L\Omega^2)^2 + R^2\Omega^2}} \qquad (g)$$

where, β is defined by

$$\cos\beta = \frac{R\Omega}{\sqrt{(\frac{1}{c} - L\Omega^2)^2 + R^2\Omega^2}}$$

$$\sin\beta = \frac{(\frac{1}{c} - L\Omega^2)}{\sqrt{(\frac{1}{c} - L\Omega^2)^2 + R^2\Omega^2}}$$

Now,

$$\frac{1}{c} - L\Omega^2 = 25000 - 1440 = 23,560 \qquad (h)$$

and,

$$R\Omega = (40)(120) = 4800 \qquad (i)$$

Since these quantities are both positive, both $\sin\beta$ and $\cos\beta$ are positive. Hence, β is in the first quadrant. Therefore,

$$\tan\beta = \frac{23,560}{4800} = 4.9083$$

Therefore,

$$\beta = 1.370 \text{ rad} = 78.48° \qquad (j)$$

Also, by Eqs. (g), (h), and (i),

$$\frac{E_o}{\sqrt{(\frac{1}{c} - L\Omega^2)^2 + R^2\Omega^2}} = \frac{120}{\sqrt{(23,560)^2 + (4800)^2}} = 0.0049908 \qquad (k)$$

Then, differentiating of Eq. (g) yields, with $i = \dot{Q}$, $\Omega = 120$ rad/s, and Eqs. (j) and (k),

$$i = 0.5989 \cos(120t + 1.370)$$

or by trigonometry,

$$i = 0.5989 \sin\left(120t + 1.370 + \frac{\pi}{2}\right)$$

Hence, $\underline{i = 0.5989 \sin(120t + 2.941)}$

Note that the phase angle is $\gamma = -2.941$ rad or $\gamma = -168.5°$ [see Eq. (20.29)]

20.57

Given: Vibration records of a refrigerator motor indicate that, in operation, the maximum vertical displacement of the motor from the equilibrium position is 0.040 in. Records show that the shaking force reaches its maximum value 120° before the maximum displacement occurs. The frequency of the shaking force is 10 Hz. The motor weighs 20 lb. The damping constant is 10% of the critical value.

(Continued)

20.57 Cont.

a) Determine the spring constant k.

b) Determine the amplitude of the shaking force.

Solution:

From the given data,

$$X_{max} = 0.040 \text{ in}$$
$$\gamma = 120° = \frac{2\pi}{3} \text{ rad}$$
$$\Omega = 10 \text{ Hz} = 20\pi \text{ rad/s}$$
$$\zeta = 0.10$$

Also, the mass of the motor is,

$$m = \frac{W}{g} = \frac{20 \text{ lb}}{384 \text{ in/s}^2} = 0.05176 \text{ lb}\cdot\text{s}^2/\text{in}$$

we can find the natural frequency of the system from the given data and Eq. (20.30). Thus, with,

$$\tan\gamma = \frac{2\zeta\left(\frac{\Omega}{\omega_0}\right)}{1-\left(\frac{\Omega}{\omega_0}\right)^2}$$

and, the given values for γ and ζ; we have,

$$\tan(2\pi/3) = -1.73205 = \frac{2(0.10)\left(\frac{\Omega}{\omega_0}\right)}{1-\left(\frac{\Omega}{\omega_0}\right)^2}$$

or,

$$\left(\frac{\Omega}{\omega_0}\right)^2 - 0.11547\left(\frac{\Omega}{\omega_0}\right) - 1 = 0$$

The real root of this equation is,

$$\frac{\Omega}{\omega_0} = 1.0594$$

Therefore,

$$\omega_0 = \frac{20\pi}{1.0594} = 59.31 \text{ rad/s}$$

From $\omega_0 = \sqrt{k/m}$, we find the spring constant as,

$$k = \omega_0^2 m = 182.1 \text{ lb/in}$$

b) With Ω, ζ, and ω_0 known, we can find the magnification factor from Eq. (20.32),

$$\frac{X_{max}}{\Delta} = \frac{1}{\sqrt{\left(1-\left(\frac{\Omega}{\omega_0}\right)^2\right)^2 + 4\zeta^2\frac{\Omega^2}{\omega_0^2}}}$$

$$\frac{X_{max}}{\Delta} = 4.0873$$

Now, the static deflection is,

$$\Delta = X_{max}/4.0873 = 0.04/4.0873 = 0.009786 \text{ in}$$

The amplitude of the forcing function is,

$$F_0 = k\Delta = (182.1)(0.009786) = 1.782 \text{ lb}$$

20.58

Given: A seismic pick up used to record vertical vibrations of the earth consists of a mass m supported on two springs. Damping is represented by a dashpot (Fig. a). When the casing of the instrument is displaced a vertical distance z_c, the mass m is displaced a vertical distance z_m from its equilibrium position (Fig. b). The relative displacement between the mass and the casing is $z_r = z_m - z_c$. Assuming that the vibration of the earth, and hence the casing, is,

$$z_c = z_c \sin\Omega t \qquad (a)$$

The differential equation for relative displacement of the mass m is,

$$m\ddot{z}_r + c\dot{z}_r + kz_r = z_c\Omega^2 m \sin\Omega t \qquad (b)$$

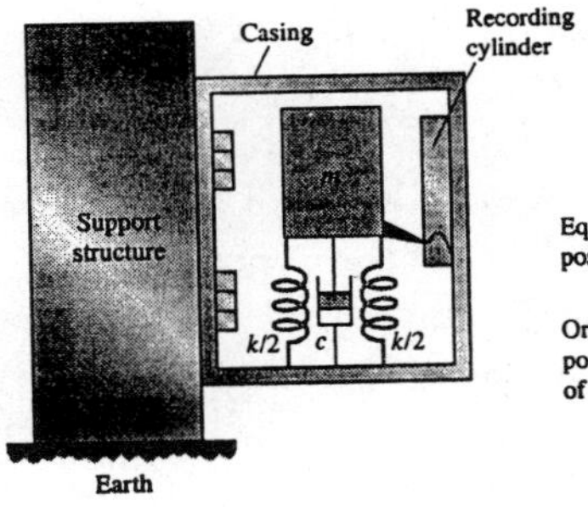

Figure a

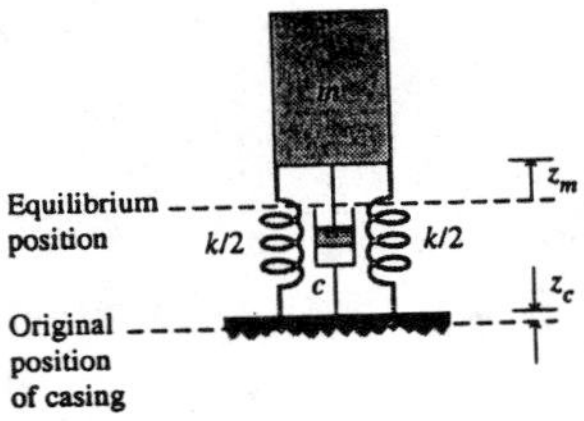

Figure b

a) Plot the ratio z_r/z_c as a function of the frequency ratio Ω/ω_0, where z_r is the amplitude of relative displacement z_r given by

$$z_r = \frac{z_c(\Omega/\omega_0)^2}{\sqrt{[1-(\Omega/\omega_0)^2]^2 + (2\zeta\Omega/\omega_0)^2}} \qquad (c)$$

Use the following values of ζ: 0, 0.10, 0.20, 0.40, 0.80, 1.0, and 2.0.

b) Discuss the behavior of z_r/z_c as a function of Ω/ω_0 and ζ

(Continued)

20.58 Cont.

Solution:

a) The function given in Eq. (c) can be coded into a spreadsheet to relate z_r/z_c to (Ω/ω_o) for the 7 values of ζ. The resulting curves are shown in Fig. c, for the frequency range $0 \le \frac{\Omega}{\omega_o} \le 3.0$

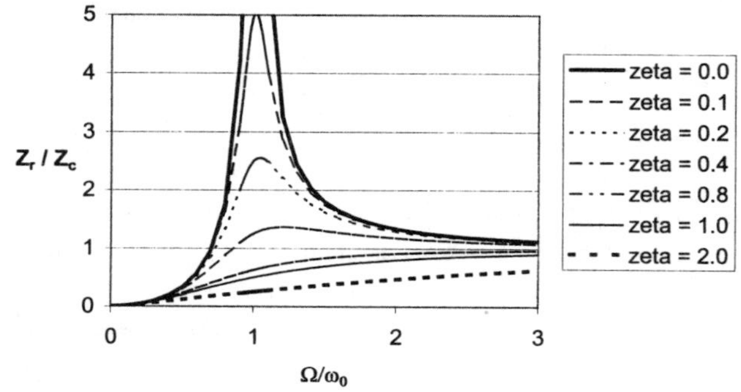

Figure c

b) The function of Eq. (c) behaves in a reverse manner, relative to the magnification factor given by Eq. (20.32). That is, the curves all start at $z_r/z_c = 0$ for $\Omega/\omega_o = 0$ and are asymptotic toward 1.0 as Ω/ω_o gets large. Otherwise the behavior is quite similar to that of Eq. (20.32), shown in Fig. 20.11. When Ω/ω_o is very small (the natural frequency of the seismic pickup is high relative to the frequency of the ground motion), the seismic pickup will perform poorly - it will record almost no relative displacement. When the seismic pickup has relatively low natural frequency, Ω/ω_o is high and the pickup will perform well, recording a broad range of ground motions. A relative damping ratio greater than about 0.2 should be provided to avoid excessive displacement of, and possibly damage to, the device.

20.59

Given: Consider the forced vibration of a spring-mass-damper system of one degree of freedom.

a) Plot the phase angle γ for forced vibrations of the system [see Eq. (20.30)] as a function of the frequency ratio Ω/ω_o. Use the following values ζ: 0, 0.02, 0.20, 0.50, 1.00, 2.00. Note that for $\Omega/\omega_o = 1$, $\gamma = \pi/2$, irrespective of ζ.

b) Discuss the behavior of γ as a function of Ω/ω_o and ζ.

Solution:

a) From Eq. (20.30) the phase angle is given by,

$$\gamma = \tan^{-1}\left(\frac{2\zeta(\Omega/\omega_o)}{1-(\Omega/\omega_o)^2}\right)$$

The resulting curves are shown in Fig. a for the frequency range $0 \le \Omega/\omega_o \le 3.0$

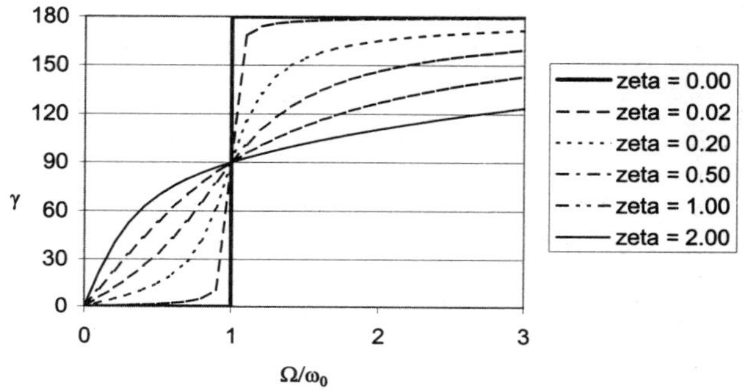

Figure a

Note: The $\tan^{-1}$ function on the spreadsheet must return a value for γ between 0° and 180° (between 0 and π rad). If instead the returned value is between -90° and 90° (between $-\pi/2$ and $\pi/2$ rad), then the negative (4th quadrant) values must be converted to equivalent positive (2nd quadrant values).

b) Since γ is always between 0 and 180°, this demonstrates that the peak displacement of the system always occurs simultaneously with or after the peak value of the forcing function. The function for γ is discontinuous at resonance if the system is undamped. As the relative damping ratio increases, the rate of change

(Continued)

20.59 Cont

in γ at resonance decreases. For highly damped systems, the phase angle increases quickly at low frequency ratios ($\Omega/\omega_o < 1.0$) but does not approach 180° until the frequency ratio is very large ($\Omega/\omega_o >> 3$).

20.60

Given: A sping/mass system is in equilibrium in the position shown (Fig. a). The mass m oscillates on the frictionless rod under the action of gravity and the springs.

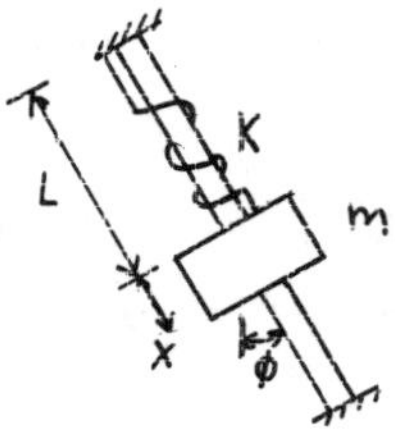

Figure a

Use the energy method to derive the frequency of small vibration.

Solution:

As the system vibrates, the kinetic energy T and the potential energy V are,

$$T = \frac{1}{2} m \dot{x}^2 \qquad (a)$$

$$V = \frac{1}{2} k x^2 \qquad (b)$$

For harmonic vibration, the displacement is,

$$x = A \sin(\omega t - \beta) \qquad (c)$$

and the velocity is,

$$\dot{x} = A\omega \cos(\omega t - \beta) \qquad (d)$$

So the maximum displacement and maximum velocity are

$$x_{max} = A$$

$$\dot{x}_{max} = A\omega$$

At maximum displacement $V = V_{max}$ and $T = 0$

At maximum velocity $T = T_{max}$ and $V = 0$

Since total mechanical energy is constant at these extremes, we have,

$$T_{max} = V_{max} \qquad (e)$$

Substitution from Eq. (a), (b), (c), and (d) into Eq. (e) yields,

$$\tfrac{1}{2} m (A\omega)^2 = \tfrac{1}{2} k A^2$$

Therefore,

$$\omega = \sqrt{\frac{k}{m}} \qquad \text{or} \qquad f = \frac{1}{2\pi}\sqrt{\frac{k}{m}}$$

20.61

Given: Two identical masses that weigh 675 N each rest on a frictionless plane (Fig. a). They are connected by a spring with stiffness 36 N/mm. Forces F are applied to pull the masses apart. Then the forces are suddenly removed.

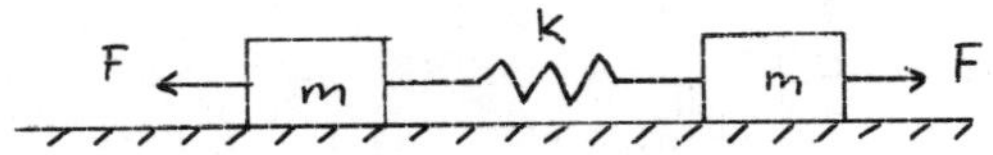

Figure a

Use the energy method to derive the frequency of small vibration.

Solution:

Let the unstretched length of the spring be 2a. Then by symmetry, each mass moves with a displacement x while the spring has an elongation of 2x.

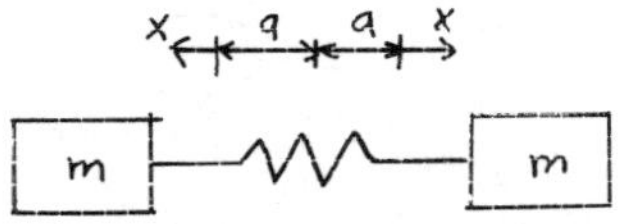

Figure b

For harmonic vibration

$$x = A \sin(\omega t - \beta) \qquad (a)$$

$$\dot{x} = A\omega \cos(\omega t - \beta) \qquad (b)$$

The maximum kinetic energy of the masses occurs at $\dot{x} = \dot{x}_{max} = A\omega$, which gives,

$$T_{max} = 2\left(\tfrac{1}{2} m (A\omega)^2\right) = m A^2 \omega^2 \qquad (c)$$

At this time $V = 0$,

The maximum potential energy of the spring occurs at $x = A$, which gives,

$$V_{max} = \tfrac{1}{2} k (2A)^2 = 2kA^2 \qquad (d)$$

and at this time $T = 0$.

(Continued)

20.61 Cont.

For conservation of energy, we can equate T_{max} and V_{max} from Eq. (c) and Eq. (d), respectively to obtain,

$$mA^2\omega^2 = 2kA^2$$

From which,

$$\omega = \sqrt{\frac{2k}{m}} \quad \text{and} \quad f = \frac{1}{2\pi}\sqrt{\frac{2k}{m}}$$

From the given data,

$$k = 36 \text{ N/mm} = 36000 \text{ N/m}$$

$$m = \frac{675 \text{ N}}{9.81 \text{ m/s}^2} = 68.81 \text{ kg}$$

so,

$$\underline{\underline{\omega = 32.35 \text{ rad/s}}}, \quad \underline{\underline{f = 5.148 \text{ Hz}}}$$

20.62

Given: The disk in Fig. a rotates about its center O. Its mass is $m = 1.5$ lb·s²/in and the constant for each spring is $k = 10$ lb/in. The disk is rotated slightly from its equilibrium position and released. Assume that the spring forces act tangentially to the disk during oscillation.

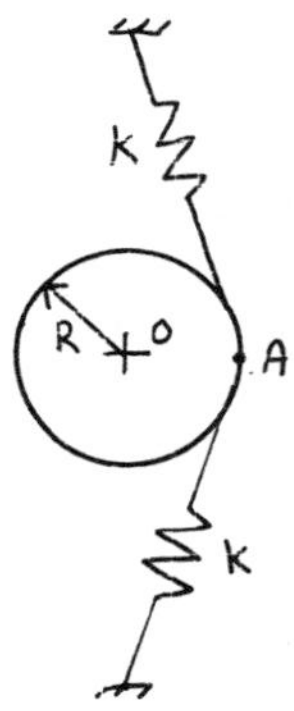

Figure a

Use the energy method to derive the frequency of small vibration.

Solution:

During rotation, point A moves a distance (see Fig. b)

$$s = R\theta$$

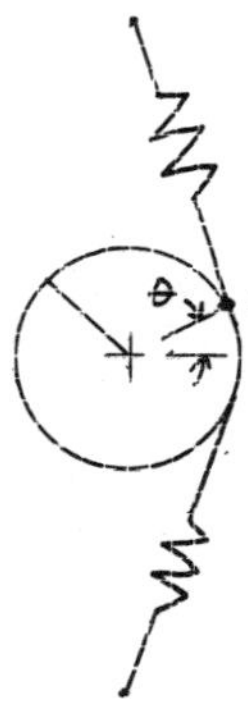

Figure b

If rotation is harmonic, then

$$\theta = A\sin(\omega t - \beta)$$

$$\dot{\theta} = A\omega\cos(\omega t - \beta)$$

The maximum kinetic energy of the disk is,

$$T_{max} = \tfrac{1}{2} I (A\omega)^2$$

At which time $V = 0$. For the disk,

$$I = \tfrac{1}{2} m R^2$$

The maximum potential energy in the spring is,

$$V_{max} = 2\left[\tfrac{1}{2}k(RA)^2\right]$$

At which time $T = 0$. By conservation of energy, we can equate T_{max} with V_{max} to obtain,

$$\tfrac{1}{2}\left(\tfrac{1}{2}mR^2\right)(A\omega)^2 = 2\left[\tfrac{1}{2}k(RA)^2\right]$$

Therefore,

$$\omega = 2\sqrt{\frac{k}{m}} \quad \text{or} \quad f = \frac{1}{\pi}\sqrt{\frac{k}{m}}$$

With $k = 10$ lb/in and $m = 1.5$ lb·s²/in, this gives,

$$\underline{\underline{\omega = 5.164 \text{ rad/s}}} \quad \text{or} \quad \underline{\underline{f = 0.822 \text{ Hz}}}$$

20.63

Given: A block of mass m hangs from a string that is wrapped around a disk and is attached to a spring with stiffness k as shown in Fig. a. The disk has moment of inertia I_O.

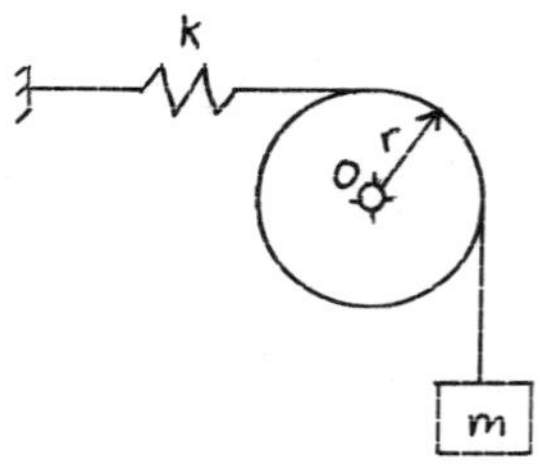

Figure a

Use the energy method to derive the frequency of small free vibration for the system. Assume that the string does not slip on the disk.

Solution:

During clockwise rotation of the disk, the spring is stretched an amount,

$$\delta = r\theta$$

If rotation is harmonic

$$\theta = A\sin(\omega t - \beta)$$

$$\dot{\theta} = A\omega\cos(\omega t - \beta)$$

(Continued)

20.63 Cont.

The maximum kinetic energy of the system occurs when V=0 and has value,

$$T_{max} = \frac{1}{2} I_o (A\omega)^2 + \frac{1}{2} m (rA\omega)^2$$

The maximum potential energy of the spring occurs when T=0 and has value,

$$V_{max} = \frac{1}{2} k (rA)^2$$

By conservation of energy, we can equate T_{max} and V_{max} to obtain

$$\tfrac{1}{2} I_o A^2 \omega^2 + \tfrac{1}{2} m r^2 A^2 \omega^2 = \tfrac{1}{2} k r^2 A^2$$

Therefore,

$$\omega = \sqrt{\frac{kr^2}{I_o + mr^2}} \quad \text{or} \quad f = \frac{1}{2\pi} \sqrt{\frac{kr^2}{I_o + mr^2}}$$

20.64

Given: The cylinder in Fig. a has mass m and is displaced from its equilibrium position. It rolls on the surface.

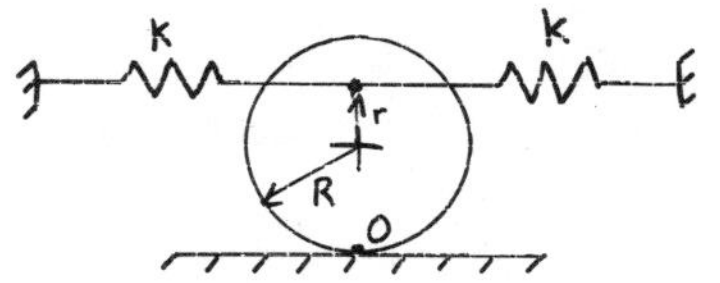

Figure a

Use the energy method to derive the frequency of small free vibrations for the system.

Solution:

As the cylinder rolls on the surface, the springs elongate/compress an amount,

$$\delta = (r+R)\theta$$

If rotation is harmonic,

$$\theta = A \sin(\omega t - \beta)$$
$$\dot{\theta} = A\omega \cos(\omega t - \beta)$$

The maximum kinetic energy of the cylinder occurs when V=0 and has value,

$$T_{max} = \tfrac{1}{2} I_o (A\omega)^2$$

The moment of inertia about point O, the point of zero velocity, is

$$I_o = \frac{1}{2} mR^2 + mR^2 = \frac{3}{2} mR^2$$

(see Appendix F and apply Steiner's Theorem)

So, $T_{max} = \tfrac{3}{4} m R^2 A^2 \omega^2$

The maximum potential energy of the spring occurs when T=0 and has value,

$$V_{max} = 2\left[\tfrac{1}{2} k (r+R)^2 A^2\right]$$

By conservation of energy, we can equate T_{max} and V_{max} to obtain,

$$\frac{3}{4} m R^2 A^2 \omega^2 = k (r+R)^2 A^2$$

$$\omega = \frac{2(r+R)}{R} \sqrt{\frac{k}{3m}} \quad \text{or} \quad f = \frac{(r+R)}{\pi R} \sqrt{\frac{k}{3m}}$$

20.65

Given: The cylinder shown in Fig. a undergoes small vibrations as it rolls on the inclined plane. k= 15 lb/in, m= 6 lb·s²/ft.

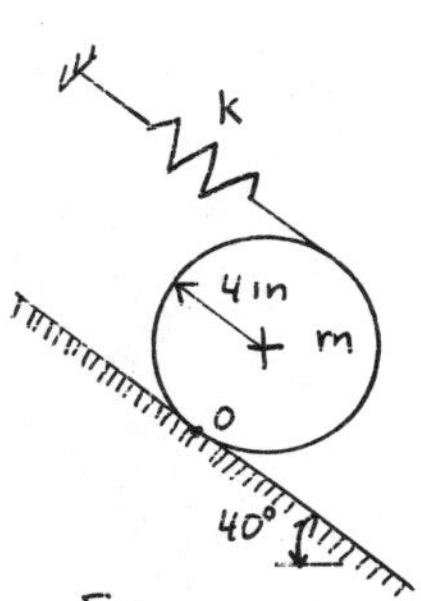

Figure a

Use the energy method to derive the frequency of free vibration for the system.

Solution:

As the cylinder rolls on the inclined plane, the spring elongates an amount

$$\delta = 2r\theta$$

If rotation is harmonic

$$\theta = A \sin(\omega t - \beta)$$
$$\dot{\theta} = A\omega \cos(\omega t - \beta)$$

The maximum kinetic energy of the cylinder occurs when V=0 and has value

$$T_{max} = \tfrac{1}{2} I_o (A\omega)^2$$

The moment of inertia about point O, the point of zero velocity, is

$$I_o = \frac{1}{2} mr^2 + mr^2 = \frac{3}{2} mr^2$$

(see Appendix F & apply Steiner's Theorem)

(Continued)

20.65 Cont.

So, we have $T_{max} = 3/4\, mr^2 A^2 \omega^2$

The maximum potential energy of the spring occurs when $T = 0$ and has value

$$V_{max} = 1/2\, k(2rA)^2$$

By conservation of energy, we can equate Tmax and Vmax to obtain,

$$3/4\, mr^2 A^2 \omega^2 = 2kr^2 A^2$$

$$\omega = \sqrt{\frac{8k}{3m}} \quad \text{or} \quad f = \frac{1}{\pi}\sqrt{\frac{2k}{3m}}$$

with the given numerical values for k and m, we obtain,

$$\underline{\omega = 8.944 \text{ rad/s}} \quad \text{or} \quad \underline{f = 1.424 \text{ Hz}}$$

20.65

Given: Consider the spring-mass system shown in Fig. a.

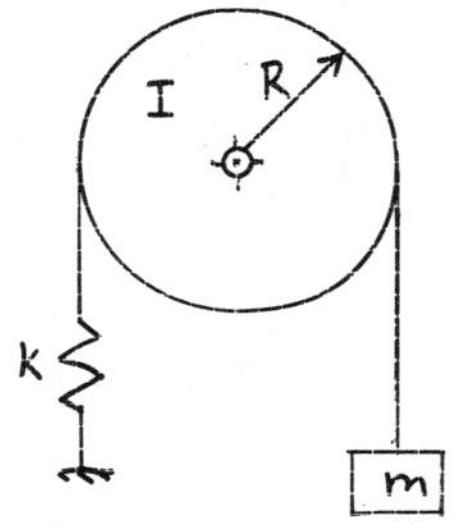

Figure a

a) Using the energy method, derive the differential equation of motion for the system shown in Fig. a. Neglect friction.

b) Calculate the natural frequency.

Solution:

a) We select the rotation angle θ as the independent coordinate. For rotation through an angle θ, the spring elongation is,

$$\delta = R\theta$$

The potential energy of the spring is

$$V = 1/2\, k(R\theta)^2$$

The kinetic energy of the system is,

$$T = \frac{1}{2} I \dot{\theta}^2 + \frac{1}{2} m (R\dot{\theta})^2$$

For conservation of energy $T + V = C$ or,

$$1/2\, kR^2\theta^2 + 1/2\, I(\dot{\theta})^2 + 1/2\, mR^2(\dot{\theta})^2 = c \qquad (a)$$

Now, we can differentiate Eq. (a),

$$kR^2\theta\dot{\theta} + I\dot{\theta}\ddot{\theta} + mR^2\dot{\theta}\ddot{\theta} = 0 \qquad (b)$$

Since $\dot{\theta} \neq 0$ for all times, Eq. (b) yields,

$$\underline{kR^2\theta + (I + mR^2)\ddot{\theta} = 0} \qquad (c)$$

b) The natural frequency is found directly from the differential equation of motion [Eq. (c)] as,

$$\underline{\omega = \sqrt{\frac{k_{eff}}{m_{eff}}} = \sqrt{\frac{kR^2}{I + mR^2}}} \quad \text{or} \quad \underline{f = \frac{1}{2\pi}\sqrt{\frac{kR^2}{I + mR^2}}}$$

20.67

Given: The uniform bar AB, shown in Fig. a, weighs 10 lb and the spring constants are $k_1 = 10$ lb/in and $k_2 = 30$ lb/in.

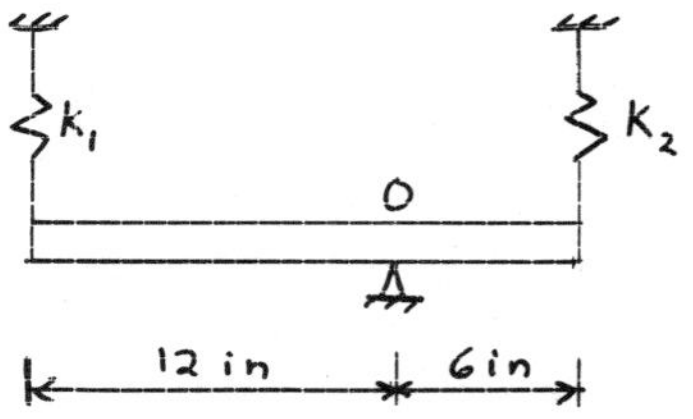

Figure a

Using the energy method, determine the frequency of small free vibrations of the system. The bar is in equilibrium in the position shown.

Solution:

For a rotation of the bar through an angle θ, the spring elongations are

$$\delta_1 = 12\theta \quad , \quad \delta_2 = 6\theta$$

The total potential energy of the springs is,

$$V = 1/2\, k_1\delta_1^2 + 1/2\, k_2\delta_2^2$$

$$V = 1/2\,(144k_1 + 36k_2)\theta^2$$

or

$$V = (72k_1 + 18k_2)\theta^2$$

The kinetic energy of the bar for rotation about O is,

$$T = 1/2\, I_o(\dot{\theta})^2$$

where $I_o = \frac{1}{12} m(18^2) + m(3^2)$

$$I_o = 36\, m$$

(see Appendix F and apply Steiner's Theorem)

(Continued)

20.67 Cont.

So,

$$T = 18\,m\,(\dot{\theta})^2$$

For harmonic vibrations

$$\theta = A \sin(\omega t - \beta)$$

$$\dot{\theta} = A\omega \cos(\omega t - \beta)$$

So using conservation of energy, we equate the maximum potential energy with the maximum kinetic energy to get,

$$(72k_1 + 18k_2)A^2 = 54m(A\omega)^2$$

$$\omega = \sqrt{\frac{72k_1 + 18k_2}{18m}}$$

Substitution of numerical values yields,

$$\omega = \sqrt{\frac{72(10) + 18(30)}{18(10/386.4)}} = 52.01 \text{ rad/s}$$

$$f = \frac{\omega}{2\pi} = 8.28 \text{ Hz}$$

20.68

Given: The uniform disk D is attached to the bar B by a frictionless pin (Fig. a). Bar B is hinged at A. A torsional spring in joint a exerts a restoring moment of 68 N·m/rad, when the bar is displaced from the downward vertical. The masses of the disk and the bar are, respectively, 7.5 kg and 15 kg. The disk D rolls on the circular track with center at A.

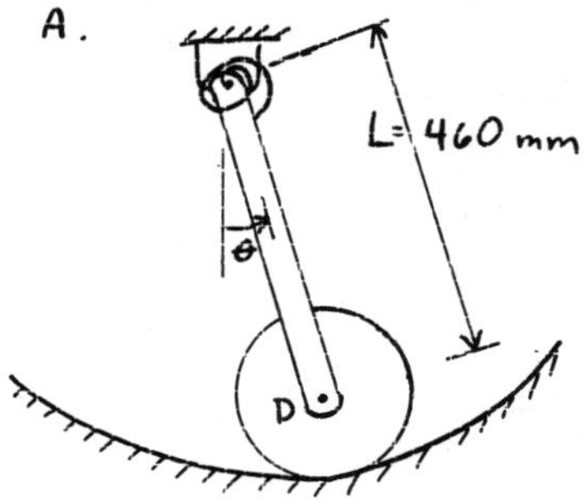

Figure a

Determine the frequency of small free vibrations of the system.

Solution:

Let θ be the position coordinate for the bar (see Fig. a). First, we determine the potential energy and the kinetic energy of the system for small vibrations.

Potential energy:

of the disk – $V_D = m_D\, g L (1 - \cos\theta)$

of the bar – $V_B = m_B\, g (L/2)(1 - \cos\theta)$

of the spring – $V_s = (1/2)\, k\theta^2$

Therefore,

$$V = \left(m_D + \frac{m_B}{2}\right) g L (1 - \cos\theta) + \tfrac{1}{2} k \theta^2$$

For small vibrations, $\cos\theta = 1 - \theta^2/2$

So, the total potential energy is approximately,

$$V = (m_D + m_B/2)\left(\frac{gL}{2}\right)\theta^2 + \tfrac{1}{2}k\theta^2 \quad (a)$$

kinetic energy:

of the disk – $T_D = \frac{1}{2} m_D v_D^2 + \frac{1}{2} I_D \omega^2$

$$T_D = \tfrac{1}{2} m_D (L\dot{\theta})^2 + \tfrac{1}{2}\left(\tfrac{1}{2} m_D r^2\right)\left(\frac{L\dot{\theta}}{r}\right)^2$$

Note that $V_D = L\dot{\theta} = r\omega$, where ω is the angular velocity of the disk and r is its radius. So, $\omega = \left(\frac{L\dot{\theta}}{r}\right)$

of the bar – $T_B = \frac{1}{2} I_A \dot{\theta}^2$

$$T_B = \tfrac{1}{2}\left(\tfrac{1}{3} m_B L^2\right)\dot{\theta}^2$$

Therefore,

$$T = \tfrac{3}{4} m_D L^2 \dot{\theta}^2 + \left(\tfrac{1}{6} m_B\right) L^2 \dot{\theta}^2$$

or

$$T = \tfrac{1}{12}(9m_D + 2m_B) L^2 \dot{\theta}^2 \quad (b)$$

Assuming that the motion is harmonic, we can write,

$$\theta = A \sin(\Omega t - \beta) \quad (c)$$

where Ω is the natural angular frequency of vibration (Note that we use ω as the angular speed of the disk)

Hence,

$$\dot{\theta} = A\Omega \cos(\Omega t - \beta) \quad (d)$$

The maximum values of θ and $\dot{\theta}$ are,

$$\theta_{max} = A, \quad \dot{\theta}_{max} = A\Omega \quad (e)$$

Now, we can equate the maximum values of potential and kinetic energy,

$$V_{max} = T_{max}$$

(Continued)

20.68 Cont.

From Eq. (a), (b), and (e), this gives

$$(m_D + m_B/2)\left(\frac{gL}{2}\right)A^2 + \frac{1}{2}kA^2 = \frac{1}{12}(9m_D + 2m_B)L^2(A\Omega)^2$$

With $g = 9.81\ m/s^2$, $m_D = 7.5\ kg$, $m_B = 15\ kg$, $L = 0.46\ m$, and $k = 68\ N\cdot m/rad$, we can find the angular frequency of vibration from,

$$67.84A^2 = 1.719\,L^2A^2\Omega^2$$

Therefore, $\Omega = 6.28\ rad/s$

$$f = \frac{\Omega}{2\pi} = 1.00\ Hz$$

Note: The frequency of vibration does not depend on the radius of the disk!

20.69

Given: A thin rectangular plate is bent into A semicircular cylinder of radius r. It rocks on a horizontal surface.

a) Show by the law of conservation of energy that the differential equation of motion is,

$$r\dot{\theta}^2\left(1 - \frac{2}{\pi}\cos\theta\right) - \frac{2g}{\pi}\cos\theta = \text{constant}$$

where θ is the angular displacement from the equilibrium position.

b) Differentiate the equation given in part a with respect to t, and linearize with the aid of the approximations $\sin\theta \doteq \theta$ and $\cos\theta \doteq 1$. Hence, derive a formula for the frequency of small oscillations.

Solution:

a) We can start with a drawing of the cylinder in its rotated position. Setting the origin O at the point of contact between the cylinder and the surface for $\theta = 0$, we have Fig. a. See Appendix D for the location of the center of mass.

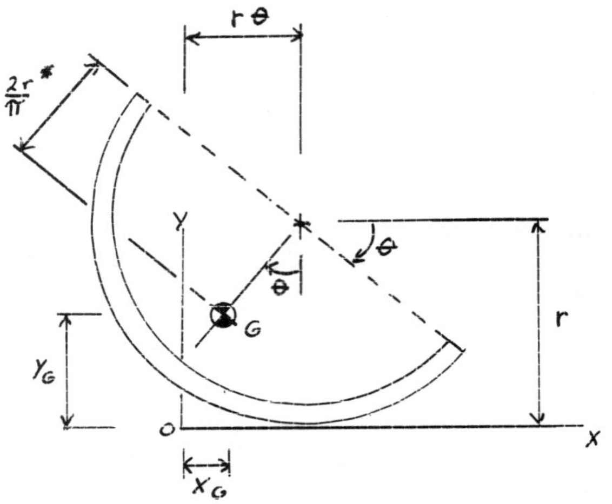

Figure a

As the cylinder rocks, the coordinates of the center of mass (x_g, y_g) change according to,

$$x_G = r\theta - \frac{2r}{\pi}\sin\theta \qquad (a)$$

$$y_G = r - \frac{2r}{\pi}\cos\theta \qquad (b)$$

The speed S_G of the center of mass is given by,

$$S_G^2 = (\dot{x}_G)^2 + (\dot{y}_G)^2 \qquad (c)$$

where, from Eq. (a) and (b),

$$\dot{x}_G = r\dot{\theta} - \left(\frac{2r}{\pi}\cos\theta\right)\dot{\theta} \qquad (d)$$

$$\dot{y}_G = \left(\frac{2r}{\pi}\sin\theta\right)\dot{\theta} \qquad (e)$$

Substitution of Eq. (d) and (e) into Eq. (c) yields,

$$S_G^2 = r^2(\dot{\theta})^2\left[\left(1 - \frac{2}{\pi}\cos\theta\right)^2 + \left(\frac{2}{\pi}\sin\theta\right)^2\right] \qquad (f)$$

Now, we write the total energy of the system in terms of Y_G and S_G.

The kinetic energy is

$$T = \frac{1}{2}I_G\dot{\theta}^2 + \frac{1}{2}mS_G^2 \qquad (g)$$

where I_G is the mass moment of inertia with respect to the center of mass G.

To find I_G, we first find I_C, the mass moment of inertia about the axial line through the center of the cylinder, see Fig. b.

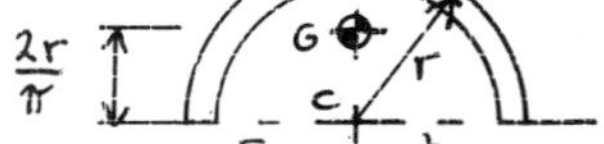

Figure b

(Continued)

20.69 Cont.

$$I_C = \int_0^{\pi} r^2\,dm$$

$$I_C = \int_0^{\pi} r^2 \rho r\,d\theta$$

$$I_C = \rho r^3 \pi$$

Also, $m = \rho \pi r$, so $I_C = mr^2$

$$I_G = I_C - \left(\frac{2r}{\pi}\right)^2 m = mr^2 - m\left(\frac{4r^2}{\pi^2}\right)$$

$$I_G = mr^2\left(1 - \frac{4}{\pi^2}\right) \qquad (h)$$

Substitution of Eq. (h) into Eq. (g) gives,

$$T = \frac{mr^2\dot\theta^2}{2}\left[\left(1-\frac{4}{\pi^2}\right) + \left(1 - \frac{2}{\pi}\cos\theta\right)^2 + \left(\frac{2}{\pi}\sin\theta\right)^2\right]$$

$$T = \frac{mr^2\dot\theta^2}{2}\left[1 - \frac{4}{\pi^2} + 1 - \frac{4}{\pi}\cos\theta + \frac{4}{\pi^2}\right]$$

$$T = mr^2\dot\theta^2\left[1 - \frac{2}{\pi}\cos\theta\right] \qquad (i)$$

Next, we write the potential energy of the cylinder as

$$V = mgy_G = mgr\left(1 - \frac{2}{\pi}\cos\theta\right) \qquad (j)$$

Conservation of energy allows us to write,

$$T + V = \text{constant.}$$

Substitution from Eq. (i) and (j) yields,

$$r\dot\theta^2\left(1 - \frac{2}{\pi}\cos\theta\right) + g\left(1 - \frac{2}{\pi}\cos\theta\right) = \text{constant}$$

Grouping all constant terms on the right side of the equation, we have

$$r(\dot\theta)^2\left(1 - \frac{2}{\pi}\cos\theta\right) - \frac{2g}{\pi}\cos\theta = \text{constant} \qquad (k)$$

b) Differentiation of Eq. (k) gives,

$$2r\dot\theta\ddot\theta\left(1 - \frac{2}{\pi}\cos\theta\right) + r(\dot\theta)^2\left(\frac{2}{\pi}\sin\theta\right)\dot\theta + \frac{2g}{\pi}(\sin\theta)\dot\theta = 0$$

This expression reduces to,

$$r\ddot\theta\left(1 - \frac{2}{\pi}\cos\theta\right) + \frac{r(\dot\theta)^2}{\pi}\sin\theta + \frac{g}{\pi}(\sin\theta) = 0 \quad (\ell)$$

For small vibrations, $\sin\theta = \theta$ and $\cos\theta = 1$, so Eq. (ℓ) simplifies to,

$$r\ddot\theta\left(1 - \frac{2}{\pi}\right) + \frac{r(\dot\theta)^2}{\pi} + \frac{g}{\pi}\theta = 0$$

The term $(\dot\theta)^2$ is high order, so we can neglect it.

Therefore,

$$r\ddot\theta\left(1 - \frac{2}{\pi}\right) + \frac{g}{\pi}\theta = 0$$

$$\ddot\theta + \frac{g}{\pi r\left(1 - \frac{2}{\pi}\right)}\theta = 0$$

$$\ddot\theta + \frac{g\,\theta}{r(\pi - 2)} = 0 \qquad (m)$$

Now, the frequency of small vibration is,

$$\omega = \sqrt{\frac{g}{r(\pi-2)}} \qquad \text{or} \qquad f = \frac{1}{2\pi}\sqrt{\frac{g}{r(\pi-2)}}$$

20.70

Given: A thin hoop of mass m and radius R hangs over a fixed horizontal shaft of radius r (Fig. a). The plane of the hoop is perpendicular to the axis of the shaft. The hoop swings in its plane without slipping on the shaft.

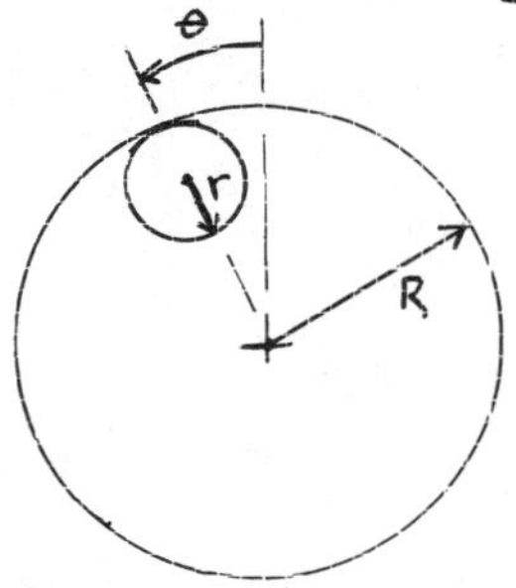

Figure a

a) By the energy principle, set up the differential equation for large oscillations of the hoop in terms of r, R, and θ.

b) Determine the frequency of small oscillations of the hoop.

Solution:

We start with a drawing of the system with the hoop in its rotated position (Fig. b). The origin O is located at the center of the hoop for θ = 0.

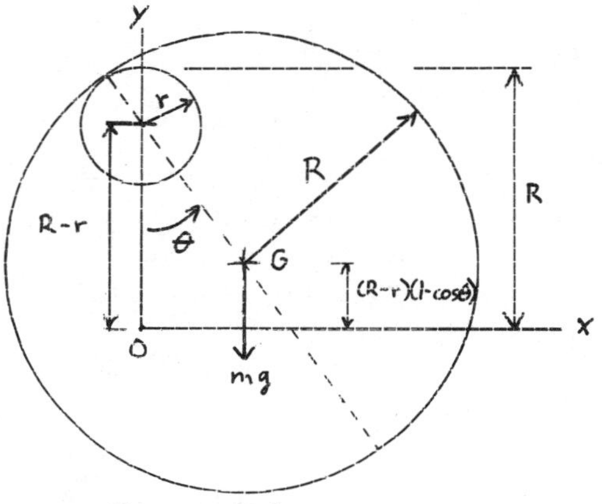

Figure b

(Continued)

20.70 Cont.

The mass center G of the hoop swings in a circular arc with radius (R-r) about the center of the shaft. As it swings, the hoop rolls on the shaft. For one complete revolution of the center of mass about the shaft, the hoop makes (R-r)/R revolutions as it rolls on the shaft. Hence, the angular velocity of the hoop is,

$$\omega = \left(\frac{R-r}{R}\right)\dot{\theta} \qquad (a)$$

The velocity of the mass center G is,

$$V_G = (R-r)\dot{\theta} \qquad (b)$$

The mass moment of inertia of the hoop is,

$$I_G = mR^2 \qquad (c)$$

From Eq. (19.12), the kinetic energy of the hoop is,

$$T = \frac{1}{2} I_G \omega^2 + \frac{1}{2} m V_G^2$$

with Eqs. (a), (b), and (c), this gives

$$T = \frac{1}{2}(mR^2)\left(\frac{R-r}{R}\right)^2 (\dot{\theta})^2 + \frac{1}{2} m (R-r)^2 (\dot{\theta})^2$$

or

$$T = m(R-r)^2(\dot{\theta})^2 \qquad (d)$$

The potential energy of the hoop, with respect to the origin O is,

$$V = mg\,(R-r)(1-\cos\theta) \qquad (e)$$

For conservation of energy

$$T + V = \text{constant}$$

Substitution from Eq. (d) and (e) yields

$$m(R-r)\left[(R-r)(\dot{\theta})^2 + g(1-\cos\theta)\right] = \text{constant}$$

Move all constants to the right side of the equation to obtain,

$$(R-r)(\dot{\theta})^2 - g\cos\theta = \text{constant} \qquad (f)$$

Differentiation of Eq. (f) gives,

$$2(R-f)\dot{\theta}\ddot{\theta} + g(\sin\theta)\dot{\theta} = 0$$

Eliminating $\dot{\theta}$ to get the equation of motion,

$$\underline{\underline{2(R-r)\ddot{\theta} + g\sin\theta = 0}} \qquad (g)$$

b) For small vibrations, $\sin\theta \simeq \theta$ and Eq. (g) reduces to,

$$2(R-r)\ddot{\theta} + g\,\theta = 0$$

which gives,

$$\omega = \sqrt{\frac{g}{2(R-r)}} \quad \text{or} \quad f = \frac{1}{2\pi}\sqrt{\frac{g}{2(R-r)}}$$

20.87

Given: A system consists of a mass m suspended from a spring with constant k.

Derive a relation between the natural frequency f and static deflection Δ of the system and sketch a graph of f versus Δ.

Solution:

For the system in Fig. a, the natural frequency is,

$$f = \frac{1}{2\pi}\sqrt{\frac{k}{m}} \qquad (a)$$

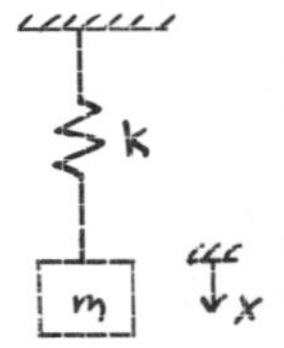

Figure a

The static deflection of the spring is,

$$\Delta = mg/k \qquad (b)$$

From Eq. (b) we have

$$\frac{k}{m} = \frac{g}{\Delta} \qquad (c)$$

Substitution of Eq. (c) into Eq. (a) gives

$$\underline{\underline{f = \frac{1}{2\pi}\sqrt{g/\Delta}}}$$

A sketch of this relation is shown in Fig. b.

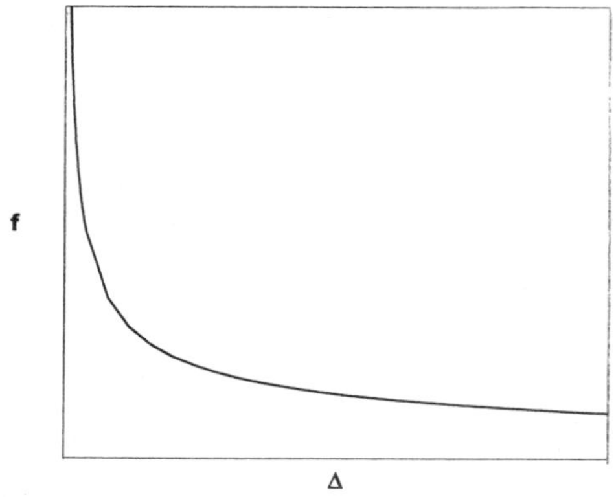

Figure b

21.1

Given: Each of the bodies shown in Fig. I is a uniform solid.

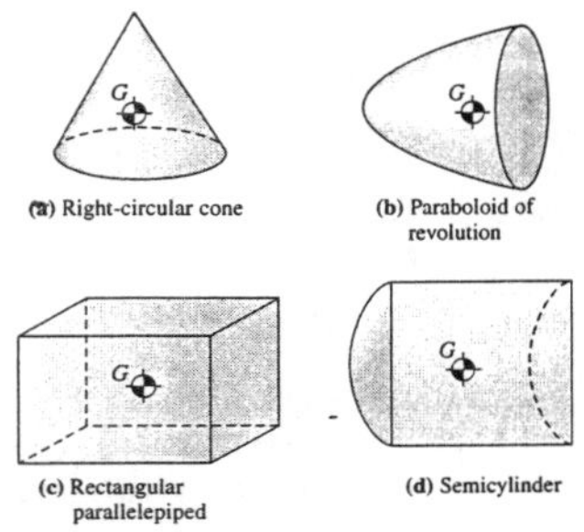

Figure I

Draw a sketch of each body showing principal axes xyz with origin at the center of mass G.

Solution: The principal axes are shown in each body directly in Fig. I.

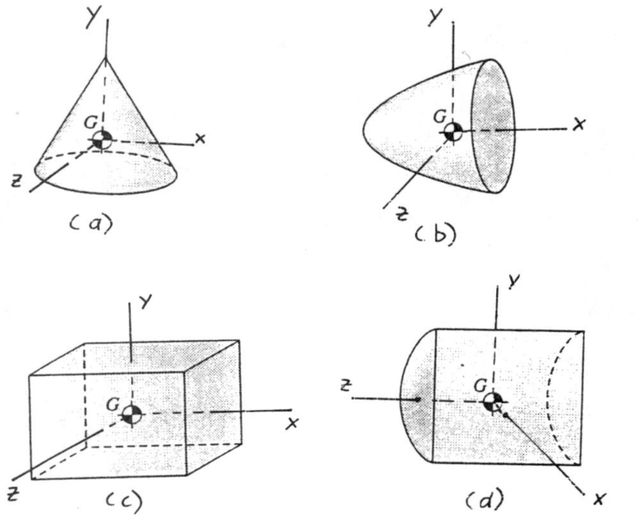

21.2

Given: Figure a is an end view of a homogeneous triangular prism of mass m.

Show that

$$I_{xy} = \frac{1}{12} mab \qquad (a)$$

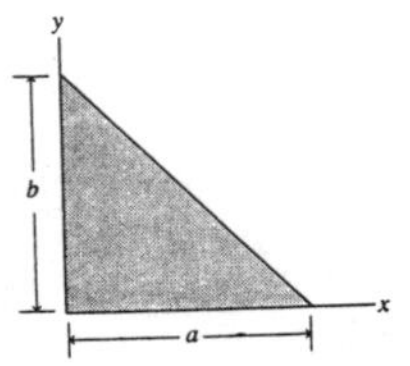

Figure a

Solution:

The diagonal line of a prism is defined by

$$y = \frac{b}{a}(a-x) \qquad (b)$$

Then since an element $dx\,dy$ of the prism has mass (Fig. b)

$$dm = \rho L\, dx\, dy \qquad (c)$$

where ρ is the mass density and L is the length of the prism, the mass m of the element is, with Eqs. (b) and (c),

$$m = \rho L \int_0^a dx \int_0^{\frac{b}{a}(a-x)} dy = \frac{1}{2}\rho abL \qquad (d)$$

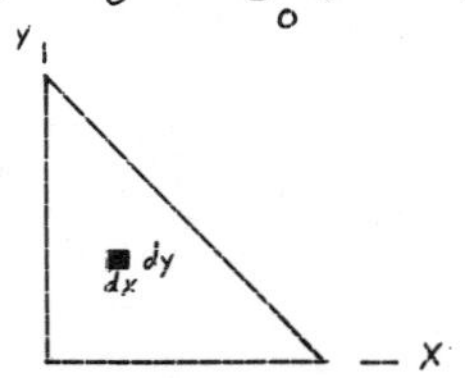

Figure b

Then, by definition of I_{xy} and Eqs. (b), (c), and (d),

$$I_{xy} = \int xy\,dm = \rho L \int_0^a x\,dx \int_0^{\frac{b}{a}(a-x)} dx = \frac{1}{2}\frac{\rho b^2 L}{a^2}\int_0^a x(a-x)^2 dx$$

Integration yields, with Eq. (d),

$$I_{xy} = \frac{1}{24}\rho L a^2 b^2 = \frac{1}{12} mab \qquad (e)$$

Equation (e) is the result required [see Eq. (a)].

21.3

Given: A homogeneous cube is situated so that one edge coincides with the negative x axis, one edge coincides with the negative y axis, and one edge coincides with the positive z axis.

Using integration, derive formulas for the moments and products of inertia with respect to xyz coordinate axes [see Eqs. (21.2)].

Solution: The cube is shown in Fig. a.

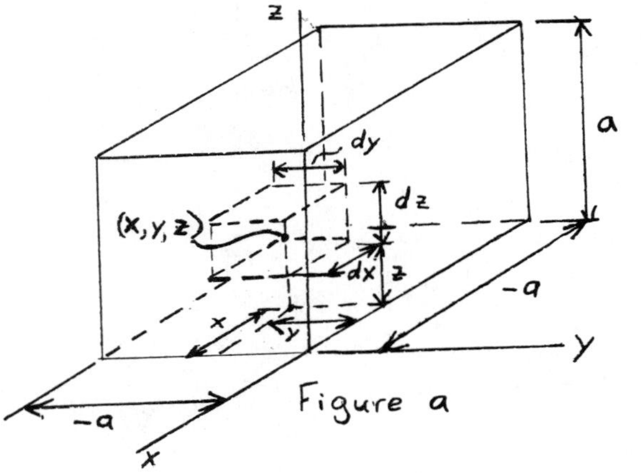

Figure a

The mass of element $dx\,dy\,dz$ is

$$dm = \rho\, dx\,dy\,dz \qquad (a)$$

(Continued)

21.3 Cont.

Hence, the mass of the cube is

$$m = \rho \int_{-a}^{0} dx \int_{-a}^{0} dy \int_{0}^{a} dz = \rho a^3 \qquad (b)$$

By definition [Eqs. (21.2)], with Eq. (a)

$$I_{xx} = \int (y^2 + z^2)\, dm$$

or $$I_{xx} = \rho \int_{-a}^{0} dx \int_{-a}^{0} dy \int_{0}^{a} (y^2+z^2)\, dz$$

Integration yields, with Eq. (b),

$$\underline{\underline{I_{xx} = \frac{2}{3} m a^2}}$$

Similarly, by Eqs. (21.2),

$$\underline{\underline{I_{yy} = \int (z^2 + x^2)\, dm = \frac{2}{3} m a^2}}$$

$$\underline{\underline{I_{zz} = \int (x^2 + y^2)\, dm = \frac{2}{3} m a^2}}$$

Also, by Eqs. (21.2) and Eqs. (a) and (b),

$$\underline{\underline{I_{xy} = \int xy\, dm = \int_{-a}^{0} dx \int_{0}^{a} dz \int_{-a}^{0} xy\, dy = \frac{1}{4} m a^2}}$$

Similarly,

$$\underline{\underline{I_{yz} = \int yz\, dm = -\frac{1}{4} m a^2}}$$

$$\underline{\underline{I_{zx} = \int zx\, dm = -\frac{1}{4} m a^2}}$$

By definition,

$$I_{XY} = \int XY\, dm \qquad (c)$$

Hence, by Eqs. (b) and (c),

$$I_{XY} = \int (X_G + x)(Y_G + y)\, dm \qquad (d)$$

$$I_{XY} = X_G Y_G \int dm + X_G \int y\, dm + Y_G \int x\, dm + \int xy\, dm$$

But since G is the origin of the xyz axes,

$$\int x\, dm = \int y\, dm = 0 \qquad (e)$$

and also,

$$m = \int dm, \quad \int xy\, dm = I_{xy} \qquad (f)$$

Therefore, Eqs. (d), (e), and (f) yield [see Eq. (a)]

$$\underline{\underline{I_{XY} = I_{xy} + m X_G Y_G}}$$

Similarly, by definition,

$$I_{YZ} = \int YZ\, dm, \quad I_{ZX} = \int ZX\, dm \qquad (g)$$

with Eqs. (b) and (g), following the same procedure as for I_{XY}, we obtain

$$\underline{\underline{I_{YZ} = I_{yz} + m Y_G Z_G}}$$

$$\underline{\underline{I_{ZX} = I_{zx} + m Z_G X_G}}$$

21.4

Given: Axes xyz are rectangular coordinate axes with origin at G, the center of mass of a body having mass m (Fig. a). Axes XYZ form a parallel coordinate system, in which G has cordinates (X_G, Y_G, Z_G).

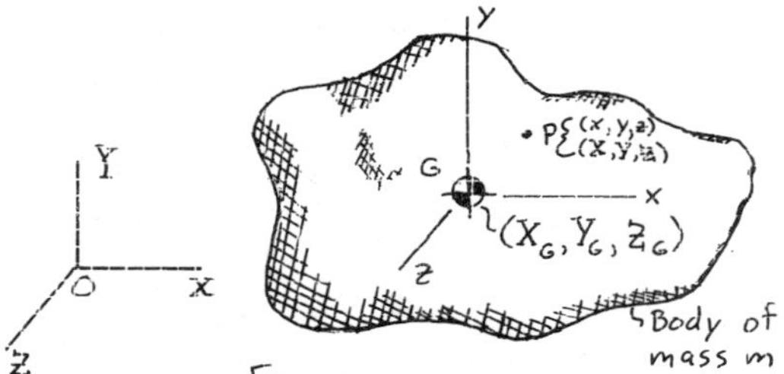

Figure a

show that

$$I_{XY} = I_{xy} + m X_G Y_G \qquad (a)$$

and that similar formulas apply for I_{YZ} and I_{ZX}.

Solution: Consider a point P in the body (Fig. a). Relative to axes xyz the cordinates of P are (x, y, z). Relative to axes XYZ, the coordinates of P are (X, Y, Z). Hence, by Fig. a,

$$X = X_G + x$$
$$Y = Y_G + y \qquad (b)$$
$$Z = Z_G + z$$

21.5

Given: An end view of a homogeneous quarter cylinder of radius a and mass m is shown in Fig. a.

Show that

$$I_{xy} = ma^2/2\pi$$

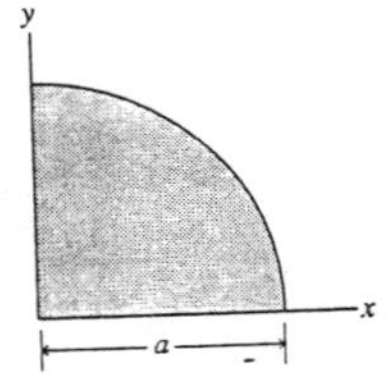

Figure a

Solution:

By definition,

$$I_{xy} = \int xy\, dm \qquad (b)$$

where (Fig. b),

$$x = r\cos\theta$$
$$y = r\sin\theta$$
$$dV = (r\,d\theta)(dr)L \qquad (c)$$
$$dm = \rho L r\, dr\, d\theta$$

ρ = mass density

L = length of the cylinder

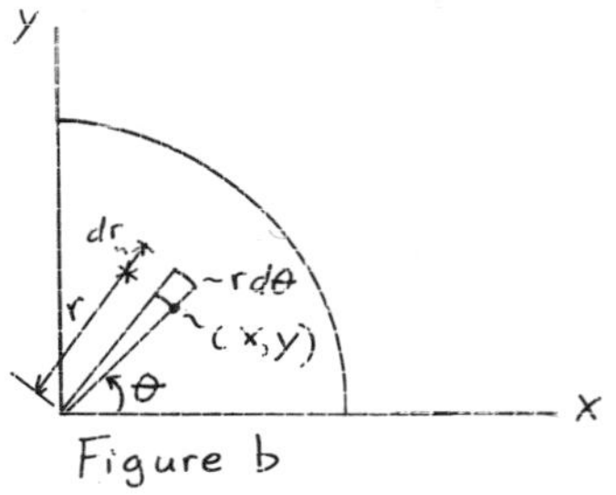

Figure b

(Continued)

21.5 Cont.

By Eqs. (b) and (c)

$$I_{xy} = \rho L \int_0^a r^3 dr \int_0^{\pi/2} \sin\theta \cos\theta \, d\theta \qquad (d)$$

Note that $\int_0^{\pi/2} \sin\theta\cos\theta \, d\theta = \int_0^{\pi/2} \frac{\sin 2\theta}{2} d\theta$

$$= -\frac{1}{4}\cos 2\theta \Big|_0^{\pi/2} = \frac{1}{2}$$

Hence, integration of Eq. (d) yields

$$I_{xy} = \frac{1}{8}\rho L a^4 \qquad (e)$$

also,

$$m = \int_0^a r\,d\theta \int_0^{\pi/2} d\theta = \frac{\pi}{4}\rho L a^2 \qquad (f)$$

Hence, Eqs. (e) and (f) yield [see Eq. (a)]

$$\underline{\underline{I_{xy} = ma^2/2\pi}}$$

21.6

Given: The diagonal of one face of a homogeneous cube is shown in Fig. a.

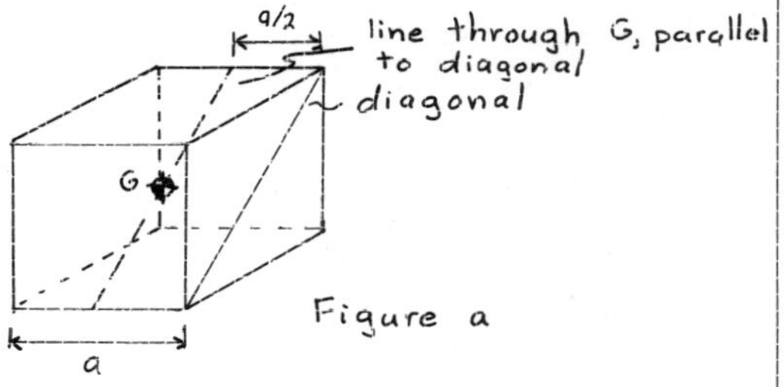

Figure a

Derive a formula for the mass moment of inertia of the cube about the diagonal.

Solution: For any axis through the mass center G (Fig. a)

$$I_G = \frac{1}{6} ma^2 \qquad (a)$$

(see Example 21.4 or Table F1, Appendix F, the rectangular parallelepiped with b=L=a)

Hence, by the Steiner theorem (parallel-axis theorem), we find,

$$I_{diagonal} = I_G + md^2 = \frac{1}{6}ma^2 + m\left(\frac{a}{2}\right)^2$$

or $\quad \underline{\underline{I_{diagonal} = \frac{5}{12} m a^2}}$

21.7

Given: A homogeneous rectangular solid has mass density $\rho = 0.001$ lb·s²/in⁴. The body is 8 in long, 4 in wide, and 3 in thick.

Determine the moment of inertia of the body about a line passing through its center and connecting two diagonally opposite corners.

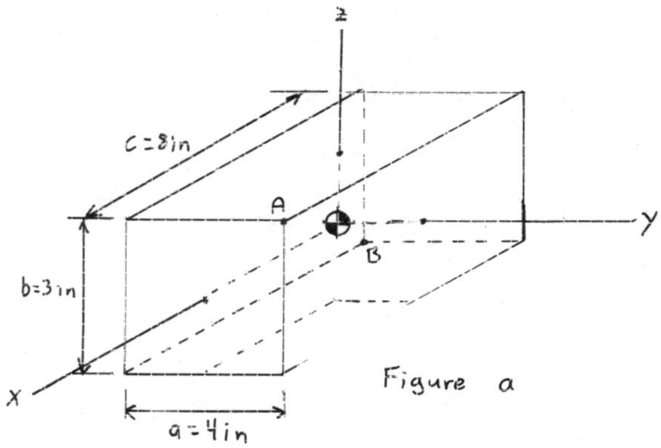

Figure a

Solution:

The rectangular solid is sketch in Fig. a. Select axes xyz with origin at the center (center of mass G) of the solid and parallel to the edges of the solid. Then, the products of inertia vanish, and the mass moment of inertia I about any line through G is given by,

$$I = I_{xx}\ell^2 + I_{yy}m^2 + I_{zz}n^2 \qquad (a)$$

where I_{xx}, I_{yy}, I_{zz} are the moments of inertia about axes x, y, z and (ℓ, m, n) are the direction cosines of the line relative to axes xyz. The moment of inertia of the body about a line passing through G and connecting any two diagonally opposite corners is the same for any two corners. So, let's select corners A and B (Fig. a)

By Appendix F, Table F_1, and Fig. a,

$$I_{xx} = \frac{1}{12} m(a^2 + b^2)$$
$$I_{yy} = \frac{1}{12} m(b^2 + c^2) \qquad (b)$$
$$I_{zz} = \frac{1}{12} m(a^2 + c^2)$$

where the mass of the body is

$$M = \rho abc = (0.001)(4)(3)(8) = 0.096 \text{ lb·s}^2/\text{in} \qquad (c)$$

Since the coordnates of corner A are x = 4 in, y = 2 in, z = 1.5 in, the direction cosines of line AB are

$$\ell = \frac{4}{\sqrt{4^2 + 2^2 + 1.5^2}} = \frac{4}{4.717} = 0.848$$

$$m = \frac{2}{4.717} = 0.424 \qquad (d)$$

$$n = \frac{1.5}{4.717} = 0.318$$

Equation (b) and (c) yield with a = 4 in, b = 3 in, and c = 8 in,

(Continued)

21.7 Cont.

$$I_{xx} = 0.200 \text{ lb}\cdot\text{in}\cdot\text{s}^2$$
$$I_{yy} = 0.584 \text{ lb}\cdot\text{in}\cdot\text{s}^2 \qquad \text{(e)}$$
$$I_{zz} = 0.640 \text{ lb}\cdot\text{in}\cdot\text{s}^2$$

Therefore, by Eqs. (a), (d), and (e)

$$I_{AB} = (0.200)(0.848)^2 + (0.584)(0.424)^2 + (0.640)(0.318)^2$$

or

$$\underline{\underline{I_{AB} = 0.3135 \text{ lb}\cdot\text{in}\cdot\text{s}^2}}$$

21.8

Given: The homogeneous solid wedge shown in Fig. a has mass m.

Determine the product of inertia I_{yz}, I_{xy}, and I_{xz}, using integration or symmetry conditions.

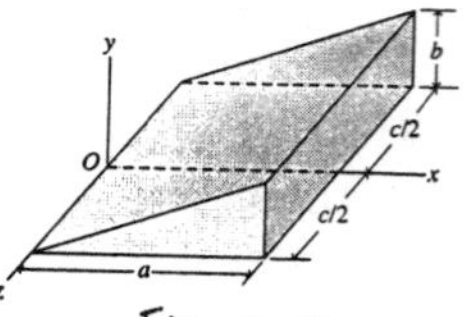

Figure a

Solution:

By Fig. a, the xy plane is a plane of symmetry. Therefore,

$$\underline{\underline{I_{xz} = I_{yz} = 0}}$$

For a volume element $dx\,dy\,dz$, the corresponding mass is

$$dm = \rho\, dx\, dy\, dz \qquad \text{(a)}$$
$$\rho = \text{mass density}$$

By definition,

$$I_{xy} = \int xy\,dm = \rho \int_0^a x\,dx \int_0^{\frac{b}{a}x} y\,dy \int_{-c/2}^{c/2} dz$$

or

$$I_{xy} = \left(\tfrac{1}{2}\rho abc\right)\left(\tfrac{ab}{4}\right) \qquad \text{(b)}$$

Since the volume of the prism is $\frac{1}{2}abc$, the mass is

$$m = \tfrac{1}{2}\rho abc \qquad \text{(c)}$$

Hence, Eqs. (b) and (c) yield

$$\underline{\underline{I_{xy} = \tfrac{1}{4} mab}}$$

21.9

Given: The homogeneous rectangular parallelepiped shown in Fig. a has mass m.

Use integration to evaluate the moments and products of inertia I_{xx}, I_{zz}, I_{xy}, and I_{yz} [lb·ft·s²] of the parallelepiped relative to the xyz axes, in terms of the mass m.

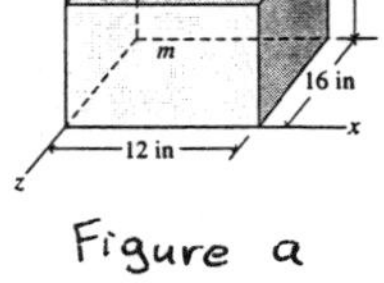

Figure a

Solution: Consider an element $dx\,dy\,dz$ of the parallelepiped. (Fig. b)

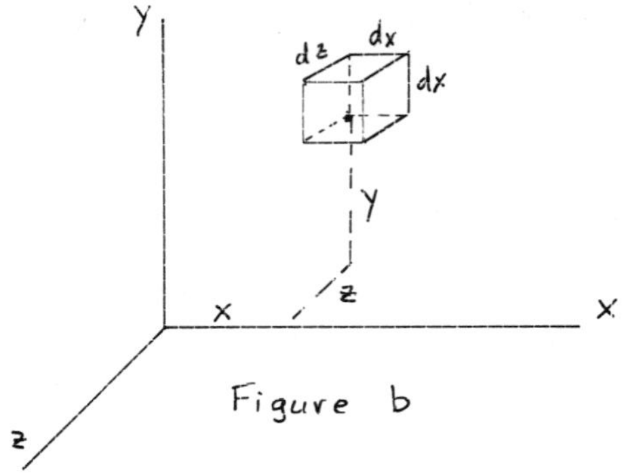

Figure b

By definition [Eq. (21.2)] and Fig. b,

$$I_{xx} = \int (y^2+z^2)\,dm \qquad \text{(a)}$$

where,

$$dm = \rho\, dx\, dy\, dz \qquad \text{(b)}$$

Equations (a) and (b) yield

$$I_{xx} = \rho \int_0^1 dx \int_0^{2/3} dy \int_{-4/3}^{0} (y^2+z^2)\,dz$$

Integration yields, with $m = \rho(1)(\frac{2}{3})(\frac{4}{3}) = \frac{8}{9}\rho$,

$$\underline{\underline{I_{xx} = \tfrac{20}{27} m}}$$

Similarly, by Fig. b,

$$I_{zz} = \int (x^2+y^2)\,dm$$

or

$$I_{zz} = \rho \int_0^1 dx \int_0^{2/3} (x^2+y^2)\,dy \int_{-4/3}^{0} dz$$

Integration yields, with $m = 8\rho/a$,

$$\underline{\underline{I_{zz} = 13/27\, m}}$$

Next,

$$I_{xy} = \int xy\,dm = \rho \int_0^1 dx \int_0^{2/3} xy\,dy \int_{-4/3}^{0} dz$$

Integration yields, with $m = 8\rho/9$,

$$\underline{\underline{I_{xy} = \tfrac{1}{6} m}}$$

Finally,

$$I_{yz} = \int yz\,dm = \rho \int_0^1 dx \int_0^{2/3} dy \int_{-4/3}^{0} yz\,dz$$

Integration yields, with $m = 8\rho/9$,

$$\underline{\underline{I_{yz} = -\tfrac{2}{9} m}}$$

21.10

<u>Given:</u> The homogeneous circular cylinder has mass m (Fig. a)

Use integration to evaluate the moments and products of inertia relative to axes xyz, in terms of m, r, and L. Then, by letting the appropriate dimension become small, derive the inertia properties for a thin disk and for a slender rod.

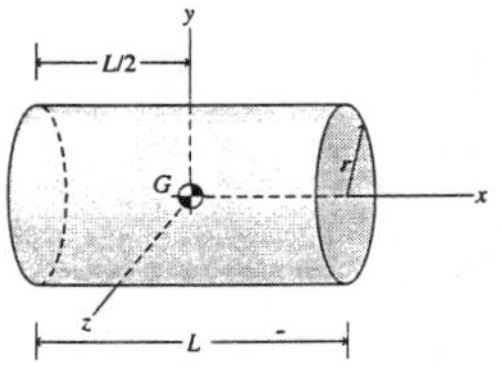

Figure a

<u>Solution:</u>

Since xyz axes are principle axes (Fig. a), $I_{xy} = I_{zx} = I_{yz} = 0$.

It is convenient to use cylindrical coordinates (Fig. b).

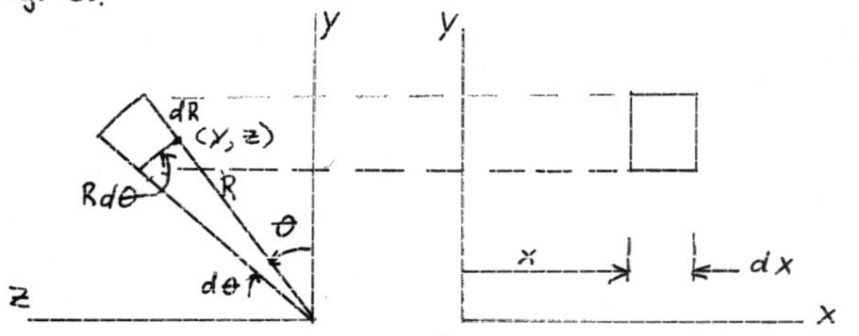

Figure b

By Fig. b, the elemental volume is

$$d_{(Vol)} = (R\,d\theta)(dR)(dx) = R\,dR\,d\theta\,dx$$

and the corresponding mass element is

$$dm = \rho\, d_{(Vol)} = \rho R\,dR\,d\theta\,dx \qquad (a)$$

By Fig. a, the mass of the cylinder is

$$m = \rho \pi r^2 L \qquad (b)$$

and,

$$y = R\cos\theta \qquad (c)$$

$$z = R\sin\theta \qquad (d)$$

By definition,

$$I_{xx} = \int (y^2 + z^2)\,dm \qquad (e)$$

$$I_{yy} = \int (x^2 + z^2)\,dm \qquad (f)$$

$$I_{zz} = \int (x^2 + y^2)\,dm \qquad (g)$$

By Eqs. (a), (c), (d), and (e) and Fig. a,

$$I_{xx} = \rho \int_{-L/2}^{L/2} dx \int_0^r R^3 dR \int_0^{2\pi} d\theta$$

Integration yields, with Eq. (b),

$$I_{xx} = \tfrac{1}{2}(\rho \pi r^2 L) r^2 = \tfrac{1}{2} m r^2 \qquad (h)$$

Similarly, by Eqs. (a), (d), and (f) and Fig. a

$$I_{yy} = \rho \iiint (x^2 + R^2 \sin^2\theta) R\,dR\,d\theta\,dx$$

$$= \rho \iiint x^2 R\,dR\,d\theta\,dx + \rho \iiint R^3 \sin^2\theta\,dR\,d\theta\,dx$$

or,

$$I_{yy} = \rho \int_{-L/2}^{L/2} x^2 dx \int_0^r R\,dr \int_0^{2\pi} d\theta + \rho \int_{-L/2}^{L/2} dx \int_0^r R^3 dR \int_0^{2\pi} \sin^2\theta\, d\theta$$

Integration yields, with Eq. (b),

$$I_{yy} = \tfrac{1}{12}[(\rho \pi r^2 L)L^2 + 3(\rho \pi r^2 L) r^2]$$

or $$I_{yy} = \tfrac{1}{12} m (3r^2 + L^2) \qquad (i)$$

By symmetry,

$$I_{zz} = \tfrac{1}{12} m (3r^2 + L^2) \qquad (j)$$

For a thin disk, $L \ll r$. Then, by Eqs. (h), (i), and (j), for a thin disk,

$$I_{xx} = \tfrac{1}{2} m r^2$$

$$I_{yy} = \tfrac{1}{4} m r^2$$

$$I_{zz} = \tfrac{1}{4} m r^2$$

For a slender rod, $L \gg r$. Then, by Eqs. (h), (i), and (j), for a slender rod,

$$I_{xx} \approx 0$$

$$I_{yy} = \tfrac{1}{12} m L^2$$

$$I_{zz} = \tfrac{1}{12} m L^2$$

21.11

<u>Given:</u> The circular disk shown in Fig. a is mounted obliquely on a shaft (Fig. b). The disk has principal moments of inertia $I_{xx} = I_{yy} = \frac{1}{4} m r^2$ and $I_{zz} = \frac{1}{2} m r^2$.

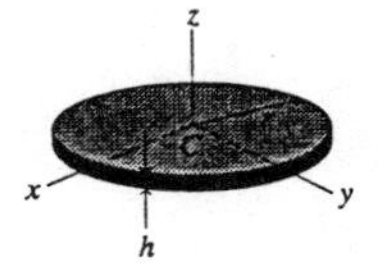

Figure a

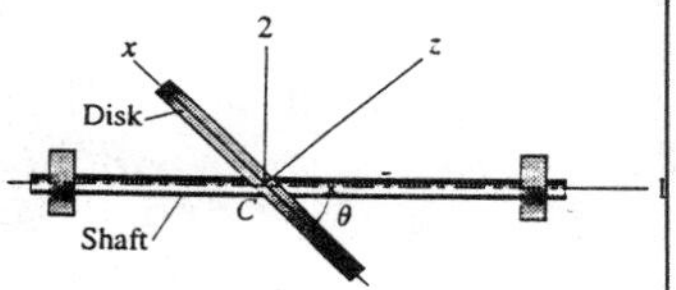

Figure b

Determine its mass moment of inertia about axis 1 in terms of m, r, and θ. Then plot I_1/mr^2 as a function of θ for $0 \le \theta \le 360°$

<u>Solution:</u>

Consider a unit vector $\hat{i}$ along the axis of the shaft (Fig. c). Its direction cosines relative to xyz axes are

$$l = -\cos\theta$$

$$m = \cos 90° = 0 \qquad (a)$$

$$n = \sin\theta$$

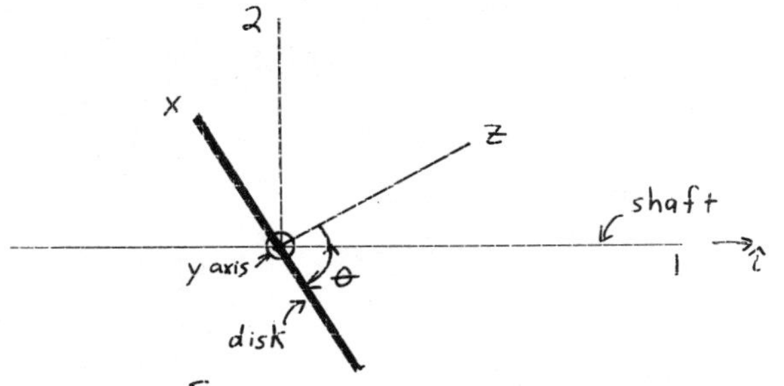

Figure c

(Continued)

21.11 Cont.

Hence, by Eqs. (21.3) and (a), with Fig. a,

$$I_1 = l^2 I_{xx} + m^2 I_{yy} + n^2 I_{zz}$$

$$I_1 = (\cos^2\theta)(\tfrac{1}{4} m r^2) + 0 + (\sin^2\theta)(\tfrac{1}{2} m r^2)$$

or

$$\underline{I_1 = \tfrac{1}{4} m r^2 (1 + \sin^2\theta)} \qquad \text{(b)}$$

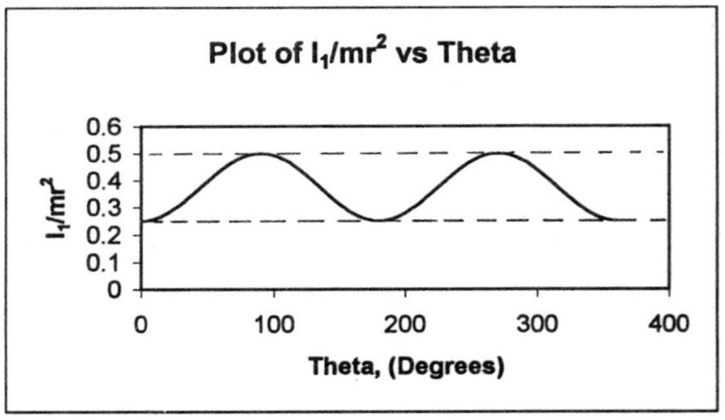

Figure d

By Fig. d, we see that the ratio I_1/mr^2 oscilates between 0.25 to 0.50, and its slope is zero at $\theta = 0°, 90°, 180°, 270°$, and 360°

21.12

Given: The wedge of Fig. P 21.8 is shown in Fig. a below.

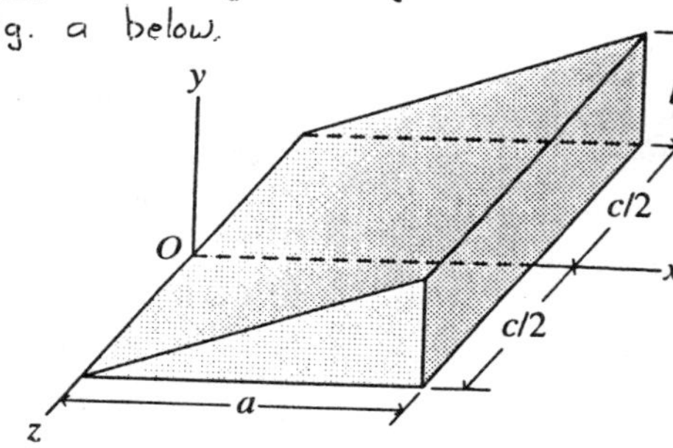

Figure a

Using integration, determine the mass product of inertia $I_{\bar{x}\bar{y}}$ of the wedge relative to $\bar{x}\,\bar{y}\,\bar{z}$ axes that are parallel to the xyz axes and have their origin at the center of mass of the wedge.

Solution:

By definition, relative to $\bar{x}\bar{y}\bar{z}$ axes,

$$I_{\bar{x}\bar{y}} = \int \bar{x}\,\bar{y}\, dm \qquad \text{(a)}$$

The center of mass of the wedge is located at $(\tfrac{2}{3}a, \tfrac{1}{3}b, 0)$ relative to xyz axes; see Figs. a and b (where a slice of the cross section of the wedge in the xy plane, $z = 0$, is shown).

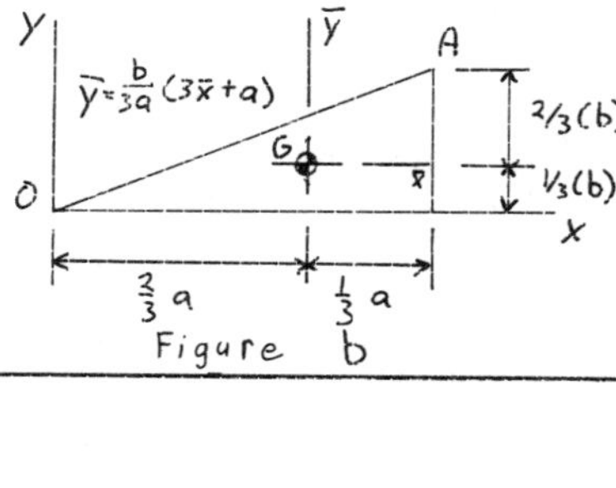

Figure b

The equation of line OA, in terms of coordinates $\bar{x}, \bar{y}$ is

$$\bar{y} = \frac{b}{3a}(3\bar{x} + a) \qquad \text{(b)}$$

and by Fig. a, the mass of the prism is

$$m = \tfrac{1}{2}\rho a b c \qquad \text{(c)}$$

where ρ is the mass density of the wedge. Also, for a volume element $d\bar{x}\,d\bar{y}\,d\bar{z}$, the corresponding element of mass is

$$dm = \rho\, d\bar{x}\, d\bar{y}\, d\bar{z} \qquad \text{(d)}$$

Hence, by Eqs. (a), (b), and (d) and Figs. a and b,

$$I_{\bar{x}\bar{y}} = \int \bar{x}\bar{y}\,dm = \rho \int_{-\frac{2a}{3}}^{a/3} \int_{-b/3}^{\frac{b}{3a}(3\bar{x}+a)} \int_{-\frac{c}{2}}^{c/2} \bar{x}\bar{y}\, d\bar{x}\, d\bar{y}\, d\bar{z}$$

or

$$I_{\bar{x}\bar{y}} = \rho c \int_{-\frac{2a}{3}}^{a/3} \frac{\bar{x}}{2}\left[\left(\frac{b}{3a}\right)^2 (3\bar{x} + a)^2 - \left(\frac{b}{3}\right)^2\right] d\bar{x}$$

Completing the integration, we find with Eq. (c),

$$\underline{I_{\bar{x}\bar{y}} = \tfrac{1}{36} m a b}$$

21.13

Given: The homogeneous solid wedge shown in Fig. a has mass m.

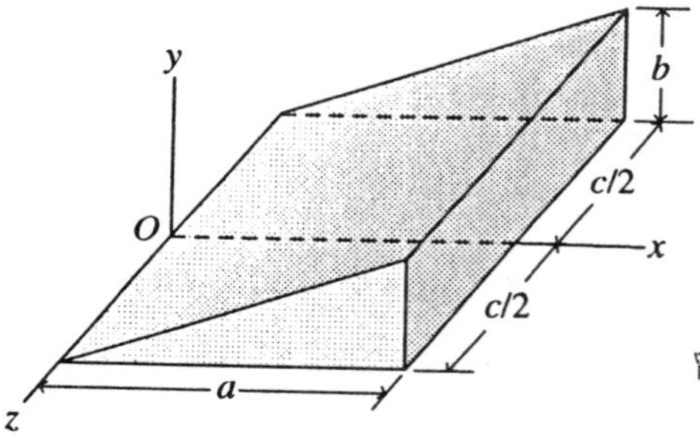

Figure a

Select $\bar{x}\bar{y}\bar{z}$ axes parallel to the xyz axes and having their origin at the wedge's center of mass G.

Use the parallel-axis theorem to show that $I_{\bar{x}\bar{y}} = \tfrac{1}{36} m a b$, given that $I_{xy} = \tfrac{1}{4} m a b$.

Solution: By Fig. a, the center of mass is located at $x = \tfrac{2}{3}a$, $y = \tfrac{1}{3}b$, and $z = 0$ (by symmetry). Consider a slice of the wedge cross section in the xy plane $z = 0$, Fig. b.

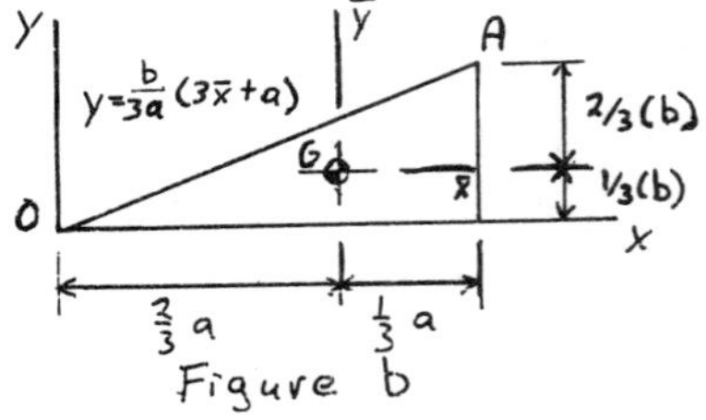

Figure b

(Continued)

21.13 Cont.

By Eq. (21.8),

$$I_{xy} = I_{\bar{x}\bar{y}} + m\,\bar{x}_o\,\bar{y}_o \qquad (a)$$

where $\bar{x}_o, \bar{y}_o$ are the coordinates of point O with respect to axes $\bar{x}\bar{y}$. Therefore, by Fig. b,

$$\bar{x}_o = -\tfrac{2}{3}a\,, \quad \bar{y}_o = -\tfrac{1}{3}b \qquad (b)$$

Hence, Eqs. (a) and (b) yield

$$I_{\bar{x}\bar{y}} = I_{xy} - m\bar{x}_o\bar{y}_o = \tfrac{1}{4}mab - m(-\tfrac{2}{3}a)(-\tfrac{1}{3}b)$$

or $$I_{\bar{x}\bar{y}} = \tfrac{9}{36}mab - \tfrac{8}{36}mab = \tfrac{1}{36}\,mab$$

21.14

Given: A homogeneous cube of dimension a has principal axes $\bar{x}\bar{y}\bar{z}$ with origin at the center of mass G. For any axis through G, the mass moment of inertia is $I_G = \frac{1}{6}ma^2$ (see Example 21.4 or Appendix F, Table F.1; the rectangular parallelepiped becomes a cube with b = L = a).

Derive a formula for the mass moment of inertia of the cube about a line that coincides with one of its edges.

Solution: Consider line b-b along an edge of the cube parallel to axis $\bar{x}$ (Fig. a).

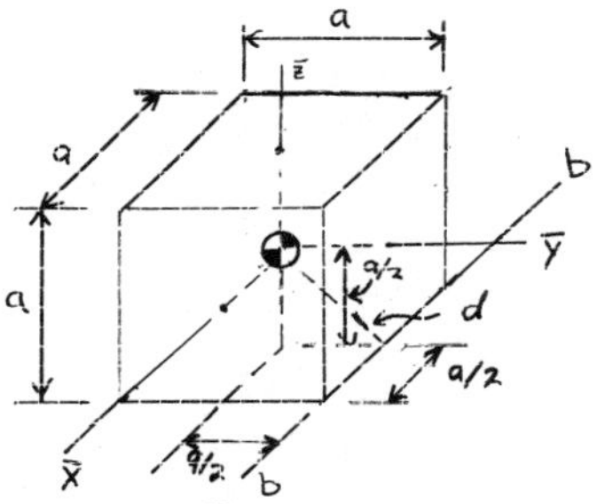

Figure a

The distance from axis $\bar{x}$ to line b-b is (Fig. a)

$$d = \sqrt{(\tfrac{a}{2})^2 + (\tfrac{a}{2})^2} = \frac{a}{\sqrt{2}} \qquad (a)$$

By Eq. (21.8),

$$I_{b\text{-}b} = I_{\bar{x}\bar{x}} + md^2 \qquad (b)$$

where $$I_{\bar{x}\bar{x}} = \tfrac{1}{6}ma^2 \qquad (c)$$

Hence, by Eqs. (a), (b), and (c),

$$I_{b\text{-}b} = \tfrac{1}{6}ma^2 + \tfrac{1}{2}ma^2 = \tfrac{2}{3}ma^2$$

21.15 Cont.

Given: An end view of a homogeneous prism of mass m is shown in Fig. a.

Using the parallel-axis theorem, determine the product of inertia $I_{\bar{x}\bar{y}}$ relative to center-of-mass axes $\bar{x}\bar{y}$, which are parallel to the xy axes.

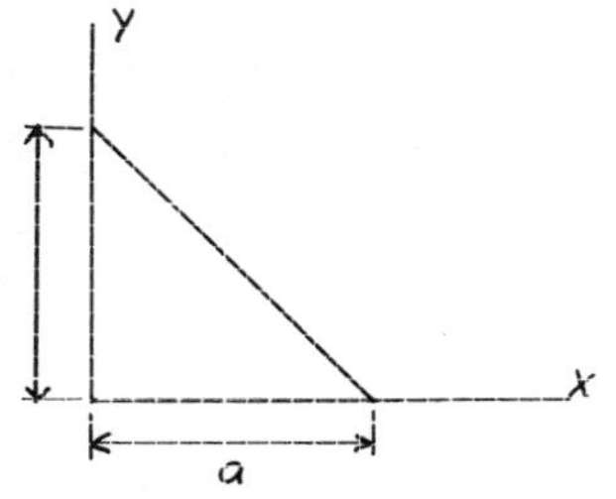

Figure a

Solution:
By the given information in Problem 21.2,

$$I_{xy} = \tfrac{1}{12}mab \qquad (a)$$

The center-of-mass axes $\bar{x}\bar{y}$ are located as shown in Fig. b.

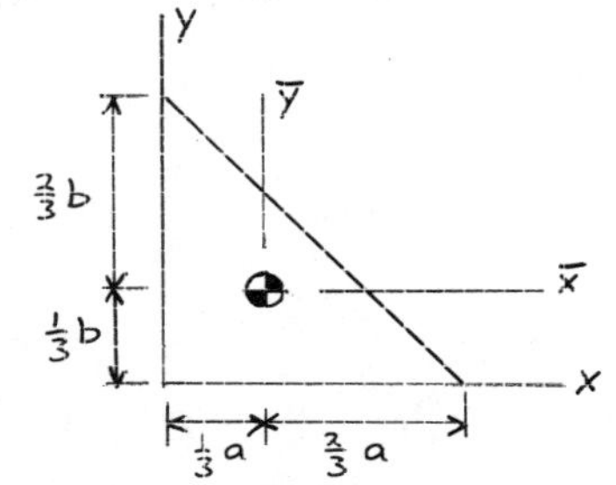

Figure b

By the first of Eqs. (21.8),

$$I_{xy} = I_{\bar{x}\bar{y}} + m\,\bar{x}_o\bar{y}_o \qquad (b)$$

where $\bar{x}_o, \bar{y}_o$ are the coordinates of the origin O of axes xy, with respect to axes $\bar{x}\bar{y}$ with origin at G. Then, by Fig. b,

$$\bar{x}_o = \tfrac{-1}{3}a\,, \quad \bar{y}_o = -\tfrac{1}{3}b \qquad (c)$$

Hence, by Eqs. (a), (b), and (c),

$$I_{\bar{x}\bar{y}} = I_{xy} - m\bar{x}_o\bar{y}_o = \tfrac{1}{12}mab - m(-\tfrac{1}{3}a)(-\tfrac{1}{3}b)$$

or $$I_{\bar{x}\bar{y}} = -\tfrac{1}{36}mab$$

21.16

Given: An end view of a homogeneous quarter cylinder of radius a and mass m is shown in Fig. a.

Using the parallel-axis theorem, determine the product of inertia $I_{\bar{x}\bar{y}}$ relative to the center-of-mass axes $\bar{x}\bar{y}$, which are parallel to the xy axes.

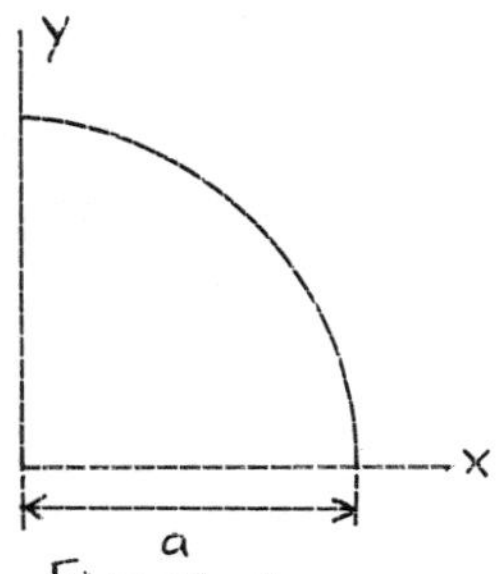

Figure a

(Continued)

21.16 Cont.

Solution: By Problem 21.5,

$$I_{xy} = \frac{ma^2}{2\pi} \qquad (a)$$

The center-of-mass axes $\bar{x}\bar{y}$ are located as shown in Fig. b.

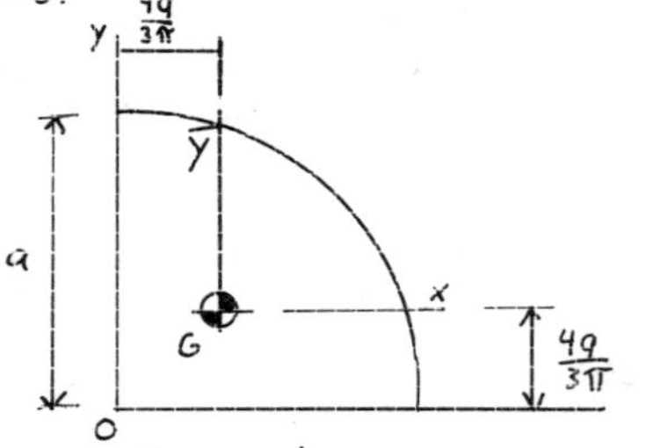

Figure b

By the first of Eqs. (21.8),

$$I_{xy} = I_{\bar{x}\bar{y}} + m\bar{x}_o\bar{y}_o \qquad (b)$$

where $\bar{x}_o, \bar{y}_o$ are the coordinates of the origin of the xy axes, with respect to the $\bar{x}\bar{y}$ axes, with origin at G. By Fig. b,

$$\bar{x}_o = -\frac{4a}{3\pi}, \quad \bar{y}_o = -\frac{4a}{3\pi} \qquad (c)$$

Hence, by Eqs. (a), (b), and (c),

$$I_{\bar{x}\bar{y}} = I_{xy} - m\bar{x}_o\bar{y}_o = \frac{ma^2}{2\pi} - m\left(-\frac{4a}{3\pi}\right)\left(-\frac{4a}{3\pi}\right)$$

or $I_{\bar{x}\bar{y}} = \frac{ma^2}{\pi}\left(\frac{1}{2} - \frac{16}{9\pi}\right) = -0.02097ma^2$

21.17

Given: The moments and products of inertia of the solid right circular cone in Fig. a relative to the X Y Z axes are

$$I_{XX} = \frac{3}{10}mr^2$$

$$I_{YY} = I_{ZZ} = \frac{3}{20}m(4h^2+r^2) \qquad (a)$$

$$I_{XY} = I_{XZ} = I_{YZ} = 0$$

Use the parallel-axis theorem to derive formulas for the moments and products of inertia of the cone relative to the xyz axes.

Solution: By the first of Eqs. (21.6),

$$I_{XX} = I_{\bar{x}\bar{x}} + md_{X\bar{x}}^2 \qquad (b)$$

where $I_{\bar{x}\bar{x}}$ is the mass moment of inertia of the cone relative to the center-of-mass axis $\bar{x}$ parallel to axis X, and $d_{X\bar{x}}$ is the distance between the X axis and the $\bar{x}$ axis (Fig. a)

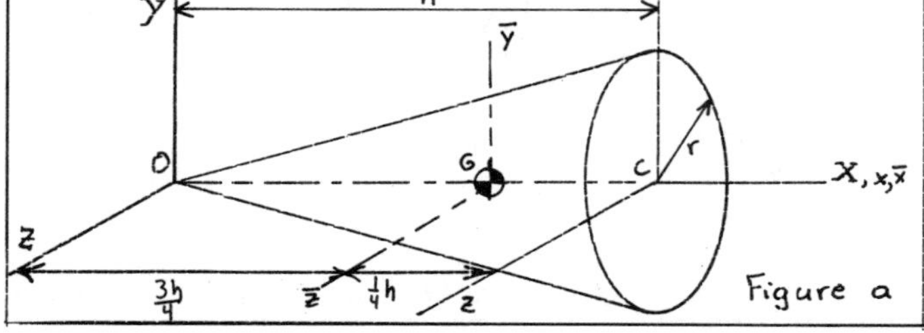

Figure a

Also, by the first of Eqs. (21.6),

$$I_{xx} = I_{\bar{x}\bar{x}} + md_{x\bar{x}}^2 \qquad (c)$$

But,

$$d_{X\bar{x}} = d_{x\bar{x}} = 0 \qquad (d)$$

since axes X, $\bar{x}$, and x coincide. Therefore, by Eqs. (b), (c), and (d), and the first of Eqs. (a),

$$I_{xx} = I_{XX} = \frac{3}{10}mr^2$$

Of course this result could have been obtained by inspection of Fig. (a).

By the second of Eqs. (21.6), we have

$$I_{YY} = I_{\bar{y}\bar{y}} + m\left(\tfrac{3}{4}h\right)^2 \qquad (e)$$

and

$$I_{yy} = I_{\bar{y}\bar{y}} + m\left(\tfrac{1}{4}h\right)^2 \qquad (f)$$

Therefore, by (e) and (f), and the second of Eqs. (a),

$$I_{yy} = I_{YY} - \frac{9}{16}mh^2 + \frac{1}{16}mh^2$$

$$= \frac{3}{20}m(4h^2+r^2) - \frac{1}{2}mh^2$$

or

$$I_{yy} = \frac{1}{20}m(2h^2+3r^2)$$

Similarly, or by symmetry,

$$I_{zz} = \frac{1}{20}m(2h^2+3r^2)$$

For products of inertia, we have by Eqs. (21.8) for I_{xy},

$$I_{xy} = I_{\bar{x}\bar{y}} + m\bar{x}_o\bar{y}_o \qquad (g)$$

where $I_{\bar{x}\bar{y}}$ is the $\bar{x}\bar{y}$ product of inertia of the cone, and $\bar{x}_o, \bar{y}_o$ are the coordinates of point O (Fig. a) relative to axes $\bar{x}\bar{y}$.

By Fig. a,

$\bar{x}_o = -\frac{3}{4}h$ and $\bar{y}_o = 0$. Hence, Eq. (g) yields

$$I_{xy} = I_{\bar{x}\bar{y}} \qquad (h)$$

Also, by the first of Eqs. (21.8), and Fig. a,

$$I_{XY} = I_{\bar{x}\bar{y}} + m\bar{x}_c\bar{y}_c = I_{\bar{x}\bar{y}} \qquad (i)$$

since $\bar{y}_c = 0$. Then, by Eqs. (a), (h), and (i),

$$I_{xy} = I_{XY} = 0$$

Similarly, since $I_{XZ} = I_{YZ} = 0$ and $\bar{y}_o = \bar{y}_E = 0$,

$$I_{xz} = I_{yz} = 0$$

Note: Since axes XYZ and xyz are principal axes, the products of inertia for these axes are zero (see Section 21.3).

21.18

Given: As in Problem 21.17, the moments and products of inertia of the solid right circular cone in Fig. a, relative to XYZ axes are

$$I_{XX} = \frac{3}{10} m r^2$$
$$I_{YY} = I_{ZZ} = \frac{3}{20} m (4h^2 + r^2) \qquad (a)$$
$$I_{XY} = I_{XZ} = I_{YZ} = 0$$

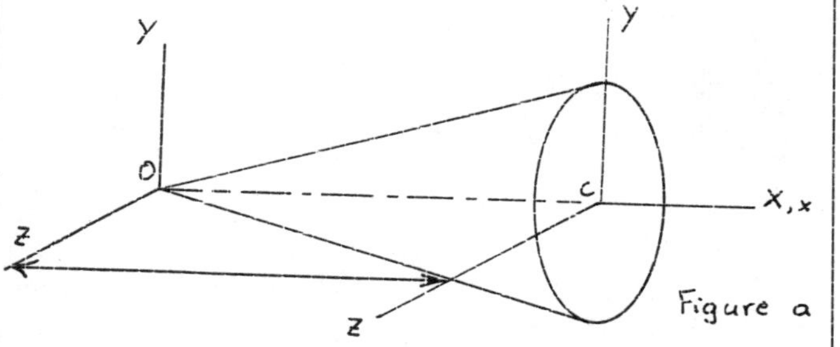

Figure a

Use the parallel-axis theorem to derive formulas for the moments of inertia relative to $\bar{x}\bar{y}\bar{z}$ axes, which are parallel to axes XYZ and have origin at the cone's center of mass. [Check your answers against the formulas in Appendix F (Table F.1)].

Solution: By the first of Eqs. (21.6),

$$I_{XX} = I_{\bar{x}\bar{x}} + m d_{X\bar{x}}^2 \qquad (b)$$

where $d_{X\bar{x}}$ is the distance between the X axis and the $\bar{x}$ axis. By Fig. b, the location of $\bar{x}\bar{y}\bar{z}$ are shown. Hence, since axes X and $\bar{x}$ coincide, $d_{X\bar{x}} = 0$. Then, by Eq. (b),

$$I_{\bar{x}\bar{x}} = I_{XX} = \frac{3}{10} m r^2$$

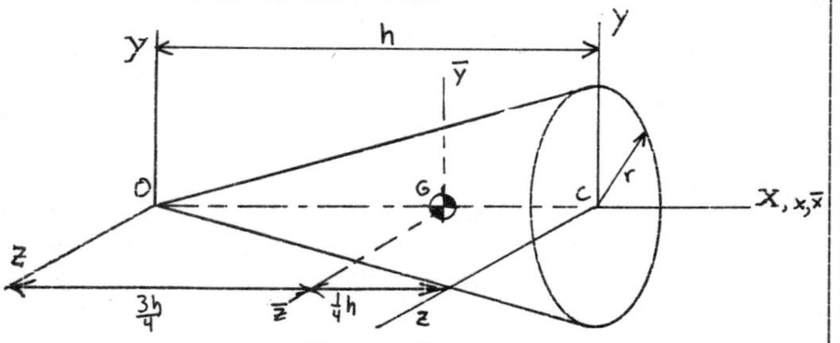

Figure b

By the second of Eqs. (21.6),

$$I_{YY} = I_{\bar{y}\bar{y}} + m d_{Y\bar{y}}^2 \qquad (c)$$

where $d_{Y\bar{y}}$ is the distance between axes Y and $\bar{y}$. By Fig. b, $d_{Y\bar{y}} = \frac{3}{4}h$. Hence, the second of Eqs. (a) and Eq. (c) yield,

$$I_{\bar{y}\bar{y}} = I_{YY} - m d_{Y\bar{y}}^2 = \frac{3m}{80}(h^2 + 4r^2)$$

Similarly, or by symmetry,

$$I_{\bar{z}\bar{z}} = \frac{3m}{80}(h^2 + 4r^2)$$

The formulas for $I_{\bar{x}\bar{x}}$, $I_{\bar{y}\bar{y}}$, and $I_{\bar{z}\bar{z}}$ agree with those in Table F.1 for the solid right circular cone.

21.19

Given: By Appendix F, Table F.1, the moments of inertia for the solid right circular cone shown in Fig. a, relative to principal axes at the center of mass G, are,

$$I_{\bar{x}\bar{x}} = \frac{3}{10} m r^2$$
$$I_{\bar{y}\bar{y}} = I_{\bar{z}\bar{z}} = \frac{3}{80} m (4r^2 + h^2) \qquad (a)$$

a) Derive a formula for the moment of inertia about a line that passes through the origin O and the point P on the base of the cone at $y = r$.

b) Derive a formula for the moment of inertia about a line that passes through the center of mass and is parallel to the line that passes through O and the point P on the base of the cone at $y = r$.

Solution:

a) The two lines described in parts a and b are shown in Fig. a.

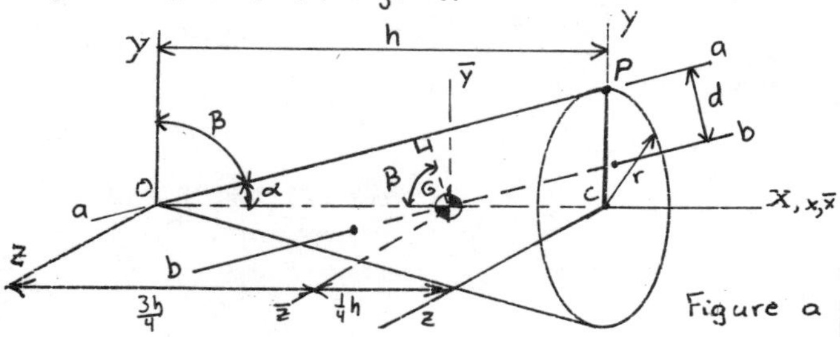

Figure a

By Eq. (21.10), the moment of inertia about line a-a is

$$I_a = I_1 \ell^2 + I_2 m^2 + I_3 n^2 \qquad (b)$$

where I_1, I_2, I_3 are principal moments of inertia at point O of the cone, and (ℓ, m, n) are the direction cosines of line a-a.

By Fig. a,

$$\cos\alpha = \ell = \frac{h}{\sqrt{h^2 + r^2}}$$

$$\cos\beta = m \quad \frac{r}{\sqrt{h^2 + r^2}} \qquad (c)$$

$$n = 0$$

By the parallel-axis theorem [Eqs. (21.6)]

$$I_1 = I_{\bar{x}\bar{x}} + m d_{x\bar{x}}^2$$
$$I_2 = I_{\bar{y}\bar{y}} + m d_{y\bar{y}}^2 \qquad (d)$$
$$I_3 = I_{\bar{z}\bar{z}} + m d_{z\bar{z}}^2$$

where $d_{x\bar{x}}, d_{y\bar{y}}, d_{z\bar{z}}$ are the distances between axes $x, \bar{x}$, $y, \bar{y}$, and $z, \bar{z}$ respectively.

(Continued)

21.19 Cont.

Since axes X and $\bar{x}$ coincide, $d_{x\bar{x}} = 0$, and by Fig. a,

$$d_{y\bar{y}} = d_{z\bar{z}} = \frac{3}{4}h$$

Hence, Eqs. (d) yield, with Eqs. (a),

$$I_1 = I_{\bar{x}\bar{x}} = \frac{3}{10}mr^2$$
$$I_2 = I_3 = \frac{3}{20}m(r^2+4h^2) \qquad (e)$$

See also, Appendix F, Table F.1.
Then, by Eqs. (b), (c), and (e)

$$I_a = \left(\frac{3}{10}mr^2\right)\left(\frac{h^2}{h^2+r^2}\right) + \frac{3}{20}m(r^2+4h^2)\left(\frac{r^2}{h^2+r^2}\right)$$

or

$$I_a = \frac{3}{20}mr^2\left(\frac{6h^2+r^2}{h^2+r^2}\right)$$

(b) Line b-b has the same direction cosines as line a-a [Eqs. (d)]. Hence, by Eq. (2.10),

$$I_b = I_1 l^2 + I_2 m^2 + I_3 n^2 \qquad (f)$$

where for point G, by Eq. (a),

$$I_1 = I_{\bar{x}\bar{x}} = \frac{3}{10}mr^2$$
$$I_2 = I_{\bar{y}\bar{y}} = \frac{3}{80}m(4r^2+h^2) \qquad (g)$$
$$I_3 = I_{\bar{z}\bar{z}} = \frac{3}{80}m(4r^2+h^2)$$

Hence, by Eqs. (c), (f), and (g),

$$I_b = \frac{3}{10}mr^2\left(\frac{h^2}{h^2+r^2}\right) + \frac{3}{80}m(4r^2+h^2)\left(\frac{r^2}{h^2+r^2}\right)$$

or

$$I_b = \frac{3}{80}mr^2\left(\frac{9h^2+4r^2}{h^2+r^2}\right)$$

21.20

Given: Refer to Example 21.5. The slider guide of mass density $\rho = 7846\ \text{kg/m}^3$ (Fig. a) may be considered as two parts (Fig. b); part 1 (400 mm × 600 mm × 600 mm) and part 2 (300 mm × 600 mm × 300 mm). Part 1 has positive mass

$$m_1 = 7846(0.400)(0.600)(0.600) = 1130\ \text{kg} \qquad (a)$$

and part 2 has negative mass

$$m_2 = -7846(0.300)(0.600)(0.300) = -423.684\ \text{kg} \qquad (b)$$

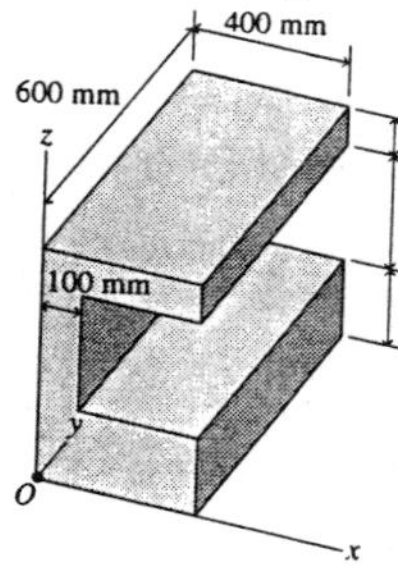

Figure a

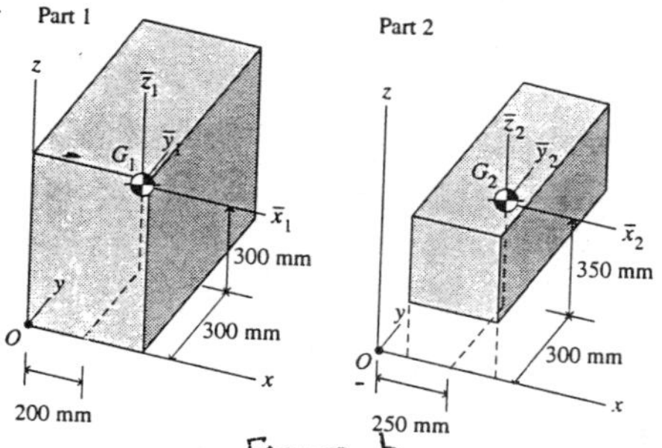

Figure b

a) Show that the center of mass of the guide is located at

$$x_G = 170\ \text{mm},\ y_G = 300\ \text{mm},\ z_G = 270\ \text{mm}$$

relative to axes xyz.

b) Determine the mass moments and products of inertia of the guide relative to axes $\bar{x}\bar{y}\bar{z}$, which are parallel to the xyz axes and have origin at the center of mass.

Solution:

a) By definition, the position vector $\bar{r}_G$ of the center of mass is given by the relation,

$$(m_1+m_2)\bar{r}_G = m_1\bar{r}_{G1} + m_2\bar{r}_{G2} \qquad (c)$$

where $\bar{r}_{G1}$ and $\bar{r}_{G2}$ are the position vectors of the centers of mass of part 1 and 2, respectively.

By Eqs. (a), (b), and (c) and Fig. b, we find,

$$(1130-423.684)\bar{r}_G = (1130)(0.2\hat{i} + 0.3\hat{k}) - (423.684)(0.25\hat{i} + 0.30\hat{j} + 0.35\hat{k})$$

or

$$706.316\,\bar{r}_G = 120.079\hat{i} + 211.8948\hat{j} + 190.7106\hat{k}$$

Hence,

$$\bar{r}_G = 0.170\hat{i} + 0.300\hat{j} + 0.270\hat{k} \quad [\text{m}]$$

or,

$$x_G = 170\ \text{mm},\ y_G = 300\ \text{mm},\ z_G = 270\ \text{mm} \qquad (d)$$

b) Since the $\bar{x}\bar{z}$ plane ($\bar{y} = 0$) is a plane of symmetry $I_{\bar{x}\bar{y}} = I_{\bar{y}\bar{z}} = 0$. Then, by Eqs. (21.6), (21.7), and (21.8),

$$I_{\bar{x}\bar{x}} = I_{xx} - md_{x\bar{x}}^2$$
$$I_{\bar{y}\bar{y}} = I_{yy} - md_{y\bar{y}}^2 \qquad (e)$$
$$I_{\bar{z}\bar{z}} = I_{zz} - md_{z\bar{z}}^2$$
$$I_{\bar{z}\bar{x}} = I_{zx} - m\bar{z}_0\bar{x}_0$$

where, with Eq. (d),

$$d_{z\bar{z}}^2 = y_G^2 + x_G^2 = (0.17)^2 + (0.30)^2 = 0.1189\ \text{m}^2$$
$$d_{x\bar{x}}^2 = y_G^2 + z_G^2 = (0.30)^2 + (0.27)^2 = 0.1629\ \text{m}^2$$
$$d_{y\bar{y}}^2 = x_G^2 + z_G^2 = (0.17)^2 + (0.27)^2 = 0.1018\ \text{m}^2 \qquad (f)$$
$$z_G x_G = (0.27)(0.17) = 0.0459\ \text{m}^2$$
$$m = m_1 - m_2 = 1130 - 426.684 = 706.316\ \text{kg}$$

By Example 21.5,

$$I_{xx} = 165.3\ \text{N}\cdot\text{m}\cdot\text{s}^2$$
$$I_{yy} = 11.2\ \text{N}\cdot\text{m}\cdot\text{s}^2 \qquad (g)$$
$$I_{zz} = 115.4\ \text{N}\cdot\text{m}\cdot\text{s}^2$$
$$I_{zx} = 30.7\ \text{N}\cdot\text{m}\cdot\text{s}^2$$

Equations (e), (f), and (g) yield

$$I_{\bar{x}\bar{x}} = 165.3 - (706.316)(0.1629) = 50.2\ \text{N}\cdot\text{m}\cdot\text{s}^2$$
$$I_{\bar{y}\bar{y}} = 111.2 - (706.316)(0.1018) = 39.3\ \text{N}\cdot\text{m}\cdot\text{s}^2$$
$$I_{\bar{z}\bar{z}} = 115.4 - (706.316)(0.1189) = 31.4\ \text{N}\cdot\text{m}\cdot\text{s}^2$$
$$I_{\bar{z}\bar{x}} = 30.7 - (706.316(0.0459) = -1.72\ \text{N}\cdot\text{m}\cdot\text{s}^2$$

21.21

Given: An ovoid surface is constructed with radius I, rather than $I^{-1/2}$ (see the discussion of the ellipsoid of inertia in Sec. 21.3).

Show that the equation of the surface is

$$(x^2+y^2+z^2)^3 = (I_{xx}x^2 + I_{yy}y^2 + I_{zz}z^2 - 2I_{yz}yz - 2I_{zx}zx - 2I_{xy}xy)^2$$

Solution:

A point P on the ovoid surface is given by

$$x = lI, \quad y = mI, \quad z = nI \tag{a}$$

where l, m, n are direction cosines of the line from origin O of the ovoid to the point P: (x, y, z). By Eqs. (21.3) and (a), we have for the moment of inertia about the line through O and P,

$$I^3 = x^2 I_{xx} + y^2 I_{yy} + z^2 I_{zz} - 2yz I_{yz} - 2zx I_{zx} - 2xy I_{xy} \tag{b}$$

Also, by Eq. (a),

$$x^2+y^2+z^2 = (l^2+m^2+n^2)I^2 = I^2 \tag{c}$$

Therefore, by Eq. (c),

$$I^3 = (x^2+y^2+z^2)^{3/2} \tag{d}$$

Equations (b) and (d) and squaring both sides of the resulting equation, we obtain the required result, namely,

$$(x^2+y^2+z^2)^3 = (I_{xx}x^2 + I_{yy}y^2 + I_{zz}z^2 - 2I_{yz}yz - 2I_{zx}zx - 2I_{xy}xy)^2$$

21.22

Given: Let XYZ and xyz be two sets of rectangular coordinate axes with common origin. Let the cosines of the angles between the axes of the two systems be denoted by letters as indicated in Table P21.22. For example, l_2 is the cosine of the angle between the y axis and the X axis.

Table P(21.22)

	X	Y	Z
x	l_1	m_1	n_1
y	l_2	m_2	n_2
z	l_3	m_3	n_3

a) Show that

$$I_{xx} = l_1^2 I_{XX} + m_1^2 I_{YY} + n_1^2 I_{ZZ} - 2m_1n_1 I_{YZ} - 2n_1l_1 I_{ZX} - 2l_1m_1 I_{XY}$$

and similar formulas exist for I_{yy} and I_{zz}. Then, show that

$$I_{xx}+I_{yy}+I_{zz} = I_{XX}+I_{YY}+I_{ZZ} = I_1+I_2+I_3$$

b) Show that,

$$I_{xy} = -l_1l_2 I_{XX} - m_1m_2 I_{YY} - n_1n_2 I_{ZZ} + (m_1n_2+m_2n_1)I_{YZ} + (n_1l_2+n_2l_1)I_{ZX} + (l_1m_2+l_2m_1)I_{XY}$$

and similar formulas exist for I_{yz} and I_{zx}.

Solution: By Table P21.22,

$$\begin{aligned} x &= l_1X + m_1Y + n_1Z \\ y &= l_2X + m_2Y + n_2Z \\ z &= l_3X + m_3Y + n_3Z \end{aligned} \tag{a}$$

By Eq. (21.3) and Table P21.22,

$$\begin{aligned} I_{xx} &= l_1^2 I_{XX} + m_1^2 I_{YY} + n_1^2 I_{ZZ} - 2m_1n_1 I_{YZ} - 2n_1l_1 I_{ZX} - 2l_1m_1 I_{XY} \\ I_{yy} &= l_2^2 I_{XX} + m_2^2 I_{YY} + n_2^2 I_{ZZ} - 2m_2n_2 I_{YZ} - 2n_2l_2 I_{ZX} - 2l_2m_2 I_{XY} \\ I_{zz} &= l_3^2 I_{XX} + m_3^2 I_{YY} + n_3^2 I_{ZZ} - 2m_3n_3 I_{YZ} - 2n_3l_3 I_{ZX} - 2l_3m_3 I_{XY} \end{aligned} \tag{b}$$

By Eqs. (b), with the relations (Table P21.22)

$l_1^2+l_2^2+l_3^2=1$, $m_1^2+m_2^2+m_3^2=1$, $n_1^2+n_2^2+n_3^2=0$, $m_1n_1+m_2n_2+m_3n_3=0$, $n_1l_1+n_2l_2+n_3l_3=0$, and $l_1m_1+l_2m_2+l_3m_3=0$

$$I_{xx}+I_{yy}+I_{zz} = I_{XX}+I_{YY}+I_{ZZ} \tag{c}$$

Also, for principal axes, $I_{YZ}=I_{ZX}=I_{XY}=0$ and $I_{XX}=I_1$, $I_{YY}=I_2$, $I_{ZZ}=I_3$. Then, Eqs. (b) yield,

$$I_{XX}+I_{YY}+I_{ZZ} = I_1+I_2+I_3 \tag{d}$$

Hence, by Eqs. (c) and (d),

$$I_{xx}+I_{yy}+I_{zz} = I_{XX}+I_{YY}+I_{ZZ} = I_1+I_2+I_3$$

as was to be shown.

b) For the products of inertia, by definition,

$$I_{xy} = \int xy\,dm, \quad I_{yz} = \int yz\,dm, \quad I_{zx} = \int zx\,dm \tag{e}$$

By Eqs. (a) and the first of Eqs. (e), we have,

$$I_{xy} = \int (l_1X+m_1Y+n_1Z)(l_2X+m_2Y+n_2Z)\,dm$$

or

$$I_{xy} = \int [l_1l_2X^2 + m_1m_2Y^2 + n_1n_2Z^2 + (m_1n_2+m_2n_1)YZ + (n_1l_2+n_2l_1)ZX + (l_1m_2+l_2m_1)XY] \tag{f}$$

Now by the second and third of Eqs. (21.2)

$$I_{YY}+I_{ZZ} = 2\int X^2\,dm + I_{XX}$$

or $\int X^2\,dm = \frac{1}{2}(-I_{XX}+I_{YY}+I_{ZZ})$

and similarly,

$$\int Y^2\,dm = \tfrac{1}{2}(I_{XX}-I_{YY}+I_{ZZ})$$
$$\int Z^2\,dm = \tfrac{1}{2}(I_{XX}+I_{YY}-I_{ZZ})$$

Hence, Eq. (f) becomes,

$$\begin{aligned} I_{xy} = {} & \tfrac{1}{2} l_1l_2(-I_{XX}+I_{YY}+I_{ZZ}) \\ & + \tfrac{1}{2} m_1m_2(I_{XX}-I_{YY}+I_{ZZ}) \\ & + \tfrac{1}{2} n_1n_2(I_{XX}+I_{YY}-I_{ZZ}) \\ & + (m_1n_2+m_2n_1)I_{YZ} + (n_1l_2+n_2l_1)I_{ZX} \\ & + (l_1m_2+l_2m_1)I_{XY} \end{aligned}$$

(Continued)

21.22 Cont.

or, since $l_1 l_2 + m_1 m_2 + n_1 n_2 = 0$,

$$I_{xy} = -l_1 l_2 I_{xx} - m_1 m_2 I_{yy} - n_1 n_2 I_{zz} + (m_1 n_2 + m_2 n_1) I_{yz} + (l_1 n_2 + l_2 n_1) I_{zx} + (l_1 m_2 + l_2 m_1) I_{xy}$$

Similarly,

$$I_{yz} = -l_2 l_3 I_{xx} - m_2 m_3 I_{yy} - n_2 n_3 I_{zz} + (m_2 n_3 + m_3 n_2) I_{yz} + (n_2 l_3 + n_3 l_2) I_{zx} + (l_2 m_3 + l_3 m_2) I_{xy}$$

$$I_{zx} = -l_3 l_1 I_{xx} - m_3 m_1 I_{yy} - n_3 n_1 I_{zz} + (m_3 n_1 + m_1 n_3) I_{yz} + (m_3 l_1 + m_1 l_3) I_{zx} + (l_3 m_1 + l_1 m_3) I_{xy}$$

21.23

Given: The principal moments of inertia of a body with respect to a fixed point O are,

$$I_1 = 5 \text{ lb}\cdot\text{ft}\cdot\text{s}^2$$
$$I_2 = 3 \text{ lb}\cdot\text{ft}\cdot\text{s}^2 \quad \text{(a)}$$
$$I_3 = 1.5 \text{ lb}\cdot\text{ft}\cdot\text{s}^2$$

The body rotates about a fixed axes through point O, with direction cosines

$$l = 0.3\sqrt{2}$$
$$m = 0.5\sqrt{2} \quad \text{(b)}$$
$$n = -0.4\sqrt{2}$$

relative to principal axes, with an angular speed of $\omega = 40$ rad/s.

Calculate the kinetic energy of the body.

Solution:

The angular velocity of the body is, with Eqs. (b) and $\omega = 40$ rad/s,

$$\bar{\omega} = 40[0.3\sqrt{2}\,\hat{\imath} + 0.5\sqrt{2}\,\hat{\jmath} - 0.4\sqrt{2}\,\hat{k}]$$

Hence,

$$\omega_1 = 40(0.3\sqrt{2}) = 12\sqrt{2} \text{ rad/s}$$
$$\omega_2 = 40(0.5\sqrt{2}) = 20\sqrt{2} \text{ rad/s} \quad \text{(c)}$$
$$\omega_3 = 40(-0.4\sqrt{2}) = -16\sqrt{2} \text{ rad/s}$$

By Eq. (21.15), the kinetic energy of the body relative to principal axes is

$$T = \tfrac{1}{2} I_1 \omega_1^2 + \tfrac{1}{2} I_2 \omega_2^2 + \tfrac{1}{2} I_3 \omega_3^2 \quad \text{(d)}$$

Therefore, by Eqs. (a), (c), and (d), we have,

$$T = \tfrac{1}{2}[(5)(288) + 3(800) + 1.5(512)]$$

$$T = 2304 \text{ ft}\cdot\text{lb}$$

21.24

Given: Before being placed in orbit, a communications satellite is tested by mounting it on a short rigid rod attached to a ball-and-socket joint at the fixed point O (Fig. a). The mass of the satellite is $m = 10$ lb·s²/ft. With respect to principal axes 1, 2, and 3 at O, its moments of inertia are

$$I_1 = 21 \text{ lb}\cdot\text{ft}\cdot\text{s}^2$$
$$I_2 = 35 \text{ lb}\cdot\text{ft}\cdot\text{s}^2 \quad \text{(a)}$$
$$I_3 = 42 \text{ lb}\cdot\text{ft}\cdot\text{s}^2$$

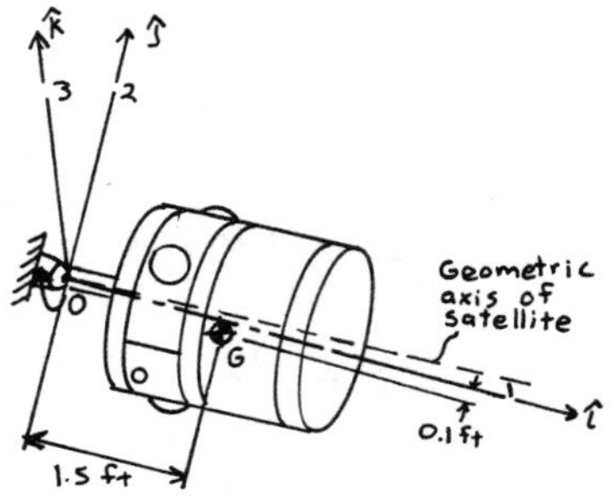

Figure a

Since the satellite is nonhomogeneous, its center of mass is eccentric from axis 1, which in turn does not coincide with the geometric axis. The satellite is spun with an angular velocity of

$$\bar{\omega} = 8\hat{\imath} - 12\hat{\jmath} + 16\hat{k} \quad [\text{rad/s}]$$

or

$$\omega_1 = 8 \text{ rad/s}, \quad \omega_2 = -12 \text{ rad/s}, \quad \omega_3 = 16 \text{ rad/s} \quad \text{(b)}$$

Determine the kinetic energy of the satellite.

Solution: Since O is a fixed point, by Eq. (21.15), the kinetic energy is,

$$T = \tfrac{1}{2} I_1 \omega_1^2 + \tfrac{1}{2} I_2 \omega_2^2 + \tfrac{1}{2} I_3 \omega_3^2 \quad \text{(c)}$$

Therefore, by Eqs. (a), (b), and (c),

$$T = \tfrac{1}{2}[21(64) + 35(144) + 42(256)]$$

$$T = 8568 \text{ ft}\cdot\text{lb}$$

21.25

Given: A uniform slender rod AB of length L and mass m is pinned at A to rod CD and rotates with angular speed Ω about the axis of CD (Fig. a). As the rods rotate, the angle θ that rod AB forms with CD varies with time t, that is, $\theta = \theta(t)$.

a) Determine the kinetic energy of rod AB.

b) Determine the kinetic energy of rod AB. Express the results in terms of $L, \Omega, \theta, \dot{\theta}$, and m.

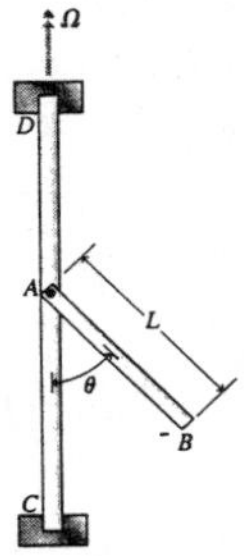

Figure a

Solution:

a) Consider the fixed principal axes 1, 2, 3, with origin at A (Fig. b).

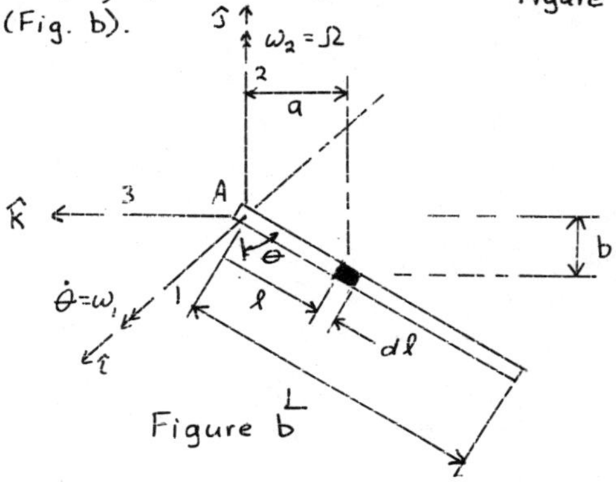

Figure b

By Fig. b, the angular velocity of rod AB is

$$\bar{\omega} = \bar{\omega}_1 + \bar{\omega}_2 = \dot{\theta}\hat{\imath} + \Omega\hat{\jmath} + 0\hat{k} \qquad (a)$$

Hence, the angular speed of AB is

$$\underline{\omega = \sqrt{(\dot{\theta})^2 + \Omega^2}}$$

b) Let ρ be the mass per unit length of the rod. Then, the moments of inertia of rod AB relative to axes 1, 2, 3 are (Fig. b)

$$I_1 = \tfrac{1}{3} mL^2$$

$$I_2 = \int a^2 dm = \int_0^L l^2 \sin^2\theta \, (\rho\, dl)$$

$$I_2 = \tfrac{1}{3}\rho L^3 \sin^2\theta$$

or

$$I_2 = \tfrac{1}{3} mL^2 \sin^2\theta \qquad (b)$$

$$I_3 = \int b^2 dm = \int_0^L l^2\cos^2\theta\,(\rho\, dl) = \tfrac{1}{3}\rho L^3\cos^2\theta$$

or

$$I_3 = \tfrac{1}{3} mL^2\cos^2\theta$$

By Eq (21.15), the kinetic energy of bar AB is

$$T = \tfrac{1}{2} I_1\omega_1^2 + \tfrac{1}{2} I_2\omega_2^2 + \tfrac{1}{2} I_3\omega_3^2 \qquad (c)$$

Also, by Eq. (a),

$$\omega_1 = \dot{\theta}, \quad \omega_2 = \Omega, \quad \omega_3 = 0 \qquad (d)$$

Hence, by Eqs. (b), (c), and (d)

$$T = \tfrac{1}{2}\left(\tfrac{1}{3} mL^2\right)(\dot{\theta})^2 + \tfrac{1}{2}\left(\tfrac{1}{3} mL^2 \sin^2\theta\right)\Omega^2$$

or

$$\underline{\underline{T = \tfrac{1}{6} mL^2 \left(\dot{\theta}^2 + \Omega^2\sin^2\theta\right)}}$$

21.26

Given: A disk A having radius 300 mm and mass 10 kg rotates at constant angular velocity $5\hat{k}$ [rad/s] in a horizontal plane (Fig. a). A second disk B having radius 150 mm and mass 5 kg mounted perpendicular to disk A at its circumference by means of a small axle. Disk B rotates about the axle with an angular velocity of $3\hat{\jmath}$ [rad/s].

Determine the kinetic energy of the system.

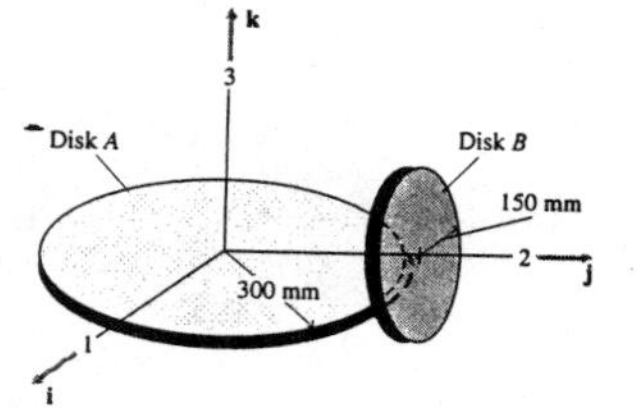

Figure a

Solution:

The angular velocity of disk A, as given, is

$$\bar{\omega}_A = 5\hat{k}, \quad \text{or} \quad \omega_{AB} = 5 \text{ rad/s} \qquad (a)$$

The angular velocity of disk B is

$$\bar{\omega}_B = 3\hat{\jmath} + 5\hat{k}$$

or

$$\omega_{B2} = 3 \text{ rad/s}, \quad \omega_{B3} = 5 \text{ rad/s} \qquad (b)$$

The moments of inertia of disks A and B, relative to axes 1, 2, 3, are

$$I_{A3} = \tfrac{1}{2} m_A r_A^2 = \tfrac{1}{2}(10)(0.30)^2 = 0.45 \text{ kg}\cdot\text{m}^2$$

$$I_{B2} = \tfrac{1}{2} m_B r_B^2 = \tfrac{1}{2}(5)(0.15)^2 = 0.05625 \text{ kg}\cdot\text{m}^2 \qquad (c)$$

$$I_{B3} = \tfrac{1}{4} m_B r_B^2 + m_B d_B^2 = \tfrac{1}{4}(5)(0.15)^2 + (5)(0.30)^2$$

or

$$I_{B3} = 0.4781 \text{ kg}\cdot\text{m}^2$$

The kinetic energy of the system is

$$T = \tfrac{1}{2} I_{A3}\omega_{A3}^2 + \tfrac{1}{2} I_{B2}\omega_{B2}^2 + \tfrac{1}{2} I_{B3}\omega_{B3}^2 \qquad (d)$$

Hence, Eqs. (a), (b), (c), and (d) yield

$$T = \tfrac{1}{2}(0.45)(5)^2 + \tfrac{1}{2}(0.05625)(3)^2 + \tfrac{1}{2}(0.4781)(5)^2$$

or

$$\underline{T = 11.85 \text{ N}\cdot\text{m}}$$

21.27

Given: A particle P of mass m slides along a hoop of radius R and mass 2m with constant speed v relative to the hoop (Fig. a). The hoop rotates with angular speed ω about an axle that coincides with the x axis. For the instant shown, the hoop lies in the xy plane, and the particle forms an angle of 60° with the axis of rotation.

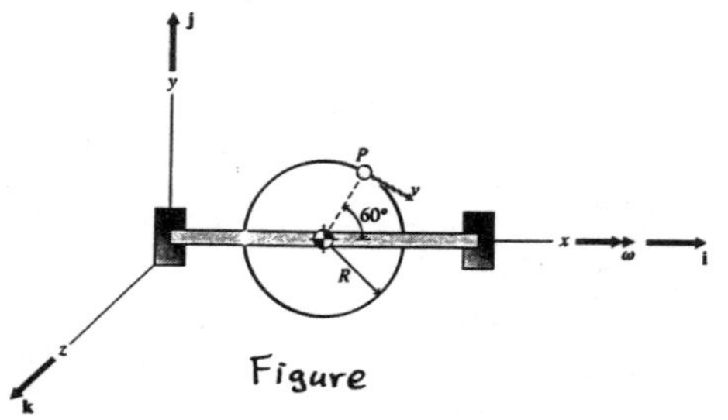

Figure

a) Determine the velocity of the particle with respect to the fixed axes xyz at this instant.

b) Determine the kinetic energy of the system at this instant.

Express results in terms of v, R, ω, and m.

Solution:

a) By Fig. a, the velocity of P is

$$\bar{v}_P = (v\cos 30°)\hat{\imath} - (v\sin 30°)\hat{\jmath} + (R\sin 60°)\omega\hat{k}$$

or

$$\bar{v}_P = \frac{\sqrt{3}}{2}v\hat{\imath} - \frac{1}{2}v\hat{\jmath} + \frac{\sqrt{3}}{2}R\omega\hat{k} \qquad (a)$$

b) The kinetic energy of the system is the sum of the kinetic energy of the ring and the kinetic energy of the particle. For the instant shown, the kinetic energy of the ring is

$$T_{ring} = \frac{1}{2}I_{xx}\omega^2 \qquad (b)$$

where, by Fig. a,

$$I_{xx} = \int y^2\, dm_{ring}$$

and

$$y = R\sin\theta$$

$$dm_{ring} = \left(\frac{2m}{2\pi R}\right)R\,d\theta = \frac{m}{\pi}\,d\theta$$

Hence,

$$I_{xx} = \frac{mR^2}{\pi}\int_0^{2\pi}\sin^2\theta\, d\theta = mR^2 \qquad (c)$$

By Eqs. (b) and (c),

$$T_{ring} = \frac{1}{2}mR^2\omega \qquad (d)$$

The kinetic energy of the particle P is

$$T_{particle} = \frac{1}{2}mv_P^2 \qquad (e)$$

By Eq. (a), $v_P^2 = v^2\left(\frac{3}{4}+\frac{1}{4}\right) + \frac{3}{4}R^2\omega^2 \qquad (f)$

Hence, by Eqs. (e) and (f)

$$T_{particle} = \frac{1}{2}mv^2 + \frac{3}{8}mR^2\omega^2 \qquad (g)$$

Then, Eqs. (d) and (g) yield

$$T = \frac{1}{2}mv^2 + \frac{7}{8}mR^2\omega^2$$

21.28

Given: A collar of mass $m_c = 8$ kg slides out along a uniform slender rod (Fig. a). The rod has mass $m_r = 12$ kg and length $L = 2$ m and rotates about fixed axes y and z with angular speeds $\omega_y = 4$ rad/s and $\omega_z = 5$ rad/s.

For the case where $\theta = 0$, $s = 1.5$ m, and $\dot{s} = 2$ m/s, determine the kinetic energy of the system.

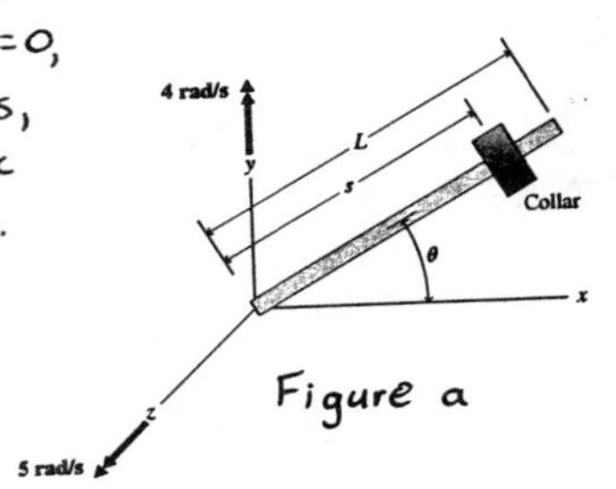

Figure a

Solution:

For $\theta = 0$, $s = 1.5$ m and $\dot{s} = 2$ m/s, the velocity of the collar is (see Fig. a)

$$\bar{v}_c = \dot{s}\hat{\imath} + s\omega_z\hat{\jmath} - s\omega_y\hat{k}$$

or with the given data

$$\bar{v}_c = 2\hat{\imath} + 7.5\hat{\jmath} - 6\hat{k} \quad [\text{m/s}] \qquad (a)$$

where $\hat{\imath}, \hat{\jmath}, \hat{k}$ are unit vectors along x, y, z axes, respectively. Therefore, the kinetic energy of the collar is, with Eq. (a),

$$T_c = \frac{1}{2}m_c v_c^2 = \frac{1}{2}(8)(2^2 + 7.5^2 + 6^2)$$

or

$$T_c = 385 \text{ N·m} \qquad (b)$$

The kinetic energy of the rod is

$$T_r = \frac{1}{2}I_y\omega_y^2 + \frac{1}{2}I_z\omega_z^2$$

where

$$I_y = \frac{1}{3}m_r L^2$$

$$I_z = \frac{1}{3}m_r L^2$$

$$\omega_y = 4 \text{ rad/s}$$

$$\omega_z = 5 \text{ rad/s}$$

Hence,

$$T_r = \frac{1}{6}m_r L^2(\omega_y^2 + \omega_z^2)$$

$$T_r = \frac{1}{6}(12)(2)^2(4^2 + 5^2)$$

$$T_r = 328 \text{ N·m} \qquad (c)$$

So, the kinetic energy of the system is, by Eqs. (b) and (c)

$$T = T_c + T_r = 385 + 328 = 713 \text{ N·m}$$

21.29

Given: The homogeneous top shown in Fig. a has the form of a cone of height h. The angle θ varies as the top spins. See Appendix F, Table F.1 for the principal moments of inertia of a cone.

Neglecting friction, write the equation based on conservation of mechanical energy ($T+V=$ constant) in terms of the projections of $\bar{\omega}$ on the principal moments of inertia.

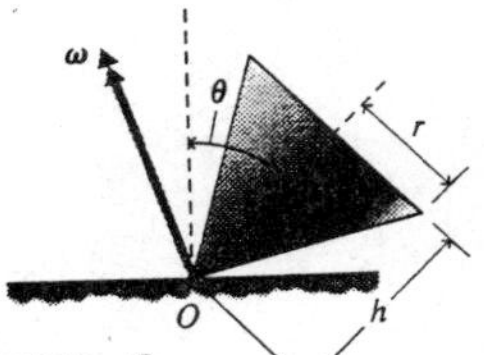

Figure a

Solution:

By Appendix F, Table F-1 and Fig. b,

$$I_{xx}=\frac{3}{10}mr^2 \qquad (a)$$
$$I_{yy}=I_{zz}=\frac{3}{20}m(4h^2+r^2)$$

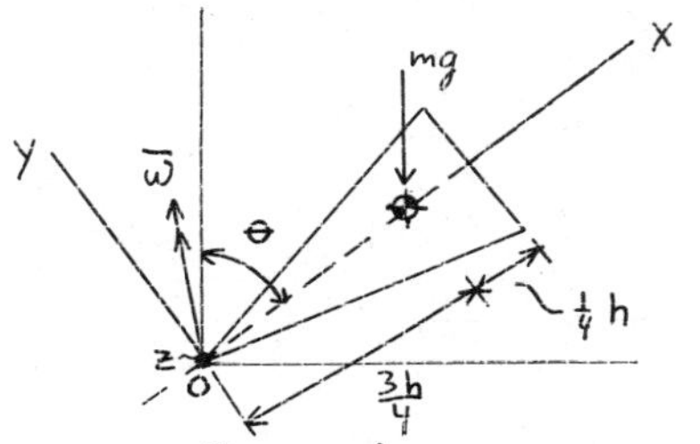

Figure b

The projections of $\bar{\omega}$ on principal axes xyz are ω_x, ω_y, and ω_z. Hence, the kinetic energy of the top is

$$T=\frac{1}{2}I_x\omega_x^2+\frac{1}{2}I_y\omega_y^2+\frac{1}{2}I_z\omega_z^2$$

or with Eqs. (a)

$$T=\frac{1}{2}\left(\frac{3}{10}mr^2\right)\omega_x^2+\frac{1}{2}\left[\frac{3}{20}m(4h^2+r^2)\right](\omega_y^2+\omega_z^2) \qquad (b)$$

The height of the center of mass of the cone relative to the potential energy datum O is $\frac{3}{4}h\cos\theta$. Hence, the potential energy of the cone is

$$V=mg\left(\frac{3}{4}h\cos\theta\right) \qquad (c)$$

Therefore, by the law of conservation of mechanical energy and Eqs. (b) and (c),

$$T+V=\frac{3m}{40}\left[2r^2\omega_x^2+(4h^2+r^2)(\omega_y^2+\omega_z^2)\right]+\frac{3}{4}mgh\cos\theta=\text{constant}$$

or simplifying,

$$10gh\cos\theta+2r^2\omega_x^2+(4h^2+r^2)(\omega_y^2+\omega_z^2)=\text{constant}$$

21.30

Given: The communications satellite of mass $m=10\ \text{lb}\cdot\text{s}^2/\text{ft}$ (see Problem 21.24) is placed aboard the space shuttle, which orbits at a speed of 17,500 mi/h ($\approx$ 25,700 ft/s). Assume that the rigid rod of Problem 21.24 has negligible length, so that the distance along axis 1 from point P, at which the rod was attached to the satellite, to the center of mass G of the satellite is 1.5 ft (Fig. a).

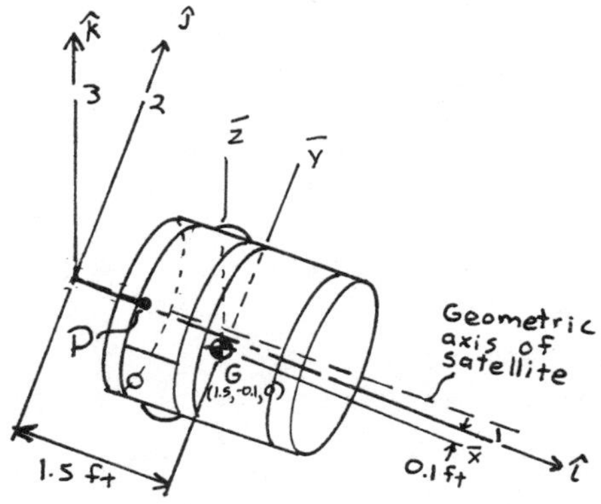

Figure a

The satellite is released into space with point P having a velocity

$$v_P=25{,}700(-0.9129\hat{i}+0.3651\hat{j}-0.1826\hat{k})$$
$$v_P=-23{,}460\hat{i}+9385\hat{j}-4692\hat{k}\ \ [\text{ft/s}] \qquad (a)$$

and the satellite having angular velocity

$$\bar{\omega}=8\hat{i}-12\hat{j}+16\hat{k}\ \ [\text{rad/s}] \qquad (b)$$

Determine the kinetic energy of the satellite at the instant it is released.

Note that the position vector $\bar{r}$ from P to G lies in the plane of axes 1 and 2 and is

$$\bar{r}=1.5\hat{i}-0.1\hat{j}\ \ [\text{ft}] \qquad (c)$$

Solution:

Either Eq. (21.13) or Eq. (21.17) may be used to solve this problem. First, we will use Eq. (21.13). Then since axes xyz (axes 1, 2, 3 in Fig. a) are principal axes, $I_{yz}=I_{zx}=I_{xy}=0$. Also, in Eq. (21.13), by Fig. a,

$$\int x\,dm=mX_G=(10)(1.5)=15\ \text{lb}\cdot\text{s}^2 \qquad (d)$$
$$\int y\,dm=mY_G=(10)(-0.1)=-1.0\ \text{lb}\cdot\text{s}^2 \qquad (e)$$
$$\int z\,dm=mZ_G=(10)(0)=0 \qquad (f)$$

(Continued)

21.30 Cont.

with Eqs. (d), (e), and (f), Eq. (21.13) reduces to (with O = P)

$$T = \frac{1}{2} m v_P^2 + 15(v_{Py}\omega_z - v_{Pz}\omega_y) - 1.0(v_{Pz}\omega_x - v_{Px}\omega_z) + \frac{1}{2}(I_{xx}\omega_x^2 + I_{yy}\omega_y^2 + I_{zz}\omega_z^2) \quad (f)$$

By the data given in Problem 21.24 and Eqs. (a) and (b)

$$I_{xx} = I_1 = 21 \text{ lb}\cdot\text{ft}\cdot\text{s}^2$$
$$I_{yy} = I_2 = 35 \text{ lb}\cdot\text{ft}\cdot\text{s}^2 \quad (g)$$
$$I_{zz} = I_3 = 42 \text{ lb}\cdot\text{ft}\cdot\text{s}^2$$

$$v_P^2 = (25,700)^2$$
$$v_{Px} = -23,460 \text{ ft/s}$$
$$v_{Py} = 9385 \text{ ft/s} \quad (h)$$
$$v_{Pz} = -4692 \text{ ft/s}$$

and

$$\omega_x = 8 \text{ rad/s}$$
$$\omega_y = -12 \text{ rad/s} \quad (i)$$
$$\omega_z = 16 \text{ rad/s}$$

Substitution of Eqs. (g), (h), and (i) into Eq. (f) yields

$$T = \frac{1}{2}(10)(25,700)^2 + (15)[(9385)(16) - (-4692)(-12)] - (1.0)[(-4692)(8) - (-23,460)(16)] + \frac{1}{2}[(21)(8)^2 + (35)(-12)^2 + (42)(16)^2]$$

or $T = \underbrace{3.30\times 10^9}_{\text{Translational energy}-\bar{v}_P \text{ effect}} + \underbrace{1.08\times 10^6}_{\text{Rotational energy}-\bar{\omega}\times\bar{r} \text{ effect}}$ (ft·lb)

Therefore,

$$\underline{T = 3.30\times 10^9 \text{ ft}\cdot\text{lb}}$$

Note: The kinetic energy due to the translational velocity $\bar{v}_P$ is much larger than that due to the rotational effect, $\bar{\omega}\times\bar{r}$.

As a check, we may use Eq. (21.17). Then, we must determine the velocity $\bar{v}_G$ of the center of mass of the satellite and the moments of inertia relative to axes parallel to axes 1, 2, 3 (Fig. a), with origin at G. By kinematics, the velocity $\bar{v}_G$ is given by

$$\bar{v}_G = \bar{v}_P + \bar{v}_{G/P} = \bar{v}_P + \bar{\omega}\times\bar{r} \quad (j)$$

where by Eqs. (b) and (c),

$$\bar{\omega}\times\bar{r} = \begin{vmatrix} \hat{i} & \hat{j} & \hat{k} \\ 8 & -12 & 16 \\ 1.5 & -1 & 0 \end{vmatrix} = 16\hat{i} + 24\hat{j} + 10\hat{k} \text{ [rad/s]} \quad (k)$$

By Eqs. (a), (j), and (k), we obtain

$$\bar{v}_G = -23,444\hat{i} + 9409\hat{j} - 4682\hat{k} \quad (\ell)$$

Now, by Eqs. (21.6), (21.7), and (21.8), where $\bar{x}_P = -1.5$ ft, $\bar{y}_P = 0.1$ ft, $\bar{z}_P = 0$ are the coordinates of the origin P of the xyz axes (axes 1, 2, 3 in Fig. a), relative to axes $\bar{x}\bar{y}\bar{z}$, the moments and products of inertia relative to axes $\bar{x}\bar{y}\bar{z}$ are, with Fig. a and data from Problem 21.24

$$I_{\bar{x}\bar{x}} = I_{xx} - md_{x\bar{x}}^2 = 21 - (10)(0.1)^2 = 20.9 \text{ lb}\cdot\text{ft}\cdot\text{s}^2$$
$$I_{\bar{y}\bar{y}} = I_{yy} - md_{y\bar{y}}^2 = 35 - (10)(-1.5)^2 = 12.5 \text{ lb}\cdot\text{ft}\cdot\text{s}^2$$
$$I_{\bar{z}\bar{z}} = I_{zz} - md_{z\bar{z}}^2 = 42 - (10)[(1.5)^2 + (0.1)^2] = 19.4 \text{ lb}\,\text{ft}\,\text{s}^2 \quad (m)$$
$$I_{\bar{x}\bar{y}} = I_{xy} - m\bar{x}_P\bar{y}_P = 0 - (10)(-1.5)(0.1) = 1.5 \text{ lb}\cdot\text{ft}\cdot\text{s}^2$$
$$I_{\bar{x}\bar{z}} = I_{\bar{y}\bar{z}} = 0$$

Hence, by Eq. (21.17), with $I_{\bar{x}\bar{z}} = I_{\bar{y}\bar{z}} = 0$, the kinetic energy is

$$T = \frac{1}{2} m v_G^2 + \frac{1}{2}(I_{\bar{x}\bar{x}}\omega_x^2 + I_{\bar{y}\bar{y}}\omega_y^2 + I_{\bar{z}\bar{z}}\omega_z^2) - I_{\bar{x}\bar{y}}\omega_x\omega_y \quad (n)$$

Hence, by Eqs. (i), (ℓ), (m), and (n), we have with m = 10 lb·s²/ft,

$$T = \frac{1}{2}(10)(6.6007\times 10^8) + \frac{1}{2}[(20.9)(8)^2 + (12.5)(-12)^2 + (19.4)(16)^2] - (1.5)(8)(-12)$$

or $$T = 3.30\times 10^9 \text{ ft}\cdot\text{lb}$$

21.31

Given: A uniform slender rod of length L and mass m rotates in the xy plane about the z axis (at point O; Fig. a). A disk of mass M and radius R rolls along the rod away from point O, with v, the speed of its center relative to the rod. For the instant shown, the disk is at a distance b from point O.

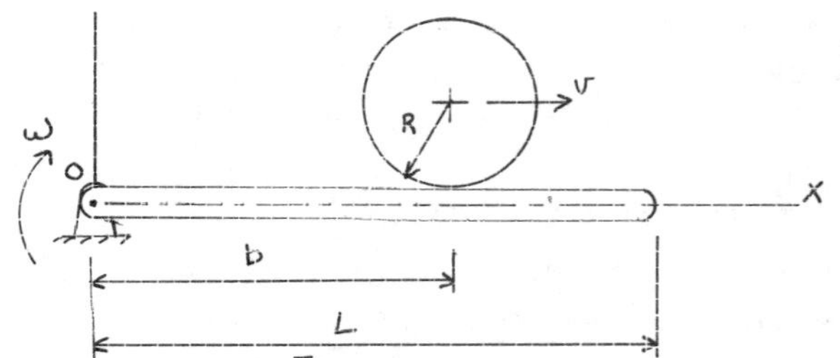

Figure a

Determine the kinetic energy of the system at this instant

Solution: Refer to Section 21.4. Let XYZ be Newtonian axes with origin at O (Fig. b). Then the kinetic energy of the rod is

$$T_{rod} = \frac{1}{2}(I_{rod})_O\,\omega^2 = \frac{1}{2}\left(\frac{1}{3}mL^2\right)\omega^2 = \frac{1}{6}mL^2\omega^2 \quad (a)$$

(Continued)

21.31 Cont.

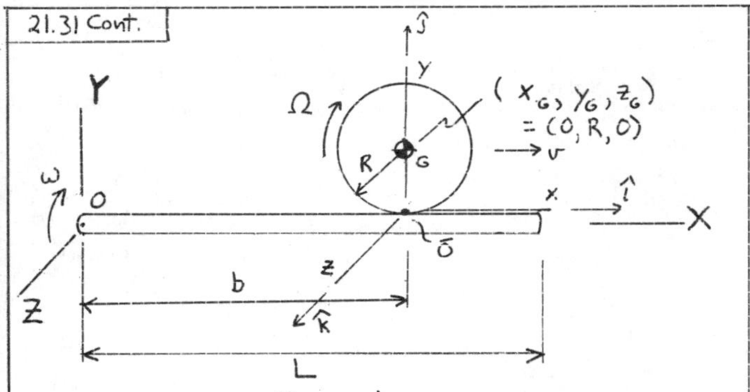

Figure b

Let xyz axes be fixed in the disk, with origin at $\bar{o}$, the point in the disk that is in contact with the rod. The velocity of point $\bar{o}$ relative to axes XYZ is

$$\bar{v}_{\bar{o}} = -b\omega \hat{J} = v_{\bar{o}y}\,\hat{J} \qquad (b)$$

Hence,

$$v_{\bar{o}x} = 0 \;,\; v_{\bar{o}y} = -b\omega \;,\; v_{\bar{o}z} = 0 \qquad (c)$$

The angular velocity of the disk relative to the rod is (Fig. b), since the disk rolls on the rod,

$$\Omega = \frac{v}{R} \qquad (d)$$

where v is the speed of the center of the disk relative to the rod. Hence, the absolute angular velocity of the disk is,

$$\bar{\omega}_{disk} = -(\omega + \Omega)\hat{k} = -(\omega + \tfrac{v}{R})\hat{k}$$

or, $(\omega_{disk})_x = 0,\ (\omega_{disk})_y = 0,\ (\omega_{disk})_z = -(\omega + \frac{v}{R})$ (e)

Therefore, by Eqs. (21.13), (b), (c), (d), and (e), the kinetic energy of the disk is,

$$T_{disk} = \tfrac{1}{2}Mb^2\omega^2 + (b\omega)(\omega + \tfrac{v}{R})\int x\,dM + \tfrac{1}{2}I_{zz}(\omega + \tfrac{v}{R})^2 \qquad (f)$$

Now, by Fig. b,

$$\int x\,dM = Mx_G = 0 \qquad (g)$$

and

$$I_{zz} = \tfrac{1}{2}MR^2 + MR^2 = \tfrac{3}{2}MR^2 \qquad (h)$$

By Eqs. (f), (g), and (h), we find

$$T_{disk} = \tfrac{1}{2}Mb^2\omega^2 + \tfrac{1}{2}(\tfrac{3}{2}MR^2)(\omega^2 + \tfrac{2\omega v}{R} + \tfrac{v^2}{R^2})$$

or,

$$T_{disk} = \tfrac{3}{4}Mv^2 + \tfrac{1}{4}M(3R^2\omega^2 + 2b^2\omega^2 + 6vR\omega) \qquad (i)$$

The kinetic energy of the system is, by Eqs. (a) and (i),

$$T = T_{rod} + T_{disk} = \tfrac{1}{6}mL^2\omega^2 + \tfrac{3}{4}Mv^2 + \tfrac{1}{4}M(3R^2\omega^2 + 2b^2\omega^2 + 6vR\omega)$$

Alternative Solution: Consider the system when the rod forms angle θ with respect to the X axis (Fig. c)

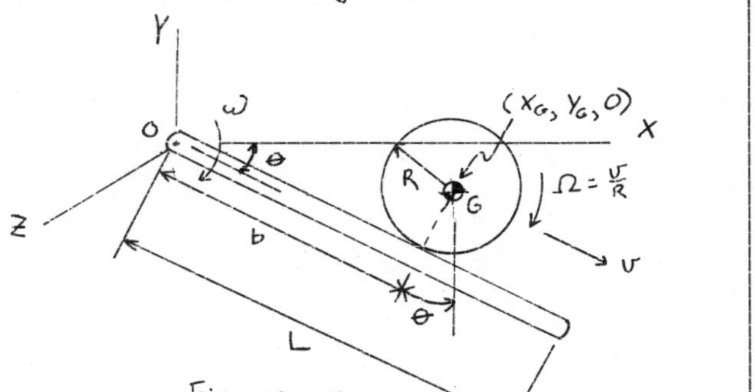

Figure c

as before,

$$T_{rod} = \tfrac{1}{2}I_o\omega^2 = \tfrac{1}{6}mL^2\omega^2 \qquad (j)$$

and

$$\omega_{disk} = \omega + \frac{v}{R} \qquad (k)$$

By the theory of a rigid body motion, the kinetic energy of the disk is,

$$T_{disk} = \tfrac{1}{2}M(\dot{X}_G^2 + \dot{Y}_G^2) + \tfrac{1}{2}I_G\omega_{disk}^2 \qquad (\ell)$$

where by Fig. c,

$$X_G = b\cos\theta + R\sin\theta$$
$$Y_G = -b\sin\theta + R\cos\theta$$

Differentiation with respect to time yields,

$$\dot{X}_G = \dot{b}\cos\theta - b\dot{\theta}\sin\theta + R\dot{\theta}\cos\theta$$
$$\dot{Y}_G = -\dot{b}\sin\theta - b\dot{\theta}\cos\theta - R\dot{\theta}\sin\theta$$

or since $\dot{b} = v$ and $\dot{\theta} = \omega$,

$$\dot{X}_G^2 + \dot{Y}_G^2 = v^2 + R^2\omega^2 + b^2\omega^2 + 2vR\omega \qquad (m)$$

also since $I_G = \frac{1}{2}MR^2$, with Eq. (k),

$$\tfrac{1}{2}I_G\omega_{disk}^2 = \tfrac{1}{2}(\tfrac{1}{2}MR^2)(\omega^2 + \tfrac{2v\omega}{R} + \tfrac{v^2}{R^2}) \qquad (n)$$

Hence, Eqs. (j), (ℓ), (m), and (n) yield

$$T = T_{rod} + T_{disk} = \tfrac{1}{6}mL^2\omega^2 + \tfrac{3}{4}Mv^2 + \tfrac{1}{4}M(3R^2\omega^2 + 2b^2\omega^2 + 6vR\omega)$$

as before

21.32

Given: A homogeneous solid sphere of mass m and radius r is rigidly mounted on a shaft whose axis intersects the center of the sphere (Fig. a). The shaft rotates with angular velocity $\bar{\omega}$ about its axis.

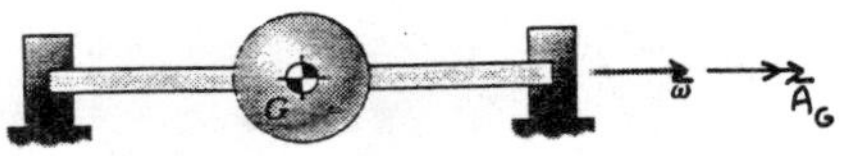

Figure a

(Continued)

21.32 Cont.

a) Determine the angular momentum $\bar{A}_G$ of the sphere about its center of mass.
b) Show $\bar{A}_G$ and $\bar{\omega}$ in a sketch.

Solution:

a) The mass moment of inertia of the sphere is (Appendix F, Table F.1)

$$I_G = \frac{2}{5} m r^2 \qquad (a)$$

Hence, the angular momentum $\bar{A}_G$ is,

$$\bar{A}_G = I_G \bar{\omega}$$

b) By Eq. (a), the direction of $\bar{A}_G$ is the same as $\bar{\omega}$ (see Fig. a).

21.33

Given: By Problem 21.33, a body rotates with angular speed $\omega = 40$ rad/s about a fixed axis through a point O, with direction cosines $(0.3\sqrt{2}, 0.5\sqrt{2}, -0.4\sqrt{2})$ relative to principal axes. The principal moments of inertia of the body relative to O are,

$$I_1 = 5 \text{ lb·ft·s}^2$$
$$I_2 = 3 \text{ lb·ft·s}^2 \qquad (a)$$
$$I_3 = 1.5 \text{ lb·ft·s}^2$$

Determine the magnitude A of the angular momentum of the body with respect to O.

Solution:

By Eqs. (21.24),

$$A_{O1} = I_1 \omega_1$$
$$A_{O2} = I_2 \omega_2 \qquad (b)$$
$$A_{O3} = I_3 \omega_3$$

where

$$\omega_1 = 40(0.3\sqrt{2}) = 12\sqrt{2} \text{ rad/s}$$
$$\omega_2 = 40(0.5\sqrt{2}) = 20\sqrt{2} \text{ rad/s} \qquad (c)$$
$$\omega_3 = 40(-0.4\sqrt{2}) = -16\sqrt{2} \text{ rad/s}$$

By Eqs. (a), (b), and (c), we find

$$A_{O1} = (5)(12\sqrt{2}) = 60\sqrt{2} \text{ lb·ft·s}$$
$$A_{O2} = (3)(20\sqrt{2}) = 60\sqrt{2} \text{ lb·ft·s} \qquad (d)$$
$$A_{O3} = (1.5)(-16\sqrt{2}) = -24\sqrt{2} \text{ lb·ft·s}$$

Hence, by Eq. (d)

$$A^2 = (60\sqrt{2})^2 + (60\sqrt{2})^2 + (-24\sqrt{2})^2 = 15{,}552$$

or

$$\underline{A = 124.71 \text{ lb·ft·s}}$$

21.34

Given: A homogeneous solid sphere of mass m and radius r is rigidly attached to a rod that passes through its center of mass G (Fig. a). The rod rotates about the x axis with angular velocity $\bar{\omega}$ and about the vertical y axis with angular velocity $\bar{\Omega}$.

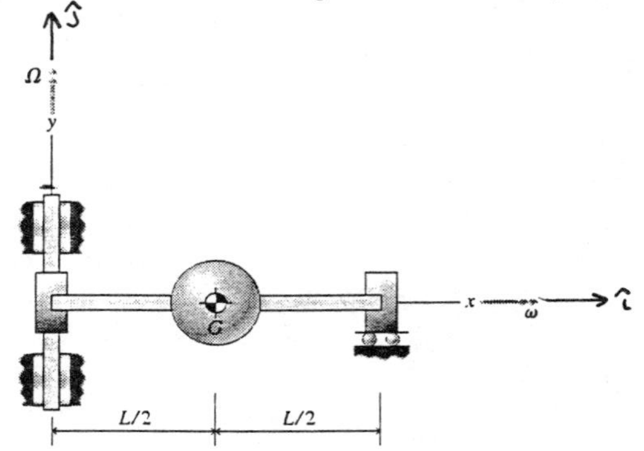

Figure a

a) Determine the angular momentum of the sphere.
b) Show the angular momentum and the angular velocity vectors for the sphere in a sketch.

Solution:

a) By Fig. a, the angular velocity of the sphere is

$$\bar{\omega}_s = \bar{\omega} + \bar{\Omega} = \omega\hat{i} + \Omega\hat{j} \qquad (a)$$

By Appendix F, Table F.1, the moments of inertia are

$$I_{xx} = \frac{2}{5} m r^2$$
$$I_{yy} = \frac{2}{5} m r^2 + m\left(\frac{L}{2}\right)^2 = \frac{1}{20} m (8r^2 + 5L^2) \qquad (b)$$

Then, with Eqs. (a) and (b), the angular momentum of the sphere is

$$\bar{A}_s = I_{xx}\omega\hat{i} + I_{yy}\Omega\hat{j}$$

or

$$\underline{\bar{A}_s = \frac{1}{20} m [8r^2\omega\hat{i} + (8r^2 + 5L^2)\Omega\hat{j}]} \qquad (c)$$

b) The vectors $\bar{\omega}_s$ and $\bar{A}_s$ [see Eqs. (a) and (c)] are shown in Fig. b,

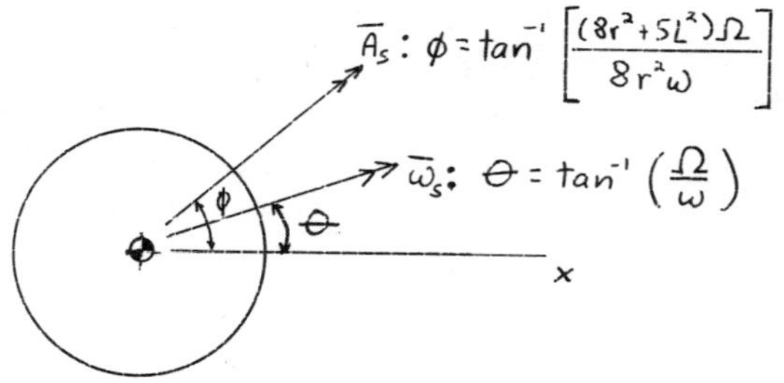

Figure b

21.35

Given: The sphere in Problem 12.34 is replaced by a homogeneous square flat plate of mass m and dimension a with its face parallel to the xy plane and its edges parallel to the x and y axes (Fig. a). The rod passes through the center of mass of the plate.

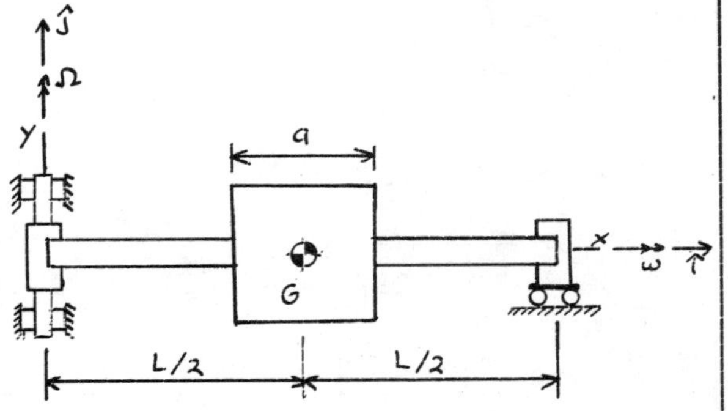

Figure a

a) Determine the angular momentum of the plate.

b) Show the angular momentum and the angular velocity vectors for the plate in a sketch.

Solution: By Fig. a, the angular velocity of the plate is

$$\bar{\omega}_p = \bar{\omega} + \Omega = \omega \hat{\imath} + \Omega \hat{\jmath} \quad (a)$$

By Appendix F, Table F.1, the moments of inertia of the plate are,

$$I_{xx} = \frac{1}{12} ma^2$$

$$I_{yy} = \frac{1}{12} ma^2 + m\left(\frac{L}{2}\right)^2 = \frac{1}{12} m(a^2 + 3L^2) \quad (b)$$

Then, with Eqs. (a) and (b), the angular momentum of the plate is

$$\bar{A}_P = I_{xx}\omega \hat{\imath} + I_{yy}\Omega \hat{\jmath}$$

or

$$\bar{A}_P = \frac{1}{12} m[a^2\omega \hat{\imath} + (a^2 + 3L^2)\Omega \hat{\jmath}] \quad (c)$$

b) The vectors $\bar{\omega}_p$ and $\bar{A}_P$ [see Eqs. (a) and (c)] are shown in Fig. b.

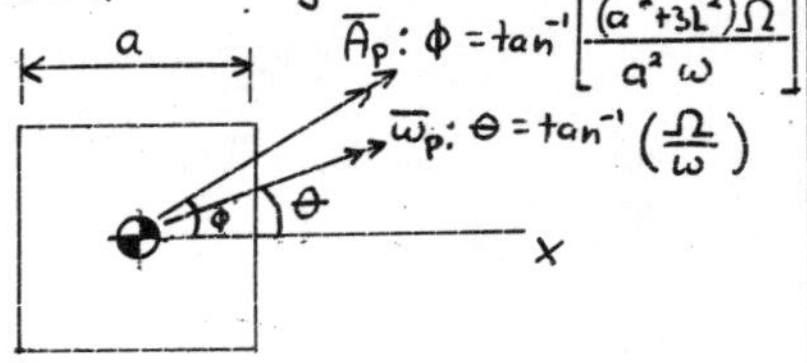

Figure b

21.36

Given: Two point masses m and 3m are rigidly attached by light rods to a shaft that rotates with angular velocity $\bar{\omega}$ about its longitudinal axis (Fig. a). The masses and the axis of the shaft lie in the xy plane, a plane of symmetry for the masses.

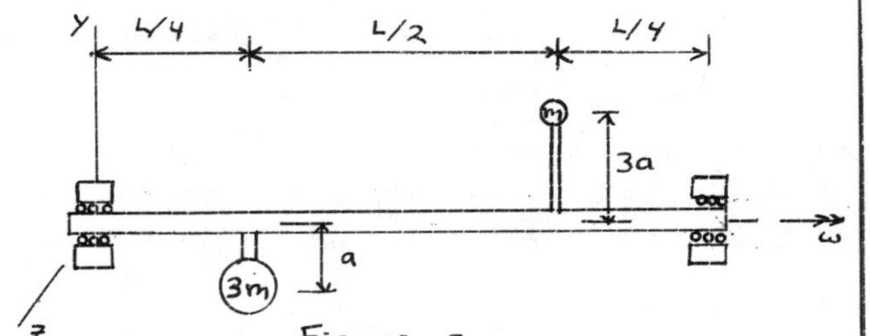

Figure a

a) Locate the center of mass G of the system of two masses with respect to the xyz axes.

b) Select axes $\bar{x}\bar{y}\bar{z}$ parallel to the xyz axes, and with origin at G. Determine the mass moments and products of inertia of the system of masses relative to the $\bar{x}\bar{y}\bar{z}$ axes.

c) Determine the angular momentum $\bar{A}_G$ of the system of masses relative to G.

d) Show $\bar{A}_G$ and $\bar{\omega}$ in a sketch.

Solution:

a) The coordinates x_G, y_G of the center of mass G are defined by the equations (see Fig. a).

$$4m x_G = (3m)\left(\frac{L}{4}\right) + (m)\left(\frac{3L}{4}\right)$$

$$4m y_G = (3m)(-a) + (m)(3a)$$

or

$$x_G = \frac{3}{8}L \quad , \quad y_G = 0 \quad (a)$$

b) Figure b shows the location of G and $\bar{x}\bar{y}\bar{z}$ axes.

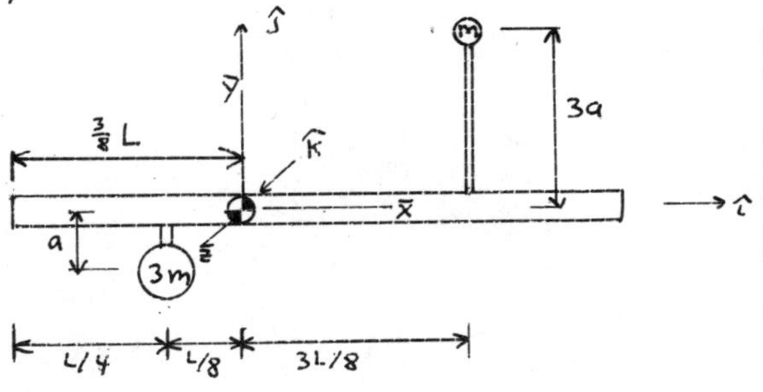

Figure b

(Continued)

21.36 Cont.

By Fig. b, since the xy plane is a plane of symmetry, the mass moments and products of inertia are, relative to $\bar{x}\bar{y}\bar{z}$ axes,

$$I_{\bar{x}\bar{x}} = (3m)(-a)^2 + (m)(3a)^2 = 12\,ma^2$$
$$I_{\bar{y}\bar{y}} = (3m)(-\tfrac{L}{8})^2 + (m)(\tfrac{3L}{8})^2 = \tfrac{3}{16}\,mL^2$$
$$I_{\bar{z}\bar{z}} = (3m)[(L/8)^2 + a^2] + (m)[(3L/8)^2 + (3a)^2] \quad (b)$$
$$= (3/16)\,m\,(L^2 + 64a^2)$$
$$I_{\bar{x}\bar{y}} = (3m)(-L/8)(-a) + (m)(3L/8)(3a) = \tfrac{3}{2}mLa$$
$$I_{\bar{x}\bar{z}} = I_{\bar{y}\bar{z}} = 0$$

c) The $\bar{x}, \bar{y}, \bar{z}$ components of the angular momentum $\bar{A}_G$ relative to G are, by Eq. (b) and Eq. (21.22), with $\omega_x = \omega$, $\omega_y = \omega_z = 0$

$$A_{G\bar{x}} = I_{\bar{x}\bar{x}}\,\omega_x = 12\,ma^2\omega$$
$$A_{G\bar{y}} = -I_{\bar{x}\bar{y}}\,\omega_x = (-3/2)\,mLa\omega$$
$$A_{G\bar{z}} = 0$$

Therefore,

$$\bar{A}_G = 12\,ma^2\omega\,\hat{\imath} - (3/2)\,mLa\omega\,\hat{\jmath}$$

or

$$\bar{A}_G = \tfrac{3}{2}m\omega a\,(8a\,\hat{\imath} - L\,\hat{\jmath}) \quad (c)$$

d) The vectors $\bar{A}_G$ and $\bar{\omega}$ are shown in Fig. c [see Eq. (c) and Fig. a].

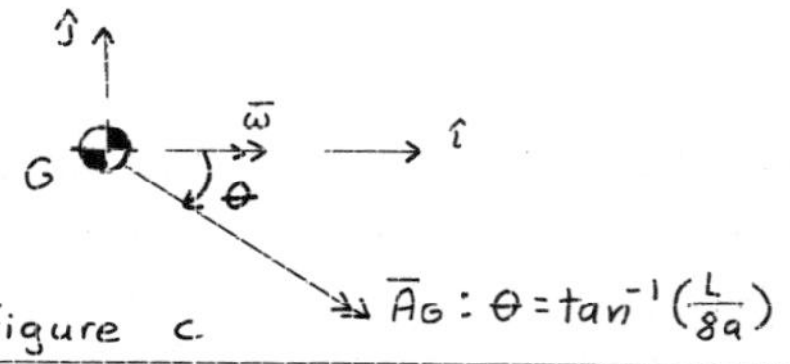

Figure c

21.37

Given: A rigid body rotates about a fixed axis with angular velocity $\bar{\omega}$. The angular moment $\bar{A}$ of the body with respect to point O on the axis of rotation forms an angle θ with $\bar{\omega}$. The principal axes of inertia I_1, I_2, I_3 through O form angles α, β, and γ with $\bar{\omega}$ (Fig. a).

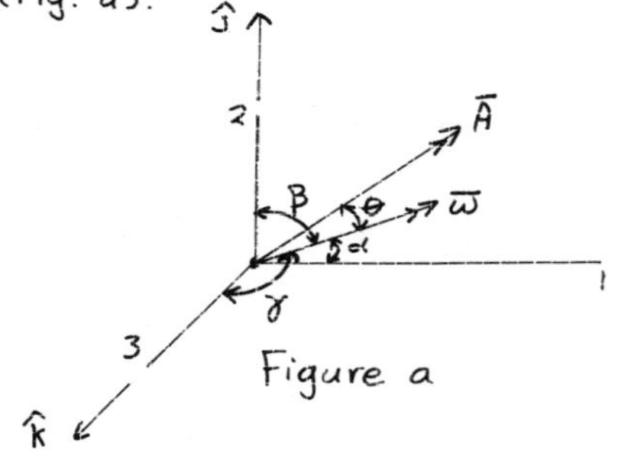

Figure a

Show that

$$\cos\theta = \frac{I_1\cos^2\alpha + I_2\cos^2\beta + I_3\cos^2\gamma}{\sqrt{I_1^2\cos^2\alpha + I_2^2\cos^2\beta + I_3^2\cos^2\gamma}}$$

Solution:

By Fig. a,

$$\bar{\omega} = \omega[(\cos\alpha)\hat{\imath} + (\cos\beta)\hat{\jmath} + (\cos\gamma)\hat{k}] \quad (a)$$

and

$$\bar{A} = A_1\hat{\imath} + A_2\hat{\jmath} + A_3\hat{k} \quad (b)$$

where

$$A_1 = I_1\omega_1\,,\quad A_2 = I_2\omega_2\,,\quad A_3 = I_3\omega_3 \quad (c)$$

Therefore, by Eqs. (a), (b), and (c),

$$\bar{A} = \omega[(I_1\cos\alpha)\hat{\imath} + (I_2\cos\beta)\hat{\jmath} + (I_3\cos\gamma)\hat{k}] \quad (d)$$

By the scalar product of vectors,

$$\cos\theta = \frac{\bar{A}\cdot\bar{\omega}}{A\omega} \quad (e)$$

Hence, by Eqs. (a), (d), and (e)

$$\cos\theta = \frac{\omega^2[I_1\cos^2\alpha + I_2\cos^2\beta + I_3\cos^2\gamma]}{\left(\sqrt{\omega^2(I_1^2\cos^2\alpha + I_2^2\cos^2\beta + I_3^2\cos^2\gamma)}\right)(\omega)}$$

or,

$$\cos\theta = \frac{I_1\cos^2\alpha + I_2\cos^2\beta + I_3\cos^2\gamma}{\sqrt{I_1^2\cos^2\alpha + I_2^2\cos^2\beta + I_3^2\cos^2\gamma}}$$

21.38

Given: A rigid body of mass m has angular velocity $\bar{\omega}$ and angular momentum $\bar{A}_G$ relative to body axes xyz, with origin at the bodie's center of mass G.

Show that the kinetic energy of the body can be written as

$$T = \tfrac{1}{2}mv_G^2 + \tfrac{1}{2}\bar{\omega}\cdot\bar{A}_G$$

where v_G is the magnitude of the velocity $\bar{v}_G$ of G.

Solution: Let $\hat{\imath}, \hat{\jmath}, \hat{k}$ be unit vectors along x, y, z axes, respectively. Then,

$$\bar{\omega} = \omega_x\hat{\imath} + \omega_y\hat{\jmath} + \omega_z\hat{k} \quad (a)$$

where $\omega_x, \omega_y, \omega_z$ are x, y, z projections of $\bar{\omega}$.

By Eq. (21.22), the x, y, z projections of $\bar{A}_G$ are

$$A_{Gx} = I_{xx}\omega_x - I_{xy}\omega_y - I_{xz}\omega_z$$
$$A_{Gy} = -I_{yx}\omega_x + I_{yy}\omega_y - I_{yz}\omega_z \quad (b)$$
$$A_{Gz} = -I_{zx}\omega_x - I_{zy}\omega_y + I_{zz}\omega_z$$

Then, by Eqs. (a) and (b),

$$\bar{\omega}\cdot\bar{A}_G = (\omega_x\hat{\imath} + \omega_y\hat{\jmath} + \omega_z\hat{k})\cdot(A_{Gx}\hat{\imath} + A_{Gy}\hat{\jmath} + A_{Gz}\hat{k})$$

or

$$\bar{\omega}\cdot\bar{A}_G = I_{xx}\omega_x^2 - I_{xy}\omega_y\omega_x - I_{xz}\omega_z\omega_x$$
$$- I_{yx}\omega_x\omega_y + I_{yy}\omega_y^2 - I_{yz}\omega_z\omega_y$$
$$- I_{zx}\omega_x\omega_z - I_{zy}\omega_y\omega_z + I_{zz}\omega_z^2$$

(Continued)

21.38 Cont.

Therefore, simplifying,

$$\frac{1}{2}\bar{\omega}\cdot\bar{A}_G = \frac{1}{2}I_{xx}\omega_x^2 + \frac{1}{2}I_{yy}\omega_y^2 + \frac{1}{2}I_{zz}\omega_z^2 - I_{xy}\omega_y\omega_x - I_{xz}\omega_z\omega_x - I_{yz}\omega_z\omega_y \qquad (c)$$

Therefore, with Eq. (c)

$$T = \frac{1}{2}mv_G^2 + \frac{1}{2}\bar{\omega}\cdot\bar{A}_G$$

is identical to Eq. (21.17).

21.39

Given: A uniform slender rod AB of length L and mass m is pinned at A to rod CD and rotates with angular speed Ω about the longitudinal axis of CD (Fig. a). As the rods rotate, the angle θ that AB forms with CD varies with time t: $\theta = \theta(t)$.

a) Determine the angular momentum of rod AB about point A in terms of $L, \Omega, \theta, \dot{\theta}$, and m.

b) Determine the time derivative of the angular momentum in terms of $L, \Omega, \theta, \dot{\theta}, m$, and $\ddot{\theta}$.

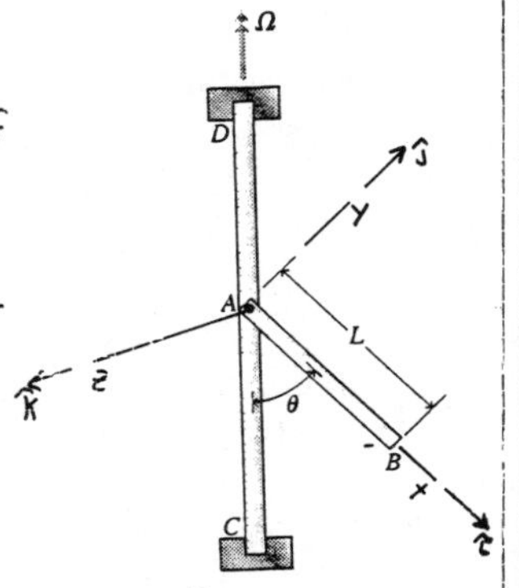

Figure a

Solution:

a) Select principal axes x y z with origin at A, and let them rotate with rod AB (Fig. a). Relative to axes x, y, z

$$I_{xx} \approx 0 \;,\quad I_{yy} = I_{zz} = \frac{1}{3}mL^2 \qquad (a)$$

By Eqs. (21.24),

$$A_{Ax} = I_{xx}\omega_x \;,\quad A_{Ay} = I_{yy}\omega_y \;,\quad A_{Az} = I_{zz}\omega_z \qquad (b)$$

where by Fig. a,

$$\omega_x = -\Omega\cos\theta, \quad \omega_y = \Omega\sin\theta, \quad \omega_z = \dot{\theta} \qquad (c)$$

or $\bar{\omega} = -(\Omega\cos\theta)\hat{\imath} + (\Omega\sin\theta)\hat{\jmath} + \dot{\theta}\hat{k}$ (d)

By Eqs. (a), (b), and (c),

$$A_{Ax} = 0, \; A_{Ay} = \frac{1}{3}mL^2\Omega\sin\theta, \; A_{Az} = \frac{1}{3}mL^2\dot{\theta}$$

or $\bar{A}_A = \frac{1}{3}mL^2[(\Omega\sin\theta)\hat{\jmath} + \dot{\theta}\hat{k}]$ (e)

b) Differentiation of Eq. (e) with respect to time t yields

$$\frac{d\bar{A}_A}{dt} = \frac{1}{3}mL^2[(\Omega\cos\theta)\dot{\theta}\hat{\jmath} + (\Omega\sin\theta)\dot{\hat{\jmath}} + \ddot{\theta}\hat{k} + \dot{\theta}\dot{\hat{k}}]$$

But

$$\dot{\hat{\jmath}} = \bar{\omega}\times\hat{\jmath} = \begin{vmatrix} \hat{\imath} & \hat{\jmath} & \hat{k} \\ -\Omega\cos\theta & \Omega\sin\theta & \dot{\theta} \\ 0 & 1 & 0 \end{vmatrix} = -\dot{\theta}\hat{\imath} - (\Omega\cos\theta)\hat{k}$$

and

$$\dot{\hat{k}} = \bar{\omega}\times\hat{k} = \begin{vmatrix} \hat{\imath} & \hat{\jmath} & \hat{k} \\ -\Omega\cos\theta & \Omega\sin\theta & \dot{\theta} \\ 0 & 0 & 1 \end{vmatrix}$$

or $\dot{\hat{k}} = (\Omega\sin\theta)\hat{\imath} + (\Omega\cos\theta)\hat{\jmath}$

Hence,

$$\frac{d\bar{A}_A}{dt} = \frac{1}{3}mL^2[(2\Omega\dot{\theta}\cos\theta)\hat{\jmath} + (\ddot{\theta} - \Omega^2\sin\theta\cos\theta)\hat{k}]$$

21.40

Given: A body is mounted on a rigid shaft through point O (Fig. a). Principal axes of the body at O form angles (α, β, γ) with the axes of the shaft. The constant speed of the shaft is ω.

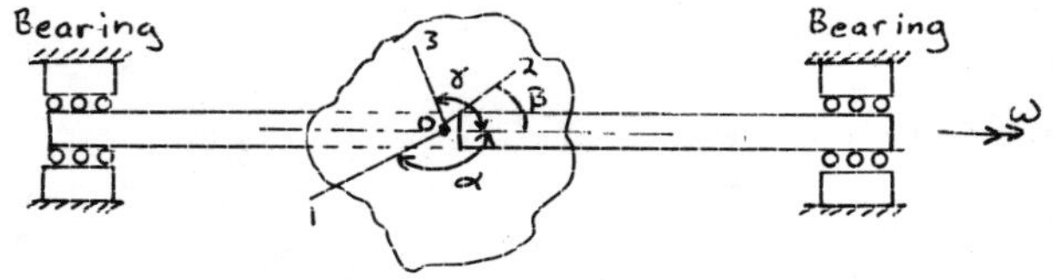

Figure a

Using the Euler equations of motion, determine the projections (M_1, M_2, M_3) of the moment $\bar{M}$ that the shaft exerts on the body.

Solution:

The projections $\bar{\omega}$ on the principal axes are (Fig. a),

$$\omega_1 = \omega\cos\alpha \;,\quad \omega_2 = \omega\cos\beta, \; \omega_3 = \omega\cos\gamma \qquad (a)$$

Since $\bar{\omega}$ is constant, $\dot{\omega}_1 = \dot{\omega}_2 = \dot{\omega}_3 = 0$. Hence, by Eqs. (21.34) and (a), with $I_{xx} = I_1$, $I_{yy} = I_2$, and $I_{zz} = I_3$,

$$M_1 = (I_3 - I_2)\omega^2\cos\beta\cos\gamma$$

$$M_2 = (I_1 - I_3)\omega^2\cos\alpha\cos\gamma$$

$$M_3 = (I_2 - I_1)\omega^2\cos\alpha\cos\beta$$

21.41

Given: Refer to the problem of Example 21.18.

Using the Euler equations, work Example 21.18; that is, determine the moment exerted by the shaft on the block.

Solution: Since the angular speed is constant and axes 1, 2, 3 are principal axes, Eqs. (21.34) yield,

$$M_1 = (I_3 - I_1)\omega_2\omega_3$$
$$M_2 = (I_1 - I_3)\omega_1\omega_3 \qquad (a)$$
$$M_3 = (I_2 - I_1)\omega_1\omega_2$$

(Continued)

21.41 Cont.

Also, by Example 21.18,

$$\omega_1 = 86.6 \text{ rad/s}, \quad \omega_2 = -50 \text{ rad/s}, \quad \omega_3 = 0 \qquad \text{(b)}$$

$$I_1 = 2.342 \text{ lb}\cdot\text{in}\cdot\text{s}^2, \quad I_2 = 0.2833 \text{ lb}\cdot\text{in}\cdot\text{s}^2, \quad I_3 = 2.208 \text{ lb}\cdot\text{in}\cdot\text{s}^2 \qquad \text{(c)}$$

Equations (a), (b), and (c) yield

$$M_1 = M_2 = 0, \quad M_3 = 8914 \text{ lb}\cdot\text{in}$$

Hence, the moment exerted by the shaft on the block is

$$M = \sqrt{M_1^2 + M_2^2 + M_3^2} = 8914 \text{ lb}\cdot\text{in}$$

or

$$\bar{M} = 8914\,\hat{\tau} \text{ lb}\cdot\text{in}$$

where τ is the unit vector along principal axis 3.

21.42

Given: A uniform rigid bar of mass m and length L is attached at one end to a shaft that rotates at constant angular speed ω (Fig. a)

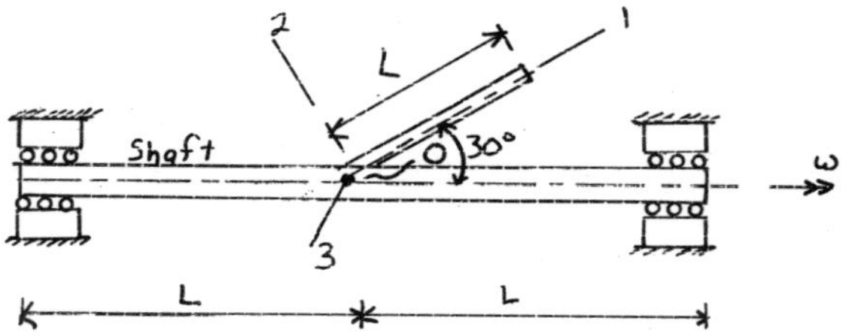

Figure a

Determine the forces exerted on the shaft by its bearings.

Solution:

First, we determine the moment exerted on the rod by the shaft. Therefore, consider principal body axes 1, 2, 3 of the rod, with origin at O (Fig. a). The axis 3 and the unit vector $\hat{k}$ are directed out of the plane of the figure. By Eq. (21.14), the angular momentum of the rod is

$$A_{o1} = I_1\omega_1, \quad A_{o2} = I_2\omega_2, \quad A_{o3} = I_3\omega_3 \qquad \text{(a)}$$

where

$$I_1 \approx 0 \quad \text{and} \quad I_2 = I_3 = \tfrac{1}{3}mL^2 \qquad \text{(b)}$$

and,

$$\omega_1 = \omega \cos 30^\circ = \frac{\sqrt{3}}{2}\omega$$

$$\omega_2 = \omega \cos 120^\circ = -\frac{1}{2}\omega \qquad \text{(c)}$$

$$\omega_3 = \omega \cos 90^\circ = 0$$

By Eqs. (a), (b), and (c), we obtain

$$A_{01} = 0, \quad A_{02} = -\tfrac{1}{6}mL^2\omega, \quad A_{03} = 0 \qquad \text{(d)}$$

Since $\bar{\omega}$ is constant and hence $\bar{A}_o$ is constant, Eq. (21.36) yields, with Eqs. (c) and (d),

$$\bar{M}_o = \bar{\omega} \times A_o = \begin{vmatrix} \hat{\imath} & \hat{\jmath} & \hat{k} \\ \frac{\sqrt{3}\omega}{2} & -\frac{1}{2}\omega & 0 \\ 0 & -\frac{1}{6}mL^2\omega & 0 \end{vmatrix}$$

or

$$\bar{M}_o = -\frac{\sqrt{3}}{12}mL^2\omega^2\hat{k} \qquad \text{(e)}$$

The moment $\bar{M}_R$ that the rod exerts on the shaft is $-\bar{M}_o$, or (see Fig. b).

$$\bar{M}_R = \frac{\sqrt{3}}{12}mL^2\omega^2\hat{k} \qquad \text{(f)}$$

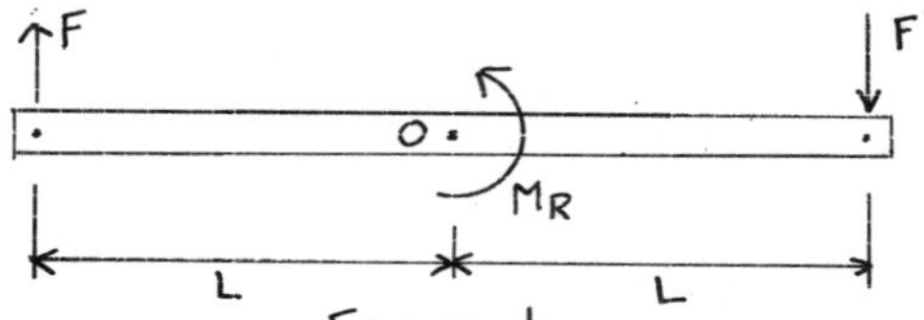

Figure b

Hence, by the equilibrium for the shaft,

$$\Sigma M_o = M_R - 2FL = 0 \qquad \text{(g)}$$

Equations (f) and (g) yield

$$F = \frac{\sqrt{3}}{24}mL^2\omega$$

Alternatively, by the Euler equations [Eqs. (21.34)],

$$M_{o1} = (I_3 - I_2)\omega_2\omega_3 = 0$$

$$M_{o2} = (I_1 - I_3)\omega_1\omega_3 = 0$$

$$M_{o3} = (I_2 - I_1)\omega_1\omega_2 = \left(\tfrac{1}{3}mL^2\right)\left(\tfrac{\sqrt{3}}{2}\omega\right)\left(-\tfrac{1}{2}\omega\right)$$

or $M_{o3} = -\frac{\sqrt{3}}{12}mL^2\omega^2$

Therefore,

$$M_o = M_{o3}\hat{k} = -\frac{\sqrt{3}}{12}mL^2\omega^2\hat{k}$$

The remainder of the solution procee des as before

21.43

Given: A uniform solid thin disk of mass m and radius r is attached at its center of mass G to a shaft that rotates about its axis at a constant speed Ω [rad/s]. The axis of the disk forms an angle θ with the axis of the shaft (Fig. a).

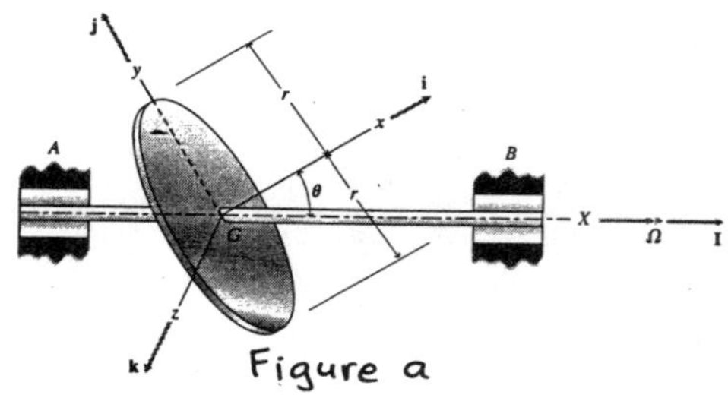

Figure a

(Continued)

21.43

Determine the couple exerted on the shaft by the bearings A and B, in terms of m, r, Ω, and θ. Neglect gravity.

Solution:

This problem may be solved using the Euler equations. Hence, by Eqs. (21.34), since Ω is constant,

$$M_{Gx} = (I_{zz} - I_{yy})\,\omega_y \omega_z$$
$$M_{Gy} = (I_{xx} - I_{zz})\,\omega_x \omega_z \qquad (a)$$
$$M_{Gz} = (I_{yy} - I_{xx})\,\omega_x \omega_y$$

where xyz are principal body axes with origin at G (Fig. a).

The angular projections $\omega_x, \omega_y, \omega_z$ are obtained from the condition.

$$\Omega \bar{I} = \Omega[(\cos\theta)\hat{\imath} - (\sin\theta)\hat{\jmath}]$$

or $\omega_x = \Omega\cos\theta\,,\ \omega_y = -\Omega\sin\theta\,,\ \omega_z = 0 \qquad (b)$

Also, by Appendix F, the principal moments of inertia for the disk are

$$I_{xx} = \tfrac{1}{2}mr^2\,,\ I_{yy} = I_{zz} = \tfrac{1}{4}mr^2 \qquad (c)$$

So, by Eqs. (a), (b), and (c),

$$M_{Gx} = M_{Gy} = 0\,,\quad M_{Gz} = \tfrac{1}{8}mr^2\Omega^2 \sin 2\theta$$

Therefore,

$$\bar{M}_G = \tfrac{1}{8}mr^2\Omega^2(\sin 2\theta)\hat{k} \qquad (d)$$

The couple $\bar{M}_G$ is exerted on the disk by the shaft. Hence, $\bar{M}_D = -\bar{M}_G$ is the couple exerted on the shaft by the disk (Fig. b)

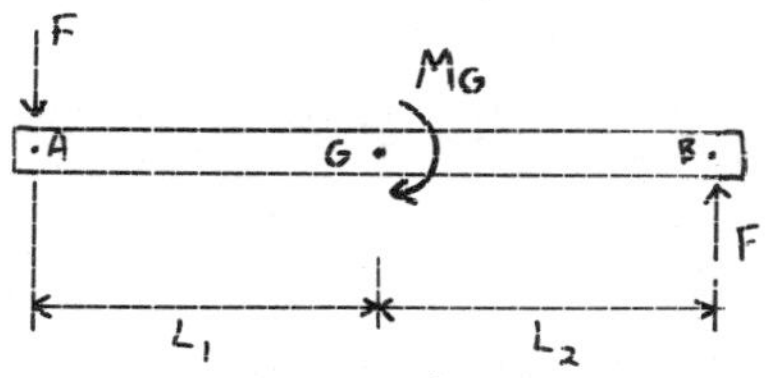

Figure b

By Fig. b,

$$\Sigma M_G = F(L_1 + L_2) - M_G = 0$$

where $F(L_1 + L_2)$ is the couple exerted on the shaft by the bearings.

Hence, $F(L_1 + L_2) = M_G$

or, the couple exerted on the shaft by the bearings is also $\bar{M}_G$.

Alternatively, by Eq. (21.36), since $d\bar{A}_G/dt = 0$,

$$\bar{M}_G = \bar{\omega} \times \bar{A}_G$$

where by Eqs. (21.24), (b), and (c),

$$A_{G1} = I_{xx}\omega_x = \tfrac{1}{2}mr^2\Omega\cos\theta$$
$$A_{G2} = I_{yy}\omega_y = -\tfrac{1}{4}mr^2\Omega\sin\theta \qquad (f)$$
$$A_{G3} = 0$$

So, by Eqs. (b), (e), and (f),

$$\bar{M}_G = \begin{vmatrix} \hat{\imath} & \hat{\jmath} & \hat{k} \\ \Omega\cos\theta & -\Omega\sin\theta & 0 \\ \tfrac{1}{2}mr^2\Omega\cos\theta & -\tfrac{1}{4}mr^2\sin\theta & 0 \end{vmatrix}$$

or $\bar{M}_G = \tfrac{1}{8}mr^2\Omega^2(\sin 2\theta)\hat{k}$

as before [see Eq. (d)]

21.44

Given: A body rotates about a fixed axis passing through its center of mass. The axis coincides with a principal axis of inertia of the body.

a) Show by Eq. (21.37) that the body exerts no inertial couples that cause reactions of the bearings on the shaft.

b) Show, in this case, that Eq. (21.37) is equivalent to the equation of motion for rotation, $M = I\alpha$, where M is the moment of the applied forces, I is the principal moment of inertia about the axis of rotation and α is the angular acceleration about the axis.

Solution:

Relative to the center of mass G of the body (Fig. a), by Eq. (21.37), the moment $\bar{M}_G$ of the external forces about G is,

$$\bar{M}_G = \bar{\omega} \times \bar{A}_G + \frac{\bar{A}_G}{A_G}\frac{dA_G}{dt} \qquad (a)$$

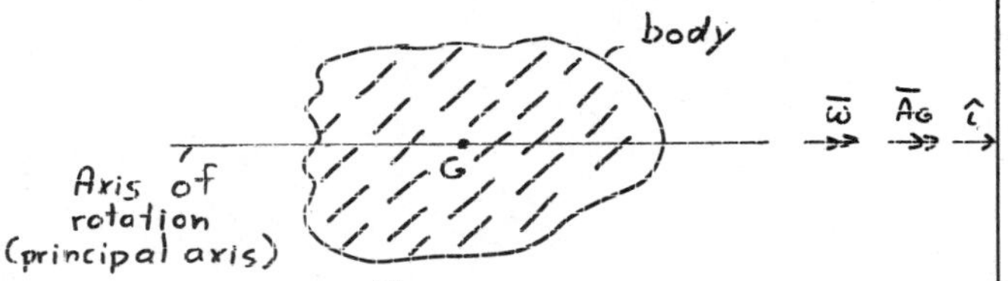

Figure a

Since $\bar{\omega}$ and $\bar{A}_G$ are parallel, $\bar{\omega} \times \bar{A}_G = 0$. Therefore, by Eq. (a)

$$\bar{M}_G = \frac{\bar{A}_G}{A_G}\frac{dA_G}{dt} \qquad (b)$$

But $\bar{A}_G/A_G$ is a unit vector $\hat{\imath}$ along the axis of rotation. Also, $A_G = I\omega$ where I is the mass moment of inertia of the body about the axis of rotation. Hence, by Eq. (b),

$$\bar{M}_G = \frac{d(I\omega)}{dt}\hat{\imath} = I\frac{d\omega}{dt}\hat{\imath} = I\alpha\hat{\imath} \qquad (c)$$

where $\alpha = d\omega/dt$

(Continued)

21.44 Cont.

So, since $\overline{M}_G$ lies along the axis of the shaft, it does not require reactions of the bearings on the shaft.

b) It is immediately seen from Eq. (c) that

$$\underline{\underline{M_G = I\alpha}}$$

21.45

Given: A homogeneous cylinder that weighs 12 lb is mounted obliquely on a light rigid shaft that rotates at 40 rad/s (Fig. a). Let $g = 384\ \text{in/s}^2$, and neglect gravity.

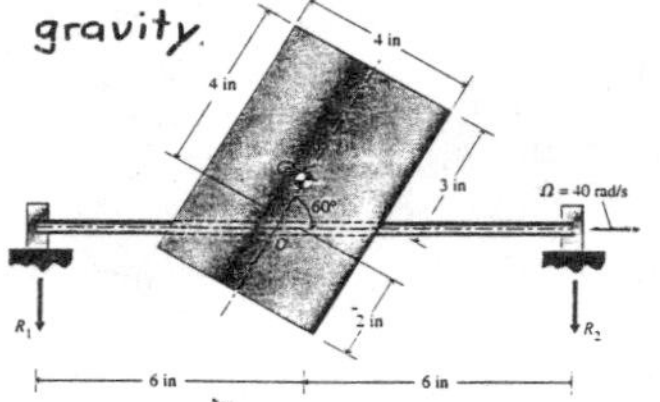

Figure a

Determine the reactions R_1 and R_2, using the Euler equations. Note that the center of mass G does not lie on the axis of the shaft.

Solution:

Since Ω is constant, Eq. (21.37) yields,

$$\overline{M}_O = \overline{\omega} \times \overline{A}_O \qquad (a)$$

Let (1,2,3) be principal axes at O (Fig. b)

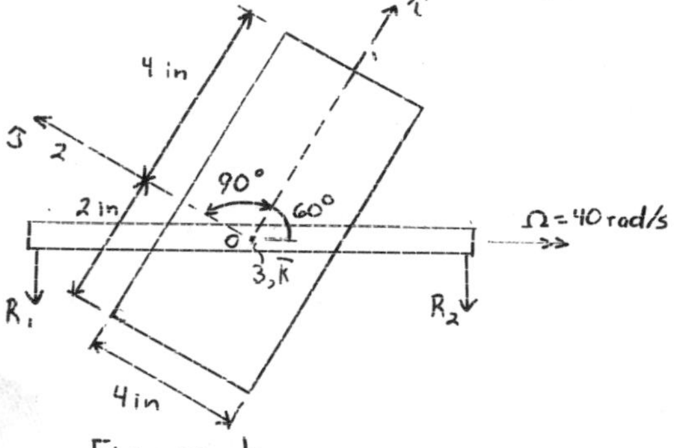

Figure b

The mass moments of inertia for axes (1,2,3) are (see Appendix F), with the parallel axis theorem,

$$I_1 = \tfrac{1}{2} m r^2$$
$$I_2 = I_3 = \tfrac{1}{12} m (L^2 + 3r^2) + m(1)^2$$

or since $m = 12/384 = 0.03109\ \text{lb}\cdot\text{s}^2/\text{ft}$

$L = 6$ in, and $r = 2$ in;

$$I_1 = 2m = 0.06218\ \text{lb}\cdot\text{in}\cdot\text{s}^2$$
$$I_2 = I_3 = 5m = 0.15544\ \text{lb}\cdot\text{in}\cdot\text{s}^2 \qquad (b)$$

The projections of $\overline{\Omega}$ on axes 1,2,3 are

$$\omega_1 = 40\cos 60° = 20\ \text{rad/s}$$
$$\omega_2 = 40\cos 150° = -34.641\ \text{rad/s} \qquad (c)$$
$$\omega_3 = 0$$

Hence, the projections of $\overline{A}_O$ on axes 1,2,3 are,

$$A_{O1} = I_1\omega_1 = 1.2435\ \text{lb}\cdot\text{in}\cdot\text{s}$$
$$A_{O2} = I_2\omega_2 = -5.3846\ \text{lb}\cdot\text{in}\cdot\text{s} \qquad (d)$$
$$A_{O3} = I_3\omega_3 = 0$$

Equations (a), (c), and (d) yield

$$\overline{M}_O = \begin{vmatrix} \hat{i} & \hat{j} & \hat{k} \\ 20 & -34.641 & 0 \\ 1.2435 & -5.3846 & 0 \end{vmatrix}$$

Therefore, the couple exerted on the cylinder by the shaft is,

$$\overline{M}_O = -64.62\,\hat{k}\ [\text{lb}\cdot\text{in}].$$

Therefore, the projections of $\overline{M}_O$ on axes 1, 2, 3 are,

$$M_{O1} = M_{O2} = 0$$
$$M_{O3} = -64.616\ \text{lb}\cdot\text{in} \qquad (e)$$

Note also that the acceleration of the center of mass G is, directed downward,

$$a_G = (1)(\sin 60°)(\Omega^2) = 1385.64\ \text{in/s}^2$$

Hence, the net force exerted on the shaft by the bearings is, directed downward,

$$F = m a_G = 43.080\ \text{lb} \qquad (f)$$

Thus, by the free-body diagram of the shaft (Fig. c),

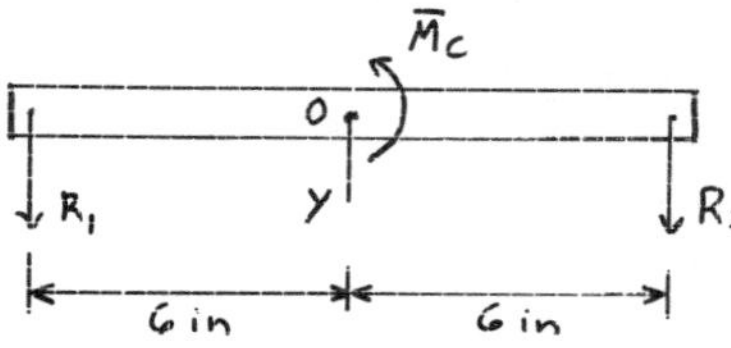

Figure c

$$\Sigma F_y = R_1 + R_2 = F = m a_G = 43.080\ \text{lb}$$
$$\Sigma M_O = 6R_1 - 6R_2 + M_c = 0 \qquad (g)$$

where $\overline{M}_c$ is the couple that the cylinder exerts on the shaft. By the law of action and reaction,

$$\overline{M}_c = -\overline{M}_O = 64.616\,\hat{k}$$

or

$$M_c = 64.616\ \text{lb}\cdot\text{in} \qquad (h)$$

Thus, by Eqs. (g) and (h),

$$R_1 + R_2 = 43.080$$
$$6R_1 - 6R_2 = -64.616$$

The solution of these equations is

$$\underline{\underline{R_1 = 16.155\ \text{lb}}}$$
$$\underline{\underline{R_2 = 26.925\ \text{lb}}}$$

21.46

Given: The Eq. (21.37) for the couple $\bar{M}$ that acts on the shaft of Fig. 21.14, section 21.8. Show that the projection of $\bar{M}$ on the axis of the shaft is equal to $I\,d\omega/dt$, where I is the moment of inertia of the body about the axis of the shaft.

Solution: By Eq. (21.37),

$$\bar{M} = \bar{\omega} \times \bar{A}_Q + \frac{\bar{A}_Q}{A_Q}\frac{dA_Q}{dt} \qquad (a)$$

By Fig. 21.14, the unit vector along the axis of the shaft is,

$$\hat{\imath} = \frac{\bar{\omega}}{\omega} \qquad (b)$$

Then, by Eqs. (a) and (b), the projection of $\bar{M}$ on the axis of the shaft is,

$$\bar{M}\cdot\hat{\imath} = (\bar{\omega}\times\bar{A}_Q)\cdot\frac{\bar{\omega}}{\omega} + \frac{\bar{A}_Q}{A_Q}\cdot\frac{\bar{\omega}}{\omega}\frac{dA_Q}{dt} \qquad (c)$$

But, $(\bar{\omega}\times\bar{A}_Q)\cdot\bar{\omega} = 0$ (d)

since $\bar{\omega}\times\bar{A}_Q$ is perpendicular to $\bar{\omega}$

also, $\bar{A}_Q\cdot\bar{\omega} = A_{Q1}\omega_1 + A_{Q2}\omega_2 + A_{Q3}\omega_3$

where,

$$A_{Q1} = I_1\omega_1,\quad A_{Q2} = I_2\omega_2,\quad A_{Q3} = I_3\omega_3 \qquad (e)$$

or,

$$\bar{A}_Q\cdot\bar{\omega} = I_1\omega_1^2 + I_2\omega_2^2 + I_3\omega_3^2 \qquad (f)$$

where I_1, I_2, I_3 are the mass moments of inertia of the body relative to principal body axes (1, 2, 3) with origin at Q and $(\omega_1, \omega_2, \omega_3)$ are the projections of $\bar{\omega}$ on axes (1, 2, 3)

By Eqs. (e),

$$A_Q = \sqrt{I_1^2\omega_1^2 + I_2^2\omega_2^2 + I_3^2\omega_3^2} \qquad (g)$$

and by Fig. 21.14,

$$\omega_1 = \omega\cos\alpha,\ \omega_2 = \omega\cos\beta,\ \omega_3 = \omega\cos\gamma \qquad (h)$$

Therefore, by Eqs. (f) and (h)

$$\bar{A}_Q\cdot\bar{\omega} = \omega^2(I_1\cos^2\alpha + I_2\cos^2\beta + I_3\cos^2\gamma)$$

or, with Eq. (21.10), $\bar{A}_Q\cdot\bar{\omega} = I\omega^2$ (i)

where $I = I_1\cos^2\alpha + I_2\cos^2\beta + I_3\cos^2\gamma$ is the mass moment of inertia of the body about the axis of the shaft.

Also, by Eqs. (g) and (h)

$$A_Q = \omega\sqrt{I_1\cos^2\alpha + I_2\cos^2\beta + I_3\cos^2\gamma} \qquad (j)$$

and therefore,

$$\frac{dA_Q}{dt} = \frac{d\omega}{dt}\sqrt{I_1^2\cos^2\alpha + I_2^2\cos^2\beta + I_3^2\cos^2\gamma} \qquad (k)$$

Then by Eqs. (j) and (k),

$$\frac{1}{A_Q}\frac{dA_Q}{dt} = \frac{1}{\omega}\frac{d\omega}{dt} \qquad (\ell)$$

Consequently, by Eqs. (c), (d), (i), and (ℓ), the projections of $\bar{M}$ on the axis of the shaft is,

$$\bar{M}\cdot\hat{\imath} = \frac{\bar{A}_Q\cdot\bar{\omega}}{\omega^2}\frac{d\omega}{dt} = I\frac{d\omega}{dt}$$

21.47

Given: Initially the cylinder/shaft of Problem 21.45 is motionless. A torque of 200 lb·in is applied suddenly to the shaft in the positive sense of Ω (Fig. a)

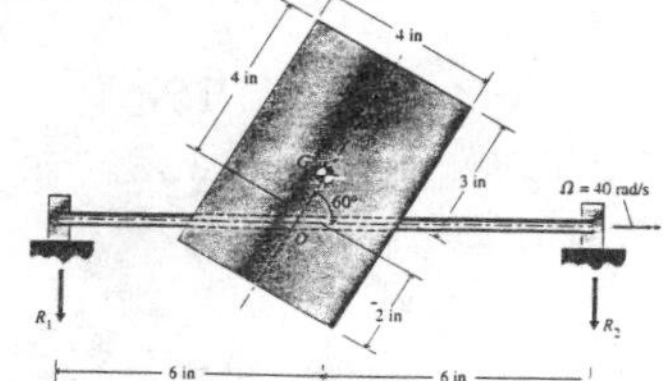

Figure a

Determine the instantaneous reactions of the bearings. Neglect gravity.

Solution:

By Eq. (21.37),

$$\bar{M}_o = \bar{\omega}\times\bar{A}_o + \frac{\bar{A}_o}{A_o}\frac{dA_o}{dt} \qquad (a)$$

Initially, $\bar{\omega} = 0$. Hence, initially

$$\bar{M}_o = \frac{\bar{A}_o}{A_o}\frac{dA_o}{dt} \qquad (b)$$

By Eq. (b), the direction of $\bar{M}_o$ coincides with the direction of $\bar{A}_o$.

Let axes 1, 2, 3 be principal body axes of the cylinder with origin O (Fig. b), and XYZ be fixed axes along and perpendicular to the shaft (axes Z is directed out of the plane of figure b)

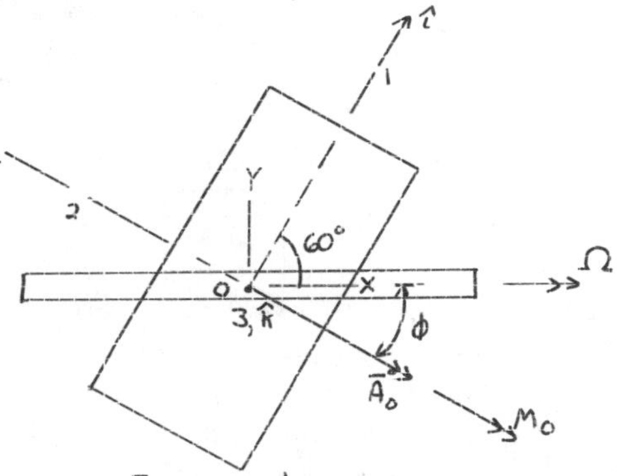

Figure b

The vectors $\bar{A}_o$ and $\bar{M}_o$ form the angle ϕ with axis X

Now, with respect to axes 1, 2, 3,

$$A_{o1} = I_1\omega_1 = I_1\Omega\cos 60° = \tfrac{1}{2}I_1\Omega$$
$$A_{o2} = I_2\omega_2 = I_2\Omega\cos 150° = -\frac{\sqrt{3}}{2}I_2\Omega \qquad (c)$$
$$A_{o3} = I_3\omega_3 = I_3\Omega\cos 90° = 0$$

where by Appendix F and the parallel axis theorem (see also Problem 21.45 solution)

$$I_1 = \tfrac{1}{2}mr^2,\quad I_2 = I_3 = \tfrac{1}{12}m(L^2 + 3r^2) + m(1)^2$$

and with $r = 2$ in and $L = 6$ in (see Problem 21.45)

$$I_1 = 2m,\quad I_2 = I_3 = 5m \qquad (d)$$

By Eqs. (c) and (d), we obtain

$$A_{o1} = m\omega,\quad A_{o2} = -\frac{5\sqrt{3}}{2}m\omega,\quad A_{o3} = 0 \qquad (e)$$

(Continued)

21.47 Cont.

Hence,

$$A_0 = \sqrt{A_{01}^2 + A_{02}^2 + A_{03}^2} = \frac{1}{2} m\omega\sqrt{79} \qquad (f)$$

and the angle ϕ (see Fig. b) is given by

$$\tan(\phi + 60°) = \frac{|A_{02}|}{A_{01}} = \frac{5\sqrt{3}/2m\omega}{m\omega} = \frac{5\sqrt{3}}{2} = 4.3301$$

or $\phi = 76.996 - 60 = 16.996°$

The x projection of $\bar{M}$ is equal to the torque 200 lb·in or

$$M_x = 200 \text{ lb·in}$$

Therefore, by Fig. b, with $\phi = 16.996°$,

$$M = \frac{M_x}{\cos 16.996°} = 209.134 \text{ lb·in}$$

Hence, the y projection of M is

$$M_y = -M \sin 16.996° = -61.131 \text{ lb·in}$$

Since M_x is along the axis of the beam. The projection M_y produces a reaction of the bearings in the z direction (Fig. c)

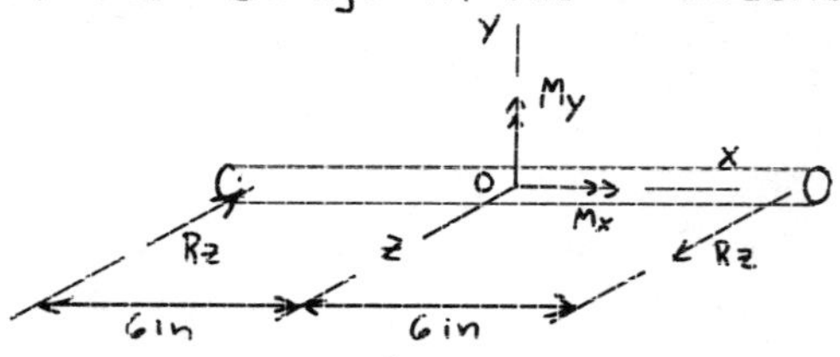

Figure c

Hence,

$$\Sigma M_0 = M_y - 12R_z = 0$$

or $R_z = \frac{M_y}{12} = 5.094$ lb

21.48

Given: A thin plate of mass m in the form of an isosceles right triangle is welded to a shaft as shown in Fig. a. An external torque $T = T_0 \sin 2t$ is applied to the shaft. The effects of the shaft on the plate may be represented by a force of magnitude F at A and a couple of moment M.

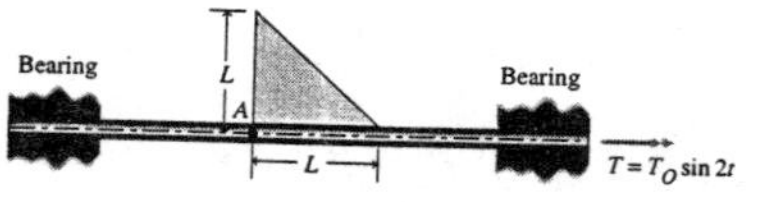

Figure a

Determine F and M at time $t = \pi/4$ s. The system is at rest when $t = 0$.

Solution:

The free-body diagram of the plate, including the inertial force, is shown in Fig. b.

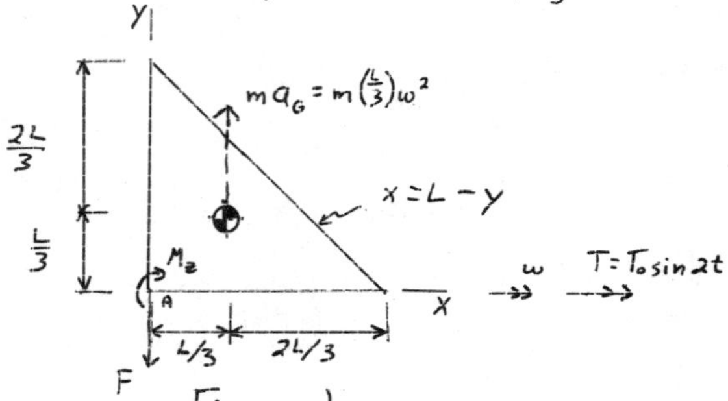

Figure b

By Fig. b,

$$\Sigma F_y = \frac{1}{3} m L\omega^2 - F = 0$$

$$\Sigma M_G = F\frac{L}{3} - M_z = 0$$

or

$$F = \frac{1}{3} m L\omega^2$$
$$M_z = \frac{1}{3} FL \qquad (a)$$

To determine ω, we note that,

$$T = I\alpha = I\frac{d\omega}{dt} \qquad (b)$$

where I is the mass moment of inertia of the plate about the axis of the shaft. Hence, with $T = T_0 \sin 2t$ and $t = \pi/4$ s, Eq. (b) yields,

$$\int_0^\omega d\omega = \frac{T_0}{I}\int_0^{\pi/4} (\sin 2t)\, dt$$

or after integration,

$$\omega = \frac{T_0}{2I} \qquad (c)$$

To determine I, by definition, we have (see Fig. c)

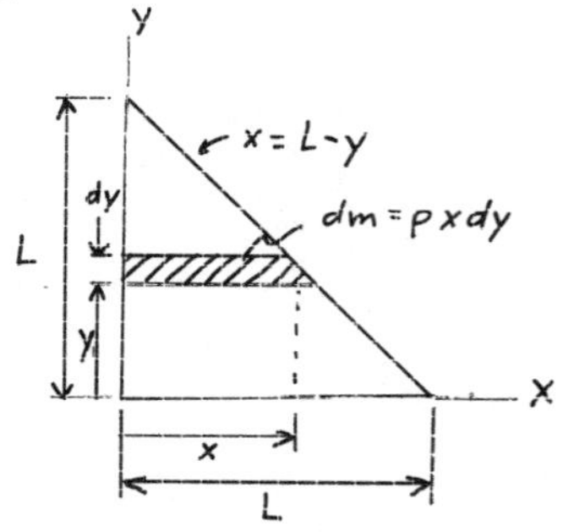

Figure c

$$I = I_x = \int y^2 dm = \int_0^L y^2 \rho x\, dy$$

so, $I = \rho\int_0^L y^2(L-y)\,dy = \frac{1}{12}\rho L^4 = \frac{1}{6} m L^2 \qquad (d)$

since $m = \frac{1}{2}\rho L^2$

(Continued)

21.48 Cont.

Therefore, by Eqs. (c) and (d),

$$\omega = \frac{3T_0}{mL^2} \qquad (e)$$

Then, by Eqs. (a) and (e)

$$F = \frac{3T_0^2}{mL^3} \text{ [in the negative } y \text{ direction (Fig. b)]}$$

$$M_z = \frac{T_0^2}{mL^2} \text{ [in the negative } z \text{ direction (Fig. b)]}$$

Note: In addition to M_z, the torque $T_0 (= M_x)$ is applied to the plate at time $t = \pi/4$ s.

21.49

Given: A uniform slender bar of mass m is welded to a shaft AB of negligible mass at an angle of 30° (Fig. a). The shaft rotates at constant angular speed Ω.

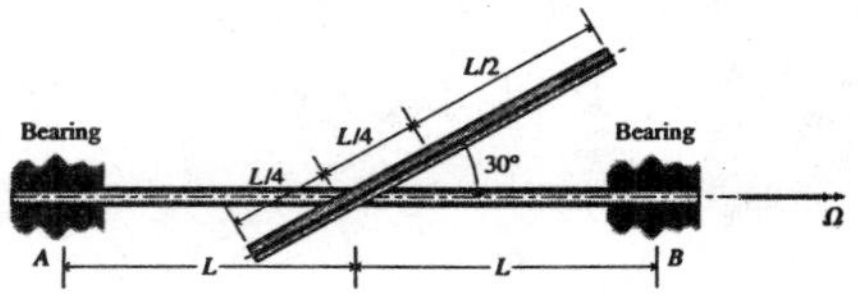

Figure a

Determine the forces exerted on the shaft by the bearings at A and B.

Solution:

Select principal body axes 1, 2, 3 (Fig. b), where axis 3 is directed outward from the page.

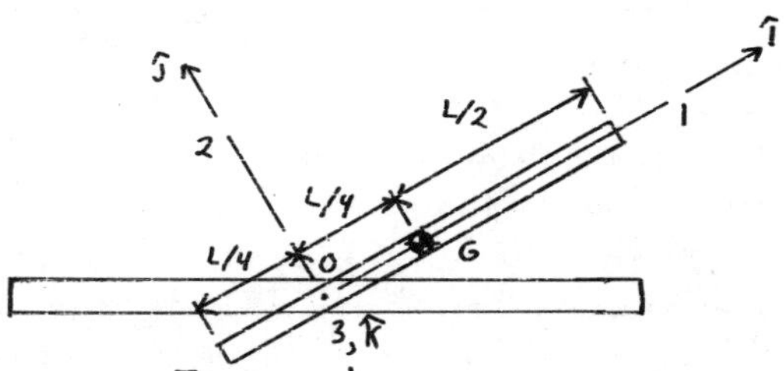

Figure b

The mass moments of inertia of the bar relative to point O are

$$I_1 \approx 0$$

$$I_2 = I_3 = I_G + m\left(\frac{L}{4}\right)^2 = \frac{1}{12}mL^2 + \frac{1}{16}mL^2 = \frac{7}{48}mL^2 \quad (a)$$

The moment of momentum is

$$\bar{A}_o = A_{o1}\hat{\imath} + A_{o2}\hat{\jmath} + A_{o3}\hat{k} \quad (b)$$

where, with respect to axes 1, 2, 3

$$A_{o1} = I_1\omega_1, \; A_{o2} = I_2\omega_2, \; A_{o3} = I_3\omega_3 \qquad (c)$$

and,

$$\omega_1 = \omega\cos 30° = \frac{\sqrt{3}}{2}\omega$$

$$\omega_2 = \omega\cos 120° = -\frac{1}{2}\omega \qquad (d)$$

$$\omega_3 = \omega\cos 90° = 0$$

The couple exerted on the bar by the shaft is given by Eq. (21.37), namely

$$\bar{M} = \bar{\omega} \times \bar{A}_o + \frac{\bar{A}_o}{A_o}\frac{dA_o}{dt} \qquad (e)$$

But since $\bar{\omega}$ = constant, $\frac{dA_o}{dt} = 0$, so Eq. (e) reduces to [using Eqs. (a), (b), (c) and (d)]

$$\bar{M}_o = \bar{\omega} \times \bar{A}_o \qquad (f)$$

or

$$\bar{M}_o = \begin{vmatrix} \hat{\imath} & \hat{\jmath} & \hat{k} \\ \frac{\sqrt{3}\omega}{2} & -\frac{1}{2}\omega & 0 \\ 0 & -\frac{7}{96}mL^2\omega & 0 \end{vmatrix} = -0.06315\, mL^2\omega^2\,\hat{k}$$

Thus, the shaft exerts the couple

$$\bar{M}_o = -0.06315\, mL^2\omega^2\,\hat{k} \qquad (g)$$

on the bar.

In addition to the couple $\bar{M}_o$, the shaft exerts a force F on the bar at O. To determine F, consider the free-body diagram of the bar (Fig. c).

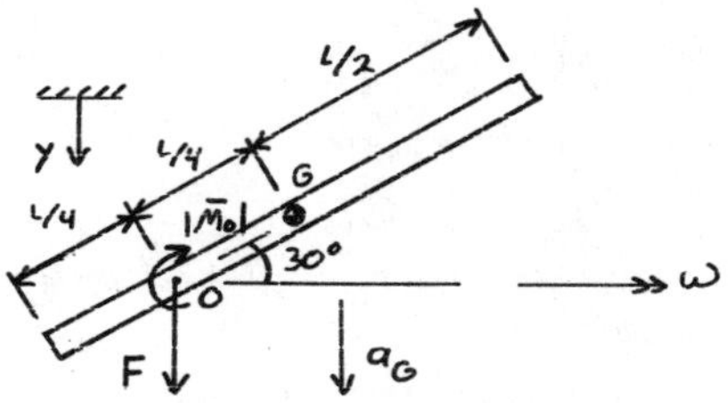

By Fig. c, and Newton's second law,

$$\Sigma F_y = F = ma_G = m\left(\frac{L}{4}\sin 30°\right)\omega^2$$

or

$$F = \frac{1}{8}mL\omega^2 \qquad (h)$$

Hence, by the law of action and reaction, the free-body diagram of the shaft is shown in Fig. d.

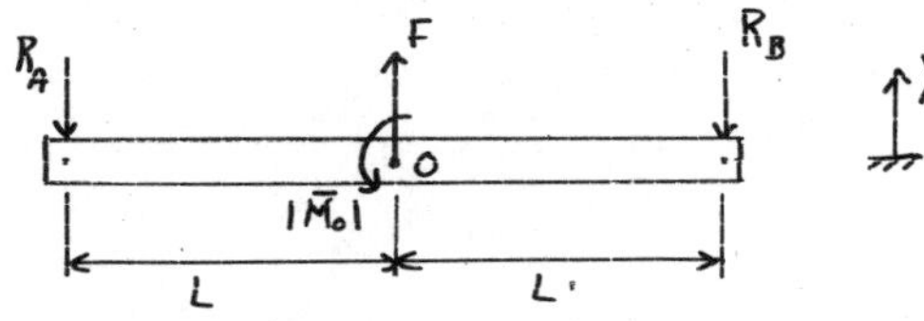

Figure d

By Fig. d,

$$\Sigma F_y = F - R_A - R_B = 0$$

or, with Eq. (h),

$$R_A + R_B = \frac{1}{8}mL\omega^2 \qquad (i)$$

also, by Fig. d,

$$\circlearrowleft + \Sigma M_O = |\bar{M}_o| + R_A L - R_B L = 0$$

or, with Eq. (g), $R_A - R_B = -0.06315\, mL\omega^2$ (j)

The solution of Eqs. (i) and (j) is

$$\underline{R_A = 0.03093\, mL\omega^2} \qquad \underline{R_B = 0.09407\, mL\omega^2}$$

21.50

Given: Figure a represents a homogeneous cylindrical body of diameter 100 mm and length 150 mm. The body is mounted obliquely on a shaft that rotates with angular speed $\omega = 60$ rad/s. The body weighs 54 N.

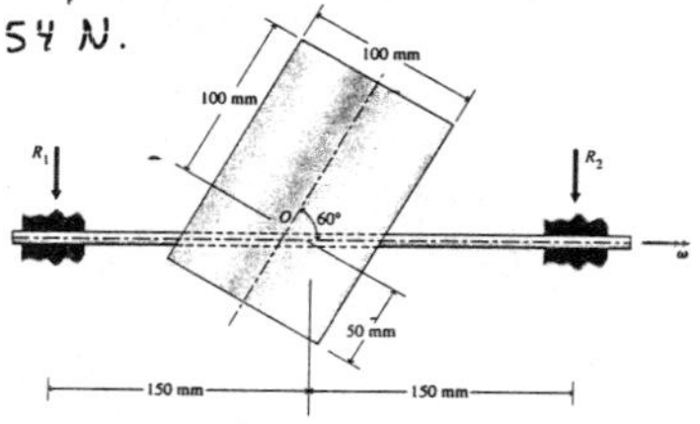

Figure a

Calculate the forces R_1 and R_2, neglecting gravity effects.

Solution:

This problem can be solved using Eq. (21.37), namely,

$$\bar{M}_o = \bar{\omega} \times \bar{A}_o + \frac{\bar{A}_o}{A_o} \frac{dA_o}{dt} \qquad (a)$$

and since $\bar{\omega}$ is constant,

$$\frac{dA_o}{dt} = 0$$

Therefore, Eq. (a) reduces to

$$\bar{M}_o = \bar{\omega} \times \bar{A}_o \qquad (b)$$

where $\bar{M}_o$ is the moment exerted on the body by the shaft.

Let (1, 2, 3) be principal body axes at O (Fig. b)

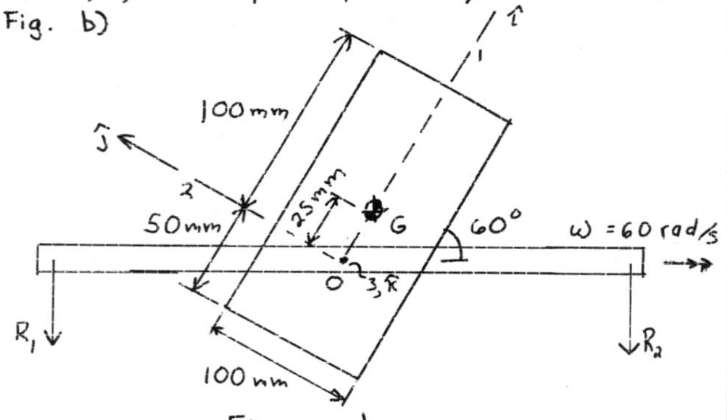

Figure b

The mass moments of inertia for axes 1, 2, 3 are by Appendix F and the parallel-axis theorem, with L = 150 mm, r = 50 mm, and m = 54/9.81 = 5.5046 kg,

$$I_1 = \tfrac{1}{2} m r^2 = 6880.8 \text{ Kg}\cdot\text{mm}^2$$
$$I_2 = I_3 = \tfrac{1}{12} m (L^2 + 3r^2) + m(25)^2 = 17{,}202 \text{ Kg}\cdot\text{mm}^2 \qquad (c)$$

The projections of $\bar{\omega}$ on axes 1, 2, 3 are

$$\omega_1 = 60 \cos 60° = 30 \text{ rad/s}$$
$$\omega_2 = 60 \cos 150° = -51.962 \text{ rad/s} \qquad (d)$$
$$\omega_3 = 60 \cos 90° = 0$$

Hence, the projections of $\bar{A}_o$ on axes 1, 2, 3 are, with Eqs. (c) and (d),

$$A_{o1} = I_1 \omega_1 = 206.42 \text{ N}\cdot\text{mm}\cdot\text{s}$$
$$A_{o2} = I_2 \omega_2 = -893.85 \text{ N}\cdot\text{mm}\cdot\text{s} \qquad (e)$$
$$A_{o3} = I_3 \omega_3 = 0$$

Then, Eqs. (a), (d), and (e) yield

$$\bar{M}_o = \begin{vmatrix} \hat{\imath} & \hat{\jmath} & \hat{k} \\ 30 & -51.962 & 0 \\ 206.42 & -893.85 & 0 \end{vmatrix}$$

or,

$$\bar{M}_o = -16.09 \times 10^3 \hat{k} \text{ [N}\cdot\text{mm]} \qquad (f)$$

The force F exerted on the body by the shaft is, by Newton's second law (see Fig. c),

$$F = m a_G = m(25 \sin 60°)\omega^2$$

or

$$F = (5.5046)(25 \sin 60°)(60)^2 = 429.03 \text{ N} \qquad (g)$$

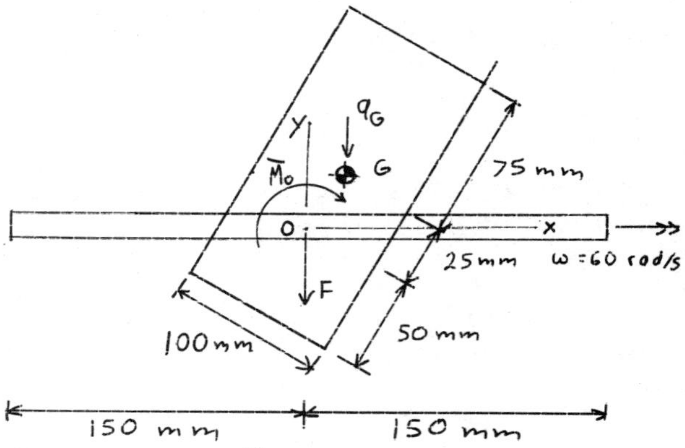

Figure c

Hence, by Eqs. (f) and (g) and Fig. c, the shaft exerts the force F and the couple $\bar{M}_o$ on the body. So, by the law of action and reation the body exerts the force −F and the couple $-\bar{M}_o$ on the shaft. Therefore, by equilibrium of the shaft (Fig. d), with Eqs. (f) and (g),

$$\Sigma F_y = F - R_1 - R_2 = 0\ ;\ R_1 + R_2 = 429.03 \qquad (h)$$
$$\circlearrowleft + \Sigma M_o = 16.09 \times 10^3 + (150)(R_1 - R_2) = 0;$$
$$R_1 - R_2 = -107.27 \qquad (i)$$

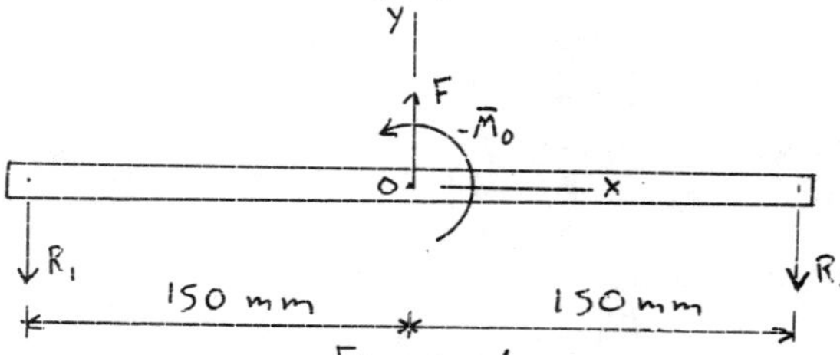

Figure d

The solution of Eqs. (h) and (i) is

$\underline{\underline{R_1 = 160.88 \text{ N}}}$ $\underline{\underline{R_2 = 268.15 \text{ N}}}$

where R_1 and R_2 are the forces exerted on the shaft.

21.51

Given: In the system described in Problem 21.50, a torque of 10 N·m is applied suddenly to the shaft in the positive direction of $\bar{\omega}$ (Fig. a).

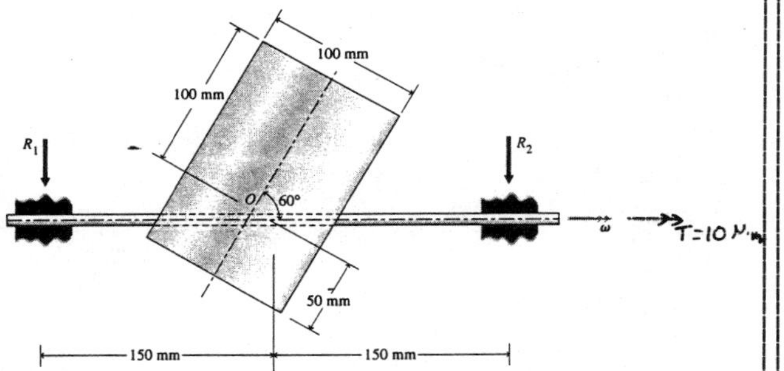

Figure a

Determine the immediate reactions of the bearings, if initially $\bar{\omega} = 0$.

Solution:

As in Problem 21.50, the couple $\bar{M}_o$ exerted on the body by the shaft is given by,

$$\bar{M}_o = \bar{\omega} \times \bar{A}_o + \frac{\bar{A}_o}{A_o}\frac{dA_o}{dt}$$

and since initially $\bar{\omega} = 0$,

$$\bar{M}_o = \frac{\bar{A}_o}{A_o}\frac{dA_o}{dt} \qquad (a)$$

By Eq. (a), the direction of $\bar{M}_o$ coincides with the direction of $\bar{A}_o$. Now the projections of $\bar{A}_o$ on principal axes 1, 2, and 3 (Fig. a) are

$$A_{o1} = I_1\omega_1 = I_1\omega\cos 60° = \tfrac{1}{2}I_1\omega$$
$$A_{o2} = I_2\omega_2 = -I_2\omega\cos 30° = -\tfrac{\sqrt{3}}{2}I_2\omega \qquad (b)$$
$$A_{o3} = I_3\omega_3 = I_3\omega\cos 90° = 0$$

where by Eq. (c) of the solution of Problem 21.50,

$$I_1 = 6880.8 \text{ kg·mm}^2$$
$$I_2 = I_3 = 17{,}202 \text{ kg·mm}^2 \qquad (c)$$

Therefore, by Eqs. (b) and (c), with $[\omega] = [\frac{rad}{s}]$,

$$A_{o1} = 3440.4\,\omega \quad [\text{N·mm·s}]$$
$$A_{o2} = -14{,}897\omega \quad [\text{N·mm·s}] \qquad (d)$$

Let $\bar{A}_o$ from the angle ϕ with respect to axis 1 (Fig. b), where axis x is along the axis of the shaft and axis y is perpendicular to the shaft.

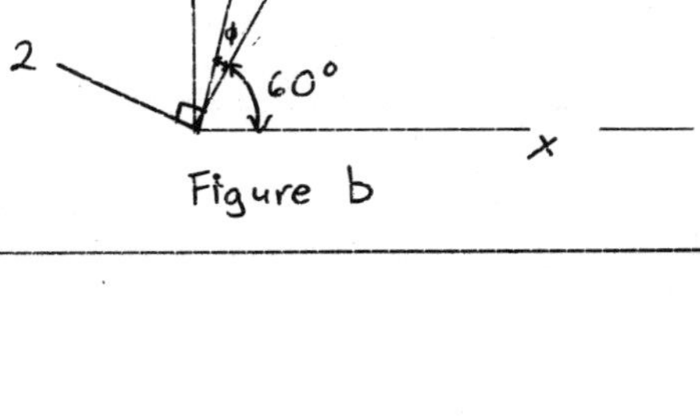

Figure b

Then, by Eqs. (d),

$$\tan\phi = \frac{A_{o2}}{A_{o1}} = -4.330$$

or

$$\phi = -77.0°$$

The angle ϕ determines the direction of $\bar{A}_o$ and $\bar{M}_o$ (Fig. c).

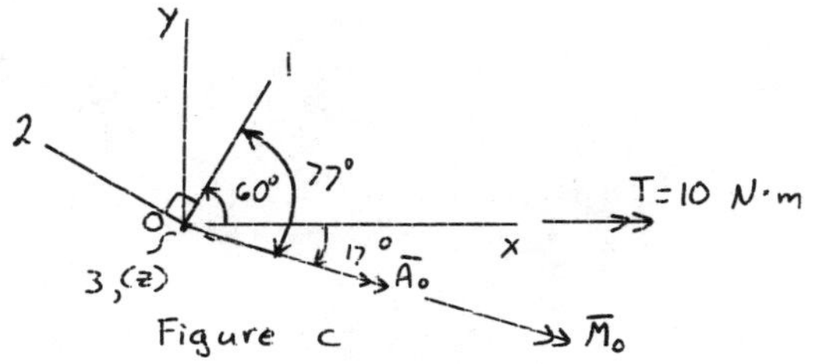

Figure c

The x projection of $\bar{M}_o$ is (Fig. c)

$$M_x = T = 10 \text{ N·m} \qquad (e)$$

Hence,

$$|\bar{M}_o| = M_o = \frac{M_z}{\cos 17°} = 10.457 \text{ N·m}$$

Then, the y projection of $\bar{M}_o$ is

$$M_y = -M_o \sin 17° = -3.057 \text{ N·m} \qquad (f)$$

The force F initially exerted on the body by the shaft is, by Newton's second law (see Fig. a),

$$F = m a_G = m(25\sin 60°)\omega^2 = 0$$

since initially $\omega = 0$. Consequently, only the projection M_y of $\bar{M}_o$ results in a reaction of the bearings (Fig. d; note that the moment exerted on the shaft by the body is $-M_y = +3.057$ N·m)

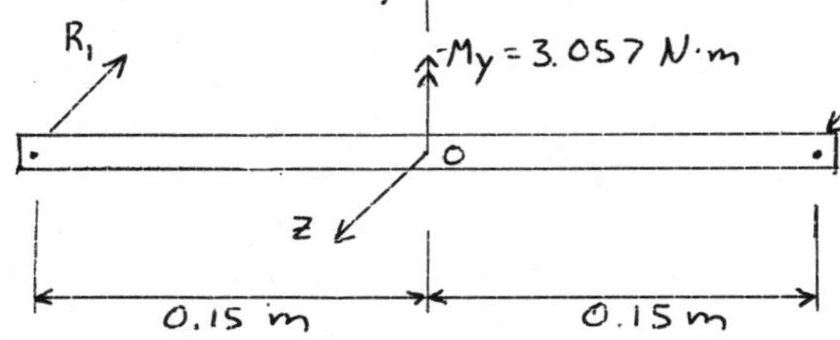

Figure d

By Fig. d,

$$\Sigma F_z = R_2 - R_1 = 0 \;;\; R_2 = R_1 = R$$
$$\Sigma M_o = 3.057 - (0.15)(2R) = 0$$

Hence, $\underline{R = R_1 = R_2 = 10.19 \text{ N}}$

21.52

Given: The shaft in Example 21.18 is subjected to the angular acceleration 30 rad/s². At the same instant its angular speed is 100 rad/s (Fig. a).

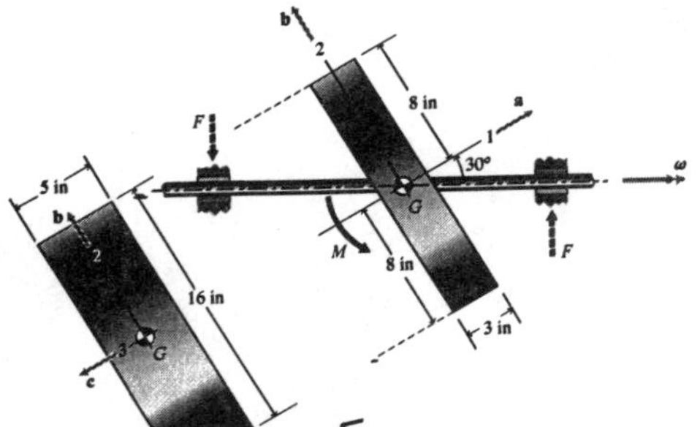

Figure a

Each bearing is 10 in from the center of the block, the shaft is horizontal, and the mass of the shaft is negligible.

a) Determine the bearing reactions including the effect of the weight of the body.

b) Determine the torque applied to the shaft.

Solution:

a) The couple exerted on the body by the shaft is given by

$$\bar{M}_G = \bar{\omega} \times \bar{A}_G + \frac{\bar{A}_G}{A_G}\frac{dA_G}{dt} \qquad (a)$$

The projections of $\bar{A}_G$ along axes 1, 2, 3 are (axis 3 is perpendicular to the plane of axes 1 and 2, Fig. a)

$$A_{G1} = I_1\omega_1;\ A_{G2} = I_2\omega_2,\ A_{G3} = I_3\omega_3 \qquad (b)$$

where by Example 21.18,

$$I_1 = 2.342\ \text{lb·in·s}^2,\ I_2 = 0.2833\ \text{lb·in·s}^2,\ I_3 = 2.208\ \text{lb·in·s}^2 \qquad (c)$$

and by Fig. a,

$$\omega_1 = \omega\cos 30° = 0.8660\,\omega$$
$$\omega_2 = \omega\cos 120° = -0.5000\,\omega \qquad (d)$$
$$\omega_3 = \omega\cos 90° = 0$$

Equations (b), (c), and (d) yield

$$A_G = \sqrt{A_{G1}^2 + A_{G2}^2 + A_{G3}^2} = 2.033\,\omega \qquad (e)$$

Hence, since $\alpha = d\omega/dt = 30$ rad/s²,

$$\frac{dA_G}{dt} = 2.033\frac{d\omega}{dt} = 2.033\alpha = 60.99\ \text{lb·in} \qquad (f)$$

For $\omega = 100$ rad/s, Eq. (e) yields

$$A_G = 203.3\ \text{lb·in·s} \qquad (g)$$

So, by Eqs. (b), (c), (d), and (g),

$$A_{G1}/A_G = 0.9976$$
$$A_{G2}/A_G = -0.06968 \qquad (h)$$
$$A_{G3}/A_G = 0$$

Therefore, by Eqs. (f) and (h), in terms of unit vectors $\bar{a}$ and $\bar{b}$ (Fig. a),

$$\frac{\bar{A}_G}{A_G}\frac{dA_G}{dt} = 60.84\,\bar{a} - 4.250\,\bar{b} \qquad (i)$$

Also, by Eqs. (b) and (d) for $\omega = 100$ rad/s,

$$\bar{\omega} \times \bar{A}_G = \begin{vmatrix} \bar{a} & \bar{b} & \bar{c} \\ 86.60 & -50.00 & 0 \\ 202.82 & -14.165 & 0 \end{vmatrix}$$

or,

$$\bar{\omega} \times \bar{A}_G = 8914.3\,\bar{c} \qquad (j)$$

Then, by Eqs. (a), (i), and (j), we have the couple exerted on the body by the shaft as,

$$\bar{M}_G = 60.84\,\bar{a} - 4.250\,\bar{b} + 8914.3\,\bar{c} \qquad (k)$$

(see Fig. b)

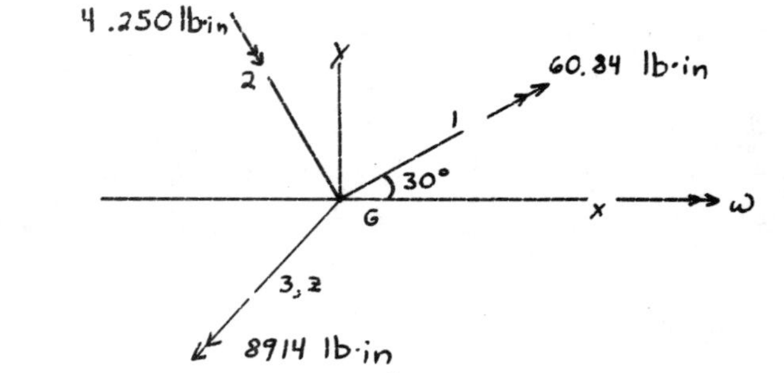

Figure b

By the free-body diagram of the shaft (Fig. c), we have

$$\Sigma F_x = 0$$
$$\Sigma F_y = R_{1y} + R_{2y} - 38.6 = 0$$
$$\Sigma F_z = R_{1z} + R_{2z} = 0 \qquad (\ell)$$
$$\Sigma M_y = 10R_{1z} - 10R_{2z} - 60.84\sin 30° + 4.250\cos 30° = 0$$
$$\Sigma M_z = 10R_{2y} - 10R_{1y} - 8914.6 = 0$$

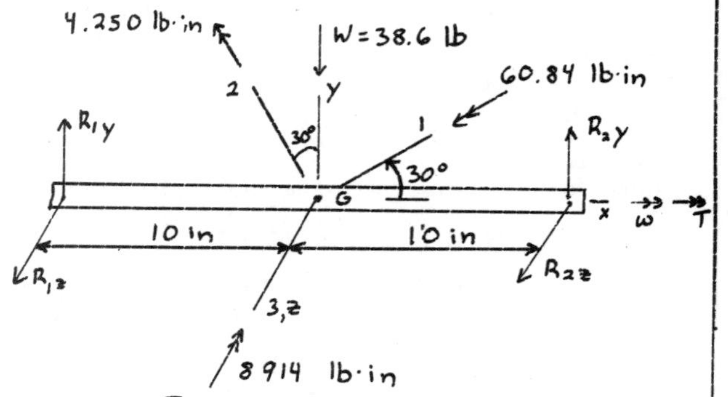

Figure c

The solution of Eqs. (ℓ) gives the forces exerted on the shaft by the bearings, namely,

$R_{1y} = -426.4$ lb , $R_{2y} = 465.0$ lb

$R_{1z} = 1.337$ lb , $R_{2z} = -1.337$ lb

(Continued)

21.52 Cont.

(b) Since the mass of the shaft is negligible, by Fig. c,

$$\Sigma M_x = T - 60.84 \cos 30° - 4.250 \sin 30° = 0$$

or

$$T = 54.81 \text{ lb}\cdot\text{in}$$

21.53

Given: A homogeneous cone that weighs 7 lb is mounted obliquely on a shaft that rotates with angular speed $\omega = 60$ rad/s (Fig. a)

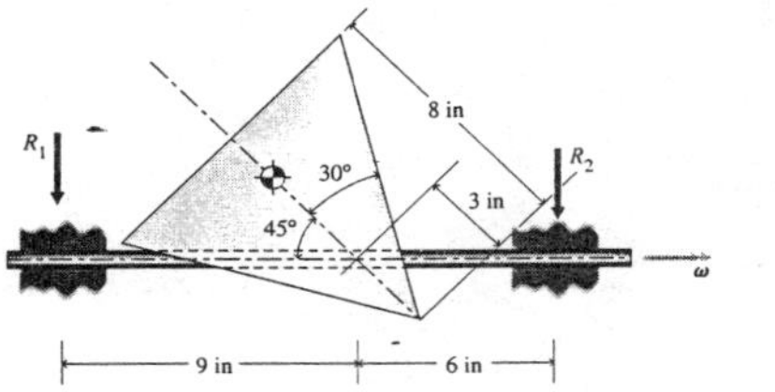

Figure a

Determine the reactions of the bearings. Neglect gravity.

Solution:

Consider principal axes 1, 2, 3 with origin O on the axis of the shaft (Fig. b)

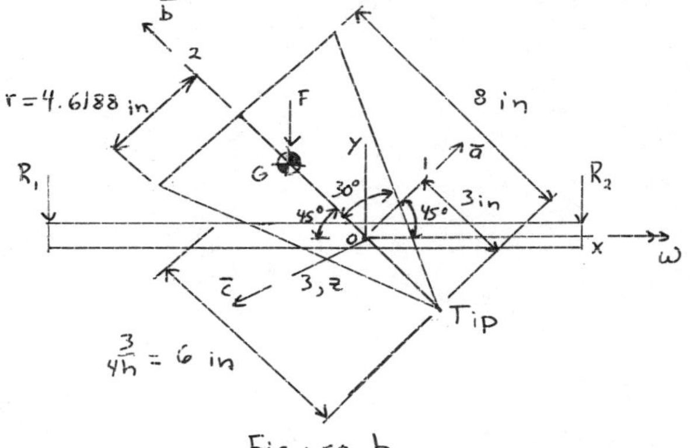

Figure b

By Fig. b and Appendix F (Table F.1), the mass moment of inertia about axis 2 is,

$$I_2 = \frac{3}{10} m r^2 \qquad (a)$$

where,

$$m = \frac{7}{386} = 0.01813 \text{ lb}\cdot\text{s}^2/\text{in}$$

and

$$r = 8 \tan 30° = 4.6188 \text{ in}$$

or

$$I_2 = 0.11603 \text{ lb}\cdot\text{in}\cdot\text{s}^2 \qquad (b)$$

For the transverse axis through the mass center G,

$$I_G = \frac{3}{80} m (h^2 + 4r^2)$$

where $h = 8$ in Hence,

$$I_G = 0.1015 \text{ lb}\cdot\text{in}\cdot\text{s}^2$$

Hence, for the transverse axes through O,

$$I_1 = I_3 = I_G + m(3)^2$$

or

$$I_1 = I_2 = 0.2647 \text{ lb}\cdot\text{in}\cdot\text{s}^2 \qquad (c)$$

Also, by Fig. b, relative to axes 1, 2, 3,

$$\omega_1 = 60 \cos 45° = 42.426 \text{ rad/s}$$
$$\omega_2 = 60 \cos 135° = -42.426 \text{ rad/s} \qquad (d)$$
$$\omega_3 = 60 \cos 90° = 0$$

Then, by Eqs. (b), (c), and (d),

$$A_{01} = I_1 \omega_1 = 11.230 \text{ lb}\cdot\text{in}\cdot\text{s}$$
$$A_{02} = I_2 \omega_2 = -4.923 \text{ lb}\cdot\text{in}\cdot\text{s} \qquad (e)$$
$$A_{03} = I_3 \omega_3 = 0$$

Therefore, the couple $\bar{M}_o$ that the shaft exerts on the cone is, since ω = constant (A_o = constant),

$$\bar{M}_o = \bar{\omega} \times \bar{A}_o + \frac{\bar{A}_o}{A_o}\frac{dA_o}{dt} = \bar{\omega} \times \bar{A}_o$$

or

$$\bar{M}_o = \begin{vmatrix} \bar{a} & \bar{b} & \bar{c} \\ 42.426 & -42.426 & 0 \\ 11.230 & -4.923 & 0 \end{vmatrix}$$

Therefore,

$$\bar{M}_o = 267.6\ \bar{c} \quad [\text{lb}\cdot\text{in}\cdot\text{s}]$$

where $\bar{a}, \bar{b}, \bar{c}$ are unit vectors along axes 1, 2, and 3 respectively (see Fig. b).

Since $\bar{M}_o$ is perpendicular to the (x, y) plane (Fig. b),

$$M_x = M_y = 0, \quad M_o = M_z = 267.6 \text{ lb}\cdot\text{in}\cdot\text{s} \qquad (e)$$

Also, by Newton's second law, the shaft exerts the force F on the body, where

$$F = m a_G = m(3 \sin 45°)\omega^2$$

or

$$F = (0.01813)(3)(0.7071)(60)^2 = 138.45 \text{ lb} \qquad (f)$$

Then, by the free-body diagram of the shaft (Fig. c)

$$\Sigma F_y = R_1 + R_2 - F = 0$$
$$\Sigma M_z = 9R_1 - 6R_2 - M_o = 0 \qquad (g)$$

with Eqs. (e) and (f), the solution of Eqs. (g) is

$$R_1 = 73.22 \text{ lb}, \quad R_2 = 65.23 \text{ lb}$$

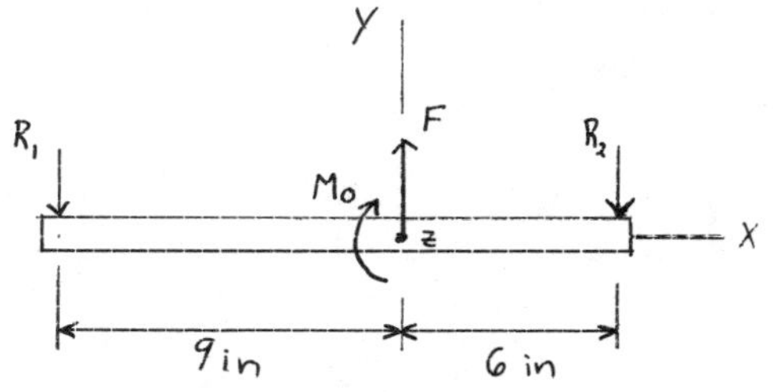

Figure c

21.54

Given: The disk and shaft (Fig. a) rotate with angular velocity $\bar{\omega}$ about the axis z of the shaft, and the shaft rotates with angular velocity $\bar{\Omega}$ about the axis y. A force P is applied to the right end of the shaft.

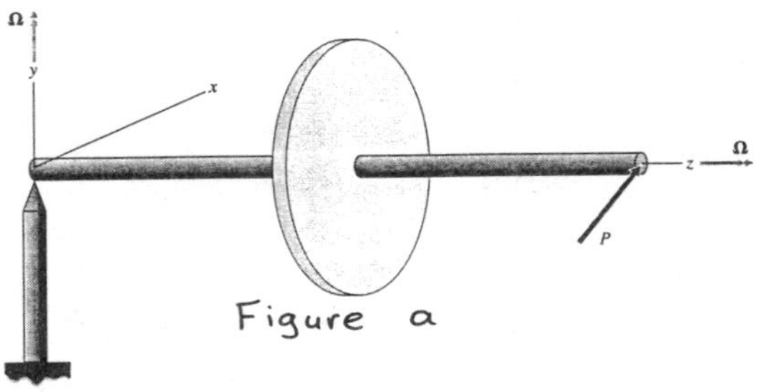

Figure a

In words, describe the response of the shaft for the following cases:

a) The force is directed in the $+x$ direction.
b) The force is directed in the $-x$ direction.
c) The force is directed in the $+y$ direction.
d) The force is directed in the $-y$ direction.

Solution:

a) The tip of the shaft will move up.
b) The tip of the shaft will move down.
c) The precessional speed Ω will be decreased.
d) The precessional speed Ω will be increased.

21.55

Given: A gyroscope consists of a homogeneous ball, with a diameter of 75 mm, that rotates at 200 rad/s about a shaft of negligible mass (Fig. a). The shaft precesses in a horizontal plane.

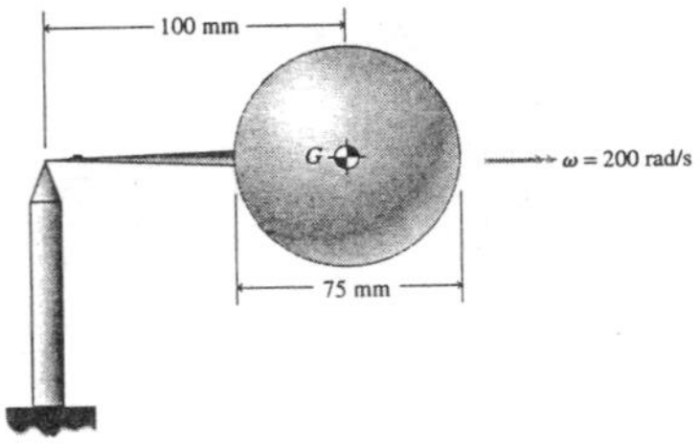

Figure a

What is the angular velocity of precession of the gyroscope?

Solution:

Since $\omega_1 = 200$ rad/s, the precession speed ω_2 is determined by Eq. (21.41), namely,

$$k_1^2 \omega_1 \omega_2 = gb \qquad (a)$$

where k_1 is the radius of gyration of the ball with respect to the axis of the shaft, and $b = 0.100$ m (see Fig. 21.17). The constant k_1 is determined by the relation (see Appendix F, Table F.1)

$$I_1 = \frac{2}{5} mr^2 = k_1^2 m$$

where m is the mass of the ball and r is its radius.

Thus, $$k_1^2 = \frac{2}{5} mr^2 \qquad (b)$$

Equations (a) and (b) yield

$$\omega_2 = \frac{5gb}{2r^2\omega_1} = \frac{5(9.81)(0.100)}{2(0.075/2)^2(200)}$$

or,

$$\omega_2 = 8.72 \text{ rad/s}$$

21.56

Given: A top consists of a disk of mass m_d and radius R welded to a slender stem of mass m_s and length L (Fig. a). The stem is placed on a peg, and the top precesses about axis 2 with angular speed p.

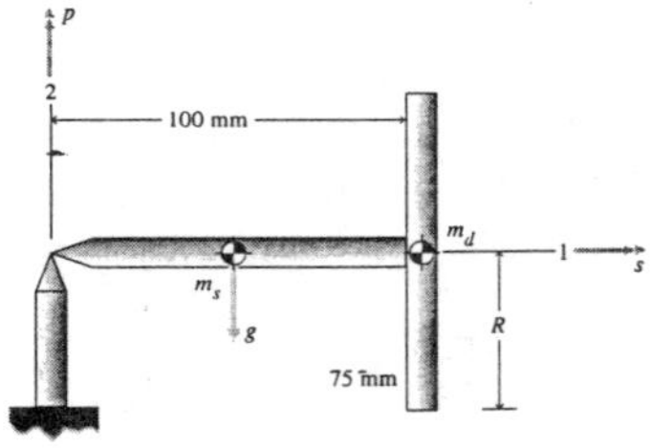

Figure a

Determine the spin s of the top about axis 1 in terms of the parameters p, m_d, m_s, L, R, and g (the acceleration of gravity).

Solution:

The spin S is determined by Eq. (21.40), namely,

$$M = I_1 \omega_1 \omega_2 = I_1 sp \qquad (a)$$

where $\omega_1 = s$ and $\omega_2 = p$

By Fig. a, the mass moment of inertia of the top with respect to axis 1 is (since the mass moment of the stem about axis 1 is approximately zero)

$$I_1 = \frac{1}{2} m_d R^2 \qquad (d)$$

(Continued)

21.56 Cont.

To determine M, consider the free-body diagram of the top (Fig. b), where we have neglected the small thickness of the disk.

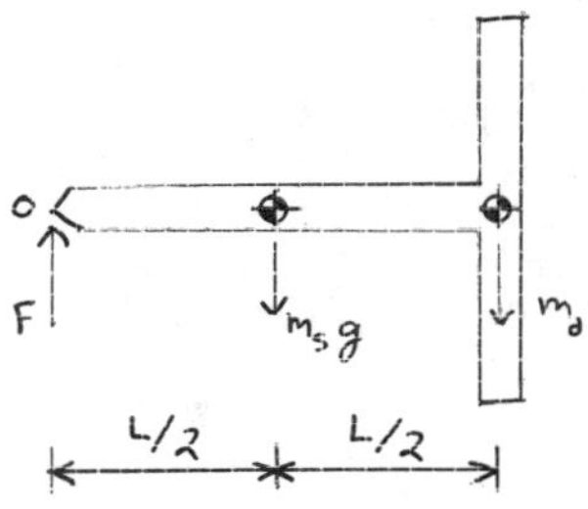

Figure b

By Fig. b,

$$\Sigma M_o = M = (m_s g)\left(\frac{L}{2}\right) + (m_d g)(L) \qquad (c)$$

Hence, by Eqs. (a), (b), and (c), we obtain

$$s = \frac{(m_s + 2m_d)gL}{m_d R^2 p}$$

21.57

Given: The top of Problem 21.56 spins about axis 1 with angular speed s and precesses at a uniform angular speed p (Fig. a)

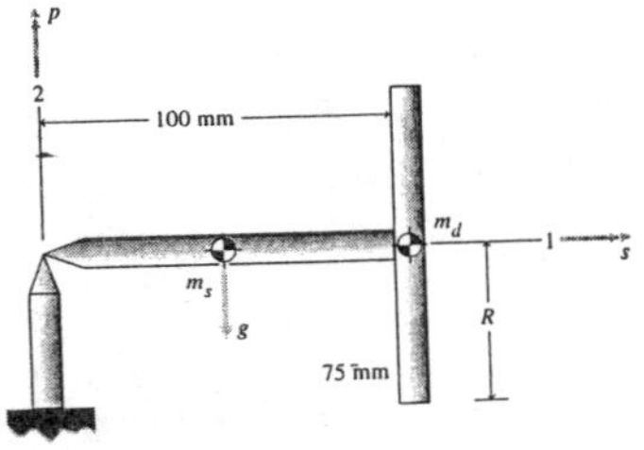

Figure a

Determine p in terms of m_d, m_s, L, R, and g (the acceleration of gravity).

Solution:

The precession speed p is determined by Eq. (21.40), namely,

$$M = I_1 \omega_1 \omega_2 = I_1 s p \qquad (a)$$

where $\omega_1 = s$ and $\omega_2 = p$

By Fig. a, the mass moment of inertia of the top with respect to axis 1 is (since the mass moment of the stem about axis 1 is approximately zero),

$$I_1 \approx \frac{1}{2} m_d R^2 \qquad (b)$$

To determine M, consider the free-body diagram of the top (Fig. b), where we have neglected the small thickness of the disk.

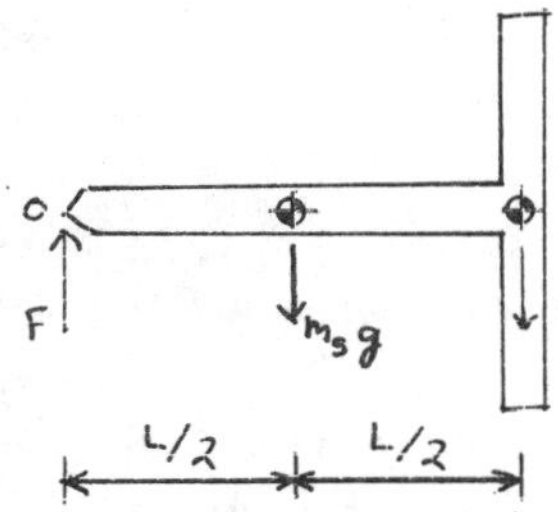

Figure b

By Fig. b,

$$\Sigma M_o = M = (m_s g)\left(\frac{L}{2}\right) + (m_d g)L \qquad (c)$$

So, by Eqs. (a), (b), and (c), we obtain

$$p = \frac{(m_s + 2m_d)gL}{m_d R^2 s}$$

21.58

Given: The propeller of an air boat rotates at 1800 rpm (Fig. a). The mass moment of inertia of the propeller about its spin axis is $I_1 = 4$ slug·ft². The boat makes a circular turn of radius 40 ft at a speed of 40 mi/h.

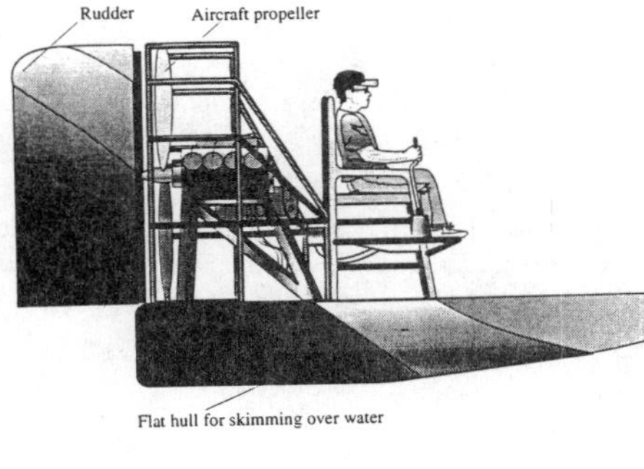

Figure a

Determine the moment exerted on the boat due to the gyroscopic effect of the wheel.

Solution: By Eq. 21.40, the moment M is given as,

$$M = I_1 \omega_1 \omega_2 \qquad (a)$$

From the given data,

$$I_1 = 4 \text{ slug}\cdot\text{ft}^2$$

$$\omega_1 = \left(1800 \frac{\text{rev}}{\text{min}}\right)\left(\frac{2\pi \text{ rad}}{\text{rev}}\right)\left(\frac{1 \text{ min}}{60 \text{ s}}\right) = 60\pi \frac{\text{rad}}{\text{s}} \qquad (b)$$

$$\omega_2 = \frac{v}{r} = \frac{(40)(88/60)}{40} = 1.467 \text{ rad/s}$$

(Continued)

21.58 Cont.

Then, by Eqs. (a) and (b), we get

$$M = (4)(60\pi)(1.467)$$

or $\underline{\underline{M = 1105.8 \text{ lb}\cdot\text{ft}}}$

21.59

Given: A satellite in orbit has mass moments of inertia $I_1 = 50 \text{ kg}\cdot\text{m}^2$ and $I_2 = 100 \text{ kg}\cdot\text{m}^2$ about axes 1 and 2, respectively (Fig. a). The angular speed about axis 1 is $\omega_1 = 2$ rad/s, and the angle between axis 1 and the angular momentum $\bar{A}$ (which is fixed in direction) is 50°

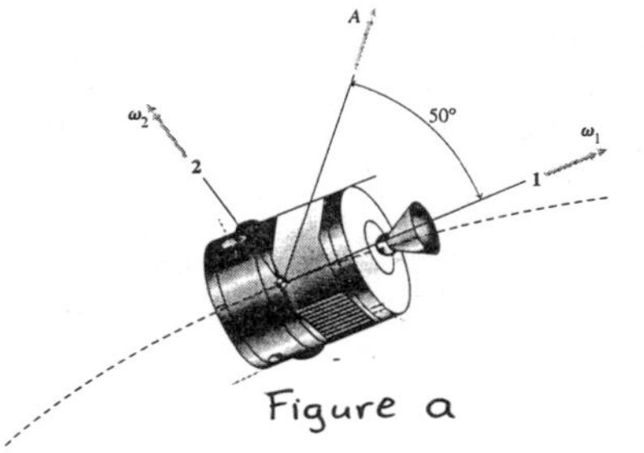

Figure a

a) Determine the angular speed of precession p of axis 1 about $\bar{A}$. Hint: The angular speed of precession of axis 1 about $\bar{A}$ is the magnitude of the projection of $\bar{\omega}_1$ on $\bar{A}$.

b) Determine the angle between the angular velocity $\bar{\omega}$ of the satellite and axis 1.

Solution:

a) Using the hint, we have,

$$\underline{\underline{p = \bar{\omega}_1 \cdot \bar{A}/A = (\omega_1)(1)\cos 50° = 2\cos 50° = 1.286 \ \tfrac{\text{rad}}{\text{s}}}}$$

b) By Fig. a and the given data

$$A_1 = I_1\omega_1 = (50)(2) = 100 \, [\text{kg}\cdot\text{m}^2\cdot\text{s}^{-1}]$$

$$A_2 = I_2\omega_2 = 100\,\omega_2 \quad [\text{kg}\cdot\text{m}^2\cdot\text{s}^{-1}]$$

Hence, $\tan 50° = \dfrac{A_2}{A_1} = \dfrac{100\,\omega_2}{100} = \omega_2$

or, $\omega_2 = 1.1918$ rad/s

Therefore, the angular velocity $\bar{\omega}$ of the satellite is,

$$\bar{\omega} = \omega_1\hat{\imath} + \omega_2\hat{\jmath} = 2\hat{\imath} + 1.192\hat{\jmath} \ [\text{rad/s}] \qquad (a)$$

where $\hat{\imath}$ and $\hat{\jmath}$ are unit vectors along axes 1 and 2, respectively.

Therefore, the angle between $\bar{\omega}$ and $\bar{\omega}_1$ is given by the scalar product and Eq. (a),

$$\bar{\omega}\cdot\bar{\omega}_1 = \omega_1^2 = 4 = \omega\omega_1\cos\theta \qquad (b)$$

where θ is the angle between $\bar{\omega}$ and ω_1 and $\omega = \sqrt{2^2 + 1.1918^2} = 2.3281$ rad/s by Eq. (a).

Hence, with Eq. (b), we find

$$\cos\theta = \frac{4}{(2.3281)(2)} = 0.85905$$

or

$$\underline{\underline{\theta = 30.79°}}$$

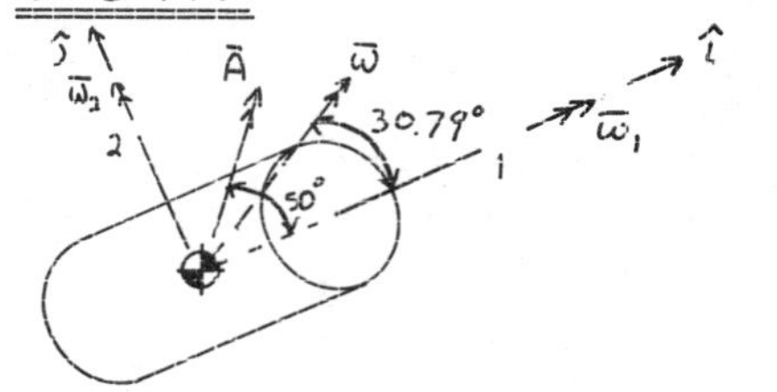

Figure b

21.60

Given: The uniform circular flywheel of a car rotates about its central axis, which is aligned with the longitudinal axis of the car. The mass moment of inertia of the flywheel about this axis is 2.7 kg·m². When the car travels at 72 km/h, the flywheel rotates at 3600 rpm. The distance between the front and rear axles is 3 m. The car turns left on a curve of radius 20 m at a speed of 20 m/s.

Determine the change in net force on the front wheels due to the gyroscopic effect of the flywheel.

Solution:

We note that 72 km/h is equal to 20 m/s. Hence,

$$\omega_1 = \left(3600\,\frac{\text{rev}}{\text{min}}\right)\left(\frac{2\pi\,\text{rad}}{\text{rev}}\right)\left(\frac{1\,\text{min}}{60\,\text{s}}\right) = 120\pi\,\frac{\text{rad}}{\text{s}} \qquad (a)$$

The angular speed due to the turn is

$$\omega_2 = \frac{v}{r} = \frac{20\text{ m/s}}{20\text{ m}} = 1 \text{ rad/s} \qquad (b)$$

Since the flywheel rotates clockwise when viewed from the driver's seat and since the car turns left, the angular velocities $\bar{\omega}_1$ and $\bar{\omega}_2$ are directed as shown in Fig. a.

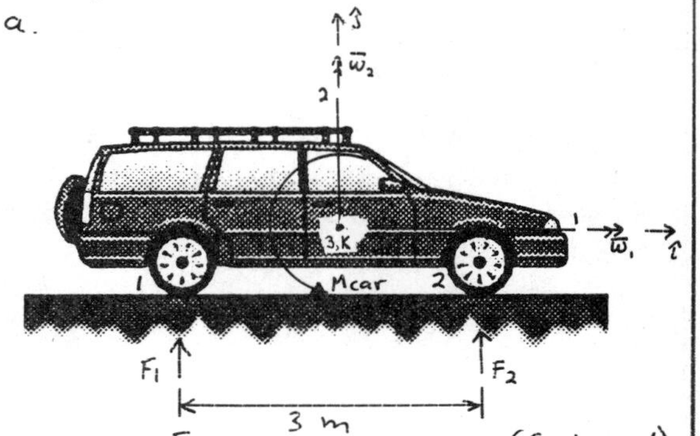

Figure a

(Continued)

21.60 Cont.

In Fig. a, F denotes the net change in force due to the gyroscopic effect of the flywheel.

By Eq. (21.42), the couple that the supports exerts on the flywheel is,

$$\bar{M} = I_1 \frac{d\bar{A}}{dt} = I_1 \bar{\omega}_2 \times \bar{\omega}_1 \quad (c)$$

where by Fig. a,

$$\bar{\omega}_1 = \omega_1 \bar{i}$$
$$\bar{\omega}_2 = \omega_2 \bar{j} \quad (d)$$

where $\bar{i}, \bar{j}$ are unit vectors along axes 1 and 2 (Fig. a). Hence, by Eqs. (c) and (d),

$$\bar{M} = I\,\omega_1\omega_2\,\bar{j} \times \bar{i} = -I_1\,\omega_1\,\omega_2\,\bar{k} \quad (e)$$

where the unit vector $\bar{k}$ is directed out of Fig. a.

By Eqs. (a), (b), and (e), with $I_1 = 2.7\ \mathrm{kg \cdot m^2}$, we find,

$$\bar{M} = -(2.7)(120\pi)(1)\bar{k} = -1017.9\,\bar{k}\ [\mathrm{N \cdot m}] \quad (f)$$

Consequently, the couple that the flywheel exerts on the supports (on the car) is (see Fig. a)

$$\bar{M}_{car} = -\bar{M} = 1017.9\,\bar{k}\ [\mathrm{N \cdot m}] \quad (g)$$

The car does not rotate about axis 3. Hence, by Fig. a,

$$\Sigma M_1 = M_{car} - 3F_2 = 0$$

or $$F_2 = \frac{M_{car}}{3} = \frac{1017.9}{3} = 339.3\ \mathrm{N}$$

Thus, the net vertical upward force on the front wheels is decreased by 339.3 N, since F_2 acts downword.

21.61

Given: A ship's gyro is a solid disk of mass 3450 slugs and radius 6 ft. The gyro rotates with an angular speed of 900 rpm about its spin axis which is directed vertically up. The ship encounters a wave that causes it to roll momentarily about its longitudinal axis at a rate of 0.1 rad/s, towards its starboard.

a) Determine the couple exerted on the ship by the gyro supports.

b) How is the motion of the ship affected by the couple?

Solution:

a) The spin of the gyro in rad/s is given by,

$$s = \left(900\ \frac{\mathrm{rev}}{\mathrm{min}}\right)\left(2\pi\ \frac{\mathrm{rad}}{\mathrm{rev}}\right)\left(\frac{1\ \mathrm{min}}{60\ \mathrm{s}}\right) = 94.248\ \frac{\mathrm{rad}}{\mathrm{s}} \quad (a)$$

Therefore, the gyro has a spin $s = 94.248\ \frac{\mathrm{rad}}{\mathrm{s}}$ and a precession $p = 0.1$ rad/s (Fig. a), where axis 2 is directed toward the bow of the ship and axis 3 is directed into the plane of Fig. a.

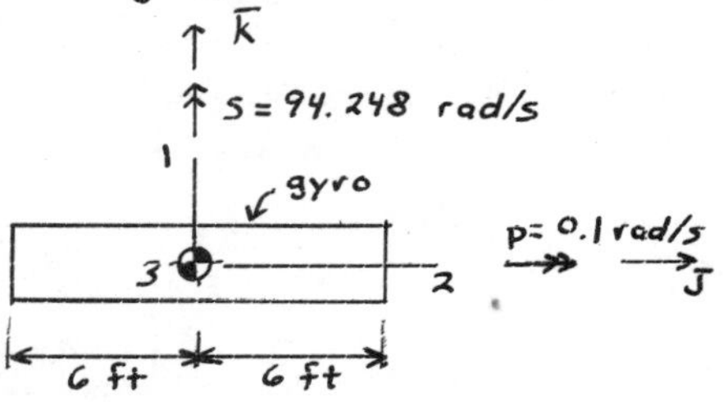

Figure a

The couple that the supports exert on the gyro is given by Eq. (21.42), namely,

$$\bar{M} = I_1 \bar{\omega}_2 \times \bar{\omega}_1 = I_1\,\bar{p} \times \bar{s}$$

Therefore,

$$\bar{M} = I_1\ (0.1\,\bar{j}) \times (94.248\,\bar{i})$$

or, $$M = -I_1\ (9.425\,\bar{k}) \quad (b)$$

where $\bar{k}$ is a unit vector directed into the plane of Fig. a, along axis 3 and

$$I_1 = \tfrac{1}{2}mr^2 = \tfrac{1}{2}(3450)(6)^2 = 62{,}100 \quad (c)$$

Hence, with Eqs. (b) and (c), we obtain the couple exerted on the gyro by the supports as,

$$\bar{M} = -5.853 \times 10^5\,\bar{k}\ [\mathrm{lb \cdot ft}] \quad (d)$$

Therefore, the couple exerted on the ship by the gyro supports is,

$$\bar{M}_s = -\bar{M} = 5.853 \times 10^5\,\bar{k}\ [\mathrm{lb \cdot ft}]$$

b) The couple $\bar{M}_s$ is directed perpendicular to the longitudinal axis (axis 2 in Fig. a) along axis 3. Hence, it will cause the ship to rotate (pitch) so that the bow goes down.

21.62

Given: The ship in Problem 21.61 momentarily pitches (its bow moves up) at a rate of 0.02 rad/s. The gyro spin is 900 rpm and there is no roll.

a) Determine the couple exerted on the ship by the gyro supports.

b) How does the couple affect the motion of the ship.

Solution:

a) The spin of the gyro in rad/s is

$$s = \left(900\ \frac{\mathrm{rev}}{\mathrm{min}}\right)\left(2\pi\ \frac{\mathrm{rad}}{\mathrm{rev}}\right)\left(\frac{1\ \mathrm{min}}{60\ \mathrm{s}}\right) = 94.248\ \frac{\mathrm{rad}}{\mathrm{s}} \quad (a)$$

(Continued)

21.62 Cont.

Therefore, the gyro has a spin $s = 94.248$ rad/s and a precession $p = 0.02$ rad/s (Fig. a), where the horizontal axis 2 is directed perpendicular to the length of the ship and axis 3 is directed toward the bow along the longitudinal axis of the ship.

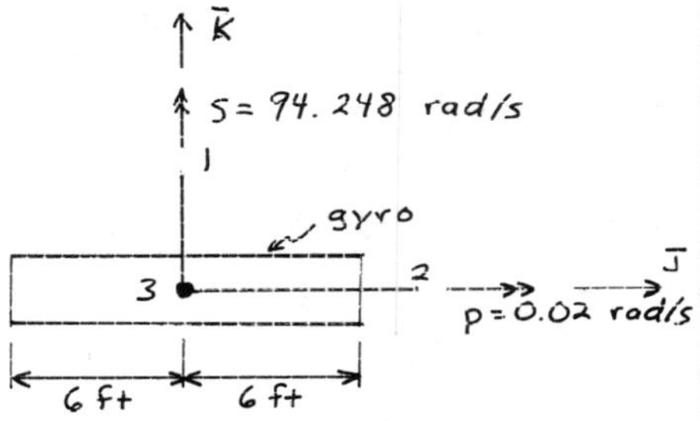

Figure a

The couple that the supports exert on the gyro is given by Eq. (21.42), namely,

$$\bar{M} = I_1 \bar{\omega}_2 \times \bar{\omega}_1 = I_1 \bar{p} \times \bar{s}$$

Therefore (see Fig. a), with $I_1 = 62{,}100$ lb·ft·s², (see the solution of Problem 21.61),

$$\bar{M} = I_1 (0.02\bar{j}) \times (94.248\bar{i})$$

or $\quad \bar{M} = -1.1706\,\bar{k}$ [lb·ft] $\qquad$ (b)

where $\bar{k}$ is a unit vector directed along the axis 3, into the plane of Fig. a. The couple $\bar{M}_s$ exerted on the ship by the gyro supports is

$$\underline{\bar{M}_s = -\bar{M} = 1.1706 \times 10^5\,\bar{k}\ [\text{lb}\cdot\text{ft}]}$$

b) The couple $\bar{M}_s$ is directed longitudinally toward the bow, along axis 3. Hence, it will cause the ship to roll toward its starboard.

21.63

Given: The ship in Problem 21.61 yaws momentarily (the bow swings toward the port side) at a rate of 0.04 rad/s. The gyro spin remains at 900 rpm and there is no roll or pitch.

a) Determine the couple exerted on the ship by the gyro supports.

b) How does the couple affect the motion of the ship?

Solution:

The gyro is subjected to its spin s and the angular speed ω due to the ship's yaw (Fig. a).

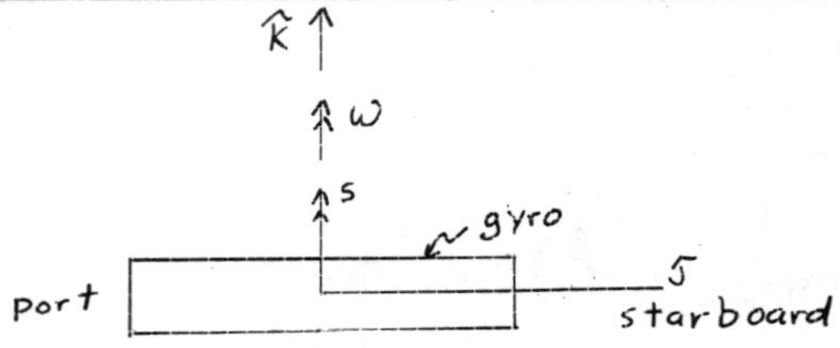

Figure a

The couple $\bar{M}$ exerted on the gyro is, by Eq. (21.42)

$$\bar{M} = I_1\,\bar{\omega} \times \bar{s}$$

So, by Fig. a,

$$\bar{M} = I_1 (\omega\bar{j}) \times (s\bar{j}) = 0$$

Since the couple exerted on the gyro by the supports is zero, the couple that the supports exert on the ship is also zero.

b) Since the supports exert a zero couple on the ship, the motion of the ship is not affected by the gyro.

21.64

Given: A solid homogeneous disk of mass m and radius r is rigidly attached obliquely to a shaft that rotates with angular speed ω (Fig. a).

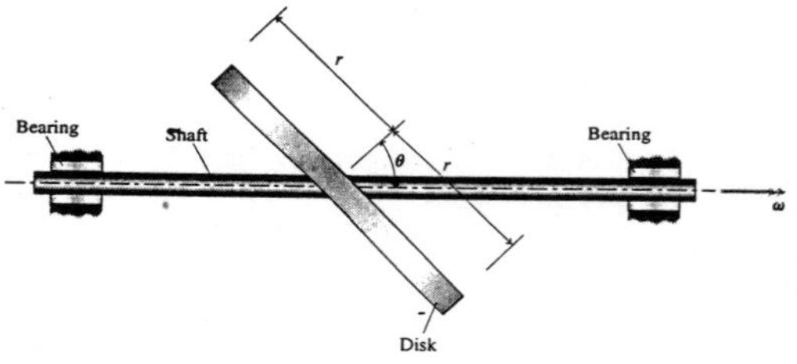

Figure a

Show that the gyroscopic couple on the bearings has magnitude

$$M = \tfrac{1}{8} m r^2 \omega^2 \sin 2\theta \qquad \text{(a)}$$

Solution: Consider axis 1 as shown in Fig. b. Axis 1 is perpendicular to the disk and axes 2 and 3 are diametral axes in the disk, with axis 3 directed out of the plane of Fig. b.

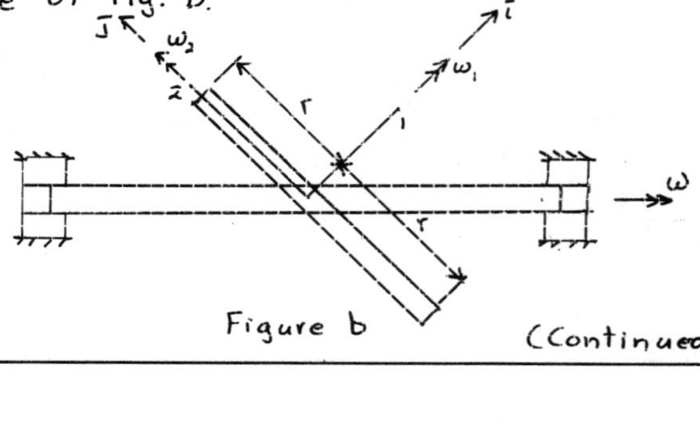

Figure b

(Continued)

21.64 Cont.

By Eq. (21.42), the couple exerted on the disk is,

$$M = \frac{d\bar{A}}{dt} \qquad (b)$$

To calculate $\bar{A}$, consider the projections of $\bar{A}$ on axes (1, 2, 3), namely,

$$A_1 = I_1\omega_1 \ , \ A_2 = I_2\omega_2 \ , \ A_3 = I_3\omega_3 \qquad (c)$$

Now, by Appendix F, Table F.1,

$$I_1 = \tfrac{1}{2}mr^2, \ I_2 = I_3 = \tfrac{1}{4}mr^2 \qquad (d)$$

and by Fig. b,

$$\omega_1 = \omega\cos\theta, \quad \omega_2 = -\omega\sin\theta \ , \ \omega_3 = 0 \qquad (e)$$

Hence, by Eqs. (c), (d), and (e),

$$\bar{A} = A_1\bar{i} + A_2\bar{j} + A_3\bar{k}$$
$$= I_1(\omega\cos\theta)\bar{i} - I_2(\omega\sin\theta)\bar{j} + I_3(0)$$

or,

$$\bar{A} = \tfrac{1}{4}mr^2\omega\left[2(\cos\theta)\bar{i} - (\sin\theta)\bar{j}\right] \qquad (f)$$

Then, Eqs. (b) and (f) yield

$$\bar{M} = \tfrac{1}{4}mr^2\omega\left[2(\cos\theta)\frac{d\bar{i}}{dt} - (\sin\theta)\frac{d\bar{j}}{dt}\right] \qquad (g)$$

Now by Eqs. (17.12a) and (17.12b)

$$\frac{d\bar{i}}{dt} = \bar{\omega}\times\bar{i} \ , \ \frac{d\bar{j}}{dt} = \bar{\omega}\times\bar{j} \qquad (h)$$

where by Eq. (e)

$$\bar{\omega} = \omega\left[(\cos\theta)\bar{i} - (\sin\theta)\bar{j}\right] \qquad (i)$$

So,

$$\frac{d\bar{i}}{dt} = -\omega(\sin\theta)(\bar{j}\times\bar{i}) = \omega(\sin\theta)\bar{k}$$
$$\frac{d\bar{j}}{dt} = \omega(\cos\theta)(\bar{i}\times\bar{j}) = \omega(\cos\theta)\bar{k} \qquad (j)$$

Substitution of Eqs. (j) into Eq. (g) yields

$$\bar{M} = \tfrac{1}{4}mr^2\omega^2\left[2\sin\theta\cos\theta - \sin\theta\cos\theta\right]\bar{k}$$

or

$$\bar{M} = \tfrac{1}{8}mr^2\omega^2(2\sin\theta\cos\theta)\bar{k} = \tfrac{1}{8}mr^2\omega^2(\sin 2\theta)\bar{k}$$

Hence, the magnitude of the gyroscopic couple that acts on the bearings is

$$M = \tfrac{1}{8}mr^2\omega^2\sin 2\theta$$

This result agrees with Eq. (a)

21.65

Given: The wheels of a car have total mass m, radius r, and radius of gyration k. The car enters a circular curve of radius R, inclined at an angle θ with the horizontal, at a speed v.

Show that, due to the angular velocities of the wheels and the car, the gyroscopic couple on the car has magnitude,

$$M = \frac{mk^2 v\cos\theta}{rR}\left(1 - \frac{r}{R}\sin\theta\right) \qquad (a)$$

Note that since $r/R \ll 1$ for most realistic situations, the magnitude of the gyroscopic couple is approximated by

$$M = \frac{mk^2 v\cos\theta}{rR} \qquad (b)$$

Solution:

Figure a is a sketch of the car in the curve, with velocity v directed out of the plane of the figure.

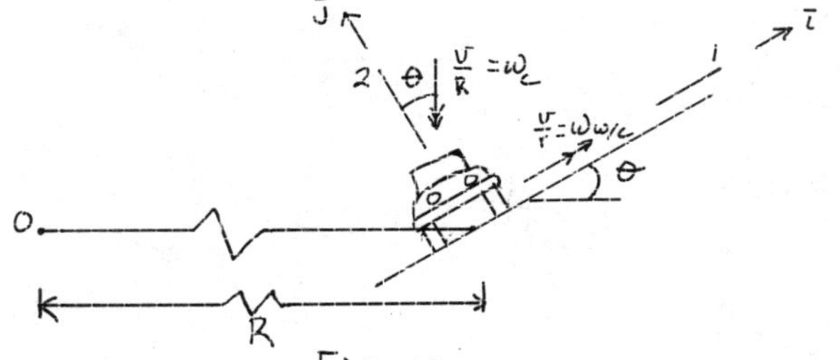

Figure a

By Fig. a, the angular speed of the car with respect to the center O of the curve is

$$\omega_c = \frac{v}{R} \qquad (c)$$

and the angular speed of the wheels relative to the car is

$$\omega_{w/c} = v/r \qquad (d)$$

Let axes 1, 2 be directed up the incline and perpendicular to the incline, respectively. The wheels of the car act like the rotor of a gyroscope. By Fig. a, the projections of the angular velocities along axes 1 and 2 are

$$\omega_1 = \frac{v}{r} - \frac{v}{R}\sin\theta$$
$$\omega_2 = -\frac{v}{R}\cos\theta$$

and the corresponding angular velocities are

$$\bar{\omega}_1 = \left(\frac{v}{r} - \frac{v}{R}\sin\theta\right)\bar{i}$$
$$\bar{\omega}_2 = -\frac{v}{R}(\cos\theta)\bar{j} \qquad (c)$$

where $\bar{\omega}_1$ is the spin vector and $\bar{\omega}_2$ is the precession vector. Hence, by Eq. 21.42, the couple acting on the wheels is

$$\bar{M} = I_1\bar{\omega}_2\times\bar{\omega}_1 \qquad (d)$$

where $\quad I_1 = mk^2 \qquad (e)$

By Eqs. (c), (d), and (e), we have

$$\bar{M} = mk^2\begin{vmatrix} \bar{i} & \bar{j} & \bar{k} \\ 0 & -\frac{v}{R}\cos\theta & 0 \\ \frac{v}{r} - \frac{v}{R}\sin\theta & 0 & 0 \end{vmatrix}$$

Therefore, the couple exerted on the wheel is

$$\bar{M} = mk^2\left(\frac{v^2}{rR}\cos\theta - \frac{v^2}{R^2}\sin\theta\cos\theta\right)\bar{k}$$

or

$$\bar{M} = \frac{mk^2v^2\cos\theta}{rR}\left(1 - \frac{r}{R}\sin\theta\right)\bar{k} \qquad (f)$$

(Continued)

21.65 Cont.

where $\bar{k}$ is a unit vector directed out of the plane of Fig. a. In turn the couple exerted on the car by the wheels is $\bar{M}_c = -\bar{M}$. It tends to tip the car. The magnitude of $\bar{M}_c$ is the same as that of $\bar{M}$. Thus,

$$M_c = M = \frac{mk^2v^2\cos\theta}{rR}\left(1 - \frac{r}{R}\sin\theta\right)$$

[See Eq. (a)] For $\frac{r}{R} << 1$,

$$M \approx \frac{mk^2v^2}{rR}\cos\theta$$

[See Eq. (b).]

21.66

Given: The Feltner rotor windmill consists of two thin uniform cylindrical shells (rotors) each 40 ft long and 8 ft in diameter, see Fig. a for an elevation view of the windmill. Each rotor weighs 1600 lb. The rotors act like the vanes of an ordinary windmill. The rotors are mounted on a concentric shaft that rotates about its center with angular speed 5 rad/s. Simultaneously, level gears cause the rotors to spin about their geometric axes at 25 rad/s.

Determine the bending moment in the shaft due to the gyroscopic couple.

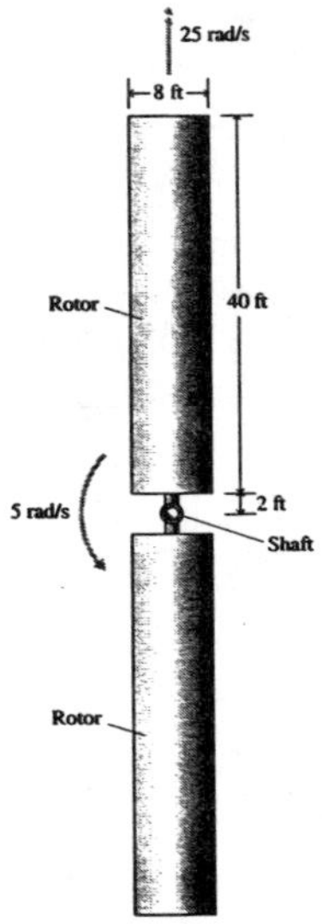

Figure a

Solution:

A side view of the top rotor is shown in Fig. b

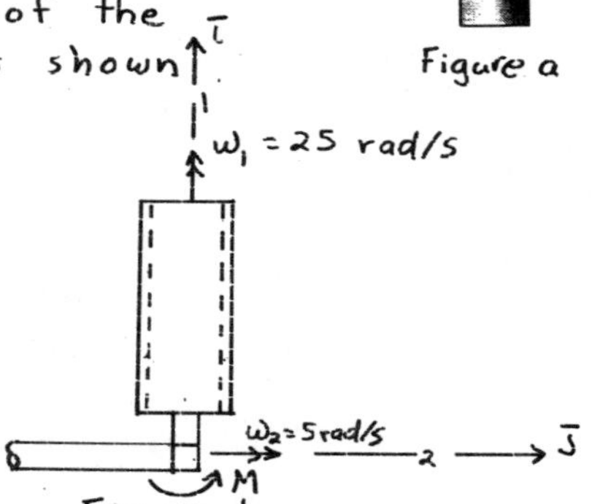

Figure b

By Eq. (21.40), the moment exerted on the rotor by the shaft is

$$\bar{M} = I_1\omega_1\omega_2 \circlearrowright \qquad (a)$$

where by Appendix F, Table F.2,

$$I_1 = mr^2 = \left(\frac{1600}{32.2}\right)(4^2) = 795.0 \text{ lb}\cdot\text{ft}\cdot\text{s}^2 \qquad (b)$$

By Eqs. (a) and (b), with $\omega_1 \doteq 25$ rad/s and $\omega_2 = 5$ rad/s, we obtain

$$\bar{M} = (795.0)(25)(5) = 99{,}380 \text{ lb}\cdot\text{ft} \circlearrowright$$

The gyroscopic couple $\bar{M}_T = -\bar{M}$, due to the top rotor, acts on the shaft; that is, it is opposite to $\bar{M}$ (Fig. b), or,

$$\bar{M}_T = 99{,}380 \text{ lb}\cdot\text{ft} \circlearrowleft \qquad (c)$$

Similarly for the bottom rotor (Fig. a),

$$\bar{M} = I_1\omega_1\omega_2 \circlearrowleft$$

or $\bar{M} = 99{,}380$ lb·ft $\circlearrowleft$

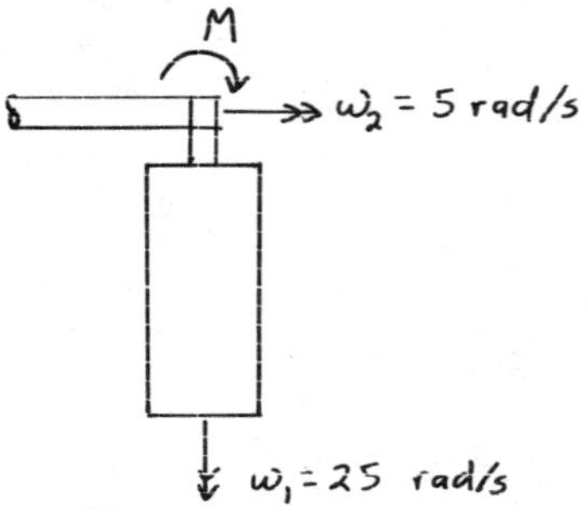

Figure c

The gyroscopic couple $\bar{M}_B = -\bar{M}$ due to the bottom rotor acts on the shaft, and is opposite to $\bar{M}$. Therefore,

$$\bar{M}_B = 99{,}380 \text{ lb}\cdot\text{ft} \circlearrowright \qquad (d)$$

The net gyroscopic couple that acts on the shaft is, by Eqs. (c) and (d),

$$\bar{M}_{Net} = \bar{M}_A + \bar{M}_B = 99{,}380 \circlearrowleft + 99{,}380 \circlearrowright = 0$$

Therefore the gyroscopic couples due to the two rotors cancel.

21.67

Given: For the gyrostabilizer of Problem 21.61, let the mass be m_0, the radius be r and the spin velocity have magnitude s and the direction cosines of its spin axis be ℓ, m, and n relative to a right-handed coordinate system xyz. Axis x is directed along the longitudinal axis of the ship toward the bow; axis y is directed vertically up; axis z is transverse to the ship, directed starboard. The ship is subjected to an angular velocity Ω with direction cosines L, M, and N relative to the xyz axes.

(Continued)

21.67 Cont.

a) Determine the couple exerted on the ship by the gyro in terms of $m_o, r, s, \Omega, \ell, m, n, L, M$, and N.

b) How does this couple affect the motion of the ship.

c) Specialize the results of part a to verify the results of Problem 21.61, 21.62, and 21.63.

Solution:

(a) Relative to axes xyz, the spin velocity is

$$\bar{\omega}_1 = s(\ell\bar{i} + m\bar{j} + n\bar{k}) \quad (a)$$

where $\bar{i}, \bar{j}, \bar{k}$ are unit vectors directed along axes x, y, z, respectively. Similarly, the angular velocity of the ship is

$$\bar{\omega}_2 = \Omega(L\bar{i} + M\bar{j} + N\bar{k}) \quad (b)$$

Hence, by Eq. (12.42), the couple exerted on the gyro by its supports is,

$$\bar{M} = I_1 \bar{\omega}_2 \times \bar{\omega}_1 \quad (c)$$

where, $I_1 = \frac{1}{2} m_o r^2$

Equations (a), (b), and (c) yield

$$\bar{M} = \tfrac{1}{2} m_o r^2 s \Omega \begin{vmatrix} \bar{i} & \bar{j} & \bar{k} \\ L & M & N \\ \ell & m & n \end{vmatrix}$$

or,

$$\bar{M} = \tfrac{1}{2} m_o r^2 s \Omega [(Mn - mN)\bar{i} + (N\ell - nL)\bar{j} + (Lm - \ell M)\bar{k}] \quad (d)$$

The couple exerted by the supports on the ship is $-\bar{M}$.

b) The couple $-\bar{M}$ produces a roll to the port due to the moment $-\frac{1}{2} m_o r^2 s \Omega (Mn - mN)$, a yaw to the starboard due to the moment $-\frac{1}{2} m_o r^2 s \Omega (N\ell - nL)$, and a pitch (the bow moves down) due to the moment $-\frac{1}{2} m_o r^2 s \Omega (Lm - \ell M)$.

c) In Problem 21.61, the ship undergoes a roll about the longitudinal axis (here the x axis) and the gyro spin axis is directed upward (here the y axis). Hence, by Eqs. (a) and (b), $\ell = n = 0$ and $M = N = 0$. Then $m = L = 1$, and the couple $-\bar{M}$ reduces to [see Eq. (d)]

$$-M = -(\tfrac{1}{2} m_o r^2 s \Omega)\bar{k}$$

and it moves the bow down (see Problem 21.61)

In Problem 21.62, the ship undergoes a pitch (the bow moves up) and the spin axis is vertically up (here the y axis). Here the pitch is a rotation about the z axis. Hence, again by Eqs. (a) and (b), $\ell = n = 0$ and now $L = M = 0$. Then, $m = N = 1$, and the couple $-\bar{M}$ reduces to, $-\bar{M} = +(\frac{1}{2} m_o r^2 s \Omega)\bar{i}$ and it causes the ship to roll toward starboard (see Problem 21.62).

In Problem 21.63, the bow swings (yaws) toward port, that is, it rotates about the y axis. Again the spin axis is directed along the y axis. Hence, by Eqs. (a) and (b),

$$\ell = n = L = N = 0$$

and $m = M = 1$. Then, the couple $-M$ reduces to zero as in Problem 21.63. It does not affect the motion of the ship.

21.68

Given: The propeller of an airplane weighs 400 lb. It rotates at 1800 rpm. When the plane pulls out of a dive at a speed of 500 mi/h, it moves on a circular path of radius 4500 ft. The mass moment of inertia of the propeller about the axis of the propeller shaft is 224 lb·ft·s². Calculate the bending moment in the propeller shaft due to the gyroscopic couple.

Solution:

The spin vector of the propeller has the magnitude (in rad/s),

$$\omega_1 = \left(1800 \frac{\text{rev}}{\text{min}}\right)\left(\frac{2\pi \text{ rad}}{\text{rev}}\right)\left(\frac{1 \text{ min}}{60 \text{ s}}\right) = 60\pi \frac{\text{rad}}{\text{s}} \quad (a)$$

and the spin vector precesses with angular speed

$$\omega_2 = v/r$$

where, $v = (500 \frac{\text{mi}}{\text{h}})(5280 \frac{\text{ft}}{\text{mi}})(\frac{\text{h}}{3600 \text{ s}}) = 733.3 \frac{\text{ft}}{\text{s}}$

$r = 4500$ ft

Hence,

$$\omega_2 = \frac{733.3}{4500} = 0.16296 \text{ rad/s} \quad (b)$$

see Fig. a,

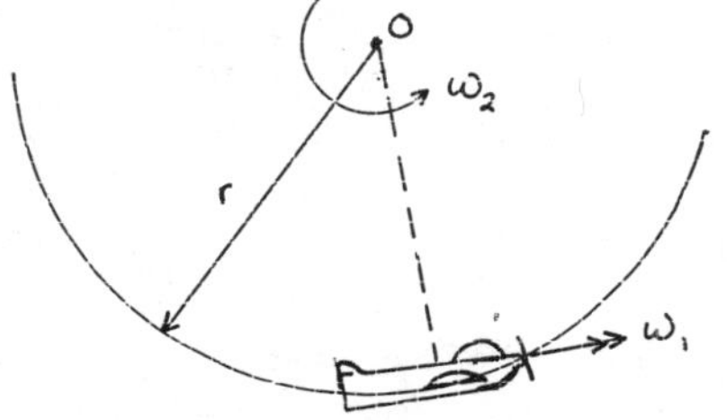

Figure a

The moment exerted on the propeller is, by Eq. (21.40), $M = I_1 \omega_1 \omega_2$ (c)

where, $I_1 = 224$ lb·ft·s²

Therefore, by Eqs. (a), (b), and (c)

$$M = (224)(60\pi)(0.16296) = 6880.8 \text{ lb·ft}$$

(Continued)

21.68 Cont.

The couple $\bar{M}$ is directed toward the center O of the circular path [see Eq. (21.42)]. The gyroscopic couple $-\bar{M}$ is exerted on the shaft by the propeller, and is directed away from O. It exerts a moment $M = 6880.8$ lb·ft on the shaft and bends the shaft toward the starboard of the plane.

21.69

Given: The wheel in Fig. a is a solid disk with diameter 100 mm. It weighs 6 N, and rotates at 3000 rpm about the uniform horizontal shaft which is pinned at A. The shaft weighs 2 N. The support at C is removed and the wheel/shaft system swings as a pendulum about the at A.

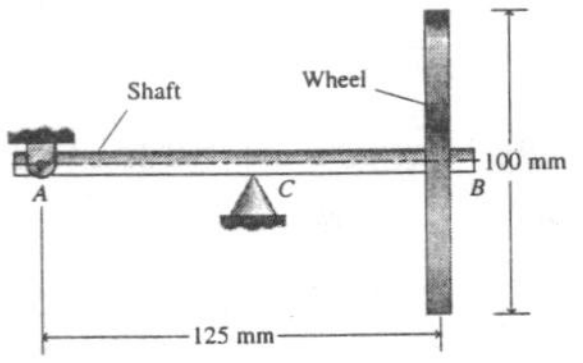

Figure a

Determine the maximum gyroscopic couple that is transmitted to the pin at A.

Solution:

Since the angular speed ω_1 of the disk is constant and the maximum angular rotation of the system about A occurs when the shaft is directed downward, the maximum gyroscopic couple occurs then. We may calculate the maximum angular speed ω_2 of the system about A by the principle of conservation of energy. Thus,

$$T_1 + V_1 = T_2 + V_2 \qquad (a)$$

where initially

$$T_1 = 0\,, \quad V_1 = \left(\frac{0.125}{2}\right)(2) + (0.125)(6) = 0.875 \text{ N·m} \qquad (b)$$

and,

$$T_2 = \frac{1}{2} I \omega_2^2 \,, \quad V_2 = 0 \qquad (c)$$

where,

$$I = \frac{1}{3} m_s L_s^2 + m_w L_s^2 + \frac{1}{4} m_w r^2$$

(the subscripts s and w denote shaft and wheel, respectively, and L_s is the length of the shaft and r is the radius of the wheel).

Here, we have taken the datum of potential energy in the vertical downward position of the system, and the datum for kinetic energy in the initial horizontal position.

Therefore,

$$I = \frac{1}{3}\left(\frac{2}{9.81}\right)(0.125)^2 + \left(\frac{6}{9.81}\right)(0.125)^2 + \frac{1}{4}\left(\frac{6}{9.81}\right)(0.050)^2$$

or,

$$I = 0.0110 \text{ N·m·s}^2 \qquad (d)$$

Hence, Eqs. (a), (b), (c), and (d) yield

$$\omega_2 = 12.613 \text{ rad/s} \qquad (e)$$

The angular spin of the wheel in rad/s is

$$\omega_1 = (3000)\left(\frac{2\pi}{60}\right) = 100\pi \text{ rad/s} \qquad (f)$$

Therefore, in the downward position, (see Fig. b) the couple exerted on the wheel [see Eq. (21.42)] is

$$\bar{M} = I_1 \bar{\omega}_2 \times \bar{\omega}_1 \qquad (g)$$

where by Fig. b,

$$\bar{\omega}_1 = \omega_1 \bar{i} \,, \quad \bar{\omega}_2 = \omega_2 \bar{j} \qquad (h)$$

axis 2 and unit vector $\bar{j}$ are directed into the page.

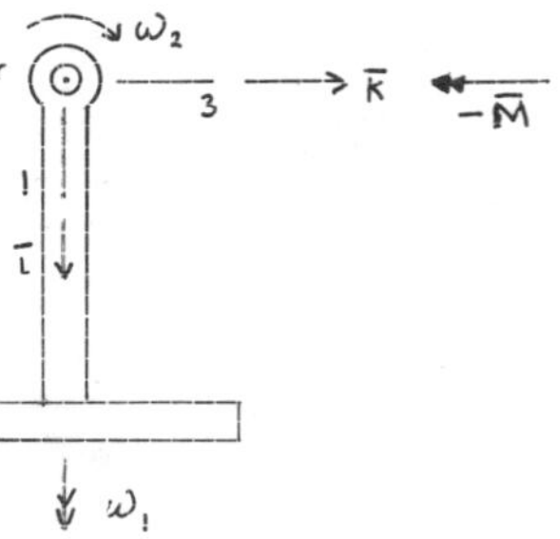

Figure b

Also, since the mass moment of the shaft about axis 1 is negligible,

$$I_1 = \frac{1}{2} m_w r^2 = \frac{1}{2}\left(\frac{6}{9.81}\right)(0.05)^2$$

or,

$$I_1 = 7.645 \times 10^{-4} \text{ N·m·s}^2 \qquad (i)$$

Hence, by Eqs. (e), (f), (g), (h), and (i),

$$\bar{M} = 7.645\times 10^{-4} \begin{vmatrix} \bar{i} & \bar{j} & \bar{k} \\ 0 & 12.613 & 0 \\ 100\pi & 0 & 0 \end{vmatrix}$$

or,

$$\bar{M} = 3.029\,\bar{k} \quad [\text{N·m}]$$

Hence, the gyroscopic couple that acts on the pin at A is

$$-\bar{M} = -3.029\,\bar{k} \quad [\text{N·m}]$$

This couple bends the pin up (see Fig. b).